42nd Conference of the International Deep Drawing Research Group

IOP Conference Series: Materials Science and Engineering
Volume 1284

Lulea, Sweden
19 - 22 June 2023

ISBN: 978-1-7138-7497-3
ISSN: 1757-8981

Printed from e-media with permission by:

Curran Associates, Inc.
57 Morehouse Lane
Red Hook, NY 12571

Some format issues inherent in the e-media version may also appear in this print version.

Additional copies of this publication are available from:

Curran Associates, Inc.
57 Morehouse Lane
Red Hook, NY 12571 USA
Phone: 845-758-0400
Fax: 845-758-2633
Email: curran@proceedings.com
Web: www.proceedings.com

TABLE OF CONTENTS

IOP Conf. Series: Materials Science and Engineering 1284 (2023) 011001 doi:10.1088/1757-899X/1284/1/011001

42nd International Deep Drawing Research Group (IDDRG) Conference

iddrg 2023

Luleå, Sweden, 19-22 June 2023

Figure 1. The city of Luleå, Sweden.

1. Introduction

The 42nd International Deep Drawing Research Group (IDDRG) conference was held in Luleå, Sweden, on 19-22 June 2023 (IDDRG 2023), Figure 1. The conference was chaired by Nader Asnafi and co-chaired by Johnny Sjöström, both associate professors at Luleå University of Technology, Sweden. The conference was organized by the Austrian Society for Metallurgy and Materials (ASMET).

The International Deep Drawing Research Group (IDDRG) was started in 1957 as an organization of national groups from all over the world devoted to the study of sheet metal forming including forming processes, materials, formability issues, tooling, tribology, and many other interesting related aspects of sheet metal forming research and industrial practice.

IDDRG conferences provide an open and scientifically highly acknowledged forum, where leading sheet metal specialists and researchers worldwide gather for fruitful discussions on challenging technical topics in an inspiring environment for knowledge exchange, interactions, and development.

IOP Conf. Series: Materials Science and Engineering 1284 (2023) 011001 doi:10.1088/1757-899X/1284/1/011001

2. Conference topics

Fossil-free steel, sustainable/advanced/smart materials, multi-material concepts/solutions, forming/formability, lightweight structures, digital transformation, digitalization, smart plants, smart press shops, cold and hot forming and trimming (incl. single and multi-step forming and trimming), tribology, simulations (incl. material models, topology optimization, tribology, springback compensation, and geometry assurance), testing, testbeds, springback, shape accuracy, geometry assurance, surface quality, tools and dies, and additive manufacturing were the topics of IDDRG 2023, the 42nd International Deep Drawing Research Group (IDDRG) conference in Luleå, Sweden.

3. Chairs

- Nader Asnafi, Associate Professor, Luleå University of Technology, Sweden, Executive chair
- Johnny Sjöström, Associate Professor, Luleå University of Technology, Sweden, Co-chair

4. Organizing Committee

- Nader Asnafi, Associate Professor, Luleå University of Technology, Sweden, Executive chair
- Michelle Grabmaier, Ms., Austrian Society for Metallurgy and Materials (ASMET), Austria
- Gerhard Hackl, Dr., Austrian Society for Metallurgy and Materials (ASMET), Austria

5. Scientific Committee

- Paul Åkerström (Sweden)
- Nader Asnafi (Sweden)
- Ton van den Boogaard (The Netherlands)
- Daniel Casellas Padró (Sweden)
- Kester Clarke (USA)
- Patrick Duroux (France)
- Gašpar Gantar (Slovenia)
- Gerhard Hackl (Austria)
- Eiji Izuka (Japan)
- Jörgen Kajberg (Sweden)
- Celalettin Karadogan (Turkey)
- Nico Langerak (The Netherlands)
- Jari Larkiola (Finland)
- Myoung-Gyu Lee (Korea)
- Mathias Liewald (Germany)
- Lars-Erik Lindgren (Sweden)
- Zsolt Lukács (Hungary)
- Mohamed Mohamed (UK)
- Marion Merklein (Germany)
- Dirk Mohr (Switzerland)
- Krishnaiyengar Narasimhan (India)
- Erik Olsson (Sweden)
- Johan Pilthammar (Sweden)
- Pasi Peura (Finland)
- Bernard Rolfe (Australia)
- Eneko Saenz de Argandoña (Spain)
- Abel Santos (Portugal)
- Mats Sigvant (Sweden)
- Sandrine Thuillier (France)
- Wolfram Volk (Germany)
- Leopold Wagner (Austria)
- Michael Worswick (Canada)
- Shi-Hong Zhang (China)

6. Venue

The conference was held at The House of Culture, Skeppsbrogatan 17, Luleå, Sweden, Figure 2.

Figure 2. The House of Culture in Luleå, the venue of the IDDRG 2023 conference.

7. Plenary speakers

Eight plenary speeches were held in the following order:

- Martin Pei, Executive Vice President & CTO, SSAB AB (Swedish Steel): *SSAB's transformation to fossil-free steel with the HYBRIT technology*
- Per-Emil Back, R&D Director at Gestamp R&D Sweden: *History of hot stamping process and hot stamping simulations*
- Mats Sigvant, Technical Leader in the area of Sheet Metal Forming Simulations, Volvo Cars, Sweden: *Sheet metal forming simulations at Volvo Cars: the past, the present, and the future*
- Jan Larsson, Business development manager for aftermarket services at AP&T: *Press hardening – history and future*
- Johan Pilthammar, Technical Expert in the area of Simulation Driven Stamping Development at Volvo Cars, Sweden: *Towards virtual tryout and digital twins at Volvo Cars*
- Bart Carleer, Corporate Technical Director, AutoForm Engineering Switzerland: *Optimal milling data preparation for draw dies*
- Mathias Liewald, Professor and Head of the Institute for Forming Technology at the University of Stuttgart, Germany: *Perspectives on AI applications in sheet metal forming*
- Vincent Grolleau, Professor, ETH, Switzerland, and Université de Bretagne Sud, France: *Novel and revisited experiments for plasticity and fracture testing of sheet metals*

8. Number of presentations, published scientific papers/first authors, and participants

At this conference, 98 presentations were held in accordance with the following:

- 8 plenary speeches.
- 1 Ph.D. candidate presentation for which no paper is published in the present IDDRG 2023 proceedings.
- 89 presentations of the (89) scientific papers that are included in the present IDDRG 2023 proceedings. Figure 3 shows the number of scientific papers by country.

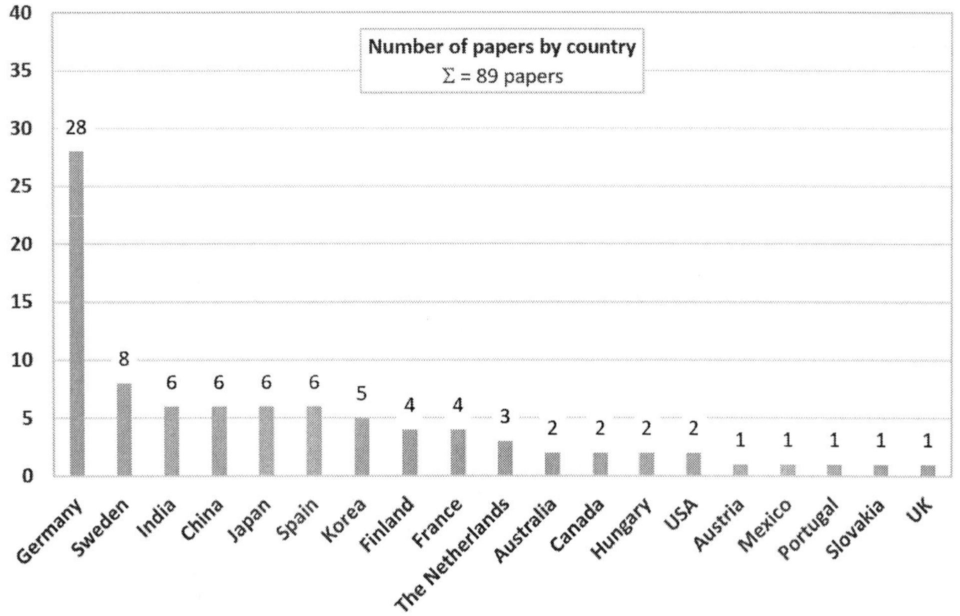

Figure 3. Number of scientific papers by country. These papers are included in the present IDDRG 2023 proceedings.

Figure 4 shows the number of participants by country. Totally 150 participants attended IDDRG 2023 (the figures are from 11 May 2023, the date this preface was prepared). 37% of the participants were from the industry, while 67% were from academia.

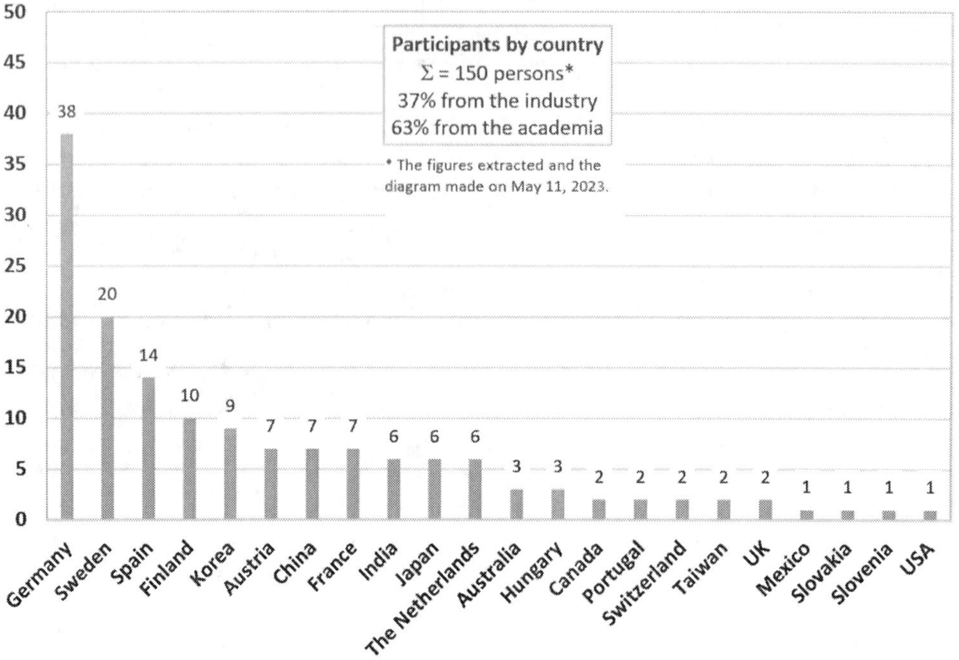

Figure 4. Number of participants by country. Totally 150 participants attended IDDRG 2023. 37% were from the industry and 63% from academia. Note that the figures are from 11 May 2023, the date this preface was prepared.

9. Technical/Industrial tours

Technical/industrial tours were arranged for 40 conference participants on 22 June 2023. These tours went to Scania Luleå and Gestamp HardTech.

10. Sponsors and exhibitors

Autoform Engineering, AP&T Group, and voestalpine were the gold, silver, and bronze sponsor respectively of this conference, Figure 5.

AUTOFORM
Forming Reality

AP&T
AUTOMATION · PRESSES · TOOLING

voestalpine
ONE STEP AHEAD.

Figure 5. The sponsors of IDDRG 2023.

IOP Conf. Series: Materials Science and Engineering 1284 (2023) 011001 doi:10.1088/1757-899X/1284/1/011001

In addition to these companies, Erichsen and Vialux were exhibitors, Figure 6.

Figure 6. In addition to the sponsors (Figure 5), these companies were exhibitors at IDDRG 2023.

11. Acknowledgments
The Chairs, Organizing Committee members, and Editors wish to express their sincere gratitude to the members of the Scientific Committee, plenary speakers, authors and presenters, participants, sponsors, exhibitors, Gestamp HardTech, Scania Luleå, and the staff at the House of Culture for their outstanding support and a highly memorable IDDRG conference.

12. Editors
- Nader Asnafi, Associate Professor, Luleå University of Technology, Sweden
- Lars-Erik Lindgren, Professor, Luleå University of Technology, Sweden

IOP Conf. Series: Materials Science and Engineering 1284 (2023) 011002 doi:10.1088/1757-899X/1284/1/011002

Peer Review Statement

All papers published in this volume have been reviewed through processes administered by the Editors. Reviews were conducted by expert referees to the professional and scientific standards expected of a proceedings journal published by IOP Publishing.

- **Type of peer review:** Single Anonymous
- **Conference submission management system:** The submission management system of ASMET*/Indico
- **Number of submissions received:** 110
- **Number of submissions sent for review:** 110
- **Number of submissions accepted:** 89
- **Acceptance Rate (Submissions Accepted / Submissions Received × 100):** 80.9
- **Average number of reviews per paper:** 2
- **Total number of reviewers involved:** 33
- **Contact person for queries:**
  Name: Nader Asnafi
  Email: nader.asnafi@outlook.com
  Affiliation: Luleå University of Technology, Sweden

IOP Conf. Series: Materials Science and Engineering 1284 (2023) 012001 doi:10.1088/1757-899X/1284/1/012001

A mechanical interlocking joint between sheet metal and carbon fibre reinforced polymers through punching

Núria Latorre[1,2], **Daniel Casellas**[1,3] **and Josep Costa**[2]

[1]Eurecat, Centre Tecnològic de Catalunya, Unit of Polymeric and Composite Processes, Av. Universitat Autònoma, 23 – 08290 Cerdanyola del Vallès, Spain
[2]AMADE, Polytechnic School, University of Girona, Av. Universitat de Girona, 4. 17003 Girona, Spain
[3]Division of Solid Mechanics, Luleå University of Technology, 971 87 Luleå, Sweden

E-mail: nuria.latorre@eurecat.org

Abstract. The joint between different lightweight materials plays a significant role in multi-material design of structural components for the automotive industry, aiming to reduce the vehicle's weight without compromising performance or safety. Yet, conventional mechanical joining technologies between metals and Carbon Fibre Reinforced Polymers (CFRP) result in either a hole being drilled in the composite material, leading to damages which reduce the load bearing capacity, or the weight of the part being increased due to the incorporation of fasteners. At the same time, alternative mechanical joining methodologies involve complex and costly processing, hindering their industrial application. This work presents a new, simple, cost-efficient and non-weight penalizing mechanical joining technology between a metal sheet and fibre reinforced polymer prepregs consisting of a single-step punching process. In this process, the metallic sheet is completely perforated, while the prepreg is not. The punch pushes the carbon fibres through the metallic hole, with no, or minimal fibre breakage, generating a mechanical interlock. The shear strength and the absorbed energy of the co-cured joint increase with the incorporation of the mechanical interlocking joint.

1. Introduction

Nowadays, concern over fuel consumption and pollutant emissions is driving transport industry efforts towards vehicles weight reduction, powering a trend known as lightweighting. Such weight decrease is a key factor to lower CO_2 emissions in fuel powered vehicles and to increase the driving range in electric ones. An interesting strategy to achieve such weight decrease is Multi-Material Design (MMD) of structural components, consisting of combining materials with high specific properties such as advanced high strength steels, aluminium alloys, or Fibre Reinforced Polymers (FRP). MMD seeks to meet the mechanical and safety requirements at affordable costs and high production volumes.

Considering this, the joint between these dissimilar materials is usually the weakest point of the structure and it determines its structural efficiency. Traditional joining strategies for metal-FRP multi-material structures include mechanical fastening, adhesive bonding, hybrid mechanical-adhesive bonding and welding [1].

Mechanical fastening strategies consist of generating a mechanical interlock between two materials, with their main advantages being low cost, simply processing, easy maintenance and non-sensitivity to the working environment [2]. Conventional mechanical joining strategies for metals and CFRP include bolted joining [3], Self-Piercing Rivets (SPR) [4] and mechanical clinching [5]. However, these conventional methodologies either increase the weight of the structure due to fasteners employment or require to drill a hole in the composite, which can cause damage such as delamination or fibre breakage, thus reducing the load bearing capacity of the structure [6].

Modifications to such conventional mechanical joining strategies have also been proposed. Some examples are the incorporation of bonded metallic inserts in the composite to reduce the stress concentration factors at the hole in the composite [7] or steering the fibres onto the tacky prepreg to match the load path to improve the stiffness and strength of the bolted joint [8]. A post-curing SPR method to join CFRP and aluminium alloy sheets has also been proposed to reduce the damage generated in the composite material [4]. Conventional clinching has also been adapted to extend its suitability to materials with low ductility, as CFRP are. For instance, using a spring die [9], Friction Assisted Clinching [10], or softening the polymeric matrix in thermoplastic composites to increase composite toughness and formability [11].

Alternative mechanical joining procedures have also been developed in order to overcome the previous drawbacks, such as pin joints [12] or loop joints [13] among other joining methodologies.

However, all these methods imply extra process steps or high complexity, which hampers their implementation in high volume production industries [14]. Aimed at boosting the deployment of MMD in the automotive industry, this work presents a straightforward, cost-efficient, non-weight penalizing mechanical joining technology for metal and FRP based on a single-step punching process of the metal sheet and the uncured prepreg. Such technology can be easily implemented in automotive production lines, where punching operations are commonly used. It does not increase the weight of the part with the addition of fasteners, on the contrary, it removes part of the metallic material, making the part even lighter. In addition, the hole created in the metallic material relives stress concentrations more efficiently than the composite due to its plasticity [15]. Therefore, damage on the composite is minimized without adding weight to the structure or complexity to the joining process. Moreover, the joint does not generate any protrusion, making it suitable for completely flat applications.

2. Materials and methods

2.1. Materials

Rolled sheets of 1.5-mm-thick AA5754 H111 aluminium alloy were used in this work. The high formability and corrosion resistance of this alloy makes it one of the most common non heat treatable aluminium alloy used in automotive industry to produce medium strength parts. Aluminium sheets were waterjet cut perpendicular to the rolling direction and joined with CFRP sheets of 0.65 mm thickness, which consisted of one layer of a Twill 2x2 prepreg of 650 g/m^2 with the MTC275 toughened epoxy resin.

2.2. Microstructural and mechanical characterization of the joint

The cross-section of the generated joint was inspected through stereomicroscopy using an Olympus SZX10.

The mechanical performance of the joint was evaluated with the Single Lap Shear (SLS) test performed at room temperature using a universal testing machine at constant crosshead speed of 1mm/min. Specimen geometry and dimensions were the ones depicted in Figure 1. A discontinuity is present in each one of the substrates, equidistant from the specimen centre and in opposite directions for each substrate, so that the area between both discontinuities is the area of the joint which is being tested. The maximum shear load, shear strength and absorbed energy were evaluated from the load-displacement curves.

The used SLS configuration introduces peeling (mode I) and non-linear geometric effects due to the slight curvature of the specimens, which is caused by the difference on the Coefficient of Thermal Expansion between aluminium and the carbon fibres. However, this mode-mixity was considered to be representative of industrial applications and no effort was devoted to avoiding it.

Two different types of specimens, with five specimens per each type, were manufactured to analyse the mechanical performance of the joint (Table 1). Co-cured specimens (CO) were used as reference specimens, and they were prepared with no mechanical interlock with the aim to characterize the joint strength given by the adhesion of the epoxy resin on the aluminium substrate when co-curing.

Mechanically interlocked (MI) specimens were prepared with the mechanical joining procedure developed in this work.

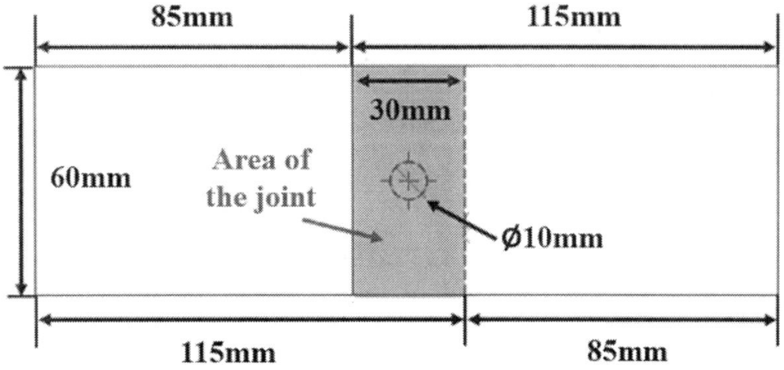

Figure 1. Dimensioned drawing of Single Lap Shear specimens.

Table 1. Types of specimens and process parameters.

	Punched materials	d_d (mm)	Clearance (%)	Stroke (mm)
Co-cured specimens (CO)	None	-	-	-
Mechanically interlocked (MI)	Al+CFRP	11.7	38	2.3

3. Developed joining procedure

The mechanical joining procedure consists of laying up uncured CFRP prepreg layers on top of an aluminium sheet and punching the whole system with the CFRP facing the punch side of the set-up. The aluminium surface was cleaned with acetone using a wipe cloth to remove any surface contamination before adding the prepreg. To avoid the prepreg from sticking to the punch, a silicon paper was placed between the uncured CFRP prepreg and the punch.

By adjusting the cutting clearance and the punch stroke, the aluminium sheet is completely punched through while the carbon fibres are not (Figure 2). Instead, the carbon fibres are pressed against the hole walls in the aluminium, generating a mechanical interlock between both materials. As joining takes place in an uncured state, delamination of the CFRP does not occur when punching.

The mechanical joints were achieved mounting a punching tooling on a universal testing machine. The specimens were cured afterwards by thermoforming. Co-curing of the composite epoxy resin on the aluminium sheet takes place and therefore, adhesive bonding between both substrates is present apart from mechanical interlocking.

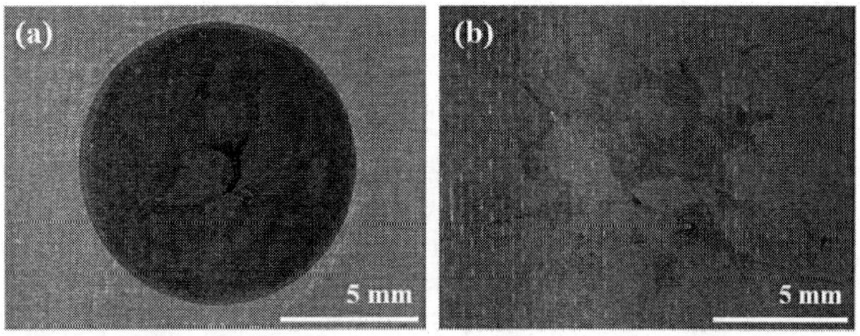

Figure 2. Mechanical interlocking joint (a) from the aluminium sheet side and (b) from the CFRP side.

The cutting clearance, c equation (1), is defined as the ratio between the punch (d_p) and the die (d_d) diameters gap and the thickness of the punched material, t [16].

$$c = \frac{d_d - d_P}{2t} \cdot 100 \qquad (1)$$

Typical clearance values in the sheet metal forming industry range from 5% to 20%, with 10% being the most common. However, higher cutting clearances have been used in this work to allow the carbon fibres to slide between the punch and the die without being completely cut. Another relevant punching parameter is the punch stroke (s), which is the penetration of the punch on the substrate being punched.

In the present work, the joints were performed at a constant punching speed of 10 mm/min. The punch diameter was 10 mm, as is commonly found in automotive body-in-white parts, and the fillet radius was 0 mm. Three die diameters were used for joint geometry characterization, 11.7 mm, 11.3 mm and 10.8 mm, which correspond to clearance values of 18%, 29% and 38%, respectively. Mechanical performance was evaluated for CO specimens and MI specimens performed with 11.7 mm die diameter (table 1).

During the punching process, load vs stroke curves were recorded (Figure 3). When punching a metallic sheet (Figure 3a), three different regimes were observed [17]. The first (A) corresponds to the deformation and strain hardening of the aluminium sheet. The second (B) corresponds to the cutting of the aluminium and the third (C) corresponds to the complete punching of the aluminium sheet.

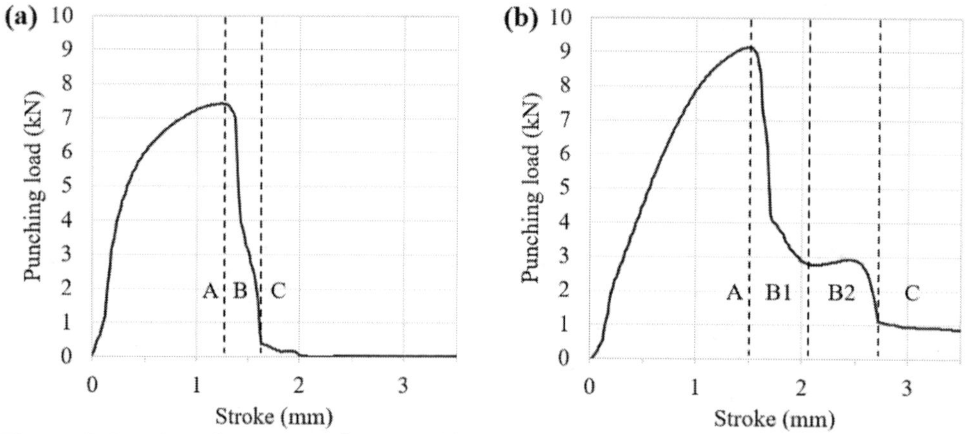

Figure 3. Load-stroke curves when punching with d_p =10 mm and d_d=11.7 mm (a) a 1.5-mm-thick aluminium sheet, and (b) a 1.5-mm-thick aluminium sheet with CFRP prepreg.

When punching a metallic sheet with a CFRP prepreg (Figure 3b), the same punching regimes are observed, but regime B is divided into two differentiated regions: B1 and B2. Region B1 corresponds to the cutting of the aluminium sheet. On the other hand, region B2 initiates with complete punching of the aluminium without fibre breakage and, as the punch stroke increases, carbon fibres progressively break until maximum fibre breakage is reached at end of regime B2.

Thus, the process parameters to completely punch the aluminium sheet while avoiding or minimizing the cutting of the carbon fibres were selected for this study. Such conditions fall into regime B2 of the punching load vs stroke curve and are the ones which allow the generation of the developed joint (Figure 3b).

4. Results

The cross-section of the joints was inspected with stereomicroscopy (Figure 4). The carbon fibres were pressed against the roll-over depth of the aluminium cut edge, while the area of the hole in contact with the fracture depth was resin rich. The carbon fibres filled the rest of the aluminium hole.

IOP Conf. Series: Materials Science and Engineering 1284 (2023) 012001 doi:10.1088/1757-899X/1284/1/012001

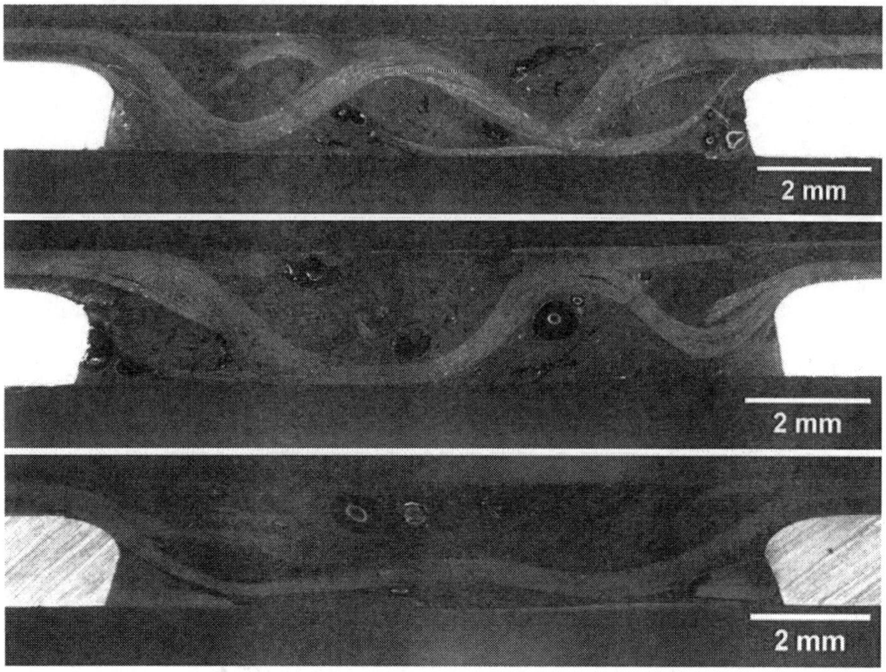

Figure 4. Cross-sections of the developed joint.

The cross-sections of punched aluminium sheets and MI specimens were also inspected (Figure 5). The morphology of the Al sheet cut edge was modified with the incorporation of the CFRP prepreg in the punching process. Punched aluminium sheets without CFRP present four different regions at the cut edge: roll-over depth, smooth-sheared depth, fracture depth and burr (Figure 5). On the other hand, when incorporating the carbon fibre prepreg in the punching step (MI specimens), the burr and the smooth-sheared depth disappear. Therefore, only the roll-over depth and fracture depth are present in the cut edge, regardless of the clearance.

Load-displacement curves obtained for the SLS tests of CO and MI specimens are shown in Figure 6. Average shear strength and absorbed energy are plotted in Figure 7a and Figure 7b, respectively.

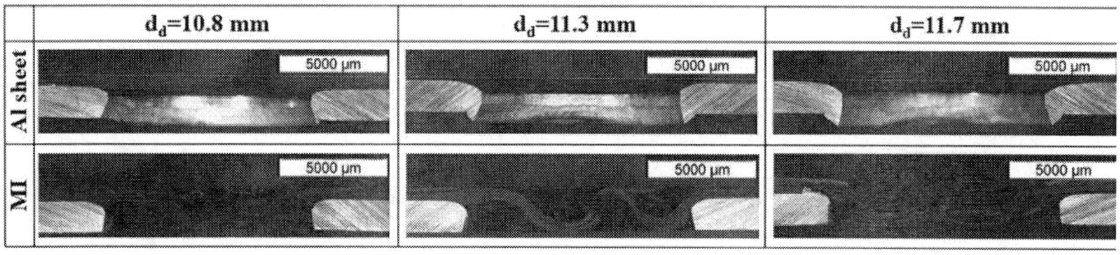

Figure 5. Cross-section of aluminium sheets and MI specimens generated with different die diameters.

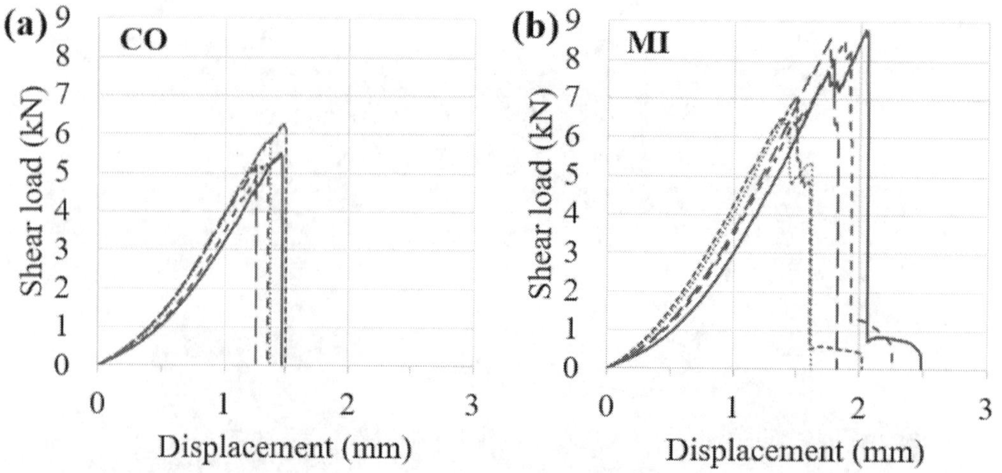

Figure 6. Load-displacement curves obtained in the SLS test for (a) co-cured samples (CO) and (b) MI specimens.

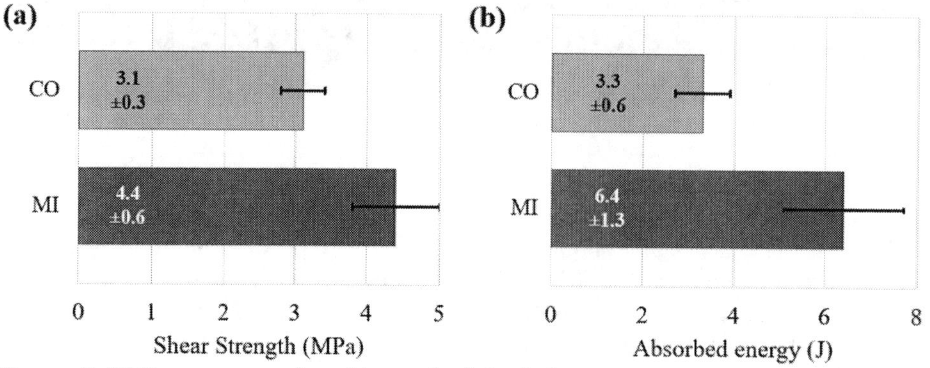

Figure 7. SLS average results with standard deviation regarding (a) shear strength and (b) absorbed energy.

CO specimens showed adhesive failure between the CFRP and the aluminium substrate (Figure 8a). On the other hand, MI specimens failed by complete pull-out or unbuttoning of the CFRP bulge from the aluminium hole (Figure 8b). Catastrophic failure occurred for CO specimens, whereas a more progressive failure was observed for the MI specimens, with several small drops before the load peak and a non-zero load queue after the main load drop (Figure 6b).

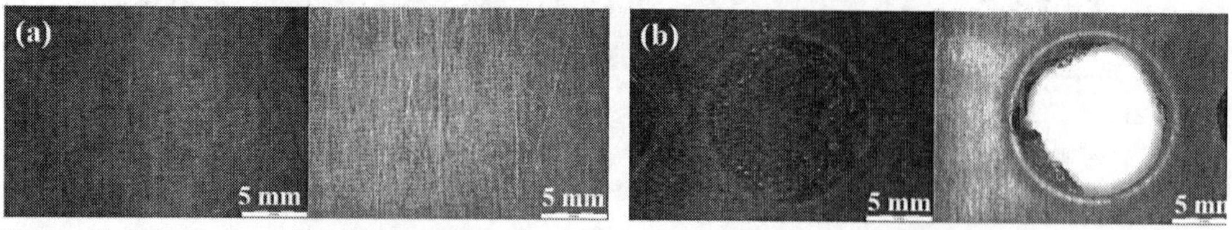

Figure 8. Joint region of SLS specimens after failure (a) CO specimens and (b) MI specimens.

5. Discussion

The present work describes a new type of joining methodology which is not intuitive, since the aluminium sheet, which is in contact with the die, can be punched without completely cutting the carbon fibres, which are in contact with the punch. This is possible because the punch penetrates the material slightly before the actual cutting, and compresses the aluminium surface near the punch inducing strain hardening on the material. Therefore, the resistance of the material to penetration rises and material fracture does not start on the edge close to the punch, but rather occurs on the soft area close to the die, which is not hardened [18].

As the punching is performed with the uncured prepreg, the polymeric resin is in a viscous flow state, and therefore delamination and CFRP tear around the joining area are reduced. When curing, the epoxy resin flows inside the aluminium hole and fills it up. Consequently, there is a resin-rich region near the fracture depth of the aluminium cut edge (Figure 4) and the fibres are less compacted inside the aluminium hole than on the rest of the specimen.

Regarding the process window, the punch strokes that allow the generation of the joint are those that fall within punching regime B2, where the aluminium is completely punched, but the CFRP is not.

When punching an aluminium sheet together with a CFRP prepreg (MI specimens), the cut edge geometry is changed with respect to when only punching aluminium sheets. The smooth-sheared depth and the burr vanish due to the presence of the carbon fibres (Figure 5), which fill the gap between the punch and the die during the punching process, reducing the effective clearance and leaving no free space for burr formation. In addition, burnishing does not take place and, therefore, the smooth-sheared depth cannot be formed.

With respect to the mechanical properties MI specimens increased the mechanical performance of the co-cured CO joints in terms of shear strength and absorbed energy. When punching the specimen to generate the mechanical interlock, the punch pushes the carbon fibres all the way down through the aluminium hole. Because of this, failure of the specimen is delayed with respect to CO specimens, thus higher shear strength and absorbed energy are obtained. In addition, MI specimens fail by CFRP bulge unbuttoning (Figure 8b), leading to a less catastrophic failure, which also contributes to the increase of the absorbed energy at failure. Another contribution to the higher mechanical properties of these specimens can be the strain hardening caused in the hole surroundings, which is due to the aluminium compression caused by the tension introduced into the carbon fibres when punching.

The mechanical behaviour of the developed joint was compared to other similar joining methodologies between composites and sheet metals found in the open literature (Figure 9) [9], [19]–[25]. Mechanical clinching and its variations give maximum shear loads ranging from 0.8 to 5.3 MPa when joining aluminium and CFRP [11]. The highest shear loads were found for hole clinching. However, this process requires to drill a hole in the composite material and then align this hole with the clinching device to generate the joint. Mechanical clinching of thermoplastic composites, where there is also adhesive bonding in addition to mechanical interlocking, show maximum shear loads ranging from 0.8 MPa to 3.9 MPa [11]. Therefore, considering the maximum joint force of 4.4 MPa, and an improvement of 1.3 MPa with respect to adhesive bonding resulting from the co-curing of the CFRP with the aluminium, the joint procedure presented in this work can be postulated as an interesting alternative to join dissimilar materials in engineering applications.

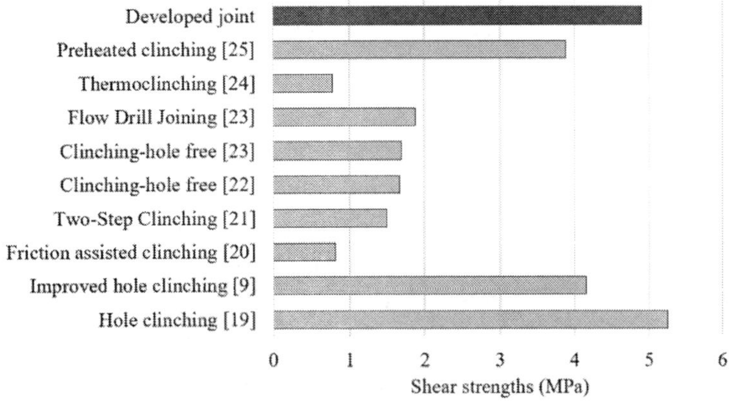

Figure 9. Shear strengths calculated from maximum shear loads and the joint area for different clinching technologies [9], [19]–[25].

6. Conclusions

A new mechanical joining methodology for a metal sheet and FRP based on a single-step punching process was developed in this work. Such technology is new, simple, cost-efficient and does not penalize in the weight of the part. In addition, it avoids damage in the composite material and does not add complexity to the joining process.

The resulting mechanical interlocking joint improved the shear strength of the co-cured joint in a 41% and the absorbed energy a 94% with respect the co-cured joint for the studied materials.

Moreover, the mechanical performance of the joint was found to be within the same range as other mechanical metal-composite joining methodologies such as clinching, indicating its potential to be used to join dissimilar materials in practical applications.

Acknowledgments
Núria Latorre is a fellow of Eurecat's "Vicente López" PhD grant program. This work was financially supported by the Catalan Government through the funding grant ACCIÓ-Eurecat (Project Flagship - Intelligent).

The authors would like to thank Aludium for providing the AA5754 H111 aluminium sheets. Gratitude is also extended to Daniel Berglund and Yvonne Aitomäki from the Research Institute of Sweden (RISE) for the fruitful discussions.

References

[1] A. Pramanik *et al.*, “Joining of carbon fibre reinforced polymer (CFRP) composites and aluminium alloys – A review,” *Compos. Part A Appl. Sci. Manuf.*, vol. 101, pp. 1–29, 2017, doi: 10.1016/j.compositesa.2017.06.007.

[2] A. Galińska, “Mechanical joining of fibre reinforced polymer composites to metals—a review. Part I: Bolted joining,” *Polymers (Basel).*, vol. 12, no. 10, pp. 1–48, 2020, doi: 10.3390/polym12102252.

[3] S. D. Thoppul, J. Finegan, and R. F. Gibson, “Mechanics of mechanically fastened joints in polymer-matrix composite structures - A review,” *Compos. Sci. Technol.*, vol. 69, no. 3–4, pp. 301–329, 2009, doi: 10.1016/j.compscitech.2008.09.037.

[4] J. Wang *et al.*, “A self-piercing riveting method for joining of continuous carbon fiber reinforced composite and aluminum alloy sheets,” *Compos. Struct.*, vol. 259, no. September 2020, p. 113219, 2021, doi: 10.1016/j.compstruct.2020.113219.

[5] F. Lambiase, “Mechanical behaviour of polymer-metal hybrid joints produced by clinching using different tools,” *Mater. Des.*, vol. 87, pp. 606–618, 2015, doi: 10.1016/j.matdes.2015.08.037.

[6] M. A. Caminero, M. Lopez-Pedrosa, C. Pinna, and C. Soutis, “Damage monitoring and analysis of composite laminates with an open hole and adhesively bonded repairs using digital image correlation,” *Compos. Part B Eng.*, vol. 53, pp. 76–91, 2013, doi: 10.1016/j.compositesb.2013.04.050.

[7] P. P. Camanho, C. M. L. Tavares, R. De Oliveira, A. T. Marques, and A. J. M. Ferreira, “Increasing the efficiency of composite single-shear lap joints using bonded inserts,” *Compos. Part B Eng.*, vol. 36, no. 5, pp. 372–383, 2005, doi: 10.1016/j.compositesb.2005.01.007.

[8] R. Li, D. Kelly, and A. Crosky, “Strength improvement by fibre steering around a pin loaded hole,” *Compos. Struct.*, vol. 57, no. 1–4, pp. 377–383, 2002, doi: 10.1016/S0263-8223(02)00105-8.

[9] C. J. Lee, B. M. Kim, B. S. Kang, W. J. Song, and D. C. Ko, “Improvement of joinability in a hole clinching process with aluminum alloy and carbon fiber reinforced plastic using a spring die,” *Compos. Struct.*, vol. 173, pp. 58–69, 2017, doi: 10.1016/j.compstruct.2017.04.010.

[10] F. Lambiase, A. Paoletti, and A. Di Ilio, “Advances in Mechanical Clinching: Employment of a Rotating Tool,” *Procedia Eng.*, vol. 183, pp. 200–205, 2017, doi: 10.1016/j.proeng.2017.04.021.

[11] B. Gröger *et al.*, “Clinching of thermoplastic composites and metals—a comparison of three novel joining technologies,” *Materials (Basel).*, vol. 14, no. 9, pp. 1–18, 2021, doi:

10.3390/ma14092286.

[12] P. N. Parkes, R. Butler, J. Meyer, and A. de Oliveira, "Static strength of metal-composite joints with penetrative reinforcement," *Compos. Struct.*, vol. 118, no. 1, pp. 250–256, 2014, doi: 10.1016/j.compstruct.2014.07.019.

[13] F. Möller, C. Thomy, F. Vollertsen, P. Schiebel, C. Hoffmeister, and A. S. Herrmann, "Novel method for joining CFRP to aluminium," *Phys. Procedia*, vol. 5, no. PART 2, pp. 37–45, 2010, doi: 10.1016/j.phpro.2010.08.027.

[14] A. Galińska and C. Galiński, "Mechanical joining of fibre reinforced polymer composites to metals-A review. Part II: Riveting, clinching, non-adhesive form-locked joints, pin and loop joining," *Polymers (Basel).*, vol. 12, no. 8, 2020, doi: 10.3390/POLYM12081681.

[15] F. Gamdani, R. Boukhili, and A. Vadean, "Tensile strength of open-hole, pin-loaded and multi-bolted single-lap joints in woven composite plates," *Mater. Des.*, vol. 88, pp. 702–712, 2015, doi: 10.1016/j.matdes.2015.09.008.

[16] D. Frómeta, A. Lara, S. Parareda, and D. Casellas, "Evaluation of edge formability in high strength sheets through a fracture mechanics approach," *AIP Conf. Proc.*, vol. 2113, no. July, 2019, doi: 10.1063/1.5112704.

[17] K. Nothhaft, J. Suh, M. Golle, I. Picas, D. Casellas, and W. Volk, "Shear cutting of press hardened steel: Influence of punch chamfer on process forces, tool stresses and sheared edge qualities," *Prod. Eng.*, vol. 6, no. 4–5, pp. 413–420, 2012, doi: 10.1007/s11740-012-0404-8.

[18] Z. Tekiner, M. Nalbant, and H. Gürün, "An experimental study for the effect of different clearances on burr, smooth-sheared and blanking force on aluminium sheet metal," *Mater. Des.*, vol. 27, no. 10, pp. 1134–1138, 2006, doi: 10.1016/j.matdes.2005.03.013.

[19] C. J. Lee, S. H. Lee, J. M. Lee, B. H. Kim, B. M. Kim, and D. C. Ko, "Design of hole-clinching process for joining CFRP and aluminum alloy sheet," *Int. J. Precis. Eng. Manuf.*, vol. 15, no. 6, pp. 1151–1157, 2014, doi: 10.1007/s12541-014-0450-6.

[20] F. Lambiase and A. Paoletti, "Friction-assisted clinching of Aluminum and CFRP sheets," *J. Manuf. Process.*, vol. 31, pp. 812–822, 2018, doi: 10.1016/j.jmapro.2018.01.014.

[21] F. Lambiase and D. C. Ko, "Two-steps clinching of aluminum and Carbon Fiber Reinforced Polymer sheets," *Compos. Struct.*, vol. 164, pp. 180–188, 2017, doi: 10.1016/j.compstruct.2016.12.072.

[22] F. Lambiase and D. C. Ko, "Feasibility of mechanical clinching for joining aluminum AA6082-T6 and Carbon Fiber Reinforced Polymer sheets," *Mater. Des.*, vol. 107, pp. 341–352, 2016, doi: 10.1016/j.matdes.2016.06.061.

[23] H. Seidlitz, L. Ulke-Winter, and L. Kroll, "New Joining Technology for Optimized Metal/Composite Assemblies," *J. Eng. (United Kingdom)*, vol. 2014, no. Mmd, 2014, doi: 10.1155/2014/958501.

[24] J. Vorderbrüggen, B. Gröger, R. Kupfer, A. Hoog, M. Gude, and G. Meschut, "Phenomena of forming and failure in joining hybrid structures - Experimental and numerical studies of clinching thermoplastic composites and metal," *AIP Conf. Proc.*, vol. 2113, pp. 1–7, 2019, doi: 10.1063/1.5112580.

[25] P. C. Lin, J. W. Lin, and G. X. Li, "Clinching process for aluminum alloy and carbon fiber-reinforced thermoplastic sheets," *Int. J. Adv. Manuf. Technol.*, vol. 97, no. 1–4, pp. 529–541, 2018, doi: 10.1007/s00170-018-1960-7.

IOP Conf. Series: Materials Science and Engineering 1284 (2023) 012002 doi:10.1088/1757-899X/1284/1/012002

Process chain for forming and consolidating fiber-reinforced thermoplastics and metallic sheets in a two-stage isothermal tool

B-A Behrens[1], A Raatz[2], S Hübner[1], E Farahmand[1], L Albracht[1], T Fünfkirchler[1], S Yarcu, C Schumann[2] and J Wehmeyer[1]

[1] Leibniz Universität Hannover, Institute of Forming Technology and Forming Machines, An der Universität 2, 30823, Garbsen, Germany.
[2] Leibniz Universität Hannover, Institute of Assembly Technology, An der Universität 2, 30823, Garbsen, Germany.

E-mail: wehmeyer@ifum.uni-hannover.de

Abstract. Due to their good mechanical properties and low structural weight, multi-material structures are a promising approach in the automotive industry to lightweight design, body construction and functionalization. Especially metal and plastic are mainly combined to achieve improved properties of the final component compared to mono-material structures. This paper describes the development of a manufacturing cell for the joint forming and heat-assisted press joining of steels and continuous fiber-reinforced thermoplastics in the form of unidirectional carbon-fiber tapes. In order to achieve shorter cycle times and to ensure economical production, a manufacturing cell, supplemented with automated handling by means of two robots and an isothermal, two forming stages tool concept was developed and tested.The composite components produced were tested with regard to their mechanical performance. The feasibility of the production was demonstrated. All composite components had a higher specific load capacity than a pure steel component. Cycle times of well under 60 seconds were achieved. An enormous reduction in process time compared to variothermal tool concepts could be achieved with the new manufacturing cell.

1. Introduction

The efficient use of resources is becoming increasingly important, particularly in view of the changes in our environment, but also from a competitive perspective. Various measures can be used to increase resource efficiency. These include, for example, more efficient product design, less raw material-intensive manufacturing, or the possibility of separating and recycling the material.

By combining metals and fiber-reinforced plastics to form hybrid components, the advantages of the two groups of materials are combined, synergy effects are exploited and lightweight construction potential is optimally exploited. The problem with implementing such material concepts is often the lack of cost-effectiveness of the associated manufacturing processes. The manufacturing of fiber-reinforced plastic-metal hybrid components requires a handling and forming process adapted to different material behaviours and a large-area connection of the two components. Two basic approaches exist in research and industry to produce such joints:

On the one hand, the two joining partners can be joined with the aid of a suitable adhesive. The use of such a joining additive ensures increased bond strengths, but it is also very cost-intensive and causes challenges in manufacturing. The use of adhesives requires additional process steps, increased occupational safety measures and an increase in process time due to the curing of the adhesive.

A one-step way of producing a large-area joint between metals and fiber-reinforced plastics is press joining with so-called variothermic, i.e. alternately warm, tools. In this process, the materials to be joined are heated in the tool. The thermoplastic is melted and pressed together with the metallic component in the heated tool. The tool is then cooled in a time-consuming process and the thermoplastic solidifies. Industrial use of these tools, which are first heated and then cooled, is neither economically and technically feasible nor energetically reasonable in series production due to the long process times and the high thermal alternating loads [1].

The majority of the process variants developed to date have been primarily geared to basic feasibility or the fulfilment of specific requirements, which have been confirmed or achieved in initial results. The next step is to focus on cost reduction in the sense of process effort and process time reduction. The challenge is to develop processes and process routes that ensure complete functionality and at the same time an economic incentive for series production.

For this purpose, a process chain based on a two-stage, isothermal tool concept with automated handling technology was developed in the scope of this research and implemented for the production of load-adapted components. The tool consists of a heating and impregnation unit and a forming and consolidation unit. Both stages are permanently heated at different temperatures. Robots will be used to stack the semi-finished products, insert them into the first die stage, transfer them between the die stages and remove the finished work pieces.

First, the sample component which is based on a structural component from the automotive industry was defined. The load-bearing capacity of the component is to be realized by reinforced areas. This is to be done by applying thin stiffeners made of fiber-reinforced plastic tapes. Based on the component geometry, the design of the impregnation and the forming tool was carried out. The multi-material design of the components and the associated different properties of the semi-finished products pose further challenges for the automated handling systems. These must be capable of reliably gripping and stacking both stiff sheet metal and dimensionally unstable plastic film in the unprocessed state and fixing it for further transfer to the mold. After commissioning, the optimal process window with regard to impregnation of the fibers and consolidation of the components was determined. The process parameters forming force, holding time and tool temperature are varied and their effects on component stability and quality were analysed. A comparison of a pure steel component with load-compliant hybrid components will highlight the lightweight construction potential of the final components.

2. State of the Art

2.1. Fiber-reinforced thermoplastics

Fiber-reinforced thermoplastics (FRP) present a promising solution to overcome the challenges mentioned in [1]. In order to achieve future lightweight construction objectives, the use of FRP has been progressively increasing in recent years and has become a crucial aspect. Semi-finished FRP products can be classified into various types, such as long fiber reinforced thermoplastics (LFT), glass mat reinforced thermoplastics (GMT), fiber reinforced unidirectional tapes (UD tapes), and organic sheets [2].

When exposed to high temperatures, thermoplastics can be shaped using stamp forming processes, which can facilitate large-scale production, depending on the matrix material [4, 5]. Apart from consolidating the matrix materials, other joining techniques like clinching can also be used at elevated temperatures [6].

Although FRPs are considered a class of materials in their own right, they basically consist of a composite of several materials. In this composite structure, the individual components have a clear division of tasks: the high-strength fibers bear the mechanical compressive forces and loads, while the matrix in which the fibers are embedded fixes them in position, supports them and transmits the mechanical stresses to the fibers [7].

2.2. Heat-assisted press joining of hybrid FRP steel components

Heat-assisted press joining enables the joining of thermoplastic FRPs and metals and offers the potential of low process times and shortened process chains. In this process, the thermoplastic matrix material of the fiber composite is melted so that a kind of fusion bonding with the metallic joining partner is realized [8]. With the help of heat-assisted press joining, the process step of bonding is omitted and structural strengths can be achieved with appropriate process control and surface pretreatment [9]. Within the framework of the AACHEN CLUSTER OF EXCELLENCE "Integrative Production Technology for High-Wage Countries", a plant technology for heat-assisted press joining of metal-plastic hybrid joints was developed. Here, the thermoplastic is melted by heat conduction due to the contact between the metallic workpiece and the tool in the joining zone [10]. The contact temperature when the metal surface is wetted by the molten plastic is of crucial importance for the formation of the bond. For a combination of polyamide 66 and steel, for example, an optimum joining temperature in the range of 260 °C – 270 °C was determined [11], whereas a bond between polyamide 6 and steel at a joining temperature of 250 °C is considered promising [12]. Other subprojects dealt with inductive heating of the tool [13] and conductive heating of the metallic joining component with the aim of reducing process time [14]. The joining process of heat-assisted press joining is divided into several substeps. First, both materials must be heated to the required joining temperature. The temperatures are set by means of thermocouples. The thermoplastic is heated, for example, by an infrared field [15]. The heating takes place superficially in the area of the joining zone, so that consolidation of the organic sheet is not necessary. The temperature of the organic sheet is controlled with a pyrometer. When both materials have reached the required joining temperatures, the metal sheet is pressed on the organic sheet with a defined force. In the process, the molten thermoplastic flows into the undercut geometries of the metal surface and a form-fit joint is formed. Furthermore, material bonds are formed between the thermoplastic and the metal surface [16]. Afterwards, the cooling process begins, for example with the aid of compressed air. Once the thermoplastic of the organic sheet has solidified, the test specimen can be removed. The necessary process parameters for heat-assisted press joining are the joining temperatures of both joining partners and the joining force. Extensive studies have already been carried out on this [15, 17].

In the IGF (Industrial Collaborative Research) project "19560N Forming thermojoining", a combined process for joint forming and heat-assisted press joining of high-strength steels and continuous fiber-reinforced thermoplastics was developed. In this project, galvannealed steels and glass fiber-reinforced polyamide 6 thermoplastics were joined directly. To investigate the joining process, tensile shear tests were carried out in which the steel was heated to 230 °C respectively 280 °C and the organic sheet to 280 °C and then joined. Tensile shear strengths of 10 MPa to 20 MPa were achieved. Furthermore, continuous fiber-reinforced plastics were used in this work, focusing on the impregnation of the FRP prepregs, the creation of local and thus load-adapted reinforcement structures, and the reduction of process time. To achieve higher composite strengths during thermal joining, various pretreatment processes of the joining partners were necessary, since the strengths are below 5 MPa without pretreatment [18]. So far, a wide variety of pretreatment processes have been considered, such as mechanical pretreatment by sandblasting [10, 19], imprinting of structures [20, 21], corundum blasting [22 , 23] or different chemical pretreatments [25], which create an activated and partly new disordered surface structure. Significant research has also been carried out in the area of undercut-optimized, targeted structuring of the surface. Fig. 1 shows examples of these approaches namely CMT pin welding [13], HyTensile [25, 26], laser structuring [20, 27] and electron beam structuring [13]. The joint strengths achieved depend significantly on the surface pretreatment and the load direction. In tensile shear tests, for example, strengths of up to 35 MPa could be achieved by laser pretreatment [27]. The strengths achieved underline the potential of press joining technique, but at the same time show the need for a material-, component- and application-specific selection of the most suitable pretreatment.

IOP Conf. Series: Materials Science and Engineering 1284 (2023) 012002 doi:10.1088/1757-899X/1284/1/012002

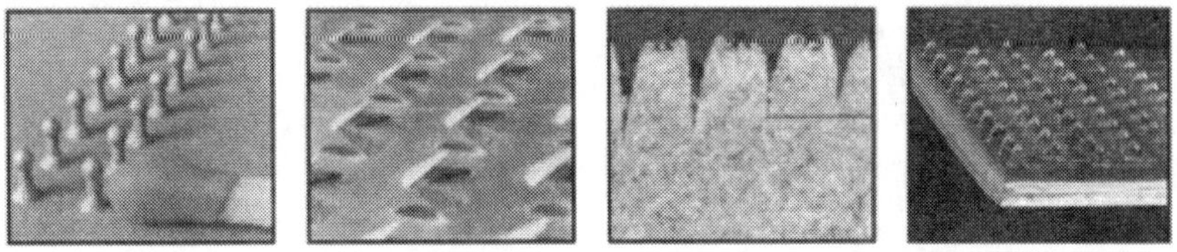

Figure 1. Surface structuring processes by CMT pin welding [13], HyTensile [25, 26], laser structuring [20, 27] and electron beam structuring [13].

In "High-volume process chains for highly integrated components made of hybrid fiber-plastic/metal composites" [28], a variothermal forming tool for the production of hat sections consisting of carbon fiber reinforced thermoplastic tapes and metal sheets with different frame angles was manufactured. After the tool was put into operation, hybrid hat profiles were produced and component properties were determined. The forming tool has a maximum heating power of 12.4 kW and is heated to 250 °C. The tool can be cooled down to a removal temperature of 150 °C within 45 seconds due to cooling channels, filled with permanently circulating water, running close to the contour.

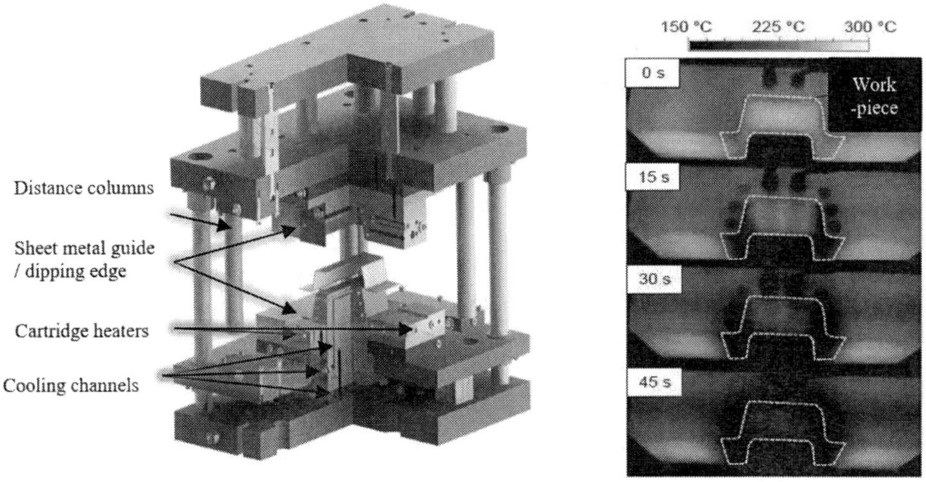

Figure 2. Sectional view of the CAD model of the variothermal forming tool (left), temperature profile of the punch during cooling (right) [28].

The CAD model of the forming tool is shown in a sectional view in Figure 2. However, several minutes are required for the reheating of the tool. This fact was the main motivation for the work presented in this paper to develop a process chain with which economically attractive production times for FRP-metal hybrids can be realized. Another goal is to save energy by avoiding the need of reheating the mold after a part has been manufactured.

3. Experimental setup

3.1. Work piece

The component developed here is a hybrid sandwich component with specially reinforced areas. It consists of metallic cover sheets on which PA6 (Polyamid 6) films are located as adhesion promoters. No other adhesives are needed to create a permanent bond. The core of the component consists of a full-surface layer of UD-CFRP tape (unidirectional carbonfiber reinforced plastic tape). The areas of the reinforcing structures are reinforced with further narrow strips of UD-CFRP tape. Figure 3 shows the schematic layer structure of the semi-finished product and a manufactured component.

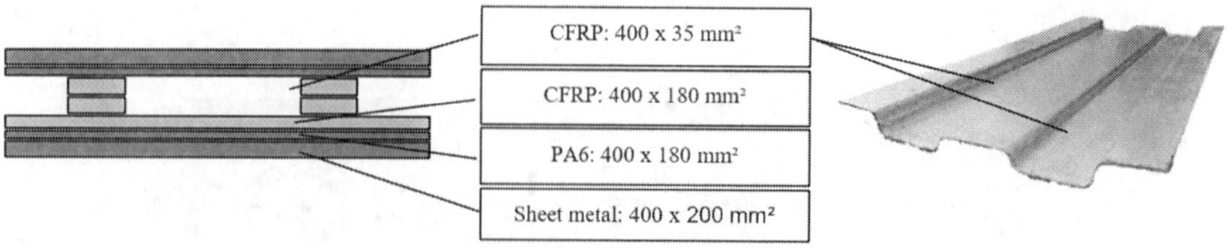

Figure 3. Layer structure of the semi-finished products (left), manufactured component (right) [29].

3.2. *Two-stage isothermal heating / impregnating and consolidation / forming* tool

In order to produce sandwich components designed to withstand bending loads and to save energy costs as well as to significantly reduce production time, a two-stage isothermal heating / impregnation tool (cf. Figure 4, left (1)) and consolidation / forming tool (cf. Figure 4, left (2)) was developed, which ensures accessibility for automated handling technology. In contrast to the variothermal tools described in the state of the art, in this concept both stages are heated permanently, but at different temperatures, by heating coils. This eliminates the energy-consuming heating process for each new component. Cooling channels are also no longer required.

In the first tooling stage, the stacked semi-finished products are heated and the still loose carbon fibres are impregnated into the PA6 plastic with a force of 325 kN and thermally joined. The temperature of this tooling stage is 270 °C (10 °C below the decomposition temperature of the PA6). This temperature is selected so that the PA6 melts and a maximum heat cushion is created for the transfer to the second tooling stage, while the plastic does not solidify before the actual forming. Contours in the die of the mold ensure that the additional layers of CFRP-UD tape for the reinforcement contour do not slip.

In the second tooling stage, the forming takes place at a temperature of 120 °C due to the deeper contours of the die. The temperature was determined so that the plastic does not solidify before the forming is completed. Cooling then takes place during holding time of average 25 s in the mold. After this time, the consolidation of the composite is completed and the still warm but completely solidified component can be removed. The CAD model of the tools and the sectional view of the impregnation / heating stage are shown in Figure 4.

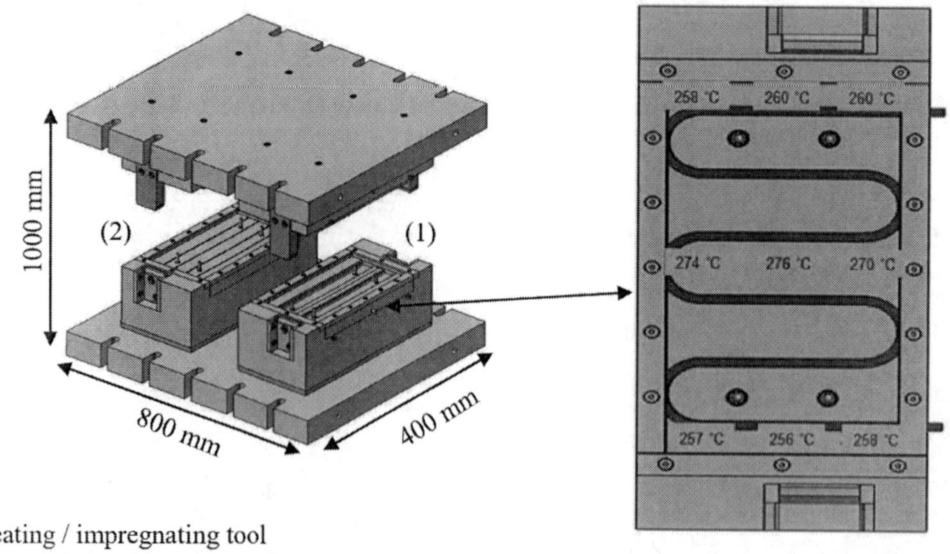

1: Heating / impregnating tool
2: Consolidation / forming tool

Figure 4. CAD model of the tools (left), sectional view of the impregnation / heating stage with measured temperatures (right) [29].

IOP Conf. Series: Materials Science and Engineering 1284 (2023) 012002 doi:10.1088/1757-899X/1284/1/012002

3.3. Manufacturing cell with automated handling technology

To further reduce the production time and to increase the repeatability of the tests, an automated stacking of the semi-finished products as well as an automated transfer into the tools using two robots was developed.

A Kuka KR6 (see figure 5) was used for stacking the semi-finished products. The challenge here was that the robot's gripper had to be able to reliably grip and place both rigid metal sheets and CFRP UD tapes (in various sizes), as well as flabby PA6 film in the unprocessed state. For this reason, a vacuum suction cup in the form of two arrays (each with four suction cups) was chosen. In addition, the gripper has two small heated punches that can be used to thermally bond the semi-finished product layers at specific points to prevent subsequent slippage during transfer to the mold. After completed stacking process, the fixed stacks of semi-finished products are deposited at a defined location for further handling.

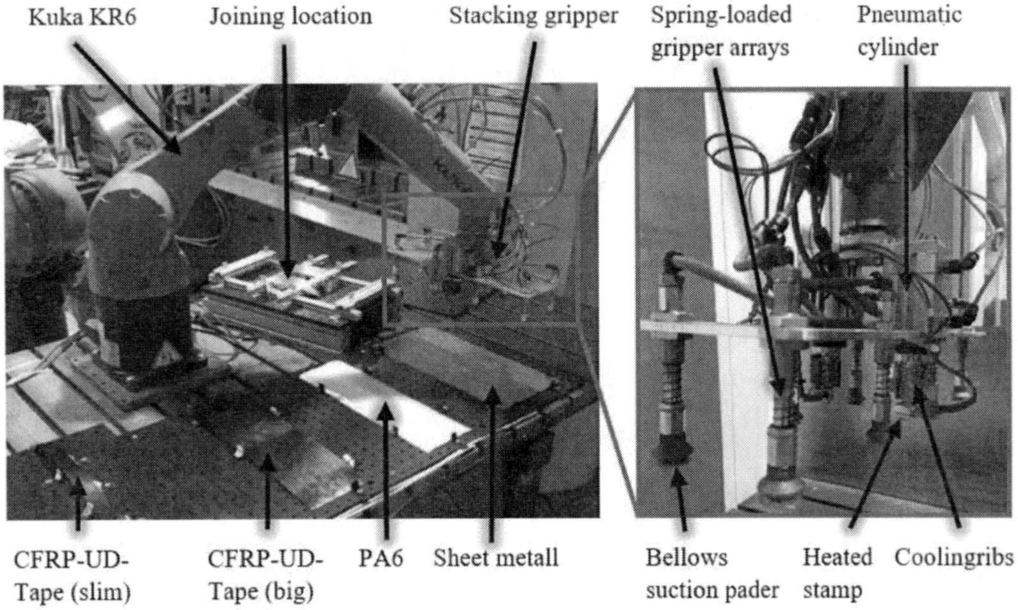

Figure 5. Stacking cell consisting of table, robot and semi-finished product storage (left), stacking gripper on robot with component identification (right) [29].

A Kuka KR60 robot was used to transfer the stacked semi-finished products to the heating stage, from the heating stage to the forming stage and from the forming stage to the deposit location. The KR60 grips the stacks of semi-finished products by means of a magnetic gripper. The axial speed of the robot was set to approximately 50 % of the possible path speed (by safety reasons) that the robot can achieve. The sequences of the stacking process and the transfer process were synchronized with each other in order to generate the best possible production times. With a heating or forming time of 20 s (+12 s opening and closing) and a transfer time of 38 s, this results in a production time per component of 70 s. As both stages of the tool are always loaded in each stroke, this value can still be divided by 2 resulting in a theoretical cycle time of 35 s. Figure 6 shows the complete manufacturing cell.

Figure 6. Production cell consisting of Kuka KR60, Kuka KR6, two-stage isothermal tool and Dunkes 250 forming press [29].

4. Conducting experiments

The following test matrix (Table 1.) shows the production parameters (forming force, holding time in the tool, temperature in the forming stage) used in this research. For the metallic cover sheets, DC03 (tensile strength Rm = 270-370 MPa) with a measured average surface roughness Ra of 0.1401 µm and ZM70 (Rm = 340-395 MPa, Ra = 0.5711 µm) were used. Furthermore, monolithic reference components were manufactured from 1 mm thick DC03 to reference the effects of the reinforcement structures designed to withstand the load. Furthermore, a special adhesion promoter film was tested. Destructive component tests in the form of three-point bending tests are carried out in order to demonstrate the lightweight construction potential of the hybrid components compared to the monolithic components.

Table 1. Test matrix for the production of the components.

Test series	Steel	Forming force [kN]	Holding time [s]	Temp. forming stage [°C]
A	DC03	250 - 400	20 - 35	100 - 140
B	ZM70	250 - 400	20	100 - 140
C*	ZM70	250 - 400	20	100 - 140
D*	DC03	250 - 400	20	100 - 140

*: Additional layer of adhesion promoter film

5. Results

The testing of the double-hat profiles, manufactured using the test setup described in chapters 3 and 4, was carried out by means of three-point bending tests following DIN 14130. The test set-up is shown in Figure 7 (left). The test components were inserted in such a way that the areas with the UD tape reinforcements were maximally far off the compression fin of the test unit and were thus loaded in tension. The specific load capacity of the composites can be calculated from the weight of the composites and the maximum average forces per test series from the three-point bending tests.

Depending on the combination, different weights are obtained. The specific load value can be determined from the maximum force values using the formula below:

$$Max. spec. load \left[\frac{N}{g}\right] = \frac{Force}{Weight\ of\ the\ unit}$$

The max. specific load capacities for the different test series are shown in Figure 7 (right).

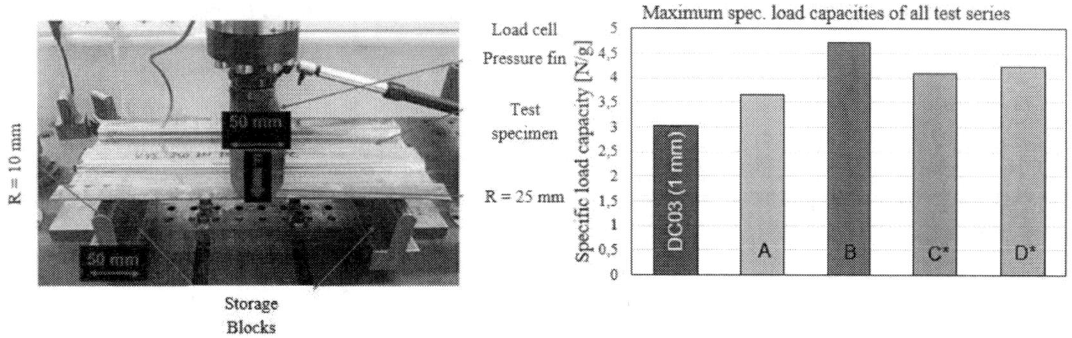

Figure 7. Test set-up three-point bending tests (left), test results (right) [29].

It can be seen (figure 7, right) that all test series show a higher specific load capacity compared to the monolithic component. In test series A, the influence of the holding time in the mold, which was varied from 20 to 35 s, was determined to be negligible, as the PA6 melted completely after 20 s. In addition, a lower temperature of 100 °C in the forming stage proved to be advantageous in all test series. This is related to the fact that the actual forming is completed after approx. 2 s and that the material composite can then solidify better.

Table 2. Parameter combination for generating the maximum specific load capacity per test series.

Test series	Steel	Forming force [kN]	Holding time [s]	Temp. forming stage [°C]	Max. spec. load [N/g]
A	DC03	400	20	100	3.66
B	ZM70	325	20	100	4.72
C*	ZM70	250	20	100	4.08
D*	DC03	250	20	100	4.23

*: Additional layer of adhesion promoter film

The best bond was achieved with the ZM70 at a forming force of 325 kN as the rough surface of the steel provides better conditions for thermal press joining than that of the rather smooth DC03. The specific load capacity was in this case 56.8 % higher than the of the 1 mm thick DC03 sheet. In the test series C* and D*, an additional adhesion promoter film was used. This still had a positive influence on the DC03 components, but proved to have a negative effect on the ZM components. The reason for this is that the tool was not designed for the additional layers on the top and bottom sheet metal. The increased amount of material led to the PA6 being squeezed out and ultimately to delamination of the composite.

6. Summary, conclusion and outlook

In the scope of this paper, a novel, two-stage, isothermal tool concept with automated handling technology consisting of two robots was presented.

This tooling system can be used to produce stacks of semi-finished products consisting of steel, PA6 film and unidirectionally reinforced carbon fibre tapes. In the final process, both tool stages are

permanently occupied and new stacks of semi-finished products are continuously stacked by the first robot. With a heating or forming time of 20 s (+12 s opening and closing) and a transfer time of 38 s, this results in a production time per component of 70 s. As both stages of the tool are always loaded in each stroke, this value can still be divided by 2 resulting in a theoretical cycle time of 35 s. Compared to a variothermic tool, the process time is reduced by a factor of 8.5 for this component, which shows the immense time-saving potential of the two stage isothermal tool concept.
Subsequently, a detailed examination of all fabricated components was carried out by means of 3-point bending tests. Four series of tests were examined in which different metals and an additional bonding agent were investigated. All test series showed higher specific load capacities up to 56.8 % in the later comparison with a single 1 mm thick DC03 sheet.

Within the scope of this project, a process was developed that enables the production of components from loosely stacked semi-finished products made of sheet metal and UD tapes in a two-part, isothermal heating and impregnation stage and a consolidation and forming stage. Due to the individual assembly of the component from individual layers and the comparatively short cycle time, this production technology offers the potential to economically manufacture components that can withstand loads in series production. Further validation tests with lower temperatures in the forming stage will be the subject of future investigations. In addition, more component testing should be carried out in terms of the alpha delta problem and fatigue strength tests.

Acknowledgements
The authors thank the Industrial Collaborative Research (IGF) and the European Research Association for Sheet Metal Working (EFB) for the financial support of the research project "Economical production of load-compliant FRP/metal composites" (IGF-Code: 19603N).

References
[1] Behrens B-A, Neumann A 2016 Ermittlung der Prozessgrenzen bei der Umformung von Musterbauteilen aus Faser-Kunststoff/MetallVerbunden; Schlussbericht zu dem IGF-Vorhaben DFG-AiF-Cluster Großserientaugliche Prozessketten für hochintegrierte Bauteile aus hybriden FaserKunststoff/Metall-Verbunden Teilprojekt 4
[2] Friedrich K, Almajid A A 2013 Manufacturing of Advanced Polymer Composites for Automotive Applications. In: Applied Composite Materials 20, pp. 107-128
[3] Gebai S, Hallal A, Hammoud M 2018 Composite Materials Types and Applications: A Review on Composite Materials. In: Mechanical Properties of Natural Fiber Reinforced Polymers. Emerging Research and Opportunities, IGI Global, pp. 1-29
[4] Behrens B-A, Hübner S, Bonk C, Bohne F, Micke-Camuz M 2017 Development of a combined Process of Organic Sheet forming and GMT Compression Molding. In: Procedia Engineering 207, Cambridge, pp. 101-106
[5] Behrens B-A, Hübner S, Neumann A 2014 Forming sheets of metal and fibre-reinforced plastics to hybrid parts in one deep drawing process. In: Procedia Engineering 81, pp. 1608-1613
[6] Behrens B-A, Hübner S, et al. 2017 Forming and Joining of Carbon-Fiber-Reinforced Thermoplastics and Sheet Metal in One Step. In: Procedia Engineering 183, pp. 227-232
[7] Schürmann H 2007 Konstruieren mit Faser-Kunststoff-Verbunden. Springer-Verlag, Berlin, ISBN 9783540721895
[8] Ageorges C, Ye L, Hou M 2001 Advances in fusion bonding techniques for joining thermoplastic matrix composites: a review, Composites: Part A, Bd. 32, pp. 839-857
[9] Heckert A, Zaeh M F 2015 Induction Heated Joining of Aluminum and Carbon Fiber Reinforced Nylon 66, Journal of Laser Applications, Bd. 27, Nr. S2
[10] Brecher C 2011 Integrative Produktionstechnik für Hochlohnländer, Berlin Heidelberg: Springer-Verlag
[11] Flock D 2011 Wärmeleitungsfügen hybrider Kunststoff-Metall-Verbindungen, Norderstedt
[12] Reig F, Steinhilper R 2018 Handbuch Konstruktion, 2. Auflage, München. Hanserverlag

[13] Scheik S, Schleser M, Reisgen U 2012 Thermisches Direktfügen von Metall und Kunststoff - Eine Alternative zur Klebtechnik?, Adhäsion
[14] Schoene J, Marx B, Schönberger J 2014 Fügen von Metall-Kunststoff-Verbunden, ISF Direkt 49
[15] Lippky K, Hartwig S, Blass D, Dilger K 2019 Bonding performance after aging of fusion bonded hybrid joints. International Journal of Adhesion and Adhesives
[16] Üzüm M 2015 Metal polymer hybrids - Multiscale adhesion behavior and polymer dynamics. Technische Universität Berlin, Dissertation
[17] Lippky K, Mund M, Blass D, Dilger K 2018 Investigation of hybrid fusion bonds under varying manufacturing and operating procedures. Composite Structures
[18] Lippky K, Gerling L, Fischer F, Dilger K 2014 „online proceedings", 37. Annual Meeting of the Adhesion Society, Bahia Resort, San Diego, CA. February 23rd until February 26th
[19] Bergmann J P, Stambke M 2012 Potential of laser-manufactured polymer-metal hybrid joints, Physics Procedia 39, pp. 84-91
[20] Ageorges C, Ye L, Hou M 2001 Advances in fusion bonding techniques for joining thermoplastic matrix composites: a review, Composites: Part A, Bd. 32, pp. 839-857
[21] Matsuzaki R, Tsukamoto N, Taniguchi J 2016 Mechanical Interlocking by im-printing micropatterns for improving adhesive strength of polypropylene. International Journal of Adhesion & Adhesives 68, pp. 124-132
[22] Mittal K, Pizzi A 1999 Adhesion Promotion Technique. Technological Applications. Basel: Marcel Dekker. ISBN: 0-8247-0239-5, pp. 2, 4-5, 19-20
[23] Velthuis R 2007 LAUDATIO - Fügetechnik - Verfahrensvergleich: Leichtbau aus Metall und Faser-Kunststoff-Verbunden, Kunststoffe : Werkstoffe, Verarbeitung, Anwendung ; Organ deutscher Kunststoff-Fachverbände, Kunststoffe / Deutsche Ausgabe, in: Kunststoffe : Werkstoffe, Verarbeitung, Anwendung, pp. 52–55
[24] Didi M, Mitschang P 2011 Diskontinuierliches Induktionsschweißen von CF/PEEK und CF/PA66 mit Aluminium, CCeV
[25] Bräunling S, Staiger E 2015 Textil-Blech-Verbund Hybride auf Basis von Kohlenstofffasern und Thermoplast durch umformende Verbundherstellung, laufendes EFB Projekt, 3. Zwischenbericht AK-Sitzung „Hybride Strukturen"
[26] Dröder K, Brand M, Gerdes A, Grosse T 2014 Grefe, H.; Lippky, K.; Fischer, F.; Dilger, K.: An Innovative Approach for Joining of Hybrid CFRP-Metal Parts by Mechanical Undercuts: Proceedings, Euro Hybrid Materials and Structures
[27] Heckert A, Zaeh M F 2014 Laser Surface Pre-treatment of Aluminium for Hy-brid Joints with Glass Fibre Reinforced Thermoplastics, Physics Procedia
[28] Behrens B-A, Neumann A 2016 Großserientaugliche Prozessketten für hochintegrierte Bauteile aus hybriden Faser-Kunststoff/Metall-Verbunden – Teilprojekt 4: Ermittlung der Prozessgrenzen bei der Umformung von Musterbauteilen aus Faser-Kunststoff/Metall-Verbunden. AiF-Abschlussbericht 17689BR, Hannover
[29] Behrens B-A, Raatz A, Wehmeyer J, Bohne F, Lorenz R, Schumann C 2021 Wirtschaftliche Fertigung belastungsgerechter FVK/Metall-Verbunde: IGF-Vorhaben 19603 N, Schlussbericht

IOP Conf. Series: Materials Science and Engineering 1284 (2023) 012003 doi:10.1088/1757-899X/1284/1/012003

Carburization behavior at elevated temperatures and mechanical properties of a hot stamped complex phase steel

J H Risse*, F Sippel and M Merklein

Friedrich-Alexander-Universität Erlangen-Nürnberg, Institute of Manufacturing Technology, Egerlandstr. 13, 91058 Erlangen, Germany

E-mail: *henning.risse@fau.de

Abstract. As a lightweight construction strategy, hot stamped parts of ultra-high-strength steels with tailored properties are increasingly used for crash relevant components in car bodies, e.g., as B-pillars. With the process of tailored carburization, parts can be reinforced locally without increasing the sheet metal thickness. Thus, this process has the potential to be applied to lightweight components for no deformation zones, e.g., in the battery housings. Standardly, carburization is performed at 950 °C, with the highest strengths achieved at long times of up to 6 h, which prolongs the total process time. By raising the temperature, the carbon diffusion increases, which enables a reduced heat treatment time and, consequently, a shorter process time. Therefore, the objective of this study is to investigate the influence of elevated carburization temperatures on the mechanical properties of a carburized and hardened complex phase steel, CP-W® 800. A single carburization step at the enhanced temperatures leads to an embrittlement of the samples. Hence, an additional diffusion-annealing step is implemented to homogenize the carbon content, without enlarging the total heat treatment time. Depending on the time allocation of carburization and diffusion, the application of the diffusion annealing step results in higher strengths and ductility compared to only carburized samples.

1. Introduction

The steadily increasing requirements for lower fuel consumption or in the case of electric vehicles further range, accompanied by higher passenger safety specifications, are calling for concepts for lightweight vehicle bodies with high protection capabilities. Two approaches to fulfill this prerequisite are either substituting the part material with lightweight materials such as aluminum alloys or utilizing ultra-high-strength steels [1]. Because of the limited formability of the latter, hot stamping is commonly used to produce ultra-high-strength steel components. Due to the restricted energy absorption of fully high-strength parts, various strategies are pursued to locally adjust the properties e.g., increasing the ductility by adjusting the microstructure via partial austenitization or partial hardening [1]. With this strategy, for instance, the energy absorption of the lower part of a B-pillar can be improved while the high-strength upper part protects the occupants from intrusion, thus enhances the crashworthiness [2].

Another strategy for locally adapting the properties of crash-relevant components is tailored carburization. In contrast to the above-mentioned approaches, the intention is to strengthen component segments by raising the carbon content in the according areas of the sheet metal [3]. The process consists of three steps, namely the coating step, the carburization step, and the hot stamping step. First, the sheet metal is coated with a carbon source in the carburized areas and with a barrier layer to prevent carbon intake in the non-carburized areas. The coated material is then heat treated at temperatures above 900 °C for several hours to enable carbon absorption and diffusion throughout the sheet thickness. Lastly, austenitizing and hot stamping are used to hot-form the sheet into the component, which can then be laser-cut to its final geometry. Since the carburization step can be performed preceding, the fabrication of tailored carburized parts could be integrated into the existing hot stamping production line. Furthermore, this technology allows tailored properties to be achieved without changing the sheet thickness, with small transition zones between the strengthened and unmodified sections and with a high complexity of the carburized zones. Consequently, tailored carburization has the potential to be utilized for crash-relevant parts where a local strengthening of complex areas without differences in sheet thickness is required [4].

Recent investigations show the feasibility of producing parts with locally graded carbon content and thus with tailored mechanical properties. The surface hardness of a low-carbon steel can be increased by approx. 100 % by carburization with a transition zone of about 3 mm to the non-carburized zone [5]. For carburized CP-W® 800, the ultimate tensile strength (UTS) and elongation at break increase with raising carburization time, reaching a maximum of 1500 MPa and 3 % at 950 °C for 6 h [3]. Additionally, with a raising carburization time the hardness distribution across the sheet thickness harmonizes and becomes homogeneous at 6 h, which induces the increase in elongation at break.

As the additional process step carburization reaches its' highest potential at long durations, this study aims to investigate higher carburization temperatures to reduce the processing time. According to the temperature dependency of the diffusion coefficient, an accelerated carbon absorption and diffusion can be achieved by increasing the heat treatment temperature.

2. Materials and Methods

In this study, the hot rolled complex phase steel CP-W® 800 with a thickness of $t_0 = 1.6$ mm is used as the sheet material. The carbon content of max. 0.14 wt.% [6] is significantly lower compared to the conventional hot stamping steel 22MnB5 with max. 0.25 wt.% [7], which enables further carburization of the material. In addition, the used sheets are uncoated to minimize the influence on the process.

The carbon source used is Sigraflex® high-purity graphite foil from SGL Carbon SE, which is bonded to the sheet using Resbond 931C graphite adhesive from Cotronics Corp. To prevent the carbon from dissipating during the carburization step, the individual samples are heat treated in a closed package of 0.051 mm thick stainless steel hardening foil from Arnold Schröder Ofen GmbH.

In this work, two different heat treatment routes are applied namely a single hardening route with a carburization and an austenitization step as well as a diffusion-annealed hardening route with an additional diffusion annealing step interposed between the other steps. **Figure 1** shows the schematic temperature-time procedure for both routes. For the single hardening route, only one combination of the carburization temperature T_{carb} and of the carburization time t_{carb} of 1050 °C and 2 h is investigated. The diffusion annealing temperature T_{diff} is kept constant at 1050 °C for the diffusion-annealed hardening route. Here, the total heat treatment time of 2 h is the same as during the single hardening but it is divided into t_{carb} and diffusion annealing time t_{diff}. These 2 h are allocated in t_{carb}/t_{diff} 0.5 h/1.5 h, 1 h/1 h and 1.5 h/0.5 h as shown in Table 1. All samples are austenitized at $T_{aust} = 900$ °C for $t_{aust} = 5$ min and immediately water quenched to simulate hardening from the hot stamping and induce martensite formation.

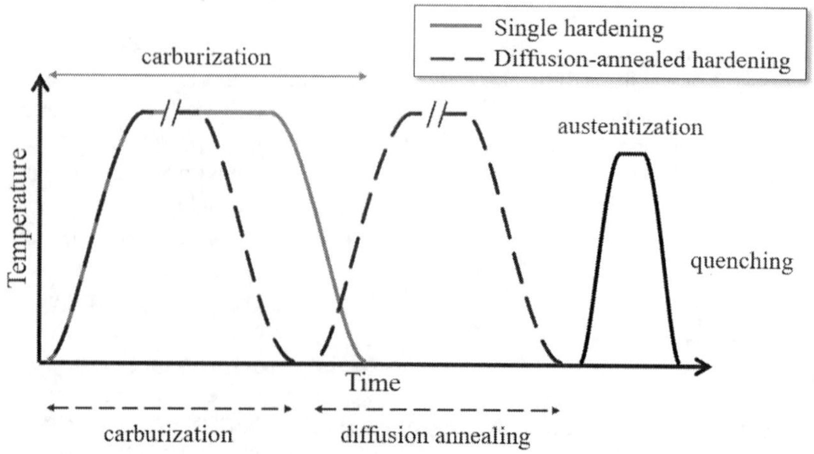

Figure 1. Schematic temperature-time course for the single hardening route (red line) and diffusion-annealed hardening route (blue dashed line).

Afterward, the mechanical properties of the carburized samples are determined via uniaxial tensile tests according to DIN EN ISO 6892-1:2019 [8] using the universal testing machine Zwick Z100 combined with the digital image correlation system Aramis from Carl Zeis GOM Meteorology. Additionally, the microhardness distribution of heat treated and cut specimens across the sheet thickness is determined by Vickers hardness HV 0.2 measurements using a Fischerscope HM2000. The measurement procedure was conducted according to DIN EN ISO 6507-1:2018 [9]. Furthermore, metallographic cuts are etched with 3 % Nital to investigate the grain structure of the non-carburized and carburized samples. Images of the polished and etched cuts were taken with a light microscope at 500x magnification.

Table 1. Heat treatment settings for the different steps of the two routes.

Route	T_{carb} in °C	t_{carb} in h	T_{diff} in °C	t_{diff} in h	T_{aust} in °C	t_{aust} in min
Single hardening	1050	2.0	/	/	900	5
Diffusion-annealed hardening	1050	0.5, 1.0, 1.5	1050	1.5, 1.0, 0.5	900	5

3. Results and discussion

Preceding investigations at different elevated carburization temperatures have shown that only the heat treatment at 1050 °C results in a homogeneous hardness distribution after two hours, which according to Horn [4] is essential for obtaining maximum ductility. Based on these findings, only these data and the additional variation of the process route with a diffusion annealing step will be discussed in this article.

3.1. Mechanical properties

Ensuing from the faster diffusion of carbon, a reduction in the processing time of carburized complex phase steel is pursued by applying higher carburization temperatures to the conventional single hardening heat treatment route. **Figure 2** shows the UTS and elongation at break of the non-carburized and carburized samples heat treated in the single hardening and diffusion-annealed hardening routes. The non-carburized specimens, which experienced the same heat treatment at 1050 °C for 2 h but without the carbon source, strengthen to 1113 ± 58 MPa in UTS. This corresponds to an improvement of about 40 % compared to the initial state [6] due to martensite formation induced by the quenching.

Concurrently, the ductility of the material decreases from 12 % to approx. 3.6 ± 0.8 %. The single hardening route shows a further decline in ductility, where all the carburized samples break within the elastic range. Consequently, the material fails without reaching its' UTS and only the maximal strength at failure can be determined of roughly 745 ± 240 MPa. The large standard deviation is caused by the fact that the failure occurring at different stresses during the elastic deformation without reaching the plastic range to align the rupture stresses. This early failure could be initiated by a high carbon content due to the elevated carburization temperature leading to a hardened but less tough material. An additional factor could also be locally dissimilar properties caused by an inhomogeneous carbon distribution or residual stresses originating from hot warpage and water quenching. Furthermore, the combination of both the low toughness and the fast cooling can initiate the formation of quench cracks at the surface leading to the samples fail prematurely in the elastic range [10].

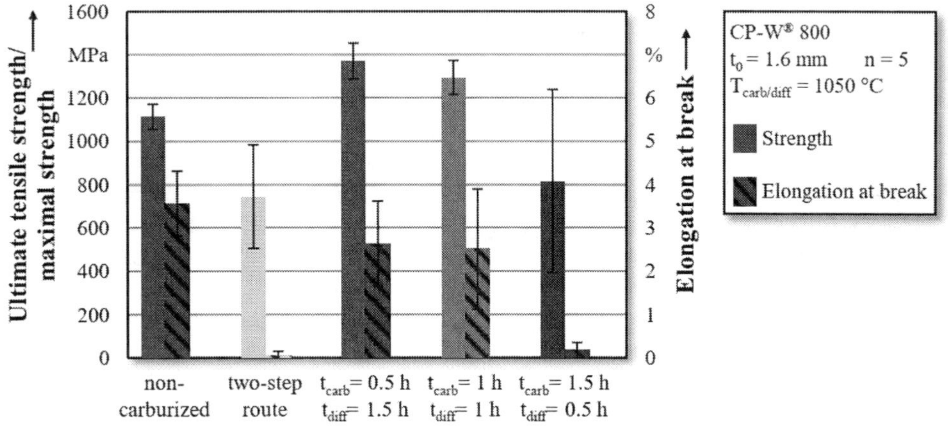

Figure 2. Comparison of the UTS/maximal strength and the elongation at break of the non-carburized state and conditions after different carburization methods.

In order to increase the homogeneity of the carbon content and thus the material properties, a diffusion annealing step is integrated after the carburization step without increasing the total heat treatment duration. So the two-hour carburization step is allocated in 0.5 h/1.5 h, 1 h/1 h, and 1.5 h/0.5 h of carburization and diffusion annealing time t_{carb}/t_{diff}, respectively. Due to the absence of the carbon source at the non-carburized condition, the tensile properties and hardness are not influenced by the allocation of the heat treatment. Therefore, only one non-carburized state is presented in this study. **Figure 2** presents the UTS resp. maximal strength and elongation at break for these combinations. In contrast to the single hardening, the 0.5 h/1.5 h allocation resulted in four out of five samples deforming plastically and those four samples reach a UTS of about 1369 ± 84 MPa. Additionally, the elongation at break increases to 2.6 ± 1.0 %. As a result, the combination of this heat treatment with partly area coated with a graphite source would enable a grading of the materials' properties via tailored carburization. Nevertheless, one sample still failed in the elastic range at approx. 350 MPa, which hampers the failure prediction of the material. By doubling the carburization time, the UTS attains about 1294 ± 79 MPa, which is an increment of roughly 15 % in comparison with the non-carburized condition. The elongation at break of 2.5 ± 1.4 % is similar to the 0.5 h/1.5 h and lower than the non-carburized state. Consequently, with this heat treatment combination a grading in the tensile strength and ductility of a component would be feasible. Furthermore, the absence of failure in the elastic region enables a better predictability of the failure. As a result of further increasing the carburization time to 1.5 h, the material shows an analogous behavior to the single hardened samples failing within the elastic range and reaching a maximal strength of approx. 816 ± 422 MPa. This indicates, that either the 1.5 h at this elevated temperature leads to an excessive carbon intake resulting in low ductility, or the 0.5 h is insufficient to homogenize the material. Here, further research is required to determine the influence of each heat treatment step on the material properties.

In comparison with the benchmark of a 6 h carburization step at 950 °C [3], the reduction of the carburization time to 2 h with a parallel increase of the temperature to 1050 °C results in a drop of the ductility and strength. However, by further reducing the carburization time to use the remaining time of the 2 h for diffusion annealing, the tensile properties can be improved to a slightly inferior level than the benchmark.

As the samples were coated with the graphite source on both the top and the bottom, the hardness distribution in **Figure 3** is presented only for half of the sheet thickness, assuming a symmetrical behavior. For reasons of clarity and comprehensibility, the data are split into two diagrams. Both the non-carburized and the single hardened samples show a homogeneous hardness distribution across the sheet thickness at approx. 400 HV0.2 for the non-carburized condition and at about 600 HV0.2 for the carburized state. These findings are in good agreement with the results of Horn and Merklein [3] showing a comparable hardness courses for non-carburized and carburized samples heat treated at 950 °C for 6 h. Consequently, a similar hardness level and distribution can be achieved by increasing the carburization temperature to 1050 °C while reducing the carburization duration to one third. In the latter case, except for the uniform hardness course, the last measuring point at the edge of the sample decreases to roughly 560 HV0.2. Since austenitizing is carried out without the hardening foil protecting the samples from the oxidative atmosphere, the edge of the material can decarburize, resulting in less martensite and a decrease in hardness. Choi and Cooman [11] report a similar hardness course for hot stamped 22MnB5 with a decarburization at the edges during austenitization. Despite the hardness decline, this improves the bendability of the material without affecting the tensile properties. Furthermore, the standard deviations of the carburized condition are significantly larger compared to the non-carburized state, as displayed in Figure 3 a). This indicates local variations in hardness and to that effect also in carbon content which may explain the very low ductility and the failure in the elastic range of the single hardened samples in the tensile test.

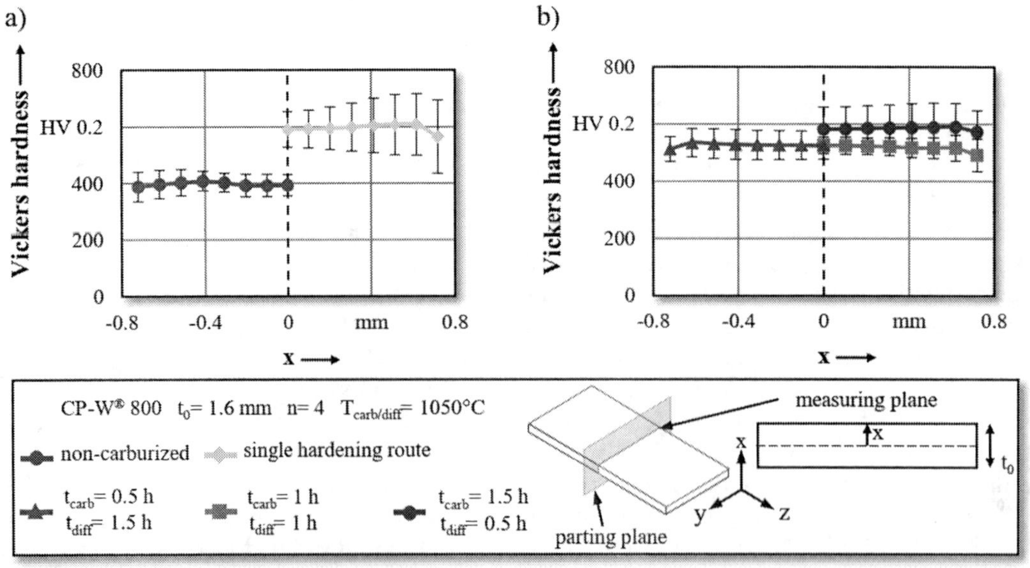

Figure 3. Hardness distribution across the sheet thickness for non-carburized and single hardened samples (a) as well as different combinations of carburization and diffusion annealing times (b).

In order to improve the material's ductility a shorter carburization time plus a diffusion annealing step is implemented for the homogenization of the material. Figure 3 b) presents the hardness distribution of the three investigated combinations. As the total heat treatment time is constant, a homogeneous hardness distribution is obtained in all cases. Both the 0.5 h/1.5 h and 1 h/1 h allocations result in a similar hardness of approx. 520 HV0.2 with a drop at the last measuring point before the edge due to the decarburization during the austenitization. Subsequently, the hardness values confirm the similarity of the tensile properties of both combinations. Regarding the hardness, on the one hand, the carbon intake induces a hardness increment of about 30 % compared to the non-carburized samples. This approves the results of the tensile test regarding the feasibility of producing locally graded parts utilizing the 1 h/1 h combination. But on the other hand, the shorter carburization time leads to a lower hardness level in comparison with the single hardening route. However, by allocating the carburization time into carburization and diffusion time, the standard deviation of the hardness measurements can be significantly reduced. This is in good agreement with the tensile test results and indicates a more homogeneous material leading to improved ductility. With an increment of the carburization time to 1.5 h, the hardness inclines to almost the same level as the single hardened samples, accompanied by an increase of the standard deviation. Consequently, the higher carbon intake produces a more inhomogeneous material wit reduced ductility, which precludes this heat treatment route for technically applicable components.

3.2. Grain structure

Additional to the hardness measurements, **Figure 4** presents the etched micrographs of the non-carburized sample, the single hardening specimen and two diffusion annealed samples. The hardened but not carburized complex phase steel has a mixed microstructure consisting of martensite and bainite, which is in good agreement with the measured hardness of 400 HV0.2. Based on the enhancement with carbon, the single hardened sample consists of a finely martensitic microstructure, which explains the increment in the hardness compared to the non-carburized sample. Since the 1 h/1 h samples are only carburized for half of the time of the single hardening route, less carbon is able to diffuse into the material, but is homogenized as a result of the diffusion annealing step. This led to a homogeneous martensitic structure as shown in **Figure 4**. The decrement of about 80 HV0.2 of the diffusion-annealed samples can be explained by the inferior carbon content in the martensite [12].

As all hardness courses decrease in the last measuring point at the edge, the microstructure of the 0.5 h/1.5 h sample in the near edge area is shown in **Figure 4** d). The microstructure composition is gradually changing from a fully martensitic in the bulk over a mixed microstructure to a ferritic layer at the edge, visible through the appearance of white grains. The bulk structure is analogous to that of the 1 h/1 h sample, which is consistent with the hardness measurements. Behrens et al. [13] show a similar microstructure appearance for 22MnB5 heat treated in a decarburizing atmosphere. The increasing ferritic amount is caused by diffusion-controlled decarburization of the edge during the austenitization in the oxidative furnace atmosphere. Due to the decreasing carbon content, the material does not form martensite during quenching but ferrite resulting in an inferior hardness in the near-edge region [14]. As visible in the hardness courses in **Figure 3**, this process is more distinctive in the carburized sample due to an increased thermodynamic driving force caused by a higher difference in carbon concentration between the surface layer and atmosphere. Moreover, in Figure 4 d) also a cinder layer on the sheet surface is observable. This cinder formed in the oxidative atmosphere during the austenitization step due to the absence of a protective layer such as AlSi or Zinc coatings, which are commonly used in hot stamping of ultra-high strength steels [1]. This type of coatings would also prevent the edges of the sheet from decarburization, but also inhibit the carbon intake and therefore, the sheets used in this study are uncoated.

IOP Conf. Series: Materials Science and Engineering 1284 (2023) 012003 doi:10.1088/1757-899X/1284/1/012003

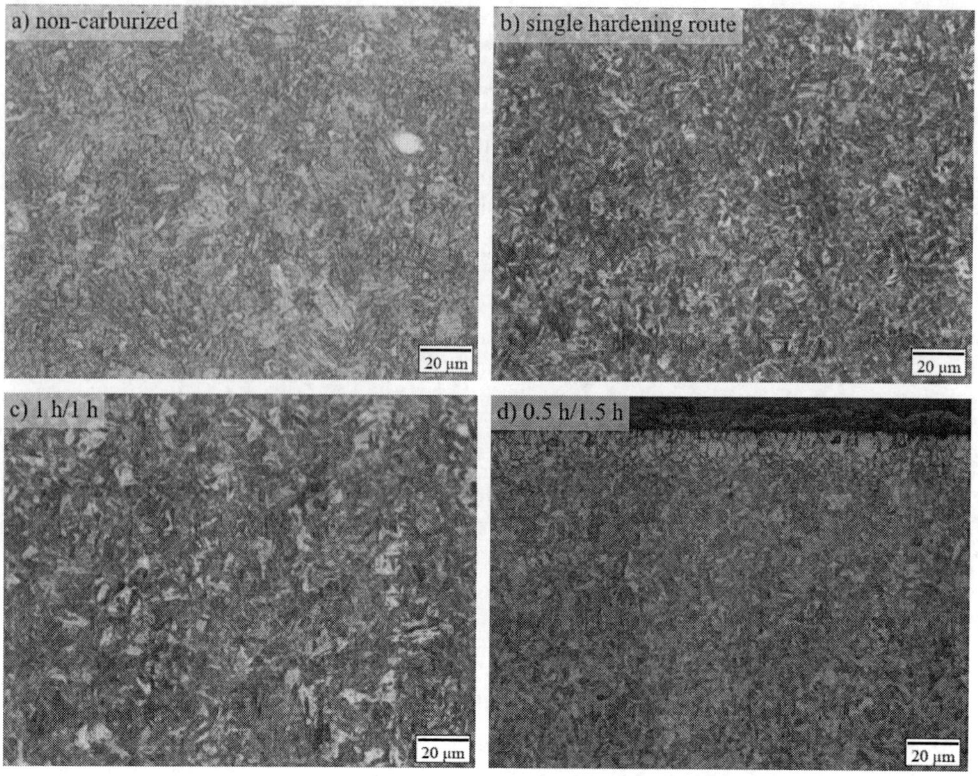

Figure 4. Etched micrographs of the sheet's center of the non-carburized sample a), the single hardened sample b), and the 1 h/1 h sample c) as well as of the edge area of the 0.5 h/1.5 h sample d).

4. Summary and Outlook

The objective of this study was to investigate the effect of elevated temperatures during the carburization of the sheet material CP-W® 800 in order to reduce the process time for the fabrication of locally carburized parts. Therefore, different heat treatment strategies were applied, namely, the established single hardening route but at 1050 °C and the route with an additional diffusion annealing step. The tensile tests showed that the investigated single hardening route parameter is not effective to produce a technically applicable material because of very low ductility. This may be caused by locally inhomogeneous properties, as indicated by excessive standard deviations at the hardness distribution which is in the mean homogeneous as required. Hence, not only does the average hardness have to be homogeneously distributed across the sheet thickness, but also the variations must also be small in order to enable ductility in the material. For homogenization of the material, a diffusion annealing step is integrated after the carburization without enlarging the heat treatment time. In this way, the standard deviation of the hardness distribution can be reduced with only a slight decrement in hardness level, and consequently, the ductility of the carburized material increases, enabling an average UTS of approx. 1294 MPa at the 1 h/1 h samples. However, the success of the new heat treatment route depends on the allocation of carburization and diffusion annealing time.

Further investigations are required to define a process window for the time allocation as well as the heat treatment temperature to achieve the tensile properties of the carburized material at 950 °C for 6 h. Moreover, the influence of the accelerated diffusion on the formation of the transition zone between carburized and non-carburized areas needs to be investigated.

Acknowledgments
The authors would like to thank the German Research Foundation (DFG) for funding the project "In-situ Charakterisierung eines local aufgekohlten Komplexphasenstahls zur Herstellung von Halbzeugen mit maßgeschneiderten Eigenschaften" (Project number 450281370).

References

[1] Taylor T and Clough A 2018 Critical review of automotive hot-stamped sheet steel from an industrial perspective *Materials Science and Technology* **34** 809–61

[2] Karbasian H and Tekkaya A E 2010 A review on hot stamping *Journal of Materials Processing Technology* **210** 2103–18

[3] Horn A and Merklein M 2020 Functional optimization of hot-stamped components by local carburization International *Journal of Lightweight Materials and Manufacture* **3** 43–54

[4] Horn A 2022 *Grundlegende Untersuchungen zur Gradierung der mechanischen Eigenschaften pressgehärteter Bauteile durch eine örtlich begrenzte Aufkohlung* Dissertation Lehrstuhl für Fertigungstechnologie, Friedrich-Alexander-Universität Erlangen-Nürnberg

[5] Rolfe B F et al 2017 Local Patchworking of Sheet Properties Combining Carburization and Hot Stamping Advanced High Strength Steel and Press Hardening *Advanced High Strength Steel and Press Hardening - Proceedings of the 3rd International Conference on Advanced High Strength Steel and Press Hardening (ICHSU2016)* 346–51

[6] thyssenkrupp Steel Europe AG 2017 CP-W® and CP-K® - Product Information for Complex-Phase Steels https://www.thyssenkrupp-steel.com/en/products/sheet-coated-products/multiphase-steel/complex-phase-steel/ (accessed 12 Oct 2017)

[7] thyssenkrupp Steel Europe AG 2022 precidur® HLB 22 / 22MnB5 - Product information for hot-rolled precision strip made in Hohenlimburg https://www.thyssenkrupp-steel.com/en/products/precision-steel-strip/product-details/boron-alloyed-special-structural-steel/ (accessed 24 Jan 2023)

[8] DIN German Insitute for Standardization 2019 Metallic materials - Tensile testing: Part 1: Method of test at room temperature (6892-1:2019) (Berlin: Beuth Verlag)

[9] DIN German Insitute for Standardization 2018 Metallic materials - Vickers hardness test: Part 1: test method (DIN EN ISO 6507-1:2018) (Berlin: Beuth Verlag)

[10] Samuel A and Prabhu K N 2022 Residual Stress and Distortion during Quench Hardening of Steels: A Review *J. of Materi Eng and Perform* **31** 5161–88

[11] Choi W S and Cooman B C de 2014 Characterization of the Bendability of Press-Hardened 22MnB5 Steel *steel research int.* **85** 824–35

[12] La Concepción V L de, Lorusso H N and Svoboda H G 2015 Effect of Carbon Content on Microstructure and Mechanical Properties of Dual Phase Steels *Procedia Materials Science* **8** 1047–56

[13] Behrens B-A, Hübner S, Holländer U, Langohr A, Pfeffer C and Albracht L 2021 Increasing the energy absorption of monolithic manganese boron steels in oxygen-free environment *IOP Conf. Ser.: Mater. Sci. Eng.* **1157** 12021

[14] Zorc M, Nagode A, Burja J, Kosec B and Zorc B 2018 Surface Decarburization of the Hypo-Eutectoid Carbon Steel C45 during Annealing in Steady Air at Temperatures T > AC1 *Metals* **8** 425

Analysis of process limits for partial hot stamping with controlled pre-cooling by radiation exchange

A Reihani[1], L Donat[1], S Heibel[1], T Schweiker[1], M Gienger[2], A Güner[2], M Winderlich[3] and M Merklein[4]

[1]Mercedes-Benz AG, Béla-Barényi-Straße 1, 71063, Sindelfingen, Germany
[2]AutoForm Engineering Deutschland GmbH, Joseph-von-Fraunhofer-Straße 13a, Dortmund, Germany
[3]GEDIA Gebrüder Dingerkus GmbH, Röntgenstraße 2-4, 57439 Attendorn, Germany
[4]Institute of Manufacturing Technology, Friedrich-Alexander Universität Erlangen-Nürnberg, Egerlandstr. 13, 91058, Erlangen, Germany

E-mail: alborz.reihani_masouleh@mercedes-benz.com

Abstract. The increasing demands on part complexity and tailored properties in form of local strength and ductility cannot be achieved with conventionally manufactured hot stamped body components. To meet the requirements of safety-relevant body parts, partial hot stamping is increasingly used in the automotive industry. One approach to achieve tailored properties on hot stamped parts is based on controlled cooling of the blank prior to forming and quenching. The related furnace technology is based on a furnace chamber in which a cooled aluminum mask protects local areas of the blank from thermal radiation while simultaneously absorbing the blank's own emitted radiation. The sheet thickness and utilized material significantly influence the controlled radiation exchange and present potential process limits. To analyse these lower process limits, blanks made of the materials 22MnB5 AS150 and 8MnB7 AS150 are partially pre-cooled and hot stamped with different sheet thicknesses. In addition, the process is simulated with AutoForm®R10 and validated by the obtained experimental results. The results show that the sheet thickness is a key influencing factor in the process and significantly affects the mechanical properties after partial hot stamping. Especially sheet thicknesses t_0 of 2.25 mm can limit the process, as the effectiveness of the radiation exchange for controlled pre-cooling and phase transformation decreases.

1. Introduction

The fundamental objective of hot stamped body parts is to combine lightweight construction with a sufficiently high crash safety level [1]. Crash-relevant parts have to comprise a high intrusion resistance and energy absorbing potential to provide for the passenger's safety [2]. However, crash performance investigations show that hot stamped components with homogeneous high-strength properties can reach their limits in highly stressed body areas [3]. In particular, the lower ductility causes a lower energy absorption. Furthermore, the thermal joints with other reinforcement parts affect the hardness in the softening zone. Studies by Mohamadizadeh et al. [4] indicate that the application of resistance spot welding on hot stamped parts made of 22MnB5 reduces the hardness in the heat affected zone from 500–520 HV to 300–320 HV. In the event of an impact, the deformation localises around the softened zone, leading to cracks and a decreased crash performance [5]. To circumvent these challenges, hot stamped parts can be adapted through the targeted integration of ductile areas. According to the current state of the art, different semi-finished products such as tailor-welded blanks, tailor-rolled blanks or patchwork blanks as well as cooling strategies exist to produce tailor-tempering parts [6]. For the production of tailor-tempering parts, three process strategies can be distinguished: Partial austenitization prior to hot stamping, partial cooling by means of different tool tempering, tool surfaces and/or tool materials as well as partial tempering after hot stamping [6]. To achieve a partially austenitized blank prior to forming and quenching, various innovative furnace technologies have been established in the automotive industry [7]. One of these furnace technologies is based on a special furnace chamber with integrated cooled aluminium masks, the so-called TemperBox®. In partial hot stamping with the TemperBox®, the blank is first homogeneously austenitized and then positioned between cooled aluminium masks protecting the covered areas from radiation and absorbing the blank's own radiation simultaneously. The combination of thermal radiation protection and absorption of the blank's own emitted radiation results in a controlled cooling below the critical cooling rate. This ensures the decomposition of the austenite into soft phase mixtures of ferrite, pearlite and/or bainite. At the same time, uncovered areas are reheated in order to remain in the austenitized state. Consequently, the blank is formed and quenched in a partially austenitized state. Parts produced by partial pre-cooling show an improved component quality resulting from a higher dimensional accuracy compared to alternative tailored tempering approaches [8]. Additionally less heat distortion occurs, as no significant temperature gradient remains in the part after hot forming. Further advantages of this process are a homogeneous diffusion between coating and substrate, the avoidance of elaborately designed tempering tools and the avoidance of an additional process step. Nevertheless, this process can have sheet thickness-dependent limitations due to the restricted effectiveness of the heat radiation exchange. This work contains the experimental investigation of partial pre-cooling with the TemperBox® for different sheet thicknesses and semi-finished product concepts with the materials 22MnB5 and 8MnB7 as well as the simulative prediction of the phase transformation and the resulting hardness in AutoForm®R10.

2. Investigation methods and materials

2.1. Materials

Besides the intention to achieve higher strengths and thus fully exploit the potential for lightweight construction, there is an increasing demand for steels with higher ductility [9]. These enable targeted energy absorption as a result of plastic crash deformation [1]. Hence, in addition to the conventional aluminium-silicon coated 22MnB5 AS150, 8MnB7 AS150 is investigated in this work. Table 1 summarises the chemical compositions of the manganese-boron steels 8MnB7 and 22MnB5 with the sheet thicknesses respectively.

Table 1. Chemical composition of 22MnB5 and 8MnB7 in wt. %

Material/sheet thickness	C (%)	Mn (%)	B (%)	Cr (%)	Si (%)	Al (%)	Ti (%)	Ni (%)	Mo (%)
22MnB5/0.80 mm	0.23	1.18	0.003	0.17	0.26	0.04	0.04	0.01	0.003
22MnB5/1.25 mm	0.22	1.18	0.003	0.17	0.26	0.03	0.03	0.01	0.003
22MnB5/1.50 mm	0.23	1.18	0.003	0.18	0.25	0.03	0.04	0.02	0.003
22MnB5/2.25 mm	0.22	1.16	0.003	0.18	0.25	0.03	0.04	0.02	0.003
8MnB7/0.80 mm	0.081	1.64	0.003	0.08	0.32	0.03	0.05	0.01	0.002
8MnB7/1.50 mm	0.078	1.60	0.003	0.08	0.31	0.04	0.05	0.01	0.001

In contrast to 22MnB5, 8MnB7 has a lower carbon content increasing the critical cooling rate which is partly compromised by the raised amount of manganese [10]. However, carbon has a dominant impact in steels as it increases hardenability and strength, even in small amounts [11]. This is based on the martensite formation in the overall microstructure. During hot forming the body-centred cubic lattice in the austenite state is transformed into a tetragonally distorted martensite lattice due to the non-diffusion of the carbon atoms while quenching [12]. Besides the carbon content, boron and manganese have a decisive influence on the hardenability of the hot stamped steel [10]. An overview of the influence of the alloying elements on the cooling-rate dependent phase transformation is shown in the CCT diagram (continuous cooling transformation) of the manganese-boron steel 22MnB5 (see figure 1). The critical cooling rate $\dot{T}_K$ describes the minimum necessary cooling rate for a fully martensitic transformation.

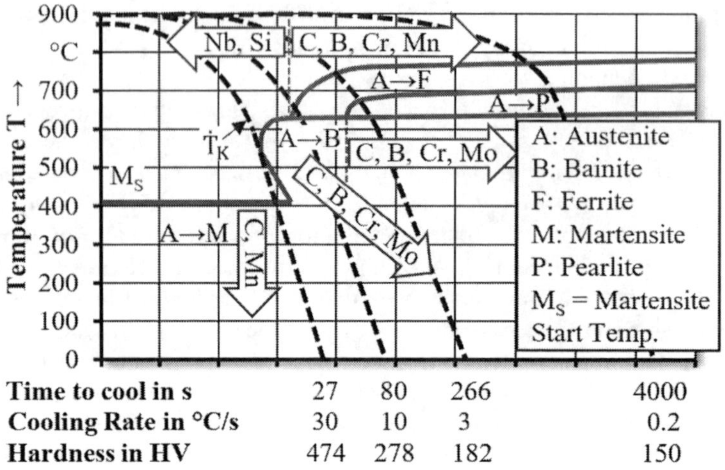

Figure 1. Influence of the alloying elements on the phase transformation in the CCT diagram of 22MnB5 [10]

2.2. Experimental setup and corresponding simulation model

In this work, blanks with the dimensions 157 mm x 496 mm are heated and pre-cooled in the prototype chamber furnace of AP&T Sweden AP and partially hot stamped with the hydraulic press Imposite 2500 from Dieffenbacher GmbH. Proceeding from the simple full-surface quenching of flat blanks with different materials and sheet thicknesses, the geometric and process complexity is extended by the investigation of patched-hat profiles. For the patched-hat profile, 8MnB7 with a sheet thickness t_0 of 1.5 mm is used as main blank and 22MnB5 with a sheet thickness t_0 of 1.25 mm is used as patch. The advantages of patchwork blanks are that the material, geometry, sheet thickness and position of the patches can be freely designed according to the component requirements. An overview of the applied sheet thicknesses and material combinations as well as the basic dimensions of the blanks are shown in figure 2a.

In the experimental trials, the furnace temperature is set at 950 °C and the heating time to obtain a homogeneously austentized blank is varied between 270 s for sheet thicknesses t_0 of 0.80 mm and 600 s for the patchwork blanks ($t_0 = 2.75$ mm). Subsequently, the austenitized blank is positioned in the TemperBox® between two cooled aluminum masks by a furnace feeder. Due to the radiative and the convective heat exchange with the ambient air, the temperature drops during the transfer to the TemperBox®. After positioning the blank, the lower aluminum mask is moving vertically upwards until the defined mask distance of 2 mm has been reached which starts the pre-cooling phase. Within this phase, the blank is locally cooled in the covered area and simultaneously heated above the eutectoid temperature in the uncovered area to maintain an austenitic microstructure. The selected partial pre-cooling time $t_{pre\text{-}cooling}$ of 75 s is chosen based on a realistic value in industrial manufacturing and to quantify differences in phase transformation and final hardness at different sheet thicknesses, materials and semi-finished product concepts. The cooling rate in the covered area must remain below the critical cooling rate to ensure the decomposition of the austenite into ferrite, pearlite and bainite. Finally, the partially austenitized blank is positioned in the tool after approx. 11 s of transfer time, where it is either quenched directly or, as in the production of the patched-hat profiles, first formed and then quenched.

The holding time during quenching is set to 25 s and a pressing force of 1250 kN is used. To ensure the reproducibility of the examination, all trials are repeated three times.
To calculate the phase transformation and the final hardness after partial hot stamping with the TemperBox®, a simulation model is created in AutoForm®R10. For this purpose, the thermo-mechanical properties and the process boundary conditions must be integrated. The material cards for 8MnB7 and 22MnB5 used for this numerical investigation combine the thermal properties, the temperature and strain-rate dependent flow behaviour as well as the phase transformation. The phase transformation and the final hardness calculation are based on experimentally determined CCT diagrams. To calculate the temperature changes on the blank and the phase transformation during partial pre-cooling with the TemperBox®, experimental temperature measurements are first carried out with thermocouples. A total of three thermocouples are welded onto the blank surface, two in the covered area and one in the uncovered area (see figure 2b). The results of the temperature measurements are then specified as boundary conditions in the simulation model. In the uncovered or heated area, the temperature is given as a fixed value based on the experimental results, while in the covered area the temperature distribution is calculated using the locally measured temperature and convective cooling. Figure 2b summarises the experimentally determined and numerically calculated temperature curves of the blank during partial pre-cooling with the TemperBox®. For the numerical calculation of the transfer phase and the quenching phase, the boundary conditions are taken from the experimental trials. The discretisation of the blank is done with elastic-plastic shell elements with 11 integration points across the thickness.

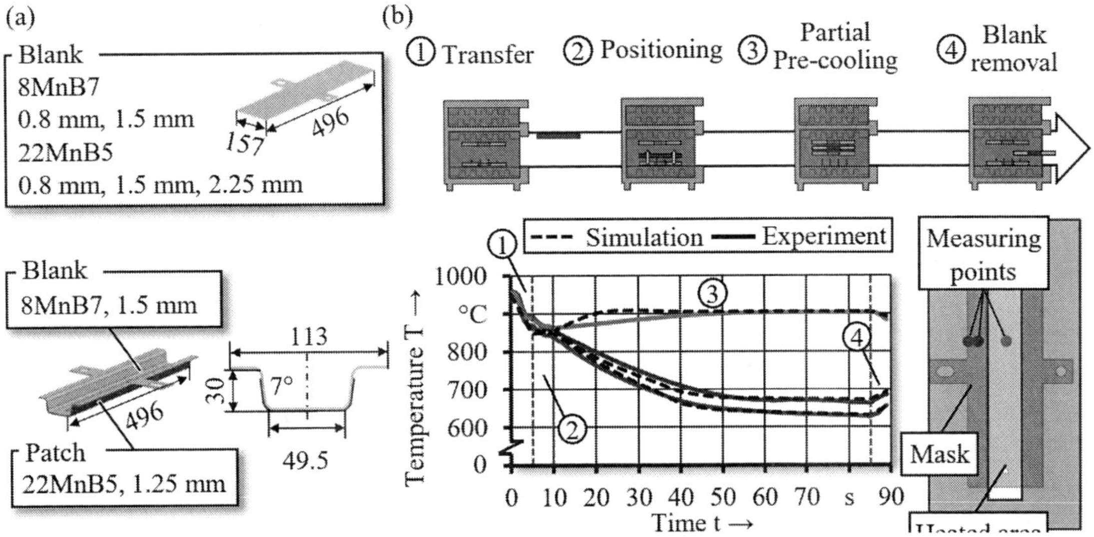

Figure 2. a) Used materials and dimensions of the components examined as wells as b) experimental and numerical time-temperature curves on the blank surface

3. Results and discussion

A significant mechanical parameter in the analysis of hot stamped components is the hardness, which enables the evaluation of the cooling-rate dependent phase transformation. For the hardness measurements, a representative sample is laser-cut for each partially hot stamped blank and separated in cross-section. After the preparation of the samples, Vickers microhardness tests are carried out with the semi-automatic hardness tester DuraScan G7 from Struers GmbH and the associated software Ecos Workflow from EMCO-TEST GmbH. The results of the numerically calculated and experimentally determined hardness curves of the blanks with sheet thicknesses t_0 of 0.80 mm and 1.50 mm are shown in Figure 3. The results are based on the measurements of all three samples for respectively each sheet thickness and material combination, which are carried out and averaged in 1 mm point intervals.

IOP Conf. Series: Materials Science and Engineering 1284 (2023) 012004 doi:10.1088/1757-899X/1284/1/012004

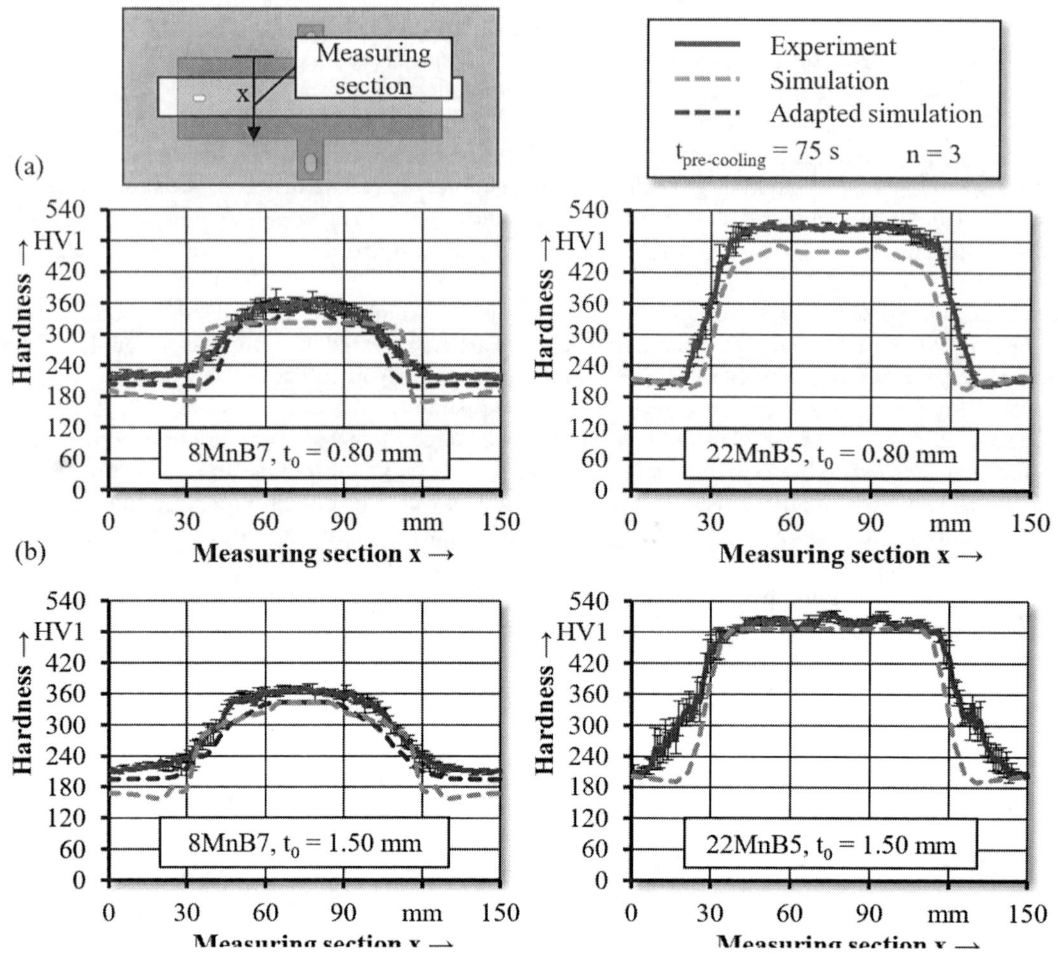

Figure 3. Numerically and experimentally determined hardness curves for 8MnB7 and 22MnB5 for the sheet thickness variants a) t_0 = 0.80 mm and b) t_0 = 1.50 mm

The analysis of the results shows that the hardness curves of 22MnB5 differ significantly from those of 8MnB7. In the uncovered area (35–115 mm), 22MnB5 has a higher hardness than 8MnB7 due to its higher carbon content and consequently lower critical cooling rate. As a result of the internal heat conduction in the material, the hardness decreases continuously from the uncovered to the covered surface. Compared to 22MnB5, bainite phases start to occur in 8MnB7 already at cooling rates of less than 50 °C/s [13]. Consequently, the decrease in hardness towards the covered area of the blank starts earlier, resulting in a smaller hard zone and larger soft zone at 8MnB7. In addition, it should be noted that depending on the sheet thickness, different cooling rates are recorded during the partial pre-cooling of the blank, which affects the phase transformation and the length of the soft, transition and hard zones. The analysis of the hardness curves shows the influence of the sheet thickness by a wider hard zone and a correspondingly shorter soft zone for sheet thicknesses t_0 of 1.50 mm compared to sheet thicknesses t_0 of 0.80 mm resulting in slower cooling rates for thicker blanks due to increased internal heat conduction. While the numerical hardness curves for 22MnB5 reproduce the experimentally determined hardness curves with good accuracy, significant differences are evident for 8MnB7. A possible reason for the deviation in the numerically calculated and experimentally determined hardness curves is that the phase transformation and the calculated hardness values are based on CCT diagrams. In partial hot stamping with the TemperBox®, a highly discontinuous cooling occurs in the partial pre-cooling phase, which cannot be accurately reproduced by a CCT diagram. Furthermore, in the simulation it is assumed that the thermo-mechanical properties such as thermal conductivity and heat capacity are constant regardless of the blank temperature. Consequently, there could be inaccuracies in the calculation of the temperature gradient in the direction of the analysed measuring section. Compared to 22MnB5, these assumptions cause inaccuracies for 8MnB7 due to the higher sensitivity of the cooling rate-dependent phase transformation. An adjustment of the simulation with additionally applied cooling zones in the transition between covered and uncovered areas of the blank leads to a good accuracy in the calculation of the hardness curve for 8MnB7. These additional cooling zones are modelled in order to oppose the assumptions of temperature-independent thermomechanical properties such as thermal conductivity and heat capacity and to ensure a successive temperature gradient at the transition between covered and uncovered blank areas.

In contrast to thermal conduction, where energy is transferred by direct excitation of the free electrons, cooling is less efficient with thermal radiation and convection [14]. Consequently, the sheet thickness can have a decisive influence on partial hot stamping with the TemperBox®, as the partial pre-cooling is mainly affected by the thermal radiation. The extent to which the phase transformation and the resulting hardness are influenced by the use of higher sheet thicknesses is shown in figure 4 using a blank with a sheet thickness t_0 of 2.25 mm made of 22MnB5. The results are based on a partial pre-cooling time of 75 s to ensure comparability.

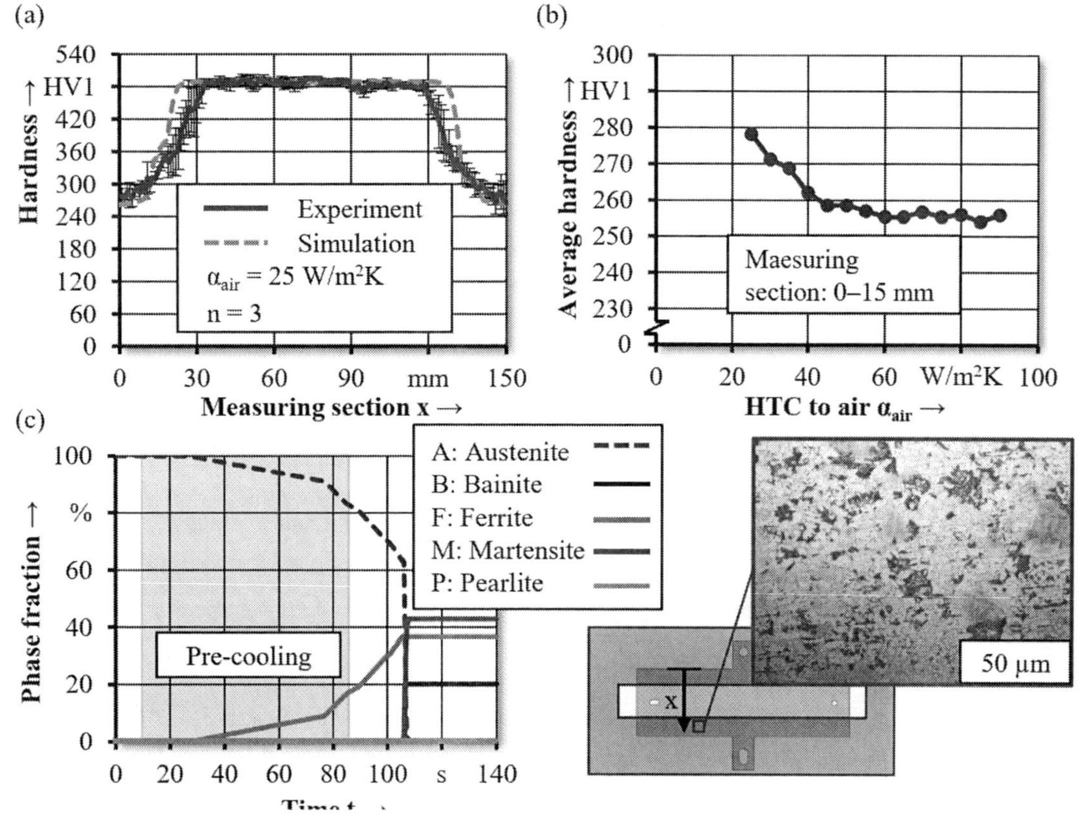

Figure 4. a) Numerically and empirically determined hardness curve, b) numerically determined correlation between hardness and heat transfer coefficient and (c) microscopically recorded microstructure and calculated phase fractions

By analysing the experimentally determined hardness curves in figure 4a, it is noticeable that the hardness in the covered area drops down to 280 HV. The hardness values and the microstructure indicate that a homogeneous ferrite-pearlite structure did not form in the covered area. The numerically calculated phase transformation process during partial hot stamping with the TemperBox® (see figure 4c) confirms that the decomposition of the austenite is not fully completed after the partial pre-cooling of the blank and thus the transfer to the tool can influence the phase transformation and the resulting hardness values. During transfer, the temperature of the blank decreases due to radiation emission and convection with the environment. To simulate the heat loss with the environment, temperature-dependent heat transfer coefficients are specified at 25 °C and 950 °C blank temperature and intermediate values are interpolated. The simulation-based heat transfer coefficients describe effective values that are composed of the heat losses through radiation and convection. An iterative approach is used in this work to determine the heat transfer coefficient by matching the calculated hardness values in the measuring section 0–15 mm (see figure 4b). With a set heat transfer coefficient of 25 W/m²K, good agreement is achieved between numerically calculated and experimentally determined hardness curves (see figure 4c). As the sheet thickness increases, the efficiency of cooling in the covered area of the blank decreases, which can lead to the transformation of hard phases such as bainite and martensite after quenching, depending on the partial pre-cooling time. The martensite and bainite phases can be seen in the recorded microstructure image and in the simulated phase fraction (see figure 4c).

An extension of the complexity with regard to process boundary conditions and component geometry is represented by the investigation of the patched-hat profile. In particular, the contact conditions caused by the part geometry of the patchwork blanks can lead to challenges in the quenching phase. Due to the patch, an undercut is created in the quenching phase (see figure 5b). This prevents the tools from exerting contact pressure on both sides of the main blank. Investigations by Güner et al [13] with a conventionally hot stamped patched B-pillar made from an 8MnB7 main blank show that the hardness drops by up to 13 % representatively for a measurement in the area of the undercut. In contrast, the experimentally determined average hardness curve for the main blank of the partially pre-cooled patched-hat profile indicates that there is no significant drop in hardness in the undercut area (17–22 mm and 128–133 mm). The hardness decreases by up to 6 % in this area. Compared to the conventional direct hot stamping process of patchwork blanks, austenitic decomposition in the undercut area starts during partial pre-cooling. Accordingly, the quenching phase does not significantly influence the austenite decomposition and the final hardness in the pre-cooled area. Numerically, the hardness curves in the main blank and in the patch are reproduced with good accuracy.

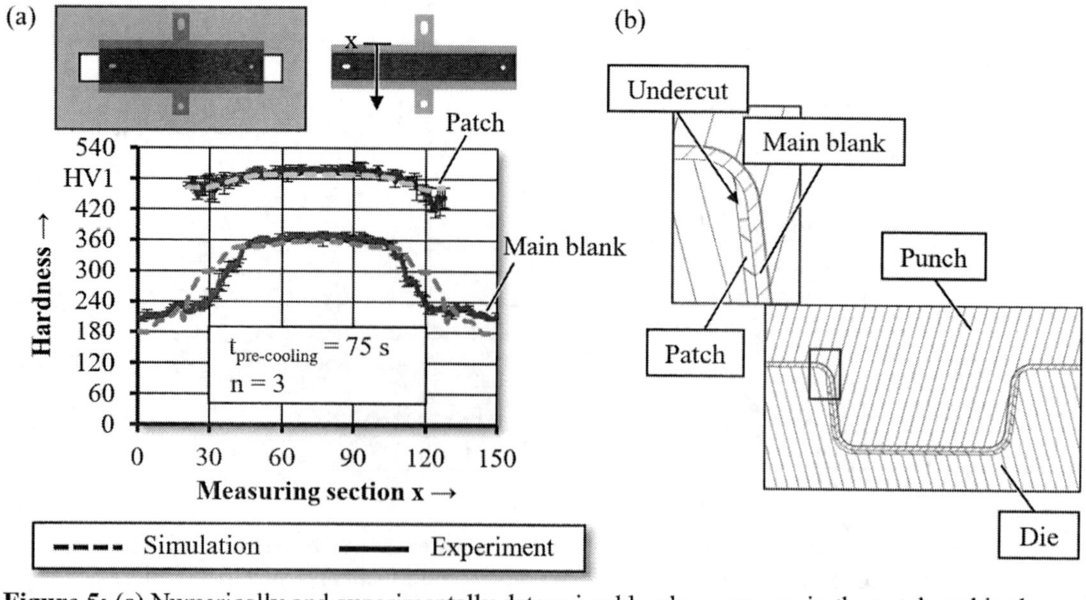

Figure 5: (a) Numerically and experimentally determined hardness curves in the patch and in the main blank, as well as (b) detailed representation of the contact conditions in the closed tool state

4. Summary and outlook

The impact of partial hot stamping with TemperBox® on the final hardness for different sheet thicknesses, materials and semi-finished product concepts was investigated in this work. In the first step, flat blanks made of 8MnB7 and 22MnB5 in sheet thicknesses t_0 of 0.80 mm and 1.50 mm were investigated. Comparing the results between 8MnB7 and 22MnB5 illustrate that the carbon content has a significant influence on the final hardness profile of the partially hot stamped blanks. Simulatively, the hardness curve can be calculated with sufficient accuracy based on the used material cards for 8MnB7 and 22MnB5. In addition, the results show that the use of higher sheet thicknesses t_0 such as 2.25 mm of 22MnB5 can lead to process limits, as a homogeneously soft microstructure with a partial pre-cooling time $t_{pre\text{-}cooling}$ of 75 s does not form in the covered area. Numerically, the final hardness after partial hot stamping can be determined with good agreement for a sheet thickness t_0 of 2.25 mm. Thereby, the heat transfer coefficient must be taken into account in the simulation, as the decomposition of the austenite is not completely finished after the partial pre-cooling. In the last step, geometrical complexity is extended by the investigation of hot formed patched-hat profiles. The results indicate that the use of partial pre-cooling for patched-hat profiles can compensate the missing both side tool contact in the quenching phase. As austenitic decomposition begins in the pre-cooling phase, the quenching phase does not affect the phase formation and the final hardness in the pre-cooled area significantly. Accordingly, homogeneous soft areas occur regardless of the quenching phase.

References

[1] Billur E 2018 Hot Stamping of Ultra High-Strength Steels (Heidelberg: Springer)

[2] Boria S 2016 Lightweight design and crash analysis of composites (*Lightweight Composite Structures in Transport* vol 1) ed J Njuguna (Sawston: Woodhead Publishing) chapter 13 pp329–360.

[3] Rosenstock D, Banik J ,Gerber T and Myslowicki S 2019 Hot stamping steel grades with increased tensile strength and ductility - MBW-K 1900, tribond 1200 and tribond 1400 IOP Conf. Ser.: Mater. Sci. Eng. **651** 012040.

[4] Mohamadizadeh A, Biro E and Worswick M 2020 Shear band formation at the fusion boundary and failure behaviour of resistance spot welds in ultra-high-strength hot-stamped steel *Sc. And Tech. of Welding and Joining* **25** 556–63.

[5] Burget S and Sommer S 2012 Characterization and modeling of fracture behavior of spot welded joints in hot-stamped ultra-high strength steels *LS-DYNA Forum*, October 5–7, Ulm, Germany.

[6] Merklein M Wieland M Lechner M Bruschi S Ghiotti A 2016 Hot stamping of boron steel sheets with tailored properties: A review *Journal of Materials Processing Technology* **228** 11–24.

[7] Lehmann H 2015 New developments in furnaces for press-hardening *5th International Conference on Hot Sheet Metal Forming of High Performance Steel*, May 31st–June 3rd Toronto, Canada.

[8] Feuser P 2012 Ein Ansatz zur Herstellung von pressgehärteten Karosseriekomponenten mit maßgeschneiderten mechanischen Eigenschaften: Temperierte Umformwerkzeuge. Prozessfenster, Prozesssimulation und funktionale Untersuchung, *Doctoral theses*, University of Erlangen–Nuremberg (Bamberg: Maisenbach Verlag).

[9] Múnera D D, Pic A, Abou-Khalil D, Shmit F and Pinard F 2009 Innovative Press Hardened Steel Based Laser Welded Blanks Solutions for Weight Savings and Crash Safety Improvements *SAE Int. Journal of Materials and Manufacturing* **1** 472–78.

[10] Mohrbacher H 2014 Martensitic Automotive Steel Sheet - Fundamentals and Metallurgical Optimization Strategies *Advanced Materials Research* **1063** 130–42.

[11] Naderi, 2007 Hot Stamping of Ultra High Strength Steels *Doctoral theses,* RWTH Aachen University (Aaachen: Publikationsserver der RWTH Aachen University).

[12] Ohring M 1995 HOW ENGINEERING MATERIALS ARE STRENGTHENED AND TOUGHENED (*Engineering Materials Science* vol 1) ed M Ohring (Cambridge: Academic Press) chapter 9 pp 431–500.

[13] Güner A, Sonntag M, Philippot C, Bittendiebel J, Sarre B, Dormegny L, Reihani A and Heibel S 2022 Experimental and numerical investigation of final product properties of Ductibor® 1000 AS under different process conditions, 8th *Int. Conf. Hot Sheet Metal Forming of High-Performance Steel,* May 30th–June 2nd, Barcelona, Spain.

[14] Baehr H and D Stephan K 2006 Heat and Mass Transfer (Heidelberg: Springer).

IOP Conf. Series: Materials Science and Engineering 1284 (2023) 012005 doi:10.1088/1757-899X/1284/1/012005

Investigation on evolution of microstructure and mechanical properties of heat-treatable 6010-S aluminium alloy during gas-based hot sheet metal forming process

T Teeuwen[1*], N K Baru[2], D Bailly[3] and G Hirt[4]

[1-4] Institute of Metal Forming (IBF), RWTH Aachen University, Intzestraße 10, 52072 Aachen, Germany

E-mail: * tobias.teeuwen@ibf.rwth-aachen.de

Abstract. Modern hot sheet metal forming processes offer the opportunity, especially in the automotive sector, to meet current demands for ultra-lightweight design. Due to the increased formability at the high process temperatures, high-strength aluminium alloys are increasingly coming into the focus of the industry. However, the complex thermo-mechanical interactions within these processes can lead to undesirable microstructural changes that could have a negative impact on the final mechanical properties. The purpose of this work is therefore to investigate the microstructural evolution of age-hardenable EN AW 6010-S alloy in the course of a modern gas-based sheet metal forming process and to quantify its influence on the resulting mechanical T6 properties. For this purpose, two different components were first formed at laboratory scale and the areas of microstructural interest were identified. Metallographic examinations and EBSD measurements were performed to visualize the influence of process temperature and deformation on the microstructure. EDX analysis helped to identify non-metallic phases. In the next step, artificial aging of the components was performed to increase the mechanical properties. Tensile tests and hardness measurements showed that the deformation and temperature have no negative influence on the final mechanical properties of the alloy.

1. Introduction

Within the context of lightweight automobile design, 6xxx (Al-Mg-Si) aluminum sheets have proven to be optimal candidates, due to their high specific strength comparable to some steels. Solution heat treatment of these alloys results in complete dissolution of the hardening Mg_2Si precipitates and alloying elements in the solid solution. Due to this reason, the material shows good hot formability and increased forming limits compared to cold forming. Rapid quenching followed by artificial ageing at warmer temperatures is then typically done to enable the segregation of that solid solution, resulting in the formation of strength-enhancing Mg_2Si precipitates. But, dynamic grain growth can take place when deformation and high temperatures are superimposed. The final strength of the 6xxx aluminum alloys after hot forming processes thus depends on the proportion and size of Mg_2Si precipitates as well as the materials grain size, which further depends on the evolution of local temperature and strain along the hot forming process chain.

For previously established hot forming processes of 6xxx alloys, the time-temperature profile is defined in order to achieve optimum mechanical properties. The process routes of modern gas-based hot forming processes such as quick plastic forming, high speed blow forming and hot metal gas forming drastically differ from the existing established hot forming processes. On one hand, the material undergoes rapid cooling beyond the solvus temperature via forced convection due to pressurized gas and immediate rapid re-heating upon contact with the tools at the elevated temperature. On the other hand, the local strain within the workpiece is significantly higher, especially due to relatively complex part geometries. Therefore, for achieving the best mechanical properties of the aluminum parts after rapid gas-based hot forming, a clear understanding of local microstructure development is mandatory. Within this study, the evolution of local microstructure and resulting mechanical properties under thermo-mechanical conditions of a laboratory-scale gas-based hot forming processes are critically analyzed for the precipitation hardening aluminum alloy EN AW-6010-S.

2. State of the Art

The EN AW 6010-S aluminum alloy was previously implemented within the gas-based hot sheet metal forming processes, and its ability to be formed into complex laboratory-scale parts is demonstrated in [1, 2]. Moreover, in the authors' previous study [3], the high temperature formability of this alloy is characterized in terms of a forming limit curve, which is further validated within a laboratory scale test. These studies mainly focused on the flow stress and formability of the 6010 alloy under gas-based hot forming conditions, rather than the microstructure and final yield strength. In another recent study, Taylor et al. [4] investigated the EN AW 6010 alloy via tensile tests to determine their optimal hot formability. Upon solutionizing at 565 °C, quenching to 500 °C, water quenching to room temperature and rapid aging at 205 °C for 1 h, the material has achieved its peak yield strength of 335 MPa. However, the influence of deformation is not considered within this measurement.

Even though 6010 alloy is not extensively investigated in terms of its microstructure under hot forming conditions, several other heat-treatable 6xxx alloys were extensively studied. He et al. [5] investigated the microstructure of 6061 alloy tube after hot metal gas forming. In this study, the initially fine equiaxial grains grew as the temperature increased. Moreover, the grains also stretched in the deformation direction. In [6], Talebi-Anaraki et al. investigated the die-less hot tube gas forming of extruded EN AW 6063 alloy tube. They concluded that the low-rate air cooling of the deformed tube from solution heat treatment temperatures resulted in an increase of the grain size. Hu et al. [7] investigated the microstructure development for an aluminum alloy with 3 % Mg_2Si via upsetting tests to create a process map. At higher temperatures and low forming rates, nucleation led to uniform, fine and recrystallized microstructure. The development progressed as a function of the strain. In addition, the grain orientation angle increased due to the transformation of small angle grain boundaries into large angle grain boundaries upon continuous dynamic recrystallization. Additionally, debonding of the precipitates and the surrounding aluminum matrix was found.

Based on these existing results, it is not yet clear how the microstructure of the 6010 alloy evolves under the temperature and deformation conditions of rapid gas-based hot forming processes. Moreover, the influence of the temperature and deformation conditions on the resulting post forming and post heat-treatment microstructure as well as mechanical properties cannot be predicted. The current study investigates these aspects with help of laboratory-scale process reproduction and a variety of established characterization methods.

3. Material and Methods

The material under investigation is cold rolled (H18) EN AW 6010-S sheet metal with a thickness of 2 mm received from Speira GmbH (previously Hydro Aluminium GmbH). The chemical composition according to the manufacturer is given in Table 1.

Table 1. Chemical composition of the investigated EN AW 6010-S sheet metal in wt.%.

Al	Si	Fe	Cu	Mn	Mg	Cr	Zn	Ti
remaining	0.94	0.30	0.23	0.53	0.76	0.01	0.19	0.024

Within this study, this alloy is processed by a gas-based hot forming process and the microstructure and mechanical properties are analyzed along the process route. The alloy is first heated conductively with hot dies to the solution heat treatment temperature (565 °C) within 5 min, annealed there for 15 min and then rapidly formed using gas starting at this temperature. Through this process sequence, two different components, shown in Figure 1 (a) and (b), are produced.

The first part is a cross-die component known from [1, 3] with a depth of 32 mm, an arm width of 32 mm and an arm length of 30 mm. Within the cross-die component, high local strains can be achieved, which can demonstrate the influence of high degrees of deformation on the microstructure. On the other hand, a round cup, known from [2], with a depth of 40 mm and an inlet diameter of 100 mm is formed. Due to its relatively simpler geometry, this component allows tensile specimens to be taken from the bottom area in order to investigate the mechanical properties after passing through the process chain. Manufacturing via the gas-based forming process is analogous to the previously published work [1–3]. After forming, the components are removed from the hot die and quenched in a water bath. The round cup is then subjected to further artificial aging at 205 °C for 45 min. The time-temperature history of the round cup measured via a K-type thermocouple is shown in Figure 1 (c).

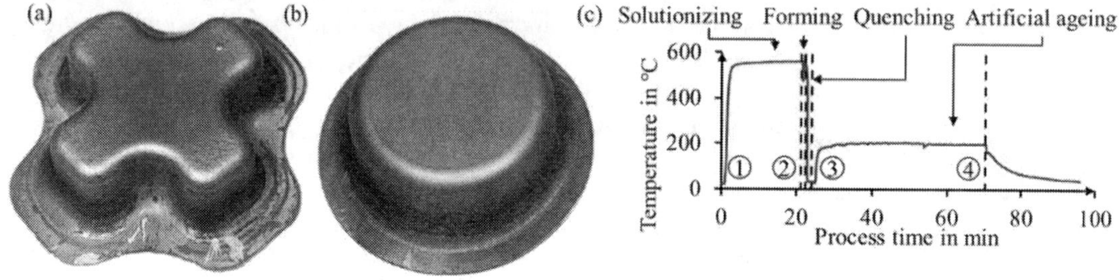

Figure 1. (a) Cross-die and (b) round cup components; (c) Time-temperature profile during gas-based hot forming and artificial ageing.

The evolution of the microstructure is observed via metallographic micrographs at four distinct time points along the process chain in Figure 1 (c): (1) Initial condition; (2) After solution heat treatment (SHT); (3) After forming; (4) After artificial ageing. The areas to be examined are cut from the components with a band saw, embedded in epoxy resin, and wet ground and diamond polished according to standard practice. Electrolytic etching according to Barker is used to display the grain structure. The optical microscopy is done on a Keyence VHX-7000 digital microscope. For in-depth analysis of non-metallic phases, Scanning Electron Microscopy (SEM) alongside with Energy-dispersive X-ray spectroscopy (EDX) is performed. Grain size measurements are done via Electron Back-Scatter Diffraction (EBSD).

Additionally, the influence of forming and artificial aging on the mechanical properties of the alloy is quantified via tensile tests on DIN 50125 (type E 2×6×20) specimens extracted from the flat bottom area of the round cup components. The tensile tests are performed quasi-statically on a Zwick/Roell Z100 tensile testing machine. Finally, the tensile strength is comparatively analysed along the component contour by means of Vickers hardness measurements.

4. Results

4.1. Initial condition

The initial condition of the aluminum alloy is H18 (cold rolled). Accordingly, a typical rolled microstructure with highly elongated grains is expected. Before considering the microstructure, the initial state is examined unetched, since Barker etching can dissolve particles out of the matrix and correspondingly reveal artifacts and pores that can lead to misinterpretation.

Figure 2 (a) and (b) show the unetched initial state. Non-metallic particles and individual pores are visible. However, in the etched state, Figure 2 (c), the expected rolling texture of the microstructure can be seen and the etchant has removed the particles from the surface (dark artifacts).

(a) (c)

Figure 2. (a) Unetched micrograph of 6010-S H18 (magnification: 500x); (b) Detailed micrograph (magnification: 2000x); (c) Etched micrograph of 6010-S H18 (magnification: 400x).

EDX analysis of the non-metallic particles shows that they are α-AlFeMnSi (bright color) inclusions and primary Mg_2Si phases (dark grey), Figure 3. Coarse particles of both phases are formed during casting [8] due to the alloy composition. Finer dispersoids of the AlFeMnSi phase are additionally formed during further processing of the cast ingots into the thin strips [9]. These finer dispersoids help stabilize the microstructure [10] during hot forming and contribute to the alloy designation "-S (superplastically enhanced)". This means that the present material will show a fine grain structure and no tendency for grain growth when held on high temperatures for longer periods. The black areas on the SEM image, Figure 3 (a), are pores formed by detachment of the matrix from the non-metallic inclusions, as already described in [7, 11].

(a) (b)

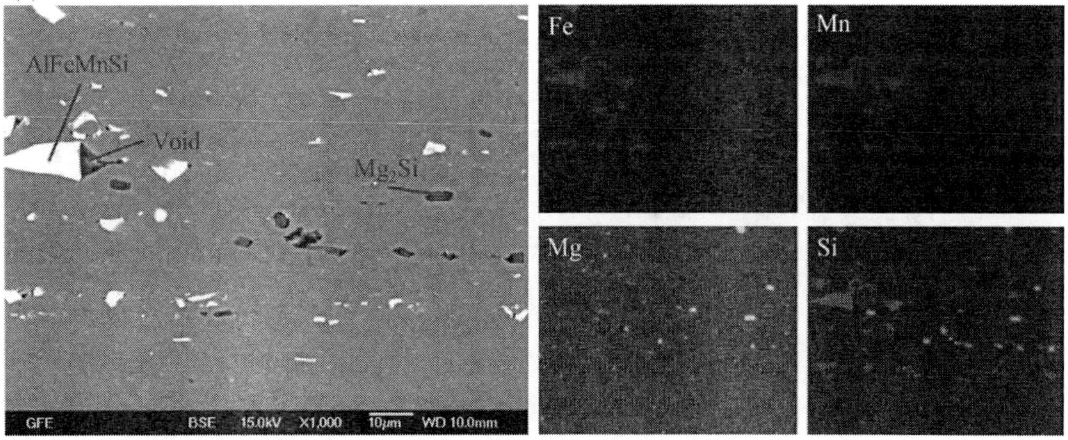

Figure 3. (a) SEM image of 6010-S H18 (magnification 1000x); (b) Element distribution (EDX).

4.2. After solution heat treatment

After solution heat treatment at 565 °C for at least 15 min, a fully recrystallized globulitic microstructure can be seen, Figure 4 (a). Since a quantitative grain size evaluation on the optical microscopy images was not possible due to weak contrast, EBSD measurements were performed. Here, the individual grains

IOP Conf. Series: Materials Science and Engineering 1284 (2023) 012005 doi:10.1088/1757-899X/1284/1/012005

are color-coded with respect to their orientation in the inverse pole figure map shown in Figure 4 (b). This enables easy evaluation of the microstructure. Overall, a fine-grained, recrystallized structure can also be seen here, consisting of smaller and larger grains. The larger grains occur more frequently in areas of greater non-metallic particle spacing, while the smaller grains are found primarily in the areas of low particle spacing. These results are in correspondence with the recrystallization observed by [7] initialized on non-metallic phases due to particle-stimulated nucleation.

The average grain size derived from EBSD data is 12.5 µm. Due to the SHT, the alloy's hardness decreased from 62.7 HV1 (initial condition) to 47.9 HV1 (solutionized).

Figure 4. (a) Etched micrograph of 6010-S after solution heat treatment (SHT) at 565 °C for 20 min (magnification: 400x); (b) inverse pole figure map from EBSD measurements.

4.3. *After deformation – cross-die component*

The cross-die component has been produced according to the process route mentioned in chapter 3. For complete form filling, comparatively high degrees of strain must be achieved in the component. Figure 5 (a) shows the final equivalent plastic strain distribution in the part determined via FE simulations. Details of the simulation model can be found in the authors' previous publications [1–3].

Based on these results, a total of four component areas have been identified whose microstructure is to be investigated after forming. Figure 5 (b) shows the positions of the four specimens taken from the cross-die component. With the help of these specimens, the influence of all strain levels and different temperature developments in the component can be visualized. However, due to similarities in the results, just the arm corner (2) will be shown in this section.

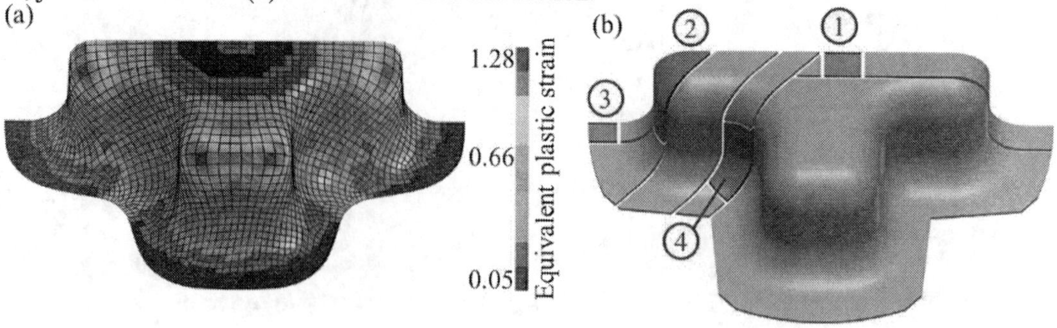

Figure 5. (a) Simulated equivalent plastic strain distribution; (b) location of specimens for metallographic analysis: (1) bottom; (2) arm corner; (3) flange; (4) concave wall region.

Since the arm corners are formed at the end of the process, the highest degree of deformation and the highest cooling occur here. As described in [2], the gaseous medium first causes cooling of the sheet via forced convection and then the sheet re-heats upon contact with the hot die surface. In the arm corners, this cooling amounts to almost 45 °C down to 520 °C. Although this results in the sheet cooling down beyond the SHT temperature (approx. 545 °C-590 °C), the cooling and reheating take place within a process time of 5 s. This time period is so short that no pre-segregation and therefore negative influence on the later Mg_2Si formation is expected. Therefore, in the following section just microstructural effects will be displayed. Figure 6 shows the microstructural evolution along the arm corner specimen.

It can be seen in Figure 6 (d) that the grains elongate in the direction of sheet stretching and in the areas with more strain, the grain elongation is more pronounced. EBSD measurements show a slightly finer-grained microstructure in this region with a mean grain size of 10.7 µm, Figure 6 (e). This can be explained by the Continuous Dynamic Recrystallization (CDRX), i. e., a continuous increase of the misorientation of sub-grains up to the transformation of Low Angle Grain Boundaries (LAGB) into High Angle Grain Boundaries (HAGB), a phenomenon that usually occurs in aluminum alloys during high temperature forming [10]. The Kernel Average Misorientation (KAM) data from the EBSD measurements support this observation by showing an increased amount of sub-grain boundaries (green lines) within individual grains, Figure 6 (f).

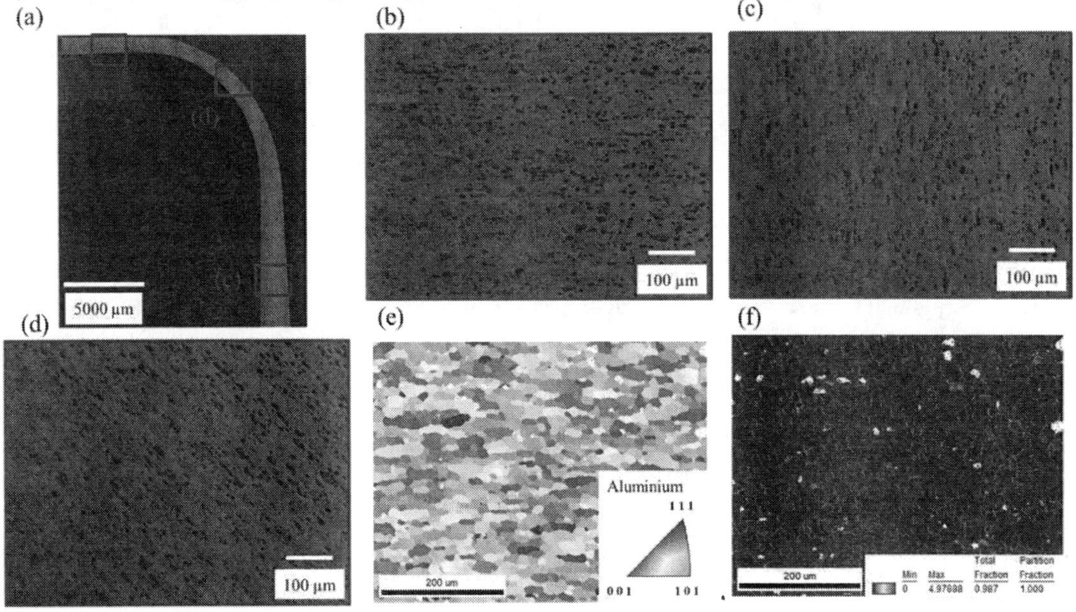

Figure 6. (a) Metallographic preparation of arm corner specimen (150x); (b) Micrograph of the bottom section; (c) wall section; (d) radius section of the arm corner (b-d: 400x); (e) EBSD analysis of the radius section; (f) KAM value derived from (e) showing LAGB formation.

These results indicate that there is a slight reduction in grain size due to CDRX during forming in areas with high degree of deformation. It needs to be investigated whether these effects influence the final strength after artificial aging. As it was not possible to take tensile specimens from the cross-die component due to its size, the round-cup component will be used for these investigations.

4.4. After artificial ageing – round cup component

The round-cup component underwent the same process route as the cross-die component. In addition, however, the round cup was also artificially aged. The associated time-temperature data can be found in chapter 3 and Figure 1. The selected artificial ageing is aimed at setting the T6 condition, whereby maximum strengths should be achieved [4]. The tensile specimens were taken from the bottom of the cup, see Figure 7 (a). To ensure that CDRX also occurs in the round cup, metallographic sections have been taken from the bottom and the radius. Figure 7 (b) shows the microstructure in those areas after forming but before artificial ageing and thus corresponds to the condition as the cross-die component.

The microstructure shows the same elongated and locally refined structure by CDRX. Artificial ageing does not affect this grain structure and it remains unchanged, see Figure 7 (c). Differences in amount and size of etched out artefacts are just due to the chosen view field and not material related.

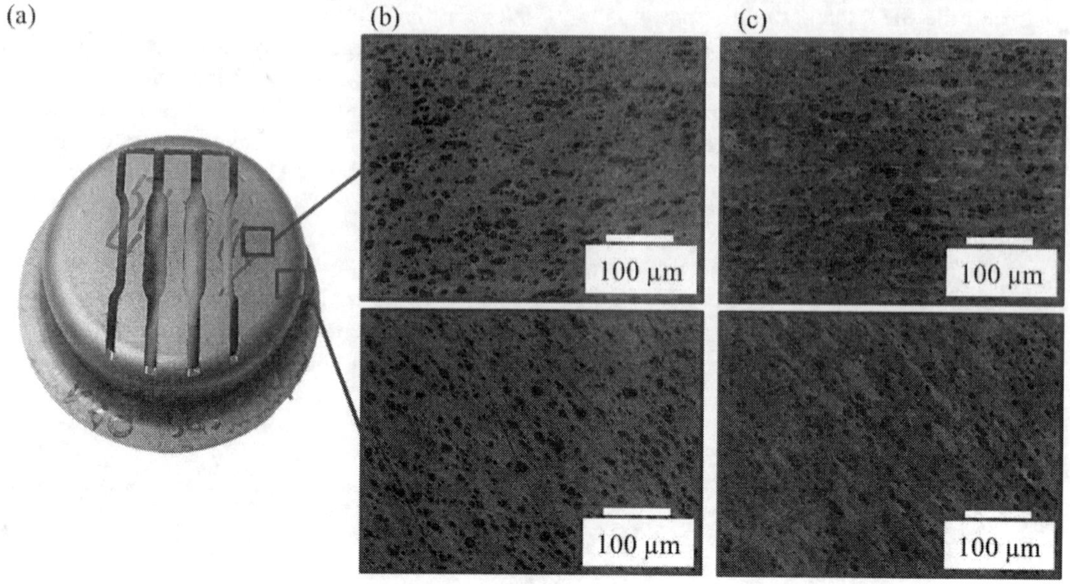

Figure 7. (a) Round-cup component with tensile specimens; (b) microstructure of cup bottom and radius after forming; (c) microstructure of cup bottom and radius after T6-treatment (b/c: 400x).

The tensile strengths achievable in the T6 state after forming are on average 353 MPa, displayed in Figure 8 (a), while the elongation at fracture is about 9.4 %. These data agree well with the results measured by Taylor et al. [4] on non-deformed material (335 MPa/11.5 %). The slightly higher strength and slightly lower elongation at fracture in the present work might be due to the shorter artificial aging.

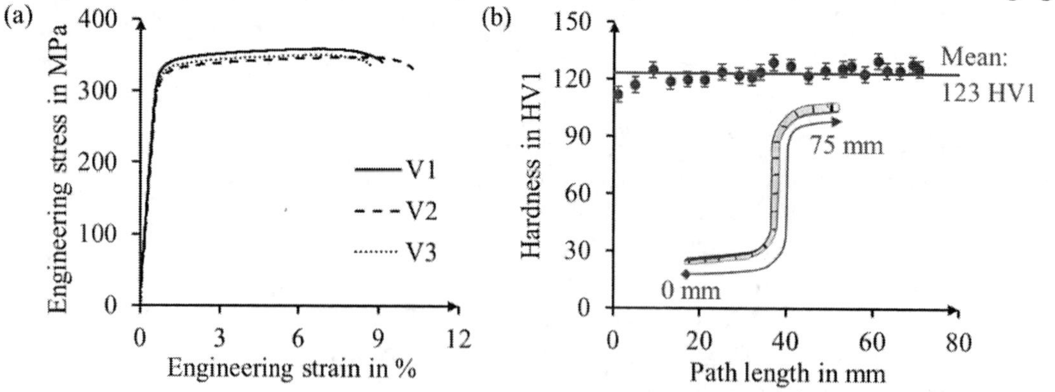

Figure 8. (a) Eng. Stress-strain curves after T6-treatment; (b) Vickers-hardness along cup contour.

Since the bottom of the cup also shows signs of CDRX but undergoes less strain compared to the radius or cross-die component, the strength will be correlated across the contour of the round cup using hardness measurements. Figure 8 (b) shows the evolution of the Vickers hardness measurement along the cup contour. It can be seen that the hardness in the bottom region is slightly lower (112 HV1) than in the flange region (126 HV1), but shows no significant fluctuations. The average hardness is 123 HV1. According to [12], a tensile strength of 354.5 MPa can be calculated from this hardness. This value corresponds to the tensile strengths measured in the tensile test, so that it can be concluded that even larger strains do not have a negative influence on the achievable tensile strengths in the T6 state.

5. Conclusion and Outlook

Within this study, the microstructure evolution of the precipitation-hardenable EN AW 6010-S alloy has been investigated over the course of a gas-based hot forming process. For this purpose, several metallographic micrographs at four different time points along the process chain have been prepared. EBSD measurements revealed the grain structure with better contrast whereas non-metallic precipitates have been identified via EDX measurements. Tensile strengths in the T6 condition have also been determined at the final stage. Finally, correlation of tensile strength with hardness measurements has helped to approximate the strength development also over the entire part contour.

In summary, the following key findings can be formulated:

- In its initial state, the selected alloy shows a large number of non-metallic inclusions of the α-AlFeMnSi phase as well as of coarse primary Mg_2Si. These inclusions serve as initiators for particle-stimulated nucleation and help to stabilize the grain structure so that a homogeneous, globulitic, fine-grained microstructure occurs after solution annealing.
- After forming, the grains elongate in the direction of sheet stretching. High elongations result in a more pronounced elongation of the grains. There is a slight reduction in grain size due to continuous dynamic recrystallization.
- T6 heat treatment resulted in high tensile strengths, which agree well with, and even slightly exceed existing literature data.
- The correlation of tensile strength and hardness proves that there is no significant fluctuation of hardness and consequently tensile strength along component contour.

Although the results suggest that the gas-based hot forming process does not have a negative influence on the selected alloy, it was already evident in the initial state that matrix separation occurs at the non-metallic inclusions. Future work should therefore critically examine how this matrix separation and thus porosity develops in dependency of strain and temperature. It should also be clarified whether this porosity influences the cyclic load-bearing capacity of the hot- formed components and how this could be reduced by process adjustments.

Acknowledgments

The authors gratefully thank Speira GmbH (previously Hydro Aluminium GmbH) for providing the aluminium sheets and LKR Ranshofen (AIT), especially Dr.-Ing. Jürgen Nietsch and Aurel Arnoldt, for providing EBSD measurements and the Central Facility for Electron Microscopy for EDX analyses.

References

[1] Baru N K, Teeuwen T, Bailly D and Hirt G 2022 *NUMISHEET 2022* ed K Inal *et al* (Cham: Springer International Publishing) pp 785–801
[2] Baru N K, Teeuwen T, Teller M, Hojda S, Braun A and Hirt G 2021 *IOP Conf. Ser.: Mater. Sci. Eng.* **1157** 12027
[3] Teeuwen T, Baru N K, Tilly K J, Bailly D and Hirt G 2022 *IOP Conf. Ser.: Mater. Sci. Eng.* **1238** 12049
[4] Taylor S, Dhara S, Slater C and Kotadia H 2023 *Metals* **13** 76
[5] He Z, Fan X, Shao F et al. 2012 *Trans. Nonfer. Met. Society of China* **22** 364–69
[6] Talebi-Anaraki A, Chougan M, Loh-Mousavi M and Maeno T 2020 *JMMP* **4** 56
[7] Hu M, Sun Y, He J, Li C, Li H, Yu L and Liu Y 2022 *Materials Characterization* **183** 111623
[8] Yang W, Ji S, Zhou X, Stone I, Scamans G, Thompson G E and Fan Z 2014 *Met and Mat Trans* A **45** 3971–80
[9] Muggerud A M F, Mørtsell E A, Li Y and Holmestad R 2013 *Mater. Sci. Eng: A* **567** 21–8
[10] Huang K and Logé R E 2016 *Materials & Design* **111** 548–74
[11] Österreicher J A, Schiffl A, Falkinger G and Bourret G R 2016 *IOP Conf. Ser.: Mater. Sci. Eng.* **119** 12028
[12] Meng C 2010 Effect of Preheating Condition on Strength of AA6060 Aluminium Alloy for Extrusion *Masterthesis* School of Engineering, Auckland University of Technology

IOP Conf. Series: Materials Science and Engineering 1284 (2023) 012006 doi:10.1088/1757-899X/1284/1/012006

Numerical analysis of the load paths and the resulting damage evolution during deep drawing of dual-phase steel

Martina Müller[1*], Tim Herrig[1] and Thomas Bergs[1,2]

[1]Laboratory for Machine Tools and Production Engineering (WZL) of RWTH Aachen University, Campus-Boulevard 30, 52074 Aachen, Germany
[2]Fraunhofer Institute for Production Technology (IPT), Steinbachstr. 17, 52074 Aachen, Germany

E-mail: martina.mueller@wzl.rwth-aachen.de

Abstract. Within the deep drawing process chain, damage develops and evolves – starting with foreign particles from the slag as well as non-metallic inclusions leading to voids, which develop, accumulate and ultimately lead to failure. This damage accumulation and evolution along the process chain decisively influences the performance of components. Accordingly, the influence on the damage accumulation and evolution offers a lever to increase the performance of components. The final damage state in the component has been shown to be dependent on the prevailing load path by means of the stress and strain states along the process route. The investigation of the stress and strain states requires numerical modelling. From the models, the load path can then be extracted and a correlation can be obtained with the damage that results. The present work investigates the load paths during deep drawing of u-shaped profiles as a function of process set-up and process parameters of dual-phase steel DP800. The resulting load paths are extracted, analyzed and correlated with the numerically predicted damage.

1. Introduction

The automotive industry has seen a shift towards the integration of lightweight measures in the product design process, driven by the need to meet the European Union's ambitious climate targets [1]. The use of lighter vehicles has been identified as an auspicious strategy for reducing CO_2 in transportation. As a result, the use of high strength materials has become a crucial aspect of the design process since other benefits such as improved vehicle performance and cost reduction goes along. Dual-phase (DP) steel is among the most used materials in the context of lightweight design and the automotive industry. DP steel is a type of high-strength, low-alloy steel that is composed of a soft ferrite matrix with a hard martensitic second phase in the form of islands. On account of the different mechanical properties of the two phases, the plastic behavior of DP steels exhibits a relatively low ratio between yield strength and tensile strength, pronounced work hardening, and exceptional overall formability. This is due to the fact that the distribution of stress and strain between the phases involved allows a large number of degrees of freedom for microstructural design [2]. This unique property profile of DP steels makes it attractive for a wide range of applications. The ability to control the microstructure through careful manipulation of the manufacturing process makes them particularly useful for applications where high strength and formability are required [3]. The use of DP steels in the automotive industry has been shown to significantly improve the strength and weight of body panels and frames, resulting in improved fuel efficiency and lower emissions. In addition, the excellent formability enables the production of complex shapes and geometries, which is essential for the design of modern cars [4]. Besides the peculiarity of the two phases and the consequent differences in the mechanical properties, DP steel also differs significantly in the development of microscopic damage compared to other types of steel. While inclusions in particular are responsible for void initiation in other steels, different mechanisms dominate in DP steels [5]. These include martensite fracture, phase boundary detachment and grain boundary decohesion, with the damage mechanisms depending in particular on the microstructure [6]. The dominant damage mechanisms, however, result from the grain sizes and the martensite content while the damage evolution itself and its amount depends on the stress-strain-state the material experiences along the process route. Nowadays, DP steel is used in particular for the production of vehicle body panels and frames by means of deep drawing. The sheets used in deep drawing typically go through a series of processes before the forming process such as steelmaking, continuous casting, hot and cold rolling, and heat treatment. These processes can all have an impact on the damage evolution of the material. The deep drawing process then further impacts the accumulated damage in the material. The damage state after deep drawing, however, greatly affects the component's properties in terms of fatigue and crash behavior.

In the past, many studies have been conducted to identify factors that influence the deep drawing process, such as friction, blank holder force, blank shape, and punch velocity [7]. Most of this research has focused on producing a defect-free component at a macroscopic level, such as preventing wrinkling, e.g. in Chen et al. [8], sheet thinning, e.g. in Preedawiphat et al. [9], or springback, e.g. in Durmaz et al. [10] or Nanu et al. [11]. In addition, a number of efforts have been made to analyze damage evolution during deep drawing. Therefore, different numerical models have been developed over the years to predict damage during deep drawing. Currently, there are two main types of models used to predict the behavior of materials during deformation: coupled and uncoupled models [12]. Coupled models take into account the effects of damage on the material's ability to withstand deformation, and are often based on micro-mechanical or phenomenological principles. Examples of coupled models include the Gurson-Tvergaard-Needleman (GTN) [13–15] model and the Lemaitre model [16,17]. Micromechanical models, such as the GTN model, aim to depict physical phenomena like void nucleation using a set of parameters, which require a significant amount of calibration to obtain accurate results. Phenomenological models, like the Lemaitre model, take a more macroscopic approach, using mathematical approach to describe the effects of damage on the material's behavior. On the other hand, uncoupled models describe the material's behavior, including fracture, without taking damage into account. Notable examples of uncoupled models include the Johnson-Cook [18] and Bai-Wierzbicki [19] models. Considering the Lemaitre model, it is based on the concept of the damage parameter D, which affects the plastic behavior of the material and reflects the formation of voids and microcracks. D is defined as the ratio of the area affected by damage ∂S_D to the damage-free area increment ∂S, as introduced by KACHANOV [20] and further utilized by LEMAITRE [16,17]. As the material is loaded with mechanical forces, the damage variable D increases and the stress distribution changes, requiring the use of an effective stress $\tilde{\sigma}$ to estimate the change in elastic and plastic behavior of the material. This coupling of D and stress σ allows for the estimation of the material's response to applied stress (equation 1).

$$\tilde{\sigma} = \frac{\partial F}{\partial S - \partial S_D} = \frac{\partial F}{(1-D)\partial S} = \frac{\sigma}{1-D} \tag{1}$$

Although a variety of models have been developed to predict damage, many have been used to forecast failure at the macro level, .e.g. Barrera et al. [21,22] or Saxena et al. [22]. Only few, however, focused on micro level damage with the understanding that *ductile damage means the formation, growth and coalescence of voids in the microstructure. These voids are formed in structural materials by decohesion at interfaces such as phase or grain boundaries and inclusions, or by the formation of new surfaces within phases or inclusions. This damage causes degradation of the performance of the corresponding component [23].* Consequently, there are knowledge deficits against the context of this comprehension of damage, also considering its potentials in manufacturing processes. The present work focuses on the influence of process parameters and process set-up on the damage evolution during deep drawing of u-shaped profiles from DP steel DP800. The overall goal is to gain a deeper understanding of the underlying mechanisms of damage evolution during deep drawing and to progress the development of more efficient and effective manufacturing processes.

2. Materials and Methods

2.1. Approach

The purpose of this study is to analyze the development of damage in the form of accumulation and distribution. To achieve this, various types of deep drawing methods including classical deep drawing (DD), multistep deep drawing (MSDD), and reverse deep drawing (RDD) as well as different drawing die radii r_D are considered. The aim is to produce identical geometries using different process routes in order to determine the impact of different load paths on damage development. This will allow for targeted adjustments to be made to the accumulation and distribution of damage in order to enhance component performance in the future. Since damage is affected by stress and strain states, the load paths are initially analyzed based on their time- and location-dependent states, which are simulated numerically in this study. Finally, the load paths are linked to the accumulation and distribution of damage to uncover causal relationships. To describe the load paths, the von Mises stress σ_v (equation 2), the triaxiality η (equation 3) and the load angle parameter $\bar{\theta}$ (Equation 5) are used. These can be calculated from the respective principal stresses σ_i and the lode angle θ.

$$\sigma_v = \sqrt{0{,}5\,[(\sigma_1 - \sigma_2)^2 + (\sigma_2 - \sigma_3)^2 + (\sigma_3 - \sigma_1)^2]} \tag{2}$$

$$\eta \frac{\sigma_1 + \sigma_2 + \sigma_3}{3\sqrt{0{,}5\,[(\sigma_1 - \sigma_2)^2 + (\sigma_2 - \sigma_3)^2 + (\sigma_3 - \sigma_1)^2]}} \tag{3}$$

$$\theta = \operatorname{arccot}\left(\frac{2}{\sqrt{3}}\left(\frac{\sigma_1 - \sigma_3}{\sigma_2 - \sigma_3} - \frac{1}{2}\right)\right) \tag{4}$$

$$\bar{\theta} = 1 - \frac{6\,\theta}{\pi} \tag{5}$$

2.2. Deep drawing process

The deep drawing process and its variants used in this study are depicted in Figure 1. The geometry to be deep drawn was a u-shaped component from DP800 dual phase steel with a sheet thickness s of $s = 1.5$ mm. The punch speed v_p was set to $v_p = 15$ mm/s and the blank holder force F_{BH} to $F_{BH} = 40$ kN.

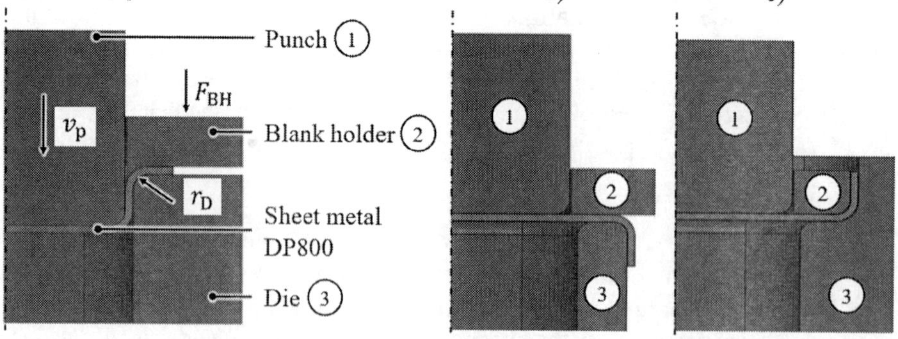

Figure 1. Schematic illustration of a) deep drawing (DD), b) reverse deep drawing (RDD), c) multi-step deep drawing (MSDD)

2.3. *Process model*

The different process variants were modelled using Abaqus/explizit FEA from Dassault Systèmes. A general contact with a constant coefficient of friction $\mu = 0.5$ was used to model the sheet metal's interactions with the tools, as determined in previous strip drawing tests. To save computation time, only one quarter of the geometries was modelled utilizing the symmetry. The sheet metal was modelled as a solid deformable body meshed with three-dimensional reduced-integration brick elements (C3D8R). The blankholder and punch were represented as shell discrete rigid bodies using a 4-node, bilinear quadrilateral mesh (R3D4). As material model a local Lemaitre damage model was used based on the work from Sprave et al. [24] The model has been successfully used in earlier work, e.g. [23,25,26].

2.4. *Reference process, position and point*

The reference process selected for the study is the DD process, with a drawing die radius r_D of r_D= 3 mm. The analysis of the load paths were conducted at a reference point located at the outer bend of the component which reflects the area of the highest bending radius. This area is considered the most sensitive area for performance in later applications. To compare different die radii r_D, the components were drawn without a flange. For comparing different process set-ups, the reference position of s = 17.6 mm was used. This punch path s corresponds to the punch path s just before the first component experiences failure.

3. Results

Figure 1 shows the course of triaxiality η, lode angle parameter $\bar{\theta}$ and von Mises stress σ_v in reliance of equivalent plastic strain φ_{pl} at the outer bend of the component in dependence of the drawing die radius r_D at the reference position. The curves of all three variables are overall similar and possess an

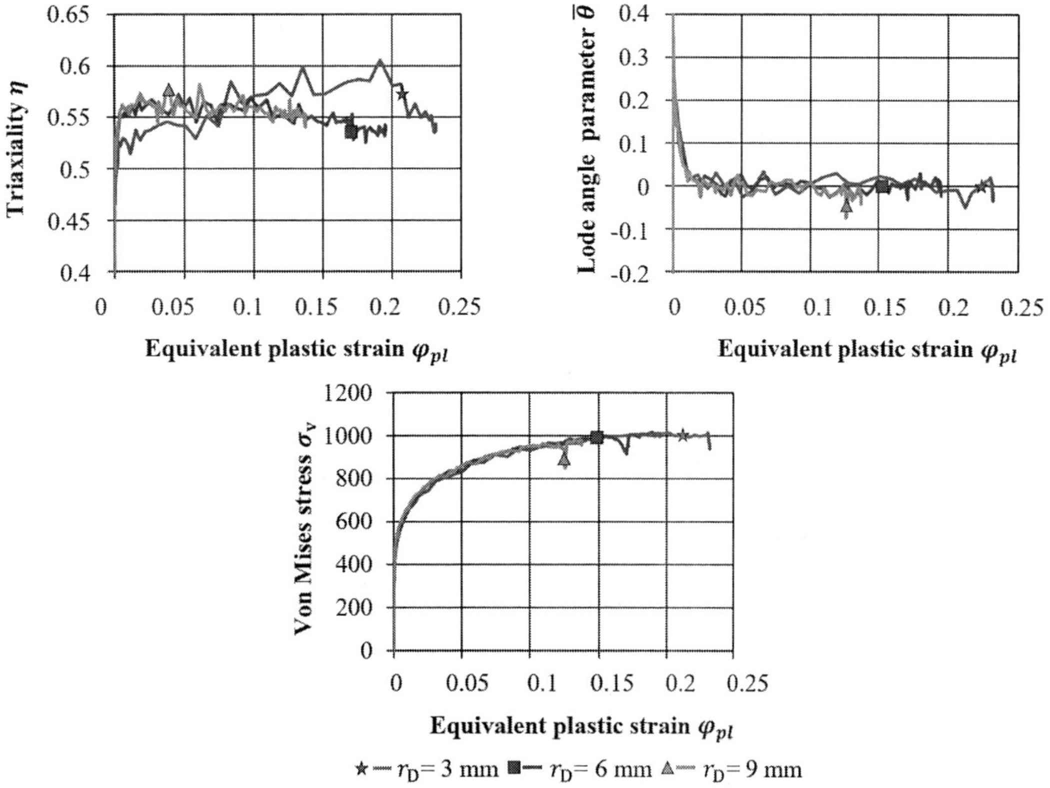

Figure 2: Course of triaxiality η, lode angle parameter $\bar{\theta}$ and von Mises stress σ_v in reliance of equivalent plastic strain φ_{pl} at outer bend of component in dependence of drawing die radius r_D

almost identical range in all areas. Only the triaxiality η course of $r_D = 3$ mm shows a small incline. The curve starts at a lower value and then crosses the other two due to higher gradient. It is also noticeable that the selected area at the reference position is subjected to a different equivalent plastic strain φ_{pl} in order to achieve an almost identical geometry.

The damage distribution of the damage parameter D is shown in figure 3a. It is evident that by changing the drawing die radius r_D not only the damage distribution can be influenced, but also the damage maximum D_{max} can be shifted. This is also visible in the course of the damage parameter D on the outside of the component (Figure 3b). When comparing the components with the different drawing die radii r_D, no damage is initially evident in the bottom for $x < 20$. In the outer bend area, however, the damage parameter D increases. First, an elevation is seen for $D_{r_D = 3\,mm}$, followed by $D_{r_D = 6\,mm}$ and $D_{r_D = 9\,mm}$. The local maximum of D in the outer bend is approximately the same for all variants. While the damage parameter D then drops again sharply for $r_D = 6$ mm and $r_D = 9$ mm, a further peak is seen for $r_D = 9$ mm. In the area of the wall, a clearly different course is shown for each individual variant. For $r_D = 3$ mm, the course of the damage parameter rises sharply and reaches the global maximum directly, only to drop again in an oscillating manner. The variant with a target drawing die radius of $r_D = 6$ mm, on the other hand, shows a later onset of increase, which, however, also rises to the global maximum.

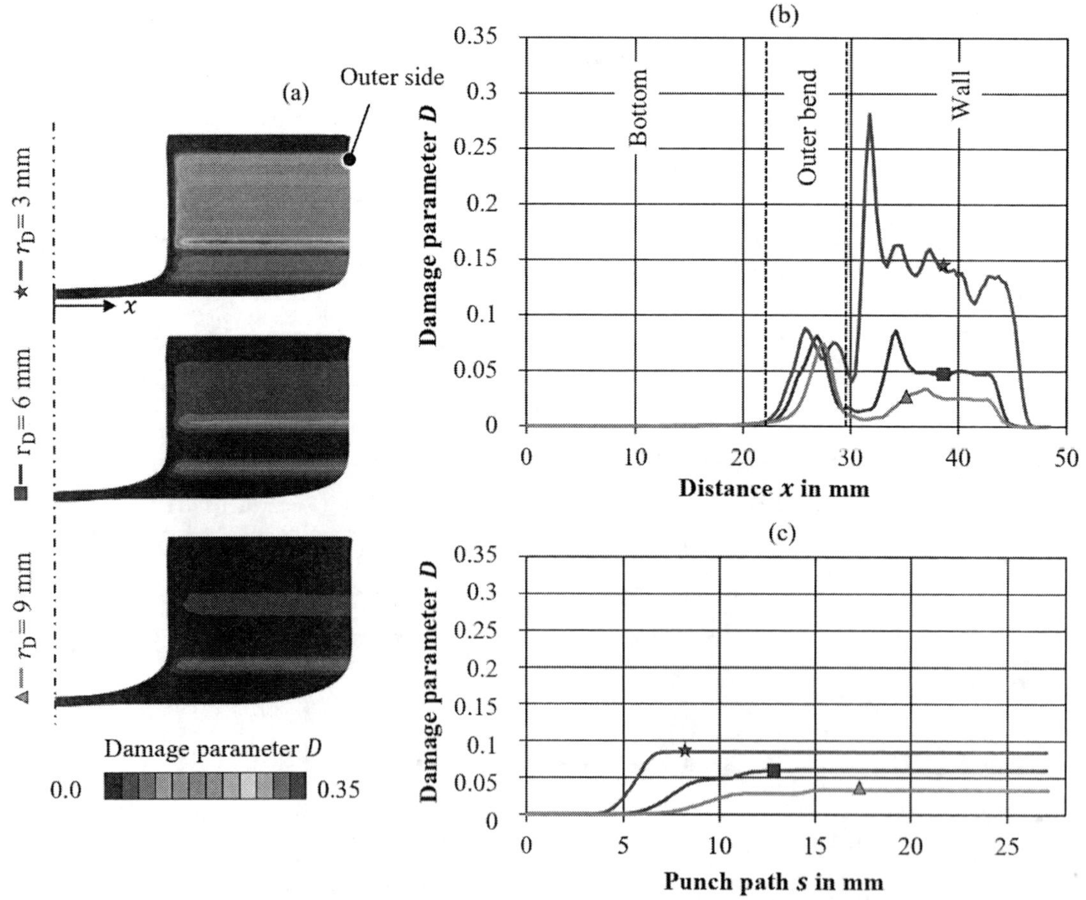

Figure 3: Damage distribution in component (a) and course of damage parameter D at outer side of components (b) and damage parameter D at reference point in reliance of punch path s (c) in dependence of drawing die radius r_D

Subsequently, the damage parameter D drops to a plateau and then to zero. The course of the damage parameter D of the target ring radius $r_D = 9$ mm mm shows an even flatter increase. In general, the amount of damage within the component decreases with an increasing drawing die radius r_D. The increase does not result in a global but in a local maximum of D for this section. A drop to zero occurs at an identical path as for $r_D = 6$ mm. The curves of the damage parameter D over the punch path s at the reference point can be seen in Figure 3c. With increasing drawing ring radius r_D, a later increase of the curve can be observed with a lower gradient. At the same time, D_{max} is smaller with increasing drawing ring radius r_D.

Figure 4 depicts the variation of triaxiality η, lode angle parameter $\bar{\theta}$ and von Mises stress σ_v in reliance of equivalent plastic strain φ_{pl} at the outer bend of the component for different process set-ups at the reference position. The curves of all three variables show remarkable resemblance and have a nearly identical range throughout. However, the course of RDD and MSDD start with $\varphi_{pl} > 0$ due to the preliminary step.

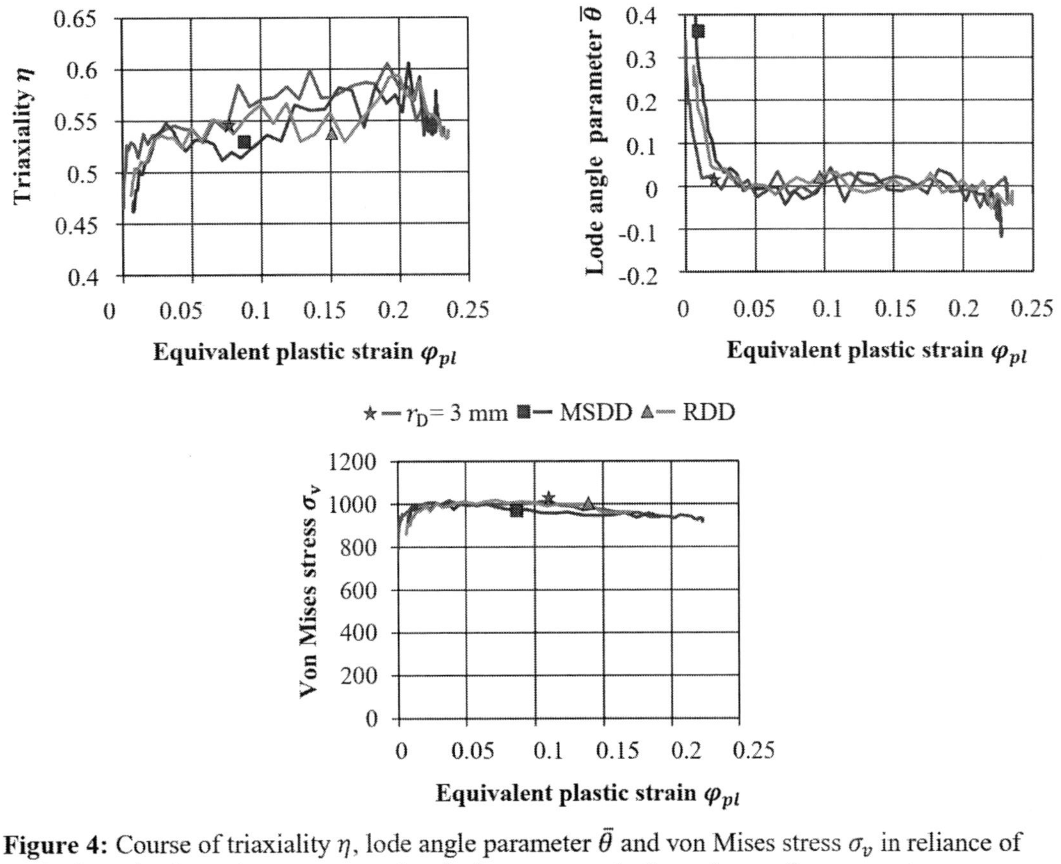

Figure 4: Course of triaxiality η, lode angle parameter $\bar{\theta}$ and von Mises stress σ_v in reliance of equivalent plastic strain φ_{pl} at outer bend of component in dependence of process set-up

Figure 5a shows the distribution of the damage parameter D in dependence of process set-up. As can be seen, changes in the process set-up not only influence the damage distribution, but also the position of the maximum damage parameter D_{max}. This is also evident in the course of the damage parameter D on the outer surface of the component shown in Figure 5b. While there is initially no damage in the bottom of all three curves, the damage parameter D increases in the outer bend, where it changes to an oscillating, decreasing curve. The course of D in the outer bend is almost identical in all three variants. Immediately after the transition from the outer bend to the wall, there is a sharp increase in D for all three courses. This is most pronounced for RDD, followed by MSDD and DD. The course from DD then drops to a plateau with an oscillating course to drop to zero at the end of the wall. The course from RDD shows two more local maxima of D to then drop to zero just before the transition to the flange followed by a peak again in the flange. The course of MSDD is very similar to the characteristics of the course, but the increases of D show clearly higher maximum values. The damage parameter over the punch path is shown in Figure 5c. The initial increase in the curves occur with an offset, with RDD increasing first, followed by $r_D = 3$ mm and MSDD. The maximum damage parameters D_{max}, however, are at a very similar level.

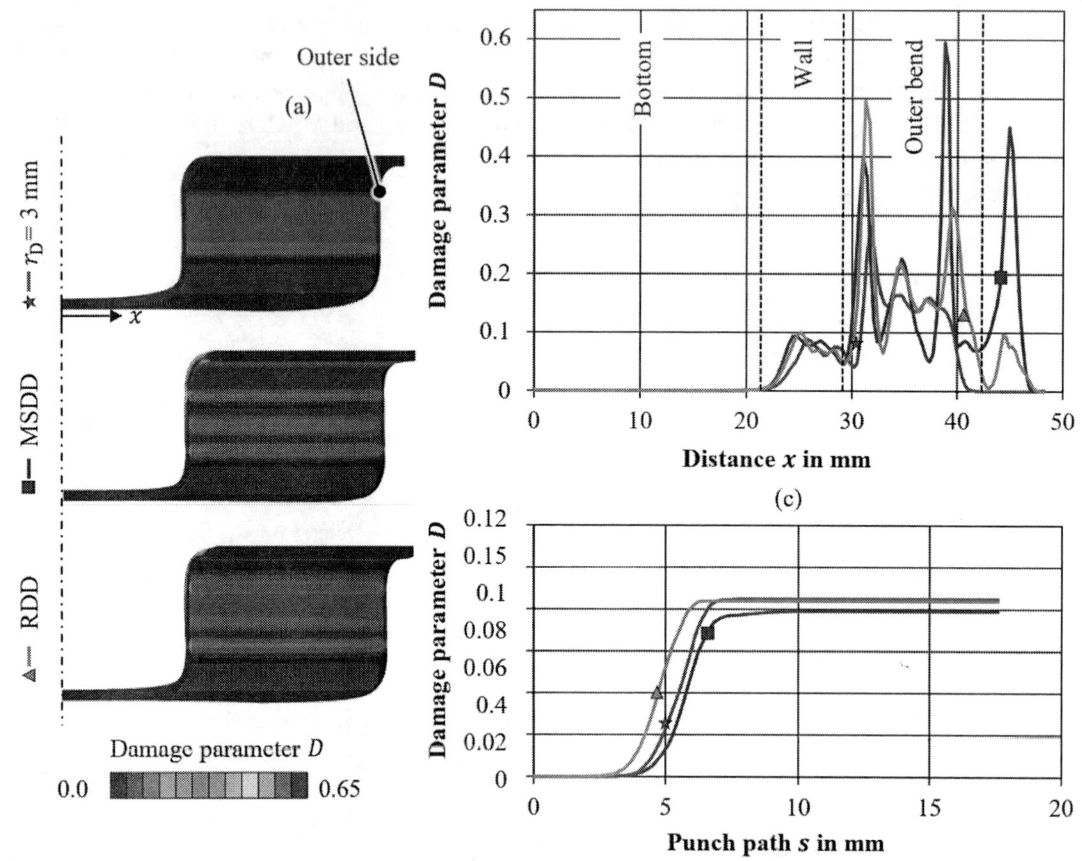

Figure 5: Damage distribution in component (a), course of damage parameter D at outer side of components (b) in dependence of process set-up and damage parameter D at reference point in reliance of punch path s (c) in dependence of process set-up

4. Discussion and summary

Considering the load paths as a function of the drawing die radii r_D in terms of triaxiality η, lode angle parameter $\bar{\theta}$ and von Mises stress σ_v as a function of equivalent plastic strain φ_{pl} at the reference position, the courses differ in particular in the maximum equivalent plastic strain $\varphi_{pl.max}$. The course of the damage parameter D, on the other hand, differs in two characteristics. On the one hand, the damage development takes place at a later stage with increasing drawing ring radius r_D. On the other hand, as the drawing ring radius r_D increases, the maximum damage parameter D_{max} decreases. This later rise in damage development is believed to be related to the time-dependent bending radius. The larger the drawing die radius r_D, the later the final bending radius is reached at the reference position around the punch radius. Accordingly, the damage development starts at a delay. With regard to the relatively lower maximum damage parameter D_{max}, a correlation is drawn with the relatively lower equivalent plastic strain φ_{pl}. Since damage development is only associated with plastic strain φ_{pl} and the plastic strain φ_{pl} is greater with a smaller drawing die radius r_D, a slightly higher value of damage development is also achieved at the reference position. With regard to the damage distribution in the entire component, a change in drawing die radius r_D shows a shift in position of the maximum damage parameter D_{max} as well as a general modification of the damage distribution. It is assumed that due to the smaller bending radius that the sheet experiences when exceeding the drawing die radius r_D and the associated lower equivalent plastic strain φ_{pl}, correspondingly less damage occurs. Similar to the change in the drawing die radii r_D, the load paths at the reference position also show similar curves when the process set-up is changed. The same applies to the maximum damage parameter D_{max} attained. However, the damage distribution and the location of the damage maximum change when the process set-up is modified. Due to the different process set-up, other plastic strains φ_{pl}, especially in the form of bending radii, coupled with altered stress-strain states result in other component areas. These are assumed to lead to different damage distribution in the component.

5. Conclusion and outlook

On the one hand, different drawing ring radii and, on the other hand, different process set-ups in the form of MSDD and RDD and the associated influence on the damage development in the component were investigated. For this purpose, the load paths were analyzed in terms of triaxiality η, lode angle parameter $\bar{\theta}$ and von Mises stress σ_v as a function of equivalent plastic strain φ_{pl}. Furthermore, the location of the maximum damage parameter D_{max} and the damage distribution were studied. Although the damage at the reference position was not significantly affected, it could be shown that by changing process parameters as well as process set-up especially the damage distribution as well as the location of the maximum damage parameter D_{max} can be influenced in the components wall. In general, the damage in the component could be reduced by changing the drawing ring radius r_D. The next step is to investigate the influence of the change in drawing ring radius r_D and process set-up on other component-immanent properties, such as springback or hardening.

Acknowledgement

This research was funded by Deutsche Forschungsgemeinschaft (DFG, German Research Foundation; Projectnumber 278868966 – TRR 188; Damage Controlled Forming Processes). The authors would also like to thank A. Erman Tekkaya, Till Clausmeyer, Alexander Schowtjak and Jan Gerlach from the Institute of Forming Technology and Lightweight Components of TU Dortmund, sub-project S01 of TRR 188, for the usage of their implementation of the LEMAITRE damage model. Simulations were performed with computing resources granted by RWTH Aachen University under project rwth0907.

References

[1] European Comission. *European Green Deal.* https://climate.ec.europa.eu/eu-action/european-green-deal_en
[2] Bieler T R, Eisenlohr P, Roters F, Kumar D, Mason D E, Crimp M A and Raabe D The role of heterogeneous deformation on damage nucleation at grain boundaries in single phase metals 2009 *International Journal of Plasticity* **9** pp 1655–1683
[3] Kalhor A, Soleimani M, Mirzadeh H and Uthaisangsuk V A review of recent progress in mechanical and corrosion properties of dual phase steels 2020 *Civ.Mech.Eng* **3**
[4] Olsson K, Gladh M, Hedin J and Larsson J Microalloyed high strength steels for reduced weight and improved crash performance in automotive applications 2006 *International Symposium on Niobium Microalloyed Sheet Steel for Automotive Application* pp 223–234

IOP Conf. Series: Materials Science and Engineering 1284 (2023) 012006 doi:10.1088/1757-899X/1284/1/012006

[5] Tasan C C, Hoefnagels J and Geers M Microstructural banding effects clarified through micrographic digital image correlation 2010 *Scripta Materialia* **11** pp 835–838
[6] Ahmad E, Manzoor T, Ali K L and Akhter J I Effect of Microvoid Formation on the Tensile Properties of Dual-Phase Steel 2000 *Journal of Materials Engineering and Performance* **3** pp 306–310
[7] Tiwari P R, Rathore A and Bodkhe M G Factors affecting the deep drawing process – A review 2022 *Materials Today: Proceedings* pp 2902–2908
[8] Chen K, Carter A J and Korkolis Y P Flange Wrinkling in Deep-Drawing: Experiments, Simulations and a Reduced-Order Model 2022 *JMMP* **4** p 76
[9] Preedawiphat P, Koowattanasuchat P, Mahayotsanun N and Mahabunphachai S Sheet thinning prediction method based on localized friction effect in deep-drawing 2020 *Advances in Mechanical Engineering* **9**
[10] Durmaz U, Heibel S, Schweiker T, Prabhakar A and Merklein M Influence of the forming process on springback 2022 *IOP Conf. Ser.: Mater. Sci. Eng.* **1** p 12074
[11] Nanu N and Brabie G Analytical model for prediction of springback parameters in the case of U stretch–bending process as a function of stresses distribution in the sheet thickness 2012 *International Journal of Mechanical Sciences* **1** pp 11–21
[12] Besson J Continuum Models of Ductile Fracture: A Review 2010 *International Journal of Damage Mechanics* **1** pp 3–52
[13] Gurson A L Continuum Theory of Ductile Rupture by Void Nucleation and Growth: Part I—Yield Criteria and Flow Rules for Porous Ductile Media 1977 *Journal of Engineering Materials and Technology* **1** pp 2–15
[14] Tvergaard V and Needleman A Analysis of the cup-cone fracture in a round tensile bar 1984 *Acta Metallurgica* **1** pp 157–169
[15] Tvergaard V Influence of voids on shear band instabilities under plane strain conditions 1981 *Int J Fract* **4** pp 389–407
[16] Lemaitre J A Continuous Damage Mechanics Model for Ductile Fracture 1985 *Journal of Engineering Materials and Technology* **1** pp 83–89
[17] Lemaitre J 1996. *A course on damage mechanics*. 2nd ed. Berlin, New York, Springer
[18] Johnson G R and Cook W H Fracture characteristics of three metals subjected to various strains, strain rates, temperatures and pressures 1985 *Engineering Fracture Mechanics* **1** pp 31–48
[19] Bai Y and Wierzbicki T A new model of metal plasticity and fracture with pressure and Lode dependence 2008 *International Journal of Plasticity* **6** pp 1071–1096
[20] Kachanov L M 1986. *Introduction to continuum damage mechanics*. Dordrecht, Springer. (Mechanics of Elastic Stability, 10).
[21] Barrera C, García-Herrera C, Celentano D J and Signorelli J W Material and Damage Characterization of the Elastoplastic Response of the EK4 Deep Drawing Steel 2022 *Metals* **5** p 720
[22] Saxena R K and Dixit P M Numerical analysis of damage for prediction of fracture initiation in deep drawing 2011 *Finite Elements in Analysis and Design* **9** pp 1104–1117
[23] Müller M, Weiser I F, Herrig T and Bergs T Numerical Prediction of the Influence of Process Parameters and Process Set-Up on Damage Evolution during Deep Drawing of Rectangular Cups 2022 *Engineering Proceedings* p 1–10
[24] Sprave L, Schowtjak A, Meya R, Clausmeyer T, Tekkaya A E and Menzel A On mesh dependencies in finite-element-based damage prediction: application to sheet metal bending 2020 *Prod. Eng. Res. Devel.* **1** pp 123–134
[25] Nick M, Feuerhack A, Bergs T and Clausmeyer T Numerical Investigation of Damage in Single-step, Two-step, and Reverse Deep Drawing of Rotationally Symmetric Cups from DP800 Dual Phase Steel 2020 *Procedia Manufacturing* pp 636–642

IOP Conf. Series: Materials Science and Engineering 1284 (2023) 012006 doi:10.1088/1757-899X/1284/1/012006

[26] Bergs T, Nick M, Feuerhack A, Trauth D and Klocke F Numerical-experimental investigation of load paths in DP800 dual phase steel during Nakajima test 2018 *Proceedings of international Esaform conference*

Oxygen-free resistance heating with nitrogen and silane as an energy-efficient heating process for hot stamping

B-A Behrens[1], S Hübner[1], U Holländer[2], A Langohr[2], L Albracht[1*], E Faramand[1], J Wehmeyer[1], S Yarcu[1], S Raumel[3] and M C Wurz[3]

[1] Institute of Forming Technology and Machines, An der Universität 2, 30823 Garbsen, Germany
[2] Institute of Materials Science, An der Universität 2, 30823 Garbsen, Germany
[3] Institute of Micro Production Technology, An der Universität 2, 30823 Garbsen, Germany

E-mail: albracht@ifum.uni-hannover.de

Abstract. Hot stamping is a process for the production of ultra-high-strength components used in the automotive industry for passenger protection. In this process, cut blanks are heated in a roller hearth furnace, which is operated with gas, to approx. 950 °C in 8 to 10 min and then formed and quenched in a water-cooled tool. This extends the yield strength of the components up to 1500 MPa. The long heating time is caused by the coating of aluminium and silicon (AlSi) with a thickness of 70 µm. During the 8 to 10 min, an intermetallic phase forms and ensures that the blank does not scale. Due to the long heating time and the energy carrier gas, the process is very energy inefficient and can only be controlled slowly. Resistance heating provides an energy-efficient alternative. Due to the direct current flow through the sheet, up to 68% of the energy can be saved compared to roller hearth furnace heating. The process is a high-speed heating process that can heat sheets to 950 °C in less than 10 seconds. The heating time is not sufficient for conventional coatings to bond sufficiently with the base material, which means that there are currently no suitable coatings for resistance rapid heating. A new approach is to suppress scale formation by reducing the oxygen during heating. By lowering the oxygen partial pressure to the XHV level, scale formation is not possible. At the same time, this absence of oxygen is an ideal condition for applying coatings. Due to the developed resistance heating device, sheets can be heated within a very short time without scale and without the need for fossil fuels.

1. Introduction

Hot stamping is used industrially to produce ultra-high-strength near-net-shape components. Due to the achievable high tensile strengths of 1500 MPa, these components are mainly used in the automotive industry for passenger protection. Hot stamping begins by heating the manganese boron steels above the austenitizing temperature to approx. 950 °C. The manganese boron steels are then quenched and cooled down to the ambient temperature. [1] For scale protection, the sheets are usually coated with an aluminum-silicon alloy (AlSi) coating. This allows only a low heating rate to ensure sufficient adhesion between the sheet material and the coating. For this reason, the sheets are heated in 8 to 10 minutes. Heating is usually accomplished by roller hearth furnaces. These are matched to the cycle time of the press and are therefore several 10 meters long, so that a hot sheet can be removed from the roller hearth furnace several times a minute to be hot-stamped. At the same time, they are operated with natural gas, which is an immense cost driver in production, especially in view of the current geopolitical situation. After heating, the sheet is removed from the furnace and placed in a press with a water-cooled tool. Cooling of at least 27 K/s creates a component with a complete martensitic microstructure. [2] Due to the forming process in combination with rapid cooling, cracks are formed in the AlSi coating, which can also extend to the base material. Since these components have active aluminum coursed corrosion protection and are also sealed by cathodic dip painting, such cracks are considered negligible [3]. The components produced in this way are usually laser-cut afterwards and integrated into the vehicle structure in the automotive sector by resistance spot welding. An alternative heating method is resistance heating. The direct flow of current through the sheet makes efficient use of the energy and allows heating rates of more than 100 K/s [4]. However, the industrially widespread AlSi coating is unsuitable for the resistance heating process because the electric field affects the coating and the short heating time is insufficient for the entire coating to react completely with the base material. Uncoated sheet scales during resistance heating in a similar manner to furnace heating, although to a lesser extent [5]. A new approach to this challenge is being researched in the Collaborative Research Center 1368 "oxygen-free production". Using a developed process chamber and the process gases nitrogen and silane, the oxygen content is reduced to an extremely high vacuum (XHV) adequate level. In this test chamber, sheets can be heated without scale forming. At the same time, the XHV adequate atmosphere and heating is used to coat the sheet. In the following, the coating formation of NiP(Cr) coatings on manganese-boron steels is investigated during rapid heating with an inductive laboratory setup in silane nitrogen atmosphere and a process window with the heating temperatures and holding times is worked out. These will then be validated by resistance heating experiments in XHV equivalent atmosphere. Some of the nickel-coated sheets will then be investigated for resistance spot welding by providing sheets with a spot weld and determining the maximum shear tensile strength in a tensile testing machine. [7]

2. Materials and Methods

2.1. Coating Materials

In this work, the conventional, low-melting NiCrP brazing alloys Ni700 (89Ni-11P, data in Ma.-%), Ni710 (76Ni-14Cr-10P) and Ni720 (65Ni-25Cr-10P) are used. According to the manufacturer, depending on the alloy, heat treatments in the temperature range between about 900 and 1000 °C are required according to their melting range, whereby the coating result also depends on the holding time.

Table 1. Chemical Properties of the coating

Lot name	Chemical composition / mass %			Melting range / °C
	Ni	Cr	P	
Ni700	89	-	11	875
Ni710	76	14	10	890
Ni720	65	25	10	950

2.2. Test materials

Uncoated 22MnB5 sheet of thickness 1.5 mm, were in-situ nickel coated under XHV-adequate atmosphere and AlSi-coated 22MnB5 of thickness 1.8 mm were used for the reference tests. The chemical properties can be found in Table 2 and the mechanical properties in Table 3.

Table 2. Chemical Properties of the 22MnB5

Element	C	Mn	Si	P	S	Al	Ti	Cr	B	Mo	Cu	Ni
[%]	0.22 – 0.25	1.20 – 1.40	0.20 – 0.30	0.020	0.005	0.02 – 0.05	0.02 – 0.05	0.11 – 0.20	0.002 – 0.0035	0.10	0.10	0.10

Table 3. Mechanical Properties of the 22MnB5 (delivery state, untreated)

Yield strength $R_{p\,0,2}$	Tensile strength R_m	Total elongation A_{80}
310 – 400 MPa	280 – 580 MPa	> 20 %

2.3. Preliminary tests for coating with Ni(Cr)P powders

The high heating rates favoured for coating production under SiH4-doped inert gas were simulated in preliminary tests by inductively heating test sheets in a quartz glass inert gas recipient. The setup with the individual components is shown in Figure 1. The Temperature-time (T-t) regimes are determined by video recordings of the melting and flow behaviour with the sheet temperature, see Fig. 2 [6]. Thus, from the combination of the obtained temperature curves from the sample substrate with the timing of the observed melting process, the actual surface temperature of the sample to be coated can be specified to be about ±3 °C. A total of around 50 individual tests were carried out with the above-mentioned coating systems, with maximum target temperatures ranging from 850 to 1120 °C. Metallographic and scanning electron microscopic analyses of the coating results are then used to evaluate the coatings and coating process parameters used with regard to their potential suitability for subsequent application in resistance heating [7].

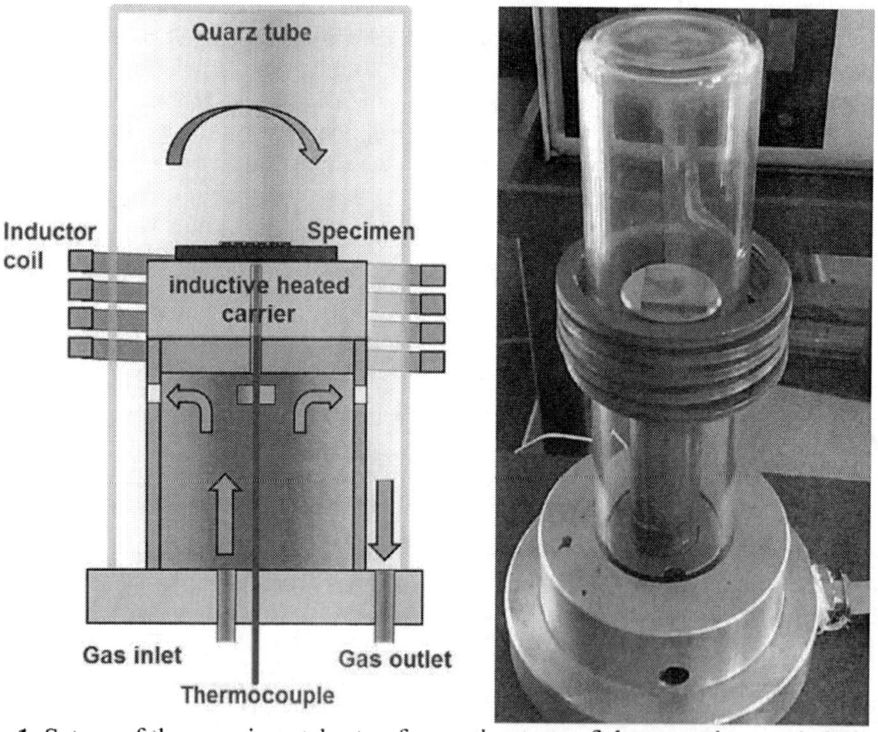

Figure 1. Set-up of the experimental setup for coating tests of sheet specimens via inductive heating. [7]

As an example, the procedure for determining the temperature profile at the specimen surface during layer formation when melting Ni700 powder coating is shown in Figure 4. The eutectic alloy has a defined melting point of 875 °C, according to which the start of melting, the end of melting and the start of solidification can be assigned. A further visually visible assignment point is the attainment of the maximum spread of the melt, which correlates with the maximum surface temperature. Due to the indirect heating via the sample carrier, there is also the boundary condition that the surface temperature must always be below the measured temperature in the sample carrier. This consideration results in the dashed curve of the surface temperature in Fig. 4. The curve clearly shows that in the present case the maximum temperature of 905 °C is reached about 10 s after the start of melting and that good wetting and layer consolidation has occurred due to the visible spreading.

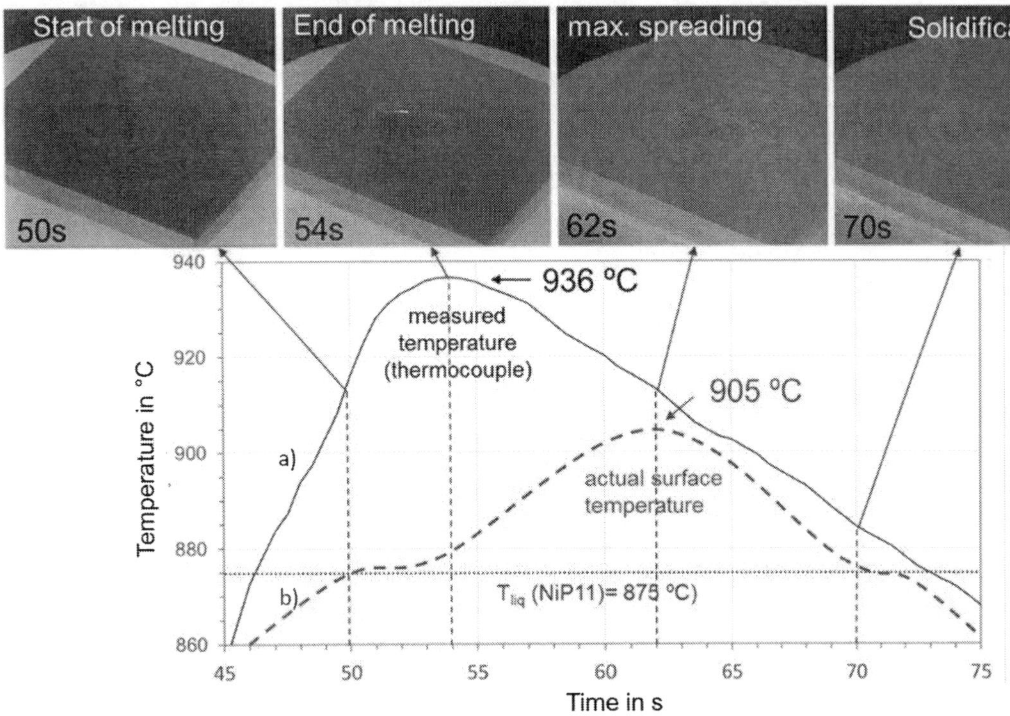

Figure 2. Correlation between the measured temperature profile, a) black line, and the observed melting and wetting behavior of the coating material with the temperature profile at the sheet surface, b) red line, shown as an example for a melting test with Ni700. [6]:

2.4. Experimental setup for resistance heating in XHV adequate atmosphere

For resistance heating with nitrogen and silane gases in XHV adequate atmosphere, a test chamber was specially constructed and integrated into an existing resistance system (see Figure 3). During heating, the blank is partially located in the test chamber and fixed between the upper and lower boxes. The contact surface between the welded housing and the sheet metal is lined with a high-temperature insulation Bra-Board®M, which can withstand temperatures of up to 1200 °C. There are two viewing windows, one at the front to inspect the process and one at the top to control the heating by means of a pyrometer. At the rear inlets are positioned for inserting the process gases nitrogen and silane, a corrugated stainless-steel tube for sampling the process atmosphere and measuring oxygen, and a 2-inch exhaust port for extracting the process atmosphere before opening the test chamber. Oxygen is measured by means of a calibrated lambda probe (MESA Industrie-Elektronik GmbH, Marl, Germany), with which automated recording and display is possible.

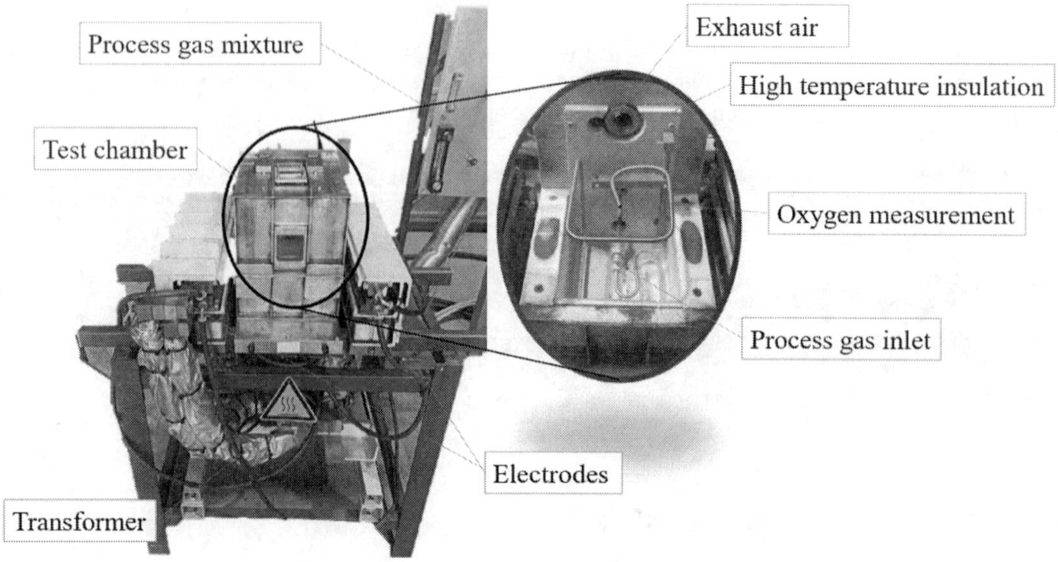

Figure 3. Design of the test chamber [7]

2.5. Process sequence of resistance heating in XHV adequate atmosphere

The process of resistance heating in XHV adequate atmosphere is shown in Figure 4 with the development of the sheet temperature and oxygen concentration over time. In the first step (1), the required process atmosphere is built up and, with approx. 260 s, the longest process step. There, pure nitrogen 5.0 is admitted to the chamber in the first 230 s and silane is additionally introduced in the last 30 s. The minimization of oxygen is greater with silane, since the silane does not merely displace the oxygen as nitrogen 5.0 does, but reacts actively with the oxygen. Subsequently, heating is carried out to the hot-stamping temperature of 950 °C (2), which is achieved within 10 s due to resistance heating. The oxygen concentration decreases analogously with temperature, since the activity of the silane increases with rising temperature and reaches a minimum of 10-21 vol. % O_2. This low oxygen level is maintained until the end of heating. The temperature level is maintained for a certain time to allow the braze coating to bond with the base material. The time above the solidus temperature of the coating material and the maximum heating temperature are crucial to compare resistance heating with inductive heating. At temperatures below 875 °C, the chamber is opened because the solidus temperature is undershot and the sheet is transported to hot stamping press. The rapid increase in oxygen content is caused by the opening of the chamber, with air flowing into it. Therefore, the oxygen value reaches the atmospheric concentration of 21 vol. %.

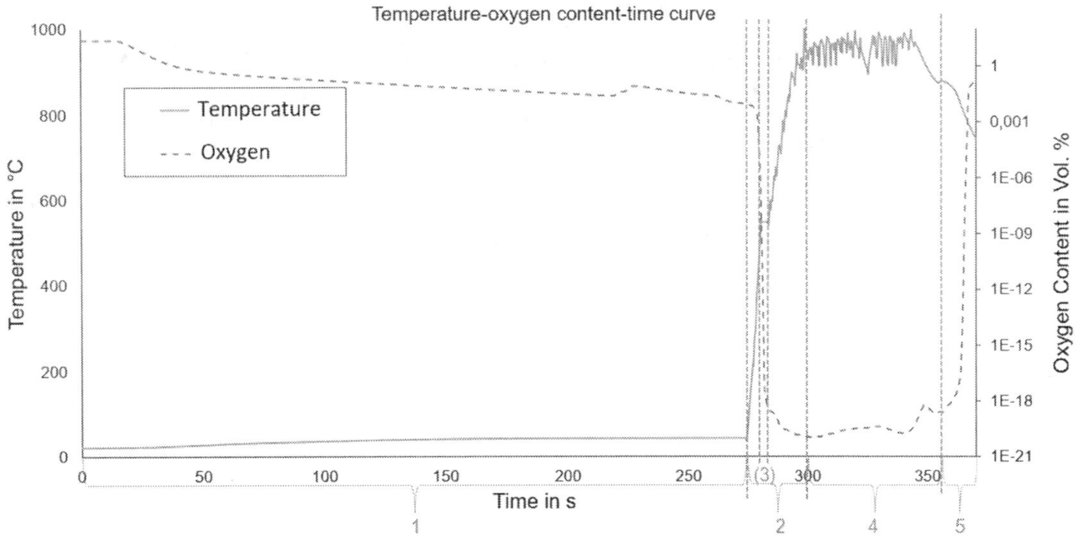

Figure 4. Process flow of oxygen-free resistance heating [7]

To investigate the suitability of nickel coatings for resistance spot welding, hot stamped plates with uniform coating application on the surface are joined to the reference plates of 22MnB5 AlSi by means of resistance spot welding (Sila PPN 253). These plates are tested in a universal tensile testing machine of the company Otto Wolpert Werke GmbH for maximum joint strength in shear tension. This tests were repeated three times each. To examine the weld lens, micrograph specimens are fabricated and examined by light microscopy. As a comparison, reference tests with standard AlSi-coated sheets are executed. The hardness of the nickel coating was investigated by nanoindentation using the Trobindenter Ti900 of Hysitron (Bruker, Minneapolis, Mn, USA). A capacity three-plate transducer and a diamand Berkovich Corner Tip (tip radius less than 150 nm) were used for the tests. The experiments were force-controlled using a trapezoidal force-displacement function. Eleven indents were performed per position with a descending force of 500 µN each starting at 6000 µN. The obtained indentation depths were between 170.6 nm and 40 nm. Prior to indentation, the surface was scanned by atomic force microscopy (AFM) and the indents were positioned in a defined array in an area of 40 x 40 µm with a spacing of 8 µm.

3. Results

3.1. Layer formation of the nickel coatings in the laboratory test rig

Light microscopic analyses of metallographic microsections provide information on the layer structure at different temperature-time regimes set for the coating processes. As an example, Figure 5 shows the resulting structures as a function of the coating material (or chromium content) and the maximum temperature reached. The states comprise the incipient material bonding between the particles of the substrate surface in the range of the solidus temperatures of the coating materials up to the necessary minimum temperature for the formation of a smooth closed and pore-free layer. For the latter - as the tests showed - superheating up to the respective specified target temperatures (905 °C for Ni 700, 920 °C for Ni 710 and 1000 °C for Ni 720) is required if the layer formation must be completed within approx. 10 s after the onset of melting. In particular, the very chromium-rich Ni 720 thus requires strong superheating for the formation of defect-free coatings. With regard to rapid resistance heating, the coating material Ni 700 is therefore particularly well suited; however, the minimum temperature required for Ni 710 is also within the target temperature for hot stamping of the sheets (approx. 950 °C).

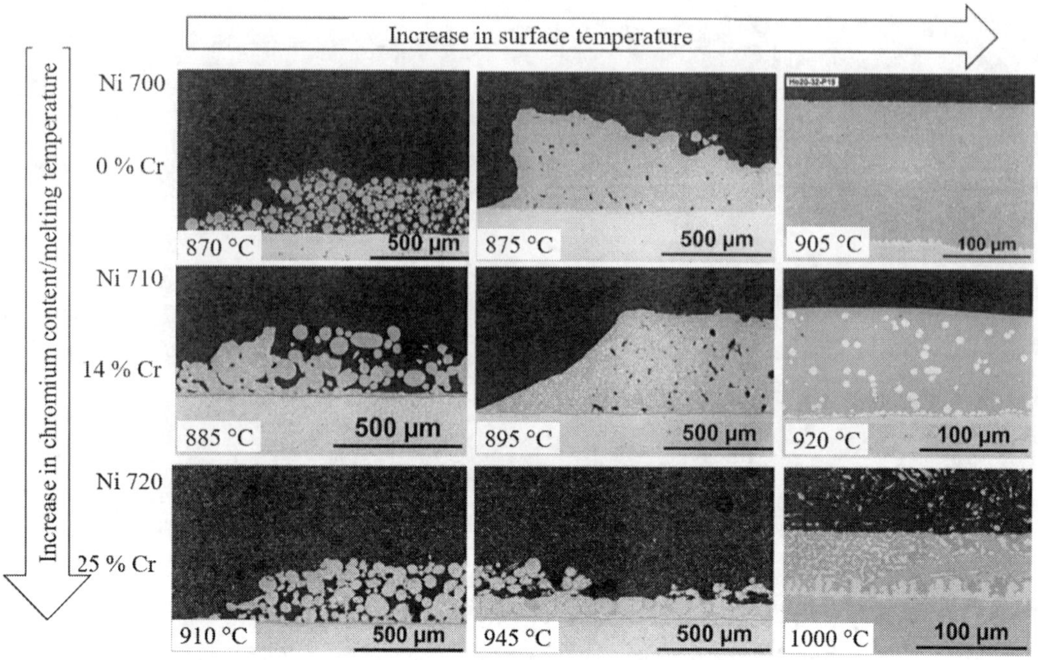

Figure 5. Compilation of selected micrographs of coated sheet specimens as a function of the coating material and the maximum surface temperature reached [8]

Coatings where the layer formation is completed exhibit a good metallurgical bonding in the entire interface as demonstrated in Figure 6 in the case of Ni700. The plotted element mappings via EDX analysis reveals a small zone of interdiffusion between the base material and the Ni700 coating.
The hardness of the nickel coating Ni700 was at 10. ± 0.8 GPa, which corresponds to 1020 HV. This value is thus significantly higher than 22MnB5 with a completely martensitic structure at approx. 480 HV01.

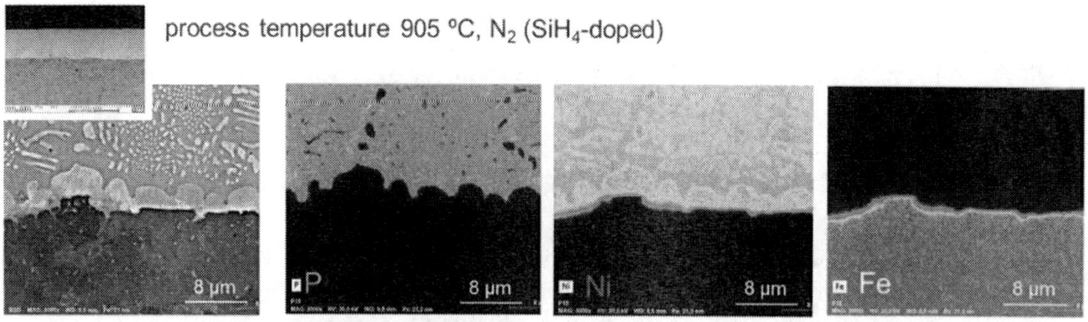

Figure 6. EDX analysis of Ni.700 coating processed at 905°C. The element mapping reveals a small diffusion zone in the interface as result of a metallurgical bonding between steel and coat

3.2. Assessment of the weldability of the nickel coatings

The following welding parameters in table 3 were used for the reference tests and for the in-situ coated tests.

Table 4. Welding settings

Electrode	Wahlenmeier Mk3x65-Z/ Dm25_SW22 Ø 8 mm
Lead time	200 ms
Welding time	250 ms
Hold-off time	200 ms
Electrode force	13 kN
Welding current	11kA

The shear tensile tests showed that the maximum strength of the reference specimens (AlSi-coated-sheet joined AlSi-coated-sheet) was on average 16.29 kN and that of the nickel braze coated specimens (Nickel-coated-sheet joined AlSi-coated-sheet) was on average 11.80 kN, which is 73 % lower than the strength of the reference specimens. An explanation for this is provided by the sections of the spot welds in Figure 6. The spot weld with the nickel braze coating is generally wider than the reference spot weld, but has several pores in the weld lens, which can explain the lower shear tensile strength.

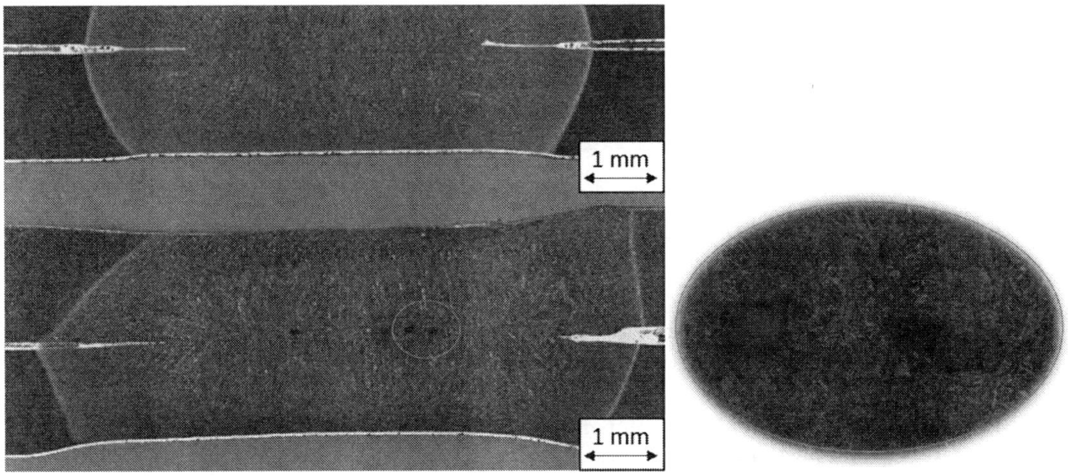

Figure 7. Micrograph of the resistance spot welding tests; top reference point bottom Ni700 coated with internal nickel coating and pores.

4. Summary and outlook

In the scope of this research, it was shown that the development of an inline coating process for manganese-boron steel sheets during resistance heating for hot stamping is possible. Powdered nickel braze metals, which are suitable for coating application during rapid heating processes were used as scale-resistant coating materials. To ensure that the powders, which melt at approx. 900 °C, form homogeneous coatings with good material bonding to the steel surface, an oxygen-free atmosphere is essential. For this purpose, a research facility was developed in which sheet blanks are resistance heated in a silane-doped XHV adequate atmosphere. The investigations showed that the process conditions required for coating formation can be realized within the desirable process window for rapid resistance sheet heating. Compared to the base material, the nickel coating has a much higher hardness (1020 HV) and could thus provide abrasive protection. At the same time, however, resistance spot welding shows that the nickel coating is rather unsuitable at this time for joining. The use of these nickel coatings for resistance spot welding is limited. Due to the voids in the weld lens, an acceptable weld spot has not yet been joined, resulting in lower shear strength compared to conventional AlSi coatings. At the same time, the conventional AlSi coatings have active corrosion protection compared to the nickel coating, which significantly improves the performance of the components. Therefore, for resistance heating under XHV-adequate atmosphere, the coating systems should be extended to aluminium and zinc to obtain active corrosion protection. However, this needs adaptations, because a large part of the heating time is used for the formation of an intermetallic phase. Thus, an adaptation of the coating systems and an adaptation of the equipment to the lower application rate of the coating is necessary.

Acknowledgements
Funded by the Deutsche Forschungsgemeinschaft (DFG, German Research Foundation) – Project-ID 394563137 – SFB 1368

References
[1] Karbasian, H., Tekkaya, A.E., A review on hot stamping Journal of Materials Processing Technology 210 (2010) 2103–2118, DOI:10.1016/j.jmatprotec.2010.07.019
[2] Behrens, B.-A.; Hübner S.; Farahmand E.; Albracht L. (2022): Resistance Heating by Means of Direct Current for Resource-Saving CO_2-Neutral Hot Stamping *(Key Engineering Materials ISSN: 1662-9795, Vol. 926)* pp 2363-2370 (Switzerland: Trans Tech Publications Ltd).
[3] Allély, C.; Dosdat L.; Clauzeau, O.; Ogle, K.; Volovitch, P. (2014): Anticorrosion mechanisms of aluminized steel for hot stamping. Surface and Coatings Technology, 238, 188-196.
[4] Wolf, L., Diekamp, M., Gretzki, Th., Nürnberger, F., Bach, F.-W., D. Rodman, J. Moritz, J. Schrödter, S. Hübner, B.-A. Behrens (2014) Hot stamping and subse-quent spray cooling: A neu manufacturing approach, Plastic Deformation of Metals (pp.36-55)
[5] Behrens, B.-A.; Hübner, S.; Schrödter, J.; Uhe, J. (2015): Conductive Heating opens up Various New Opportunities in Hot Stamping. Proceedings 5th Interna-tional Conference on Accuracy in Forming Technology, ICAFT 2015, pp. 157-173.
[6] Jacobson, D. M.; Humpston, G. (2005): Principles of brazing. Chap. 2: Brazes and their Metallurgy, ASM International, Materials PARK, OH 44073-0002, USA.
[7] Albracht, L., Hübner, S., Holländer, U., Behrens, B.-A. (2022): Coating Materials Under Oxygen-Free Silane Atmosphere for Hot Stamping, In: Behrens, B.-A., Brosius, A., Drossel, W.-G., Hintze, W., Ihlenfeldt, S., Nyhuis, P. (Hg.): Produc-tion at the Leading Edge of Technology. Proceedings of the 11th Congress of the German Academic Association for Production Technology (WGP), Dresden, Sep-tember 2021. Cham: Springer International Publishing, pp. 3–10. DOI: 10.1007/978-3-030-78424-9_1
[8] Behrens, B.-A.; Hübner, S.; Farahmand, E.; Holländer, U.; Langohr, A. (2023): Konduktive Erwärmung und Beschichtung in sauerstofffreier Umgebung durch Stickstoff und Monosilan. 5. Symposium der Materialtechnik Clausthal.

IOP Conf. Series: Materials Science and Engineering 1284 (2023) 012008 doi:10.1088/1757-899X/1284/1/012008

Study of Microstructural Evolution of Press Hardening Steels using Dilatometer and In-situ Studies for a Simulated Hot Stamping Condition

Hong Zhu, Hongbin Yin and Sriram Sadagopan

ArcelorMittal USA Research LLC, 3001 E. Columbus Drive, East Chicago, IN 46312, USA

hong.zhu@arcelormittal.com

Abstract. Press hardening grades are widely used in automotive industries for safety-critical structural parts due to their unique combination of high strength, excellent formability, and crash performance. Considering various scenarios on thermo-mechanical profiles in different hot stamping lines, achieving the targeted strength and ductility / fracture strain in the hot-stamped parts is still challenging to some hot stampers for some grades. In this investigation, a dilatometer study for Usibor®1500 and two emerging grades Usibor®2000 and Ductibor®1000 under a given hot stamping condition was conducted with consideration of the entire hot-stamping processes (i.e., austenitization, blank transfer, forming and final quenching from 700°C) to understand the differences in critical cooling rates, and evolution of microstructures. Influence of large forming strain (15%) on final properties is also examined for Ductibor®1000 and Ductibor®500 by DIL 805 A/D dilatometer under tensile deformation mode. In-situ observation of microstructural evolution during hot stamping process for Usibor®1500 is explored using Confocal scanning laser microscope to uncover some physical phenomena for further refinement of hot stamping practices.

1. Introduction

Press hardening steel (PHS) grades are widely used in the automotive industry for structural parts such as bumper beams, door beams, A-pillars, B-pillars, roof rails, cross members as well as door-rings to reduce vehicle's weight and meet vehicle's safety requirements during crash tests. The entire hot-stamping process includes: (a) austenitization, to generate fully uniform austenite phase in steel blank; (b) blank transfer in air with cooling; (c) stamping in a temperature range to take advantage of the excellent formability of austenite; (d) die-quenching the stamped parts to obtain targeted properties such as strength and ductility / fracture strain through microstructure control (mostly martensite microstructure). Besides wrinkling, double metal, thermal-distortion, another challenge for hot stamping process development is targeted part property achievement which is linked with microstructure evolutions (solid phase transformations and recrystallization) during entire thermal-mechanical history of hot stamping process.

Temperature histories of blank during hot stamping process is extremely complex. Conventional continuous cooling transformation (CCT) curves are generated through a set of constant cooling rates from 0.5°C/s to 100°C/s after austenitization temperature to room temperature by dilatometer tests to determine final microstructures. However, these curves are not sufficient for realistic hot stamping process development as the influences of blank transfer time, die-closing time, forming strain, strain rate and deformation temperature are not included. These knowledge and data gaps for optimizing hot stamping processes triggered many investigations on 22MnB5, the most common grade used for hot stamping with tensile strength on the order of 1500 MPa such as Usibor®1500. Nikravesh et al. (2015) constructed deformation influenced continuous cooling transformation (DCCT) curves at a deformation temperature of 850°C, compressive true strain of 0.4, strain rate of 7/s using Baehr 805 deformation dilatometer simulator. When comparing with no deformation case, they found that in the die-quenching range (20°C/s to 100°C/s), the region of austenite decomposition was translated to the left, bainite region was expanded in CCT diagram, and the final amount of martensite due to hot deformation was decreased. Fan et al. (2010) used a Gleeble3500 thermo-mechanical simulator to conduct high temperature tensile tests with 0.1 and 0.3 engineering strain in the temperature range of 600°C to 800 °C with a strain rate of 0.5/s, followed by a quenching process with a 60°C/s cooling rate. It was found that with increasing plastic strain, Ms temperature is reduced, and final hardness is increased at 800°C deformation temperature due to fully martensitic microstructure with grain refinement. The hardness and tensile strength at deformation temperatures of 700°C and 650°C show significant reduction due to the increased volume of ferrite and bainite, with plastic strain having no influence on hardness due to complex phase microstructures at the deformation temperature of 600°C. Min et al. (2012, 2013) conducted similar studies, in the same deformation temperature range of 600°C to 800 °C with tensile strain from 0.1 to 0.3 under strain rates of 0.1/s and 0.5/s performed on a Gleeble 3800 simulator. They found that the largest volume of ferrite (>50%) was generated at a deformation temperature of 700°C at a cooling rate of 30°C/s. This large volume of ferrite is generated through deformation induced ferrite transformation (DIFT), which is nucleation dominated and can be finished in a very short time. The nucleation, growth, and impingement mechanisms at fine length scales are linked to the evolution of various dislocations and their substructures, such as increased dislocation density and decreased mean free path of dislocations due to austenite deformation (Barcellona and Palmeri, 2009). Severe deformation causes mechanical stabilization of the austenite because highly dense dislocation substructures interfere with the motion of glissile transformation interfaces (Bhadeshia, 1992). Different dislocation structures can be generated from different austenite deformation temperatures (Gong et al. 2013). Local carbon depletion may occur in austenite region nearby dislocation which is carbon enriched (Dhara et al. 2022).

Usibor®2000 and Ductibor®1000 are two emerging PHS grades developed by ArcelorMittal with different combinations of strength and ductility to provide complete solutions for weight reduction while maximizing crash performance. Among the limited published studies for these two grades, Güner et al. (2021) attempted to predict part properties of Ductibor®1000 using FEM. Samadian et al. (2022) performed microstructure-based modeling of flow and fracture behavior of tailored microstructures of Ductibor®1000 at different die quenching temperatures. Palmieri et al. (2022) performed tensile tests for Usibor®2000 at 850°C, 800°C and 750°C under different strain rates and studied tailored hot stamping process for B-pillar.

Prior austenite grain size (PAGS) is an important factor for final part property control. PAGS influences the density of potential grain boundary nucleation sites for further transformation and strength of austenite (Dharra et al. 2022). The packet size and block width of a lath martensitic structure are also PAGS-dependent. Horvath and Wang (2019) reported an over-baked Usibor®1500 door beam with large PAGS fractured during a side impact test while properly processed door beam with normal PAGS did not fracture. However, current PAGS's measurement is post-mortem based without any evolution information during austenitization process.

This paper summarizes results from several investigations conducted at ArcelorMittal to evaluate PHS behavior under simulated hot stamping cycles, more specifically the stamping temperature and forming strain. Results from dilatometer studies, microstructural characterization of Ductibor® 1000 and Ductibor® 500 grades under large tensile deformation at 700°C followed by quenching, and in-situ observation of microstructural evolution of Usibor®1500 grade using a confocal scanning laser microscope (CSLM) are presented in this paper. From the investigations it can be concluded that actual hot stamping conditions will have a big impact on the transformation behavior of the PHS products.

2. Dilatometer studies to simulate hot stamping cycles for Ductibor®1000, Usibor®1500, and Usibor®2000

Chemical compositions for the PHS grades from ArcelorMittal are shown in Table 1. Thicknesses of the materials used for this study are: 1.4 mm for Ductibor®1000, 1.4 mm for Usibor®1500 and 1.6 mm for Usibor®2000.

Table 1.1. Chemistry of the emerging PHS grades used in this study (max. wt%)

Grades	C	Mn	Si	Al	Ti	B	Nb	Cr
Usibor®1500	0.25	1.4	0.4			0.005		
Ductibor®1000	0.10	1.8	0.6	0.01-0.1	0.05	0.005	0.10	0.20
Usibor®2000	0.36	0.8	0.8	0.01-0.06	0.07	0.005	0.02-0.07	0.50

This portion of the study was conducted to study the phase transformation of Ductibor®1000, Usibor®1500 and Usibor®2000 under realistic hot stamping thermal history generated using a DIL 805L Dilatometer. Associated dilatometer curves are generated for these three grades under specific thermal conditions. Microstructures and hardness are analyzed and measured at the end of the process.

Conventional CCT curves were generated through a set of constant cooling rates after austenitization temperature to room temperature by dilatometer tests as shown in Figure 1.1(a). These curves can't be directly used for CAE inputs since they don't consider transfer time and die-closing time. To fill these data and knowledge gaps for this specific study, one OEM provided following thermal history which is shown below and Figure 1.1 (b) & 1.1(c): furnace temperature at 920°C; heating rate: 5°C/s and 10°C/s; dwell time (total): 6 minutes; transfer time: 10 seconds (corresponding temp drops from 920°C to 820°C); die closing and forming process: 6 seconds (corresponding temp drops from 820°C to 700°C); final quenching rates (from 700°C): 10°C/s, 20°C/s, 30°C/s, 40°C/s, 50°C/s and 60°C/s.

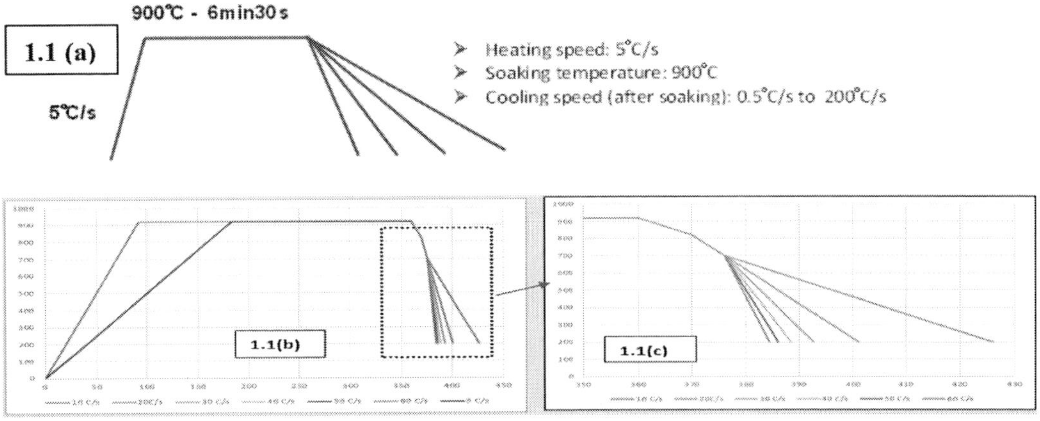

Figure 1.1. (a) Schematic of thermal profile for conventional CCT curves, (b) current thermal history provided by OEM to simulate stamping condition; (c) Cooling phase of specified thermal history. Y-axis is the temperature (°C) and X-axis is the time (in sec).

Dilatometer Test Results

The DIL 805L quenching dilatometer was employed to perform the tests under the specified thermal histories using 4mm x 10mm specimens. Changes in temperature and length with time in the specimens during the tests are recorded. Specimens after tests are kept for microstructure analysis and hardness measurement. From early testing, it was found that initial heating rate has a negligible effect on the quenching portion of dilatometer curves (length change vs temperature) for these three grades. So, only dilatometer curves with 10°C/s heating rate were selected for further analysis.

Quenching portions of expansion-temperature curves for Ductibor®1000, Usibor®1500 and Usibor®2000 are shown in Figures 1.2a to 1.2c. Inflection points on these curves indicate potential start and finish temperatures of phase transformation. For Usibor®2000, it is shown that Martensite transformation occurred from 10°C/s to 60°C/s (confirmed by microstructure analysis and hardness measurement in the next section). For Usibor®1500, it is observed that fully Martensitic transformation occurred from 20°C/s to 60°C/s (confirmed by microstructure analysis and hardness measurement in the next section). For Ductibor®1000, the situation is more complex and only 60°C/s shows fully Martensite transformation potential.

The associated cooling curves quenching from 700°C for these three grades are plotted in Figure 1.3a to 1.3c with estimated Martensite transformation temperature from the expansion-temperature curves from Figure 1.2. Martensite starting (Ms) temperature for Ductibor®1000 is significantly higher than Usibor®1500 and Usibor®2000. This difference may influence final part mechanical properties and geometric tolerance for tailored parts (e.g., door ring) when an existing PHS grade is replaced by a new PHS grade.

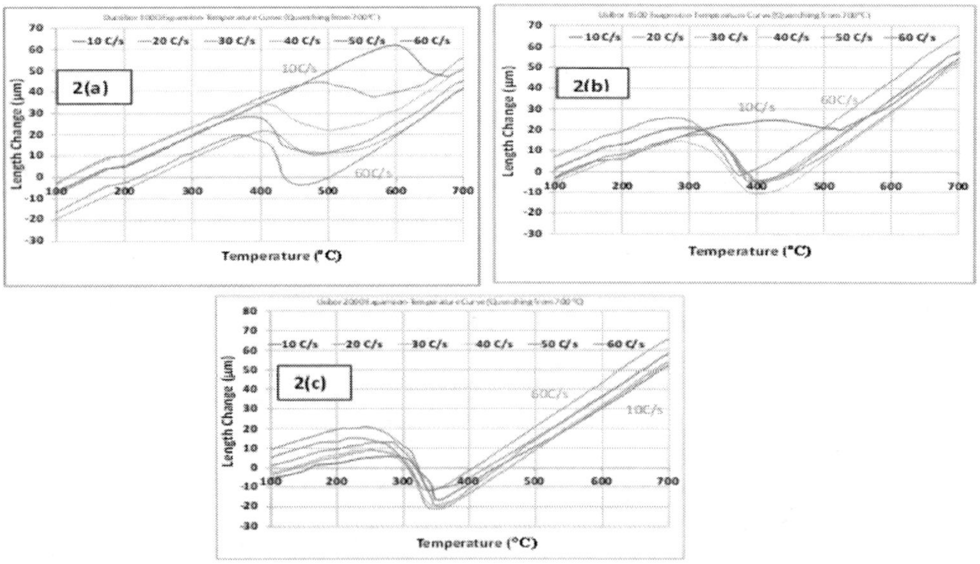

Figure 1.2. Temperature–expansion curves for (a) Ductibor®1000 (b) Usibor®1500 (c) Usibor® 2000.

IOP Conf. Series: Materials Science and Engineering 1284 (2023) 012008 doi:10.1088/1757-899X/1284/1/012008

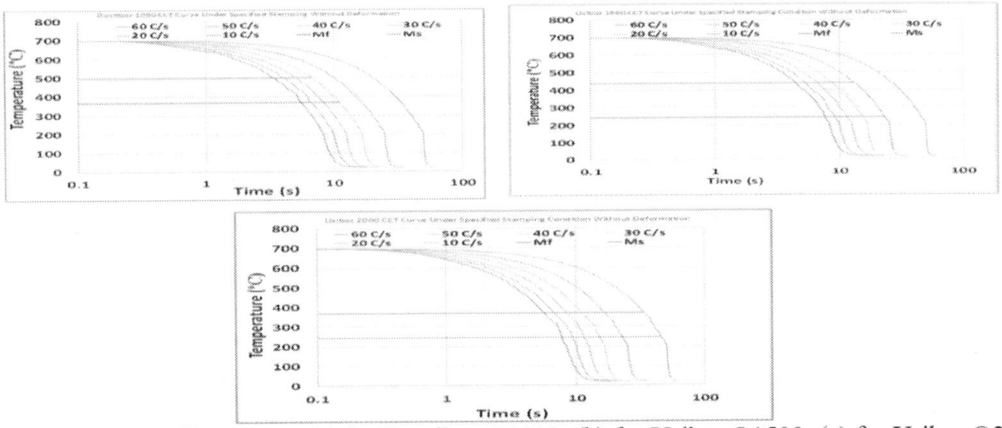

Figure 1.3. Cooling curves (a) for Ductibor ®1000; (b) for Usibor ®1500; (c) for Usibor ®2000.

Ductibor®1000

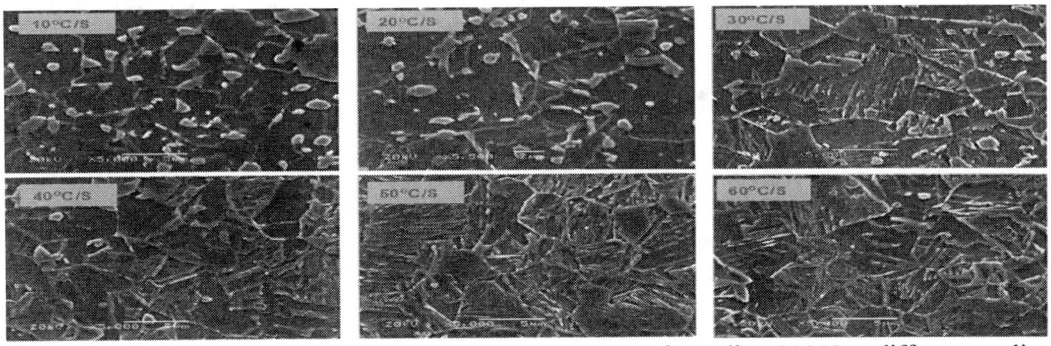

Figure 1.4. Representative steel microstructure (@ 5000x) of Ductibor®1000 at different cooling rates (from 700°C).

The microstructures of Ductibor®1000 changed significantly with increasing cooling rate as shown in Figure 1.4. At lower cooling rates, 10-20°C/sec, it is predominantly ferrite with martensite. At 30-40°C/s, higher fractions of martensite and bainite were observed. Finally, at 50°C/sec or higher, microstructure mainly consists of martensite with some bainite.

Usibor®1500

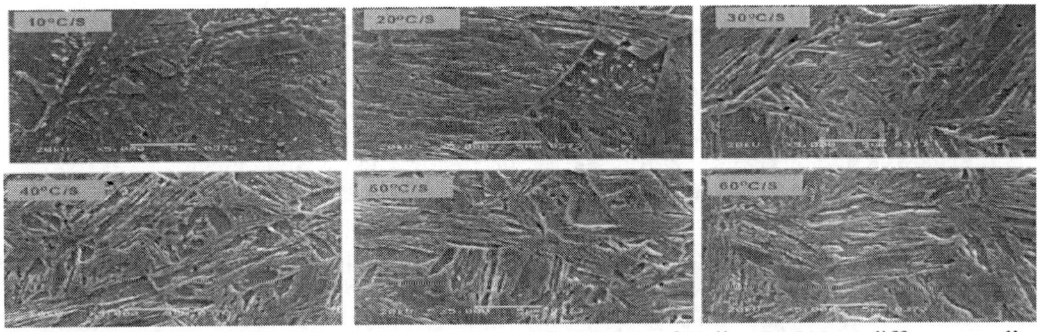

Figure 1.5. Representative steel microstructure (@ 5000x) of Usibor®1500 at different cooling rates (from 700°C).

IOP Conf. Series: Materials Science and Engineering 1284 (2023) 012008 doi:10.1088/1757-899X/1284/1/012008

As seen in Figure 1.5, compared to Ductibor®1000, for a similar thermal profile, microstructure of Usibor®1500 was predominantly martensite except at 10°C/s.

Usibor®2000

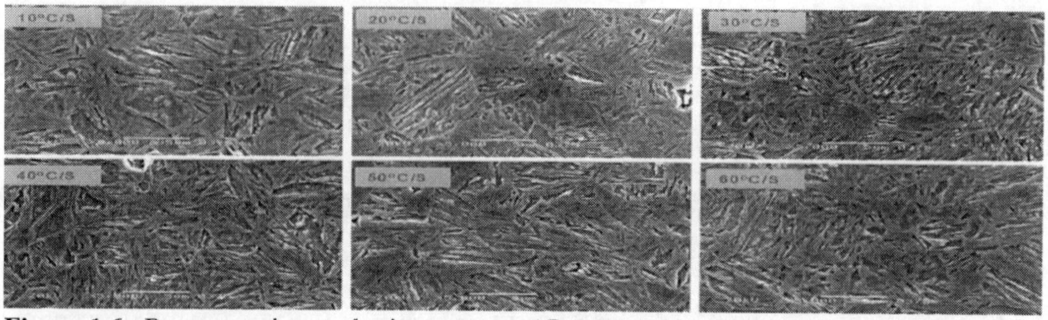

Figure 1.6. Representative steel microstructure (@ 5000x) of Usibor®2000 at different cooling rates (from 700°C).

As shown in Figure 1.6, given the higher carbon content and other alloying elements, the microstructure of Usibor®2000 was predominantly martensite even with lower cooling rate of 10^{0}C/sec.

Summary of Hardness with different quenching rates for these three PHS grades is shown in Figure 1.7.

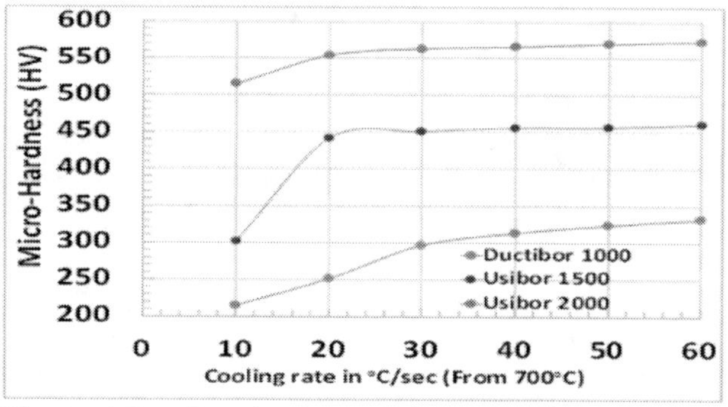

Figure 1.7. Micro-hardness with cooling rate for the selected PHS grades

3. Characterization of Ductibor®1000 and Ductibor®500 grades with and without deformation at 700°C during a simulated hot-stamping cycle

The properties of the final part are greatly influenced by hot stamping process conditions which vary between part suppliers. To investigate the associated mechanisms behind these observations, the impact of tensile strain on final properties of Ductibor®1000 and Ductibor®500 was investigated using a DIL 850 A/D, the latest TA instrumental Dilatometer designed for sheet under tension (Abad, 2017 and 2018). Microstructures and hardness are analyzed at the end of the thermomechanical testing.

The following thermo-mechanical profile is described below and shown in Figures 2.1 (a) & 2.1(b), is used for this specific study: 920°C furnace set temperature, 7.5°C/s heating rate, 6 minutes dwell time, 10 seconds transfer time: 10 seconds (corresponding temp drops from 920°C to 820°C), 6 seconds die closing and forming process (corresponding temp drops from 820°C to 700°C), tensile deformation 15% for Ductibor® 1000 and 8% and 15% for Ductibor® 500 (both at 700°C with strain rate 1.0 /s), final quenching rates (from 700°C) for Ductibor® 1000 are 20°C/s, 40°C/s, 60°C/s, 80°C/s; for Ductibor® 500 are 20°C/s, 40°C/s, 60°C/s, 80°C/s. The above thermal-mechanical cycle was designed to simulate severe conditions occurring in hot stamping practice in terms of quench-start temperature (from 700°C) and 15% tensile deformation at 700°C. Except that, cooling rate from 20°C/s to 80°C/s is within the normal die-quenching range, strain rate is selected at 1.0 /s, close to the upper limit in stamping practice and favor to suppress diffusion phase transformation.

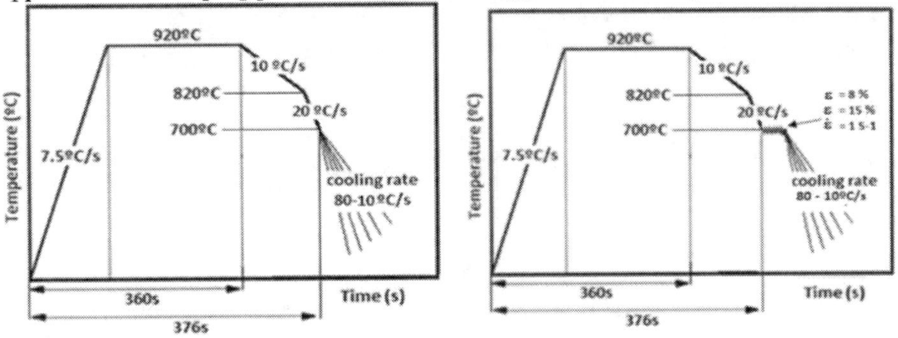

Figure 2.1. (a) Thermal history used to simulate stamping condition, (b) thermal-mechanical history to simulate stamping condition with tensile strain 8% and 15%.

The DIL 805 A/D quenching dilatometer was employed to perform the tests under the specified thermo-mechanical histories using tensile-type specimen described in Figure 2.2. The maximum cooling rate is 80°C/s with this specimen design under this equipment. Temperature, length change, and force in the specimens during the tests are recorded with time for further data processing. Specimens after tests are kept for microstructure analysis and hardness measurement.

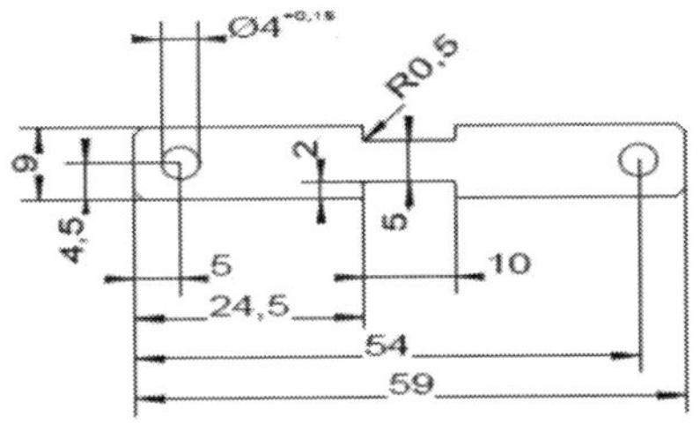

Figure 2.2. Specimen geometry for dilatometer testing (from Abad, 2017).

Ductibor®1000

Under typical cooling rates, microstructures and hardness with and without tensile deformation are compared in Figures 2.3 (a) and 2.3 (b). It is found that without deformation, satisfactory mechanical properties were achieved at a cooling rate ~ 40 °C/s. Without deformation, the microstructures after 40°C/s are predominantly martensite with some evidence of bainite with the hardness in the range of 310 HV. Martensite transformation can be observed in Figure 2.4 in the temperature range of 490°C to 380°C for cooling rate after 40°C/s. Whereas, with 15% tensile strain, microstructures after 40 °C/s are predominantly bainite and ferrite with some evidence of martensite islands with the hardness is ~ 229 (HV). Bainite transformation can be seen clearly in Figure 2.5 in the temperature range of 685°C to 530°C for cooling rate after 20°C/s. Ferrite transformation is not explicitly showed in the cooling phase in Figure 2.5, which should have occurred during deformation stage at 700°C from DIF (deformation-induced ferrite) transformation. These hardness results are 26% lower than the initial target 310 (HV) corresponding to 1000 MPa tensile strength for this grade. These phenomena are similar to the findings for 22MnB5 grades with large tensile strain at 700°C (Fang et al, 2010; Min et al, 2012), where significant increasing dislocation density provides more nucleation sites for bainite and DIF transformations, although there are significant differences among their chemical compositions and other thermo-mechanical parameters such as strain rate, et al. The simple solution to achieve the initial strength target is increasing deformation-start temperature above 700°C by reducing blank transferring time or increasing furnace temperature.

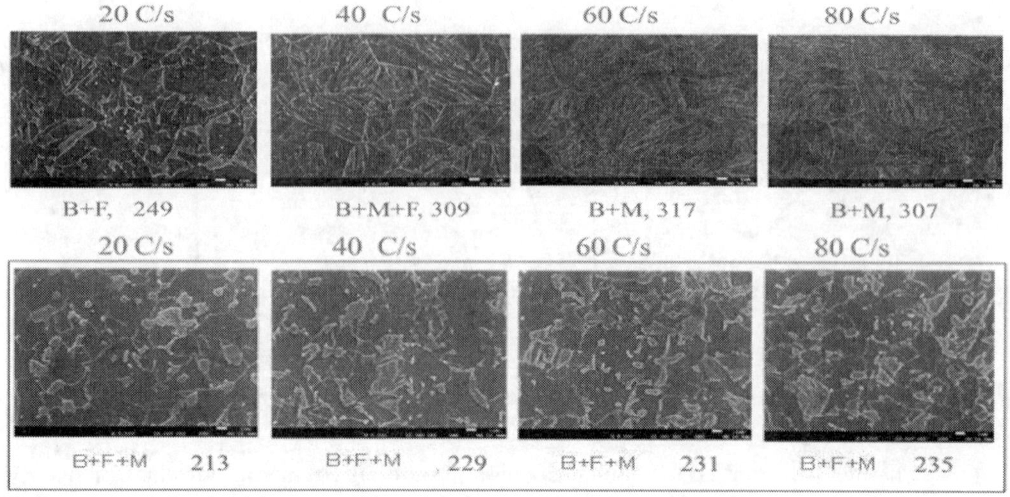

Figure 2.3 (a) (top) with 0% strain and **Figure 2.3 (b)** (bottom) with 15% strain. Representative steel microstructure (@ 5000x) and hardness of Ductibor®1000 at different quenching rates (deformed at 700°C).

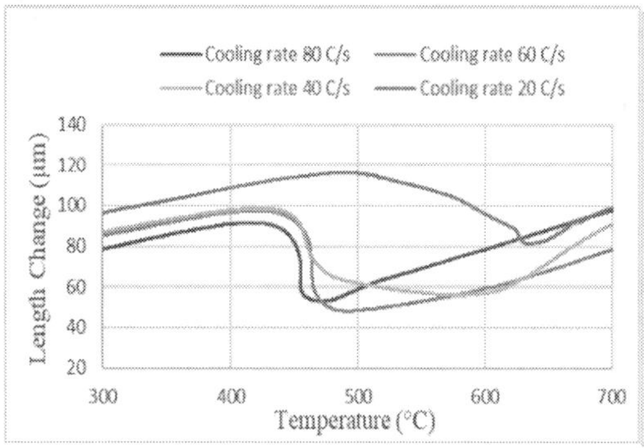

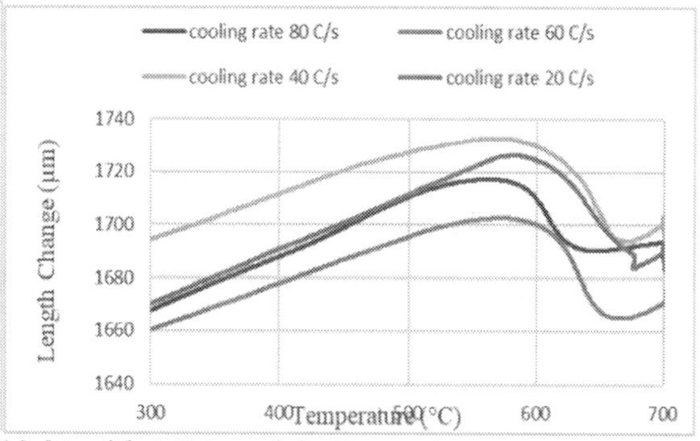

Figure 2.4 (left with 0% strain) and **Figure 2.5** (right with 15% strain). Expansion-temperature curve for Ductibor®1000 during quenching phase (from 700°C)

From a crash perspective, one could argue that the introduction of ferrite in the soft regions of tailored hot stampings (due to inevitable or intentionally imposed deformation) can be beneficial rather than harmful (Bardelcik et al, 2014) due to ductility and fracture strain increasing. For Ductibor®1000, Samadian et al. (2022) reported that with decreasing cooling rate from 243 °C/s to 21°C/s, UTS reduces by 26% reduction, and the fracture strain increases by 40–60% increase for most of loading paths. It was also observed small amount of RA is existed on Ductibor®1000 after 15% tension strain at 700°C followed by quenching from an independent XRD study. Metastable retained austenite (RA) was also found in the final microstructures of a 38MnB5 grade (Taylor et al, 2018). Ding et al (2022) used an in-line quenching and partitioning process for a newly developed Cr-alloyed PHS (Fe-0.23C-1.1Mn-(Cr + Si) < 5.0, wt%) to achieve 6 % nano-sized metastable austenite into the final microstructure of the prototype parts to improve parts bendability.

Ductibor®500

Cooling curves under different quenching rates from 700 °C with 15% pre-strain are shown in Figure 2.7 for comparison. Compared to Ductibor®1000, no obvious solid phase transformation is observed during cooling phase after 700°C. Most of the austenite may transform into ferrite during die closing corresponding temperature drop from 820°C to 700°C, as shown in Figure 2.8.

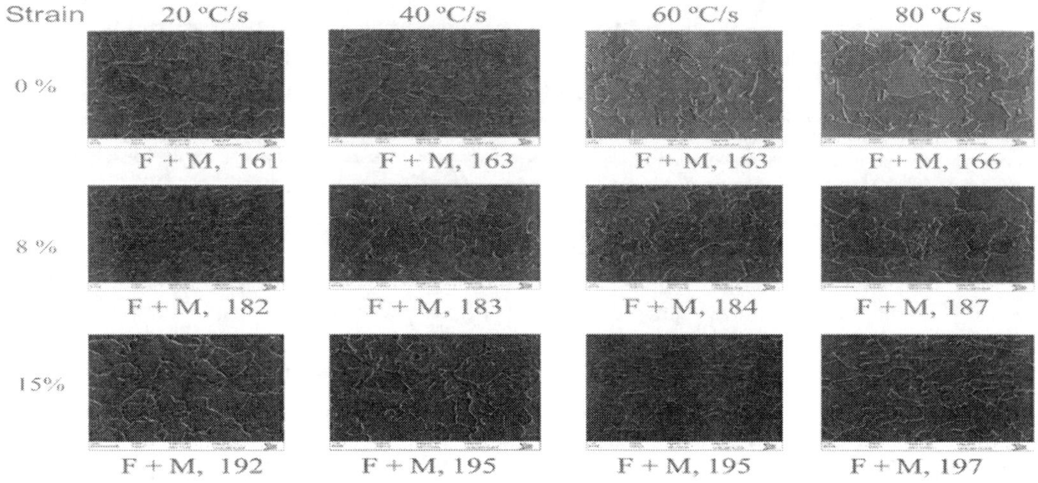

Figure 2.6. Representative steel microstructure (@ 5000x) and hardness variation of Ductibor® 500 at different cooling rates and tensile strain (top row: 0% strain; middle row, 8%, bottom row: 15% tensile strain) from 700°C.

Microstructural characterization and hardness data with different cooling rates under different pre-strain conditions are shown in Figure 2.6. Irrespective of deformation mode, the final microstructure is predominantly ferrite with some evidence of martensite. Subtle increase in hardness values were observed with tensile strain, from 163 at 0%, to 183 at 8%, to 195 at 15% with 40 °C/s, corresponding to tensile strength 518 MPa, 582 MPa, and 624 MPa, which can be explained as ferrite strain hardening during deformation stage at 700°C. Maximum hardness is ~ 197 (HV) at 80 °C/s, corresponding to 650 MPa.

Samadian et al (2021) reported interesting results for Ductibor® 500 under air-cooled, die-cooled and water-cooled conditions respectively. Our hardness results with pre-tensile strain at 700°C forming temperature are among their air-cooled results (167 HV) and die-cooled results (248 HV). For the simulated hot-stamping cycle we adopted, initial air transfer and cooling in die closing for Ductibor®500 play a critical role on final ferrite volume due to its super lower Boron content. Besides Ductibor®500 grade, pre-forming-strain caused final part hardness increase is also observed for a 38MnB5 grade (2000 MPa), which is explained by RA and finer martensite lath size correlated to plastic deformation, dislocation density and cooling rate (Taylor et al, 2018).

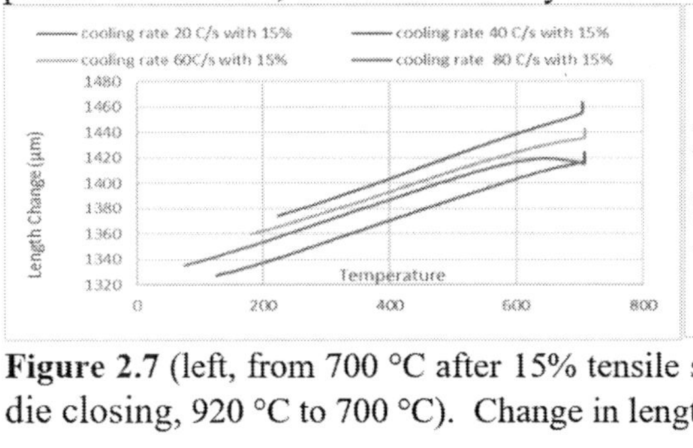

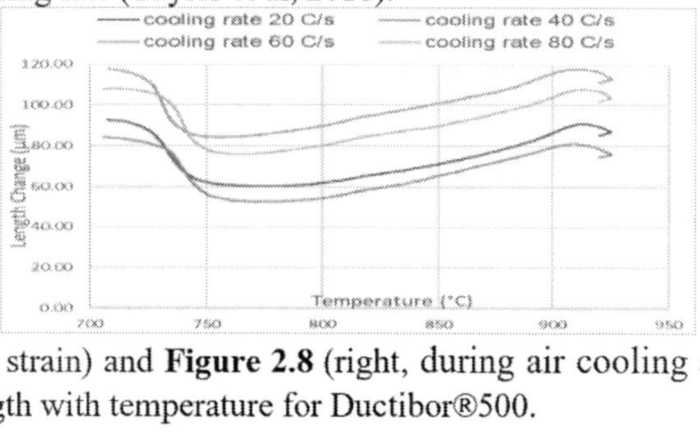

Figure 2.7 (left, from 700 °C after 15% tensile strain) and **Figure 2.8** (right, during air cooling and die closing, 920 °C to 700 °C). Change in length with temperature for Ductibor®500.

4. In-situ observation of microstructural evolution of Usibor®1500 during a simulated hot stamping cycle using confocal scanning laser microscope

High temperature confocal scanning laser microscope (HT-CSLM) is a powerful tool for visualization of elevated temperature phenomena (up to 1650°C) at 200 ~ 2000x magnifications. The laser scans the sample surface and detects any surface depth variation at around 1 micron and displays the scanned

IOP Conf. Series: Materials Science and Engineering 1284 (2023) 012008 doi:10.1088/1757-899X/1284/1/012008

area at maximum 60 fps (frames per second). This system was explored for in-situ observation of microstructure evolution during the entire hot stamping process (heating, soaking, cooling and quenching) to uncover hidden mechanisms / physics, and extract critical process parameters such as PAGS, lower bound of forming temperature etc. to help OEM to optimize their hot stamping practices.

This study was only conducted on Usibor®1500 (1.4 mm). Specimens is round-shaped with a diameter of 7mm diameter. Specimen surface was polished to 1 μm roughness. The thermal profile for a simulated hot stamping is as follows: (a) 5°C/s heating rate from 110°C to 920°C after slow pre-heating to 110°C, (b) 300 seconds soaking at 920°C, (c)10°C/s cooling rate from 920°C to 820°C, (d) 20°C/s cooling rate from 820°C to 700°C, (e) ~20°C/s quenching rate from 700°C to room temperature. This process was successfully performed in a HT-CSLM system and videos for microstructure evolution were recorded. Typical pictures extracted from the recorded videos are showed in Figure 3.1 to 3.4 to shown austenite grain boundary evolution and martensite transformation during austenization and subsequent cooling phases.

The observed PAGS is ~ 20 μm to 50 μm from the picture of 920°C-end in Figure 3.1. It is interesting to see martensite phase inception in the temperature range of 502°C to 462°C at mesoscale in Figure 3.3. Martensite start temperature Ms, defined as 1% martensitic volume fraction in the microstructure, is 420°C for this grade. These results imply that martensite phase transformation occurs much early at mesoscale and microscale than Ms.

Yang and Bhadeshia (2009) reported that Ms is increases with the austenite grain size increasing and calculated a fundamental martensite– start temperature for an infinitely large austenite grain size, which can be more than 100°C high than Ms. The temperature for martensite phase inception at microscopic and this fundamental martensite share a similarity: much high than Ms.

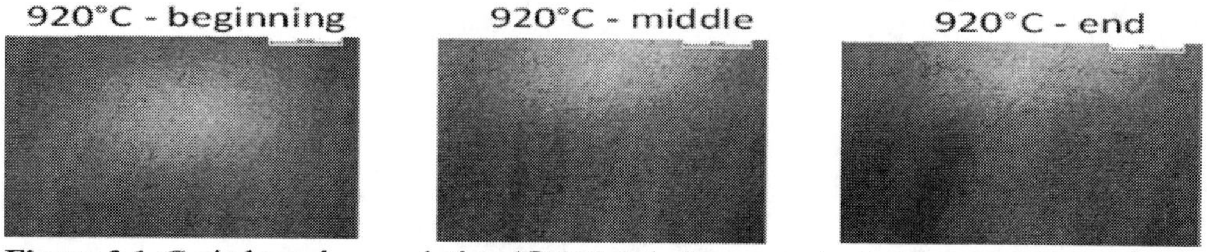

Figure 3.1. Grain boundary evolution (@ 200x) during austenization process (920°C for 5 minutes)

Figure 3.2. Grain boundary evolution (@ 200x) during cooling phase

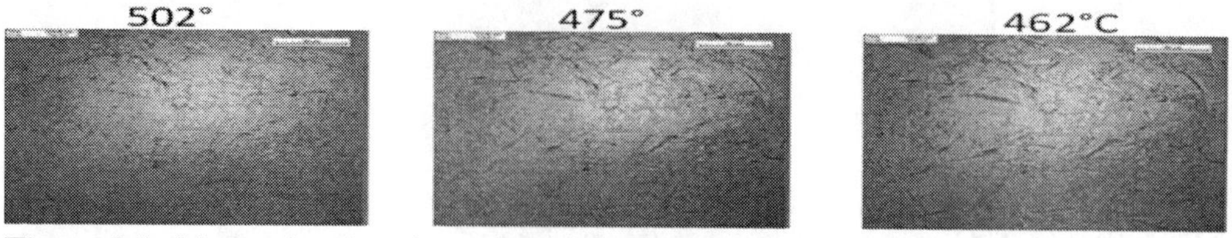

Figure 3.3. Martensite phase inception (@ 200x) at meso-micro scale

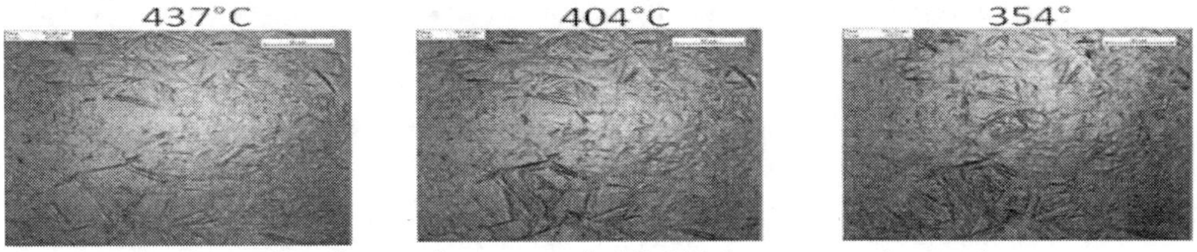

Figure 3.4. Martensite phase volume growth (@ 200x) at meso-micro scale

5. Conclusions / Summary

- Increase in hardness was observed with increase in cooling rate for tested grades. Higher the cooling rate, the higher the hardness of the material.
- For similar thermal profile, significant differences in critical cooling rates were observed between tested grades. These differences should be considered in fine tuning of hot stamping process when existing PHS grades are replaced by new grades to achieve targeted mechanical properties in hot stamped parts with good dimensional tolerance.
- For Ductibor®1000 and Ductibor®500 grades, selected tensile deformation during hot stamping has significant influences on final properties (i.e. hardness / microstructure).
- For severe conditions of deformation start temperature (i.e. from 700°C) and deformation (tensile strain ~ 15%), Ductibor® 1000 exhibited more variation in properties compared to Ductibor® 500. Ductibor®1000 behavior is similar to the published results for 22MnB5 grades (1500 MPa).
- Lower hardness values of Ductibor® 1000 with large strain irrespective of deformation mode can be explained as follows. Considering solid phase transformation of steel, where significant increase in dislocation density after large strain (at 700 °C) will provide more nucleation sites for Bainite and SIF transformations which delay Martensite transformation during subsequent cooling. This may result in deviation of final phase fraction and hardness of specimens from initial targeted microstructure and hardness at tested thermal-mechanical profile.
- HS-CSLM is successful used for observation of Usibor®1500 microstructure evolution during austenization and cooling phase on a simulated hot stamping condition. This system can be used in situ measuring PAGS and martensite phase transformation observation.
- Martensite transformation may occur at relative higher temperature than Ms at mesoscale and microscale. This phenomenon may need to be considered for process development to avoid early strain partitioning caused damage during hot stamping process.

References

[1] Abad, C., 2017. Thermomechanical Study of DUCTIBOR 1000 Sheet Steel by Means of Dilatometry. CTM Technical report 583SATM17 for ArcelorMittal, unpublished.

[2] Abad, C., 2018. Thermomechanical Study of DUCTIBOR 500 Sheet Steel by Means of Dilatometry. CTM Technical report 133SATM18 for ArcelorMittal, unpublished.

[3] Barcellona, A., Palmeri, D., 2009. Effect of Plastic Hot Deformation on the Hardness and Continuous Cooling Transformations of 22MnB5 Microalloyed Boron Steel. Metall and Mat Trans A, vol. 40, 1160–1174.

[4] Bardelcik, A., Worswick, M.J., Wells, M.A., 2014. The influence of martensite, bainite and ferrite on the as-quenched constitutive response of simultaneously quenched and deformed boron steel – Experiments and model. Materials and Design 55, 509–525.

[5] Bhadeshia, H.K.D.H., 1992. Bainite in Steels. The Institute of Metals. Book Number 504.

[6] Chen,Y.,et al 2014. Meso-scale FEA modeling to simulate crack initiation and propagation in Boron steel. 13th International LS-DYNA Users Conference.

[7] Ding, W., Gong, Y., Liu, Q., Wang, J., Wang, Z., Li, W., Jin, X., 2022. Improve bendability of a Cr-alloyed press-hardening steel through an in-line quenching and non-isothermal

partitioning process. Journal of Manufacturing Processes, 481-493.
[8] Dhara, S., Bohemen, S.M.C., Santofimia, M.J., 2022. Isothermal decomposition of austenite in presence of martensite in advanced high strength steels: A review. Materials Today Communications 33, 104567.
[9] Fan, D.W, Park, R.B, Cho, Y.R, Cooman, B.C de, 2010. Influence of Isothermal Deformation Conditions on The Mechanical Properties of 22MnB5 HPF Steel. Steel Research Int., vol. 81, 292–298.
[10] Gong, W., Y. Tomota, Y. Adachi, A.M. Adachi, et al., 2013. Effects of ausforming temperature on bainite transformation, microstructure and variant selection in nanobainite steel. Acta Materialia 61, 4142-415.
[11] Güner, A., Sonntag, M., Philippot, C., Bittendiebel, J., Sarre, B., Dormegny, L., Reihani, A.,Heibel, S., 2021. Assuring Final Product Properties in BIW Applications Using Ductibor ®1000 AS. In Proceedings of the 16th Erlanger Workshop Warmblechumformung, Erlangen, Germany.
[12] Horvath, C., and J. Wang, 2019. Microstructural Optimization of AHSS and Press Hardened Steel, 2019 Great Design in Steel.
[13] Min, J., Lin J., Min, Y., Li, F., 2012. On the ferrite and bainite transformation in isothermally deformed 22MnB5 steels. Materials Science and Engineering A, vol. 550, 375–387.
[14] Min, J., Lin, J., Min, Y., 2013. Effect of thermo-mechanical process on the microstructure and secondary-deformation behavior of 22MnB5 steels. J. Materials Processing Technology, 213, 818-825.
[15] Nikravesh, M., M. Naderi, G.H. Akbari, and W. Bleck, 2015. Phase Transformations in a Simulated Hot Stamping Process of the Boron Bearing Steel. Materials and Design, vol. 86, pp 18-24.
[16] Palmieri, M.E.; Galetta, F.R.; Tricarico, L., 2022. Study of Tailored Hot Stamping Process on Advanced High-Strength Steels. J. Manuf. Mater. Process. 6, 11.
[17] Samadian, P., C. Butcher, M.J. Worswick, 2021. A mean-field homogenization approach to predict fracture in as-quenched microstructures of Ductibor® 500-AS steel: characterization and modelling. International Journal of Solids and Structures 229, 111137.
[18] Samadian, P., A. Abedini, C. Butcher, M.J. Worswick, 2022. Microstructure-based Modelling of Flow and Fracture Behavior of Tailored Microstructures of Ductibor® 1000-AS Steel. Metals 2022, 12, 1770.
[19] Taylor, T., McCulloch, J., 2018. Effect of Part/Die Boundary Conditions on Microstructural Evolution during Hot Stamping 2000 MPa Class Boron Steel. Steel Res. Int., 89, 1700495.
[20] Yang, H-S., and Bhadeshia, H.K.D.H., 2009. Austenite Grain Size and the Martensite–Start Temperature. Scripta Materialia 60. 493–495.

ACKNOWLEDGMENTS

Thanks to support from management teams of ArcelorMittal Global R&D. Special thanks go to L. Huang @ Ford, J. Dykman @ Honda, P. Belanger @ Gestamp, Dr. D. Casellas @ CTM and AM Global R&D colleagues (current and formal) Mr. N. Ramisetti, Dr. J. Chintamani, Dr. M. Kuo, Dr. H. Seyedrezai, Dr. C. Philipps, Dr. L. Dormegny, Ms. Z. Wang, Dr. D. Fan, Mr. S. Lepre, Mr. F. Nietzel, E. Kekeis, L. Laus and M. Alvarez, for their various inputs and support. CTM (Spain) work on using DIL 805 A/D quenching dilatometer for Ductibor®1000 and Ductibor®500 is highly appreciated.

Development of geometry variants and forming technology for metal diaphragms to increase the efficiency of diaphragm compressor

M Solovev[1], A Leonhardt[1], A Graf[1], A Kunke[1] and V Kräusel[2]

[1]Institute for Machine Tools and Production Processes, Chemnitz University of Technology, Chemnitz, Germany
[2]Fraunhofer Institute for Machine Tools and Forming Technology IWU, Chemnitz, Germany

E-mail: mikhail.solovev@mb.tu-chemnitz.de

Abstract. This article outlines the development of an elastically foldable, bistable diaphragm geometry that enables an increase in the compression volume of diaphragm compressors compared to the use of flat diaphragms. Using a Finite Element Model (FEM) in ABAQUS/Standard, several membrane geometries (truncated cone, cone, and dome) were calculated and analyzed with regard to their elastic deformation. Selected geometries were produced using Single Point Incremental Forming (SPIF) and tested by applying full-area pneumatic pressure to determine the diaphragm's lifetime.

1. Introduction

The utilization of hydrogen is critical to the development of an environmentally friendly, low-emission, clean, and sustainable energy system. [1-2]. Fuel cell vehicles (FCVs) that use hydrogen as fuel have become a major focus of research because they are considered zero-emission vehicles and do not strain existing power grids. As technology and systems continue to advance, FCVs are expected to be more competitive than pure electric vehicles, which is why many automobile manufacturers are interested in hydrogen fuel cells. FCVs are among the most promising vehicles for the future, due to the development of related technologies. Hydrogen refueling stations, which provide fuel for FCVs, are essential components and critical infrastructure for operating these vehicles. [3].

Hydrogen gas is compressed to a high pressure of 350 bar to 700 bar in a hydrogen refueling station, and is subsequently stored in high-pressure containers prior to being dispensed to FCVs. [4-6]. The most desirable approach to compressing hydrogen gas is to employ a diaphragm compressor, as its metallic diaphragm effectively separates the hydraulic oil and gas systems, thus guaranteeing gas purity and preventing any contamination of the gas. [7]. Diaphragm compressors are a specialized piece of equipment for hydrogen refueling stations due to their advantages of low power consumption, minimal cooling requirements, and suitability for handling pure gases. Typically, a hydrogen diaphragm compressor with a hydraulic drive consists of a driven piston, a hydraulic oil system, a diaphragm set, and a compression chamber, as shown in figure 1. During operation, when the piston approaches the bottom dead center, the gas suction process is completed as the diaphragms are at their closest position to the perforated plate. The piston then moves upward toward the cavity, pushing the diaphragm to compress the gas using the hydraulic oil. Once the gas pressure reaches the discharge pressure, the discharge valve opens and the diaphragm makes contact with the cavity surface until the piston reaches the top dead center and the oil pressure reaches its maximum. The piston then moves downward, causing the diaphragm to deform and start the suction process again. [8]. Two disadvantages of current diaphragm compressors are a low flow rate due to cavity profiles as well as a short diaphragm life. [8-9]. The latter is influenced by radial stresses causing cracks in the membrane. [8].

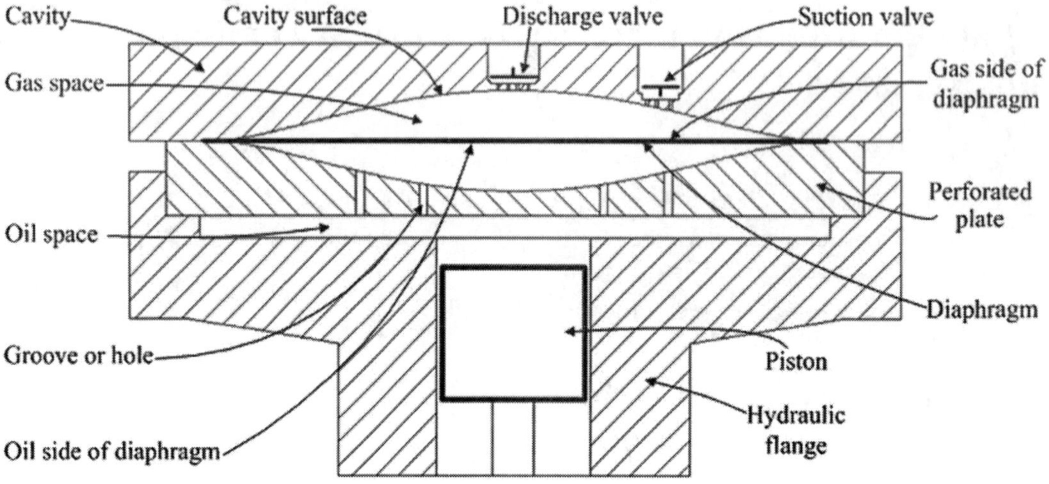

Figure 1. The schematic of a diaphragm compressor with a hydraulic drive [10].

The purpose of this paper is to investigate the potential of using a new diaphragm geometry to improve the cavity volume and therefore the efficiency of a hydrogen compressor. The goal is to develop a membrane with a larger displacement volume that can compress more hydrogen in a shorter amount of time. By adjusting the geometry of the membrane, the objective is to enhance the efficiency of the hydrogen compressor and lower the energy needed for the compression procedure. The paper aims to provide a solution for the challenge of compressing hydrogen gas, which is essential for the storage and transportation of hydrogen in fuel cell systems. Ultimately, the goal of this research is to contribute to the development of more efficient and cost-effective hydrogen compression technology, which could help accelerate the adoption of hydrogen as a clean energy source.

2. Calculation of geometries with elastic folding using Abaqus

2.1. Simulation model setup

The simulation of membrane folding was conducted using a 3D Abaqus simulation model. To reduce calculation time, only one quarter of the model was considered. The elastic-plastic material properties of the austenitic stainless spring steel 1.4310 in the simulation model were described using a yield curve obtained from tensile testing. As a preliminary step, the folding of the diaphragm was simulated using different strength levels of 1.4310 (1100-1300, 1300-1500, 1500-1700 and 1900-2100 MPa) and different sheet thicknesses of 0.05, 0.1, and 0.2 mm. The geometries with a sheet thickness of ≤ 0.1 mm provided better folding properties. Although the folding of the diaphragm improves with increasing strength grade of the 1.4310, a minimum elongation is needed for the forming of the membrane. The optimal combination of strength and elongation is shown in table 1.

Table 1. Mechanical properties of 1.4310 with 0.1 mm sheet thickness.

Strength grade	Yield strength (MPa)	Tensile strength (MPa)	Elongation (%)
F15-1700	1471	1651	10.3

The simulation employed shell elements and utilized an implicit approach, with a mesh size of 1 mm. This allowed for accurate computation of the membrane's behaviour during the folding (compression process). The simulation of the folding was divided into two steps. In the first step, the diaphragm is pressurized to 100 bar at the top side and to 99.9 bar at the bottom side. In the second step, the pressures are changed to 99.9 bar at the top side and 100 bar at the bottom side (figure 2).

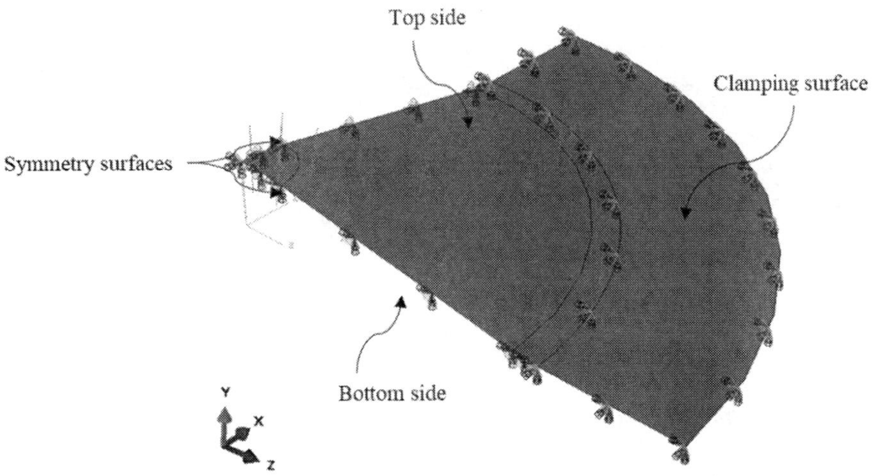

Figure 2. Abaqus 3D simulation model of a diaphragm.

2.2. Calculated geometries

A cone, a truncated cone and a dome geometry were selected as geometry variants for diaphragms. Figure 3 shows the selected geometries. The outer diameter D of the three geometries is constant and set to 140 mm. The other four parameters (height h, cone angle α, radii R_1 and R_2) are variable. The Abaqus software was used to simulate the folding behaviour of the geometry variants. The aim of the simulation was to find an optimal combination of geometric parameters in which the folding of the diaphragm occurs in both directions without plastic deformation and a maximum volume is compressed.

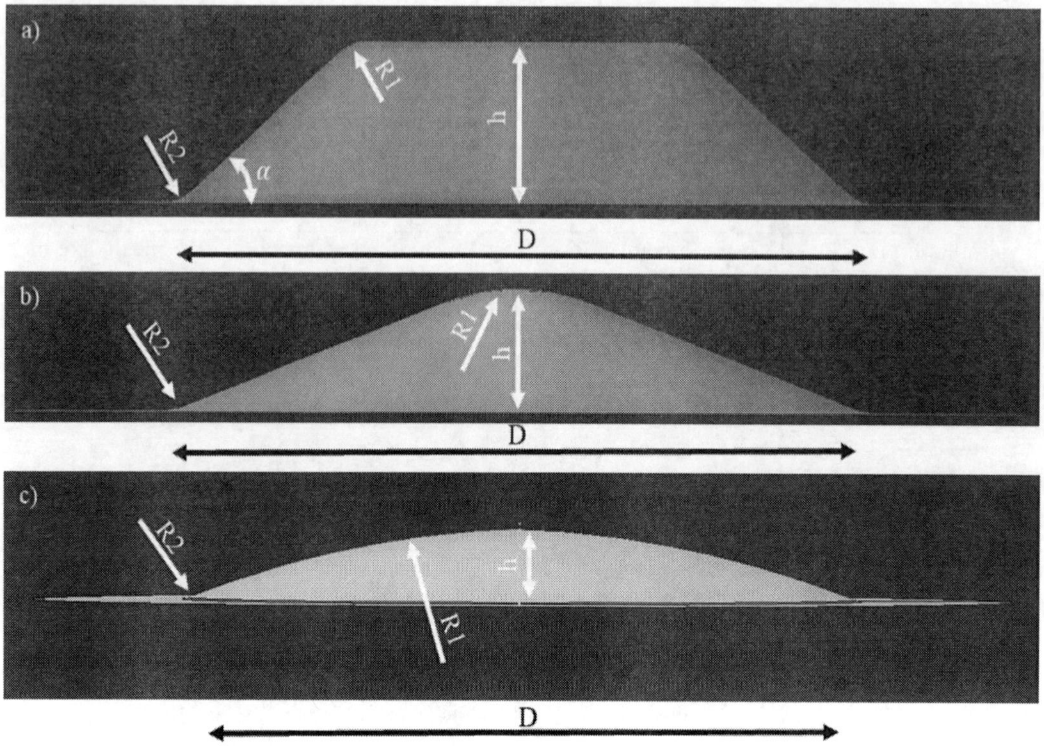

Figure 3. Selected diaphragm geometries: a) cone, b) truncated cone and c) dome.

2.3. Evaluation of simulation results

First, the volume of a flat diaphragm with an outer diameter of 140 mm was determined. To calculate the volume of the flat diaphragm, a series of pneumatic load tests were performed to measure the diaphragm deflection using the test rig (figure 7) and a dial gauge. A flat diaphragm made of 1.4310 with strength grade F1500-1700 and a sheet thickness of 0.1 mm was subjected to a pressure of 0.1 bar. The deflection of the diaphragm was recorded and compared with simulation results. Both the experiment and the simulation yielded the same deflection value of 3.2 mm. In the next step, the diaphragm contour from Abaqus was imported into a CAD programme (Creo) and the volume of the diaphragm was determined (table 2). By comparing the compressed volume of the diaphragm geometries listed in table 3 with the volume of a flat diaphragm, an increase of 2 to 3 times can be achieved.

Table 2. Volume of the flat diaphragm.

Diameter (mm)	Thickness (mm)	Pressure (bar)	Deflection (mm)	Volume (cm^3)
140	0.1	0.1	3.2	24.568

In evaluating the simulation results for the folding of the non-flat membrane geometries, two quantities, AC YIELD and Mises stress, were analyzed. AC YIELD provides a yes/no flag (1/0) indicating if the material is currently yielding or not. The maximum Mises stress can be compared to the flow curve. If AC YIELD is 1 and the maximum Mises stress is above the flow stress, the folding of the membrane is not elastic but plastic and not suitable for application in a diaphragm compressor. Figure 4 shows an example of the evaluation of the cone geometry. Table 3 summarizes the calculated and evaluated diaphragm geometries.

IOP Conf. Series: Materials Science and Engineering 1284 (2023) 012009 doi:10.1088/1757-899X/1284/1/012009

Table 3. Listing of the calculated geometries.

Geometry	h (mm)	α (°)	R_1 (mm)	R_2 (mm)	Volume (cm^3)	Max. Mises (MPa)	AC YIELD
Cone	10	9.03	60	30	51.542	743	0
Cone	11	10.05	60	30	56.752	1079	0
Cone	12	11.08	60	30	61.979	1207	0
Cone	13	12.14	60	30	67.226	1363	0
Cone	14	13.23	60	30	72.495	1423	0
Cone	15	14.34	60	30	77.787	1523	1
Truncated cone	5	15	30	30	51.071	1049	0
Truncated cone	5	15	60	30	50.466	1214	0
Truncated cone	6	15	60	30	57.044	1502	1
Dome	8	-	400	30	51.251	943	0
Dome	9	-	400	30	55.560	875	0
Dome	10	-	400	30	59.898	1242	0
Dome	11	-	400	30	64147	1474	1

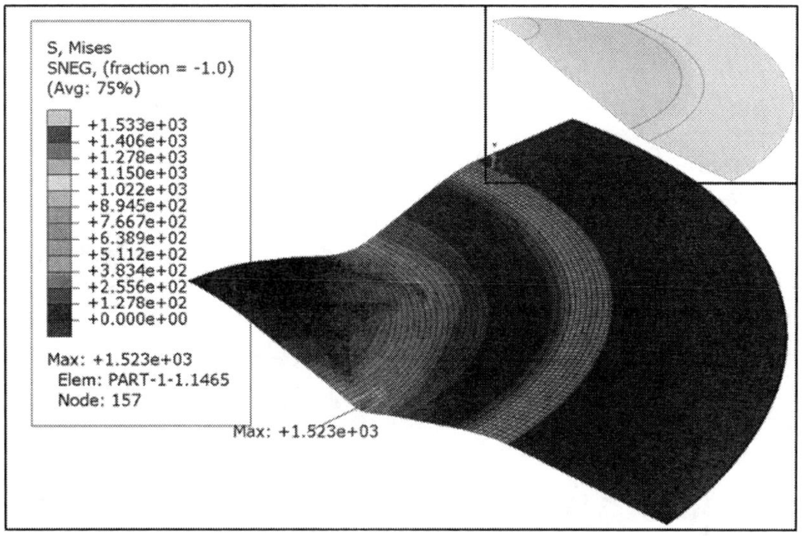

Figure 4. An example of the calculated geometries: h15 $R_1$60 $R_2$30 with Mises stress distribution.

3. Manufacturing of membrane prototypes with single point incremental sheet metal forming

3.1. Selected geometries and experimental setup for the forming tests

Four membrane geometries were produced using SPIF: a cone with depths of 10, 12, and 15 mm, $R_1$60 mm and $R_2$30 mm, as well as a truncated cone with a depth of 5 mm, 15° cone angle, $R_1$30 and $R_2$30 mm.

To perform the SPIF process, initial square blanks with dimensions of 200 mm x 200 mm were cut from the employed sheet. The SPIF of the diaphragm prototypes was performed on a conventional 3-axis CNC milling machine, using a hemispherical head forming tool made from cold working tool steel with a diameter of 10 mm and 30 mm. A spiral tool path with step width of 1 and 2 mm and a feed rate of 2000 mm/min without spindle rotation were used. To reduce friction between the sheet and the forming tool, hydraulic oil was utilized as a lubricant. Figure 5 shows the experimental setup.

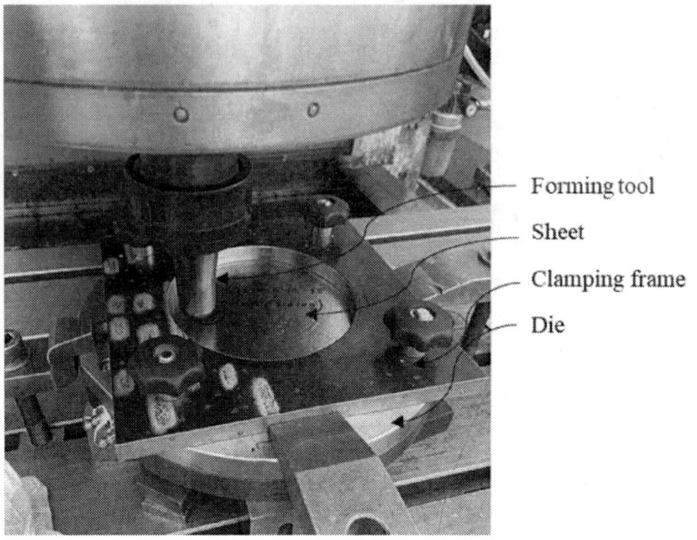

Figure 5. Experimental setup for the SPIF of the diaphragm geometries.

3.2. Results of SPIF tests
The SPIF process was carried out in two steps. In the first step, the diaphragm geometry was pre-formed with a forming stylus with a diameter of 30 mm and 2 mm step width. In the second step, it was formed with a 10 mm forming tool and 1 mm step width increment. A formed cone and truncated cone geometry are shown in figure 6. During the SPIF of both geometries local springback of the material occurs due to its high strength grade when the forming stylus moves across the sheet surface. In the case of the truncated cone this springback causes wrinkles which are still visible after the forming process. As these wrinkles cause local plastic deformation in the folding process, this geometry is not further investigated.

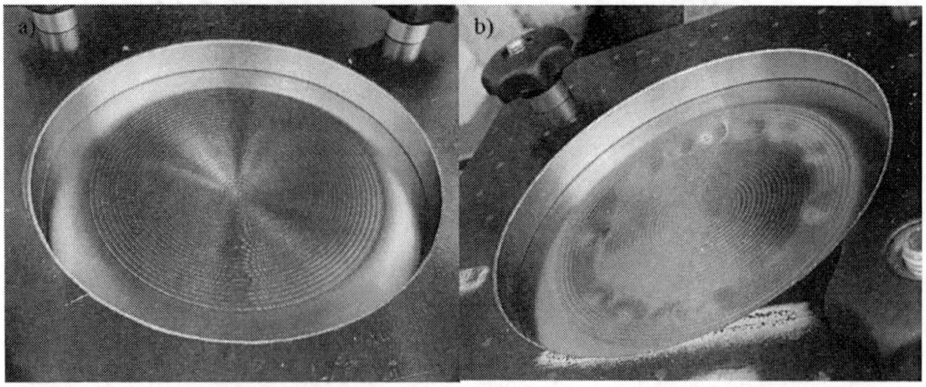

Figure 6. The membranes after SPIF (clamped): a) Cone h10 $R_1$60 $R_2$30 and b) truncated cone h5 α15° $R_1$30 $R_2$30.

4. Folding tests of selected geometries

4.1. Folding properties of the formed geometries

For the investigation of the folding behaviour and assessment of the service life of formed membranes, a test rig was developed and provided by Kyros Hydrogen Solutions GmbH (figure 7). The frequency controller is used to control the 5/2-way pneumatic valve and set the switching frequency. The actuation frequency was set to 2 Hz. The pressure sensor serves to monitor the pressure and detect any decreases in pressure which indicates a cracking of the membrane. The counter was utilized to register the switching of the 5/2-way pneumatic valves and consequently count the number of folding cycles.

Figure 7. Test rig for investigating the folding properties of a membrane.

The results of the folding tests are shown in figures 8-10. The membranes were folded 1000 times and visually inspected for wrinkles or bulges. The best folding behaviour was observed for the 10 mm cone geometry, with no wrinkles visible on the membrane surface. The 12 mm and 15 mm cone geometries showed plastic deformation of the membrane during folding, which was clearly visible on the membrane surface. The 15 mm cone geometry failed after only a few folding cycles and developed a crack.

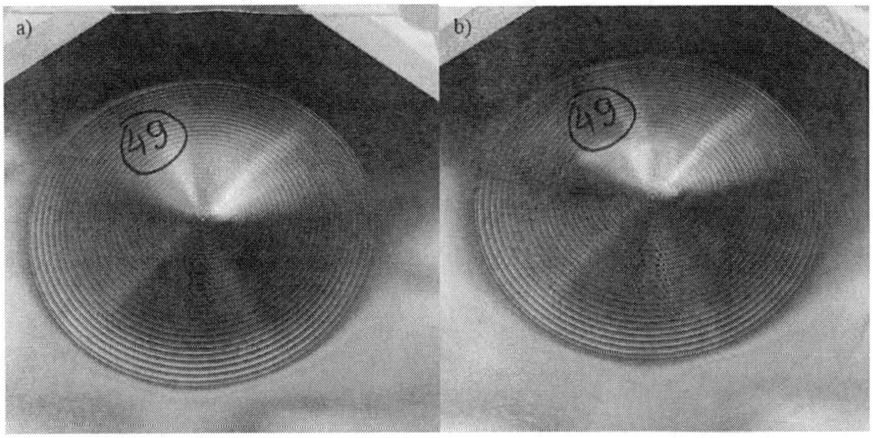

Figure 8. Cone h10 $R_1$60 $R_2$30: a) after SPIF and b) after 1000 folding cycles.

IOP Conf. Series: Materials Science and Engineering 1284 (2023) 012009 doi:10.1088/1757-899X/1284/1/012009

The folding tests of the 10 mm and 15 mm cones show good agreement with the simulation results. According to the simulation results, the 10 mm cone can be folded purely elastically, while the 15 mm cone undergoes plastic deformation.

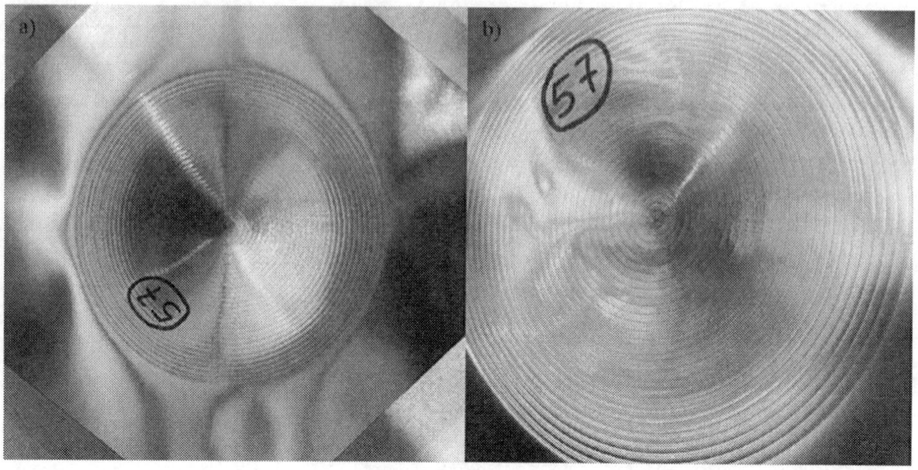

Figure 9. Cone h12 $R_1$60 $R_2$30: a) after SPIF and b) after 1000 folding cycles.

The simulation results suggested that the 12 mm cone should also fold elastically, but after multiple folding cycles in the experiment, wrinkles occurred due to plastic deformation. The reason for this is that the previous forming process of the membrane during SPIF was not taken into account in the performed simulations of the membrane folding. The occurring strain hardening and inhomogeneous stress and strain distribution in the sheet influence the folding properties of the diaphragm.

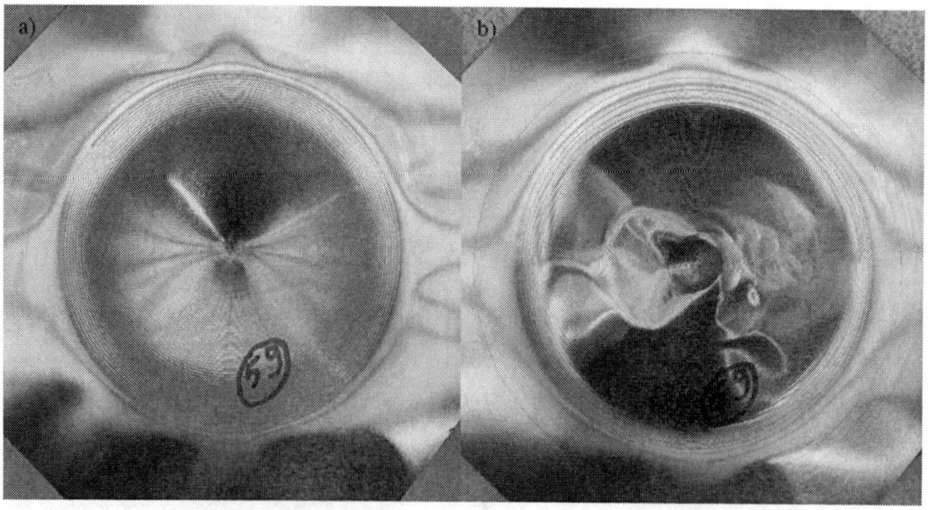

Figure 10. Cone h15 $R_1$30 $R_2$30: a) after SPIF and b) after a few folding cycles.

4.2. Lifetime of the diaphragms formed with SPIF

As a short diaphragm life is a disadvantage of current diaphragm compressors with a flat membrane, the lifetime of the formed membranes was tested within the test rig. The number of folding cycles was counted until a first crack appeared in the membrane.

A list of the tested membranes is presented in table 4 indicating that the achieved number of cycles for the formed membranes are still too small for industrial application. During the tests on the membrane's lifespan, it was found that all of the diaphragms fracture near to the membrane tip. When examining the fracture site, it was detected that the crack always follows the contact path between the forming tool and the sheet. The reason for this is that the surface of the membrane was strain hardened and damaged by the forming tool.

Table 4. Listing of the formed diaphragms.

Geometry	h (mm)	α (°)	R_1 (mm)	R_2 (mm)	Volume (cm^3)	Max. Mises (MPa)	Number of cycles
Cone	10	9.03	60	30	51.542	743	5628
Cone	12	11.08	60	30	61.979	1207	420
Cone	15	14.34	60	30	77.787	1523	<100

5. Conclusion and prospects

As part of this project, a membrane geometry was developed using simulation tools that exhibits purely elastic folding behavior and allows for an increase in compressed/cavity volume of a diaphragm compressor compared to a flat membrane. The calculated geometries were produced using SPIF and tested for their folding behavior and operational lifetime. It was found that the manufacturing process of the membrane has a significant impact on its durability.

Further investigations will involve testing other forming methods, such as hydroforming and stretch forming. These methods, especially hydroforming, do not affect the surface quality during the forming process. Additionally, the simulation model will be optimized through coupled forming and folding simulations to better understand the influence of forming on folding behavior. Furthermore, the geometry of the membranes after forming and folding will be measured to capture the global springback and any changes in geometry that occur.

Acknowledgments

The authors would like to thank German Federal Ministry of Education and Research (BMBF) for funding the research project “PEM4Heat - Entwicklung eines PEM Hochdruckelektrolyse Stacks mit Prozesswärmeauskopplung, eines H2-Hochdruckverdichters und eines Kreislaufmotors am WIR!-Ausgangspunkt Sonneberg” as well as Kyros Hydrogen Solutions GmbH for the provision of testing facility used in the presented investigations.

References

[1] Balat M 2008 Potential importance of hydrogen as a future solution to environmental and transportation problems *International Journal of Hydrogen Energy* 33(15) pp 4013-4029

[2] Winter C J 2009 Hydrogen energy-Abundant, efficient, clean: A debate over the energy-system-of-change *International Journal of Hydrogen Energy* 34(14) pp S1-S52

[3] Wilberforce T, El-Hassan Z and Khatib F N et al 2017 Developments of electric cars and fuel cell hydrogen electric cars *International Journal of Hydrogen Energy* 42(40) pp 25695-25734

[4] Kang J E, Brown T and Recker W W et al 2014 Refueling hydrogen fuel cell vehicles with 68 proposed refueling stations in California: Measuring deviations from daily travel patterns *International Journal of Hydrogen Energy* 39(7) pp 3444-3449

[5] Yu W, Dianbo X and Jianmei F et al 2010 Research on sealing performance and self-acting valve reliability in high-pressure oil-free hydrogen compressors for hydrogen refueling stations *International Journal of Hydrogen Energy* 35(15) pp 8063-8070

[6] Weinert J X, Shaojun L and Ogden J M et al 2007 Hydrogen refueling station costs in Shanghai *International Journal of Hydrogen Energy* 32(16) pp 4089-4100
[7] Reddi K, Elgowainy A and Sutherland E 2014 Hydrogen refueling station compression and storage optimization with tube-trailer deliveries *International Journal of Hydrogen Energy* 39(33) pp 19169-19181
[8] Li J, Jia X and Wu Z et al 2014 The cavity profile of a diaphragm compressor for a hydrogen refueling station *International Journal of Hydrogen Energy* 39(8) pp 3926-3935
[9] Wang T, Tang Z and Jia X 2019 Study on the stress and deformation of a diaphragm compressor cylinder head under extreme conditions *International Conference on Compressors and their Systems* 604(1) pp 012029
[10] Jia X, Chen J, Wu H und Peng X 2016 Study on the diaphragm fracture in a diaphragm compressor for a hydrogen refueling station *International Journal of Hydrogen Energy* 41(15) pp 6412-6421

Part quality improvement in flexible roll forming of a combined variable width and depth component

B Abeyrathna*[1], B Rolfe[2], R Taube[3], B Oxley[4] and M Weiss[1]

[1] Institute for Frontier Materials, Deakin University, Waurn Ponds, Pigdons Rd. VIC. 3216, Australia
[2] School of Engineering, Deakin University, Waurn Ponds, Pigdons Rd. VIC. 3216, Australia
[3] Ford Motor Company of Australia Ltd, 1735 Sydney Rd, Campbellfield VIC. 3061
[4]SHAPE CORP., 1835 Industrial Park Dr, Grand Haven, MI 49417 USA

buddhika.abeyrathna@deakin.edu.au

Abstract. Flexible Roll Forming (FRF) is an advanced manufacturing process that can be used to form long and complex part families from hard to form materials for the automotive and truck industry. However, depending on the part complexity and material strength shape defects such as end flare, springback and twist can occur. In this work a high strength automotive part (Sled runner) is produced from 3 high strength steels; the complex component combines shape variations in width and depth. The optimum forming sequence is identified to produce the part without any flange wrinkles. In addition to this, a new over forming technique is introduced to overcome end flare and springback. The experimental results are evaluated and discussed with numerical analysis and suggest that the presented forming technology enables the manufacture of high strength components in accordance with automotive specifications and quality standards.

1. Introduction

Flexible Roll Forming (FRF) can be used to form complex automotive components from Advanced High Strength Steel (AHSS) and Ultra High Strength Steel (UHSS) for structural and crash applications. Due to the limited formability of AHSS and UHSS, shape defects such as springback, end flare are inevitable in part manufacturing. However, the FRF process provides the required tool flexibility to overcome such defects [1].

The most common types of shape defects in FRF are web warping, wrinkling, springback and end flare. Web warping occurs if the flange does not undergo the required plastic deformation. A blank holder solution has been introduced to hold the web in place during FRF to achieve the required plastic deformation in the flange and to eliminate web warping [2, 3]. A similar approach was introduced in [4] which ~~web is~~ constrained the web by a set of rolls. Heat assisted FRF was also introduced in [5] to soften the critical flange region and facilitate the required longitudinal compression deformation.

Flange wrinkling is another common shape defect in FRF and occurs if the longitudinal compressive stress within the flange exceeds the critical wrinkling limit. This defect is difficult to eliminate as the state of stress is hard to control. A one step model was introduced in [6] which used geometric and material parameters to determine the likelihood of wrinkling initiation. For that, the maximum longitudinal compressive stress is compared with the critical wrinkling limit of the flange. A different study [7] compared the compressive strain in the flange predicted with FEA with the theoretical

IOP Conf. Series: Materials Science and Engineering 1284 (2023) 012010 doi:10.1088/1757-899X/1284/1/012010

longitudinal strain required in the flange. Wrinkling was defined to occur when the FEA strain was lower than the analytical strain. A combined experimental and theoretical approach was used to identify the occurrence of wrinkling in [8]. For that a wrinkling limit plot together with the evolution of the effective strain - stress triaxiality in the critical flange region was used. Wrinkles were observed if the stress triaxiality showed a wave type deviation changing from positive to negative and vice versa. Some studies have shown that wrinkling in FRF can be reduced with additional tool support and by the use of small angle increments [1, 9]

Springback and end flare are the other two common shape defects in FRF that need to be controlled. Springback occurs due to the elastic recovery of the material after forming and can be successfully compensated for by overbending [1]. However, end flare in combination with springback does not allow for a uniform overbending approach. Therefore variable overbending was proposed in [10] by altering the tool path to generate different degrees of overbending along the length of the part.

In this study a complex automotive part with combined variable width and depth is formed for the first time. The effect of the forming sequence on wrinkling and the part quality is investigated and a flexible overbending approach is proposed to overcome springback and end flare.

2. Material and part geometry

Three Dual Phase (DP) steels a DP600, a DP780 and a DP1000 were tested. Standard tensile tests were performed to obtain the material properties in the rolling direction in an Instron 5967 machine with a 30kN load cell according to ASTM E8/E8M [11]. A crosshead speed of $0.025mms^{-1}$ was used giving a strain rate of 0.001 s^{-1}. The average true stress-strain curves are shown in **Figure 1**(a). The material parameters were obtained by fitting the Hollomon's power equation (Equation (1)) to the true stress - effective plastic strain obtained from the tensile curves as shown in **Figure 1**(b) and the corresponding material parameters are given in Table 1.

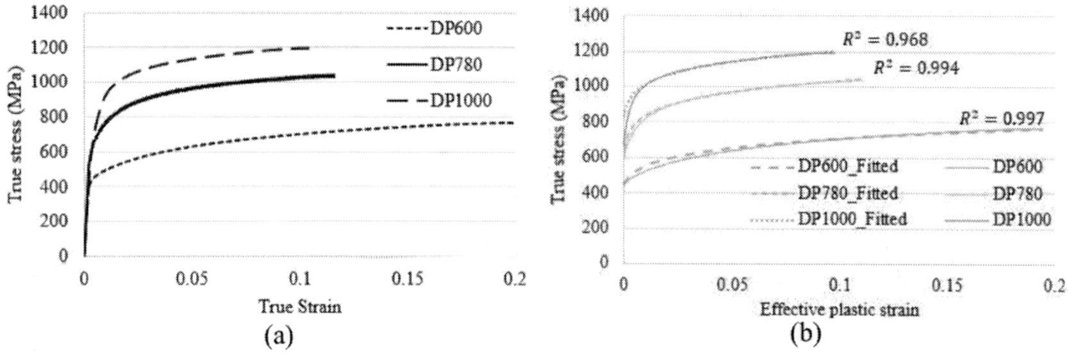

Figure 1. (a)Averaged true stress strain curves determined along the rolling direction for the DP steels (b) true stress- effective plastic strain curves and fitted Hollomon's power law

$$\sigma = k\varepsilon_{eps}{}^{n} \qquad (1)$$

Where σ is the true stress, ε_{eps} is the effective plastic strain, n and k are hardening exponent and strength coefficient respectively.

Table 1. Material properties determined with the fitted Hollomon's law for the DP600, the DP780 and the DP1000

Material	Yield Strength (MPa)	Ultimate tensile strength (MPa)	Elastic limit	n	K (MPa)
DP600	446.5	767.7	0.002	0.14	981
DP780	594	960	0.003	0.09	1284
DP1000	764.1	1194.3	0.004	0.08	1447

The part considered for the investigation is a Sledrunner design as shown in **Figure 2**(a) which was provided by the Ford Motor Company. The cap-like closed in geometry at the back end of the part has to be removed to enable part manufacture with FRF (see **Figure 2**(b)). Finally, the part was simplified as shown in **Figure 2**(c) by removing the surface depressions of the original part. The dimensions of the front cross section of the simplified part are shown in **Figure 2**(d). This is a complex part which has a varying cross section in both width and depth direction along the length.

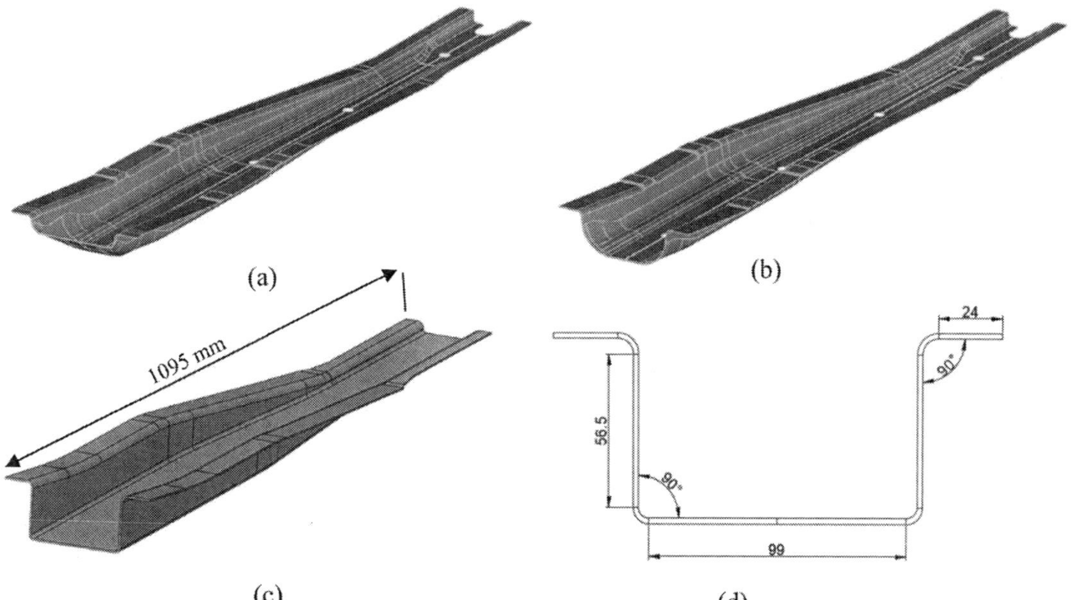

(a) (b)

(c) (d)

Figure 2. (a) Sledrunner component design provided by Ford (b) Sledrunner after front cap removal (c) simplified Sledrunner (d) cross section dimensions of the front end.

3. Flexible roll forming trials

The flexible roll forming experiments were performed with the FRF facility at Deakin University [12]. The operating principle of Deakin's FRF machine is different to that of a conventional flexible roll forming machine and its operation principle is explained in [12]. Three different forming sequences were used to form the part to understand the effect of the forming sequence on the part quality. In all 3 forming sequences, the flange is formed to 57° in 4 passes as shown in Figure 3(a). In the first forming sequence (FS1), the flange was formed to 90° in the following 2 passes and then top hat was formed in 4 subsequent passes as shown in Figure 3(b). In the second forming sequence (FS2), the top hat profile was fully formed from pass 5 to 8 and then the remaining flange was formed to 90° in the following two passes as shown in Figure 3(c). In the third forming sequence (FS3), the top hat profile was partially formed in passes 5 and 6, and then both the flange and the top hat were formed to 90° in pass 7.

The outer surfaces of the formed parts were scanned with a HandySCAN 3D [13] and three cross sections were considered for the part shape evaluation as shown in **Figure 4**(a). Angle α and β were measured to get a quantitative measure of the springback and end flare (see **Figure 4**(b)). Finally, an overbending operation was performed using two cylindrical rolls as shown in **Figure 5** to compensate for springback and the end flare defect.

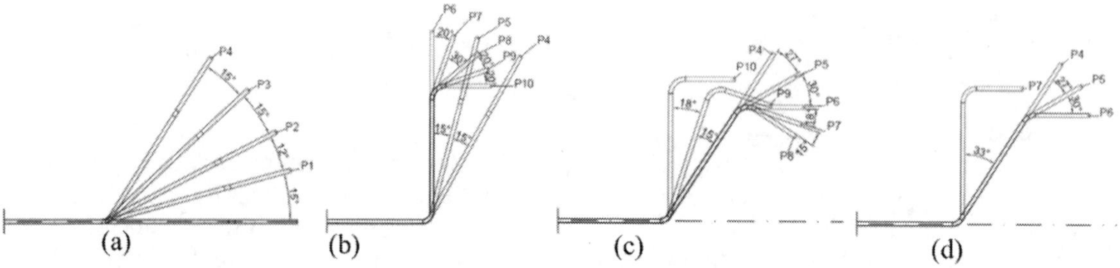

Figure 3. Forming sequence of (a) pass 1-4 used for all forming sequences (b) pass 4-10 in FS1 (c) pass 4-10 in FS2 (d) pass 4-7 in FS3

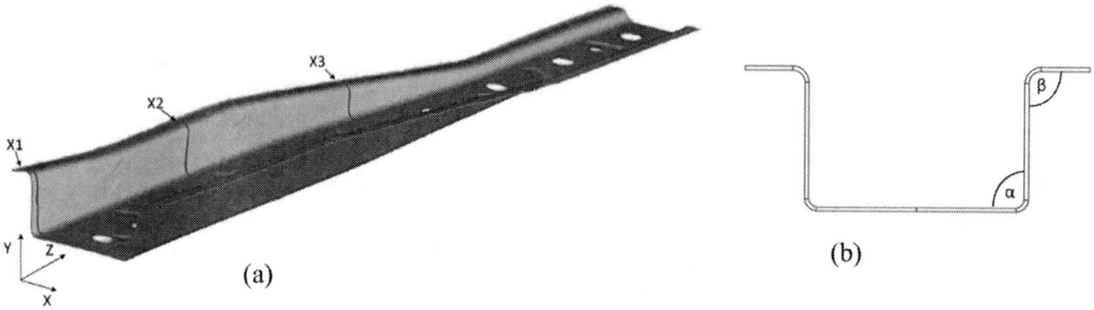

Figure 4. (a)Locations of different cross sections considered for the part shape evaluation (b) measured angles of each cross section

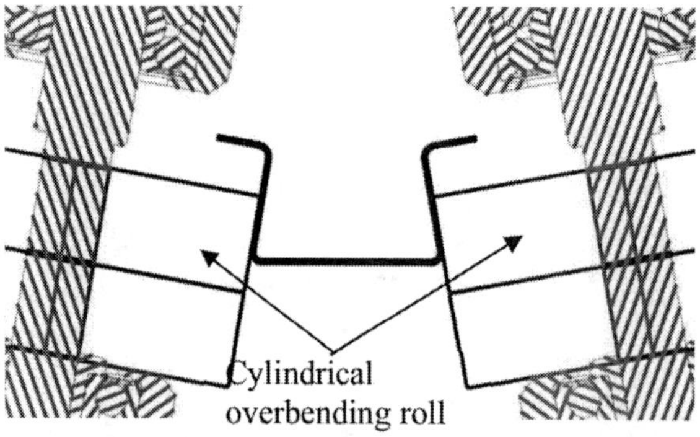

Figure 5. Overbending tooling

4. Finite element analysis (FEA)

The finite element simulation of the FRF process was performed with Copra RF FEA [14] which uses an implicit solver. In the experimental forming process the top and bottom clamps that carry the pre-cut sheet move forward while the forming rolls change their position and angle to form the flange incrementally. However, in the FEA, the pre-cut sheet is assumed stationary, and the forming rolls are moved forward along a pre-defined tool path to form the part as shown in Figure 6(a). The forming tools and clamps were designed as rigid bodies and the pre-cut sheet was discretised as full integration, hexahedral, type 7 arbitrarily distorted brick elements [15]. A symmetric half of the pre-cut sheet was modelled and two elements through the thickness were used. A Young's modulus of 200GPa and a Poisson's ratio of 0.3 were used to define the elastic part of the material properties [12], and the von Mises yield criteria in combination with the plastic part of the true stress strain curves shown in Figure 1 were applied to define the plastic material properties. Three boundary conditions were used on the pre-cut sheet as shown in Figure 6(b). An X lock boundary condition was applied on all nodes in the symmetric line which restricts the nodes in the X direction to represent symmetry. A Y lock boundary condition was used on 6 leading nodes on the symmetric line to keep the pre-cut sheet in place before clamping and after unclamping. A Z lock boundary condition was used on 6 nodes on the leading edge to avoid material movement in the Z direction while forming. The longitudinal stress along the flange edge was evaluated in the FEA model to understand the effect of forming sequence on the part quality (see Figure 6(c)).

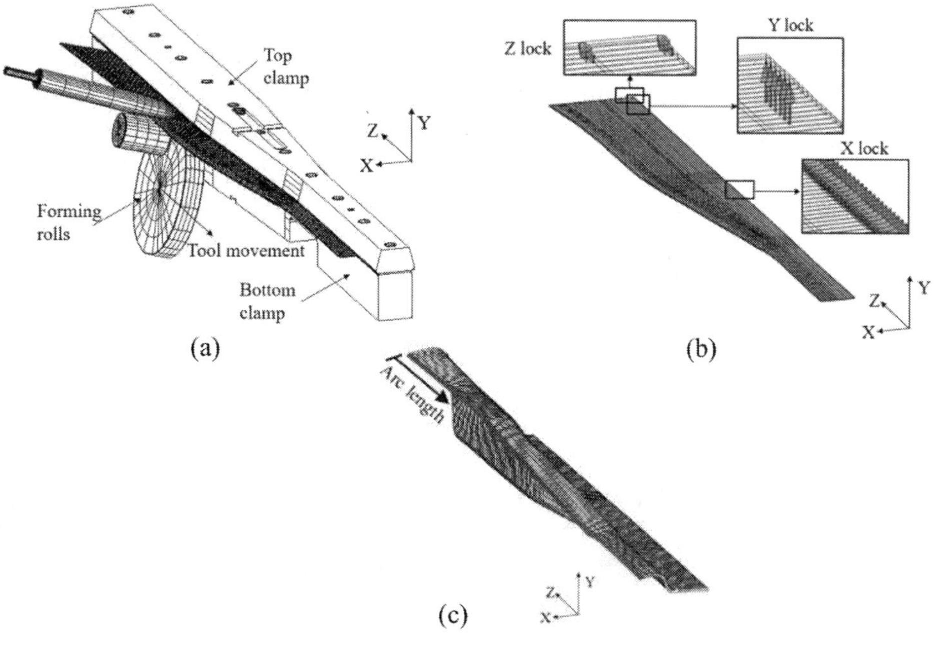

Figure 6. (a) Finite element model assembly (b) boundary conditions on the pre-cut (c) the node path considered for the analysis of longitudinal stress

5. Results and discussion

The part shape in the FEA is investigated at different stages of the three forming sequences. The first 4 forming passes are the same for all 3 forming sequences. The part shape after pass 4 is shown in **Figure 7** and indicates the formation of one localised buckle. The corresponding longitudinal stress distribution for the first four passes in FS1, FS2 and FS3 is shown in **Figure 9**(a). The stresses are mainly compressive for the first 4 passes and the waviness in the stress distribution after pass 4 indicates the development of a buckle. That buckle further propagates and creates a wave when the flange is fully formed to 90° in pass 6 in FS1 (see **Figure 7**). According to **Figure 9**(a) and (b), the stresses in the critical region are mainly compressive for FS1 and the waviness of the plots confirms the presence of wrinkles. In FS2, the top hat forming starts after pass 4 and the buckle is partially formed out during the top hat forming operation (see **Figure 8**(b)). The longitudinal stress plots for passes 5-10 of this condition are shown in **Figure 9**(c) and indicate a longitudinal tensile stress in the last two passes, P9 and 10 when the top hat is formed. Such tensile stresses cannot be achieved with FS1 due to the high compressive stresses that accumulate during the flange forming operation which already creates some severe wrinkling. The buckle observed after pass 4 is partially formed out in FS2 after the top hat forming operation due to the introduced tensile longitudinal stress. In FS3, the first 6 forming passes are like the FS2 condition. However instead of completing the top hat forming operation, the partially formed flange and top hat are formed to the final shape in pass 7 in forming sequence FS3. This results in a perfectly formed flange as shown in **Figure 8**(c). It shows a better flange quality than the part formed with FS2 (compare **Figure 8**(b) with (c)). According to longitudinal stress diagram in **Figure 9**(c) and (d), after the last pass in both FS2 and FS3, the same level of tensile longitudinal stress is achieved. However, in FS2, pass 7 and 8 do not generate any tension while completing the top hat formation. Therefore, a small depression is locked in on the flange as shown in **Figure 8**(d). On the other hand, in FS3, the top hat forming is completed at the last pass while creating a tensile stress on the top hat. Therefore, a clean flange can be achieved with FS3 without creating any localised depression in the critical region of the flange.

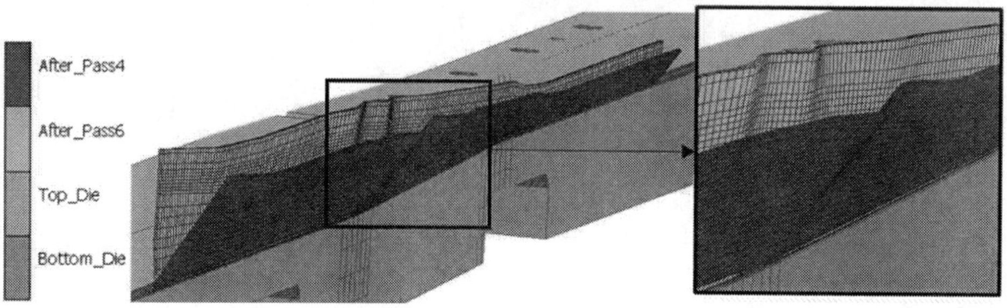

Figure 7. Part quality after pass 4 for FS1, FS2 and FS3 and after pass 6 in FS1

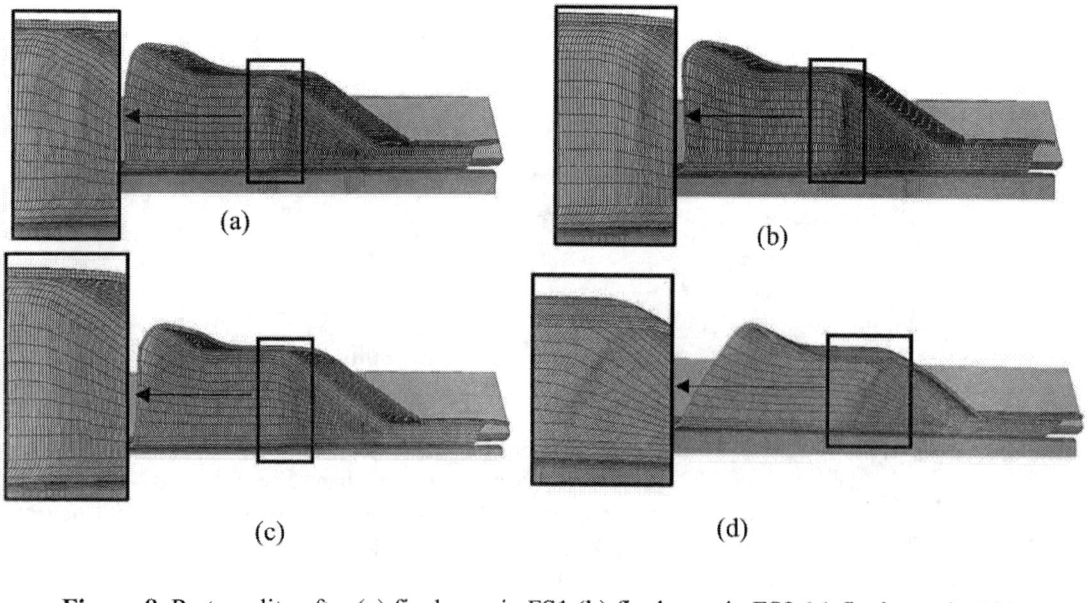

Figure 8. Part quality after (a) final pass in FS1 (b) final pass in FS2 (c) final pass in FS3 (d) pass 8 in FS2

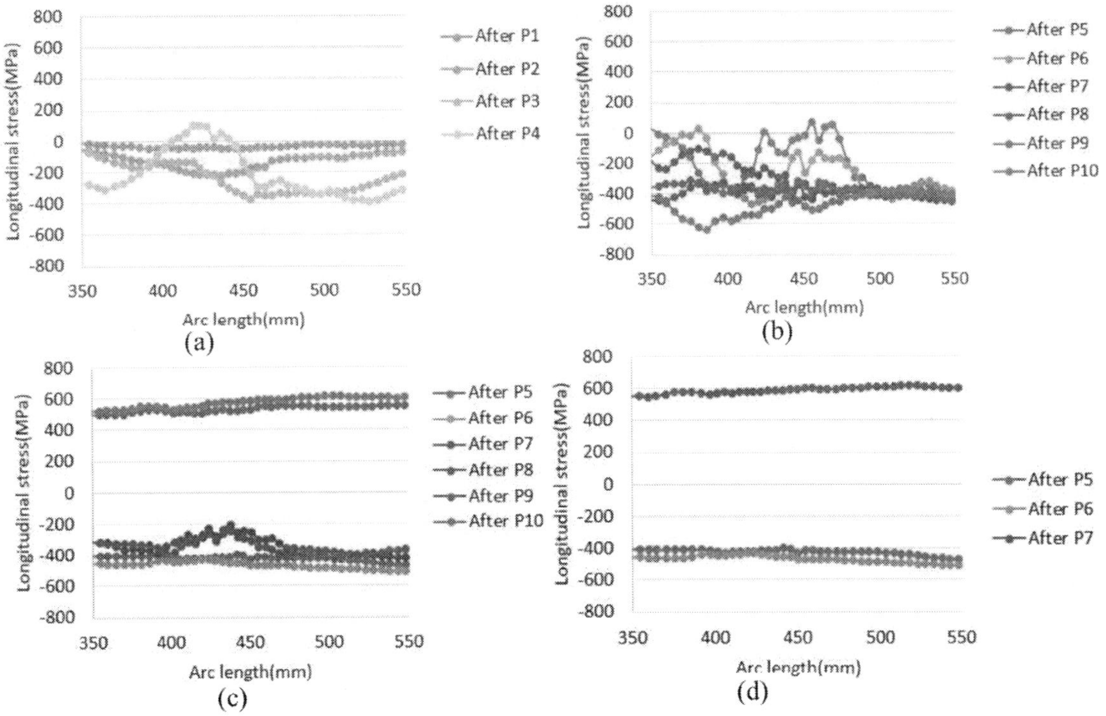

Figure 9. The longitudinal stress distribution in the critical region for (a) first 4 passes in FS1, FS2 and FS3 (b) pass 5-10 in FS1 (c) pass 5-10 in FS2 (d) pass 5-7 in FS3

Based on the initial FEA analysis it was concluded that FS3 will give the best part quality combined with a minimum number of forming passes. Therefore, FS3 was used to perform the experimental FRF prototype forming trials with the 3 DP steels (Table 1). The part quality after the last flange forming pass (pass 4) is shown in **Figure 10** (a) and (b) for the FEA and the experiments respectively. Both, the FEA and the experiments show some waviness after forming pass 4 for all three DP steels. However, in the following 3 forming passes the waviness is completely removed and a a wrinkle free part produced as it is shown in **Figure 11**. This confirms that the waviness in the flange can be formed out by introducing a top hat forming operation. The top hat introduces tensile stresses that pull out the wrinkles.

.

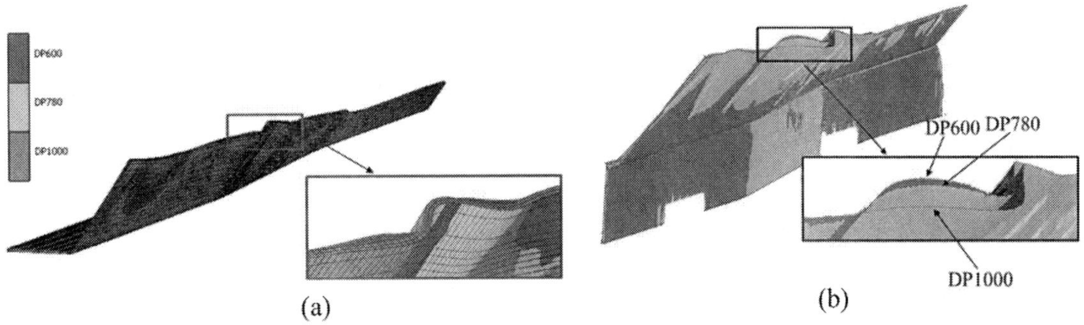

Figure 10. Part shape comparison between FEA and experiments after pass 4 for (a) FEA (b) experiments

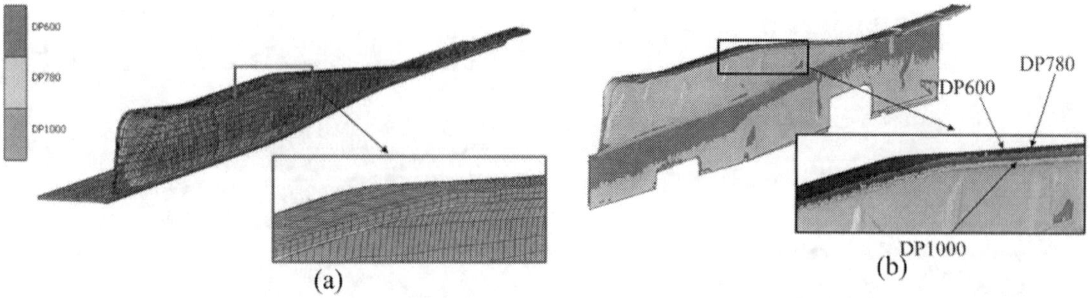

Figure 11. Part shape comparison between FEA and experiments after pass 7 for (a) FEA (b)experiments

The cross-section shape was analysed in three locations as explained in section 3. and the springback angles are presented in **Table 2**. The springback angle increases with material strength. In addition, forming angle α shows higher springback than angle β. That could be due to the tool configuration of the last pass where angle β is fully supported by a top and a bottom roll while angle α is only supported by the bottom roll as indicated in **Figure 12**(a). The cross section X1 (refer to Figure 4) shows higher springback than X2 and X3 for angle α; this indicates that there is end flare.

Table 2. Springback angle of different cross sections

Cross section	X1		X2		X3	
Material	$\Delta\alpha$	$\Delta\beta$	$\Delta\alpha$	$\Delta\beta$	$\Delta\alpha$	$\Delta\beta$
DP600	6.28	-1.42	-0.88	-1.21	-0.54	-3.74
DP780	9.53	1.10	1.60	0.76	2.04	-1.05
DP1000	14.07	2.13	5.38	2.55	6.93	1.07

A new approach was introduced to compensate for end flare and springback by overbending. For that a pair of cylindrical rolls was used as shown in **Figure 5**. The overbending angle and the distance over which the flange was overformed was determined based on the maximum springback angle and its spread over the length of the part, **Figure 12**(b). A two-step overbending approach was used for the DP1000 which showed the highest variation of springback over the length of the part. For the DP 1000 14° and 7° overbending steps were applied over 110mm and 315mm lengths respectively (see **Figure 12**(b)). The springback angles after compensation are given in **Table 3**. There is a clear improvement in springback after the 2nd overbending step for the DP1000, especially in X2 and 3. Images of the physically formed parts before and after overbending are shown in **Figure 13**; this indicates a significant part quality improvement after overbending for all 3 materials.

IOP Conf. Series: Materials Science and Engineering 1284 (2023) 012010 doi:10.1088/1757-899X/1284/1/012010

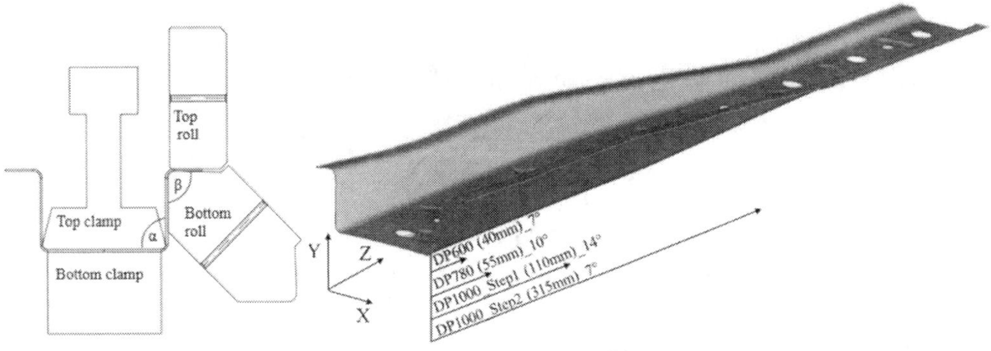

(a) (b)

Figure 12. (a) Tool configuration of the last pass (b) overbending procedure for different materials

Table 3. Springback angle of different cross sections after overbending

Cross section	X1		X2		X3	
Material	$\Delta\alpha$	$\Delta\beta$	$\Delta\alpha$	$\Delta\beta$	$\Delta\alpha$	$\Delta\beta$
DP600	2.20	-0.55	-0.92	-1.26	-0.13	-3.07
DP780	-0.56	1.30	2.41	-2.42	2.35	-2.38
DP1000_Step1	-0.86	6.29	4.57	3.45	7.53	0.915
DP1000_Step2	0.51	1.07	1.16	0.95	-0.49	-4.16

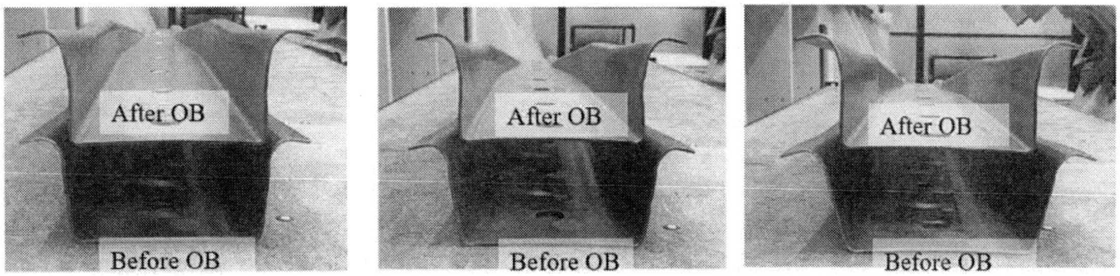

(a) (b) (c)

Figure 13. Formed parts before and after overbending for (a) DP600 (b) DP780 (c) DP1000

6. Conclusion

- A complex automotive part with combined variation in width and depth was successfully formed from 3 high strength steels.
- Wrinkles can be formed out by a subsequent top hat forming operation; however, the top hat forming step needs to start before the wrinkles in the flange get too severe.
- In FRF of combined variable width and depth components, localised flange wrinkles can be locked in if the top hat is fully formed on a partially formed flange.
- A new stepwise overbending approach was introduced to compensate for springback and end flare in FRF.

Acknowledgment
The authors would like to thank ARC Training Centre in Lightweight Automotive Structure (project number IC160100032) and Ford Motor Company for the financial support. The authors further acknowledge the technical and software support provided by data M Sheet Metal Solutions GmbH.

References

[1] Abeyrathna B, Ghanei S, Rolfe B, Taube R and Weiss M 2022 Optimising part quality in the flexible roll forming of an automotive component *The International Journal of Advanced Manufacturing Technology* **118** 3361-73
[2] Berner S, Storbeck M and Groche P 2011 A Study on flexible roll formed products accuracy by means of FEA and experimental tests. In: *The 14th International ESAFORM Conference on Material Forming,* p 345
[3] Abee A, Berner S and Sedlmaier A 2008 Accuracy improvement of roll formed profiles with variable cross sections. In: *Proceeding of 9th International Conference of Plasticity (ICTP),*
[4] Park J C, Yang D Y, Cha M H, Kim D G and Nam J B 2014 Investigation of a new incremental counter forming in flexible roll forming to manufacture accurate profiles with variable cross-sections *International Journal of Machine Tools and Manufacture* **86** 68-80
[5] J. Larrañaga S B, L. Galdos, P. Groche 2011 Geometrical accuracy improvement in flexible roll forming lines. In: *International Conference on Advances in Materials and Processing Technologies.,*
[6] Groche P, Zettler A, Berner S and Schneider G 2011 Development and verification of a one-step-model for the design of flexible roll formed parts *International Journal of Material Forming* **4** 371-7
[7] Kasaei M M, Naeini H M, Abbaszadeh B, Mohammadi M, Ghodsi M, Kiuchi M, Zolghadr R, Liaghat G, Tafti R A and Tehrani M S 2014 Flange Wrinkling in Flexible Roll Forming Process *Procedia Engineering* **81** 245-50
[8] Kasaei M, Naeini H M, Liaghat G, Silva C, Silva M and Martins P 2015 Revisiting the wrinkling limits in flexible roll forming *The Journal of Strain Analysis for Engineering Design* **50** 528-41
[9] Abeyrathna B, Rolfe B, Pan L, Ge R and Weiss M 2016 Flexible roll forming of an automotive component with variable depth *Advances in Materials and Processing Technologies* **2** 527-38
[10] Abeyrathna B, Ghanei S, Rolfe B, Taube R and Weiss M 2020 Springback and end flare compensation in flexible roll forming. In: *IDDRG 2020,* (IOP Publishing) p 012048
[11] ASTM 2011 Standard test methods for tension testing of metallic materials. In: https://www.astm.org, (ASTM International)
[12] Abeyrathna B, Rolfe B, Harrasser J, Sedlmaier A, Ge R, Pan L and Weiss M 2017 Prototyping of automotive components with variable width and depth. In: *IDDRG 2017,* (IOP Publishing) p 012092
[13] 2020 HandyScan3D.
[14] Solutions d S M 2014 COPRA® RF - Software for Roll Forming.
[15] 2020 Marc® User Manual. (MSC Software Corporation)

IOP Conf. Series: Materials Science and Engineering 1284 (2023) 012011 doi:10.1088/1757-899X/1284/1/012011

Development of rapid die cooling technology using the Groove Pressing method in hot stamping process

J M Park[1], K J Park[1], J Y Kong[1], S C Yoon[1], J H Rhee[1], J S Hyun[1]

[1] Application Engineering Group, Hyundai-Steel R&D Center, Dangjin, South Korea

E-mail: zigsism@hyundai-steel.com

Abstract. The demand for hot stamping parts in the automobile industry is increasing due to the growth of eco-friendly vehicles with improved crash safety performance and lighter weight. However, the use of ultra-high-strength steel in cold stamping is challenging due to its high yield strength and low elongation. Hot stamping is a widely used process to overcome this challenge, but the long die quenching time reduces its productivity. In order to address this issue, a Groove Pressing method was applied to the sidewall of the die to improve cooling performance without requiring additional devices within the die. The study investigated the required contact pressure between the sheet and the die for the implementation of the groove shape on the sidewall. Computational analysis was performed to design optimal groove shape for the sidewall, and the results showed that Case 3 was the most suitable in terms of cooling performance and contact pressure. The groove shape was incorporated into the die making, and experiments were conducted on aluminized coated 22MnB5 1.2 mm sheet. The results showed that the implementation of the groove shape on the sidewall improved cooling performance, dimensional quality, and mechanical properties without affecting the corrosion resistance of the component. The study demonstrates the potential of the Groove Pressing method as a practical solution for improving the productivity of the hot stamping process in the automobile industry.

1. Introduction

In recent years, there has been increasing demand for hot stamping parts in the automobile industry, driven by the need for eco-friendly vehicles with improved lightening and crash safety performance. Ultra-high strength steel is an excellent material for achieving these goals, but it is difficult to cold stamp due to its high yield strength and low elongation. In order to address this issue, the hot stamping process has become widely applied in the car body manufacturing process. This process involves simultaneously stamping and quenching a high-temperature heated steel sheet to produce ultra-high-strength components of 1.5GPa or more. However, it currently takes more than 10 to 15 seconds to achieve the required dimensional quality and mechanical properties during die quenching. As a result, the hot stamping process has a disadvantage in terms of productivity, with press SPM (Stroke per Minute) of only 3 to 4, which is less than that of cold stamping. To improve the efficiency of the hot stamping process, the industry needs to research technologies that can shorten the die quenching time [1~5].

The die quenching time is dependent on the shape of the component being produced. Generally, horizontal parts in the direction of press operation experience lower contact pressure between the die and sheet compared to vertical parts, leading to slower cooling rates. Specifically, the sidewall and flat parts of the component correspond to the horizontal parts, with the thickness of the sidewall decreasing significantly during draw forming, resulting in a significant reduction in cooling rate. In cases where die quenching time is inadequate, dimensional quality and mechanical properties may be compromised, leading to degraded component performance. Therefore, it is crucial to determine the appropriate die quenching time based on the cooling performance of the sidewall [6~7]

Conventionally, direct water cooling has been studied to improve the cooling rate of the sidewall. Maeno et al. improved the cooling rate of the component sidewall by filling the lower die with water, which allowed water to flow to the sidewall during forming [8]. Narayanasamy et al. studied a cooling system that sprays water onto the surface of the component through flow paths connected to the cooling channel [9]. Eiichi et al. prevented local wall-thinning by spraying water or air on the sidewall of the sheet before forming and improved the cooling rate of the component [10]. However, direct water cooling and air spray require additional devices to be equipped in the die and have difficulty in being applied to mass production due to poor workability.

In this study, the Groove Pressing method was applied to the sidewall of the die to improve cooling performance without requiring additional devices within the die. The method involved engraving groove shape on the sidewall using a pressing technique, which increased the contact pressure between the sheet and the die during quenching, resulting in an improved cooling rate, as shown in Figure 1.

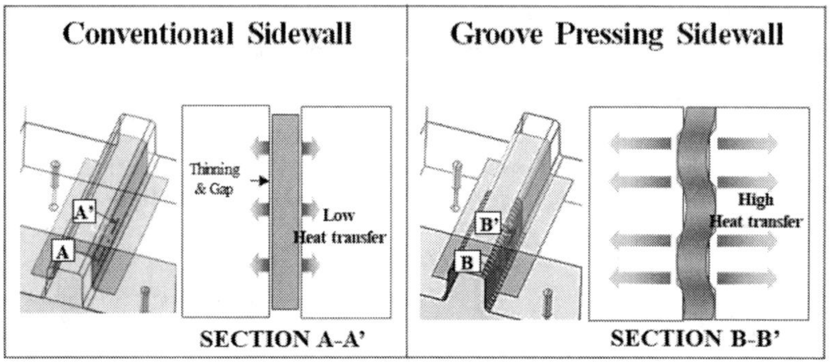

Figure 1. Comparison of conventional and the Groove Pressing methods

2. Experimental Procedure

The required contact pressure between the sheet and the die for the implementation of the groove shape on the sidewall was investigated. The die cooling behavior under various contact pressure conditions was analyzed through a numerical heat transfer model, and the results are represented in Figure 2. The improvement in cooling rate was observed to be proportional to the contact pressure between the die and the sheet. However, if the contact pressure is too high on the sidewall, it may negatively affect productivity due to scratches on the component surface and problems with jamming. Thus, it is important to design the groove shape to achieve adequate contact pressure on the sidewall.

In order to design optimal groove shape for the sidewall, contact pressure and cooling performance were compared through FEM (Finite Element Method) analysis of different shapes as represented in Table 1. In this study, computational analysis was performed on a HAT-type component using a draw method to induce thickness reduction. The analysis revealed that the contact pressure on the top, which is a flat part of the component, was approximately 15 MPa, while the contact pressure on the Conventional Sidewall (CS) was significantly lower due to thickness reduction. In order to induce a cooling rate in the sidewalls similar to that of the TOP, contact pressure of approximately 15 MPa seemed appropriate, and among the conditions tested, Case 3 of Groove Pressing Sidewall (GPS) appeared to be the most suitable in terms of cooling performance and contact pressure. In contrast, although Case 2 improved cooling performance, the contact pressure was too high. Therefore, the groove shape of Case 3 was incorporated into the die making in this study.

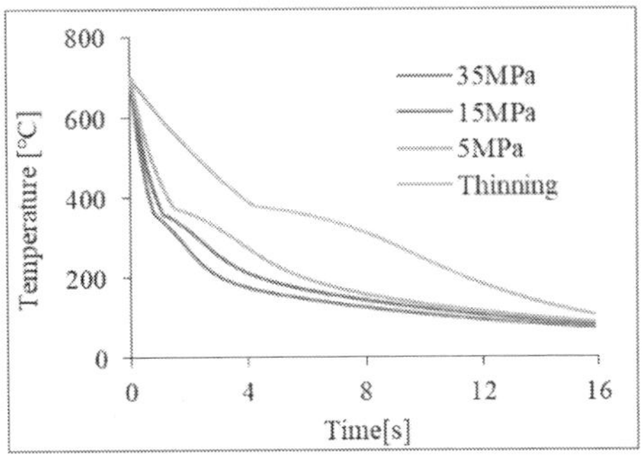

Figure 2. Effect of contact pressure on cooling behavior

Table 1. FEM analysis results according to the groove shape conditions

Shape		Depth (mm)	Width (mm)	Pitch (mm)	Temp. at 6 sec. (°C)	Pressure (MPa)	Schematic of The Groove Shape
TOP		-	-	-	71	15	
CS		-	-	-	350	-0.01	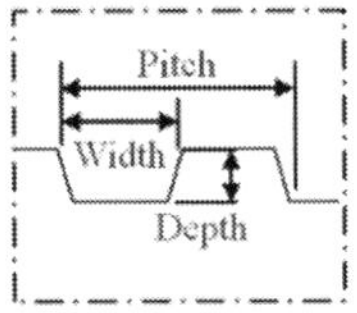
GPS	Case 1	0.1	2.5	4.0	319	1	
	Case 2	0.2	2.5	4.0	129	31	
	Case 3	0.15	2.5	5.0	143	16	
	Case 4	0.15	3.0	4.0	186	11	

The die machining method was performed using a conventional 5-axis machining center. The experimental sheet, aluminized coated 22MnB5 1.2 mm, was heated to 950 °C for 4 minutes and then stamped at 650 °C. The sidewall temperatures were measured using a thermal camera for each quenching time of 6, 8, 10 and 16 seconds to capture the temperature behavior during die cooling. The dimensional quality was measured using a Checking Fixture (C/F) to measure the flush of the component flange. In addition, tensile specimens were collected from the flat part and the sidewall of the component, and the strength was evaluated by comparison to analyze the mechanical properties. Finally, Corrosion Cycle Test (CCT) evaluation was performed to verify the corrosion resistance due to the groove shape. The test conditions followed the evaluation standards of Hyundai Motor Company and Kia.

3. Results and discussion

After the initial visual inspection of the component, no problems with jamming to the sidewall were detected. Figure 3 illustrates the appearance of the component immediately after forming, demonstrating the successful implementation of the groove shape on the sidewall.

Table 2 presents the results of comparison between the temperatures of the die sidewalls at different quenching times. The CS cooled slowly to 281 °C within 6 seconds, while the GPS cooled rapidly to 95 °C. At 8 seconds, the temperatures of CS and GPS were 247 °C and 90 °C, respectively. The slower cooling of CS was due to a decrease in thickness, which caused gaps between the die and the sheet. In contrast, GPS showed a significantly improved cooling rate compared to CS, attributed to the increased contact pressure resulting from the groove shape. At 10 seconds, CS and GPS had temperatures of 179 °C and 76 °C, respectively, with the difference in cooling rate still evident. At 16 seconds, CS and GPS had temperatures of 107 °C and 76 °C, respectively, with a significant reduction in temperature difference compared to shorter quenching times. Assuming a post-cooling component temperature of approximately 100 °C, GPS demonstrated a cooling speed that was faster than CS for at least 10 seconds.

The effect of the groove shape on flange dimensions was analyzed, and the results are represented in Table 3. Flange dimensions were measured for both CS and GPS at each quenching time using a dial gauge with a C/F of HAT-type component as the reference to measure the flange flush surface. At the 6 seconds quenching time, a dimension change of -0.29 mm was measured for CS, while GPS showed a smaller change of -0.06 mm. GPS demonstrated significantly less dimensional deformation compared to CS. At 8 seconds, changes of -0.12 mm and -0.01 mm were observed for CS and GPS, respectively, with GPS still demonstrating superior dimensional quality compared to CS. Moreover, at 10 seconds, while a change of -0.08 mm was observed for CS, no change was observed in the pattern application area for GPS, resulting in superior dimensional quality compared to CS. CS and GPS both demonstrated satisfactory changes in dimensions of -0.05 mm and -0.02 mm, respectively, at the 16 seconds quenching time. However, a spring-back tendency was observed in CS due to thermal deformation caused by slower cooling compared to the flange or upper flat part when the quenching time was insufficient. Meanwhile, GPS demonstrated improved dimensional stability as the cooling rate on the sidewall was similar to that of the flat part.

In order to investigate the effect of the groove shape on mechanical properties, the tensile strength of each part was evaluated at different quenching times. Table 4 shows the tensile strength results of the TOP, CS, and GPS parts for each die quenching time. At 6 seconds, the GPS exhibited approximately 8 % higher tensile strength (1634 MPa) compared to CS (1518 MPa). At 8 seconds, the GPS showed approximately 7 % higher tensile strength (1609 MPa) compared to CS (1504 MPa). At 10 seconds, the GPS showed approximately 6 % higher tensile strength (1603 MPa) compared to CS (1517 MPa). At 16 seconds, the GPS exhibited approximately 3 % higher tensile strength (1603 MPa) compared to CS (1552 MPa). Overall, at all quenching times, GPS showed higher tensile and yield strength than CS, and improved to the tensile strength level of the flat part of the component, TOP. This was achieved by increasing the cooling rate of the groove part to the level of the TOP, resulting in uniform mechanical properties throughout the component. However, due to the rapid cooling rate of GPS, its elongation slightly decreased.

The CCT was conducted to validate the corrosion resistance of the parts where the groove shape applied and to compare CS and GPS for corrosion resistance based on a 6 seconds quenching time. The results evaluated using 2 cycles (1 cycle = 12 hours) of the CCT are represented in Figure 4, which visually shows that the degree of discoloration for CS and GPS was similar. This suggests that the Al-Si coating layer on the GPS area is intact, and its corrosion resistance level can be considered equivalent to that of CS.

In the future, this study plans to investigate the impact of the groove shape applied components on vehicles crash safety performance. Through this research, valuable information is expected to be provided for improving both productivity and vehicles crash safety.

Figure 3. HAT-type component with the Groove Pressing method

Table 2. Temperatures of the components after forming (°C)

Parts	Quenching time			
	6 sec.	8 sec.	10 sec.	16 sec.
CS	281	247	179	107
GPS	95	90	76	76

Table 3. Flush inspection results of the components (mm)

Parts	Quenching time			
	6 sec.	8 sec.	10 sec.	16 sec.
CS	-0.29	-0.12	-0.08	-0.05
GPS	-0.06	-0.01	0	-0.02

Table 4. Mechanical properties of the components

Quenching time	Parts	Yield Strength (MPa)	Tensile Strength (MPa)	Elongation (%)
6 sec.	**TOP**	1122	1614	8.0
	CS	1003	1518	8.7
	GPS	1133	1634	7.3
8 sec.	**TOP**	1116	1608	8.3
	CS	995	1504	8.5
	GPS	1116	1609	7.7
10 sec.	**TOP**	1121	1615	8.1
	CS	1015	1517	8.4
	GPS	1127	1603	7.8
16 sec.	**TOP**	1126	1599	8.5
	CS	1064	1552	7.5
	GPS	1094	1603	7.8

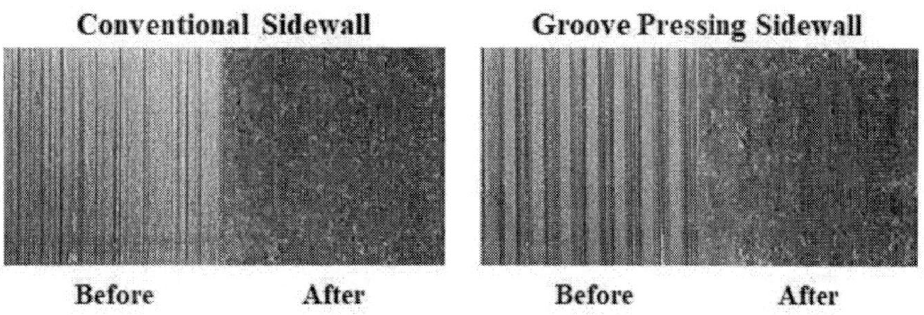

Figure 4. Results of the Corrosion Cycle Test

4. Conclusions

This study presents a novel method to improve the cooling rate of hot stamped component side walls using the Groove Pressing in areas where cooling is insufficient. The study evaluates the effects of this method on the productivity and quality of the components and draws the following conclusions:

- The Groove Pressing significantly improves the temperature of components, with a cooling rate of 95 °C in 6 seconds compared to the CS temperature of 281 °C. Despite a reduction in thickness due to draw forming, the contact pressure between the die and sheet increases, leading to a higher cooling rate. This results in a quenching time reduction of more than 10 seconds, assuming a part cooling target temperature of approximately 100 °C.

- The improved cooling rate of the sidewalls leads to excellent dimensional quality of the GPS, even with short quenching times. Additionally, the tensile strength is increased as the level of the top flat surface of the components.

- The effect of the groove shape on corrosion resistance is insignificant, similar to that of the CS. Thus, applying the Groove Pressing method improves both the productivity and quality of the parts.

References

[1] Karbasian H, Tekkaya A E 2010 J. *Mater. Process. Technol.* **210** no. 15 pp. 2103-2118
[2] Merklein M, Lechler J *J. Mater. Process. Technol.* **177**, no. 1-3 pp. 452-455
[3] Naganthan A, Penter L 2012 *"Sheet Metal Form. : Proc. & Appl.* (ASM Int.) pp. 133-142
[4] Kusumi K, Nomura N, Yamamoto S, Nakata M, Abe M, Suehiro M 2014, *Procedia Eng.* **81** pp. 1719-1724.
[5] Hu P, He B and Ying L 2016 *Appl. Therm. Eng.* **96** pp. 338-351
[6] Ying C, Shujuan L, Xiaodong L, Cunyu W, Ping H, Kunmin Z 2016 *Appl. Therm. Eng.* **99** pp. 419-428
[7] Kim N H, Kang C G 2010 *Trans. Mater. Process.* **19** pp. 378-386
[8] Maeno T, Mori K I, Fujimoto M 2015 *CIRP Annals* **64** pp. 281-284
[9] Narayanasamy P S, Singh A K, Narasimhan K 2020 *IOP Conf. Ser.: Mater. Sci. Eng.* **967** 012059
[10] Eiichi O, Yasuhiro Y, Takamichi I, Noritoshi I 2018 *R&D Rev. Toyota CRDL* **49** no.2 pp. 15-23

Industrial Internet of Things and Big Data Techniques for the Smart Press Shop 4.0 Development in Automotive Industry

Ivan Peinado-Asensi[1,2], N Montés[1] and E García[2]

[1] Mathematics, Physics and Technology Sciences Department, University CEU Cardenal Herrera, C/ San Bartolomé 55, 46115, Alfara del Patriarca, Valencia, Spain
[2] Ford Spain, Polígono Industrial Ford S/N, 46440, Almussafes, Valencia, Spain

E-mail: ivpeias@gmail.com

Abstract. In recent years, one of the main challenges in industry has been to obtain work parameters in real time for later analysis to understand the process better. The purpose is to obtain substantial information about the process for better decision-making and to find out what happens during manufacturing to know both the state of the equipment and the product, looking for savings in maintenance costs and quality controls among others. This has not been an exception for the automotive industry. These challenges have been associated with the concept of Industry 4.0, specifically with Big Data and IIoT (Industrial Internet of Things) techniques. Following this concept, many of the companies and research teams develop their algorithms and ideas based on the installation of new sensors without taking into account the implications that this proposal has in real factories. The cost of the sensor, the installation setup, maintenance, machine modifications and replication for hundreds of machines make many of the ideas fail when innovation is implemented in the companies. For that reason, in our previous work we have introduced a novel concept in Industry 4.0, the development of algorithms by using exclusively the information that is available in the equipment. Machines, as well as presses and general automated systems, include sensors for their automated normal production. These sensors are installed and the information they provide is almost free. Following this philosophy, the present paper proposes the Criterion-360 for real press shops. With this criterion, it is possible to measure a lot of variables with no cost, not only the tonnage, but also, many variables such as the press speed, Counterbalance and overload pressure, Cushion Pressure & Position, lubrication, etc., and therefore with this amount of information many possibilities arise. As the new developed system does not need new sensors and additional installation setup, it is easy and cheap to replicate it in other presses and plants.

This article presents two tools that have been developed from the information obtained with this criterion, these tools are currently in operation and have been used to modify work processes and get the press to work in optimal conditions.

1. Introduction

It is a fact that the world is increasingly being digitized for the advantages that this entails, in addition to the easy access that there is today to technologies that can make this transition possible. In the technological section and more specifically in that of manufacturing, great benefits can be achieved by applying these technologies of Industry 4.0. It is easy to see this trend with the increase in publications that have been made in recent years [1] regarding the different areas that the manufacturing industry encompasses today. Adopting IoT technologies to improve production quality and reduce downtime by digitalization in manufacturing [2] and combining IoT and Cloud, enable equipment to be adjusted, thus allowing energy consumption to be monitored in real-time and activating service-oriented energy [3] and in addition, production data can also be used to increase the automation of certain crucial activities such as resource allocation and scheduling [4]. In the automotive world this is no exception and two of the great pillars that are pursued with this transition are, on the one hand, obtaining greater knowledge and control over the process that gives us an advantage in decision-making and on the other, it seeks to achieve economic savings by minimizing possible losses that may happen as it is in the case of unexpected stops of the equipment due to breakdowns with the aim of having a plant that improves competitiveness compared to other manufacturers. In the stamping process different techniques are applied to achieve this saving as is the case in modifying the material used, such as using thinner sheets, which results in a reduction of the amount of material used. Also, regarding the use of lighter materials, in which these two factors lead us to a lower weight of the finished chassis of the vehicle. Another process is the use of optimization techniques in the design of both the sheets to be cut and the final part in order to reduce the amount of material that can be discarded in the stamping process. In this paper we are going to show two tools developed that we put into operation in the Press Shop of Ford Valencia which has allowed us to obtain new information about the process and the advantages that this entails, therefore continuing with the purpose of digitization pursued by the company but also ensuring a reduction in costs. Here is where the real challenge lies, since on many occasions there are very promising projects or improvement proposals but due to the high cost involved it is difficult to amortize the investment and therefore the realization of these projects in the industry may not be considered viable.

Since the beginning of the so-called Industry 4.0 this trend is increasingly on the rise and numerous methods of detection, prediction and correction of anomalies have been developed [5] in many processes of all kinds thanks to the connectivity and technology available today. In general, by installing sensors in certain critical parts of a machine we have been able to obtain more information about the process, having the ability to know in real time what is happening and this is very good, but most of these tools are developed in laboratories in specific equipment that works very well so the problem comes when trying to apply these solutions to the real industry and this is caused by several factors. The first issue is the cost and return time of these systems, since it is a new technology with high cost, that is why from the clients' point of view this return is not considered profitable. The second issue is the behaviour of certain equipment in real production that sometimes differs from that tested in laboratories or previous tests carried out so that sometimes you can obtain biased or even erroneous information of the process under the appearance of false negatives and the third issue is the difficulty of implementing this whole system in the large number of devices that can be found in a factory such as Ford. For example, we may have a solution that could be applied to find errors in gears by vibration but when we try to apply this to the large number of gears that we can have in the factory and the complexity of the installation needed, the cost may increase prices that the company will hardly accept. To this we must add the bad situation the automotive world is going through today where the cost savings go in parallel with the modernization of the plants in order to be more competitive against rivals and therefore the development of these systems becomes more complex.

This paper is divided into 4 sections, in the following section we will explain the work done previously and how the tool that allows us to digitize information from the stamping process has been developed. The monitored process variables and examples of these data will be shown in section 3 and finally the tools developed from the information obtained and the conclusions will be described in section 4 and 5.

2. Previous Work

Under the need to obtain process information to be able to anticipate breakdowns and unexpected stops during production, the Mini-Terms standard has been implemented in Ford Valencia in order to get information directly from the equipment to be monitored. As an initial hypothesis, it is proposed that the later a mechanical element performs the action in the process, the closer a fault is to appearing in that component. We have found behaviour patterns that validated this hypothesis in sub-elements of the work cells such as a hydraulic cylinder that actuates the movement of a welding clamp, or the movement time of a claw that is assembling a part of the chassis. We have been able to validate that a change in trend with positive slope in the operating time of the equipment indicated a wear in it and shortly after a breakdown happened. To date, the standard has been extended throughout the entire factory and there are already more than 25,000 items monitored with the Miniterms system.

The main objective of the Miniterms philosophy is to obtain the maximum possible information available in industrial equipment. This means that information is obtained from the sensors already found in the equipment, without the need to install new ones. We can take advantage of the existing communication of these sensors to extract the information from the PLC and store it in a database, and thus have access to them later. From this, we have found out new possibilities in which tools can be developed to provide additional and more detailed information about the process. Therefore, with the new paradigm of Mini-terms [6] developed by the co-authors of this paper at Ford plant where the time of action of subcomponents in the different assembly plants has been measured, we have been able to understand in more detail the operation of the majority of the equipment. We have therefore been able to predict a large number of faults and reduce repair and maintenance work since the system is in operation [7]. To be able to obtain this information from any PLC and make this system easy to implement in other plants, we have standardised the routines created for programming the PLC. Therefore, from now on the data available in the PLC can be extracted and stored in a database which can be accessed from any device with an internet connection, something that we had not been able to obtain until the implementation of this tool was carried out. This has given a great power of analysis to the staff at the time of addressing a breakdown or performing a maintenance task.

In the stamping plant the same need arises since the monitoring parameters of the stamping process do not follow the same pattern as in the processes followed in the assembly plant. This is because unexpected stops and equipment failures are caused by different reasons than those taking into account the initial hypothesis of the Mini-Terms. It should be clarified that there is equipment in the stamping process which does not directly affect the result of the process and therefore we have been able to know its status through the Mini-terms standard, this equipment is currently implemented and monitored on the platform. Examples of this type of equipment are: the actuation time of the brake and the clutch of the presses, the die fasteners and, if any, the automatic tables for changing the dies. But the scope sought in this case is to understand what happens during the stamping process both in the press and in the die at the time of performing the stamping process. Thus, we will be able to anticipate major breakdowns and also predict whether we will manufacture the parts that may be below quality standards and may be classified as scrap during the production process.

For this purpose, the so-called Criterion 360 has been developed based on the Mini-Terms standard. The initial approach is to obtain a value from the sensors installed in a press and available in the PLC for each position of the press cycle. It is known that, in a press cycle, the position of the movable part referred to as slide, where we have the upper die tied, is defined between a range from 0 to 359 depending on the rotational position of the main shaft of the eccentric gear motion transmission system. By implementing a routine designed from Mini-terms in the PLCs of the presses in order to obtain 360 data per cycle, we have been able to monitor a large amount of process information and subsequently store it on the servers. In the next point we will describe the digitized variables and show several examples of the visualization of the data that are stored on the server.

3. Data available through Criterion 360

More than 40 different types of variables are involved in the stamping process. The ideal would be to be able to digitize them all in real time but with the technology we have at our disposal today this becomes a very complex challenge. In the future the vast majority of these variables, if not all, may be able to be measured in real time. So far, the working parameters shown in the following table are being obtained from the PLC of a factory plant press.

Table 1. Monitored variables of a stamping press.

Process Value	Sensors qty.	Units	Criteria
Tonnage Force	4	Tm	360
Cushion Cylinder Pressure	8	bar	360
Cushion Cylinder Position	2	mm	360
Counterbalance Pressue	1	bar	360
Overload Pressure	2	bar	360
Energy consumpion	1	W/h	360
Clutch activation time	1	ms	Miniterm
Brake activation time	1	ms	Miniterm
Slide Position	1	mm	Miniterm
Pres speed	1	hit/min	Miniterm
Die clamp travelling time	12	ms	Miniterm
Lubrication mapping*	1	l/s	-

In addition to this information we also have the properties of the coil provided by the supplier with their respective tolerances, such as thickness, minimum and maximum yield, the minimum % of stretch, the min R and the type of material, among others. Another important information of the process and considered as a critical variable is the mapping* of the amount of oil for the blanks, we have found out this information directly from the pc of the Coil-Fed Lubricant Brush Washer. The great advantage of the data collection system presented is that it allows us to have a large amount of free information in each cycle and thus being able to store up to 5400 data per cycle, that is, every 8-10 seconds in which the acquisition of the data is done in about 4 seconds and the sending of the packets to the servers is done during the idle operation.

With the large amount of information available to us, new opportunities have arisen for development and creation of tools that allow us to achieve the requirements pursued by the company through the digital transformation. An example is the web tool developed for the visualization and analysis of information in real time from any device of the company, where we can check the process variables mentioned in table 1. In the figures shown below two signals obtained from the press can be seen. The first one is from a tonnage sensor, where we can see the force applied by the press in a cycle, and the second is the pressure sensor of the cylinder of a press cushion, where we can see the force exerted by the drawbed to control the material flow inside the die.

In the next section we will explain two of the tools that have been developed taking advantage of the amount of data that we have at our disposal which allow us to have a greater knowledge of the process and obtain an improvement of both the equipment health and savings.

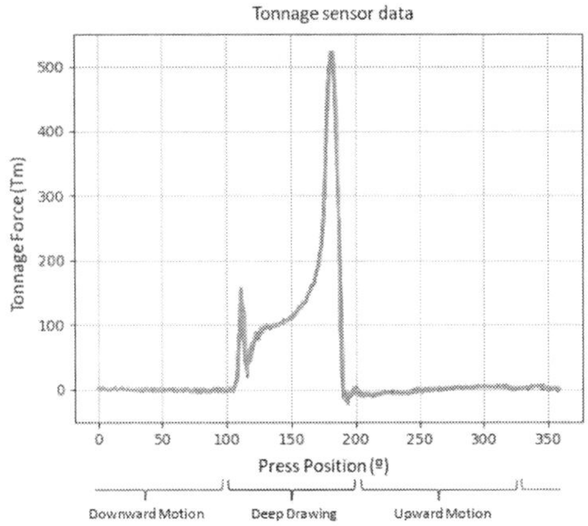

Figure 1. Tonnage data for 1 cycle.

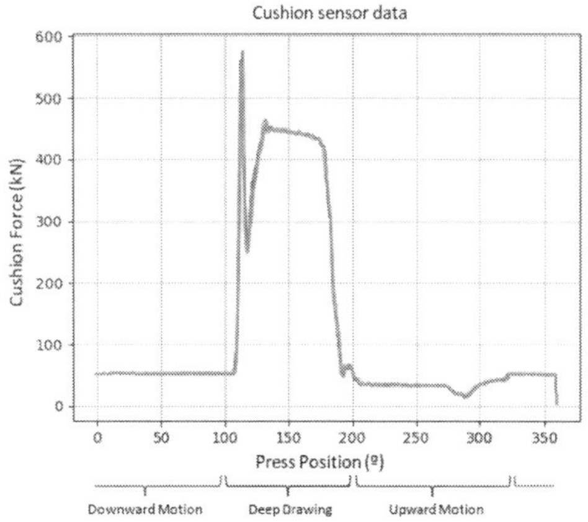

Figure 2. Cushion data for 1 cycle.

4. Developed tool and real cases

Of the tools developed, the first one to be presented allows to know the state of the stroke during the stamping process [8]. This tool is developed under the need to know the state of the different mechanical elements that make up the kinematic drive system of the press. With the passage of time the mechanical elements may deteriorate, such as for example the lower part of the connecting rod, the gears of the eccentric system or even cracks may appear in the structure of the press. If these faults are not detected, the press will work under unfavourable conditions and this may trigger a major breakdown. This type of damage has a great economic impact for the company. So, it is important to anticipate the fault or adjust the press when it is working in forced conditions in order to prevent this type of breakdown from happening. Apart from the major faults, we may find other problems such as press slide unbalance which may affect the wear of the die surface and therefore the quality of the manufactured part.

To have greater control over the process and avoid what has just been discussed in the previous paragraph the Gravity Centre Monitoring tool has been developed with which the value of the gravity centre of the slide can be monitored for each press position within a cycle. The theory states that the more centred the gravity centre of a press is during a work cycle, the better it will work. Therefore, by monitoring this parameter we are able to find out when it begins to unbalance or even detect anomalies in operation with the amount of information available.

We have been able to get this information from the data of the tonnage sensors that we have available in the press where this tool has been developed. With the monitoring of the tonnage data we can find out the gravity centre of the slide in each cycle, where we obtain 360 values of each sensor, calculating the gravity centre for each of the 360 data obtained we have,

$$GC_j = \frac{\sum T_i D_i}{\sum T_i}, \quad i = 1, \ldots, 4, \quad j = 0, \ldots, 359. \tag{1}$$

With T being the value of the tonnage of each sensor and D the distance from the point of the forces to the centre of the slide. Under the hypothesis of the more centred the stroke, the better the health of the press, some restrictions have been defined for the working limit of the press which are provided by the manufacturer, therefore being able to identify when there are work values outside what is defined as normal. Several tests have been performed to verify that the developed tool works correctly. By forcing the unevenness between the corners of the slide with values between 1 and 2 mm of difference, we were able to check that the system can detect this imbalance. The test was successful and the tool was implemented in production including an alarm system that reports when work patterns happen outside the desired conditions. With the system in production, we have detected different real cases of parts in which the press was working under overstress conditions. By making the appropriate adjustments, we have reduced the overstress and ensured the correct functioning of the equipment.

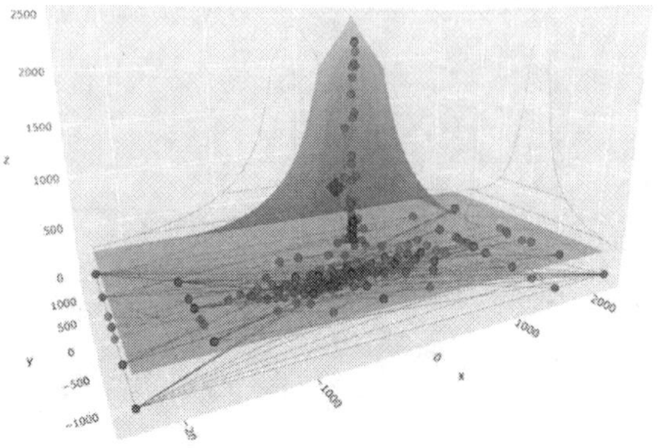

Figure 3. Gravity Center 1 Cylce.

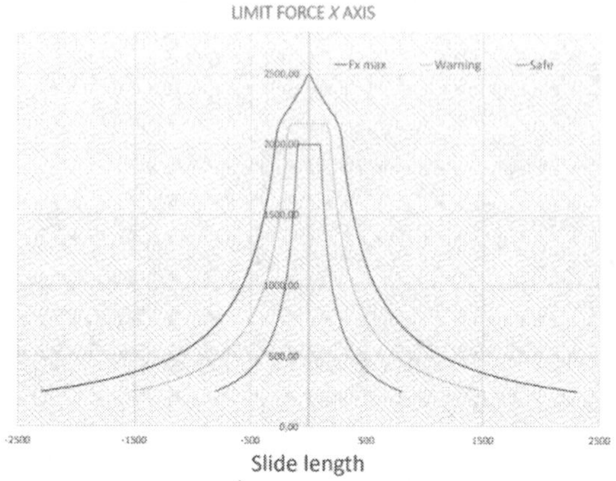

Figure 4. Gravity Center Limits Monitoring

In figure 3 we can see the calculated values of the gravity centre with the data obtained from the tonnage sensors for each of the positions of a cycle. In figure 4 we can see the safety limits defined to classify the alarms. Depending on the type of level exceeded by the value of gravity centre and the number of repetitions found, we can decide when to track and review what is happening in the equipment or intervene in it.

Continuing with the premise of developing new tools that cover the needs demanded by the process, an automatic adjustment of parameters is proposed to ensure the minimum consumption carried out by the company. This tool [9] has given us the following advantages. Energy efficiency: optimizing consumption by modifying the working parameters of the press. Improving sustainability in manufacturing: reducing consumption contributes to reducing the company's carbon footprint and therefore, achieving a reduction in polluting emissions and contributing to a sustainable future. And also, increasing the competitiveness of the company: with the savings obtained they can become more competitive compared to other manufacturers in the market. Background information indicates that a reduction in energy can be obtained by the press through the modification of working parameters such as pressure compensation and slide adjustment. An in-depth study of the behaviour of the consumption of the electric motor is carried out with respect to the change of the values of these parameters without affecting the quality of the material. A common denominator is observed, that is, for a certain weight of the upper die there is a non-linear relationship for the value of the compensation to be configured provided by the manufacturer that does not adjust to the day-to-day in plant since, as production progresses, each technician responsible for the line performs his/her own adjustment procedure for a problem that has been developing over time and thus taking into account his/her experience acquired. This affects both the compensation and many other parameters that are modified through tests in order to carry out a production according to demand. Another factor difficult to control that was discovered at the time of the study is that there are presses with compensation pressure values above the value advised by the manufacturer, which is different to the value that ensures a minimum consumption of the press. This is mainly due to the existence of small leaks in the pneumatic circuit, where a higher pressure is required to ensure the optimal work in raising the slide within the press cycle, resulting in minimum consumption. After carrying out this, a pattern is obtained that is repeated in all the dies regardless of their weight. And it has a similar consumption curve with a minimum point for a specific compensation pressure value. This can be extrapolated to the different items from the study carried out with the equation obtained that describes the pattern we have just defined. Therefore, a polynomial approximation of the behaviour of the measured electrical consumption of the engine has been made,

$$p(x) = a_0 + a_1 x + a_2 x^2 + \ldots + a_n x^n. \quad (2)$$

Where the function obtained $p(x)$ is an approximation to the real function $f(x)$ of the consumption made by the press, thus $p(x) \simeq f(x)$. And to find the minimum point within our function automatically we use the expression of the gradient descent,

$$x_{n+1} = x_n - \alpha \nabla f(x_n). \quad (3)$$

In this way we are able to find the proper working value of the compensation pressure to ensure a minimum consumption by the main motor of a press during production, ensuring its optimal operation.

These are two tools developed by the potential of obtaining the amount of information from the Criterion 360 presented in this article. They are currently operating in a press at Ford factory and are giving good results since working parameters have been modified under certain unfavourable conditions found thanks to these systems.

5. Conclusions

In this article we have seen how we are able to obtain a large amount of information from the process and store a large amount of data for analysis without the need to make a large economic investment and therefore, improving the knowledge of the process and decision making. Following the philosophy of Miniterms developed in the factory years ago makes it possible to obtain great results by using Industry 4.0 techniques, meeting the needs of a large company such as the Ford factory in Valencia, improving its efficiency, reducing costs and being able to manufacture better products. It can also help the company stay competitive in a rapidly changing market.

Acknowledgments

This study was supported by the Universidad CEU Cardenal Herrera, Ford Spain S.L. and Fundación para el Desarrollo y la Innovación (FDI), Spain, which the authors gratefully acknowledge.

References

[1] Zheng T, Ardolino M, Bacchetti A and Perona M 2021 *International Journal of Production Research* **59** 1922–1954
[2] Saqlain M, Piao M, Shim Y and Lee J Y 2019 *Journal of Sensor and Actuator Networks* **8** 25
[3] Ocampo-Martinez C *et al.* 2019 *Journal of Manufacturing Systems* **52** 131–145
[4] Mourtzis D and Vlachou E 2018 *Journal of manufacturing systems* **47** 179–198
[5] Canizo M, Conde A, Charramendieta S, Minon R, Cid-Fuentes R G and Onieva E 2019 *IEEE Access* **7** 52455–52466
[6] Garcia E and Montes N 2019 *IFAC-PapersOnLine* **52** 165–170
[7] Garcia E, Montés N, Llopis J and Lacasa A 2022 *Sensors* **22** 6222
[8] Peinado-Asensi I, Montes N and García E 2022 *Key Engineering Materials* **926** 853–861
[9] Peinado-Asensi I, Montes N and García E 2022 In situ calibration algorithm to optimize energy consumption in an automotive stamping factory process *Proceedings of the 19th International Conference on Informatics in Control, Automation and Robotics - Volume 1: ICINCO,* INSTICC (SciTePress) pp 169–176 ISBN 978-989-758-585-2

On-line quality control and tool wear evaluation in trimming process by data analytics techniques

E. Garcia-Llamas[1*], J. M. Gonzalez Castro[2], G. Ramirez[1], J. Pujante[1]

[1] Eurecat, Centre Tecnològic de Catalunya, Unit of Metallic and Ceramic Materials, Plaça de la Ciència 2, 08243 Manresa, Spain
[2] Eurecat, Centre Tecnològic de Catalunya, Unit of Applied Artificial Intelligence, Av. Universitat Autònoma, 23, 08290 Cerdanyola del Vallès, Spain

E-mail: *eduard.garcia@eurecat.org

Abstract. Sheet metal is an extremely common form of use in broad engineering sectors and represents the main consumption of steel and aluminum in Europe. All production processes where sheet metal is used require cutting operations. Therefore, it is an extremely relevant procedure, common and of great economic importance in material-intensive sectors such as automotive, construction and engineering. In a competitive environment where the demand for quality increases continuously and using harder to process high-performance materials, online monitoring and quality control become an attractive prospective. In this regard, application of data analytics techniques offers a promising toolset for fast and efficient analysis. This work presents a pilot demonstration of an instrumented sheet metal cutting line that integrates online quality control and predictive maintenance concepts, namely detection of wear in tools as well as problems arising from the raw material and process. Automated sequential cutting was performed using a straight trimming die mounted on a hydraulic press to cut High-Mn TWIP steel strip. Three piezoelectric force sensors were located in the trimming tool, collecting cycle data. Tests performed included studying the effect of blunting on the M2 (1.3343) tool, as well as the influence of correct lubrication. The resulting data were analyzed through the extraction of several parameters from each cycle to correlate die wear-induced force response evolutions with the shear-induced deformation in High-Mn TWIP steel sheets. Extracted parameters were the maximum and minimum force of each cycle and the impulse (integrated force over time) was computed and segmented by plastic deformation and crack contributions. Our analysis show that advanced algorithms can be used to analyze sensor data on an automated installation, revealing OK/NOK conditions and anomalous events.

1. Introduction

Nowadays, the transportation sector is challenged to increase the presence of the advanced high-strength steels (AHSS) in their newest models, due to their great combination of mechanical properties and lightweight which is a paramount option for safety components, structural parts of the car body and chassis. For these reasons, AHSS has been extended in automotive industry to fulfill the demanding safety and fuel consumption legislations [1].

Advanced high-strength steel materials are part of the family of multi-phase steels. The typical AHSS materials are Complex-Phase (CP) steels, Dual-Phase (DP, ferritic-martensitic) steels, TRIP (Transformation Induced Plasticity) steels and TWIP (Twinning Induced Plasticity) steels.

The high-Mn twinning plasticity (TWIP) steels are exceptional candidates for the development of lightweight vehicles due to their combination of strength and ductility. Their excellent ability to form complex shapes by stamping overcome the others AHSS with equivalent strength levels [2,3].

In general terms, AHSS sheet manufacturing processes added a new challenge to control the dimensional accuracy of blanks, the wear in the dies, and subsequently the wear induced damage in the finished part [4, 5]. The cutting and trimming processes have become a key aspect due to their economic impact, as it is necessary to manufacture large quantities of sheet metal blanks with a high-quality standard like the cut surface characteristic values [6,7]. At the present, the methods for controlling aspects such as sheared edge quality and die wear are costly and time consuming, because these analyses are necessary performed post-operation and partial disassembly of the tools [8]. However, there is an unquestionable interest in performing these quality checks online, optimizing process and quality control and introducing industry 4.0 concepts into production. In this regard, some authors have been working in non-destructive techniques, such as audio signals and acoustic emissions sensors [9,10].

This work presents a pilot demonstration of an instrumented sheet metal close cutting line to trim High-Mn TWIP steel strip that integrates online quality control and predictive maintenance concepts. The objective behind is to correlate the force responses obtained from the sensorized trimming press with the blunting effect on the M2 (1.3343) tool, as well as the influence of correct lubrication, and the sheared edge quality. Moreover, the analysis of the sensor data by advanced algorithms shows the potential to reveal OK/NOK conditions and anomalous events. The identification of these phenomena in the data will allow future developments of data-driven artificial intelligence models for on-line identification of tool wear in the trimming tool.

2. Experimental methodology

2.1. Material

In this study, a commercial High-Mn TWIP steel strip was used; this material was acquired as cold rolled, uncoated 1.45 mm thick sheet, cut into into 23 mm wide strips. The yield and tensile strengths of the material are 500 MPa and 930 MPa, respectively. Chemical composition was determined by means of spectrometry; Table 1 shows the most relevant alloying elements on of the studied material.

Table 1: Chemical composition of High-Mn TWIP steel.

Element	C	Mn	Al	Si	Cr	P	S	Fe
Weight %	0.39	20.08	1.35	0.28	1.35	0.020	0,011	Bal.

The cutting tests were conducted with a blade made of uncoated M2 (HSS) 1.3343 steel with the critical dimension of 100x30x40 mm and an average hardness of 61± 1 HRC.

2.2. Tooling and cutting process

Automated sequential cutting was performed using a straight trimming die mounted on a 150 ton hydraulic press. Each stroke was performed with a 22 mm/s upper die displacement rate. The TWIP strip was automatically feed by a decoiler and a pneumatic feeder, both synchronized. The cutting tool consists of three main parts: upper die, blank-holder plate and lower die. Within upper die six guide pressure springs apply the necessary forces to hold the sheet between the die and blank-holder plate during the cutting process. The instrumented blade with the three sensors were located in the upper die. Four guide pillars are attached to the base plate (lower die) to accommodate the shearing forces (Figure 1). The relative cutting clearance was 10% of the strip thickness.

The trimming tool was instrumented with three piezoelectric sensors for force measurement in the z-direction, each one a capacity of 50 kN. These ring force transducers contain a quartz sensor element which measures the force directly transferred to the quartz element located within the sensor. The sensors were bolted to the upper die with a pre-tension of 20 N·m torque. The sensors were connected to laboratory charge amplifier LabAmp 5165A with data acquisition, charge, IEPE, voltage, 1 or 4 channels. This amplifier was used for the data acquisition and storage. Data treatment would be performed offline by advanced algorithms, after the test campaign and data collection had finished.

The first 400 cutting strokes were performed using blades with as-machined sharp cutting edges. Next 300 strokes were performed with an artificial blunt edge in the upper blade. The artificial wear was done machining the edge by a radial saw. In this way, it was possible to compare the as-machined sharp tool edge and the artificial worn edge (blunt blade) in terms of: 1) the shared strip steel edge, and 2) the force variation measured by sensors.

A second batch of tests was performed applying lubricant on the tool, to compare its effect against the tool without lubrication. This test was performed with sharp blades and blunt blades and included 200 strokes with new blades and 200 strokes with blunt blades. The effect of the lubricant was checked by using it in half of the strokes for each blade.

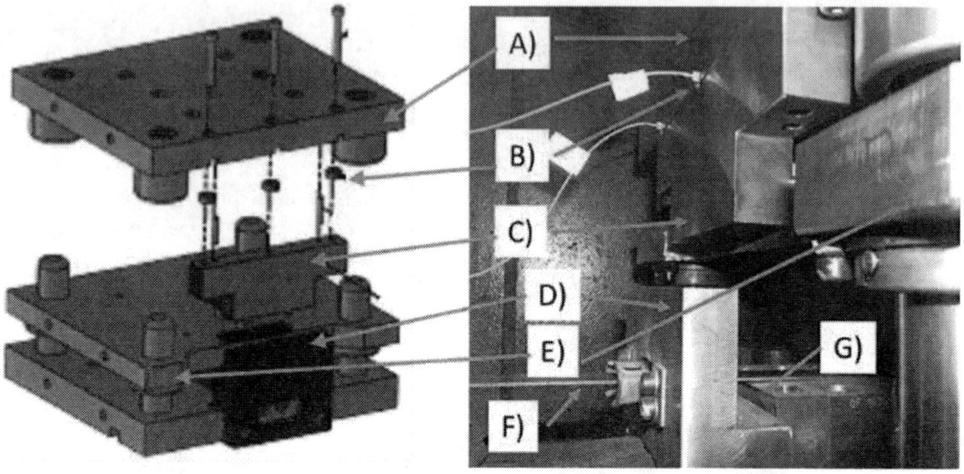

Figure 1. At the left, cutting tool exploded view and at the right, detailed cutting tool assembled in the trimming line. The labelling indicates: A) the upper die, B) the piezoelectric sensors force. C) blade, D) blade-holder to guarantee no bending, E) blank-holder F) High-Mn TWIP steel strip, and G) bed blade.

2.3. Wear damage measurement of tools

Wear damage of tools were evaluated using following criteria: (1) the surface roughness in lateral face of blade, 2) the quality of the trimmed strip, and (3) cutting edge radius of tool.

The surface roughness parameters were calculated from topographic images acquired by infinite focus microscopy (Alicona Infinite Focus SL, Graz, Austria). The three-dimensional image acquisition of surfaces, classified as optical profilometer measurement, are made without contacting the sample. Images from different depths are acquired point-by-point and reconstructed generating a complete 3D topography of the surface. Roughness was expressed in terms of Ra and Rsk (μm) and calculated according to UNE EN ISO 4228:1998 and UNE EN ISO 3274:1998 standards by means of the SensoMap topography analysis software. The Arithmetic Average Roughness (Ra) is the absolute average relative to the base length. Skewness, noted as Rsk, is a measure of the average of the first derivative of the surface, the departure of the surface from symmetry.

On the other hand, the quality of trimmed was analyzed taking into account the roll over, shear zone, fracture zone, burr and fracture angle. In this regard, cross-sectional and frontal-sectional images of trimmed strips were acquired using the infinite focus microscope. Finally, the cutting edge of blades were also measured by means of optical profilometer.

2.4. Data engineering

Data collected by the sensors described in section 2.2 were analyzed through the extraction of several parameters from each cycle, with the intent to correlate die wear-induced force response evolutions with

IOP Conf. Series: Materials Science and Engineering 1284 (2023) 012013 doi:10.1088/1757-899X/1284/1/012013

the shear-induced deformation in High-Mn TWIP steel sheets. Extracted parameters were the maximum and minimum force of each cycle, and the impulse (integrated force over time, defined as $\vec{J}$) was computed (Figure 2 as an example). To integrate the force, curve an algorithm was developed to automatically detect the times when the blade started the contact with the steel strip and when was completely fractured. These times labeled as t_0 and t_f for each stroke were selected following these rules:

- t_0: For each stroke, it was the first instant in which the derivative was larger than one-fifteenth of the maximum derivative. This value was assess experimentally comparing with different threshold the capacity to identify the initial deformation point of the measured force curves.
- t_f: Located at the first minimum after the maximum force.

Once the segmentation for each stroke was selected, the force was integrated to obtain the impulse.

$$\vec{J} = \int_{t_0}^{t_f} \vec{F} \cdot dt$$

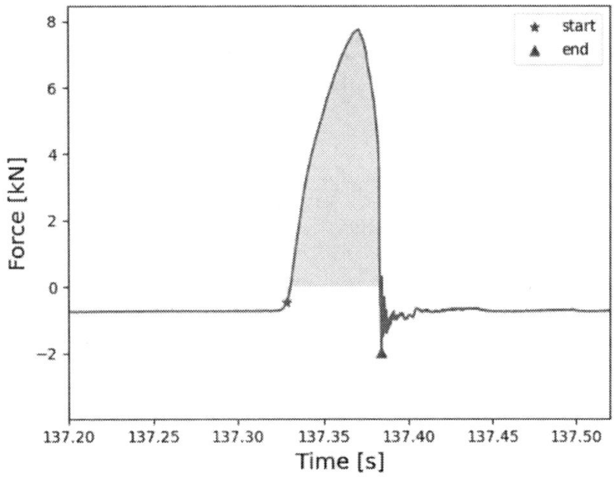

Figure 2. Example of automatic segmentation to compute the area under the curve corresponding to the impulse.

3. Results

3.1. Tribological characterization

3.1.1. Blade wear. The characterization of the initial and blunt blade is shown in Figure 3. Initial radius of the fresh edge was 54 μm. The blunt blade aims to reproduce a wear effect by machining the radius with a radial saw; the blunt edge radius measured was 628 μm. Moreover, it is shown the change of the initial and blunt roughness quantified by Ra_1 = 0.391± 0.046, Rsk_1 = -0.0697 and Ra_2 = 0,39 ± 0,059 and Rsk_2 = 3,19, respectively. The Ra parameter did not change before and after the tests which means that the average surface roughness for the length of the measurement performed remained constant. As a reliable parameter to detect change on the surface during the test was found the parameter Rsk. The initial Rsk_1 had a negative value which indicates that the surface is made up of valleys, while after the test Rsk_2 has positive value which means that the surface contains mainly peaks and asperities due to the adhered material from the TWIP strip.

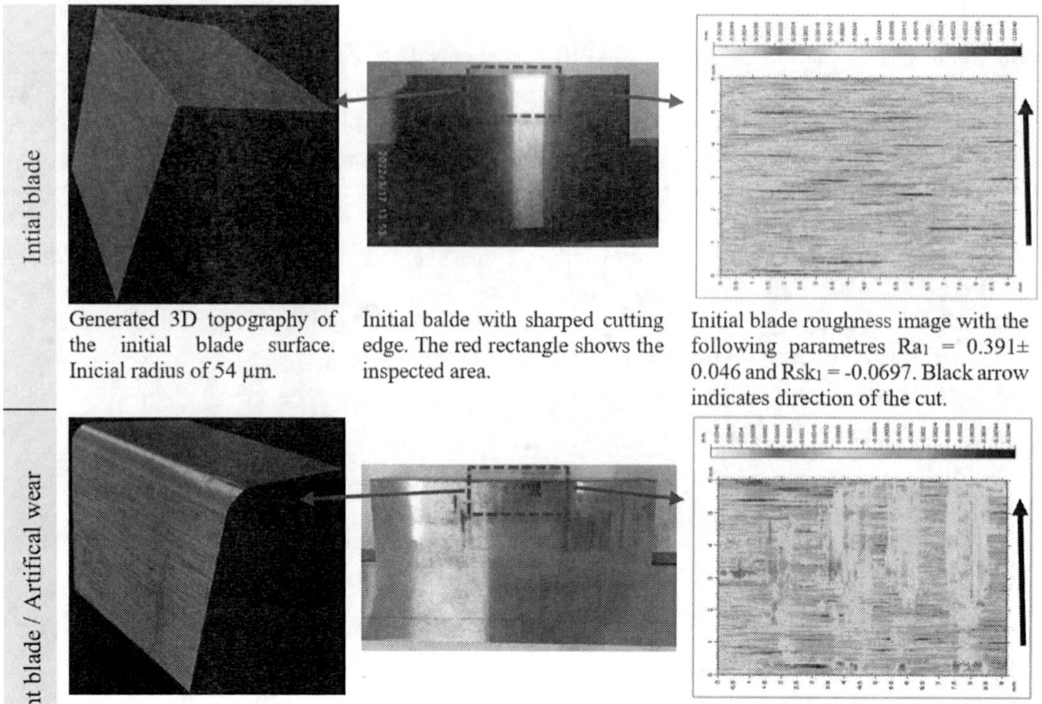

Generated 3D topography of the initial blade surface. Inicial radius of 54 μm.

Initial balde with sharped cutting edge. The red rectangle shows the inspected area.

Initial blade roughness image with the following parametres Ra_1 = 0.391± 0.046 and Rsk_1 = -0.0697. Black arrow indicates direction of the cut.

Generated 3D topography of the blunt blade surface. Blunt radius of 628 μm.

Blunt balde with the cutting edge modified with artificial wear. The red rectangle shows the inspected area.

Blunt blade roughness image with the following parametres Ra_2 = 0,39 ± 0,059 and Rsk_2 = 3,19. Black arrow indicates direction of the cut.

Figure 3. 3D images of cutting edge and roughness of the initial and blunt blade by Alicona.

A schematic side profile of a trimmed strip and the most relevant cutting surface characteristics are presented in Figure 4. The analogous sheared zones are shown in Figure 5, where the evolution at different strokes of the cross-sectional and front view of the trimmed edges of High-Mn TWIP strip steel are discerned in terms of fracture and shear zone. Measurements taken from these zones are presented in the graph of Figure 6.

The following phenomena were discerned: In the first 150 strokes, a wear-in and stabilization of the cutting system was produced (i.e., slight natural wear of cutting edge, blade and holder accommodation, first adhesion transferred between blade and strip, etc.), resulting in a decrease in the shear depth and at the same time, an increase in the fracture depth. Once the cutting system stabilized after these 150 strokes), the shear depth tended to decrease and fracture depth increased due to the natural wear of tool. Thus, natural worn of cutting edge induced an increment in force, which can be traduced in an increment of energy required to produce the final fracture of the sample.

This phenomenon was even more evident when the edge was artificially blunted (Figure 6, starting at 450 strokes), the shear and fracture depth increased compared to the new edge values.

The degradation of the sheared edge is primarily due to the deformation of the material layers adjacent to the die edge caused by sliding contact at the interface between the strip and the cutting tool.

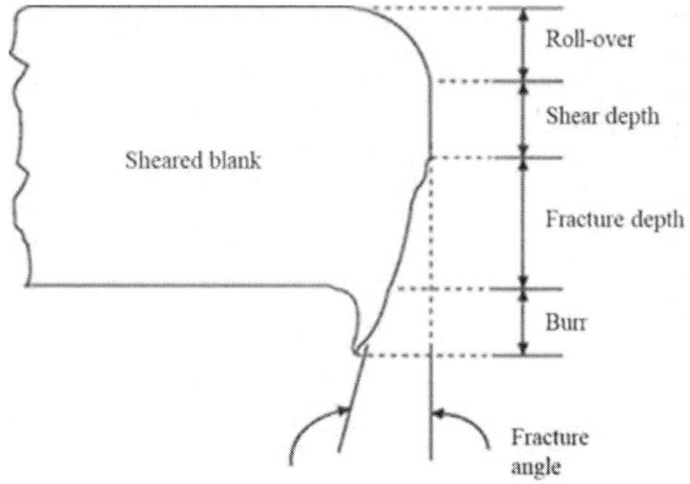

Figure 4. General schematic of the sheared edge and principal deformation zones.

	Strokes	Cross-section view	Sheared surface / Front section view
Sahrp edge – Natural wear	50	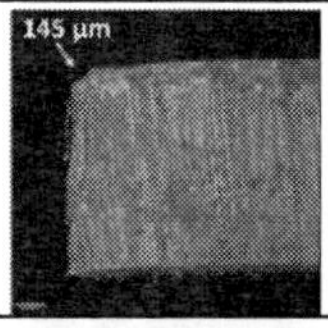	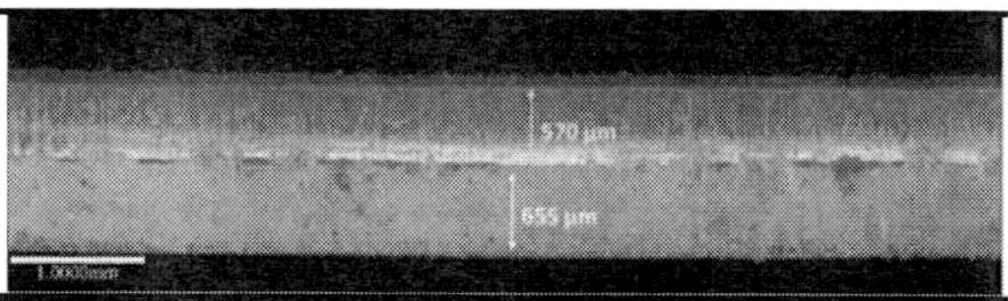
	100	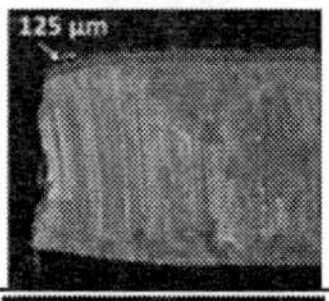	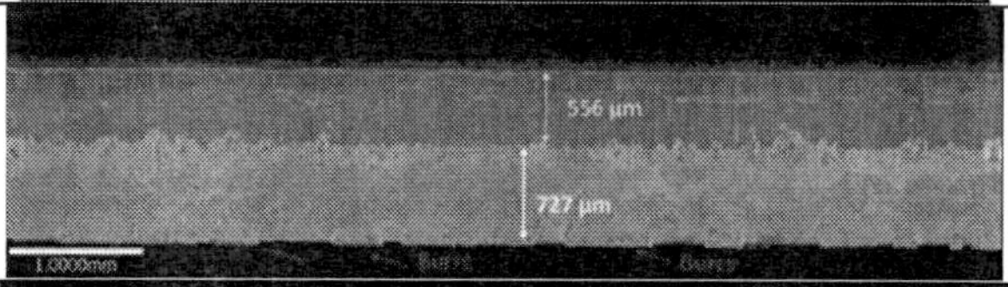
	150	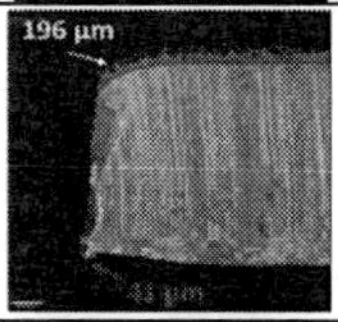	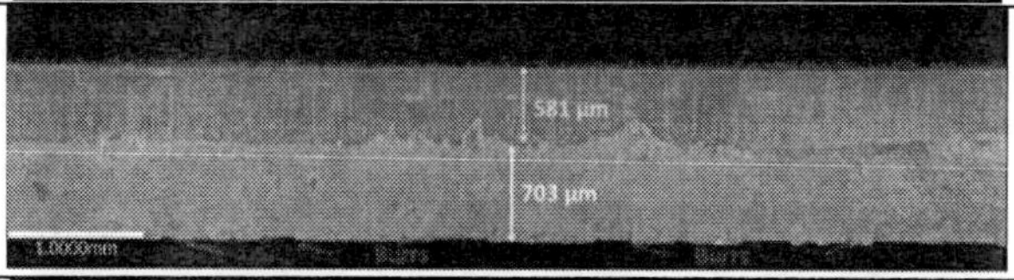
	200	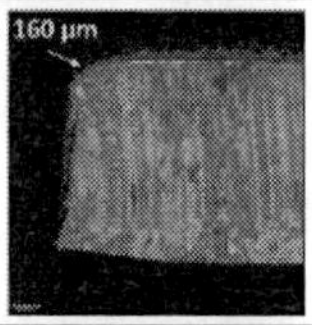	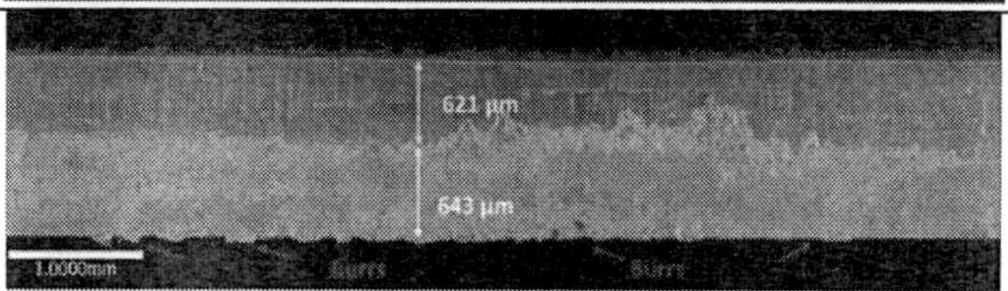
	400	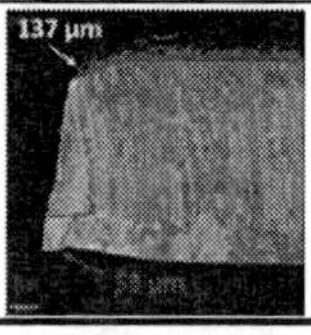	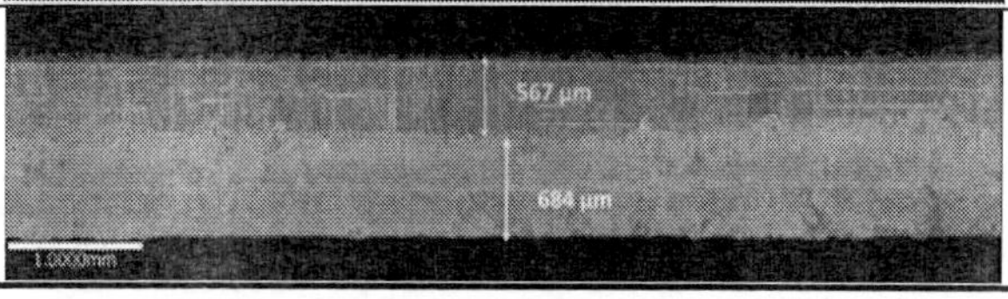

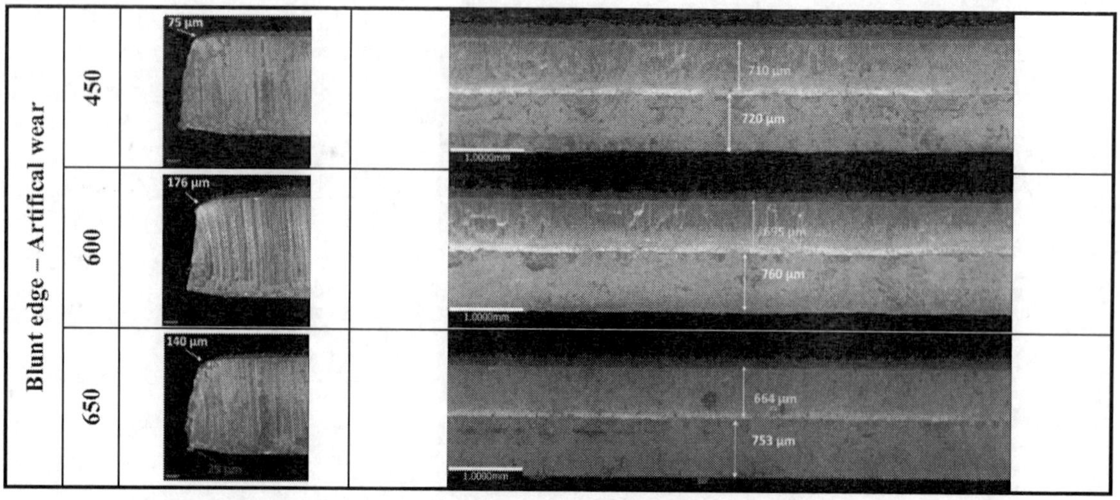

Figure 5. Alicona images of cross and front section of the shared edges, obtained using both new and blunt edge at different strokes.

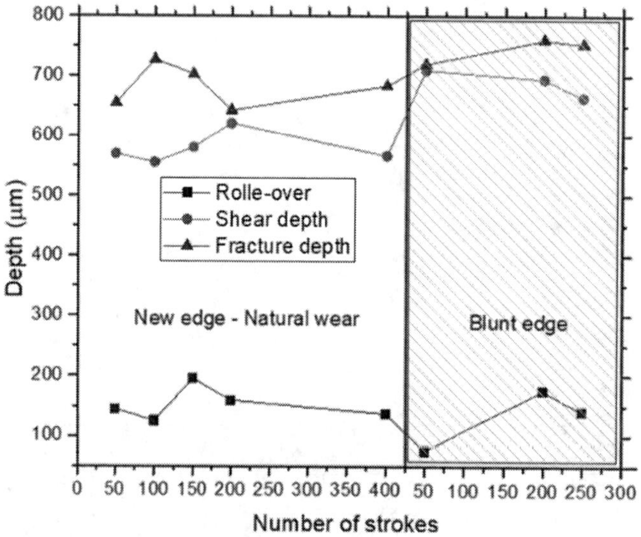

Figure 6. Evolution of the roll-over, shear dept and fracture depth versus stroke number for new edge and blunt edge.

3.1.2. Effect of the Lubricant. The second batch of tests pretended to discern the effect of the lubricant. In this work no remarkable differences could be found when comparing cuts performed with or without lubricant on either new edge or blunt edge; as can be seen in Figure 7. However, it must be noticed that although no clear lubricant effect cannot be observed by edge analysis, this parameter will become relevant in the analysis of the force recorded by a piezoelectric sensor in the following section,

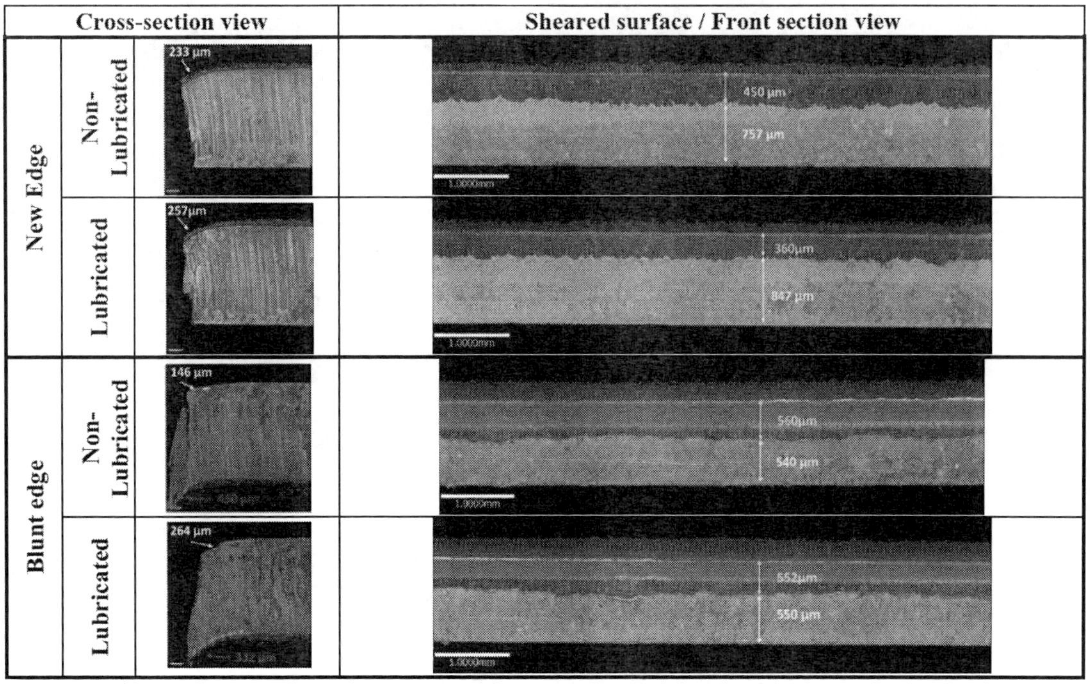

Figure 7. Alicona images show the cross and front sections of the shared edges, obtained using both new and blunt edges, with and without lubricant.

3.2. Data analysis

3.2.1. Blade wear. The force measured at each stroke was integrated following the methodology explained in Section 2.4 to obtain the impulse for each stroke. To identify changes in the blade status, for each of the two experiments, the measurements were grouped into stacks of 50 measurements: these stacks or groups are visualized in boxplots. This visual tool allows to identify the distribution of the data showing the middle 50% of the data (interquartile range) and the median of each group. The boxplot also defines outliers as values 1.5 times beyond the interquartile range.

The result of the first experiment with a fresh blade and then artificially blunted is shown in Figure 8. The most important behavior is observed in the central force sensor, which is measuring the force above the steel strip. Over the complete experiment, the cutting impulse (total force integrated over time) required to cut the steel has different regimes.

The first 400 strokes correspond to the test run initiating with a new tool and allowing natural wear to set in. For the first 150 strokes, the total impulse reduced slightly due to the stabilization of the cutting system, as mentioned in the previous section. After that, impulse starts to increase along the series. This increment is tied to the wear increment of the edge discussed in section 3.1.

After the first 400 strokes, the blade edge was blunted using a radial saw, to artificially generate a geometry corresponding to an worn tool. This wear produced an increment in the impulse for the central sensor. When the worn blade continued cutting the steel, the impulse decreased, and after 200 strokes its level was similar to the impulse before the blade was mechanized. It is likely that the initial geometry of the blunt edge was not optimized for the trimming process and for this reason the initial impulse detected was the one with the highest value. Once the blunt edge adapted to the interaction blade/sheet the force reduced until it reached similar impulse value than the one before the edge was blunted.

The importance of the central sensor is stressed in Figure 9 which contains the mean values and the variance of the measurements by each 50 strokes measured in the central sensor. Here it is possible to see that natural wear resulted in an approximate 5 % increment in impulse before the blade was blunted at 400 cycles. After machining, this severe change in geometry produced an increment of 54% on the impulse compared with the sharp-edged blade.

In the case of the left and right sensors, the changes are smaller in each regime. This happens because the sensors are located outside the surface of the steel strip and the blade wear is not modified due to the use. Notably, when the blade was blunted (beyond 400 strokes), the impulse was incremented for both sensors, reflecting the increment of the total work required to cut the steel strip.

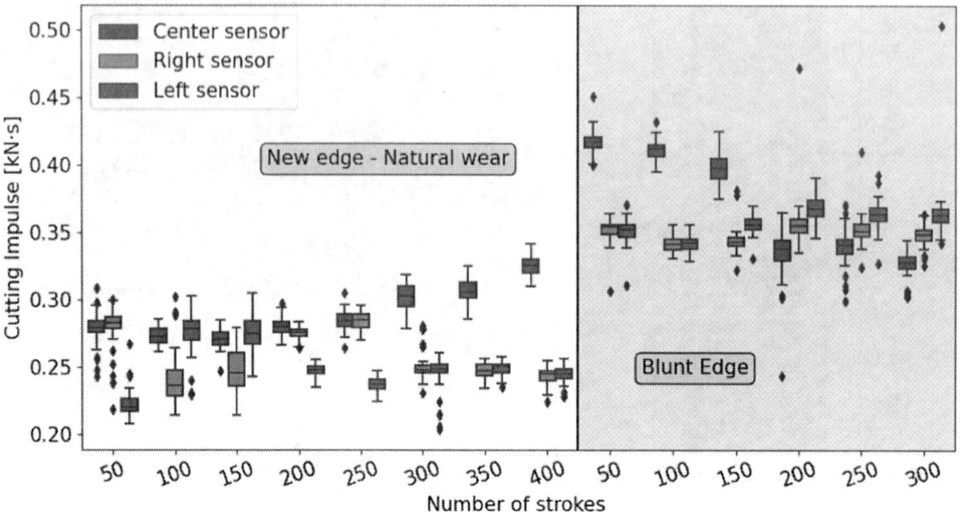

Figure 8. Impulse aggregated in groups of 50 strokes. Until 400 strokes the edge was fresh, and the degradation follows the natural use. After 400 strokes artificial wear was induced by machining the edge.

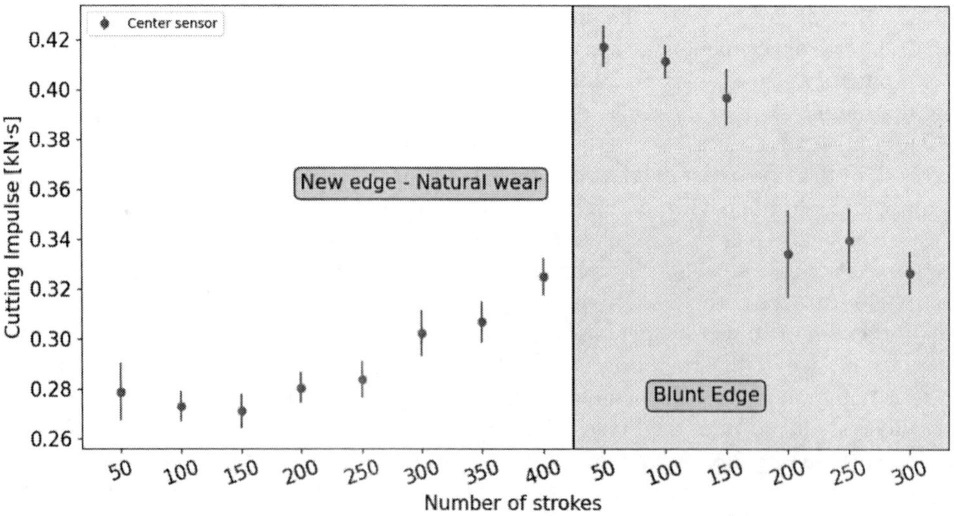

Figure 9. Mean and standard deviation of the impulse by each 50 strokes.

3.2.2. Effect of the Lubricant. The second experiment consisted in comparing the effect of using lubricant on the measured force signals. The effect was assessed in new blades and blunt blades. This analysis is available in Figure 10. In this analysis, it is possible to see the effect of the lubricant in the stroke impulse, but a clearer understanding is seen when the reduction in the mean impulse is measured for each blade (Table 2).

In this case, the lubricant reduced the total impulse by more than 8% in new and blunt blades, but the effect was higher in the blunt blade with a larger than 12% for the sensor above the cutting area.

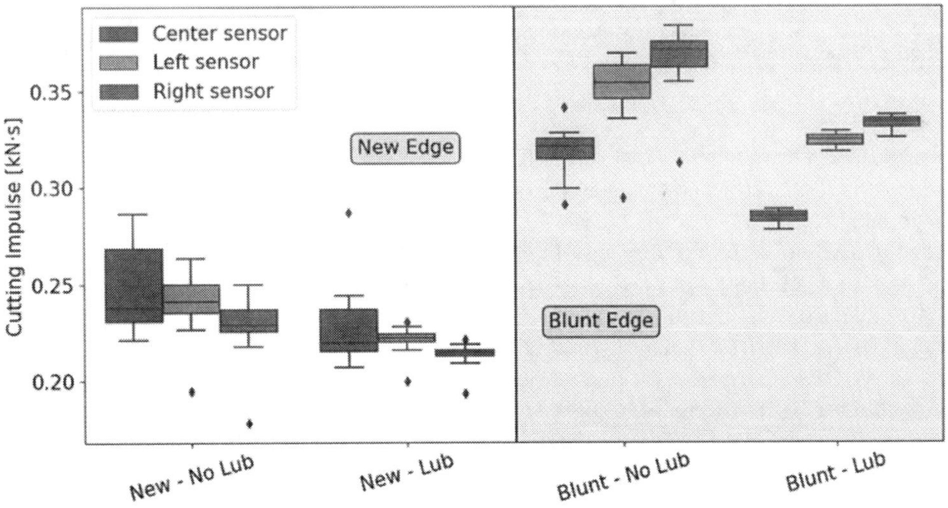

Figure 10. Boxplots of impulse aggregated by experiment considering the use of lubrication and the blade state.

Table 2: Percentual reduction of cutting impulse for each sensor due to the lubrication of the cutting tool.

Edge state	Left sensor	Central sensor	Right sensor
New	-8,3%	-8%	-8.7%
Blunt	-8.5%	-12.5%	10.8%

4. Conclusions

In this work, a pilot cutting line was used to evaluate the capability of relatively simple and robust force sensors to be used for control of quality and stability of a cutting/trimming process. The following conclusions could be drawn from the obtained results:

- Our analysis show that advanced algorithms can be used to analyze sensor data on an automated installation, revealing OK/NOK conditions and anomalous events.
- The relatively simple setup was sensitive enough to identify force changes due to wear, as well as the effect of lubrication on the cutting process.
- Average trimming impulse (integrated force over time) increased due to die edge blunting during the natural wear by 18 %. On the other hand, 8% and 12.5% force decrease were measured with the lubricant applied on the cutting tool for new and blunt edge, respectively. The identification of these phenomena in the data will allow future developments of data-driven artificial intelligence models for on-line identification of tool wear in the trimming tool.
- In this study, force-time curves were used for the analysis, as opposed to force-displacement curves required to obtain physical measurements of work. Results using this approach were successful, demonstrating that this can be a cheap alternative since we do not need expensive linear encoders.
- The relation between the shear depth and fracture depth gives a correlation to the force measured by the sensors.

- Central sensor is more reliable due to the optimization localization just on the top of the cutting area. Therefore, it may be possible to obtain relevant data with fewer, well positioned sensors.
- Initially, artificial wear gives higher impulse value but once the system gets stabilized the sensors showed similar force than with natural wear after 400 strokes.

Acknowledgements
This work was financially supported by the Catalan Government through the funding grant ACCIÓ-Eurecat (Project PRIV-INTELCUT).

References

[1] Steel Market Development Institute 2020 *2020 Steel Industry Technology Roadmap for Automotive* URL https://www.steel.org/wp-content/uploads/2021/01/2020-AISI- Automotive-Roadmap.pdf
[2] Bouaziz O, Allain S, Scott C P, Cugy P and Barbier D 2011 *Curr. Opin. Solid State Mater. Sci.* **15** 141–168 URL https://doi.org/10.1016/j.cossms.2011.04.002
[3] De Cooman B C, Estrin Y and Kim S K 2018 *Acta Mater.* **142** 283–362 URL https://doi.org/10.1016/j.actamat.2017.06.046
[4] Ingarao G, Di Lorenzo R and Micari F 2009 Mater. Des. **30** 4421-4433 URL https://doi.org/10.1016/j.matdes.2009.04.001
[5] Wu-Rong W, Chang-Wei H, Zhong-Hua Z and Xi-Cheng W 2011 *Mater.Des.* **32** 3320-3327 URL https://doi.org/10.1016/j.matdes.2011.02.021
[6] Mori K -I, Abe Y and Suzui Y 2010 J. Mater. Process.Technol. **210** 653–659 URL https://doi.org/10.1016/j.jmatprotec.2009.11.014
[7] Wu X, Bahmanpour H and Schmid K 2012 J. Mater. Process. Technol. **212** 1209–1224 URL https://doi.org/10.1016/j.jmatprotec.2012.01.006
[8] Cui Z, Bhattacharya S, Green D E and Alpas A T 2018 *IOP Conf. Ser.: Mater. Sci. Eng.* **418** 012066 URL https://doi.org/10.1088/1757-899X/418/1/012066
[9] Ubhayaratne I, Periera M, Xiang Y and Rolfe B 2017 *Mech Syst Sig Process* **85** 809-826 URL https://doi.org/10.1016/j.ymssp.2016.09.014
[10] Yoon J, Han H, Kang B and Kim Y 2019 *Key Eng Mater* **794** 285-294 URL https://doi.org/10.4028/www.scientific.net/KEM.794.285

IOP Conf. Series: Materials Science and Engineering 1284 (2023) 012014 doi:10.1088/1757-899X/1284/1/012014

Application of a neural network for predicting cutting surface quality of punching processes based on tooling parameters

A Schenek, M Görz, K R Riedmüller and M Liewald

Institute for Metal Forming Technology, University of Stuttgart, Holzgartenstraße 17, 70174 Stuttgart, Germany

E-mail: adrian.schenek@ifu.uni-stuttgart.de

Abstract. Punching represents one of the most frequently used manufacturing processes in the sheet metal processing industry. Thereby, one of the most important quality criteria for such punching processes constitutes the geometric shape of the cutting surface. High cutting surface qualities are usually characterized by the lowest possible edge draw in, a high clean cut, as well as a small fracture surface and a low burr height. In this respect, conventional punching processes can only produce clean cut proportions (CCP) up to 20-50% of the sheet thickness. By means of the so called "concave punch nose design" developed at the Institute for Metal Forming Technology, the geometry of conventional punches could be optimized in such a way that the clean-cut proportion along the cutting surface is significantly increased. The findings reported about in this paper show that the quality parameters of the sheared component edges thereby show highly nonlinear relationships to the tooling and sheet metal material parameters used. In order to quantify these effects, the data from several numerical punching simulations is used to pretrain an artificial neural network (ANN). Input data for the neural network included features of the punch geometry, the size of the clearance, the sheet thickness and the material data of the semi-finished product. The output of the neural network is precise predictions of the achievable cutting surface quality parameters. The experimentally validated findings presented in this paper show that the process design for novel punching processes such as cutting with concave punch nose design is possible, when using a modern machine learning approach.

1. Introduction and state of the art

Almost every sheet metal component has to be trimmed or punched along its production process [1]. To evaluate the quality of trimmed and punched components, several geometric features along the sheared sheet metal component edge are considered. These features include the edge draw-in height (h_E), the clean cut height (h_S), the fracture surface height (h_B), as well as the burr height (h_G) and are illustrated in figure 1 [2]. In order to comparatively quantify the quality of sheared sheet metal component edges, these parameters are often specified in relation to the thickness (t) of the sheet metal material used. As shown in figure 1, these relative values can be referred to as edge draw-in height proportion (EPD), clean cut proportion (CCP), fracture surface height proportion (FSP), and burr height proportion (Burr). A technical challenge in designing shear cutting processes is to select tooling parameters in such a way that the clean cut portion is maximized and at the same time the other cutting surface characteristics are minimized [3], [4]. Here, special attention must be paid to the tooling parameters clearance and edge geometry of punch and die, since these parameters significantly influence the punching process respectively the quality of cutting surfaces [5]–[7]. For the process design of conventional shear cutting processes, industrial companies often rely on experimentally determined tabular data in this respect. A more systematic, precise, and rapid approach for designing shear cutting processes involves the training of machine learning models via experimentally or numerically determined data. For this purpose, the geometric features of cutting surfaces can be combined to an output vector as shown in figure 1.

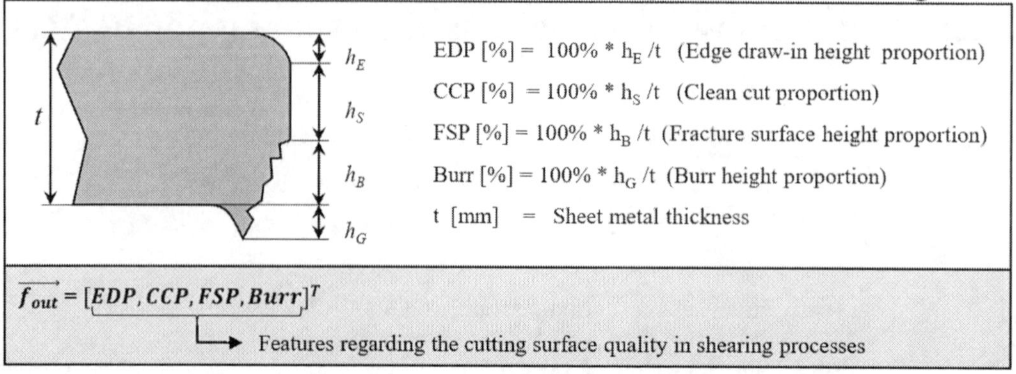

Figure 1. Cutting surface parameters of sheared sheet metal components [2] summarized as output-vector

In previous studies in this context, Djavanroodi et al. [8] and Stanke et al. [9] predicted edge draw-in on fine-blanked workpieces based on an artificial neural network model assessing the process influence of the V-ring geometry, the blank holder force, and the counter punch force. Here, numerical simulation data was used to train the neural network model, as this requires a considerable amount of process data which is difficult to obtain by experiments. Al-Momani et al. [16] also employed artificial neural networks and multiple regression analysis to develop a model for predicting burr height in a shearing process. Mao et al. [10] successfully predicted clean cut heights of fine-blanked parts also using an artificial neural network trained on the basis of numerically determined punch and counterpunch forces. Furthermore, machine learning models have been used several times to quantify tool wear in shear cutting processes [11]–[15]. In summary, these studies stated that data-driven models have been successfully used as design aids and monitoring methods for conventional and already well-known shearing and fine blanking processes. However, none of the mentioned studies addressed the prediction of the entirety of the cutting surface parameters presented in figure 1. Furthermore, the continuously increasing demands on cutting surface quality and production costs require the development of new shearing processes having additional process parameters. Until today, it has not been clarified whether machine learning models are also suitable for predicting cutting surface parameters for newly developed punching processes. One of these processes is shear cutting with the so-called "concaved punch nose design" (see figure 2), which was initially proposed by Liewald and Senn in 2018 [16]. The idea of the concave punch nose design is to optimize the geometry of conventional punches in order to enlarge clean cut proportion along the produced cutting surface due to superimposed compressive stresses within the shear affected zone.

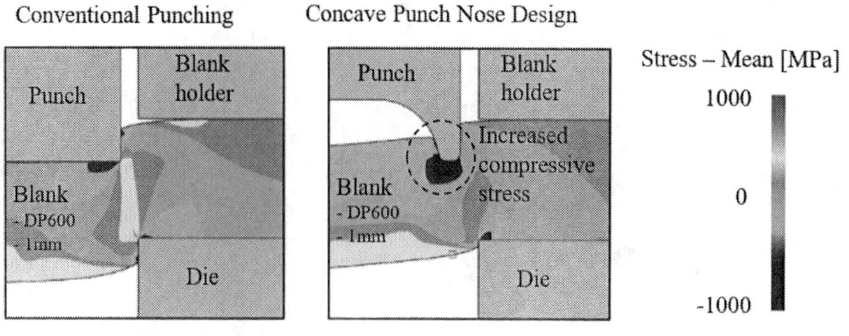

Figure 2. Conventional punching process (left) and concaved punch nose design (right) [17]

The major advantage of using a concave punch nose design as opposed to other precision blanking processes (e.g. fine-blanking) is that cutting surface quality can be improved by minor changes of the punch geometry, thus maintaining the high production rate of conventional shear cutting processes [18]. In this context, the objective of the present paper involves the development of a predictive model for the cutting surface geometries achievable by this novel punching process. Since even small changes in the demonstrated punch geometry have a significant effect on the cutting surface parameters of sheared sheet metal component edges [17], a concaved punch nose design appears to be a promising approach for answering the following research question:

„Is a numerically pre-trained neural network able to predict the cutting surface parameters (output) of sheared sheet metal edges based on given tooling and semi-finished product parameters (input)?"

The research approach applied in this context is presented below and involves the pre-training of a neural network based on numerically determined data. Subsequently, the predictive accuracy of the neural network is evaluated and compared with experimentally determined data.

2. Investigated sheet metal materials and simulation model setup

Three different sheet metal materials (DC03, DP600, and DP800) were analyzed within the study presented here. The yield curves of these materials were determined through experimental tensile tests and then extrapolated using the Hockett-Sherby approach. Table 1 provides a summary of the Hockett-Sherby fitting parameters and the determined ultimate tensile strengths of the investigated sheet metal materials

Table 1. Investigated sheet metal materials and Hockett-Sherby extrapolation parameters

	$k_f(\varphi) = k_{f,s} - (k_{f,s} - k_{f,0}) \cdot e^{-m\varphi^P}$		
	DP600	**DP800**	**DC03**
$k_{f,s}$ **[MPa]**	794.013	1015.087	457.886
$k_{f,0}$ **[MPa]**	311.541	398.240	197.799
m **[-]**	8.076	7.726	4.508
P **[-]**	0.692	0.706	0.784
R_m **[MPa]**	628.3	810.1	317.6

A numerical simulation model was used to investigate the effect of different tooling parameters on the cutting surface and in order to generate a data set for training a corresponding neural network. The numerical investigations were performed using the simulation program DEFORM 2D, which is well-suited for accurate numerical calculations of cutting and punching processes [19]. To simulate material separation, damage modeling was implemented using the "Normalized Cockroft Latham" approach, which is often used for material separation tasks [5], [19]. Table 2 summarizes the numerically investigated parameter variations and figure 3 shows the geometric setup of the simulation model.

Table 2. Parameter variation used for numerical punching simulations

Parameter		**Variation and total number of variations**
Punch Diameter	D	10mm
Web width	SB	0.1mm, 0.2mm, 0.3mm, 0.4mm, 0.5mm, 5.0mm [a]
Web angle	SW	0° [a], 45°, 55°, 65°, 75°
Punch radius	R_S	50µm
Die radius	R_M	20µm
Clearance	u	5%, 10%, 15%
Sheet thickness	t	0.5mm, 1.0mm, 1.5mm
Sheet metal	WS	DC03, DP600, DP800
[a] Conventional punching process.		

The die and the blank holder were modeled as rigid bodies to reduce calculation time. For meshing the blank, mesh windows with element sizes specifically adapted to the process were utilized, resulting in a minimum element edge length of 6 micrometers within the shear affected zone. A constant punch diameter of 10mm was used for all numerical and experimental investigations, while web width, web angle, clearance, sheet metal thickness, and sheet metal material were varied. Cutting edge radii were kept constant at 50 micrometers in all simulations and experiments, in accordance with current industrial recommendations for punching high-strength sheet metal materials [20]. The sheet metal parameters regarding the ultimate tensile strength and sheet metal thickness as well as the tooling parameters regarding the clearance, the web width and the web angle were defined as input vector for our further investigations.

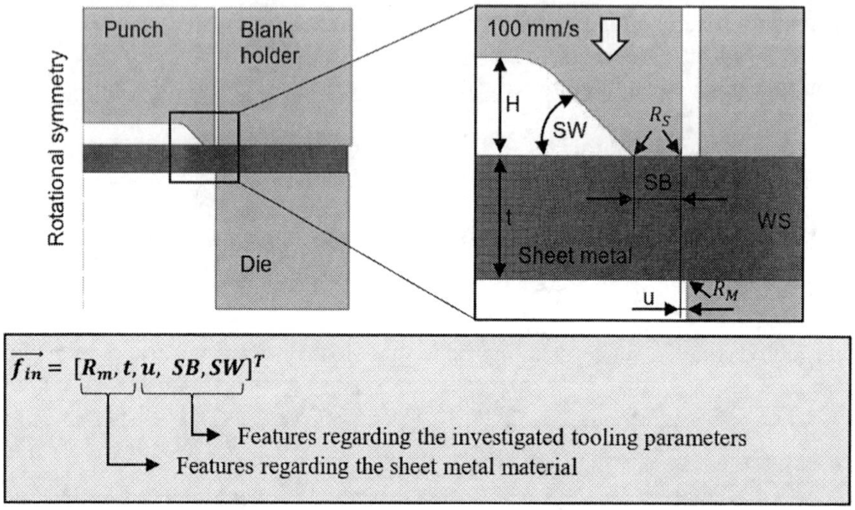

Figure 3. Investigated tooling parameters and sheet metal data summarized as input-vector

A full-factorial experimental design was employed for the numerical investigations. With five different investigated web widths (SW), four different web angles (SW), and in each case three different clearances (u), sheet thicknesses (t) and sheet materials (WS), the total number of simulations for the concave punch nose design was 540 (= 5·4·3·3·3). Additionally, 27 simulations were performed for a conventional punching process, so that the final data set consisted of 567 simulation results. In the following subchapter, relationships between the calculated cutting surface parameters and the investigated process parameters will be discussed. The relationships were obtained by using a pearson correlation analysis.

3. Data exploration of numerically determined data

Data exploration is an important step in data-driven projects and investigations. It involves examining and analyzing data to gain a better understanding of the characteristics and structure of considered datasets. Furthermore, data exploration is also important for identifying patterns and trends, detecting outliers and errors, and guiding the selection of appropriate statistical models. The Pearson correlation analysis is one of the most commonly used methods of analyzing datasets with multiple features. Here, Pearson's correlation coefficient (r) varies between -1 and +1 with 0 implying no correlation. Correlation values of -1 or +1 imply an exact linear relationship between two variables. Positive correlation values mean that as variable 1 increase, variable 2 also increases. Negative correlations imply that as variable 1 increases, variable 2 decreases. Figure 4 shows the correlation matrix for the results numerically determined in the present study. Regarding the investigated features of the sheet metal material (R_m, t) and the numerically varied tooling parameters (u, SB, SW), the following conclusions can be drawn from the correlation matrix:

- Higher tensile strength R_m of the sheet metal material leads to a decrease in edge draw-in height proportion EDP (r = -0.42), clean cut proportion CCP (r = -0,74) and burr height proportion Burr (r = -0,48). The fracture surface height proportion FSP increases as the tensile strength of the sheet metal material R_m increases (r = 0.81)
- As the sheet metal thickness t increases, the edge draw-in proportion EDP increases (r=-0.59)
- An increasing size of the clearance u leads to an increased edge draw-in height proportion EDP (r=0.44)
- Decreasing the web width (SB) leads to higher clean cut proportions CCP (r = -0.53) and smaller fracture surface height proportions FSP (r = 0.38)
- Almost no linear correlation can be found for the relations between the web angle SW and any of the investigated cutting surface parameters.

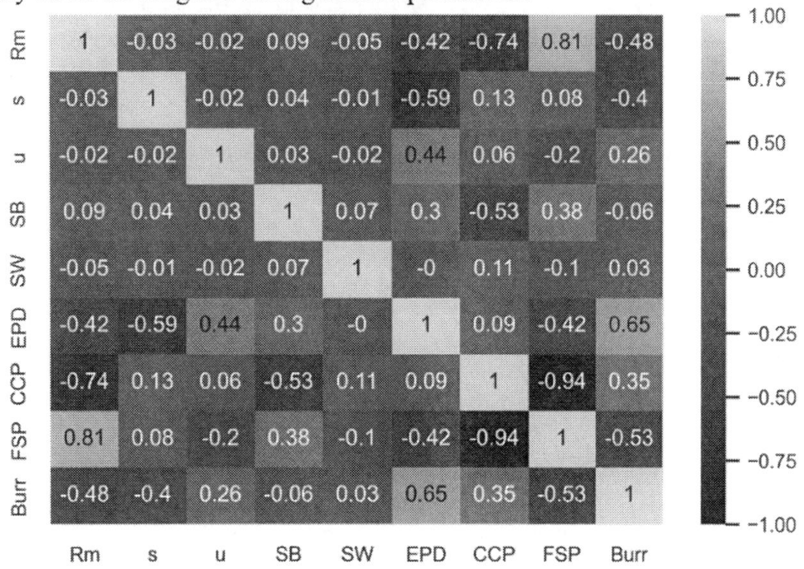

Figure 4. Pearson correlation matrix for analysing the numerically determined results [21]

Since none of these correlations indicate high linear relationships (abs(r) > 0.9), analytical modelling or a simple linear regression model seem not to be promising approaches for predicting cutting surface features based on tooling and sheet metal parameters. For example, training a linear regression model resulted in a relatively large deviation between the numerically determined and the model predicted cutting surface features, with a mean absolute error (mae) of 3.0% (see figure 5).

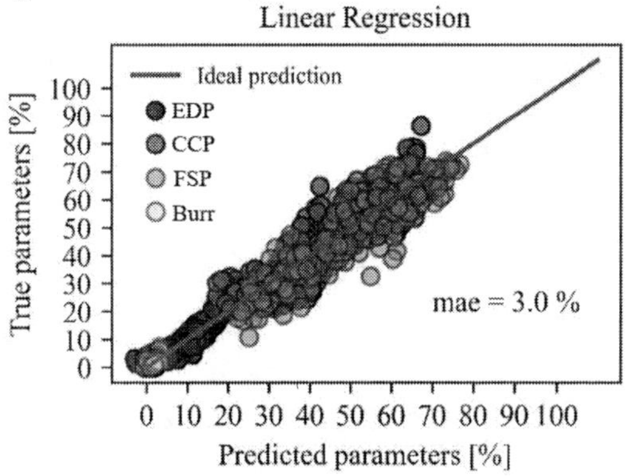

$$\overrightarrow{f_{out}} = \mathbf{A} \cdot \overrightarrow{f_{in}} + \vec{b}$$

$$\begin{pmatrix} EDP \\ CCP \\ FSP \\ Burr \end{pmatrix} = \begin{bmatrix} A_{11} & \cdots & A_{14} \\ \vdots & \ddots & \vdots \\ A_{51} & \cdots & A_{54} \end{bmatrix} \cdot \begin{pmatrix} R_m \\ t \\ u \\ SB \\ SW \end{pmatrix} + \begin{pmatrix} b_1 \\ b_2 \\ b_3 \\ b_4 \\ b_5 \end{pmatrix}$$

Figure 5. Prediction result when training a multiple linear regression model on the numerically determined data

IOP Conf. Series: Materials Science and Engineering 1284 (2023) 012014 doi:10.1088/1757-899X/1284/1/012014

In contrast, the results described in the following section will show that an artificial neural network is much better suited to predict cutting surface features from tooling and sheet metal parameters. Here, the presented linear regression model serves as a reference model to evaluate the predicting performances of the neural network.

4. Neural network setup, model training and evaluation

A fully connected feed forward neural network (FFNN) was used to examine the predictability of an ANN with regard to the cutting surface features based on sheet metal parameters and tooling features. The FFNN was implemented in Python using the machine learning libraries tensorflow and keras. Figure 6 provides a summary of the neural network setup employed, consisting of one input layer, four hidden layers and one output layer. The sheet metal parameters (R_m, s) as well as the tooling features (u, SB, SW) served as inputs for the neural network, with the output layer consisting of four neurons representing the cutting surface features EDP, CCP, FSP, and Burr. The rectified linear activation function (ReLU) was applied to each layer. The activation function, the number of hidden layers and the number of neurons for each hidden layer were determined via systematic hyperparameter tuning using keras hyperband tuner.

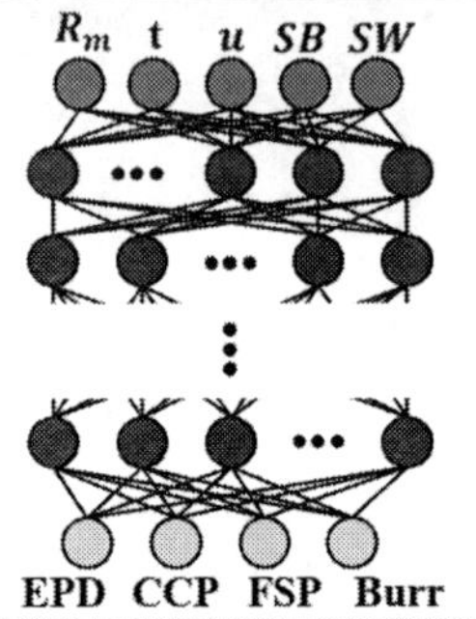

Inputs	Hidden Layers	Activation Function	Outputs
R_m t u SB SW	Layer 1: 15 Neurons Layer 2: 50 Neurons Layer 3: 45 Neurons Layer 4: 45 Neurons	reLu	EDP CCP FSP Burr

Figure 6. Neural network setup and activation function

In order to train the FFNN, 80% of the numerically determined data (0,8·567 = 453) was used as the training dataset. 10% of the data was used for validation purposes to improve the model's performance during 200 training epochs. The remaining 10% was used for testing the neural network on unseen data (test data). The mean squared error (mse) was selected as the loss function, and the Adam optimization algorithm was used to incrementally improve the randomly initialized weights of the ANN based on the errors calculated during the training cycles. The progression of the loss function and mean squared error throughout the 200 epochs as well as the final prediction results are presented in figure 7. The model did not exhibit overfitting during the entire training process, as proved by the non-divergent blue and black curves in the upper diagrams in figure 7. As shown by the lower diagrams, the neural network reveals good adaptation to the training dataset, achieving a mean absolute error (mae) of 1.2%. This in general is considered as a satisfactory result for a machine learning task. Moreover, the mae of 1.5% for the test dataset, which was not used in the training process, indicates good generalization ability of the numerically pre-trained neural network. Furthermore, the trained neural network shows much better fitting capabilities (mae = 1.2%) to the numerically determined data than the initially proposed linear regression model (mae =3.0%, see figure 5). To evaluate the "real" predictive quality of the numerically pre-trained neural network on experimental data, the network's response was finally tested on experimentally determined data.

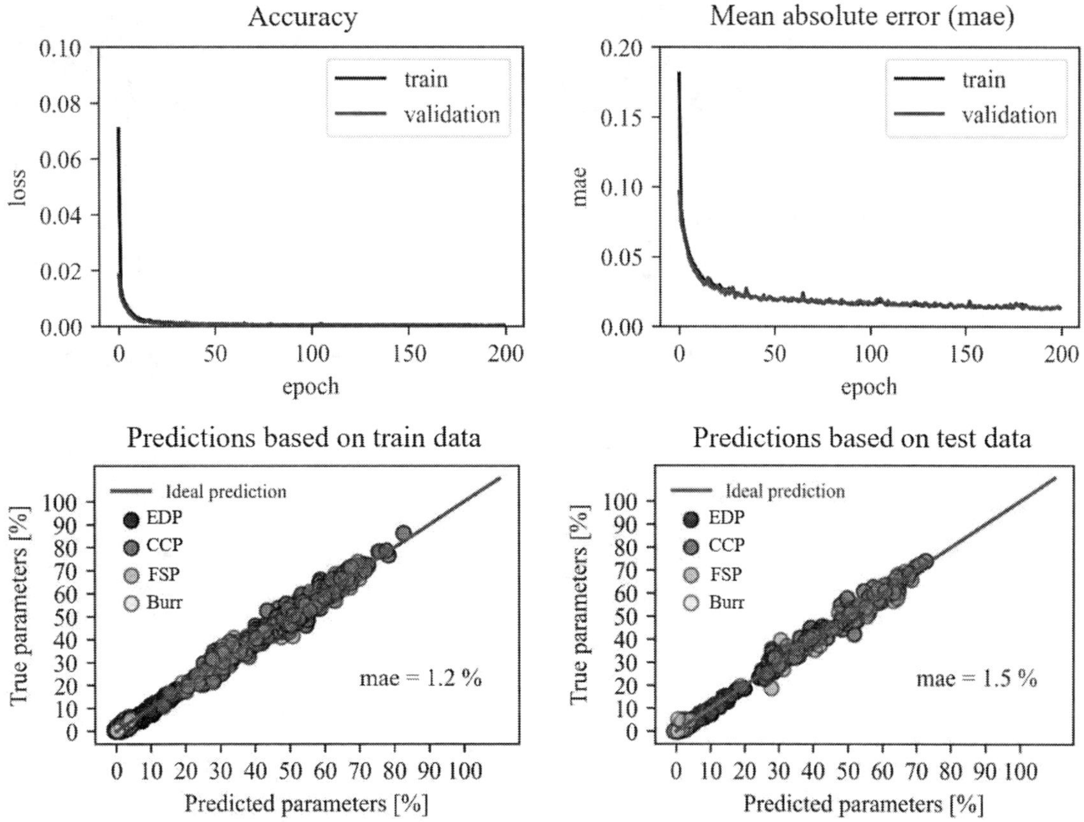

Figure 7. Learning process and prediction results for train and test data

5. Experimental validation

Validating a numerically pretrained model involves two crucial steps. The first step is to ensure that the numerically determined results provide good agreement with the experimentally determined results. If there is a significant deviation between the cutting surface features obtained through numerical and experimental methods, the preprocessing parameters of the simulation setup may have to be adjusted. This adjustment involves modifying the finite element size or the damage values in order to improve the accuracy of the numerical results. To demonstrate the validity of the simulation model used in the present study, subchapter 5.1 presents a direct comparison between numerically and experimentally determined cutting surfaces. The second step of the validation process involves comparing the neural network predictions with experimentally determined cutting surface features. This step is covered in subchapter 5.2, which presents a comparison between the neural network predictions obtained and the experimentally determined cutting surface features. This validation process ensured that the numerically pretrained model was suitable for the intended application.

5.1. Experimental validation of the simulation model accuracy

The accuracy of the simulation model used was verified by comparing the numerically predicted cutting surfaces with experimentally determined cross-sectional contour scans. The contour scans were recorded using an Alicona Digital Microscope Infinite Focus G5 on samples cut during the experiments. A total of six experimental parameters shown in figure 8 were used for validating the established simulation model.

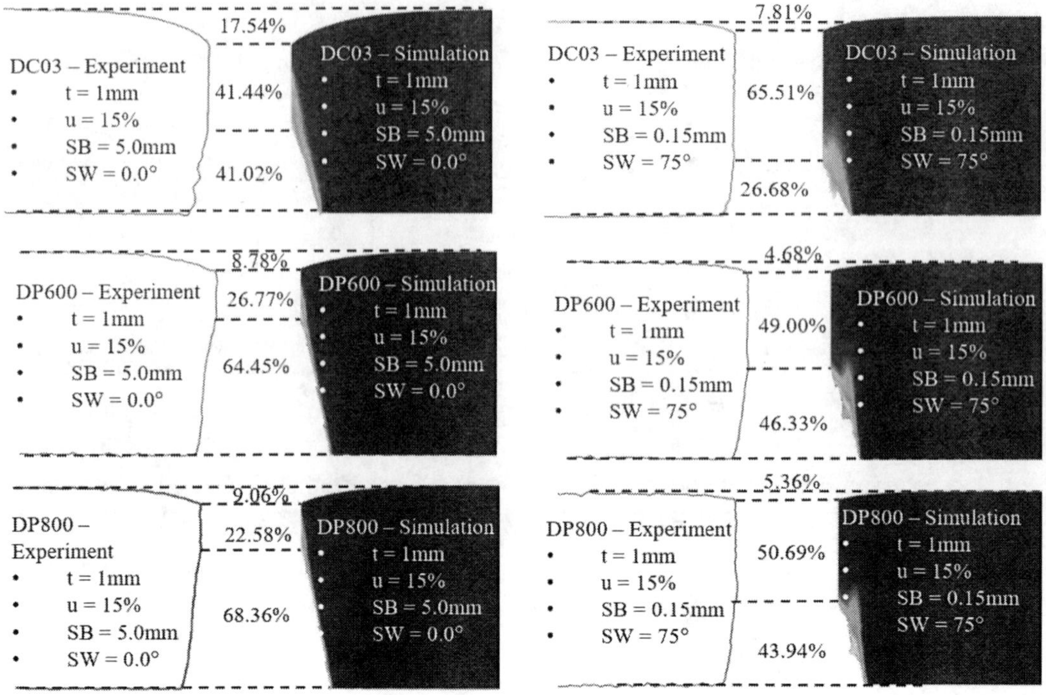

Figure 8. Accuracy of the simulation model for six experimentally determined cutting surfaces [21]

The results depicted in figure 8 reveal a high level of agreement between the numerically calculated and the experimentally determined cross-sectional contours. Consequently, an iterative refinement of the simulation model was deemed unnecessary. In the following subsection, the experimentally determined cutting surface features shown in figure 8 will finally be compared with the predictions of the numerically pre-trained neural network.

5.2. Experimental validation of the neural network predictions

The applicability of the numerically pre-trained network to experimentally measured cutting surface parameters was finally verified for the experimental data described in figure 8. For this purpose, the corresponding sheet metal data (R_m, t) as well as the tooling parameters (u, SB, SW) (see figure 8) were used as input for the neural network. The predicted cutting surface features were subsequently compared with the experimentally determined EDP, CCP, FSP and Burr. Figure 9 finally summarizes the result of this comparison.

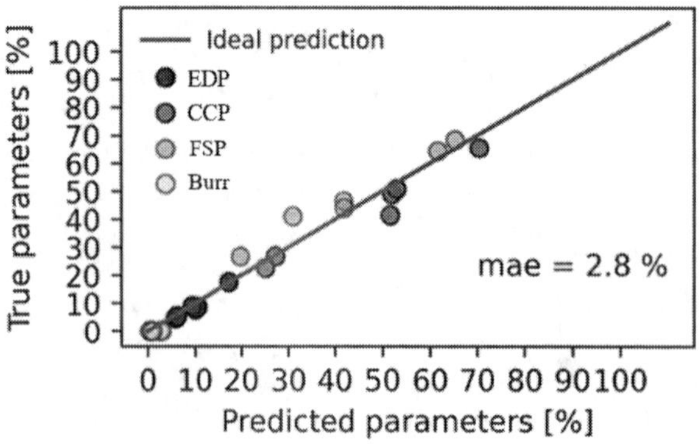

Figure 9. Prediction accuracy of the numerically pre-tained neural network for experimentally determined cutting surfaces

The mean absolute error for the predictions of the pretrained neural network for experimentally determined cutting surfaces is 2.8%. Considering the research question posed at the beginning, numerically pretrained neural networks are well suited for predicting cutting surface features based on sheet metal data (R_m, s) and tooling parameters (u, SB, SW). In order to further improve the prediction accuracy of the presented approach, additional experimental investigations are planned to extend the existing data set for the training of the neuronal network.

6. Summary and further research

In sheet metal processing, punching is a widely used manufacturing process, with the quality of the cutting surface being a crucial factor. High-quality cutting surfaces are characterized by low edge draw-in, a clean cut, small fracture surface, and low burr height. However, conventional punching processes are limited to clean cut proportions of only 20-50% of the thickness of the sheet metal material. To address this issue, the "concave punch nose design" was developed at the Institute for Metal Forming Technology in order to optimize the geometry of conventional punches for significantly increasing the clean-cut proportion along the cutting surface. In this paper, investigations on the highly nonlinear relationships between the quality parameters of the sheared component edges and the tooling and sheet metal material parameters used were presented. To quantify these effects, an artificial neural network was pre-trained using data from several numerical punching simulations. Input features included punch geometry, clearance size, sheet thickness, and material data. The output of the neural network provided precise predictions of achievable cutting surface quality parameters. The experimentally validated findings demonstrate that modern machine learning approaches can be used to predict cutting surface parameters using the punch geometry and properties of the sheet metal material as inputs for an artificial neural network. The mean absolute error for the predictions of the pretrained neural network for experimentally determined cutting surfaces was determined by 2.8%, which is a satisfying result for the investigated machine learning task. Future research concerns a transferability analysis of the presented content for further sheet metal materials and process parameters (e.g. open cut line).

References

[1] Demmel P Golle R Hoffmann H and Petry R 2015 *Blechumformung – Verfahren, Werkzeuge und Maschinen* ed K Siegert (Berlin Heidelberg: Springer) chapter 6 pp 223-241

[2] Verein Deutscher Ingenieure 1994 *VDI 2906 – Blatt2: Schnittfächenqualität beim Schneiden, Beschneiden und Lochen von Werkstücken aus Metall* (Berlin: Beuth) p 2

[3] Klocke F 2017 *Fertigungsverfahren 4: Umformen* (Berlin Heidelberg: Springer) chapter 5 pp 487-544

[4] Kolbe M 2020 *Stanztechnik – Grundlagen, Werkzeuge, Maschinen* (Wiesbaden: Springer Fachmedien) chapter 5 pp 23-52

[5] Schenek A and Liewald M 2021 Scherschneiden – Auslegung von Schneidkantenradien *WT Werkstattstechnik* **10** pp 712-717

[6] Feistle M Golle R and Volk W 2016 Determining the Influence of Shear Cutting Parameters on the Edge Cracking Susceptibility of High-strength-steels Using the Edge-fracture-tensile-test *Procedia CIRP* **41** pp 1078-1083

[7] Deutsches Institut für Normung e. V. 2011 *DIN 9830: Schnittgrathöhen an Stanzteilen* (Berlin: Beuth) pp 1-6

[8] Djavanroodi F Pirgholi A and Derakhshan E 2010 FEM and ANN Analysis in Fine-Blanking Process *J. Materials and Manufacturing Processes* **25** pp 864-872

[9] Stanke J Feuerhack A Trauth D Mattfeld P and Klocke F 2018 A predictive model for die roll height in fine blanking using machine learning methods *Procedia Manuf.* **16** pp 570-577

[10] Mao H Chen H Liu Y and Ji K 2022 A Novel Force Variation Fine-Blanking Process for the High-Strength and Low-Plasticity Material *Metals* **12** p 458

[11] Jin J and Shi J 2000 Diagnostic Feature Extraction From Stamping Tonnage Signals Based on Design of Experiments *J. Manuf. Sci. Eng.* **122** pp 360-369
[12] Klingenberg W and Singh U P 2004 *Proceedings of the 34th International MATADOR Conference* ed S Hinduja (London: Springer) pp 169-174
[13] Lee W B, Cheung C F, Chiu W M and Chan L K 1997 Automatic supervision of blanking tool wear using pattern recognition analysis *Int. J. Mach. Tools Manuf.* **37** pp 1079-1095
[14] Breitling J Pfeiffer B Altan T Siegert K 1997 Process control in blanking *J. Mater. Process. Technol.* **71** pp 187 192
[15] Asahi A Karadogan C Tamura S Hayamizu S and Liewald M 2021 Process data based estimation of tool wear on punching machines using TCN-Autoencoder from raw time-series information *IOP Conf. Ser. Mater. Sci.* Eng. **1157** pp 1-9
[16] Senn S and Liewald M 2018 Numerical investigation of a new sheet metal shear cutting tool design to increase the part quality by superposed compression stress *J. Phys. Conf.* Ser. **1063**
[17] Schenek A Senn S Liewald M 2022 *Production at the Leading Edge of Technology* ed M Liewald et al. (N.N.: Springer International Publishing) pp 179-188
[18] Senn S Liewald M 2019 Investigation of a new sheet metal shear cutting tool design to increase the part quality by superposed compression stress *IOP Conf. Ser. Mater. Sci. Eng.* **651**
[19] Uhlmann E von der Schulenburg M G 2007 Finite Element Modeling and Cutting Simulation of Inconel 718 *CIRP Ann.* **56** pp 61-64
[20] Thyssenkrupp AG 2023 *DP-W und DP-K Produktinformation für Dualphasenstähle* product brochure vol. 05/23
[21] Schenek A Görz M Liewald M Riedmüller K R Prediction of Cutting Surface Parameters in Punching Processes Aided by Machine Learning *TMS 2023 152nd Annual Meeting & Exhibition Supplemental Proceedings* pp. 607-619

Formability evaluation based on a constitutive model with stress rate direction dependency considering bifurcation modes

T Oya[1], K Ito[2], G Uemura[2] and N Mori[2]

[1] Faculty of Science and Technology, Keio University, 3-14-1 Hiyoshi, Kohoku-ku, Yokohama 2238522, Japan
[2] M & M Research, Inc., Tokyo, Japan

E-mail: oya@sd.keio.ac.jp

Abstract. Accurately predicting the fractures that occur in materials during forming not only improves quality, but also reduces the amount of material that is discarded. Therefore, as a key technology for digital transformation to achieve carbon neutrality, it is necessary to improve the accuracy of forming analysis and establish an efficient method for predicting forming limits. Formability assessment is typically performed using ductile fracture criteria or experimental forming limit diagrams (FLDs), which require prior fracture testing. The authors have established a forming evaluation method based on three-dimensional local bifurcation theory as a versatile method that does not require destructive testing. This method can follow abrupt changes in the stress field, such as localized necking, by using a flow rule that takes into account the directionality of the stress rate. In this study, the proposed method is applied to the tensile testing and the forming analysis of a square cup drawing to analytically verify the location and onset time of local bifurcation modes that may lead to fracture.

1. Introduction

When computer simulation is applied to press forming, prototype studies can be carried out on a computer, which can reduce costs, shorten development times and save resources. For this reason, computer models for metallic materials are being actively developed and researched. For example, accurate prediction of the fractures that occur in materials during forming not only improves quality, but also reduces the amount of material that is discarded. Therefore, as a key technology for digital transformation towards carbon neutrality, it is necessary to improve the accuracy of forming analysis and establish efficient methods for predicting forming limits. However, the results of forming simulations using computer models do not always match the actual behavior, especially in drawing processes, which show complex stress states, and there can be differences between forming simulations and the actual process. Therefore, there is a strong need to establish a more accurate method of predicting forming limits using forming simulation.

There are various types of forming failures, of which fracture is one of the most difficult to predict. As forming fails if the material fractures, appropriate conditions must be found when designing the die geometry and forming process to achieve the desired forming geometry without fracture. Instead of time-consuming and costly experiments, it would be useful if forming simulation could be used to design an optimum forming process that prevents fracture. However,

IOP Conf. Series: Materials Science and Engineering 1284 (2023) 012015 doi:10.1088/1757-899X/1284/1/012015

as fracture involves material separation, it is inherently difficult to handle and various methods have been devised. There are many examples of combining ductile fracture criteria[1][2] with finite element analysis, but as it is a phenomenological model, it lacks generality and versatility as many criteria have been proposed and parameters have to be determined experimentally. The experimental method of forming limit diagrams (FLDs) also has a wide range of applications, but requires a lot of experimentation to produce FLDs and is not user-friendly. A major method for evaluating the formability of sheets was proposed by Marciniak and Kuczynski[3], a numerical method introducing imperfections of sheets known as the M-K model. Many successor methods have been developed [4][5], because of their ability to predict forming limits through relatively simple calculations. Due to the introduction of virtual defects in the computational model, various proposals have been made for their numerical values and determination methods. A different theoretical approach is the modified maximum force criterion (MMFC) [6][7], which focuses on the hardening state after the diffused necking, and has yielded good prediction results for several types of sheets. These approaches have their own challenges, such as requiring burdensome fracture testing, models and parameters being phenomenological and lacking generality, and including numerical ambiguities such as the introduction of defects. In addition to that, they do not seem to address fracture in the shear zone, although shear fracture can be observed in a variety of forming processes. Theoretical FLD creation based on localized necking theory [8] is more versatile because it is based on mechanical principles and is considered most useful if it can be applied to any material or forming configuration. For the forming limits of sheets, the localized necking theory of Hill has been extended by Stören and Rice [9] to take into account the location of the deformation modes. This theory is hereafter referred to as S-R theory. It should be noted that the mechanical modeling of S-R theory is based on assumptions such as plane stress state throughout the deformation process and uniform stress distribution before localized necking occurs. Therefore, S-R theory does not take into account deformation modes that occur obliquely in the thickness direction and may exclude deformation modes that may lead to actual fractures. Based on S-R theory, Ito et al. [10] treated the localized necking problem from the viewpoint of 3D bifurcation analysis to classify possible bifurcation modes, and also proposed a new analysis scheme to estimate FLD. Although localized necking is not a failure itself, in the case of thin sheets, necking in the thickness direction can almost be regarded as a failure. Based on the stress rate direction-dependent plastic constitutive equation, the authors developed a forming limit prediction technique [10][11] that can be applied to arbitrary materials and forming types. Although the proposed method has been developed as a subroutine and can be used in general-purpose finite element analysis software, there are still not sufficient examples of its usefulness to be investigated. In this study, the proposed method was applied to tensile tests and forming analyses of square cup drawings to analytically verify the location and onset time of local bifurcation modes that may lead to failure.

2. Theoretical models

This section provides brief descriptions of the theoretical and analytical models[10][11] used in this study.

2.1. Stress rate direction-dependent plastic constitutive equation

The hypothesis that extends the information obtained from uniaxial tensile tests to general stress conditions and gives a concrete expression of the flow rule is the normality law shown below.

$$d\varepsilon^p = \frac{\partial g(\sigma)}{\partial \sigma} d\lambda. \tag{1}$$

This indicates that the plastic strain increment $d\varepsilon^p$ occurs only in the direction outward normal to the current stress point on the plastic potential function $g(\sigma)$. Normality flow rules are classified into two types: associated flow rules and non-associated flow rules. In the associated flow rule, the yield function $f(\boldsymbol{\sigma})$ is the plastic potential. In non-associated flow rules, on the other hand, the plastic potential and the yield function are generally different. In the normality law, the direction of the plastic strain increment $d\boldsymbol{\varepsilon}^p$ is uniquely determined in the direction normal to the current stress point on the plastic potential surface, independent of the direction of the stress increment $d\boldsymbol{\sigma}$. This, in other words, denies that the direction of the plastic strain increment follows changes in the direction of the stress increment. Therefore, it is not applicable to bifurcation phenomena in which the direction of the strain increment changes instantaneously, such as wrinkling or localized necking.

Although the normality law can be established as a reasonable approximation and gives generally valid results for many problems, a more accurate plastic flow law must take into account the dependence on the direction of stress increment. In other words, to make the plastic constitutive equation applicable to instability and bifurcation problems, the normality law must be rejected and the direction of plastic strain increment must change in accordance with the change in the direction of stress increment. With an awareness of these issues, the authors proposed the following stress rate direction-dependent plasticity constitutive model:

$$d\boldsymbol{\varepsilon}^p = \Lambda \left(\boldsymbol{n}_F : d\boldsymbol{\sigma}'\right) \left(K_C \boldsymbol{l} + (1 - K_C)\boldsymbol{n}_{\boldsymbol{N}}\right), \tag{2}$$

$$\boldsymbol{n}_F = \frac{\nabla\bar{\sigma}(\boldsymbol{\sigma})}{|\nabla\bar{\sigma}(\boldsymbol{\sigma})|}, \quad \boldsymbol{n}_{\boldsymbol{N}} = \frac{\nabla\bar{\sigma}^*(\boldsymbol{\sigma})}{|\nabla\bar{\sigma}^*(\boldsymbol{\sigma})|}, \quad \boldsymbol{l} = \frac{d\boldsymbol{\sigma}'}{|d\boldsymbol{\sigma}'|}, \tag{3}$$

where Λ is a scalar constant and $0 < K_C < 1$ is a material parameter that expresses the degree of dependence on the stress increment direction; when $K_C = 0$, it is consistent with the normality law. In addition, $\bar{\sigma}(\boldsymbol{\sigma})$ is the equivalent stress related to the yield function, $\bar{\sigma}^*(\boldsymbol{\sigma})$ is the equivalent stress related to the plastic potential function, and $d\boldsymbol{\sigma}'$ is the deviatoric stress increment. Thus, $\boldsymbol{n}_F$ means the normal direction of the yield surface, and $\boldsymbol{n}_N$ is the outward unit normal of the plastic potential function, which determines the direction of strain. And $\boldsymbol{l}$ is the direction of current deviatoric stress.

2.2. Forming limit prediction based on the 3-dimensional bifurcation theory

As Hill has formulated in the general bifurcation theory, bifurcation occurs at when the following condition is satisfied.

$$I[\Delta\boldsymbol{v}] = \int \Delta\boldsymbol{L} : \Delta\dot{\boldsymbol{S}} dV = 0, \tag{4}$$

where $\Delta\boldsymbol{v}$ is velocity field, $\boldsymbol{L}$ and $\dot{\boldsymbol{S}}$ are velocity gradient tensor and 1st Piola-Kichhoff stress tensor rate, respectively. $\dot{\boldsymbol{S}}$ can be represented by the Cauchy stress tensor as

$$\dot{\boldsymbol{S}} = \boldsymbol{D} : \dot{\boldsymbol{\varepsilon}} + \boldsymbol{\omega} \cdot \boldsymbol{\sigma} - \boldsymbol{\sigma} \cdot \boldsymbol{\omega} - \boldsymbol{L} \cdot \boldsymbol{\sigma} = \boldsymbol{A} : \boldsymbol{L}, \tag{5}$$

where $\dot{\boldsymbol{\varepsilon}}$, $\boldsymbol{\omega}$, $\boldsymbol{D}$ are strain rate tensor, spin tensor, and tangent stiffness tensor, respectively. And $\boldsymbol{A}$ is a fourth rank tensor that relates the nominal stress rate and velocity gradient tensor $\boldsymbol{L}$. To characterize the bifurcation mode, the velocity gradient tensor is allowed to be discontinuous when the velocity gradient tensor crosses the bifurcation border Γ. $\boldsymbol{L}$ can be represented by the normal vector $\boldsymbol{n}$ on the bifurcation border and the local deformation mode vector $\boldsymbol{m}$ that is normal to $\boldsymbol{n}$ as

$$\boldsymbol{L} = \boldsymbol{m} \otimes \boldsymbol{n}. \tag{6}$$

The mode vector $\boldsymbol{m}$ can be composed of two different vectors in the Γ plane; namely, $\boldsymbol{m}_{\mathrm{SH}}$ and $\boldsymbol{m}_{\mathrm{SV}}$ are the vectors in horizontal and vertical direction, respectively. These vectors are expressed with three angle parameters.

Substituting Eq. (5) and (6) into (4), the following bifurcation criterion is obtained:

$$I[\boldsymbol{m},\boldsymbol{n};\boldsymbol{\sigma}] = hH[\boldsymbol{m},\boldsymbol{n};\boldsymbol{\alpha}] - \sigma\Sigma[\boldsymbol{m},\boldsymbol{n};\boldsymbol{\alpha}], \tag{7}$$

where the first and second term of this functional are described as

$$H[\boldsymbol{m},\boldsymbol{n};\boldsymbol{\alpha}] = \boldsymbol{m}\cdot\boldsymbol{n}\cdot\overline{\boldsymbol{D}}(\boldsymbol{\alpha})\cdot\boldsymbol{n}\cdot\boldsymbol{m}, \quad \text{where} \quad \overline{\boldsymbol{D}}(\boldsymbol{\alpha}) \equiv \boldsymbol{D}(\boldsymbol{\alpha})/h, \tag{8}$$

$$\Sigma[\boldsymbol{m},\boldsymbol{n};\boldsymbol{\alpha}] = \frac{1}{2}\left[\boldsymbol{m}\cdot\boldsymbol{\alpha}\cdot\boldsymbol{m} - \boldsymbol{n}\cdot\boldsymbol{\alpha}\cdot\boldsymbol{n}\right]. \tag{9}$$

Here, the current stress is represented by

$$\boldsymbol{\sigma} = \sigma\boldsymbol{\alpha}, \quad \sigma = \sqrt{\boldsymbol{\sigma}:\boldsymbol{\sigma}}, \quad \boldsymbol{\alpha} = \boldsymbol{\sigma}/\sigma, \tag{10}$$

where σ and $\boldsymbol{\alpha}$ are the norm of the current stress tensor and normalization tensor which gives stress ratios for each stress component, respectively. In an elasto-plastic material subjected to large strain, ignoring elastic deformation, the tangent stiffness tensor can be assumed to be proportional to the hardening rate h.

Based on these relations, finally have the following local bifurcation criterion.

$$\left(\frac{\sigma}{h}\right)_{\mathrm{cr}} = \min\left(\frac{H[\boldsymbol{m},\boldsymbol{n};\boldsymbol{\alpha}]}{\Sigma[\boldsymbol{m},\boldsymbol{n};\boldsymbol{\alpha}]}\right) \tag{11}$$

The bifurcation criterion represented by Eq.(11) indicates that the local bifurcation, which is specified by the mode vectors $\boldsymbol{m}$ and $\boldsymbol{n}$ that are based on the current stress ratio tensor $\boldsymbol{\alpha}$, should be identified by the ratio σ/h that means the ratio of the stress level to the work-hardening. Mechanically, stress σ means the intensity of fracture initiation, and the hardening coefficient h means the material's resistance against fracture. Therefore, the formability represented in the σ/h plane is free from the strain-path-dependency that is usually observed in a typical FLD represented in strain space. Thus, because exhibiting forming limits on the σ/h plane is mechanically reasonable, this new expression is called SHFLD.

The fracture limits in the SHFLD can show 3D local bifurcation limits. The probable fracture would be lie between the lower bound represented by the S-R limit[9] and the upper bound represented by the 3D local bifurcation limit[10].

3. Results

3.1. Simulation settings

As a material model, the stress rate direction-dependent model described in the previous section was developed as a user subroutine and introduced into the commercial FEM software LS-DYNA. This model is referred to as Ito-Goya model in the following sections. The algorithm is implemented by an implicit method based on the return mapping. The sheets are modeled by solid elements, and three elements are used in the thickness direction. In the following simulations, aluminum alloy A6061 sheet was used as the material under consideration. The material constants are shown in Table 1.

IOP Conf. Series: Materials Science and Engineering 1284 (2023) 012015 doi:10.1088/1757-899X/1284/1/012015

Table 1. Material constants of A6061 sheet used in the simulations.

Property	Value
Mass density (g/cm^3)	2.70
Young's modulus (GPa)	70
Poisson's ratio	0.33
Initial yield stress (MPa)	276

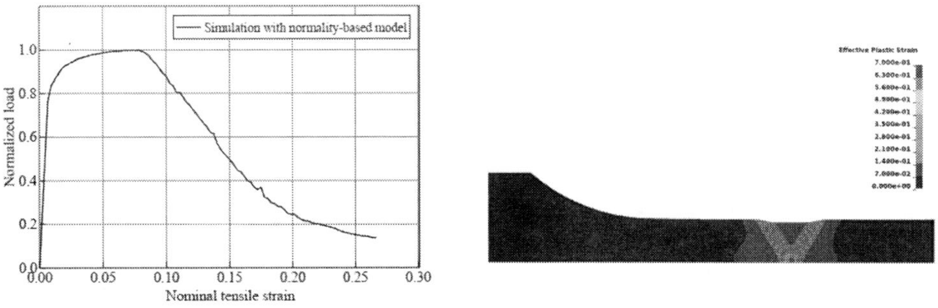

Figure 1. A result of tensile analysis using the normality law model, (left) tensile load history, (right) equivalent plastic strain distribution of specimen model immediately after maximum load.

3.2. Tensile analysis

The Ito-Goya model, implemented by a user subroutine, was applied to the analysis of tensile tests. The purpose of this analysis is to clarify the effect of the stress rate direction-dependent parameter K_C on the expression of the localized necking. A comparison was made with an elasto-plastic model based on normality law, which is included in the LS-DYNA material model library. This model was compared with the Ito-Goya model with the stress rate direction-dependent parameter K_C set to 0. When K_C is set to 0, the dependence on the stress rate direction disappears, making the model essentially equivalent to a normality law, and localized necking should essentially not occur.

First, as shown in Fig.1, a tensile analysis was performed on the normality law model. Immediately after the maximum tensile load was reached, a sudden decrease in load and localization of deformation were observed. The figure of the FE model on the right shows the equivalent plastic strain distribution immediately after the maximum load is reached (tensile strain of about 0.1). In actual tensile tests, the deformation mode shifts from uniform deformation to diffuse necking after the maximum load is reached, but in this analysis, it appears that a localized necking-like state is reached directly without such a process. Although this phenomenon cannot be explained by the normality law model, the strong deformation constraint in the tensile test may have caused this unrealistic abrupt transition of deformation modes.

Then, as shown in Fig.2, tensile analyses were performed by varying the K_C value. Although the K_C value is actually determined based on the consistency of the uniaxial tensile test results with the predicted fracture limit results, our previous studies have generally obtained reasonable results with a K_C value of 0.2. Therefore, in the numerical experiments conducted here, $K_C = 0.2$ was used as the standard value, and cases with values around 0.2 were considered. The

IOP Conf. Series: Materials Science and Engineering 1284 (2023) 012015 doi:10.1088/1757-899X/1284/1/012015

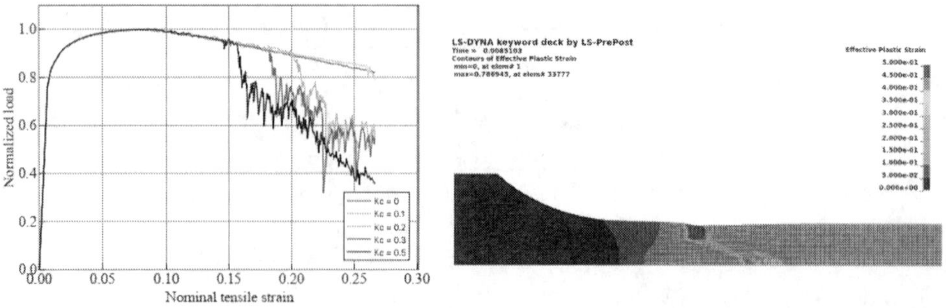

Figure 2. Results of tensile analysis using the Ito-Goya model, (left) tensile load histories, (right) example of equivalent plastic strain distribution after localized necking initiation ($K_C = 0.2$).

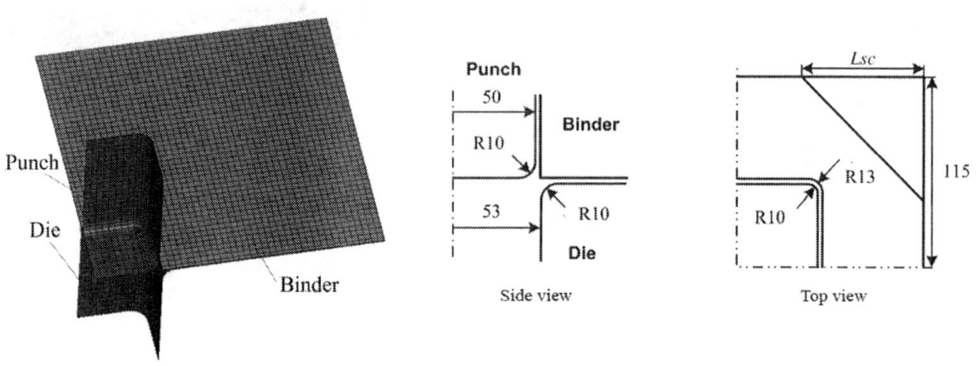

Figure 3. Finite element model and dimensions for square cup drawing analysis.

history of the tensile load is shown. In the present analysis, no localized necking was observed for $K_C = 0$ and $K_C = 0.1$. The case of $K_C = 0$ is similar to the case of the comparison with the normality law model described above, but in the case of $K_C = 0.1$, it is considered that a larger deformation is required before the occurrence of localized necking due to the smaller dependence on the stress rate direction. In the cases of $K_C = 0.2$, 0.3, and 0.5, the timing of the onset of localized necking accelerated with increasing value of K_C. This means that the larger the stress rate direction-dependent plastic strain is, the more likely deformation localization is to occur, which is a physically valid result. In conclusion, the analysis using the Ito-Goya model and its user subroutine shows the validity of the expression of the localized necking and confirms the usefulness of the proposed model.

3.3. Drawing analysis

Next, a square cup deep drawing analysis was performed using the Ito-Goya model. The finite element model of the square cup deep drawing was constructed as shown in Fig.3 referring to the work of Nishimura et al.[12] and Oliveira et al.[13].

Forming limits were determined using NXT Defect Evaluator 5, developed by M&M Research,

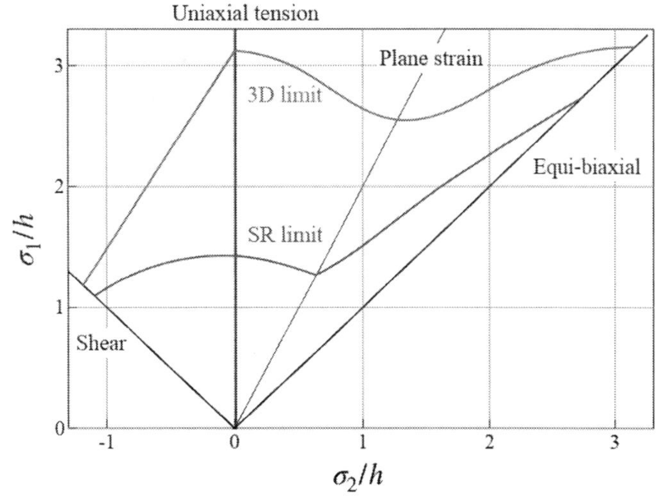

Figure 4. Schematic illustration of a SHFLD.

which implements the Ito-Goya model and SHFLD. In our previous studies and in NXT Defect Evaluator 5, the onset of localized necking is referred to as the S-R limit and the onset of local shear band is referred to as the 3D limit. These represent the lower and upper limits at which fracture occurs, since fracture actually occurs later than the lower limit (S-R limit) and before the upper limit (3D limit). That is, below the lower limit (S-R limit), there is no concern about fracture. Fracture is caused by the force, or stress, applied to the material, not by strain. Stress alone, however, is not sufficient; work-hardening of the material also affects fracture susceptibility. The three-dimensional local bifurcation theory states that the lower and upper limits of fracture (S-R limit and 3D limit) can be uniquely expressed as the ratio of the stress σ to the work-hardening rate h of the material, i.e., σ/h. Therefore, the fracture limit can be expressed without excess or deficiency by expressing the limit in the form $(\sigma/h)_{\mathrm{cr}}$. Since $(\sigma/h)_{\mathrm{cr}}$ takes different values depending on the stress ratio, a forming limit diagram can be constructed with σ_1/h and σ_2/h as axes. By comparing the stress of each element obtained from forming simulation results with SHFLD, which is represented by the stress/work-hardening rate plane, the work allowance can be evaluated. This method does not depend on strain history, and thus provides an objective evaluation independent of forming conditions. A schematic diagram of SHFLD is shown in Fig.4. In the color band display shown later, areas where the normalized formability evaluation value exceeds 2.0, i.e., where fracture is considered to have occurred, are indicated in red.

Drawing analysis was carried out by varying the K_C value to 0.1, 0.2 and 0.3. In each case, the SHFLD and forming results when fracture occurs are shown in Fig.5. It was confirmed that increasing the K_C value reduces the maximum drawing height as the forming limit is reached earlier. In all cases, fracture is predicted to occur in the first quadrant of the SHFLD, i.e. near the equi-biaxial stress state, and the location of the fracture is near the punch corner. This fracture occurs near the junction of the punch shoulder corner and the side wall, where a significant reduction in thickness tends to occur, and is the most commonly observed fracture phenomenon in the drawing process. The SHFLD also shows that there are different trends for different K_C values on the second quadrant side. This indicates a risk of shear mode failure. To verify the shear mode of failure, an analysis was performed for $K_C = 0.2$ with a 70 mm corner cut in the sheet. The results are shown in Fig.6. In general, a corner cut can reduce the forming

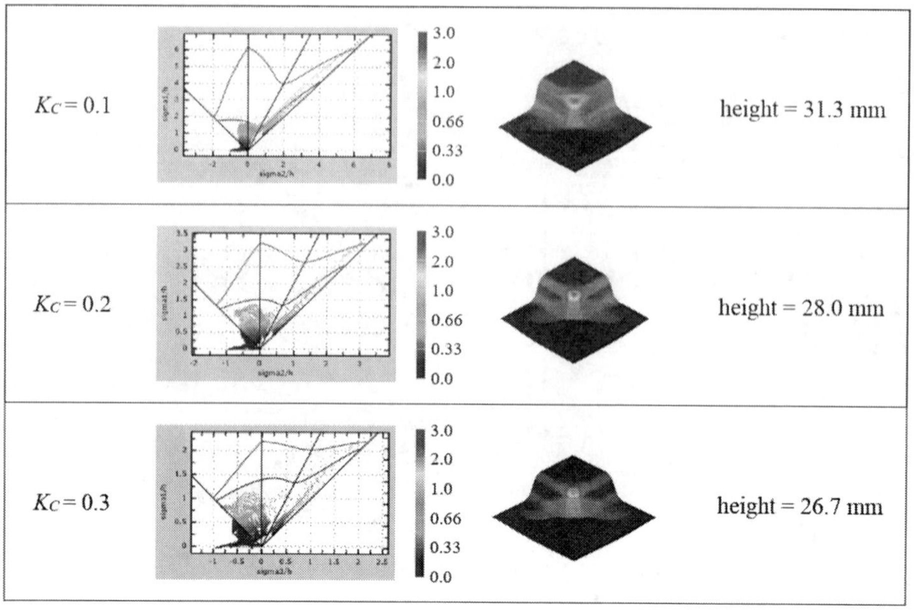

Figure 5. Forming limit prediction results for different K_C values, (left) SHFLDs, (right) corresponding FE models.

load and therefore improve the drawing formability. However, it is known that too much cutting can cause fracture near the cup wall. Wall fracture is a phenomenon that occurs mainly on the side walls of the corner section in the later stages of deep drawing of square cups and is difficult to predict and to develop preventive techniques. The non-uniform inflow of material can cause shear deformation of the sidewalls, resulting in larger than expected deformations. Therefore, this type of failure is generally difficult to predict because it is caused by a significant loss of material ductility. Although there is a considerable margin on the equi biaxial side, Fig.6 shows that shear fracture occurs at the cup wall rather than at the punch corner. Shear fracture cannot be predicted by general forming limit diagrams, but the proposed method can predict wall fracture, which means that the forming analysis and forming limit prediction method using the Ito-Goya model and SHFLD give physically valid results. Since the analysis and evaluation of formability is based on theory and does not require destructive testing, the proposed method can be used to design an optimum forming process by simulation only.

4. Conclusions

In this study, the stress rate direction-dependent plasticity constitutive equation was introduced into the material model of the FEM analysis solver LS-DYNA, and after a basic study in tensile test analysis, a square cup deep drawing process was analyzed. In particular, we focused on the influence of a parameter that expresses the dependence of plastic deformation on the stress rate direction, called the K_C value, and studied it numerically; when the K_C value is small, localized necking is less likely to occur because it approaches the normality law model, while when the K_C value is large, localized necking is more likely to occur and formability is deteriorated. Although these results are theoretically predictable, it is significant that the validity of the model has been confirmed in practical forming analysis. Future work includes theoretical and analytical

Forming height	Without corner-cut	With corner-cut
28.0 mm	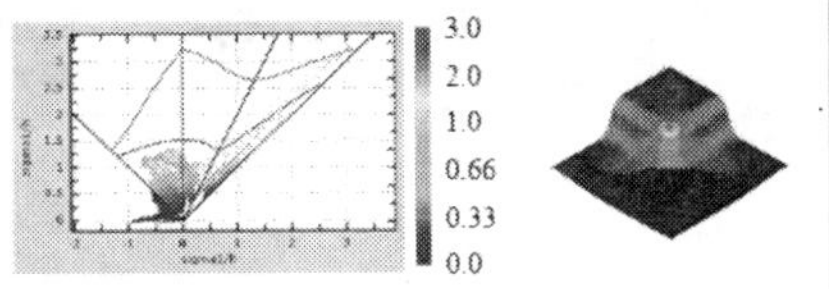	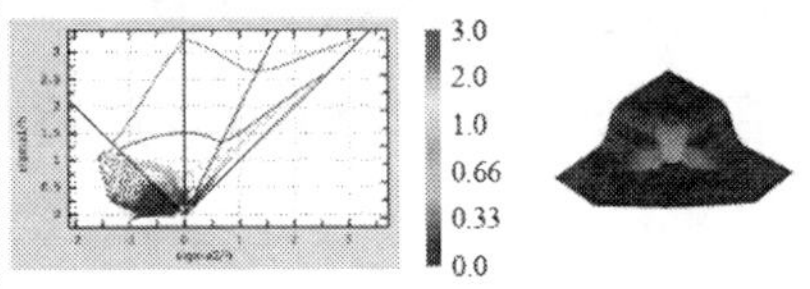
29.3 mm		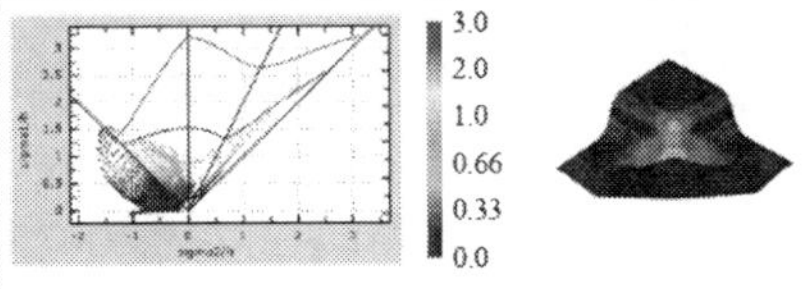

Figure 6. Effects of having corner-cuts in the case of $K_C = 0.2$.

parameter studies and experimental validation. In particular, it is important to improve the method of determining K_C values to increase usefulness and reliability.

References

[1] Cockcroft, M.G., Latham, D.J.: Ductility and the workability of metals. J. Inst. Met. 96, 33–39(1968).
[2] Zheng, L., Wang, Z., Meng, B., Wan, M.: A unified ductile fracture criterion suitable for sheet and bulk metals considering multiple void deformation modes. Int. J. Plast. 164, 103572(2023).
[3] Marciniak, Z., Kuczynski, K.: Limit strains in the processes of stretch-forming sheet metal. Int. J. Mech. Sci., 9 (9), 609–620(1967).
[4] Erfanian, M., Hashemi, R.: A comparative study of the extended forming limit diagrams considering strain path, through-thickness normal and shear stress. Int. J. Mech. Sci.,148, 316–326(2018).
[5] Banabic, D., Kami, A., Comsa, D-S., Eyckens, P.: Developments of the Marciniak-Kuczynski model for sheet metal formability: A review. J. Mater. Process. Technol. 287, 116446(2021).
[6] Hora, P., Tong, L., Berisha, B.: Modified maximum force criterion, a model for the theoretical prediction of forming limit curves. Int. J. Mater. Form. 6, 267–279(2013).
[7] Pham, Q. T., Islam, M. S., Sigvant, M., Caro, L. P., Lee, M-G., Kim, Y-S.: Improvement of modified maximum force criterion for forming limit diagram prediction of sheet metal. Int. J. Solids Struct. 273, 112264(2923).
[8] Hill, R.: Some basic principles in the mechanics of solids without a natural time. J. Mech. Phys. Solids, 7, 209(1959).
[9] Stören, S. and Rice, J. R.:Localized necking in thin sheets. J. Mech. Phys. Solids, 23, 421(1975).
[10] Ito, K., Sato, K., Goya, M. and Yoshida, T.: Prediction of limit strain in sheet metal-forming processes by 3D analysis of localized necking. Int. J. Mech. Sci., 42, 2233–2248 (2000).
[11] Oya, T., Yanagimoto, J., Ito, K., Uemura, G., Mori, N.: Material model based on stress-rate dependency related with non-associated flow rule for fracture prediction in metal forming, Procedia Engineering, 207, 227–232(2017).
[12] Nishimura, R., Ito, K., Nakazawa, Y., Uemura, G., Mori, N.: Evaluation of forming limit by the 3 Dimensional local bifurcation theory, Proceedings of the 9th International Conference on Numerical Methods in Industrial Forming Processes, 1433–1438, (2007).
[13] Oliveira, M.C., Alves, J.L., Menezes, L.F., Ito, K., Mori, N.: Local bifurcation and instability theory applied to formability analysis. Int. J. Mater. Form., 4, 347–356 (2011).

Process Monitoring using Digital Twin-Based Sensors integrated in Tool Clamping Surfaces

Robin Kurth*[1], Mohaned Alaluss[1], Martin Wagner[1], Robert Tehel[1], Matthias Riemer[1] and Steffen Ihlenfeldt[1,2]

[1] Fraunhofer Institute for Machine Tools and Forming Technology IWU, 09126 Chemnitz, Germany
[2] Institute of Mechatronic Engineering, Dresden University of Technology, 01062 Dresden, Germany

E-mail: robin.kurth@iwu.fraunhofer.de

Abstract. Process monitoring in forming technology can serve as a basis for transparent manufacturing and closed-loop control systems. Particularly, sensor systems associated with tool clamping interfaces offer great potential for robust inline process monitoring but are hardly investigated. In this paper, a concept to monitor a sheet metal forming process combined with a digital twin-based approach for virtual sensors is introduced. Here, sensors integrated into the T-slots of tool clamping surfaces in forming machines are enhanced by a mathematical representation of the elasto-mechanical behavior of the tool clamping interfaces. Moreover, the potential of the certain sensor system for inline process monitoring is highlighted. The resulting approach is substantiated by simulative analysis and by experimental investigation.

1. Introduction and motivation

The monitoring of tryout processes on forming tools, forming processes as well as machine behavior at the tool interface is an essential enabler for increased agility and efficiency in forming technology. Key parameters for assessing the process quality as well as for digitizing the tryout process entail the process force distribution and the elastic deflection of the tool clamping surfaces of press table and slide. In various research projects and works, solutions and methods to monitor the quality of formed components in production, during tool development and tryout have been developed and investigated [1-3].

Particularly, the challenge consists in an accurate, sensitive, and robust measurement and its proximity to the forming process during the production process. Existing monitoring approaches can be distinguished into tool integrated monitoring systems and into machine internal sensors. For instance, Skåre et al. [4] describe a method to identify wear-related changes using vibration sensors. By integrating strain sensors like strain gauges or piezoelectric sensors [5], force changes can be recorded locally and can be used for trend detection, wear monitoring or quality prediction. Moreover, quality problems like cracks or unfavorable process parameters can be identified by means of (mostly optical) sensors that monitor the sheet metal draw-in during a drawing process [6]. The mentioned monitoring systems measure directly and thus are very sensitive and often close-to-application. Usually, they are only used for one individual measuring task in a forming tool. Contrary, machine internal sensors like pressure, force and displacement sensors integrated in drive trains of modern forming machines, are often used for both process monitoring and machine analysis. These sensors can be integrated robustly. However, the achievable sensitivity regarding process fluctuations is severely limited due to the distance between sensor and active surfaces of the forming tool.

A promising approach is to use the tool clamping surfaces as a sensor location for inline monitoring [7]. This enables the possibility of combining a measurement that is close to the forming process and thus potentially sensitive with the advantages of a tool-independent integration. Alaluss et al. [7] analyzed underlying interdependencies and mechanism of action. An assessment of the various sensors systems [2, 3, 8-10] associated with tool clamping interfaces regarding different applications requirements (capability for inline monitoring, complexity, robustness, adaptability and state of development) is documented. Here, these sensor systems can be divided into three main groups depending on the mechanism of action and the target parameters:

- Sensor systems based on optical, fiber-optic or eddy current-based measurement sensors, which are arranged either on the rear sides of the clamping plates or in the T-slots of the press [3, 8, 9]: This measuring method promises high sensitivity concerning small process fluctuations, since components subjected to reaction forces are sensed. In addition, a local resolution of the deflection (and thus also of the acting force distribution) can be achieved. Nevertheless, these systems are hardly used industrially, because either the robustness or the adaptability is too low, and the integration effort is immense.
- The elastic deformation behavior of clamping devices due to changes in the acting press forces during the forming operation is used as a basis for strain measurement [10]: By evaluating the strains as target parameter, process trends and machine reaction can be continuously recorded. However, clamping elements are often located outside the path of process force, so that small process changes are difficult to detect. In addition, some of the measurement results cannot be clearly assigned to mechanisms of action e.g., the effect of deflection of press table and slide, the elastic deformation of the forming tool or the frictional slip behavior between forming tool and machine.
- A sensor system for force measurement between forming tool and machine using segmented measuring plates [2] consisting of an intermediate plate equipped with many strain sensors, which is additionally arranged between forming tool and machine: Although force distributions and changes can be measured sensitively, the described sensor system influences machine properties, in particular the machine stiffness and geometric parameters (tool installation height). Thus, a widespread use in production cannot take place without great expense.

It can be clearly stated that sensor systems associated with tool clamping interfaces offer great potential for robust inline monitoring. However, the evaluation of sensor data is hardly investigated for complex process monitoring applications. Here, evaluation concepts using a digital twin representation as described in [11] can help to unleash the full potential of these sensor systems. Therefore, in this paper the utilization of a certain sensor system integrated in the T-slots of the tool clamping surfaces of a press table or press slide is discussed. Furthermore, the capability for inline process monitoring is demonstrated by experimental investigations as well as the potential for a digital twin-based data evaluation regarding elastic deflection of press table and slide is described and investigated.

2. Inline monitoring approach using the tool clamping surfaces

The elastic behavior of tool clamping surfaces in presses is characterized by a large number of influencing variables and reactions. Possible sensor systems that utilize this behavior and the interaction between the forming tool and the press are presented in [7]. As shown, the use of elastic T-slot deformation as a reaction to acting process forces is particularly promising in the context of inline monitoring and hence represents the basic monitoring principle for the investigation within this work.

Figure 1a exemplarily depicts the elastic deflection of the table of a tryout press due to a process force [7]. The resulting elastic T-slot deformation $\Delta S_{x,r,sim}$ is demonstrated for T-slot 4. As displayed, the values of T-slot deformation depend on the evaluation position and change in accordance to the geometrical specifications of the clamping plate (e.g. cushion pin holes, ribs). In [12] a novel sensor system is introduced to measure the T-slot deformation related to process conditions. The sensor concept follows the principle shown in Figure 1b and 1c. The sensor consists of a base body with an attached elementary sensor (strain gauge or piezo) and is clamped inside the T-slot using a clamping screw and a clamping body with slope geometry. The measurement direction is orthogonal to the longitudinal direction of the T-slots.

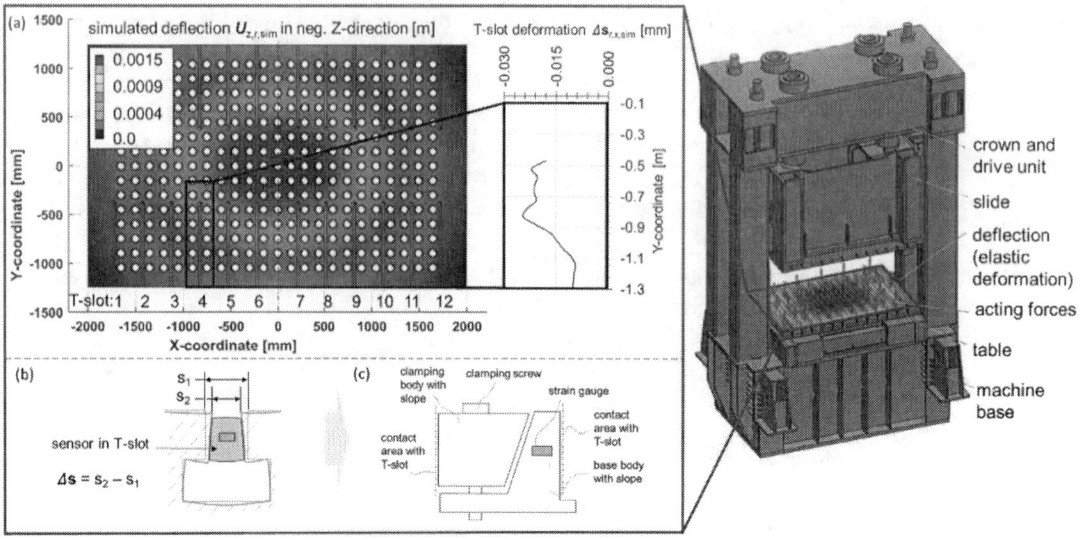

Figure 1. Schematic illustration of a tryout press with (a) exemplary FEA of the elastic deflection of the tool clamping surface (press table) due to process load and resulting elastic T-slot deformation, (b) principle T-slot deformation shape and sensor location, (c) schematic illustration of sensor design and measuring principle.

As shown in [12], with this sensor setup and the arrangement in the T-slots of the presses, process force changes that lead to changes in the elastic reaction behavior of the T-slot can be detected by continuous measurement. In the remainder of this paper, the described sensor setup is used to investigate the aptitude and potential for inline monitoring taking the example of a multi-stage sheet metal forming process.

3. Potential of process monitoring using sensors integrated in tool clamping surfaces

3.1. Experimental setup

Figure 2 shows the experimental setup to demonstrate the functionality and potential of the sensor system for process monitoring. In detail, the experimental setup consists of a servo forming press with screw drive and mounted multi-stage forming tool. This allows to manufacture a complex component as illustrated in Figure 2c. Using a feeder and a decoiler, this setup can perform continuous manufacturing process producing multiple parts. For conducting the experimental investigations, the used material was AA5754 H111 with a thickness of 1.0 mm. The 14 T-slots of press slide and table have a flank distance of 28 mm and are continuously integrated in the clamping plate. The described sensor system is integrated in the 10th T-Slot of the press table (counted in direction of material flow) under the last section of the forming tool. For the experimental investigation the sensor base body is equipped with a piezoelectric sensor element (instead of a strain gauge) due to its high sensitivity and the expectation of small elastic T-slot deformation due to a stiff press table and small acting process forces.

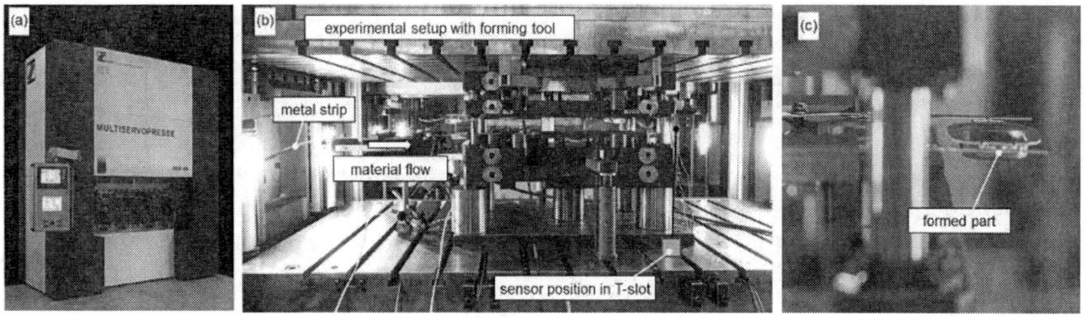

Figure 2. Test setup for experimental investigations with (a) servo forming press, (b) forming tool equipped with sensor systems and (c) formed part.

The multi-stage forming tool comprises 3 forming operations starting with two cutting stages followed by the deep drawing operation of the metal workpiece. A detailed illustration of the forming tool and the method plan is shown in Figure 3a. The resulting part geometry is detailed in Figure 3b. Using this forming tool and material it is possible to provoke cracks in the parts by changing the amount of oil lubrication for the sheet metal strip. Without or insufficient oil lubrication, a characteristic crack occurs in the risk zone at the bottom area of the part (Figure 3c). This crack is caused due to high restrain forces within the deep drawing stage of the forming tool.

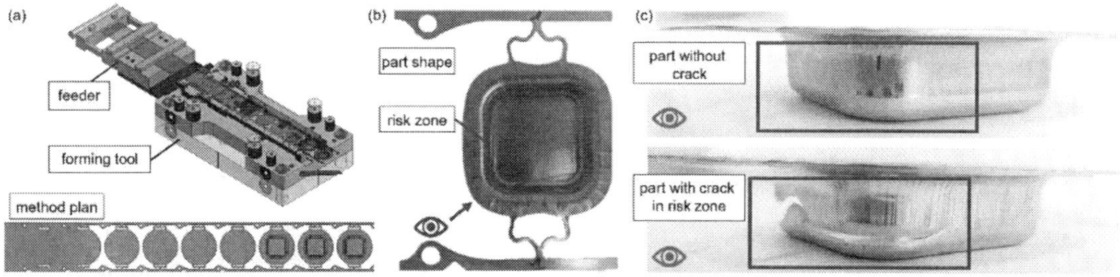

Figure 3. CAD model of the forming tool (lower part) and the method plan (a), geometric shape of the formed part (b) and potential production error (crack) in the risk zone of the part (c).

3.2. Experiment execution and results

During the experimental investigation the oil lubrication of the sheet metal strip is varied in a way that parts with and without crack in the risk zone are produced. If the occurrence of this defect pattern during production leads to changes in the process force curve of the deep-drawing stage, potential detection of the crack in the sense of an anomaly detection is possible on the basis of the continuously recorded sensor signal of the elastic T-slot deformation.

For performing inline process monitoring and anomaly detection different mathematical methods are known. Depending on the reaction of the measured values to the process changes, a suitable analysis method has to be selected.

Based on the resulting sensor signal characteristic, significant T-slot deformation can be detected during production. Since the sensor follows the principle of strain measurement and a conversion into Δs is not necessary for anomaly detection. The strain signals $\boldsymbol{\varepsilon}_{std}$ are visualized in the diagrams in Figure 4. For better comparability, the measurement values are standardized to overall maximum value. Here, a great temporal change of the sensor signal can be observed. One reason lies in the used signal amplifier and A/D converter, which has a high sensitivity but also a low intrinsic impedance. Thus, the charge in the piezoelectric element caused by strain can already dissipate during the measurement.

In Figure 4a, a comparison between a sensor signal curve representing a forming process of crack-free part and a sensor signal characterizing a forming process resulting a part with a crack is given. The comparison shows differences in form of three features. Especially in feature 2 and 3 clear deviations can be determined, which pose great potential for an automated detection of such deviations. Thereby, this automated detection can be implemented based on a local evaluation of the sensor signal values by means of reference value analysis or by means of a global envelope analysis. Both approaches require to identify (train) the reference values or the envelope limits on the basis of a set of curves from experiments with crack-free parts. Within the experimental analyses, 6 curves from experiments with crack-free parts were used to identify the upper $\boldsymbol{env}_{upper,i}$ and lower $\boldsymbol{env}_{lower,i}$ limit values for a global envelope analysis. The calculation of the limit values for each time step i is given in equation (1). Herein, the limits are identified as positive or negative sum of the mean value of the 6 signals in the relevant time step, the standard deviation $\boldsymbol{\sigma}_i$ and a constant value c.

$$\begin{bmatrix} \boldsymbol{env}_{upper,i} \\ \boldsymbol{env}_{lower,i} \end{bmatrix} = \begin{bmatrix} \overline{\boldsymbol{\varepsilon}_{std,\iota}} + 5\boldsymbol{\sigma}_i + \mathrm{c} \\ \overline{\boldsymbol{\varepsilon}_{std,\iota}} - 5\boldsymbol{\sigma}_i - \mathrm{c} \end{bmatrix} \tag{1}$$

The resulting envelope is depicted in Figure 4b. In order to increase the robustness of the calculation, the raw signal curves have been smoothened by means of a moving average filter. By application of constant value of 3% of the value range of $\boldsymbol{\varepsilon}_{std}$ a robust envelope analysis of the production data from 36 manufactured forming parts (mixed with and without crack in the risk zone) is possible.

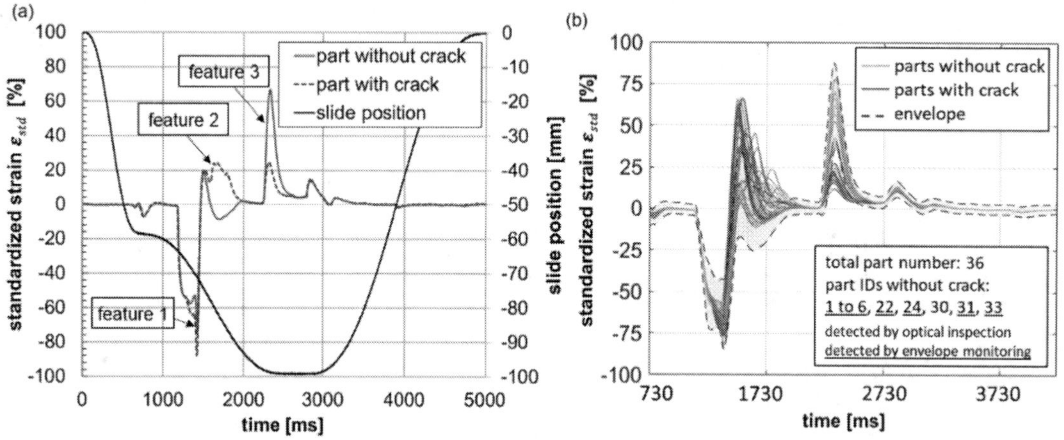

Figure 4. (a) Diagram of the experimental sensor data over one stroke producing a crack-free part and a part with crack, (b) diagram of envelope monitoring during production of 36 parts mixed with and without crack in the risk zone.

To correlate the part quality with the strokes and signal cures $\boldsymbol{\varepsilon}_{std}$, the parts have been marked by numbers (ID) and time stamp. The quality inspection was performed manually at the end of line by optical inspection. As results indicate in Figure 4b, 11 parts out of 36 were produced without and 25 parts with crack in the risk zone. With the described sensor system and the envelope analysis it was possible to detect all parts with characteristic crack. Only one part (ID 30), which did not have a crack in the risk zone, was incorrectly detected as part with a crack (false positive).

The results from the experimental analysis show that a sensor system as introduced in [7] and [12] can be used for inline process monitoring and anomaly detection. The high sensitivity of the measurement setup used within the investigations allows inline identification of deviations in the forming process (cracks in the risk zone of the produced parts). In further investigations, it must be shown how a more detailed signal analysis and/or more extensive measurement system can reduce the existing deviations (misidentification). In addition, the applicability for further quality variations (e.g. sheet thinning, cracks, wrinkles) need to be investigated.

4. Potential for digital twin-based data evaluation

The procedure described above for signal evaluation of sensor data provides limited information due to its principle. Model-based data analysis (digital twin-based virtual sensors) offers a possibility to expand the information content of sensor data [11]. This requires a mathematical model describing the considered system, which can be used for sensor data evaluation. Since a large number of the sensors used in this paper can be integrated into the tool clamping surfaces for the detection of the elastic T-slot deformation, a correlated sensor data evaluation in the sense of digital twin-based virtual sensors seems promising. As already mentioned in [12], it is possible to measure the elastic deflection of the clamping surfaces of press table and slide, which has a decisive influence on the resulting quality of formed parts (especially large sheet metal parts) and is a major parameter during forming tool tryout. The principal concept for digital twin-based virtual sensors is depicted in Figure 5 and adopts to the aim of monitoring the elastic deflection of the clamping surfaces on press table and slide using the mentioned sensor system measuring the elastic T-slot deformation at multiple sensor positions in the T-slots. In this regard, the digital twin describes the elastic deformation reaction at the tool clamping surfaces as a function of the acting forces.

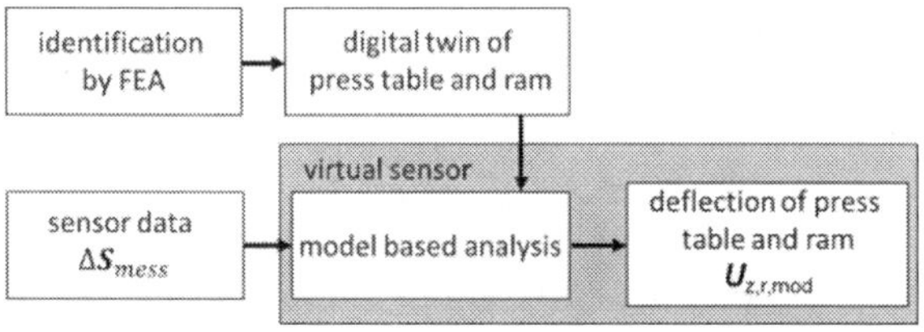

Figure 5. Principal concept of digital twin-based virtual sensors in context of deflection measurement utilizing elastic T-slot deformation.

This concept is used to analyze the potential of determining elastic deflection of press table and slide. For this purpose, a virtual analysis is performed to investigate the regarding detection potentials. A simplified plate model with T-slot geometry is used in numerical finite element analyses (FEA). The detailed geometry and constrains used in the FE simulations are shown in Figure 6a.

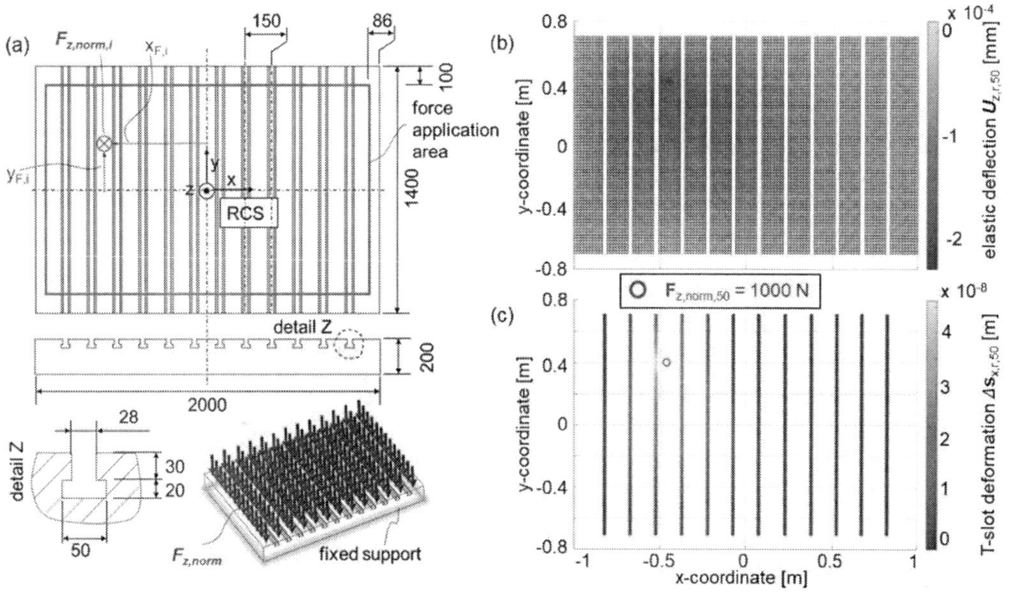

Figure 6. (a) Geometry, load scenarios and constrains for FEA, (b) simulated elastic deflection of the plate and (c) simulated elastic T-slot deformation under exemplary base load case.

With the use of this simplified plate geometry, the elastic deformation reactions (plate deflection $U_{z,r,norm}$ and T-slot deformation $\Delta S_{x,r,norm}$, both as fields in x- and y-direction) under acting base load cases (equation (2)) are determined by FEA. Simulation results for an exemplary base load case are depicted in Figure 6b and c. Each base load case consists of an acting single force $\boldsymbol{F}_{z,norm}$ in negative z-direction of 1,000 N applied at defined position $x_{F,i}$ and $y_{F,i}$. The deflection $\boldsymbol{U}_{z,r}$ and the T-slot deformation $\Delta\boldsymbol{S}_{x,r}$ for complex load cases can be determined analytically by a linear combination (equations (3) and (4)) of the i base load cases and the associated elastic deformation reactions using the scale factor a_i.

$$\begin{bmatrix} \boldsymbol{U}_{z,r,norm} \\ \Delta\boldsymbol{S}_{x,r,norm} \end{bmatrix} = f(\mathbf{F}_{z,norm}) \tag{2}$$

$$\Delta\boldsymbol{S}_{x,r} = \sum_i(\Delta\boldsymbol{S}_{x,r,norm,i} \cdot \mathbf{F}_{z,norm,i} \cdot a_i) \tag{3}$$

$$\boldsymbol{U}_{z,r} = \sum_i(\boldsymbol{U}_{z,r,norm,i} \cdot \mathbf{F}_{z,norm,i} \cdot a_i) \tag{4}$$

Thus, the elastic deformation behavior can be described efficiently. In order to realize the intended monitoring solution with the available sensor system in the T-slots, it is necessary to assign the sensor data $\Delta\boldsymbol{S}_{x,r,mess}$ to a holistic shape of the elastic T-slot deformation of the mathematical model $\Delta\boldsymbol{S}_{x,r}$. This can be achieved by a mathematical optimization, which solves equation 3 for varying scaling factors a and compares the result with the values at the measuring points in an objective function.

$$\boldsymbol{R} = \left|\Delta\boldsymbol{S}_{x,r,mess} - \sum_i(\Delta\boldsymbol{S}_{x,r,norm,i} \cdot \mathbf{F}_{z,norm,i} \cdot a_i)\right| \tag{5}$$

For optimal parameters $a_{mess,i}$ the residuum R reaches its minimum. With the identified parameter the complex acting load scenario $\boldsymbol{F}_{z,mess}$ can be determined by equation (6). Thus, it is possible to determine the deflection $\boldsymbol{U}_{z,r,mod}$ and total deformation of the T-slots $\Delta\boldsymbol{S}_{x,r,mod}$ related to this linear combination.

$$\boldsymbol{F}_{z,mess} = [\boldsymbol{F}_{z,norm,i} \cdot \boldsymbol{a}_{mess,i}] \tag{6}$$

Figure 7 summarizes the previously described method.

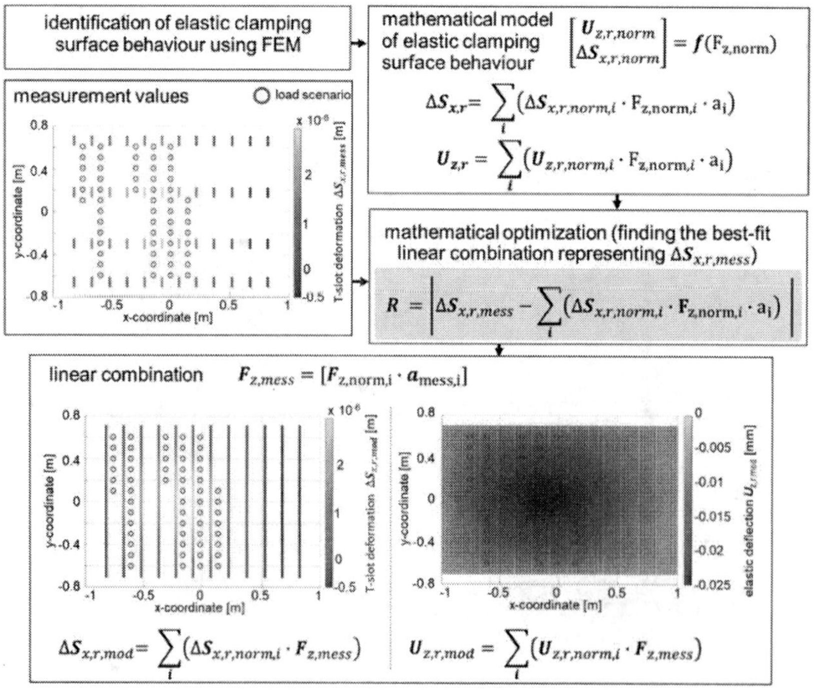

Figure 7. Method and simulation results for digital twin-based detection of elastic press table and slide deflection under complex load scenario by measuring the elastic T-slot deformation.

The method is demonstrated on a complex load scenario with two load application areas and force values of 5,000 N for each single point load (defined by a_{origin}). This load scenario leads to a complex

T-slot deformation pattern. Since the number of sensors and their possible integration positions during real application is limited, only T-slot deformation values $\Delta S_{x,r,mess}$ at selected locations (representing the measuring points) are used as input information for the mathematical optimization. Despite this limitation, a suitable parameter set a_{mess} is found, which approximates the actual load scenario very well. The deviations err_U and $err_{\Delta S}$ of the values for deflection $U_{z,r,mod}$ and total deformation of the T-slots $\Delta S_{x,r,mod}$ determined with a_{mess} from the values $U_{z,r,origin}$ and $\Delta S_{x,r,origin}$ resulting from the original load scenario are used as error parameters. According to equation (7) the error values with $err_{\Delta S} = 4 \cdot 10^{-7}$ and $err_U = 5 \cdot 10^{-5}$ are one magnitude below the evaluation values.

$$\begin{bmatrix} err_U \\ err_{\Delta S} \end{bmatrix} = \begin{bmatrix} \| U_{z,r,mod} - U_{z,r,origin} \| \\ \| \Delta S_{z,r,mod} - \Delta S_{z,r,origin} \| \end{bmatrix} \tag{7}$$

Thus, the method for a model-based evaluation of sensor data representing the elastic T-slot deformation to identify the elastic deflection of a press table or slide has been demonstrated and confirmed by FEA taking the example of a simplified plate.

5. Conclusion and future work

The presented investigations show the high potential of sensor systems integrated in tool clamping surfaces for inline monitoring during forming operation. The elasto-mechanical behavior of the tool clamping interface due to changes in the acting forming forces can be detected by the discussed sensor systems without affecting the forming tool or the press characteristics. The obtained sensor data achieves good results for an inline process monitoring. Two different analysis methods and procedures were demonstrated. On the one hand, it could be shown that inline quality monitoring and anomaly detection is possible by direct measurement and envelope analysis. Using an envelope curve analysis, it was possible to detect cracks on parts with the proposed sensor system.

In future work the investigations will focus on the accuracy of the detection method by applying other mathematical analysis methods based for example on artificial intelligence, sensor data fusion with other (available) sensors or improving the sensor system by utilization of multiple measurement points inside the T-slots of press table and slide.

By validation on simulative level, a novel digital twin-based data evaluation approach was proven to be accurate enough to determine the elastic surface deflection behavior of the press table and slide based on reduced information of real sensors integrated in the T-slots of the tool clamping surface. In this context it was shown that a mathematical model representing the elastic deformation behavior could be used in a specific method to build virtual sensors and increase the information provided by the sensor system. In further work, the focus should be shifted on strengthening the simulatively obtained results by validating them with experimental investigations within untrained scenarios. Moreover, the digital twin-based data evaluation should be developed further to include different heterogeneous data from the forming machine environment. Additionally, investigations regarding the minimum necessary number of sensors and measurement positions should be performed.

Acknowledgements

This work was supported by the Fraunhofer High-Performance Center for Smart Production and Materials and by the Fraunhofer Cluster of Excellence Cognitive Internet Technologies CCIT.

References

[1] Cao J, Brinksmeier E, Fu M, Gao R, Liang B, Merklein M, Schmidt M and Yanagimoto J 2019 *CIRP Annals* **68** 605-629

[2] Drossel W G, Zorn W and Hamm L 2019 *CIRP Annals* **68/1** 309-312

[3] Pilthammar J *et al.* 2021 *IOP Conf. Ser.: Mater. Sci. Eng.* **1157** 012076

[4] Skåre T and Krnatz F 2003 *Wear* **255** 1471-1479

[5] Biehl S, Rumposch C, Paetsch N, Bräuer G, Weise D, Scholz P and Landgrebe D 2016 *Microsystem Techn. Res. Journal* **22** 1757–1765

[6] Micro-Epsilon 2020 Available online: https://www.micro-epsilon.com/applications/branch/Automotive/Blecheinzug-Pressen/?sLang=en *(Acessed on 09 March 2023)*

[7] Alaluss M, Kurth R, Tehel R, Wagner M and Ihlenfeldt S 2022 *Journal of Machine Eng.* **22** 116-131

[8] Pilthammer J, Sigvant M, Hansson M, Palsson E and Rutgersson W 2017 *IOP Conf. Series. Journal of Physics* **896** 012-068

[9] Gesellschaft für Optische Messtechnik 2010 Availabe online: https://www.gom.com/fileadmin/ user_upload/ industries/forming_machine_EN.pdf. *(Acessed on 09 March 2023)*

[10] Kurth R, Tehel R, Päßler T, Putz M, Wehmeyer K, Kraft C and Schwarze H 2019 *Proc. Manufacturing* **27** 65–71

[11] Kurth R, Alaluss M, Tehel R, Reichert W and Ihlenfeldt S 2022 *Engineering Proc.* **26** (1):10

[12] Kurth R, Bergmann M, Tehel R, Dix M and Putz M 2021 *CIRP Annals* **70/1** 235–238

Local and global forming potential of present dual-phase steel developments

Thorsten Beier, Stefan Woestmann and Sascha Sikora

thyssenkrupp Steel Europe AG, Eberhardstraße 12, 44145 Dortmund, Germany

thorsten.beier@steeleurope.com

Abstract. Modern car bodies are constantly relying higher proportions of high and ultra-high strength steels. In the past, the increase in material strengths led to a conflict regarding the complexity of component geometries. Limits of material formability are reached and compromises had to be made in part design. By integrating application-oriented problems into the material development, these limits are expanded with respect to the given requirements. Within the present study, for several dual-phase steels with a tensile strength ≥800 MPa, practical laboratory tests are presented and their benefits in press shop application. In particular influencing factors on stretch flange formability are presented and discussed for example cutting parameters, anisotropy, time between cutting and forming or the orthogonal strain gradient.

1. Introduction

Modern car bodies are constantly relying on higher proportions of high- and ultra-high-strength steels. In the past, the increase in material strength led to a conflict regarding the complexity of the component geometries. Limits of material formability were quickly reached, and compromises must be made in component design. Through new approaches in material development and the integration of application-oriented problems, these boundaries are now significantly expanded. However, to select materials according to the requirements, the special features must be measurable and the associated benefits in component design and production must be implemented in the development process.

As a rule, the elongation or the forming limit curve is an essential feature for the formability of a material. Hereby the global formability or the resistance to instability is represented. In feasibility analyses using Finite-Element-Method (FEM), these characteristics are constantly used. In the press shops, the simple rule with higher elongation values more complex part geometries or higher process stability can be ensured. However, practical experience shows that local forming behavior often limits process reliability. Thus, the stretch flangeability of the blank edges is significantly influenced by the previous mechanical cutting process. These local properties can be still not sufficiently considered for the part design in FEM simulation. The interaction of the influencing parameters is complex and can only be estimated by statistical considerations.

The aim of the present study is to work out extended material characteristics of dual-phase steels from 800 MPa up to 1000 MPa tensile strength for an efficient selection and secure processing in the press shop. Therefore, existing information's are gathered and enhanced by additional material testing. In a further chapter it is discussed how to transfer these laboratory results to process design for press shop application.

2. Test procedure and used material

In the present study different DP-steels grades of the 800 MPa and 1000 MPa class are used. Dual-phase steels are the classic universal grades when it comes to high and ultra-high strengths combined with good forming and joining properties. Their share in multi-phase lightweight construction steels for cold forming has been rising for years and accounts currently for about 90 percent [1]. Variants with higher ductility and increased strengths now open new potential for customized lightweight solutions with steel. Table 1 shows the representative values in rolling direction for hot dip galvanized material. Beside the standard, classic grade also grades with higher elongation (DH), higher hole expansion ratios (HHE), higher yield strength (HY) or 3rd generation types (jetQ®) [1, 2] are investigated in this work regarding their local and global characteristics.

Table 1: Representative mechanical properties of DP steel grades; DIN EN ISO6892-1, rolling direction, hot dip galvanized

		t [mm]	Y.S. [N/mm²]	U.E. [%]	T.E. [%]	r_m	n
800 MPa class	DP-K® 440Y780T	1.5	468	12.1	19.1	0.85	0.117
	DP-K® 440Y780Y-DH	1.5	480	13.6	20.3	0.89	0.144
	DP-K® 440Y780T-HHE	1.5	537	10	17.9	0.90	0.099
1000 MPa class	DP-K® 590Y980T	1.5	621	9.5	14.8	0.77	0.098
	jetQ® 980	1.5	839	8.8	13.6	0.917	0.098
	DP-K® 700Y980T-HY	1.5	765	6.1	13.1	1.00	0.063

Looking at the yield strength or at uniform and total elongation the differences regarding their global properties are obvious. These global characteristics are derived from standard tensile tests according to DIN EN ISO6892-1 [3]. To investigate the local characteristics the used tensile specimen are surveyed by measuring the fracture surface. Figure 1 shows schematically the procedure and the equations to calculate local ductility which can be referred to the whole surface (TFS) or to the point of minimum thickness (eps_{3f}). First basic investigations have been made by Hance [4].

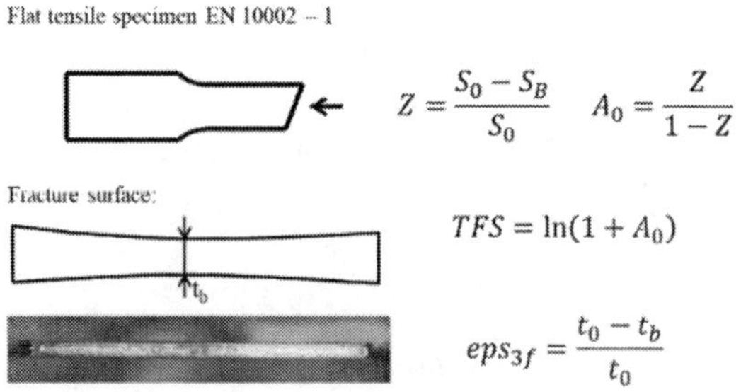

Figure 1: Evaluation of local ductility from fracture surface of tensile specimen

One important and frequently discussed local characteristic of multiphase steels is stretch flange formability which is often specified by hole expansion tests (HET). The various influencing factors on the hole expansion ratio (HER) are quantified within the material development inside tkSE. The comprehensive test program is given in table 2.

To work out the influence of material scatter on HER accompanying to serial production different batches (up to 50) are tested. The effects of aging, cutting parameter or rolling direction are investigated in further test runs for each material. Additional to the described factors the load situation at the hole edge is studied by varying the punch shape and die dimension for HET. Dimensions of the different punch shapes and test equipment is summarized in table 3.

Table 2: Overview test program

Test run	Influencing factors	Investigation target, test procedure & boundaries
1	Material scatter	Statistics on HER for in-plane hole expansion test determined from several coils using standard test parameters (hole ∅20 mm, cutting clearance 11-13%, time between cutting & forming >48h)
2	Orthogonal strain gradient	HER resulting from different punch shapes/ opening angles α (50°, 120°, 180°) for standard test parameters
3	Aging of the cutting edge	HER for in-plane hole expansion test with time lag <10min and >48h using standard cutting parameters
4	Cutting parameters	HER for in-plane hole expansion test for standard time between cutting & forming >48h, in addition various cutting clearances and cutting punch radii
5	Rolling direction	Maximum edge strain for in-plane hole expansion test using bisected specimen 0° & 90° rolling direction using standard cutting parameters

Table 3: Geometry and dimensions HET tools

Test	Punch type	Punch dimension	Hole	Die	Evaluation criterion
HET_A (ISO16630)	conical 60°	Ø 40 mm	Ø 10 mm	Ø 55 mm; R5	
HET_B	conical 50°	Ø 100 mm		Ø 110 mm; R5	Global HER [%]
HET_C	conical 120°	Ø 100 mm	Ø 20 mm	Ø 110 mm; R5	
HET_D	flat bottom	Ø 100 mm; R15		Ø 140 mm; R5	

By using different punch shapes the strain distribution close to edge is changing. The load condition can be quantified and differentiate by the orthogonal strain gradient. The definition of the gradient is introduced in [5] and it's definition is illustrated in figure 2 and evaluated for a developed length of 5 mm. With the help of this definition it is possible to use limits from HET for FEM analysis to predict edge crack. In laboratory during testing the diameter of the hole expansion is measured on the punch side. For a more accurate comparability it is transferred to membrane layer and logarithmic strain.

For the quantification of the impact of anisotropy on the maximum edge strain punched specimen (HET_D) are bisected. These two halves are tested at the same time in one test run to prevent tilting of the punch. During forming the strain is measured via digital image correlation. The test procedure is described in more detail in [6].

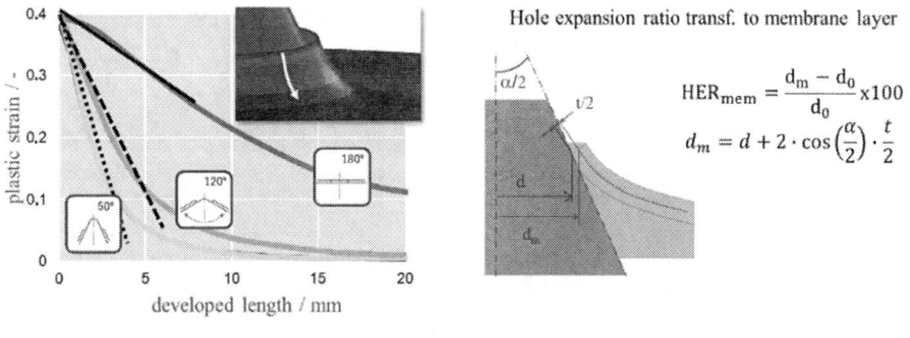

Figure 2: Definition orthogonal strain gradient and transformation to membrane layer

3. Test evaluation and analysis

3.1. Global properties

As can be expected from UE in table 1, the three variants show differences in the work hardening behavior. Figure 3b shows the hardening behavior measured according to DIN EN ISO 6892.

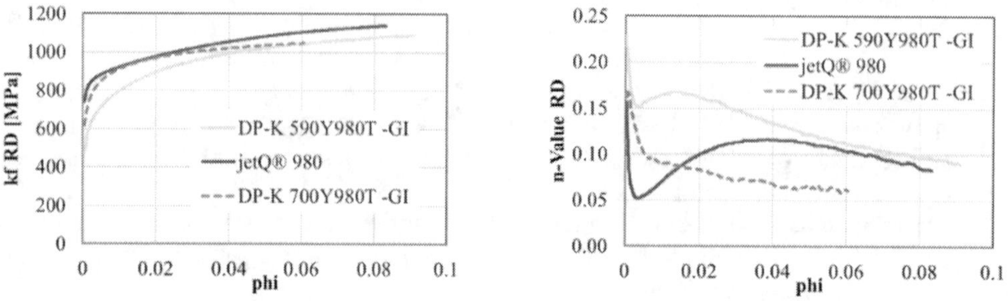

(a) (b)

Figure 3: a) quasi static yield curve b) evolution of hardening coefficient

To evaluate forming limit curves Nakajima experiments are done during the previous material development according to DIN EN ISO 12004 [7]. For a better understanding of the different hardening behavior of the material concepts these Nakajima experiments are analyzed more deeply. By looking at the strain evolution from digital image correlation (DIC) for the critical facet it is obvious that the DP-K 590Y980T reaches a higher drawing depth until fracture but a lower strain level of ~25% (figure 4a) for a biaxial strain path. The strain distribution for 25mm punch travel shown in figure 4b shows are more homogenous distribution with a higher quantity of facets of lower strains at the pole of the specimen. This correlates to the different n-value evolution in figure 3b.

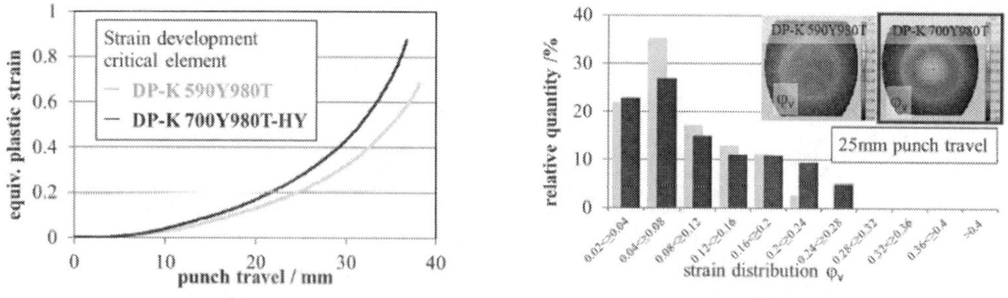

(a) (b)

Figure 4: Nakajima test a) strain development critical facet b) strain distribution from DIC 25 mm punch travel [8]

3.2. Local Ductility

A first orientation regarding the formability properties is represented by local elongation at fracture (TFS) vs. uniform elongation (UE) according to Hence [9]. In the present study the local elongation at fracture (eps_{3f}) instead of the true logarithmic fracture strain (TFS) according to [10] is used, as given in figure 1. Whereby the local elongation at fracture (eps_{3f}) refers to the maximum sheet thickness reduction over the fracture surface of the tensile specimen. For a larger quantity of material batches, the eps_{3f} values and the corresponding UE is given in figure 5.

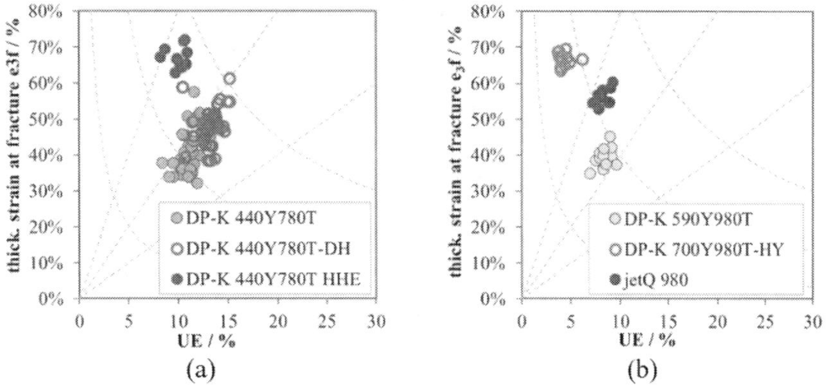

Figure 5: Local and global forming potential of DP800 (a) and DP1000 (b) steel grades (test direction longitudinal)

For both strength classes the classic and the DH or jetQ grade show higher global ductility. Whereby the DP-K 700Y980T-HY or the DP-K 440Y780T HHE grade differentiates itself by a higher local ductility. But also can be seen that especially for the DP-K 440Y780T the scatter in the material leads sometimes equal global ductility. The jetQ 980 offers a good combination of local and global properties (figure 5b).

3.3. Stretch flange formability

To evaluate the stretch-flangeability of cutting edges, additional hole expansion tests with varying test conditions compared to the ISO16630 [11] hole expansion are carried out to describe the failure as comprehensively as possible.

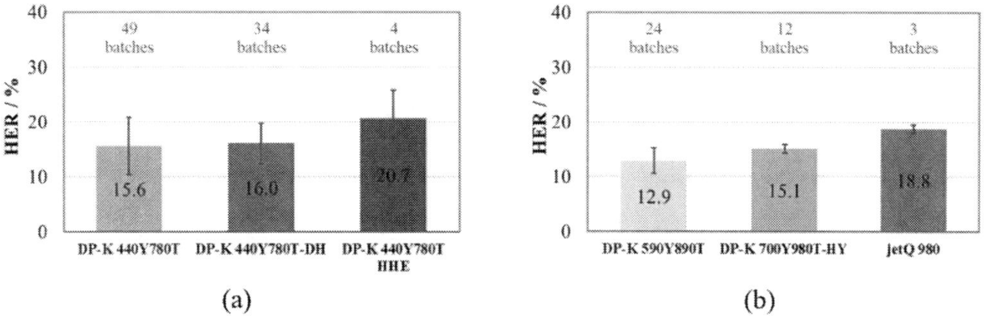

Figure 6: In-plane hole expansion test; ∅20 mm holes, cutting clearance 11-13 %, testing >48 h after punching; typical mean values HER from several batches

Primarily the standard test is an in-plane hole expansion (HET_D table 3) using a flat bottom punch. In test run 1 the material side scatter due to the local micro structural differences is considered by a high number of different batches. For the 800 MPa DP-steels (figure 6a) the scatter is significant higher compared to the 1000 MPa class (figure 6b). As could be assumed the higher strength material shows reduced stretch flange formability. In this case for DP-K 700Y980T-HY and jetQ 980 it must be pointed out that the necking close to edge is the cause for failure. So these two materials are not sensitive to edge cracks but of course limited due to their lower global forming properties represented by the forming limit curve.

The mean value of the hole expansion minus the standard deviation can determined as the material-specific limit. This limit is recommended for process and part design for example by FEM. Experiences on further boundary conditions in FEM have been discussed in [6].

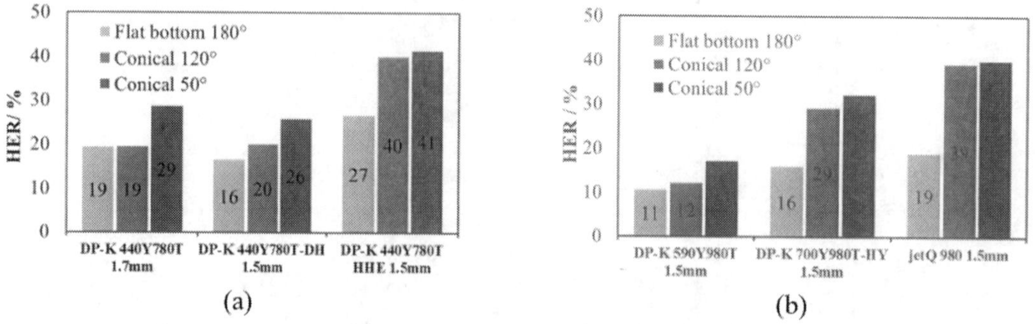

Figure 7: Hole expansion ratio using different punch geometries; ∅20 mm; cutting clearance 10-11%, testing >48 h after stamping; hole expansion converted to the sheet mid layer after figure 2

In typical applications of cold-rolled multi-phase steels, the critical edge strain is often in the flange area with a low orthogonal strain gradient [5, 12]. To evaluate the influence of the orthogonal strain gradient, in test run 2 additional hole expansion tests with different conical punches with opening angles of 120° and 50° (figure 7) have been carried out for one characteristic batch. The mechanical cutting conditions are kept identical. The higher hole expansion ratio for the HHE and jet Q concept is obvious in all test conditions. Looking at the influence of the orthogonal strain gradient at the cutting edge the higher HER values for conical punch geometries has been detected. By using conical punches, a higher orthogonal strain gradient is achieved [5, 12]. A higher strain gradient is also typical for the standard hole expansion according to ISO16630 which is the most popular test usually for comparing edge crack sensitivity.

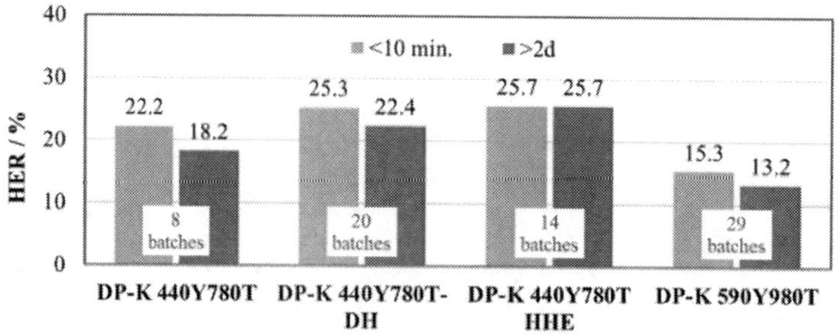

Figure 8: In-plane HET ∅20 mm holes; testing <10 min. and >48 h after punching, cutting clearance 11-13%; mean value from several batches

Cutting edges with local high strain levels have a high aging potential, so that the time interval between the punching process and the following forming operation has a significant influence on the failure level [13]. Extra tests are carried out directly after punching the holes (<10 min.) to estimate the effect of ageing on the hole expansion ratio. The higher HER values for a smaller time interval could be observed in test run 3 (figure 8). In the standard procedure for the hole expansion tests in product development, the critical case is assumed inside tkSE, so that a minimum time of 48 h between punching and hole expansion is always used.

By this test run the reduced HER caused by aging effects except for the HHE grad is getting obvious. This influencing factor has also kept in mind for a robust process design. When a production directly from the coil is not possible a lower potential for stretch flange ability must be considered.

Another important aspect are the parameters of the cutting process itself. They have been investigated in test run 4. Here the standard cutting clearance of 11-13 % (referred to sheet thickness) comparable to ISO hole expansion is compared to different punching parameters as shown in figure 9. The influence of tool wear is also investigated by using defined tool edge radius of 250 µm. This radius represents the worn cutting tool and leads also to a significant loss of edge formability.

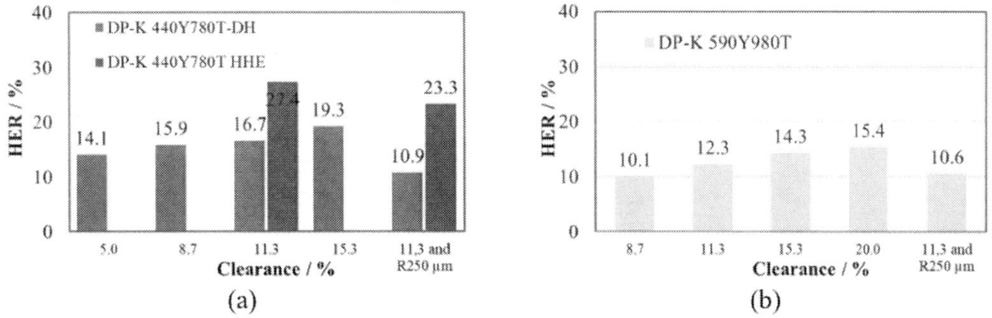

Figure 9: In-plane HET ∅20 mm holes; testing >48 h after stamping; HER using the different cutting parameter clearance [%] and tool radius [µm]

The last parameter considered in test run 5 is the effect of the rolling direction on stretch flange formability. On real part geometries, the strain at the blank edges is aligned in a certain orientation to the rolling direction. To measure the effect on the failure limits, special halved samples with a 1 mm grid for strain measurement via DIC are used. The middle-punched hole expansion sample is split in the designated test direction (L/D/T) before the test, so that there is a strain localization in the two concave areas during forming. For DP-K 700Y980T-HY and jetQ 980 no edge crack was observed using the in-plane HET. Here failure was driven by necking beside the edge. For the case of DP K 440Y780T and DP K 440Y780T-DH the fracture is located at the edge in the middle of the crescent-shaped cutout. The reduced forming potential for transverse load is given in figure 10. This effect driven by the anisotropy of the material is noticeable for the displayed grades DP-K 440Y780T and DP-K 440Y780T-DH. This fact must be considered with respect to the nesting position of the blanks on the coil.

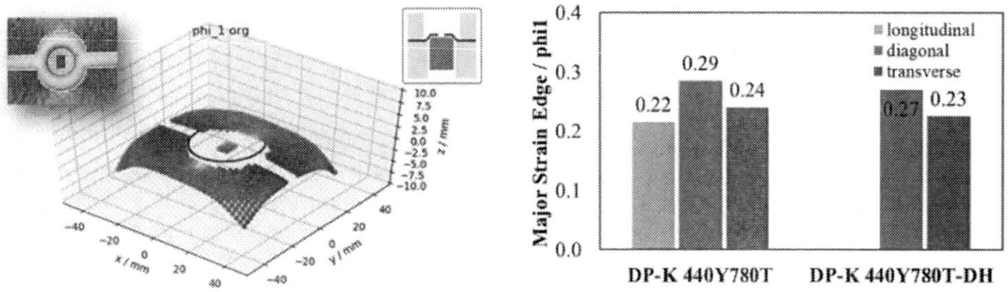

Figure 10: HET with halve split blanks and strain measurement; clearance 11%; testing >48 h after stamping

4. Conclusions and outlook

The aim of these study is to evaluate the impact of a higher strength level on stretch-flange formability. Therefore DP-steels of the strength class 800 MPa and 1000 MPa are investigated by several hole expansion test. Table 4 gives an overview of the interaction of the individual effects which can be transferred to critical areas in real geometries. The result is a recommendation for the design of the

IOP Conf. Series: Materials Science and Engineering 1284 (2023) 012017 doi:10.1088/1757-899X/1284/1/012017

forming processes in the form of a maximum edge strain (major strain, phi_1). So each factor has an impact on the forming potential at the edge and the whole has to be considered in sum. In table 4 in the rightmost column the impact is given as scale factor for the classic grades. Here the impact for the example of a process with challenging flange areas, stamping directly from coil, worn cutting tools and the main load transverse to rolling direction is summarized.

It is getting obvious, that especially the 800 MPa strength class shows a high sensibility on shear cutting. The 1000 MPa strength class show nearly the same maximum edge strain with respect to process design. For critical parts or processes the derivatives HHE or jetQ can help to achieve a robust process in the press shop and offer more degrees of freedom in part design. The impact of the investigated influence factors on stretch-flange formability show the same ratio for both strength classes.

Table 4: Consideration of process boundary conditions to the stretch-flange formability of cutting edges

Influential factor	Customer process(e.g.)	**Recommendation max. edge strain (phi_1)** DP800		DP1000		Scale factor
		Classic	HHE	Classic	jetQ	Classic
Scatter (MV-SD) in-plane HET	Different coils	0.1	0.15	0.1	0.17	basis
Strain gradient	Flange areas (low gradient)	0.1	0.15	0.1	0.17	x 1
Aging cut edge	Stamping directly from coil	0.12	0.15	0.12	0.17	x 1.15
Cutting parameters	worn	0.1	0.15	0.1	0.17	x 0.85
Rolling direction	Transverse (e.g)	0.1	0.15	0.1	0.17	x 1

* HHE & jetQ show instability (FLC) next to the edge before cracking (in-plane HET)

Experience in recent past show that the process reliability of ultrahigh-strength dual-phase steels is mainly influenced by the local formability and the forming capacity of the cutting edges. Until today it is difficult to make an accurate failure prediction by FEM for the cutting edges [14]. In the described studies, the most important influencing variables on the complex failure on cutting edges and a pragmatic approach for estimating the forming potential were shown. New optimized material concepts can help to design the forming processes more robustly. Here, it is necessary to weight between global and local forming potential. Press trials using a demonstrator tool representing a A-pillar reinforcement [6] show that the named product advantages are directly applicable. Tests on the demonstrator part show that the failure level in this individual case is assessed too conservatively and that there is still a need for further research about edge failure criterium.

References

[1] https://www.thyssenkrupp-steel.com/en/products/sheet-coated-products/multiphase-steel/dual-phase-steel/

[2] N.N. VDA 239-100; Flacherzeugnisse aus Stahl zur Kaltumformung / Sheet Steel for Cold Forming (Version 06/2016)

[3] N.N. DIN EN ISO 6892-1; Zugversuch, Prüfverfahren bei Raumtemperatur für metallische Werkstoffe

[4] B.M. Hance; Advanced High Strength Steel: Deciphering Local and Global Formability; International Automotive Body Congress 2016

[5] Iizuka, E.; Urabe, M.; Yamasaki, Y.; Effect of strain gradient on stretch flange deformation limit of steel sheets – Prediction method of cut edge failure; Umformen im Karosseriebau, Bad Nauheim, Deutschland 2015

[6] Beier, T; Linnepe, M; Woestmann, S; Sikora, S; Forming Potential of Current Dual-Phase Steel Developments with Strengths of 800 MPa; 7th International Conference on Accuracy in Forming Technology ICAFT, Chemnitz, Germany, 2–3 November 2022

[7] N.N. DIN EN ISO 12004; Metallic materials – Sheet and strip – Determination of forming-limit curves – Part 1: Measurement and application of forming-limit diagrams in the press shop (ISO 12004-1:2008)

[8] Beier, T; Gerlach, J; Roettger R; Kuhn R; Possibilities for specific utilization of material properties for an optimal part design; IDDRG 2017; Munich; Germany; 2017

[9] B.M. Hance; T.M. Link, Effects of Fracture Area Measurement Method and Tension Test Specimen Type on Fracture Strain Values of 980 Class AHSS; IDDRG 2019, IOP Conf. Series: Materials Science and Engineering 651 (2019) 012061

[10] Iizuka, E.; Hashimoto, K.; Kuwabara, T.; Ishiwatari, A.; Inazumi, T.; Effects of anisotropic yield functions on the accuracy of forming simulations of hole expansion; IDDRG 2009, Golden, CO, USA, 2009

[11] N.N. 2017-09 International Standard ISO 16630 Metallic Materials – Sheet and Strip – Hole Expanding Test

[12] T. Beier, S. Woestmann; Berücksichtigung von schergeschnittenen Blechkanten zur Auslegung von Formgebungsprozessen höherfester Stahlwerkstoffe in der FEM-Umformsimulation; 14. LS-DYNA Forum, Bamberg, Deutschland 2016

[13] E. Atzema, P. Seda; Effect of zinc coating and time on edge ductility; 5th International Conference on Steels in Cars and Trucks; Amsterdam, Netherlands 2017

[14] Braun, M. et al.; Umformsimulation unter Einbeziehung einer verminderten Umformbarkeit schergeschnittener Stahlblechkanten – Wirksamkeit von 3 verschiedenen Kantenriss-Kennwerten und 3 Software-Systemen im Realversuch.; Umformen im Karosseriebau, Bad Nauheim, Deutschland 2016

Numerical and experimental investigation of hole flangeability of AA6061 alloy

V Kumar[1*], D Ravi Kumar[1]

[1]Department of Mechanical Engineering, Indian Institute of Technology Delhi, New Delhi, India-110016

E-mail: vinay.kumar@mech.iitd.ac.in

Abstract. Hole flangeability is a measure of edge formability or local formability, and it is important for forming processes like stretch flanging. Hole flangeability of sheet metals is evaluated by determination of hole expansion ratio (HER) using a standard hole expansion test (HET). In the present work, numerical and experimental investigation of hole flangeability has been carried out on 1.6mm thick AA6061 alloy sheets. The material model plays an important role in the accuracy of predicted results in numerical simulations. The simulations have been carried out using two different work hardening laws (Swift and Voce). The failure has been predicted using the experimentally determined forming limit curve (FLC) of AA6061 alloy sheets. HER and thickness have been predicted along the edge of the expanded hole by simulation. The predicted HER and thinning from FE simulation have been validated experimentally by performing hole expansion tests according to ISO 16630-2009 using a conical punch. The predicted results have been found to be in good agreement with the experimental results.

1. Introduction

The application of aluminum alloys is growing in the automotive industry to meet the increasing requirement for lightweighting. High strength to weight ratio offered by aluminum alloys combined with good formability makes them suitable for manufacturing complex sheet metal parts for light weight vehicles. Aluminum alloys of 5xxx and 6xxx series are particularly more important for applications in automotive and aviation industries for which formability of sheet metals is critical.

In the assessment of formability of sheet metals, stretch flangeability or hole flangeability is an important aspect. It is the measure of local formability at the edge of the hole in a stretch flanging operation. The hole flangeability of sheet metals is normally quantified by hole expansion ratio (HER). The hole expansion ratio is the increase in hole diameter to the original hole size when the edge cracks appear [1] and is regarded as a relative "local formability" parameter with application specific to edge-cracking analysis and prediction [2]. The hole expansion test (shown in Fig. 1) is designed specifically for characterizing the deformation of sheets in the flange. The hole expansion ratio $\boldsymbol{\lambda}$ is usually expressed in percentage as given by Equation 1:

$$\lambda(\%) = \left[\frac{D_h - D_0}{D_0}\right] * 100 \quad (1)$$

where D_0 is the initial hole diameter and D_h is the average hole diameter when the crack appears.

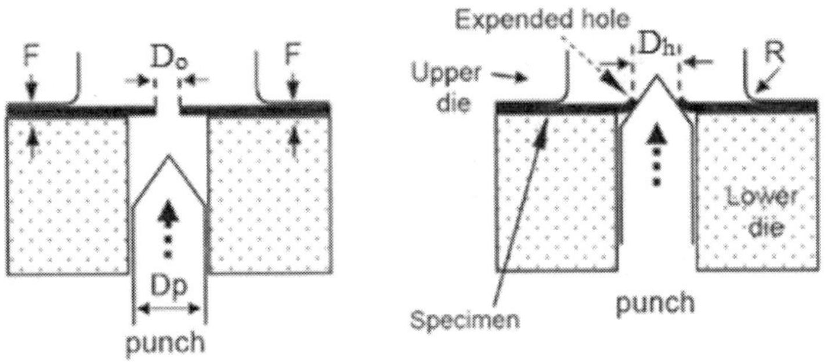

Figure 1. Schematic diagram of hole expansion test

The prediction of hole flangeability of sheet metals is always a concern, especially in steels [2] and aluminum alloys [3] due to edge cracking or edge fracture of sheet metals in hole expansion. In the past, a few attempts have been made to study the stretch flangeability of aluminum alloys. The thinning behavior in the hole expansion test of AA6016-O has been predicted by FE simulation using three different yield criteria [4]. In hole expansion tests, hole flangeability of Al alloys depends on various factors such as punch shape, hole preparation technique, burr position and sheet thickness [2,3,5,6]. In simulation of hole expansion forming processes of Al alloys, it is essential to use suitable material models for accurately predicting the behavior of the material, formability, thinning and load. This is crucial for optimizing the process and minimizing defects in the parts for automotive industry [7]. In the material model, the hardening laws are important for defining the material behaviour during plastic deformation and forming limit curve for prediction of failure [8].

In this work, the hole flangeability of AA6061 alloy sheets is predicted numerically by FE simulation using Dynaform software with LS-Dyna solver. Two different work hardening laws were used to define the material model. The forming limit curve of AA6061 alloy sheets was determined experimentally and it has been used to predict the failure in the simulation. The predicted HER and thinning from FE simulations have been validated by conducting the hole expansion tests.

2. Alloy composition and properties

6xxx series aluminium alloys have various applications mainly in the automotive and aircraft industries due to their superior mechanical properties, corrosion resistance and machinability. AA6061-O Al alloy sheets of thickness 1.6 mm have been used in the present study. The chemical composition of the alloy (in weight%) is given in Table 1.

Table 1. Chemical composition of the AA6061 aluminium alloy (by weight %)

Grade	Cu	Mg	Si	Fe	Mn	Zn	Cr	Al
6061-O	0.20	0.92	0.81	0.54	0.11	0.20	0.19	remaining

The mechanical properties of the alloy were determined using uniaxial tensile test as per ASTM E8M specification and the properties are given in Table 2. From the experimental data of engineering stress and engineering strain, the log true stress - log true strain values were plotted from which the n and K values were determined. To characterize anisotropy, the plastic strain ratio values (R_0, R_{45} and R_{90}) were determined as per ASTM E517 specification. The values along with average plastic strain ratio $\bar{R}$ are also reported in Table 2.

Table 2. Mechanical properties of AA 6061-O alloy sheets

Grade	Thickness	Orientation w.r.t RD	YS (MPa)	UTS (MPa)	n	K (MPa)	R	$\bar{R}$
		0 deg	60.1	124.1	0.272	245.1	0.598	
6061-O	1.6 mm	45 deg	59.5	122.8	0.263	236.2	0.658	0.642
		90 deg	60.0	124.3	0.267	244.6	0.656	

3. Methodology

3.1 Prediction of HER by simulation

3.1.1 Modelling of blank and tools. The simulation of hole expansion test was conducted using Dynaform software with LS-Dyna solver. The tools (die, blank holder, and conical punch) and the blank were modelled in Solidworks software as shown in Fig. 2. The blank modelled is of dimension 100 mm X 100 mm with a circular hole of 10 mm diameter in the center of blank. The die and blank holder with appropriate clearance have been used. The blank was meshed with Belytschko-Tsay shell elements of size 2.0 mm with five integration points. Refining mesh option with a maximum refinement level of 5 was used for better accuracy in the predicted results. A blank holding force of 40 kN was used to hold the blank between the dies and a constant punch velocity of 200 mm/min was applied. The contact between the tool surface and the blank was defined using the 'forming one way surface to surface' option. A constant friction coefficient value of 0.125 was defined at the tool-blank interface.

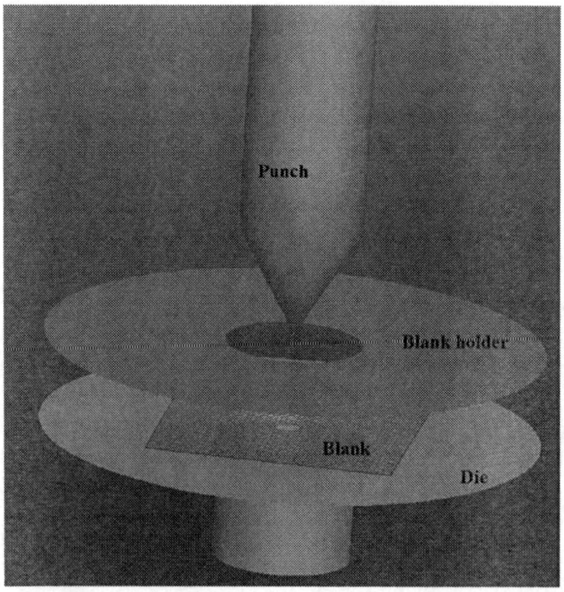

Figure 2. Model of tools and blank for hole expansion test simulation

3.1.2 Material model. In the material model, 3-paramter_Barlat 89 yield criterion was used to model the yielding behaviour of the alloy by taking the yield exponent to be 8 as suggested for FCC materials [9]. Two different hardening laws (Swift and Voce) were used to model the hardening behaviour of the alloy during plastic deformation. Both the hardening laws are empirical relationships that describe dependence of flow stress on plastic strain.
Swift law of hardening is given by

$$\sigma = K\ (\varepsilon_0 + \varepsilon_p)^n \tag{2}$$

where σ is true stress, ε_0 is initial elastic strain, ε_p is plastic strain, K is strength coefficient, and n is strain hardening exponent.

The Voce law of hardening is given by

$$\sigma = A\text{-}\ B\ e^{-C\varepsilon_p} \tag{3}$$

where A, B and C are material constants and ε_p is plastic strain.
The material constants in the Voce law of hardening for the AA6061-O alloy are taken from Kacem et al. [10]. These are corrected by a linear piecewise relation after necking as shown in Table 3.

Table 3. Material parameters of Voce hardening law for the A6061-O alloy

Before necking ($0 < \varepsilon_p < 0.095$)			After necking ($\varepsilon_p > 0.095$)						
A (MPa)	B (MPa)	C	ε_p	0.1	0.2	0.3	0.4	0.5	0.6
154	106	25	σ(MPa)	146	166	180	192	202	210

The flow curves predicted by Swift law and Voce law are shown in Fig. 3. It can be observed that flow stress predicted by Swift law is lower than that of Voce law at smaller plastic strains, but it is higher at larger strains (> 0.6).

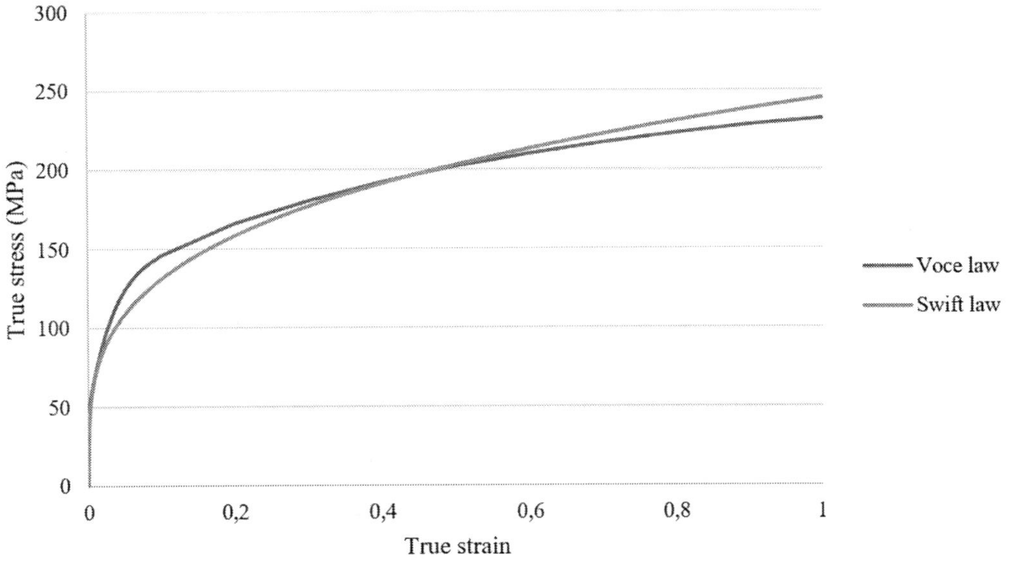

Figure 3. Comparison of flow curves predicted by the hardening law

3.2 FLC determination for failure prediction

Forming limit curve is a convenient and useful tool for design and manufacturing analysis of sheet metal products and can be used to evaluate the formability of sheet metals under various strain paths. The forming limit curve of the alloy was experimentally determined by means of the Nakazima method[11]. The blanks were laser marked with a circular grid of 5mm diameter. After the testing the samples (Fig. 4) under different strain paths up to necking/failure, the major and minor strains were plotted. Some of the tested samples are shown in Fig. 4. The experimentally determined FLC of this alloy is shown in Fig. 5. The plane strain intercept value (FLD_0) is 27.1%.

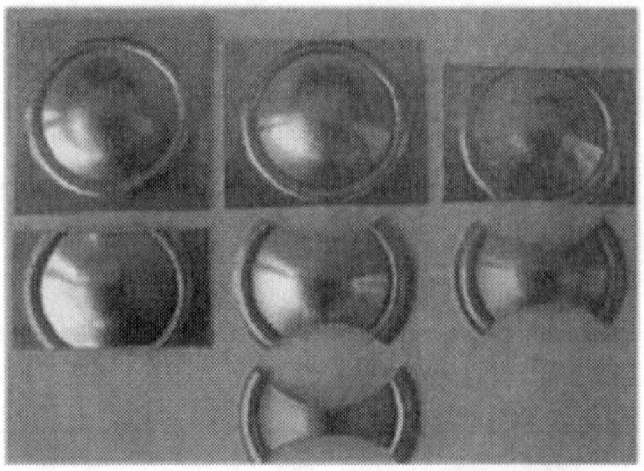

Figure 4. FLC test specimens after deformation

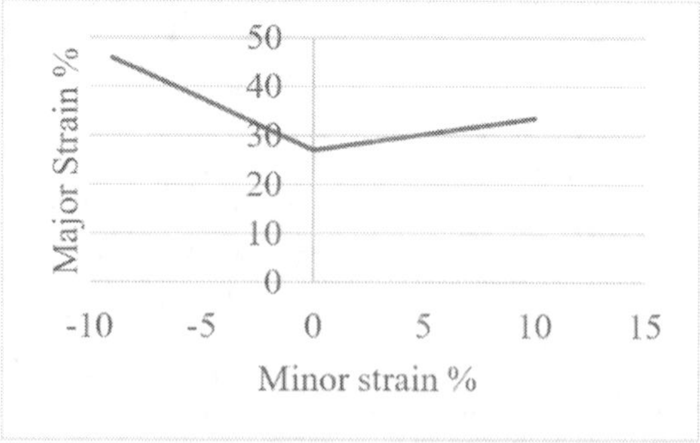

Figure 5. Experimentally determined forming limit curve of AA6061-O sheets

3.3 Experimental validation by hole expansion test

The hole expansion tests were performed as per ISO standard 16630: 2017 [1]. Sheet samples of dimensions 100 mm X 100 mm with a circular hole of 10 mm diameter in the center of the sample were used. The holes were made by punching on a hydro-pneumatic punching press of 15-ton capacity. The initial hole size D_0 was measured in two directions, perpendicular to each other. A burr is normally visible on the bottom side of the sample after punching the hole. Though the burrs were removed by polishing the surface, the samples were used with burr surface at the top during the hole expansion test [6]. The hole expansion tests were performed on a 100-ton capacity double action hydraulic press using a 50mm conical punch of 60° angle as shown in Fig. 6. The experiments were carried out with a blank holding force of 40 kN and the punch speed was kept constant at 0.5 mm/sec. The test was stopped immediately when a crack at the edge appeared which was detected with the help of a camera placed above the die setup. A tested sample is in Fig. 7. The final hole size D_h was measured both physically by a vernier caliper and by using an image analyser tool on the images acquired from the test. The average final thickness was determined by measuring thickness at different locations along the edge of the expanded hole.

Figure 6. Conical punch used in HET

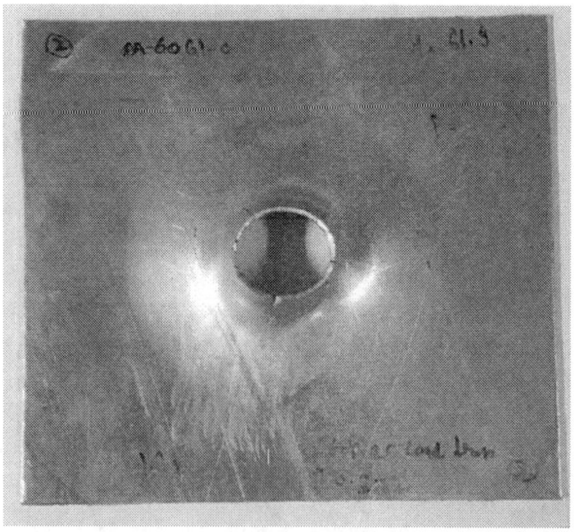

Figure 7. Hole expansion test sample after testing

4. Results and discussion

The predicted HER, thinning and peak load from simulation are compared with experimental results and discussed in this section.

4.1 HER

The prediction of failure in HET simulation using the experimentally determined FLC is shown in Fig. 8(a) and Fig. 8(b) using Swift and Voce laws, respectively. The figures clearly indicate the onset of failure in tension-compression region on the left-hand side of the FLC. The state of stress at the failure location on the hole edge is almost uniaxial tension. The predicted values of HER with Swift hardening law and Voce hardening law are 55.1% and 48.8%, respectively. The experimental value of HER is 67.2%. The discrepancy could be due to the additional punch displacement in experiments until a macroscopic crack was detected for analysis by the image analyzer tool. It could have lead to larger HER. Two or three small cracks were observed in the experiments when the test was stopped, and the average diameter was taken for determination of HER.

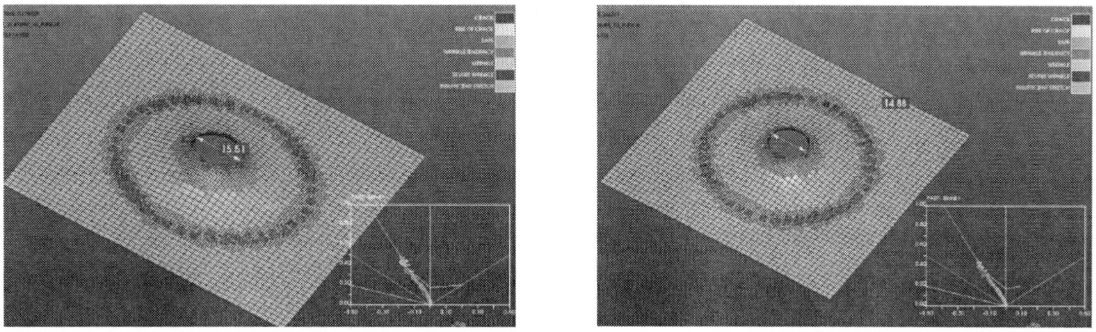

(a) (b)

Figure 8. Prediction of hole expansion ratio in HET simulation using (a) Swift law and (b) Voce law

4.2 Thinning prediction

The predicted thinning variation in the blank in simulation at the onset of crack is shown in Fig. 9 (a) and 9 (b) for both the hardening laws. The maximum thinning has been found to be 26.55% and 26.036% for Swift law and Voce law, respectively. The predictions using both the laws agreed well with experimentally measured maximum thinning which is 24.8%. The predicted plastic strain at the location of failure has been found to be slightly with Swift law (0.69) when compared to Voce law (0.66). It can be observed from Fig. 3, flow stress predicted by Swift is marginally higher than Voce at failure strains. Due to this, higher thinning is predicted in the simulation when Swift law is used.

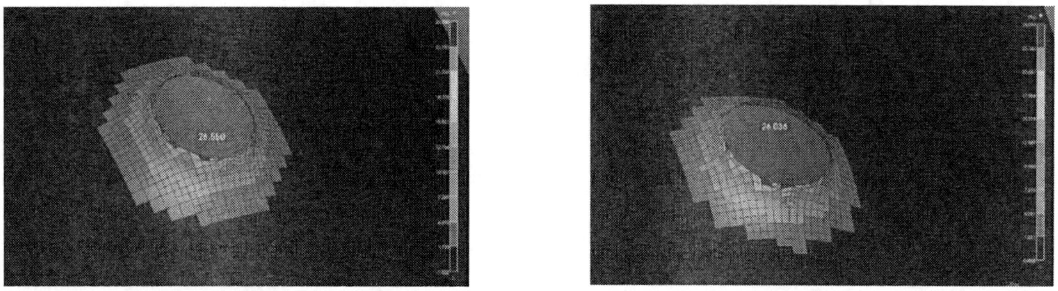

(a) (b)

Figure 9. Prediction of thinning variation in HET simulation using (a) Swift law and (b) Voce law

4.3 Comparison of load

The peak load predicted in simulation using both Swift hardening law and Voce hardening law matched well with the experimental peak load with very small discrepancy. This shows the suitability of both the hardening laws for load prediction in hole expansion test simulation.

The predicted and experimental values of HER, maximum thinning and peak load are summarized in Table 4.

Table 4. Comparison of predicted and experimental results of HET

Parameter	Simulation		Experiment
	Swift law	Voce law	
% HER	54.2	48.4	67.2
Max % thinning	26.6	26.0	24.8
Load (in kN)	10.32	10.30	11.28

5. Conclusions

In this work, two different hardening laws (using Swift and Voce) were used for the numerical simulation of HET of AA6061-O alloy. The following are the main findings of this investigation:

1. The predicted HER using Swift law has been found to be in better agreement with the experimental value than Voce hardening law.
2. Prediction of thinning and peak load in HET using both the laws agreed well with experimentally measured maximum thinning and peak load indicating suitability of both the laws for load and thinning prediction in the hole expansion test simulation of this alloy.
3. The predicted plastic strain at the location of failure in HET has been found to be slightly higher with Swift law and also, the flow stress predicted by Swift law is higher at the failure strain in HET.

Acknowledgement

The author would like to acknowledge the Ministry of Human Resource Development (MHRD), Government of India for receiving the assistantship for this research work.

References

[1] ISO 16630:2017 Metallic materials-sheet and strip-hole expanding test
[2] Hance B M 2017 *SAE Int J Engines* **10** pp 247–57
[3] Stanton M, Bhattacharya R, Dargue I, Aylmore R and Williams G 2011 *AIP Conference* Proceeding 1353 pp 1488-93
[4] Akşen T A and Firat M 2021 *SN Appl Sci* **3** p 320
[5] Li J, Wang H and Yan D-W 2016 *Int conference AMTIA*
[6] Paul S K 2020 *Materialia (Oxf)* **9**
[7] Kuwabara T, Mori T, Asano M, Hakoyama T and Barlat F 2017 *Int J Plast* **93** pp 164–86
[8] Sing W M and Rao K P 1997 *Journal of Materials Processing Technology,* **63(1-3***)* pp 105–10
[9] Barlat F and Lian K 1989 *Int J Plast,* **5(1***)* pp 51-66
[10] Kacem A, Krichen A, Manach P Y, Thuillier S and Yoon J W 2013 *Eng Fract Mech* **99** pp 251–65
[11] Nakazima K, Kikuma T and Hasuka K 1968 *Yawata Tech Rep* **264** pp 8517-30

IOP Conf. Series: Materials Science and Engineering 1284 (2023) 012019 doi:10.1088/1757-899X/1284/1/012019

Study on the ultrasonic-assisted spinning of thin-walled cylindrical parts with longitudinal and transverse ribs

Shuangli Li[1], Yixi Zhao[1], Zhongqi Yu[1] and Shuhui Li[1]

[1] Shanghai Key Laboratory of Digital Manufacture for Thin-Walled Structures, Shanghai Jiao Tong University, Shanghai, 200240, China.

E-mail: yxzhao@sjtu.edu.cn

Abstract. Current studies have shown that the application of composite energy fields in the plastic forming process can help reduce the forming load and promote material flow. In this paper, a new method of ultrasonic-assisted spinning integral forming for thin-walled cylindrical parts with outer ribs was proposed, which could not only improve the manufacturing efficiency of ribbed cylindrical parts, but also further improve the forming height of outer ribs. A simulation model of ultrasonic-assisted spinning for thin-walled cylindrical parts with longitudinal and transverse outer ribs was established, the influence of ultrasonic amplitude on the forming accuracy of longitudinal and transverse outer ribs was obtained, and the mechanism of ultrasonic vibration improving the forming accuracy of ribs was revealed. On this basis, ultrasonic-assisted spinning equipment was built to verify the feasibility of this process. The results showed that with the increase of ultrasonic amplitude, the forming height of longitudinal and transverse outer ribs gradually increases, and the effect of raising the forming height of transverse ribs is higher than that of longitudinal ribs. In addition, as the ultrasonic amplitude increases, the uniformity of the forming morphology of the outer rib is not obviously improved. The spinning of the outer rib is mainly realized by the material in the form of integral filling into the groove. The ultrasonic vibration improves the material flow properties, and promotes the filling of the material into the rib groove, thereby increasing the spinning forming height of the outer ribs. The forming height of the outer rib was measured with or without ultrasonic vibration. When the rolling reduction is 1 mm, the average forming height of the outer ribs is increased by 50% compared with that without ultrasound, which provides a new research idea for the integral forming of the spinning of thin-walled cylindrical parts with longitudinal and transverse ribs.

1. Introduction

With the rapid development of space industry, carrier rockets are developing towards the trend of light weight and heavy load. As the main structure of carrier rocket, the thin-walled cylindrical parts with ribs plays the role of bearing mechanical load and providing installation support for other components. The structure of longitudinal and transverse ribs can make the cylinder in the axial and radial direction to bear greater pressure, which is the more common structure of the thin-walled cylindrical parts with ribs. However, this kind of component is geometrically characterized by "high ribs and thin walls", and its high-precision integral manufacturing faces technical challenges.

In recent years, spinning process has received certain attention in the integral manufacturing of thin-walled components with ribs, which can effectively improve the material utilization rate, reduce the forming defects of the components and improve the structural reliability [1]. However, due to the influence of adverse factors such as structural constraints and mandrel friction resistance, the flow of materials into the groove is limited, and the filling height of ribs is insufficient, which restricts the development of spinning process to a certain extent. The present study found that the auxiliary energy field on the plastic forming process can promote the flow of the material in the deformation zone and improve the forming accuracy of the ribs. The auxiliary energy fields can be divided into electricity [2], sound, heat[3] and light[4]. Among them, ultrasonic has the advantages of concentrated energy and strong propagation direction. As an auxiliary means, ultrasonic has been widely used in welding [5], upsetting [6] and surface rolling process [7]. The application of ultrasonic vibration in the process of metal forming can effectively reduce the yield stress and flow stress of materials [8], decrease the interface friction between mandrel and blank [9], and optimize the surface roughness of formed components [10].

Based on the above research, in order to improve the filling of thin-walled cylindrical parts with longitudinal and transverse ribs, an ultrasonic-assisted spinning process was proposed and its forming principle was studied. The finite element model of 2219-O aluminum alloy was established, and the influence of ultrasonic amplitude and rolling reduction on the forming height and uniformity of longitudinal and transverse ribs was studied. The experimental platform of ultrasonic-assisted spinning was established to verify the accuracy of finite element model and feasibility of the process.

2. Forming principle and modeling of ultrasonic-assisted spinning process

2.1. Forming principle of ultrasonic-assisted spinning process

The thin-walled cylindrical parts with longitudinal and transverse ribs were formed by internal spinning method. The side wall of the cylinder is the plastic forming area, and its forming principle is shown in Figure 1(a). The mandrel is connected with the spindle of the spinning machine to drive the thin-walled cylindrical part to rotate synchronously. The spinning roller first carries out circumferential feed to form a certain amount of rolling reduction. With the increase of circumferential feed, the friction between the thin-walled cylindrical part and the spinning roller gradually increases, so the spinning roller will be driven to rotate, as shown in Figure 1(b). The plastic flow of the material is realized through the axial feed of the rotary wheel. The purpose of the tail jacking is to limit the axial flow of the material and thus better promote the radial flow of the material into the rib groove, as shown in Figure 1(c). The effect of ultrasonic application is mainly reflected in the acoustic softening effect when ultrasonic vibration acts on materials, which is manifested as reducing the yield stress and flow stress of materials and effectively promoting the flow of materials. Ultrasonic device is usually used to transfer ultrasonic energy to the spinning roller, as shown in the Figure 1(a).

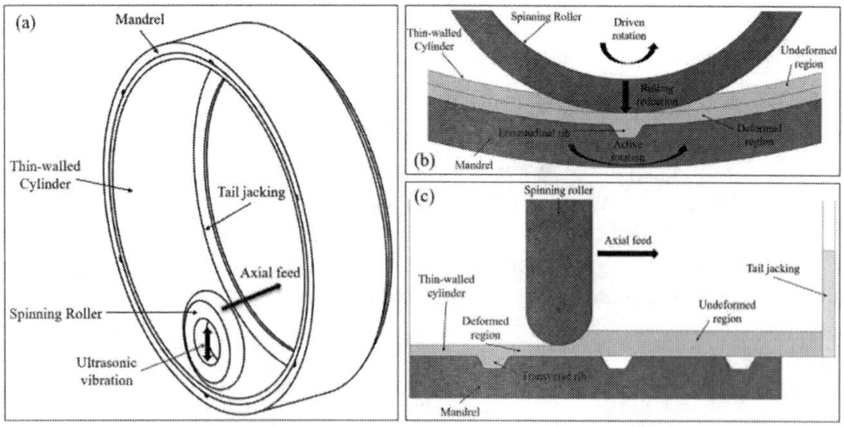

Figure 1. Forming principle of ultrasonic-assisted spinning process: (a) Forming process; (b) Forming longitudinal rib; (c) Forming transverse rib.

2.2. *Finite element modeling of ultrasonic-assisted spinning process*

In this study, the ABAQUS/Explicit software was used to simulate the ultrasound-assisted spinning of thin-walled cylindrical parts with longitudinal and transverse outer ribs. The finite element model, as shown in the Figure 2, is mainly composed of mandrel, spinning roller, thin-walled blank and constraint ring. The mandrel and spinning roller are set as discrete rigid bodies, and the thin-walled blank is set as variable bodies. The face to face contact is adopted between each part. The friction coefficient between different contact surfaces is set by penalty friction model to simulate the friction force, and the friction coefficient between spinning roller and workpiece and between mandrel and workpiece is set as 0.05 and 0.15, respectively. The restraint ring is bound with the thin-walled blank to the rotation of the blank driven by the spindle. The simulation process is mainly divided into three analysis steps: (I) the mandrel and blank rotate synchronously to the specified speed, (II) the spinning roller radial feed to the set distance, (III)the spinning roller axial feed at the set speed.

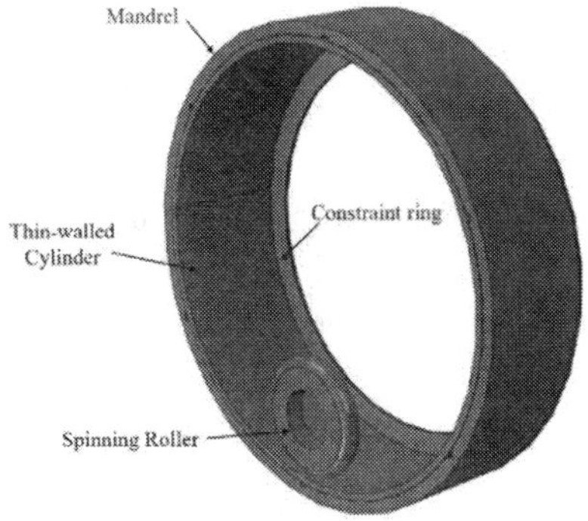

Figure 2. Finite element model of ultrasound-assisted spinning process.

The blank material used is 2219-O aluminum alloy. In the simulation process, the material is assumed to be isotropic, continuous and uniform, and the thermal effect, gravity and inertia between the spinning roller and the blank are ignored. The relevant mechanical property parameters of the blank are shown in the Table 1.

Table 1. Material properties of 2219-O aluminum alloy.

Parameters	Density (kg/m3)	Young's modulus (GPa)	Poisson's ratio
AA2219-O	2840	73.1	0.33

Stress-strain data of 2219-O aluminum alloy under different ultrasonic amplitudes (UTA) were obtained by ultrasound-assisted uniaxial tensile test, as shown in the Figure 3.

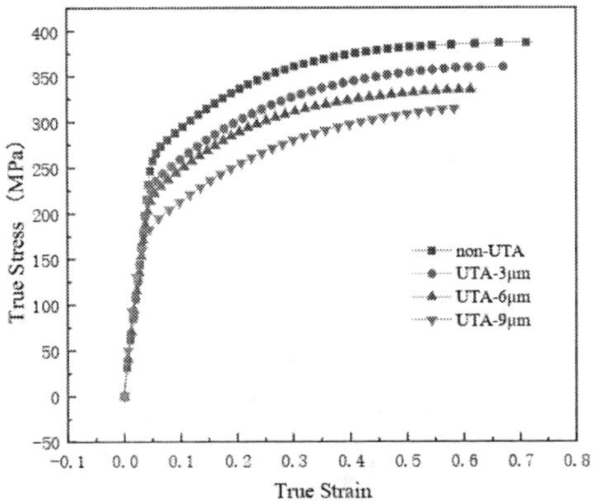

Figure 3. Uniaxial tensile true stress- true strain curves with different ultrasonic amplitudes.

Different stress-strain data are used to simulate the spinning process under different ultrasonic amplitudes, and the parameter settings are shown in the Table 2.

Table 2 Finite element simulation cases with parameter settings.

Simulation case	Rolling reduction (R/mm)	Ultrasonic amplitude (UTA/μm)
#1~#4	0.5	0, 3, 6, 9
#5~#8	1	0, 3, 6, 9
#9~#12	1.5	0, 3, 6, 9

3. Ultrasonic-assisted spinning experimental platform

3.1. Structural design of mandrel

The mandrel with rib groove mainly adopts trapezoidal slider structure, its core structure includes fixed slider and forming slider, other structures include end plate, slider base, restraint plate and regulating plate, as shown in Figure 4. The function of the fixed slider is to connect the slider base and the forming slider, the forming slider is connected with the fixed slider, and its surface is processed with longitudinal and transverse rib grooves. Each forming slider corresponds to a fixed slider, and 12 forming sliders are combined to form a structure with six longitudinal grooves and three transverse grooves. The end plate plays the role of connecting the spindle of the spinning machine and the slider base. The restraint plate has the function of restricting the axial movement of the blank. The regulating plate can adjust the tightness between the forming sliders. When assembling the blank, the gap between the forming sliders can be increased by adjusting the regulating plate, which is convenient for the assembly of blank. Before forming, the regulating plate is used to make the forming sliders contact closely, which ensures the forming accuracy of the component, and the interference fit between the slider and the blank ensures the synchronous rotation of the spindle and blank. After forming the thin-walled cylindrical parts with longitudinal and transverse ribs, the gap between the forming sliders can be increased by adjusting the regulating plate for the convenience of demoulding.

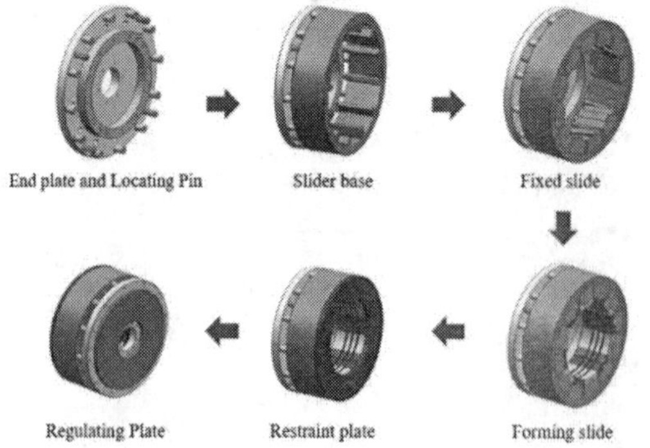

Figure 4. Mandrel structure and assembly process.

3.2. Ultrasonic vibration device

Ultrasonic vibration device needs to be designed according to the introduction of ultrasonic energy. At present, the application of ultrasonic energy can be divided into two types: introduction from the spinning roller and direct transfer of ultrasonic energy to the blank. In this study, the ultrasonic vibration device is directly contacted with the blank. Ultrasonic vibration device mainly includes ultrasonic generator, ultrasonic transducer, ultrasonic horn and ultrasonic tool head, as shown in Figure 5. The ultrasonic vibration frequency is 20kHz, and the ultrasonic amplitude can be changed by adjusting the power of ultrasonic generator.

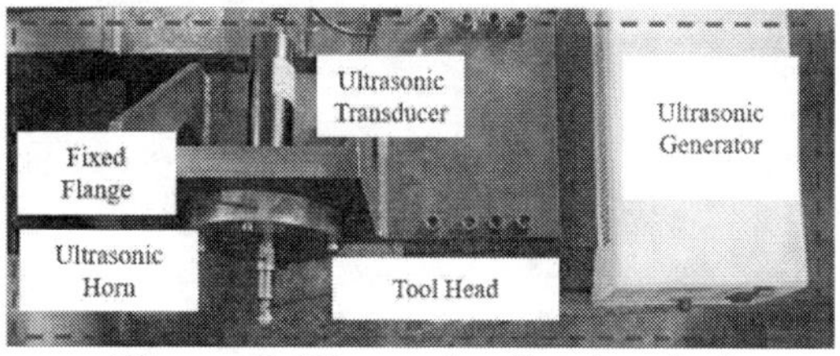

Figure 5. Ultrasonic vibration device.

3.3. Ultrasonic-assisted spinning platform

The mandrel is assembled on the Okay800 CNC spinning machine, as shown in Figure 6(a). The ultrasonic vibration device is assembled in conjunction with the spinning roller, and its structure schematic diagram is shown in Figure 6(b). The trajectory of spinning is programmed using Siemens numerical control system.

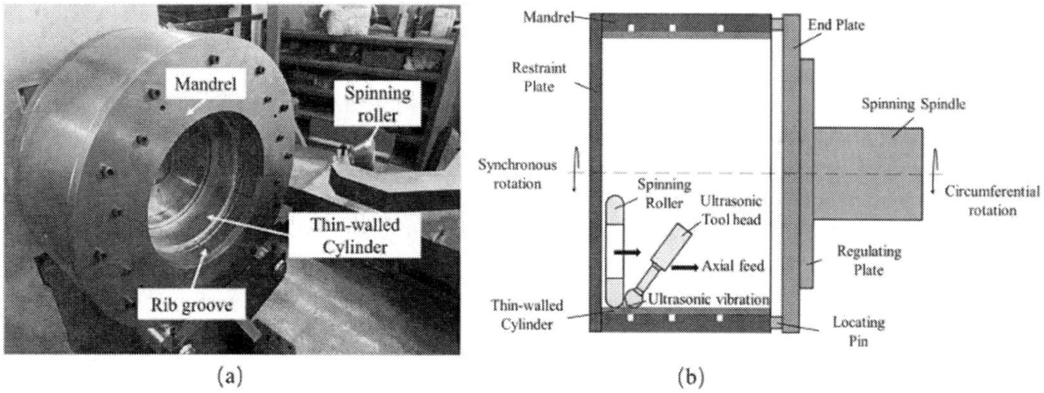

Figure 6. Ultrasonic-assisted spinning platform: (a) Spinning machine with mandrel; (b) Structure schematic diagram.

The spinning roller is first fed radial, then fed axial to realize the forming of the longitudinal and transverse outer ribs. The ultrasonic device and the spinning roller maintain the same axial feed speed. When the spinning roller is fed axially, the ultrasonic energy can soften the material, promote the material flow and improve the forming accuracy of the ribs. However, due to the performance limitation of the spinning machine, the maximum rolling reduction can only reach 1mm at present.

4. Result and Discussion

4.1. Effect of ultrasonic amplitude on spinning force

Figure 7 shows the variation of simulated spinning force with the rolling reduction at different ultrasonic amplitudes. It can be observed that with the increase of ultrasonic amplitude, the spinning force presents a trend of gradual decrease. And with the increase of the rolling reduction, this decreasing trend is more significant.

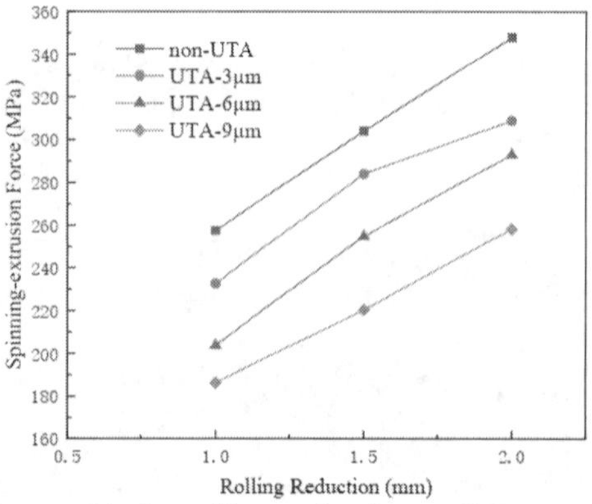

Figure 7. Spinning force with the rolling reduction at different ultrasonic amplitudes.

This is because the acoustic softening effect of ultrasonic vibration in the plastic forming process will reduce the spinning force, and with the increase of ultrasonic amplitude, the effect of acoustic softening will become more obvious, so the spinning force decreases gradually with the increase of ultrasonic amplitude.

4.2. Stress-strain state

Figure 8 shows the plastic stress-plastic strain distribution of thin-walled cylinder with outer ribs varying with the rolling reduction at ultrasonic amplitude 9μm. It can be seen from Figure 8(a) that in the initial forming stage, the material flows circumferentially and fills into the longitudinal rib groove under the action of the spinning roller, the formed longitudinal ribs show a large stress distribution. And with the increase of the rolling reduction, the longitudinal rib will present a larger stress state. After the formation of the transverse rib, the stress of the longitudinal rib is relatively smaller, and the stress is mainly concentrated in the root of the longitudinal rib. Aiming at the stress distribution of the transverse rib, the stress at the root of the transverse rib is quite different from that at the top, and with the increase of rolling reduction, the difference will be gradually obvious. In the spinning process, due to the existence of intersection position, the formation of longitudinal rib is regarded as discontinuous, while the formation of transverse rib is regarded as continuous under the action of axial feed of spinning roller. The continuous formation of transverse rib makes its stress distribution more regular.

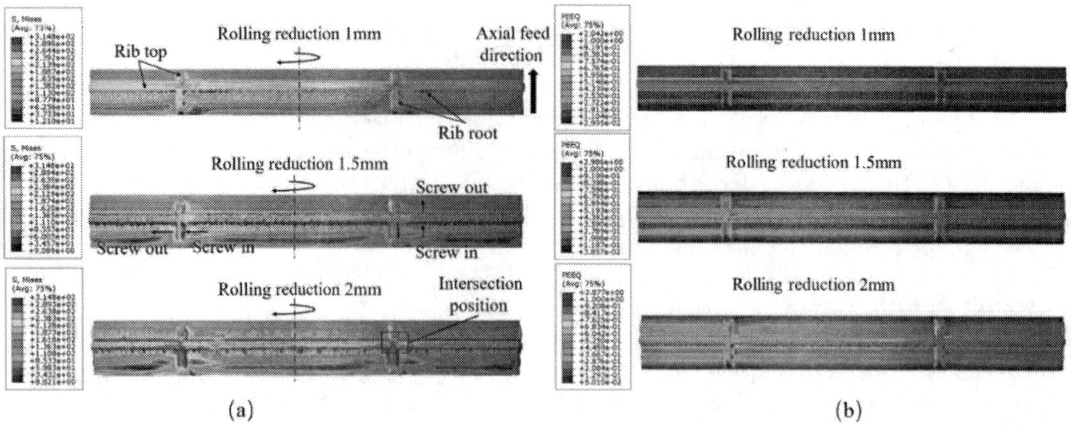

Figure 8. Plastic stress-plastic strain distribution of thin-walled cylinder with outer ribs varying with the rolling reduction when ultrasonic amplitude is 9μm: (a) Stress distribution; (b) Strain distribution.

Based on the analysis of the strain distribution of the transverse rib, as shown in Figure 8(b), the strain at the root of the transverse rib is large and evenly distributed, while the strain at the top of the rib is relatively small. The strain at the root of the longitudinal rib is smaller than that at the root of the transverse rib, but the overall strain distribution is the same as that of the transverse rib. This indicates that the deformation of the material is large only at the position of the screw-in end and screw-out end of the rib groove, while which is small at the position of the rib groove. The materials come into the groove mainly in the form of integral filling, and a small part of the material through the form of flow.

4.3. Forming height of outer ribs

It can be seen from Figure 9 that with the increase of the rolling reduction, the forming height of the outer ribs gradually increases. In addition, under the same rolling reduction, the forming height of outer ribs increases with the increase of ultrasonic amplitude.

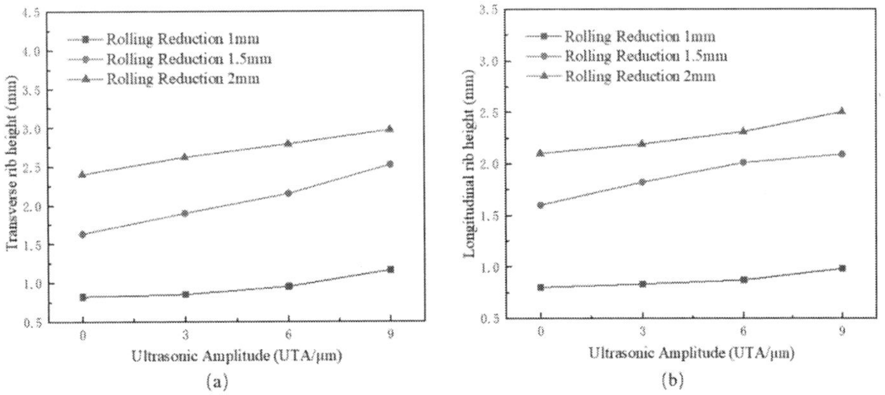

Figure 9. Forming height of the outer ribs with the change of the rolling reduction at different ultrasonic amplitudes: (a) Transverse rib height; (b) Longitudinal rib height.

When the rolling reduction is small(1mm), the axial feed of the spinning roller causes less material flow, resulting in less material filling at the rib groove, so the rib height is less than the rolling reduction. Ultrasonic vibration will promote the material flow and improve the filling of materials. With the increase of ultrasonic amplitude, the filling amount of materials will increase, making the rib height gradually improve.

With the increase of rolling reduction, the material flow accumulation driven by the spinning roller increases rapidly, making the overall filling amount of material in the rib groove increases significantly. Therefore, when the rolling reduction is 1.5mm and 2mm respectively, the rib height will be slightly greater than the rolling reduction. On this basis, under the action of ultrasonic softening effect, the material accumulation driven by the spinning roller will further increase, promoting more material filling in the rib groove. Thus, the formed rib height is higher than that without ultrasound, as shown in Figure9.

Under the same rolling reduction, the forming height of transverse rib is greater than that of longitudinal rib under different ultrasonic amplitudes. According to the stress-strain distribution of the outer ribs in 4.2, it can be seen that due to the large deformation of the transverse rib, more materials are filled in the transverse rib groove. In addition, as shown in Figure 1(c), when the spinning roller is under a certain amount of rolling reduction, the area of contact between the side wall of the spinning roller and the material is larger, so more materials will be driven to flow axially in the axial feed direction. Since the formation of the transverse rib is caused by the material flowing axially to the transverse rib groove for radial filling, more axial flow of material means more radial filling of the transverse groove. In contrast, the circumferential flow of the material caused by the spinning roller is less, resulting in less radial filling of the material in the longitudinal groove. Thus, compared with the longitudinal rib, the forming height of the transverse rib is higher.

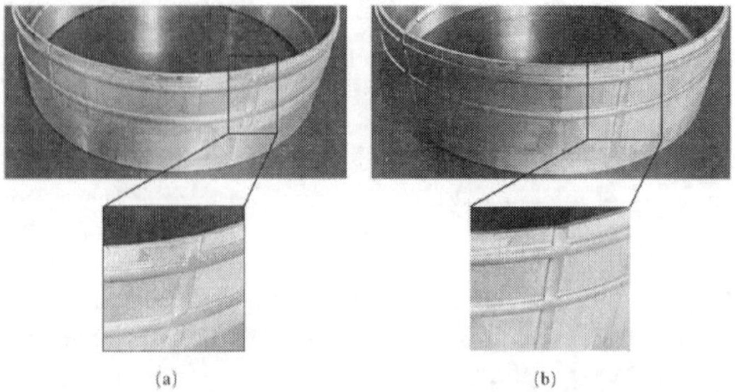

Figure 10. Forming height of the outer ribs formed by spinning test: (a) without ultrasound; (b) with ultrasonic amplitude of 9um.

The forming height of the outer ribs formed by spinning test was measured, as shown in Figure 10. When the rolling reduction was 1mm, the forming height of transverse ribs without ultrasound and with ultrasonic amplitude of 9um were 0.76mm and 1.12mm, and the forming height of longitudinal ribs were 0.70mm and 1.04mm, respectively. The forming height of the outer ribs obtained by simulation, as shown in Figure9, is in good agreement with the experimental results, which verifies the correctness of the simulation results. This also indicates that the introduction of ultrasound can effectively improve the forming height of the outer ribs.

4.4. Forming uniformity of outer ribs

The forming uniformity analysis of the outer rib mainly includes the forming morphology at the transverse rib, the longitudinal rib and the intersection of the two kinds of ribs. Figure 11 shows the forming morphology of the outer ribs under different rolling reduction when ultrasonic amplitude is 9μm.

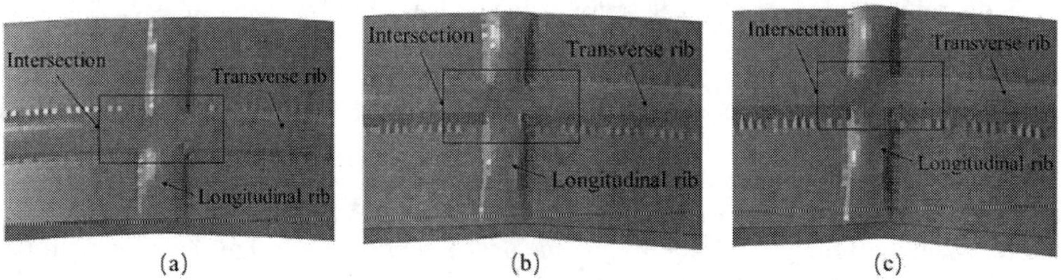

Figure 11. Forming morphology of outer ribs: (a) Rolling reduction 1mm; (b) Rolling reduction 1.5mm; (c) Rolling reduction 2mm.

It can be found that the forming morphology of transverse rib always maintains a uniform state, while that of the longitudinal rib shows a trend of increasing inhomogeneity with the increase of rolling reduction. In addition, the forming morphology at the intersection position shows a trend of nonuniformity. Based on the analysis in Section 4.3, axial flow of materials is easier than circumferential flow. In the forming process of longitudinal rib, the existence of transverse groove has a guiding effect on the axial flow of material. The closer the position is to the transverse groove, the more significant this effect is, resulting in less circumferential flow of the material. Thus, radial filling of longitudinal rib groove near transvers rib grooves is less, which leads to the uneven morphology of longitudinal rib. In the circumferential annular groove, due to the axial flow of more materials, the radial filling amount in the transverse groove is larger, and the influence of the longitudinal groove is small, which can be regarded as continuous filling. So the forming morphology of the transverse rib is more uniform. At the intersection position, the axial flowing material should be filled radially in the transverse rib groove, but the longitudinal rib groove that has not been filled plays the role of axial guidance at this time, so that part of the material that should be filled radially changes into axial flow, which leads to the insufficient filling of the material.

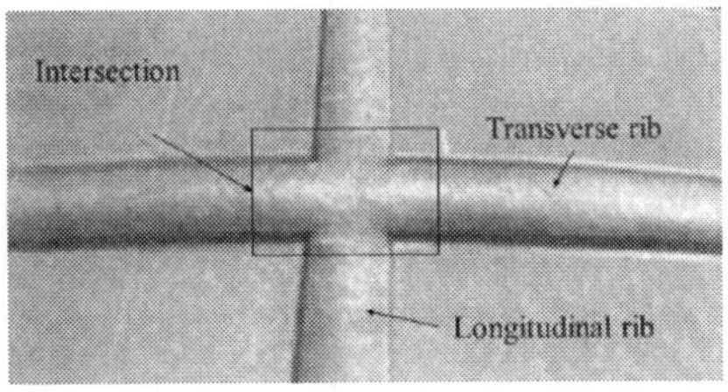

Figure 12. Forming morphology of the outer ribs formed by spinning test.

The forming morphology of the outer ribs formed by spinning test with ultrasonic amplitude of 9μm was observed, as shown in Figure 12. The results showed that the uniformity distribution of the outer ribs obtained by the experiment is in good agreement with the simulation results, which further verifies the correctness of the simulation analysis. Meanwhile, it can be concluded that the forming uniformity of the outer ribs has not been significantly improved under the action of ultrasound.

5. Conclusion

In order to solve the problem that the filling height of outer ribs is not ideal in the spinning process of thin-walled cylinder with outer ribs, an ultrasonic-assisted spinning process is proposed in this paper. Based on ABAQUS/Explicit software, the principle and forming characteristics of ultrasonic-assisted spinning process of 2219-O aluminum alloy thin-walled cylinder with outer ribs has been systematically studied. The experimental platform of ultrasonic-assisted spinning was established to verify the accuracy of finite element model and feasibility of the process. The conclusions are as follows:

(1) Ultrasonic vibration can effectively reduce the spinning force in the spinning process, and with the increase of ultrasonic amplitude, the reduction of spinning force becomes more obvious.

(2) The strain at the root of the outer ribs is large, while the strain at the top of the outer ribs is relatively small, which means that the material is mainly in the form of integral filling into the groove, and only a small part of the material through the form of flow.

(3) The forming height and uniformity of the transverse rib are better than those of the longitudinal rib, which is closely related to the axial and circumferential flow of material.

(4) The introduction of ultrasound can significantly improve the forming height of the outer ribs, which has a good application prospect in the formation of the ribbed components.

Acknowledgment

The authors much acknowledge the financial support from the National Key Research and Development Program of China (No. 2020YFA0711102). The authors also thank the reviewers of IDDRG for the suggestive comments which improve the technical content of this paper.

References

[1] M.S. Domack and J.A. Wagner 2015 *Innovative manufacturing of cylinders with integral stiffeners* (NASA Langley Research Center Report)
[2] Zhao Y X, Wan X M and Gao L T 2018 *J. Manuf. Sci. Eng.* **140** 101012
[3] Liang W, Guan L, Lv Q Y and Xing Z G 2019 *Adv. Mater. Sci. Eng.* **2019** 8930134
[4] Klocke F and Brummer C M 2014 *Procedia Eng.* **81** 2385-2390
[5] Zhong Y B, Wu C S and Padhy G K 2017 *J. Mater. Process. Technol.* **239** 273-283
[6] Liu Y, Suslov S and Han Q 2012 *Mater. Lett.* **67** 52-55
[7] Liu Y, Zhao X H and Wang D P 2014 *Mater. Sci. Eng., A* **600** 21-31
[8] Yang C, Shan X and Xie T 2016 *Int. J. Adv. Manuf. Tech.* **83** 645-655
[9] Dawson G R, Winsper C E and Sansome D H 1970 *Met. Form.* **37** 254–261
[10] Vahdati M, Mahdavinejad R and Amini S 2015 *Proc. Inst. Mech. Eng., Part B* **231** 1-12

An evaluation method for experimental necking detection of automotive sheet metals

Quoc Tuan Pham[1], Md Shafiqul Islam[1], Alexander Barlo[1], Mats Sigvant[1,2]

[1] Department of Mechanical Engineering, Blekinge Institute of Technology, Karlskrona, Sweden

[2] Volvo Cars, Department 81110 Strategy Development, Olofström, Sweden

E-mail: quoc.tuan.pham@bth.se

Abstract. In sheet metal stamping, the occurrence of strain localization in a deformed sheet is considered a failure. As so, sheet metal's formability is conventionally evaluated using the Forming Limit Diagram (FLD), which separates the principal strain space into safety and unsafety regions by a Forming Limit Curve (FLC). This study presents an evaluation method for detecting strain localization based on Digital Image Correlation (DIC) during the experiment. The commercial DIC software ARAMIS is adopted to monitor the strain-field distribution on the deformed specimen's surface. A detailed analysis of the proposed method is presented considering Nakajima tests conducted for two automotive sheet metals: AA6016 and DP800. The identified FLC based on the proposed method is compared with that of well-established methods such as ISO 12004:2-2008 and time-dependent methods. For both investigated materials, the proposed method presents a lower FLC than the others.

1. Introduction

Evaluation of the material formability or the forming limit is a crucial demand in sheet metal stamping processes. The forming limit is conventionally determined at the moment of strain localization on the deformed specimen surface. For a particular material, combining the forming limit with attainable strain states formulates the Forming Limit Curve (FLC). Accurate determination of the FLC is important not only for designing parts but also for validating the formed products.

Different methods have been presented in the literature for detecting strain localization. The ISO 12004:2-2008 standard is a well-established method for this purpose based on an inverse parabola fitted to the strain distribution along a section on the deformed surface. Volk and Hora [1] introduced an alternative by comparing the thinning rate of the necking area observed in the pre-necking and post-necking stages. The linearity of strain acceleration is quantified in the correlation methods (Merklein et al. [2] Hotz et al. [3]) in order to determine the strain localization. Martínez-Donaire et al. [4] suggested examining the strain rate evolution of a point located at the boundary of the necking region. Furthermore, the geometrical variations during a test were proposed to use for detecting the strain localization as well, see Wang et al. [5], and Min et al. [6].

The summarized methods successfully determined the FLC of several automotive sheet metals in the corresponding studies. Although their preliminary results are promising, examining their applications for a wide range of materials is necessary to validate their effectiveness. However, it is obvious that all of these methods are sensitive to the interpretations and prejudices of the practitioners. That always affects the robustness of a selected method.

This study presents an evaluation method for experimentally detecting the strain localization appeared on the specimen's surface. The proposed criterion is based on the evolution of the Coefficient of Variation (CoV) of the major strain rate observed in a region of interest (ROI). The proposed method is applied to investigate the FLC of two sheet metals, DP800 and AA6016, that are widely used in the automotive industry. The derived FLCs are compared to those of other common methods to verify the effectiveness of the proposed method.

2. Experiment procedure

2.1. Investigated materials

The tested materials in this study are Dual-Phase (DP) 800 and Aluminum Alloy (AA) 6016. Both of them have a thickness of 1.2 mm. They are well-known to be medium-to-high strength materials, which are used in making vehicle frames and chassis. Both materials used in this study were supplied by Volvo Cars (VCBC).

2.2. Nakajima test

To determine their FLC, a series of Nakajima tests are conducted for the investigated materials following the ISO 6892-1:2009 standard. Seven tests are carried out for each material to characterize the necking strains under different stress states, ranging from uniaxial tension to equi-biaxial tension. Fig. 1 shows the blank geometries used in these tests.

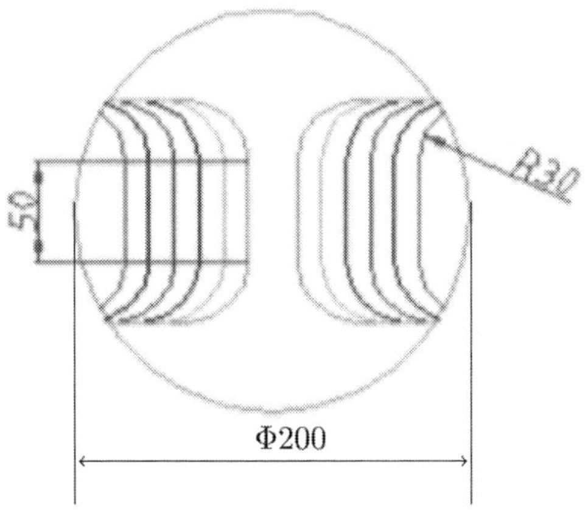

Figure 1. Geometry of the specimens used in Nakajima tests. Specimen widths (W): 25, 50, 75, 100, 125, 150, 200. Unit in mm.

A hemispherical punch with a diameter of 100 mm is used to deform the blank during each test. To reduce the friction acting on the contact area, full lubrication of mild oil is spread out, and a thin polymer sheet is used to cover the punch. Moreover, a drawbead is used to prevent the unnecessary drawing of the blank edges. These tests are conducted under a constant ram-forming speed of 25 mm/s until rupture. During the tests, a commercial Digital Image Correlation (DIC) system ARAMIS is used to monitor the strain evolution on the top surface of the deformed specimens.

3. Evaluation method

3.1. ISO standard method

The evaluation method presented in ISO 12004-2:2008 is generally accepted to determine the FLC of sheet metals. In this method, the major and minor strains along sections perpendicular to the crack profile are interpolated by inverse parabola functions. The pole of these fitted functions determines the limiting strains. The method is implemented in the ARAMIS built-in function, which makes evaluating more user-friendly. Fig. 2 shows an example of selecting sections on the surface of a deformed AA6016-W100 specimen. Fig. 3 shows the identified necking strains of the considered specimen.

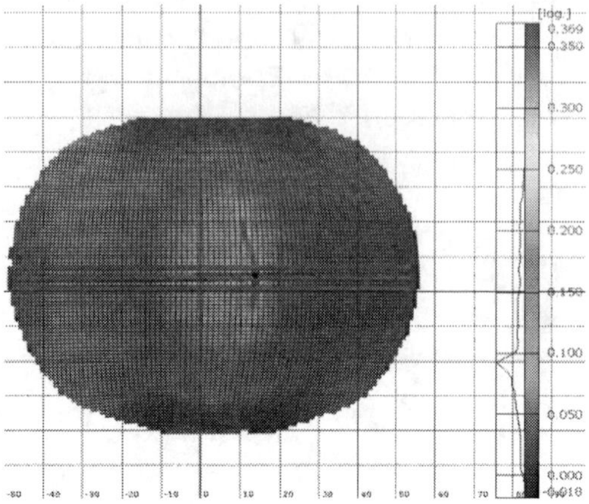

Figure 2. Selection of three sections for FLC calculation in ARAMIS.

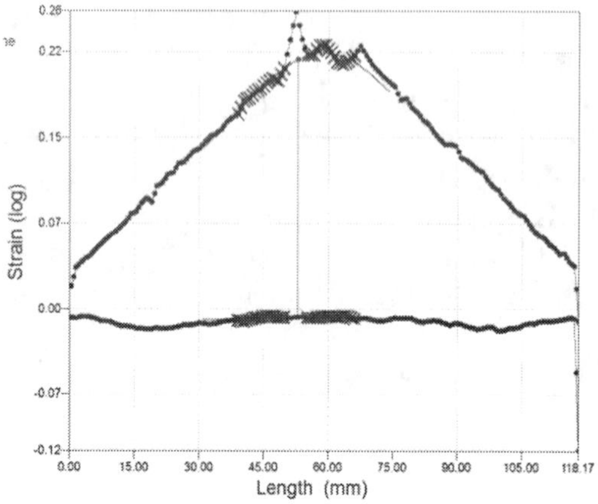

Figure 3. ARAMIS build-in function for FLC determination based on the ISO 12004-2:2008 standard.

3.2. Time-dependent method

Besides its benefits, the ISO standard method produces higher FLC than the forming limits observed in the production line of various automotive sheet metals [3, 7]. To overcome the drawback, Volk and Hora [1] presented a time-dependent method that takes into account the evolution of the thinning rate of a necking zone. The intersection of two linear curves, fitted to the necking zone's thinning rate evolution in the stable and unstable periods of deformation, indicates the instability.

This study adopts the time-dependent method (labeled "LBF method") to evaluate the FLC of the tested materials. In each specimen, a rectangular region of $2 \times 10\ mm^2$ in the necking zone is constructed in the second last image before visible fracture, which is equivalent to the recommendation of 5 elements for a 2 mm grid in the original article [1]. The arithmetic means of the thinning rate evaluated over the constructed necking zone are calculated for all stages of deformation. Based on the calculation, the necking strains of the corresponding specimen are determined following the procedure described in the cited work [1]. Fig. 4 presents the identified necking stage 81th of the AA6016-W100 specimen based on the time-dependent method. It is noticeable that the thinning rates of the last four stages before visualized crack were ignored in the calculation since these values are higher than one.

3.3. Volvo Cars method

An alternative method was developed and extensively used at Volvo Cars to determine the FLC of sheet metals. The basic idea of this method was presented in [8]. In deep, the minimum and maximum major strain rates (denoted by "Min-$\dot{\varepsilon_1}$" and "Max-$\dot{\varepsilon_1}$", respectively) observed in a ROI are compared to the lower and upper lines, which are constructed based on the mean (μ) and variation (σ) of the strain rates evaluated on the ROI. Crossing the boundary by either the Min-$\dot{\varepsilon_1}$ or Max-$\dot{\varepsilon_1}$ indicates the stage of necking initiation. Fig. 5 presents an example of the

VCBC method applied to determine the necking stage of the AA6016-W100 specimen, of which the derived result is stage 78th.

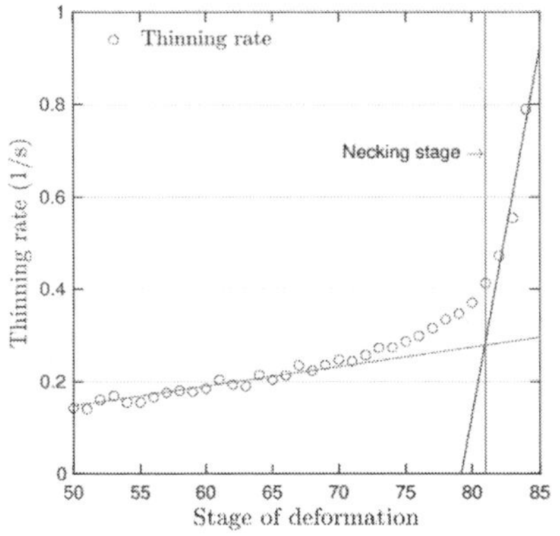

Figure 4. Evaluation of the linear best fitted method for AA6016-W100 specimen.

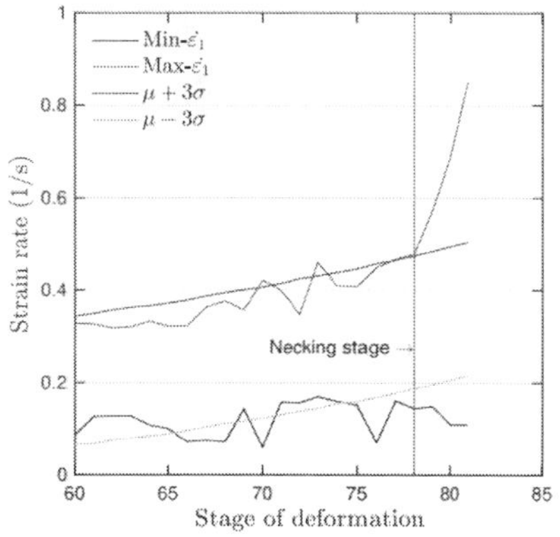

Figure 5. Evaluation of the VCBC method for AA6016-W100 specimen.

Implementation of this method in practice requires experienced experts who deeply understand the sensitivity of the method on the selection of ROI as well as the adjustment of the necking stage. In this study, the FLCs of two tested materials are evaluated by VCBC's experts and reported subsequently (denoted by "VCBC-FLC") for comparison purposes.

3.4. New method for necking detection

In statistics, mean, μ, is the average of the data set, while variation, σ, measures the spread of the data in the set. The Coefficient of Variation (CoV) of which the formulation is expressed in the equation below measures the dispersity of these values distribution.

$$CoV = \frac{\sigma}{\mu} \tag{1}$$

Motivated by the Volvo Cars method, it is suggested that the CoV of the major strain rate observed in a ROI can provide information regarding the necking initiation on the surface of the deformed specimen. Unlike a rectangular ROI preference in the previous studies, this study proposes to use a circular area located at the center of the necking region for this purpose. Therefore, a 5 mm diameter circle (denoted by "D5") is constructed on the surface of the deformed specimen at the last stage before the crack visualization for further investigation. The center of the circle is selected by the point that shows the largest equivalent plastic strain on the specimen surface at the last forming stage.

Fig. 6 depicts the evolution of mean and variation of the major strain rates evaluated on the ROI of the deformed DP800-W50 specimen. In addition, the corresponding CoV evaluated in this area is also plotted in this figure. A magnified view of this figure is presented in Fig. 7 focused on the identified necking stage. These figures show that μ always increases as the deformation continues. In contrast, the evolution of σ can be classified into two periods or ranges: (i) varying around a fixed value in the stages of small deformation and (ii) increasing significantly at later stages of deformation. As a result, the CoV decreases in the first period and increases in the latter. It is reasonable to assume that the deformation of the whole ROI is homogeneous in the first period, while strain localization occurs in the second period. Thus, the

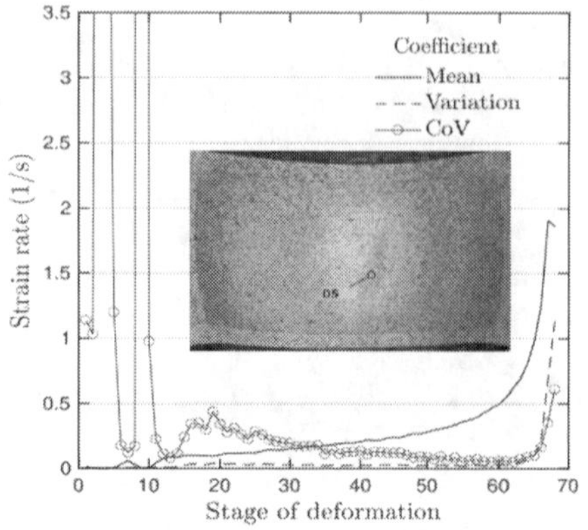

Figure 6. Statistic coefficients monitored at the necking region.

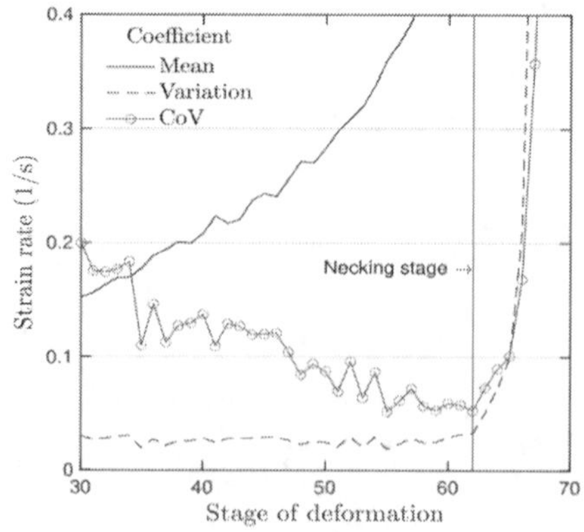

Figure 7. A magnified view of Figure 6 with the identified necking stage.

global minimum of CoV evolution indicates the strain localization established on the deformed specimen.

As seen in Fig. 7, the CoV curve exhibits various bumps due to inevitable noises contained in the measured data. One may consider a smoothed curve, which is a numerical approximation for the measured one, to identify the global minimum and the relevant necking stage. Although investigating the smoothing method and its influence on the derived results is omitted in this work, that will be the scope of a forthcoming study. To deal with the bumpiness of the curve, it is assumed that the last local minimum before a steep increase of the CoV indicates the necking stage, as shown in Fig. 7. An in-house Matlab code is developed to determine the necking stage based on the proposed method.

In order to study the sensitivity of ROI selection to the derived results, two additional ROIs (denoted "D3" and "D10") are constructed on the specimen surface, which have the same center as the D5 but with different diameters of 3 mm and 10 mm, respectively. In addition, a "D5-sub" region is constructed with the same diameter of 5 mm and 5 mm away from the other's center point, as shown in Fig. 8.

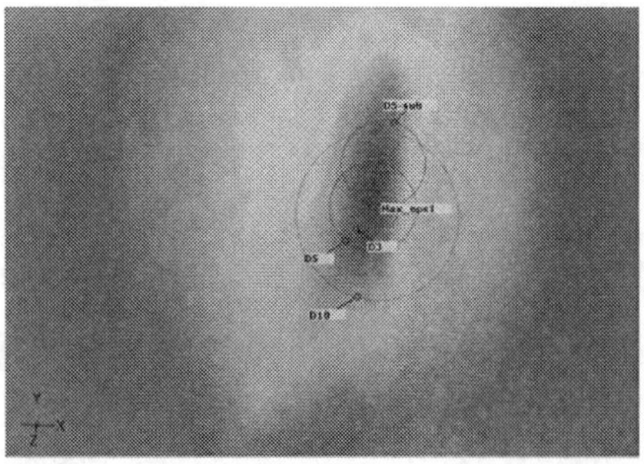

Figure 8. Location of different ROIs for sensitive study.

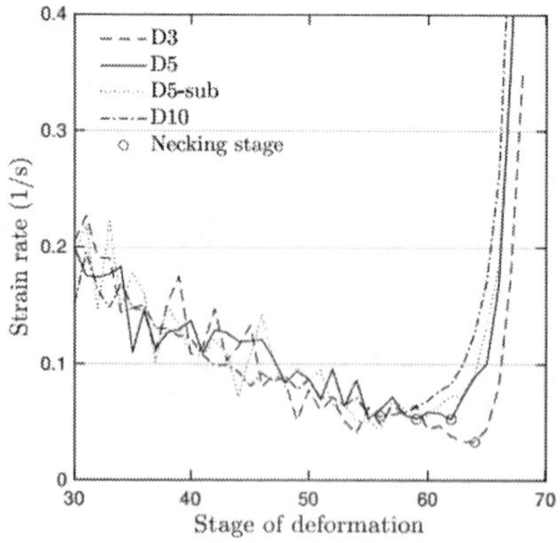

Figure 9. Evolution of the CoV observed in different ROIs.

Fig. 9 compares the evolution of CoV of the major strain rate observed in these ROIs. In addition, the calculated necking stage based on these ROIs is depicted in this figure. These CoV curves appear in a similar tendency in the decreasing period. However, their distinction is obviously displayed in the increasing period where the smaller ROI produces the later stage of increasing line. Consequently, adopting a smaller ROI yields a later necking stage, which is explainable as follows. Increasing CoV indicates an increase of the inhomogeneous strain distribution over the ROI due to the concentrated deformation on the necking area. In other words, the minimum of CoV indicates the last stage of homogeneous deformation over the ROI. As a result, the size of ROI is directly related to the size of the "identified" necking area or the "necking band". Since a size of 2 mm to 3 mm of the necking band is commonly reported for automotive sheet metals [9, 10], a circular area with a 5 mm diameter of ROI is recommended in this study.

4. Comparison

Fig. 10 and Fig. 11 show comparisons between the FLCs of AA6016 and DP800 sheets evaluated based on different methods. As seen in these figures, the ISO standard method provides the highest FLC for both tested materials, while the results of the VCBC method are always the lowest curves. It is noticed that the VCBC-FLCs are considered the ground trust since they are actually used in a production line. The time-dependent method results in an intermediate FLC for both tested materials. The difference between the perceived curves in Figs. 10-11 demonstrates the significance of developing a new evaluation method instead of the ISO 12004-2:2008 standard.

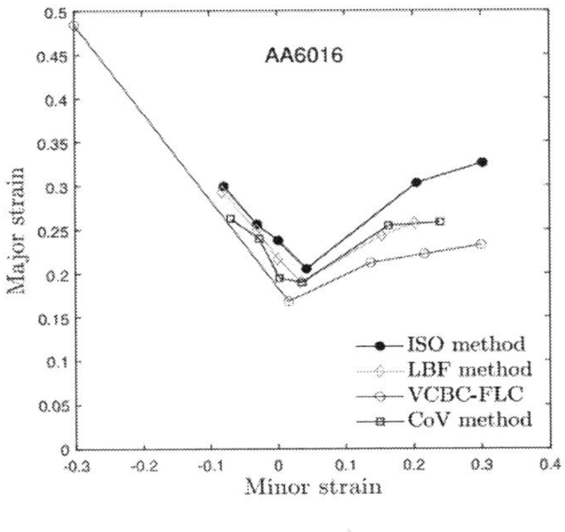

Figure 10. Comparison between predicted FLCs of AA6016 sheet based on different methods.

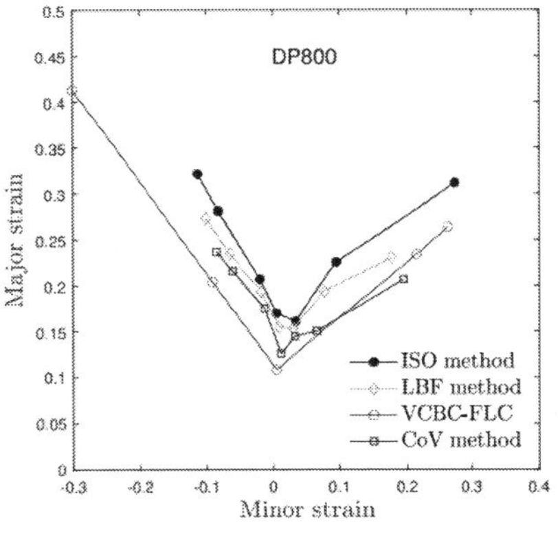

Figure 11. Comparison between predicted FLCs of DP800 sheet based on different methods.

Compared to the time-dependent method, the proposed CoV method derives lower FLC for two examined materials, except for the right side of the FLC of AA6016. Although the evaluated FLC of the proposed method is higher than the VCBC-FLC, the comparison demonstrates its potential in determining the FLC of sheet metals. Moreover, implementing the CoV method is programmable since it seeks a minimum event. That is a substantial benefit in the current era of big data.

Many factors could influence the derived result of the proposed method, which will be considered in a forthcoming study. Some of the foreseeing factors can be listed as the region of interest (i.e., the location, shape, and size of the investigating area), the mesh size of the ROI, and the different sources of quantification (i.e., the major strain rate and the thinning rate). Adopting acoustic emission measurement techniques is a promising approach for validating the accuracy of the proposed method [11, 12], which deserves further investigation. It is necessary to investigate the applications of the proposed method for different materials and various test batches to verify its robustness and effectiveness.

5. Conclusion

A new evaluation method for experimental detecting strain localization of sheet metals is proposed in this study. The method seeks the minimum of the CoV observed in the major strain rate field spreading in a region of interest, which makes its implementation programmable. The proposed method is applied to determine the FLC of AA6016 and DP800 sheets, of which the derived FLCs are lower than that of the well-established methods, such as, ISO 12004:2-2008 standard and the time-dependent method [1]. Compared to an industrial used curve, the VCBC-FLC, the proposed method presents closer results than the others.

Acknowledgments

Open Access funding was provided by the Blekinge Institute of Technology, Sweden. This study was also funded by VINNOVA in the Sustainable Production sub-program within Vehicle Strategic Research and Innovation (FFI) program (grant number 2020-02986).

References

[1] Volk W and Hora P 2011 *Int. J. Mater. Form.* **4** 339–46
[2] Merklein M, Kuppert A and Geiger M 2010 *CIRP Annals* **59(1)** 295–98
[3] Hotz W, Merklein M, Kuppert A, Friebe H and Klein M 2013 *Key Eng. Mater.* **549** 397–404
[4] Martínez-Donaire A J, García-Lomas F J and Vallellano C 2014 *Mater. Des.* **57** 135–45
[5] Wang K, Carsley J E, He B, Li J and Zhang L 2014 *J. Mater. Process. Technol.* **214(5)** 1120–30
[6] Min J, Stoughton T B, Carsley J E and Lin J 2017 *Exp. Mech.* **57** 521–35
[7] Min J, Stoughton T B, Carsley J E and Lin J 2016 *Int. J. Mech. Sci.* **117** 115–34
[8] Sigvant M, Mattiasson K and Larsson M 2008 *Proc. IDDRG 2008 Int. Conf., June 16-18, Olofström, Sweden* ed Nader Asnafi
[9] Bao C, Francois M and Joncour L L 2016 *Strain* **52(3)** 244–60
[10] Zhang R, Shi Z, Shao Z, Dean T A and Lin J 2021 *Int. J. Mech. Sci.* **189** 105997
[11] Petit J, Montay G and François M 2018 *Exp. Mech.* **58** 743-58
[12] Baral M, Al-Jewad A, Breunig A, Groche P, Ha J, Korkolis Y P and Kinsey B L 2022 *J. Mater. Process. Technol.* **310** 117758

Edge fracture of the first and third-generation high-strength steels: DP1000 and QP1000

Z Li[1], Y Chang[2], J Rong[1], J Min[3] and J Lian[1]*

[1] Department of Mechanical Engineering, Aalto University, 02150 Espoo, Finland
[2] Steel Institute, RWTH Aachen University, Intzestraße 1, 52072 Aachen, Germany
[3] School of Mechanical Engineering, Tongji University, Shanghai 201804, China

E-mail: junhe.lian@aalto.fi

Abstract. Advanced high-strength steels (AHSS) have shown profound progress in improving tensile ductility or global formality in the last decades over three generations. For a complete assessment of both the global and local formability, this study aims to characterize and compare the tensile and edge fracture behavior of the first and third-generation AHSS (dual-phase steel and quenching & partitioning steel) with the same nominal strength level of 980 MPa. Uniaxial tensile tests are performed to characterize the tensile properties. Hole expansion tests are conducted with two edge conditions based on separated preparation techniques (waterjet with polishing and punching) to investigate the edge fracture for both materials. The hole expansion ratios and edge fractures are compared between two materials and two edge conditions. It is concluded that the investigated QP1000 has promoted global formability while the DP1000 shows better local formability due to its damage-tolerant and crack-resistant responses.

1. Introduction

Advanced high-strength steels (AHSS) were developed in the automotive industries to improve crashworthiness and lightweight structure [1]. The first-generation (1st G) AHSS elevates the strength level beyond 800 MPa from traditional high-strength steels, whereas the superior strength comes with the sacrifice of ductility. Among the widely adopted grades, dual-phase (DP) steels are the most popular and developed ones, due to their balanced properties of high strength, good strain hardening, and low production cost [2]. Those steels are commonly used to produce complex parts from numerous manufacturing processes, in which the formability of the material becomes vital to be evaluated. Formability is assessed from two aspects, global and local. Global formability is sufficient to describe the necking limit of the sheet material in the global forming modes, such as stretching, drawing, and plane-strain tension, where the deformation of the material is distributed evenly over a relatively large region. Whereas the local formability is determined by the fracture limit when concentrated deformation occurs locally in the local forming modes including bending, collar forming, and hole expansion. Among all global and local failures, edge fracture tends to be extremely critical since the edge formability is depending on both material properties and the edge quality.

Numerous studies have been conducted to characterize both the local and global formability [3-8] of DP steels. In the meanwhile, the third-generation (3rd G) steels were proposed [9] with specific processes. One typical example is the Quenching and Partitioning (QP) steels, offering a superior combination of strength and ductility, which is significantly promoted from the 1st G. QP 1000 grades and relatively novel with improved global formability [10]. It is still under evaluation for many applications, e.g., in manufacturing structural parts such as bumpers and B-pillars, where the edge area undergoes large deformation, usually at uniaxial tension, plane-strain tension, or rather complex stress states. However, when it comes to the local formability, especially the stretch-flange-ability of the edge, the investigation is quite limited.

To accurately characterize the stretch-flange-ability of the edge, a variety of experimental methods have been reported [11]. Among them, the hole expansion test (HET) is a commonly used method to effectively evaluate the stretchability of sheet metal and its edge fracture [12-14]. It is verified that HET gives good results in characterizing the edge crack sensitivity of AHSS steels [15]. Different influencing factors have been studied for DP and complex-phase (CP) steels to find out the most critical ones [16]. As reported in a large number of studies, the severe pre-damage introduced to the hole edge by the blanking process in the shear-affected zone (SAZ), places a dominating role in the edge stretch-flange-ability. It is verified that HET gives good results in characterizing the edge crack sensitivity of AHSS steels [15]. Different influencing factors have been studied for DP and complex-phase (CP) steels to find out the most critical ones [16]. As reported in a large number of studies, the severe pre-damage introduced to the hole edge by the blanking process in the shear-affected zone (SAZ), places a dominating role in the edge stretch-flange-ability [17-19]. The edge effect therefore must be taken into consideration for edge formability. Madrid et al. [20] reported QP1000 is not sensitive to the punch geometry. Goshert et al. [21] studied four edge conditions with angular stretch bend tests for QP980 and DP600, which emphasized the influence of edge effects. Nevertheless, there are few comparative studies regarding the local formability between the 1st and 3rd Gen. steels [22], and particularly no data is available for QP1000 and DP1000 in the current literature.

Therefore, in the current study, uniaxial tensile and hole expansion tests are conducted to investigate the tensile and edge fracture behavior of QP1000 and DP1000. Two conditions for smooth and blanked edges based on different forming methods including waterjet cutting with polishing and punching, are prepared. After hole expansion tests, the edge fracture patterns are then characterized, followed by a damage mechanism analysis for both materials.

2. Materials and experiments

The sheet metals chosen for this study are as-received high-strength DP steel (DP1000) with a measured thickness of 1.46 mm and quenching & partitioning steel (QP1000) with a thickness of 1.36 mm. The chemical composition of the DP1000 and QP1000 material is performed via a BELEC LAB 3000s optical emission spectrometer. The analysis results are summarized in Table 1. The microstructure was obtained from scanning electron microscopy (SEM). The micrographs of both materials are shown in Figure 1. DP1000 is composed of ferrite and martensite phases, with an approximate martensite phase fraction of 45% [23]. QP1000 contains metastable retained austenite (RA), tempered martensite, secondary martensite, as well as non-martensitic constituents (NMC) such as bainitic ferrite apart from ferrite.

Table 1. Chemical composition of DP1000 and QP1000 (mass content in %).

Material	C	Si	Mn	Al	P	S	Ti + Nb	Cr + Mo
DP1000	0.07	0.30	2.55	0.05	<0.040	<0.010	<0.15	<1.0
QP1000	0.15	1.67	2.42	0.074	<0.002	<0.002	0.011	<0.03

The tensile properties were tested using a Zwick/Roell Z020 screw-driven tensile machine. Uniaxial tensile tests of the smooth dog-bone (SDB) specimens were performed at room temperature (RT, 25°C) under quasi-static (QS, 10^{-4} s^{-1}) loading along rolling direction (RD). The testing setup is shown in Figure 2. A stereo digital image correlation (DIC) system was used to measure the displacement during the deformation, with the detecting camera resolution of 50 pixels/mm. To ensure the image quality, the surface of the specimen was lacquered with an even layer of white paint and then sprayed to form speckle patterns with a speckle diameter of 5-10 pixels. A parallel filter was placed before the spotlight, which can minimize the reflections from the material surface and therefore, improve the lighting condition.

Observations were recorded from Hikvision cameras with two separate signals (Left-0, Right-1) from both sides.

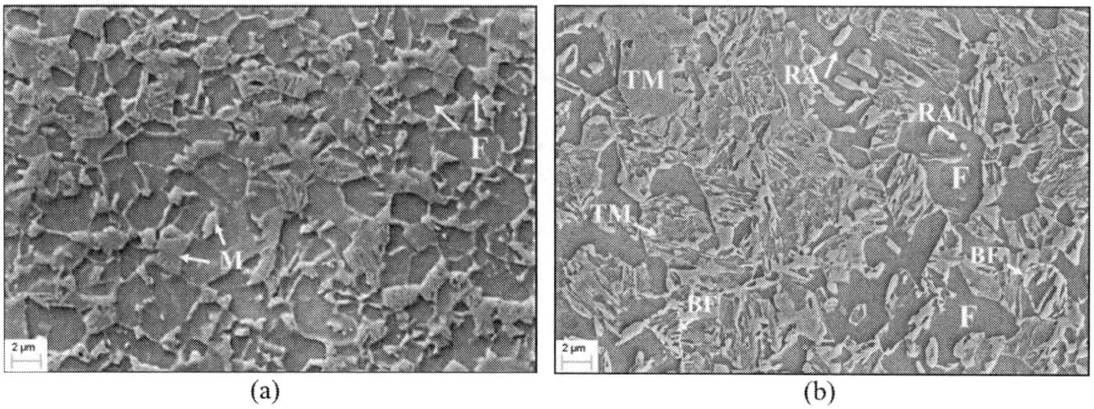

Figure 1. Microstructure of (a) DP1000 [24] with martensite (M) and ferrite (F) phases indicated, and (b) QP1000 with retained austenite (RA), tempered martensite (TM), bainitic ferrite (BF) phases indicated.

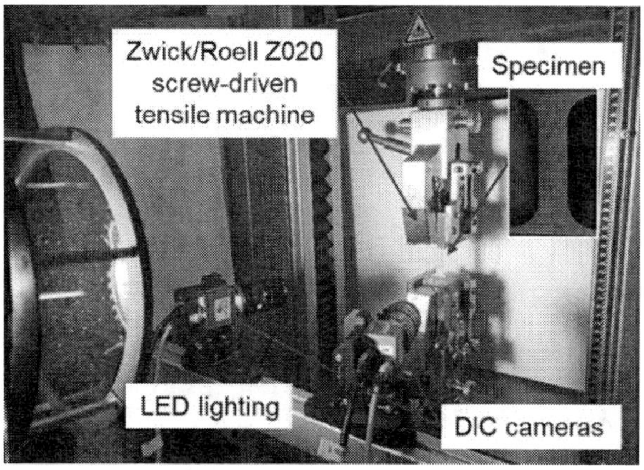

Figure 2. Uniaxial tensile testing set-up with stereo-DIC system and lighting.

To characterize the edge fracture, circular specimens with a diameter of 90 mm were cut as shown in Figure 3 (a). Central holes with separated edge conditions were prepared based on different manufacturing techniques. The smooth edge condition was introduced by waterjet cutting and then ground to 10 mm with a smooth finish. The blanked hole is punched from the hole blanking process as schematically illustrated in Figure 3 (b). Then these two materials under both edge conditions were conducted in the HET following the ISO/16630:2017-09 standard. The set-up of the HET is depicted in Figure 3 (c). The specimen was clamped in between the die and blank holder under a holding force of 235 kN. Then the specimen was expanded by a conical punch at a 5 mm/min expansion speed until the first through-thickness crack (1st TTC) was detected on the edge of the hole. The tracking of the first through-thickness crack easily got overshot with bare eyes. Therefore, a camera was introduced to record the images and observe the deformation at each frame. The hole expansion ratio (HER), as a criterion to characterize the resistance to edge fracture, was then calculated as defined in Eq. (1), where d_h is the inner hole diameter after expansion and d_0 is the original hole diameter before the test. Both were measured by digital caliper and averaged across three repeats. The results after expansion were further verified with the measured values from the images captured by the camera, to make sure the diameters were taken at the exact time when the first through-thickness crack occurred.

$$HER = \frac{d_h - d_0}{d_0} \times 100\% \quad (1)$$

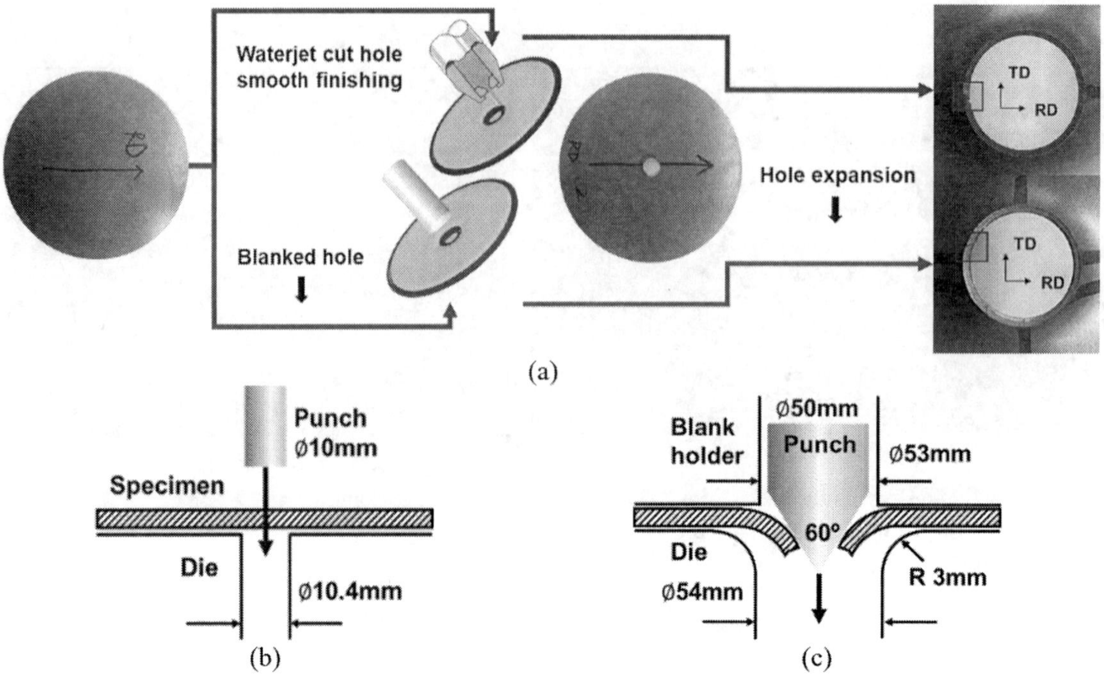

Figure 3. Schematic diagram of (a) the edge fracture characterization approach, (b) cross-section view of the hole blanking, and (c) cross-section view of the hole expansion process.

3. Results and discussion

3.1. Tensile properties

The engineering stress-strain, true stress-strain, r-value, and forming limit curves from the uniaxial tensile tests along RD for both materials are shown in Figure 4. The mechanical properties of the DP and QP steels such as yield strength (YS), ultimate tensile strength (UTS), percentage uniform elongation (% UE), percentage total elongation (% TE), and Lankford coefficients (r-value) at UE are listed in Table 2. DP1000 shows a slightly higher YS value, while QP1000 surpasses UTS. From the results, QP1000 reaches yield strength sooner before DP1000, while it takes much longer elongation to reach UTS and then to fracture. The UTS is improved by 10% from DP1000 to QP1000. What is significantly enhanced is the ductility, the UE of QP1000 is more than three times higher than the one in DP1000, and the fracture elongation is also increased by more than 100%. There is no significant difference between the two materials in terms of the r-value and their evolution pattern. Further investigation can be made by taking into account the loading effect for the anisotropic properties, Further investigation can be made by taking into account the loading effect for the anisotropic properties, which can be found in detail in a follow-up study [25].

Table 2. Tensile properties of DP1000 and QP1000.

Material	YS, MPa	UTS, MPa	UE, %	TE, %	R-value at UE, -
DP1000	770.00	983.38	5.22	10.97	0.73
QP1000	752.20	1090.20	18.09	23.57	0.75

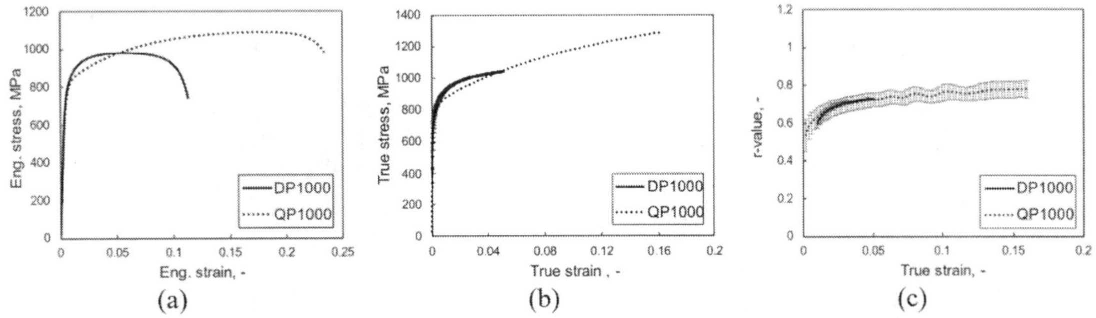

Figure 4. The tensile properties from uniaxial tensile tests, (a) engineering stress-strain curves, (b) true stress-strain curves, and (c) r-value curves of QP1000 in comparison with DP1000.

3.2. Edge fracture behavior

3.2.1. HER. The HER values for the blanked and machined holes of DP1000 and QP1000 are compared in Figure 5 (a). Sensitivity to the edge condition for both materials was characterized by comparing the HER on each condition. Both steels demonstrate strong sensitivity to the edge condition. The HER value for DP1000 is higher than QP1000 for both conditions, i.e., DP1000 shows higher edge fracture resistance than QP1000. For the DP1000 machined hole, the HER is 92%, which is almost one and a half times the value measured for the machined hole of QP1000, 62%. A similar trend exists for the blanked hole of DP1000, which has an HER of 29%, while the HER for the blanked hole of QP1000 is 21%, giving a slightly reduced difference in between. The confidence interval has a rather small value for the blanked holes, suggesting there is no significant variation for the pre-damage introduced to the blanked edges. Whereas the error for machined holes is comparatively high, which indicates the ideal smooth condition can be easily breached, adding up the fact that it is often difficult to accurately track down the moment when the first through-thickness crack is forming. The measured HER includes error from overshoot of the punch head. As shown in Figure 5 (b) with the blue and red bars, both materials can be clearly differentiated between the edge conditions with formability reductions of 66-68%. The variations for machined holes in DP1000 are larger than in QP1000 due to the fact that the specimen preparation process is slightly different, with QP1000 cut directly from electrical discharge machining (EDM) with higher surface quality. The formability reduction for DP1000 is slightly higher than in QP1000 can be explained by the study [26] that for blanked edges, a higher density of voids can be found at ferrite-martensite interfaces. The excellent damage tolerance and crack resistance of DP1000 got depressed, making it more vulnerable to fracture. The formability distinguishes between QP1000 and DP1000 are around 27% for smooth edges as shown in a green bar, and 33% for blanked edges plotted in an orange bar.

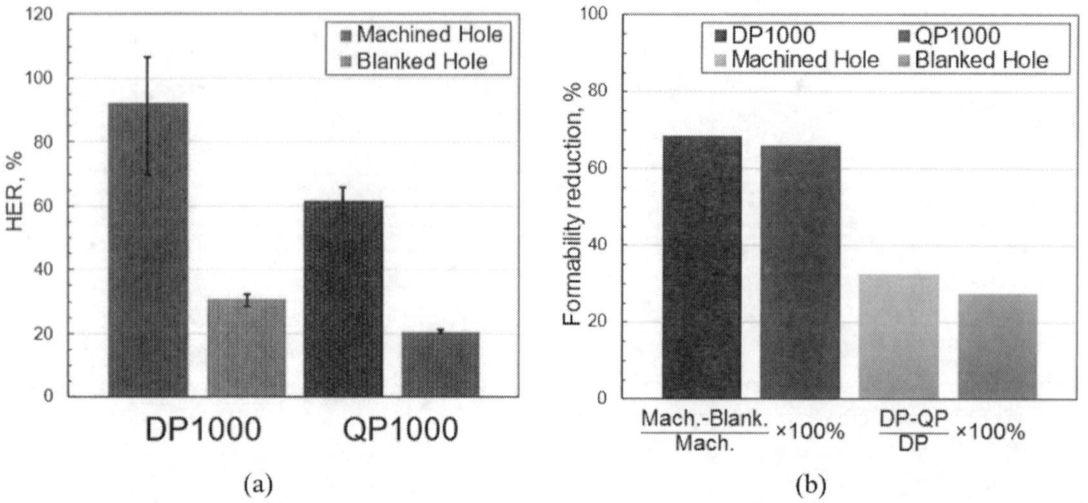

Figure 5. HER values analysis (a) of the punched and machined holes for DP1000 and QP1000, (b) the formability reduction from smooth edge to punched edge for each material, and from DP1000 to QP1000 for each edge condition.

3.2.2. Edge fracture on smooth holes. The edge crack distribution and pattern for DP1000 and QP1000 on smooth holes are shown in Figure 6 and Figure 7, respectively. The localized necking is clearly observed at the hole edge close to RD in DP1000 as marked in Figure 6 (a), whereas no obvious thickness reduction can be found on the hole circumference for QP1000. As shown in Figure 6 (b), the first through-thickness crack locates near RD, with fractures prone to a slant angle of conical punch during deformation, then rapidly propagating through the thickness to form the first major crack. If the test is not halted immediately, the crack would expand further in the radial direction following a serrated path as sketched in Figure 6 (c). For QP1000, after the first major crack, a limited number of minor cracks can be observed initiating from the bottom layer, as blue triangles annotated in Figure 7. One out of all the minor cracks can also develop further to form another through-thickness crack, whereas the same behavior is not observed on DP1000. Fracture is prone to develop into one or at most two major cracks in the smooth edge region for both materials.

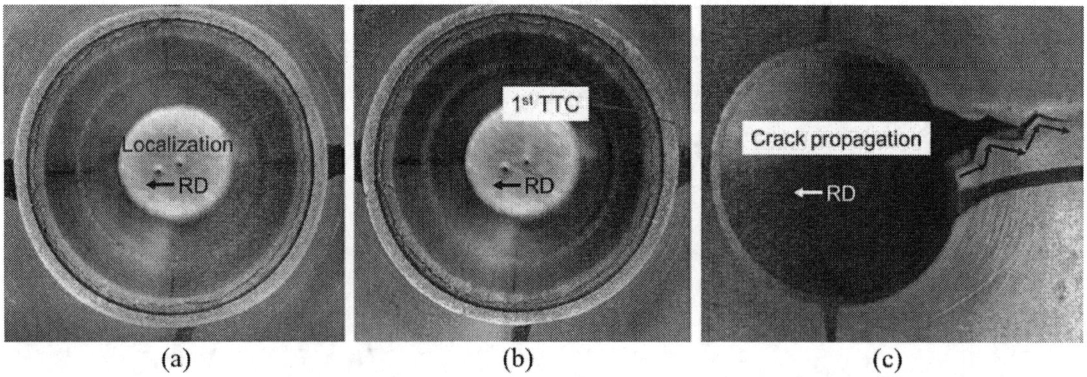

Figure 6. Edge fracture pattern and locations on smooth edges for DP1000 (a) after strong localization, (b) the moment when the first through-thickness crack occurs, and (c) crack propagation with blue marks.

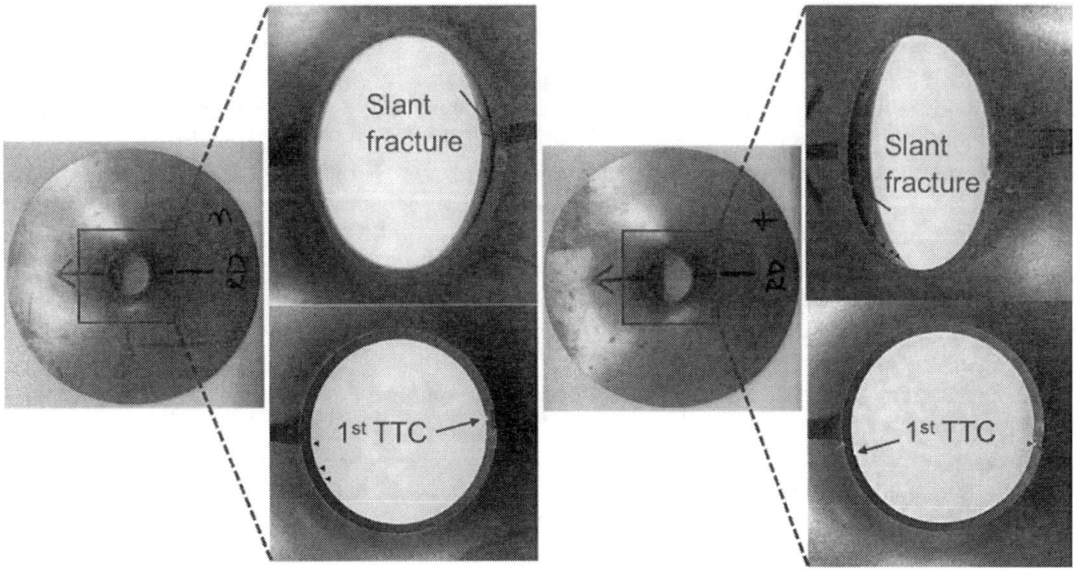

Figure 7. Edge fracture pattern and locations on smooth edges for QP1000 after first through-thickness crack, and blue marks for minor cracks.

3.2.3. Edge fracture on punched holes. The edge crack distribution and pattern for DP1000 and QP1000 on punched holes are shown in Figure 8 (a) and (b), respectively. No noticeable localization can be witnessed for either material. The first through-thickness crack has the same location close to RD as in smooth edges, while the fractures were prone to a slant angle of 45°. Different from the pattern on the smooth edge, the crack initiates not only from the bottom layer but also from the top surface of the punched hole. The number of minor cracks on punched edges is distinctively larger than in the smooth condition for both materials, while DP1000 collects more minor cracks than QP1000. When the force is applied continuously after the first major crack appears, the hole gets expanded further. For DP1000, the number of major cracks grows to four. The location of these four major cracks at the open-up circumference is along RD and transverse direction (TD), as marked with red circles in Figure 8 (c). For QP1000, the number of major cracks ended into six as shown in Figure 8 (d). The locations are not only at RD, and TD but also along the diagonal direction (DD).

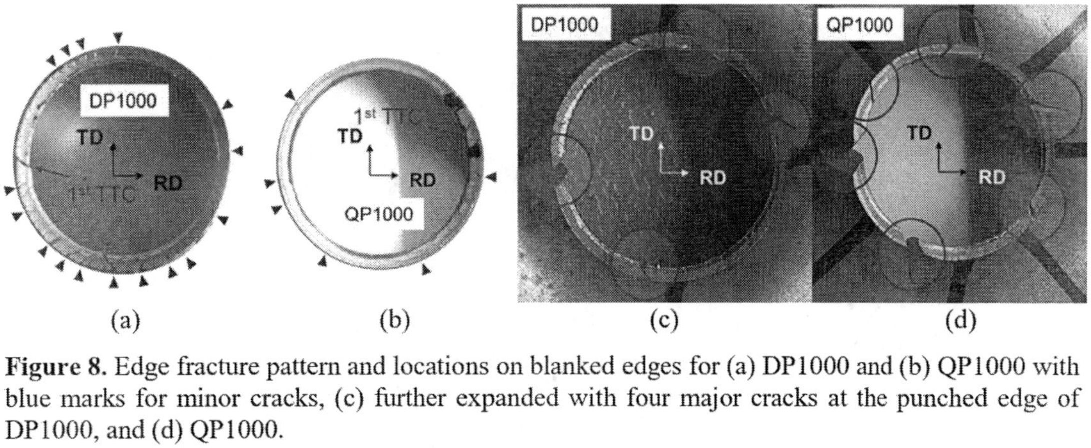

Figure 8. Edge fracture pattern and locations on blanked edges for (a) DP1000 and (b) QP1000 with blue marks for minor cracks, (c) further expanded with four major cracks at the punched edge of DP1000, and (d) QP1000.

3.2.4. Damage mechanism analysis of the fractured specimens.

The SEM fractography of the local critical area in the section close to the fracture surface after uniaxial tensile tests are shown in Figure 9 (a), with the red quadrilaterals and blue circles indicating microcracks and voids, respectively. The number of defects, in terms of voids and micro-cracks, is distinctively larger in QP1000 than in DP1000, especially for voids. It is noticeable that both ferrite and martensite grains are extremely elongated in the severe deformation zone in DP1000, as previously reported [27]. For DP steels, three void nucleation damage mechanism modes including martensite brittle cracking, ferrite-martensite interface decohesion, and ferrite-ferrite grain boundary decohesion are commonly found. Since both phases are relatively fine microstructures, it is difficult to distinguish between microcracks and voids from the phase boundaries. However, as the ferrite phase shows generally higher ductility and deformed more easily compared to martensite, phase boundary debonding is the main failure type under tensile loading at room temperature [27], while for higher temperatures, more complex damage mechanisms take place due to dynamic strain aging effects [28]. The current investigated QP grade contains both tempered martensite and secondary martensite, arising the fraction of the harder and more brittle phases in general. In addition, the interaction between RA and martensite in QP steels exacerbates the occurrence of phase boundary debonding during uniaxial tension, leading to the formation of voids and cracks more easily to reach rupture.

The deformation history of both materials, including the plastic localization, damage initiation, and fracture can be found in the schematic drawing from Figure 9 (b). QP1000 shows a higher uniform deformation strain than DP1000. DP1000 yields a large amount of plastic localization before damage occurs followed by a fast damage evolution stage until the final fracture. While QP steels exhibit low resistance to fracture due to their sensitivity to damage development, resulting in lower local formability. In DP1000, damage concentrates on localized regions. QP1000 on the other hand has a more homogeneous deformation with more uniformly distributed damage. However, with a larger amount of voids and micro-cracks, the rather uniform deformation and uniform damage distribution lead to a sudden fracture with the cracks growing along. In contrast, the high level of strain localization and corresponding concentrated damage of DP1000 leads to a fracture pattern with significant necking.

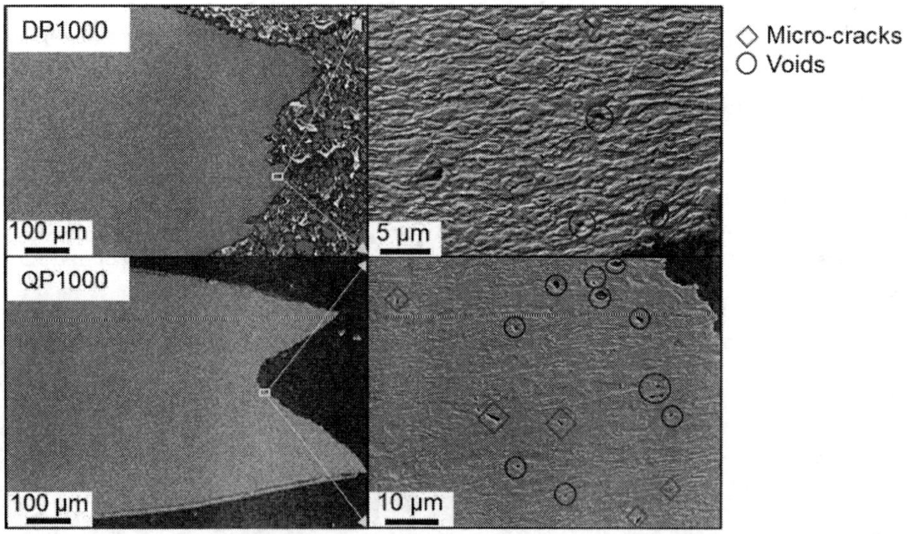

(a)

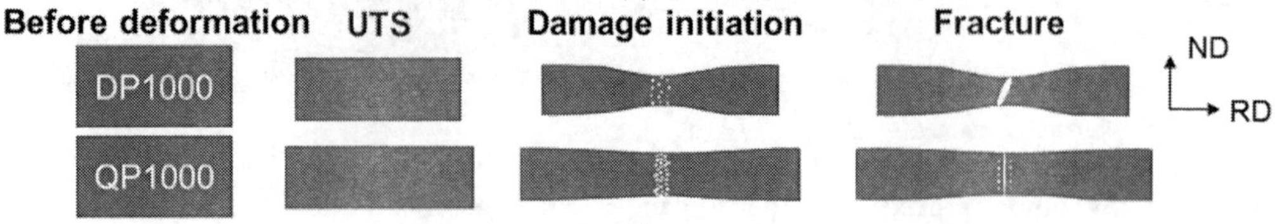

(b)

Figure 9. Microstructure-level damage mechanism analysis: (a) SEM fractography of fractured specimens with the red and blue indicators pointing out the micro-cracks and voids, respectively, and (b) schematic drawing of the relationship pattern during uniaxial tension for DP1000 and QP1000.

4. Conclusions

(1) The QP1000 steel in this study has an enhanced global formability with yield strength over 750 MPa and a superior total elongation of about 25%, while the local formability is conservatively proved by the hole expansion ratio of 62% for the smooth edge and 21% for the blanked edge.
(2) The DP1000 steel in this study shows satisfied tensile properties with a yield strength of 770 MPa and a moderate total elongation of 11%, while the local formability is promoted with a relatively high hole expansion ratio of 92% for the smooth edge and 29% for the blanked edge.
(3) The surface condition plays a dominant role in the edge formability and the formability reduces severely from machined hole to blanked hole for both materials in 66-68%.
(4) DP1000 demonstrates superior local formability with higher damage tolerance and edge fracture resistance due to fewer micro-cracks and local voids generated along deformation, while, despite the superior global formability and strain hardening rate, QP1000 exhibits strong sensitivity to damage with more micro defects during deformation due to its more complex microstructure, therefore limiting the local formability.

References

[1] Bouaziz O, Zurob H and Huang M 2013 Driving force and logic of development of advanced high strength steels for automotive applications *Steel Res Int* **84** 937-47
[2] Rana R and Singh S B 2016 *Automotive steels: design, metallurgy, processing and applications*: Woodhead Publishing)
[3] Tsipouridis P, Werner E, Krempaszky C and Tragl E 2006 Formability of High Strength Dual-phase Steels *Steel Res Int* **77** 654-67
[4] Lian J, Liu P and Münstermann S 2012 Modeling of damage and failure of dual phase steel in Nakajima test *Key Engineering Materials* **525-526** 69-72
[5] Lian J, Sharaf M, Archie F and Münstermann S 2013 A hybrid approach for modelling of plasticity and failure behaviour of advanced high-strength steel sheets *International Journal of Damage Mechanics* **22** 188-218
[6] Wu-Rong W, Chang-Wei H, Zhong-Hua Z and Xi-cheng W 2011 The limit drawing ratio and formability prediction of advanced high strength dual-phase steels *Materials & Design* **32** 3320-7
[7] Hance B 2016 Advanced high strength steel: Deciphering local and global formability. In: *Proc. International Automotive Body Congress, Dearborn, MI,* p 2
[8] Westhäuser S, Schneider M, Teschner M and Denks I 2019 Local ductility–key parameter for predicting formability of AHSS. In: *IOP Conference Series: Materials Science and Engineering*: IOP Publishing) p 012049
[9] Speer J, Matlock D K, De Cooman B C and Schroth J 2003 Carbon partitioning into austenite after martensite transformation *Acta Mater* **51** 2611-22
[10] Chen X, Niu C, Lian C and Lin J 2017 The evaluation of formability of the 3rd generation advanced high strength steels QP980 based on digital image correlation method *Procedia Engineer* **207** 556-61
[11] Feistle M, Golle R and Volk W 2022 Edge crack test methods for AHSS steel grades: A review and comparisons *J Mater Process Tech* **302** 117488
[12] Leu D-K, Chen T-C and Huang Y-M 1999 Influence of punch shapes on the collar-drawing process of sheet steel *J Mater Process Tech* **88** 134-46
[13] Huang Y-M and Chien K 2002 Influence of cone semi-angle on the formability limitation of the hole-flanging process *The International Journal of Advanced Manufacturing Technology* **19** 597-606
[14] Xu L, Chen L, De Cooman B, Steglich D and Barlat F 2010 Hole expansion of advanced high strenth steel sheet sample *International Journal of Material Forming* **3** 247-50
[15] Larour P, Freudenthaler J, Kerschbaum M and Dolzer D 2020 Edge crack sensitivity versus tensile local ductility of AHSS sheet steels. In: *IOP Conference Series: Materials Science and Engineering*: IOP Publishing) p 012080

[16] Pathak N, Butcher C and Worswick M 2016 Assessment of the critical parameters influencing the edge stretchability of advanced high-strength steel sheet *Journal of Materials Engineering and Performance* **25** 4919-32

[17] Chiriac C and Chen G 2008 Local formability characterization of AHSS–Digital camera based hole expansion test development. In: *IDDRG 2008 Conference,* pp 16-8

[18] Wang K, Luo M and Wierzbicki T 2014 Experiments and modeling of edge fracture for an AHSS sheet *Int J Fracture* **187** 245-68

[19] Mu L, Wang Y, Zang Y and Araujo Stemler P M 2017 Edge Fracture Prediction Using Uncoupled Ductile Fracture Models for DP780 Sheet *Journal of Failure Analysis and Prevention* **17** 321-9

[20] Madrid M, Van Tyne C, Sriram S, Pavlina E, Hu J and Clarke K 2018 Hole expansion ratio in intercritically annealed QP 980/1180 steel grades as a function of testing condition. In: *IOP Conference Series: Materials Science and Engineering*: IOP Publishing) p 012083

[21] Goshert B, Terrazas O, Matlock D and Van Tyne C 2018 Sample edge effects on tensile properties and sheet formability. In: *IOP Conference Series: Materials Science and Engineering*: IOP Publishing) p 012064

[22] Shen F, Wang H, Liu Z, Liu W, Könemann M, Yuan G, Wang G, Münstermann S and Lian J 2022 Local formability of medium-Mn steel *J Mater Process Tech* **299** 117368

[23] Liu W, Lian J, Aravas N and Münstermann S 2020 A strategy for synthetic microstructure generation and crystal plasticity parameter calibration of fine-grain-structured dual-phase steel *International Journal of Plasticity* **126** 102614

[24] Chang Y 2021 *Evidence and effects of heterogeneities in complex phase and dual phase steels* Dissertation RWTH Aachen University Aachen, Germany

[25] Li Z, Chang Y, Liu W and Lian J 2023 Predicting edge fracture in dual-phase steels: significance of anisotropy-induced localization *under review*

[26] Misra R, Thompson S, Hylton T and Boucek A 2001 Microstructures of hot-rolled high-strength steels with significant differences in edge formability *Metallurgical and Materials Transactions A* **32** 745-60

[27] Liu W, Lian J and Münstermann S 2019 Damage mechanism analysis of a high-strength dual-phase steel sheet with optimized fracture samples for various stress states and loading rates *Engineering Failure Analysis* **106** 104138

[28] Liu W and Lian J 2021 Stress-state dependence of dynamic strain aging: Thermal hardening and blue brittleness *International Journal of Minerals, Metallurgy and Materials* **28** 854-66

IOP Conf. Series: Materials Science and Engineering 1284 (2023) 012022 doi:10.1088/1757-899X/1284/1/012022

Strain evolution during hole expansion testing of 800 MPa tensile strength hot-rolled steels

P Plosila[1*], V Kesti[2], A Kaijalainen[1], R Vierelä[3], P Rautio[3] and J Kömi[1]

[1]Materials and Mechanical Engineering, University of Oulu, P.O. Box 4200, FI-90014 University of Oulu, Finland
[2]SSAB Europe, Rautaruukintie 155, P.O. Box 93, FI-92101 Raahe, Finland
[3]Arctic Steel and Mining, Lapland University of Applied Sciences, Tietokatu 1, FI-94600 Kemi, Finland

E-mail: pekka.plosila@oulu.fi

Abstract. One limitation of the standardized ISO 16630 hole expansion test is that it provides only one result: limiting hole expansion ratio (HER). In practice, steels with similar HER values can have different cut edge forming behavior due to possible differences in strain localization tendencies. Digital image correlation (DIC) strain measurement during formability testing allows more in-detail analysis of strain-state near the cut edge. In this paper, strain evolution during hole expansion testing was investigated for three 800 MPa tensile strength grade hot-rolled strip steels. The steels were selected to have differences in microstructures and anisotropies of mechanical properties. Two different hole expansion test methods with DIC strain measurement were utilized to investigate different edge loading scenarios: in-plane stretching with a flat-top punch and more out-of-plane stretching with a hemispherical punch. Test holes were prepared according to ISO 16630 standard. In order to examine strain evolution and localization during testing, strains were measured with circle-shape sections around the hole in various distances from the cut edge. Strain localization behavior was investigated in different sheet directions and the effect on the hole expansion ratio was evaluated. Results show considerable differences in the cut edge forming behavior between the investigated materials.

1. Introduction

Utilization of advanced high-strength steel grades has been increasing in the automotive industry resulting from increasing requirements for weight reduction and safety improvement. Production of components for the applications often include shearing, which introduces damage to the cut edge and promotes crack formation during forming [1]. Sheared edge formability has been generally considered to be more serious issue in high-strength steels compared to conventional mild steel grades [2].

The most widely used method to estimate edge-crack sensitivity of steel sheets is the ISO 16630 hole expansion test [3], where a 10 mm diameter punched hole is expanded with a conical tool. One limitation of the test is that it provides only one result: limiting hole expansion ratio (HER), which is determined from the ratio between the initial hole diameter and the diameter after testing. The test does not consider the effects of localization of strain to certain directions in the edge. In practice, even steels with similar HER values can have widely different cut edge forming behavior.

Flat-top punch tests have been shown to have considerably lower HER compared to the conical punch ISO test depending on the material [4,5], whereas the results of hemispherical test set-ups have been reported being closer to the values of ISO HER [2,5]. In addition to punch shape and hole edge condition, microstructure of the material is also known to have effects on the hole expansion. For example, high hardness difference between ferrite and martensite phases [6] and secondary constituents, such as inclusions and cementite [7], have been reported to decrease HER results.

Measurement of strains during hole expansion testing can be enabled by utilizing digital image correlation (DIC). Similarly to the ISO test, the hemispherical and flat-top punch tests create an initially uniform strain distribution around the hole, however, the different punch shapes lead to differences in strain gradients in the edges [8]. The strain localization tendency of each material also causes the strain to concentrate in certain directions around the edge, as the punch and specimen geometry in these tests, ideally, should not force the strain to localize. Different steel grades have been shown to have differences in the major strain distributions at failure around the hole edge of hemispherical punch test specimens [9].

In this paper, the research aim was to investigate strain evolution during hole expansion testing for 800 MPa tensile strength grade hot rolled steels. To enable more precise strain data collection around the hole edge, as on-line strain measurement is challenging with a conical punch, hemispherical and flat-top punch test set-ups were instead selected for the hole expansion testing. Previously, localized thinning was observed in the electrical discharge machined ISO hole expansion test specimen edges in the same hot-rolled materials [10]. One motivation for the present research was to confirm with DIC measurements if edge failure is preceded by localization of strain in the investigated materials also with punched hole condition.

2. Experimental

2.1. Research materials

The research materials in this study were industrially hot-rolled low-carbon steel strips with 3 mm thickness. The steels were selected to have similar tensile strength grade while simultaneously consisting of different microstructures. The three investigated steels were: a multi-phase steel consisting of ferrite, bainite and martensite (CP), essentially a single-phase precipitation-hardened steel consisting primarily of polygonal ferrite (F), and essentially a single-phase precipitation-hardened steel consisting primarily of quasi-polygonal ferrite (QF). The overall grain size was considerably finer in CP compared to the single-phase steels. The grain morphology was elongated in the rolling direction in CP and mostly equiaxed in F and QF. Some amount of banding was observable in the grain structure of F, whereas the grain structure of QF was more randomly orientated. Typical microstructures at quarter thickness are presented in Figure 1.

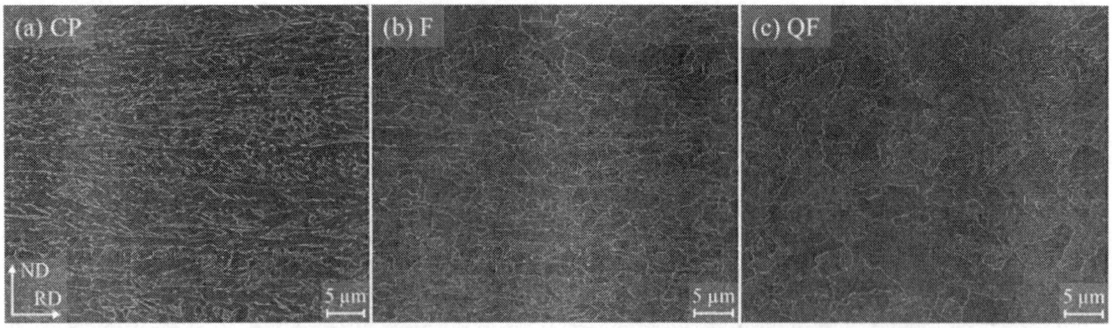

Figure 1. Typical quarter thickness microstructures of the investigated materials: (a) CP, (b) F, and (c) QF, scanning electron microscope images after 2% Nital etching.

2.2. *Mechanical testing*

Tensile tests were performed in accordance with ISO 6892-1:2019 [11] at room temperature. Flat specimens had parallel length of 120 mm and width of 20 mm. The tests were performed to an angle of 0° (RD), 45° (DD), and 90° (TD) relative to the rolling direction with two repeated tests. A GOM Aramis optical strain measurement system (DIC) was utilized to obtain strains for determination of plastic strain ratios (r). The r values, normal anisotropies, and planar anisotropies were determined according to ISO 10113:2020 [12] from 2%-A_g plastic strain range.

Two different hole expansion test methods were utilized to investigate different edge loading scenarios: hole expansion with a hemispherical punch (Nakajima hole expansion test, NHE) and hole expansion with a flat-top punch (Kaiser Wilhelm Institut test, KWI). The test holes for NHE and KWI tests were prepared similarly to the ISO 16630:2017 hole expansion standard [3] by punching a 10 mm diameter hole with 11.7% cutting clearance in an Erichsen universal sheet metal testing machine. Blank diameter for the round NHE specimen was 215 mm, and for KWI tests, 150 x 500 mm specimen size was utilized with 150 mm spacing between the holes. The burr side of each specimen was painted with a stochastic paint pattern to enable DIC measurements.

Hole expansion testing was performed with an Erichsen universal sheet metal testing machine utilizing 600 kN blank holder force and 0.5 mm/s punch speed. In the NHE tests, the hemispherical punch diameter was 100 mm, and multipurpose grease lubrication was used on the punch side. In the KWI tests, the flat-top punch diameter was 50 mm, and Teflon foil with 0.1 mm thickness and multipurpose grease on both sides of the foil was used as lubricant. The burr side was set to face away from the punch in both test set-ups, and three repetitions were performed for each test and material combination. For comparison purposes, ISO 16630 conforming conical punch hole expansion tests without DIC measurement were conducted with four repeated tests for all materials. The different tool geometries and typical specimen shapes after testing are presented in Figure 2a-c.

NHE and KWI testing was recorded with a GOM Aramis DIC system. Utilized frame rate in the beginning of each test was 1 Hz and was increased to 10 Hz close to the end of the test. Strain analysis was conducted as close to the edge as possible with a facet size of 2.1 mm x 2.1 mm. A typical strain map during NHE testing is presented in Figure 2d. Local point strains from the start of the test up to the moment before through-thickness crack appearance were measured for three different edge strain components accordingly to Schneider et al. [8]: major strain (circumferential strain), minor strain (radial strain), and thickness strain, illustrated in Figure 2d. Strains were measured with points along three circle-shape sections around the hole in various distances from the cut edge (Figure 2e): the first section was as close to the edge as possible (approx. 2 mm) to obtain coherent data around the edge, the second 1 mm outward of the first section, and the third 4 mm from the first section. In this paper, only the major strain measurements are discussed. From the measurement data, strain evolution and localization during testing was evaluated.

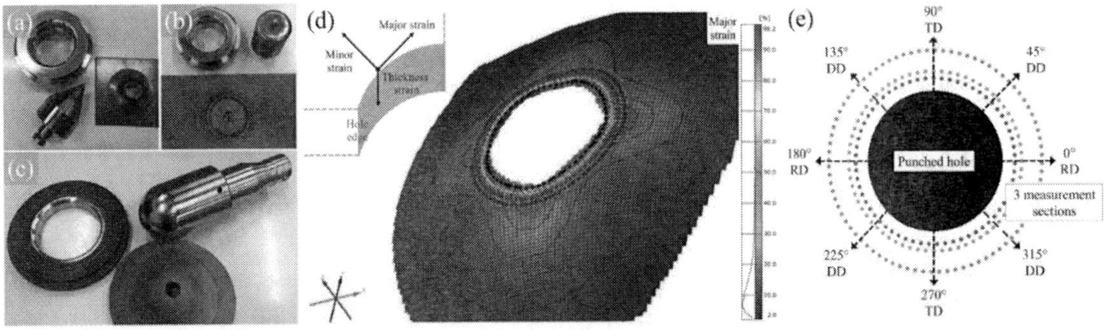

Figure 2. Utilized tools and typical specimen shapes after (a) ISO, (b) KWI, and (c) NHE hole expansion tests. (d) Illustration of strain components near the cut edge and a typical DIC strain map of NHE test, and (e) illustration of strain measurement circle sections and directions around test holes.

Hole expansion ratios were determined from NHE and KWI tests with pixel line measurements of hole diameters from the image at the beginning of the test and the last image before through-thickness crack appearance. From ISO 16630 HE-tests, HER was determined by caliper measurements according to the standard before and after the testing.

3. Results and discussion

3.1. Tensile tests

Longitudinal tensile properties were mostly similar between the investigated materials. Yield strengths ($R_{p0.2}$) were almost identical and ultimate tensile strength (R_m) was slightly lower in QF. F and QF had similar uniform elongation (A_g) and total elongation (A_{80}) levels, whereas in CP both elongation values were lower. Tensile testing and plastic strain measurement results are summarized in Table 1.

Table 1. Mean tensile test results measured in the rolling direction. Determined plastic strain ratios, normal-, and planar anisotropies.

Material	$R_{p0.2}$ (MPa)	R_m (MPa)	A_g (%)	A_{80} (%)	r_{RD}	r_{DD}	r_{TD}	r_m	Δr
CP	751	846	5.8	11.5	0.46	1.20	0.64	0.87	-0.65
F	757	845	9.1	16.4	0.58	1.20	0.75	0.93	-0.53
QF	751	828	9.1	17.4	0.70	1.01	0.89	0.89	-0.22

The highest plastic strain ratios were obtained in the diagonal direction and the lowest ones in the rolling direction in each material, indicating weakest thinning resistance in the rolling direction. Determined normal anisotropies (r_m) were below 1.0 and essentially similar between the materials. All planar anisotropy (Δr) values were negative and had more variation compared to r_m. The least planar anisotropy was observed in QF. Tensile properties and anisotropy measured by tensile testing will be discussed more comprehensively in another publication [10].

3.2. Hole expansion ratio

Hole expansion ratio measurement results are summarized in Table 2. HER was observed to be the highest in QF material regardless of the testing method. The flat-top punch test produced lower HER values compared to the hemispherical and conical punch test set-ups. ISO test HER was similar between CP and F and significantly higher in QF. Hemispherical punch test set-up produced the highest HER in CP and F, whereas in QF NHE, HER was lower compared to the ISO HER value.

Table 2. Mean hole expansion test results. Standard deviations of the measurements are given.

	Pixel measurement		Caliper measurement
Material	NHE HER (%)	KWI HER (%)	ISO HER (%)
CP	53 ± 6	28 ± 3	38 ± 7
F	35 ± 1	23 ± 1	33 ± 6
QF	76 ± 4	48 ± 2	106 ± 14

The different measurement methods could have had effects on the determined HER values. In the ISO standard method, the diameter of the expanded hole is measured after the testing when the load of the expanding punch is removed. The pixel line measurements were done from camera images taken during the testing from the last image before crack appearance where the specimen was in loaded condition. In high-strength steels, springback can cause the edge to withdraw after the load removal leading to a decrease in the hole diameter. Additionally, due to the angles of the cameras and the punch potentially covering bottom edges, the line measurements were done on the top edge of the hole. These two factors could have overestimated the HER values measured from the NHE and KWI specimens compared to the standard ISO HER measurements.

3.3. Major strain distribution along the edge

In all NHE and KWI specimens, the highest local major strain peaks in the measurement circle sections around the hole edge were observed to appear in the sheet rolling direction, maximum value being close to either 0° or 180° angle relative to the rolling direction. Secondary major strain peaks were seen near the sheet transversal direction (90° and 270°), and major strain minimums were seen close to the sheet diagonal directions (45°, 135°, 225°, and 315°). The findings are consistent with the tensile test results, where anisotropy was observed in the materials as the diagonal direction had higher thinning resistance compared to the other sheet directions. Through-thickness cracks were observed after the testing for both NHE and KWI specimens in the rolling and transversal directions where the major strain peaks were located. Unfortunately, in the present tests, we were unable to confirm whether the first through-thickness crack in the edge was always appearing in the location with the maximum major local strain or possibly in other strain peak directions.

Different major (true) strain measurement results from the closest section to the hole edge are summarized in Figure 3. Maximum local major strain, mean local major strain, and minimum local major strain along the circle section were determined from each specimen and mean values from the three repeated tests were calculated. Maximum local major strains along the section were observed to be highest in QF and lowest in F. CP showed higher measured strain values compared to F despite its worse elongation properties measured by the tensile testing. The maximum strain levels were similar in NHE and KWI tests, however, scatter between the specimens was high. Mean local major strains along the section were also highest in QF and lowest in F. In the mean local major strain, there was a difference between the test methods as KWI showed lower mean strains compared to NHE for all materials. Minimum local strains along the section were similarly highest in QF. The minimum strain levels were lower in KWI tests compared to NHE tests. Scatter between the specimens was considerably lower for mean local and minimum local major strains.

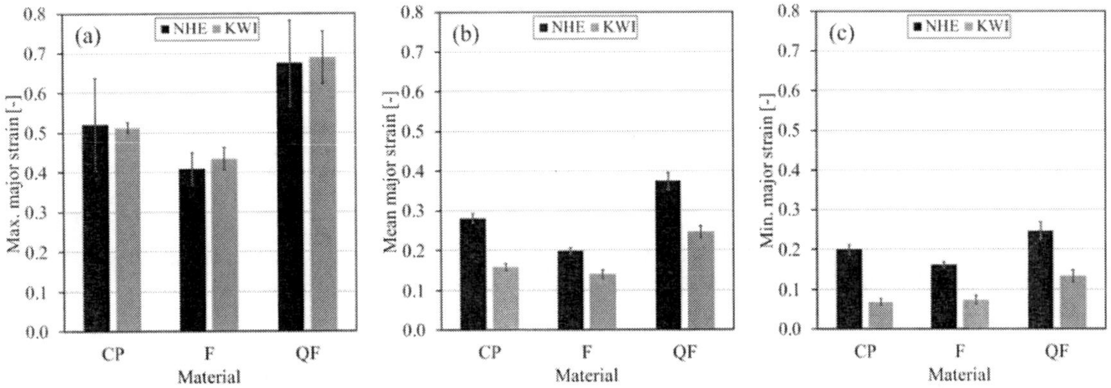

Figure 3. Calculated means of different major strain measurement results in the closest section to the hole edge: (a) maximum local major strain, (b) mean local major strain, and (c) minimum local major strain along the circle section. Error bars represent standard deviation of the measurements.

In Figure 4, hole expansion results are plotted against different local major strain measurement results closest to the hole edge for both NHE and KWI tests. The results show excellent correlation between HER and mean local major strain along the hole edge, and the correlation was slightly lower for minimum local major strains. The maximum major strain results had more scatter even compared to the hole expansion ratios, and the correlation was found to be weak.

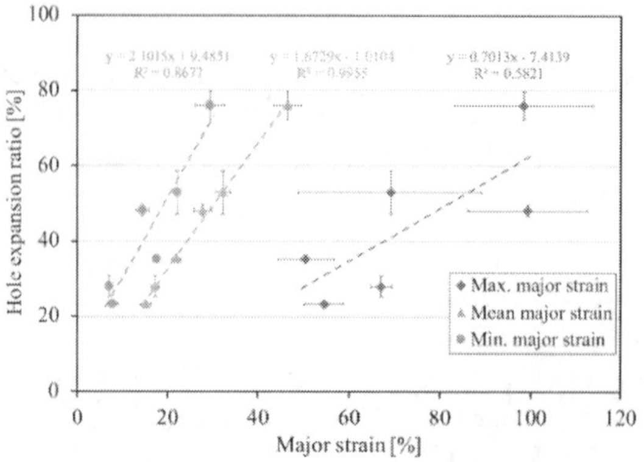

Figure 4. Correlations between means of different local major (engineering) strain measurement results closest to the hole edge and hole expansion ratio. Error bars represent standard deviation of the measurements.

For the more precise investigation of strains, representative specimens were selected from each test and material combination. Figure 5 shows polar maps of major strains close to the edge from the last image before cracking occurs. The polar maps show the directions of the maximum local major strain along the hole edge as an angle relative to the rolling direction. NHE tests show higher peak and uniform strains around the hole compared to KWI tests. The peak strains are located in the rolling and transversal direction, similarly to some of the polar maps presented by Denks et al. [9].

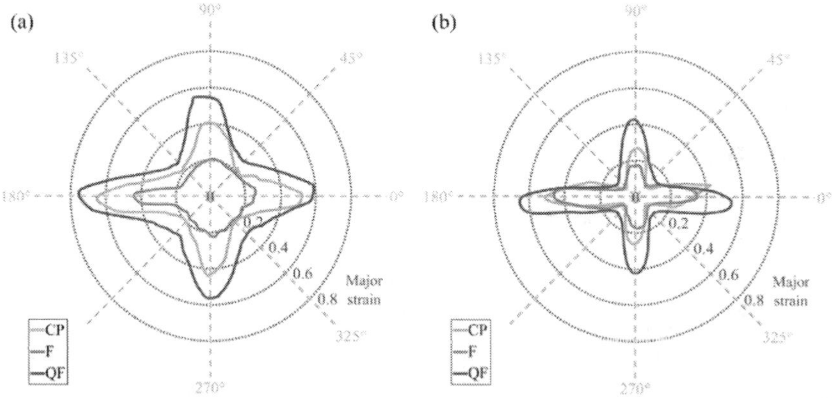

Figure 5. Polar maps of local major strains in closest section to the hole edge of (a) NHE and (b) KWI specimens before the through-thickness fracture.

Major strain distributions in the closest section to the hole edge are presented in a different form in Figure 6a-f, where the four dotted lines are representing different time points during the tests. The first line was selected to be common between the specimens (15 s), where the major strain distribution is still uniform around the edge. The second line is at the time where the strain distribution starts to show some signs of localization. The third line shows the distribution before the start of pronounced necking, and the final line is presenting the strain distribution in the moment before through-thickness fracture appearance. In NHE test F shows the most difference between the strain distribution shapes compared to two other materials. In KWI tests the distribution shapes are more similar. The strain peaks are located in rolling and transversal directions, and minimums in diagonal directions. QF has the highest strain levels before strain localization as well as the highest local peak strains. The levels of uniform strain are similar in F and CP, but the withstood local stains are comparatively smaller in F. This could indicate that F has local ductility issues that could be related to fracture toughness, as steels with higher fracture toughness can display higher strains at fracture [13].

IOP Conf. Series: Materials Science and Engineering 1284 (2023) 012022 doi:10.1088/1757-899X/1284/1/012022

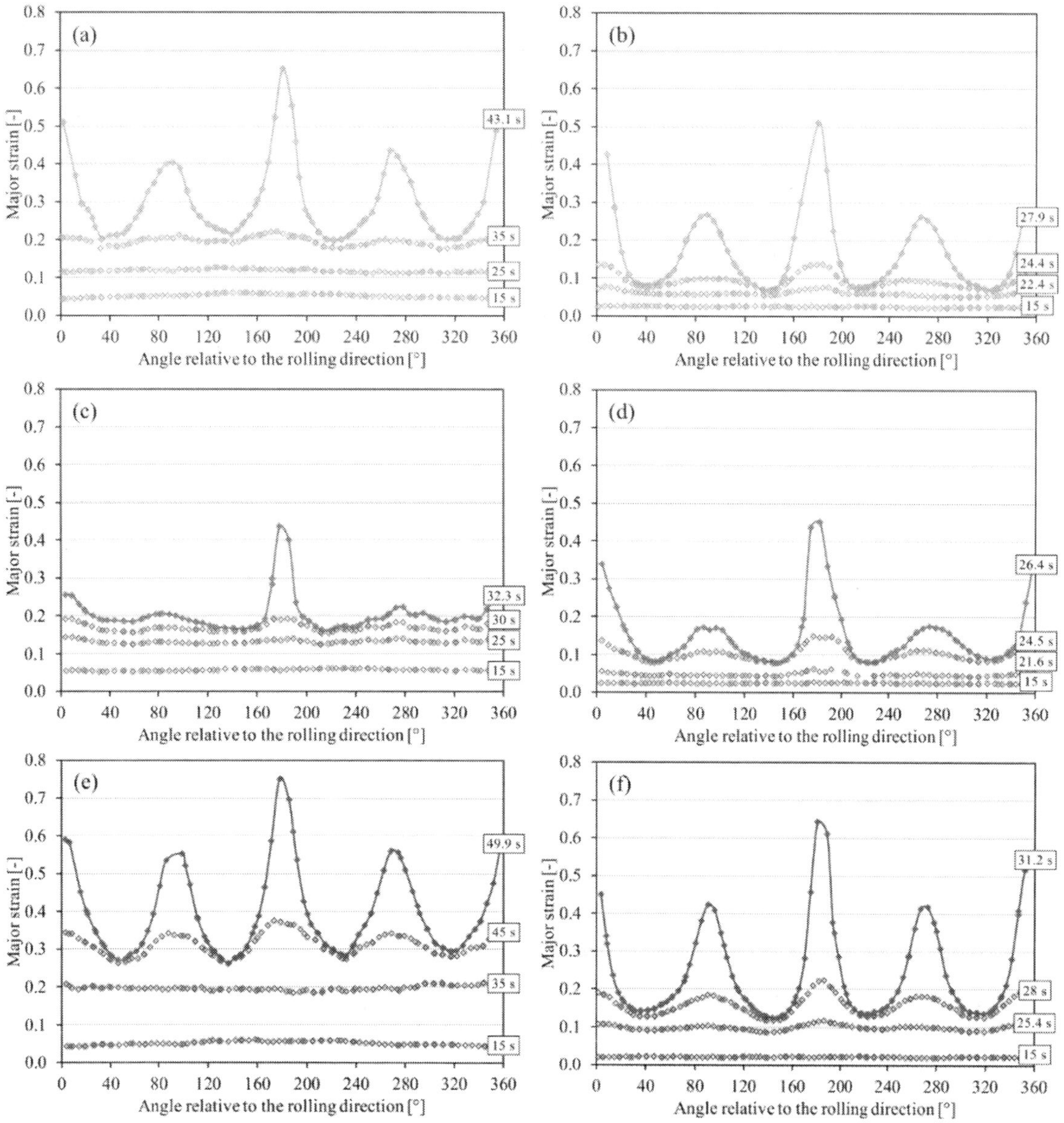

Figure 6. Major strain distribution in the section closest to the hole edge at different time intervals during testing of (a) CP - NHE, (b) CP - KWI, (c) F - NHE, (d) F - KWI, (e) QF - NHE, and (f) QF - KWI.

Higher HER results may be possible by increasing the mean major strain before the fracture occurrence around the hole edge. This could possibly be achieved by increasing the level of uniform strain in the edge by delaying the onset of localization of strain, or by increasing the tolerance to the peak strains. From these, the uniform strain level seems to be more effective as seen with the difference between QF and CP NHE test results. However, the material also should have tolerance to strain peaks as otherwise fracture would occur as soon as the strain starts to localize. High tolerance to localized strains could be increasingly important in applications where localization is forced by the specimen geometry or tool set-up.

3.4. Major strain evolution during testing

In Figure 7, evolution of the major strain in the section closest to the edge is presented in different sheet directions for the selected specimens: rolling direction (RD), diagonal direction (DD), and transversal direction (TD).

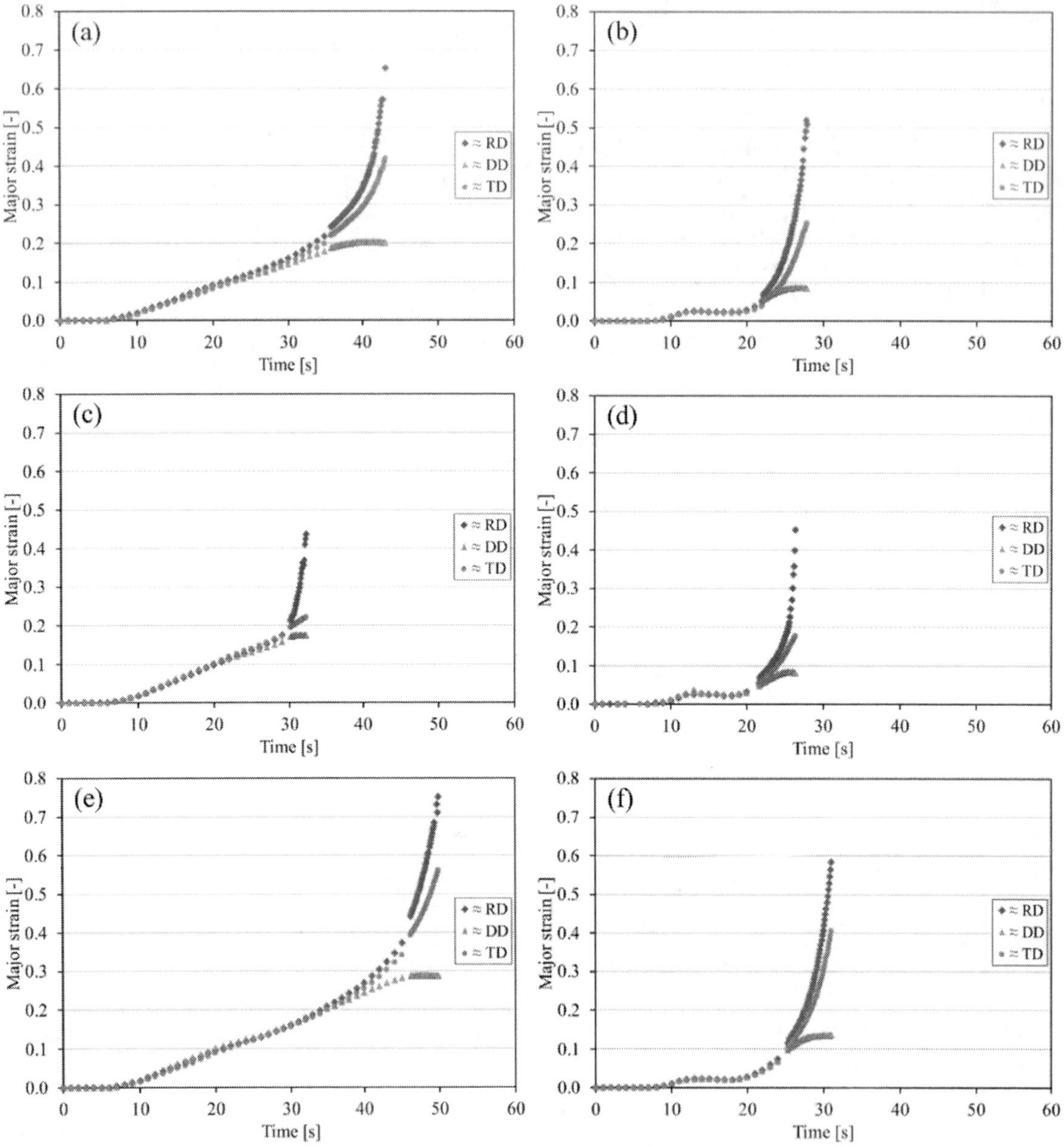

Figure 7. Evolution of major strain in the section closest to the hole edge in different approximate sheet directions during testing of (a) CP - NHE, (b) CP - KWI, (c) F - NHE, (d) F - KWI, (e) QF - NHE, and (f) QF - KWI.

The sheet directions are approximate, as the measurement points along the circle sections were not always exactly in the desired directions, for example in 180°, but rather within a few degrees. At the beginning of each test, strain level is zero as punch must reach the specimen for deformation to start. In NHE specimens, strains increase linearly until start of localization and all materials seem to follow approximately the same strain rate. In KWI specimens, there is a low plateau of strain as the deformation is focusing on the edges of the flat-top punch before the strains in the edge starts to increase. Localization to RD and TD occurs quicker and in lower strain levels compared to NHE specimens. In common for both test set-ups, the major strain stops to increase in DD when pronounced necking starts. Strains in TD are following those of RD in behind but fail to reach as high levels. In both tests set-ups QF shows delayed localization of the strain compared to the other materials.

3.5. Major strain distribution inwards from the edge

In Figure 8, major strains measured in the different circle sections around the hole edges of NHE and KWI tests are presented for different sheet directions. Distances of the points from the edges were determined from the coordinate data and assuming ideal 10 mm diameter for the initial hole. The distance is not the same at the end of the test as the material close to the edge contracts in the radial direction as the test proceeds. However, the results should estimate the strain gradients at end of the tests. Due to challenges in DIC measurements close to the edge and placement of the measurement sections, the measurement distances from the edges have differences between the specimens.

Based on the gradients, it can be inferred that the strains are considerably higher closer to the edge compared to the results from the first measurement section. Strain gradients close to the hole edge have been shown in simulations to be affected by the punch shape and to have proportional ratio to logarithmic hole expansion test results [5]. The major strain levels in the radial direction were generally higher in NHE specimens, the one exception being KWI F specimen. In addition to the differences resulting from the punch shapes, the strain levels also had differences in the sheet directions. Strain gradients in most specimens were steeper in RD and TD compared to DD resulting from the localization of the strain.

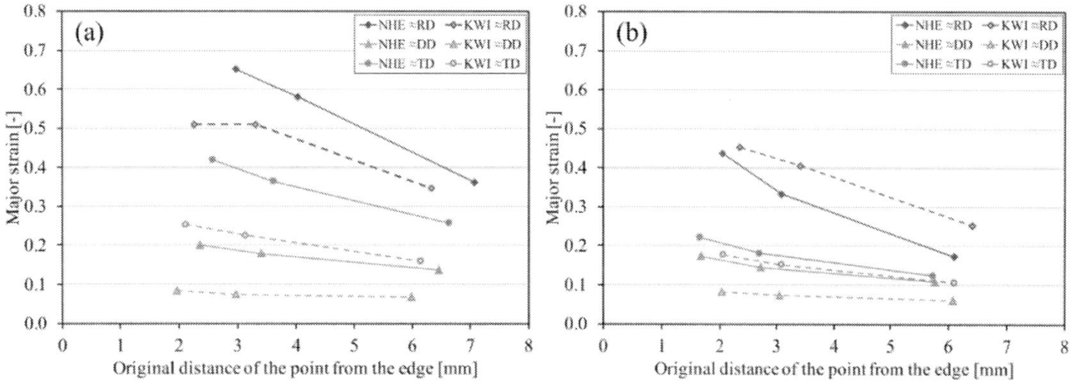

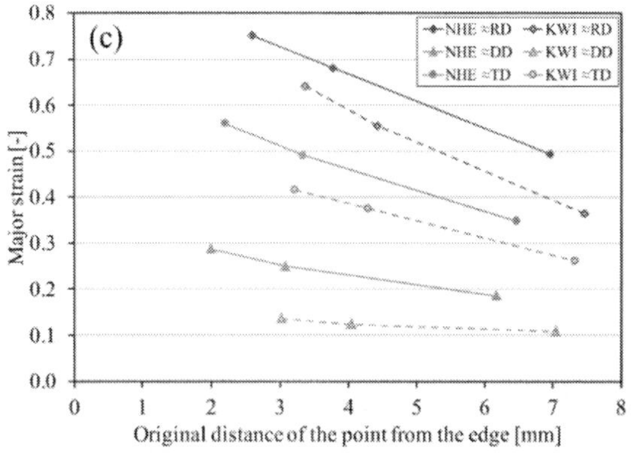

Figure 8. Major strains of NHE and KWI tests of (a) CP, (b) F, and (c) QF specimens in different approximate sheet directions and the distance of circle section measurement points from the hole edge at the start of the test.

4. Conclusion

Strain evolution and localization during hole expansion testing was investigated for three 800 MPa tensile strength grade hot-rolled steels with different microstructures. Hemispherical punch and flat-top punch test set-ups were utilized to obtain different edge loading scenarios. Different local strain components, most importantly major strain (circumferential strain), around the expanding hole were measured close to the hole edge by utilizing digital image correlation, and tensile testing was conducted to determine plastic strain ratios.

In all hole expansion tests, major strain was detected to localize in the directions with the least amount of thinning resistance: the rolling and transversal directions. Through-thickness cracks in the edges were observed to appear in these strain peak directions. Significant localization of the major strain was occurring earlier in flat-top punch tests compared to hemispherical punch tests. The material with the least amount of planar anisotropy was observed to have highest peak and mean local major strains, and the localization of strain was appearing comparably later. Hole expansion ratios showed excellent correlation with measured mean local major strain along the edge in both hemispherical and flat-top punch test set-ups.

Results suggest that in applications, where the strain is not forced to localize in certain directions by the specimen geometry, such as in hole expansion tests, superior sheared edge formability for high-strength hot-rolled steels could be obtained by improving the mean major strain component along the edge. This could be achieved by combining factors of delayed localization of strain promoted by lesser degree of anisotropy, and by improving tolerance of the material to withstand localized strains. However, due to the limited number of investigated materials, more work is needed to further validate the findings.

Acknowledgments

Financial support for the research work by Business Finland project FOSSA – Fossil-Free Steel Applications is highly appreciated.

References

[1] Levy B S and Van Tyne C J 2012 *J. Mater. Eng. Perform.* **21** 1205–13
[2] Konieczny A and Henderson T *2007 SAE Tech. Paper 2007-01-03*
[3] ISO 16630:2017 Metallic materials - Sheet and strip - Hole expanding test
[4] Pathak N, Butcher C and Worswick M 2016 *J. Mater. Eng. Perform.* **25** 4919–32
[5] Larour P, Hinterdorfer J, Wagner L, Freudenthaler J, Grünsteidl A and Kerschbaum M 2022 *IOP Conf. Ser.: Mater. Sci. Eng.* **1238** 012041
[6] Pushkareva I, Allain S, Scott C, Redjaïmia A and Moulin A 2015 *ISIJ International* **55** 2237–46
[7] Kaijalainen A, Kesti V, Vierelä R, Ylitolva M, Porter D and Kömi J 2017 *J. Phys.: Conf. Ser.* **896** 012103
[8] Schneider M, Geffert A, Peshekhodov I, Bouguecha A and Behrens B A 2015 *Materwiss. Werksttech.* **46** 1196–217
[9] Denks I A, Schneider M, Westhäuser S and Lesch C 2019 *Steel Res. Int.* **90** 1–16
[10] Plosila P, Kesti V, Kaijalainen A, Hannula J and Kömi J 2023 Effect of anisotropy of mechanical properties on hole expansion of 800 MPa tensile strength hot-rolled steels (in preparation)
[11] ISO 6892-1:2019 Metallic materials - Tensile testing - Part 1: Method of test at room temperature
[12] ISO 10113:2020 Metallic materials - Sheet and strip - Determination of plastic strain ratio
[13] Frómeta D, Lara A, Parareda S and Casellas D 2019 *AIP Conf. Proc.* **2113** 160007

Accuracy of stretch flange formability prediction by considering strain gradient for actual car body parts

E Iizuka[1], A Okada[2]

[1]JFE Steel Corporation, 3-28-12, Meieki, Nakamura-ku, Nagoya, Aichi Pref., 450-6427, Japan
[2]Bellsonica Corporation, 630-18, Yamaguchi, Kosai-city, Shizuoka Pref., 431-0043, Japan

E-mail: e-iizuka@jfe-steel.co.jp

Abstract. The effect of the strain gradient on the stretch flange deformation limit of steel sheets was investigated by hole expansion tests under various forming conditions and by FEM forming simulations. The hole expansion ratio was changed by changing the initial hole diameter and punch shapes. This study clarified the fact that deformation limit strain increases, as it does not depend on the initial hole diameter or punch shape, and thus the strain gradient increases. The deformation limit thickness strain of the stretch flange portion could be changed by changing the strain gradient. These results suggest that the formability of the stretch flange deformation area should be determined from not only the maximum principal strain but also the strain gradient. Then, the accuracy of stretch flange formability prediction by considering the strain gradient for actual car body parts in mass production was investigated. The maximum principal strain and the strain gradient of the parts were calculated by FEM or measured in the press part. These results suggest that the formability of the stretch flange deformation area for actual car body parts could not be predicted by not only the maximum principal strain or the thickness reduction, but also the conventional FLD. This investigation verified the accuracy of stretch flange formability prediction by considering the strain gradient for actual car body parts.

1. Introduction

Demand for high strength steel sheets is increasing with the aims of realizing larger vehicle weight reductions and enhanced crash performance. One of the largest problems in press forming of high strength steel sheets is fracture, which often occurs in stretch flanging areas, not only in inner panel parts but also in frame and underbody parts. Therefore, there has been strong demand for improvement of the predictive accuracy of the stretch flange formability of high strength steels by finite element method (FEM) forming simulations [1].

To improve the predictive accuracy of stretch flange formability by FEM analysis, estimation of the deformation limit as well as estimation of the deformation behavior of the material is generally important. Regarding stretch flange deformation behavior, Kuwabara et al. [2,3] investigated the influence of the anisotropy yield function. Regarding the stretch flange deformation limit, many studies have shown the influence of the shearing edge condition [4] and material properties [5].

In most studies of the influence of the strain gradient on the stretch flange deformation limit, the stretch flange deformation limit was basically evaluated by the results under shaving edge conditions. However, no studies have reported quantitative estimations of the influence of the strain gradient on the stretch flange deformation limit for sheared edges. Thus, there will also be no subsequent studies which verify stretch flange formability considering the influence of the strain gradient of the stretch flange deformation portion in FEM forming simulations.

In this situation, Iizuka et al. [5-7] investigated the effect of the strain gradient on the stretch flange deformation limit strain of steel sheets. The results clarified the fact that the stretch flange deformation limit strain basically depends on the strain gradient in the radial direction. As conclusions, it was suggested that the formability of the stretch flange deformation area should be determined from not only the maximum principal strain or the thickness reduction, but also the strain gradient and blanking clearance. Then, the accuracy of stretch flange formability prediction by considering the strain gradient for actual car body parts in mass production was investigated. The maximum principal strain and the strain gradient of the parts were calculated by FEM or measured in the press parts. These results suggest that the formability of the stretch flange deformation area for actual car body parts could not be predicted by not only the maximum principal strain or the thickness reduction, but also the conventional FLD. These investigations verified the accuracy of stretch flange formability prediction by considering the strain gradient for actual car body parts.

2. Experimental procedure

2.1. Target part

To verify the accuracy of stretch flange formability prediction, a target part was selected from parts in which stretch flange fracture is a concern in mass production. The material grade of this part in mass production is JAC440W, and the thickness is 1.4 mm. Figure 1 show the part shape and seven alarm portions for stretch flange fracture. The seven alarm portions were selected based on the thickness reduction calculated by FEM analysis or measured in the press trial part.

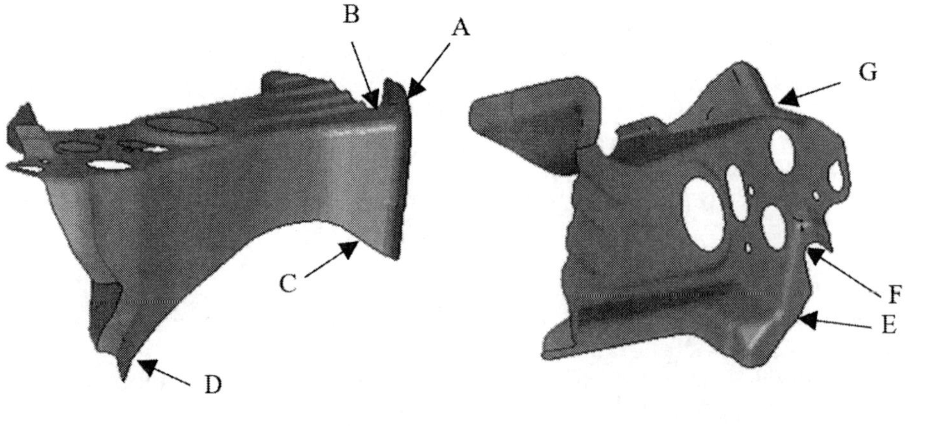

Figure 1. Part shape and alarm portions for stretch flange fracture.

2.2. Test material

The mechanical properties of the material in the hole expansion test are shown in Table 1. Yield strength YS, tensile strength TS, and elongation El were obtained by uniaxial tensile tests in the rolling direction.

Table 1. Thickness and mechanical properties of material.

Thickness/mm	YS/MPa	TS/MPa	El/%
1.4	320	466	36

2.3. Hole expansion test procedure

Hole expansion tests were carried out to evaluate the stretch flange deformation limits of the test material. Two punch shapes (conical, cylindrical) and three initial hole diameters (10 mm, 25 mm, 50 mm) were selected to investigate the effects of the strain gradient on the stretch flange deformation limit. Table 2 shows the experimental conditions for the hole expansion tests by conical punches and table 3 shows the experimental conditions for the hole expansion tests by flat-bottomed punches in this study. To make fracture occur from a hole edge intentionally by a flat-bottom hole expansion test, the punch diameter was changed suitably. All initial holes of the test materials were punched out with a clearance between the punch and die of 12.5%. In all hole expansion tests, anti-rust oil was used as the lubrication condition, burr was formed on the die side, and the number of samples was five.

Table 2. Experimental conditions for hole expansion tests by conical punches.

Diameter/mm			Shoulder radius/mm
Initial hole	Punch	Die	Die
10	50	59.6	5
25	98	150	5
50	140	160	5

Table 3. Experimental conditions for hole expansion tests by flat-bottomed punches.

Diameter/mm			Shoulder radius/mm	
Initial hole	Punch	Die	Punch	Die
10	50	59.6	8	5
25	80	160	5	5
50	100	160	10	5

3. Results

3.1. Stretch flange deformation limit of test material

Figure 2 shows the effect of the strain gradient on the deformation limit strain of the test material. The deformation limit strain was calculated by the average and minimum hole expansion ratio in each test as the average deformation strain of the hole edge. Figure 2 show the schematic diagram of strain gradient. The strain gradient was defined as the average gradient of the maximum major strain in the radial direction (direction crossing the maximum major strain direction at right angles) from the hole edge, and was analyzed by an FEM forming simulation. The deformation limit strain increased linearly, mostly with increases in the strain gradient, and did not depend on the punch shape and initial hole diameter, which was the same as the previous results [6]. In Figure 3, the minimum line is approximately 15% lower than the average line.

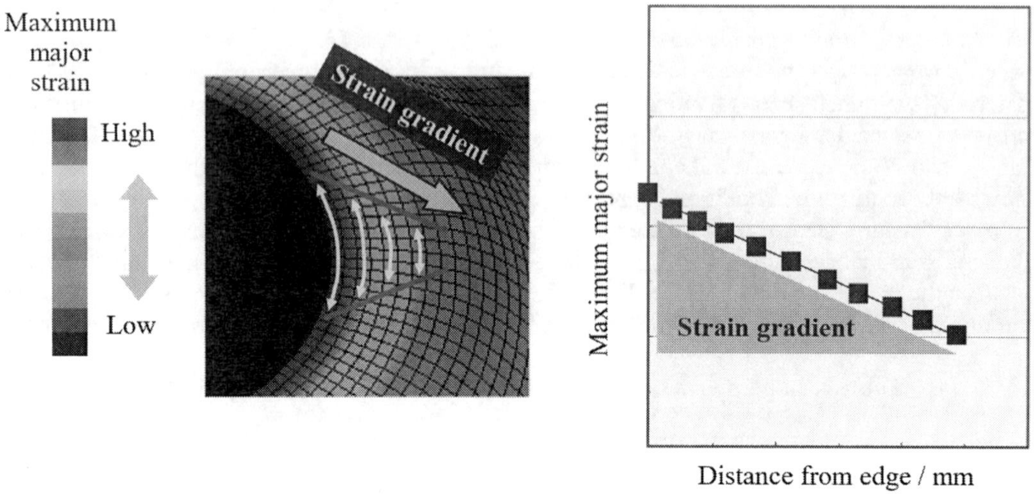

Figure 2. Schematic diagram of strain gradient.

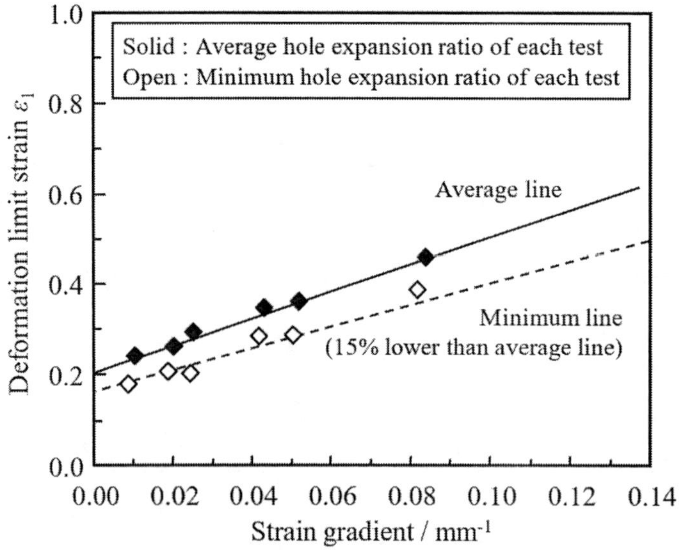

Figure 3. Effect of strain gradient on deformation limit strain of test material.

3.2. Stretch flange formability of target part in mass production

Table 4 shows the stretch flange formability predictions for the alarm portions by FLD calculated by FEM or thickness reduction by press trial before mass production and the results of stretch flange formability in mass production.

Table 4. Stretch flange formability predictions and results of stretch flange formability.

Result	Alarm portion						
	A	B	C	D	E	F	G
FEM prediction by FLD	Fracture	Fracture	Fracture	Formable	Formable	Formable	Fracture
Thickness reduction by press trial	Fracture	Formable	Fracture	Formable	Formable	Formable	Formable
FEM prediction by considering strain gradient	Fracture	Fracture	Fracture	Formable	Fracture	Fracture	Formable
Mass prodction	Fracture	Fracture	Fracture	Formable	Fracture	Fracture	Formable

In mass production, stretch flange fracture appeared sporadically at alarm portions A, B, C, E, and F. The stretch flange formability predictions of the alarm portions by FLD or thickness reduction did not correspond with the results of actual stretch flange formability in mass production at portions B, E, F and G.

4. Discussion

Figure 4 shows the results of stretch flange formability prediction by considering the strain gradient. The maximum principal strain and the strain gradient of the parts were calculated by FEM or measured in the press part. The maximum principal strain calculated by FEM corresponded with the measurement result of the press part.

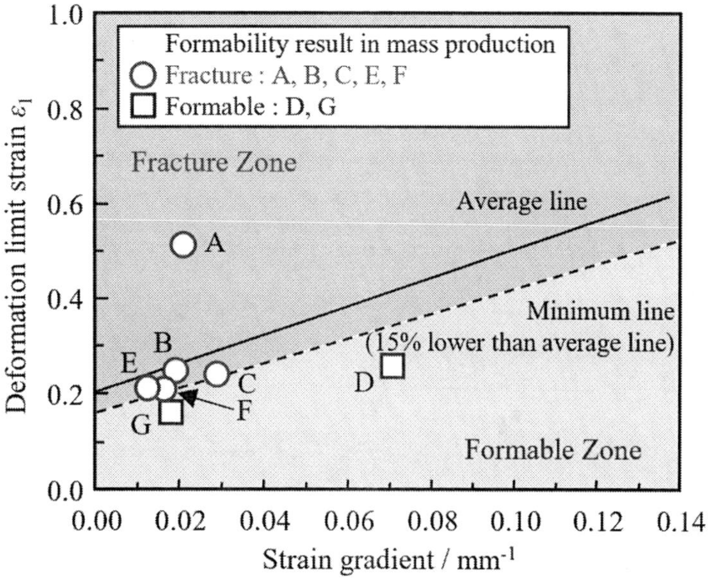

Figure 4. Results of stretch flange formability prediction by considering strain gradient.

Stretch flange formability in mass production can be predicted exactly by the minimum line, which is approximately 15% lower than the average line. This result suggests that stretch flange formability should be judged by considering the strain gradient, and the forming allowance, which includes the forming and material tolerance in mass production, should be set with a margin of approximately 15%.

5. Conclusions

The accuracy of stretch flange formability prediction by considering the strain gradient for actual car body parts in mass production was investigated. As the results, the following conclusions were obtained.

(1) The formability of the stretch flange deformation area for an actual car body part in mass production did not correspond with the FEM prediction result by the FLD or the thickness reduction by the press trial.
(2) Stretch flange formability in mass production can be predicted exactly by the minimum line, and that line is approximately 15% lower than the average line.
(3) Stretch flange formability should be judged by considering the strain gradient, and the forming allowance, which includes the forming and material tolerance in mass production, should be set with a margin of approximately 15%.

These investigations verified the accuracy of stretch flange formability prediction by considering the strain gradient for actual car body.

References

[1] Iizuka E 2017 *Journal of the JSTP* **58** 110–114.
[2] Kuwabara T, Hashimoto K, Iizuka E, Yoon J W 2009 *Journal of the JSTP* **50** 925–930.
[3] Worswick M J, Finn M J 2000 *Int. J. of Plasticity* **16** 701-720.
[4] Iizuka E, Hira T, Yoshitake A 2005 *Journal of the JSTP* **46** 625–629.
[5] Iizuka E, Urabe M, Yamasaki Y, Hiramoto J 2010 *Journal of the JSTP* **51** 700–705.
[6] Iizuka E, Urabe M, Yamasaki Y, Hiramoto J 2017 *J. of Physics: Conference Series* **896**, 012008.
[7] Iizuka E, Higai K, Yamasaki Y 2019 *J. of Physics: Conference Series* **651**, 012017.

Effect of blank size on impact hydroforming of 2195 Al-Li alloy half-tube part with curvature

H L Zhu[1,2], Y Xu[1,*], S F Chen[1] and S H Zhang[1]

[1] Shi-changxu Innovation Center for Advanced Materials, Institute of Metal Research, Chinese Academy of Sciences, Shenyang 110016, People's Republic of China
[2] School of Materials Science and Engineering, University of Science and Technology of China, Shenyang 110016, People's Republic of China

E-mail*: yxu@imr.ac.cn

Abstract. Aluminum-Lithium (Al-Li) alloy has excellent properties such as low density, high elastic modulus and high specific strength, and it is considered as the most ideal structural material in aerospace. However, due to the low plasticity of Al-Li alloy at room temperature, it is easy to crack during cold forming, so its application and development have been limited. In order to explore the possibility of improving the plasticity of 2195 Al-Li alloy at room temperature, a half-tube part with curvature and deep cavity structure was formed by using impact hydroforming (IHF) technology. In this paper, the process route of half-tube part was optimized and four sizes of blank were designed. Through finite element numerical simulation and analysis, it was found that the F area was prone to wrinkle defects when there was less material at the inner arc position, and the most serious thinning occurred in the design margin area. After analysis and comparison, the blank size with drawing ratio of 2.22 was finally selected. Based on the verification experiment, it was found that due to the material properties of Al-Li alloy and the characteristics of IHF, one-step forming was difficult, and the forming of half-tube part was prone to insufficient drawing depth or large area fracture; Although the part obtained by two-step forming also had cracks, the overall forming quality of the part would not be affected because they were located in the design margin area to be removed later. In addition, it was observed that the surface quality of the part removed the design margin area were free of wrinkles and cracks, which meets the objective design requirements.

1. Introduction

As a kind of aluminum alloy, aluminum-lithium (Al-Li) alloy has been improved after three generations and it has many excellent properties such as lower density, higher elastic modulus and higher specific strength, etc. According to statistics, replacing conventional high-strength aluminum alloy with Al-Li alloy can reduce the structural mass by 10%-20% and increase the rigidity by 15%-20%[1]. Therefore, Al-Li alloy is considered as the most ideal structural material in aerospace and it has broad application prospects[2]. However, due to its poor plasticity at room temperature, obvious anisotropy and easy cracking during cold forming, the forming of Al-Li alloy is difficult[3]. Although the warm and hot forming methods can improve the formability of Al-Li alloy[4,5], it has not yet established a complete process specification and industry standard in the aerospace field, which limits the wide application of Al-Li alloy in complex structural parts[6].

Traditional sheet forming processes such as skin stretch forming, rubber forming and drop hammer forming, etc. are adopted to form Al-Li alloy, which often have problems such as uniformity of product, various processes, serious spring-back, difficult to control residual stress, uneven deformation, wrinkling and cracking[7]. Therefore, the key technology and development direction of Al-Li alloy widely used in the future new generation of launch vehicles, large civil aircraft, national defense industry and other important fields is to solve and realize the room-temperature forming process of complex thin-walled components, and constantly develop new technologies to improve the forming accuracy, efficiency and quality on the basis of traditional technologies.

Impact hydroforming (IHF) is a novel plastic forming technology with high strain rate, which can combine the flexible characteristics of hydroforming with the high-speed characteristics of IHF, and it has been attracted wide attention. Broomhead[8] studied the fracture behaviour of the sheet under biaxial tensile stress at high strain rate by using the test method of high-speed drop hammer hitting the liquid chamber to transfer the load to the sheet. Kosing[9] designed and manufactured a high-speed hydraulic punch press with a drop hammer hitting the liquid to obtain the pulse peak, and he carried out a study on the forming limits of sheet and tube. In addition, scholars[10] from Ukraine, Germany and Australia have cooperated to carry out simulation analysis and experimental verification on the free bulging process of tube under the IHF. Lang[11] developed an impact liquid-filled composite forming process, and the feasibility of forming some local small features through high-speed loading was verified by experiments. Yang[12-14] studied the deformation law of tube during IHF and observed the actual effect of improving the formability of tube. In recent years, Zhang[15] et al. first proposed the IHF technology to realize sheet forming only under the impact hydraulic loading. Based on the demonstration and application of different brands of aluminum alloy by Ma, Chen and El-Aty [16-17] et al. , it is proved that IHF technology can significantly improve the formability and precision quality of aluminum alloy. The research of many scholars above has fully proved that the formability of most materials can be improved under high strain rate, and some sheet and tube parts with typical simple structural characteristics can be formed by using IHF technology. However, there are few reports on the forming experiments and formability of low plasticity metal materials such as Al-Li-Cu alloy at room temperature.

Therefore, the paper aims to explore the feasibility of the IHF technology with high strain rate to improve the ductility of 2195 Al-Li alloy at room temperature. A half-tube part with curvature and deep cavity structure was taken as an example, and the forming experiment was carried out by IHF. The model of half-tube parts was optimized, four kinds of blank sizes and outlines were designed, and the best scheme of size and outline was selected based on numerical simulation and experimental research. The margin area that can be removed later was designed, which can improve and ensure the forming quality of half-tube parts. Through experimental research, it was found that due to the material characteristics of Al-Li alloy and the characteristics of IHF technology, one-step forming was difficult, and two-step forming can form a half tube that meets the objective requirements, which fully shows that the IHF technology with high strain rate can effectively improve the formability of 2195 Al-Li alloy at room temperature.

2. Objective part and analysis of forming principle

2.1. Objective part

The objective part is a half-tube part with curvature, whose geometric dimensions and outline are shown in Figure 1 (a), and the 3D model established by SolidWorks is described in Figure (b). The radius R_1 of the tube is 45.8 mm, and the arc radius R_2 of the integral half-tube part is 200 mm. The half-tube part with curvature has the characteristics of deep cavity and large size, and the material is aluminum-lithium (Al-Li) alloy with lower plasticity at room temperature, and it is difficult to form by conventional forming methods, so it is necessary to try to use a new advanced technology to form the half-tube part. The sheets used in the experiment are 2195 Al-Li alloy with a thickness of 1.1 mm provided by a Chinese research institutions. As shown in Table 1, the chemical composition was analysed by plasma emission spectrometer, and the main elements are Al, Li, Cu and Mg. Figure 2 describes the engineering stress-strain curve of 2195 Al-Li alloy obtained by uniaxial tensile experiment.

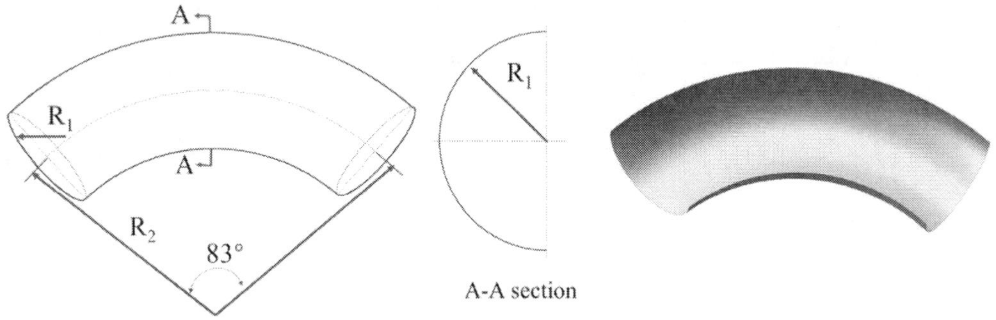

Figure 1. Drawing of half-tube part with curvature (a) Dimension and outline (b) 3D model

Table 1. Chemical composition of 2195 Al-Li alloy (Wt.%)

Element	Li	Cu	Mg	Zr	Ag	Al
Min. [18]	0.80	3.70	0.25	0.08	0.25	Bal
Max. [18]	1.20	4.30	0.80	0.16	0.60	Bal
This work	1.06	3.90	0.54	0.12	0.35	Bal

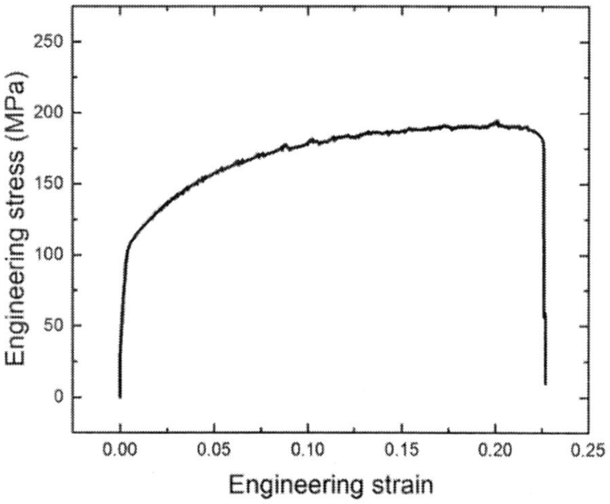

Figure 2. Engineering stress-strain curve of 2195 Al-Li alloy

2.2. Impact hydroforming technology

Impact hydroforming (IHF) is a novel high strain rate forming technology, which has the characteristics of dynamic impact loading and liquid flexibility, and it effectively avoids some shortcomings of traditional high strain rate forming. IHF has many advantages, such as large impact energy, convenient adjustment of initial energy, high forming efficiency and wide application range. As shown in Figure 3 (a) and (b), The basic principle of IHF is to apply the energy of the high-speed impact projectile on the liquid medium to produce the high-energy liquid that can replace the static high-pressure effect of the supercharger output in the traditional hydroforming, and it can make the workpiece quickly attach to the die through one or more impacts[19], and it can form the sheet or tube parts with the specified shape, Figure 3 (c) is the self-developed IHF equipment used in this experiment.

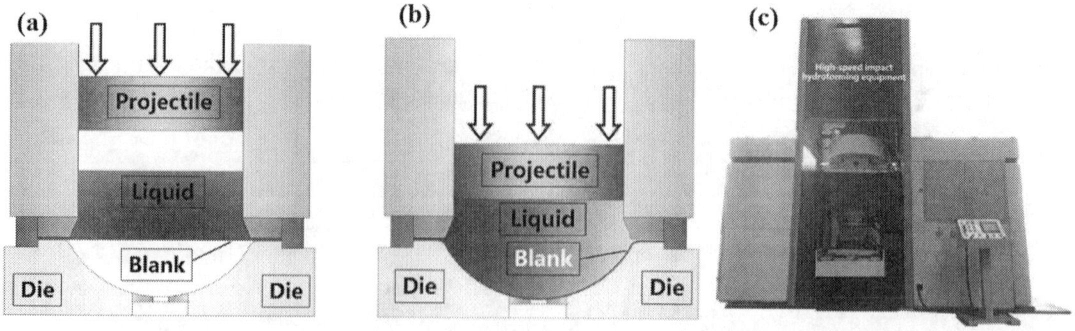

Figure 3. Principle and equipment of IHF (a) Before forming (b) After forming (c) Equipment

2.3. Route of process and size of blank

There are two ideas about the process route of IHF of the half-tube part: first, it is necessary to add the design margin at the large section and both sides of the part, then the IHF process is used to form the half-tube part, and the Wire cut Electrical Discharge Machining (WEDM) is adopted to remove the design margin, and finally the objective part is obtained; Second, according to the principle of equal volume, the initial blank size of the half-tube part without design margin (Figure 5 (a)) is directly designed, and then the IHF is performed. However, in the early stage, through the numerical simulation of the part without design margin, it was found that even if the initial position of the blank was changed and the process parameters were optimized for many times, there were still folding defects that were difficult to eliminate, as shown in Figure 5 (b). Therefore, the second idea was abandoned and the first idea was adopted. The part model with design margin is shown in Figure 4. The optimized process route of the IHF of part is shown in Figure 6. The large section of part was supplemented and the arc-shaped margin area was designed at both sides. The quality of the design margin after deformation will not affect the overall forming quality of the part, because the part or area will be cut off by WEDM or other mechanical methods after forming. Therefore, to some extent, this ensures the forming quality and precision of the objective part.

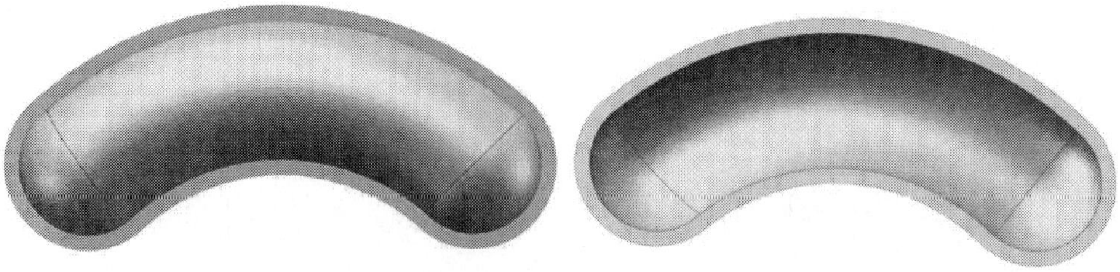

Figure 4. Model of part with design margin

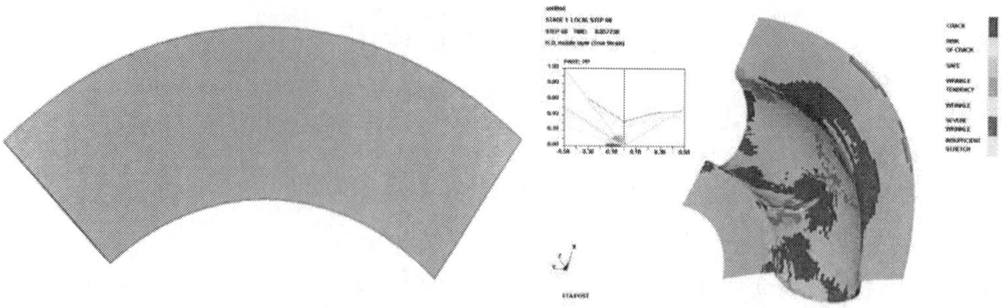

Figure 5. Simulation of part without design margin (a) Initial blank size (b) Simulation results

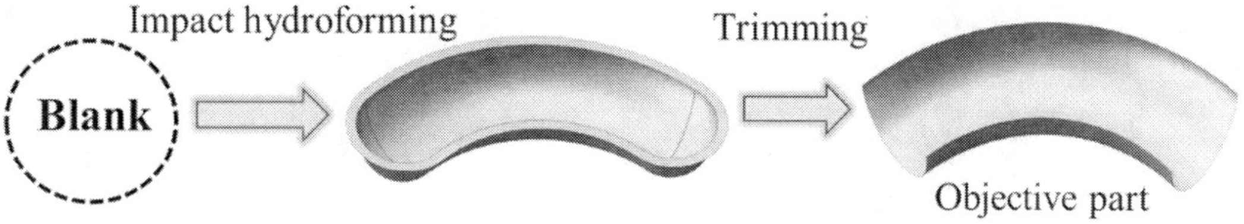

Figure 6. Optimized process route of IHF of part

Based on the blank engineering module of Dynaform-3D software, the part model with design margin was imported, and the outline and size of the blank were obtained through inverse simulation, and four kinds of outline and size were designed. Figure 7 shows the initial outline and size design details of the blank. The formula (1) was used to measure and name four kinds of blanks, for example, the blank with a drawing ratio of 2.08 was called in the first blank size scheme. According to the simulation results in Figure 8, it is found that fold defects occur in area F when the drawing ratio is 2.08 and 2.12, while there are basically no defects at the drawing ratio of 2.17 and 2.22. The reason is that the flange in area H of the former is relatively small, and the contact area with the upper die is small, so the friction force is small , that is, the tensile stress of the flange edge in the H area to the F area is low during the forming process, which leads to the wrinkle in the F area. At the same time, it is also seen that the thinning of D and E area is the most serious, and there is a risk of cracking later, but because the two areas are located in the design margin, it does not affect the forming quality of half-tube part. In summary, it can be considered that the designed process route above is feasible.

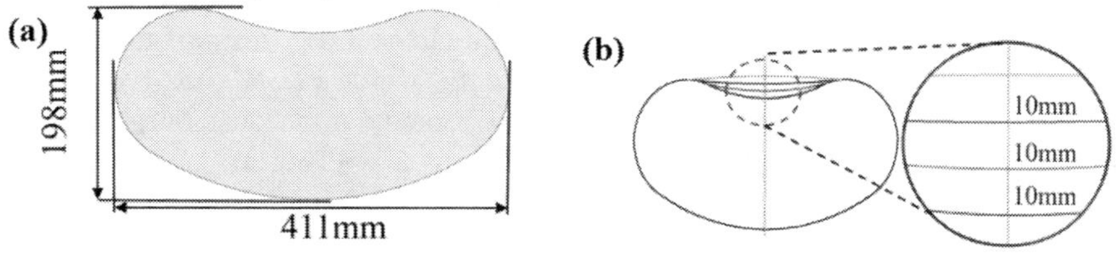

Figure 7. Size and outline of the blank with design margin (a) Size and outline (b) Design details of size

$$D_R = \frac{A_0}{A} \tag{1}$$

Where D_R represents the drawing ratio, A_0 is the area of blank, A represents the vertical projected area of the die cavity.

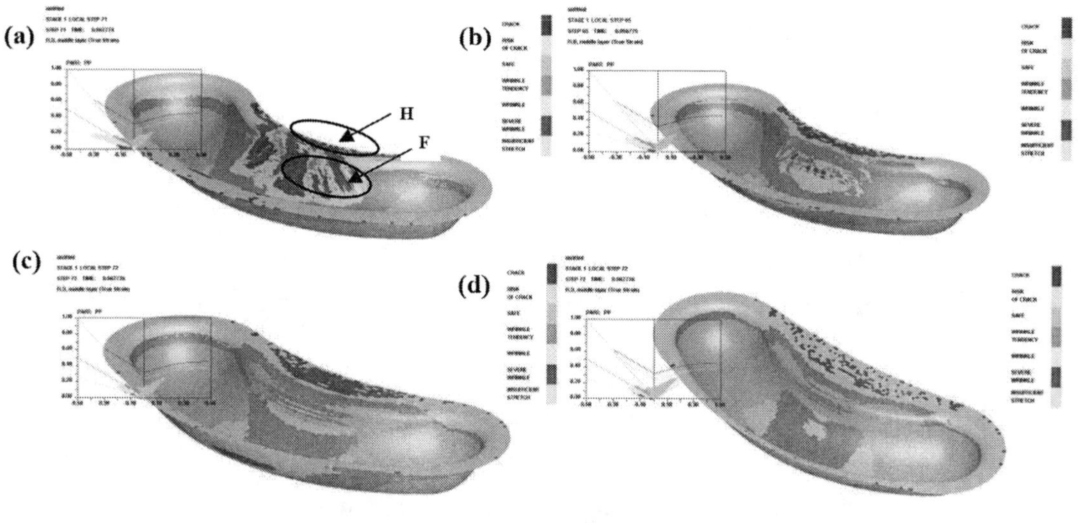

(e) (f)

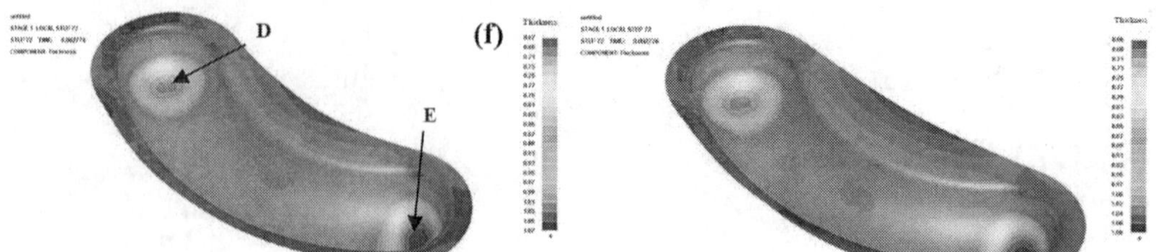

Figure 8. Simulation results after blank forming with different drawing ratios (a) FLD with D_R of 2.08 (b) FLD with D_R of 2.12 (c) FLD with D_R of 2.17 (d) FLD with D_R of 2.22 (e) Thickness with D_R of 2.17 (f) Thickness with D_R of 2.22

3. Verification experiment

In order to verify the reliability of numerical simulation analysis and explore to improve the formability of 2195 Al-Li alloy at room temperature, the self-designed IHF equipment was used for forming experiment. Based on the previous research, a drawing ratio of 2.22 was selected as the blank and it was cut by WEDM. The IHF process of half-tube part is specifically divided into one-step forming and two-step forming. The impact pressure parameters of one-step forming were 8.5 MPa and 10 MPa, and the impact pressure parameters of two-step forming is 8.5+10 MPa, and the pressure of blank holder are all set to 2 MPa. The formed objective part can be seen in Figures 9 and 10. When the pressure is 8.5 MPa, the depth of half-tube part cannot reach the objective depth, as shown in Figure 9 (a); When the pressure is 10 MPa, the cracking phenomena of large area occurs due to excessive impact energy and material characteristics of Al-Li alloy, as shown in Figure 9 (b). Figure 10 (a) shows the half-tube part after two-step forming, and the depth of part reaches the objective depth, which means that the part and the die basically fit. Although the areas of D and E at both ends are cracked, but because they are located in the design margin, the two area should be cut off by WEDM, which will not affect the overall forming quality of the half-tube part, as seen in Figure 10 (b). The surface quality of the obtained parts is good, without wrinkles and cracks, which indicates that the IHF technology with high strain rate can effectively improve the formability of 2195 Al-Li alloy at room temperature.

(a) (b)

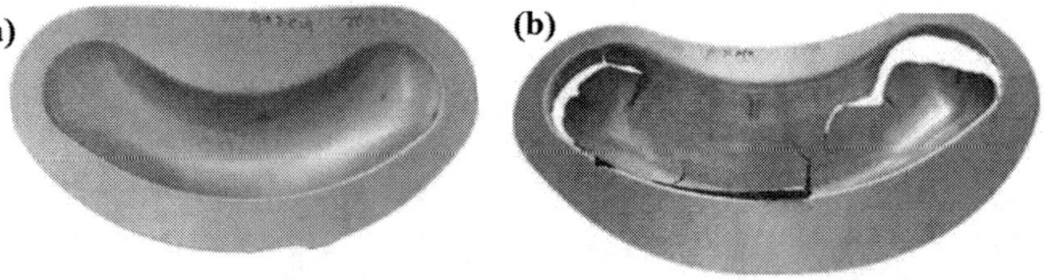

Figure 9. One-step forming of IHF (a) Impact pressure of 8.5 MPa (b) Impact pressure of 10 MPa

(a) (b)

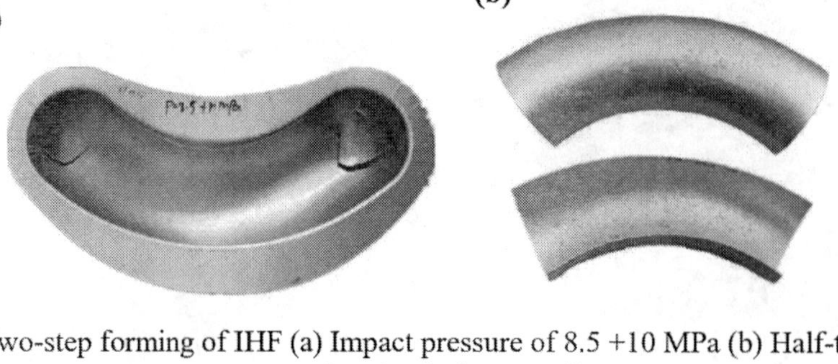

Figure 10. Two-step forming of IHF (a) Impact pressure of 8.5 +10 MPa (b) Half-tube part after removing the design margin

IOP Conf. Series: Materials Science and Engineering 1284 (2023) 012024 doi:10.1088/1757-899X/1284/1/012024

4. Conclusions

The process route and size outline of the 2195 Al-Li alloy half-tube part with curvature were designed and optimized. The forming experiment was completed by IHF technology. The main conclusions can be summarized as follows:

(1) The large section of part was supplemented and the arc-shaped margin area was designed at both sides, which ensured the forming quality and precision of the objective part.

(2) Four kinds of the blank size and outline were designed, and numerical simulation and analysis were carried out. It was found that the D and E areas were the most thinned and have the risk of cracking, but they were located in the design margin, which does not affect the forming quality of half-tube parts. Finally, the drawing ratio of 2.22 was selected as the outline and size of the blank.

(3) Due to the material characteristics of Al-Li alloy and the characteristics of IHF technology, the one-step forming of half-tube part is difficult, the part after two-step forming have a good surface quality without wrinkles and cracks, which indicates that the IHF technology can effectively improve the formability of Al-Li alloy at room temperature.

Acknowledgments

The authors would like to acknowledge the support of the National Natural Science Foundation of China (Grant NO: 51875548 and 52111530293) and Youth Innovation Promotion Association CAS (Grand NO: 2019195).

References

[1] Williams J C and Starke E A 2003 *Acta Mater.* **51** 5775-5799
[2] Dursun T and Soutis C 2014 *Mater. Des.* **56** 862-871
[3] Li H Y, Yi T, Zeng Z D, Zheng Z Q and Zheng F 2008 *Mat. Sci. Eng. A* **498** 314-320
[4] Gao H X, Weng T X, Liu J, Li C, Z Li Q and Wang L L 2016 *Manuf. Rev.* **3** 9
[5] Ma G S, Wang M and Wu X D 2008 *Chin. J. of Nonferrous Met.* **18(6)** 980-984
[6] Sun Z G, Guo X, Liu H B, Bao P L and Chen H 2012 *Aviation Manuf. Technol.* **5** 60-63
[7] Liu T J, Wang Y J and Wu J J 2012 *Aviation Manuf. Technol.* **17** 40-44
[8] Broomhead P and Grieve R J 1982 *J. Eng. Mater. Technol.* **104(2)** 102
[9] Kosing O E and Skews B W 1998 *Int. J. Impact Eng.* **21(9)** 801-816
[10] Khodkoa O, Zaytseva V, Sukaylob V and Verezubc N 2015 *J. Manuf. Process.* **20** 304-313
[11] Lang L H, Wang S H, Yang C L and Chao Y 2014 *Forg. Stamping Technol.* **7** 1-5
[12] Li K and Yang L F 2017 *Forg. Stamping Technol.* **42(6)** 77-81
[13] Liu J W, Liu X Y, Yang L F and Liang H P 2016 *P. I. Mech. Eng. B-J Eng.* **230(1)** 111-119
[14] Liu J W, Yang N J and Wu M J 2017 *Int. J. Mach. Tool. Manu.* **45(7)** 81-84
[15] Ma Y , Xu Y and Zhang S H 2018 *Cirp. Ann-Manuf. Techn.* **67** 281-284
[16] Chen D Y, Xu Y, Zhang S H, Ma Y and El-Aty A A 2020 *J. Mater. Process. Tech.* **287** 116533
[17] Xu Y, El-Aty A A and Zhang S H 2019 *IOP Conf. Ser.: Mater. Sci. and Eng.* **651** 1-9
[18] Xu J H, Huang L, Xu Y K, Xie B X, Zhao M J, Su H L, Wang Y and Li J J 2022 *Mater. Charact.* **187** 111872
[19] Chen D Y, Zhang S H and Xu Y 2020 *IOP Conf. Ser.: Mater. Sci. Eng.* **967** 012028

IOP Conf. Series: Materials Science and Engineering 1284 (2023) 012025 doi:10.1088/1757-899X/1284/1/012025

Effect of pre-strain on edge formability of high strength steels for chassis applications

A Payà[1]*, A Lara[1], D Frómeta[1], M Marés[1], I Gil[2] and C Arenillas[2]

[1] Eurecat Centre Tecnològic de Catalunya, Unit of Metallic and Ceramic Materials, Plaça de la Ciència, 2, Manresa 08243, Spain

[2] Gestamp, Technical and Innovation Office, AIC-Automotive Intelligence Center, Parque Empresarial Boroa, Parc 2A-4, Amorebieta Etxano – Bizkaia 48340, Spain

E-mail: anna.paya@eurecat.org

Abstract. There is a high level of uncertainty in the prediction of edge cracking during the design and manufacturing of high strength sheet metal products. Edge ductility is a complex parameter to measure since it is not an intrinsic property of the material, but it also depends on the trimming strategy/parameters that have been applied to the edge (edge preparation method, cutting/punching clearance, etc.). Furthermore, most forming processes involve more than one operation where the material can be deformed in a first stage, then trimmed or punched and further deformed in a final forming step to obtain the desired shape. Therefore, in order to design a robust and safe forming process able to predict the risk of edge cracking, it is crucial to understand the influence of pre-strain on the residual formability of trimmed edges. This paper analyses the effect of different pre-strain levels and paths (uniaxial tension, plane strain and equi-biaxial stretching) on the sheared edge ductility of a Complex Phase (CP) steel and a high strength low alloy (HSLA) steel used for automobile chassis applications. To this purpose, rectangular specimens are cut out from pre-strained samples and punched. Hole Tension Tests (HTTs) in combination with Digital Image Correlation (DIC) are carried out to determine the edge fracture strain of undeformed and pre-strained samples. The results are discussed in terms of major strains obtained from DIC. The proposed methodology allows describing the edge cracking sensitivity of high-strength steel sheets as a function of the pre-deformation state and can be useful to predict the risk of edge cracking more accurately in forming simulations.

1. Introduction

Stamping of automotive chassis components entails many challenges related to the forming process and material characteristics. One of the main concerns of automotive stampers is the occurrence of edge cracks in cold-formed components during serial production [1,2]. This phenomenon, known as edge cracking, is one of the most important failure modes of high-strength steels for body-in-white and chassis applications and its prediction still being a challenge for automotive product designers. The main reason behind that is the inherent complexity of this fracture mode, which does not only depend on material properties but is greatly affected by the edge quality, and the absence of robust experimental and numerical approaches for edge crack prediction.

Currently, the definition of manufacturing processes is mainly based on the results predicted by numerical simulation software, which relies on material constitutive models and failure criteria calibrated using specific experimental tests. In sheet metal forming, strain limits are accurately defined by well-established formability criteria, such as Forming Limit Diagrams (FLDs), for global deformation modes where strain localization occurs due to the application of a uniform deformation in a relatively large area (uniaxial tension, plane strain, equi-biaxial tension). However, there is no quorum on the definition of the best-suited experimental test to assess the residual formability of sheared or trimmed edges.

The strain limit of a sheared edge is usually lower than the defined by the Forming Limit Curve (FLC) for edge cracking sensitive materials [2,3]. This is due to the damage introduced in the edge during the trimming, shearing or punching process. It is well known that mechanical cutting induces high plastic strains locally in the edge, which results in a highly work-hardened area that reduces the posterior ductility of the material in comparison to an undamaged area [4,5]. Another factor that compromises the integrity of a sheared edge and limits its formability is the presence of defects (microcracks, voids, surface irregularities, etc.) in the fracture surface, which act as fracture initiation sites and lead to the propagation of macroscopic cracks during the forming process [6,7].

To better understand this particular failure mode and assess the edge cracking sensitivity of metallic sheet materials, multiple tests have been developed lately. So far, the only standardized test to characterize edge ductility is the Hole Expansion Test (HET) according to ISO 16630 [8]. However, the HET has been an object of criticism in the research community due to the large data scattering and the poor repeatability observed in many research works and round-robin tests [9,10]. Furthermore, it only represents a specific case of edge formability; a closed edge cut with a 12 ± 1% cutting clearance and expanded with a conical tool of 60°. To overcome such limitations, alternative edge crack tests based on optical strain measurements and Digital Image Correlation (DIC) techniques covering a wide range of loading cases and strain gradients have been proposed in the literature [3,11-13]. All these test configurations provide relevant strain data that may help to define edge forming limits and improve the accuracy of forming simulations.

Most research works dealing with the edge cracking sensitivity of metallic materials are focused on the characterization of the material in undeformed conditions. However, manufacturing processes usually involve a sequence of trimming and forming operations. During that sequence, it is usual that a specific part region passes through different deformation path changes. These regions can be trimmed in a subsequent operation and further deformed in a later forming step. Therefore, understanding the influence of pre-strain level and deformation history on the residual formability of cut edges is critical to predicting potential edge cracking risks.

Up until now, only a few studies have been conducted on this topic [14-16] and there are still some aspects that remain unclear. In this work, the effect of different forming mode sequences on the punched edge formability of two automotive steel grades for chassis applications is investigated through Hole Tension Tests (HTT) in combination with DIC. Different pre-strain levels and deformation paths (uniaxial tension, plane strain and equi-biaxial stretching) are considered for investigation. The objective is to analyse the combined influence of pre-strain and trimming on the edge ductility of high-strength steels used in chassis components.

2. Materials and Experimental Procedure

2.1. Materials

The materials used for this study are two hot-rolled steel grades, a 2.7 mm thick complex phase steel (CP800) and a high strength low-alloy steel (S550MC) with a thickness of 3.0mm. The mechanical properties of the investigated materials are shown in Table 1.

Table 1. Mechanical properties of the investigated steels in the transverse direction. Results from uniaxial tensile tests according to ISO6892-1 (Type 1 specimens; gauge length: 50mm, width: 12.5mm).

Material	Thickness (mm)	Yield Strength (MPa)	Ultimate Tensile Strength (MPa)	Total Elongation (%)
CP800	2.7	761	809	14.7
S550MC	3.0	588	601	18.4

2.2. *Experimental Procedure*

To reproduce the multi-stage forming sequence and analyse the influence of pre-strain on edge ductility, a three-step testing approach was implemented. First, different deformation conditions were produced in flat samples using uniaxial, plane strain and equi-biaxial pre-strain using Marciniak stretching tests. After pre-straining, testing samples were cut out from pre-strained samples and punched. Finally, Hole Tension Tests (HTTs) were carried out to measure edge fracture strains. The experimental details for the different steps are detailed below.

2.2.1. Pre-strain: Marciniak Test. Marciniak stretching tests were used to generate a controlled pre-strain in flat undeformed samples. Three deformation paths were selected for the investigations, uniaxial (ε_1=-2ε_2), plane-strain (ε_2=0) and equi-biaxial (ε_1=ε_2). The specimen geometries used to obtain these strain paths are shown in Figure 1a. Specimens were machined with the shaft transverse to the rolling direction. Three pre-strain levels per deformation mode were pre-defined, 5%, 10% and 15% for uniaxial and plane strain, and 3%, 5% and 10% for equi-biaxial tension.

The tests were carried out in a double-acting hydraulic press using a punch speed of 1.0 mm/s. The punch had a diameter of 100 mm and the press holder was equipped with a circular rib to avoid material sliding. The deformation was monitored by using a 3D DIC equipment (GOM ARAMIS) integrated into the press (Figure 1b and c) [17].

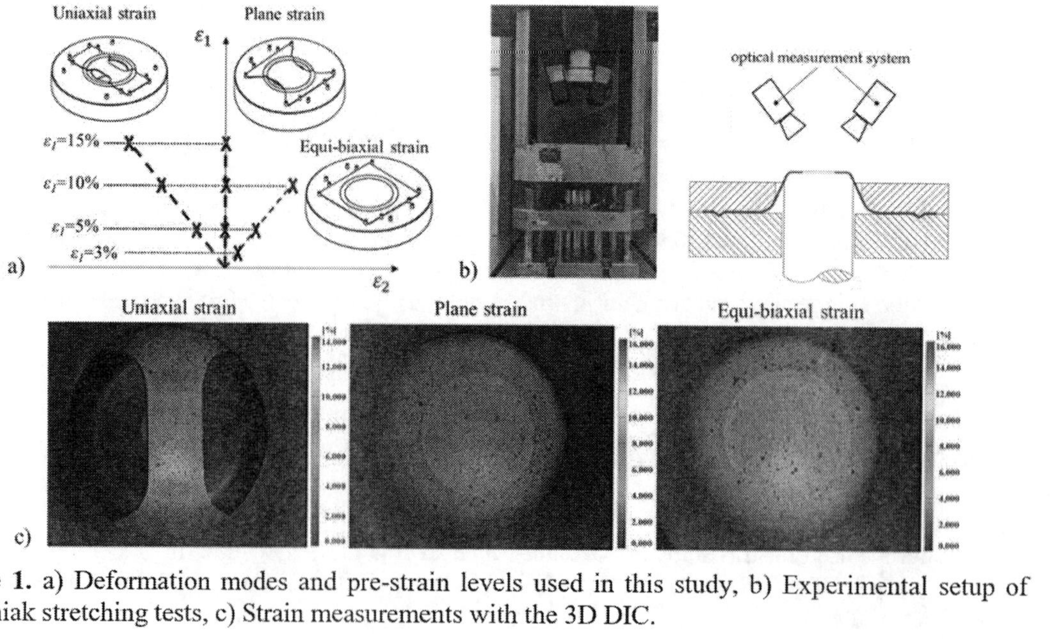

Figure 1. a) Deformation modes and pre-strain levels used in this study, b) Experimental setup of Marciniak stretching tests, c) Strain measurements with the 3D DIC.

2.2.2. Mechanical punch. Rectangular specimens for HTTs were cut from pre-strained samples by spark erosion and a 10 mm diameter hole was punched in the centre (Figure 2a). Figure 2b shows the punching tool used for hole piercing. Punch-to-die clearances of 12% and 10.7% were used for CP800 and S550MC, respectively. The punching speed was 10 mm/min. Undeformed (no pre-strained) HTT specimens were also prepared using the same punching parameters. The dimensions of the HTT specimens are shown in Figure 2c.

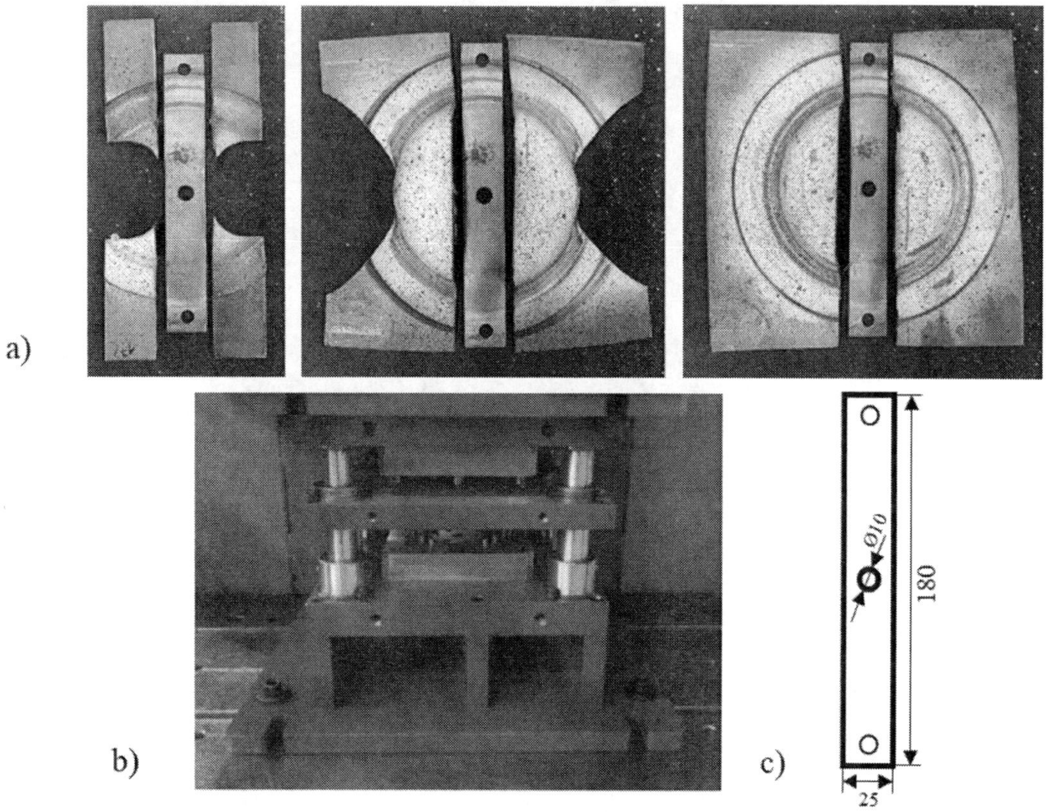

Figure 2. a) HTT specimens extracted from pre-strained samples after Marciniak stretching tests, b) Punching tool, c) Dimensions of the HTT specimen.

2.2.3. Hole Tension Test. Hole Tension Tests were conducted on a universal testing machine (Figure 3a). The specimens were loaded up to fracture at a constant displacement rate of 2.88mm/min. A minimum of two replicates per testing condition were used.

Edge fracture strains were obtained by means of the 3D DIC equipment. To do so, a speckle pattern was applied to test specimens by spray painting (Figure 3b) and the tests were recorded using the digital cameras of the GOM ARAMIS system. The image acquisition rate was set to 10Hz. The last image just before fracture initiation was used to determine the fracture strain, corresponding to the point of maximum deformation in the hole edge (Figure 3c). True logarithmic major strain values were used to characterize the edge fracture resistance of the investigated steel grades. A facet size and a step size of 11 and 9 pixels were used for full-field strain analysis respectively. The strain reference length was 0.91 mm.

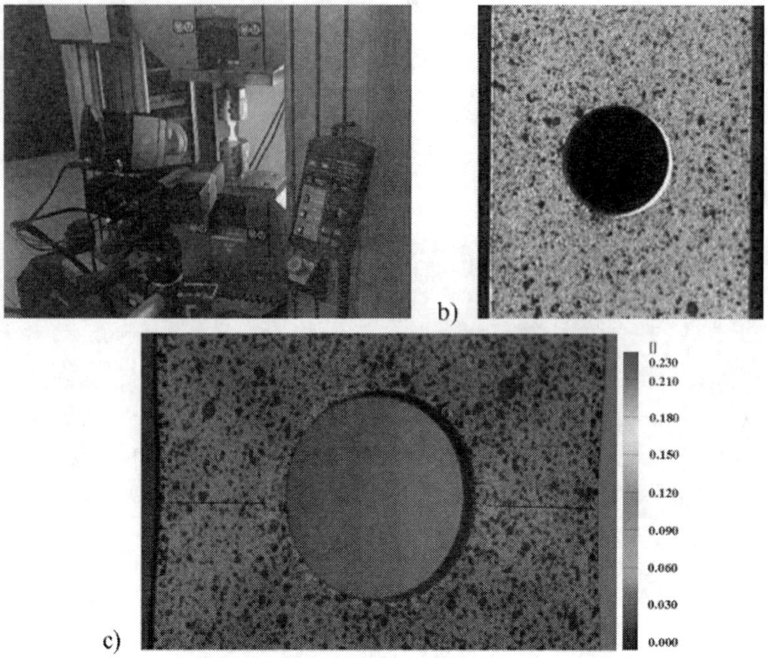

Figure 3. a) Universal testing machine with DIC system GOM Aramis, b) Sample painted with a speckle pattern, c) Image just before fracture initiation with true logarithmic major strain represented.

3. Results and discussion

Figure 4 shows the true major strain at fracture as a function of the pre-deformation level and deformation path for CP800 and S500MC. In general, little or no effect of pre-strain can be observed in either of the two steel grades investigated. For CP800, an almost constant true major strain of around 0.31 was obtained. Although small differences are observed for different pre-strain levels, no clear trend can be established in any of the deformation modes. Such differences might be motivated by experimental deviations and the large data scattering observed for this steel. In the case of S550MC, the scattering is much lower, as indicated by the error bars, but the trend is quite similar to the one observed in CP800, i.e. the true major strain remains nearly constant (≈0.23) and seems to be independent of the pre-strain level and deformation path. A very slight decrease of the true major strain with increasing pre-strain can be perceived in plane strain and biaxial strain conditions. However, the differences are practically negligible.

These results suggest that in a multistage forming process in which the sequence forming + trimming/punching + edge forming is given, the deformation induced in previous forming operations will not affect the residual formability of the cut edge. These observations are in good agreement with the findings of Chiriac et al. [14,15], who reported similar hole expansion behaviour for dual phase (DP) steels, DP780 and DP980, with different pre-deformation histories. They also showed that edge fracture strains, measured by DIC, in hole expansion tests with conical and flat punches were unaffected by (biaxial) pre-strain [15]. Similar results were reported by Feistle et al. for a DP780 using edge fracture tensile tests with uniaxial, plane strain and biaxial pre-strained specimens [15]. All these works lead to the same conclusion: forming operations prior to punching or trimming do not affect the residual formability of the sheared edge. Such behaviour may be attributed to the large strains generated locally at the hole edge during the punching process, where the material reaches its fracture limit, making the influence of previous pre-deformation on edge formability irrelevant.

Nevertheless, it is important to remark that sheared edge forming limits may depend on several factors, including cut edge quality and loading scenario (stretch flanging, bending, in-plane stretching, open or closed trim line, etc.) [11, 13]. Therefore, the results obtained in this study must be considered only representative of the specific load case investigated. Further investigations should be carried out to analyse the combined effect of pre-strain, edge condition and close-to-the-edge strain gradients on the edge stretching performance of high-strength steel sheets, depending on the loading mode involved in the manufacturing process.

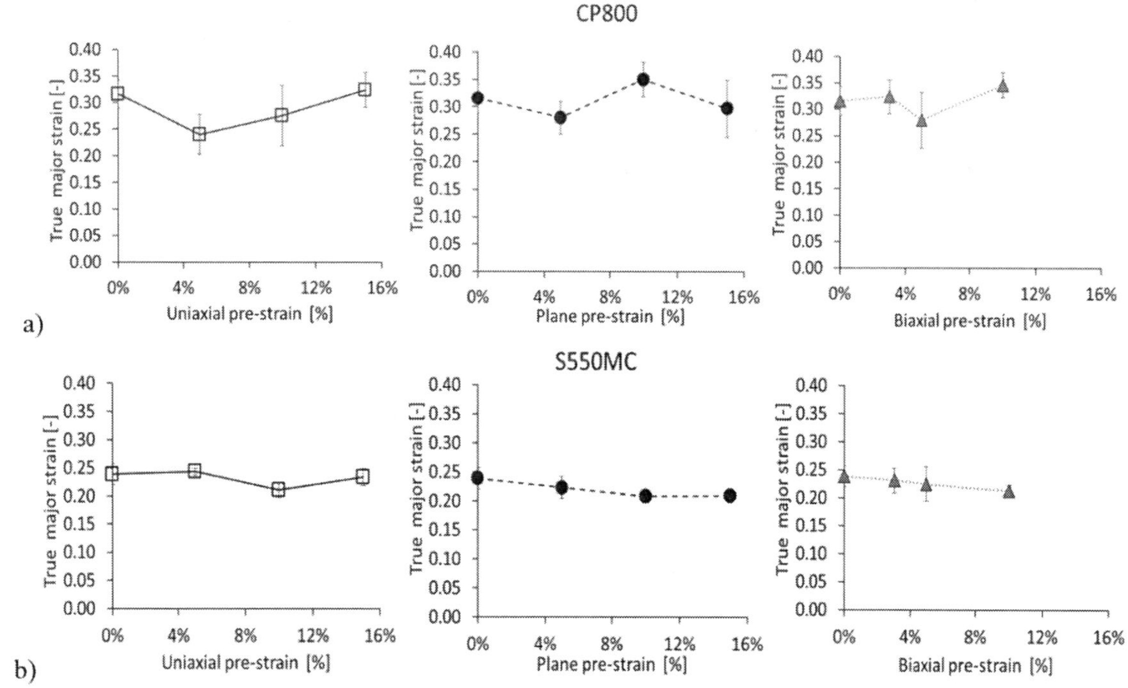

Figure 4. True major strain at fracture as a function of the pre-strain level for the different deformation paths investigated, a) CP800, b) S550MC. The error bars represent the standard deviation.

4. Conclusions

In the present work, the effect of deformation before edge punching on edge fracture resistance was studied in a CP800 and an S550MC for automotive chassis applications. From the analysis of the results obtained from Hole Tension Tests with samples pre-strained at different deformation levels and following different deformation paths (uniaxial tension, plane strain and equi-biaxial tension), it is concluded that forming operations carried out before punching have no significant influence on the residual formability of punched edges. For the studied steels, it was observed that the true major strain at fracture remained practically constant regardless of the pre-strain level and deformation mode. The results are considered relevant for the definition and design of multistage forming processes for chassis parts. Nevertheless, it is important to note that cut edge formability of high-strength steels may also depend on the stretch flanging configuration and strain gradients combinations. Therefore, in order to obtain more accurate predictions, appropriate edge crack test configurations should be used to represent the specific load case present in the industrial forming process.

Acknowledgments

The research leading to these results has been funded by the EU Horizon 2020 programme under grant agreement H2020-EU.2.1.3. – 814517 (FormPlanet).

References

[1] Konieczny A and Henderson T 2007 On formability limitations in stamping involving sheared edge stretching *SAE Technical Paper* 2007-01-0340.

[2] Silva C R M, Silva F J G and Gouveia R M 2018 Investigations on the edge crack defect in Dual Phase steel stamping process *Procedia manufacturing* **17** pp 737-745.

[3] Larour P, Freudenthaler J, Grünsteidl A and Wang K 2014 Evaluation of alternative stretch flangeability testing methods to ISO 16630 standard *Int. Conf. Proc. IDDRG2014* (Paris, France, June 01-04).

[4] Levy B S and Van Tyne C J 2012 Review of the shearing process for sheet steels and its effect on sheared-edge stretching *J. of Mater. Eng. and Perform.* **21** pp 1205–1213.

[5] He J, Li S and Dong L 2020 Experiments and FE simulation of edge cracking considering prehardening after blanking process *Int. J. Mater. Form.* **13** pp 547–56.

[6] Takahashi Y, Kawano O and Tanaka Y 2009 Fracture mechanical study on edge flangeability of high tensile-strength steel sheets *Proceedings from the MS&T Conf.* (Pittsburgh, Pennsylvania, October 25-29) pp 1317-1328.

[7] Yoon J I, Jung J, Joo S H, Song T J, Chin K G, Seo M H, Kim S J, Lee S and Kim H S 2016 Correlation between fracture toughness and stretch-flangeability of advanced high strength steels *Matter. Lett.* **180** pp 322-326.

[8] ISO16630 2017 Metallic materials – Method of hole expanding test

[9] Schneider M and Eggers U 2011 Investigation on punched edge formability *Int. Conf. Proc. IDDRG2011* (Bilbao, Spain, June 05-08).

[10] Atzema E, Borsutzki M, Braun M, Brockmann S, Bülter M, Carlsson B, Larour P and Richter A 2012 A European round robin test for the hole expansion test according to ISO 16630 *New Development in Sheet Metal Forming, Int. Conf.* (Fellbach, Germany, May 23-24) pp 171-184.

[11] Larour P, Schauer H, Lackner J and Till E T 2016 Edge crack simulation with the modular "smiley" forming tool *Int. Conf. Proc. IDDRG2016* (Linz, Austria, June 12-15) pp 457–477.

[12] Atzema E and Seda P 2015 SETi, sheared edge tensile test improved *Int. Conf. Proc. IDDRG2015* (Shanghai, China, May 31-June 03) pp 984–91.

[13] Kesti V, Folmerz M, Vierelä R, Rautio P, Ruoppa R, Plosila P and Kaijalainen A 2022 An approach to describe edge ductility *IOP Conf. Ser. Mater. Sci. Eng.* 1238 012035.

[14] Chiriac C, Shi M F, and Link T M 2012 The prestrain effect on the sheared edge flangeability of Dual Phase 780 steels *SAE Technical Paper* 2012-01-0533.

[15] Chiriac C and Shi M 2015 Experimental study of edge stretching limits of DP980IBF steel in multistage forming process *SAE Int. J. Mater. Manf.* 8*(2).*

[16] Feistle M, Kindsmüller A, Pätzold I, Golle R and Volk W 2022 Influence of sheet metal preforming on edge crack sensitivity using an AHSS steel grade *Int. J. Mater. Form.* 15*, 50.*

[17] Galanulis K, Adolf S and Friebe H 2015 Optical 3D metrology for optimization of sheet metal forming processes *Key Engineering Materials* **639**, Trans Tech Publications Ltd. pp 3-11.

Forming of a cylindrical cup in 7075-T6 under warm temperature conditions

S Royne[1], H Laurent[1] and A Maillard[2]

[1] Univ. Bretagne Sud, UMR CNRS 6027, IRDL, F-56100 Lorient, France
[2] Cetim, F-60300 Senlis, France

E-mail: sylvain.royne@univ-ubs.fr

Abstract. The high mechanical properties of aluminum alloys are very interesting for reducing the weight of parts in the automotive sector. However, a formability issue prevents the use of these alloys at room temperature especially for complex shapes where part failure occurs before full drawing. The use of a warm forming process, which consists of heating the part between 150°C and 250°C for aluminium alloys, allows to increase formability and therefore to perform complex shapes. In this study, the forming of a cylindrical cup in 7075-T6 is carried out at 200°C in two steps by two successive reductions in the diameter of the blank. Numerical simulations of these two forming steps are performed with Abaqus. The influence of friction coefficient is analysed in these two forming steps by comparing the evolutions of punch force.

1. Introduction

High-strength aluminum alloys, including the 7xxx series, have emerged as a promising alternative to conventional steels in the automotive and aerospace sectors due to their lightweight and strong characteristics. However, their poor formability is a major obstacle to their widespread use in the production of complex lightweight components. Increasing the temperature between 150°C and 250°C is a solution to significantly improve the formability of these aluminum alloy sheets. However, most existing studies have focused on one-step production processes and used heating methods that are not compatible with current industry standards [1, 2, 3].

Most of the studies on high temperature forming on the 7075-T6 alloy, consists in forming the sheets by carrying out heat treatments under the tools as for hardening steels [4, 5]. Only a limited number of studies have examined the multi-stage warm forming of sheets to assess the improvement in formability of these aluminum alloys [6, 7]. Additionally, it is crucial to preserve the T6 state of these alloys, which has been found to exhibit optimal performance. In our previous study [8], we showed that warm temperature forming should last between 10s and 20s at 200°C to maintain the T6 state. Prolonged heating can lead to the formation of precipitates which do not maintain the T6 state and then require additional heat treatment costs to recover them.

The aim of this study is to investigate the warm formability of a 7075-T6 alloy using a new process for forming a cylindrical cup in two successive steps by approaching industrial stamping conditions. Numerical simulations of these two-step forming processes at 200°C are performed in Abaqus. The punch force is compared with the experimental results during these two stages by changing the friction coefficient.

2. Material and methods

2.1. Material

Aluminum alloy sheet AA7075-T6 with a thickness of 0.8,mm1 mm is considered in this work. Table 1 presents the chemical composition of this alloy and its mechanical properties in the T6 receiving state is listed in table 2.

Zn	Mg	Cu	Si	Cr	Fe	Mn	Ti	Al
5.1-6.1	2.1-2.9	1.2-2.0	0.4	0.18-0.28	0.5	0.3	0.2	Balance

Table 1. Chemical composition of AA7075-T6 sheet in %wt.

Conditions	$Rp_{0.2\%}$ [MPa]	R_m [MPa]	$A\%$
Receiving	500	557	14.6

Table 2. Mechanical properties at RT of the AA7075-T6 measured in the receiving condition.

2.2. Warm forming and reforming of a cylindrical cup

The warm forming and redrawing tests are performed successively on a press with two specific tools developed by the CETIM. Figure 1 shows the layout of the two stamping tools.

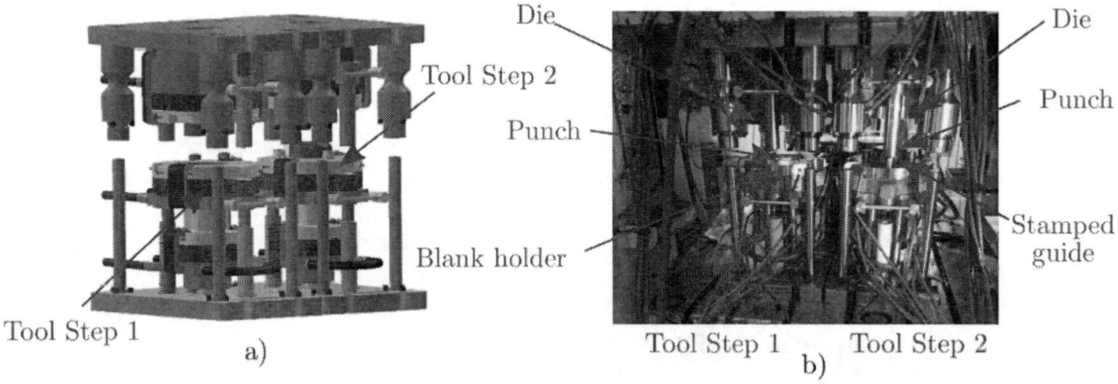

Figure 1. a) Schematic representation of the tools and b) their arrangements in the press.

An initial step of heating of the blank with a circular diameter 157mm is performed by a tool located behind the tools 1 and 2 (not shown in figure 1) in order to reach a temperature of 200°C in 0.45s. The blank is then manually transferred under the first stamping tools. Then, the cup is again manually transferred to the second stamping operation. The dimensions of the tools of the first and second steps are shown in figures 2-a and 2-b, respectively. Due to the proximity of the tools, the transfer time is reduced and the two forming steps are assumed to be performed under isothermal conditions. The total time (heating, first and second forming) is less than 20s to guarantee the mechanical characteristics of the T6 state.

Both stamping tools are heated to the target temperature using a heating cartridges. The tools temperature is controlled by K-type thermocouples located in each element that composes the tools. In addition, the temperature of the blank is monitored during the two stamping steps with a pyrometer located at a constant distance of 20mm from the blank. A clamping force of 41kN is applied on the first stamping step to the blank holder and no clamping force is applied on the second forming step. The reaction forces during both drawing processes are measured by a force sensor located under the punch.

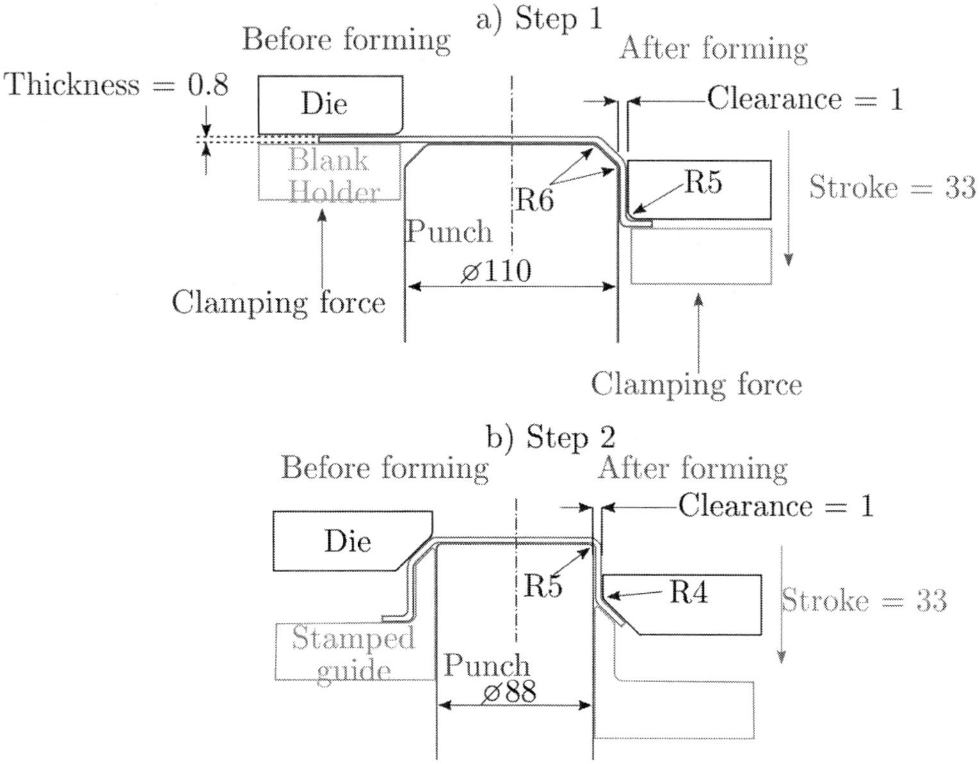

Figure 2. Dimensions and movements of tools (in mm) during a) the first and b) the second step of forming.

Forming tests are performed at RT and 200°C with a JELT 5411 lubricant. At RT, forming is impossible because the sheet breaks under the tools during the first step, so the results at RT are not preseented here. At 200°C, four tests are performed to check the reproducibility but the average results is used as a representative. Figure 3 presents the theoretical dimensions of the part after the two stamping operations and pictures of the parts obtained at the end of these two operations at 200°C.

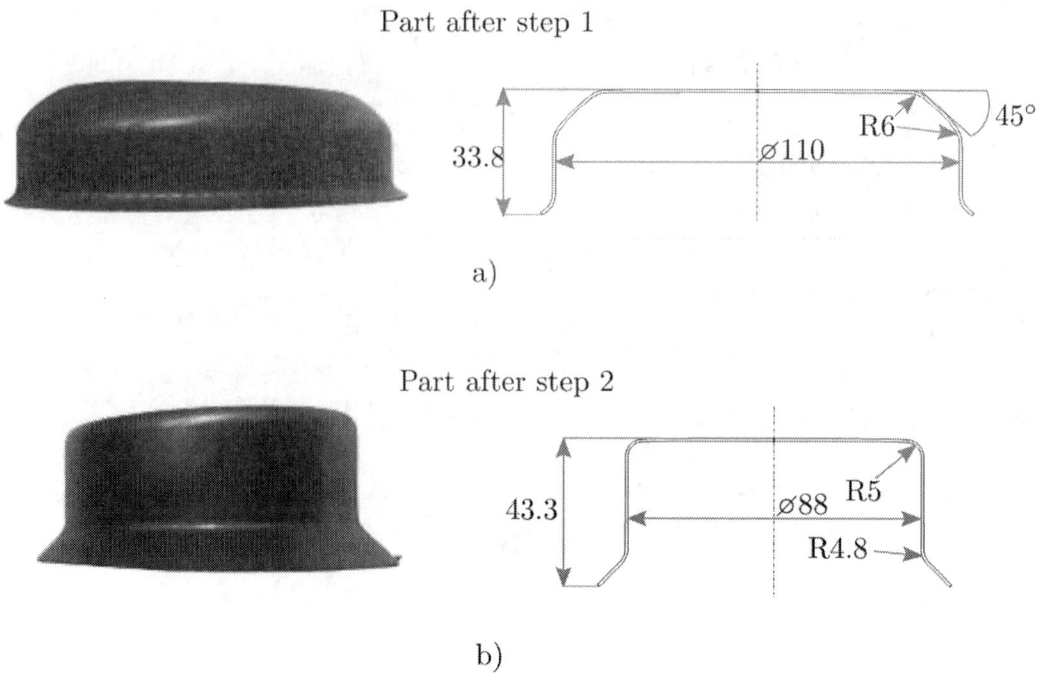

Figure 3. a) Dimensions (in mm) of the part after step 1 a) and b) after step 2.

2.3. Numerical simulations of the warm forming process

The two forming processes are simulated with Abaqus/Standard assuming isothermal conditions. A 2D axisymetric model is used with tools modeled as rigid surfaces. A mesh size (2D quadratic quadrilateral elements CAX4) of approximately 3.5mm is used in the punch zone before the punch radius whereas the size decrease to approximately 0.5mm after this radius. Moreover, 11 elements are used through the thickness. A strain rate dependent elastic-plastic behavior of the blank is simulated with a Swift law:

$$\bar{\sigma} = K(\varepsilon_0 + \bar{\varepsilon})^n \tag{1}$$

K, ε_0, n are material parameters and $\bar{\varepsilon}$ is the equivalent plastic strain.

From results presented in [8], these material parameters are identified by tensile tests at RT and 200°C for strain rates of 2×10^{-4}, 2×10^{-3}, 2×10^{-2} and $0.2s^{-1}$.

3. Results and discussion

The numerical and experimental results of the reaction force of the punch of the first stamping stage are presented in figure 4 with different friction coefficient as well as the temperature of the blank measured by the pyrometer during the stamping at 200°C. The blank temperature during this first step is 194.8±1.47 which is acceptable compared to the target temperature of 200°C. The error bar on the force curve at 200°C gives the variation in force over the four tests performed. The small force jump obtained numerically at about 16mm displacement is due to the contact of the blank on the second punch radius. At the end of the forming, the force is not zero when the press reaches bottom dead center but it is gradually cancelled when the press is raised. Only the forming sequence is analyzed here and thus the force after the bottom dead center is not simulated.

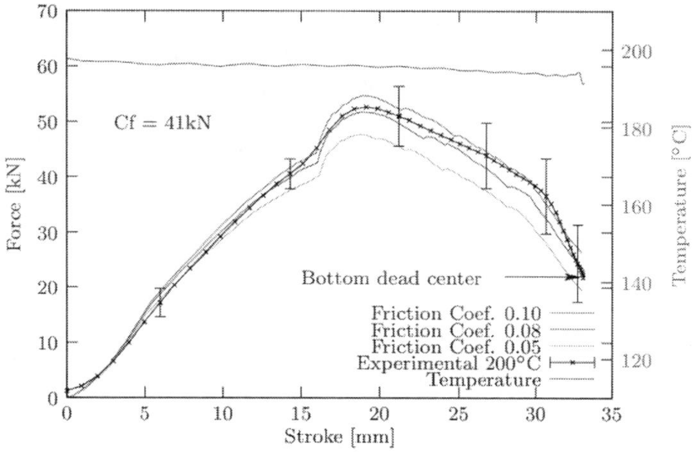

Figure 4. Experimental and numerical punch force as a function of punch displacement at 200°C and influence of friction coefficient and temperature of the blank during forming measured by pyrometer for the target temperature of 200°C.

Different friction coefficients between 0.05 and 0.1 are tested. A friction coefficient of 0.08 seems the closest to the experimental results for this first stamping step but due to the variation of the experimental reaction force a coefficient of friction of 0.10 can also be acceptable.

The numerical and experimental results of the punch force in function of the die stroke during the second step are given in figure 5. The temperature of the blank measured by the pyrometer during the stamping at 200°C is also shown in this figure.

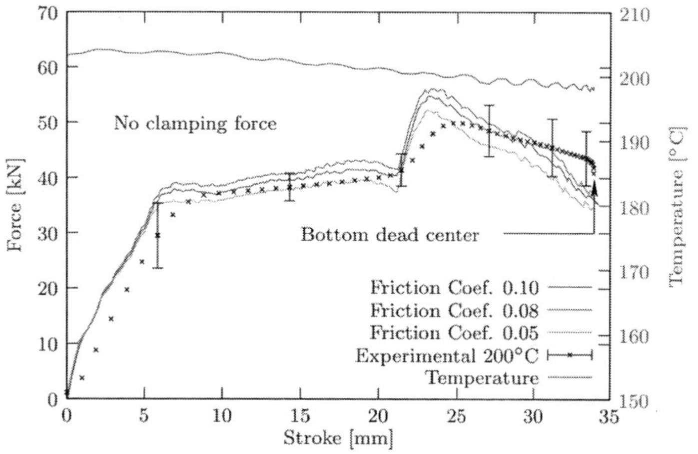

Figure 5. Experimental and numerical punch force as a function of punch displacement for the second step of stamping at RT and 200°C and influence of friction coefficient.

The blank temperature during this second step is 200.2±2.16 indicating that the forming is performed at the desired temperature. The slight variations in the temperature measurement may be due to the vibrations of the press and the lubricant which generate slight bubbles that may disturb the measurement. The stabilization of punch force at about 6mm of displacement is due to the the wrapping of the blank around the punch radius. The increase in force observed at about 23mm is due to the contact with the lower radius of the blank obtained in the first drawing step that comes into contact with the die during the second step.

A friction coefficient of 0.08 also seems the closest to the experimental results but due to the variation of experimental effort a coefficient of 0.05 or 0.10 might also be acceptable. It appears that determining a friction coefficient on the figures 4 and 5 is difficult due to the experimental variation of the reaction force of the punch.

It is also important to note the presence of wrinkles in the radius die after first step stamping as shown in figure 3-a.

4. Conclusion

Warm forming in two successive forming steps of a cylindrical cup part of 7075-T6 alloy were performed at RT and 200°C with a dedicated process. The punch force in function of die stroke obtained with Abaqus were compared with the experimental results and the influence of friction coefficient was tested. It was observed that a friction coefficient of 0.08 was closest to the experimental results of the first and second stamping steps but the other friction coefficients of 0.05 and 0.10 also gave results in the same order of magnitude as the experimental results which can suggest that the friction coefficient probably evolves during the two forming phases.

Acknowledgments

The authors gratefully acknowledge the financial support of the Brittany Region (France) via the program RB EMBHOTAL.

References

[1] Wang H, Luo Y b, Friedman P, Chen M h and Gao L 2012 *Transactions of Nonferrous Metals Society of China* **22** 1–7 URL `https://www.sciencedirect.com/science/article/pii/S100363261161131X`

[2] Kumar M, Sotirov N and Chimani C M 2014 *Journal of Materials Processing Technology* **214** 1769–1776 URL `https://www.sciencedirect.com/science/article/pii/S0924013614001150`

[3] Huo W, Hou L, Zhang Y and Zhang J 2016 *Materials Science and Engineering: A* **675** 44–54 URL `http://www.sciencedirect.com/science/article/pii/S0921509316309698`

[4] Jiang Y F, Ding H, Cai M H, Chen Y, Liu Y and Zhang Y S 2019 *Materials Characterization* **158** 109967 URL `https://www.sciencedirect.com/science/article/pii/S1044580319312069`

[5] Zheng K, Dong Y, Zheng J H, Foster A, Lin J, Dong H and Dean T A 2019 *Materials Science and Engineering: A* **761** 138017 URL `https://www.sciencedirect.com/science/article/pii/S0921509319307920`

[6] Panicker S S, Prasad K S, Sawale G, Hazra S, Shollock B and Panda S K 2019 *Materials Science and Engineering: A* **768** 138445 URL `https://www.sciencedirect.com/science/article/pii/S0921509319312316`

[7] Günzel J, Hauß J, Gaedigk C, Bergmann J and Groche P 2022 *IOP Conference Series: Materials Science and Engineering* **1238** 012014 URL `https://iopscience.iop.org/article/10.1088/1757-899X/1238/1/012014`

[8] Royne S, Laurent H and Maillard A 2022 *IOP Conference Series: Materials Science and Engineering* **1238** 012011 URL `https://dx.doi.org/10.1088/1757-899X/1238/1/012011`

IOP Conf. Series: Materials Science and Engineering 1284 (2023) 012027 doi:10.1088/1757-899X/1284/1/012027

Determination of Edge Fracture Limit Strain for AHSS in the ISO-16630 Hole Expansion Test

A Barlo[1,*], M Sigvant[1,2], V Kesti[3], M S Islam[1], Q T Pham[1], J Pilthammar[1,2]

[1] Dept. of Mechanical Engineering, Blekinge Institute of Technology, Karlskrona, Sweden
[2] Volvo Cars Dept. 81110 Strategy Development, Olofström, Sweden
[3] SSAB Europe Oy, Raahe, Finland

E-mail: Alexander.Barlo@bth.se

Abstract. With the increased demand for application of sustainable materials and lightweight structures, the sheet metal forming industry is forced to push existing materials to the limits. One area where this is particular difficult is when it comes to assessing the formability limit for sheet edges. For decades, the ISO-16630 Hole Expansion Test (HET) has been the industry standard for expressing the edge formability of sheet metals through the Hole Expansion Ratio (HER). However, in recent years, this test has been criticized for its high scatter in results for repeated experiments. This scatter has been suspected to be caused by the operator-reliant post-processing of the test, or variations in the cutting conditions for the different test specimens. This study investigates the impact of shifting the evaluation point of the test from the through-thickness crack to the onset of surface failure on the reported scatter, as well as performs inverse modeling of the Hole Expansion Test to obtain an edge limit strain value.

1. Introduction

With the increased focus on the development and application of sustainable materials, the sheet metal forming industry faces more challenges than ever before. Demands for lightweight structures and complex component geometries forces the industry to push the capabilities of the materials to their limits, why it is more than ever important to be able to accurately determine failure limits for said materials.

One area where this historically has been difficult is on the topic of edge fracture. To describe the edge formability of sheet metals, the ISO standardized (ISO-16630) Hole Expansion Test (HET) [1] is most commonly used. This test has however over the years been heavily criticized for the high scatter in results and operator-reliant post-processing [2–5]. In an effort to reduce the reported scatter, Barlo et al [6] proposed a modification to the Hole Expansion Test setup to include draw beads (modified setup presented in Figure 1), thereby changing the boundary conditions of the test. In this study 63 repetitions of the Hole Expansion Test was performed off a DP800 AHSS material and yielded a mean Hole Expansion Ratio of $\bar{\lambda} = 31.184$ [%] with a standard deviation of $\sigma = 4.2910$. Even though the study did not succeed in reducing the reported scatter, it raised a valid question - when should the test be evaluated. In the ISO-standard, it is specified that the test should be evaluated at the time of appearance of a through-thickness crack, however, with the introduction of a 3D-DIC system in the modified setup, this opens up for evaluation at i.e. the onset of fracture, which aligns better with industrial needs.

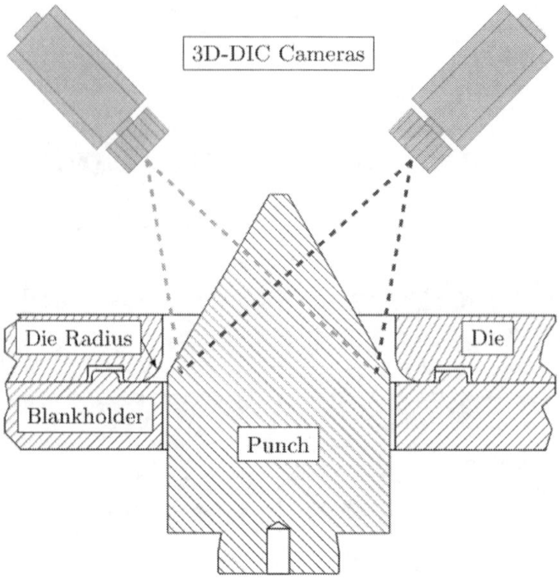

Figure 1. Modified experimental setup in fully closed configuration with draw beads and 3D DIC [6].

The application of the 3D-DIC system is however not as straight forward as usual. Due to the collar formation in the Hole Expansion Test, the DIC system fails to generate the strain field in the region of interest (RoI) around the hole edge. This is exemplified in Figure 2 where it is shown that the generated strain field does not cover the RoI. Therefore, this study will use the results from the 3D-DIC system to calibrate a Finite Element model for inverse modeling of the principal strains at the onset of edge fracture.

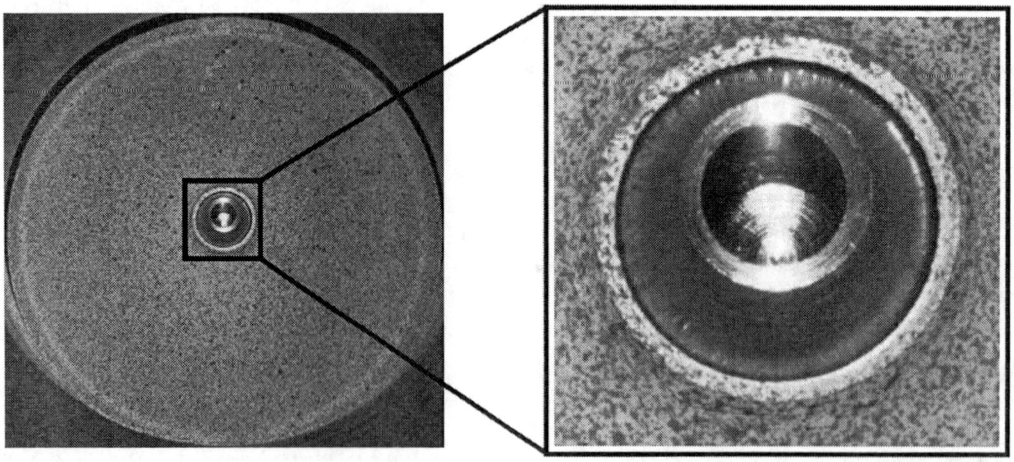

Figure 2. 3D-DIC strain field overlain experimental geometry with the RoI highlighted.

The number of experimental repetitions dealt with in this study is however significantly lower than the one presented in [6]. In this study, only 11 repetitions will be used (Test No. 10-20) as this test window shows a somewhat stable condition. In the original study, Test No. 5, 21, and

36 yielded no usable results as the test was terminated due to limitations by the test equipment, and from test 48 and onward, a downwards trend in results suspected to be caused by wear of the cutting tool, or changes in friction conditions between tool and blank. The selected range of results can be seen in Figure 3.

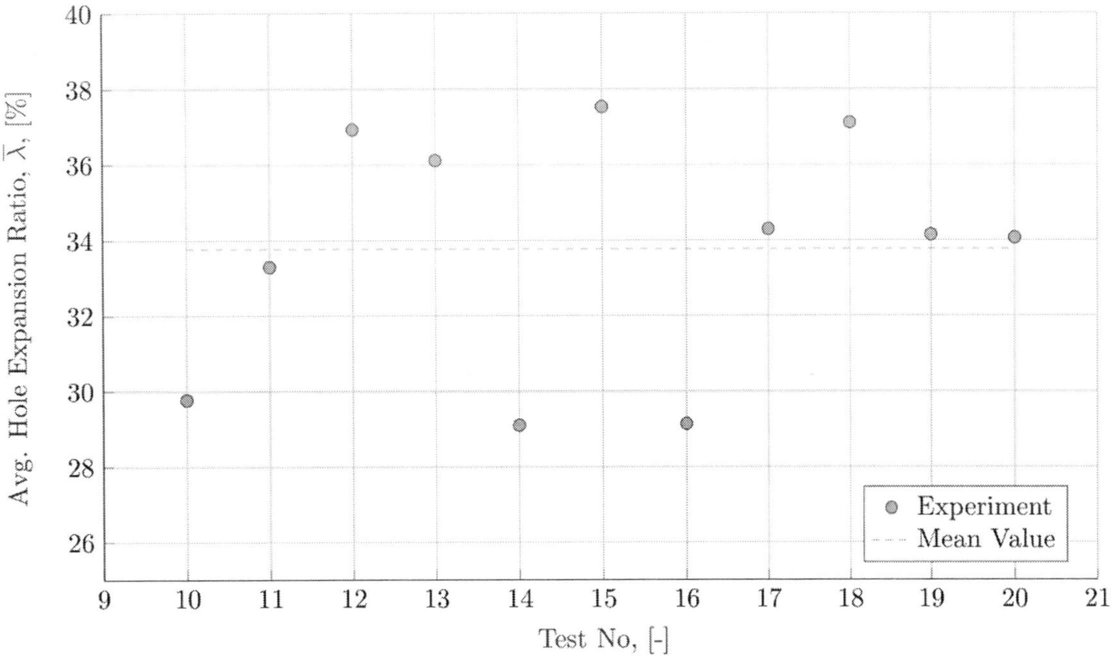

Figure 3. Results reported in [6] for test number 10-20. The test range has a mean Hole Expansion Ratio $\overline{\lambda} = 33.771$ [%] with at standard deviation of $\sigma = 3.175$.

2. Calibration of Finite Element Model

For the calibration of the Finite Element model, the commercial FE-code AutoForm™ R10 was used. For the description of the material behaviour, the BBC2005 material model was used, where the input parameters are presented in Table 1 and the hardening curve and yield surface can be found in Figure 4.

σ_0	σ_{45}	σ_{90}	σ_b	r_0	r_{45}	r_{90}	r_b	M
379.5	382.5	385.2	377.6	0.759	0.975	0.860	1.002	7.6

Table 1. Material parameters used as input to the BBC2005 material model in AutoForm™ R10. The exponent $M = 2k$ value has been obtained through inverse modeling to capture the principal strain predictions more accurately than with the standard values based on crystallographic structure [7].

For the modeling of the friction condition in the experimental setup, a combination of static Coulomb and dynamic TriboForm friction models were used. The dynamic TriboForm friction model was applied in the die radius (see Figure 1) where high tool pressures and sliding velocities were expected. The full friction model setup is outlined in Table 2.

For the exemplification of the calibration, the experimental results from Test No. 12 is used. For the setup of the model, an initial master element size of 32 [mm] with an allowed refinement

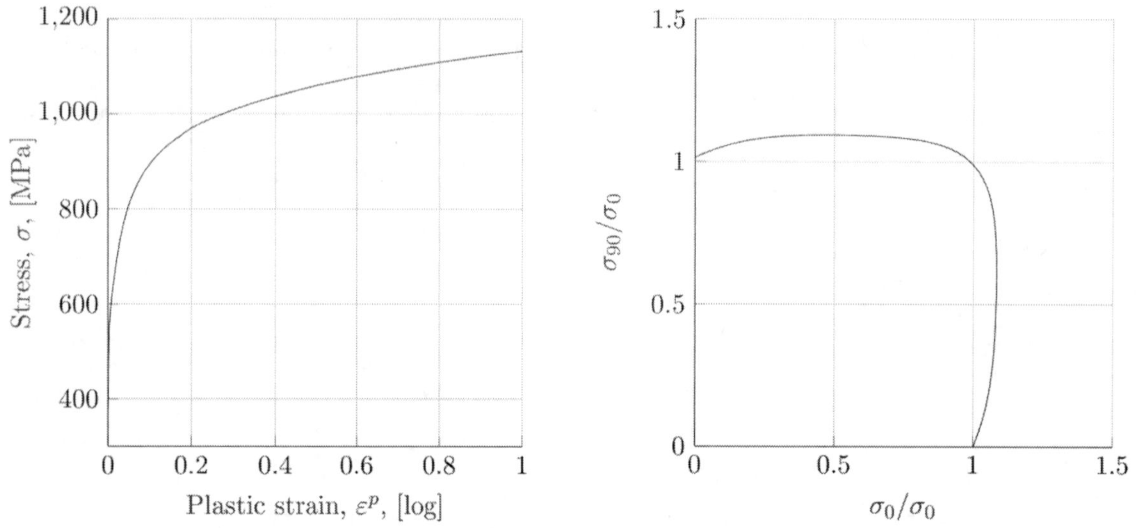

Figure 4. Hardening Curve and Yield Surface of the DP800 AHSS material.

Tool Part	Friction Model	Friction Coefficient, μ, [-]
Die	Coulomb	0.15
Binder	Coulomb	0.15
Punch	Coulomb	0.01
Die Radius	TriboForm	-

Table 2. Friction model setup of Finite Element Model.

level of 7 (allowing for an element size of 0.25 [mm]) was used. To avoid fusion and fission of elements in the RoI during the simulation, an initial refinement area (circular, radius 22.5 [mm]) was created with a constant element size of 0.5 [mm]. The initial and final mesh can be seen in Figure 5.

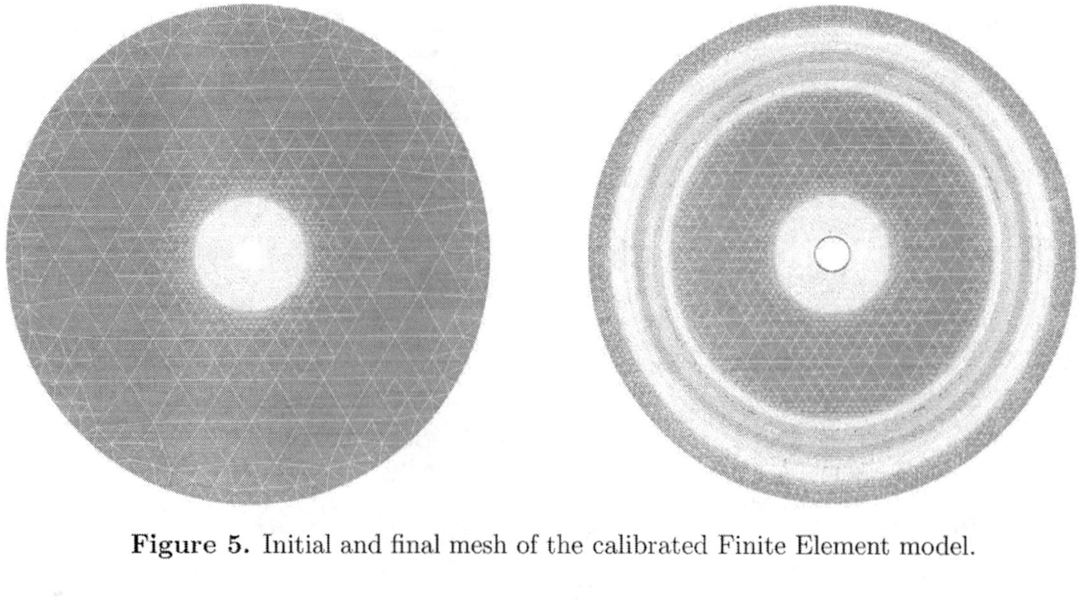

Figure 5. Initial and final mesh of the calibrated Finite Element model.

The Finite Element model is set up to dump data for every 0.5 [mm] punch displacement, why the results will be placed into these bins during evaluation. To check the accuracy of the model, initially the force-displacement response of the experiment and simulation was compared. This comparison can be found in Figure 6. As can be seen, there is a fairly good correspondence between the simulation and experiment except between roughly 6.5 and 8.5 [mm] of punch displacement. Here an unexpected decrease of forces is seen, that is assumed to be due to the numerical solution using more iterations before converging. The two markers in Figure 6 marking the failure illustrates how the simulated results are binned, where the difference in between simulation and experiment is approximately 0.2 [mm].

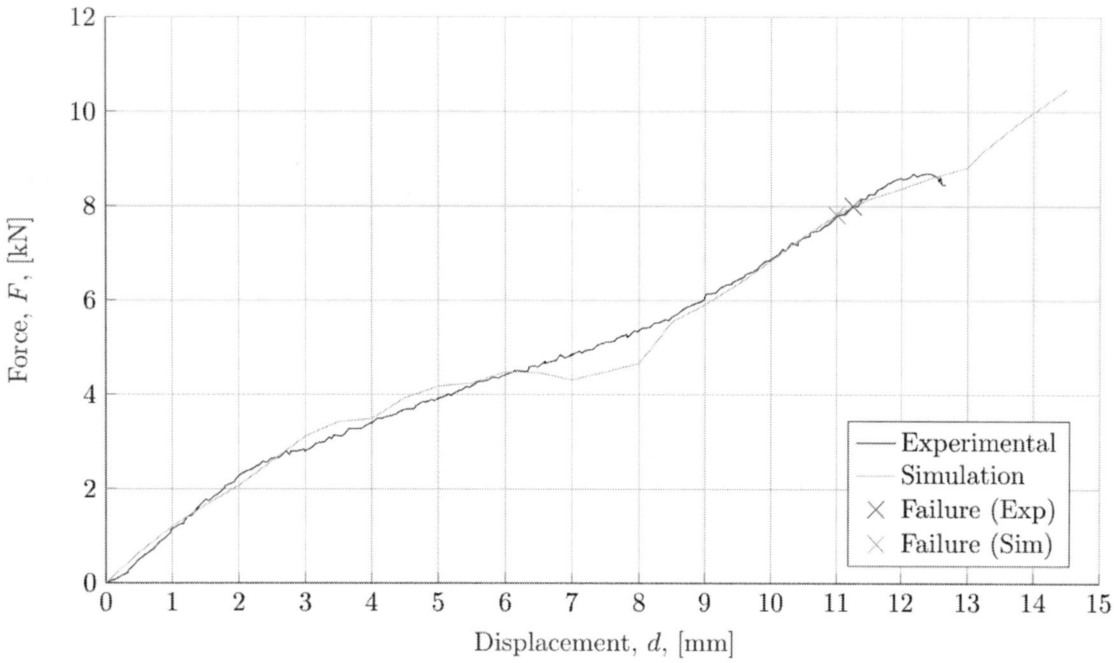

Figure 6. Comparison of experimental and simulated force-displacement response. The simulated failure is located approximately 0.2 [mm] before the experimental due to binning of results from the simulation.

Next, the deviation between the experimental and simulated geometries are checked. Two sections are created aligning with the x- and y-plane of the experimental measurements. The observed deviation along the two sections ranges between -0.1 and 0.32 [mm] in the Z-direction, yielding an acceptable geometrical fit between the experimental and simulated geometry. The section definition and deviation developments can be found in Figure 7.

Finally, the principal strain predictions from the simulation is compared to the experiment. For this, the sections defined in Figure 7 are used. It is expected that the maximum major and minor strain values will be significantly higher in the simulation due to the lack of information in the RoI in the measurement (see Figure 2), why the general development of the major and minor strains are of interest. Figure 8 presents the comparison of the principal strain. As can be seen, the principal strain profiles from the simulation matches the experimental quite well both for Section 1 and 2. As expected, the simulated maximum major strain is significantly higher than the experimental for both sections, indicating that the inverse modeling is justified.

With the calibration procedure presented, the 11 Hole Expansion Tests can be evaluated.

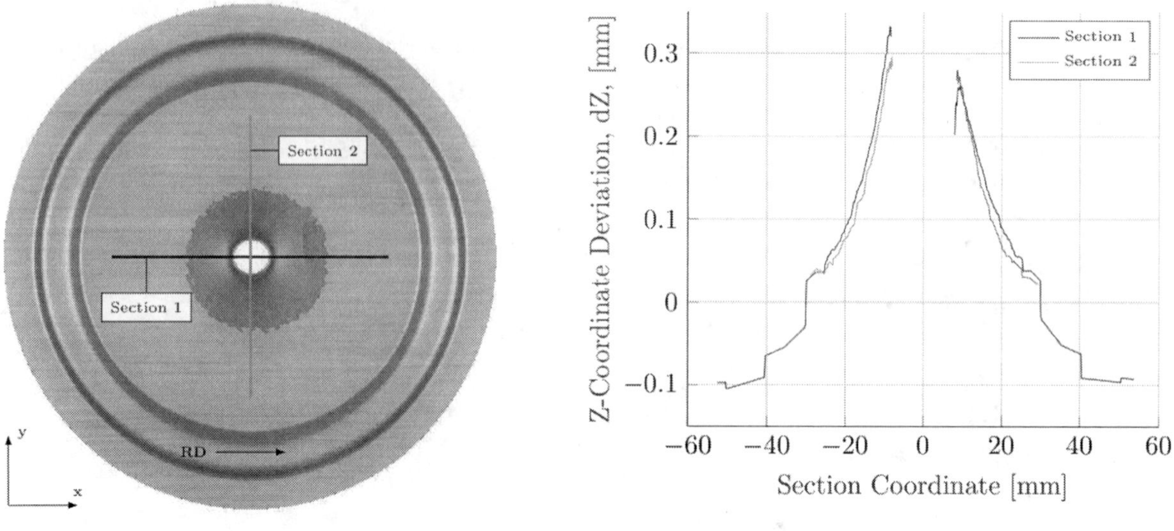

Figure 7. Z-coordinate deviation between experimental and simulated geometry across two sections.

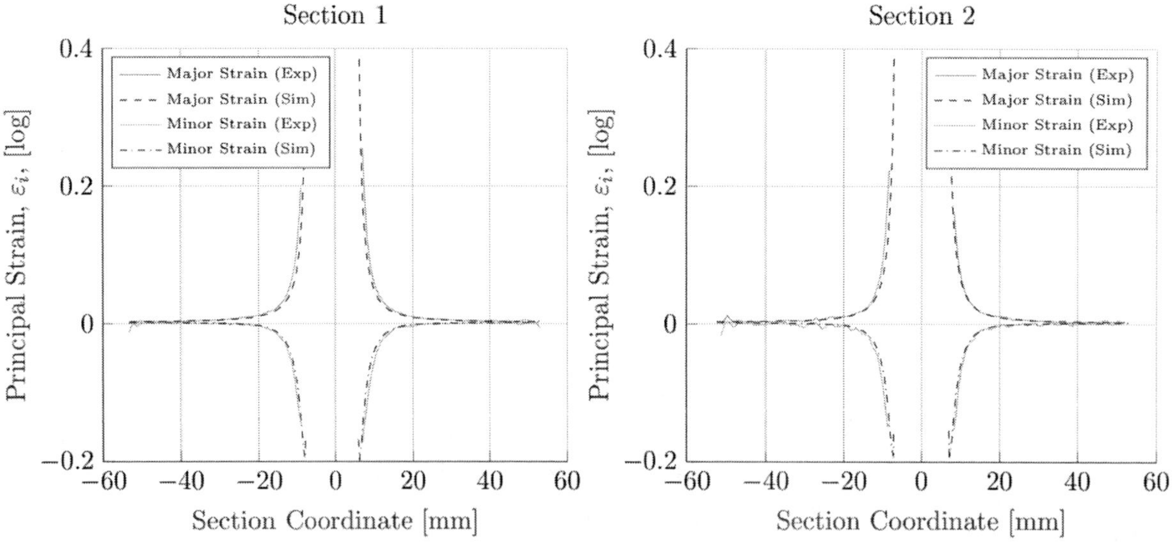

Figure 8. Comparison of principal strain profiles obtained from experiments and simulation for Section 1 and 2.

3. Results

With the shift of evaluation from through-thickness crack to the onset of surface fracture (illustrated in Figure 9), the impact of the crack propagation can now be investigated. Figure 10 presents the difference in punch depth for the 11 tests. As can be seen, there is variation between the change in punch displacement, indicating that the propagation of the crack through the surface could have an impact on the scatter of the determined Hole Expansion Ratios.

Using the calibrated FE-model, the failure strain can be identified for both the surface and membrane layer. Table 3 presents the identified punch displacement along with the maximum major strain (and the corresponding minor strain) for both the surface and membrane layer

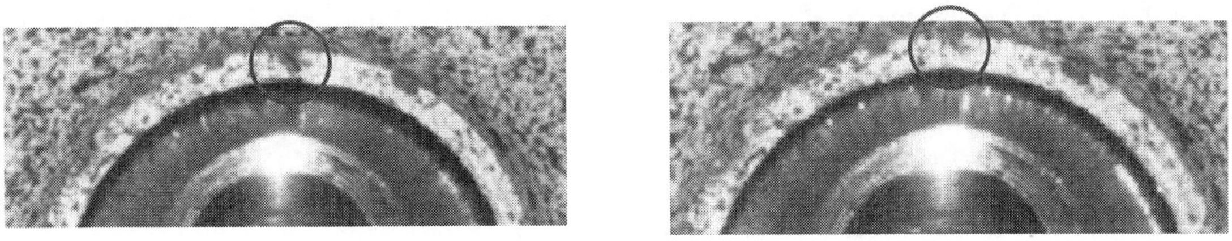

Figure 9. Illustration of the shift in evaluation point from through-thickness crack (left) to onset of surface crack (right).

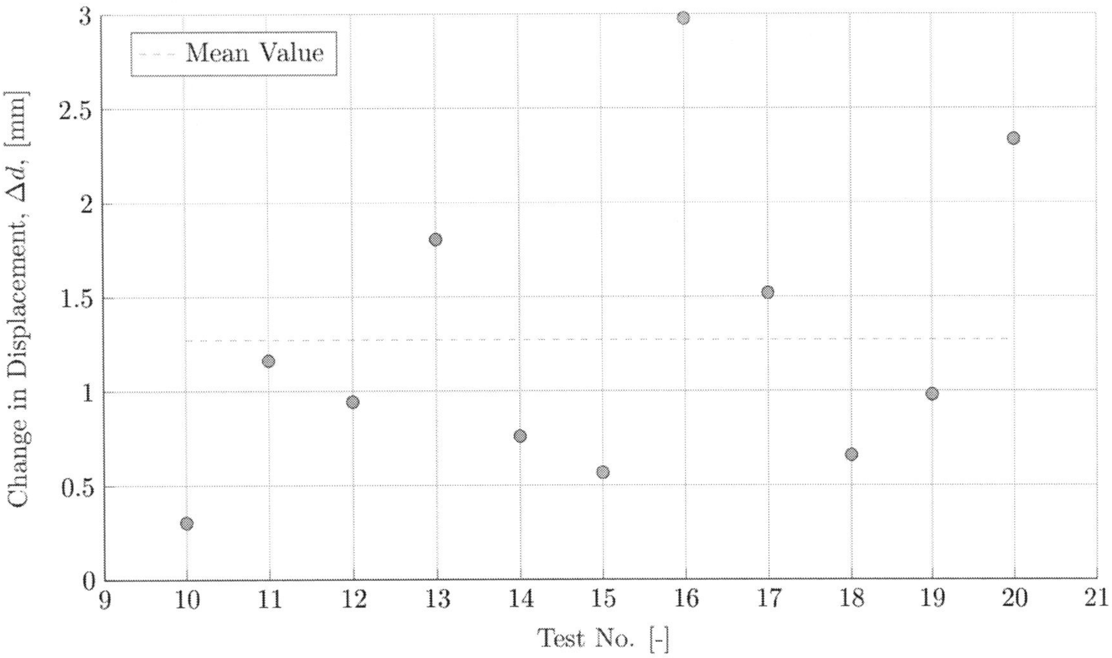

Figure 10. Difference in punch displacement between failure stages identified at through-thickness fracture [6] and onset of surface fracture. The differences have a mean value of 1.2734 [mm] with a standard deviation of $\sigma = 0.8158$

in the Finite Element model. As presented, the 11 tests investigated yielded seven different punch depths at the onset of surface fracture, with the failure strain at the surface layer ranging between 0.259 and 0.466, and the failure strain at the membrane layer ranging between 0.194 and 0.400.

Looking at the failure strains from a standard formability point of view, Figure 11 presents the seven failure bins at both the surface and membrane layer. Since the 11 experimental tests are divided into 'failure bins' the Forming Limit Diagram only presents the bins, why the number of occurrences of each bin can be found in Figure 12. It should here be noted that this figure presents the data for the surface layer, in terms of failure strain, but the number of occurrences are identical for the membrane layer.

		Surface Layer		Membrane Layer	
Test No	Displacement	Minor Strain	Major Strain	Minor Strain	Major Strain
10	11.0	-0.171	0.420	-0.096	0.356
11	10.5	-0.159	0.405	-0.085	0.332
12	11.0	-0.171	0.420	-0.096	0.356
13	10.0	-0.151	0.381	-0.075	0.308
14	10.0	-0.151	0.381	-0.075	0.308
15	12.0	-0.185	0.466	-0.116	0.400
16	7.5	-0.109	0.259	-0.040	0.194
17	10.0	-0.151	0.381	-0.075	0.308
18	11.5	-0.177	0.445	-0.107	0.380
19	10.5	-0.159	0.405	-0.085	0.332
20	9.5	-0.140	0.357	-0.071	0.285

Table 3. Identified failure strain points in surface and membrane layers for the 11 Hole Expansion Tests.

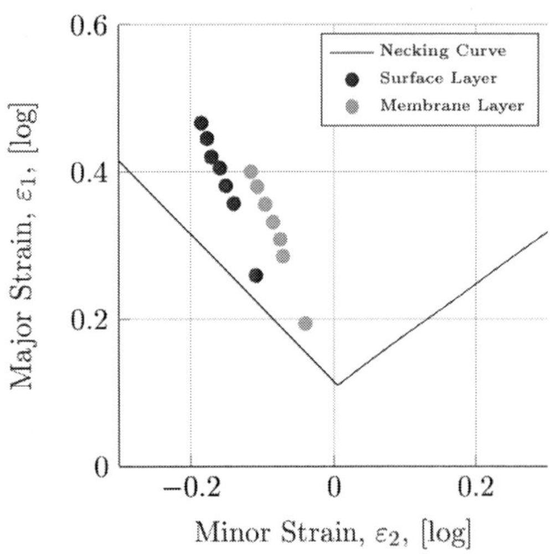

Figure 11. Failure points plotted against the necking curve for the DP800 material.

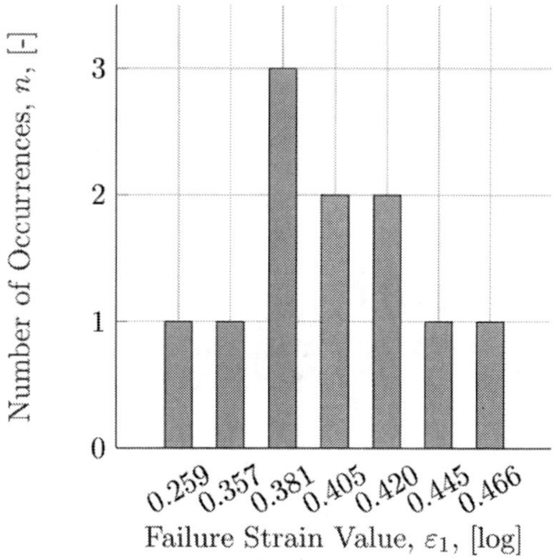

Figure 12. Number of occurrences for each failure strain for the surface layer.

4. Discussion

In the presented work 11 Hole Expansion Tests were analysed, and failure strain values were obtained through inverse Finite Element modeling. As presented in Table 3 and Figure 11 some degree of scatter in the result is still present. Interesting is the distribution of occurrences of the various failure strains presented in Figure 12, where a normal distribution is nearly obtained. It should here be noted, that since the binning is performed in intervals of 0.5 [mm] of punch displacement, small changes in displacement would move a result from one bin to another. In the case of the results from the tests with a failure strain of 0.381 at the surface level, the three tests are 0.3, 0.06, and 0.05 [mm] away from being placed in the next bin. If an assumption of normal distribution is made based on this, it could be an indication of some stochastic material behaviour contributing to the scatter in the test, however one should be aware of the low number of data points perhaps not being sufficient for the theoretical distribution of the sample mean to be distributed roughly normal [8]. Therefore, a more in-depth study, taking more samples into account, should be performed to understand how e.g. how stochastic hardening influences simulation models.

5. Conclusion

The presented study attempted to eliminate the scatter in the ISO-16630 Hole Expansion Test by moving the point of evaluation from through-thickness crack to onset of surface fracture, along with identifying a failure limit strain through inverse Finite Element modeling. The study did not succeed in eliminating the scatter, but was able to obtain limit strain values from calibrated Finite Element Models. The study further uncovered a near normal distribution of failure strains for the 11 experimental tests investigated, indicating that stochastic material behaviour could contribute to the scatter in results.

Acknowledgements

Open Access funding was provided by Blekinge Institute of Technology. This study was also funded by VINNOVA in the Sustainable Production sub-program within Vehicle Strategic Research and Innovation (FFI) program (grant number 2020-02986).

References

[1] International Standard Organization 2018 Metallic Materials - sheet and strip - hole expanding test (ISO 16630).
[2] Schneider, M. *et al* 2015 Overview and comparison of various test methods to determine formability of a sheet metal cut-edge and approaches to the test results application in forming analysis *Mat.-wiss u. Werkstofftech* **46**(12) pp 1196-1217
[3] Larour P *et al* 2014 Evaluation of alternative stretch flangeability testing methods to iso 16630 standard *IDDRG 2014 Conference Proceedings, June 1-5, Paris, France.*
[4] Schneider, M. and U. Eggers 2011 Investigation on punched edge formability *IDDRG 2011 Conference Proceedings, June 5-8, Bilbao, Spain.*
[5] Hance B. M. 2017 Practical Application of the Hole Expansion Test *SAE Int. J. Engines* **10**(2)
[6] Barlo, A. *et al* 2022 A Study of the Boundary Conditions in the ISO-16630 Hole Expansion Test *IOP Conf. Ser.: Mater. Sci. Eng.* **1238**, p. 012031.
[7] Pilthammar, J. *et al.* 2018 BBC05 with non-integer exponent and ambiguities in nakajima yield surface calbration *Int. J. Mater. Form.* **14**, pp. 577-592.
[8] Ayyub, B. M. and R. H. McCuen 2011 Probability, statistics, and reliability for engineers and scientists *CRC Press*, 3rd edn.

IOP Conf. Series: Materials Science and Engineering 1284 (2023) 012028 doi:10.1088/1757-899X/1284/1/012028

Hot-forming of a 980 MPa third generation advanced high strength steel

Abdelbaset R H Midawi[1*], C Tolton[1], R George[1], M Subramanian[2], T Skszek[3], C Butcher[1] and M Worswick[1]

[1]Forming and Crash lab, MME, University of Waterloo, Waterloo, ON, Canada
[2]Magna International, Troy, MI, United States,
[3]Pacific Northwest National Laboratory (PNNL), Richland, WA, United States

E-mail: amidawi@uwaterloo.ca

Abstract. This paper considers the effect of processing route on the final microstructure hardness and tensile behavior of a 980 MPa third generation advanced high strength steel (3G-AHSS). Of particular interest is the use of a hot stamping thermal schedule that mimics a quench and partition (Q&P) heat treatment process normally used to produce Q&P grade steel sheet. For comparison purposes, samples were also produced using a quench and temper (Q&T) processing history. The experiments were performed using a Gleeble3500 thermo-mechanical simulator system. Large coupons were produced using both processing routes from which microstructure, hardness, and tensile samples were extracted. Mechanical properties results were compared with the as-received material. Both processing routes resulted in strength levels that were close to that of the as-received material; however, the total elongation of the Q&P processed samples was 52% higher than that of the Q&T processed material.

1. Introduction

Third generation advanced high strength steels (3G-AHSS) offer an attractive combination of high strength and ductility due to their multiphase microstructures [1–3] containing retained austenite (RA) that transforms into martensite during deformation. 3G-AHSS are currently seeing adoption for use in a weight reduction of structural automotive components, such as A- and B-pillars, roof rails, and front and rear rails [2]. Key challenges in conventional room temperature stamping of these alloys include high springback, tool wear, and press tonnage. One approach to alleviate these issues, and the focus of the current paper, is to hot stamp 3G-AHSS using a process similar to that used for press-hardenable 22MnB5 grades [4–6]. Unfortunately, austenization followed by rapid quenching of 3G-AHSS to room temperature will result in a fully martensitic microstructure without the RA needed to enable the transformation induced plasticity (TRIP) effect that imparts such desirable mechanical properties to 3G-AHSS. In this paper, the use of a two-step hot stamping process is considered an effort to mimic a quench and partition (Q&P) thermal treatment process [7,8] during hot stamping. In this approach, complex structural parts can be hot formed at lower press tonnages with reduced springback while the Q&P treatment recovers the 3G-AHSS microstructure with its high work hardening rate for use in secondary forming stages or crash energy absorption.

Previous studies have examined the multi-step thermal processing of AHSS. Rana et al. [9] considered a quench and partition (Q&P) two-step process for a medium Mn (10 Wt%) AHSS to stabilize a high-volume fraction of RA, reporting ultra-high strength (1150-1400 MPa) and high elongation (20-40%). Zhang et al. [10] studied an alternative processing route corresponding to a quench and temper (Q&T) process, in which an AHSS was quenched from 910 °C to room temperature and then tempered at temperatures in the range of 400-600 °C for 45 minutes. They reported that the impact energy increased with tempering temperature, which was attributed to the change in the martensite morphology from fine to coarse laths and a reduction in the dislocation density; however, the strength decreased as the tempering temperature increased.

The current study focuses on hot stamping of a commercially available 980 MPa grade Q&P 3G-AHSS. A Gleeble thermo-mechanical simulator is used to simulate what is termed a Q&P hot stamping process in which the material is first subjected to an intercritical anneal (ICA), followed by quenching to a temperature above the martensite finish (M_f) temperature. The material is immediately heated to a partitioning temperature to stabilize the retained austenite. Microstructural examination, microhardness measurements, and tensile tests are used to characterize the samples following the Gleeble processing. For comparison purposes, a Q&T heat treatment process is also examined to contrast the mechanical behaviour of the resulting microstructures. Finally, hot stamping trials were performed on the 3G-AHSS material to examine the potential benefit of reducing press tonnage relative to conventional room temperature stamping.

2. Experimental

2.1. Material

The material used in this study is a 1.6 mm galvanized (GI) zinc-coated 3G-980 AHSS, with the nominal chemical composition shown in table 1. The as-received microstructure is shown in Figure 1 and consists of martensite islands in a ferrite matrix, bainite laths, and retained austenite (RA). This unique microstructure is usually produced using a quenching and partitioning heat treatment process. The volume fraction of 15.4% RA was calculated using X-Ray diffraction according to the methodology of [11].

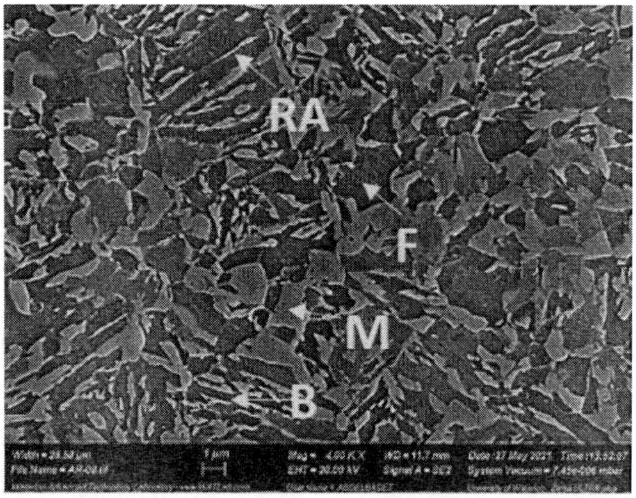

Figure 1. Secondary electron micrograph showing the as-received material microstructure, with the different phases present (M: martensite, F: ferrite, B: Bainite, RA: retained austenite).

The carbon equivalent of 0.68 was calculated using the Yurioka et al. [12] formula. The critical transformation temperatures and continuous cooling transformation (CCT) diagram, shown in Figure 2, were predicted using JmatPro [13] based on the chemical composition of the as-received material shown in Table 1. The critical transformation temperatures A_{C1} and A_{C3} were estimated to be 700 °C and 798 °C respectively, while the martensite start (M_S) and finish (M_f) temperatures were predicted to be 320 °C and 190 °C, respectively.

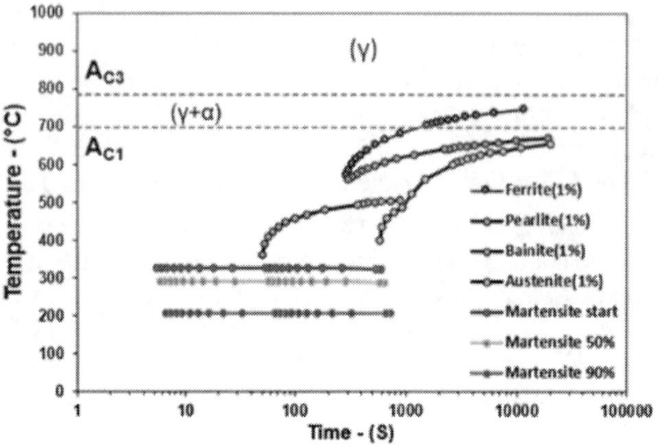

Figure 2. The predicted CCT diagram for the 3G-980 AHSS material. (γ: Austenite, α: Ferrite).

Table 1. Nominal chemical composition (Wt%) of the as-received material.

Material ID	C	Mn	Si	Cr	Ni	Mo	CE_N
3G-980	0.24	2.4	0.62	0.04	0.01	0.006	0.68

2.2. Sample preparation and testing

The thermal processing was performed using a Gleeble®3500 thermo-mechanical simulator. Rolling direction (RD) Gleeble specimens (Figure 3-a) were cut from the as-received sheet using a waterjet cutter. Metallurgical coupons were machined from the as-received and as-processed Gleeble samples. The metallurgical samples were cross-sectioned, hot-mounted in conductive bakelite, and prepared according to ASTM E3-11 [14]. The microstructure samples were prepared using a 5 second 5% nital etchant solution, followed by cleaning using alcohol. A field emission scanning electron microscope (FESEM, UltraZeiss) was used to image the microstructure. Microhardness tests were performed according to ASTM-E384 [15] using a Clemex automated Vickers microhardness tester. The indentation load was 0.3kgf, dwell time was 15 seconds, and spacing between indents was 500 μm. The hardness analysis was performed on a polished surface without etching.

Miniature tensile coupons (Figure 3-b) were machined from the central region of the Gleeble specimens to ensure the reported mechanical properties represent the as-processed microstructure. The microhardness was measured along the gauge length of the Gleeble specimens. The mini-tensile coupons featured a small taper (600 mm radius in Figure 3-b) along the edge of the gauge region to promote localization at the specimen center where the temperature was controlled by the Gleeble. The tensile tests were conducted using a 100 kN electromechanical MTS tensile frame. The Vic 3D R9.1 3D DIC software, from Correlated Solutions Inc., was used to measure the strain field.

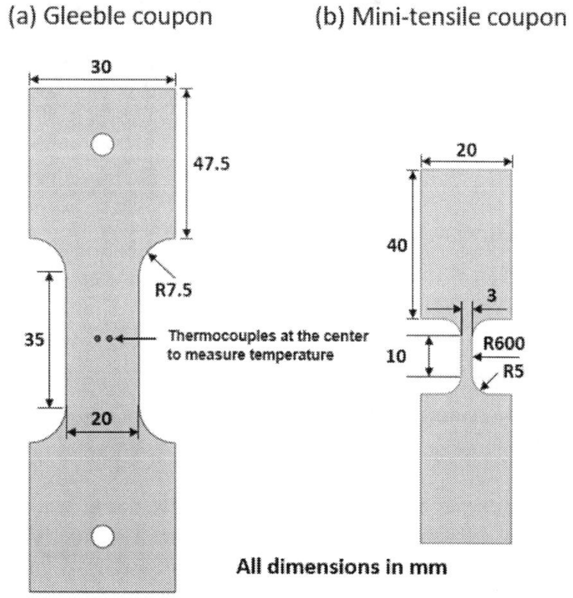

Figure 3. Schematic shows the Gleeble coupons and the miniature tensile test coupon (note: the minimum gauge width at the center of the miniature tensile coupon was approximately 2.98 mm).

2.3. Gleeble simulation method – thermal processing histories

The thermal-processing tests were used to assess the response of the microstructure and resulting mechanical properties to two processing schedules corresponding to either Q&P or Q&T hot stamping. Both processing routes started with an ICA treatment at a temperature of 750 °C. The heating and cooling rates were set to approximate rates that can be achieved using industrial furnaces and cooled tooling. Therefore, the specimens were heated to 750 °C at a slow heating rate of (2.5 °C/s), to represent conventional furnace heating and then held at a temperature for 300 s. The specimens were subsequently processed with either a quench and partition (Q&P) or a quench and temper (Q&T) route.

The schematic in Figure 4 shows the thermal history for each processing route. The intercritical Q&P history, shown in Figure 4-a, used a quench from 750 °C to 250 °C at 50 °C/s. Immediately following the quench, the sample temperature was further reduced to 230 °C over a period of 10 s to mimic the convective heat loss during the transfer of the sample from the hot stamping die to the partitioning furnace. To model the partitioning operation, the sample was heated to a partitioning temperature of 400 °C over a period of 5 min, after which the samples were held at the partitioning temperature for 120 seconds and then air cooled. The quench and partitioning temperatures were determined based on the martensite start (M_S) and martensite finish (M_f) temperatures, shown in the CCT diagram Figure 2. The schematic displayed in Figure 4-b shows the thermal history of the quench and temper specimen (Q&T). The specimen was quenched from the ICA temperature of 750 °C to room temperature (25 °C) and then heated at a rate of 1.25 °C/s to the tempering temperature of 500 °C and held for five minutes to allow the martensite tempering reaction to take place, followed by air cooling.

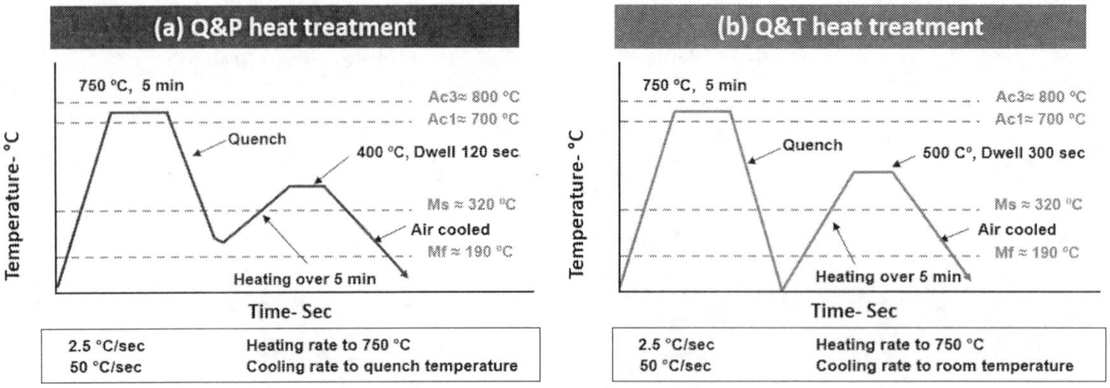

Figure 4. Schematic showing the heat treatment schedule used in the Gleeble experiments for (a) the ICA Q&P hot stamping process and (b) the Q&T hot stamping process.

2.4. Hot stamping

Hot stamping formability tests were performed using an automated hot stamping press system to assess the reduction in press tonnage achievable by hot stamping the 3G-AHSS. The tooling comprised a stationary 100 mm diameter hemispherical Nakazima [16] punch mounted on the base of a hydraulic forming press, as shown in Figure 5. The specimen geometry comprised a 230 mm square blank intended to produce a near equal-biaxial strain path for conventional room temperature Nakazima testing. A water-cooled binder surrounded the punch and was mounted on pneumatic springs to provide up to 355 kN of clamping force on the blank. The binder featured spring-loaded lifter fittings that enabled the heated blank to be accurately positioned and prevented initial contact with the binder, thereby delaying quenching prior to the forming stroke. The water-cooled die was installed on a 600-ton hydraulic actuator that created the downstroke to form the sample. Prior to hot stamping, the blanks were solutionized in a conventional furnace at 750 °C for a period of 5 min, after which they were robotically transferred to the tooling and immediately hot stamped with a punch speed of 50 mm/s. Since the focus of these experiments was on the reduction in press tonnage using hot stamping, the primary instrumentation was the die displacement versus punch load history. For the same reason, a partitioning operation was not considered in the current hot stamping experiments, and the samples were quenched to room temperature without subsequent thermal treatment.

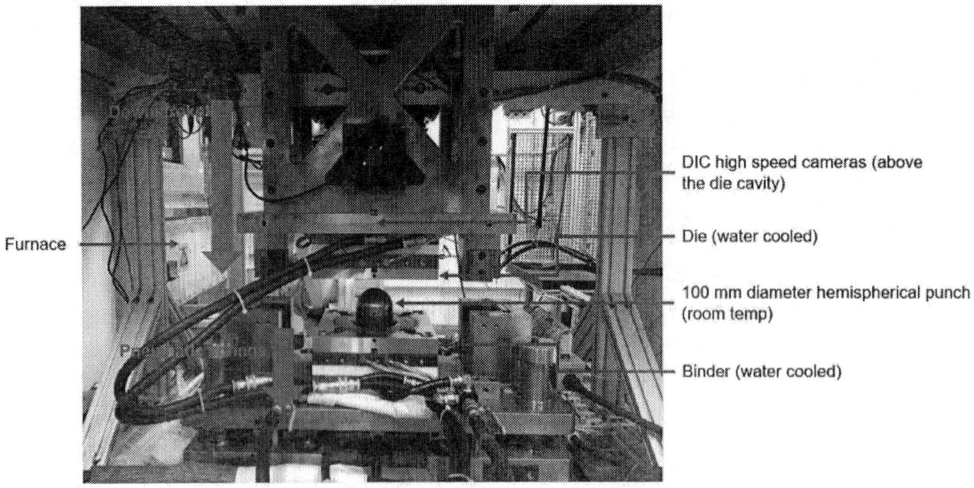

Figure 5. Photograph of the Nakazima tooling used in the hot stamping experiments.

3. Results and discussion

3.1. AR and as-processed microstructure

Scanning electron micrographs (SEM) representing the microstructure of the various thermal process conditions are shown in Figure 6. The as-received (AR) microstructure of the 3G-980 AHSS sheet (Figure 6-a) consists of multiple phases, including ferrite (F), martensite (M), bainite (B), and retained austenite (RA).

The microstructure of the Q&P heat treatment is shown in Figure 6-b. Qualitative comparison between the as-received and Q&P microstructures reveals that both conditions are similar in terms of the microstructural phases present. In addition, the retained austenite volume fraction measurements for the Q&P process were 20% RA compared to 15.4% for the AR condition. This increase in the RA phase could improve the mechanical properties by promoting a stronger TRIP effect during deformation.
The as-processed Q&T microstructure is shown in Figure 6-c. This processing route results in a dual-phase microstructure (martensite and ferrite) following quenching to room temperature with essentially no RA. The tempering reaction results in the decomposition of the martensite phase to form ferrite and carbide (Fe3C). The tempering process is a function of time and temperature [17]. The relatively short tempering time of 5 minutes results in partial tempering, with some martensite islands in the final microstructure.

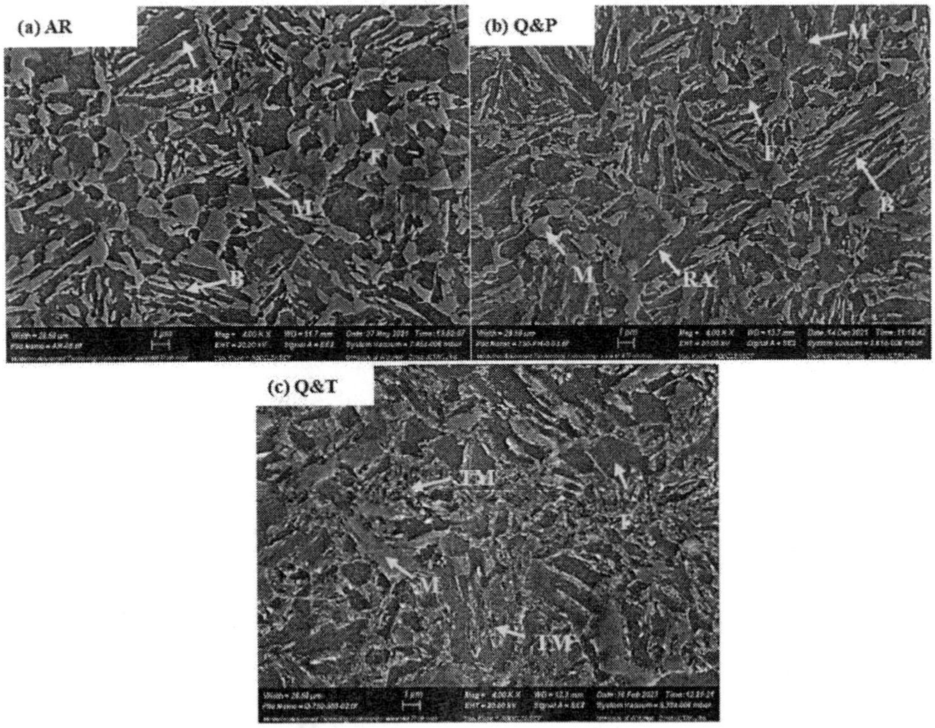

Figure 6. Secondary electron micrographs show the microstructure of the (a) AR condition; (b) Q&P processing route, and (c) Q&T processing route.

3.2. Microhardness measurements

The microhardness measurements were performed on the cross-section at the middle of the Gleeble coupons. The average hardness of 15 indents (and standard deviation) on each condition is shown in Figure 7 for the AR and the two thermally processed conditions. The hardness of the as-received material was 304 ± 7 HV. The Q&P and Q&T thermal-processing routes resulted in hardness levels of 301 ± 4

HV and 313 ± 6 HV. Thus, despite the striking differences in as-processed microstructures (Figure 6), the two processing routes produce similar hardness levels.

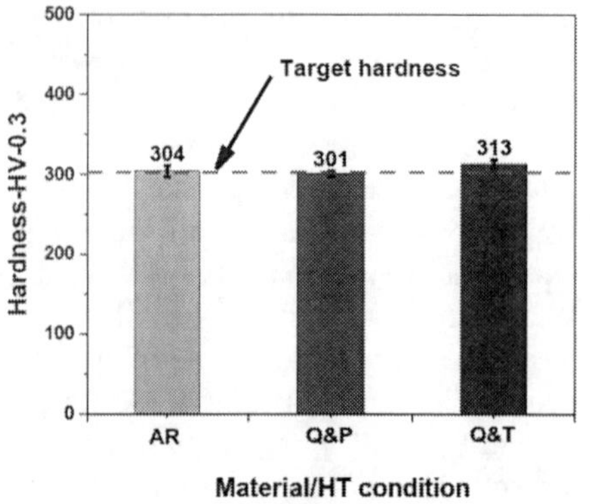

Figure 7. Comparison of the average hardness of the as-received 3G-980 AHSS with the hardness obtained in different heat treatment routes.

3.3. *Tensile behaviour*

The engineering stress-strain response of the AR and two as-processed microstructures are shown in Figure 8-a, while Table 2 summarizes the mechanical properties. The as-received ultimate tensile strength (UTS) was 1033 MPa. The Q&T sample exhibited a UTS of 988 MPa, which is within 5% of the AR UTS; however, the total elongation of the Q&T sample was only 19%, a 36% drop relative to the AR sample. Furthermore, yield point elongation behaviour was observed in the Q&T stress-strain curve that was attributed to the presence of a tempered martensite microstructure. The Q&P condition showed an 8% lower UTS compared to the AR condition; however, the total elongation increased from 27% to 32%, a relative improvement of 17%. Additionally, the area under the stress-strain curve, which represents a quantitative measure of the material toughness (energy absorption), is summarized in Figure 8-b. The Q&P condition showed a 7% improvement in toughness compared to the AR material and was 45% higher than the toughness of the Q&T condition. This loss in elongation and toughness of the Q&T processed material, compared to the Q&P and AR conditions which both contain high levels of RA, can be attributed to the more brittle phases such as martensite and tempered martensite present in the Q&T microstructure, and loss of RA due to the full quenching process. Finally, it is acknowledged that the UTS of the Q&P processed material fell somewhat below that of the AR material. This is not viewed as an issue since the UTS can be corrected by adjusting the ICA temperature prior to hot stamping and is left for future work.

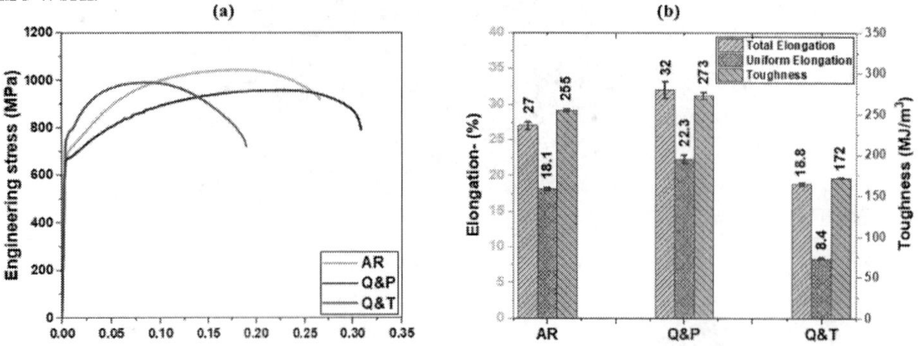

Figure 8. (a) Engineering stress-strain curves for selected conditions and (b) elongation and area under the stress-strain curve (toughness).

Table 2. Summary of the mechanical properties and hardness for the tested conditions. The extensometer gauge length was 9 mm.

Material/HT Process	YS [MPa]	UTS [MPa]	Uniform El. [%]	Total El. [%]	Hardness [HV]	RA [%]
AR	677±4	1033±8	18.1±0.20	27±0.50	304±7	15.40
Q&P	663±1	953±11	22.3±0.60	32±1.20	301±3	20.50
Q&T	773±13	988±2	8.4±0.10	18.8±0.2	313±5	-

3.4. Hot stamping Nakazima results

The Nakazima punch force versus stroke curves are shown in Figure 9. The results reveal a significant reduction of approximately 78% in the peak press force for the elevated-temperature tests relative to the room temperature stamping. This outcome indicates significant promise for press tonnage reduction using a Q&P hot stamping process. Note that there was no significant difference in the peak press force between the tests using initial blank temperatures of 750 °C and 900 °C.

The Nakazima data does reveal a reduction in the dome height of 34% for the Q&P process relative to the room temperature forming. This drop in dome height is attributed to austenite stabilization and suppression of the TRIP reaction at elevated temperature, a necessary consequence of retaining the austenite in the as-formed microstructure, which results in highly desirable as-formed properties. The drop in dome height may also be due to the adoption of a hemispherical punch geometry which tends to quench the centre of the sample first, thereby forcing localization towards the periphery of the blank under plane strain conditions, as described by He et al. [18].

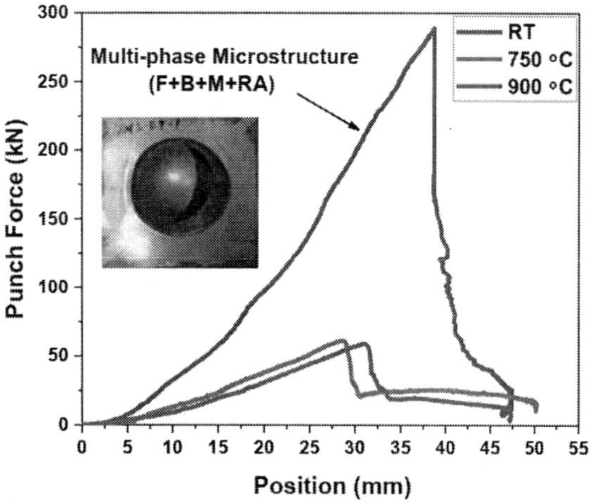

Figure 9. Punch force versus displacement curves for the as-received material at different initial forming temperatures in the Nakazima dome tests.

4. Conclusion

The results of the present study demonstrate that the Q&P thermal cycle can be used within a hot stamping process to essentially recover the AR microstructure of the 980 MPa 3G-AHSS, offering exceptional strength, ductility, and energy absorption after forming. From a process perspective, additional complexity to perform a Q&P hot stamping operation lies in the need to use a heated die to arrest the quench step above the M_f temperature as well as an additional furnace heating stage to perform the partitioning operation. Future work will consider the influence of deformation during the hot forming stage on the resulting RA along with evaluation of the formability and fracture limits of the processed Q&P route with an emphasis upon crash performance.

References

[1] E. De Moor, J.G. Speer, D.K. Matlock, D.N. Hanlon, Effect of Retained Austenite on Tensile Behavior of AHSS Revisited, Mater. Sci. Technol. (2011) 568–579.
[2] M. Davenport, Third-generation advanced high-strength steel emerges., Stamp. J. (2017). https://www.thefabricator.com/stampingjournal/article/stamping/third-generation-advanced-high-strength-steel-emerges.
[3] D. Krizan, K. Steineder, S. Kaar, T. Hebesberger, Development of third generation advanced high strength steels for automotive applications, 19th Int. Sci. Conf. Transf. (2018).
[4] M. Sajan, M. Amirthalingam, U. Chakkingal, A novel method for the spring-back analysis of a hot stamping steel, J. Mater. Res. Technol. 11 (2021) 227–234. https://doi.org/10.1016/j.jmrt.2021.01.017.
[5] K. Ikeuchi, J. Yanagimoto, Valuation method for effects of hot stamping process parameters on product properties using hot forming simulator, J. Mater. Process. Tech. 211 (2011) 1441–1447. https://doi.org/10.1016/j.jmatprotec.2011.03.017.
[6] M. Merklein, J. Lechler, Investigation of the thermo-mechanical properties of hot stamping steels, 177 (2006) 452–455. https://doi.org/10.1016/j.jmatprotec.2006.03.233.
[7] J. Speer, D.K. Matlock, B.C. De Cooman, J.G. Schroth, Carbon partitioning into austenite after martensite transformation, Acta Mater. 51 (2003) 2611–2622. https://doi.org/10.1016/S1359-6454(03)00059-4.
[8] David K. Matlock and John G. Speer, Third Generation of AHSS: Microstructure Design Concepts, in: S. Suwas, Arunansu Haldar, D. Bhattacharjee (Eds.), Microstruct. Texture Steels Other Mater., springer, 2009: pp. 185–205. https://doi.org/https://doi.org/10.1007/978-1-84882-454-6_11.
[9] R. Rana, C.H. Carson, J.G. Speer, Hot Forming Response of Medium Manganese Transformation Induced Plasticity Steels, in: 5th Int. Conf. Hot Sheet Met. Form. High-Performance Steel, 2014.
[10] Y. Zhang, J. Yang, D. Xiao, D. Luo, C. Tuo, H. Wu, Effect of Quenching and Tempering on Mechanical Properties and Impact Fracture Behavior of Low-Carbon Low-Alloy Steel, Metals (Basel). 12 (2022) 2–13. https://doi.org/10.3390/met12071087.
[11] G.P. Medina, H.L. Ferreira, P.Z. Robledo, A.M. Pérez, F.A.R. Valdés, Microstructural Development in a TRIP-780 Steel Joined by Friction Stir Welding (FSW): Quantitative Evaluations and Comparisons with EBSD Predictions, Soldag. e Insp. 21 (2016) 146–155. https://doi.org/10.1590/0104-9224/SI2102.04.
[12] N. Yurioka, Carbon equivalents for hardenability and cold cracking susceptibility of steels., 1990.
[13] N. Saunders, Z. Guo, A.P. Miodownik, J.-P. Schillé, The Calculation of TTT and CCT diagrams for General Steels, JMatPro Softw. Lit. (2004) 1–12.
[14] ASTM, E3-11 Standard Guide for Preparation of Metallographic Specimens 1, ASTM Copyright. i (2011) 1–12. https://doi.org/10.1520/E0003-11.2.
[15] ASTM Standard, E384 -17 Standard Test Method for Microindentation Hardness of Materials, ASTM Int. (2017) 1–40. https://doi.org/10.1520/E0384-17.
[16] K. Nakazima; T. Kikuma; and K. Hasuka, Study on the formability of steel sheets. Yawata Technical Report, No. 264, 1968.
[17] D.C. Ramachandran, A.R.H. Midawi, M. Shojaee, O. Sherepenko, H. Ghassemi-Armaki, E. Biro, A Comprehensive Evaluation of Tempering Kinetics on 3 rd Generation Advanced High Strength Steels, Acta Materialia SSRN. J. (2022) 101644.

https://doi.org/10.2139/ssrn.4141388.
[18] M. He, R., George, R., DiCecco, S., Butcher, C., Chiriac, C., Shi, C., Boettger, J., Worswick, Formability Characterization of Al-Si Coated Ultra-High Strength Steel in Hot Stamping through Nakajima Dome Testing, in: CHS2, Barcelona, 2022.

The Investigations on Formability of Tantalum RO5200 Thin Foils: Bio-Material.

Omkar Kulkarni[1] and Ganesh Kakandikar[2]

[1]Ph. D. Research Scholar and Assistant Professor, School of Mechanical Engineering, Dr. Vishwanath Karad MIT World Peace University, Pune, Maharashtra, India
[2]Professor, School of Mechanical Engineering, Dr. Vishwanath Karad MIT World Peace University, Pune, Maharashtra, India

E-mail: kakandikar@gmail.com

Abstract. Tantalum RO5200 (TaRO5200) is an excellent and new emerging Bio-material with wide applications. The most important property of TaRO5200 is Osseointegration which helps in integration of direct structural and functional connection between living bone and the surface. The process of forming in at actual is a plastic deformation where in the material is transforming in permanently is some shape without occurrence of any type of defect of failure. Few different types of failure enlisted are local necking, wrinkling, earing etc. In recent days transforming from macro level to micro level is now trending part of life. The most difficult part in moving towards micro level forming which means forming sheets under 100 microns thickness, is control over process. Hence this gives significance and rise for the investigations over forming at micro level and prediction of mechanical and formability properties at micro level. This research work posited in this paper focusses on the plotting of Forming limit curve (FLC) which is one of the methods to investigate formability of ultra-thin foil of TaRO5200 with sheet thickness 80 microns (µm). The experiments performed for FLC plotting were based on the Nakkajima test following the ASTM 2218-14 standards. The FLCs are plotted experimentally and via numerical simulation as well which in comparison shows a good agreement. The variance obtained between experimental and numerical simulation is up to 15%. Which in another words can be stated as numerical simulation is 15% safer in design than the experimental work. The novelty of the work lies in testing of the bio material and defining FLC by performing experiments via newly designed tool for the specific sheet thickness of 80 µm following the ASTM standards. Microstructural study is also performed on the specimen prior and after forming test to analyse and understand the physics along with mechanism of the material in test. The material behaviour is also explained in the microstructural study section.

1. Introduction

In the fields of medical, precision and electronic information equipment, the demand for thin metal parts is growing, and micro forming [1] processing technology is attracting attention. However, reducing the material size also has some problems, such as reduced ductility of the material and non-uniform deformation; therefore, there is a need to develop materials and processes suitable for micro forming [2]. Today, the use of micro components in technical systems and devices is a decisive value-adding factor in many industries such as medical technology, healthcare, mobility or communications.

Therefore, the micro molding techniques used to manufacture these components play a key role in creating corresponding innovative and affordable products[3].

The growing trend of miniaturization has been the driving force for the development of various new manufacturing techniques and the simplification of traditional processes over the past few years. This is especially the case when conventional approaches to silicon-based microelectromechanical systems (MEMS) components reach their limits in terms of materials to be processed, aspect ratios, and more. The forming process is a promising method due to its numerous advantages. The corresponding microscopic details are produced from metallic materials. Micro forming[4]–[6] can provide near-net shape properties, improved formation of complex geometries, the ability to produce miniature parts with relatively high stiffness and strength, and the ability to use a wide variety of materials. Furthermore, in general, the molding process provides the high-volume production of micro components required today. Micro components play an important role in a variety of industrial products in economically important markets. Micro forming-based products can find applications in medicine, mechanical engineering, electrical and electronic systems, optoelectronics, chemical systems, sensors, and thermal control[7].

The current research aims and focusses on the ability of the Tantalum RO5200 Biomaterial to be formed with respect to one investigation and analysis technique[8] i.e FLC. The current research focussed on the material thickness of 80 μm. the methodology utilized in this research work is by the use of Nakkjima test with standard specimen with following the ASTM-2218-14 standard which finally end up in comparison of simulation and numerical approach. Simulation is carried out by means of M-K model criterion. A novel tool was designed for the experimentation purpose specification of which are with hemispherical head of dia 4mm. The microstructural analysis with the help of Optical microscope and Scanning electron microscope (SEM) was performed to study the behaviour of the material which explains the formability of material. Energy dispersive X-ray spectroscopy (EDS) was performed to verify the elements which are present in material which shows a good agreement with material. This research finally attempts to define and compare FLC of material Ta RO5200 by experimentally according to Nakkajima test and by simulation. The microstructural study performed with the help of SEM supports the behaviour of the material.

2. Formability & Forming Limit Curve

The Forming limit curve (FLC)[9]–[11] describes the maximum formability of a sheet metal material. The stratigraphic limit map is the key combination of two major surface strains (large and small) above which local necking instability exists. The Forming Limit Curve (FLC) provides an excellent guide for material, tooling and lubrication conditions. Furthermore, it depends heavily on the parameters of the material. The idea of forming limit maps was first proposed by Keeler when he noticed that the maximum local elongation was not sufficient to determine the possible deformation rate of the sheet[12], [13]. He determined that, in the same figure, the construction of the main deformation at failure of ε1 gives a curve: ε2 forms a limit curve along both axes. This curve is first restricted to the region where $\varepsilon 2 > 0$, done by Goodwin when $\varepsilon 2 < 0$. This curve is interesting because it divides the plane into two areas. The success area below the shape constraint curve and the failure area above it are used for drawing operations. The current criterion for rejecting slender parts is the presence of localized necking.

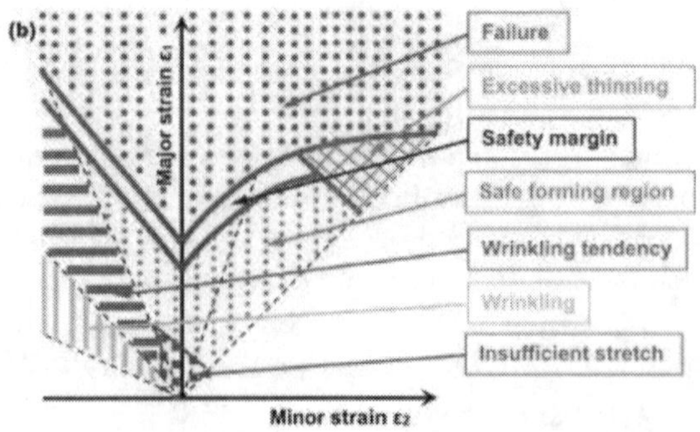

Figure 1. Schematic diagram of Failure limit curve FLC[14]

Figure 1 indicates the schematic diagram of FLC in which horizontal and vertical axis are minor and major strain respectively. All the zones are indicated in diagram with different colour combination. The red zone on top indicates failure and green zone indicates safe zone for forming. The blue, purple light green and brown zones are actually zones of defects where in formed object in this zone ends in some or the other kind of defect. This methodology is used to predict formability of Ta RO5200.

3. Tantalum RO5200

The discovery of tantalum[15]–[17] is often credited to the discovery of the mineral by the Swedish chemist Anders Gustav Eckenberg in 1802 in the mines of Iterby, Sweden. Furthermore, many authors believe that he has only discovered the allotrope of niobium, which is chemically similar to tantalum[5], [14], [18]. It wasn't until 1866 that Swiss chemist Jean Charles Galissard de Marignac proved that tantalum and niobium were two different elements. Ta Ro5200 material is considered for testing in this research. This material consists of elements such as Tantalum, Carbon, Oxygen, Niobium etc. this material shows properties such as excellent resistance to corrosion and also resistance to reaction to organic acids and osseointegration. Which actually defines this material as bio material. This material has a vast range of application in the medical field. Tantalum RO5200 is a biocompatible material that has been used in various medical applications, such as orthopedic and dental implants, cardiac pacemakers, and aneurysm clips[18]. It provides superior bone ingrowth and fixation, good biocompatibility and osseointegration, and excellent mechanical properties. Due to its stability, reliability, and resistance to corrosion, Tantalum RO5200 is a valuable material for medical devices that require long-term implantation [15], [16], [19], [20]. Table 1 indicates the element present in the current material considered these elements are predicted by energy dispersive X-ray spectroscopy (EDS) analysis. The Figure 2 indicates the results of EDS analysis of Ta RO5200[21], [22] which defines the energy levels of the elements present. The mechanical properties of the Ta RO5200 are notified in Table 2.

Table 1. Chemical composition of Ta RO5200

Elements	% by weight
Ta	85.47
C	8.1
O	6.30
Nb	0.1

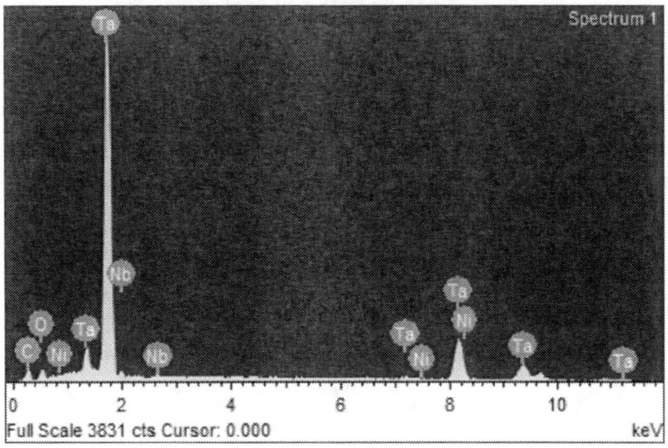

Figure 2. EDS analysis of Ta RO5200

Table 2. Mechanical properties of the Ta RO5200

Tensile Strength	**498.75 Mpa**
Elongation	**2.12%**
Thickness	**80 μm**

4. Specimen Preparation

The test specimens were made ready by firstly performing a circular grid marking by screen printing process. The stencil used for the circle grid marking was of 1mm dia hole and 2mm distance between 2 holes. The grid marking was performed and is shown in Figure 3. These circular grids marked will get stretched while the experimentation in forming process into ellipse which then will have major and minor axis. From these major and minor axis, the major and minor strain are evaluated according to Equation 1. This eventually will end up in plotting of FLC for the said material. The experimentation and measurement are performed for each specimen from a set. The cutting of specimens from the sheet are done by Wire cut EDM process. The specified shape required for the experimentation was performed in accordance with ASTM 2218-14 standard, Standard is a standard which is set to determine the FLC of any material with some particular specimen shapes and sizes. This is a specific standard which is mostly used by researches to predict and define formability of any material with specified test samples. Anil Mashalkar et.al.[2] posited a research work on thin brass foils, he followed the same ASTM standard with sheet thickness considered as 100 μm. gyan patel et.al.[9], [23] pursued research on plotting of FLC for thin brass foil with following the same ASTM standard. The same research work was continued with some microstructural aspects on Zirconium material with sheet thickness of 50 μm[24] [25]. specimen dimeter considered for experimentation was 10mm for the current research sample was prepared with respect to Nakajima test as prescribed in figure 4. Various strain paths were predicted as following to such as uniaxial, uniaxial intermediate, plane, biaxial intermediate and biaxial[22], [26].

$$Major/minor\ strain = \frac{Major/minor\ dia - initial\ dia}{initial\ dia} \qquad (1)$$

IOP Conf. Series: Materials Science and Engineering 1284 (2023) 012029 doi:10.1088/1757-899X/1284/1/012029

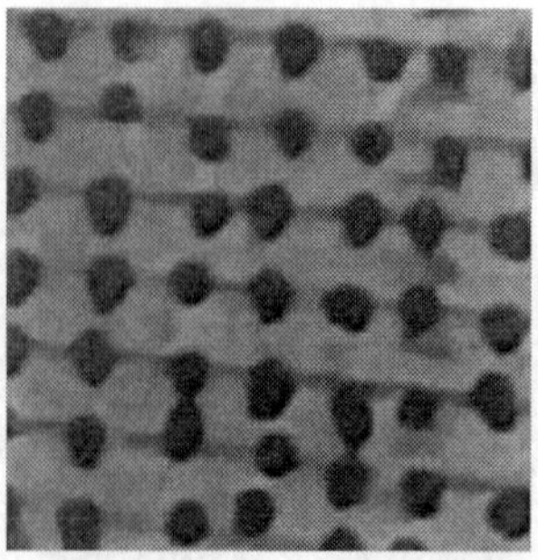

Figure 3. Grid marked specimen

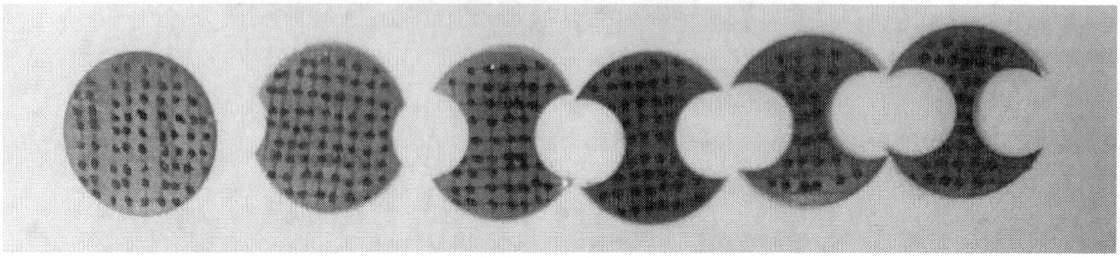

Biaxial Strain	*Intermediate Biaxial Strain*	*Plain Strain*	*Intermediate Uniaxial Strain*	*Uniaxial Strain*

Figure 4. Nakajima test specimen preparation.

5. Development of Tool Setup

A special purpose tool as a fixture was designed and developed for the purpose of investigations in micro forming of the material Ta RO5200 as shown in Figure 5. The designed tool used as a special fixture consists of two set of dies as upper die and lower die, in between these two dies two friction plates are used to develop friction. The tool is specifically designed with the hemispherical punch head of 4mm dia. The 4 holes present on upper die are having internal threading via which 4 bolts are used to fasten and develop pressure on friction plates which ultimately will develop pressure in blank. Blank is hold in between the two pressure plates.

This entire setup is mounted on the Universal Testing Machine (UTM) of 5 KN capacity for the purpose of performing Nakajima test. The UTM for the experimentation was run with a crosshead travel speed of 0.4mm/min to 1.2mm/min.

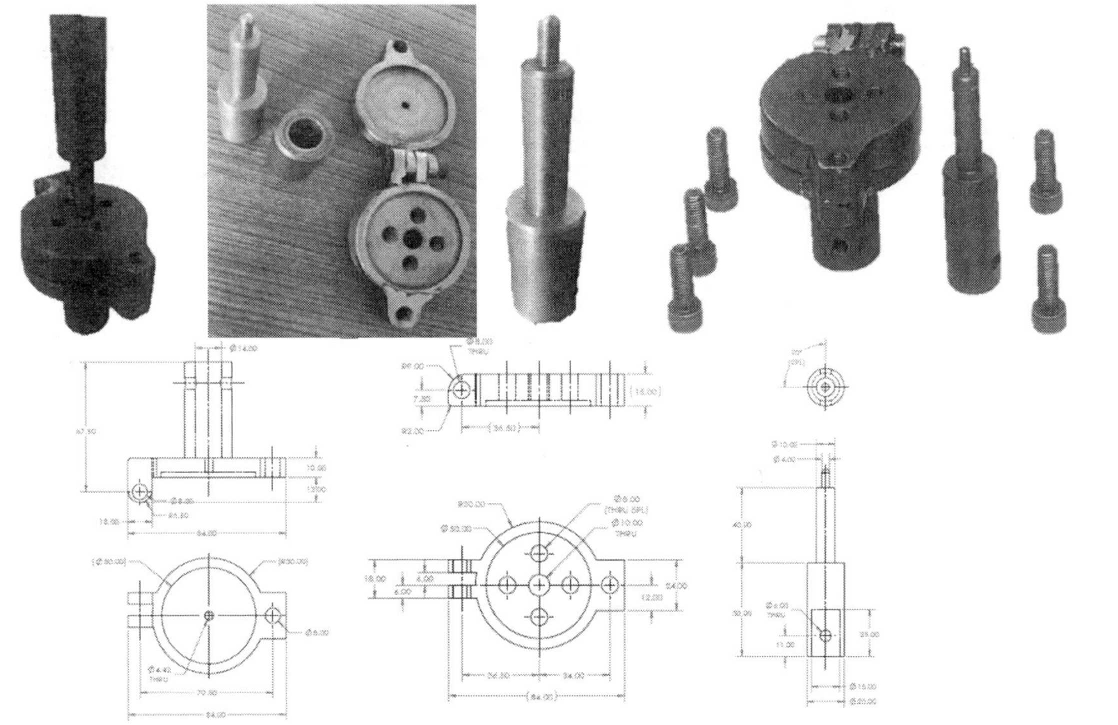

Figure 5. Design and development of tool for micro forming investigations.

6. Experimental Investigation

The experimental tests performed for the investigation were as per the ASTM 2218-14 standard via Nakajima test performed on 6 different samples of a set. There was total 3 sets of experiment performed for the purpose of precision the specimens utilised, were already printed with circular grid via screen printing as a shown in figure. Every specimen for every set was processes for a single crack development just after necking. Where ever the formation of crack is started the measurement of major and minor axis is performed over every sample. The measurement is performed via vision measuring system to increase the accuracy. The prediction of Major and Minor strains is performed by the measurement of major and minor axis near the necking area occurrence. Following the equation 1 the strain rates are calculated which is ratio change in length to original length. This ultimately secures the FLC plots for the material, in all 6 strain paths are plotted which is shown in Figure 6.

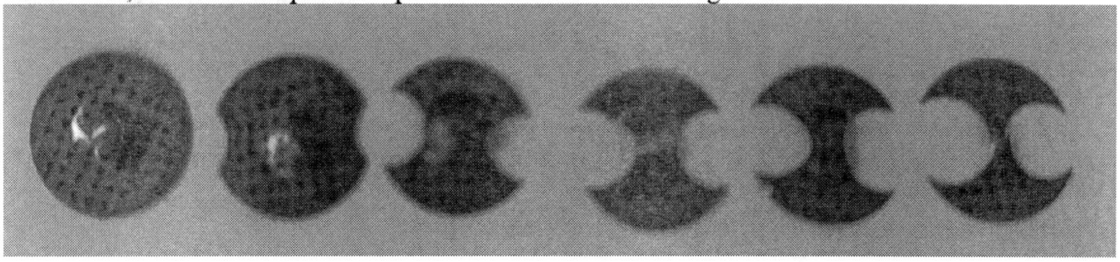

Biaxial Strain	*Intermediate Biaxial Strain*	*Plain Strain*	*Intermediate Uniaxial Strain*	*Uniaxial Strain*

Figure 6. Specimen after experimentation

IOP Conf. Series: Materials Science and Engineering 1284 (2023) 012029 doi:10.1088/1757-899X/1284/1/012029

7. Theoretical Prediction of FLC by M-K Criterion

In order to obtain the FLC for any material this following method is widely used which is accurate and fast respectively. It's based on the theory of plasticity and plastic deformation. In Swift's Model relation of rms values along with shear stress margin increments are used for the calculation. Marchiniak et.al proposed an M-K Model, which utilised initial thickness and defects depth. This model was widely used due to its simplicity and accuracy. Furthermore, LI X et.al improved the existing model of M-K by adding the factor, angle between the groove and principal direction which rectified the M-K model in a better way. A schematic figure for M-K model is shown in Figure 7

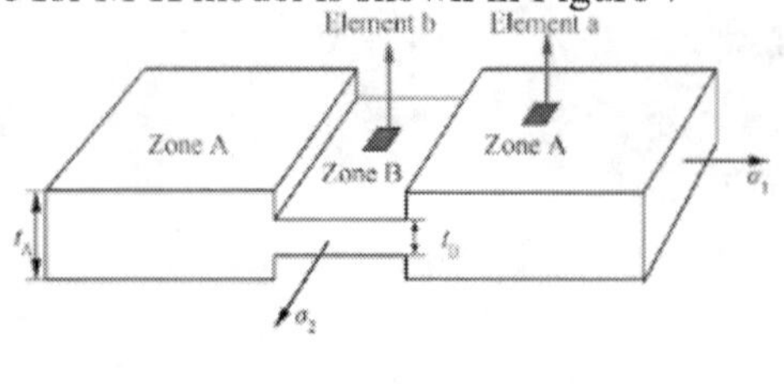

Figure 7. Schematic diagram for M-K model

7.1. Tensile Test

Tensile test is performed to get a record of properties of the material specifically mechanical properties. The important property to me noted and mentioned is Ultimate tensile strength, elongation, and angle at which it gets cracked which needs to be measured. This specific data is required to perform simulation trials on M-K model. The mechanical properties noted are specified in Table 2. For the tensile test to be performed the specimens need to be prepared as per standards which is indicated in Figure 8.

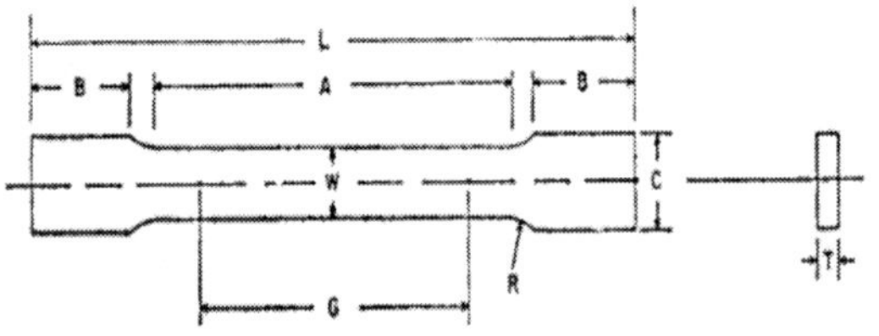

	Subsize Specimen
	6 mm [0.250 in.] Wide
	mm [in.]
G—Gauge length (Note 1 and Note 2)	25.0 ± 0.1 [1.000 ± 0.003]
W—Width (Note 3 and Note 4)	6.0 ± 0.1 [0.250 ± 0.005]
T—Thickness (Note 5)	
R—Radius of fillet, min (Note 6)	6 [0.250]
L—Overall length, min (Note 2, Note 7, and Note 8)	100 [4]
A—Length of reduced section, min	32 [1.25]
B—Length of grip section, min (Note 9)	30 [1.25]
C—Width of grip section, approximate (Note 4 and Note 9)	10 [0.375]

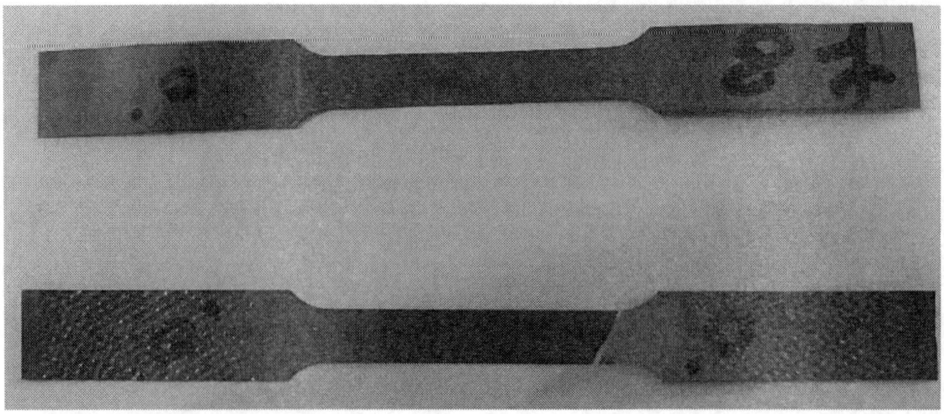

Figure 8. Standard tensile test specimen for thin foil

The specimen for the tensile test were done by the wire cut EDM process as it was used in order to reduce the stress concentration. The test was carried out on a specific 5KN UTM machine of low capacity with high accuracy with the specific fixture designed. The test resulted in the mechanical properties of the material along with the engineering stress strain curve.

8. Results and Discussions

The experiments were performed on 6 specimens and from the measurement of strain the FLC was plotted. The experimental strains i.e. major and minor strains are measured on vision measuring system for all specimens and are plotted on a graph for FLC to be plotted as shown in Figure 9. The strain path measured are mentioned in Table 3. ABAQUS software is used to perform simulation trials over all the measure paths following the M-K model. The results from the simulation are also presented in Table 4. The FLC for Material Ta RO5200 was plotted with the help of experiments and average from those experiments along with simulation from M-K model for the comparison which actually shows a good agreement.

Table 3. Strain Measurements in Six Strain Paths (Experimental Data)

Sr.no.	Specimen Type	Minor Strain	Major strain
1	Biaxial	0.08	0.16
2	Biaxial Intermediate	0.06	0.12
3	Plane	-0.01	0.07
4	Uniaxial Intermediate 1	-0.06	0.09
5	Uniaxial Intermediate 2	-0.1	0.12
6	Uniaxial	-0.16	0.19

Table 4. ABAQUS Software (Simulation data)

Sr.no.	Specimen Type	Minor Strain	Major strain
1	Biaxial	0.08	0.11
2	Biaxial Intermediate	0.05	0.09
3	Plane	-0.01	0.05
4	Uniaxial Intermediate 1	-0.06	0.07
5	Uniaxial Intermediate 2	-0.09	0.09
6	Uniaxial	-0.14	0.13

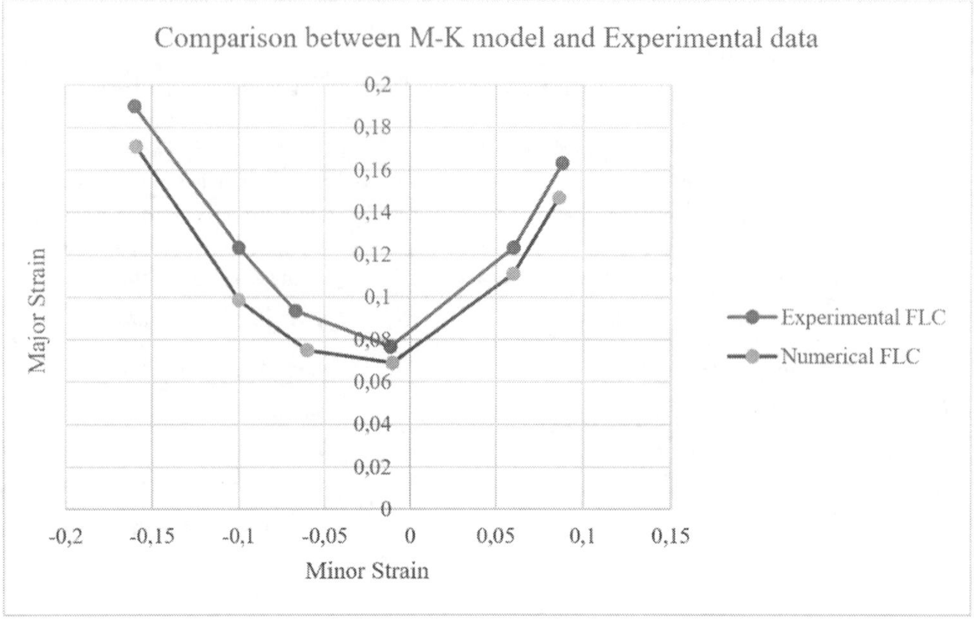

Figure 9. Comparison of Experimental and M-K model FLC

8.1. Micro Structural Study of Tantalum RO5200

An optical metallurgical microscope was used for microstructural analysis of the Tantalum 80-micron sheet. In microstructure at 100X, two phases were observed. One phase was present near about 39% and the second phase was present near about 61%. These phases affect the properties of the Tantalum. Image analyser software was used for the phase analysis.

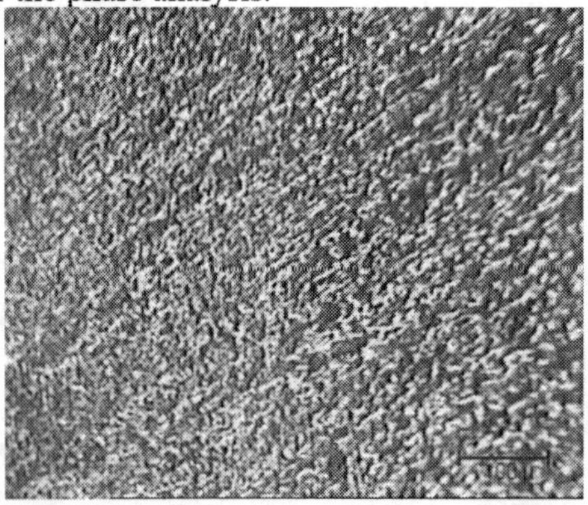

Figure 10. Microstructure of Ta RO5200

To ensure the chemical composition of the 80-micron sheet of specimen EDS method was used. EDS characterization method can rapidly give information about the chemical composition of a specimen, including Which elements are present, their distribution. Energy Dispersive Spectroscopy (EDS) analysis has been done on the given specimen. Ni, Nb, O, C elements were present in this specimen other than Tantalum[16]. Microstructure of Ta RO5200 is shown in Figure 10.

Figure 11. Unformed SEM Image of Ta RO5200

Figure 12. Formed SEM Image of Ta RO5200

The Field Emission Scanning Electron Microscope (SEM) and Energy Dispersive X-ray Spectrometer (EDS) analysis were performed on machine named as FEI Nova NanoSEM 450 and Bruker XFlash 6130 respectively. FESEM images are captured on Unformed and formed samples of Ta RO5200 which are represented in Figure 11 and

Figure 12 respectively. The formed surface after the Nakajima test was analysed to study the forming mechanism of sample. Figure 11 show the unformed SEM image of Ta RO5200 sheet of thickness 80 μm and Figure 12 shows the formed SEM image of Ta RO5200. From the SEM images it can clearly be observed that the unformed Ta RO5200 image does not show any crack developed on the surface whereas the formed Ta RO5200 image shows a significant number of cracks developed. In the unformed Ta RO5200 image it can clearly be seen the presence of two different phases i.e., alpha and beta phase. Due to the stretching the grain structure in formed Ta RO5200 got distorted. After forming this sheet due to stretching the structure get elongated and due to which the micro cracks in Nano meter level are developed which are also seen clearly in the formed Ta RO5200 images.

In the unformed sample it can be clearly observed no voids or stretchiness of material is present. There are patches which are present which represents the phases in the material. As compared to the formed sample the image shows a clear stretchy pattern on the sample. This represents the material sample is being stretched and ended in to failure. At the top side of the image small crack formation is also observed. These observation in the image represent that the failure of the material is due to expansion in material of forming of material and not due to the voids present or any material irregularities.

9. Conclusions

The Micro forming investigations was performed on Ta RO5200 foil: Bio Material, via Forming limit curve. The foil sheet thickness is considered as 80 μm. The experimental and simulation analysis was performed the experimental method was performed by help of Nakajima test and simulation was carried out by the help of M-K model in ABQUS software. The plotted FLC from experimentation and simulation method showed a good agreement which in at actual sets a fundamental benchmark for researches for the purpose of designing and simulating any another component in the specified micro forming domain focusing on the application in the field of biomedical. The fixtures, tools and specimen were designed and developed in accordance to the ASTM standards. To support the behaviour of the material the microstructural study is performed on Ta RO5200. This microstructural study is performed to study and analyse the formability behaviour of the material. The Ta RO5200 80 μm sheet was observed with the deformations and stretching as per the load applied while performing the Nakajima test in microstructural studies. The Ta Ro5200 was tested with EDS analysis which shows the presence of Tantalum, Niobium, Oxygen, Carbon content in it.

References

[1] N. Srinivasan, D. Rajenthirakumar, and R. Sridhar, "Micro Forming-Key Issues and Research Opportunities," *Adv. Nat. Appl. Sci.*, vol. 11, no. 4, pp. 557–579, 1998, [Online]. Available: http://www.aensiweb.com/http://creativecommons.org/licenses/by/4.0/.

[2] A. Mashalkar, G. Kakandikar, and V. Nandedkar, "Micro-forming analysis of ultra-thin brass foil," *Mater. Manuf. Process.*, vol. 34, no. 13, pp. 1509–1515, 2019, doi: 10.1080/10426914.2019.1655158.

[3] F. Jia, J. Zhao, L. Luo, H. Xie, and Z. Jiang, "Experimental and numerical study on micro deep drawing with aluminium-copper composite material," in *Procedia Engineering*, 2017, vol. 207, pp. 1051–1056, doi: 10.1016/j.proeng.2017.10.1129.

[4] A. Dhal, S. K. Panigrahi, and M. S. Shunmugam, "Investigation into the micro deep drawing capabilities of a specially engineered refined aluminium alloy," in *MATEC Web of Conferences*, 2018, vol. 190, doi: 10.1051/matecconf/201819010001.

[5] A. Teyfouri, M. Ahmadi, E. S. Lori, and S. Sorooshian, "A Review on Micro Formings," *Mod. Appl. Sci.*, vol. 9, no. 9, 2015, doi: 10.5539/mas.v9n9p230.

[6] T. C. Chen, J. C. Lin, and R. M. Lee, "Analysis of deep drawing process for stainless steel micro-channel array," *Materials (Basel).*, vol. 10, no. 4, 2017, doi: 10.3390/ma10040423.

[7] J. Zhu, Y. Lin, S. Liu, X. Ma, and G. Wang, "Plasticity and size effects of micro-forming sheet processed by electropulsing," *Mater. Manuf. Process.*, vol. 35, no. 10, pp. 1146–1155, 2020, doi: 10.1080/10426914.2020.1772482.

[8] N. Ayachi, N. Guermazi, C. H. Pham, and P.-Y. Manach, "Development of a Nakazima Test Suitable for Determining the Formability of Ultra-Thin Copper Sheets," *Metals (Basel).*, vol. 10, no. 9, p. 1163, Aug. 2020, doi: 10.3390/met10091163.

[9] G. Patel, K. Ganesh M, and O. Kulkarni, "Experimental and numerical investigations on forming limit curves in micro forming," *Adv. Mater. Process. Technol.*, vol. 00, no. 00, pp. 1–12, 2020, doi: 10.1080/2374068X.2020.1793268.

[10] K. I. Manabe, "Metal micro-forming," *Metals*, vol. 10, no. 6. MDPI AG, pp. 1–3, 2020, doi: 10.3390/met10060813.

[11] C. Maier, "Failure prediction in sheet metal forming using FEA of Nakazima test," *Machines, Technologies, Materials*, no. 7. SCIENTIFIC PROCEEDINGS INTERNATIONAL SCIENTIFIC-TECHNICAL CONFERENCE "TECHNICS. TECHOLOGIES. EDUCATION. SAFETY '13", pp. 9–12, 2013.

[12] M. Geiger, M. Kleiner, R. Eckstein, N. Tiesler, and U. Engel, "Microforming," *CIRP Ann. - Manuf. Technol.*, vol. 50, no. 2, pp. 445–462, 2001, doi: 10.1016/S0007-8506(07)62991-6.

[13] B. Kuhfuss *et al.*, "Micro Forming Processes BT - Cold Micro Metal Forming: Research Report of the Collaborative Research Center 'Micro Cold Forming' (SFB 747), Bremen, Germany," 2020, pp. 27–94.

[14] A. Groth, E. Schedin, and R. Moshfegh, "Investigation of different techniques to evaluate the FLC from experiments and FE simulations," in *Key Engineering Materials*, 2014, vol. 611–612, pp. 41–48, doi: 10.4028/www.scientific.net/KEM.611-612.41.

[15] G. Maccauro, P. R. Iommetti, F. Muratori, L. Raffaelli, P. F. Manicone, and C. Fabbriciani, "An Overview about Biomedical Applications of Micron and Nano Size Tantalum," *Recent Patents on Biotechnology*, vol. 3. pp. 157–165, 2009.

[16] B. R. Levine, S. Sporer, R. A. Poggie, C. J. Della Valle, and J. J. Jacobs, "Experimental and clinical performance of porous tantalum in orthopedic surgery," *Biomaterials*, vol. 27, no. 27, pp. 4671–4681, 2006, doi: 10.1016/j.biomaterials.2006.04.041.

[17] N. Patil, K. Lee, and S. B. Goodman, "Porous tantalum in hip and knee reconstruetive surgery," *Journal of Biomedical Materials Research - Part B Applied Biomaterials*, vol. 89, no. 1. John Wiley and Sons Inc., pp. 242–251, 2009, doi: 10.1002/jbm.b.31198.

[18] D. Cristea, I. Ghiuţă, and D. Munteanu, "TANTALUM BASED MATERIALS FOR IMPLANTS AND PROSTHESES APPLICATIONS," *Bulletin of the Transilvania University of Braşov* •,

vol. 8, no. 57. 2015.

[19] D. Cristea *et al.*, "Properties of tantalum oxynitride thin films produced by magnetron sputtering: The influence of processing parameters," *Vacuum*, vol. 98, pp. 63–69, 2013, doi: https://doi.org/10.1016/j.vacuum.2013.03.017.

[20] R. C. L. Robertson, "Repair of Cranial Defects with Tantalum," *J. Neurosurg.*, vol. 1, no. 4, pp. 227–236, 1944, doi: 10.3171/jns.1944.1.4.0227.

[21] Y. Wang and C. Zhang, "The determination of friction coefficient in Nakazima test and its application in predicting FLCs with modified M-K model," in *IOP Conference Series: Materials Science and Engineering*, 2020, vol. 967, no. 1, doi: 10.1088/1757-899X/967/1/012056.

[22] D. Lumelskyj, J. Rojek, and M. Tkocz, "NUMERICAL SIMULATIONS OF NAKAZIMA FORMABILITY TESTS WITH PREDICTION OF FAILURE," *Ro. J. Techn. Sci. − Appl. Mechanics*, vol. 60. pp. 184–194, 2015.

[23] G. Patel and G. Kakandikar, "Investigations on effect of thickness and rolling direction of thin metal foil on forming limit curves in microforming process," *Modern Manufacturing Processes*. Elsevier, pp. 145–155, 2020, doi: 10.1016/b978-0-12-819496-6.00007-5.

[24] O. Kulkarni and G. Kakandikar, "Formability Assessment with Microstructural Investigations for Zirconium 702 Thin Foils: Bio-Material Applications," *Adv. Mater. Process. Technol.*, pp. 1–11, Feb. 2022, doi: 10.1080/2374068X.2022.2044131.

[25] O. Kulkarni and G. Kakandikar, "Novel product design of tool for investigating formability with microstructural study of bio-material titanium grade-II thin foils," *Int. J. Interact. Des. Manuf.*, 2022, doi: 10.1007/s12008-022-00903-3.

[26] R. Sridhar and S. ShankarC, "Review On Micro Deep Drawing Process For Thin Foil Materials," *Int. J. Innov. Eng. Technol.*, vol. 11, 2018, doi: 10.21172/ijiet.113.03.

IOP Conf. Series: Materials Science and Engineering 1284 (2023) 012030 doi:10.1088/1757-899X/1284/1/012030

Effect of the intercritical annealing in the formability and microstructure of a medium Mn steel sheet

J L Hernández-Rivera[1], A C González-Castillo[2], M O Ramos Azpeitia[3] and C G Garay-Reyes[4]

[1] CONACYT- Instituto de Metalurgia, Universidad Autónoma de San Luis Potosí, Sierra Leona 550, Lomas 2ª Sección, San Luis Potosí 78210, México.
[2] Instituto de Metalurgia, Universidad Autónoma de San Luis Potosí, Sierra Leona 550, Lomas 2a Sección, San Luis Potosí 78210, México.
[2] Facultad de Ingeniería, Universidad Autónoma de San Luis Potosí, Dr. Manuel Nava 8, Zona Universitaria, San Luis Potosí 78290, México.
[4] Laboratorio Nacional de Nanotecnología, Centro de Investigación de Materiales Avanzados (CIMAV), Miguel de Cervantes 120, Chihuahua, 31136, México.

luis.rivera@uaslp.mx

Abstract. A medium Mn steel (MMS) was cast by employing a vacuum induction furnace. After that, the steel was hot rolled (HR) in order to achieve a final thickness of 1.4 mm. The microstructure of the sheet was found to be composed of martensite and a little amount of austenite, the Ultimate Tensile Strength (UTS in this processing condition was close to 1600 MPa and a negligible elongation was found. An intercritical annealing (IA) heat treatment was applied to the steel to promote the austenite reversion and increase its amount and stability. Previous thermodynamic simulations and experimental results were used to determine the temperature and time parameters of the IA. After the steel was subjected to this heat treatment, it exhibited an elongation close to 35% and an UTS close to 1100 MPa. Bendability testing was carried out in that condition in order to correlate it with the microstructural changes in the sheet. It was found that the critical bending angle was higher in the IA condition in comparison with the HR state.

1. Introduction

Medium Mn steel is one group belonging to the third generation of advanced high strength steels (AHSS) that have received much attention in the last years [1, 2]. The motivation for the intensive research regarding these steels is that they exhibit a very interesting combination of mechanical properties that are required to fulfill the actual necessities in the automotive industry regarding weight reduction and safety aspects. Typically, the thermomechanical condition for the optimal use of MMS comprises a hot or cold rolling step and after that, the application of an intercritical annealing on which there is promoted the presence of ferrite, austenite and in some cases martensite. The determination of the most convenient temperature and time values for this heat treatment (in terms of austenite stability and quantity), has been determined both by thermodynamic simulation and by experimental trials. MMS could exhibit the Transformation Induced Plasticity (TRIP) and/or Twinning Induced Plasticity (TWIP) phenomena which makes them an attractive material since significant improvement has been reported in the literature regarding ductility and formability [3] which allowed them to be shaped into a more complex geometry parts without failure.

The resistance of a sheet metal to strain localization and fracture under complex stress/strain conditions is called formability. Yi et al. [4] have mentioned that the actual trend for developing

AHSS that exhibit more than 30% of total elongation is with the purpose to ensure a successful plastic forming and also maintain enough "residual ductility" in the part for a crash event. Recently, Zhang et al. [5] reported the effect of different heat treatments on the formability of a Fe-10Mn-0.2C-2Al-0.1V (in wt.%) steel. They concluded that IA produced better formability in terms of strain hardening in comparison with the IA-Q&P treatment, it was associated to an enhanced TRIP effect produced in the first treatment. In other publication Sugimoto et al. [6] found that the maximum stretch height during stretch forming tests was achieved by a steel with 5wt.% of Mn in comparison with other steel that contain 1.5 and 3 wt.% of Mn, respectively. Also the tensile elongation was higher for the 5wt.% of Mn. These superior properties were attributed to a higher austenite quantity and stability. Despite the above studies, there is still scarce literature about the evaluation of the effective formability of MMS by another means such as Forming Limiting Diagrams or Bending tests, this is the reason why the actual study was carried out.

2. Experimental section

Homogenized cast ingots with a nominal composition of 0.15C-7Mn-1.5Al-2Si-0.1Nb-Fe (wt.%) with 30 mm in thickness were first heated to 1100 °C for 45 min to achieve a complete austenitization and then they were HR up to 1.4 mm in thickness. Each cycle consisted of three continuous rolling passes followed by reheating at 1100 °C and then a new cycle of three rolling passes was applied again until getting the desired thickness, finally the sheet was air cooled. The thermomechanical cycle was established to prevent that sheet temperature did not drop below 850 °C to avoid workability problems.

Samples of 10 mm in width and 60 mm in length were cut from as-rolled steel in the transverse and longitudinal direction. Subsequently, IA heat-treatment was performed at 700°C by 1 h in a Thermo-Scientific/FB1415M Muffle Furnace. After that, samples were quenched in brine. Characterization of the steel in the HR and in the intercritical annealed condition was performed by a metallographic analysis by optical (OM) and scanning electron microscopy (SEM) in an Olympus GX51 and a JEOL-6610LV operated at 30 kV, respectively. The etching process of the steel consisted of two stages: 1 to 2 s pre-etching in Nital solution, followed by 20 s etching in sodium metabisulfite solution. The phases present in both conditions were determined using a Bruker D8 Advance high-resolution X-ray diffractometer, with a Cu-Kα radiation source (λ=1.5405 A). The scan range and step size were 40 to 110° and 0.02°, respectively. MAUD software (based on Rietveld refinement method) was used for quantitative analysis using the JCPDS-2/Release 2016 database.

The air bending tests were performed according to the ASTM E-290-14 "Standard test methods for bend testing of material for ductility", employing a sharp punch (r = 0.6 mm radius), which moves downwards at a speed of 10 mm/min between two pivot-mounted rolls with a radius of 15 mm. The roll spacing was adapted to 2 times the thickness of the sheet + 0.5 mm. The bendability of the materials was characterized by the bending angle at maximum force, which is denominated critical bending angle (CBA) in this work. The average value of the CBA was determined by two tests per sheet condition. Samples cut from the transverse section of the bended sheet were metallographically prepared, etched and observed in the same way as it was explained in past paragraphs. Tensile testing was carried out according to ASTM E8- "Standard test methods for tension testing of metallic materials" and r parameter was determined according to E-517- "Standard test method for plastic strain ratio r for sheet metal". The uniaxial strain, plane strain, intermediate biaxial strain and equibiaxial strain points were predicted according to the work of Abspoel et al. [7] by using experimental values obtained from tensile testing, after that the forming limit diagram was built.

3. Results

3.1. Microstructure before and after IA

Figure 1 shows the X-ray spectrum of HR and IA samples. As can be seen in this figure, there were intense peaks corresponding to martensite/ferrite phases in the case of HR condition while for the IA sample, the peaks intensity of austenite showed a significant increment. It is important to note that it was not possible to distinguish between the individual contribution of ferrite and martensite to the total peaks intensity since both phases diffracted at the same 2θ positions, so corresponding peaks in the spectrum were indexed as ferrite/martensite. The increased intensity observed in austenite peaks after IA was related to a higher stability of this phase which was developed during annealing temperature because of enrichment of Mn as has been reported before [8]. The amount of retained austenite achieved at 700 °C was about 26 wt.%, whereas in the HR condition, there was a negligible content on this phase. Regarding other works on similar steels, Y. Zou et al. [9] reported about 22wt.% of retained austenite in a MMS with nominal composition of 0.005C-5.0Mn-0.5Si-1.4Ni-0.12V (wt.%) which was processed by hot rolling in the non-recrystallization zone and IA at 630 °C for 5 h, and finally air cooled to RT. In another report, Liu et al. [10] studied a MMS with a nominal composition of 0.12C-5.0Mn-1Al-0.2Mo-0.05Nb (wt.%) which was processed by hot-rolling, IA and tempering process. The HR sheets were first annealed at 600, 625, 650, and 690 °C for 30 min followed by quenching and tempering at 200 °C for 15 min, and finally, the samples were cooled to RT. As a result, they obtained a minimum of 19wt.% of retained austenite at 600 °C while a maximum of 29wt.% was obtained at 650 °C.

In order to better elucidate how the IA modify the microstructure of the steel, figure 2 shows the OM micrographs of HR and IA samples at 700 °C. As can be seen, there was an homogenous acicular phase in dark and light blue color in the former condition which was correlated with the martensite (close to 50 HRC hardness) while in the latter, the microstructure consisted of a mix of ferrite and austenite (34 HRC in average) that exhibited two different morphologies, plate-like and equiaxed. The former phase appear in a brown while the latter in white color [11], respectively. Li et al. [12] have proposed that the origin of these morphologies is the zone where the nucleation takes place, the fact that austenite and ferrite grow along the interface of the martensite laths in the initial microstructure produces the plate-like morphology while the equiaxed one is caused when the ferrite and austenite nucleation occurred in the intersection of the martensite laths. They also argued that as the IA time is increased, the reduction of the interfacial energy caused the coarsening of the equiaxed zone and the transformation of the plate-like zone to an equiaxed morphology. This effect was not evident in the IA sample as the treatment time was not enough for this effect to appear.

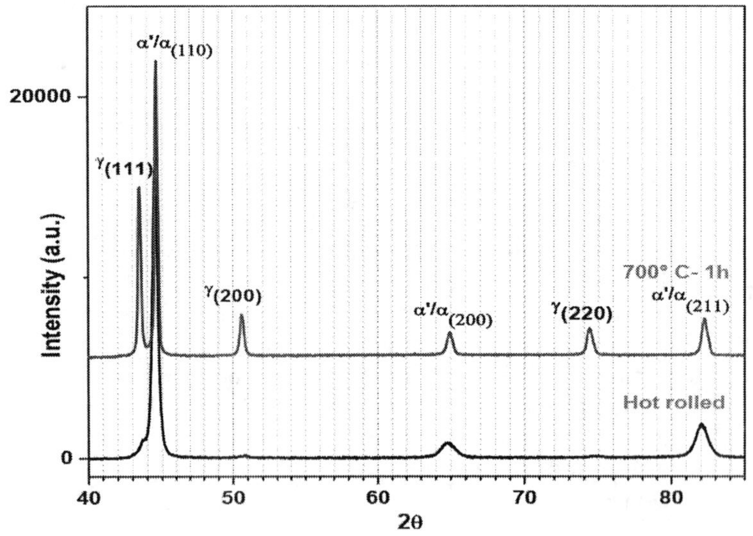

Figure 1. X-ray spectrum of the HR and IA sample.

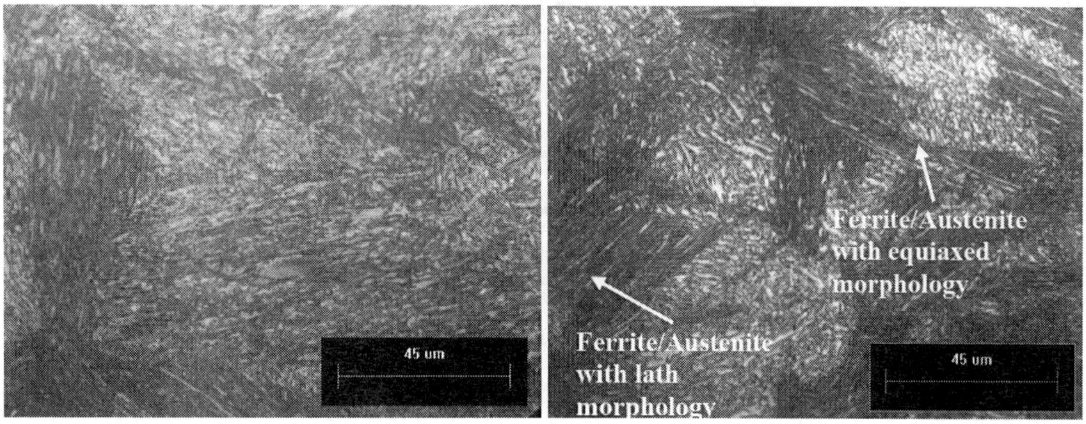

Figure 2. Optical micrographs showing the microstructure exhibited by HR and IA steel.

3.2. Bending testing

Table 1 shows the bending properties obtained from the air bending of steel sheets in the transverse and longitudinal direction. As the bending severity was maintained fixed, the more important parameter to take into account was the critical bending angle (CBA). As was expected, according to the martensitic microstructure observed in the HR condition, CBA in both directions was lower than 90°. However, this parameter was very small in the transverse direction in comparison with longitudinal one, a fact that can be caused by the strong crystallographic texture that steel acquired during the hot rolling. This result is consistent with the one published by Kaijalainen et al. [13] since they argued that bendability in the transverse direction in a conventional steel was affected by the presence of martensite-austenite plates aligned in the rolling direction and the predominantly shear texture {112}<111> component. Regarding the CBA of the IA samples, it was observed that the highest value was achieved in the longitudinal direction (103.1°), however in this case the difference between longitudinal and transverse was only 8°, which indicates that IA treatment improved markedly the bending properties of the steel sheet. The obtained CBA values are comparable and even in some cases are superior to the ones exhibited by other AHHS grades with similar thickness reported in the literature. Suppan et al. [14] have reported critical bending angles of 62 and 80° for two dual phase (DP) steels while 89 and 102° for complex phase (CP) steels. The observed results indicated that the steel studied in this work is promising to be used in stamping operations for the production of automotive parts since in the IA condition, both directions of the sheet displayed CBA higher than 90° for a very small tool radius. Regarding the relation r/t, the observed value of 2.3 is very similar to a commercial DP 700/1000 (2.0) while is better than martensitic grades of comparable UTS such as MS 950/1200 and MS 1030/1200 which have r/t values of 3.5, respectively [15]. Concerning the morphological aspects of fracture, the HR condition showed a fracture that crossed the entire thickness of the steel with some branching cracks while in the case of IA samples, they exhibited only a main crack without going across all the thickness and also without branching cracks (figure 3 and 5). In all samples the main crack followed a trajectory near to 45° respect to the surface of the sheet, this lead to the assumption that the fracture occurred along the maximum shear stress. In figure 4 it is represented the failure stages that occur typically during bending of AHSS [16]. It can be said that in the case of IA samples (figure 5), there were not observed neither ductile cuts nor fine cracks in the external surface on the bending geometry, only the unstable cracks were evident. The latter fact in combination with the absence of thickness thinning or flattening of the bended steel, could mean that the failure mechanism of bending was not the typical plastic instability observed in the majority of steels subjected to this process, but another mechanism such as decohesion of phases interface ought to the different mechanical properties and load carrying ratio of ferrite in comparison with austenite and also cleavage microcracks in the martensite formed by TRIP effect. The presence of TRIP effect in the external bending radius should be considered because the strain experimented in all samples is higher than 0.1 proposed by Sugimoto [17], a value that promoted that almost all the austenite transformed to martensite in two TRIP steels.

Table 1. Bending properties obtained from the air bending testing of sheets in the transverse and longitudinal direction.

Thermomechanical condition of steel	Bending severity (r/t)	Critical bending angle (°)	Radius in the sheet (mm)	Maximum strain in the external bending radius
HR-Transverse	2.3	37.7	4.5	0.14
HR-Longitudinal	2.3	84.3	2.7	0.22
IA-Transverse	2.3	95.4	1.3	0.35
IA-Longitudinal	2.3	103.1	1.1	0.39

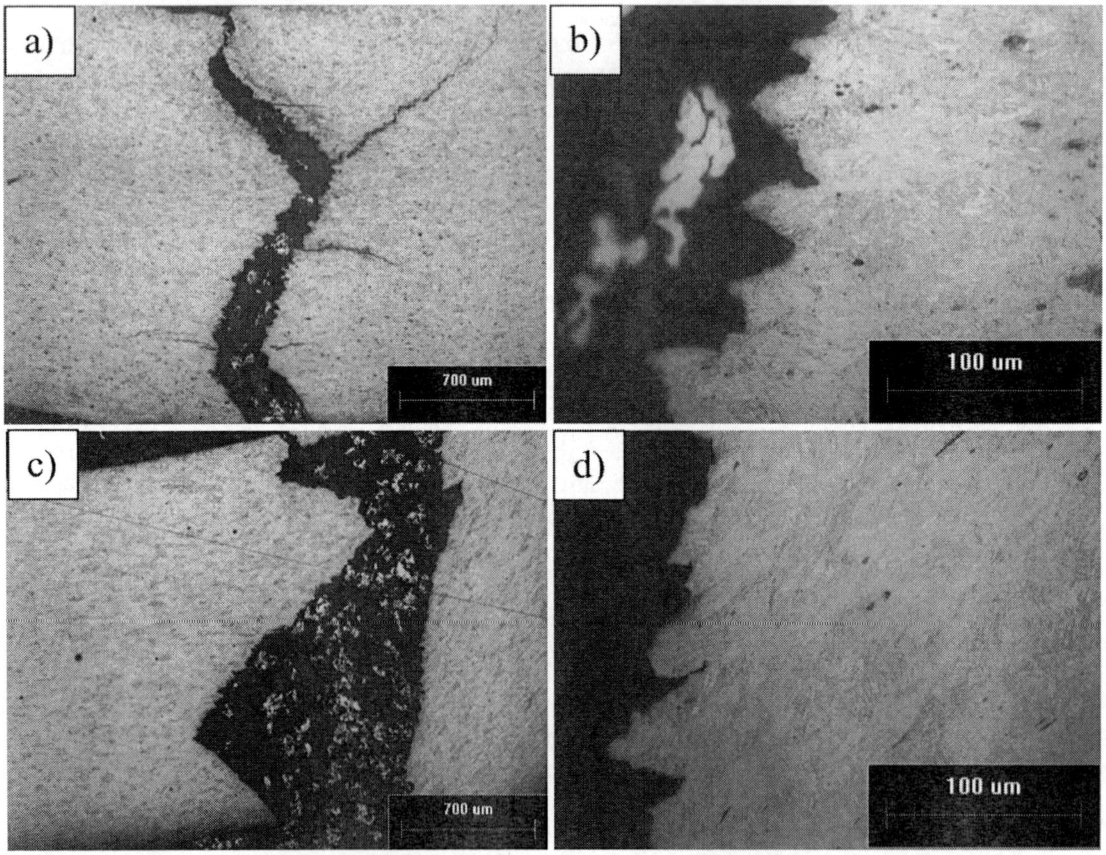

Figure 3. OM micrographs showing the fracture of the sheet in the HR condition a),b) transverse direction and c),d) longitudinal direction.

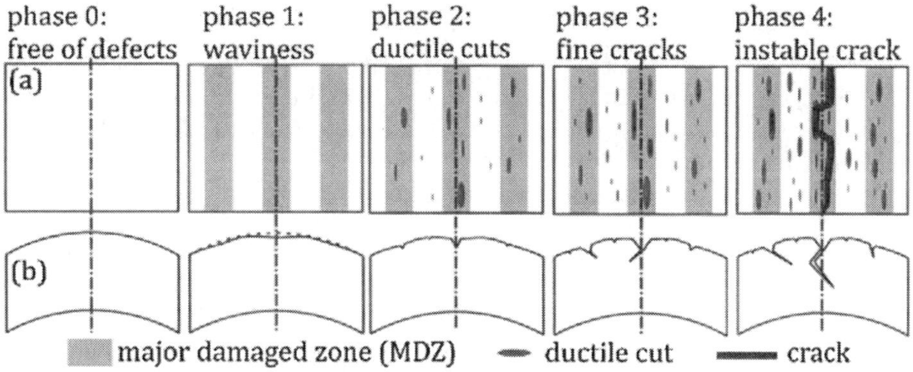

Figure 4. Failure steps during bending a) top view, b) cross section view.

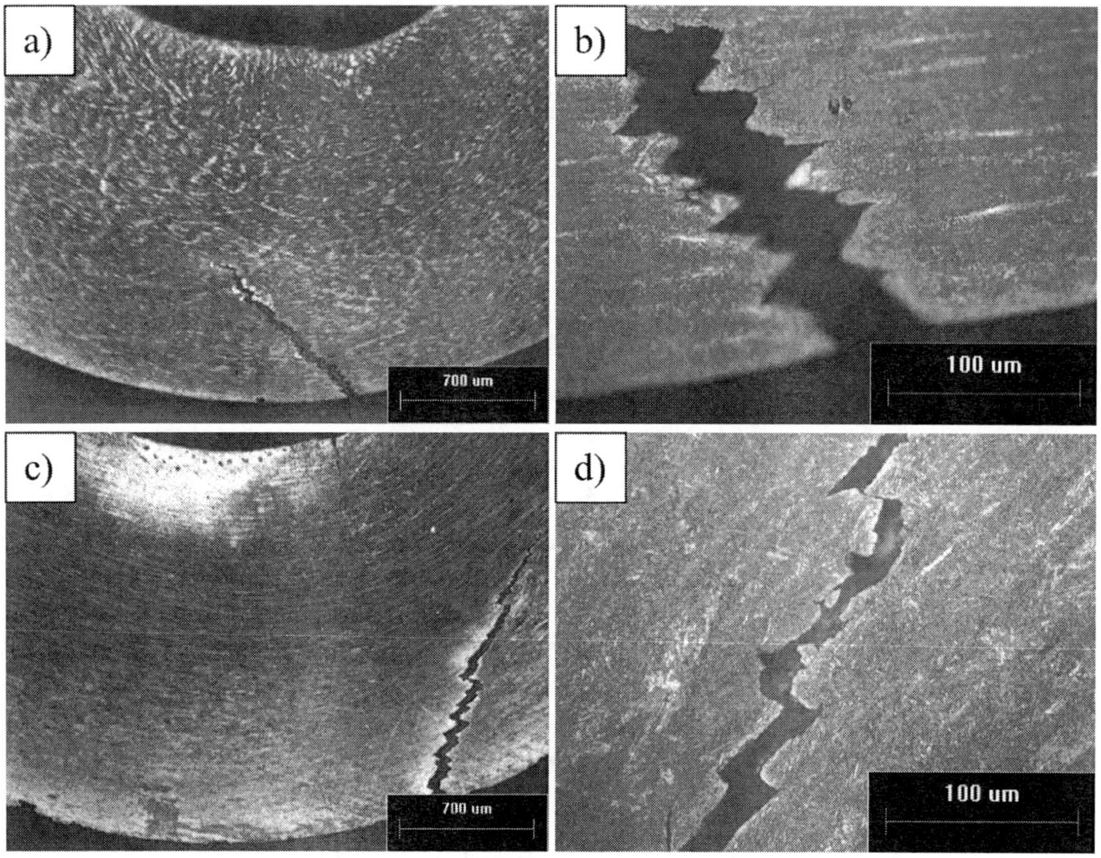

Figure 5. OM micrographs showing the fracture of the sheet in the IA condition a),b) in the transverse direction and c),d) in the longitudinal direction.

3.3. Prediction of FLD by tensile properties

Figure 6 presents the engineering stress-strain curves for both thermomechanical conditions. It is evident that the HR condition exhibited a negligible elongation value lower than 8%, however the UTS (UTS) was high (close to 1600 MPa), a value which is comparable with those typical of hot stamped steels. On the other hand, IA steel showed a very interesting combination of mechanical properties, 35% in total elongation and 1100 MPa in UTS. This combination of tensile properties allowed to classify this steel as a third generation one and that is why the formability properties are expected to be competitive with 1st. and 2nd. generation of automotive steels.

In table 2 it is presented the r values that this steel exhibited in the 0, 45 and 90° of rolling. As can be seen, the behavior of the material can be described as nearly isotropic on which the width strain and thickness are almost similar during deformation. This value promote an acceptable formability and avoid problems such as cracking or wrinkling during forming operations. The results obtained in our study are consistent with the ones reported by Chang et al. [18] since they found that a third generation automotive steel with similar Mn content (5wt.%) had an isotropic behavior according to the plastic strain ratio they found. Unfortunately, in that reference they did not report the thermomechanical processing the steel was subjected to and only described its microstructure as a mix of equiaxial grains of ferrite and austenite. Figure 7 presents the forming limit diagram (FLD) predicted by expressions established by Abspoel et al. [7] on which tensile properties are taken into account to predict the 4 main points of the FLD. Regarding the plane strain point (PS) which is considered the worse forming condition in the FLD, it can be said that its value was higher (0.31) in comparison with the same point in the experimental results reported by Gutierrez at al. [19] (0.24). It is considered that the comparison could be valid since the thickness is the same (1.4 mm) and also the UTS is very close in both steels.

At present, experimental trials are being developed in order to compare the predicted FLD with real values. It is important to point out that the initial objective for developing this steel grade was to promote its classification as third generation steel although the strength reduction was markedly reduced after IA. The focus on the promising use of this steel is not replace hot stamped steels but also other steels such as TRIP steels or even compete with commercial third generation steels which nowadays are offered by some companies. Finally, after all the results presented, the developed steel can be considered as an encouraging alternative material which fulfill the lightweight guidelines that are required in the automotive industry.

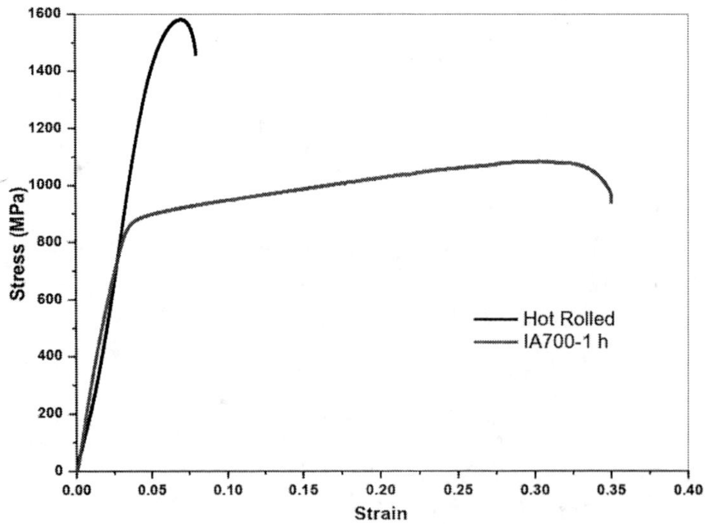

Figure 6. Engineering stress-strain graph showing the tensile mechanical properties displayed by steel sheet.

Table 2. Plastic strain ratios measured by tensile testing at different orientation of the sheet.

Sample direction	r	r_m	Δr
0°	0.96		
45°	1.14	1.06	-0.16
90°	1		

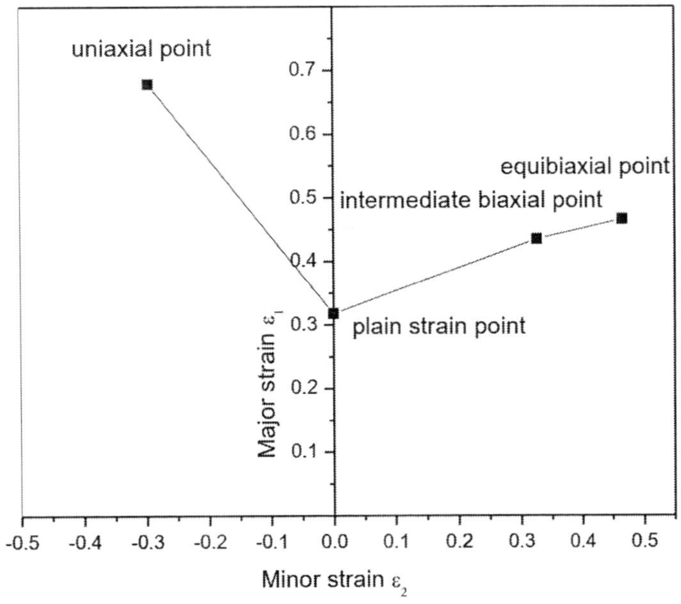

Figure 7. Predicted forming limit diagram of the MMS.

4. Conclusions

It was demonstrated that the application of IA heat treatment promoted higher amount of austenite and ferrite in the microstructure of a HR medium Mn steel, this fact caused that total elongation increased from 8 to 35%; both phases exhibited a mix morphology: equiaxed and acicular. It was found that higher critical bending angle was displayed by heat treated steel in comparison with HR state since the value of this parameter increased from 37 to 95° in the transverse direction and from 84 to 103° in the longitudinal direction. The similar values of critical bending angles that were observed in the IA condition suggested an isotropic behavior in the steel, the evaluation of plastic strain ratio also supported this argument. The prediction of the principal points in the forming limit diagram allowed to consider this steel as a promising candidate material in actual automotive components on which stamping operations are involved. When compared with other steels belonging to the third generation, the plane strain point of the predicted forming limit diagram exhibited a higher limiting value.

References

[1] Kwok TWJ, Gong P, Rose R and Dye D 2022 *Mater. Sci. Eng. A* **855** 1-13
[2] Tao Z, Yao C, Gao C, Duan D, Misra RDK, Sun C, Wu H, Gao X and Du L 2023 *J. Mater. Res. Technol.* **22** 3085-3100
[3] Zheng G, Li X, Chang Y,Wang C and Dong H 2017 *J. Mater. Eng. Perform.* **27** 530-40
[4] Yi HL, Sun L and Xiong XC 2018 *J. Mater. Sci. Technol.* **34** 1112-17
[5] Zhang B and He B 2023 *Materials* **16** 258
[6] Sugimoto K, Tanino H and Kobayashi J 2021 *Metals* **11** 1371
[7] Abspoel M, Scholting EM and Droog MM 2013 *J. Mater. Process. Technol.* **213** 759-69

[8] Cai M, Li Z and Hodgson P 2014 *Metall. Mater. Trans. A* **45** 5624-34
[9] Zou Y, Xu YB, Hu ZP, Chen SQ, Han DT, Misra RDK and Wang GZ *Mater. Sci. Eng. A* **707** 270-79
[10] Liu C, Peng Q, Xue Z, Wang S and Yang C 2018 *Materials* **11** 2242
[11] Zakerinia H, Kermanpur A and Najafizadeh 2009 *Int. J. ISSI* **6** 14-8
[12] Li J, Song R, Li X, Zhou N and Song R 2019 *Mater. Sci. Eng. A* **745** 212-20
[13] Kaijalainen A, Hautamaki I, Kesti V, Pikkarainen T, Tervo H, Mehtonen S, Porter D and Komi J, 2019 *Steel Res. Inter.* **90** 190059
[14] Suppan C, Hebesberg A, Pichler A, Rehrl J and Kolednik O 2018 *Mater. Sci. Eng. A* **735** 89-98
[15] Palaniswamy H, Al-Nasser A, Forming of advanced High strength Steels *Sheet Metal Forming-Processes and Applications* volu 1 ed. Altan T, Tekkaya (ASM International) chapter 6 107-32
[16] Kauper M and Merklein M 2013 *CIRP Ann. Manuf. Technol.* **62** 247-50
[17] Sugimoto KI, Misu M, Kobayashi M and Shirasawa H 1993 *ISIJ Int.* **33** 775-82
[18] Chang Y, Wang M, Wang N, Li X, Wang C, Zheng G, Ren D and Dong H 2018 *Mater. Sci. Eng. A* **721** 179-88
[19] Gutierrez JE, Noder J and Butcher C 2020 *Materials* **10** 902

Experiments and simulations of the drawing envelope of commercially pure aluminum

Adrian J. Carter, Kelin Chen and Yannis P. Korkolis

Dept. of Integrated Systems Engineering, The Ohio State University, Columbus, OH, 43210, USA

E-mail: korkolis.1@osu.edu

Abstract. The range of blank-holding force (BHF) for which deep drawing of AA1100-H24 is successful is investigated using a combination of experiments and simulations. The experiments involve circular blanks of three different diameters: 35 mm, 37 mm and 40 mm. The thickness of the blanks is 0.51 mm. These are drawn with a punch of 20 mm diameter, i.e., at drawing ratios of 1.75, 1.85 and 2.0, respectively, using a custom, modular forming apparatus where the BHF can be controlled at will, between 0 and 2,400 N. The experiments are performed in an Instron 8872 servo-hydraulic frame that allows precise measurements of the punch force and displacement. The working envelopes are determined experimentally in the space of BHF and draw depth, which can be categorized into three regimes: wrinkling, safe and tearing. It is found that grey zones exist between the three regimes. These experiments are then simulated in Abaqus/Standard, using two types of elements: axisymmetric and shell. In the latter case, plastic anisotropy is introduced, using the Yld2000-2D anisotropic yield function. Wrinkling and tearing failure are triggered by inducing suitable geometric imperfections. It is shown that the models are able to reproduce the experiments well. This serves as a verification of the modelling framework, which can then be used for the simulations of more complex forming processes.

1. Introduction

Prediction of failure is crucial for successful virtual design of sheet metal forming processes such as stamping and deep-drawing. Failure usually occurs in the forms of tearing or wrinkling, and prediction of such failures using typical trial-and-error is time-consuming and expensive. Finite Element (FE) models are used to quickly make such predictions without using as much material and time. In deep-drawing processes, as the material is drawn into the die, the flange of the blank develops circumferential compressive stresses which will cause wrinkling if the flange is left unconstrained. Thus, the blank is usually held down by a blankholder that provides a constraining force (BHF). If the BHF is too small the flange may still show wrinkling due to the circumferential compressive stresses; if it is too large, the increased frictional resistance may prevent the flange from flowing into the die and the material may show tearing near the bottom of the cup. Because the BHF has a direct impact on the results of the cup-drawing process, it must be set in a proper range, leading to the workability envelope of the material. In this paper, we describe experiments and simulations regarding that envelope, for AA1100-H24. We start by presenting the experimental setup, that allows an infinite variation of BHF at will. In this way, the workability envelope of the material is established. We then present a finite element (FE) model that, depending on the BHF, reproduces both wrinkling and tearing, and thus can predict the forming envelope.

IOP Conf. Series: Materials Science and Engineering 1284 (2023) 012031 doi:10.1088/1757-899X/1284/1/012031

2. Experiments

2.1. Tooling

The material of this study is commercially-pure aluminum, AA1100. Sheets are received half-hardened at -H24 temper and 0.51 mm thickness. To understand the effect of BHF on cup formability, cup-drawing experiments are conducted using a modular set up. The tooling includes a flat-ended punch, a die and a blank holder. Each part is made of A2 tool steel, heat treated to 58 (die) and 44 (punch) HRC. The dimensions of the tooling can be seen in Figure 1. Experiments are run using an Instron 8872 servohydraulic testing machine with a load capacity of 25 kN. During testing, measurements of the punch force and displacement are provided by the machine. The BHF is varied by using 3 pneumatic cylinders of 38 mm bore, and varying the air pressure between 0 and 700 kPa. An experiment is also performed with no blankholder. The experimental setup was described in earlier publications (1).

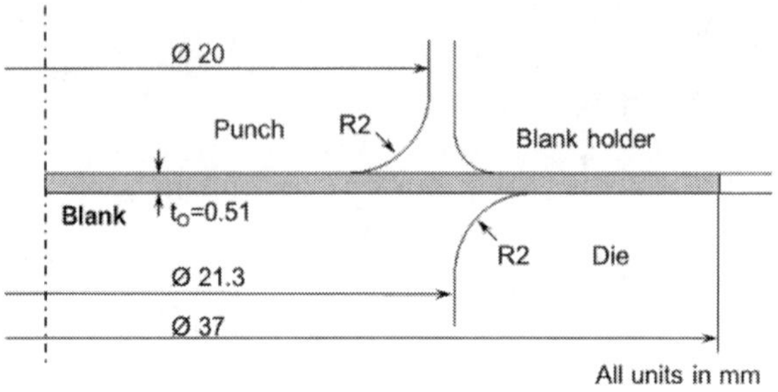

Figure 1. Geometry of the cup-drawing tooling.

2.2. Experimental results

2.2.1. Formed Cups. Figure 2 shows a set of cups with diameter of 37 mm drawn with different levels of BHF that ranges from 0 to 2,426 N. It can be seen that the results are affected the BHF. For the relatively low BHF, i.e., the 0 and 42 N cases, the two cups show severe wrinkling in the flange. As BHF increases, the number of wrinkling waves on the flange of the cup increases. For the moderate BHF cases, i.e., the 563 and 1,863 N cases, the wrinkling is suppressed successfully and the cups are drawn ideally (2). The two successful cups show earing, as is expected for this material due to its anisotropy. As the BHF is further increased to 2,184 and 2,426 N, the corresponding two cups are torn near the bottom, where the maximum thinning is expected (3,4).

Figure 2. Cups formed with different BHF's, indicating wrinkling, success or tearing.

2.2.2. Punch force-displacement. The punch force-displacement curves of these drawing experiments can be gathered from the load cell and position transducer in the Instron machine, see Fig. 3. Each curve shows a climbing slope at the beginning of the forming process, with different responses that depend on the BHF. Fully-drawn cups, e.g., the 1,836 N BHF case, reach a maximum force and which then drops to a number near to zero as the cup is fully-formed. Wrinkled cups, e.g., the 42 N BHF case, reach a maximum force significantly lower than the maximum force of a fully-drawn cup. For torn cups, the cups reach a maximum force, then the force suddenly drops toward zero. This occurs due to the sudden drop of resistance the blank is giving the punch. The punch displacement corresponding to tearing is the one corresponding to the maximum force, while the location of wrinkling is associated to the point of deviation from the fully-drawn case (1).

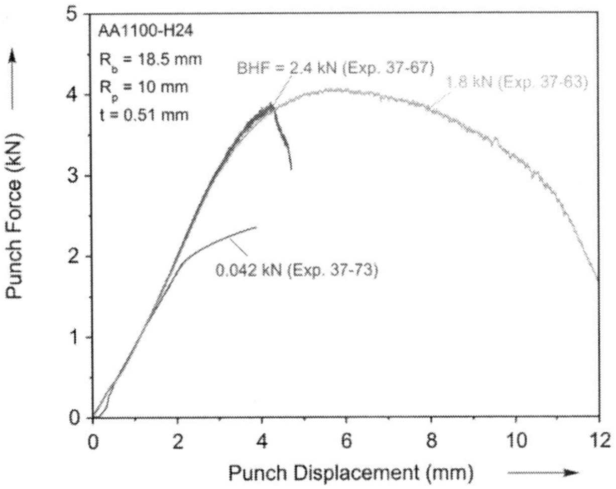

Figure 3. Punch force vs punch displacement.

2.2.3. Workability envelope. The workability envelope of AA1100-H24 can be established based on the relationship between the critical cup height (i.e., where failure, or success, occurred) and BHF. Using this envelope, it can be quickly shown whether a given BHF will lead to wrinkled, fully-drawn, or torn cups, see Fig. 4. At low BHFs wrinkling is seen; at intermediate BHFs successful cups are formed; and at high BHFs torn cups are produced. However, between the successful and tearing regions, there is often an ambiguous region where similar BHFs may result in a successful or torn cup. This may be due to the slightly different friction conditions (i.e., lubrication) in these drawing cases.

3. Finite Element Simulations

3.1. Model Description

Two FE models are created using ABAQUS/Standard ver. 6.18 (implicit) to simulate the wrinkling and tearing phenomena in our cylindrical cup-drawing experiments. The first model is for wrinkling. Figure 5a shows the layout of the model, where the punch, die and blank holder are modeled as rigid bodies. The blank is discretized with reduced integration shell elements (S4R). A total of 3,123 elements are used, arranged as shown in Fig. 5b. A finer mesh is used in areas where thinning or wrinkling is expected. A geometric imperfection is introduced to the blank, to trigger wrinkling of the flange. An out-of-plane coordinate offset of the nodes is introduced on the flange of the blank to trigger wrinkling in the model, see Fig. 5c. Each node has a random such deviation from flatness assigned to it. Since circumferential stresses vary throughout the flange area, a full blank is modelled. A through-thickness imperfection (i.e., localized thinning) equal to 3% of the undeformed sheet thickness is placed between the punch and die radii to encourage localized thinning. Contact between the blank, blank holder, punch, and die is modeled as soft contact with exponential pressure-overclosure relationships (5). The contact parameters are 30 MPa at 0 mm clearance and zero pressure at 0.02 mm of clearance. The BHF is kept constant throughout the entire process, as is the case in the experiments.

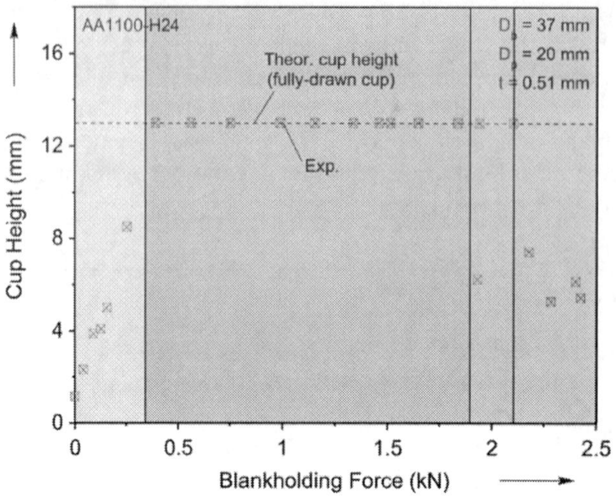

Figure 4. Workability envelope for AA1100-H24.

During simulations it was found that this shell model is not able to reproduce tearing for large BHFs, with the fidelity that would be required for all cases. The computational nature of S4R elements prevents through-thickness stress computations and this, along with the implicit FE solver is believed to be causing the predictive deficiency observed. Therefore, an additional FEA model was established, using axisymmetric CAX4R elements. A snapshot of that mesh is shown in Fig. 5d. As for the shell model, the 3% through-thickness imperfection is introduced on the blank just beyond the die radius to trigger necking and tearing.

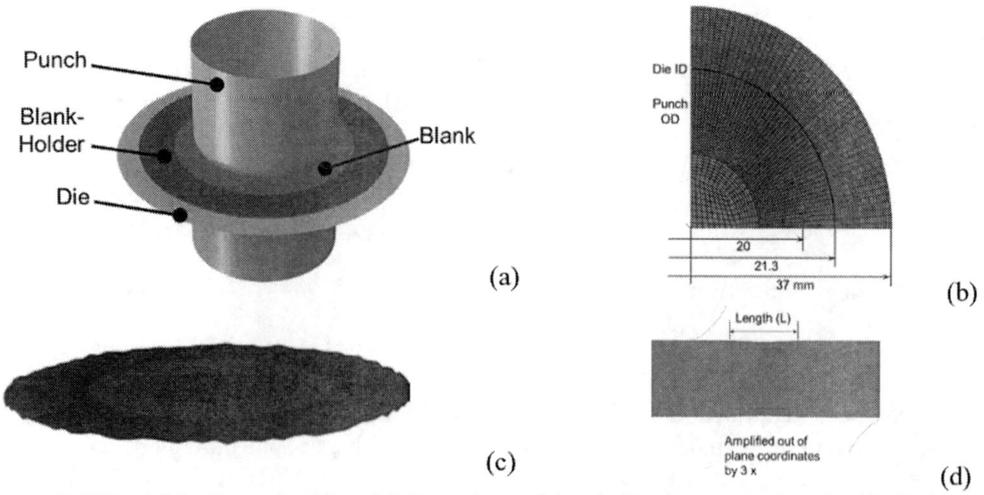

Figure 5. FE model of cup-drawing. (a) Overview of the shell element model. (b) Shell element mesh (only 1/4 is shown for clarity). (c) Isometric view of the shell element model, showing the geometric imperfection for simulating wrinkling. (d) Cross-sectional view of the axisymmetric model, showing the geometric imperfection used for simulating tearing.

Both models were run with a material hardening curve that follows the Voce law, and the following parameters:

$$\sigma = 133 + 14\,[1 - e^{(-14*\varepsilon^p)}] \quad \text{in [MPa]} \tag{1}$$

where σ is the flow stress and ε^p the equivalent plastic strain. The Yld2000-2D (6–9) anisotropic, non-quadratic yield function is used to incorporate anisotropy in the shell model. The anisotropy coefficients can be found in previous work by Ha (10).

3.2. Simulation results

The friction coefficient is inversely identified by selecting a variety of coefficients and comparing the corresponding punch force-displacement responses with known experimental results (2,3,11). A friction coefficient of 0.17 is found to perform best for a variety of BHFs at all blank diameters, see Fig. 6. Solid lines represent experimental data while dashed lines indicate FEA predictions. However, in some cases it was found that reducing the friction coefficient between the blank and punch to 0.04 provides better prediction of tearing (by the axisymmetric model).

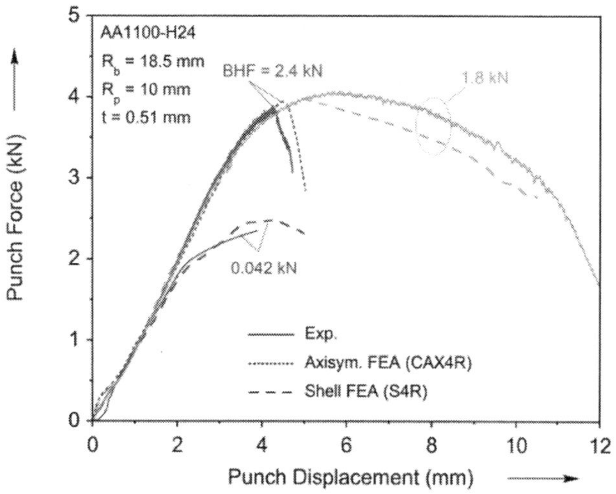

Figure 6: Punch Force-Displacement, Experiment vs FEA

Examples of formed and predicted cups are shown in Fig. 7, for 3 BHFs. It can be seen that the models reproduce the three outcomes of these experiments (wrinkling, success, tearing). It should be clarified that the first two predictions in Fig. 7 are from the shell model, while the latter is from the axisymmetric one.

Following these results, all the cases considered are simulated, and the results are added to the experimental workability envelope, see Fig. 8. That envelope includes results from all three simulation cases: shell elements with spatially-uniform friction and axisymmetric with uniform and non-uniform friction. For these results, the onset of tearing is once again determined by finding the maximum value that the force reaches before it suddenly drops (see Fig. 6) and then finding the corresponding punch displacement. Wrinkling is determined by simulating the blank with and without wrinkling imperfection and measuring the punch and blank holder displacements, as in our earlier work by Chen et al (1).

Similar workability envelopes are determined for the 35 and 40 mm blank diameters.

Figure 7. Experimental (left column) and simulated (right column) cups (37 mm blank dia.). Top row, 42 N BHF, middle row 1,836 N, bottom row 2,426 N. The contours shown are equivalent plastic strain

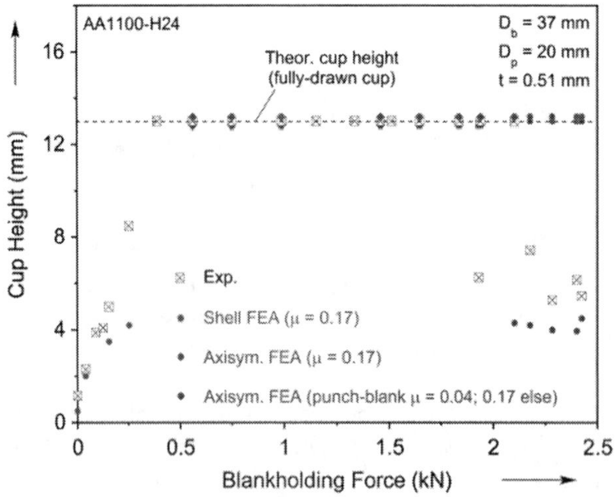

Figure 8: Punch Force-Displacement, Experiment vs FEA

3.3. Wrinkling imperfection sensitivity analysis

The exact wrinkling imperfection used has little to no effect on the punch force-displacement curve, or the number of wrinkles. This is investigated for a few cases; a 37 mm cup drawn under a BHF of 128 N is shown in Figs. 9 and 10. Three different maximum imperfection amplitudes were used, as in Figs. 9a, c and e. These are 1%, 2% and 4% of the undeformed blank thickness. As can be seen in Figs. 9b, d, f, the resulting wave numbers are very similar (ranging from 13-15). It should be noted that even experimentally, repeating the same testing condition produces cups with slightly different number of waves (by 1 to 3).

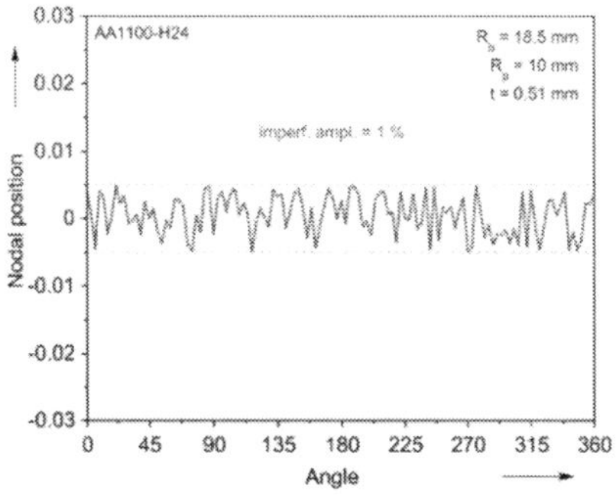

(a)

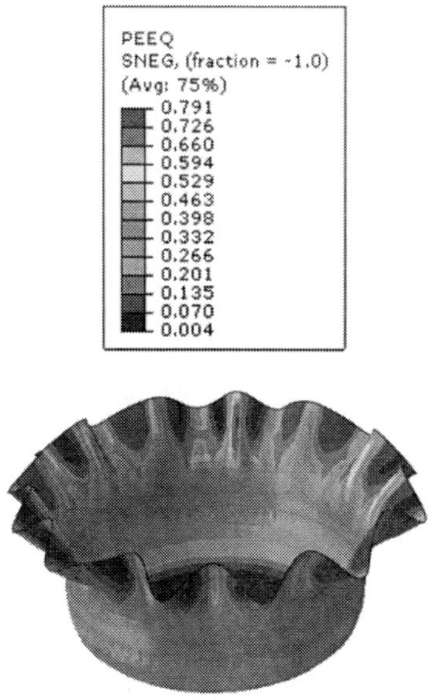

(b)

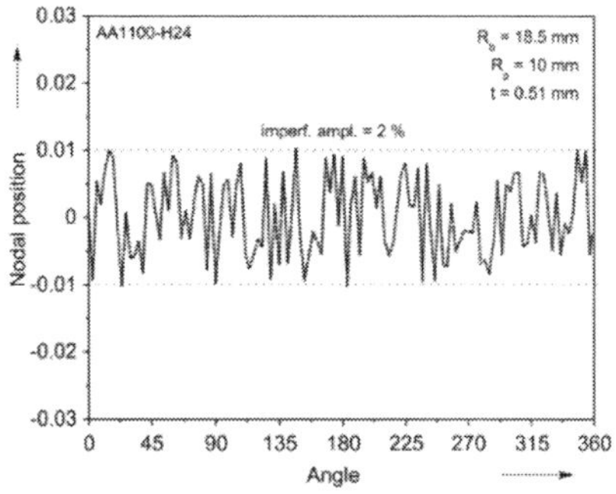

(c)

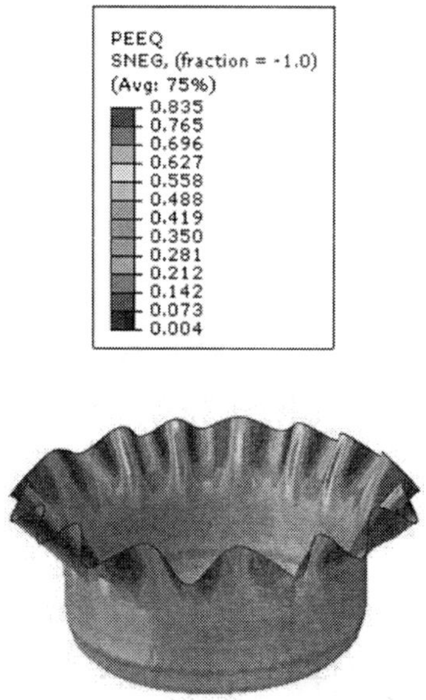

(d)

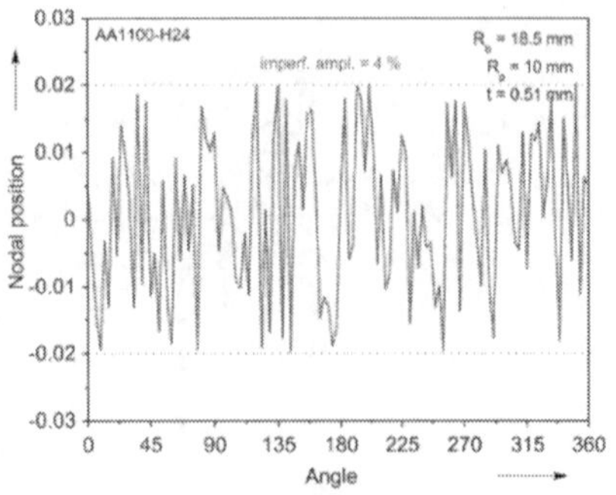

(e)

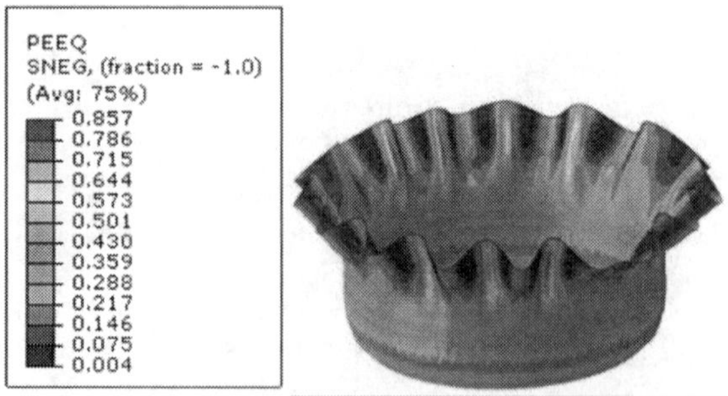

(f)

Figure 9: (a), (c), (e) Wrinkling imperfections with max. amplitudes of 1, 2 and 4% of the undeformed blank thickness. (b), (d), (f) Corresponding cup configurations, indicating the same number of waves predicted.

The punch force-displacements for these 3 different imperfections are compared to the experiment in Fig. 10. These results also show negligible effect of the imperfection amplitude on the onset of wrinkling. These findings are in agreement with our previous work (1), where a series of wrinkled cases from experiments and FEA with geometric imperfections were compared. In every case, the agreement is favorable.

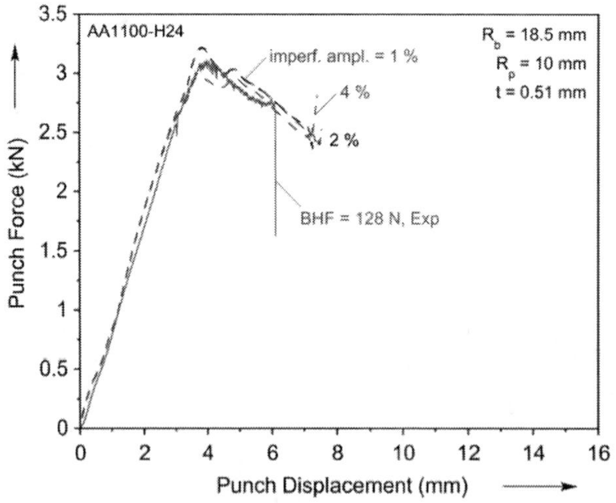

Figure 10: Punch Force-Displacement, Experiment vs FEA, for different wrinkling imperfection max. amplitudes, indicating no sensitivity of the results.

We have rigorously shown in our previous work (12, 13) that for laterally force-restrained "buckling" problem, the problem is not a conventional eigenvalue buckling problem, because the lateral force provides an energy barrier that prevent conventional bifurcation buckling. The buckled response and configuration must be obtained from post-buckling analysis. This is also true for the flange wrinkling problem under lateral BHF restraint. Therefore, the wrinkling simulation cannot be performed with eigen analysis unless BHF=0.

4. Summary and Conclusions

Flange wrinkling and tearing are both common modes of failure in the deep-drawing process, and can be predicted using these FEA models. The workability envelope of AA1100-H24 was determined using a custom experimental cup-drawing setup where the BHF can be varied at-will. Subsequently, shell and axisymmetric FE models were built in Abaqus/Standard, that succeeded in predicting the envelope. Currently, anisotropy, placement of the thickness imperfection and contact parameters are being investigated to try to combine the wrinkling and tearing attributes to a single shell model that predicts wrinkling, success and tearing with sufficient accuracy.

Acknowledgments

This research is supported by the U.S. National Science Foundation under award CMMI-1727490. This support is acknowledged with thanks. We also thank Matt Esser, Alexander Breunig, Prof. Peter Groche, Prof. Brad Kinsey and Prof. Jinjin Ha for the fruitful discussions and their engagement in this project

References

[1] Chen K, Carter AJ, Korkolis YP. Flange Wrinkling in Deep-Drawing: Experiments, Simulations and a Reduced-Order Model. Journal of Manufacturing and Materials Processing [Internet]. 2022;6(4). Available from: https://www.mdpi.com/2504-4494/6/4/76
[2] Tian H, Brownell B, Baral M, Korkolis YP. Earing in cup-drawing of anisotropic Al-6022-T4 sheets. International Journal of Material Forming. 2017;10(3):329–43.
[3] Chen K, Breuning A, Ha J, Kiney BL, Groche P, Korkolis YP. Robustness of deep-drawing finite element simulations to process variations. International Journal of Material Forming. 2022;15.
[4] Johnson W, Mellor PB. Engineering Plasticity. Van Nostrand Reinhold Company; 1978.
[5] Simulia DS. ABAQUS 6.13 User's manual. Dassault Systems, Providence, RI. 2013;305:306.
[6] Barlat F, Brem JC, Yoon JW, Chung K, Dick RE, Lege DJ, et al. Plane stress yield function for aluminum alloy sheets—part 1: theory. Int J Plast [Internet]. 2003 [cited 2017 Jun 4];19(9):1297–319. Available from: http://www.sciencedirect.com/science/article/pii/S0749641902000190
[7] Yoon JW, Barlat F, Dick RE, Chung K, Kang TJ. Plane stress yield function for aluminum alloy sheets—part II: FE formulation and its implementation. Int J Plast. 2004 Mar;20(3):495–522.
[8] Barlat F, Kuwabara T, Korkolis YP. Anisotropic Plasticity and Application to Plane Stress. In: Altenbach H, Öchsner A, editors. Encyclopedia of Continuum Mechanics [Internet]. Berlin, Heidelberg: Springer Berlin Heidelberg; 2018. p. 1–22. Available from: https://doi.org/10.1007/978-3-662-53605-6_225-1
[9] Banabic D. Sheet metal forming processes: constitutive modelling and numerical simulation. Springer Science & Business Media.; 2010.
[10] Ha J, Fones J, Kinsey BL, Korkolis YP. Plasticity and formability of annealed, commercially-pure aluminum: Experiments and modeling. Materials. 2020;13(19).
[11] Ha J, Coppieters S, Korkolis YP. On the Expansion of a Circular Hole in an Orthotropic Elastoplastic Thin Sheet. Int J Mech Sci. 2020;105706.
[12] Chen K, & Korkolis YP. A simplified model of elastic column buckling under constant lateral force restraint. Archive of Applied Mechanics, 2021; 91, 2817-2832.
[13] Chen K., & Korkolis YP. Buckling and post-buckling of an elastica under a lateral restraining force. International Journal of Solids and Structures, 2021; 233, 111178.

Influence of elevated surface temperature on the formability in cold forming of aluminum alloy sheets

E Hodžić, A Shafiee Sabet, Z Silvayeh, P Auer, A Drexler, C Sommitsch and J Domitner

Graz University of Technology, Institute of Materials Science, Joining and Forming, Research Group of Lightweight and Forming Technologies, 8010 Graz, Austria

E-mail: arash.sabet@tugraz.at, josef.domitner@tugraz.at

Abstract. In recent decades, deep drawing has been widely used in the automotive industry for producing lightweight car body components of 5xxx and 6xxx aluminum alloys. Although sheets of these alloys are usually deep-drawn at room temperature, heat generated by friction and plastic deformation may locally increase the surface temperature of the drawing tools during the serial production of components. Therefore, this work investigates the influence of elevated surface temperatures of the tool on the formability of commercial hotmelt-lubricated 1.5 mm-thick EN AW-5182 and EN AW-6016-T4 sheets. Deep drawing experiments were performed at different constant surface temperatures between room temperature (RT) and 80 °C using a cross-shaped tool with open die. With increasing surface temperature, the maximum drawing depth – that was considered as indicator for the formability – decreased by about 21 % and 28 % for EN AW-6016-T4 and EN AW-5182, respectively. Tribological experiments performed using a pin-on-plate tribometer confirmed this trend. The results clearly showed that the coefficient of friction (COF) between the sheet and the tool significantly increase at elevated surface temperature; the most notable increase of the COF occurred between 40 °C and 60 °C.

1. Introduction

Aluminum alloy sheets of the 5xxx and 6xxx alloy series with mill finished (MF) or electrical discharge textured (EDT) surfaces are widely used in the automotive industry for producing lightweight car body components [1]. The particular material properties, the specific formability behavior and the complex tribological system make forming aluminum alloy sheets more challenging than forming traditional steel sheets. Controlling the tribology in sheet metal forming is of utmost interest, as severe contact pressures, high relative velocities and elevated temperatures may occur between the sheet and the tool surfaces [2]. These contact conditions influence the friction, change the surface topography and cause different types of wear. In particular, adhesive wear (so-called galling) is a big issue in deep drawing of aluminum alloy sheets, as it affects the surface quality of both the tool and the formed components [3–8].

Thus, lubricants are necessary for reducing direct contact between the asperities of the sheet and of the tool, which enables controlling the friction and decreasing the wear. In forming of aluminum alloy sheets, water-free dry-film lubricants (so-called hotmelts) are typically used in small amounts of about 0.5-1.5 g/m^2 [9–11]. Compared to conventional water- or oil-based liquid lubricants, dry-film lubricants (DFL) possess superior characteristics in terms of environmental compatibility, cleanliness, corrosion protection, processability and lubricity. However, the ability of DFL to remove wear debris [4] and their performance at high forming speed, contact pressure or temperature [11–13] are limited. Moreover, the surface texture of the sheets (e.g., MF or EDT) influences the effectiveness of the DFL [14,15].

Even in deep drawing at room temperature (RT) heat generated by friction and by plastic deformation may notably increase the local surface temperature of the tool. The temperature-induced degradation of the lubricant was identified to increase the frictional forces at the contact surface between the sheet and the tool [16–19]. This effect on the tribological system may cause deviations and defects, especially in the serial production of deep-drawn aluminum alloy components.

The present work studies the influence of the lubricant degradation on the tribological system and on the formability of two commercial aluminum alloys as typically used for deep drawing of automotive components. For different surface temperatures from RT to 80 °C (i) the coefficient of friction (COF) between the steel tool and the aluminum alloy sheets was determined using a pin-on-plate tribometer, and (ii) the formability of the sheets was experimentally investigated using a cross-shaped tool. The results of the investigations highlight the importance of the thermal stability of hotmelt lubricants on the formability of aluminum alloy sheets.

2. Experimental methods

2.1. Materials

The formability of commercial 1.5 mm-thick sheets of aluminum alloys EN AW-5182 and EN AW-6016-T4 with EDT surface finish and hotmelt lubrication was investigated. Table 1 contains the tensile properties of these aluminum alloys, including yield strength (*YS*), ultimate tensile strength (*UTS*) and uniform strain (A_{80}). Moreover, the roughness of the sheets (R_a) and the amount of lubricant applied on the sheet surfaces (m) are provided in Table 1.

Table 1. Specifications of aluminum alloy sheets according to the supplier.

Aluminum alloy	*YS* [MPa]	*UTS* [MPa]	A_{80} [%]	R_a [µm]	m [g/m^2]
EN AW-5182	131	266	25.2	0.9	0.9
EN AW-6016-T4	102	215	26.9	0.8	1.0

2.2. Friction testing

A pin-on-plate tribometer was used to investigate the influence of elevated surface temperatures on the coefficient of friction (COF). As schematically illustrated in Figure 1, a cylindrical steel pin with dimensions of Ø 6 mm × 12 mm and with the edge radius of 0.5 mm was pressed with constant normal load against the heated aluminum alloy plate. The nominal contact pressure was 10 MPa, the sliding velocity of the pin was 1 mm/s, the sliding distance was 40 mm, the sliding direction was parallel to the rolling direction of the plate and the surface temperature of the plate was RT, 40 °C, 60 °C or 100 °C. For each of the tests the COF was calculated as ratio between the applied normal load (F_N) and the measured sliding force (F_T). Three tests were performed at each contact condition, and the mean COF was calculated based on the results of these three tests. Details about the methodologies of friction testing and lubricant analysis are provided in [19].

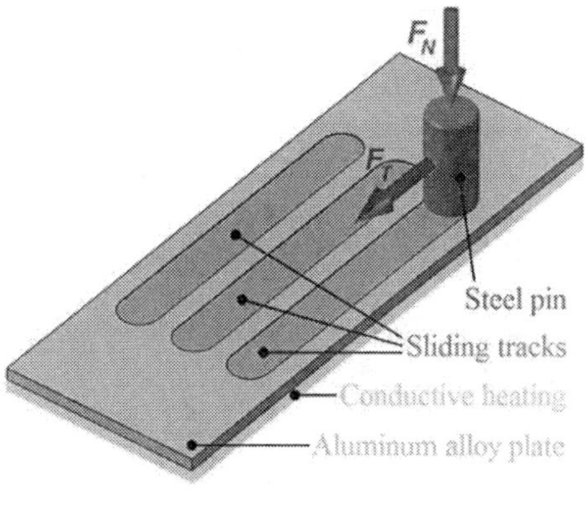

Figure 1. Schematic illustration of the setup of the pin-on-plate tribometer, adapted from [19].

IOP Conf. Series: Materials Science and Engineering 1284 (2023) 012032 doi:10.1088/1757-899X/1284/1/012032

2.3. Deep drawing
In order to investigate the effect of elevated surface temperatures on the formability of aluminum alloys, the cross-shaped tool shown in Figure 2 (a) was attached to a 400 tons hydraulic press. The punch velocity was 10 mm/s and the blankholder force was 400 kN. Before starting each series of deep drawing experiments, 15 preliminary experiments were performed at identical process parameters to achieve a steady-state lubrication condition on the tool surface. A distance sensor, a load cell and an accelerometer were used for monitoring the deep drawing process. As reported in [20] and illustrated in Figure 2 (b), the maximum drawing depth D_{max} was detected at the moment of crack initiation using the accelerometer and the load cell. D_{max} was considered as an indicator for the formability; the higher D_{max}, the better the formability.

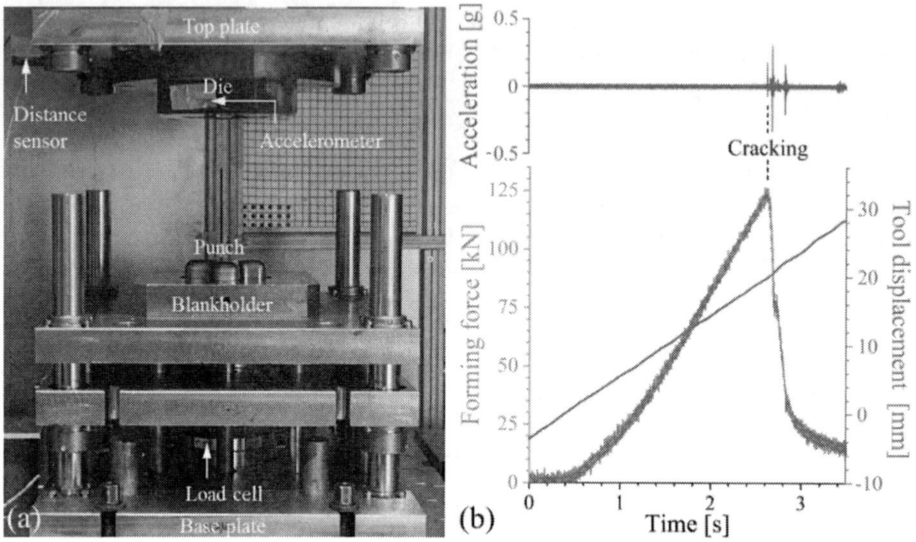

Figure 2. (a) Main components of the cross-shaped tool and (b) determination of cracking based on the detected acceleration and forming force, adapted from [20].

The deep drawing experiments were performed at four surface temperatures of the tool, namely at RT, 40 °C, 60 °C and 80 °C. For this purpose, a flexible electrical heating mat with dimensions of 500 mm × 500 mm × 1.4 mm was placed between two rectangular cover blanks. As shown in Figure 3 (a,b), this blank/mat/blank stack was clamped between the blankholder, the punch and the die. Both cover blanks were required to prevent removal of the hotmelt lubricant from the tool surface. As shown in Figure 3 (c), thermocouples TC1, TC2 and TC3 were spot-welded onto the blankholder to measure the temperature distribution on the tool surface. The thermocouples were located outside the circular contact area between the blank and the tool. The temperature distribution was very homogeneous, as the difference between the temperatures monitored by the thermocouples was just about ± 1 °C. Therefore, electrical heating was only controlled based on the temperature monitored by TC2. As exemplarily illustrated in Figure 3 (d), the heating rate was independent from the predefined maximum temperatures of 40 °C, 60 °C or 80 °C.

After the predefined maximum temperature had been achieved, the blank/mat/blank stack was removed. The surface temperatures of the blankholder, the die and the punch were manually measured using a tactile sensor to ensure that the temperature of the relevant contact surfaces was almost identical prior to deep drawing. Immediately after the final temperature check, three blanks of Ø 280 mm were deep-drawn for each set of process parameters. A mean value was finally calculated from the values of D_{max} that were measured in the three deep drawing experiments. Since the heat was stored inside the massive steel tool, the surface temperature did not notably decrease during the three deep drawing experiments.

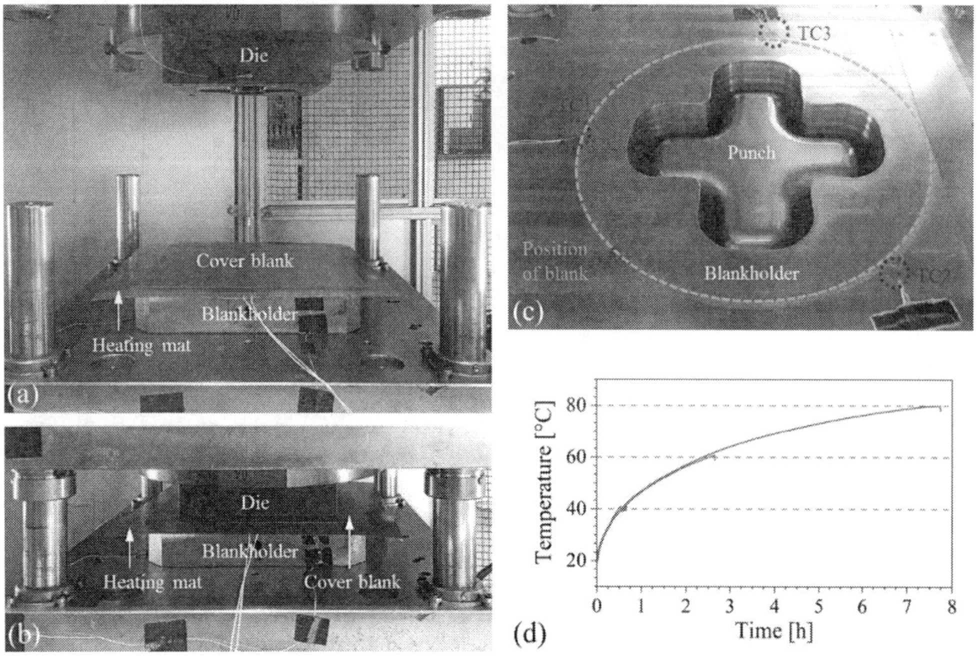

Figure 3. Setup used for heating the deep drawing tool (a) before and (b) during the heating procedure, (c) positions of the three thermocouples for monitoring the surface temperature of the tool and (d) typical increase of the surface temperature during the heating procedure as measured by one of the thermocouples.

3. Results and discussion

3.1. Friction testing

Figure 4 (a) shows the evolution of the COF over the sliding distance for each of both aluminum alloys and for each of the four surface temperatures. In general, the COF determined for EN AW-6016-T4 was lower than the COF determined for EN AW-5182, and it tended to increase with increasing temperature of the contact surface between the steel pin and the aluminum alloy plate. This trend was confirmed in Figure 4 (b), which shows the mean COF values that were calculated by averaging three of the measured curves as exemplarily shown in Figure 4 (a).

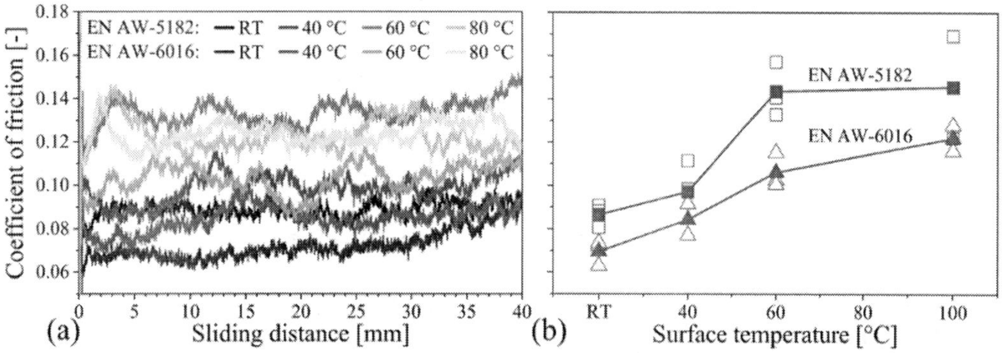

Figure 4. (a) Evolution of the COF over the sliding distance and (b) effect of the surface temperature on the COF, adapted from [19].

It is evident that the mean COF generally increased with increasing surface temperature. The COF moderately increased between RT (≈ 20 °C) and 40 °C. However, the most notable increase occurred between 40 °C and 60 °C, which is the temperature range that can easily be reached during deep drawing of aluminum alloy sheets by friction- and deformation-induced heating. Beyond 60 °C the increase of the COF was only moderate. This unfavorable behavior is mainly attributed to the partial degradation of the hotmelt lubricants at elevated temperatures and contact pressures [19].

3.2. Deep drawing

Figure 5 (a) illustrates the increase of D_{max} during the preliminary deep drawing experiments at constant blankholder force of 400 kN and at constant surface temperature (RT) of the tool. At experiment no. 1 that was performed with completely lubricant-free and clean tool surfaces D_{max} was 21.2 mm and 22.1 mm for EN AW-5182 and EN AW-6016-T4, respectively. With increasing number of experiments D_{max} increased for both alloys, since the hotmelt lubricant was transferred from the aluminium alloy blanks to the tool surfaces. The increase of D_{max} was more notable at the initial experiments, whereas D_{max} did not further increase after experiments no. 8 and no. 13 for EN AW-6016-T4 and EN AW-5182, respectively. Hence, a steady-state lubrication condition in which the hotmelt was well distributed on the tool surface was achieved. At the preliminary experiment no. 15, D_{max} was 28.4 mm and 30.5 mm for EN AW-6016-T4 and EN AW-5182, respectively. Hence, during the 15 preliminary experiments D_{max} increased by about 29 % and 44 % for EN AW-6016-T4 and EN AW-5182, respectively.

This illustrates that the build-up of a sufficient layer of lubricant on the tool surfaces was generally crucial, as it directly affected the formability of both alloys investigated. However, the increase of D_{max} and thus the improvement of the formability due to lubricant transfer from the blank surface to the tool surface was even more significant for EN AW-5182 than for EN AW-6016-T4. That was likely due to the slightly lower amount of lubricant and to the higher surface roughness of the EN AW-5182 sheets [21]. Thus, the actual number of deep drawing experiments required to achieve a steady-state lubrication condition on the tool surface depends on the process parameters, the type and amount of lubricant, and the surface topography of the sheets and of the tool. It is important to note that the results of the deep drawing experiments were only reliable, since a steady-state lubrication condition had been achieved.

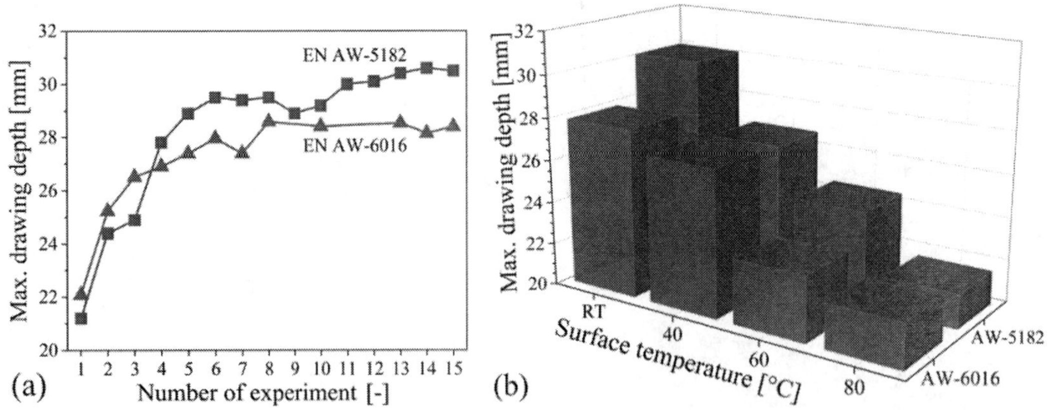

Figure 5. (a) Increase of the maximum drawing depth during the preliminary deep drawing experiments at room temperature and (b) effect of the tool surface temperature on the maximum drawing depth for alloys EN AW-5182 (grey) and EN AW-6016-T4 (blue).

As shown in Figure 5 (b), increasing the surface temperature of the tool significantly decreased the formability of both alloys as characterized by the decrease of D_{max}. When the surface temperature increased to 80 °C, D_{max} decreased to from 30.2 mm to 21.6 mm and from 27.9 mm to 22.1 mm for EN AW-5182 and EN AW-6016-T4, respectively. Hence, the formability of both alloys was almost identical at the highest surface temperature of 80 °C investigated, even though EN AW-5182 showed better formability at room temperature. As the reduction of D_{max} was just about 21 % for EN AW-6016-T4 compared to about 28 % for EN AW-5182, the hotmelt deposited on the EN AW-6016-T4 sheet was thermally more stable than the hotmelt deposited on the EN AW-5182 sheet. This basically agrees with the results of a previous study on the temperature-induced degradation behaviour of hotmelts [19]. The degradation of the lubricant increases both the COF and the adhesive wear and, thus, it decreases the formability even in deep drawing at RT.

Figure 6 shows the forming force that was measured directly at the punch during deep drawing of (a) EN AW-5182 and (b) EN AW-6016-T4 blanks. The forming force monotonically increased. At the maximum force, crack initiation occurred at the deep-drawn blank/component. During subsequent crack propagation, the forming force rapidly decreased from its maximum until deep drawing was stopped. The rate of increase and the maximum force were different for each of the investigated alloys and surface temperatures. Increasing the surface temperature from 40 °C to 80 °C decreased the maximum force, because cracking of the formed blank/component was initiated earlier (i.e., D_{max} decreased). As the COF tended to increase with increasing surface temperature, the forming force increased more rapidly. This general relationship between the COF and the forming force was also observed in a recent study [18].

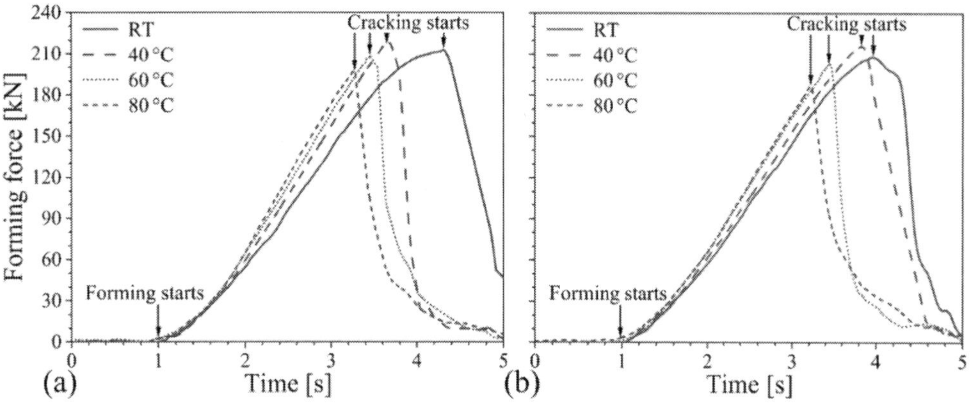

Figure 6. Influence of the surface temperature of the tool on the forming force for alloys (a) EN AW-5182 and (b) EN AW-6016-T4.

4. Summary and conclusions

The influence of elevated surface temperatures of the deep drawing tool (up to 80 °C) on the formability of commercial hotmelt-lubricated 1.5 mm-thick EN AW-5182 and EN AW-6016-T4 sheets was investigated. The following conclusions were drawn based on the results of (i) tribological pin-on-plate tests and of (ii) deep drawing experiments using a cross-shaped tool:

1. In the tribological tests the COF between the steel pin and the aluminum alloy plate increased with increasing surface temperature of the plate. The most notable increase of the COF was observed for surface temperatures between 40 °C and 60 °C, which may typically occur in serial production of deep-drawn components at room temperature. The main reason for the increase of the COF was the temperature-induced degradation of the hotmelt lubricants.
2. Preliminary deep drawing experiments at room temperature were performed to achieve a steady-state lubrication condition on the tool, which was mandatory for the subsequent tests at different surface temperatures. The formability characterized by the maximum drawing depth D_{max} increased with the build-up of the lubricant layer on the tool surface. After a certain number of preliminary experiments, the steady-state lubrication condition was achieved and, thus, D_{max} did not further increase.
3. With increasing temperature of the tool surface, the formability (i.e., D_{max}) of both aluminum alloys significantly decreased. As already indicated by the results of the tribological tests, the main reason for the decrease of the formability was the temperature-induced degradation of the lubricant, which increased the COF between the tool and the formed blank. Moreover, the decrease of D_{max} reduced the maximum forming force.

4. Even in deep drawing of aluminum alloy blanks at room temperature heat generated by friction and plastic deformation may locally increase the surface temperature of the tools (die, blankholder and punch). Since this effect significantly influences the formability of the blanks, it should particularly be considered in the design of forming tools that are used in the serial production of aluminum alloy components.

Acknowledgement
The authors would like to thank Cosma Engineering Europe GmbH for the financial support of this work and AC2T research GmbH for the assistance with the pin-on-plate tests.

References
[1] Bloeck M 2012 Aluminium sheet for automotive applications *Advanced Materials in Automotive Engineering* ed Rowe J (Cambridge: Woodhead Publishing Limited) 85–108
[2] Klocke F, Trauth D, Shirobokov A and Mattfeld P 2015 FE-analysis and in situ visualization of pressure-, slip-rate-, and temperature-dependent coefficients of friction for advanced sheet metal forming: development of a novel coupled user subroutine for shell and continuum discretization *Int. J. Adv. Manuf. Technol.* **81** 397–410
[3] Pujante J, Pelcastre L, Vilaseca M, Casellas D and Prakash B 2013 Investigations into wear and galling mechanism of aluminium alloy-tool steel tribopair at different temperatures *Wear* **308** 193–198
[4] Domitner J, Silvayeh Z, Shafiee Sabet A, Öksüz K I, Pelcastre L and Hardell J 2021 Characterization of wear and friction between tool steel and aluminum alloys in sheet forming at room temperature *J. Manuf. Process.* **64** 774–784
[5] Zabala A, Galdos L, Childs C, Llavori I, Aginagalde A, Mendiguren J and Saenz de Argandoña E 2021 The interaction between the sheet/tool surface texture and the friction/galling behaviour on aluminium deep drawing operations *Metals* **11** 979
[6] Dohda K, Yamamoto M, Hu C, Dubar L and Ehmann K F 2021 Galling phenomena in metal forming *Friction* **9** 665–685
[7] Groche P and Nitzsche G 2007 Influence of temperature on the initiation of adhesive wear with respect to deep drawing of aluminum-alloys *J. Mater. Process. Technol.* **191** 314–316
[8] Groche P, Nitzsche G and Elsen 2008 A Adhesive wear in deep drawing of aluminum sheets *CIRP Ann. Manuf. Technol.* **57** 295–298
[9] Meiler M and Jaschke H 2005 Lubrication of aluminium sheet metal within the automotive industry *Adv. Mater. Res.* **6** 551–558
[10] Losch A 2014 Sheet metal forming lubricants *Encyclopedia of Lubricants and Lubrication* vol 1 ed Mang T (Heidelberg: Springer) 1747–1769
[11] Bay N, Azushima A, Groche P, Ishibashi I, Merklein M, Morishita M, Nakamura T, Schmid S and Yoshida M 2010 Environmentally benign tribo-systems for metal forming *CIRP Ann. Manuf. Technol.* **59** 760–780
[12] Chandrasekharan S, Palaniswamy H, Jain N, Ngaile G and Altan T 2005 Evaluation of stamping lubricants at various temperature levels using the ironing test *Int. J. Mach. Tools Manuf.* **45** 379–388
[13] Kim H, Sung J H, Sivakumar R and Altan T 2007 Evaluation of stamping lubricants using the deep drawing test *Int. J. Mach. Tools Manuf.* **47** 2120–2132
[14] Meiler M, Pfestorf M, Geiger M and Merklein M 2003 The use of dry film lubricants in aluminum sheet metal forming *Wear* **255** 1455–1462
[15] Ju L, Mao T, Malpica J and Altan T 2015 Evaluation of lubricants for stamping of Al 5182-O aluminum sheet using cup drawing test *J. Manuf. Sci. Eng.* **137** 051010
[16] Hazra S, Williams D, Roy R, Aylmore R and Smith A 2011 Effect of material and process variability on the formability of aluminium alloys *J. Mater. Process. Technol.* **211** 1516–1526
[17] Schell L, Emele M, Holzbeck A and Groche P 2022 Investigation of different lubricant classes for aluminium warm and hot forming based on a strip drawing test *Tribol. Int.* **168** 107449

[18] Shafiee Sabet A, Domitner J, Öksüz K I, Hodžić E, Torres H, Rodríguez Ripoll M and Sommitsch C 2021 Tribological investigations on aluminum alloys at different contact conditions for simulation of deep drawing processes *J. Manuf. Process.* **68** 546–557

[19] Shafiee Sabet A, Domitner J, Ristić A, Öksüz K I, Rodríguez Ripoll M and Sommitsch C 2023 Effects of temperature on friction and degradation of dry film lubricants during sliding against aluminum alloy sheets *Tribol. Int.* **180** 108205

[20] Hodžić E, Domitner J, Thum A, Shafiee Sabet A, Müllner N, Fragner W and Sommitsch C 2023 Influence of alloy composition and lubrication on the formability of Al-Mg-Si alloy blanks *J. Manuf. Process.* **85** 109–121

[21] Hodžić E, Domitner J, Thum A, Shafiee Sabet A, Müllner N, Fragner W and Sommitsch C 2023 Influence of natural aging on the formability of Al-Mg-Si alloy blanks *J. Manuf. Process.* **94** 228–239

IOP Conf. Series: Materials Science and Engineering 1284 (2023) 012033 doi:10.1088/1757-899X/1284/1/012033

Experimental characterization and prediction of forming limit diagrams of PHS1800 during hot stamping

Pedram Samadian[1*], Ruijian He[1], Ryan George[1], Clifford Butcher[1] and Michael J Worswick[1]

[1] Department of Mechanical and Mechatronics Engineering, University of Waterloo, 200 University Avenue West, N2L 3G1, Waterloo, Ontario, Canada

E-mail: pedram.samadian@uwaterloo.ca

Abstract. The use of press-hardening steels (PHSs) in automotive bodies creates the opportunity of producing thinner, higher-strength components. PHS1800, an Al-Si coated PHS grade with ultimate tensile strength of around 1.8 GPa after hot stamping, is a candidate material for vehicle anti-intrusion applications. The current study aims to investigate the formability of this steel during the hot-stamping process. A custom Marciniak punch test and *in situ* digital image correlation (DIC) techniques were used to determine the in-plane forming limits of this steel during quenching from an austenitic temperature. The carrier blank thickness and geometry were exploited to quench the surrounding material of the specimens while promoting localization in their central regions. Approximately linear strain paths ranging from uniaxial to biaxial stretching were obtained in the tests while avoiding friction and out-of-plane bending. The forming limit curves (FLCs) of the material under various hot-stamping conditions were then predicted using the Marciniak-Kuczyński (MK) model, taking into account the temperature and strain-rate histories. The predicted limit strains were in good accord with the measured data.

1. Introduction

Ultra-high strength steels (UHSSs) are candidate materials to reduce vehicle weight and reinforce automotive body-in-white (BIW) structural components. Press-hardening steels, like 22MnB5, reach the strength level of UHSSs after the hot-stamping process and are often used to fabricate anti-intrusion members [1]. While their formability during hot stamping is relatively high, it remains a limiting factor for parts of higher geometrical complexity. The formability of sheet metals is often described in terms of forming limit diagrams (FLDs), which define the *locus* of maximum in-plane principal strains that the material withstands during deformation without the occurrence of localized necking. To measure the limit strains of materials, the Nakazima and Marciniak tests are commonly utilized [2]. Li *et al.* [3] evaluated the formability of 22MnB5 during hot forming in a quasi-isothermal condition using the Nakazima tests and showed an overall improved formability with an increased temperature. Georgiadis *et al.* [4] reported a decreased level of hot formability in 22MnB5 with an increased sheet thickness during a non-isothermal Nakazima testing. Dahan *et al.* [5,6] evaluated the position of localized necking within the non-isothermal Nakazima and Marciniak testing of the 22MnB5 blanks and observed centered failure only in the Marciniak tests. Pellegrini *et al.* [7] studied the influences of deformation, forming speed, and microstructural evolution on the formability of 22MnB5 steel during hot stamping.

One of the commonly used modelling approaches to predict the forming limits of materials is the Marciniak-Kuczynski (MK) model [8]. In this method, a thickness inhomogeneity perpendicular to the orientation of major principal stress is considered within the plane of the sheet, and the onset of localized necking is predicted once the equivalent plastic strain increment within the groove exceeds three or more times that outside the groove. Hutchinson *et al.* [9] improved the predictions of the MK model by considering the angle between the orientations of the groove and major principal stress to be arbitrary. Based on this technique, Li *et al.* [10] investigated the effects of the strain rate, temperature, and cooling rate on the FLD of 22MnB5 during hot stamping.

PHS1800 (32MnB5) is a new press-hardening steel that offers a strength level and energy-absorbing capacity higher than 22MnB5 in the hot-stamped condition. The open literature lacks information about the hot formability of PHS1800, which is essential in designing the hot-stamping tooling and process conditions. The current research aims to evaluate the hot formability of PHS1800 in strain states ranging from uniaxial to biaxial tension using a non-isothermal Marciniak test based on the ISO12004-2 standard [2]. The MK model is then used to predict the entire FLC. The effects of the process parameters of temperature, strain rate, and cooling rate on the FLC of PHS1800 are also examined numerically using the MK analysis.

2. Experiments

Al-Si coated PHS1800 sheet with a nominal thickness of 1.6 mm and chemical composition shown in Table 1 was used in this study.

Table 1. The chemical composition of PHS1800 material used in this study.

C	Mn	Ti	Al	Cr	Si	P	B	Fe
0.32	0.60	0.021	0.05	0.31	0.57	0.012	0.002	balance

To characterize formability under hot-stamping conditions, the non-isothermal Marciniak tests were performed using the specimen geometries shown in Figure 1, with shaft widths of 50.8 mm, 101.6 mm, 114.3 mm, and 228.6 mm corresponding to the ISO12004-2 specifications [2]. The Marciniak test equipment (Figure 2) consisted of a stationary 100 mm diameter flat-top punch fixed to the bolster of a 900 ton hydraulic forming press. The punch profile radius was 12.7 mm. The die was mounted on the moving press slide, while the binder was actuated by a hydraulic cushion to provide a 40 ton clamping force. The tooling was water-cooled to maintain a temperature of 7°C. Mild steel carrier blanks of 2 mm thickness with a 32 mm diameter central hole (Figure 1) were placed between the punch and specimens. Three layers of Teflon film were also located between the punch and carrier blanks. To ensure that the central regions of the blanks experience the highest temperature, the carrier and hot test blanks were clamped for 3 seconds in the die prior to the forming stroke. This step was key to promoting localization at the specimen center above the carrier blank hole such that the limit strains were obtained under in-plane, frictionless conditions.

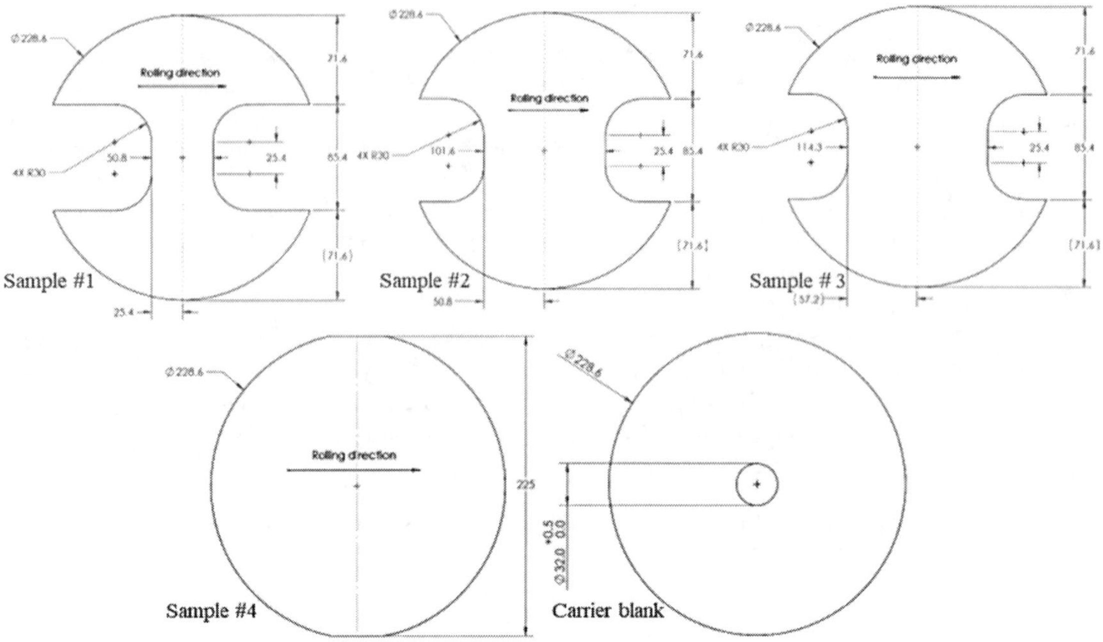

Figure 1. The geometries of the test specimens and carrier blank. Dimensions are in millimeters.

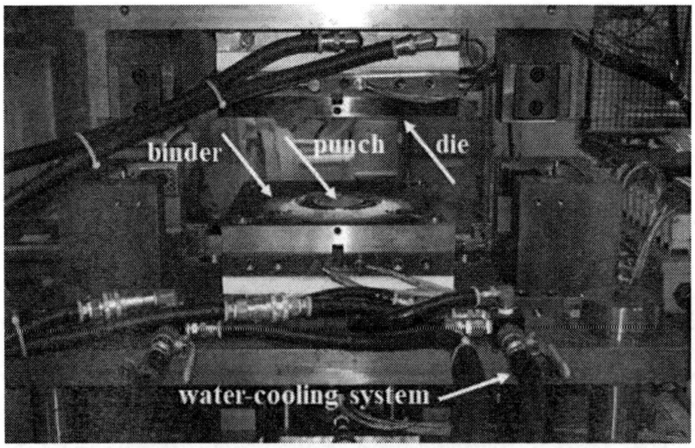

Figure 2. The Marcinial test equipment.

Stereoscopic digital image correlation (DIC) techniques were used to measure the surface strain fields of the samples. The limit strains were determined according to the analysis procedure in the ISO12004-2 standard [2]. The thermal histories of the samples were recorded using an *in situ* pyrometer. As suggested by Lu *et al.* [11], a surface-alloying heat treatment was applied to the specimens before testing to prevent the melting of the Al-Si coating during austenitization and the subsequent loss of the DIC paint speckle pattern. The surface alloying comprised heating the test blanks in a furnace at 700°C for 10 min, followed by die-quenching for 10 s between a pair of water-cooled flat dies. Lu *et al.* [11] reported that the PHS1800 hardening behavior is not significantly influenced by this heat treatment. The surface-alloyed samples were then subjected to the hot-stamping process. The austenitization step was conducted at 930°C for 5 min. The samples were then robotically transferred from the furnace to the tooling in the air over a period of 9 s. The forming-quenching step was then triggered using the Marciniak tooling at a nominal punch speed of 45 mm/s. In these tests, the temperature at the start of

IOP Conf. Series: Materials Science and Engineering 1284 (2023) 012033 doi:10.1088/1757-899X/1284/1/012033

forming (T_0), average (equivalent) strain rate ($d\bar{\varepsilon}/dt$), and cooling rate (dT/dt) were measured 770°C, 1/s, and 25°C/s, respectively.

3. MK Modelling

To predict non-isothermal FLCs for PHS1800, the MK analysis with the modifications applied by Hutchinson *et al.* [9] was used. The numerical algorithm by Butuc *et al.* [12] was implemented to calculate the forming limits for different loading conditions. The imperfection factor (*i.e.* the ratio of initial thicknesses inside and outside the groove) in the MK model was calibrated based on the measured major limit strain for the geometry that provided the closest data to the FLC_0 (*i.e.* the plane strain condition corresponding to the major strain at zero minor strain). PHS1800 was assumed to be isotropic and adhere to the von Mises yield criterion. A temperature-strain rate dependent constitutive model in the form of Equation (1) and based on the hardening response of PHS1800 measured at different temperatures and strain rates by Lu *et al.* [11] was developed to calculate the material flow behavior at large strains (Figure 3).

$$\bar{\sigma} = K(\bar{\varepsilon}_p, \dot{\bar{\varepsilon}}, T)\dot{\bar{\varepsilon}}^{m(T)} \quad (1)$$

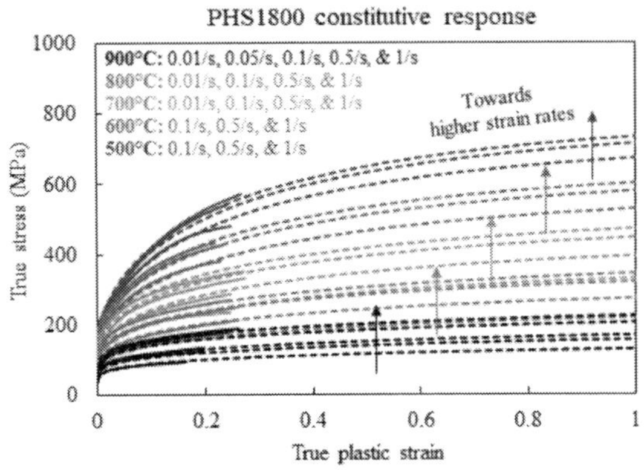

Figure 3. The hardening curves of PHS1800 at different temperatures and strain rates. The solid lines are the measured data reported by Lu *et al.* [11], and the dashed lines are the constitutive fits.

4. Results and Discussion

Figure 4 shows the measured principal strain paths of the different blank geometries within the applied hot-stamping process until the limit strains were reached. The strain histories are relatively linear, as expected in the Marciniak testing. The narrowest (50.8 mm) and widest (228.6 mm) geometries provided strain paths on the draw and stretch sides of the FLD, respectively. The intermediate specimen widths of 101.6 mm and 114.3 mm yielded strain paths close to the plane strain state (FLC_0). The major limit strain decreases from the drawing strain state (negative minor strain) towards the plane-strain state and then more gradually to the stretching strain state (positive minor strain). Li *et al.* [3] reported a similar trend in limit strains for 22MnB5 steel at 800°C and attributed it to the face-centered cubic (FCC) crystal structure of austenite.

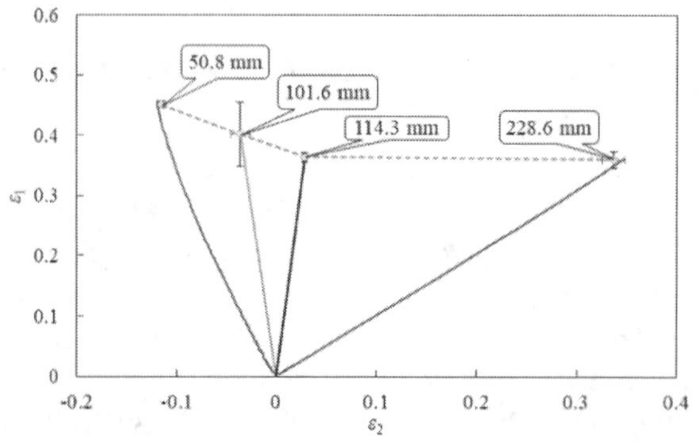

Figure 4. Measured strain paths from samples of differing widths up to necking during the applied hot stamping ($T_0 = 770°C$, $d\bar{\varepsilon}/dt = 1/s$, and $dT/dt = 25°C/s$). The ISO forming limit strains for each strain path are shown by the symbols. The error bars denote the variation in the measured limit strain data.

Figure 5 exhibits the predicted FLC of PHS1800 using the MK model for the applied hot-stamping condition (*i.e.* $T_0 = 770°C$, $d\bar{\varepsilon}/dt = 1/s$, and $dT/dt = 25°C/s$), together with the measured limit strains. An imperfectin factor of 0.97 was used in the MK model, which was calibrated using the forming limits of the 114.3 mm geometry (near the FLC_0). It is noteworthy that there is very good agreement between the prediction and measured limit strains on the draw and stretch sides that were not used in the MK model calibration. In particular, the predicted FLC captured the dependency of the major limit strain on the minor limit strain in the measure data rather well.

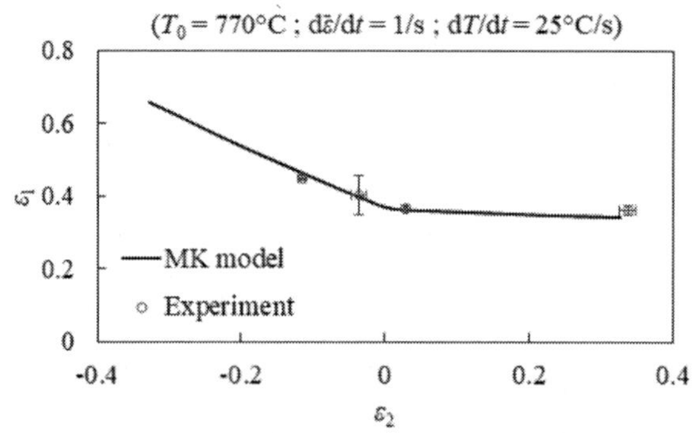

Figure 5. The predicted FLC of PHS1800 using the MK model with an imperfection factor of 0.97 for the applied hot-stamping condition (*i.e.* $T_0 = 770°C$, $d\bar{\varepsilon}/dt = 1/s$, and $dT/dt = 25°C/s$). The markers represent the measured limit strains using the ISO neck-detection method, and the error bars denote the variation in the measured data.

Given the good accordance of the MK predictions and measured formability for the calibration condition in Figure 5, the model was employed to predict the limit strains during hot stamping with other forming temperatures, strain rates, and cooling rates. Figure 6 displays the predicted FLCs of PHS1800 during hot stamping at a constant strain rate of 1/s, a constant cooling rate of 25°C/s, and different temperatures at the start of forming. The FLCs are plotted in both strain and stress space. The results show that as the initial forming temperature increases, the material softens (Figure 6b), and its formability increases (Figure 6a). An increased temperature corresponds to higher strain-rate sensitivity and a lower work-hardening rate. Given the direct relationship of formability with strain-rate sensitivity and work-hardening rate [10], the competition between the effects of these two factors determines the overall influence of the temperature on the material formability. The MK model predicted an overall positive effect of temperature on the PHS1800 formability.

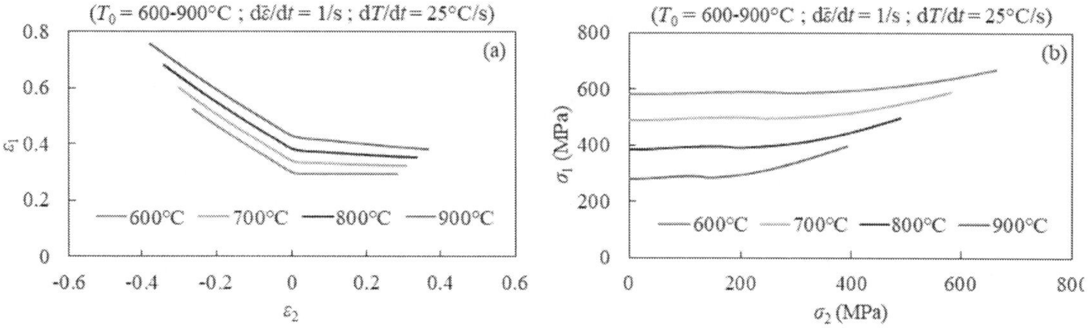

Figure 6. The predicted FLCs in (a) strain and (b) stress space using the MK model for PHS1800 during hot stamping with a constant strain rate of 1/s, a constant cooling rate of 25°C/s, and various temperatures at the start of forming. The imperfection factor used in the MK model was 0.97;

Figure 7 serves to examine the predicted effect of strain rate on the PHS1800 formability during hot stamping with a temperature of 800°C at the start of forming and a constant cooling rate of 25°C/s. The model predicts that the material formability increases at higher strain rates (Figure 7a). When the strain rate increases, the material work-hardening rate increases according to the positive strain-rate sensitivity of PHS1800 (Figure 3). On the other hand, the forming temperature also increases at higher strain rates. These two factors both improve the material formability. Figure 7b shows that the limit stresses increase by increasing the strain rate from 1/s to 10/s and by decreasing the strain rate from 1/s to 0.1/s. The increased limit stresses within the window of 1/s to 10/s are attributed to the increase in the work-hardening rate of the material due to its positive strain-rate sensitivity. However, such increases in limit stresses while reducing the strain rate from 1/s to 0.1/s are associated with the increased work-hardening rate as a result of larger temperature drops (*i.e.* lower forming temperatures), which prevail the negative influence of the strain-rate drop.

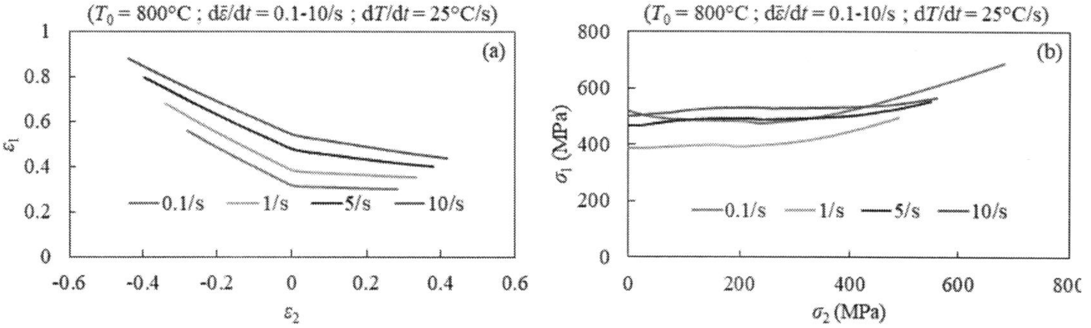

Figure 7. The predicted FLCs in (a) strain and (b) stress space using the MK model for PHS1800 during hot stamping with a temperature of 800°C at the start of forming, a constant cooling rate of 25°C/s, and various strain rates. The imperfection factor used in the MK model was 0.97.

The predictions of the MK model for the effect of cooling rate on the PHS1800 formability during hot stamping with an initial forming temperature of 800°C and a constant strain rate of 1/s are shown in Figure 8. It is evident that a higher cooling rate leads to a slight reduction in formability (Figure 8a). This response is due to the decrease in the forming temperatures. The predicted second-order effect of cooling rate on the PHS1800 formability is attributed to the relatively high strain rate employed such that the temperature drop during deformation is minimized. Figure 9a shows that reducing the strain rate to 0.1/s results in a more noticeable impact of cooling rate on the forming limits. Figures 8b and 9b demonstrate that the limit stresses increase with the increased cooling rate, which is due to larger reductions of the forming temperature. The cooling rate has a higher influence on the limit stresses at a lower strain rate of 0.1/s. This is related to the fact that the deformation prior to the necking point lasts longer at a lower deformation speed, so a higher temperature drop occurs during the forming.

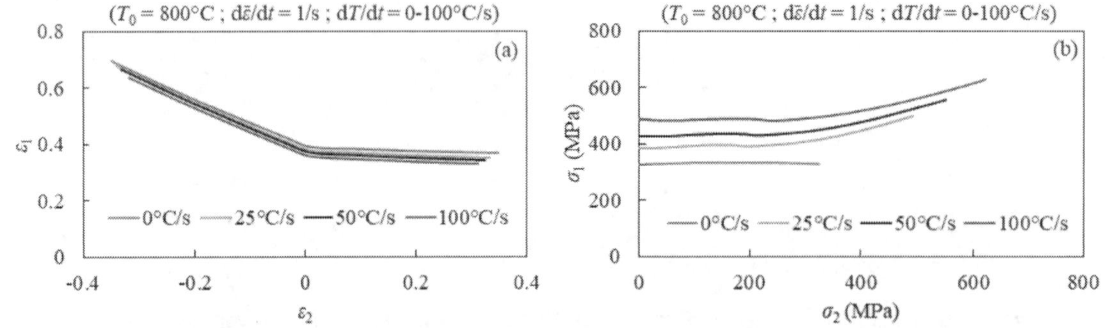

Figure 8. The predicted FLCs in (a) strain and (b) stress space using the MK model for PHS1800 during hot stamping with a temperature of 800°C at the start of forming, a constant strain rate of 1/s, and various cooling rates. The imperfection factor used in the MK model was 0.97.

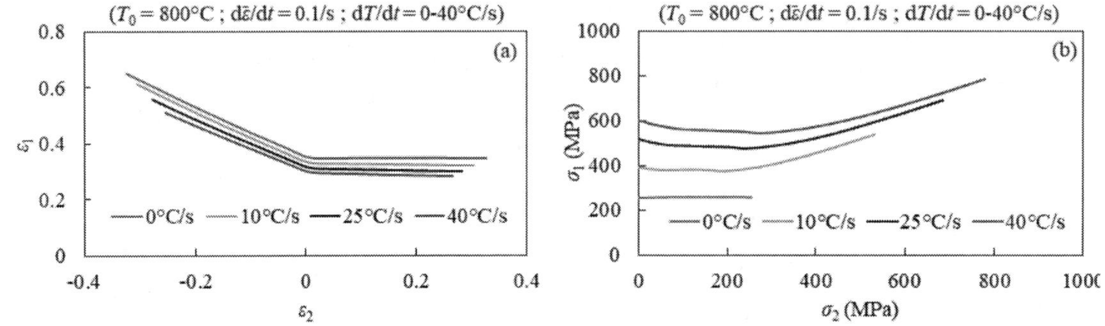

Figure 9. The predicted FLCs in (a) strain and (b) stress space using the MK model for PHS1800 during hot stamping with a temperature of 800°C at the start of forming, a constant strain rate of 0.1/s, and various cooling rates. The imperfection factor used in the MK model was 0.97.

5. Conclusion

This work investigated the formability of the PHS1800 sheet during the hot-stamping process experimentally and numerically. The measured major limit strains of this steel showed a decreasing trend with the increased minor limit strains. The FLD_0 of PHS1800 was about 0.37. The conducted MK analysis with the incorporated effects of the forming temperature, strain rate, and cooling rate predicted the non-isothermal FLC of PHS1800 during hot stamping in line with the measured limit strains. Parametric studies using the MK model showed that the formability of PHS1800 is enhanced with an increased initial forming temperature and strain rate and decreased cooling rate. Future work will examine the accuracy of these predictions.

Acknowledgments

The authors gratefully acknowledge the financial support of Ford Canada, Promatek Research Centre, the Natural Sciences and Engineering Research Council of Canada (NSERC), Ontario Centres of Innovation (OCI), and the Ontario Advanced Manufacturing Consortium.

References

[1] Merklein M and Lechler J 2006 *J. Mater. Process. Technol.* **177** 452–5
[2] ISO12004-2:2008(E): 2008 *Metallic Materials — Sheet and Strip — Determination of Forming-Limit Curves* (Geneva: ISO)

[3] Li FF, Fu MW, Lin JP and Wang XN 2014 *Int. J. Adv. Manuf. Technol.* **71** 297–306
[4] Georgiadis G, Tekkaya AE, Weigert P, Horneber S and Aliaga Kuhnle P 2017 *Int. J. Mater. Form.* **10** 405–19
[5] Dahan Y, Chastel Y, Duroux P, Hein P, Massoni E, and Wilsius 2006 *Proc. Int. Conf. of Deep Drawing Research Group* (Porto)
[6] Dahan Y, Chastel Y, Duroux P, Wilsius J, Hein P and Massoni E 2007 *Proc. Int. Conf. of Deep Drawing Research Group* (Győr)
[7] Pellegrini D, Lechler J, Ghiotti A, Bruschi S and Merklein M 2009 *Key Eng. Mater.* **410–411** 297–304
[8] Marciniak Z and Kuczyński K 1967 *Int. J. Mech. Sci.* **9** 609–20
[9] Hutchinson JW, Neale KW and Needleman A 1978 *Mechanics of Sheet Metal Forming* (New York: Plenum Press)
[10] Li H, Wu X and Li G 2013 *J. Mater. Eng. Perform.* **22** 2131–40
[11] Lu S, DiCecco S, Worswick M, Chiriac C, Luckey G, Tjong J, Boettger J and Shi C 2021 *IOP Conf. Ser. Mater. Sci. Eng.* **1157** 012012
[12] Butuc MC, Gracio JJ and Barata da Rocha A 2003 *J. Mater. Process. Technol.* **142** 714–24

IOP Conf. Series: Materials Science and Engineering 1284 (2023) 012034 doi:10.1088/1757-899X/1284/1/012034

Crashworthiness evaluation of press hardened steels with different lath-like microstructures.

V P Aroca[1], C Philippot[2], J Pujante[3], D Frómeta[3], F G Caballero[1], C Capdevila[1]

[1] National Center for Metallurgical Research (CENIM-CSIC), Gregorio del Amo 8, 28040 Madrid, Spain.
[2] ArcelorMittal Global R&D – Product center, Voie Romaine, 57280 Maizières-lès-Metz, France.
[3] Eurecat Centre Tecnològic de Catalunya, Unit of Metallic and Ceramic Materials, Placa de la Ciencia, 08243 Manresa, Spain.

E-mail: ccm@cenim.csic.es

Abstract. The deployment of Press Hardened Steels (PHSs) in the Body-In-White structures that took place during the two last decades is at the origin of a breakthrough weight saving on vehicles while guaranteeing the safety requirements of the automotive industry. The production of parts with complex shapes and a high strength between 1500-2000MPa was made possible by taking advantage of the hot rheology of austenite combined with the fast cooling obtained by in-die quenching leading to auto-tempered martensitic microstructures. In the present work, PHS1500 was hot stamped with different thermomechanical processes to promote different lath-like microstructures (bainite, tempered martensite…). An innovative approach is proposed to link the complex microstructures formed between thermoregulated dies to the mechanical behaviour and local ductility of these materials during crash-like solicitations. The microstructure has been characterised through a multi-characterization technique approach including light optical microscopy, scanning electron microscopy and X-ray diffraction. The local ductility is assessed with a combination of crack initiation tests (bending, notch tensile test) and crack propagation resistance tests (Essential work of fracture). Despite similar microstructural crystallographic features, lower bainite and auto-tempered martensite can lead to very different strength – local ductility compromise.

1. Introduction

It is already known that PHS steels are widely used in the passive safety of vehicles, as they have been developed since the 1980s and have been in widespread use since 2005. We also know that society needs a safer planet, so the automotive industry and steel manufacturers are working closely together to become more efficient in the manufacturing of their materials. For this reason, new production routes continue to be investigated and new thermo-mechanical processes are being explored to produce PHS1500 steel with different functionalities [1]. 22MnB5 is known to achieve a combination of high strength and fair local ductility when conventionally hot stamped. Its simplicity justifies that its continuous growth on the automotive market even if all its potential is still deepened with new hot stamping processes.

This steel belongs to the family of advanced high strength steels (AHSS), where a first generation started with a ferrite microstructure, followed by a second generation with substantial amounts of austenite and now we are immersed in the third generation. This generation tries to find a reasonable

IOP Conf. Series: Materials Science and Engineering 1284 (2023) 012034 doi:10.1088/1757-899X/1284/1/012034

compromise between the steels of the first and second generation to obtain microstructures (usually lath-like ferrite) with significantly improve ductility properties (low in the first generation) at a lower price than the second generation and weldable while maintaining the high strength of the first generation [2]–[4]. This is possible by exploring new forming routes or designing new alloys that produce desired microstructures.

The production of advanced high strength steel alloys has introduced new challenges in the forming process, which can be solved by using the hot stamping process. However, these alloys have proven to have ductile fracture issues in the final parts (such as energy absorption parts) due to their limited resistance to crack formation, especially when plastic strain localization takes place during crash. Martensitic and bainitic steels have been shown to have good local ductility properties [5]. Although ductility decreases with increasing strength, the microscopic uniformity of the alloy is a major determinant of its ductility[6]. In addition, the bending capacity is similar to the local ductility, as large local deformation can lead to fracture. This capacity can be adjusted by the hardness heterogeneity in the microstructure, i.e., the distribution of hard and soft phases.

The hot stamping process for these alloys, generally, can be classified into two methods, direct and indirect. In the direct method, a hot blank is transferred to a press where forming and cooling takes place, while in the indirect method, a cold-formed blank is heated and then transferred to a cooled tool for final forming and cooling of the shape. The resulting component can have tensile strength values ranging from 1500 to 2000 MPa, allowing the production of parts with different mechanical properties [7]. Although in this study we focus on steel blanks, PHS can also be produced [8] into profiles, bars and tubes; where there are a multitude of possible routes to manufacture these products.

The aim of this work is to study the mechanical properties of PHS steel under different thermo-mechanical processes. The direct hot stamping method, also known as press hardening, will be used with varying parameters to produce different lath-like ferrite microstructures.

2. Materials and experimental procedure

2.1. *Material*

In this study a PHS steel with a chemical composition as shown in Table 1 was used. For this purpose, after a laboratory casting process, the steel was first hot rolled and then cold rolled in several passes down to a thickness of 1.2 mm. This process was performed as close to an industrial process as possible. The initial state of the samples was in sheet form.

Table 1. Chemical composition. %wt.

Steel	%C	%Mn	%Si	%Al	%Cr	%B
PHS-1500	0.22	1.2	0.3	0.04	0.3	0.003

2.2. *Determination of critical temperatures*

To develop this study and to ensure the production of different lath-like ferritic microstructures in the processing, the critical temperatures were previously determined. For this purpose, continuous cooling and isothermal maintenance experiments were carried out in a DIL 805 A/D dilatometer. The experiments consisted of monitoring the change in length of a rectangular specimen (12x4 mm) during the full heat treatment. In the case of continuous cooling, the induction coil heats (7.5 °C/s) the specimen to the selected austenisation temperature, 930 °C in this case, and maintains it for 3 minutes. Finally, it is continuously cooled to room temperature at cooling rates between -100°C/s and -0.5°C/s. In the case of isothermal holding, the same austenisation cycle was applied, and specimen were cooled down at -100°C/s to the different temperatures (from 300 to 600 °C in intervals of 50 °C) and maintained at this temperature for 30 minutes. When this time is over, it is quenched to room temperature. All processes are under vacuum conditions and the temperature is controlled by K-type thermocouples [9]. From this experiment, the following are obtained by analysing the dilatometric curves and the microstructure: the beginning and end of the transformation from ferrite-to-austenite (Ac1 and Ac3) during heating, the beginning and end of the transformation from austenite-to-ferrite (Ae1 and Ae3) during cooling, the beginning and end temperatures of the bainitic transformation (Bs and Bf) and the beginning and end temperatures of the martensitic transformation (Ms and Mf).

2.3. Material processing
On a semi-industrial scale, a Press Hardening process was carried out on this steel [10]. For this purpose, the blanks were placed in a furnace until they achieved a reaustenisation temperature of 910 °C. The total heating time and time spent in the furnace was 5 minutes. Then, a robotic system takes the steel blanks to the press in less than 7 seconds. There they are die quenched in tools thermoregulated at different temperatures (20, 200, 300 and 400 °C). After being in the press for 10 to 30 seconds, they are removed until they cool down to room temperature. In order to carry out this investigation, different parts of the blanks were cut by laser cutting to perform tensile, notched tensile, bending and essential work of fracture (EWF) tests. Three specimens were cut for the tensile and bending tests and eight specimens were cut for the EWF.
It is worth mentioning that the treatment at 300 °C was only performed with a holding time of 30 seconds.

2.4. Tensile tests
Tensile tests were performed on standardised specimens (80x20x1.2mm) under quasi-static conditions and at room temperature according to ISO6892-1 [11]. From this test, the yield strength (YS), the ultimate tensile strength (UTS), uniform elongation (UEl) and total elongation (TEl) were obtained.

To find the strain at failure, also under quasi-static conditions and at room temperature, standard 50x12.5x1.2mm specimens with a notch diameter of 6.25 mm were used. This measurement shows the amount of deformation that a material can undergo before it fractures or breaks under a given load or stress. For this purpose, it is tested in a universal tensile machine until rupture. It is calculated by the equation 1 [12], where $t0$ corresponds to the initial thickness and t to the minimum thickness after the test. This thickness was measured by optical microscopy.

$$\varepsilon_F = \frac{2}{\sqrt{3}} \cdot \ln\left(\frac{t}{t_0}\right) \quad (1)$$

2.5. Bending test

The bending tests were done in accordance with VDA 238-100 standard [13]. In this test, the specimen is placed on two support points and a punch bends it in the middle. The force exerted and the displacement of the punch are recorded until the force drops to 60N, at which point the test is completed. The bending angle is measured as a result of this test.

2.6. Essential work of fracture
Crack propagation resistance was evaluated through the EWF methodology. EWF tests were performed according to CWA 17793:2021 [14]. Eight rectangular sharp-notched Double Edge Notched Tension (DENT) specimens of 200 x 55 mm (length x width) with ligament lengths from 8 to 16 mm were used for this purpose. Sharp notches were prepared following the rapid notching procedure described in [15]. The DENT specimens were tested up to fracture at a constant displacement rate of 1 mm/min and the load versus the load-line displacement was recorded. The load-line displacement was measured by means of a video extensometer using initial extensometer marks separated 50 mm. All the tests were performed at room temperature.

For each specimen, the total work of fracture (W_f) was obtained by integration of the area under the load-displacement curve. Then, the specific work of fracture (w_f) was obtained by dividing the total work of fracture by the cross-section area ($wf=Wf/l_o{\cdot}t$, where l_o is the initial ligament length and t is the initial sheet thickness). The specific essential work of fracture, w_e, was determined by extrapolation to ligament zero of the w_f vs l_o data [14].

2.7. Microstructural and fracture characterisation

For the characterisation of the microstructures, the undeformed part of the post-mortem tensile specimens (grips) was used. Pieces of 10x4 mm, the long length being perpendicular to the rolling direction, were cut and mounted in bakelite to observe the cross section to the rolling direction. After metallographic preparation: sandpaper (no. 320, 600 and 1200), polishing with diamond paste (3-1 μm) and colloidal silica (0.04 μm), and etching with 5% Nital; the microstructures were observed by optical microscopy and scanning electron microscopy (SEM). A sample of each treatment was also mounted on transparent bakelite following the same preparation process for X-ray diffraction. This experiment was performed on a Bruker AXDS D8 diffractometer with Co-Ka radiation. The composition of the retained austenite was calculated using the ratio between the lattice parameter and the chemical composition and assuming that only carbon diffuses during the transformation [16].

In addition, the hardness (HV10) of each heat treatment was also measured to assist in phase identification.

Fracture characterisation was performed on the notched tensile specimens by scanning electron microscopy using a Cold-FEG Hitachi S4800 microscope. They were performed on the specimens with the same conditions as those tested to obtain the essential work of fracture.

3. Results

After analysing the microstructures in SEM, Figure 1 is shown, in which there is a representative sample of the microstructures with a dwell time of 30s. Only these are presented, because after SEM examination of the microstructure it does not appear that the microstructure is sensitive to the forging time. We can observe a martensitic microstructure at 20, 200 and 300 °C, which will have a higher amount of tempered martensite as the temperature of the quenching dies is increased. At 400 °C we observe a microstructure of lower bainite almost entirely with some fraction of tempered martensite.

As for the austenite content, this is not significant, as only values of up to 2.7 % have been obtained, with a measurement error of ± 3%.

The mechanical properties found can be consulted in Table 2. This table shows that as the quenching die becomes hotter, the hardness decreases, decreasing YS and UTS and keeping UEl and TEl constant. Furthermore, Figure 2a shows the tensile tests where the evolution of the mechanical properties

Table 2 Mechanical Properties.

Tool temperature (°C)	Tool time (s)	HV10	RA (± 3%)	YS (MPa)	UTS (MPa)	UEl (%)	TEl (%)	BA (°)	ε_F	w_e (kJ/m2)
20	10	495 ± 3	1.6	1098 ± 17	1515 ± 11	3.9 ± 0.5	6.7 ± 0.6	59	0.43	169 ± 18
	30	488 ± 8	1.5	1088 ± 17	1510 ± 10	3.9 ± 0.1	6.3 ± 0.5	58	0.39	Not tested
200	10	479 ± 6	2.6	1033 ± 8	1453 ± 4	4.7 ± 0.1	7.4 ± 0.2	63	0.40	97 ± 34
	30	477 ± 3	2.7	1023 ± 4	1445 ± 3	4.5 ± 0.1	6.9 ± 0.3	65	0.46	116 ± 14
300	10	Not tested								
	30	448 ± 1	2.1	962 ± 3	1368 ± 2	4.3 ± 0.1	7.0 ± 0.3	67	0.43	111 ± 25
400	10	354 ± 6	0	817 ± 12	976 ± 11	4.3 ± 0.2	7.2 ± 0.4	113	0.83	245 ± 29
	30	398 ± 2	0	954 ± 11	1158 ± 11	4.1 ± 0.1	6.7 ± 0.3	93	0.66	108 ± 27

described previously can be clearly observed. Regarding the bending angle (BA), it increases with the die temperature, as well as the strain at failure (ε_F) does. For these properties, a dependence on the quenching time can be observed. This difference can be seen in the processing carried out at 400 °C,

IOP Conf. Series: Materials Science and Engineering 1284 (2023) 012034 doi:10.1088/1757-899X/1284/1/012034

where the bending angle increases by 22% when the quenching time is 10 s and the strain at fracture increases by 26%.

To further investigate this result, the dilatometry test of the isothermal treatment at 400 °C was studied more deeply. Therefore, the dilatation of the specimen was plotted as a function of time during the isothermal treatment. As shown in Figure 3a, at 10 seconds bainite is forming and at 30 seconds a second slope begins and then relaxes. This increase can be related to a secondary hardening process. This hardening could be the cause that the local ductile properties decrease with increasing die quenching time and both hardness and strength increase. To explore this hypothesis, it will be necessary in the future to characterise the precipitation state using different techniques such as TEM and the distribution of strain at bainite / martensite interface with nanohardness.

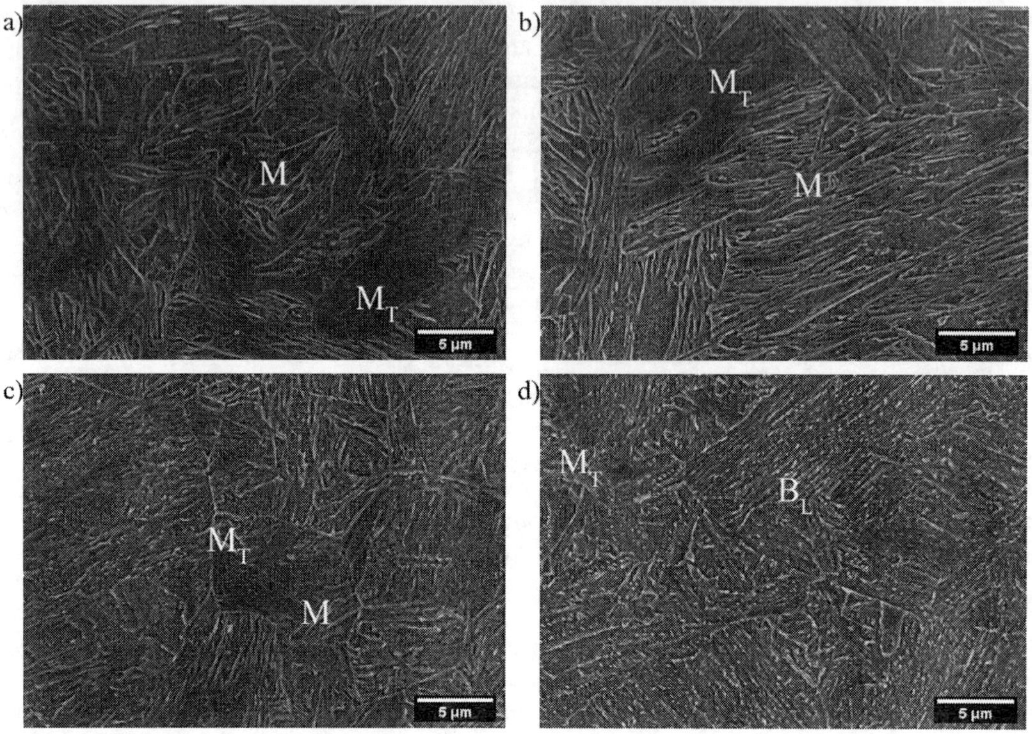

Figure 1 Microstructures after Press Hardening process. For 30s at: a)20 °C, b)200 °C, c)300 °C, d)400 °C. M: Fresh martensite, MT: Tempered martensite, BL: Lower bainite.

Looking at the EWF results (w_e), it is observed that at 400 °C during 10 s, the highest fracture toughness is obtained. There is a significant difference with the other two treatments tested, 31% at 20 °C and 60% at 200 °C. For the samples treated during 30 s, the results show that fracture toughness is not affected by the quenching temperature and the final microstructure. However, an overall comparison shows that there is a considerable improvement in the bainitic microstructure, as other authors have found [17].

Comparing the strain at fracture in the notched specimen and bending and comparing the strain at fracture in the notched specimen and EWF, Figure 2b, indicate that both bending and notched traduce the same notion of ductility into plane strain and the ability of the material to sustain a large amount of local deformation. While EWF is different, as EWF represents the energy consumed during the process of ductile crack propagation of a pre-existing crack. In contrast to samples notched by bending and plane strain, it describes the behaviour of the material after failure initiation. This is relevant when crack propagation has an important contribution to the overall fracture process, as in the collision failure mode of AHSS and PHS [18].

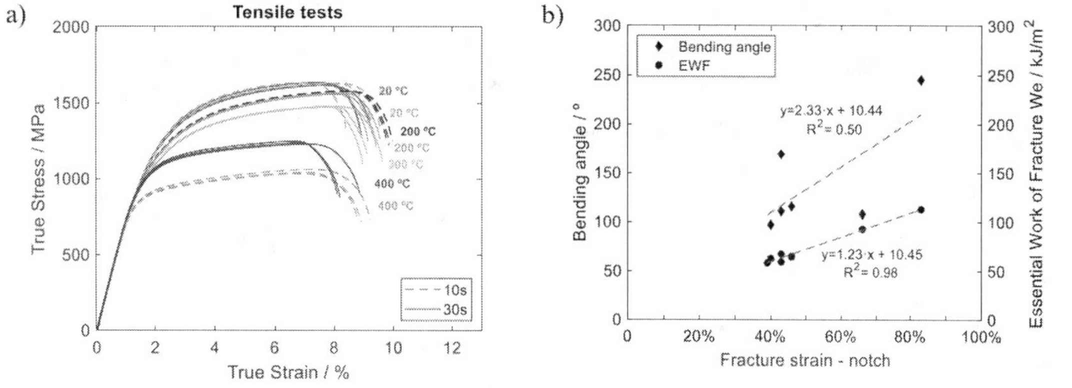

Figure 2. Mechanical properties. a) Tensile tests of the samples after Press Hardening, the difference in mechanical properties between martensitic and bainitic microstructures is noticeable. b) Representation of bending angle and EWF as a function of strain at fracture.

The fractography of the three conditions (20, 200 and 400 °C - 10s), Figure 4, reveals a completely dimple and void surface, which means that the fracture was totally ductile [19]. It can be seen that there are two types of dimples: primary and secondary. The primary ones are the larger ones, and the ones that formed first; and the secondary ones are the smaller ones. There are also visible tears associated with hard phases defragmentation, but once the mechanism of coalescence of microcavities has started, as they are observed in them. Precipitates that could be crack initiators are marked in Figure 4a and 4c.

As more primary dimples are observed at a superficial examination in the 400 °C treatment, a statistical analysis of these has been executed. Using the intersection method, the dimples were measured. It is shown in Figure 4d that the fracture surface with the largest dimples corresponds to the treatment at 400 °C, followed by the treatment at 20 °C and the one at 200 °C being the one with the lowest number of large dimples (for 10 s). This follows the same pattern as the EWF. This is not the case in the 30 s treatment where no change in EWF evolution is observed.

Finally, with the critical temperatures found, the CCT of this steel can be drawn, Figure 3b. When the thermal path suffered by the plates during the forming process is drawn, we can see that the treatments are very close to Ms, being possible to have formed a little additional martensite in the 30 seconds treatment, which gives it a higher hardness, decreasing the mechanical properties related to ductility with respect to the 10 seconds treatment.

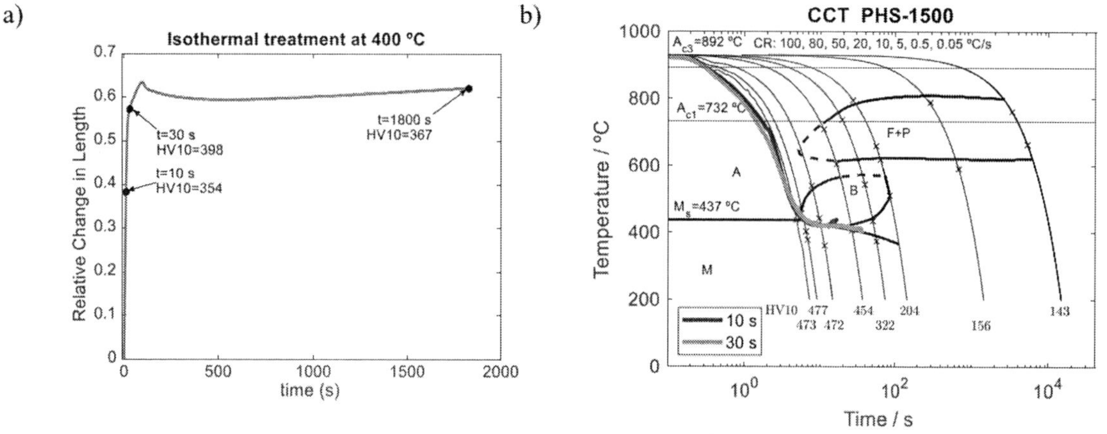

Figure 3. Dilatometric analysis. a) Representation of the isothermal treatment at 400°C. b) CCT diagram of PHS-1500.

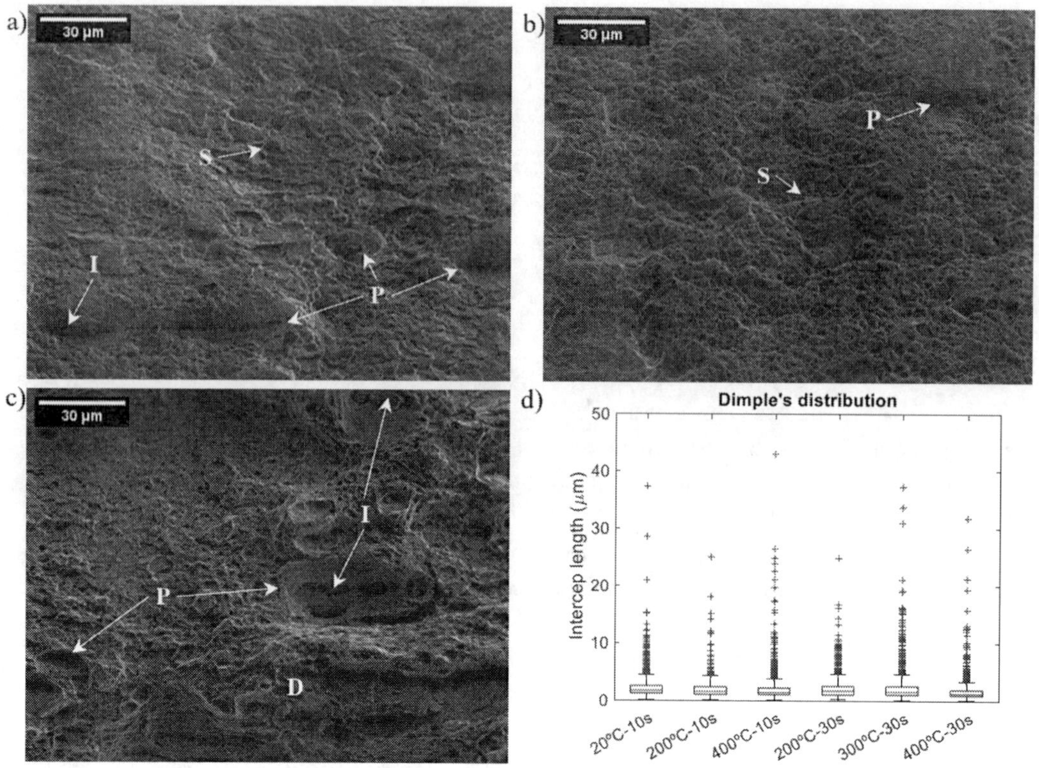

Figure 4. Fractography of samples performed at: a)20 °C, b)200 °C and c)400 °C during 10s. P: primary, S: secondary, I: inclusion, D: defragmentation. d) Dimple's distribution measurement by intercept method where a lower mean of the intercepts but a higher dispersion of the intercepts is observed, which corresponds to the treatments at 400 °C.

4. Remarks

From this investigation it is shown that the different morphologies of the lath-like ferrite and the distribution of carbon considerably modify the mechanical properties. It is experimentally verified that the higher the content of lower bainite, the higher ductile properties are obtained, which is an advantage compared to other microstructures, since having an advanced high-strength steel of simple composition, interesting properties for bending and fracture toughness are obtained.

Future work in this study should focus on the link between the amount of each phase having a ferritic lath-like structure and crack initiation/crack propagation in heterogeneous ductile microstructures. Without a quantitative evaluation of each microstructural component, it is difficult to estimate the local strain / stress triaxiality at the crack tip and understand the crack propagation. The first challenge is to be able to identify the microconstituents in a reliable way. The second challenge is to follow the crack propagation in-situ or ex-situ as the damage generated in the two microstructures is extremely similar in term of fracture features.

Funding: This research was funded by the European Research Fund for Coal and Steel under the contract RFCS2019-899268.

Acknowledgments

Electron microscopy laboratory from CENIM-CSIC.

References

[1] Aroca V P, Caballero F G, Philippot C, Germain F and Capdevila C 2022 *Proc. Int. Conf on Hot Sheet Metal Forming of High-Performance Steel (Barcelona)* Mats Oldenbury, Jens Hardell and Daniel Casellas pp 35-42.

[2] Vuorinen E P, Özügüler A G, Ion J C, Eriksson K, Somani M C, Karjalainen L P, Allain S and Caballero F G 2019 *Hot Forming of Ultra-Fine-Grained Multiphase Steel Products Using Press Hardening Combined with Quenching and Partitioning Process* Metals 9 no 3:357.

[3] Rementeria R and Caballero FG 2017 *Cold Rolling and Annealing of Advanced High Strength Steels* (Boca Raton: CRC Press) pp 482-522.

[4] Edmonds D V, He K, Rizzo F C, De Cooman B C, Matlock D K and Speer J G 2006 *Quenching and partitioning martensite-A novel steel heat treatment* Mater. Sci. Eng. A vol 438-440 pp 25-34.

[5] Philippot C, Cobo S, Aroca V P, Capdevila C, Allain S, Macchi J, Pujante J, Åkerström P, Edberg J, Bergman, G and Romero-Ruiz L 2022 *Proc. Int. Conf on Hot Sheet Metal Forming of High-Performance Steel (Barcelona)* Mats Oldenbury, Jens Hardell and Daniel Casellas pp 273-280.

[6] Caballero F G, Allain S, Cornide J, Puerta Velasquez J D, Garcia-Mateo C and Miller M K 2013 *Design of cold rolled and continuous annealed carbide-free bainitic steels for automotive application* Mater. Des., vol 49, pp 667-80.

[7] Wróbel I, Skowronek A and Grajcar A 2022 *A Review on Hot Stamping of Advanced High-Strength Steels: Technological-Metallurgical Aspects and Numerical Simulation* Symmetry 14 no 5 969.

[8] Shimada N, Tomizawa A, Kubota H, Mori H, Hara M and Kuwayama S 2014 *Development of Three-dimensional Hot Bending and Direct Quench Technology* Procedia Eng. vol 81 pp 2267-72.

[9] Zorgani M, Garcia-Mateo C and Jahazi M 2021 *Effects of ausforming temperature on carbide-free bainite transformation and its correlation to the transformation plasticity strain in a medium C- Si-rich steel* Mater. Charact. vol 176 111124.

[10] Nanda T, Singh V, Singh G, Singh M and Kumar B R 2021 *Processing routes, resulting microstructures, and strain rate dependent deformation behaviour of advanced high strength steels for automotive applications* Arch. Civ. Mech. Eng. 21 7.

[11] ISO 6892-1 *Metallic materials - Tensile testing - Part 1. Method of test at room temperature* 2019.

[12] Dietsch P, Tihay K, Bui-Van A and Cornett D 2017 *Methodology to assess fracture during crash simulation: Fracture strain criteria and their calibration* Metall. Res. Technol. vol 114 607.

[13] VDA 238-100 *Plate bending test for metallic materials* 2017.

[14] CWA 17793 *Test method for determination of the essential work of fracture of thin ductile metallic sheets*. 2021.

[15] Frómeta D, Lara A, Parareda S, Grifé L and Casellas D 2020 *New tool to evaluate the fracture resistance of thin high strength metal sheets* IOP Conf. Ser. Mater. Sci. Eng. 967 012088.

[16] Garcia-Mateo C, Caballero F G, Miller M K and Jimenez J A 2012 *On measurement of carbon content in retained austenite in a nanostructured bainitic steel* J. Mater. Sci. vol 47 pp 1004-10.

[17] Golling S, Frómeta D, Casellas D and Jonsén P 2019 *Influence of microstructure on the fracture toughness of hot stamped boron steel* Mater. Sci. Eng. A vol 743 pp 529-39.

[18] Frómeta D, Lara A, Molas S, Casellas D, Rehrl J, Suppan C, Larour P and Calvo J 2018 *On the correlation between fracture toughness and crash resistance of advanced high strength steels* Eng. Fract. Mech. vol 205 pp 319-32.

[19] Barnwal V K, Lee S Y, K J H and Barlat F 2019 *Failure characteristics of advanced high strength steels at macro and micro scales* Mater. Sci. Eng. A, vol 754 pp 411-27.

IOP Conf. Series: Materials Science and Engineering 1284 (2023) 012035 doi:10.1088/1757-899X/1284/1/012035

Strength enhancement and uniform strain distribution through cross route-constrained groove pressing

Sunil Kumar

Department of Mechanical Engineering, S V National Institute of Technology Surat, Surat, Gujarat-395007, India

E-mail: Sunil30111988@gmail.com

Abstract. Constrained groove pressing (CGP) is one of the severe plastic deformation techniques for sheet metals to impose large plastic strain and produce ultrafine-grained (UFG) structures without any change in dimensions. Cross route-constrained groove pressing (Cross-CGP), an advanced version of the conventional CGP process, involves two CGP cycles with an equivalent strain of 2.32 per cycle where the square sample is rotated by 90° after performing every cycle of CGP. The strain path in the sheet during conventional CGP and cross-CGP significantly affects the homogenization of mechanical properties. Sheets of low carbon steel are processed through conventional-CGP and cross-CGP routes. The ability to sustain the maximum strain has been increased in the cross-CGP process due to uniform strain distribution. The tensile strength and hardness in the sheet processed through cross-CGP are higher than that of conventional CGP, especially at higher pass numbers. Tensile behavior and hardness in two different orientations (parallel and perpendicular to the groove direction) have been found to be almost the same in the cross-CGP processed sheet.

1. Introduction

In recent years, the ultrafine-grained (UFG) materials have been the subject of extensive research characterized by a polycrystalline structure with grain sizes ranging from submicrometer (100~1000 nm) to nanometer scale (less than 100 nm) [1]. Due to their high fraction of grain boundaries and unique structure, UFG materials exhibit enhanced mechanical properties such as increased strength, toughness, superplasticity, hydrogen storage, fatigue life, radiation resistance, and wear resistance [2–4]. Consequently, UFG materials find promising applications in various industries including aerospace, transportation, medical devices, electronics, and defense [5,6].

Severe plastic deformation (SPD) is a top-down processing technique that can produce bulk metals with submicron-sized grains at room temperature. The gas condensation method in the presence of inert gas is first used by Gleiter and his colleagues [1,7] to demonstrate the formation of nanostructured materials. Valiev and his colleagues also worked on UFG development through SPD [8] and reviewed several SPD processes [9] in parallel with Gleiter's work. The metal undergoes significant plastic strain in all SPD processes without altering its shape. Equal channel angular pressing (ECAP) [9], accumulative roll-bonding (ARB) [10], high-pressure torsion (HPT) [9], repetitive corrugation and straightening (RCS) [11], multi-axial forging (MAF) [12], asymmetric rolling (ASR) [13], and constrained groove pressing (CGP) [14] are among the SPD techniques available. ECAP, HPT, and MAF are capable of producing small-size samples through a discontinuous route in the processing of

IOP Conf. Series: Materials Science and Engineering 1284 (2023) 012035 doi:10.1088/1757-899X/1284/1/012035

bulk metals. However, RCS, ASR, ARB, and CGP are more appropriate among SPD techniques for the production of bulk sheet metal. The degree of grain refinement depends on the SPD process used for sheet metal and CGP process has largest ability to impose strain on the sheet without changing the dimensions.

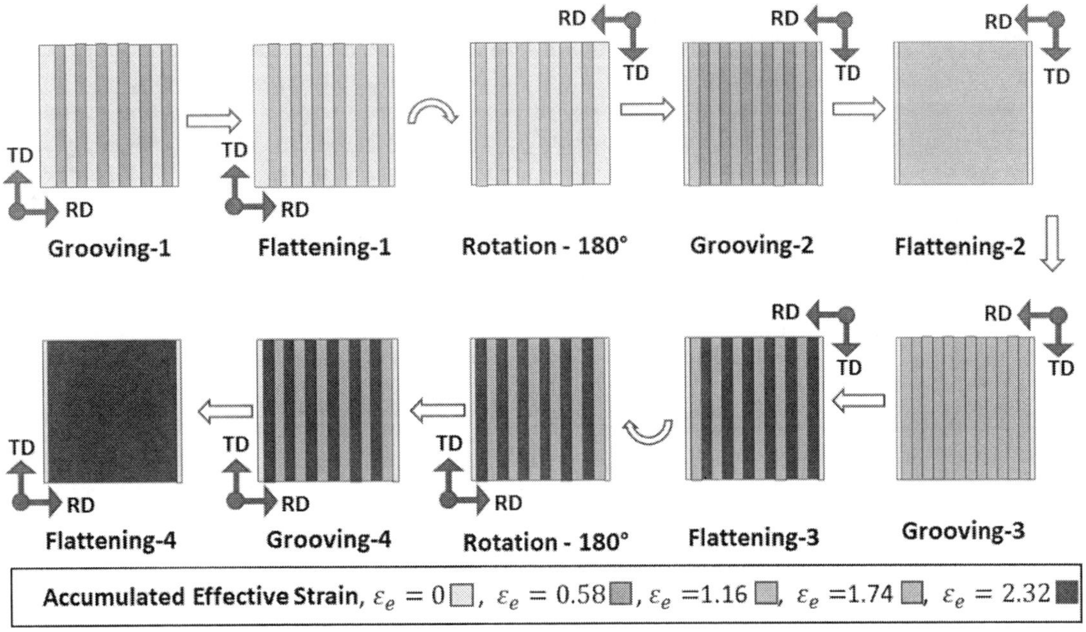

Figure 1. The stages and accumulated effective strain of conventional-CGP process (Route-1).

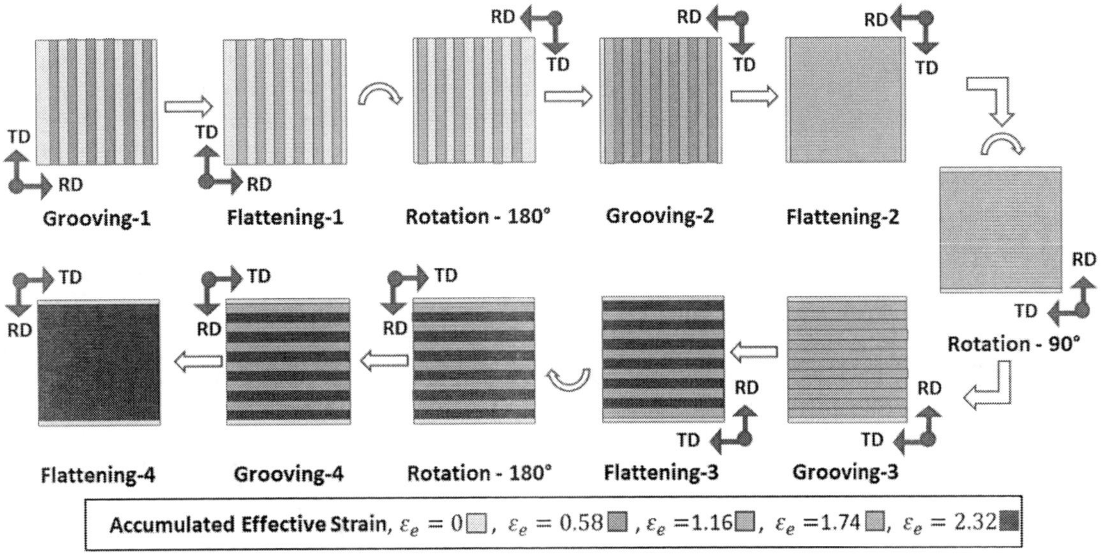

Figure 2. The stages and accumulated effective strain of cross-CGP process (Route-2).

Lightweight sheet metals with uniform grain refinement and unique properties processed through CGP technique are crucial in industrial applications to promote energy conservation and improve fuel economy. Hence, the change in strain path via different routes of the CGP process is an intriguing and promising method that can have a significant impact on the microstructure, mechanical properties, and texture of a material. Three distinct routes, namely Route-1, Route-2, and Route-3, are often used to demonstrate different ways of altering the strain path. Route-1 and Route-2 refer to the conventional-CGP and cross route-CGP (cross-CGP) processes as shown in Figure 1 and Figure 2 respectively. Route-2 involves cross-CGP, an advanced version of the conventional-CGP process, where two conventional-CGP cycles are performed with an equivalent strain of 2.32 per pass, and the square sample is rotated by 90° after each pass of conventional CGP. In reference [15], Route-3 comprises 90-degree rotations after each flattening stage for further deformation and this route is not discussed in this work.

Guler et al. [16] examined five different routes, including the three mentioned above, to improve mechanical properties and anisotropy. Aside from the significant improvement in tensile strength and hardness, they noted a noticeable improvement in material anisotropy through Route-1. Route-2 is an effective method for producing UFG structures in AZ31-O samples, and Route-3 imparts the same amount of strain in fewer passes and produces higher thinning of the sample [15]. Although the strain path slightly improves the grain refinement, it modifies the texture of the processed sample [17]. The distribution of grain size and band contrast maps from electron backscatter diffraction (EBSD) microstructural analysis shows a bimodal structure in the UFG magnesium alloy due to dynamic recrystallization in all CGP routes. The uniform grain sizes that developed after processing through Route-1, Route-2, and Route-3 are 5.04 μm, 5.69 μm, and 5.18 μm, respectively. The cross CGP process (Route-2) induces a delay in crack nucleation due to the imposition of uniform strain distribution in the sheet [18], resulting in a homogenous microstructure and mechanical properties throughout the sample compared to conventional CGP (Route-1) [19]. The variation of grain size with strain in the sample processed through both conventional and cross CGP is investigated in [20] and found that the cross CGP processed samples can sustain more strain and cause slightly finer grains before crack nucleation.

In the present paper, numerical studies are performed to understand the influence of strain path on strain distribution and mechanical properties in the sample processed through conventional-CGP and cross-CGP. The strain path-property correlations are established for uniform grain refinement and mechanical properties.

2. Finite element simulation

Commercial ABAQUS/Standard software is utilized to perform finite element analysis of the conventional and cross-CGP processes on low carbon steel material. Since the sample of low carbon steel exhibits asymmetry, a section of whole sample with three dimensions of 26×26×2 mm^3 is only analyzed. The properties used for the chemical composition (in wt%) of low carbon steel is 0.22% of Mn, 0.04% C, 0.04% Al, 0.01% P and balance Fe. The procedure of CGP process is discussed in many literature [5,6,19,20]. As per experiments, low carbon steel sample is successfully processed for five CGP passes but the surface defects like micro cracks and unevenness are observed after three passes onwards. Hence the FE analysis is carried out on the samples processed up to two passes of conventional CGP process and one pass of cross-CGP.

Isotropic material properties are assumed to analyze the conventional-CGP and cross-CGP processes, and the input stress-strain relation as $\sigma = 540\varepsilon^{0.25}$, mass density as 7870 Kg/mm^3, young modulus as 200 GPa, poison ratio as 0.3, yield strngth as 215 MPa and ultimate tensile strength as 300 MPa. The simulations include two conventional-CGP passes of all four stages each and one cross-CGP pass of eight stages, with grooved and flattened dies with a 45-groove angle and a 2 mm groove width. Material deformation is studied at room temperature and assumed to be strain rate independent, thus making the stress and strain results in the material independent of tool velocity during the simulation. A relatively high tool velocity is employed during the simulation, and displacement-based constraints are used to define the tool motion. The sheet is modeled as a deformable part and meshed using C3D8R (8-node linear brick, reduced integration, hourglass control) element type, consisting of a total of 86528 elements. The tools are modeled as analytically rigid parts, and the coefficient of friction between the sheet and die surface interface is assumed to be 0.1 throughout the analysis.

IOP Conf. Series: Materials Science and Engineering 1284 (2023) 012035 doi:10.1088/1757-899X/1284/1/012035

3. Results and discussion

The anticipated outcome of homogenous strain after second pass of conventional-CGP and first pass (eight stages) of cross-CGP is 2.32. However, as a result, Figure 3 depicts the fluctuation of equivalent plastic strain distribution contour and material response of low carbon steel in all the stages.

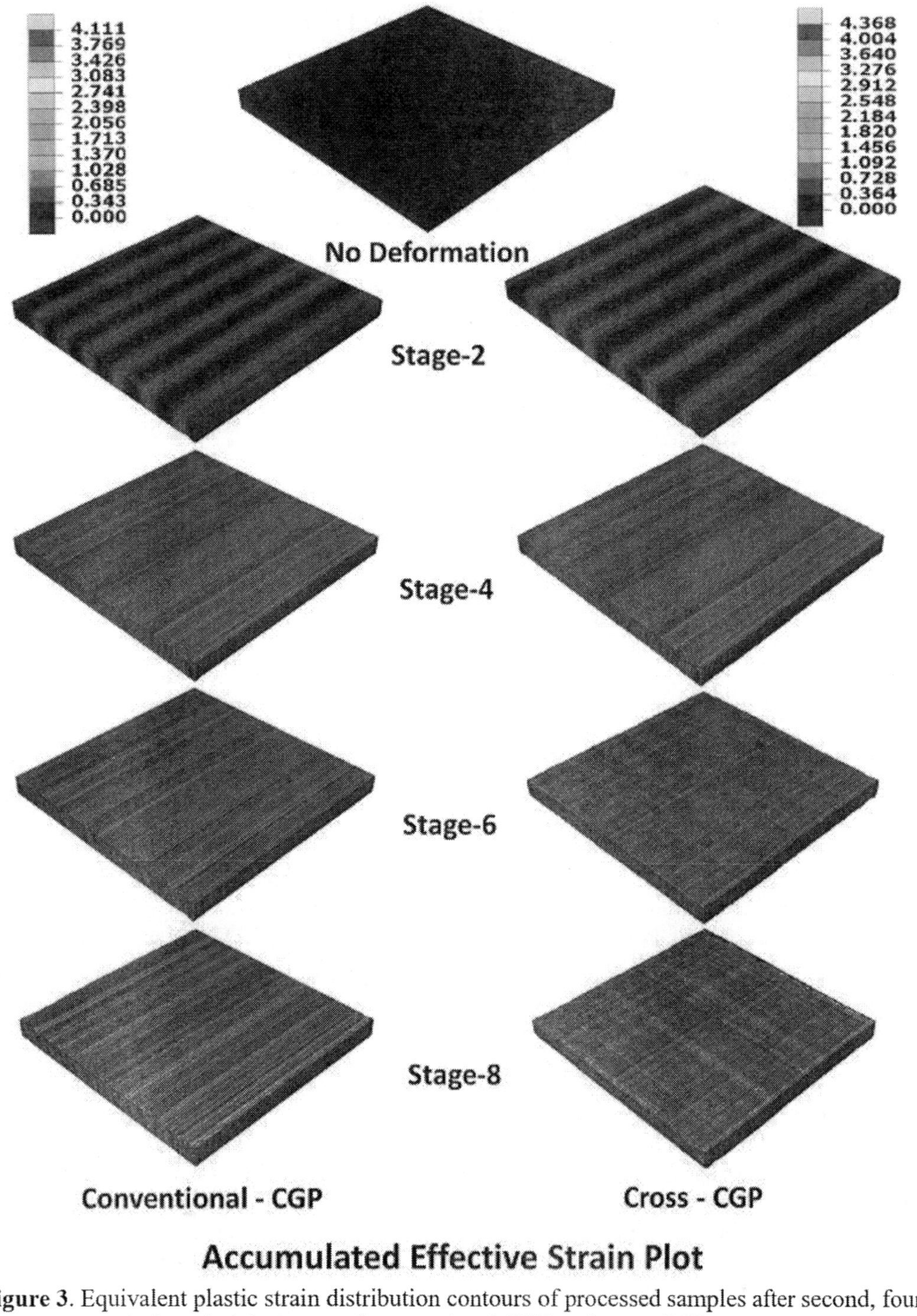

Figure 3. Equivalent plastic strain distribution contours of processed samples after second, fourth, sixth, and eight stages of conventional-CGP and cross-CGP processes.

IOP Conf. Series: Materials Science and Engineering 1284 (2023) 012035 doi:10.1088/1757-899X/1284/1/012035

The non-uniform distribution of the strain is apparent, which may lead to uneven grain refinement. In other word, a strain of 2.32 is sufficient for maximum grain refinement in conventional-CGP and cross-CGP processes albiet the distribution of grain sizes is not uniform. Table 1 illustrates the comparison of equivalent strain and inhomogeneity factor of strain distribution from conventional CGP and cross-CGP processes. As mentioned earlier, inhomogeneity of strain distribution results in non-uniformity of ultrafine grains and mechanical properties. Inhomogeneity factor as in equation (1) is used to quantify the non-uniformity in strain distribution.

$$\text{Inhomogeneity factor} = \frac{\sqrt{\sum_{i=1}^{n} \frac{(S_i - S_{avg})^2}{n-1}}}{S_{avg}} \times 100 \qquad (1)$$

Where, S_i= strain value at ith point, S_{avg} = average strain value, n=number of measurements in single sample.

Table 1: Average equivalent strain and inhomogeneity factor from the simulation of conventional-CGP and cross-CGP processes.

Process	Condition	Average equivalent strain	Inhomogeneity factor
Conventional-CGP	1 pass (4 stages)	1.29	21.34
	2 passes (8 stages)	2.53	19.94
Cross-CGP	1 pass (4 stages)	1.29	21.39
	1 pass (8 stages)	2.89	10.34

The strain distribution, which is non-uniform, shows an average value which is equivalent to the analytically calculated strain of 1.16 and 2.32 after first and second pass. As the number of stages or passes increases, the imposed strain continuously increases from 1.29 after stage-4 to 2.53 after stage-8 in conventional-CGP. While, the imposed strain increases from 1.29 after stage-4 to 2.89 after stage-8 in cross-CGP process. After eighth stage, the imposed strain is higher in cross-CGP compared to that of conventional-CGP. As per literature [21–23], the higher grain refinement is achieved through cross CGP because higher changes in shear strain path in successive passes. It also reduces the mutual dislocations that annihilate during forward and reverse shear loading compared to conventional CGP. According to Satheesh & Raghu [22], although there was a slight reduction in grain refinement after cross-CGP compared to conventional CGP, the material exhibits superior mechanical properties. The maximum of high-angle grain boundaries is in-between the peak around 15–30°, indicating that grain sub-boundaries and strain-driven misorientation are the dominant grain refinement mechanisms [22].

The variation in imposed strain throughout the sample is a result of considering not only pure shear deformation but also the effects of bending, friction, and stretching in the simulation. Thereby, the larger value of inhomogeneity factor indicates significant inhomogeneity in grain refinement and material properties. The values of inhomogeneity factor are significantly increased after fourth stage in both the processes. In conventional-CGP process, the non-uniformity of the strain distribution remains in the processed sample, albeit slightly reduced. In contrast to conventional CGP, the cross-CGP processed sample exhibits a homogeneous strain distribution due to the 90-degree symmetric rotation after CGP one pass. The values for inhomogeneity factor are decreased to 10.34 after eigth stage from 21.39 after fourth stage. Since the sample processed through cross-CGP has lower inhomogeneity factor after eighth stage, it may be inferred that the sample has higher uniformity in grain refinement and mechanical properties.

Similar to conventional CGP, the cross CGP processed aluminum samples exhibit lower strain hardening ability and higher strain rate sensitivity [22]. However, the formation of the UFG structure leads to significant increases in strength and hardness [19]. The strength initially increases after the first pass and then slightly decreases after the second pass, while hardness continuously increases with pass number. Tensile strength and hardness in the cross CGP processed sheet are higher than in conventional CGP, especially at higher pass numbers, and the anisotropy of the sample is reduced. The parallel and perpendicular orientations exhibit similar tensile behavior and hardness, indicating improved strain homogeneity at higher pass numbers as evidenced by a decrease in the IF.

4. Conclusions

The formation of the UFG structure through the imposed strain in conventional-CGP and cross-CGP processes leads to significant increases in strength and hardness. Cross-CGP process generates the best and uniform grain refinement due to uniform imposed equivalent strain compared to that of conventional-CGP. It is expected that the strain path of cross-CGP modifies the basal texture of the processed material. Moreover, in conclusion, cross-CGP processed sample has a higher capacity for strain accumulation due to change of strain path and delay in the initiation of microcracks. The strain path change in cross-CGP process contributes to improving the uniformity of strain distribution, grain refinement efficiency, and mechanical properties.

References

[1] Gleiter H 1989 Nanocyrstalline Materials *Prog. Mater. Sci.* **33** 223–315
[2] Shahmirzaloo A, Hosseini S M, Siahsarani A, Rahmatabadi D, Hashemi R and Faraji G 2020 Influences of the constrained groove pressing on microstructural, mechanical, and fracture properties of brass sheets *Mater. Res. Express* **7** 116526
[3] Valiev R 2004 Nanostructuring of metals by severe plastic deformation for advanced properties *Nat. Mater.* **3** 511–6
[4] Valiev R ., Islamgaliev R . and Alexandrov I . 2000 Bulk nanostructured materials from severe plastic deformation *Prog. Mater. Sci.* **45** 103–89
[5] Soon K, Jen M, Chua W, Atsushi D, Fong K S, Tan M J, Chua B W and Atsushi D 2015 Enabling wider use of magnesium alloys for lightweight applications by improving the formability by groove pressing *Procedia CIRP* **26** 449–54
[6] Kumar S, Hariharan K, Ravi D and Kumar S 2019 Accounting Bauschinger effect in the numerical simulation of constrained groove pressing process *J. Manuf. Process.*
[7] A.H. Chokshi, A. Rosen, J. Karch and H. Gleiter 1989 On the validity of the Hall-Patch relationship in nanocrystalline materials *Scr. Mater.* **23** 1679–89
[8] Valiev R Z, Krasilnikov N A and Tsenev N K 1991 Plastic deformation of alloys with submicron-grained structure *Mater. Sci. Eng. A* **137** 35–40
[9] Valiev R, Estrin Y, Horita Z, Langdon T G, Zehetbauer M J and Zhu Y T 2006 Producing Bulk Ultrafine-Grained Materials by Severe Plastic Deformation *Jom* **58** 33–9
[10] Saito Y, Utsunomiya H, Tsuji N and Sakai T 1999 Novel ultra-high straining process for bulk materials-development of the accumulative roll-bonding (ARB) process *Acta Mater.* **47** 579–83
[11] Huang J Y, Zhu Y T, Jiang H and Lowe T C 2001 Microstructures and dislocation configurations in nanostructured Cu processed by repetitive corrugation and straightening *Acta Mater.* **49** 1497–505
[12] Cherukuri B and Srinivasan R 2006 Properties of AA6061 processed by multi-axial compressions/forging (MAC/F) *Mater. Manuf. Process.* **21** iop
[13] Lee S H and Lee D N 2001 Analysis of deformation textures of asymmetrically rolled steel sheets *Int. J. Mech. Sci.* **43** 1997–2015

[14] Kumar S 2022 Developing methods of constrained groove pressing technique : A review *Proc. Inst. Mech. Eng. Part L J. Mater. Des. Appl.* **237** 1319–46

[15] Fong K S, Danno A, Jen T M and Chua B W 2016 Effect of Deformation and Temperature Paths in Severe Plastic Deformation Using Groove Pressing on Microstructure, Texture, and Mechanical Properties of AZ31-O *J. Manuf. Sci. Eng.* **137** 051004

[16] Guler Z and Yapici G G 2021 Application of Novel Constrained Groove Pressing Routes on Austenitic Stainless Steel *Trans. Indian Inst. Met.* **74** 2599–608

[17] Wang Z, Wang T, Guan Y and Zhu L 2021 Effect of strain path on microstructure and mechanical properties of AZ31 magnesium alloy sheets processed by constrained groove pressing *Mater. Sci. Eng. A* **804** 140794

[18] Peng K, Zhang Y, Shaw L L and Qian K W 2009 Microstructure dependence of a Cu-38Zn alloy on processing conditions of constrained groove pressing *Acta Mater.* **57** 5543–53

[19] Moradpour M, Khodabakhshi F, Mohebpour S R, Eskandari H and Haghshenas M 2019 Finite element modeling and experimental validation of CGP classical and new cross routes for severe plastic deformation of an Al-Mg alloy *J. Manuf. Process.* **37** 348–61

[20] Khodabakhshi F, Abbaszadeh M, Mohebpour S R and Eskandari H 2014 3D finite element analysis and experimental validation of constrained groove pressing-cross route as an SPD process for sheet form metals *Int. J. Adv. Manuf. Technol.* **73** 1291–305

[21] Khodabakhshi F, Abbaszadeh M, Eskandari H and Mohebpour S R 2013 Application of CGP-cross route process for microstructure refinement and mechanical properties improvement in steel sheets *J. Manuf. Process.* **15** 533–41

[22] Satheesh Kumar S S and Raghu T 2015 Strain path effects on microstructural evolution and mechanical behaviour of constrained groove pressed aluminium sheets *Mater. Des.* **88** 799–809

[23] Moradpour M, Khodabakhshi F and Eskandari H 2018 Microstructure–mechanical property relationship in an Al–Mg alloy processed by constrained groove pressing-cross route *Mater. Sci. Technol.* **34** 1003–17

IOP Conf. Series: Materials Science and Engineering 1284 (2023) 012036 doi:10.1088/1757-899X/1284/1/012036

Evaluation of hole expansion ratio of ultra-high strength martensitic steels produced with various processing routes

J. Hannula[1*], P. Plosila[1], A. Kaijalainen[1], P. Suikkanen[2], J. Kömi[1]

[1] Materials and Mechanical Engineering, Centre for Advanced Steels Research, FI-90014 University of Oulu, Finland
[2] SSAB Europe Oyj, Rautaruukintie 155, P.O. Box 93, FI-92101 Raahe, Finland

E-mail: jaakko.hannula@oulu.fi

Abstract. Hole expansion ratio of hot rolled ultra-high strength martensitic steels with tensile strength around 1000 MPa was evaluated. Steels were produced with direct quenching and traditional reheat and quenching processes with final thickness of 3 mm. Achieved yield strength values ($R_{p0.2}$) varied between 918 – 1068 MPa depending on the processing route. Mean hole expansion ratios (HER) in direct-quenched (DQ) and direct-quenched and tempered (DQT) conditions were between 22 – 36 %, and no clear improvement was seen after tempering treatment compared to quenched variant. HER values for reheat and quenched (RQ) and reheat, quenched and tempered (RQT) variants were between 31 – 49 %, and the highest values were achieved with RQT steels, which were tempered at 600 °C. Based on the field emission scanning electron microscope with electron backscatter scanning diffraction (FESEM-EBSD) analysis/characterization of martensite grain size, more uniform grain structure was discovered in RQ steels, which could be the reason for improved HER properties. HER values were also compared to tensile test results (uniform elongation, true thickness strain), and formability maps were constructed. However, only minor correlation was found between HER and true thickness strain values.

1. Introduction

In the automotive industry, the use of hot-rolled steels with improved balance of strength and ductility ensures high cost-effectiveness and better fuel economy used for chassis, underbody- and structural components. High strength hot-rolled sheet steels with excellent combinations of strength and ductility as measured e.g., globally by the elongation to fracture or locally by the hole expansion test are used to increase for example the performance of various chassis parts. However, in the assembly, stamping operations such as drawing, stretching, bending, blanking, trimming, flanging produces various stress states to the formed steel parts and may initiate manufacturing defects either in subsequent processing steps or later in use. Therefore, microstructural features that improve the damage tolerance of the material are desirable.

To achieve hot-rolled steels with tensile strength in the range 1000 – 1200 MPa, different microstructures are possible, such as martensite, tempered martensite or multiphase steels with combinations of martensite/bainite and austenite. To achieve these microstructures, correct compositions and processing parameters are needed, such as cooling rates and quenching and coiling temperatures. It is well known that microstructural characteristics such as volume fraction and morphology of the hard and soft phase, strength difference between the soft and hard phases and the volume fraction of retained austenite have an influence on hole expansion capability. Many studies have shown that hole expansion ratio of multiphase steels, such as complex phase (CP) or dual phase (DP) steels is highly dependent on the fraction of martensite, and the hardness difference between the soft ferrite and hard martensite/bainite [1–4]. Also, some research has been conducted regarding the stretch-flangeability of medium manganese and quenched and partitioned steels, where retained austenite has a dominant role on improving the mechanical properties. Those studies has been shown that these steels do not have relatively good hole expansion capability, as the hole expansion ratio is decreased with higher amount of retained austenite [5,6]. However, regarding single phase hot-rolled martensitic steels, there is not much research reported from the hole expansion ratio of martensitic steels with tensile strength over 1000 MPa. Therefore, the purpose of the study was to produce single phase martensitic microstructure to avoid the hardness differences between different phases, and to investigate the hole expansion capability of single-phase martensite produced in different processing routes.

2. Experimental

Table 1 shows the steel compositions investigated in this study. Compositions have been planned to achieve tensile strength of 1000 – 1200 MPa. Two different types of steel concepts have been included: category A steels are designed as martensitic alloys in quenched condition with carbon content of 0.1 wt.% and suitable amounts of Mn, Cr, Mo and Si to achieve sufficient hardening. Category B steels are designed as martensitic alloys suitable for high temperature tempering with equal carbon content of 0.1 wt.%, but with higher microalloying (Nb, V, Mo) for improved tempering resistance. A1 and B1 variants are alloyed with titanium and boron, whereas A2 and B2 steel are Ti- and B-free compositions.

Table 1. Chemical compositions of investigated steels [in wt.%].

Steel	C	Si + Mn + Cr	Mo	B	Ti	Nb + V
A1	0.10	2.4	0.15	0.0025	0.015	0.03
A2	0.10	3.25	0.15	-	-	0.04
B1	0.10	2.4	0.3	0.0025	0.015	0.13
B2	0.10	2.95	0.3	-	-	0.14

Laboratory scale hot rolling trials were executed at Outokumpu Research Centre in Tornio. Slabs in size of ~ 35 x 170 x 350 mm^3 were heated up to 1250 °C for one hour soaking before hot rolling was executed. The final thickness of hot rolled plates were approx. 3 mm and target finish rolling temperature (FRT) were 900 °C. The used pass schedule can be seen in Table 3. After last pass, plates were water cooled with water sprays to room temperature with cooling rate of approximately 30 – 60 °C/s to have martensitic microstructure.

Table 2. Pass schedule of hot rolling trials.

Pass	Thickness [mm]	Temperature [°C]*	Reduction [%]	Total reduction [%]
0	35	-		93.5
1	24	-	31	
2	16.6	1058	31	
3	11.5	1026	31	
4	8	979	30	
5	5.6	951	30	
6	4	901	29	
7	3	908	25	

* one example of temperatures.

Reheat and quench (RQ) variant were also produced in laboratory scale at the University of Oulu. Hot rolled samples were reheated up to 900 °C for 15 minutes prior to water quenching. Additionally, for B steels in DQ- and RQ conditions, tempering carried out by tempering temperature of 600 °C with 30 minutes holding at peak temperature.

For the microstructure characterization, field emission scanning electron microscopy combined with electron backscatter diffraction (FESEM-EBSD) was carried out. For EBSD analysis, an accelerating voltage of 20kV, and a step size of 200 nm were used with total scanned area of ~123 x 83 microns. For SEM micrographs, samples were etched using 2 % Nital solution for approximately 15-30 seconds to reveal the microstructure.

Tensile tests were performed using 3 mm thick flat specimens (three specimens for each steel) in longitudinal and transverse direction with respect to rolling direction according to the standard ISO 6892-1:2016 [7]. Fracture surfaces of the tensile specimens were investigated with a digital microscope. Due to difficulties in focusing on the whole fracture surface, only the thickness reduction at the middle of the fracture surface was measured for both specimen halves. The true thickness strain at fracture (ε_{3f}) was calculated by $\varepsilon_{3f} = ln(t_0/t_f)$, where t_0 is the original thickness of the specimen and t_f is the thickness in the middle of the fracture surface.

Hole expansion (HE) tests (four tests for each steel variant) were carried out in Lapland University of Applied Sciences with an Erichsen universal sheet metal testing machine (Erichsen 145-60) according to ISO 16630:2017 [8]. The HE test consists of two steps. The first step is punching a pre-hole at 10 mm/min punch speed using a tool with the correct clearance (inside diameter 10.7 mm) according to the standard. The second step is the forcing of a conical expanding tool into the pre-punched hole at 15 mm/min until any crack extends through the test piece thickness. The limiting hole expansion ratio is obtained by measuring the expanded hole in two perpendicular directions and expressing the increase in hole diameter as a percentage of the original hole diameters described in the standard.

3. Results and discussion

EBSD analysis was carried out to examine the differences in microstructures between steels in DQ- and RQ conditions. Measured effective- (high-angle grain boundaries, > 15°) and $d_{90\%}$ grain sizes are presented in Table 3. Based on effective grain sizes, all steels have relatively similar values. However, comparing the $d_{90\%}$ values, which refers to the largest grains in the grain size distribution, steels in RQ- and RQT conditions produced significantly smaller $d_{90\%}$ grain sizes. This implies that after reheat- and quenching, the grain structure is finer and more homogeneous compared to DQ variants. Some earlier research indicate that finer grain size can improve the overall stretch flangeability of high strength steels. Yoon et al. showed that smaller grain size improved the HER of TWIP steels, and their overall conclusion were that stretch flangeability is dominated by fracture toughness, and all factors that improve fracture toughness, i.e. finer grain size, improves HER [9]. However, there is no relevant research regarding the correlation between grain sizes of martensite (effective grain size, $d_{90\%}$ etc.) and hole expansion ratio.

Figure 1 presents the SEM images and inverse pole figures based on the EBSD data of A1 and B1 steels in different conditions. The morphology difference in prior austenite grains can be easily seen after DQ- and RQ processes, as DQ variants have pancaked austenite with combination of relatively large and small grains, as RQ variants have equiaxed grains with more homogenous grain structure. Also, the more homogeneous microstructure is evident in RQ- and RQT steels, which was seen in comparing the $d_{90\%}$ values. Based on EBSD data, misorientation distributions (2.5° – 63°) were also constructed, as seen in figure 2. Results show typical distributions for martensitic steels (peaks at ~7.5°, ~16°, ~52.5° and ~59° [10,11]) without any major differences.

Table 3. Average grain sizes based on EBSD analysis.

Steel	ECD >15° [μm]	$d_{90\%}$ [μm]
A1 DQ	1.31	12.3
A2 DQ	1.4	14.5
A1 RQ	1.5	10.4
A2 RQ	1.53	9.1
B1 DQT 600 °C	1.43	19.4
B2 DQT 600 °C	1.36	15.5
B1 RQT 600 °C	1.49	10.7
B2 RQT 600 °C	1.37	7.4

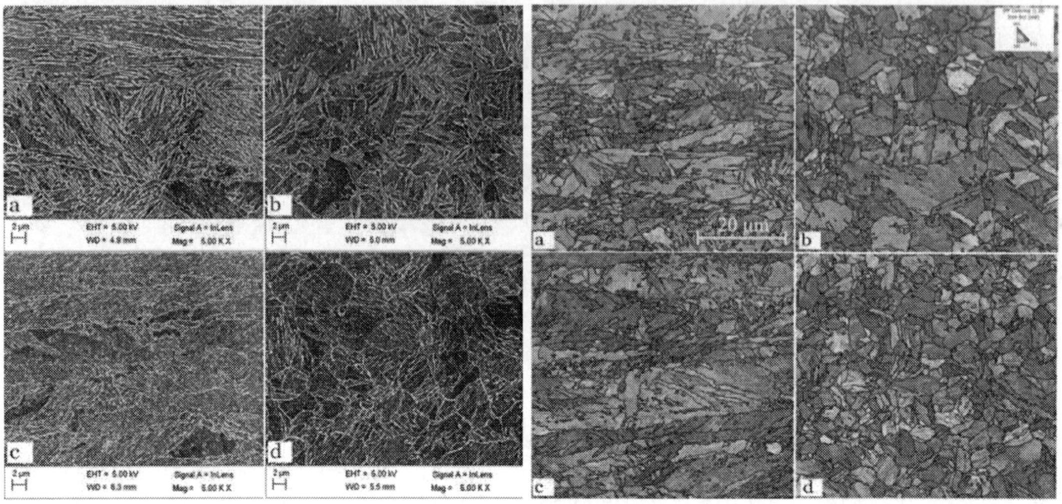

Figure 1. Left: SEM images, right: Inverse pole figures (IPF) of a) A1 DQ, b) A1 RQ, c) B1 DQT, d) B1 RQT steel.

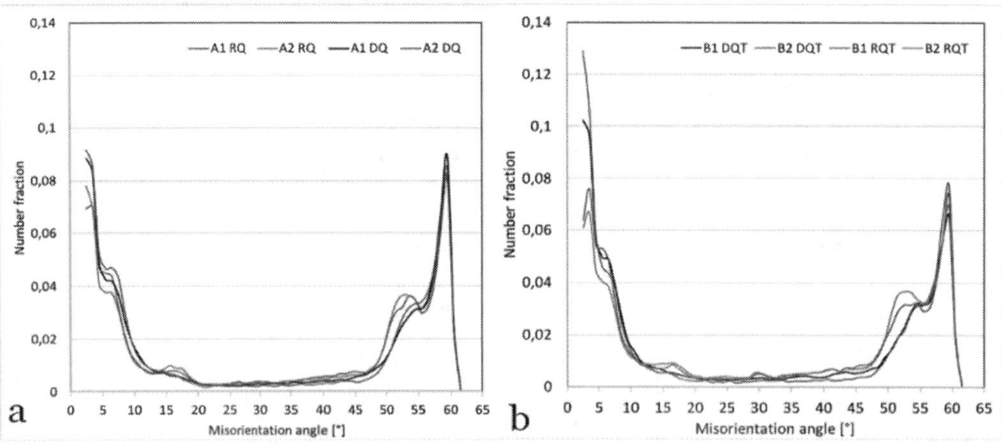

Figure 2. Misorientation distributions of a) A1-A2, b) B1-B2 steels.

Table 4 presents the mechanical properties of A1-B2 steels with different conditions. Achieved tensile strength values were between 916 – 1251 MPa (target 1000 – 1200 MPa). Lowest tensile strength values, which were all below 1000 Mpa, was achieved with B steels after high tempering temperature of 600 °C, which could be expected as tempering generally reduces the strength. Generally, there was no clear difference in mechanical properties comparing steels with and without Ti- and B alloying (variants 1 and 2). Results show that DQ process produced increase in strength with similar elongation values compared to RQ variants. However, in tempered steels, tempering at 600 °C for RQ steels produced clear increase in elongation, which is typical phenomenon comparing tempered and untempered steels. For some reason the same improvement in toughness was not seen in DQT steels, as can be seen comparing results in Table 4. When comparing Rm x A values, clearly the highest values were achieved for B steel in RQT conditions, which indicate the best overall tensile properties among the tested steels.

Table 4. Longitudinal tensile test results of A1 – B2, average of 3 tests.

Steel	$R_{p0.2}$ [MPa]	R_m [MPa]	A_g [%]	A_{50} [%]	$R_{p0.2}$ / R_m	R_m x A_{50} [MPa%]
A1 DQ	971	1207	3.0	7.5	0.80	9053
A2 DQ	974	1251	2.8	6.7	0.78	8382
A1 RQ	887	1134	3.1	7.8	0.78	8845
A2 RQ	836	1101	2.4	5.8	0.76	6386
B1 DQT	1008	1015	3.8	8.6	0.99	8729
B2 DQT	916	920	3.2	7.8	1.00	7176
B1 RQT	918	935	5.5	12.1	0.98	11314
B2 RQT	930	946	5.2	11.1	0.98	10501

In order to evaluate the global and local formability properties of A1-B2 steels, formability maps were constructed based on true uniform strains (ε_u) determined from the tensile test results and the true thinning strain at fracture measurement results. Figure 3 presents the general illustration of formability map adapted from Hance [12], and describes, how to interpret the results. The dashed lines represent limits for local, balanced and global formability classifications based on the $/\varepsilon_u$-ratio and the curves represent limits for general formability ratings. It is noticeable that formability maps presented in figure 4 have been zoomed to smaller area. Based on constructed formability maps, all steels are in category of "Fair" formability and have better local formability than global formability. Especially the RQT variants have better formability properties and are close to in category of "Good" general formability. Based on formability maps, it could be estimated, that although investigated steels have better local formability than global formability, and hole expansion test itself describes the capability of local formability, the values are mainly still in category "fair", which could indicate that hole expansion ratio would not be very high for investigated steels.

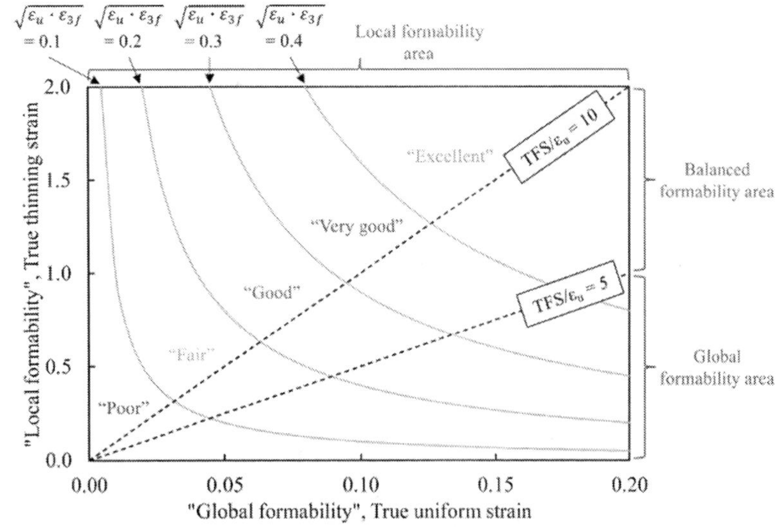

Figure 3. Illustration of formability map and the interpretation of results.

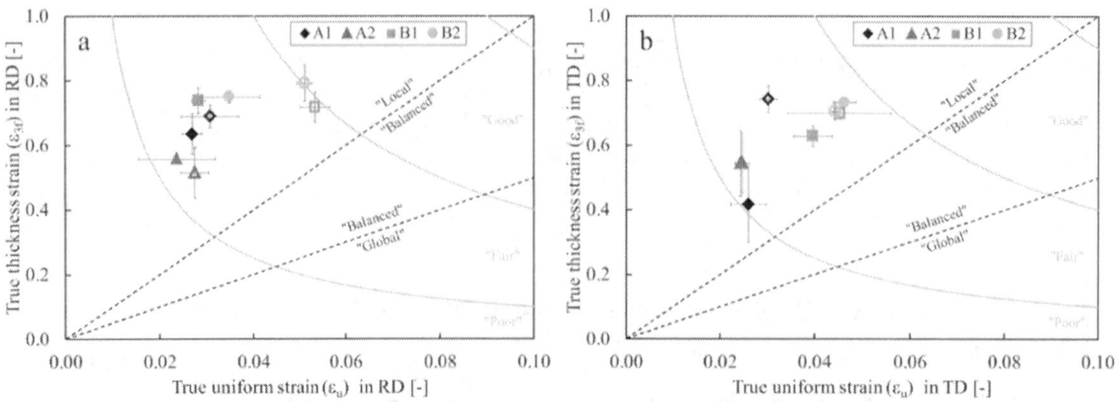

Figure 4. Formability maps of investigated DQ, DQT (solid marks), RQ and RQT (hollow marks) samples in a) rolling (RD) and b) transverse direction (TD).

Hole expansion tests were also carried out to evaluate the local formability properties of investigated steels and results are presented in Figure 5. Average hole expansion ratio (HER) values were between 24 – 49 %, and as a general conclusion, RQ steels have higher HER values compared to DQ steels. Especially B1 and B2 steels after reheat, quenching and tempering treatment achieved average HER values of 45 – 50 %. Interesting results is that tempering did not improve the HER values of DQ steels, whereas in RQ steels, tempering did improve the hole expansion. HER values indicate that, although the martensitic matrix softens during tempering, it is not the determinate component to affect the hole expansion capability especially in DQ conditions. Comparing DQT- and RQT results, it can be concluded that DQ processing affect the microstructure in certain way that will be more dominating factor affecting the HER than softening during tempering. It is possible that higher $d_{90\%}$ grain size is the reason for poorer HER of DQ variants, which cannot be affected during subsequent tempering treatment. As mentioned earlier, finer and more uniform grain size can improve the HER, as it improves the fracture toughness, which have shown to correlate with improved HER [9].

Figure 5 also presents the examples of expanded holes of A1 and B1 steels. Multiple cracks appeared simultaneously in several directions, which indicates that crack propagation is not dependent on the rolling direction, or that crack propagation is not different between DQ- and RQ variants, as DQ steels have stronger rolling texture.

Figure 6 presents the correlation between different mechanical properties and HER, and it shows that there is now correlation between these values. Some earlier research have shown that especially tensile strength and HER would have correlation [13,14]. Chen et al. showed that after 780 MPa tensile strength, HER values were constantly between 30 – 40 % [13]. However, it must be noted that many of these studies have focused on multiphase steels, where the fractions of martensite and softer ferrite are essential. Similar research has not been conducted with single phase martensitic structures. Despite different microstructures, our research has similar results indicating that with high amount of martensite and tensile strength around 1000 MPa, HER values are around 30 – 50 % in all cases. It is possible that high dislocation density and poor local formability of martensite has its limits on stretch-flangeability properties. Also some research have indicated that $R_{p0.2}$ / R_m ratio closer to value 1 will improve the hole expansion, as strain hardening would be less in these steels [15]. In our studies, all tempered steels had $R_{p0.2}$ / R_m almost 1, however steels in DQT conditions did not have any improved HER values, which does not support the correlation between $R_{p0.2}$ / R_m and HER.

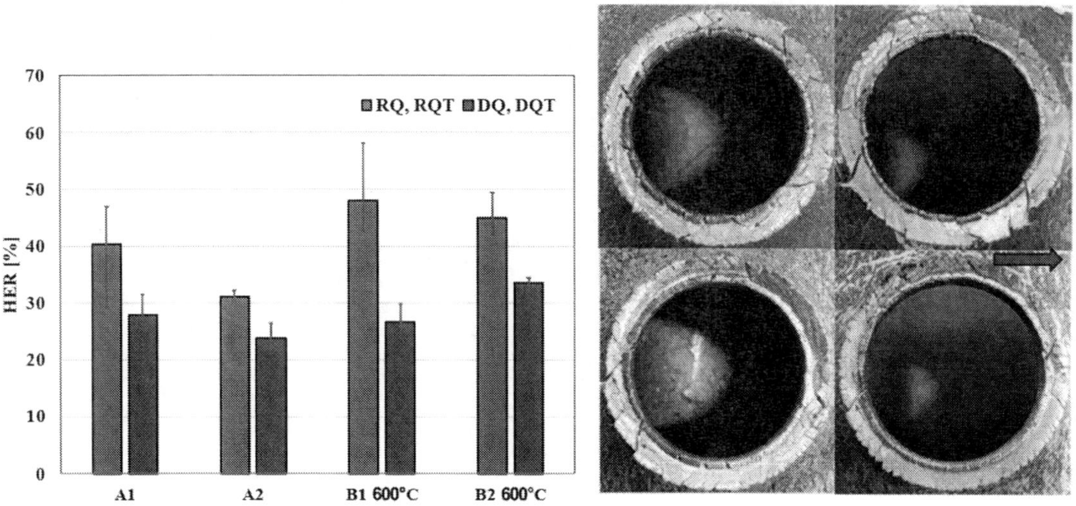

Figure 5. Left: Hole expansion ratio (HER) values of A1 – B2, four tests each, right: expanded holes, a) A1 DQ, b) A1 RQ, c) B1 DQT, d) B1 RQT.

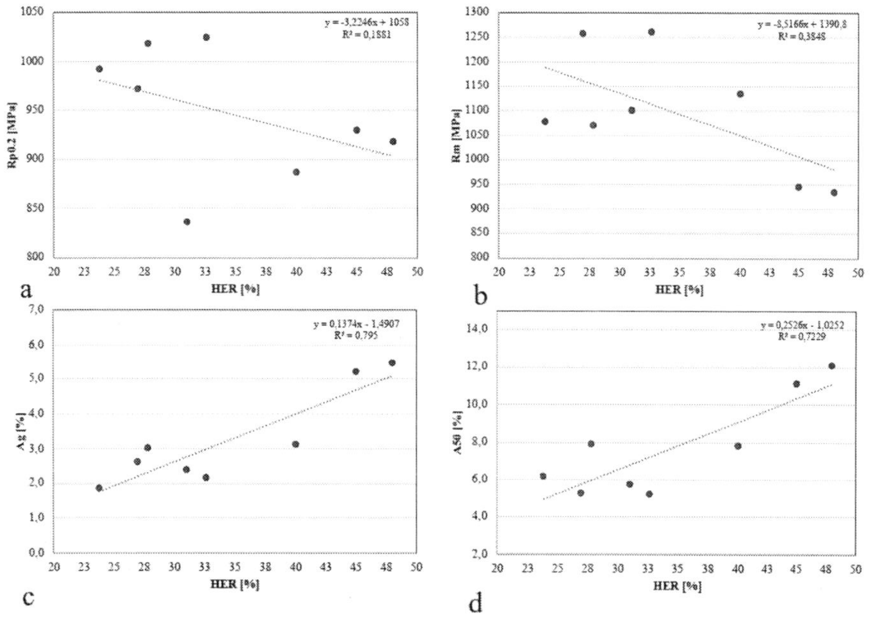

Figure 6. Correlation between HER and a) yield strength ($R_{p0.2}$), b) tensile strength (R_m), c) uniform elongation (A_g), d) total elongation (A_{50}) in longitudinal direction.

Figure 7 shows the correlation of HER and true thickness strain at fracture values. Based on the R^2 - values, only moderate correlation can be found especially comparing HER and true thickness strain in TD direction. Overall, the correlation is weak and true thickness strain does not explain the differences between HER values at least in our investigation. For example, Larour et al. have found good correlation between HER and true fracture and thickness strain measurements for a wide selection of 600 – 1200 MPa cold rolled steel sheets [16]. However, Wagner et al. showed that the specimen geometry has a large role on determining the true thickness- or true fracture strain [17]. They claimed that the width to thickness ratio (w/t) of tensile specimens should be large enough to produce shear fracture across the specimen width and thus produce consistent fracture shape regarding all tested specimens. In our study w/t ratio of tensile specimens was only 4 (12 mm / 3 mm), which can be too low for consistent true thickness fracture results based on Wagner et al. investigation. Considering other tensile results, Larour et al. showed that other values from tensile tests does not correlate with HER, as seen also in our studies [16]. However, the best HER values were achieved with steels (B steels in RQT condition) that had best overall formability based on formability maps, which indicates that formability map data can produce beneficial information regarding the HER.

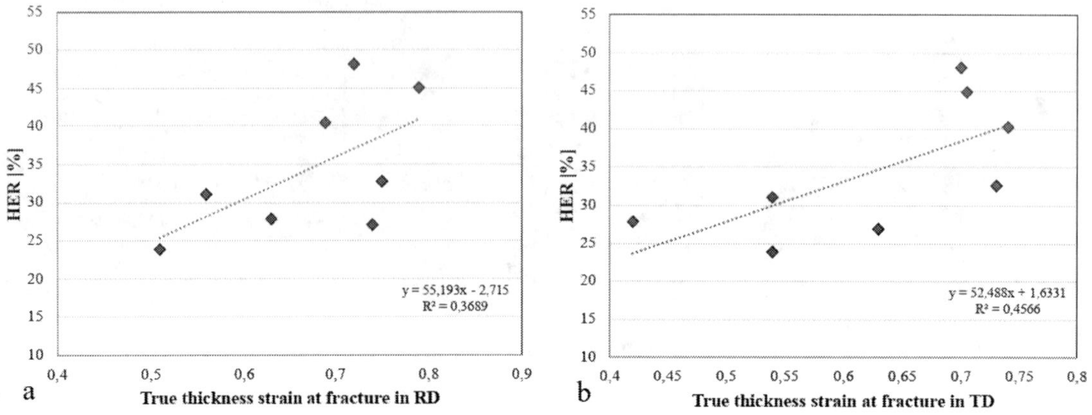

Figure 7. Correlation between HER and true thickness strain at fracture of A1-B2 steels in different conditions in a) rolling (RD) and b) transverse direction (TD).

4. Summary and conclusions

Investigated steels were hot rolled to ~3 mm thickness in several conditions including direct quenching (DQ), direct quenching and tempering (DQT) as well as reheat and quenching variants (RQ, RQT). Tensile strength values were generally above 1000 MPa, however B2 steel with 600 °C tempering had less than 1000 MPa. Elongation values were relatively low, and even after direct quenching and tempering, total elongation values were less than 10 % in DQ- and DQT steels. RQ produced lower strength values than DQ variants, although improvement in elongation was discovered after tempering at 600 °C (A_{50} ~ 11 – 12 %).

Hole expansion ratios (HER) for A1 - B2 in DQ – and DQT conditions were between 22 – 36 %, and no clear improvement was seen after tempering. Hole expansion ratios (HER) for RQ- and RQT variants were between 31 – 49 %, and the highest values were achieved with RQT samples (tempering at 600 °C). Based on the EBSD analysis of martensite grain size, more uniform grain structure was discovered in RQ steels, which could be the reason for improved HER properties. HER values were compared to tensile test results (uniform elongation, true thickness strain) and formability maps were constructed. However, only poor or moderate correlation were found between HER and true thickness strain values, although best HER values were achieved with steels that had best overall formability properties based on formability maps. Overall, it can be concluded that for martensitic steels, it was difficult to find correlation between HER and measured parameters in order to further improve the hole expansion of martensitic steels. It is possible that high dislocation density and poor local formability of martensite will be the limitation factors. However, it can be noted that hole expansion ratio of ~50 % for 1000 MPa steel is relatively good result, and the required properties, such as sufficient hole expansion ratio, are related to possible applications for this type of steel.

Acknowledgements

The financial support of CBMM (Companhia Brasileira de Metalurgia e Mineracao) and FOSSA (Fossil free steel applications) is gratefully acknowledged.

References

[1] Taylor M D, Choi K S, Sun X, Matlock D K, Packard C E, Xu L and Barlat F 2014 Correlations between nanoindentation hardness and macroscopic mechanical properties in DP980 steels *Mater. Sci. Eng. A* **597** 431–9

[2] Terrazas O R, Findley K O and Van Tyne C J 2017 Influence of Martensite Morphology on Sheared-Edge Formability of Dual-Phase Steels *ISIJ Int.* **57** 937–44

[3] Hasegawa K, Kawamura K, Urabe T and Hosoya Y 2004 Effects of Microstructure on Stretch-flange-formability of 980 MPa Grade Cold-rolled Ultra High Strength Steel Sheets *ISIJ Int.* **44** 603–9

[4] Madrid M, Van Tyne C J, Sadagopan S, Pavlina E J, Hu J and Clarke K D 2018 Effects of Testing Method on Stretch-Flangeability of Dual-Phase 980/1180 Steel Grades *JOM* **70** 918–23

[5] Kim J H, Seo E J, Kwon M-H, Kang S and De Cooman B C 2018 Effect of quenching temperature on stretch flangeability of a medium Mn steel processed by quenching and partitioning *Mater. Sci. Eng. A* **729** 276–84

[6] Im Y-R, Kim E-Y, Song T, Lee J S and Suh D-W 2021 Tensile Properties and Stretch-Flangeability of TRIP Steels Produced by Quenching and Partitioning (Q&P) Process with Different Fractions of Constituent Phases *ISIJ Int.* **61** 572–81

[7] Anon 2016 Metallic materials - Tensile testing - Part 1: Method of test at room temperature *ISO 6892-12016*

[8] Anon 2017 Metallic Materials - Sheet and Strip - Hole Expanding Test *ISO 166302017*

[9] Yoon J I, Lee H H, Jung J and Kim H S 2018 Effect of grain size on stretch-flangeability of twinning-induced plasticity steels *Mater. Sci. Eng. A* **735** 295–301

[10] Sun X, Li Z, Yong Q, Yang Z, Dong H and Weng Y 2012 Third generation high strength low alloy steels with improved toughness *Sci. China Technol. Sci.* **55** 1797–805

[11] Zajac S, Schwinn V and Tacke K H 2005 Characterisation and Quantification of Complex Bainitic Microstructures in High and Ultra-High Strength Linepipe Steels *Mater. Sci. Forum* **500–501** 387–94

[12] Hance B 2018 Advanced High Strength Steel (AHSS) Performance Levels *SAE Technical Papers*

[13] Chen X, Jiang H, Cui Z, Lian C and Lu C 2014 Hole Expansion Characteristics of Ultra High Strength Steels *Procedia Eng.* **81** 718–23

[14] Fang X, Fan Z, Ralph B, Evans P and Underhill R 2003 The relationships between tensile properties and hole expansion property of C-Mn steels *J. Mater. Sci.* **38** 3877–82

[15] Kumar Akela A, Satish Kumar D and Balachandran G 2022 Hole Expansion Test and Characterization of High-Strength Hot-Rolled Steel Strip *Trans. Indian Inst. Met.* **75** 625–33

[16] Larour P, Freudenthaler J, Kerschbaum M and Dolzer D 2020 Edge crack sensitivity versus tensile local ductility of AHSS sheet steels *IOP Conf. Ser. Mater. Sci. Eng.* **967** 012080

[17] Wagner L and Larour P 2018 Influence of specimen geometry on measures of local fracture strain obtained from uniaxial tensile tests of AHSS sheets *IOP Conf. Ser. Mater. Sci. Eng.* **418** 012074

Material variability effects on automotive part production process

A R Chezan[1], E H Atzema[1,2], J Pilthammar[3,4] and M Sigvant[3,4]

[1]Tata Steel R&D, P.O. Box 10.000, 1970 CA IJmuiden, The Netherlands
[2]Faculty of Engineering, University of Twente, P.O. Box 217, 7500 AE Enschede, the Netherlands.
[3]Volvo Cars, Dept 81110 Strategy Development, 293 80 Olofström, Sweden
[4]Department of Mechanical Engineering, Blekinge Institute of Technology, Valhallavägen, 371 41 Karlskrona, Sweden

toni.chezan@tatasteeleurope.com

Abstract. The current efforts to reduce the carbon footprint throughout the chain in the automotive industry by increased use of recycled materials poses new challenges for materials production and their use. The increase of steel scrap fraction in the current primary steel making processes, used for producing steel sheet metal for automotive components, possibly affects the material properties variability beyond the limits observed in the materials produced today despite mitigating actions in steel production. In this paper material variability increase was modelled by selecting deterministic values outside the range of the material grade used to design and manufacture an automotive part. The values were selected from an experimental data set representing the cold rolled mild steels material class range. The effects were studied numerically on a reverse engineered model of an existing automotive part production process. It was found that the manufacturing feasibility in this particular case is mainly affected by the weighted average plastic strain ratio and less by the degree of planar anisotropy.

1. Introduction

The decarbonization of steel sheet metal production process introduces new challenges for the automotive part production process. Especially, the recyclability component, the effort to increase the steel scrap in the primary steel making process produces increased variability intervals for certain residual elements outside the levels currently found in the chemistry of iron ore-based liquid steel making. It is expected that increased chemistry variability will increase the mechanical properties variability of the sheet metal products like plastic work hardening, anisotropy and total elongation, although extra efforts in the steel production will mitigate these effects. Consequently, the need of adding a material variability evaluation [1] as a feasibility step in the automotive part process development becomes more and more relevant.

Currently, the development of industrial forming processes in practice is based on a unique set of representative material properties of the selected material grade and the forming process robustness analysis is done by evaluation of variations of forming process parameters. Incorporation of material scatter as part of feasibility evaluation of the forming process robustness is a topic of scientific research. There are not yet established methodologies on how to perform such an analysis. Identification of the most relevant material parameters and their numerical implementation and automated variation is still scientifically questioned, see e.g. [2]. Further the availability of representative data sets of material properties is limited especially due to the mismatch between the number of parameters included in international norms for product specifications and the parameters used to model the material behaviour of sheet metal materials. The materials included in this study are part of the cold rolled mild steels class with mechanical properties specified in international standards in transverse to rolling direction [3]. These materials have significant processing related anisotropy of the mechanical properties and numerous previous studies have proven that accurate predictions of their behaviour require models incorporating parameters determined at least in two more directions (parallel and diagonal with respect to rolling direction) [4].

In this paper a study of material variability effect on part make-ability of an automotive part component is presented. A cold rolled mild steel material data set is used with the material parameters measured in the three main directions mentioned in the previous paragraph. First, the material parameters variability is discussed. Next, a methodology of defining output parameters of a forming process is introduced. A forming process working window-based approach is used to evaluate and compare quantitatively material and process parameter effects. This method was chosen over the bounding approach on strain distribution presented in [1] to align better with how press shops tend to evaluate robustness which is by establishing a process window.

2. Materials, experimental methods and finite element models

Standard uniaxial tensile testing (ISO 6892-1) was used to determine the material parameters included in this study. Each entry in the data base represents a set of tensile tests performed in the three main directions, parallel, diagonal and transverse to rolling direction, machined from a sheet metal blank of approx. 500 x 500 mm^2. Each sheet metal blank was cut from one cold rolled hot dip galvanized coil. The coils were produced over several years at Tata Steel (IJmuiden steelworks) integrated steel plant. Material card parameters were generated with the procedure describe in reference [4]. The material cards are composed of plastic work hardening curves with strain rate hardening dependence, Vegter 2007 yield locus and Tata Steel forming limit curve models.

The finite element model was reverse engineered from measurements of an existing automotive part manufacturing process. The surfaces of the tools of the finite element model of the door inner automotive panel [5] were based on 3D tool surface digitization using a HandySCAN 700 from Creaform. The trapped blank and draw panels were stamped at the stamping plant of Volvo Cars in Olofström. The model incorporates the tooling kinematics of the used mechanical press. The draw in measurements were performed using the same HandySCAN 700. The tribology system was modelled using the TriboForm friction model. All finite element simulations were performed using AutoForm^{plus} R10. Model accuracy was verified using strain measurements similar to the procedure presented in [6].

3. Results

Figure 1 illustrates material parameter requirements for the cold rolled mild steel grades CR1 to CR5 as specified by [3] and experimental distributions for the tensile strength (a) and the plastic strain ratio (r-value) at 20% strain. The parameters are specified and measured in a direction transverse to rolling direction. The distributions are constructed using experimental data collected over several years of material production. The data set contains 1000 points for the CR5 and CR4, 200 points for CR3, 135 points for CR2 and 83 points for CR1. The presented data was modelled in order to allow for direct comparisons of data sets with different population sizes. The data in Fig.1 illustrates that the material variability of the mild steels is significant, but falls well within the specification limits.

Data normalization was used in order to facilitate direct comparisons of the individual material parameters variability. An example of the data normalization procedure for the two material parameters from Fig. 1 and the CR4 material grade is presented in Fig. 2. Data from Fig. 2a and 2b was normalized relative to the norm specification values – a mean tensile strength of 300 MPa respectively the minimum r-value specification of 1.9 (Fig. 2c). Normalizing the data relative to mean values of each parameter from the data set, 292 MPa mean tensile strength and 2.5 mean r-value, is presented in Fig. 2d. The relative variations (total range i.e. min-max) of the tensile strength and r-value are respectively 12% and 45% of the mean. The min-max range is based on isolated outliers presenting worst case scatter and selected with the purpose of enforcing maximum variability.

The variability of tensile test data parameters required for determination of the material card parameters [4] is presented in Fig. 3. Determination of the parameters of the Vegter 2017 planar anisotropic yield locus requires the material tensile strength (Rm) the uniform tensile elongation (Ag) and r-value (r) in the three main directions relative to the rolling direction. Subscripts 00, 45 and 90 indicate respectively the parallel, diagonal and transverse orientations of the tensile test specimens relative to the material rolling direction. The Forming Limit Curve (FLC) model requires the total elongation (A80) and the r-value. Fig. 3 illustrates that the relative variation of the r-values is significantly larger as compared to the relative variations of the other parameters. This corresponds to earlier work [1] where scatter in anisotropy was shown to have a larger effect on strains than scatter in hardening.

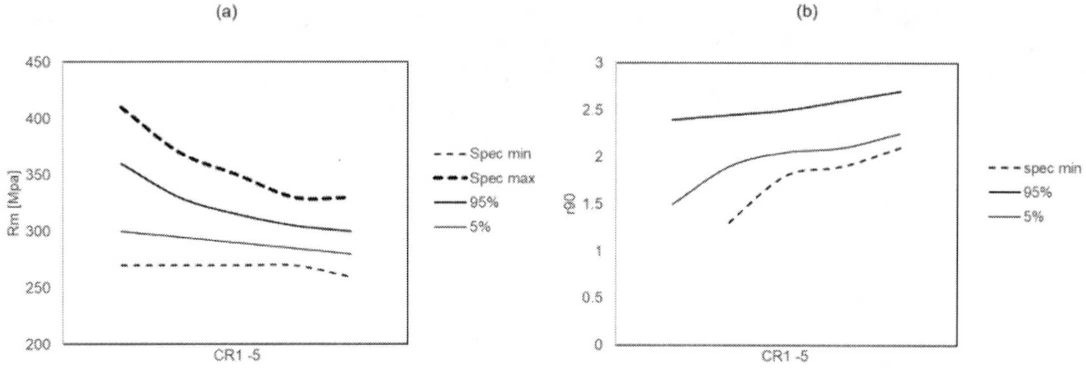

Figure 1. Graphical representation of material parameters specification limits – min and max for tensile strength R_m (a) and, respectively, min for plastic strain ratio $r_{90/20}$ (b) according to VDA-239-100 and stylised examples of measured parameters distributions represented as distributions with limits at 5 and 95% respectively.

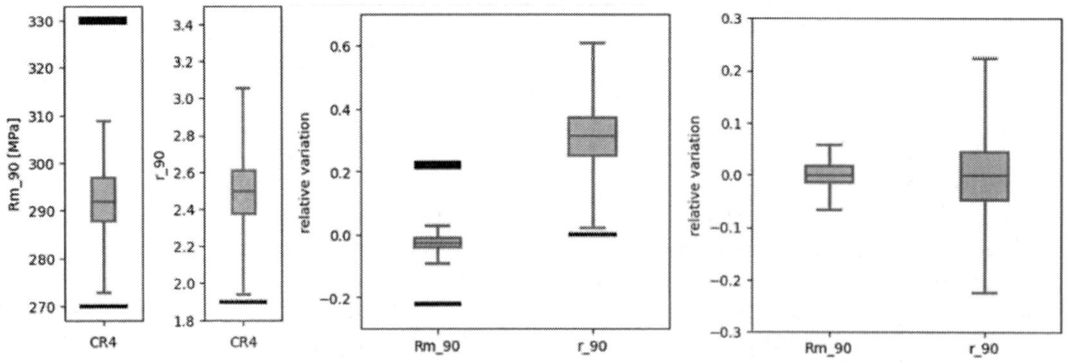

Figure 2. CR4 transverse direction tensile strength (a) and r-value (b) variations on absolute scales, (c) the variations relative to the material specification values and (d) relative to mean experimental values.

The determination of the output parameter of the FE model is described in Fig. 4. Local draw in was calculated as the local distance between the trapped blank edge and formed part edge (blue and red lines respectively in Fig. 4a) in a direction normal to the trapped blank edge. The corresponding draw in distribution along the trapped blank contour is presented in Fig. 4b. The draw in is not uniform. The local variations are due to the combined effect of blank and part geometries, material flow restraining and material plastic deformation.

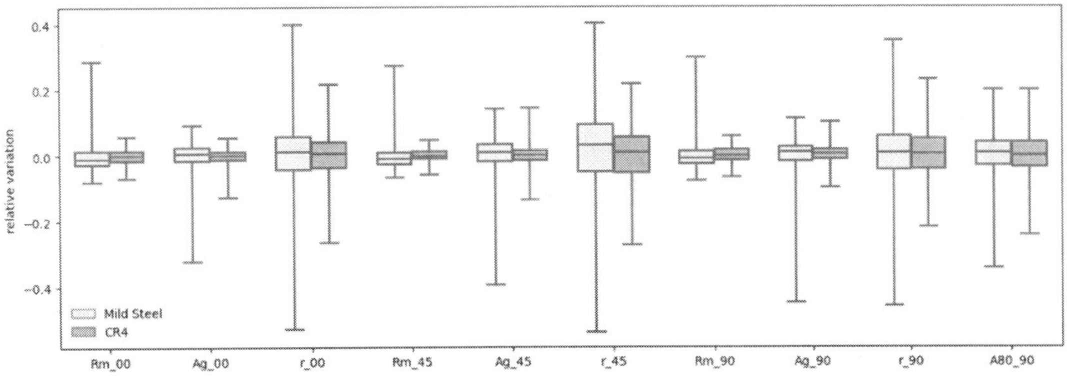

Figure 3. Mild steel vs. CR4 relative variations of the experimental material parameters values used as input for the Tata Steel models for material card for FEM.

As forming process output parameter an interval along the trapped blank contour (black line in Fig. 4a) between 1000 and 2500 mm in Fig 4b was used. The selected region concentrates on the experimental and modelled response to experimental and model parameters variations in the upper right area of the part with the orientation as in Fig. 4a. The mean draw in was calculated for this interval for various experimental and numerical conditions. In Fig. 5a is presented the experimental and predicted mean draw in variations as a function of the applied blank holder force. The two sets of experimental points labelled as Exp 1 respectively Exp 2 in Fig. 5a. were calculated from panels from two series production runs. There was a 4 year time interval between the two sets of data. The material batches were different. The material parameters used for the model represent mean CR4 values. A good agreement between measured and predicted draw in can be observed in a blank holder force interval between 1200 and 2000 kN. Linear regression (dashed lines) was used to estimate the experimental and modelled mean draw in variation sensitivity to global blank holder force variation. The two experiments (Exp 1 and 2) have 10 and respectively 7 mm / 1000 kN decrease rate of the draw in with increasing blank holder force. The FE model predicts a higher blank holder force effect of 13 mm / 1000 kN decrease.

A material parameter variability study was further performed using the model settings from Fig. 5a. A qualitative working window is suggested as in Fig. 5b. Too low blank holder force results in high draw in. The part geometrical quality deteriorates below a certain value of the applied blank holder force. The material experiences insufficient tension during part drawing and consequently the material is locally compressed and insufficiently stretched resulting in buckling and geometrical deviations (Fig. 5c). Too high blank holder force prevents normal material flow into the die cavity resulting in splitting (Fig. 5d). Situations with both increased risks of buckles and splits might occur. It is expected that parts without defects are realized for the combination of blank holder force and draw in within the polygon represented with the green continuous line in Fig. 5b. Obviously the draw in is not an input parameter like blank holder force. Furthermore, it is (partly) governed by the blank holder force, but it does represent a process average of strain level and distribution for mild steel.

Material parameters effects on draw in and the corresponding blank holder force effect were studied for four sets of r-values. They are represented as points in Fig. 6 together with several domains of the weighted average plastic strain ratio (r_m) [7] and degree of planar anisotropy parameter (Δr) [7] pair plot of r-value data from Fig. 3. The CR4 point is in the centre of the CR4 domain and represents mean r-values of this material grade. The maximum Δr is a point at the border of the mild steel material domain, outside the CR4 domain. The r_{m1} represents the mean values of the CR1 domain and the r_{m2} represents the set of r-values corresponding to the minimum weighted average plastic strain ratio r_m of the mild steel. The draw in of the max Δr variant is practically equal to that of the mean CR4 grade at the same blank holder force settings. At 1800 kN significantly lower draw in as compared to CR4 are predicted for the r_{m1} and r_{m2} cases. The simulations of these two situations predict part splitting. The simulation with a blank holder force of 1500 kN predicts a part without defects for the r_{m1} case. The draw in increases with reduction of the blank holder force with a rate similar to that of CR4. The simulation with a blank holder force of 1200 kN for r_{m2} predicts both splitting and buckles. The rate of decrease of the draw in with increasing blank holder force is 10 mm / 1000 kN, lower as compared to the CR4 case. No blank holder force working window was found for the r_{m2} case.

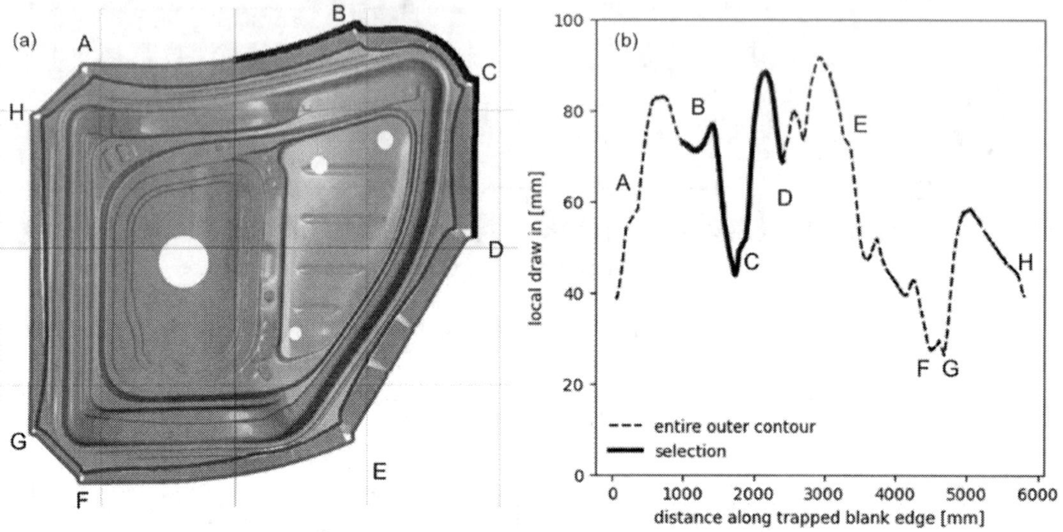

Figure 4. (a) Example of a formed panel and its outer contour (red line), the trapped blank contour (blue line) and the draw in region (light green area). Thick black line represents focus area. (b) The draw in distribution along the trapped blank contour. Capital letters indicate positions of sharp features on the trapped blank edge in (a) and the corresponding positions along the contour length in (b).

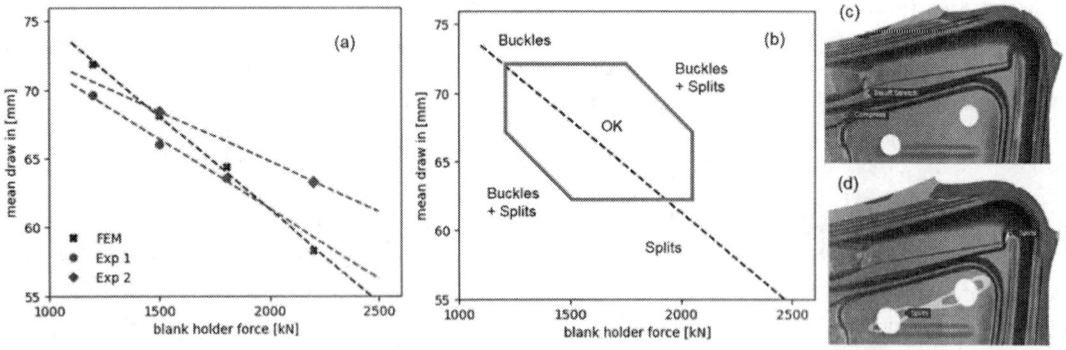

Figure 5. (a) Experimental (Exp 1 and 2) and modelled (FEM) draw in variation as a function of blank holder force. (b) Suggested working window. (c) Example of buckles. (d) Example of splits.

4. Discussion

The data presented in Fig. 1 illustrate general trends for produced material variability while comparing the material grades from the mild steel material class individually. The median of the material strength increases from CR5 (the most soft and formable material grade) towards CR1 (a general purpose mild steel). At the same time the measured interval, i.e. the difference between the maximum and minimum values, increases, the strength variability of CR1 is larger as compared to the strength variability of CR5. Similarly to material strength a gradual variation of the median and of the maximum value are visible for the experimental values of the plastic strain ratio (the r-value). The minimum values for the CR4 and CR5 grades are nearly equal to the corresponding grade specifications suggesting that the observed variability is representative while increasing the number of observations. For the CR2 and CR3 the minimum values are significantly larger as compared to the material specification. Whether this observation is representative or biased due to relatively reduced sets of data as compared to much larger data sets for the CR4 and CR5 requires further investigations.

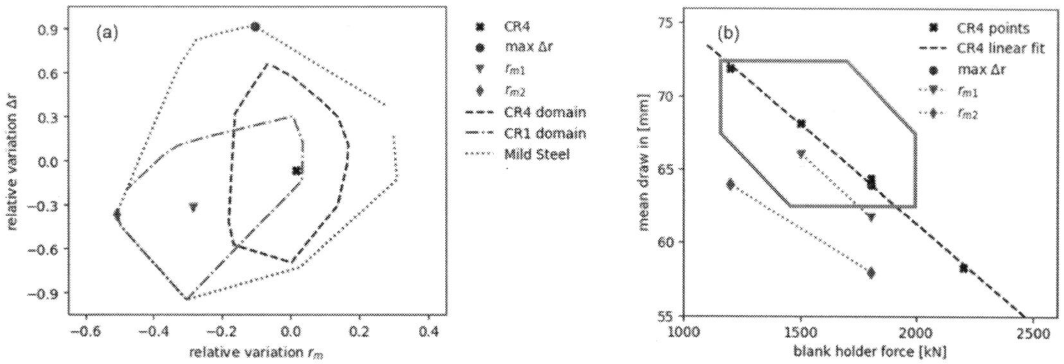

Figure 6. (a) Pair plot of the relative variations of the weighted average plastic strain ratio (r_m) and degree of planar anisotropy parameter (Δr) and the points selected for the analysis. (b) Predicted draw in and blank holder force variation of the draw in for the labelled points from (a).

It is important to realize that the material properties are specified in transverse direction while the material behaviour of mild steels is anisotropic. As a consequence, the anisotropy variability also needs to be considered as suggested by the data from Fig. 6. The relatively large component, the outer edge draw in of the door inner automotive part, appears to be less sensitive to variations of the degree of planar anisotropy parameter (Δr). However, for industrial forming applications where the tolerances of the positions of the edges of circular apertures are important it is expected that Δr plays a significantly more important role as high values of Δr are associated with significant modifications of the shape of round apertures during forming.

The current way of specifying material properties for anisotropic material grades for industrial applications practically allow further increase in produced material variability, both for the parameters in the specified direction (transverse to rolling direction in the case of mild steels) as well as for the properties in other directions. The current philosophy used in producing materials while maintaining properties within specified ranges results in essence in a normal distribution. The focus of material manufacturing process control is on limiting the distribution variance while maintaining the mean of the distribution at a cost-effective level and preventing production of non-conforming material and the associated scrap. Material chemistry modifications associated to increased recycled content will most probably need thermomechanical processing adjustments to keep the parameters specified in the norms within the specified intervals. Further work is ongoing aimed at evaluation of variability of specified material properties and associated anisotropy.

The heuristic approach used in this paper to select extreme values (somewhat akin to [1]) for the input data for the yield locus model variability study can be further extended to a systematic based study. The working window suggested in Fig. 5b could be determined by systematic variations of the 10 parameters in Fig. 3. The number of independent parameters can further be reduced by performing pair correlation studies of the experimental data set while accounting for the significant differences in the size of the populations of the mild steel material grades available data sets. At the same time such future studies need to consider practical aspects related to limitations on current implementation of automated individual and correlated parameter variation available in the finite element software. The data presented in this paper suggests that the door inner model has relative low sensitivity to variations of the degree of planar anisotropy parameter (Δr) while it is highly sensitive to the weighted average plastic strain ratio (r_m). The authors are further investigating whether this is a general or part location or part specific conclusion.

The differences of draw in variation due to modification of the blank holder pressure between experimental data and FE model (Fig. 5a) illustrate that in general the real process has lower process control capability than the virtual FE model. The origin of this difference is primarily the way the blank holder pressure is applied during the forming operation. The FE model used in this study has initially a uniform distribution of blank holder pressure while in the real process this distribution usually deviates from uniformity due to elastic effects associated to tool material properties and geometry and die and press construction [8]. The consequence is that the working window of the real process is expected to be lower as compared to the working window predicted by a FE analysis, the real process has lower capability to correct material variability via a global control. The process control strategy for both the experiment and FE model presented in this paper is based on global control parameter variation, the total blank holder force. However, the real processes have possibilities of local restraining control. The balance blocks used in the real die construction allow local reduction of contact pressure and might be used to solve local splitting issues. At the same time these blocks cause local insensitivity to blank holder force thus reducing global controllability of the process. When adding deformable tools in the FE model this could also be simulated and the simulated trend should be closer to the experimental one. Process control capabilities evaluation including local control is a subject for future studies. When pressure varies strongly over the die surfaces the importance of pressure sensitive friction models increases [6,8,9].

The experimental validation of the material performance and the corresponding working window qualitatively proposed in this study could be realized by monitoring the edge position of the formed part in-line during series production with an appropriate optical measurement system. Once the correlation between the draw in and part quality is experimentally validated the draw in could be used for automated process control of forming processes. Much literature on this is available, e.g. [10][11]. The absolute values of the draw in magnitude (20 – 90 mm range) and draw in variation (~10 mm range) presented in Fig. 4b and Fig. 5a respectively can further be used for defining requirements for the blank and part edge monitoring systems. From a dimension monitoring point of view measuring linear displacements in a range up to 100 mm with 0.1 mm accuracy would provide practically useful data sets for both evaluation of experimental working windows as well as for FE modelling validation purposes.

5. Conclusions

In this paper material variability effects were modelled deterministically by selecting values inside and outside the range of the CR4 material grade used to design and manufacture an automotive part. The values were selected from an experimental data set representing the cold rolled mild steels material class range. It was found that the manufacturing feasibility is mainly affected by the weighted average plastic strain ratio and less by the degree of planar anisotropy. The results illustrate that variability extending outside the range of the material grade used to design and manufacture the component can affect the performance of the automotive part manufacturing process. A qualitative working window accounting for variability is proposed, that could be numerically quantitative determined by performing a systematic material parameter variation study. Defining the strategy for parameter selection and variation requires further investigations aimed at independent parameter reduction using correlation studies while aligning with available implementations in FE packages for systematic material parameters variation.

The observed significantly lower blank holder force control effect on draw in of the experiments as compared to the finite element model response is expected to be representative for forming process as it reflects the difference between ideal and real processes. The results presented in this paper illustrate that the process control capability of real processes is reduced as compared to the virtual FE model and consequently the actual working window is expected to be smaller. Further comparisons of other real processes and their FE models could provide empirical correction factors that could be used during future process activities aimed at developing robust forming processes.

References

[1] Atzema E H, Scholting M E and Abspoel M 2021 Approaches to analysing scatter in forming simulations: from fundamental to pragmatic *IOP Conf. Ser.: Mater. Sci. Eng.* **1157** 012091
[2] Wienbenga JH 2014 Robust design and optimization of forming processes, PhD thesis University of Twente.
[3] VDA 239-100 (05/2016)
[4] Abspoel M, Scholting M E, Lansbergen M and Neelis B M 2017 Accurate anisotropic material modelling using only tensile tests for hot and cold forming *J. Phys.: Conf. Ser.* **896** 012049
[5] Tatipala S, Pilthammar J, Sigvant M, Wall J and Johansson C M Introductory study of sheet metal forming simulations to evaluate process robustness 2018 *IOP Conf. Ser.: Mater. Sci. Eng.* **418** 012111
[6] Sigvant M, Pilthammar J, Hol J, Wiebenga J H, Chezan T, Carleer B and van den Boogard T 2019 Friction in sheet metal forming: influence of surface roughness and strain rate on sheet metal forming simulation results *Procedia Manufacturing* 29 p512–519
[7] ISO 10113:2020
[8] Pilthammar J, Wall J and Sigvant M 2017 Framework for Simulation-Driven Design of Stamping Dies Considering Elastic Die and Press Deformations, *PROCEEDINGS OF ESAFORM 2017 / [ed] Brabazon D.,Ul Ahad I.,Naher S., American Institute of Physics (AIP)* Vol. 1896
[9] Sigvant M, Pilthammar J, Hol J, Wiebenga J H, Chezan T, Carleer B and van den Boogard A H 2018 Friction in Sheet Metal Forming Simulations: Modelling of New Sheet Metal Coatings and Lubricants *IOP Conf. Ser.: Mater. Sci. Eng.* **418** 012093
[10] "Challenges and application fields in the digital process control of sheet metal forming processes"; keynote Prof. Dr. Pavel Hora, in: MS 09: Robust process design and adaptive control; 40th IDDRG conference, 21-6 / 2-7-2021, University of Stuttgart, online.
[11] Kraft M, Liewald M 2015 Voraussetzungen und Randbedingungen zur Onlineüberwachung des Platineneinlaufs von Karosseriekomponenten für Regeleingriffe in den Tiefziehprozess

Research on toolpath optimization of robot-assisted incremental flanging

X Cheng[1], Y X Zhao[1*], Z Q Yu[1], and W Chen[1]

[1] Shanghai Key Laboratory of Digital Manufacture for Thin-Walled Structures, Shanghai Jiao Tong University, Shanghai, 200240, China

E-mail: yxzhao@sjtu.edu.cn

Abstract. Robot-assisted incremental flanging (RAIF), a cost-effective and promising forming technology for producing large-size open-edge flanged parts in multi-variety and small batches, is expected to be widely used in aerospace, automotive and other technological industries. As a multi-pass incremental forming technique, the forming quality of the flanged parts is dependent on the movement of the roller controlled by the robot directly. The toolpath of a certain pass is determined by the rotation velocity (to adjust the flanging angle), the revolution velocity (to complete the flanging process of the whole blank sheet) and the matching relationship between two velocities (synchronously or asynchronously). In this paper, the process principle of RAIF technology is illustrated by using the example of shrink flanging firstly, then the effects of asynchronous motion and synchronous motion on the forming quality of a 60° flanged part made of the 5A06 aluminum alloy are compared while the rotation and revolution velocity of the roller are set to 1°/s and 1.8°/s, respectively. Results show that the forming quality is improved significantly that the maximum height of edge waves is decreased from 2.1 mm to 0.19 mm when the roller is set to rotate and revolve synchronously. Based on the above comparison results, the rotation velocity is set as 1°/s, 1.33°/s, 2°/s and 4°/s respectively to further investigate the effect of different toolpaths on the forming quality under synchronous motion conditions. It can be found that the maximum stress gradually increases from 575.4 MPa to 621.6 MPa when the rotation velocity decreases from 4°/s to 1°/s.

1. Introduction

Flanging has extensively applications in industry to provide stiffness and rigidity or make the features on the edge to further assembly with other sheet metal parts. Lots of investigations have been conducted to various flanging process. Xu et al. [1] investigated the influence of several geometrical parameters on the formability during stretch curved flanging process. Chen et al. [2] combined rubber forming with straight flanging and analyzed the springback by rubber straight flanging with different die radii and blank thickness. Su et al. [3] revealed the influence mechanism and the inhomogeneous deformation behaviors of oblique hole-flanging parts during the electromagnetic flanging process. To obtain the main principles and methods of die design, Li et al. [4] analyzed the characteristics of the deformation, the stress properties, the stress distribution and various factors in shrink flanging. Asnafi [5] studied the stress state/deformation mode in fluid shrink flanging as well as the fracture limit in fluid stretch flanging.

The mentioned sheet metal flanging processes, such as traditional stamping, electromagnetic hole flanging and rubber flanging, are formed with the specific punch and dies for mass production. To meet the requirements of small batch flanged parts, incremental flanging (IF) has been studied considerably in view of its advantages of low cost and flexibility for making various sheet metal parts in recent years. Incremental flanging process was first proposed by Cui and Gao [6] in 2010 to optimize the formability of a hole flange. López-Fernández et al. [7] presented a systematic analysis of the deformation and wrinkling mechanisms of incremental shrink flanging and stablished the practical process window for different flange radii, spindle speed and tool diameter. Makwana et al. [8] studied the single-stage single point incremental square hole flanging of AA5052 sheet material and obtained a process-specific fracture forming limit line.

The toolpath of incremental flanging has an important influence on the geometrical accuracy, surface roughness and formability of the formed parts. Leem et al. [9] proposed a new toolpath strategy called regional plastic incremental bending (RPIB) to manufacture near-net shape corrugated structures using rectangular sheets by double-sided incremental forming (DSIF) process. Chang et al. developed a novel point-contact tool path (PCTP) method free of circumferential friction and a quadratic spiral tool path (QSTP) generation algorithm to improve surface quality and geometric accuracy, reduce forming force and restrain thickness thinning. However, when the slender sheet is used to form the open-edge flanged part with high flanging height, the defects such as springback and warpage are obvious in the forming process, which is difficult to meet the requirements of dimensional accuracy. Therefore, combining the advantages of industrial robots and incremental flanging, robot-assisted incremental flanging (RAIF) is proposed. However, as a novel incremental flanging process, the toolpath optimization of RAIF is rarely investigated.

The present work mainly focuses on the toolpath optimization of the large-size open-edge flanged parts in the RAIF process.

2. Scheme of RAIF technology

The RAIF forming system is mainly composed of a servo turntable, a backing plate, a blank holder, a roller, and an industrial robot. The forming principle of RAIF technology is shown in Figure 1. During the RAIF process, the roller rotates gradually to adjust the flanging angle until it completely contacts the flanging section of the blank sheet, and then the roller revolves around the workpiece to form the flange on the entire flanging area of the sheet metal. The rotation and revolution of the roller are controlled by a industrial robot. Firstly, the forming toolpath is designed according to the characteristics of the target flanged part, and then the trajectory is programmed using the robot language. The programmed toolpath is defined by the translation and rotation of the global Cartesian coordinate system. The system controller can control the motion of the robotic arm in six degrees of freedom to drive the wheel for flanging forming. The flow chart of the entire flanging forming process is shown in Figure 2.

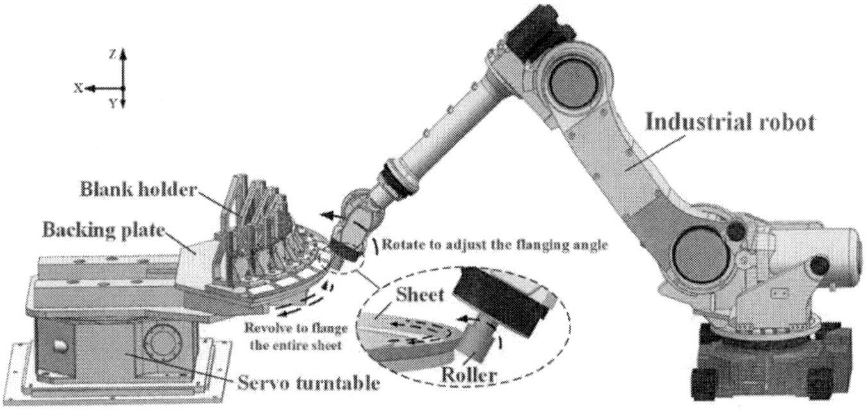

Figure 1. Schematic of RAIF technology.

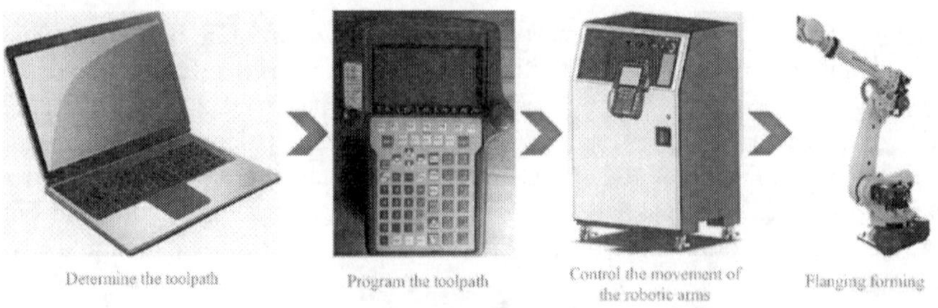

Figure 2. The flowchart of RAIF process.

3. Simulation analysis of RAIF process

3.1. Mechanical properties of the blank sheet

In order to optimize the toolpath of RAIF process, a scaled flanged part is designed and the dimension of the corresponding blank sheet is calculated, as shown in Figure 3. The geometry of the blank sheet is a truncated fan with thickness of 2 mm and the corresponding center angle (θ) is 18°, while the inner radius (R_i) and outer radius (R_o) of the truncated fan are 1003 mm and 1139 mm, respectively. The blank sheet is gradually deformed during the RAIF process, and the flanging angle (α) and the flanging height (H) of the target flanged part is 60° and 30 mm, respectively. Moreover, the inner transition radius (R_t) is set as 3 mm in this paper, which is equal to the radius of the die. The 5 key points are selected uniformly counterclockwise along the circumferential direction of the blank sheet to systematically study the deformation characteristics of the whole RAIF process. The five feature points are labeled as A, B, C, D, E, as shown in Figure 3(b).

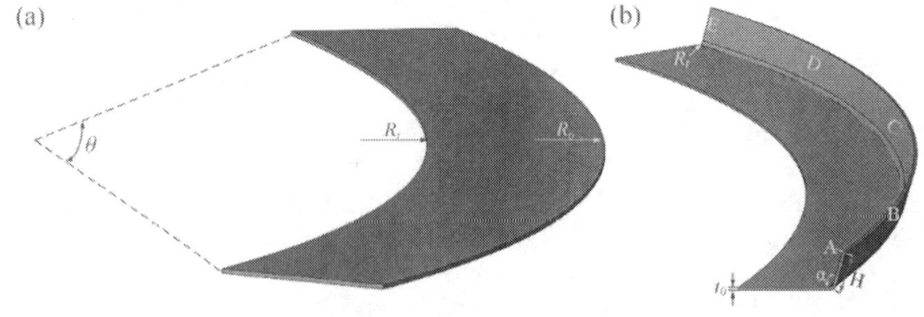

Figure 3. Ideal dimensional parameters: (a) blank sheet; (b) the target scaled flanged part.

5A06-O aluminium alloy sheet is applied as the blank material in this paper, and its chemical composition is introduced in Table 1. Mechanical properties of 5A06-O aluminium alloy sheet obtained from a series of the standard tensile tests according to ASTM E8/E8M in terms of engineering stress-strain curve is shown in Figure 4. Moreover, true stress-strain curve depicted in Figure 5 is fitted using the Hollomon constitutive equation ($\sigma = K\varepsilon^n$). The mechanical properties of 5A06-O sheet are listed in Table 2.

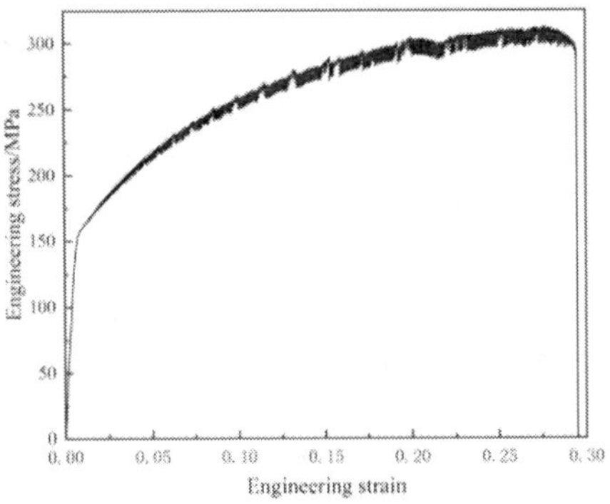

Figure 4. Engineering stress-strain curve of the 5A06-O.

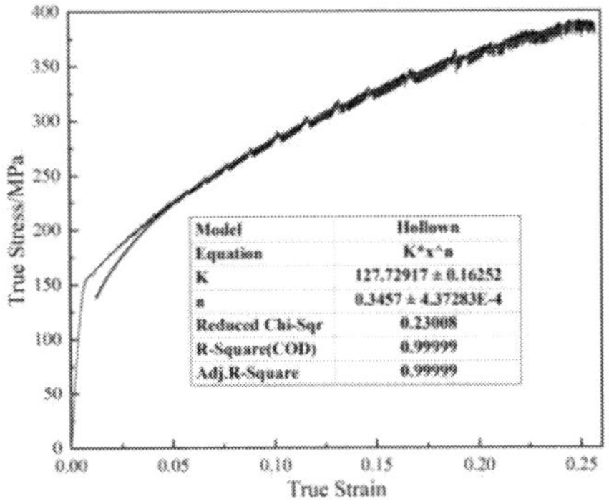

Figure 5. True stress-strain curve of the 5A06-O.

Table 1. Chemical composition of 5A06 aluminum alloy (wt.%).

Al	Mg	Si	Mn	Zn	Cu	Fe
BaL	5.8-8.8	0.4	0.5-0.8	0.2	0.1	0.4

Table 2. Mechanical properties of 5A06-O sheet.

Mechanical Property	Magnitude	Unit
Young's modulus E	70	GPa
Poisson ratio v	0.33	nil
Density ρ	2.64	g/cm^3
Elongation ratio δ	27.36	%
Yield stress σ_s (MPa)	158.8	MPa
Strength coefficient k	127.73	MPa
Hardening exponent n	0.35	nil

3.2. Toolpath generation

As described in Figure 2, the toolpath of RAIF process is determined by the rotation velocity (V_{rt}), the revolution velocity (V_{rv}) and the matching relationship between two velocities (synchronously or asynchronously). In this paper, two different toolpaths were generated to studied the effects of asynchronous motion and synchronous motion on the forming quality while V_{rt} and V_{rv} of the forming roller are set to 1°/s and 1.8°/s, respectively, as shown in Figure 6.

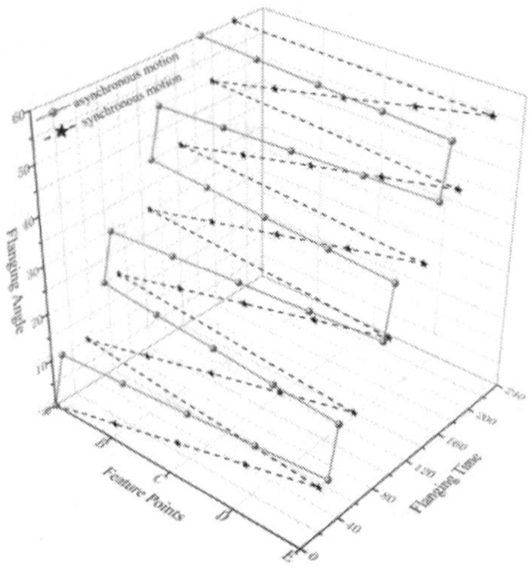

Figure 6. Toolpaths of RAIF process.

3.3. *Numerical modeling*

To investigate the effects of different toolpath on the quality of the flanged parts, the ABAQUS/Implicit solver was employed to conduct FE simulation of RAIF process, which involves complex nonlinear mechanics. The analysis step was adjusted to static implicit, and the contact among the blank sheet and dies were set to general contact. The classical Coulomb friction model was used to describe the tangential friction between the workpiece and die, and the roller. The friction coefficient was identified as 0.2.

The streamlined FE model of the entire RAIF forming system is composed of the blank sheet, die, and the roller, which are simplified based on the RAIF forming principle, as depicted in Figure 7. The blank sheet is defined as a deformable body, and the 8-node linear brick elements (C3D8) are selected. The blank sheet is divided into the flanging section (S1) and the clamping section (S2), which have varying degrees of deformation during the forming process. The S2 is only subjected the clamping force applied by the blank holder during the forming process, enabling the mesh to be coarsened to 3 mm * 4 mm * 0.5 mm (circumferential * radial * thickness direction). Conversely, S1 deforms severely due to the local continuous loading of the roller, so the mesh size is refined to 3 mm * 1.5 mm * 0.5 mm. The die is described as a discrete rigid body and the deformation that occurs during the flaning process is ignored. The mesh of die corresponds to that of the blank sheet. The blank sheet is fixed by the clamping region (D2), which contacts S2 and exerts clamping force, allowing the mesh to be coarsened to 19 mm * 19 mm (circumferential * radial direction). Moreover, the mesh needs to be refined to 19 mm * 0.5 mm since the transition zone (D1) contacts S1. Four-node 3D bilinear rigid quadrilateral discrete rigid elements (R3D4) are selected to describe the die while the roller is regarded as an analytical rigid body.

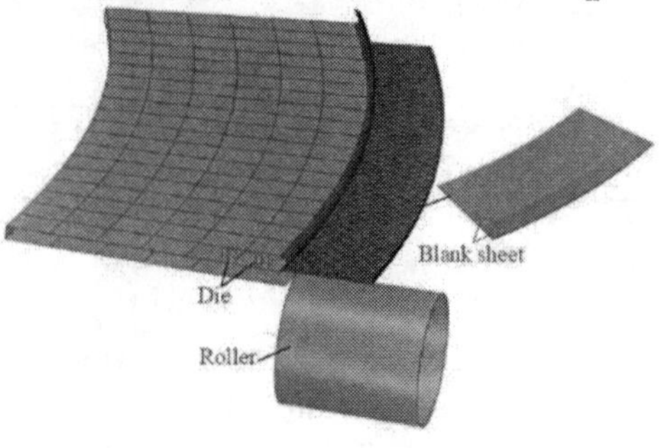

Figure 7. FEM model of shrink flanging by RAIF.

4. Results and discussion

4.1. Comparison of asynchronous motion and synchronous motion

The FE simulation results under different conditions are shown in Figure 8. It can be seen that the maximum Mises stress occurs at the feature point A in both cases. Moreover, when the roller motion is adjusted from asynchronous rotation to synchronous rotation, the Mises stress increases from 406.9 MPa to 621.6 MPa. Compared with asynchronous motion, the forming quality is improved significantly that the maximum height of edge waves is decreased from 2.1 mm to 0.19 mm when the roller is set to rotate and revolve synchronously. This can be attributed to the fact that when the roller moves synchronously, the sheet gradually bends to the target flanging angle, and the stress is continuously rebalanced, which inhibits the generation of edge waves.

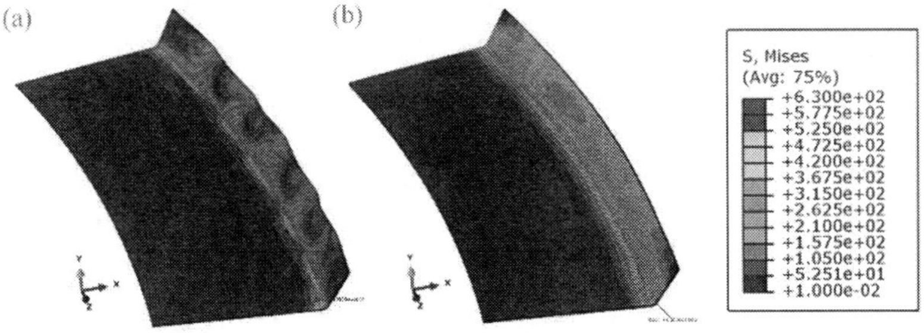

Figure 8. Simulation results: (a) asynchronous motion; (b) synchronous motion.

The springback angle distribution under different conditions is shown in Figure 9. It can be seen that when the roller rotates asynchronously, the springback angle changes greatly due to the obvious edge wave in the flanging section, and the maximum springback angle reaches 7.2°. When the synchronous motion of the roller is adopted, the springback is mainly concentrated at both ends of the sheet metal, and the maximum springback angle is 1.6°.

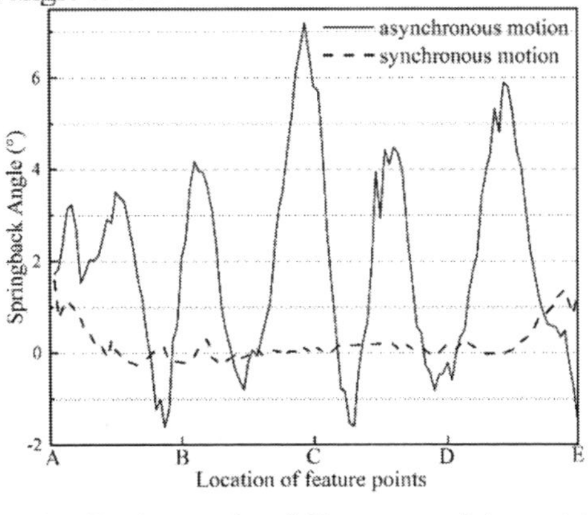

Figure 9. Springback angle distribution under different conditions: (a) asynchronous motion; (b) synchronous motion.

4.2. The effect of rotation velocity on springback angle

To further investigate the effect of different toolpaths on the forming quality under synchronous motion condition, V_{rt} is set as 1°/s, 1.33°/s, 2°/s and 4°/s while V_{rv} is defined as a constant as 1.8°/s. The generated trajectory is shown in Figure 10.

The simulation results with different toolpaths are shown in Figure 11. It can be seen that the RAIF can complete the forming of the target flanged part without defects such as edge waves and wrinkles

IOP Conf. Series: Materials Science and Engineering 1284 (2023) 012038 doi:10.1088/1757-899X/1284/1/012038

under different rotation velocities. It also can be found that the maximum stress gradually increases from 575.4 MPa to 621.6 MPa when the rotation velocity decreases from 4°/s to 1°/s.

The springback angle distribution under different conditions is shown in Figure 12. It can be found that the largest deformation occurs at the free end and then decreases sharply, which can be attributed to the large collision force generated when the roller contacts the sheet. As the rotation velocity increases, the springback angle also increases gradually, which is mainly attributed to the fact that the stress in the flanging area cannot be fully rebalanced as the velocity increases.

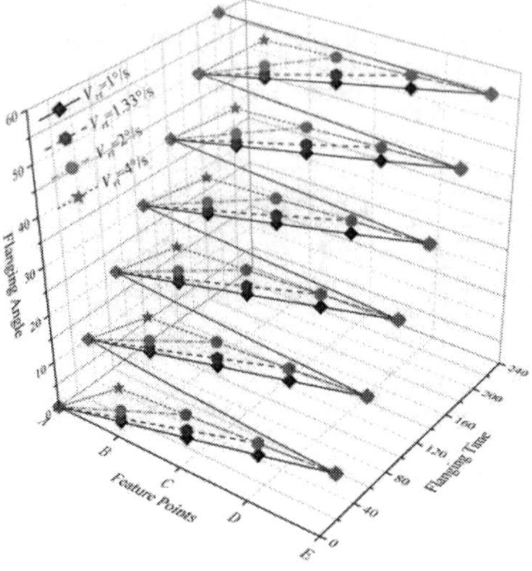

Figure 10. Toolpaths under different rotation velocities.

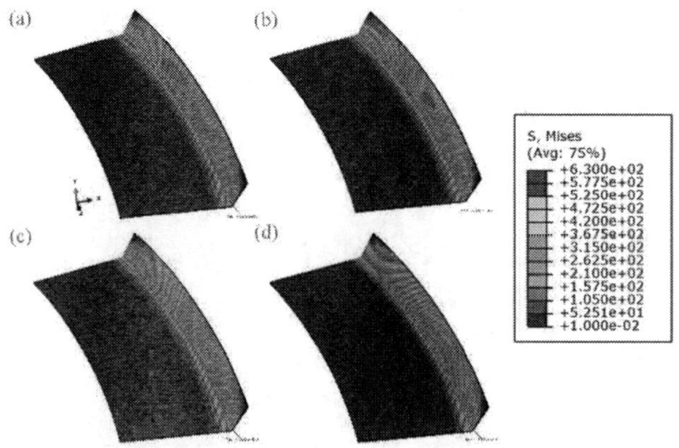

Figure 11. Simulation results under different rotation velocities.

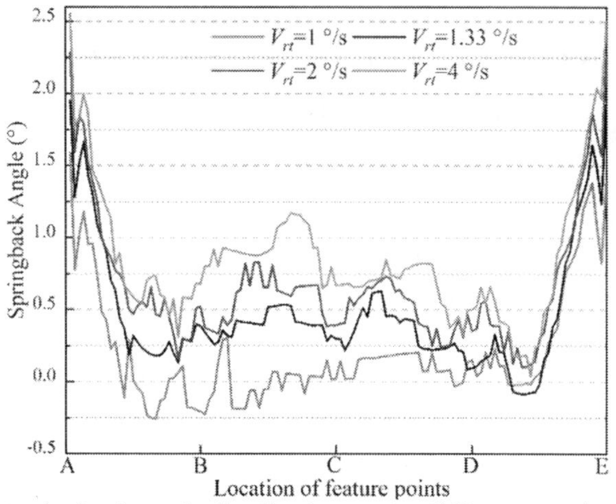

Figure 12. Springback angle distribution under different rotation velocities.

5. Conclusions

In this paper, the toolpath optimization of RAIF process was was investigated. The conclusions are summarized as follows:

Based on the matching relationship between the two velocities, two toolpaths were generated, namely synchronous motion (the roller rotates and revolves synchronously) and asynchronous motion (the roller rotates and revolves synchronously), to study the influence of different toolpaths on the forming quality, while V_{rt} and V_{rv} of the roller are set to 1°/s and 1.8°/s, respectively. Compared with asynchronous motion, the forming quality can be improved significantly when the roller is set to rotate and revolve synchronously due to the stress can be continuously rebalanced to inhibit the generation of edge wave.

As the increase of rotation velocity increases from 1°/s to 4°/s, the maximum stress gradually decreases from 621.6 MPa to 575.4 MPa, and the springback angle increases gradually.

Acknowledgment

The authors much acknowledge the financial support from the National Key Research and Development Program of China (No. 2020YFA0711102). The authors also thank the reviewers of IDDRG for the suggestive comments which improve the technical content of this paper.

References

[1] Xu F, Lin Z Q, Li S H and Xu W L 2004 *J. Mater. Process. Tech.* **145** 93-8.
[2] Chen L, Chen H Q, Wang Q Y and Li Z H 2015 *Mater. Des.* **65** 505-10.
[3] Su H L, Huang L, Li J J, Ma F, Ma H J, Huang P, Zhu H and Feng F 2020 *J. Manuf. Process.* **52** 1-11.
[4] Li C F, Yang Y Y and Li S B 1995 *J. Mater. Process. Tech.* **51** 164-70.
[5] Asnafi N 1999 *J. Mater. Process. Tech.* **96** 198-214.
[6] Cui Z and Gao L 2010 *J. Manuf. Sci. Tech.* **2** 124-8.
[7] López-Fernández J A, Centeno G and Vallellano C 2023 *Int. J. Mech. Sci.* **240** 107930.
[8] Makwana R *et al* 2023 *Mater. Manuf. Process.* **38** 680-91.
[9] Leem D, Liao S H, Bhandari S, Wang Z W, Ehmann K and Cao J 2022 *J. Mater. Process. Tech.* **308** 117727.

IOP Conf. Series: Materials Science and Engineering 1284 (2023) 012039 doi:10.1088/1757-899X/1284/1/012039

Partial resistance tempering of hot-stamped components made of 22MnB5 for subsequent bending

B A Behrens[1], S Hübner[1], S Herbst[2], L Albracht[1], J Wehmeyer[1], S Yarcu[1], P Müller[1] and E Farahmand[1]

[1]Leibniz Universität Hannover, Institute of Forming Technology and Forming Machines, An der Universität 2, 30823 Garbsen, Germany
[2]Leibniz Universität Hannover, Institute of Materials Science, An der Universität 2, 30823 Garbsen, Germany

E-mail: farahmand@ifum.uni-hannover.de

Abstract. The manganese-boron alloy 22MnB5 is particularly used for structural and safety-relevant parts in the automotive industry. Parts made from this alloy are usually produced using the hot forming process. Here, the sheet is heated to over 950 °C using an industrial roller hearth furnace. The heated sheet is then simultaneously formed and quenched in a cooled tool with a temperature gradient of more than 27 K/s. This leads to the formation of a martensitic microstructure with a hardness value of over 450 HV10 and an elongation at break of less than 6%. The small strain potential of such components makes them difficult to form after hot-stamping. Due to the high temperature gradients of resistance heating, a sheet can be heat-treated locally without a large temperature transition zone. This can be used to locally soften already hot-stamped components for subsequent operations such as bending. Within the scope of this paper, resistance heating is used to soften a hot-stamped 22MnB5+AlSi sheet stripe of 3 mm width. The sheet could consequently be bent over an angle of 90° without cracking the substrate.

1. Introduction

In order to produce ultra-high-strength sheet metal components for the automotive body sector, the industrially established process of hot stamping is used. This process combines both forming and heat treatment of a sheet in one process step [1]. In addition to high forming accuracy (no springback), it is possible to produce ultra-high-strength parts for crash-relevant applications in the body structure using this forming process which achieve very high tensile strengths of up to approx. 1500 MPa. At the same time, however, the elongation at break of the material drops from over 20% in the as-delivered condition to around 6% in the shape-hardened condition. Hot stamping can be divided into direct and indirect hot stamping [2]. In direct hot stamping, the coils are uncoiled and then the sheets are subsequently cut to size. The sheets are heated to 950 °C to austenitize the steel material. Afterwards, the heated sheet is inserted into the water-cooled tool and formed. During the contact of the heated sheet with the cold tool, the formed austenite is converted into martensite. This microstructural transformation results in components of highest strength [2]. During hot stamping, cracks occur in the FeAl phases of the coating which extend to deeper diffusion layers (increased Al and Si content in the steel) (Figure 1) [3]. Due to the subsequent cathodic dip coating, these cracks are not critical from a corrosion protection point of view. In indirect hot stamping, cold preforming takes place before the actual hot stamping step. Thus, it is possible to achieve higher strain levels with the aid of a two-stage process [2]. Direct hot stamping of

22MnB5 with an aluminum-silicon coating (AlSi) is the more common variant with a market share of over 76% in EU countries including Turkey [4]. The aluminum-silicon coating primarily serves as scale protection [5].

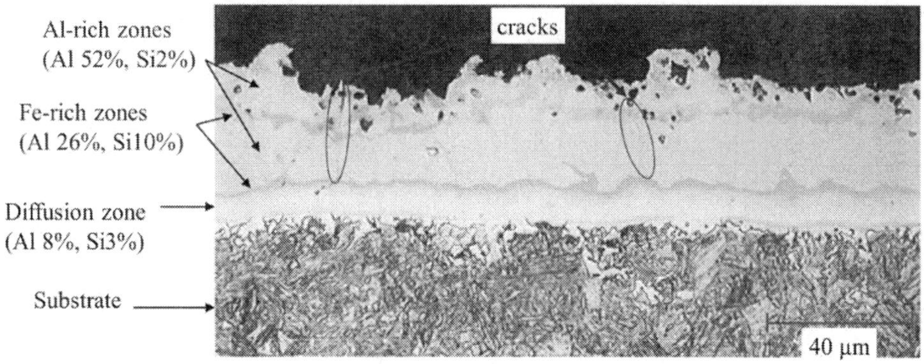

Figure 1: Cracks in the coating of a formed hot-stamped part (re-created after [3], [6])

The manganese-boron alloy 22MnB5 offers a wide range of possible material microstructures by means of which the mechanical properties can be adjusted in a targeted way [7]. During hot stamping of the aluminum-silicon coated sheets, they must be heated slowly at a maximum rate of 12°C/s to allow sufficient time for diffusion between the AlSi coating and the base material (Figure 1). During this process, the AlSi coating melts and forms new intermetallic FeAl phases with the iron of the base material resulting in a higher melting temperature at the surface [8] [9]. Since direct hot stamping is carried out in a single process step, the production of undercuts is technically difficult to implement. Subsequent bending of the sheet is only possible to a limited extent due to the low elongation at break (<6%) and, in particular, in the case of AlSi-coated sheets due to the brittle behavior of the FeAlSi alloy [10]. Subsequent forming of hot-stamped parts is basically difficult to realize due to the low elongation at fracture of the martensitic microstructure. However, the possibility of subsequent forming of such components can allow greater design freedom. For example, bending of hot-stamped components is in some cases necessary to ensure accessibility for subsequent joining operations. If subsequent forming of hot-stamped parts is indispensable, they are usually produced with locally soft areas (Figure 2).

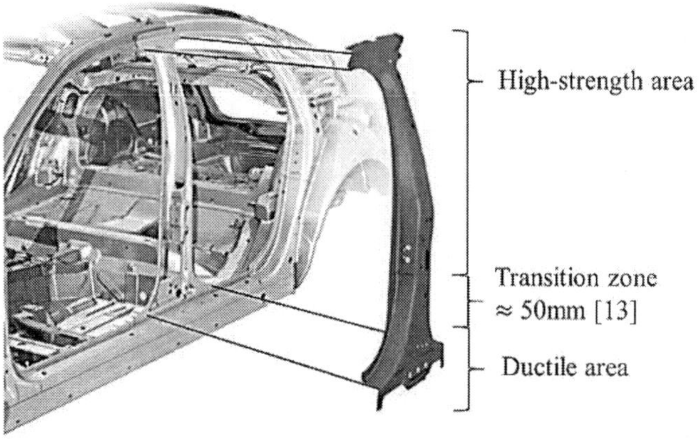

Figure 2: Hot-stamped and partially tempered B-pillar (re-created after [12] [13])

In order to generate a soft bending line with sufficiently high elongation at break in a hot-stamped part, it is possible to apply approaches similar to those used for the production of tailored properties. With such approaches, hot-stamped components with defined areas of different mechanical properties can be produced in a single operation and from a single sheet [11]. Components of this type, such as B-pillars, are frequently used in crash-relevant car body areas. In such components, areas with lower strength and higher ductility are specifically created for increased energy absorption (Figure 2). The exemplary part illustrated in Figure 3a shows a structural car body component. In this particular application, the highlighted bending part could not already be bent in the hot-stamping tool due to the necessary accessibility for the subsequent operations. Since due to its low strain potential also the hot-stamped part could not be bent, this problem was solved by producing the part from two separate parts and joining them in the final process step. However, this solution leads to an increased use of presses, requires additional forming and joining tools, reduces the lightweight quality of the car body and thus increases production costs as well as the cycle time of the production chain. Subsequent bending of an already hot-stamped component may also be necessary, if undercuts have to be created that cannot be integrated into the hot-stamping step. Figure 3b shows a demonstrator part used in the electromobility sector to protect a battery box made by BENTELER Automobiltechnik GmbH. Here, the bending after hot stamping could be realized by retaining ductile regions after hot stamping [14]. The sheet was completely austenitized. Subsequently, partial hot stamping is carried out in a partially heated tool. By further forming of the non-hardened area, the hollow part as shown in Figure 3 is formed. Using this approach, wider than necessary ductile regions were created for the production of this geometry by a bending operation. As a result, areas of the wide bending line remain softened even after the cold bending operation and represent a weakened point.

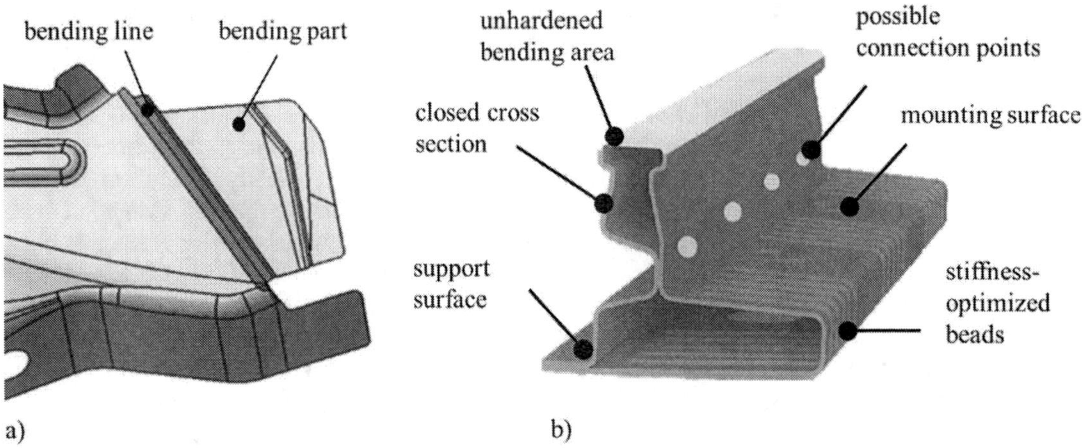

Figure 3: a) Car body structural part as an example of the need for a subsequent bending step due to the required accessibility for subsequent joining operations; b) Demonstrator part, hollow profile for the external protection of a battery box, hot-stamped and subsequently bent (re-created after [14]).

Beyond the widespread furnace heating, alternative technologies exist which can be used to create tailored properties such as induction, laser and resistance heating. For example, a sheet can be heated locally and rapidly to austenitizing temperature using a partially shielded induction coil [15, 16, 17]. Tempering of inductively hardened bulk parts is a widely used industrial process. Local inductive heat treatment of hot-stamped parts, on the other hand, has not yet been investigated and is not advisable due to the expected wide temperature transition zones and the distortion caused by rapid heating. The material can also be partially softened by means of targeted laser heating [18, 19]. In this case, the tool or the furnace technology can be kept independent and simple. However, the disadvantages are the distortion caused in the sheet metal, the high costs for the laser and the precise process conditions that must be satisfied [20, 21].

Alternatively, a sheet can be partially or completely heated by means of resistance heating or conductive heating. Here, electrical power is converted into heat inside the sheet due to current flow and as a result of ohmic resistance [22, 23]. The heat generated is proportional to the current flow and the voltage applied and can be defined according to Joule's law as follows (Equations 1 and 2).

$$P = U \cdot I \quad \text{or} \quad P = I^2 \cdot R \tag{1}$$

$$W = \int_{t1}^{t2} U \cdot I dt \tag{2}$$

Partial resistance heating by means of so-called bypass electrodes is also a possible heating option [24]. Since the current always follows the path of least resistance, the contact of a material with higher electrical conductivity can bypass the current flow in such a way that locally less heat is generated in the sheet [25]. However, geometric requirements strongly limit the use of the electrical current as well as the bypass method as a heating source. Therefore, uniform heating is essentially only applicable for rectangular geometries where all current paths are of equal length and resistance (Figure 4) [25]. Copper alloys (e.g., CuBe2) are particularly suitable as materials for current transmission as well as bypass electrodes, since their electrical conductivity ($\sigma_{CuBe2} \approx 16$ MS / m) is significantly higher than that of 22MnB5 ($\sigma_{CuBe2} \approx 1.4$ MS / m). Furthermore, local heating can also be realized by applying the electrodes specifically to the area to be heated [27]. Thus, the area between the two electrodes is heated up by the current flow, while the rest of the material remains cold. To avoid unwanted heating of the electrodes, they are usually water-cooled [25].

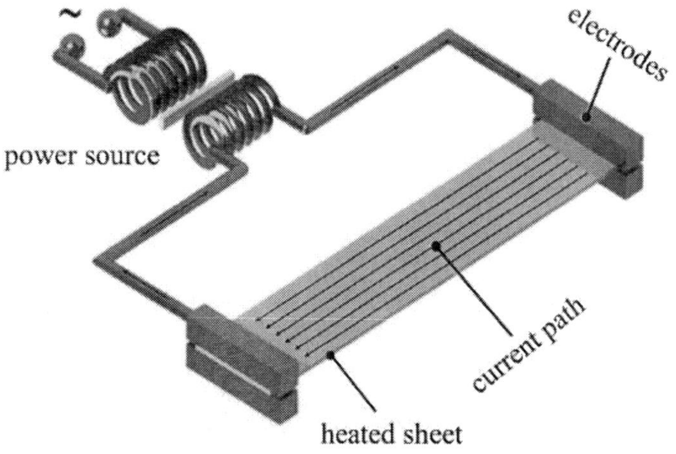

Figure 4: Principle of resistance heating of a rectangular sheet metal [26]

2. Experimental setup

To investigate the effect of local and short-time tempering of an already martensitic microstructure, the material 22MnB5 with an aluminum-silicon coating was used (see Table 1). The alloy 22MnB5 has a hardness of 200 HV10, a strength of 600 MPa, and an elongation at break A80 of more than 20% in as received condition [28]. The sheets used have a length of 220 mm, a width of 55 mm and a thickness of 1.8 mm. All sheets were first heated to 950 °C by conventional furnace heating and austenitized (cf. Figure 9a). Subsequently, the heated sheet was quenched and hardened in a water-cooled tool at a cooling rate higher than the critical cooling rate of 27 K/s. This transforms the austenitic microstructure into martensite. After hot stamping, the hardness increases to over 450 HV10, the strength to 1500 MPa and the elongation at break to 6% (Figure 5). [29]

Table 1: Chemical composition of 22MnB5 [28]

	C	Si	Mn	P	S	Al	N	Cr	Ti	B
min. in %	0.21	0.15	1.1					0.1	0.015	0.0015
max. in %	0.25	0.4	1.35	0.023	0.01	0.08	0.01	0.25	0.045	0.004

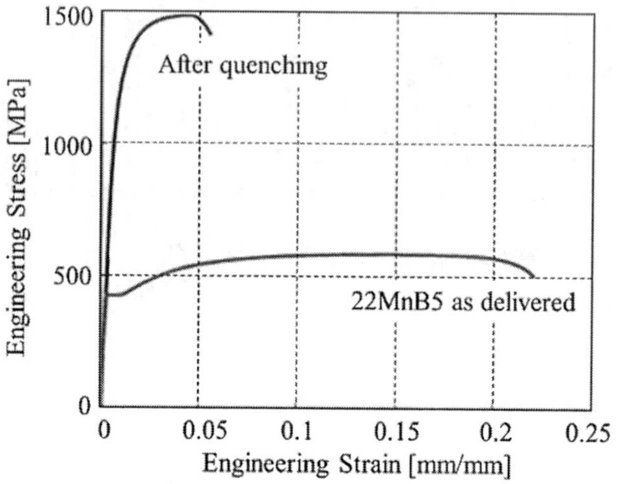

Figure 5: Engineering stress–strain curves of 22MnB5, in as-delivered conditions and after quenching (re-created after [29])

Figure 6a illustrates the principal design of the setup used to carry out the experiments. The overall setup is shown in Figure 6b. It consists of two electrodes on the upper and lower sides of the sheet. To enable contact with the sheet and to provide the necessary contact pressure for current transmission into the sheet, a blank holder was placed on the opposite side to each of the electrodes. The upper blank holder and the upper electrode can be moved vertically and ensure contact with the sheet during the current flow. This creates a fixed position by which distortion of the sheet due to rapid heating is suppressed.

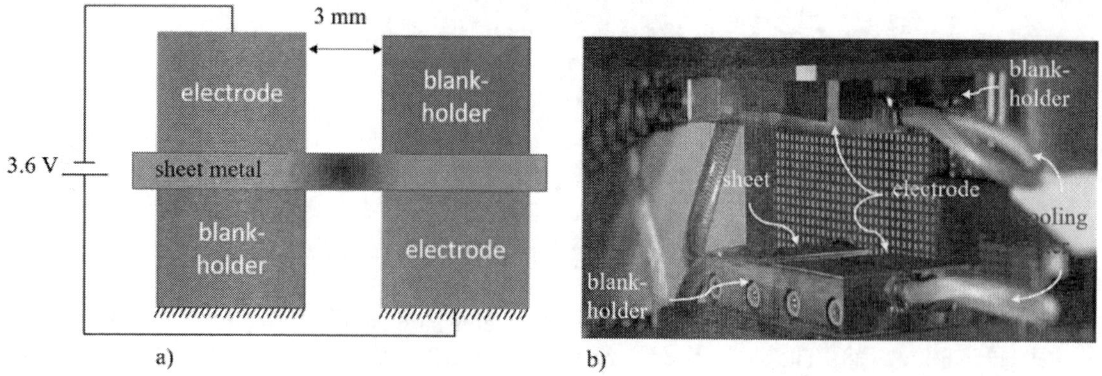

Figure 6: a) Principal illustration of local resistance tempering and b) the tool designed for local softening of hot-stamped sheets

The horizontal distance between electrode and blank holder is 3 mm. For a longer current path and improved heating, the electrodes were arranged diagonally. Furthermore, the electrodes and the blank holders were water-cooled to keep the temperature transition zone as small as possible. Alternating current of 4,160 A at a voltage of 3.6 V for a duration of 8 s was used for heating. After heating, the sheet is removed and subjected to cold bending in a bending tool as shown in Figure 7. The radius of the bending punch is 3 mm and the die has an opening angle of 90°.

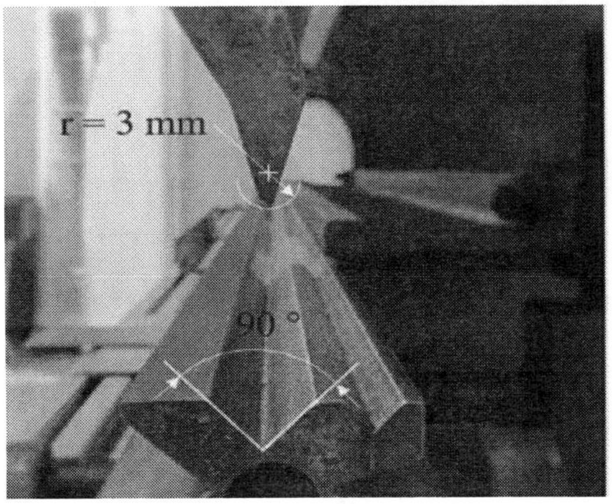

Figure 7: Bending machine of the company Wilhelmsburger Fabrik

3. Execution of experiments

Figure 8 illustrates the progression curve of heating and cooling temperatures for the hot-stamped sheet. The sheet was placed in the tool as shown in Figure 6b. The respective area was heated up to 800 °C within 3 seconds. After heating, the sheet is held under tool constraint for one more second and subsequently cooled in ambient air. The sheets were then bent in a bending machine to an angle of 90° and a radius of 3 mm (Figure 7). Under these conditions, a springback of 3° was observed on the bent part Figure 9b. The bending radius was deliberately chosen to be very sharp in order to show the limits of bendability of a resistance-tempered component.

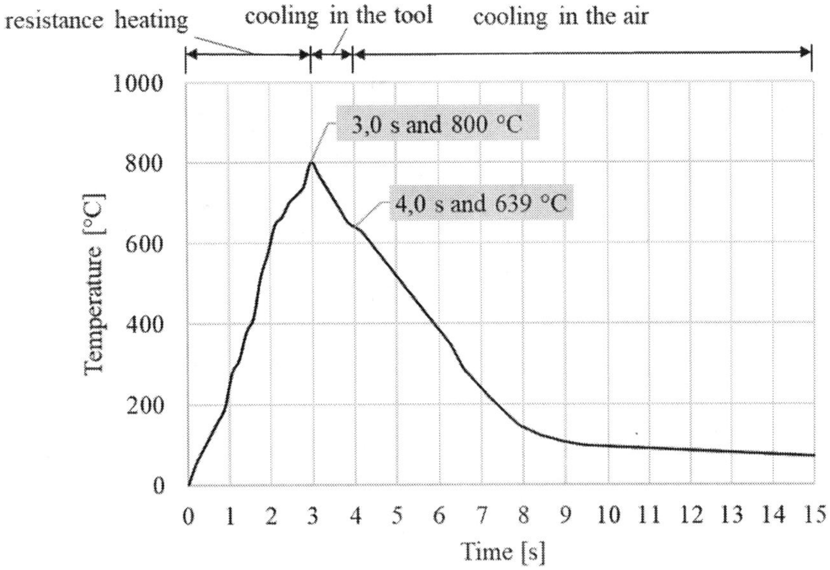

Figure 8: Temperature curve of resistance tempering

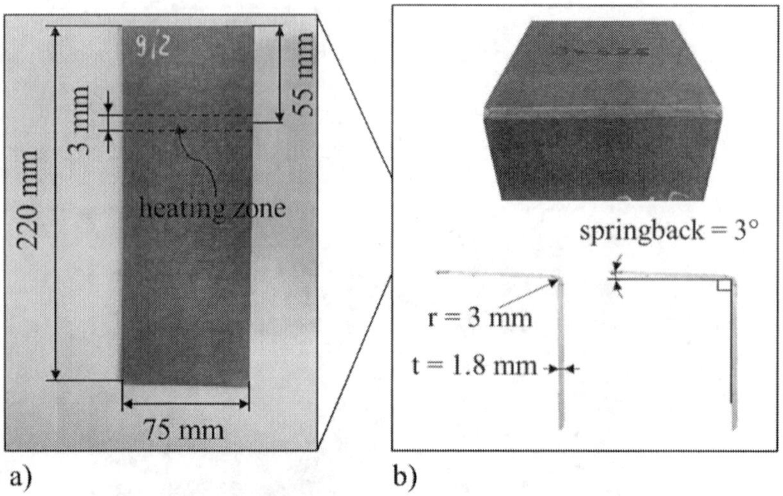

Figure 9: a) hot-stamped 22MnB5 with a martensitic microstructure, b) springback at the resistance-tempered and bent part

As a reference, a hot-stamped sheet with a fully martensitic microstructure was bent without local resistance tempering. As shown in Figure 10b, the short-time tempered sheet could be bent without any visible crack, while the reference sheet showed clear macroscopic cracks on the bending line (Figure 10a). Micrographs of the microstructures of the two sheets before bending showed a martensitic microstructure in the hardened state and tempered martensite after short-time resistance tempering. As shown in Figure 10b, no visible cracks occurred in the tempered and bent sheet. However, the bent region exhibited a changed appearance compared to the rest of the sheet.

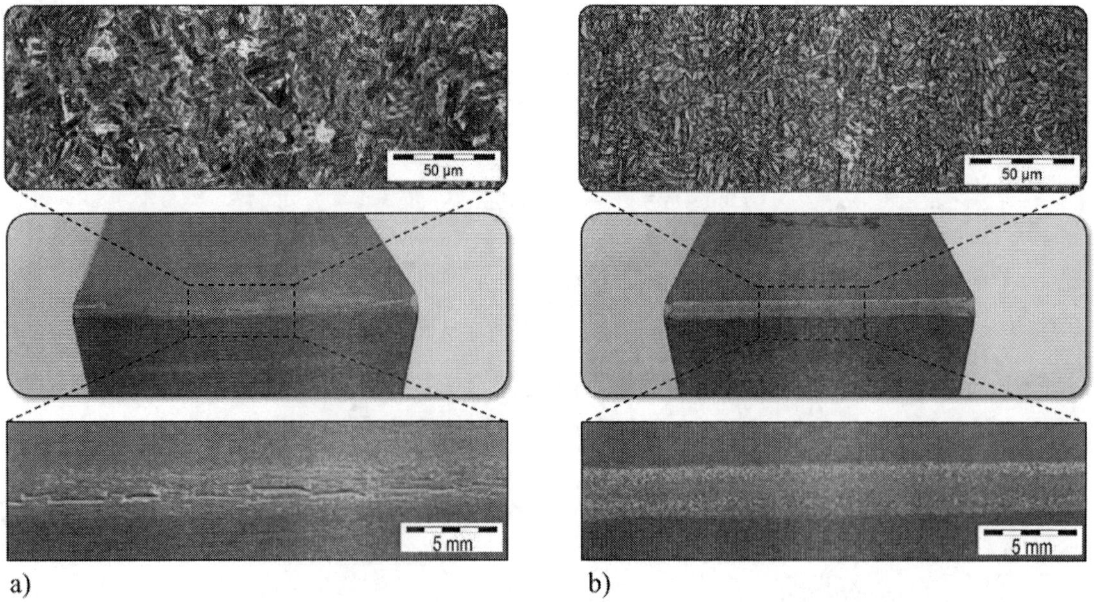

Figure 10: Results of the bending experiments a) of the hot-stamped sheet without post-treatment and b) of the additionally locally resistance-tempered sheet.

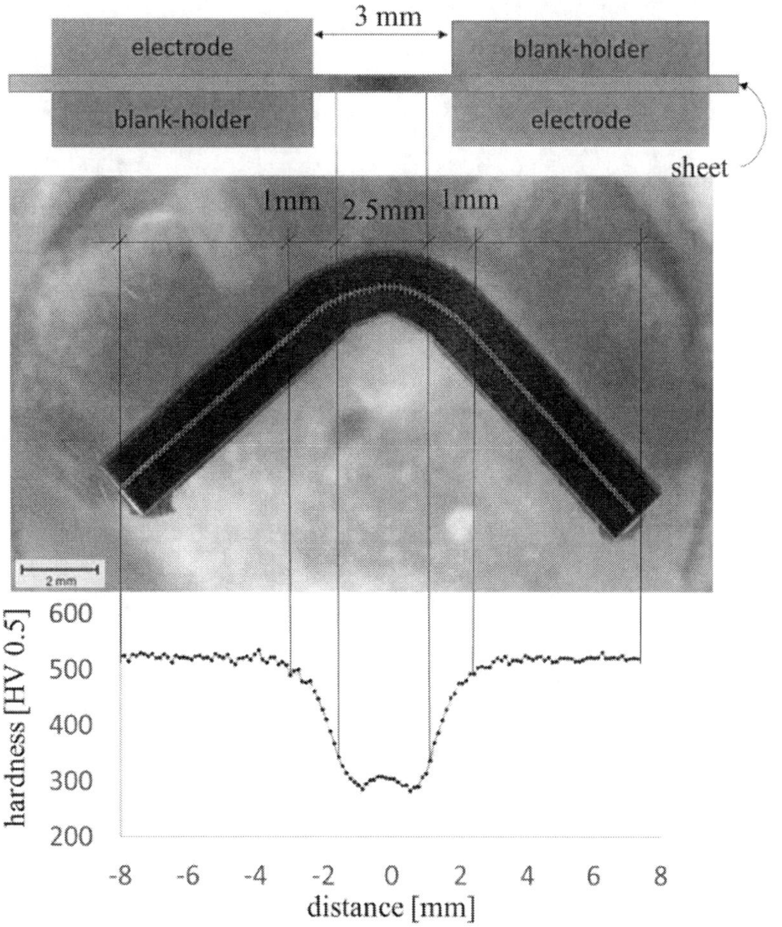

Figure 11: Hardness gradient over the tempered and bent zone

The analysis of the hardness achieved in the bent area showed a very sharp hardness gradient from the tempered zone to the hot- stamped zone. The hardness of the material could be reduced to 300 HV. The hardness increases to 500 HV outside the tempered area within a transition zone of 1 mm (Figure 11). This sharp increase in hardness is caused by the water-cooled tools, which prevent a temperature increase in the neighboring areas of the heated zone. In order to rule out possible damage to the coating, this area was comprehensively prepared and then examined in a scanning electron microscope combined with a focused ion beam (Zeiss AURIGA). As can be seen in Figure 12a and b, an area on the outer radius of the bent sheet was selected for this purpose in which a reduced layer thickness of the intermetallic FeAl coating was suspected (valley-like structure in Figure 12b). At a representative position for this area, a section perpendicular to the sample surface was prepared by means of the focused ion beam (FIB) (Figure 12c). Figure 12d shows the results of an energy-dispersive X-ray spectroscopy (EDS) analysis. Although cold bending caused partial spalling of the coating, it only occurred up to the diffusion layer enriched with aluminum and silicon. Such cracks are very common in hot-stamped parts. These cracks in the intermetallic coating, which occurred after hot stamping, already extend into this diffusion layer (Figure **1**). This means that no poorer corrosion properties are to be expected as a result of the presented cold-bending method.

IOP Conf. Series: Materials Science and Engineering 1284 (2023) 012039 doi:10.1088/1757-899X/1284/1/012039

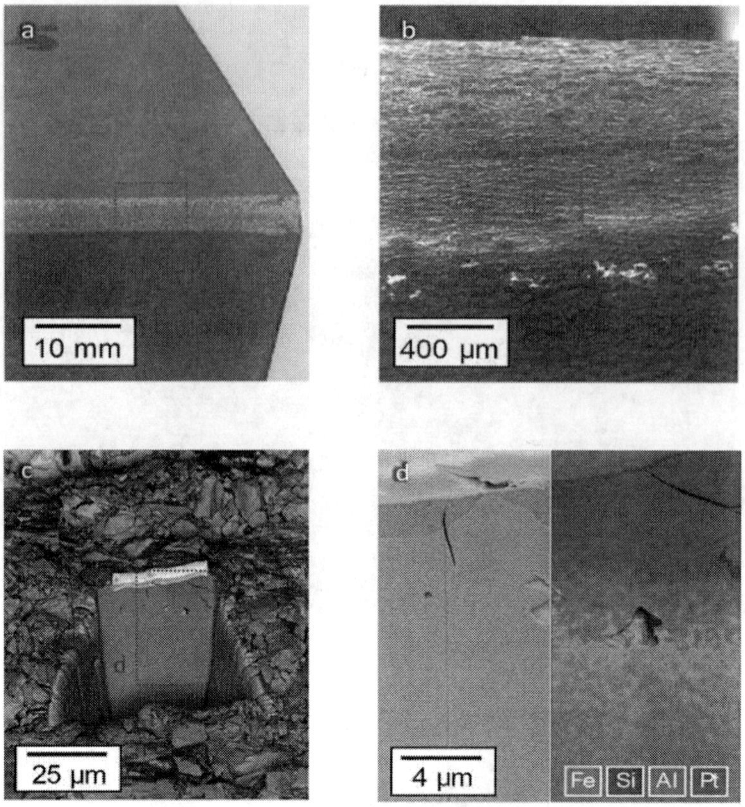

Figure 12: a) bent sheet after short-time tempering; b) SEM image (secondary electron contrast) of the area highlighted in a; c) FIB-cut in the part's surface at the position highlighted in b; d) backscattered electron contrast image (left) with half-sided overlaid elemental distribution (EDS mapping) on the cut surface shown in c.

4. Summary

Within the scope of this research, a tool was developed for the local resistance tempering of a previously hot-stamped sheet made of 22MnB5+AlSi. Using a local current flow in the tool, it was possible to soften a narrow line in an already hot-stamped sheet for a subsequent bending operation. The tool used for fixing the sheet was simultaneously used for current transfer. Since the tool was kept cold by means of cooling water and since very high heating rates were realized, the heat-affected zone remained very small. As a result, the mechanical properties could be changed exclusively in the bending line in such a way that the sheet could be cold-bent without developing visible cracks. Therefore, the mechanical properties of the neighboring areas remained unchanged. Microscopic examination of the bent sheet demonstrated that this approach can be used to produce defect-free components that do not exhibit any deep cracks extending to the substrate. This approach can essentially be used to expand the spectrum of hot-stamped components by creating new design possibilities.

Acknowledgments

We acknowledge the *Zentrale Innovationsprogramm Mittelstand* (ZIM), which is a funding program of the Federal Ministry for Economic Affairs and Climate Action in Germany for providing financial support *(Förderkennzeichen: KK5211401GM0)*.

References

[1] Bardelcik A and Worswick M J et al 2012 A strain rate sensitive constitutive model for quenched boron steel with tailored Properties (*International Journal of Impact Engineering Volume 50)* pp 49-62

[2] Doege E and Behrens B A 2018 *Handbuch Umformtechnik*, (Berlin, Heidelberg: Springer-Vieweg)

[3] Allély C and Dosdat L et al 2014 Anticorrosion mechanisms of aluminized steel for hot stamping *(Surface and Coatings Technology Volume 238)* pp 188-196

[4] Dormegny, L 2017 Efficient lightweighting with new press hardenable steels (*AMS Webinar*)

[5] Couto C and Costa I 2018 Characterization of 22MnB5 steel metallic coated with either hot-dip AlSi or electroplated ZnNi before and after hot stamping (*SAE Technical Paper Series, 27th. SAE Brasil Inte*)

[6] Arcelormittal, Steels for hot stamping - Usibor® and Ductibor®, Retrieved from https://automotive.arcelormittal.com/products/flat/PHS/usibor_ductibor date of access. 14.03.2023

[7] Meza-García E and Mosel A et al 2014 Partielles Presshärten – Grundlagenuntersuchungen zur Prozess- und Werkzeuggestaltung *Conf. 3. ICMC 2014/ 3. eniPROD Kolloquium*, eniPROD Kolloquium, Innovationen einer nachhaltigen Produktion für grüne Mobilität / Energieeffizienztechnologien in der Produktion (Chemnitz, Germany)

[8] Veit R and Hofmann H. 2011 Investigation of the Phase Formation of AlSi-Coatings for Hot Stamping of Boron Alloyed Steel (*AIP Conference Proceedings*)

[9] Kolleck R and Veit R 2009 Investigation on induction heating for hot stamping of boron alloyed steels *(CIRP Annals - Manufacturing Technology*) pp 275–278

[10] Billur E 2019 *Hot Stamping of Ultra High-Strength Steels: From a Technological and Business Perspective* (Switzerland: Springer Nature)

[11] Merklein M and Wieland M et al 2016 Hot stamping of boron steel sheets with tailored properties: A review *(Journal of Materials Processing Technology Volume 228*) pp 11-24

[12] Wilde H D and Schromm M. 2010 *Intelligente Leichtbaukarosserie: Audi A6 Karosserie* ATZextra, Bd. 15, Nr. 11, pp 166-177

[13] Sikora S and Banik J 2011 Tailored Tempering – adjustment of local mechanical properties in hot formed parts *(HTM Journal of Heat Treatment and Materials)* pp 323-328

[14] Grenz J and Buse C 2019 Pressgehärtete Hohlprofile zur Crashverstärkung in Elektrofahrzeugen *(Tagungsband zum 14. Erlanger Workshop Warmblechumformung 2019*)

[15] Löbbe C and Hering O 2016 Setting Mechanical Properties of High Strength Steels for Rapid Hot Forming Processes *(Materials 9, no. 4: 229)*

[16] Bao L and Liu W 2019 Experimental Investigation on Partition Induction Heating of High-Strength Boron Alloyed Steel Blanks *(Journal of Materials Research and Technology Volume 9)* pp 13963-13976

[17] Lee T M and et al 2011 A Demonstration of Local Heat Treatment for the Preform Annealing Process *(SAE International Journal of Materials and Manufacturing , Vol. 4)* pp 835-843

[18] Merklein M and Böhm W et al 2012 Tailoring material properties of aluminum by local laser heat treatment *(Physics Procedia Volume 39)* pp 232-239

[19] Böhm W and Merklein M Laser-assisted property gradients in multi-layered aluminum *(Physics Procedia Volume 56)* pp 1419-1426

[20] Sen N 2019 Experimental investigation of the formability of ultrahigh-strength sheet material using local heat treatment *(Ironmaking & Steelmaking Processes, Products and Applications Volume 47*)

[21] Bergweiler G 2013 Lokale Wärmebehandlung mit Laserstrahlung zur Verbesserung der Umform- und Funktionseigenschaften von hochfesten Stählen *(PhD Thesis Scientific Publisher of the IIF at the RWTH Aachen)*

[22] Field N N and Di Ciano M et al 2019 Tailoring by Direct Contact Heating During Hot Forming/Die Quenching *(Metallurgical and Materials Transactions A, Volume 50)* pp 3705-3713

[23] Liang W and Liu Y. et al 2015 Conduction Heating of Boron Alloyed Steel in Application for Hot Stamping (International Journal Of Precision Engineering And Manufacturing Vol. 16) pp 1983-1992

[24] Mori K and Bariani P F et al 2017 Hot stamping of ultra-high strength steel parts *(CIRP Ann Manuf Technol Volume 66)* pp 755–777

[25] Behrens B A and Hübner S et al 2015 Conductive heating opens up various new opportunities in hot stamping *(Proceedings of 5th International Conference on Accuracy in Forming Technology)* pp 157–173

[26] Behrens B A and Hübner S et al 2022 Resistance Heating by Means of Direct Current for Resource-Saving CO_2-Neutral Hot Stamping *(Key Engineering Materials ISSN: 1662-9795, Vol. 926)* pp 2363-2370 (Switzerland: Trans Tech Publications Ltd).

[27] Albracht L and Hübner S et al 2021 Hot stamping by resistance heating with coating application in oxygen-free atmosphere *(WT WerkstattstechnikVolume 111)* pp 893 – 898

[28] Salzgitter Flachstahl 2014 22MnB5 Colled-rolled quenched and tempered steel, boronalloyed. Retrieved from https://www.salzgitter-flachstahl.de/fileadmin/mediadb/szfg/informationsmaterial/produktinformationen/kaltgewalztes_feinblech/eng/22mnb5.pdf date of access. 14.03.2023.

[29] Eller T K and Greve L et al 2014 Plasticity and fracture modeling of quench-hardenable boron steel with tailored properties *(J. Mater. Process. Technol. Volume 214)* pp 1211–1227

Evaluation of hardening laws coupled with shear modified Lemaitre damage model for fracture prediction in incremental sheet forming

Abhishek Kumar, K Narasimhan

Department of Metallurgical Engineering and Materials Science, Indian Institute of Technology Bombay, Mumbai, India, 400076

E-mail: nara@iitb.ac.in

Abstract. Incremental sheet forming (ISF) is an emerging technology for manufacturing sheet metal components with fewer tools and equipment compared to conventional sheet forming. In this process, forming is generally performed without dies and punches and a simple hemispherical shaped tool can be used for shaping the final component. Forming limit in incremental sheet forming can be defined by the maximum formable wall angle. Finite element analysis can be used for fracture prediction by using a damage model. Also, to define the material flow behaviour hardening law should be incorporated into the finite element model. In this study four different hardening laws were used with shear modified Lemaitre damage model for fracture prediction during ISF of AA1050 alloy. The effectiveness of Hollomon, Johnson-Cook, Ludwigson and Swift hardening laws were compared for fracture prediction. Further, the predicted fracture wall angles were compared with the experimental observations. The results were analysed in terms of damage and strain evolution for the predicted fracture wall angles based on different hardening models.

1. Introduction

The incremental sheet forming (ISF) is a promising and advantageous sheet metal forming technology compared to conventional sheet metal forming. In this process sheet metal components can be formed with hemispherical or ball ended tool on a computer controlled machine. Different geometries of sheet metal can be formed without component specific tools and dies. Thus, the ISF process can result in cost effective manufacturing for batch size sheet metal components. Indeed, the process shows many advantages over conventional sheet metal forming process, but one of the major research challenges in ISF process is accurate fracture prediction.

The numerical analysis of ISF process provides the stress and strain evolution which can be used to understand the deformation mechanism of the forming process. Fracture during ISF can be predicted by implementing the damage model in finite element analysis. Bharti et al. [1] reported the effect of various uncoupled damage model during fracture prediction of ISF process. The work of Malhotra et al. [2] used damage model proposed by Xue [3]. It was found that localized deformation effect has major role on the formability of the ISF process. They also reported the effect of shear strain on the deformation mechanism during the ISF process. To incorporate the effect of shear mode, a shear modified Gurson-Tvergaard-Needleman (GTN) damage model was used by Gatea et al. [4]. Their results show good accuracy in terms of fracture depth prediction for different components. Mirnia and Shamsari [5] used modified Mohr-Coulomb (MMC3) [6] fracture criterion combined with non-linear damage evolution proposed by Xue [3]. In this work fracture depth was predicted with an error of 10%. The GTN model and its modified versions were explored to predict the fracture wall angle during ISF process by Guzman et al. [7]. Different void coalescence models were used and it was found that the Thomson void coalescence criterion predicts fracture wall angle with better accuracy as compared to the original model. The effect of different hardening criteria coupled with Lemaitre damage model was analyzed by Ben Said et al. [8]. They used Hill'48 yield criterion and mixed hardening model (isotropic-kinematic) and found thickness distribution and damage evolution prediction more realistic compared to the isotropic hardening model. A shear modified Lemaitre damage model can give good results for different forming processes as reported by Singh et al. [9] including ISF process [10]. However, very few studies were performed on the effect of hardening laws on the fracture prediction during ISF process. A comparative study of various hardening laws on failure prediction was reported by the Barnwal et al. [11] . They performed various tensile and shear test to study the failure characteristic using a damage model and various hardening laws. The aim of the present study is to analyze the effect of various hardening laws on fracture prediction coupled with shear modified Lemaitre damage model [10]. Various ISF experiments were performed on AA1050 alloy sheet and these results were compared with the simulation results to check the effectiveness of the hardening models on the fracture prediction.

2. ISF Experiments

The ISF experiment were performed on 1 mm thick AA1040 alloys and truncated cone geometry was selected as the component. A 3-axis LMW CNC machine was used to form the components. Further, a spiral tool path was considered for the various experiments with pitch of 0.5 mm. The tool rotational speed and travel speed were kept as 1000 RPM and 1000 mm/min, respectively. The experiments were performed to find the maximum formable wall angle. To achieve this various cones were formed from 60° to 75° wall angle. It was found that cone cones at 60°, 65° and 70° successfully formed up to the designed depth (40 mm), however cone at 71° wall angle fractured at depth of 15.13 mm.

3. Modeling approach

3.1. Finite element model of ISF

The finite element analysis of ISF process was performed in Abaqus/Explicit software. The Lemaitre damage model and various hardening laws were implemented in Abaqus through VUMAT. To model the forming operation, the clamping and tool were considered as analytical rigid part. The sheet was modelled as deformable part and it was discretised using linear eight-node solid elements with reduced integration formulation. The element size was considered as 0.5 mm and four elements through the sheet thickness were considered during the finite element modeling. As the tool slides on the sheet surface during forming, the contact condition between the sheet and the tool was modelled as Coulomb friction law with friction coefficient value as 0.1.

3.2. Constitutive model

The constitutive models used in the present study are discussed in this section. The section discusses about the various hardening models and the Lemaitre damage model.

3.2.1. Elastic Plastic model. The material is assumed to be isotropic, for which von-Mises yield criterion was considered. The stress-strain behaviour was modelled by isotropic Swift, Johnson-Cook, Holloman and Ludwigson law. The constitutive equations of the various hardening laws are given in the Table 1.

IOP Conf. Series: Materials Science and Engineering 1284 (2023) 012040 doi:10.1088/1757-899X/1284/1/012040

Table 1: Different hardening laws used in the present study.

S. No.	Hardening law	Formulation
1	Hollomon	$\sigma_f = k\, \varepsilon_p{}^{n_1}$
2	Johnson-Cook	$\sigma_f = \left(k_1 + k_2\, \varepsilon_p{}^{n_2}\right)(1 + C * ln\frac{\tilde{\varepsilon}}{\tilde{\varepsilon}_0})(1 - T^m)$
3	Swift	$\sigma_f = k_3\,(\varepsilon_0 + \varepsilon_p)\,n_3$
4	Ludwigson	$\sigma_f = k_4\,\varepsilon_p n_4 + exp(k_5 + k_6\varepsilon_p)$

Here, k k_2 k_3 k_4 k_5 and k_6 are the hardening model coefficients, k_1 is the yield stress, ε_0 is the strain at yield stress, and n_1 n_2 n_3 and $n_4$are the strain hardening exponent for different hardening models. For Johnson Cook (JC) model only strain effect was considered and the effect of strain rate and temperature were neglected to compare the results with other models.

3.2.2. Damage model. In the present study, Lemaitre damage model used to predict the fracture during the ISF process. This model is based on the continuum damage mechanics with some internal state variables. In the continuum damage mechanics, the model is based on the macroscopic frame work of the damage evolution. The damage in any material can be due to cracks or nucleation and growth of voids during the deformation process. Lemaitre damage model is based on the internal damage variable D, which can be considered as corrected surface density of the damage (voids and cracks). The damage is incorporated in the constitutive equation by the concept of effective stress:

$$\tilde{\sigma}_{ij} = \frac{\sigma_{ij}}{1 - D} \tag{1}$$

where $\tilde{\sigma}_{ij}$ represents the effective stress tensor and σ_{ij} is the stress tensor of undamaged material. The relationship between damage strain energy release rate *Y* and damage variable *D* can be defined as:

$$Y = \frac{\sigma_{eq}^2}{2\mathrm{E}(1 - D)^2}\left[\frac{2}{3}(1 + \nu) + 3(1 - 2\nu)\left(\frac{\sigma_H}{\sigma_{eq}}\right)^2\right] \tag{2}$$

where σ_{eq} is the von Mises equivalent stress, E is Young's modulus, ν is Poisson's ratio, σ_H is the hydrostatic pressure.

The damage evolution ($\dot{D}$) in Lemaitre model is a function of plastic strain and stress triaxiality (η). Another important variable associated with damage evolution is the third invariant (Lode parameter, $\bar{\theta}$) of the deviatoric stress tensor which was proposed by Bai and Wierzbicki [12]. The Lode parameter considers the effect of the shear mode of deformation on the damage evolution. As per the Lemaitre damage model the damage evolution ($\dot{D}$) is also dependent on the parameters known as denominator of the damage *S* and exponent of damage *s*. To incorporate all these factors, Singh et al. [9] proposed a modified Lemaitre damage evolution model which incorporates the stress triaxiality and Lode parameter in the denominator of damage as:

$$\dot{D} = \frac{\dot{\lambda}}{1 - D}\left(\frac{Y}{S_0 + (S_t - S_0)\bar{\theta}^2}\right)^s = \dot{\bar{\varepsilon}}^p\left(\frac{Y}{S_0 + (S_t - S_0)\bar{\theta}^2}\right)^s \tag{3}$$

where S_0 and S_t represents the denominator of damage under pure shear and tensile loading conditions, respectively. The model become pure tensile condition of the value of $\bar{\theta}$ = 1 then the denominator (in Eq. 3) becomes equal to S_t and if the value of $\bar{\theta}$ = 0 it reduces to S_0 and it becomes pure shear model.

The above model was implemented in Abaqus/Explicit through user subroutine. More details about the model can be found in the previous work by Singh et al [9].

3.2.3. Material parameters. The AA1050 alloy of 1 mm thickness sheets were used for the experiments and the material model parameters were determined through different mechanical testing. In order to get the hardening model parameters, a uniaxial tensile test was performed on MTS universal testing machine. The tensile test was performed at constant strain rate of 0.001 s^{-1} on an ASTM E8 tensile specimen. The other mechanical properties from the tensile test are given in Table 2.

Table 2: Mechanical properties of the AA1050 alloy.

Young's modulus (MPa)	Poisson's ratio	Yield stress (MPa)	Ultimate tensile strength (MPa)	Plastic strain at failure (%)
69000	0.33	112	120	5.57

Table 3: Material parameters for different hardening laws.

Hollomon		Johnson-Cook			Swift				Ludwigson		
k	n_1	k_1	k_2	n_2	k_3	ε_0	n_3	k_4	k_5	k_6	n_4
161	0.06	112	372	0.76	140	0.031	0.003	248	4	-33	0.24

To determine the damage parameters loading-unloading tests were performed on AA1050 alloy sheet. The test was performed on the MTS universal testing machine at 0.001 s^{-1} strain rate and room temperature. In order to calculate the damage parameter, elastic modulus of the each unloading cycle was measured and with help of Eq. 4 damage parameter for each cycle was calculated. In Eq. 4 the E represents the elastic modulus of undamaged material while $\tilde{E}$ is the elastic modulus of damaged material.

$$D = 1 - \frac{\tilde{E}}{E} \tag{4}$$

Table 3 shows the Lemaitre damage model parameters for AA1050 alloy, estimated with help of various tensile and shear tests. More details about the damage parameters estimation and calibration can be found in previous work [10]. Here D_c is the critical damage parameter at fracture, ε_D is the strain value at damage initiation and ε_R is the strain value at rupture. These parameters were determined from the load-unload test. The S_t and S_0 are the parameters estimated from the calibration.

Table 4: Calibrated Lemaitre model damage parameters.

D_c	ε_D	ε_R	S_t	S_0
0.36	0.009	0.1	6	1.2

4. Results and discussion

The experimental results of the ISF are shown in Figure 1, here it can be observed that the cones at 65° and 70° wall angle successfully formed up to designed depth (40 mm). Further, the cone formed at 71° wall angle fractured at 15.13 mm depth (Figure 1 (c)). It was found that, as the wall angle increases from 70° to 71°, the cone fractured and multiple experiments were performed to verify this observation.

Simulations were performed at different wall angles, however simulation results of 65° and 71° wall angle cones are discussed in this work.

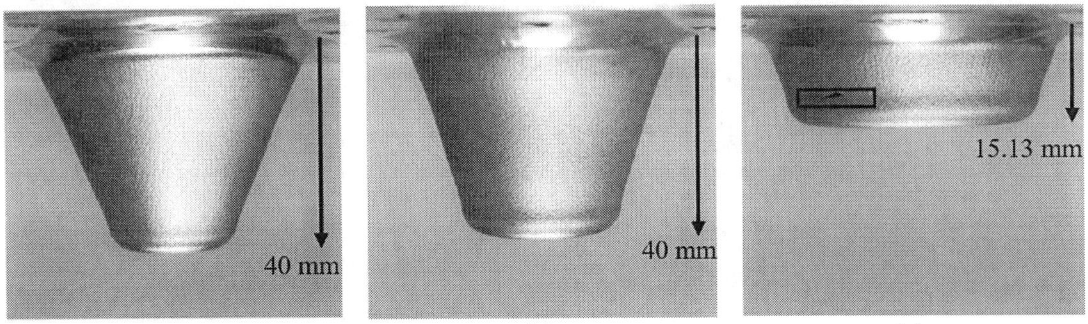

(a) (b) (c)

Figure 1: Experimental results of truncated cones formed at wall angle of (a) 65°, (b) 70° and (c) 71°.

The simulation results of 65° and 71° wall angle cones based on different hardening laws are shown in Figures 2. These results show forming depths of cones at 65° and 71° wall angle. From experimental results it can be observed that the cone at 71° wall angle fractured at 15.13 mm. The simulations show that all hardening laws predict the fracture at 71° wall angle. In Figure 1 (a), fracture can be observed for the both wall angle cones (65° and 71°), however the depth of forming (~22 mm) is more in the cone at 65° wall angle as compared to the 71° wall angle cone (~13.67 mm). As the cone at 65° wall angle also fractured, it can be concluded that FEA results based on Hollomon hardening law under predicts the fracture wall angle. Similarly, the simulation results of Johnson-Cook and Ludwigson also predicts fracture for both cones, as shown in Figure 2 (b) and (c). However, the fracture depth prediction for both wall angles (65° and 71°) based on Johnson-Cook and Ludwigson hardening models remains almost similar. These results also show under prediction of the fracture wall angle as well as depth of forming. Further, the Figure 2 (d) shows the simulation results of model based on Swift hardening law. Here, the cone at 65° wall angle successfully formed and the cone at 71° wall angle fracture at forming depth of ~ 14 mm. This shows the effectiveness of the Swift hardening model combined with Lemaitre damage model to predict the fracture.

As the fracture prediction based on different hardening models; results were further analyzed by considering the equivalent plastic strain (PEEQ) and damage variable (D) evolution. Figures 3 (a) and (b) show the damage evolution in the cones formed at 65° and 71° wall angle. It can be observed that the damage value at 65° wall angle for the models based on Hollomon, Johnson Cook and Ludwigson hardening laws reaches the critical damage value (0.36) and fracture occurs. On the other, hand, the damage value of the element (considered at the same level of forming) based on Swift hardening law remains below the critical damage value and no fracture was observed. However, at 71° wall angle, fracture was observed in all the models. It can be observed in Figure 3 (b) where damage reaches critical damage value for all models. Similarly, the PEEQ evolution for both wall angles (65° and 71°) are shown in Figure 3 (c) and (d), respectively. In Figures 1 (b) and (c) the Johnson-Cook and Ludwigson model predict similar fracture depth for both the wall angles, which can be correlated to the evolution of the plastic strain. Here, the strain levels reach similar values in both cases (65° and 71°), thus similar level of fracture depth prediction was observed. Considering the equivalent plastic strain evolution of Hollomon model, the maximum value of strain for 65° wall angle increases at a slower rate compared to the 71° wall angle cone. Moreover, the damage value also increases in similar way for both wall angles, due to this reason higher fracture depth at 65° wall angle was observed for the Hollomon model as compared to the 71° wall angle cone. Furthermore, Swift hardening model predicts fracture for 71° wall angle and no fracture was predicted for 65° wall angle. The strain evolution also shows similar trend as damage evolution which can be observed in Figure 3. A higher equivalent plastic strain can be observed in the cone at 71° wall angle compared to the 65° wall angle.

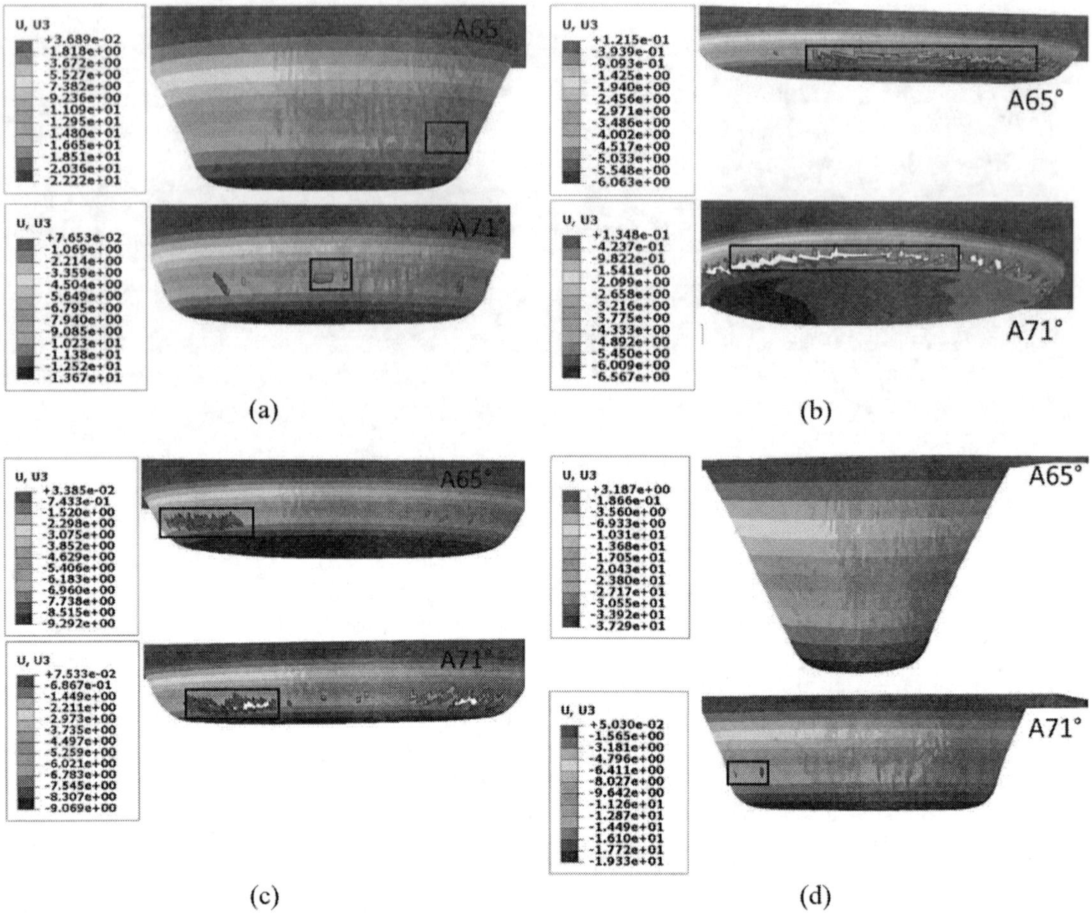

Figure 2: Simulation results of the cones formed at 65° and 71° wall angle based on (a) Hollomon (b) Johnson-Cook (c) Ludwigson and (d) Swift hardening law.

The present study was focused on the analyzing the effectiveness of the various hardening laws coupled with shear modified Lemaitre damage model to predict the fracture during ISF of AA1050. It was found that the Swift hardening model predicts the maximum formable wall angle and fracture depth with lesser difference from experimental observation as compared to the other models. The Johnson Cook and Ludwigson models under predicts the maximum formable wall angle and depth of forming. The Hollomon model predicts fracture depth for 71° wall angle with good accuracy, however this model also under predicts the maximum formable wall angle.

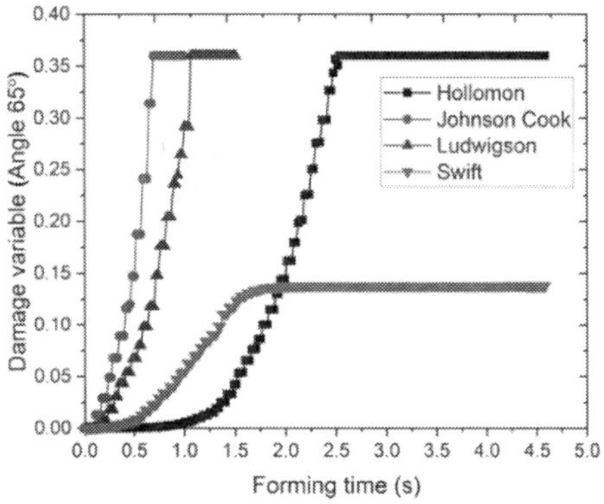

(a)

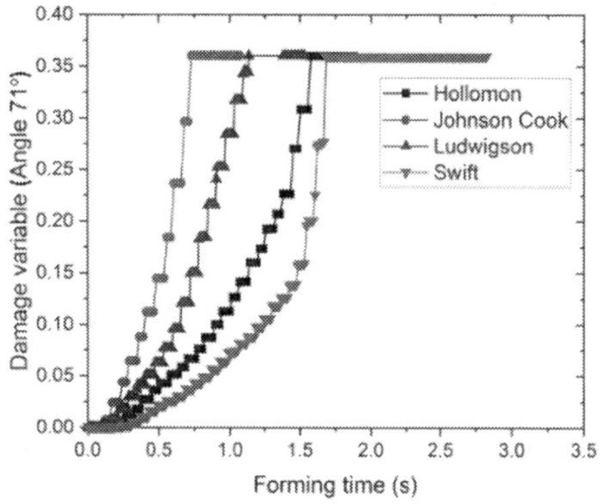

(b)

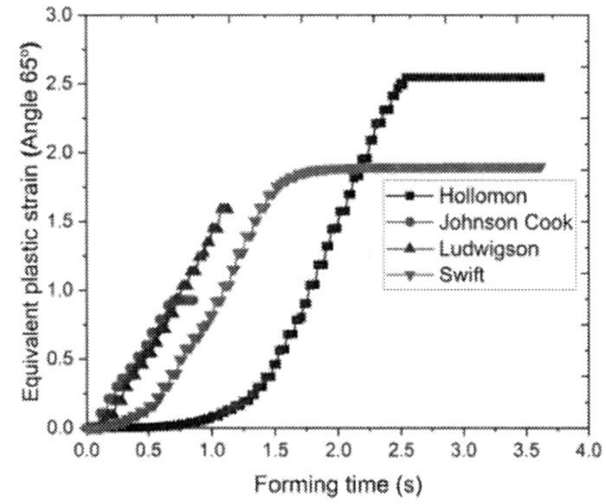

(c)

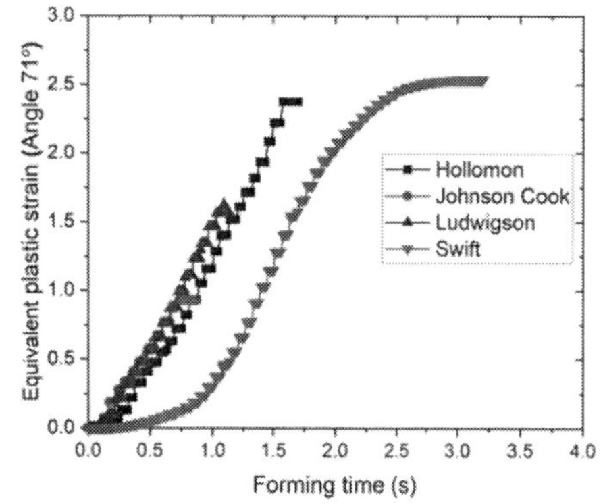

(d)

Figure 3: Simulation results of different hardening model showing evolution of (a) damage variable for 65° (b) damage variable for 71° (c) equivalent plastic strain for 65° and (d) equivalent plastic strain for 71° wall angle cone.

5. Conclusions

Four different hardening laws coupled with Lemaitre damage model were used for fracture prediction during ISF of AA1050 alloy. The experimental observations were compared with the simulation results to analyze the effectiveness of the different models in terms of maximum formable wall angle and fracture depth prediction. Two cones were simulated (65° and 71°) using different models and the following conclusions are drawn from this study:

- All models predict fracture for the 71° wall angle cone, however, Swift and Hollomon law-based model predict more accurate depth of forming as compared to the Johnson Cook and Ludwigson law.
- Swift hardening law predicts maximum formable wall angle with more accuracy as compared to the other hardening laws.

References

[1] Bharti S, Gupta A, Krishnaswamy H, Panigrahi S K and Lee M G 2022 Evaluation of uncoupled ductile damage models for fracture prediction in incremental sheet metal forming *CIRP J. Manuf. Sci. Technol.* **37** 499–517

[2] Malhotra R, Xue L, Belytschko T and Cao J 2012 Mechanics of fracture in single point incremental forming *J. Mater. Process. Technol.* **212** 1573–90

[3] Xue L 2007 Damage accumulation and fracture initiation in uncracked ductile solids subject to triaxial loading *Int. J. Solids Struct.* **44** 5163–81

[4] Gatea S, Ou H, Lu B and McCartney G 2017 Modelling of ductile fracture in single point incremental forming using a modified GTN model *Eng. Fract. Mech.* **186** 59–79

[5] Mirnia M J and Shamsari M 2017 Numerical prediction of failure in single point incremental forming using a phenomenological ductile fracture criterion *J. Mater. Process. Technol.* **244** 17–43

[6] Bai Y and Wierzbicki T 2010 Application of extended Mohr-Coulomb criterion to ductile fracture *Int. J. Fract.* **161** 1–20

[7] Guzmán C F, Yuan S, Duchêne L, Saavedra Flores E I and Habraken A M 2018 Damage prediction in single point incremental forming using an extended Gurson model *Int. J. Solids Struct.* **151** 45–56

[8] Said L Ben, Mars J, Wali M and Dammak F 2017 Numerical prediction of the ductile damage in single point incremental forming process *Int. J. Mech. Sci.* **131–132** 546–58

[9] Singh A K, Kumar A, Narasimhan K and Singh R 2021 Understanding the deformation and fracture mechanisms in backward flow-forming process of Ti-6Al-4V alloy via a shear modified continuous damage model *J. Mater. Process. Technol.* **292** 117060

[10] Kumar A, Kumar A, Shrivastava A and Mishra S 2022 Shear modified Lemaitre damage model for fracture prediction during incremental sheet forming *Int. J. Solids Struct.* **252** 111822

[11] Barnwal V K, Lee S Y, Kim J H and Barlat F 2019 Failure characteristics of advanced high strength steels at macro and micro scales *Mater. Sci. Eng. A* **754** 411–27

[12] Bai Y and Wierzbicki T 2008 A new model of metal plasticity and fracture with pressure and Lode dependence *Int. J. Plast.* **24** 1071–96

IOP Conf. Series: Materials Science and Engineering 1284 (2023) 012041 doi:10.1088/1757-899X/1284/1/012041

The effect of scrap originating trace elements on the properties of low alloyed steels

Shahroz Ahmed[1*], Ali Sabr[1], Ari Peltola[2], Olli Oja[2], Sanna Järn[2], Antti Kaijalainen[3], David Penney[4], Pasi Peura[1]

[1] Metals Technology Research Group, Tampere University, Tampere, Finland
[2] Product Development, SSAB Europe Oy, Hämeenlinna, Finland
[3] Materials and Mechanical Engineering, Oulu University, Oulu, Finland
[4] Metals Coating Group, Swansea University, Swansea, Wales, United Kingdom

E-mail: shahroz.ahmed@tuni.fi

Abstract. The present intention to reach fossil-free steel manufacturing will inevitably result in an increase in the use of steel scrap as a raw material for steel production. Consequently, the amounts of elements, seen as impurities, will increase in steels. This has already been seen in electric arc furnace (EAF) processed steels, where the Cu and Sn levels have doubled in some cases after 1980's. This may cause problems, as it is well-known, that some impurity elements have harmful effects on the properties of steel. This has been widely studied in low-alloy steels containing chromium and molybdenum which are widely used in components for the petroleum and electrical power generation applications. However, limited number of studies have been performed on formable steel grades, and the published reports/articles have mostly concentrated on the effects of P and B. Thus, there is still a need to understand the roles of other impurity elements. In the present study, a formable C-Mn steels containing additions (either individually or in combination) of Cu and Sn is investigated. The samples were cold rolled and annealed following typical time-temperature profiles of modern continuous annealing lines. Mechanical and forming properties (incl. bending and cupping tests) are determined as well as elemental profile analysis is conducted. The results identify that minor additions of impurity elements, in this case Cu and Sn, does not affect the mechanical and forming properties of low alloyed formable steel grades considerably.

1. Introduction

In order to reduce the carbon footprint from steel making, there is a great interest in transforming to 'Fossil Free Steels'. The term 'Fossil Free Steels' refers to producing steels by utilizing renewable sources of energy i.e., in practice using direct reduction with hydrogen and Electric Arc Furnace (EAF) route instead of blast furnace. Steel making through EAF requires increased use of scrap as a raw material, and as a result of this requirement the amount of trace or tramp elements, such as P, As, Cu, Sn etc, in steels will be increasing over the time. These trace or tramp elements can segregate to grain boundaries and may effect the mechanical, forming and coating properties [1], [2]; however, in some

IOP Conf. Series: Materials Science and Engineering 1284 (2023) 012041 doi:10.1088/1757-899X/1284/1/012041

cases it has been reported that some elements alone or in combination may have positive impact on properties [3].

Prior studies conducted on the topic of 'segregation of trace elements' identify the conditions for segregation of solute atoms, for example, Seah [4] studied grain boundary segregation and presented the relationship of atomic size and enthalpy of trace elements in Fe. He reported that elements with large atomic size have a higher tendency to segregate on grain boundaries and the driving force for segregation is related with elastic mismatch energy of solute and solvent atoms, and the differences in atomic dimensions. Gibson et al. [5] presented an elemental map by studying the ratio of 'solute/solvent surface energy (γ)' and 'non-dimensionalized heat of mixing ($\Delta H^{mix}/RT$)'. They also showed that the tendency of embrittlement, solute enrichment and potential increase in cohesion can be predicted by calculating the surface energy and heat of mixing at equilibrium temperature.

The effect of different trace elements, such as P, As, Sb, Sn etc., on the mechanical and forming properties have been previously investigated. For example, Rege et al. [6] studied the behaviour of P in Ti and Ti+Nb stabilized high strength IF steel during different stages of thermomechanical processing. They reported that P starts to segregate during coiling and the amount of segregation was reduced in the case of Nb alloying. They also added that cold work embrittlement resistance improved when P content was lowered from 0.06 to 0.014. Pilkington et al. [7] conducted creep and charpy impact tests to investigate the effect of P, As, Sb, Sn on the creep and impact strength of heat treated Cr-Mo steel. They reported that each element behaves differently under different testing conditions. It was found from their study that As do not segregate to grain boundaries while Sb and Sn show segregation.

Copper and tin are trace elements difficult to remove from steel. Copper is also one of the alloying elements which increase strength, hardenability and improves corrosion resistance but it should be kept to 0.2% to avoid hot shortness [1]. The effect of increasing Cu content, from 0.02 to 0.8%, was studied by Houpert et al. [8] during different stages of thermomechanical treatments and they showed that hot shortness began to appear in the steel containing more than 0.15% Cu. Study by Imai et al. [9] showed that cracking due to hot shortness occurred in mild steels during high temperature deformation; they also showed that adding Ni to mild steels can reduce hot shortness to a certain extent. Yamada et al. [10] investigated extra low carbon steel with different Cu contents and reheating temperatures. Increasing Cu was reported to increase tensile strength by solid solution strengthening and decrease elongation and anisotropy parameter.

Sn is common trace element and it's content is generally below 0.03 % because Sn can segregate to grain boundaries and cause hot shortness, temper embrittlement and decrease forming characteristics [1]. Stephenson [11] reported that increasing Sn content in steels will result in rise of ductile brittle transition temperature and reduction in shelf energy. Bruscato [12] showed that increasing Sn content in Cr-Mo steels, increases the susceptibility to temper embrittlement. Peng et al. [13] studied the effect of Sn (0.004-0.045%) on hot ductility of 20CrMnTi steel (0.02 Cu). They reported that samples with less than 0.045% Sn, show no Sn-rich phase at scale/steel interface and concluded that the reduction in hot ductility caused by Sn segregation at austenite grain boundaries reduced the cohesion of grains combined with the formation of thin ferrite film on austenite grain boundaries.

There are limited studies conducted on steels that shows the combined and individual effect of Cu and Sn on the forming and coating properties, specially in low alloyed steels. Sato et al. [14] predicted that the content of Cu and Sn in steels produced by EAF will increase in the future, as shown in Figure 1. The data from Figure 1 indicates that, thorough studies are required on steels grades that contain more than 0.2% Cu and 0.025% Sn, in order to produce fossil free steels for future applications.

In this study, the individual and combined effect of 0.3% Cu and 0.03% Sn on the properties of low alloyed steel is investigated; this work mainly concentrates on continuously annealed steels and does not consider the effects of trace elements during hot deformation. The results from this work would indicate the possibility of using low alloyed steel grades with higher content of trace elements for future applications, specially in the forming industry.

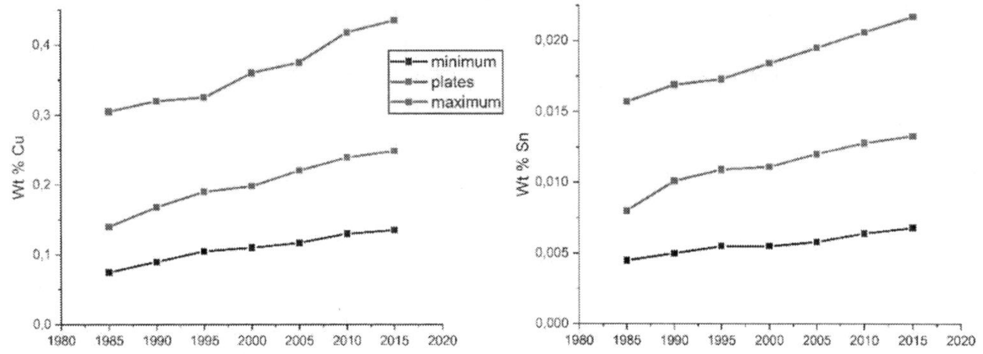

Figure 1: Increment in the content of Cu and Sn in steels over the years (predicted by Sato et al. [14]), Figure modified from [14].

2. Methods and Materials

The aim of this work was to investigate the individual and combined effect of Cu and Sn on the properties of low carbon steels. In order to study the effect of Cu-Sn on steels; four low carbon-manganese steels containing Cu-Sn were designed, the chemical composition of the steels is given in Table 1. From Table 1, it can be seen that each steel is distinct from the other; steel 1 contains no traces of Cu and Sn, steel 2 contains Cu but no Sn, steel 3 contains only Sn and steel 4 contains both Cu and Sn. The steel melts were first vacuum casted as slabs, then the slabs were subjected to hot rolling followed by cold rolling to thickness 0.8 mm. Micrographs (not shown here) from optical microscopy indicated that there are differences in the grain sizes after hot rolling. The grain size data in ASTM are 11.3, 12.3, 10.6 and 12.4 for steel 1, steel 2, steel 3 and steel 4 respectively.

Table 1: Chemical composition of the investigated steels.

Steel type	**Elemental composition in wt.%**		
	C	**Cu**	**Sn**
Steel 1	0.05	-	-
Steel 2	0.05	0.3	-
Steel 3	0.05	-	0.03
Steel 4	0.05	0.3	0.03

2.1. Annealing treatment with Hot Dip Galvanizing Simulator and Gleeble

The cold rolled 0.8 mm thin steels were cut into 125 × 210 mm rectangular sheets for annealing. The annealing cycles were conducted in a Hot Dip Galvanizing Simulator (HDS) without dipping to molten zinc; Figure 2 shows the annealing cycle, conducted in HDS. Gleeble 3800® thermomechanical simulator was also used to conduct the annealing cycle for one rectangular sheet from each steel.

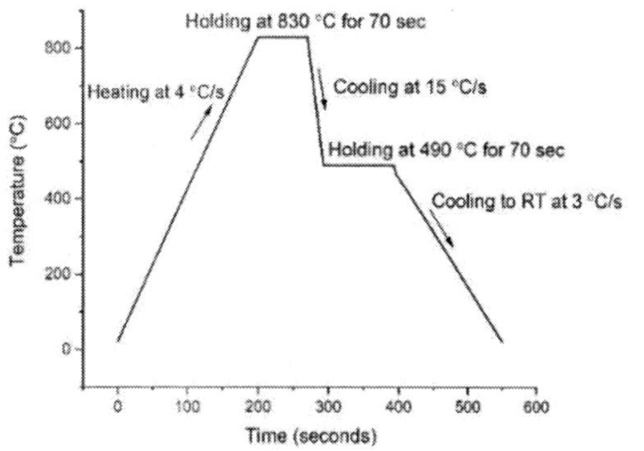

Figure 2: Heat treatment cycle for the steel samples 1 – 4 conducted in HDS and Gleeble.

2.2. *GDOES measurements*

A GDA750 glow discharge optical emission spectroscope Spectruma Analytik GmbH GDOES analysator was used to measure elemental depth profiles from surface of annealed samples down to 35 μm. The measurements were performed in a high sputter rate mode using constant current (20mA/700V).

2.3. *Mechanical and Forming tests*

After annealing cycles, the samples were investigated for their mechanical and forming properties by using uniaxial tensile tests, three point bending tests and Erichsen cupping tests.

2.3.1. Uniaxial tensile tests. The tensile tests were conducted with Instron 8800 uniaxial servo hydraulic materials testing machine. Tensile specimens with a gauge length of 25 mm and a gauge width of 9.5 mm were prepared from the annealed sheets. The tensile test was conducted according to the standard E8/E8M-22; the strain rate in the tests was 1×10^{-3} s^{-1} and the direction of loading was transverse to the direction of rolling. The strain was calculated by using a LaVision SMC 5M-140 optical extensometer with Davis software for Digital Image Correlation (DIC).

2.3.2. Three point bending tests. The three point bending tests were conducted to study the formability on samples that were annealed in Gleeble. After annealing, the specimens were cut to standard dimensions of 50 x 50 mm to conduct the tests. The test was performed using a universal testing machine with a low-friction roller carrier and a 0.2 mm radius punch, connected to a servo-hydraulic materials testing machine (specifically Instron 8800) in accordance with the specifications of standard VDA 100-238. The sample was loaded in the transverse direction of rolling at a predetermined punching speed of 20 mm/min until failure occurred. The bending angle at maximum force Fmax, referred to as α°, was determined using equation 1 (taken from VDA 100-238 standard), the equation considers the test setup and the force-displacement data.

$$\alpha^\circ = 2\times\left(-Arctan\left(\frac{\sqrt{(R+a)^2-\left(\left(\frac{-\sqrt{(h^2-4g.i-h)}}{2g}\right)+\left(R+\frac{L}{2}\right)\right)^2}-(R+a-S)}{\frac{-h-\sqrt{h^2-4g\times i}}{2g}}\right)\times\frac{180}{PI}\right) \qquad 1$$

where *g*, *h*, and *i* are test parameters defined depending on roll diameter (R), sample thickness (a), the distance between rollers (L), and displacement value (S) corresponding to maximum force.

2.3.3. Erichsen cupping tests. Erichsen cupping tests were performed with Erichsen 142-40 machine and according to the standard SFS-EN ISO 20482 standard.

3. Results and Discussions

3.1. *GDOES*

The results from GDOES are given in Figures 3 (a-d). For the minor elements (<1 wt.%), the accuracy of GDOES is low and typically within ±5-20%. From Figures 3 b and d, it can be seen that copper content is fairly the same throughout the depth of the specimen which means that there is no enrichment of Cu in the near surface region. GDOES has its limitations for the detection of minor elements and that is why the Sn data shown in Figures 3 should be considered with caution. Determining absolute Sn content needs more studies.

For reliable data about the segregation of elements to outermost surface and grain boundaries, auger electron spectroscopy (AES) or atom probe tomography (APT) studies are required.

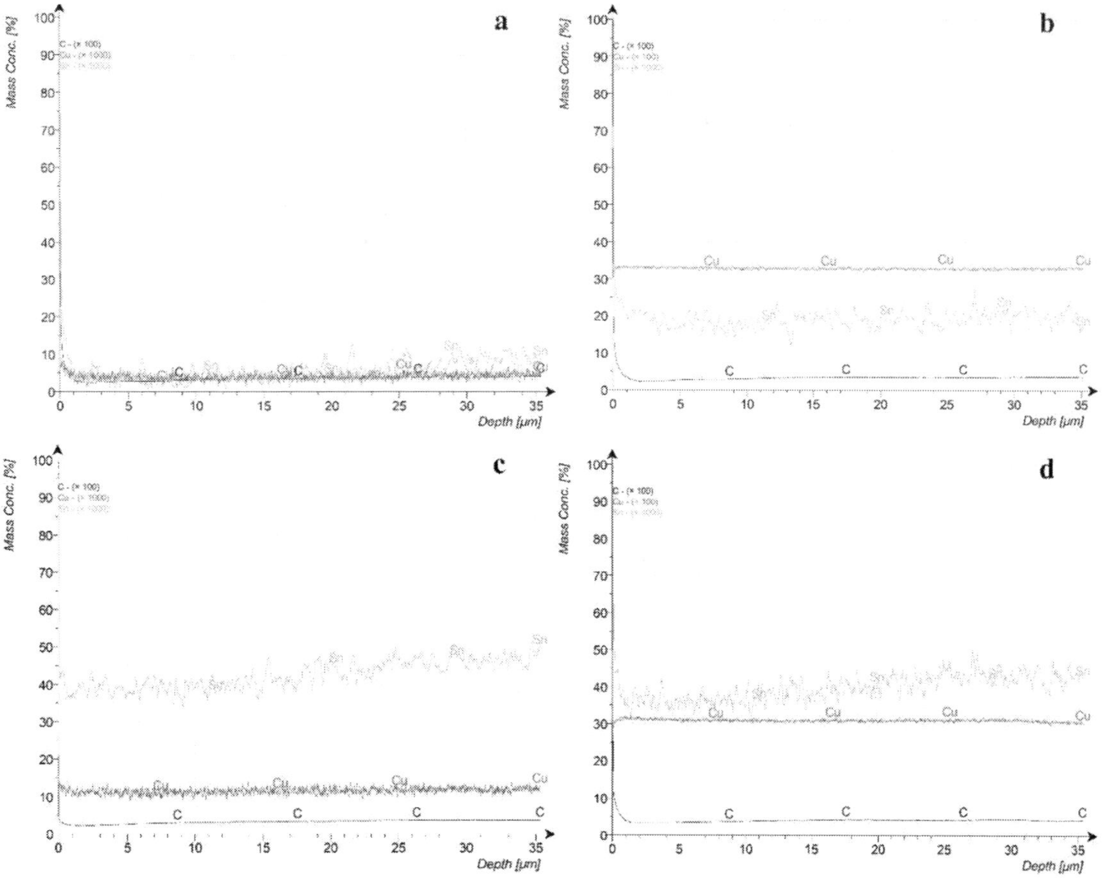

Figure 3: GDOES results from annealed steel samples (a) steel 1, (b) steel 2, (c) steel 3, (d) steel 4. Note different enhancement factors for Cu.

3.2. Tensile tests

The stress-strain data obtained from uniaxial tension tests are shown in Figures 4 (a and b) and the mechanical properties of samples are presented in Table 2. The samples were not subjected to skin pass rolling that is why the upper and lower yield points and yield point elongation are visible in Figure 4a. From the stress-strain curves (Figure 4a) it can be seen that minor additions of alloying in steel 2 and 3 suppressed the elongation slightly but the strength levels were almost the same in all of the samples. The total and uniform elongation of steel 4 (shown from stress-strain curve in Figure 4a and Table 2) are almost the same as steel 1, which might mean that minor additions of Cu and Sn together can be beneficial for low alloy steels but the combined effect of both Cu-Sn on the mechanical properties on low alloy steels needs more investigation. From Figure 4b, it can be seen that instantaneous strain hardening exponent curves looks similar for each steel and there are no major differences.

From the mechanical tests, it was seen that individual alloying of Cu and Sn (steel 2 and 3) slightly reduced the elongation but the combination of Cu-Sn alloyed steel (steel 4) had almost the same elongation. There were indications (Figure 4 a and b) that Cu alloyed steel had slightly higher strength.

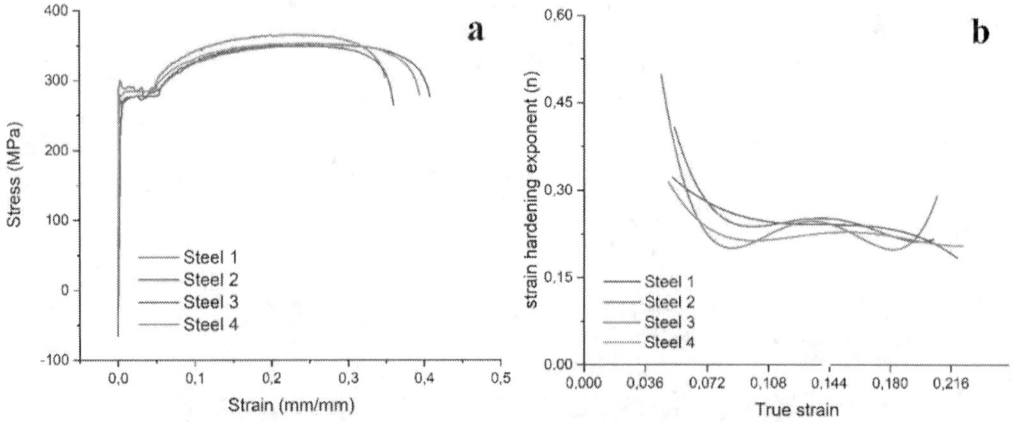

Figure 4: (a) stress-strain curves of samples 1 - 4 obtained from tensile tests (b) work hardening rate values plotted against true strain.

Table 2: Mechanical properties of samples.

Steel type	**Upper yield point (MPa)**	**Lower yield point (MPa)**	**Uniform elongation (%)**	**UTS (MPa)**	**Total elongation (%)**	**Average n value for 10-20 % elongation**
Steel 1	277	270	24.7	352	40.7	0.24
Steel 2	289	273	23.0	365	34.9	0.22
Steel 3	278	276	22.8	349	36.0	0.23
Steel 4	284	282	24.9	353	39.3	0.22

3.3. Three point bending tests

The bendability of steel samples using three-point bending tests (VDA 100-238) after annealing in a Gleeble simulator following the annealing cycle. The α° values from all four steel samples are presented in Table 3 and Figure 5 shows force vs displacement curves. The obtained α° values indicated high formability for all the steels. However, the inclusion of Cu and Sn in the steel compositions reduced the bendability to a slight degree as compared to the steel with no alloying. Steel 1, without any trace elements, had a bending angle of 115.6 degrees, while Steel 2, containing Cu, exhibited a slight decrease of 3 degrees. Steel 3, alloyed with Sn, had comparable bendability to steel 1, but Cu and Sn alloyed steel had slight lower bending angles than Steels 1, 2, and 3. The lowest bending angle achieved in three point bending test was from Cu-Sn alloyed steel but the difference in the bending angle was small.

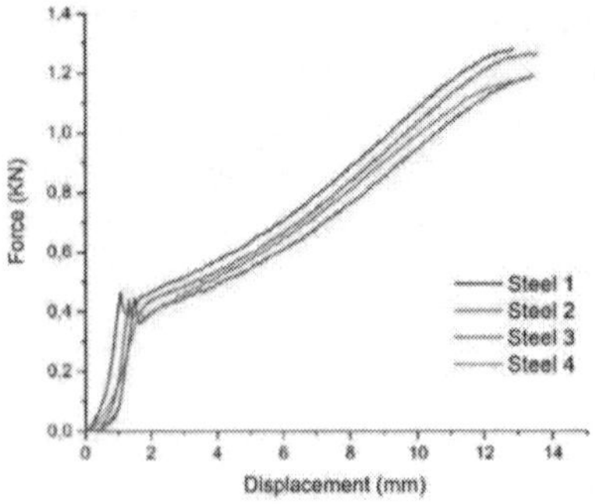

Figure 5: Force displacement curves from three point bending tests.

The bending test results show that minor additions of Cu and Sn in low alloy grades does not significantly affect the forming and mechanical properties of low alloyed grades.

Table 3: α° values obtained from bending tests.

Sample	Displacement (mm)	Force (kN)	α°(α Fmax) (degrees)
Steel 1	12.84	1.28	115.6
Steel 2	13.54	1.27	112.7
Steel 3	13.45	1.19	115.5
Steel 4	13.16	1.18	105.5

3.4. Erichsen cupping tests

The forming results from Erichsen tests are presented in Table 4. These results indicate that the change in elemental composition on low alloy formable grades had minimal changes in the cupping tests.

Table 4: Results from Erichsen cupping tests.

Sample	Thickness (mm)	Drawing speed (mm/min)	Holding force (kN)	Cup height (mm)
Steel 1	0.8	10	10	11.1
Steel 2	0.8	10	10	11.1
Steel 3	0.8	10	10	11.0
Steel 4	0.8	10	10	11.1

4. Conclusion

The individual and combined effect of addition of Cu and Sn on the forming properties of low carbon formable grades were investigated. The results suggest that there are negligible changes in the forming characteristics when Cu and Sn is alloyed with these grades. There were some indications that the combined effect of Cu and Sn might be favourable but this phenomenon needs more investigation. From this investigation, it can be concluded that minor additions of Cu and Sn does not affect the forming properties to a large extent.

Acknowledgement
The authors would like to acknowledge the efforts of Juha Uusitalo from Oulu University for conducting Gleeble treatments. The authors would also like to acknowledge Kyle Carter and Dan Britton from Swansea University for their work with Hot Dip Galvanizing Simulator. The authors are also grateful to Business Finland for funding the project 'Fossil Free Steel Applications'.

IOP Conf. Series: Materials Science and Engineering 1284 (2023) 012041 doi:10.1088/1757-899X/1284/1/012041

References

[1] Leroy V, Defourny J, and D'Haeyer R 1995 *Effects of tramp elements in flat and long products*
[2] Xu S, Brown J R and Tyson W R 2006 *Steel Res. Int.* **77** 825–835,
[3] Rod O, Becker C and Nylen M 2006 *Jernkontorets Fornkning* IM-2006-124
[4] Seah M P 1980 *Acta Metall.* **28** 955–962
[5] Gibson M A and Schuh C A 2015 *Acta Mater.* **95** 145–155
[6] Rege S J, Hua M, Garcia C I and Deardo A J 2000 *ISIJ Int.* **40** 191–199
[7] Pilkington R, Dicken R, Peura P, Lorimer G W, Allen G C, Holt M and Younes C M 1996 *MSEA* **212** 191-205
[8] Houpert C, Lanteri V, Jolive J M, Guttmann, Birat M, Jallon J P and Confente M **1996** *32nd conference of energy conservation USA*
[9] Norio I, Nozomi and Kazutoshi K 1997, *ISIJ Int.* **37** 224–231
[10] Yamada T, Oda M, and Akisue O 1995, *ISIJ Int.* **35** 1422–1429
[11] Stephenson E T 1980 *Metall. Trans. A.* **11** 517–524
[12] Bruscato R 2013 *Conference proceeding of AWS annual meeting Cleveland*
[13] Peng H, Chen W Chen L and Guo D 2014 *High Temp. Mater. Process* **33** 179–185
[14] Sato S, Takeuchi M, Mizukami Y, and Birat J P 1996, *Rev. Métallurgie* **93** 473–483

Development of a novel in-situ measurement method for thermo-mechanically coupled material characterization of high-strength aluminum alloys

N Rigas[1], T Garcin[2], C-M Kuball[1], M Merklein[1]

[1] Institute of Manufacturing Technology, Friedrich-Alexander University Erlangen-Nürnberg, Egerlandstr. 13, 91058 Erlangen, Germany
[2] Tecnar Automation Ltée, 1321 Rue Hocquart, Saint-Bruno-de-Montarville, J3V6B5 QC, Canada

E-mail: nikolaos.rigas@fau.de

Abstract. High-strength aluminum alloys are more and more used for safety and crash-relevant components due to their advantageous density-to-strength ratio. By using thermal-supported forming strategies, failure-free deep drawing is possible for these materials. In this context, the hot form quench process (HFQ®) has shown to be a promising thermally-assisted forming strategy. By using locally tempered forming tools, the production of tailored components with different mechanical properties is possible. However, material characterization and process design with conventional characterization methods are quite challenging in this context. A new approach is the use of ultrasound-assisted material characterization in combination with a thermo-mechanical simulator. Thereby it is possible to perform an in-situ material characterization. So far, this technology is mainly used for steel materials. In this contribution, the potential of an adapted measurement strategy for high-strength aluminum alloys is presented. For this purpose, the influence of different measurement parameters on modified samples and the resulting measurement signal is investigated. With the help of this study, new standards are set for time-bound material characterization under thermo-mechanical stress. As a result, thermally-assisted forming methods can be designed more efficiently and faster in the future.

1. Introduction, state of the art and goal

The use of alternative materials and improved manufacturing technologies are essential to satisfy changing mobility and powertrain concepts. Considering steadily growing environmental and resource requirements, lightweight construction is one method to fulfill these demands. Because of their advantageous density-to-strength ratio, bending stiffness and recyclability, high-strength aluminum alloys are highly attractive as lightweight materials for automotive construction in future [1]. However, the formability of high-strength aluminum alloys is limited at room temperature. For this reason, thermal-assisted forming techniques have become a focus of development in recent years.

Several thermally assisted forming operations exist for the failure-free forming of high-strength aluminum alloys. This includes the forming of sheets, which are previously completely [2] or partially heat-treated [3] and then formed at room temperature. In addition, precipitation-hardenable aluminum alloys can be formed at elevated temperatures [4] as well as under quench conditions [5]. In the past, it has already been demonstrated that the use of these thermally assisted processes can significantly reduce forming forces and springback and increase formability. The interaction-free modeling of the material and microstructure changes is essential for the process simulation. Especially for non-isothermal forming techniques, the characterization of mechanical properties and microstructural changes is challenging. Various conventional characterization methods already exist for this purpose. Mechanical material properties are usually determined by means of tensile tests. In this way, conclusions regarding the flow behavior can be made. With the help of thermally supported uniaxial tensile tests, it is also possible to determine the flow behavior at elevated temperatures. By using Differential-scanning-calorimetry-based (DSC) [6] investigations, the temperature-dependent transformation of different microstructural elements can be detected with the aid of heat flow measurements. The use of scanning electron microscope (SEM) [7] and transmission electron microscopy (TEM) [8] examinations as well as atom probe measurements enable the detection of grain and precipitation structures of precipitation-hardenable aluminum alloys. The characterization methods described, enable the precise determination of the microstructure and material properties. As a result, the temperature- [9], time- [10] and stress-dependent [11] material and precipitation behavior has already been investigated in the past. However, all of these test methods require time-consuming sample preparation.

Because age hardenable aluminum alloys already tend at room temperature to precipitation generation, the time-consuming sample preparation is a challenge for the interaction-free detection of material changes. In addition, these established measurement methods allow only the determination of microstructural or material characteristics at room temperature and at a specific point in time. For this reason, the time-related recording of structural changes is necessary in order to identify interaction-free property changes. Therefore, in this paper a non-contact in-situ measurement method for the investigation of high-strength aluminum alloys at elevated temperatures is investigated to provide a process-reliable basis for the use of suitable characterization methods.

2. Non-destructive, contact-free measurement technique

Using ultrasonic measurement the non-destructive detection of material defects or microstructural changes is possible [12]. This method, based on the principle of propagation of ultrasonic waves through the objects to be inspected, is used mainly for weld joint inspection [13], thickness measurement [14] and sheet thickness measurement [15].

The laser ultrasonic measurement is an improved method for the non-destructive contact-free characterization of materials and uses a laser to generate ultrasonic pulses in the material. For this reason, this measurement method is suitable for characterization at varying temperatures and stress conditions. In this way, it is possible to record microstructural changes, phase transformations, crystallographic textures and grain size changes under consideration of thermo-mechanical material loads. To perform these investigations, a thermomechanical simulator of the type Gleeble 3500 (Dynamic Systems Inc., USA) is used in combination with a **L**aser **U**ltrasonic **Met**allurgical Measurement (LUMet) system (Tecnar Automation Ltée, Canada). Using the thermo-mechanical simulator, the individual mechanical and thermal treatment of specimens is possible. The necessary generation laser as well as the optics are attached to the back of the Gleeble testing chamber.

To generate an ultrasonic signal in the aluminum sample, a short-pulsed, double-frequency Nd:YAG generation laser with a wavelength of 532 nm and a pulse duration of 10 ns is used. Because of the high pulse energy of 71 mJ, the laser causes a high local pressure and temperature difference on the surface of the sample (see Figure 1 a)). The surface pressure generates an acoustic initial pressure pulse inside the material, which travels through the material as an ultrasonic wave. The ultrasonic pulse is reflected by the back wall and oscillates between the front and the back of the specimen (see Figure 1 b)) [16]. Every time the ultrasonic pulse reaches the surface of the sample, it causes a small and short-term displacement of the surface. A second detection laser with a radiation duration of 50 µs and an energy of 20 mJ (on the target), which is guided by the same laser optics, enables the detection of the surface displacement on the sample surface by using a laser interferometer (see Figure 1 c)). Thus, the position change detected by a photodetector is reproduced in the form of a voltage signal and is proportional to the surface displacement. In this way, with a repetition rate of up to 50 Hz, the change of the ultrasonic signal in an aluminum sample can be detected. Afterward, the resulting ultrasound signal is analyzed with the help of the CTOME evaluation software (V2.33.1, 2022, CTOME Software & Consulting Inc., Canada). Via a fourier transformation, the echoes of the generated ultrasonic signal are detected from the unfiltered measurement signal of the position change (see Figure 1 d)).

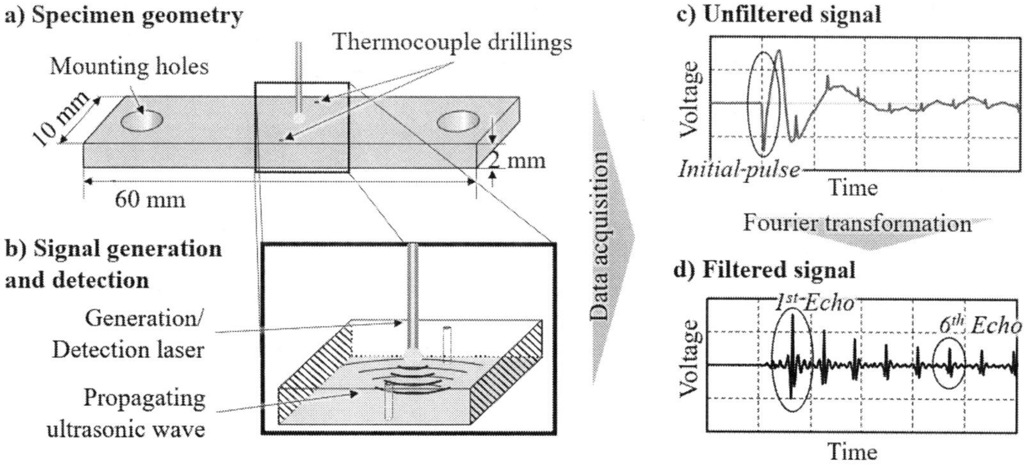

Figure 1. Overview of the a) used sample geometry, the b) signal generation and detection as well as the detected c) unfiltered signal and d) filtered signal.

By measuring two or more sequential echoes, the change in the ultrasonic signal is recorded in the form of the propagation velocity as well as the sound attenuation. On the one hand, the velocity of the ultrasonic signal depends on the elastic properties, the material temperature, the precipitation behavior and the density of the aluminum alloy. By recording the time difference of two or more successive echoes and the sample thickness, it is possible to determine the ultrasonic velocity. The attenuation, on the other hand, can be detected based on the amplitude level of one defined echo for different measurement points and is caused by divergence losses due to internal friction effects (absorption) and ultrasound scattering. In polycrystalline materials, these occur primarily at grain boundaries. Thus, in metallic materials such as aluminum alloys, conclusions can be drawn about grain size changes due to absorption [17] and precipitation effects due to velocity variation [18].

3. Experimental tests

In this study, the age-hardenable aluminum alloy AA7075 in the delivered condition T6 with a sheet thickness of $t_0 = 2.0$ mm was used. Because aluminum alloys tend to undergo microstructural and precipitation changes as a function of the respective material temperature, the tests are carried out at 450 °C, which is near the melting point. Type J thermocouples are used for precise temperature measurement. The thermocouples are inserted and twisted into previously drilled holes (see Figure 1 a)). This also avoids uncontrolled fall-off of the thermocouple wires due to their insufficient weldability on aluminum.

3.1 Adaption of different laser parameters

To generate a high-energy short-pulse generation laser, a Quality-switch (Q-switch) is used. At an optimum time of the Q-switch point, this provides the compression of energy, resulting in a short laser pulse (10 ns) with a high energy density. A delayed or early opening of the Q-switch results in a lower amount of power being emitted. Typically, the laser ultrasonic measurement is performed at a high power level of the short-pulse generation laser in order to generate a clear ultrasonic signal in the specimen. However, preliminary investigations with aluminum demonstrate that because of the low melting temperature and the strong short-term local heating of the specimen, the material is not only heated and vaporized by the generation laser but also melted. As a result, the formation of a keyhole occurs. Because a keyhole causes a significant change in sheet thickness and thus generates major measurement inaccuracy, for the successful investigation of aluminum it is essential to perform laser ultrasonic measurements on these materials without generating significant sample or signal changes. For the successful use of LUMet measurements on aluminum alloys, it is mandatory to adjust the laser power for this reason. By modifying the LUSGenCtrl software (V1.1.3.0, Tecnar Automation Ltée, Canada), it is possible to adjust the laser power by changing the timing of the Q-switch opening. In this way, the controlled reduction of the laser power of the high-energy generation laser is possible.

To measure the laser power as a function of the laser settings, a powermeter of type PowerMax USB PM10 (Coherent, USA) is positioned outside the focal area of the laser in the test chamber of the Gleeble. This enables the determination of the laser power of the generation laser.

3.2 Optical component measurement

To determine the influence of different laser powers, aluminum samples (see Figure 2 a)) are clamped in the test chamber of the Gleeble and heated to 450 °C. The samples are then tested at varying laser powers for 20 seconds each at a scanning frequency of 50 Hz. As a result, a cumulative amount of 1000 measurement pulses at isothermal temperature results on each test specimen. The influence zones of the laser on the samples have been measured in an optical profilometer of the type InfiniteFocusXL200 G5 (Bruker Alicona, Austria) with a 10x objective lens (see Figure 2 b)). Subsequently, the surface deviation of the components was measured using the GOM Inspect 2019 software (GOM GmbH, Germany). As a result, the geometric surface changes in terms of reduction of sheet thickness and surface condition have been studied (see Figure 2 c)).

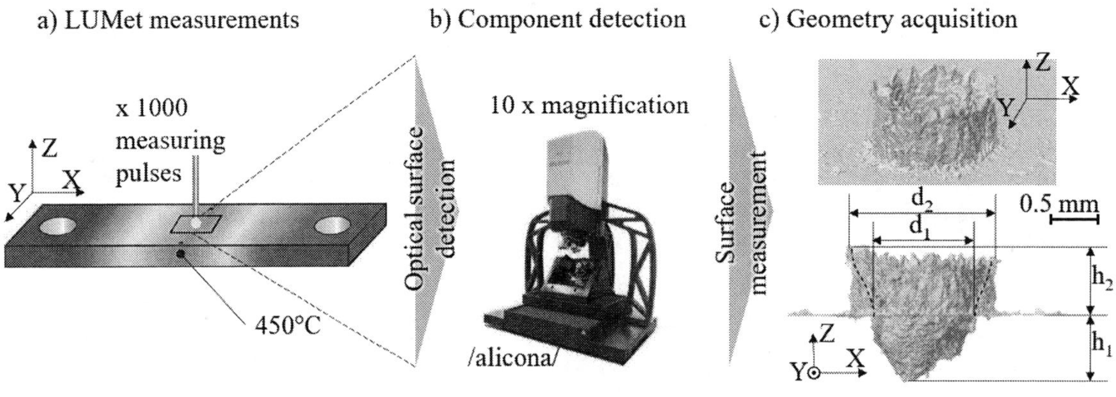

Figure 2. Component measurement via LUMet and optical 3D measurement to record signal and surface changes.

4. Results and discussion

4.1 Signal change in dependence of radiation parameters

To determine the influence of the Q-switch adjustment on the resulting generation laser power, the power of the generation was measured over a test period of approx. 10 seconds using a powermeter. For reasons of comparability, the laser power was determined in each case for a test period between six and nine seconds. A scanning frequency of 50 Hz was set for the generation laser. Figure 3 shows the determined laser power curves and the resulting laser energy as a function of the Q-switch-delay. With increasing Q-switch delay, it appears that the generation laser power decreases. Since the initial laser performance is known, the identification of the power of each generation pulse is possible. The results illustrate that by adjusting the Q-switch, a reduction of the laser power by up to 70 % is possible.

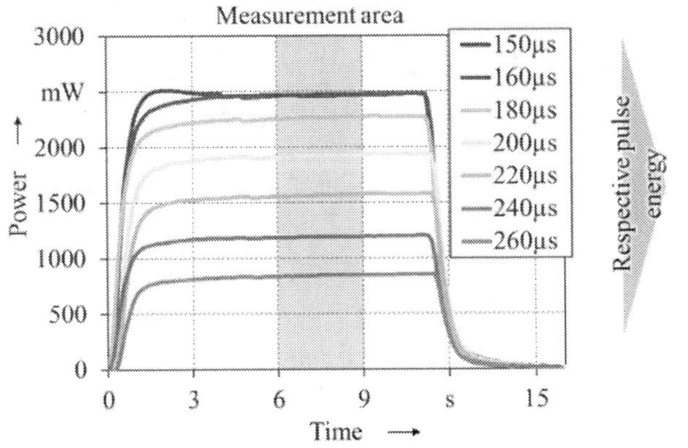

Q-Switch [µs]	Measured Power [mW]	Generation Pulse Energy [mJ]
150	2474 ± 15	49.5
160	2446 ± 15	48.9
180	2237 ± 14	44.7
200	1896 ± 13	37.9
220	1542 ± 11	30.8
240	1178 ± 7	23.6
260	827 ± 9	16.6

Figure 3. Measured laser power depending on different Q-Swichting and the resulting pulse energy.

In the next step, the influence of varying laser power on the signal quality of high-strength aluminum samples was investigated. For this purpose, aluminum samples were heated up to 450 °C and then 1000 measuring points each were applied with a measuring frequency of 50 Hz. For statistical assurance, each experiment was repeated and evaluated three times. Subsequently, the ultrasonic signal was examined in terms of signal quality. For this purpose, the ratio of measurement noise to signal quality was investigated. In Figure 4, the ultrasonic signal for one measuring point is shown schematically. The second signal echo was selected as reference amplitude in each case. To determine the signal-to-noise ratio, the amplitude level of the second echo Y_{Signal} was divided by the amplitude of the measurement noise Y_{Noise}. In the following evaluation, this ratio was determined and compared for all measuring points.

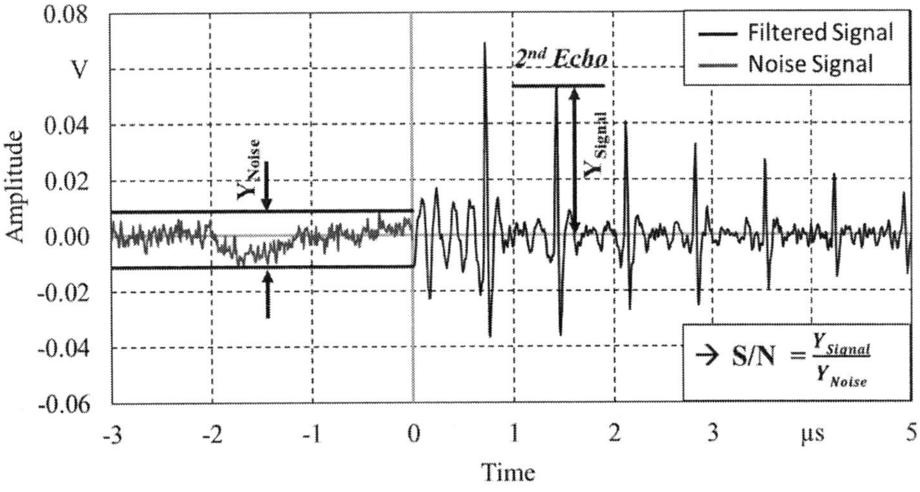

Figure 4. Determination of the signal-to-noise ratio for the detected ultrasonic signal.

The signal-to-noise ratio as a function of the respective Q-switch delay is shown in Figure 5. The results show that at high laser powers (Q-switch: 150 µs, 160 µs) at the beginning of the measurement the signal has 90 times the amplitude of the noise. As the test duration increases, the signal quality continuously decreases. The reason for the reduced signal quality can be attributed to surface defects, which occur due to local sample heating and vaporization. The decreasing signal behavior is also evident for medium outputs (Q-switch: 180 µs, 200 µs, 220 µs). However, the curves shift the signal to noise ratio downward. This is caused by the reduced local surface pressure, which also leads to a reduced sample displacement due to the lower laser power and thus the interferometer only detects lower signal amplitudes of the ultrasonic signal. Thus, it can be assumed that a lower signal quality occurs due to the reduced laser energy. However, a change of the sample surface still takes place, which leads to a successive decrease of the signal quality over the number of pulses. With comparatively low laser powers (Q-switch: 240 µs, 260 µs), a small difference between signal and measuring noise is basically evident. The reason for this also lies in the lower sample displacement and the resulting reduced signal intensity. However, at low laser powers, the signal quality decreases only slightly or not measurably with increasing measuring pulses. As a result, it can be seen that low laser powers for heated aluminum have a smaller effect on the signal quality over longer test durations.

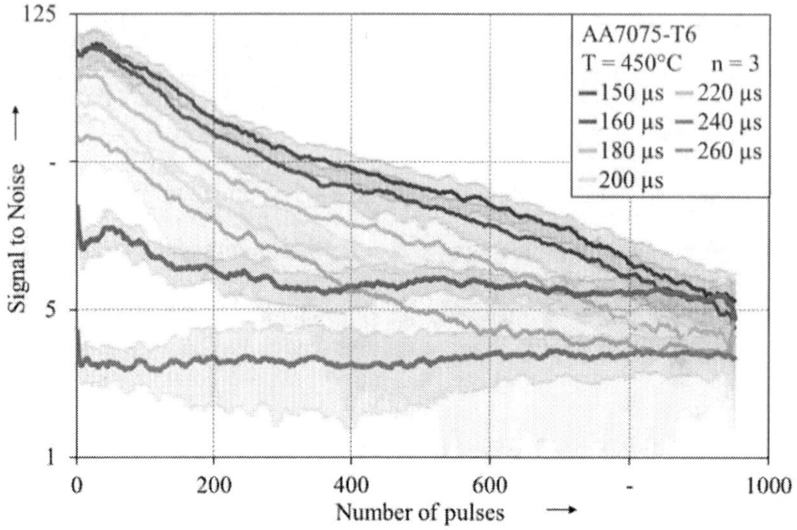

Figure 5. Signal-to-noise change as a function of the laser power and applied laser pulses.

4.2 Geometrical component penetration

The used samples have been optically measured in the area of the laser influence zone after the laser-ultrasound-assisted measurements. Figure 6 a) shows the side profile of a measured specimen. A surface change is visible both below and above the surface of the sheet. For this reason, both the penetration depth h_1 and material heightening h_2 as well as the diameter in the sheet plane d_1 and in the area of the heightening d_2 of the circular surface structures are recorded for geometrical identification. Subsequently, the geometric determination of the material reduction V_{red} in the area of the sheet thickness as well as the material accumulation above the sheet surface V_{acc} can be determined (see Figure *6* b)), whereby V_{acc} approximately corresponds to the volume of a cylinder with diameter d_2 minus the volume of a truncated cone with diameters d_1 and d_2.

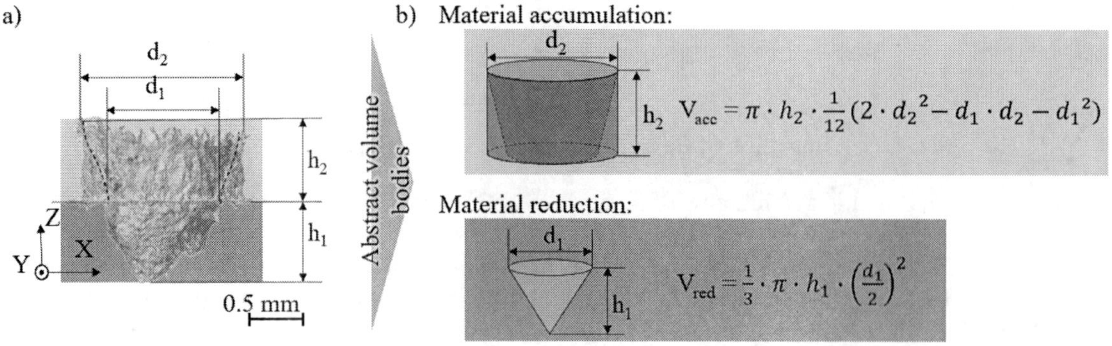

Figure 6. a) Lateral view of the surface profile, b) method for the evaluation of the geometric surface change.

In Figure 7, the influence of different laser powers on the geometric surface change is given both as volume and height differences as well as visually illustrated in each case by means of an example surface. The results show that there is almost no surface change for the lowest laser power. After 1000 pulses, there is an average height difference of 0.06 ± 0.01 mm and no measurable volume change. With a Q-switch delay of 240 µs, the height difference and the change in volume increase measurably. However, the height difference is only about 0.23 ± 0.04 mm in both directions and the volume difference in the reduced and accumulated area is 0.007 ± 0.002 mm³ after 1000 measuring points. With a further increase in laser power, a significant increase in surface deformation is evident. Height differences of up to 0.588 mm and volume differences of more than 0.1 mm³ are achieved. Due to the strong deformation of the surface, the real sheet thickness of the components to be measured is reduced significantly. This decreases the distance between the upper and lower surfaces of the sheet, resulting in an error variable concerning the ultrasonic velocity. In addition, at high laser powers, the strong surface deformation makes it difficult to clearly detect the surface deflection due to the surface reflections that occur. Consequently, it can be seen from this that, at high laser powers, the resulting surface deformation causes a measurement inaccuracy that makes it impossible to detect structural changes that are difficult to measure.

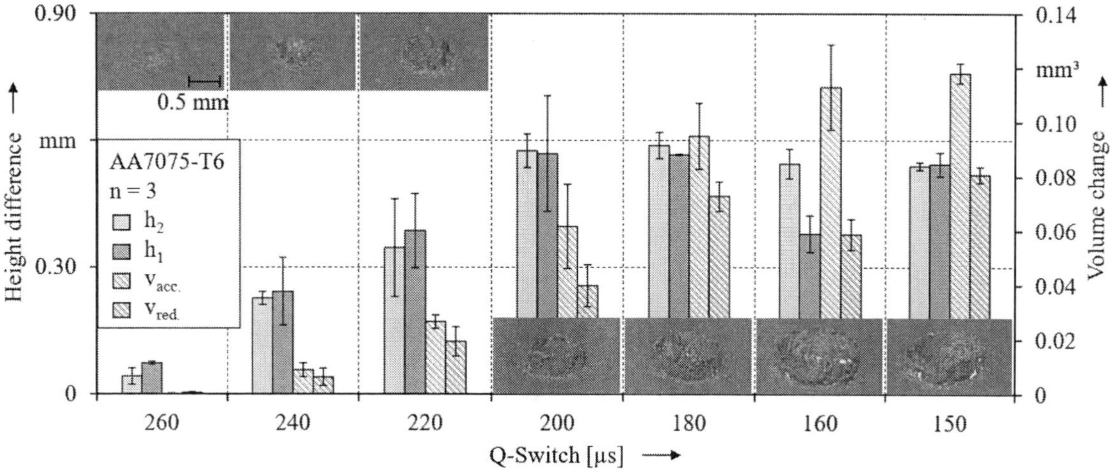

Figure 7. Influence of the laser power on the sample surface.

5. Summary and Outlook

In this study, it was demonstrated that by adjusting the Q-switch delay, it is possible to adjust the laser power of a laser ultrasound system. Using ultrasound measurements, it was also demonstrated that the laser power has an influence on the signal quality via the number of measurement points. It became evident that excessively high laser power produces unusable measurement signals with increasing measurement duration. Subsequent surface measurements validated that at high laser powers in combination with high sample temperatures a deformation of the sample surface takes place in the form of a keyhole. As a result, this study demonstrated that by adjusting the laser power, a continuous detection of microstructural changes is possible with a laser ultrasound-based measurement method at high measurement frequencies and temperatures without degrading the measurement quality over time.

Further studies will use this knowledge basis for the time-related determination of microstructural and precipitation changes of aluminum alloys at elevated temperatures. Thus, for the first time, the in-situ detection of temperature- and time-dependent precipitation processes at elevated temperatures becomes measurable.

Acknowledgments
The authors would like to thank the German Research Foundation (DFG) for supporting the present investigations, which were performed within the scope of the research project ''Manufacturing of tailored aluminum parts by controlling the local cooling rates in a combined forming, quenching and hardening process (Tailor Quenched Forming)'' (DFG 456415464) and funding of the laser-ultrasound sensor within the research project associated with the "Gottfried Wilhelm Leibniz-Preis". The authors would also like to thank the student Jakob Kölbl for his support.

References

[1] Kammer C 2021 *Aluminium Taschenbuch 1 - Grundlagen und Werkstoffe* (Berlin: Beuth)
[2] Argandoña E, Galdos L, Ortubay R, Mendiguren J and Agirretxe X 2015 Room temperature forming of AA7075 aluminum alloys *Key Engineering Materials* 199–204
[3] Geiger M, Merklein M and Vogt U 2009 Aluminum tailored heat treated blanks *Prod. Eng. Res. Devel.* **3** 401–10
[4] Huo W, Hou L, Zhang Y and Zhang J 2016 Warm formability and post-forming microstructure/property of high-strength AA 7075-T6 Al alloy *Materials Science and Engineering: A* **675** 44–54
[5] Garrett R P, Lin J and Dean T A 2005 An investigation of the effects of solution heat treatment on mechanical properties for AA 6xxx alloys: experimentation and modelling *International Journal of Plasticity* **8** 1640–57
[6] Lloyd D J, Chaturvedi M C 1982 A calorimetric study of aluminium alloy AA-7075 *Journal of materials science* **17** 1819–24
[7] Milkereit B, Wanderka N, Schick C and Kessler O 2012 Continuous cooling precipitation diagrams of Al–Mg–Si alloys *Materials Science and Engineering: A* **550** 87–96
[8] Liang Z 2012 *Clustering and Precipitation in Al-Mg-Si Alloys* (Berlin: Helmholtz-Zentrum Berlin)
[9] Milkereit B, Starink M J, Rometsch P A, Schick C and Kessler O 2019 Quench Sensitivity of Aluminium Alloys: Analysis of the Kinetics and Nature of Quench-Induced Precipitation *Materials* ***12*** 4083
[10] Kolar M, Pedersen K O, Gulbrandsen-Dahl S and Marthinsen K 2012 Combined effect of deformation and artificial aging on mechanical properties of Al–Mg–Si Alloy *Transactions of Nonferrous Metals Society of China* **8** 1824–30
[11] Feng Z, Liu C, Ma P, Yang J, Chen K, Li G, Chen L and Huang Z 2021 Initial holding time dependent warm deformation and post-ageing precipitation in an AA7075-T4 aluminum alloy *Journal of Materials Processing Technology* **294** 117111
[12] Krautkrämer J, Krautkrämer H 2013 *Werkstoffprüfung mit Ultraschall* (Berlin, Heidelberg: Springer)
[13] Norm European standard EN 23279 2017 *Non-destructive testing of welds — Ultrasonic testing* (Berlin: Beuth)
[14] Norm European standard EN 15317 2014 *Non-destructive testing – Ultrasonic testing* (Berlin: Beuth)
[15] Norm European standard EN 10160 1999 *Ultrasonic testing of steel flat product of thickness equal to or greater than 6 mm (reflection method)* (Berlin: Beuth)
[16] Schiebold K 2015 *Zerstörungsfreie Werkstoffprüfung - Ultraschallprüfung* (Berlin, Heidelberg: Springer Vieweg)
[17] Garcin T, Ueda K and Militzer M 2017 Reverse Austenite Transformation and Grain Growth in a Low-Carbon Steel *Metall and Mat Trans A* **2** 796–808
[18] Zhu Z, Peng H, Xu Y, Song X, Zuo J, Wang Y, Shu X and Yin A 2021 Characterization of Precipitation in 7055 Aluminum Alloy by Laser Ultrasonics *Metals* **2** 275

Characterization of the Mechanical Properties of Selectively Embossed Sheet Metal Materials Under Multi-Axial Loads

Pascal Heinzelmann, David Briesenick, Mathias Liewald

Institute for Metal Forming Technology, University of Stuttgart, Holzgartenstraße 17, 70174 Stuttgart, Germany

E-mail: Pascal.Heinzelmann@ifu.uni-stuttgart.de

Abstract. The application of embossed structures close to the sheet surface allows for work hardening to be introduced into sheet metal materials, leading to an improvement in their mechanical properties. So far, this was demonstrated in previous research work via tensile tests with differently embossed dual-phase steel specimens. Since tensile tests only reproduce uniaxial loading of the sheet metal material, no conclusions can yet be drawn about the property modification that can be achieved by near-surface embossing concerning real-life, multi-axial loading scenarios. To get deeper knowledge about the deformation behaviour of embossed components, it is important to consider multi-axial load cases. In the investigations reported in this paper, a multi-axial load case was realized by testing embossed bending specimens of DP600 and DP800 using a three-point bending device. The tests showed an overall improvement in the maximum bending force needed for the three-point bending test due to embossing of the materials investigated. Further, the tests indicated that lower-strength materials get more influenced by embossing than higher-strength materials. By simulating the process of embossing the bending specimens and the subsequent three-point bending test, an even deeper understanding of the material property modification could be generated.

1. Introduction

Global warming since years poses one of the biggest challenges mankind is facing in the early 21st century. It is largely caused by greenhouse gas emissions such as CO_2, which are produced, for example, as a result of motorized traffic. Both the US and the EU have therefore committed themselves to reducing CO_2 emissions, especially from road traffic. [1] Here, in addition to filtering and storing emitted gases, the aim of reducing emissions can also be achieved by saving fuel. In this regard, automotive manufacturers are focusing on lightweight construction as one approach to this problem. Thus, for example, weight-optimised aluminium alloys are increasingly being used in today's car body construction instead of heavier steel alloys, while retaining comparable strength. [2]

Another possibility to save weight is to selectively adjust material properties in relevant component areas, e.g. through selective embossing of the sheet metal components. In this process, indentations are embossed into the sheet material close to the surface, whereby material properties are changed in these areas due to induced work hardening. [3] Initial tests were carried out in 2006 by Namoco et al. using work-hardened aluminium alloys A6061-T4 and A5052-H34 [4].

Furthermore, Abel et al. investigated the formability of deep-drawn sheet metal components and found a positive effect for formability by selective embossing and reforming [5]. Walzer et al. performed tensile tests with embossed tensile specimens made of the dual-phase steel (DP) DP600 and found an increase in tensile strength due to the embossing [6]. First bending tests with embossed DP500 specimens were carried out by Briesenick et al. using a three-point bending device. It was shown that embossed bending specimens require a higher maximum bending force than unembossed bending specimens by performing a three-point bending test. [7]

The mentioned works have presented first results on property changes caused by induced work hardening. Aluminium, transformation-induced plasticity (TRIP) and DP steels were investigated. For general verification of the material properties tensile strength, yield strength and elongation at break, tensile tests were carried out by Walzer et al. In the three-point bending tests carried out by Briesenick et al., the maximum bending force was used as the measured variable to compare the embossed with the unembossed state. For dual-phase steels, property changes worth mentioning were found in this work. Dual-phase steels with tensile strength up to 600 MPa were used in the tests. In the three-point bending tests carried out, dual-phase steel with a tensile strength of 500 MPa was used.

To assess the performance of embossed sheet metal components, e.g. in the automotive context, material parameters determined in the three-point bending test under multi-axial loading, such as the maximum bending force, are more suited than uniaxial tensile test parameters. Regarding the multi-axial material properties of the embossed sheet metal materials DP600 and DP800, no experimental results have been published so far. Only aluminium, TRIP steels, and a DP500 steel were investigated. This paper aims to contribute to closing this gap, by performing three-point bending tests with DP600 and DP800 sheet metals.

Furthermore, the design of sheet metal components and the associated simulation of embossing is an essential part of process simulation. Therefore, the entire process will be simulated for this paper. For this purpose, the embossing of the bending specimens and performing the three-point bending test will be simulated and compared to the experimental process.

2. Method

For the experiments, bending specimens with dimensions 60 mm × 60 mm (see figure 1) were prepared and tested according to DIN EN ISO 7438 with the universal testing machine Roell + Korthaus RKM100. To quantify the property changes due to embossing, three-point bending tests were carried out on embossed and unembossed specimens made from DP600 and DP800 sheet metals, aligned the rolling direction of the specimens with the support and the bending punch. All experiments were performed with the embossed side facing the bending punch. The experiments were designed to quantify the maximum bending force needed in three-point bending tests.

Table 1. Maximum bending forces measured during three-point bending tests with unembossed DP600 and DP800 bending specimens in various sheet thicknesses.

Material	Sheet metal thickness t [mm]	Maximum bending force [N]
DP600	1.5	4837.7
DP600	1.77	6725.5
DP800	1.5	6529.8
DP800	1.85	10,461.4

In addition to the two different materials DP600 and DP800, two different sheet thicknesses were also considered in the investigations carried out. The sheet thicknesses $t = 1.5$ mm and $t = 1.77$ mm were used for DP600 and $t = 1.5$ mm and $t = 1.85$ mm for DP800. As a reference

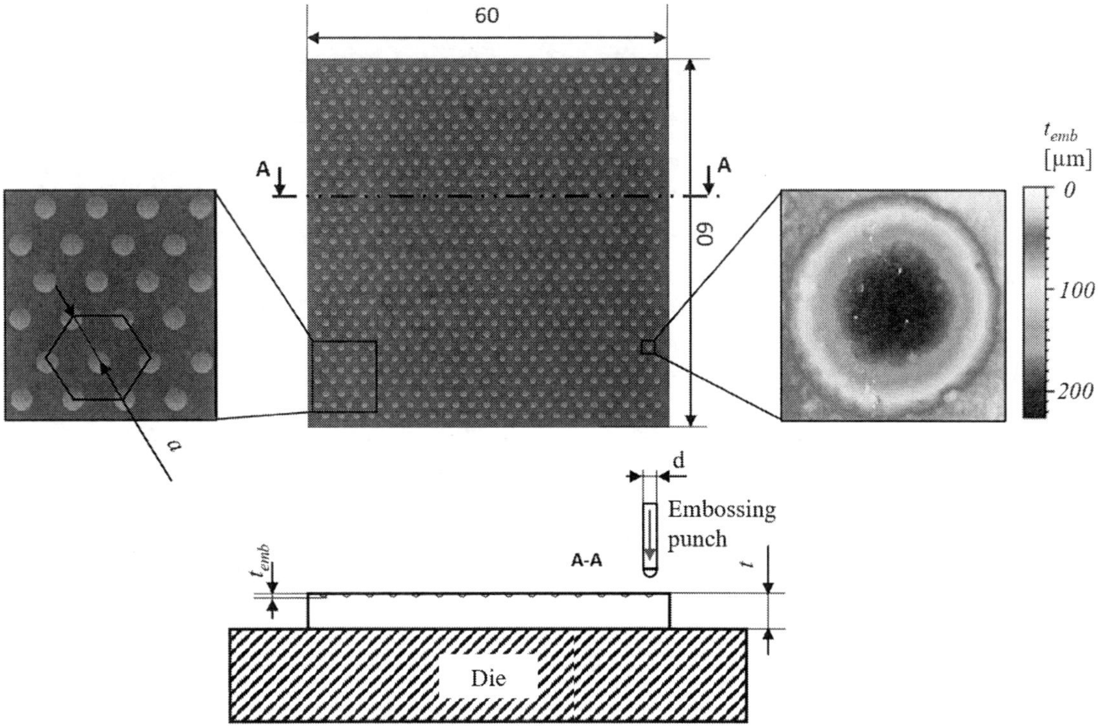

Figure 1. Overview of the embossing process, showing embossing geometry variables and a measuring example of the embossing depth t_{emb}.

for further tests, unembossed specimens were first made of the two materials investigated with the different sheet thicknesses. The corresponding results from the three-point bending tests are presented in table 1.

For the embossed specimens, six different embossing patterns were used, when producing the embossed bending specimens. The embossing patterns shown in figure 1, used can be described by the parameters a, t_{emb}, and d. The parameter a represents the distance of an embossing to its direct neighbours in a regular hexagonal pattern. So, the embossings represent the corners of a regular hexagon. The second parameter is the embossing depth t_{emb}. It indicates how deep the embossing punch penetrates the sheet metal material during the embossing process. For the embossing punch, a stamping punch was milled with the head turned into a hemisphere. The diameter of the punch, and thus of the hemisphere, is described by the parameter d.

All embossing patterns were produced using the same embossing punch with a diameter of $D = 4\,\text{mm}$. The embossing distance a was varied between 1 mm, 1.5 mm and 2 mm and the embossing depth t_{emb} between 100 µm and 200 µm. Combined with the two sheet metal materials and the two different sheet thicknesses in each case, 24 specimen configurations were thus tested. For statistical purposes, four replicate tests were carried out for each configuration. The bending specimens were produced with a CNC punching machine TRUPunch 5000R. This punching machine allows embossings to be applied on the sheet metal material freely in their position at up to 1400 strokes per minute. The embossed and unembossed bending specimens were punched by the CNC punching machine and subsequently milled to the final geometry. To prove the accuracy of the embossing depth, embossed bending specimens were randomly subjected to a depth measurement with the µsurf confocal microscope. In figure 1 the depth measurement of

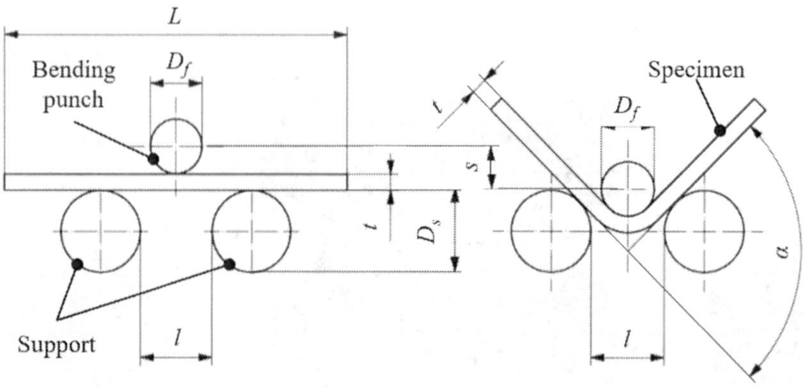

Figure 2. General illustration of a bending device with two supports and a bending punch for three-point bending tests. Representation based on [8] with adjustments regarding the variables.

embossing is shown as an example. Here, the depth in the z-direction is visualized by a colour gradient from red (0 µm) to blue (200 µm).

Embossing on the edges and in the centre was measured. The fluctuations in embossing depth between the measured embossings were small.

Only specimens having embossings depths within the range of (100 ± 10) µm and (200 ± 10) µm, respectively, were used for the bending tests.

The specimens were placed on the rotatably mounted supports and are loaded linearly by the translational movement of the bending punch. The required force and the displacement of the punch were recorded during the tests. The test setup is shown schematically in figure 2.

The supports of the test setup had a diameter D_s of 12 mm and the bending punch D_f diameter of 10 mm. The displacement s of the bending punch was fixed to 10 mm in all experiments. The distance l between the supports was set depending on the used sheet thickness t and was determined by

$$l = (D_f + 3a) \pm \frac{a}{2}. \tag{1}$$

To simulate the process of embossing numerically the bending specimens and processing the three-point bending tests, two independent simulations were performed with the LS-DYNA simulation package. The results (stress, strain, and geometry) of the embossing process were used for the further simulation of the three-point bending test. The simulation setup (as shown in figure 3) of the embossing process consisted of the three parts embossing punch (rigid), die (rigid), and bending specimen. To model the embossing punch, shell elements were used as the surface of a spherical segment. A plane surface of shell elements was used for the die. The bending specimen was modelled as a cuboid with dimensions according to figure 1. Solid elements with an element edge of length 0.5 mm were used to model it.

The edges of the bending specimen were translationally and rotationally locked in all spatial directions. Movements of the punch and the die were synchronized in the X and Y directions and opposed in the Z direction. The embossing punch penetrated the bending specimen 100 µm respectively 200 µm, depending on the embossing pattern. The die prevented the displaced material from forming on the opposite sheet surface, thus ensuring high work hardening. The contacts embossing punch and bending specimen, as well as die and bending specimen, were modelled as "forming oneway surface to surface". The materials were modelled as "MAT 024" with the flow curves given by previous work. Strain rate effects were not considered. On average, one simulation took 5 hours on an Intel® Core™ i7-7700 processor (8 cores @ 3.6 GHz).

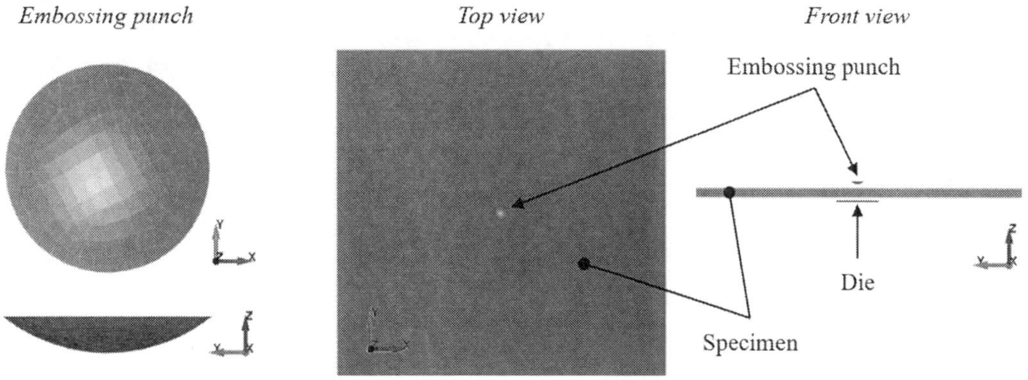

Figure 3. Simulation of the embossing process for a three-point bending specimen.

Gained data from the embossing process simulations were used in the following as an input for the simulations of the three-point bending tests of embossed bending specimens. Concretely, the geometry, the stresses and the strains calculated from the embossing process simulation were used for the further bending simulations with the "dynain" file. The simulation of the three-point bending test, shown in figure 4, consists of the bending punch (rigid), the support (rigid), and the bending specimen. There were no boundary conditions used for the bending specimen in any direction of translation or rotation. In the simulation, the bending punch moves in a negative Z direction by 10 mm. The contacts between both the support and the bending specimen, as well as the bending punch and the bending specimen, were modelled as "oneway surface to surface". The simulations of the three-point bending tests have been carried out with an Intel® Core™ i7-7700 processor (8 cores @ 3.6 GHz) for an average of 1.5 hours and was performed with explicit time integration.

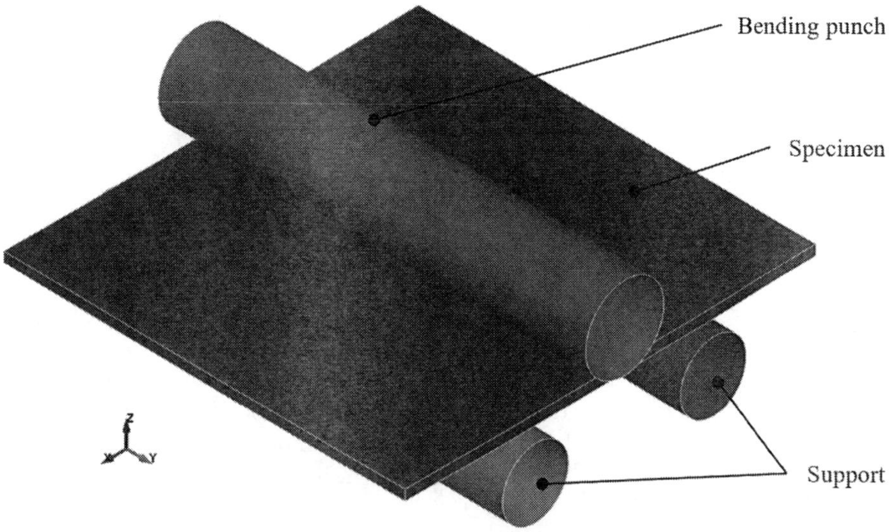

Figure 4. Simulation setup of the three-point bending test with an unembossed bending specimen.

3. Results and Discussion

This section is divided into two parts. The results of the three-point bending tests were discussed first, afterwards the simulative illustration of the embossing process and the three-point bending test second. Determined maximum bending forces measured by the experimental three-point bending tests are shown in table 2.

Table 2. Results of the three-point bending tests.

Material	Sheet metal thickness t [µm]	Embossing depth t_{emb} [µm]	Embossing distance a [mm]	Maximum bending force [N]
DP600	1.5	Reference	Reference	4837.7
DP600	1.5	100	1	5914.0
DP600	1.5	100	1.5	5497.5
DP600	1.5	100	2	5210.0
DP600	1.5	200	1	6139.8
DP600	1.5	200	1.5	6056.9
DP600	1.5	200	2	5801.0
DP600	1.77	Reference	Reference	6725.5
DP600	1.77	100	1	8186.4
DP600	1.77	100	1.5	7905.9
DP600	1.77	100	2	7431.2
DP600	1.77	200	1	8535.8
DP600	1.77	200	1.5	8184.7
DP600	1.77	200	2	7484.2
DP800	1.5	Reference	Reference	6529.8
DP800	1.5	100	1	7678.8
DP800	1.5	100	1.5	7363.6
DP800	1.5	100	2	6983.6
DP800	1.5	200	1	7449.4
DP800	1.5	200	1.5	7257.1
DP800	1.5	200	2	7098.3
DP800	1.85	Reference	Reference	10,461.4
DP800	1.85	100	1	11,673.5
DP800	1.85	100	1.5	11,156.0
DP800	1.85	100	2	11,125.4
DP800	1.85	200	1	12,215.2
DP800	1.85	200	1.5	11,915.3
DP800	1.85	200	2	11,599.5

In figure 5, required maximum bending forces detected during the performed three-point bending tests are shown depending on the embossing depth t_{emb} of the respective test configuration. Depicted results show that near-surface embossing generally leads to an increase in the maximum bending force required. Specifically, for the same material and same sheet thickness, the maximum bending force required increases when reducing the embossing distance a and increasing the embossing depth t_{emb}.

Furthermore, the bending force values recorded in the three-point bending tests were used for the validation of the embossing simulation described above. For this purpose, the experimentally determined force-displacement curves were compared with the correspondingly

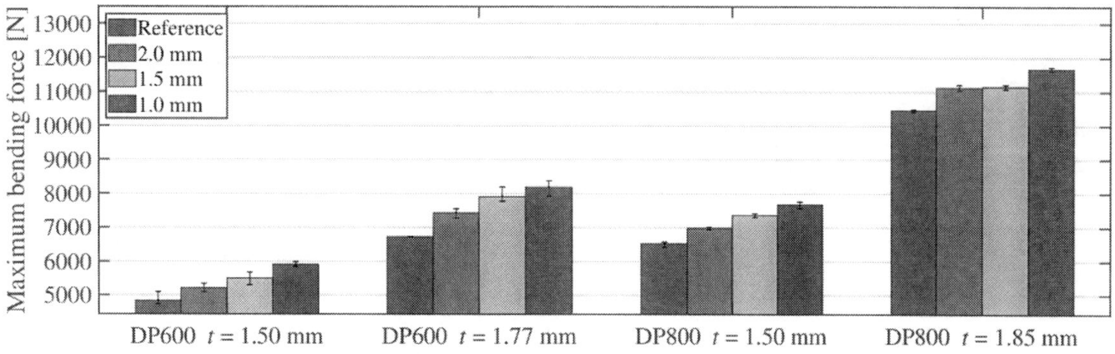

(a) Results for the embossing depth $t_{emb} = 100\,\mu$m.

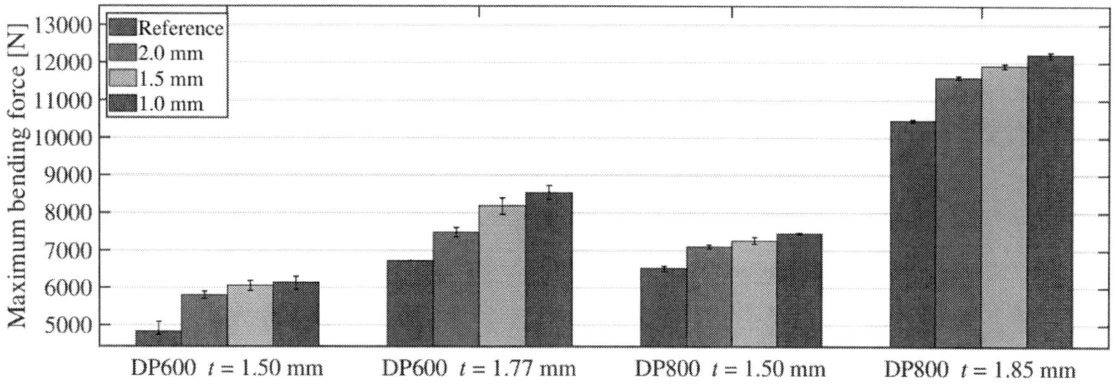

(b) Results for the embossing depth $t_{emb} = 200\,\mu$m.

Figure 5. Maximum bending force of experiments.

calculated simulation results. This comparison was carried out for both the bending tests on the unembossed and the embossed specimens. In this regard, figure 6 exemplarily shows the comparison between the experimentally determined and the numerically calculated bending force curves for the unembossed specimen made of DP800 with a sheet thickness t of 1.5 mm.

Qualitatively, the two curves in figure 6 are similar, but there are slight deviations between these two curves. The small deviations are presumably caused by the fact that the supporters and the bending punch were assumed to be rigid in the simulation, and thus the elastic effects from the universal testing machine were not considered for the calculations. These deviations could also be due to dynamic effects from the time integration or not considered manufacturing processes of the sheet metal material. Nevertheless, the simulated results are sufficiently accurate compared to the experimental data.

Figure 7 show the force-displacement curves obtained from DP800 specimens with a sheet thickness t of 1.85 mm and different embossing configurations are shown. Qualitative deviations can also be observed in these tests, which again can be explained by the elasticity of the punch and supporters not being considered in the simulations. Here, the deviations of the curves are small. They can be explained by the lack of elasticity in the simulation.

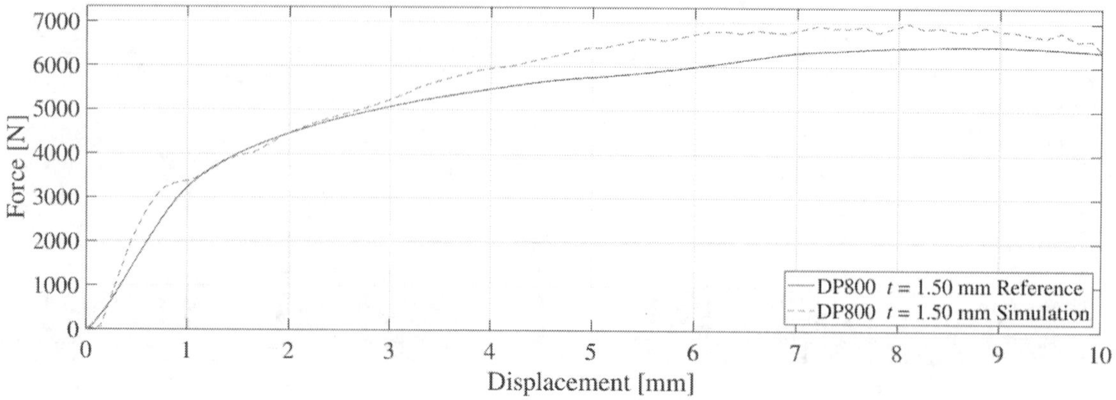

Figure 6. Comparison of experiment and simulation force and displacement results.

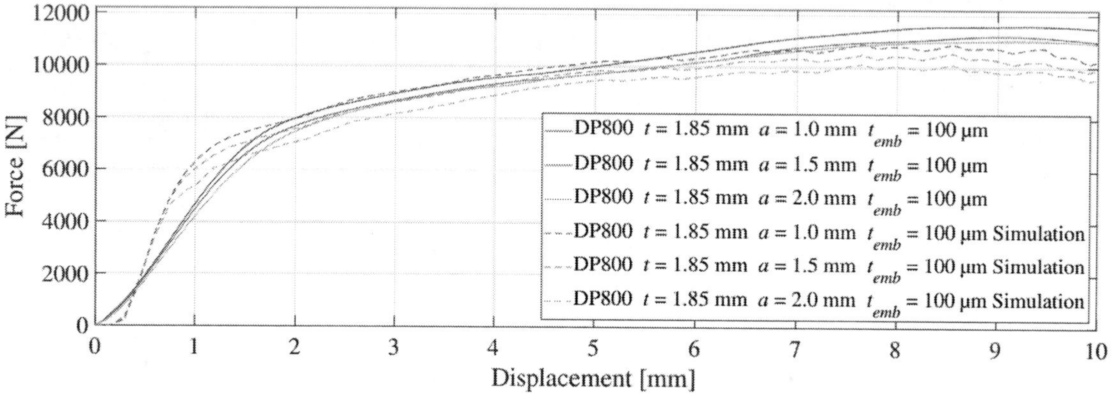

Figure 7. Comparison of experiment and simulation force and displacement results.

4. Conclusion

In the study presented in this paper, the influence of near-surface embossing on the bending properties of sheet metal components was investigated both experimentally and numerically. Experimentally, six embossing patterns with different distances ($a = 1\,\text{mm}$, $1.5\,\text{mm}$ and $2\,\text{mm}$) between the indentions and different embossing depths ($t_{emb} = 100\,\mu\text{m}$ and $200\,\mu\text{m}$) were tested. The embossing patterns were applied to bending specimens made of two materials with two different sheet thicknesses (DP600: $t = 1.5\,\text{mm}$ and $1.77\,\text{mm}$ and DP800: $t = 1.5\,\text{mm}$ and $1.85\,\text{mm}$). As a result of the experiments, it was determined that for the covered parameter range, embossing close to the surface leads to an increase in the maximum required bending force in a three-point bending test. It was also observed that increasing the embossing depth t_{emb} further encourages this effect. When considering the distances between the embossings, as a result, it could be stated, that the maximum bending force required increases with a reduction of the embossing distance in the given parameter range.

In addition to quantifying the property changes achievable by the embossings, both the embossing and the bending process could be modelled relatively accurately by means of FE simulations. The focus of the embossing process simulation was the realistic representation of the embossings introduced into the bending specimens. For the comparison between a simulated and a real embossing process, the three-point bending test was simulated to be able to make qualitative statements about the embossing simulation. To evaluate the simulation results, the force-displacement curves of the respective simulation and the experiments were compared. Here, slight deviations between the different curves were detected. This could be due to the elastic effects from the universal testing machine, the bending punch, and the support. These effects were not taken into account for the simulations. To simulate the embossing process for the design of sheet metal components, the simulated embossing process provides sufficiently accurate results. So, simulative embossed parts can be used for further simulations along the simulative manufacturing process.

Acknowledgements

The research project "Herstellung crashoptimierter Blechbauteile mittels Prägen" of the Europäische Forschungsvereinigung EFB e. V. is carried out in the framework of the industrial collective research programme (IGF no. 21940N). It is supported by the Federal Ministry for Economic Affairs and Climate Action (BMWK) through the AiF (German Federation of Industrial Research Associations e. V) based on a decision taken by the German Bundestag. Additionally, we would like to thank our project partners DYNAmore GmbH, Eckold GmbH & Co. KG, synchropress GmbH, Engineering System International GmbH, Karosseriewerke Dresden GmbH, Zeller + Gmelin GmbH & Co. KG, Novelis Switzerland SA, Tata Steel Europe, STRACK NORMA GmbH & Co. KG, F+K Werkstoffprüfung und Labor GmbH, and KIRCHHOFF Automotive GmbH.

References

[1] Shanmugam K, Gadhamshetty V, Yadav P, Athanassiadis D, Tysklind M and Upadhyayula V K 2019 *ACS Sustainable Chemistry & Engineering* **7** 4951–4963
[2] Friedrich H E 2017 *Leichtbau in der Fahrzeugtechnik* (Wiesbaden: Springer Fachmedien Wiesbaden)
[3] Namoco C S, Iizuka T, Sagrado R C, Takakura N and Yamaguchi K 2006 *Journal of Materials Processing Technology* **177** 368–372
[4] Namoco C S, Iizuka T, Narita K, Takakura N and Yamaguchi K 2007 *Journal of Materials Processing Technology* **187-188** 202–206
[5] Abe Y, Mori K i, Maeno T, Ishihara S and Kato Y 2019 *Production Engineering* **13** 589–597
[6] Walzer S and Liewald M 2019 *AIP Conference Proceedings* **2113** 160006
[7] Briesenick D, Walzer S and Liewald M 2021 Study on the effect of embossing on the bending properties of high-strength sheet metals *Forming the Future* Springer eBook Collection ed Daehn G, Cao J, Kinsey B, Tekkaya E, Vivek A and Yoshida Y (Springer International Publishing and Imprint Springer)
[8] DIN EN ISO 7438: Metallische werkstoffe - biegeversuch

IOP Conf. Series: Materials Science and Engineering 1284 (2023) 012044 doi:10.1088/1757-899X/1284/1/012044

Performance of stringer sheet parts under static and dynamic load

S Biffar and P Groche

Institute for Production Engineering and Forming Machines (PtU), Technical University of Darmstadt, Otto-Berndt-Straße 2, 64287 Darmstadt, Germany

E-mail: simon.biffar@ptu.tu-darmstadt.de

Abstract. The stringer sheet forming process chain enables the cost-effective production of components with significantly increased stiffness for mass markets such as the automotive sector. In the first process step, a stringer is added to a conventional sheet by laser welding. In the subsequent process step, the stringer sheet parts are formed by a deep drawing operation. Previous investigations of component performance focused in particular on the static stiffness of the stringer sheets. The behavior under dynamic load, such as in crash events, has not been systematically investigated to date. Static testing focused on the stringer height as the only influence on part performance. This study characterizes the crash performance of stringer sheet components by means of a drop hammer test. The energy absorption of the components is evaluated in the experiment based on the rebound height of the drop hammer, whereas the structural integrity is characterized based on the maximum dynamic as well as permanent deformation. Furthermore, a quasi-static compression test is carried out to obtain the static stiffness of the parts. In these tests, the geometric design of the stringer sheets as well as two forming process parameters are varied individually. The influence of these parameters on the static stiffness, energy absorption and structural integrity is assessed and discussed on their own and with regard to their influence on the part weight.

1. Introduction

Lightweight construction plays a crucial role in the pursuit of resource efficiency and environmentally conscious product design. For this purpose, it is necessary to transform the material properties into product properties by means of, but also limited by, the manufacturing process. Such a limitation is encountered, for example, in the production of flat stiff sheet metal products, such as those in car body construction, machine enclosures and panels of white goods. In addition to the introduction of beads [1–3], the stiffness of conventionally deep-drawn components can in general only be increased by raising the sheet thickness.

To enable the reinforcement of such components by means of stiffening elements (e.g. stringers) attached orthogonally to the sheet plane, the process chain of stringer sheet forming was developed at the PtU. In a first step, stringers are added to conventional sheet metal by laser welding. For the subsequent deep drawing process, a modified tool is used, which is equipped with slots in the punch and blank holder that receive the stringers in the forming process and support them laterally to prevent buckling. A detailed description of the process chain, which is depicted in figure 1, can be found in [4].

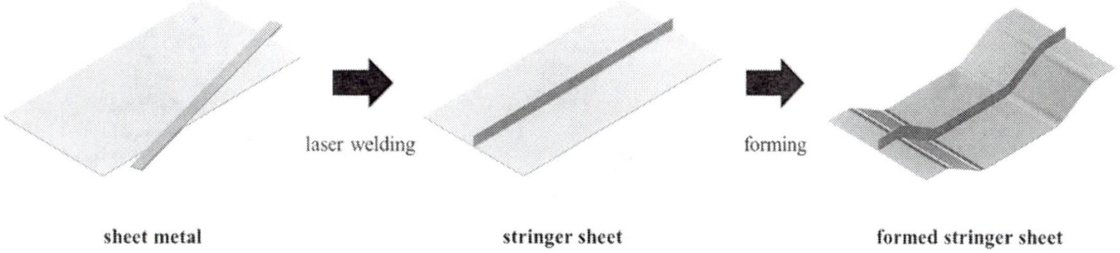

Figure 1. Stringer sheet forming process chain

As described by [5], the stiffness of the components can thus be increased significantly, which in turn allows for a reduction of the sheet thickness. In an example car body part, 22.9% of material could be saved by switching to a stringer sheet, while the stiffness of the part remained at the same level [6]. However, static stiffness is only one of several relevant parameters, especially for car body components. Behavior under dynamic load, i.e. in the event of a crash, is also of major importance in several contexts of application. This behavior has not yet been investigated for stringer sheets; previous publications refer exclusively to stiffness.

The dynamic behavior of components can be described in terms of structural integrity and energy absorption [7]. Energy absorption refers to the amount of kinetic energy that is absorbed by the component; this essentially occurs through plastification [8]. Structural integrity is not as clearly defined [9] – in the course of this paper, it is described by the elastic and permanent deformation of the component under dynamic load. Using these parameters, it is possible to make generalized statements about the behavior of the component in the event of a crash, even though this always depends on the components immediately surrounding it, the impact angle and the impact speed.

Previous studies focused exclusively on the height of the stringer as an influencing factor on the performance of stringer sheet parts. Additional variables like sheet thickness or forming process parameters were not investigated. However, these can also have significant effect on the component properties. For example, the stiffening effect of the stringers is based on the local increase of the area moment of inertia of the component. The height of the stringer is included in this cubically (in addition to the increase in the Steiner component due to the displacement of the centroid of the surface of the stringer), but the width of the stringer is incorporated linearly as well.

To asses both the dynamic behavior and the influence of other parameters than stringer height, the following scientific questions are to be answered by the present paper:

1) Are the described methods suitable for the evaluation of the parameters stiffness, structural integrity and energy absorption of stringer sheet components?
2) Which input parameters have a significant influence on the stiffness, structural integrity and energy absorption of stringer sheet components?
3) How does this influence relate to the mass of the part?

2. Method

Static and dynamic behavior of components are not independent of each other, but basically result from the same geometry and material properties of the specimen under investigation [7]. In order to better understand the relationships between these quantities, the present paper therefore measures the static quantity stiffness in addition to the dynamic quantities energy absorption and structural integrity. The compression test established by Köhler [10] is used to measure the stiffness of the components. As shown in figure 2 on the left, the components are subjected to a compression test in a tension-compression testing machine. The forward and backward stroke is 2 mm. This results in a slight plastic deformation, which is necessary to compensate for the warpage resulting from the manufacturing process. The stiffness is calculated from the slope of the force-displacement curve in the elastic return stroke.

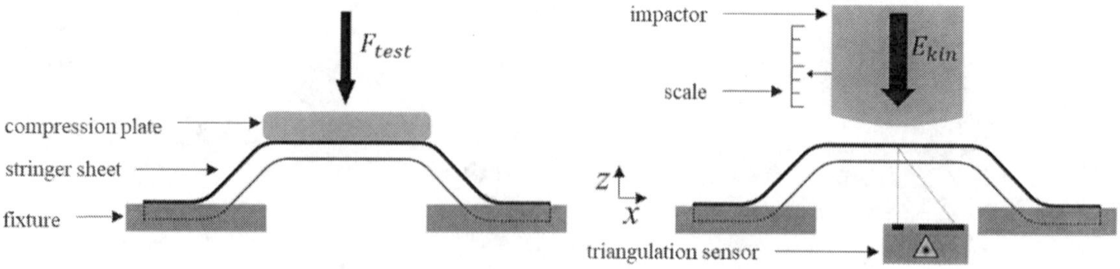

Figure 2. (left) Compression test; (right) Drop hammer test

The measurement of energy absorption and structural integrity is carried out by means of a drop hammer test rig. This is shown schematically in figure 2 on the right and consists of a holder in which the specimen is placed, and an impactor. The latter is guided laterally by rollers on profiles and thus only has one translational degree of freedom in the Z direction. In the test, the specimen is brought to a height of 1469 mm above the component surface and then dropped. On impact, the kinetic energy of the impactor is transferred to the specimen causing it to deform. From the springback of the elastic deformation component, energy is then transferred back to the impactor so that it rebounds.

A measuring mark is applied to the impactor, which points to a fixed measuring scale in the z-direction, on which the instantaneous height of the test specimen can be taken. This is done during the tests with a high-speed camera (240 frames/second). Underneath the test specimen is a fixture for the component to be tested, which is equipped with a slot to accommodate the stringers. The component can be centered to the impactor by positioning grooves, but is not fixed further during the test. Also attached to the fixture is a laser triangulation sensor, which continuously measures the deformation of the component in the z-direction from below. The measuring point is located in the center of the component in the x-direction and 3 mm off-center in the Y direction to prevent the beam from overlapping with the stringers.

The experimental setup allows the rebound height of the impactor to be evaluated from the videographic recording of the scale. By comparing the potential energy at the beginning of the free fall and that at the apex of the rebound, the energy absorbed in the experiment can be calculated in principle according to the following formula:

$$E_{abs} = E_{pot,0} - E_{pot,1} = mg(h_0 - h_1)$$

where E_{abs} is the absorbed energy, E_{pot} is the potential energy, m is the mass of the impactor, g is the acceleration due to gravity, and h is the rebound height measured on the scale. Index 0 denotes the initial state of the system, index 1 denotes the state at the time the drop hammer reverses motion during rebound. However, the energy absorbed by the component can be calculated as shown above only under the assumptions that the friction of the drop hammer with the air and the guide rails is zero, and that the component support and the substrate are ideally rigid. Both assumptions do not apply in reality, so that the calculation of energy absorption is not permissible under the given conditions. However, it can be assumed that the friction in the system and the elastic behavior of the component support do not change significantly between the tests. It is therefore possible to use the rebound height to make a comparison between the components in terms of their energy absorption without knowing their numerical value.

The deformation of the profile in the Z-direction, which is measured by the laser triangulation sensor, is used to assess the structural integrity. It is assumed that the displacement of the measuring point in the y-direction by 3 mm has a negligible influence on the measured variable. With ideal alignment of the impactor to the laser sensor, the drop hammer radius of 125 mm results in a systematic error of 0.036 mm. This value is one order of magnitude below the typical standard deviation of the measured values. In addition to measuring the deformation in the test, the permanent deformation in the center of the component is also recorded using the laser sensor.

In addition to the stringer height, further input parameters are investigated in this study. With regard to the forming process, these are the lubrication of the sheet (where maximum lubrication refers to a substantial amount of lubricant that reliably prevents oil film break) and the blankholder force used. With regard to the semi-finished product, the stringer height, the stringer width, the deviation of the stringer profile, the sheet thickness in the base sheet and the width of the base sheet were varied individually. The parameters are shown in table 1 and illustrated in figure 3. Both the stiffness and drop hammer tests were repeated three times. All specimens are made of 1.0338 steel.

Table 1: Parameter set

Parameter			**Standard config.**		
Stringer Height, *h* in mm		7	14	20	
Stringer Width, *t* in mm			1	1.5	2
Sheet Thickness, *d* in mm			1	1.5	2
Sheet Width, *b* in mm		50	150		
Blank Holder Force in kN	50	75	100		
Lubrication		Pre-Lube	Maximum		
Stringer Profile		Negative	None	Positive	

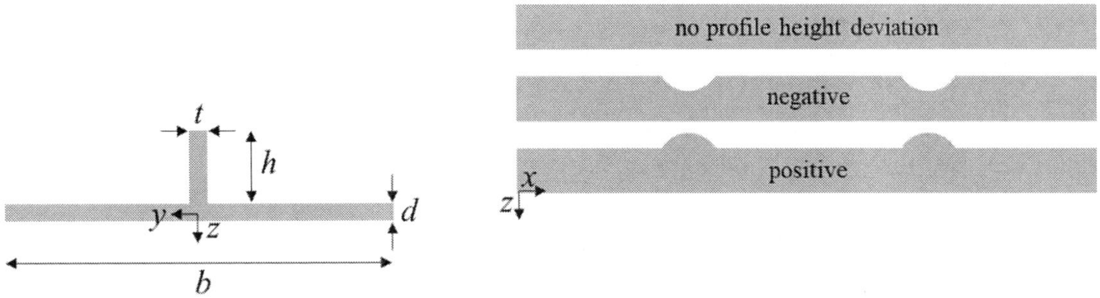

Figure 3. (left) geometric parameters of the stringer sheet; (right) stringer profile height deviation

3. Results

The results of the tests are shown in figure 4 together with their estimated standard deviation. The drop hammer test on the components with 7 mm high stringers failed. As expected, the component absorbed significantly less energy than the standard configuration. As a result, the drop hammer punched through to the fixture. Therefore, the parameters under investigation could not be reasonably calculated. However, it can be seen that the component has the lowest energy absorption and worst structural integrity of all configurations investigated.

The measured values of the dynamic and permanent deformation in the drop hammer test differ little from each other among the individual tests, which is evident from the low standard deviation. The measurement of the rebound height also produces reproducible measured values, but with a larger scatter. When measuring stiffness, there is greater scatter in the measured values. Stiffness is a parameter obtained as the first derivative of the component's force response to the deformation path in the stiffness test. Therefore, the parameter is more susceptible to variations in the measured data. The quality of the data is sufficient within the scope of the investigation to be able to make basic statements.

Stiffness, structural integrity and energy absorption are considered separately in the following. In each case, the mean values of the test results are normalized to the mean value of the test results of the standard configuration in order to obtain a simple metric for the influence of an input parameter on the respective component property.

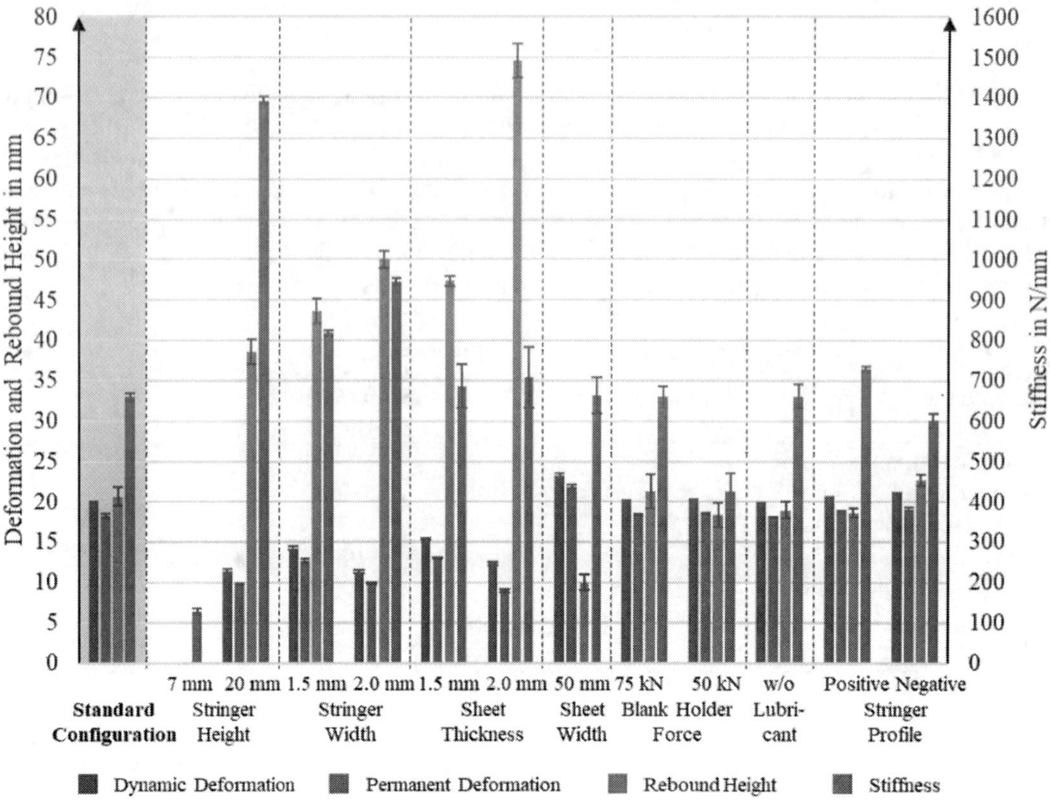

Figure 4. Results of the experiments

3.1. *Stiffness*

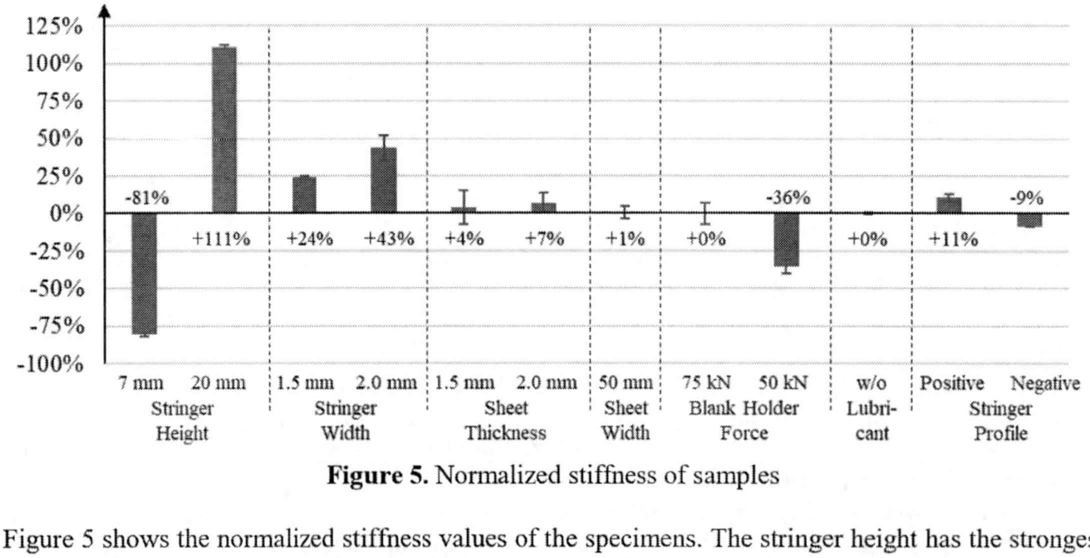

Figure 5. Normalized stiffness of samples

Figure 5 shows the normalized stiffness values of the specimens. The stringer height has the strongest effect on the stiffness. The stringer width also has a considerable influence. Increasing the thickness of the base sheet has only a minor effect, which, moreover, cannot be interpreted reliably due to the large scatter of the measured values. The width of the base sheet has a negligible effect on the stiffness. A moderate reduction in the blankholder force shows no influence on the stiffness, while a significant drop was measured when it was halved to 50 kN. The presence or absence of lubricant shows no change in stiffness. The stringers with positive height variation show a slightly increased stiffness, while those with negative height variation show a slightly decreased stiffness.

3.2. *Structural Integrity*

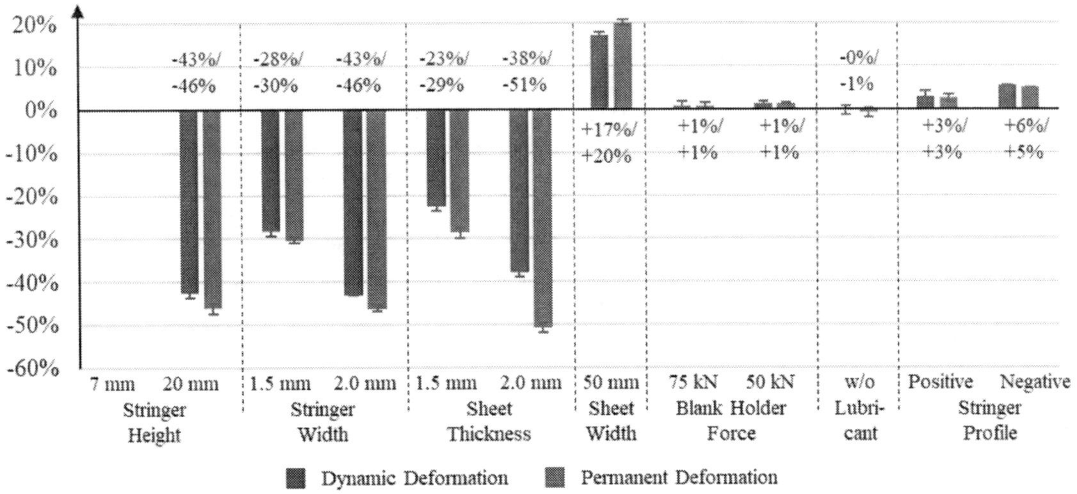

Figure 6. Normalized dynamic and permanent deformation of samples

The dynamic and permanent deformation of the specimens normalized to the respective value of the initial specimen are depicted in figure 6. In general, it can be assumed that a reduction of both values is advantageous with regard to the intended use. The strongest effect on dynamic deformation is due to the height and width of the stringers. The permanent deformation is most affected by the thickness of the base sheet. Reducing the width of the base sheet has a negative effect on both variables. Changing the blank holder force as well as the lubricant quantity has no influence on either quantity. Both the positive and the negative stringer profile deviations worsen the component behavior minimally.

3.3. *Energy absorption*

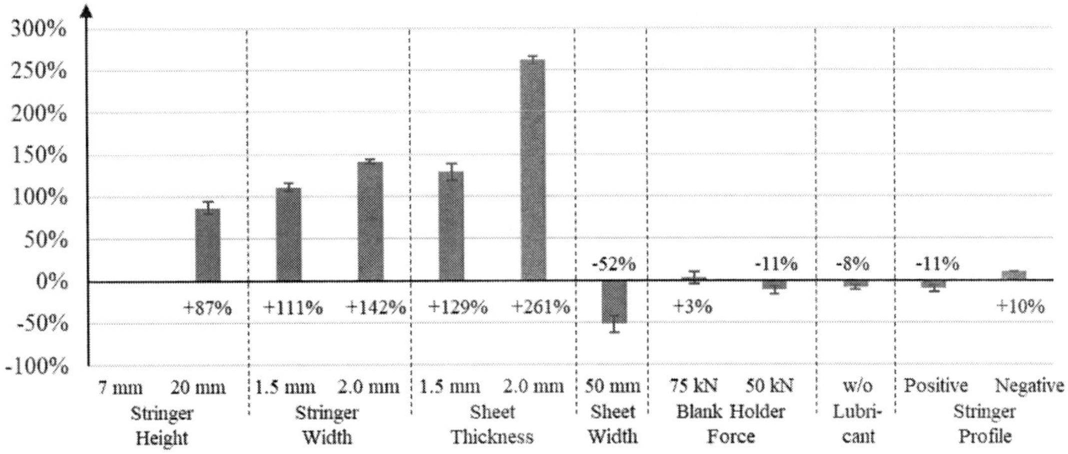

Figure 7. Normalized rebound height

Figure 7 shows the normalized rebound height as a measure of energy absorption. The thickness of the base sheet has the strongest effect on the rebound height. The stringer height, stringer width and width of the base sheet also have an influence on the rebound height. The other variables have a small to negligible effect.

4. Disscussion

4.1. Suitability of measuring methods

The applied measurement methods are basically well suited for measuring stringer sheet parts with respect to the quantities stiffness, structural integrity and energy absorption. The data allow to differentiate the influence of different input parameters. They are also consistent in themselves (with the exception of the stiffness at a hold-down force of 50 kN, see below).

However, there are some limitations, especially with regard to the measurement of energy absorption. Since the friction and energy absorption components of the fixture are not known, no calculation of the actual energy absorption can take place. Only a qualitative comparison regarding the rebound height is possible. With regard to friction, a gyroscopic sensor can be integrated into the drop hammer, which allows the speed of the drop hammer to be measured immediately before and after the impact. With this, the kinetic energy may be evaluated instead of the potential energy at the beginning of the test and at the reversal point of the rebound movement, which means that the friction losses in the guide no longer influence the measured values. In addition, the millimeter scale used in the test proved to be too coarse; the use of a gyroscopic sensor would provide data with a finer resolution. With regard to the energy absorption of the device, a more rigid design is required. Here, for example, a machine bed made of polymer concrete or gray cast iron can be used.

Another limitation is the comparatively high scatter of the stiffness data. Because stiffness is obtained as the first derivative of the force-displacement curve in the return stroke, it is sensitive to variations in the data. To counter this, for example, an increase in the number of specimens tested, a different evaluated parameter such as the maximum force achieved in the compression test, or clamping of the specimens to compensate for warpage from production should be considered.

4.2. Influence of the input parameters on the investigated variables

To begin with, it can be stated that the stringer height and the stringer width represent significant influencing parameters on all the component characteristics investigated. The base sheet thickness and the base sheet width also have an influence on the structural integrity and the energy absorption, in the selected design limits this even has a stronger effect than the properties of the stringers. There is a strong correlation between the two quantities of structural integrity (r=0.987). There is correlation between structural integrity and energy absorption (r= 0.758 and r=0.840 between dynamic and permanent deformation and rebound distance, respectively). There is a slight correlation between stiffness and structural integrity (r=0.474 and r=0.404, respectively), and there is no correlation between stiffness and rebound displacement (r=0.119).

The process parameters of the forming process essentially have no relevant influence on the component properties. One exception is the blank holder force with regard to stiffness. Here, a significant drop can be observed at 50 kN. Neither a significant geometric deviation of the components before testing nor a similarly pronounced effect in the other measured variables was found. It is possible that the decrease in blank holder force influenced the strain hardening and residual stresses in the stringer by less tensile stress being exerted on it during forming. Further investigations must be carried out to be able to make a conclusive statement.

The stiffness of the components depends mainly on their area moment of inertia with respect to the y-axis (as defined in figure 3). The area moment of inertia is calculated as follows:

$$I_y = \frac{bd^3}{12} + \frac{th^3}{12} + th\left(\frac{d+h}{2}\right)^2$$

where *h* is the stringer height, *t* the stringer width, *d* the thickness of the base sheet and *b* the width of the base sheet. The gradient of I_y is calculated as

$$\nabla I_y = \begin{pmatrix} \frac{\partial I_y}{\partial h} \\ \frac{\partial I_y}{\partial t} \\ \frac{\partial I_y}{\partial d} \\ \frac{\partial I_y}{\partial b} \end{pmatrix} = \begin{pmatrix} th^2 + \frac{td^2}{4} \\ \frac{h^3}{3} + \frac{hd^2}{4} \\ \frac{bd^2}{4} + \frac{thd}{2} \\ \frac{d^3}{12} \end{pmatrix}; \qquad \nabla I_y(h_0, t_0, d_0, b_0) = \begin{pmatrix} 196.50 \\ 918.17 \\ 44.50 \\ 0.08 \end{pmatrix}$$

and the numerical values at the point of standard configuration $h_0 = 14\,mm, t_0 = 1\,mm, d_0 = 1\,mm, b_0 = 150\,mm$ are presented on the right. The highest numerical value is calculated for the slope in the dimension of the stringer width. However, the stringer width is varied within much narrower limits than the stringer height, which can explain the apparently greater influence of the stringer height on the stiffness in the measured stiffness values. The thickness of the base sheet is varied within the same limits as the stringer thickness, but contributes to a much lesser extent to the gradient. The base sheet width has almost no relevance with regard to the stiffness.

An explanation of the relationships with regard to energy absorption and structural integrity is not readily accessible. The reason for this is the significant plastification together with the high rate of deformation. The duration of the impact is only about 12 μs in the case of the standard configuration. For both dynamic quantities, the dimensions of the base sheet become significantly more important. One possible explanation is that the propagation of the applied impulse initially occurs in all directions simultaneously and thus the transverse direction of the component initially also contributes significantly to the absorption of the impulse. As a result, both the thickness and the width of the base sheet become more relevant. A more detailed investigation by means of a numerical analysis of the impact process as well as a high-speed recording of the test with a frame rate greater than 1000 fps may contribute to the understanding of this matter in the future.

4.3. Influence of geometry parameters normalized on mass

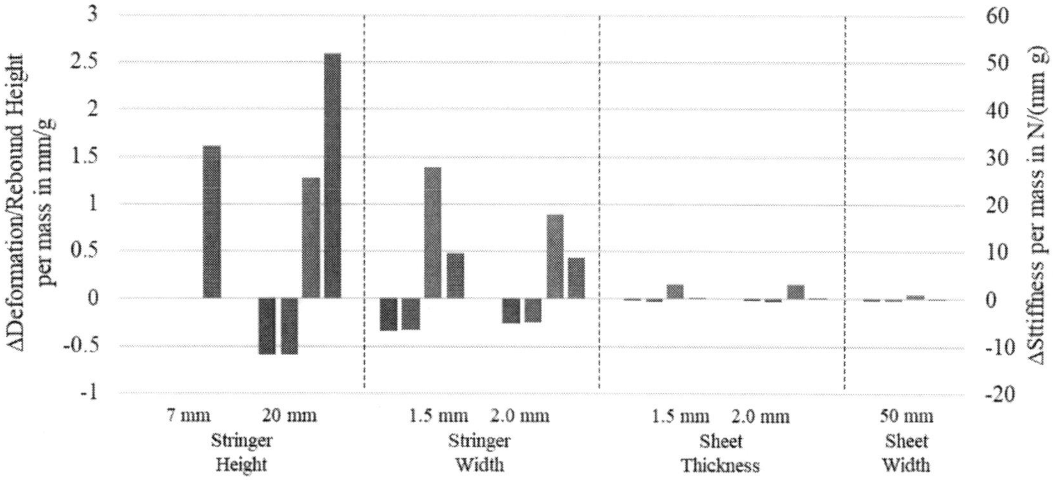

Figure 8. Change in dynamic and permanent deformation, rebound height and stiffness per mass

The graph in figure 8 represents how the measured quantities change with mass. The difference between the measured value and the corresponding value of the standard configuration is shown divided by the difference in masses. It can be seen that the geometric sizes of the stringers have the best relationship in terms of weight efficiency. The thickness and width of the base sheet require a relatively large amount of material to achieve similar effects. The values for the stringer profile deviations are not shown. Due to the very small mass differences from the standard configuration, very large values are obtained here for all component characteristics. The design space for such elements is much more limited than for the remaining influencing variables, which is also shown by the low absolute values achieved. Therefore, a comparison is only meaningful to a limited extent.

5. Conclusion

The presented measurement methods for the acquisition of static and dynamic properties of stringer sheet parts is suitable within the discussed limitations. The assumption of previous works of the stringer height as the most relevant geometric parameter for the component behavior has been confirmed within the scope of this investigation. However, in the dynamic case, the stringer width and the width and thickness of the base sheet are of similar importance. When considering the mass efficiency, it becomes apparent that the geometric dimensions of the stringer have the greatest influence per added mass on all characteristics of the component. This is particularly important in view of the argument of resource and material efficiency raised at the beginning of this article.

The present investigation was confined to the separate variation of the input parameters. This means that the interactions between these variables could not be assessed. The results of this paper can now be used to define the experimental matrix for a combined investigation of the parameters confirmed as relevant. In parallel, a simulation methodology can be developed and validated with the results obtained. The simulation of the drop hammer test can also be expected to provide further insight into the mechanisms at impact.

In the discussion of the parameters influencing stiffness, the area moment of inertia of the cross-sectional profile of a stringer sheet was used as the basis for explaining the observed effects. A more detailed modeling, which also includes the longitudinal shape of the stringer sheet part, would make it possible to achieve an analytical calculation of the stiffness. This allows the behavior to be estimated during component design without the need for complex numerical or experimental investigations.

Overall, it has been shown that stringer sheet parts allow a significant increase in performance not only in static but also in dynamic load cases. This is particularly true when it comes to designing material- and resource-efficient components, as significant improvements in component behavior can be achieved through minimal additional weight.

Acknowledgements
The research project IGF 21721/P1441 "Tailored Stringer Sheet Forming – Maßgeschneiderte Stegblechbauweisen" from the Research Association for steel Application (FOSTA), Düsseldorf, is supported by the Federal Ministry of Economic Affairs and Climate Action through the German Federation of Industrial Research Associations (AiF) as part of the programme for promoting industrial cooperative research (IGF) on the basis of a decision by the German Bundestag. The project is carried out at the Institute for Production Engineering and Forming Machines of TU Darmstadt.

References

[1] Majic N, Albers A, Kalmbach M et al. (2013) Development and statistical evaluation of manufacturing-oriented bead patterns. *Advances in Engineering Software* 57:40–47. https://doi.org/10.1016/j.advengsoft.2012.11.018
[2] Birkert A, Dreiseitel F, Hartmann B et al. (2021) Adapted part design methods for springback minimization of stamped sheet metal car body components. *IOP Conf Ser.: Mater Sci Eng* 1157:12038. https://doi.org/10.1088/1757-899X/1157/1/012038
[3] Hofmann D, Liewald M (2016) Deep drawing of one-sided, stiffness increasing sheet metal composites. *KEM* 716:879–890. https://doi.org/10.4028/www.scientific.net/KEM.716.879
[4] Groche P, Köhler S, Kern S (2018) Stamping of stringer sheets. *Journal of Manufacturing Processes* 36:319–329. https://doi.org/10.1016/j.jmapro.2018.10.025
[5] Köhler S, Groche P, Baron A et al. (2016) Forming of stringer sheets with solid tools. *AMR* 1140:3–10. https://doi.org/10.4028/www.scientific.net/AMR.1140.3
[6] Heckmann M, Köhler S, Biffar S et al. (2019) Stegblechumformung für den Karosseriebau. *ZWF* 114:554–558. https://doi.org/10.3139/104.112155

[7] Sahr C, Berger L, Lesemann M et al. (2010) Systematische Werkstoffauswahl für die Karosserie des Superlight-Car. *ATZ Automobiltech Z* 112:340–347. https://doi.org/10.1007/BF03222166
[8] Lee S-J, Lee H-A, Yi S-I et al. (2013) Design flow for the crash box in a vehicle to maximize energy absorption. *Proceedings of the Institution of Mechanical Engineers, Part D: Journal of Automobile Engineering* 227:179–200. https://doi.org/10.1177/0954407012451545
[9] Diermann V, Middendorf P (2019) Automatic evaluation of structural integrity in crashworthiness simulations using image analysis. *Int.J Automot Technol* 20:65–72. https://doi.org/10.1007/s12239-019-0006-y
[10] Köhler S, Rohnert C, Groche P (2018) Extension of geometric limits in drawing of stringer sheets. *Procedia Manufacturing* 15:693–700. https://doi.org/10.1016/j.promfg.2018.07.302

IOP Conf. Series: Materials Science and Engineering 1284 (2023) 012045 doi:10.1088/1757-899X/1284/1/012045

Micro incremental forming of thin SS304 foils and its microstructural study

Peeyush Mahajan[1], Jeet Patil[2], Sushil Mishra[1]

[1]Department of Mechanical Engineering, IIT Bombay
[2]Pillai College of Engineering, New Panvel

E-mail: sushil.mishra@iitb.ac.in

Abstract. Micro-forming of thin foils is challenging due to the size effect, friction effect, and tool design. The conventional macro-forming techniques cannot be directly applied to the thin foils. The incremental forming process (ISF) can be used for thin foils since it possesses higher formability compared to the conventional process. Hence, μ-ISF (micro incremental forming process) experiments were conducted on thin foils of SS304 stainless steel having 50, 100, and 200 μm thickness. As-received foils were further annealed at 950°C with a holding time of 20 mins to achieve strain-free microstructure. The Taguchi L18 orthogonal was applied to optimize the initial processing parameters for, μ-ISF experiments. Using the initial optimized parameters μ-ISF experiments were performed for all the thickness sheets up to the maximum wall angle. It was observed that the higher tool diameter and the lower step depth give the minimum roughness value in the deformed sheet. The tearing mode of failure was observed due to the formation of strain-induced martensite in the deformed sheet. The microstructure and bulk texture analysis relate the orientation and twin fraction with the deformation behaviour. From the texture evolution, it was observed that the brass {110} <112>, Cube {100} <001>, and goss {110} <001> texture component has a greater influence on the formability of SS304 stainless steel thin foils.

1. Introduction

Micro-forming is unique manufacturing processes, where one dimension of the component is in the micrometres. Among all Micro-forming process, micro-forming sophistically produces micro-scale features on thin metal sheets. Unique material behaviour at micro-scale makes the process challenging. One of the main challenges is the size-dependent deformation behaviour of the material, i.e., mechanical properties of the material are significantly affected by its size. This limits the direct application of traditional metal forming process parameters to micro-forming process. This necessitated evolution of new techniques and tools that must be developed to produce the desired micro-features. Frictional contact conditions in miniaturized component fabrications is another challenge in micro-forming that need to address more rigorously [1]. At the micro-scale, the contact area between the tool and the workpiece is very small, and friction can become a dominant factor in the deformation process. This can result in high surface roughness and can affect the dimensional accuracy of the produced micro-features. Furthermore, fewer grains in the thickness direction in the case of thin sheet could affect the material behaviour and the deformation process. Therefore, it is imperative to formulate new approaches to control the grain structure and deformation behaviour of the material during micro-forming [2][3][4].

To overcome these challenges, various techniques such as micro-stamping, micro-extrusion, and micro-drawing have been developed [5]. These techniques involve specialized tools and equipment that are designed to operate at the micro-scale and produce precise micro-features. Researchers have explored the size effect of sheet on the deformation behaviour during processing. Vollertsen et al. [6] studied the size effect in manufacturing the metallic components. They observed that with the decrease in the size of the metallic components the flow stress gets affected significantly. Moreover, the formability decreases with the decrease in the size of the component making micro forming process difficult to control. Sahu et al. [7] also investigated the size effect on formability and microstructural evolution in thin SS304 foils. They observed with the decrease in the foil thicknesses, the formability and the strength decrease. Palai et al. [8] studied the mechanical and the tribological behaviour of the vacuum heat treated ultra-thin SS304 foils. They observed that the with the vacuum heat treatment of the SS304 foils at 1100°C the yield strength dropped by 10% and the percentage elongation decreased by 70%.

Deep drawing, extrusion and the hydroforming are some of the processes being practise by manufacturer to produce large scaled products. Die design, tool design and process instability are prime topic of concerns when this process scaled down to the micro levels. Micro-incremental forming is one of the forming processes which is scaled down from macro level to micro level. Incremental forming process was invented by the Edward [9] in 1964 on a lathe machine. They demonstrated process advantages in terms of better formability and low deformation forces. Several studies have been conducted to study the process parameters [10][11] and the material formability [12][13]. Jeswiet et al. [14] and Petek et al. [15] developed a setup to measure the forming forces in the three principal directions. They observe that the forming forces in case of incremental forming process are comparatively low as compared conventional forming processes. In the year 2001, Saotome and Okamoto[13] proposed an ISF technique for the fabrication of the miniaturized components by hammering technique. To compensate the higher amplitude to hammering, backing plate was used for the forming. Obikawa et al. [16] developed an table top CNC in order to eliminate the need for the backing plate in case of μ-ISF. They observed that the tool rotation and step depth have a greater effect on the formability. Beltran et al. [17] studied the size effect in case of μ-ISF and concluded that the formability is significantly lower than the conventional incremental forming process. Moreover, the tearing and as well as the buckling mode of failure was observed in micro incremental forming process. Shi et al. [18] investigated the water jet incremental forming with and without the supporting of the dies. They observed that the tool path trajectory is extremely important in case of micro-incremental forming process. In addition, the buckling mode of failure was observed in case of water jet micro incremental forming process.

It has been evident that with experimental investigation, the microstructure, and the texture evolution have played a key role in understanding the deformation mechanism in case of polycrystalline material. [19]. Zhang et al. [20] studied the texture and the microstructure evolution in case of CP titanium on cold deep drawing. He found that the rolling texture components fraction changes in the stretching, drawing region and flange region. Moreover, recrystallized texture component disappears in the flange region due to the formation of twinning. Gurao et al. [21] studied the deformation behaviour of CP titanium at an extreme strain rates. They found that the texture of the deformed samples is weaker at higher strain rates compared to the deformation done at lower strain rates. In addition, strain hardening is more complicated at higher strain rates because of increased twinning. Mishra et al. [22] gives the insight about the crystallographic texture, microstructure evolution during the incremental sheet metal forming. He found that the tool diameter and the step depth have the minimal effect on texture evolution in Double sided incremental forming (DSIF). Bishoyi et al. [23]studied the microstructural and texture evolution in the uniaxial study of CP titanium. He found that the volume fraction of tensile and compressive twins has been reduced with the increase in the percentage of elongation irrespective of the deformation temperature. Barnwal et al. [24] studied the microstructure and the texture evolution on the forming behaviour of Al-6061alloy. He found that the critical texture component such as cube and brass influence the forming behaviour significantly.

From the literature it was found that the micro-incremental forming of ultra-thin sheets have not been studied. A detailed microstructural and the bulk texture analysis need to be done. With respect of mentioned work, stainless steel (grade 304) sheets with 50, 100, and 200 μm thickness have been chosen for the study. Tearing mode of failure have been observed on a 50μm sheet with 1 and 1.5mm tool diameter is due to the formation of stress induced martensite. A detailed investigation is needed in this section.

2. Experimental methodology

In the presented work, a μ-ISF experimental setup was modified from conventional incremental forming setup. It contains a hemispherical ball tool which gradually deforms the material through small incremental. Thin sheets have slippage issue during deformations that has been addressed with a new design of a backing plate. Backing plate has an extruded portion to provide clamping force. The sheet is held in between the backing plate and upper plate with four bolts. Design not only controlled the unwanted deformation during clamping but also the bending during handling was avoided. The schematic and the experimental setup of μ-ISF is shown in the Figure 1.

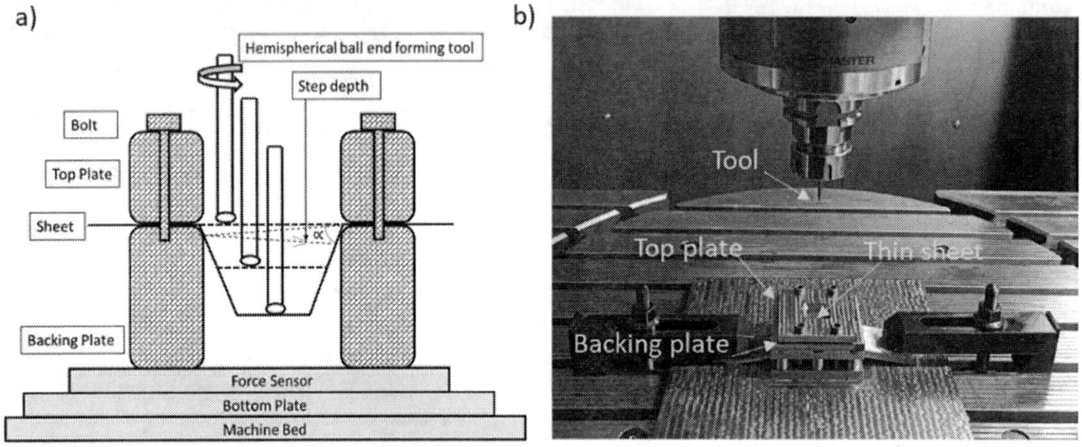

Figure 1. a) Schematic of the μ-ISF b) Experimental setup of μ-ISF

The experiments were conducted using a 5-axis numeric controlled machine. The fixture was mounted on a bottom fixture plate mounted on the machine bed.

2.1 Tensile testing

To study the size effect, uniaxial tensile tests were performed on an INSTRON® 3345 vertical column type machine with a 5KN capacity load cell. Due to the limited elongation of the sheet, the material's toughness was improved through annealing at 950°C for 20 minutes. Further, thickness to grain size measurement were carried out and it was found that 4, 8, and 15 was the ratio for sheet thicknesses of 50, 100, and 200μm, respectively. Tensile samples were prepared as per ASTM 345 standard, specimen design is shown in figure 2 (d). Wire-cut electric discharge machining (EDM) was used to cut the tensile samples in the rolling, transverse, and diagonal directions for all sheet thicknesses. Test were performed at a low strain rate of 0.5mm/min to eliminate strain rate effects. To ensure reproducibility, three tests were conducted for each condition. Figure 2 (a, b &c) depicts the true stress-strain sheet behaviour for each thickness.

2.2 Design of experiments

To understand the behaviour of process parameters in the micro incremental forming process DOE was formulated. An L18 orthogonal array with five factors and three levels was chosen. The DOE included parameters step depth, tool diameter, wall angle, and feed rate. Overall DOE is tabulated in the Table 1.

Roughness measurements for the deformed samples lower, middle, and upper regions for different forming parameters. Force measurements were performed using a mini Kistler dynamometer. Taguchi optimization revealed that step depth had the most significant impact, followed by feed rate and wall angle. The experiments were successful using a 0.75mm tool diameter, however fractures occurred with 1 and 1.5mm tool diameters for wall angles greater than 50°.

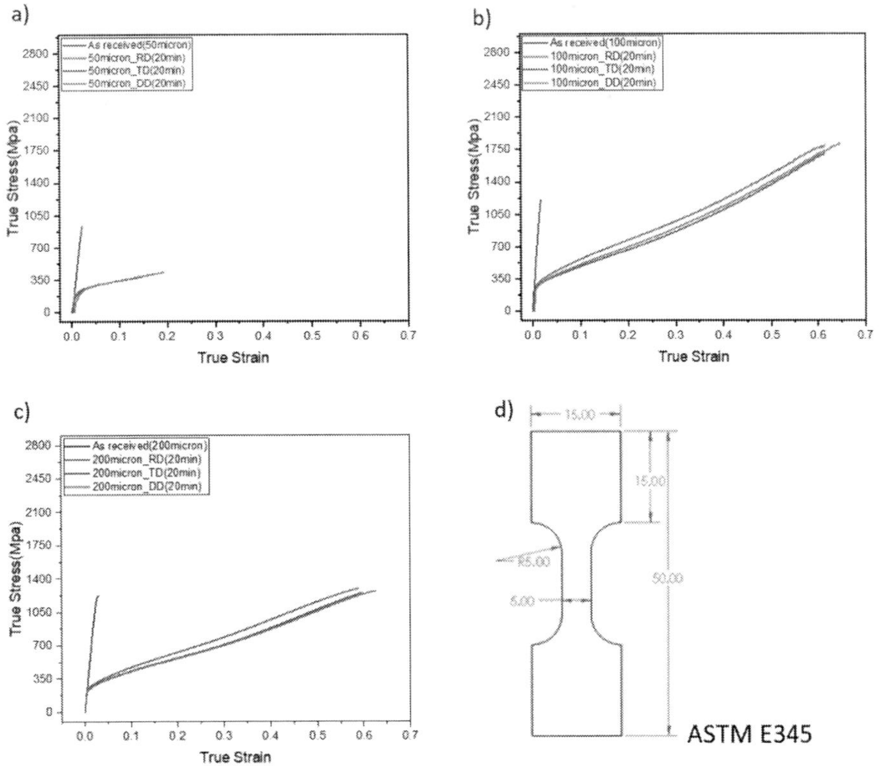

Figure 2. Tensile plots for the different sheet thicknesses i.e., 50μm (a), 100μm (b) 200μm(c) and tensile test design (ASTM E345) (d).

Table 1. Design of experiments (DOE) for μ-ISF.

Sr. No.	Sheet thickness(μm)	Tool diameter(μm)	Step depth (μm)	Feed rate(mm/min)	Wall angle(°degree)	Ra(Lower region) (nm) (Surface roughness)	Ra(centre region) (nm) (Surface roughness)	Ra(Upper region) (nm) (Surface roughness)	RMS value (Z) (N) (Root mean square value)
1	50	750	15	150	45	42	36	31	1.19
2	50	1000	25	250	50	49	39	35	1.26
3	50	1500	35	350	55	Fractured	…………	…………	…………
4	100	750	15	250	50	107	98	91	3.04
5	100	1000	25	350	55	201	165	110	4.53
6	100	1500	35	150	45	104	96	84	6.86
7	200	750	25	150	55	103	94	89	12.32
8	200	1000	35	250	45	70	39	21	16.16
9	200	1500	15	350	50	82	73	64	17.78
10	50	750	35	350	50	62	44	31	1.38
11	50	1000	15	150	55	Fractured	…………	…………	…………
12	50	1500	25	250	45	78	59	51	2.51
13	100	750	25	350	45	71	56	42	5.79
14	100	1000	35	150	50	62	77	98	5.24
15	100	1500	15	250	55	49	43	37	4.08
16	200	750	35	250	55	139	125	98	14.31
17	200	1000	15	350	45	63	59	54	12.46
18	200	1500	25	150	50	61	46	21	15.22

2.3 Surface roughness measurements
To determine the quality of the formed part, surface roughness was measured along the wall of the deformed part at three distinct locations: upper, centre, and lower region. ALICONA white light interferometer with a lateral and depth resolution of 200nm was used for surface roughness measurement.

2.4 Microstructure and texture evolution
The microstructure characterization was performed using the EBSD technique. The ZEISS Gemini 300 Field Emission Scanning Electron Microscope (FE-SEM) equipped with the OXFORD fast charge-coupled device (CCD) detector was used. A specimen was cut from the centre of the deformed sample's side wall. Sample was prepared with mechanical polishing with subsequent electropolishing. Sample was mechanically polished with 2500 Silicon carbide (SiC) polish paper. Whereas, electropolishing was performed at 20 V for 20 seconds with an electrolyte composed of 70% methanol, 20% perchloric acid, and 10% butanol. The same specimens were also used for the 2theta and bulk texture measurement. XRD studies were carried out on a PANalytical XRD machine outfitted with a GaliPIX-3D detector. The PANalytical, X-ray diffraction system has been used to study bulk texture. Further, to determine the maximum intensity of the peaks for the undeformed and deformed samples 2, analysis was performed for the theta range 30° to 110°. The bulk texture was then measured on a beam collimator area of 2mm × 2mm in X and Y directions, which covers a large area for texture measurements. The β-nickel filter was used to eliminate the β-filter radiation. Using the MTEX 5.1 software, four incomplete pole figures were used to calculate the orientation distribution function (ODF). Texture intensity has been calculated using the formula $\int \{f(g)\}^2\, dg$ where f(g) is the ODF intensity, representing anisotropy.

3. Results and discussion
Miniaturized conical shapes have been fabricated with different wall angles and aspect ratios using the micro-incremental sheet metal forming(μ-ISF). As discussed in the above section design of experiments, three different hemispherical ball end tools, such as 0.75, 1 and 1.5mm, have been used for the study. It was observed that the smaller tool diameter helps in improving the formability of the sheet.
A series of experiments have been done to find out the optimized forming parameters. These optimized parameters have been determined from the ANOVA analysis. It was found that step depth followed by wall angle and feed rate has a more significant impact in the case of micro incremental forming. These optimizations have been done considering the surface roughness. Based on the optimized parameters the constant feed rate that is 250mm/min and a constant wall angle of 50° have been chosen for the further study. With these optimized parameters the effect of three tool diameter with three different step depth for all the sheet thicknesses have been investigated.

3.1 Surface roughness measurements
With optimized parameters, the effect of the different tool diameters with different step depths is done for all the sheet thicknesses. Surface roughness measurements have been done in the three regions of the samples, the lower, centre, and upper region, as shown in the figure 3. The minimum roughness for all sheet thicknesses can be achieved by using a higher tool diameter, as it facilitates uniform contact and reduces waviness in the tool-sheet interface, resulting in a better surface finish. On the other hand, increasing the step depth leads to a higher surface roughness (Ra). With large step depth, localized sheet deformation increases which caused waviness in the deformed sheet and also the distance between consecutive steps increases tool marks on the deformed sheet, leading to a poorer surface finish. Therefore, it is recommended to use a higher tool diameter and a lower step depth to obtain a better surface finish.

It can be noted that the higher tool diameter and lower step depth that is 1.5mm and 15μm step depth gives us good surface finish and lower tool diameter that is 0.75mm at the same step depth gives us the poorer surface finish. Thus, for further investigation these cases were used for microstructure and texture evolution for 200 μm sheet thickness.

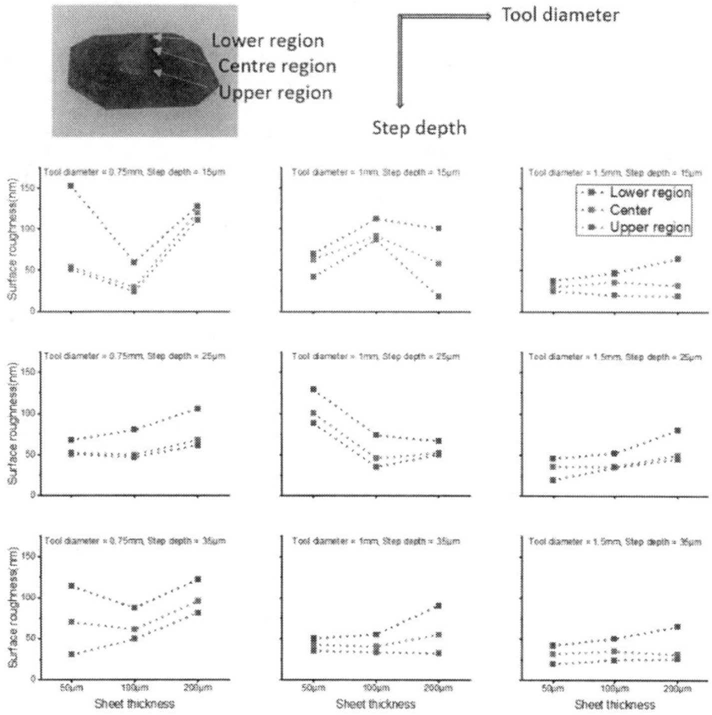

Figure 3. Surface roughness measurements plots for all the sheet thickness at different step depths with a different tool diameter

3.2. Microstructure and texture evolution

3.2.1 Microstructure evolution during μ-ISF.: It was found that as received sheets have elongated grains due to the high cold rolling. In order to maintain the thickness-to-grain size ratio the annealing of the sheets at 950°C for a holding time of 20min were performed. The Inverse pole figure of the as received material is shown in the figure 4 (a).

Figure 4 (b) shows the IPF, phase fraction chosen cases i.e., 200μm deformed sheet at 0.75mm and 1.5mm tool diameter. It was found that the phase transformation is there in the micro incremental forming process. A further investigation is needed in this section. From the IPF the grain average misorientation (GAM) value has been obtained from the software for the undeformed as well as the for the deformed sheet. The GAM value for the as received sheet is 0.72° which increases with the deformation of the sheet. For the same sheet thickness and same step depth with 0.75mm tool diameter the gam value increases to 2.1° whereas for the 1.5mm tool diameter the value increases to 2.45°. Therefore, we can conclude that the gam value increases with the increase in tool diameter for the same sheet thickness and step depth.

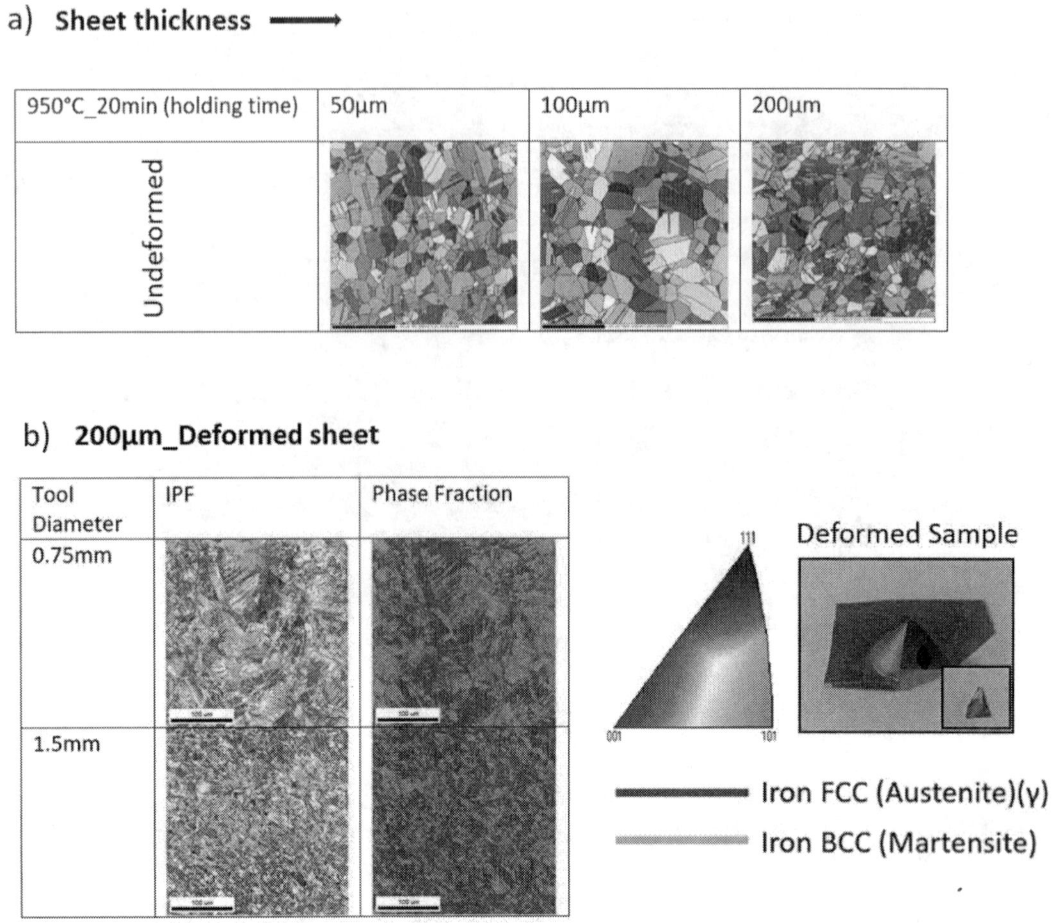

Figure 4. Inverse pole figure a) Undeformed sheet annealed at 950°C with holding time 20min b) 200μm deformed sheet at 0.75mm and 1.5mm tool diameter.

3.2.2 Texture evolution before and after μ-ISF.: The detailed bulk texture analysis, before and after deformation for chosen cases were carried out. Certain critical texture components, including Brass and cube, had a significant impact on the alloy's forming behaviour. From the literature, it was found that the in case of single point incremental forming, cube {001} <100> component is dominant deformation texture [25]. This texture assists shear deformation during the SPIF. Further, it is found that the {001} <100> cube texture component present in SPIF is undesirable, as it results in a higher stress value than the other texture component. In case of the metals with a SFE (surface free energy) less than 25mJ/m^2, their texture transition tends to shift towards an alloy-type texture known as fibre texture. This type of texture is characterized by a higher fraction of Brass {110} <112> and Goss {110} <001> components and a lower fraction of S {123} <634> and Copper {112} <111> components. The α-fibre in Euler space spans from the {011} <100> Goss texture component to the {011} <211> Brass (Bs) texture component. The standard texture component for the FCC material is given in the figure 5. As in the above section from the microstructural analysis it was found that the phase transformation is present in the micro incremental forming process.

For the same samples the 2-theta measurement also done using the X-ray technique and phase fraction has been calculated for the samples using the Rietveld analysis. The martensite fraction in the sample deformed with 0.75mm tool diameter is 82% whereas the samples deformed with 1.5mm tool diameter is 60%. Similarly, the orientation distribution function plots and volume fraction of the texture component have been determined using the MTEX 5.4.0 and MATLAB software. The volume fraction of the texture component has been calculated for both the austenite as well as for the martensite phase. As incremental forming is a plane strain deformation process [26] and it was clearly observed form the figure 5(c) that the fraction of cube component in the martensite phase increases which makes the material difficult to deform. Both goss and inverse goss texture component has been formed in the micro incremental forming which makes the material deformable.

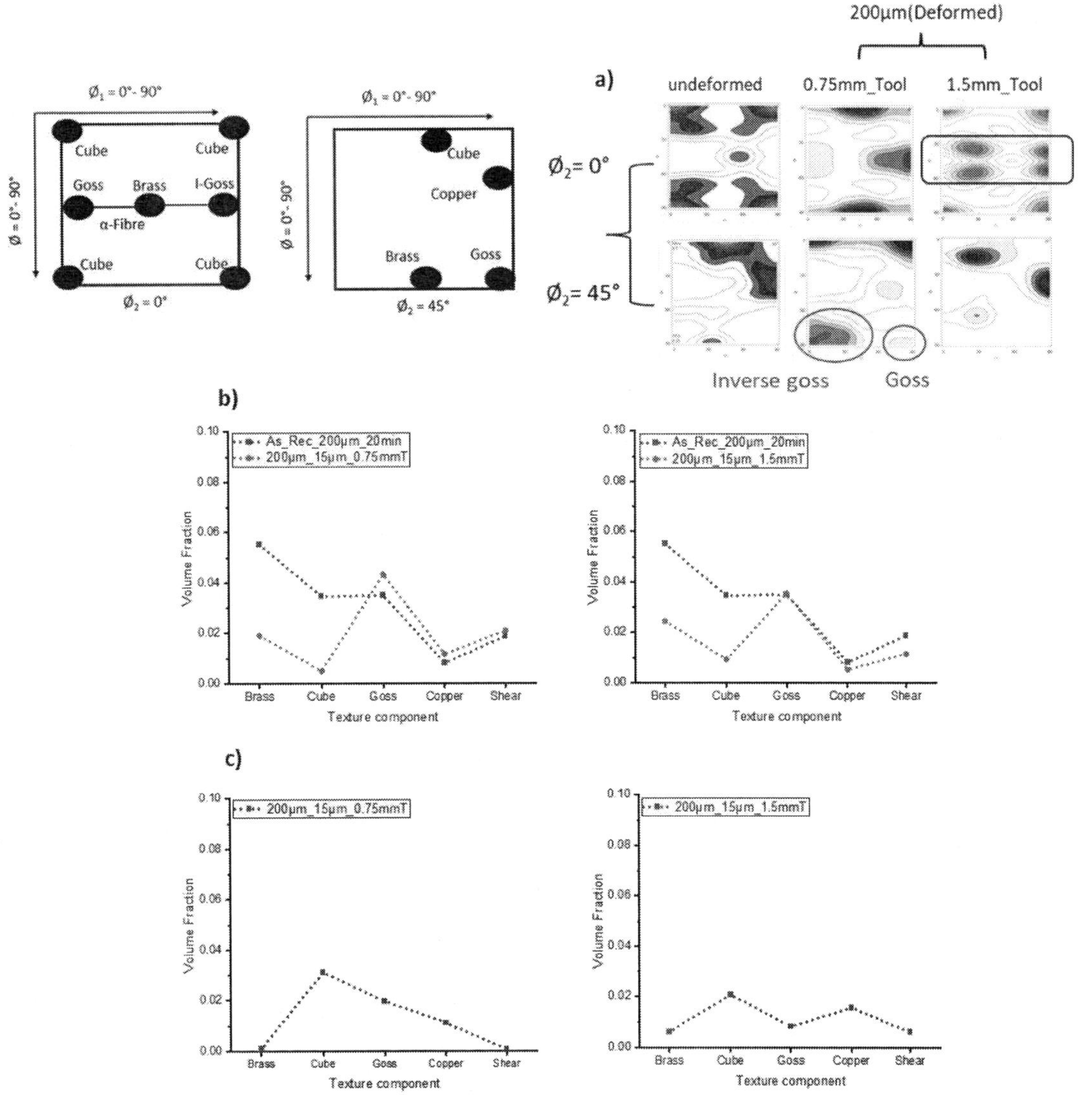

Figure 5. Texture evolution a) Orientation distribution function (ODF) for undeformed and deformed samples b) Volume fraction of texture components for austenite phase c) for martensite phase.

4. Conclusions

1. Tearing mode of failure has been observed in the 50μm sheet at higher step depth with 1mm and 1.5mm tool diameter due to the formation of stress induced martensite. A detailed investigation is needed to understand this formation of martensite in the micro-incremental forming process.
2. It was observed that the surface roughness decreases with the increase in tool diameter. The 1.5mm tool diameter gives us the minimum roughness for all the sheet thicknesses and step depth. It can be concluded that the higher tool diameter and lower step depth gives us the minimum roughness of the sheet.
3. From the microstructure analysis it was found the grain average misorientation increases with the increase in the tool diameter. For 1.5mm tool diameter there will be a more area of contact as compared to the 0.75mm tool diameter.
4. From the texture evolution it was found that the α-fibre texture has been formed after deformation which is true in the case of plain strain deformation.
5. The fraction of the cube component has been decreased in the austenite phase and increase in the martensite phase which makes the deformation difficult.

References

[1] Geiger, M. Kleiner, M. Eckstein, R. Tiesler, N. Engel U 2001 Microfroming *CIRP Ann. - Manuf. Technol.* **50** 445–62
[2] Hetz P, Kraus M and Merklein M 2022 Investigation of the Influence by Size Effects on the Material Characterization of the Uniaxial Compression Stress State *Key Eng. Mater.* **926 KEM** 955–62
[3] Hommel M and Kraft O 2001 Deformation behavior of thin copper films on deformable substrates *Acta Mater.* **49** 3935–47
[4] Liu Q and Hansen N 1995 Deformation microstructure and orientation of f.c.c. crystals *Phys. Status Solidi* **149** 187–99
[5] Razali A R and Qin Y 2013 A review on micro-manufacturing, micro-forming and their key issues *Procedia Eng.* **53** 665–72
[6] Vollertsen F, Biermann D, Hansen H N, Jawahir I S and Kuzman K 2009 Size effects in manufacturing of metallic components *CIRP Ann. - Manuf. Technol.* **58** 566–87
[7] Sahu J, Chakrabarty S, Raghavan R and Mishra S 2018 Investigations of size effect on formability and microstructure evolution in SS304 thin foils *J. Strain Anal. Eng. Des.* **53** 517–28
[8] Porwal D, Esther A C M and Dey A Microstructural, thermo-optical, mechanical and tribological behaviours of vacuum heat treated ultra thin SS304 foils *Mater. Res. Express* **3** 1–13
[9] Edward L, inventor; Edward, Leszak, assignee. Apparatus and process for incremental dieless forming. United States patent US 3,342 051. 1967 Sep 19. 1967 ÉHalli 4. -3 6 1964–6
[10] Kim Y H and Park J J 2002 Effect of process parameters on formability in incremental forming of sheet metal *J. Mater. Process. Technol.* **130** 42–6
[11] Ham, M., jeswiet J 2006 Single point incremental forming and the forming criteria for AA3003
[12] Seong D Y, Haque M Z, Kim J B, Stoughton T B and Yoon J W 2014 Suppression of necking in incremental sheet forming *Int. J. Solids Struct.* **51** 2840–9
[13] Saotome Y and Okamoto T 2001 An in-situ incremental microforming system for three-dimensional shell structures of foil materials *Journal of Materials Processing Technology* vol 113 pp 636–40
[14] Jeswiet J, Duflou J R and Szekeres A 2005 Forces in Single Point and Two Point Incremental Forming *Adv. Mater. Res.* **6–8** 449–56
[15] Petek A, Kuzman K and Kopač J 2009 Deformations and forces analysis of single point

incremental sheet metal forming *Arch. Mater. Sci. Eng.* **35** 107–16
[16] Obikawa T, Satou S and Hakutani T 2009 Dieless incremental micro-forming of miniature shell objects of aluminum foils *Int. J. Mach. Tools Manuf.* **49** 906–15
[17] Beltran M, Malhotra R, Nelson A J, Bhattacharya A, Reddy N V. and Cao J 2013 Experimental Study of Failure Modes and Scaling Effects in Micro-Incremental Forming *J. Micro Nano-Manufacturing* **1** 1–15
[18] Shi Y, Zhang W, Cao J and Ehmann K F 2019 Experimental study of water jet incremental micro-forming with supporting dies *J. Mater. Process. Technol.* **268** 117–31
[19] Suwas S, Ray R K and Satyam Suwas R K R 2014 *Crystallographic Texture of Materials*
[20] Zhang X H, Tang B, Zhang X L, Kou H C, Li J S and Zhou L 2012 Microstructure and texture of commercially pure titanium in cold deep drawing *Trans. Nonferrous Met. Soc. China (English Ed.* **22** 496–502
[21] Gurao N P, Kapoor R and Suwas S 2011 Deformation behaviour of commercially pure titanium at extreme strain rates *Acta Mater.* **59** 3431–46
[22] Mishra S, Yazar K U, Kar A, Lingam R, Reddy N V., Prakash O and Suwas S 2021 Texture and Microstructure Evolution During Single-Point Incremental Forming of Commercially Pure Titanium *Metall. Mater. Trans. A Phys. Metall. Mater. Sci.* **52** 151–66
[23] Bishoyi B, Vinjamuri R, Sabat R K, Patro S K, Suwas S and Sahoo S K 2022 Cold Drawing of Commercially Pure Titanium and Its Effect on Microstructure and Texture Evolution *Metall. Mater. Trans. A Phys. Metall. Mater. Sci.* **53** 1845–58
[24] Barnwal V K, Chakrabarty S, Tewari A, Narasimhan K and Mishra S K 2018 Forming behavior and microstructural evolution during single point incremental forming process of AA-6061 aluminum alloy sheet *Int. J. Adv. Manuf. Technol.* **95** 921–35
[25] LIU N, WANG Y, HE W jun, LI J, CHAPUIS A, LUAN B feng and LIU Q 2018 Microstructure and textural evolution during cold rolling and annealing of commercially pure titanium sheet *Trans. Nonferrous Met. Soc. China (English Ed.* **28** 1123–31
[26] Jeswiet, J. , Micari, F. , Hirt, G. , Bramley, A. ,Duflou, J. , Allwood J 2005 Asymmetric Single Point Incremental Forming of Sheet Metal *CIRP Ann. - Manuf. Technol.* **54** 88–114

IOP Conf. Series: Materials Science and Engineering 1284 (2023) 012046 doi:10.1088/1757-899X/1284/1/012046

Multi-functional SMC-aluminum battery tray to drive lightweight design

R Schmerler[1], C Drebenstedt[1], M Machhammer[2], G Gong[3], P Wolf[4]

[1] Fraunhofer IWU, Reichenhainer Straße 88, 09126 Chemnitz, Germany
[2] AP&T Sweden AB, Rönnåsgatan 3A, 523 38 Ulricehamn, Sweden
[3] RISE Research Institutes of Sweden, Fibevägen 2, 943 33 Öjebyn, Sweden
[4] FiberCheck GmbH, Technologie-Campus 1, 09126 Chemnitz, Germany

Email: michael.machhammer@aptgroup.com

Abstract. Functional integration and lightweight design are important tasks especially for modern electric vehicles. Development of multi-functional assemblies for the battery box of the future is one of the challenges in the field of electric mobility. The housing must perform many tasks: structural stiffening and sealing (waterproofness) as well as crash protection and thermal management for the batteries. In this paper results from a new developed multi-functional assembly of a SMC-aluminum battery tray will be presented and discussed. The outer housing shell is manufactured by forming SMC directly on an aluminum sandwich component in one stroke compression molding. Furthermore, passive temperature control functions are integrated into the core of the aluminum sandwich realized with switchable air-cooling and phase-change-material integrated into the foam. Additionally, sensors for structural health monitoring, temperature and humidity are integrated. Finally the process design and manufacturing tests will be explained.

1. Introduction and motivation

With the electrification of individual mobility battery systems and their components became of high interest for research and development. An essential part in the powertrain is the battery housing containing the battery cells and with this the electrical energy storage unit. The development of safe, multi-functional battery boxes and there manufacturing is one of the current and future challenges in the field of electric mobility. These housings have to meet several specifications, such as structural stiffening, crash protection, thermal management of the battery, waterproofness, fire safety, as well as noise, vibration and harshness requirements. The goal of the project presented was to develop a multi-functional lightweight battery housing and its manufacturing process that provides a higher safety to passengers based on lightweight construction methods and materials. Results of an ongoing two-year German-Swedish joint project in electric mobility applications, lightweight design and process technology for an innovative lightweight battery housing are presented

2. Requirements and design

2.1. Requirements

Battery housings have to meet several requirements, as they contain the safety critical battery cells and power electronics. General specifications include low weight, reasonable costs and scale-ability. Additionally mechanical specifications, like crash and intrusion protection, vibration and harshness requirements and the battery module fixation have to be fulfilled. Regarding thermal specifications the thermal management of the battery cells and power electronics, fire safety and thermal runaway or rather thermal propagation are important issues. Thereby the environment has to be protected from potential battery failure effects. The battery cells themselves need to be safe against environmental impacts like water, electromagnetic influences, or fire. With focus on the production low costs and cycle times in combination with a high resource efficiency are desired. For passenger as well as environment protection a reliable and early detection of system failures is crucial. Hence a well-designed monitoring system with sensors and data processing is important.

Various assumptions had to be made when designing the technology carrier. With the goal of performance comparison a reference battery housing was chosen. In this specific case, the battery housing of the VW e-Up from Volkswagen was defined as the demonstration structure. The e-Up housing was of current interest at the time of the project start as the best-selling electric car model in the first five months of 2021 in Germany [1]. For material and technology demonstration the lower tray of the housing was defined as the reference structure. It is very well suited for technology demonstration, as it provides most of the housing functionalities and represents a complex curved structure out of one material type; in this case steel. The design of the housing lid was not adjusted. Boundary conditions including design space, interfaces for assembly on the car body and positioning of power electronics were adopted from the reference.

In the following relevant design aspects and requirements within the SafEBat project will be explained. Due to defined project goals and scope a prioritization was indispensable.

2.2. Structural design

The structural and mechanical design of the SafEBat battery housing was developed taking the four load cases into account that are illustrated schematically in Figure 1. For the quasi-static load case no deflection of the housing structure was allowed while being loaded by the weight of the battery modules, which could be achieved. Furthermore the acoustical behavior of the SafEBat design in terms of the first bending and torsion natural frequency should be at least as high as for the reference. Two dynamic load cases completed the structural requirements: Two critical dynamic load cases were chosen for evaluating the mechanical performance: The side pole impact test according to the Euro NCAP regulations [2] and the bollard test, which is an original equipment manufacturer (OEM) test concerning intrusion protection from underneath the vehicle.

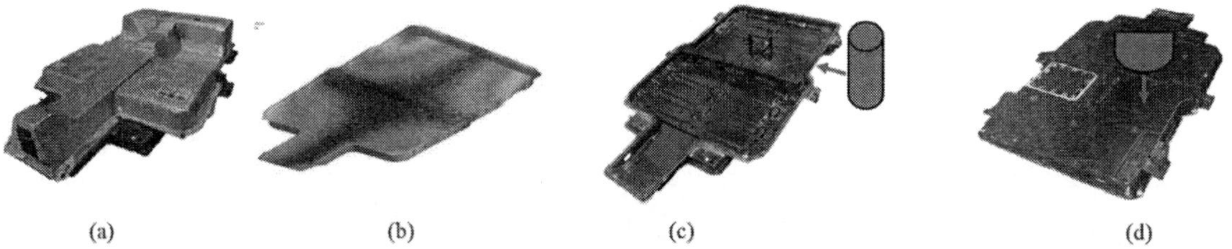

(a) (b) (c) (d)

Figure 1. Schematically illustration of the defined mechanical load cases: Quasi-static load case (a), acoustical behavior (b), side pole test (c) and bollard test (d).

Before presenting the performance comparison between the reference and SafEBat structure for these load cases, the component structure will be explained. The reference lower tray is based on a deep drawn 1.5 mm thick steel sheet stiffened with welded steel bars and mounting elements.

In contrast to the multi-unit reference, the SafEBat design consists out of less parts, illustrated in Figure 2. Its outer shell is a carbon fiber reinforced sheet molding composite (SMC) with a thickness of 1.5 mm. The SMC contains 50 wt. % of 25 mm 12K carbon fibers (CF). An aluminum foam core of 15 mm thickness with integrated aluminum hollow profiles in combination with a 2 mm thick aluminum top sheet completes the sandwich. The foam density is 0.6 g/cm³. With this multi-material lightweight design, the weight of the reference lower tray could be reduced by 15 %.

As mentioned before the housing lid for both variants is the same, consisting mainly of fiber reinforced plastics (FRP). The central crash bar was kept as well in the new design.

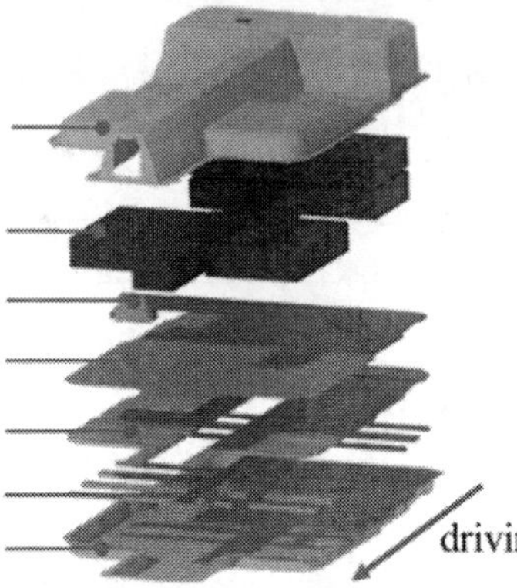

Figure 2. The multi-material SafEBat battery housing design.

A natural frequency analyses with a free-free boundary condition was simulated. For this setup of the lower tray the first bending natural frequency was 63 % higher than for the reference. Besides the first torsion natural frequency was increased by 26 %. These values meet the requirements with achieving at least the first natural frequencies as the reference and indicate a higher global bending and torsion stiffness.

With regard to the dynamic load case of the side pole impact, a bollard with a diameter of 254 mm was impacting under an angle of 15° from the side und a velocity of 32 km/hr at the weakest position of the battery. For all mechanical simulations ANSYS Workbench and for dynamic load cases the extension of LS-Dyna was used. In a real car crash for this scenario the main energy is absorbed through the sill that was not available and hence not considered. Since, with the goal of comparing the SafEBat with the reference design, the impact energy of the pole was limited to 2.3 kJ. An unstiffened area on the left side of the tray was defined as a local structural weak-point and thus chosen as the impact area.

The simulation results show a deflection decrease of 75 % for the SafEBat structure, going ahead with a maximum force peak increase of 192 %. Consequently, the intrusion into the battery housing could be reduced significant. With focus on damage the new design behaves more brittle than the more ductile steel reference. Most likely this effect can be reduced through sill load distribution, thickening, and supporting the flange area for instance with ribs.

One of the main project goals was to increase the under-body protection for a higher battery safety. With the help of the bollard test this should be investigated. In this load scenario a bollard hits the battery housing at an angle of 90° from underneath. The impact energy for design comparison was set to 150 J according to a vehicle weight of 1,600 kg. The local, structurally weak impact area is indicated in Figure 1. Under the defined boundary conditions the maximum deformation of 35 mm for the reference could be reduced by 73 % using the new sandwich design. Taking into account the required design space in z- or rather thickness direction, the intrusion was reduced by 28 %.

2.3. *Thermal Design*

Beside structural aspects a proper thermal design of the battery structure is essential for secure and efficient operation. The battery cells, dissipating heat while charging or discharging, needs to be kept in a defined temperature range. For the often used lithium ion batteries the upper temperature limit is in the range of 50 - 60 °C [3]. Exceeding these temperatures is a high risk for passengers as it can lead to thermal runaway and propagation [4].

The approach of the SafEBat design is a passive temperature control using the phase change material paraffin with a melting area around 40 °C, storing or rather releasing a high amount of energy during their phase change liquid to solid and reverse. This thermal energy storage material (PCM) is integrated into the structural aluminum foam sandwich core. The closed cell aluminum that is produced via powder metallurgical rout has micro cracks and partly small holes in the cell walls enabling the infiltration. The foam acts as a thermal conductive matrix, increasing the low thermal conductivity of PCM with 0.2 W/(m·K) by factor 85 for the PCM-aluminum foam composite. By integrating around 5 kg of the PCM a heat of circa 360 W·hr can be stored. Additionally, to the maximum temperature limitation, thermal load peaks should be flattened and the temperature homogeneity increased. Integrating and using the PCM is a design feature with the purpose of using high power battery cells that most likely dissipate more heat than high capacity cells.

Depending on the environmental conditions a heat exchange with the environment is desired. In Figure 3 two scenarios are sketched. In case of an environmental temperature that is above the necessary battery temperature condition the thermally low conductive SMC has an isolating function, see Figure 3 (a). In the second scenario a heat exchange with the colder environment is desired for battery cell cooling, see Figure 3 (b). In Figure 3 the foam and sandwich top layer are hidden for better visualization of the concept. In this case flaps positioned in the front of the central hollow profiles open enabling a flow through the channels and with this dissipating heat from the battery system.

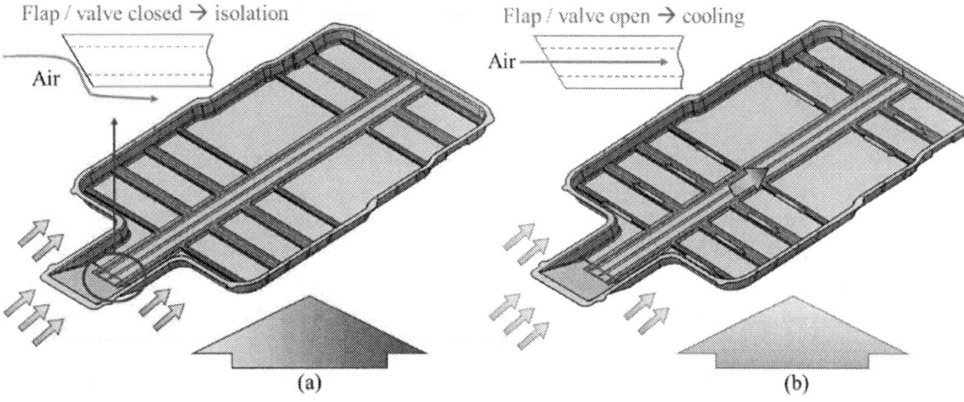

Figure 3. Scenario-depending temperature control of the SafEBat design: Isolation (a) and cooling (b).

2.4. Sensor Design
As an intended technological advancement, the existing embroidered strain sensor was taken as a base and further developed according to the requirements of the battery tray. Conventional sensors of this type have so far only been manufactured for wind power applications. These sensors formed the starting point for this project. The layout of the sensors was optimized iterative.

This included a variation of the area of the sensors and the resistance value that can be applied to them. Sensors that are too large form an antenna structure and are therefore susceptible to electromagnetic influences. Sensors that are too small can no longer be evaluated electrically and therefore do not display a measured value. As a final solution for battery case monitoring, the following characteristics were identified for the sensors. The basic resistance is 120 ohms, so that the best compromise between antenna and electrical properties is chosen. In addition, the sensors with this resistance value are industry-compatible, so that evaluation via existing control units is easily possible. About 0.5 m of wire was embroidered for the sensor. With the necessary embroidery thread spacing of 1 mm, this creates a sensitive area of 4.2 cm^2 with side dimensions of 21 x 20 mm.

3. Production technology
In this project a FRP-aluminum hybrid battery housing in a one stroke compression molding will be manufactured, which has high potential for being scaled up for high volume production. In order to reach the target of short pressing cycling time, sufficient mechanical properties and good mold-ability, chopped carbon fiber reinforced molding compound (CF-SMC) is used for manufacturing of the FRP part. Specifically, a commercial CF-SMC product, AMC®85590, from Lyondellbasell, is selected among a large number of CF-SMC candidates in order to realize low density, good mechanical properties, sufficient flow-ability and coefficient of thermal expansion (CTE) close to aluminum. The aluminum sandwich will be inserted in the tool and heated up together to the temperature for processing the CF-SMC. When the tool and the aluminum sandwich are thermally equilibrated, the CF-SMC charge will be placed on top of the aluminum sandwich following the placement recommendation from process simulation results. The tool will then be closed to reach the preset pressure. When the CF-SMC is cured it will be adhesively bonded to the aluminum sandwich.

3.1. Process development

3.1.1. Overall process. First an overview of the overall process is described, followed with detailed explanation in the further sub chapters. The structural design of the reference consists of a high number of parts accompanied with many joining process steps. The approach of the SafEBat design was to reduce these. Hence, the number of parts for the lower tray was reduced to two semi-finished components: The aluminum foam component and the SMC part. These are combined in an integral forming, joining and SMC curing process.

In the first process steps the aluminum sheet and profiles are cut into shape for foaming. Afterwards they are assembled in position and welded together. Subsequent the assembly is placed in a foaming mold. The foam-able precursor material is placed in between the profiles into the mold. After closing the mold and fixed in position by a mold carrier, it is placed in an oven. At temperatures around 600 °C the foaming process takes place and with this the aluminum foam structure is generating, thereby joining metallically with the aluminum sheet and the profiles. After the deforming and cooling of the aluminum part it is placed upside down in the tool of a press. Subsequently the SMC material is positioned on the aluminum part. While closing the tool the SMC is formed and simultaneously joined with the aluminum part. Parallel the SMC is curing. The next process steps after opening the tool are deforming of the part from the tool and a trimming operation.

3.1.2. Metal foam process.
The precursor for the foam consists of the aluminum alloy AlSi10 that was mixed with titanium di-hydride (TiH_2) in portion of about 0.8 weight-% and further additives. TiH_2 serves as the foaming agent. The TiH_2 splits into titanium and hydrogen beginning at temperatures around 400 °C [5]. At temperatures around 600 °C the released hydrogen in combination with the molten aluminum builds up the pore structure. The mixed powder can be pressed into required shapes with extrusion or rolling, like the flat aluminum band shown in Figure 4. After the precursor cutting into desired portions it is placed into a foaming mold of graphite or steel. The mold is closed in the next step and positioned on a supporting carrier structure, before it is positioned in an oven, where the foaming is processed under temperature.

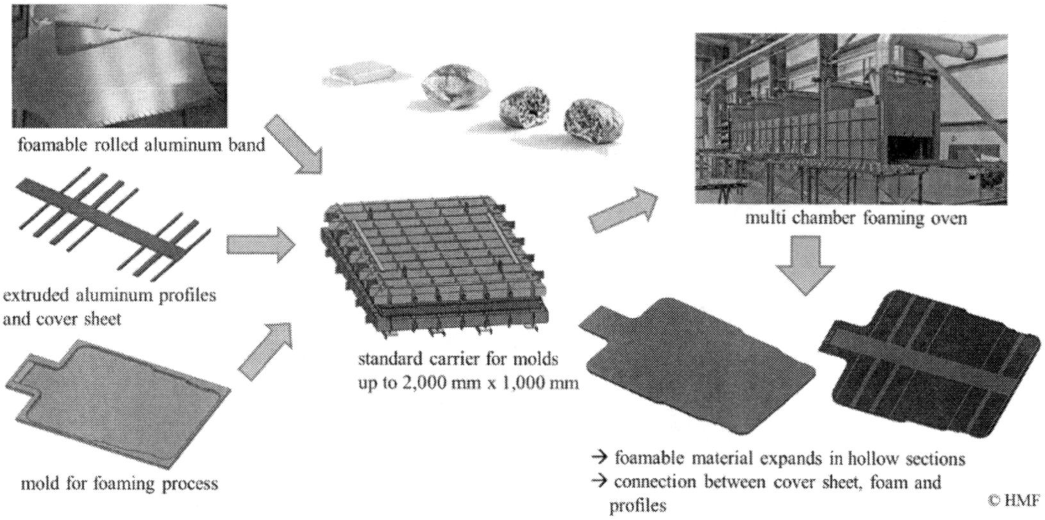

Figure 4. Process overview for the aluminum foam part production.

3.1.3. Manufacturing verification and process simulation. A testing tool (Figure 5) with the same profile and components as in the final demonstrator was designed and manufactured in order to:

1) test how aluminum sandwich will withstand the compression molding conditions,
2) test how the selected CF-SMC will flow in the tool,
3) test the tolerances of the tool,
4) test the influence of CTE difference between the selected CF-SMC and aluminum sandwich on structural integrity,
5) test the method of integrating sensor in the manufacturing and structure without being damaged, and
6) facilitate the process simulation.

Based on this testing tool, a process simulation using 3D Timon was performed. The process parameters include a mold temperature of 140 °C and a clamp force of 300 tons. According to the simulation results, the CF-SMC charge should be placed on top of the aluminum sandwich, covering almost 100 % of the sandwich's surface, in order to avoid any weld lines and no resin rich areas in critical parts of the geometry, as well as enable a fairly even fiber distribution. Simulation results in Figure 6 show a strong alignment of the fibers in z-direction at the side walls.

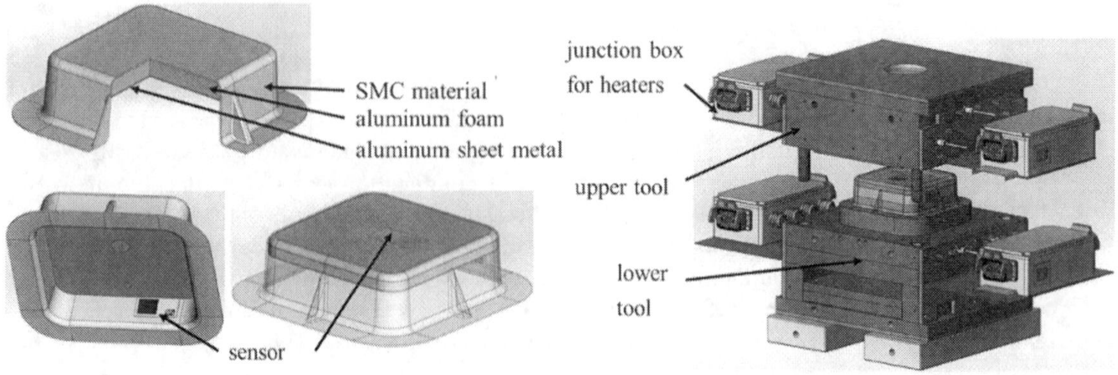

(a) (b)

Figure 5. Test part (a) and testing tool (b) for verification of manufacturing concept and also used for process simulation.

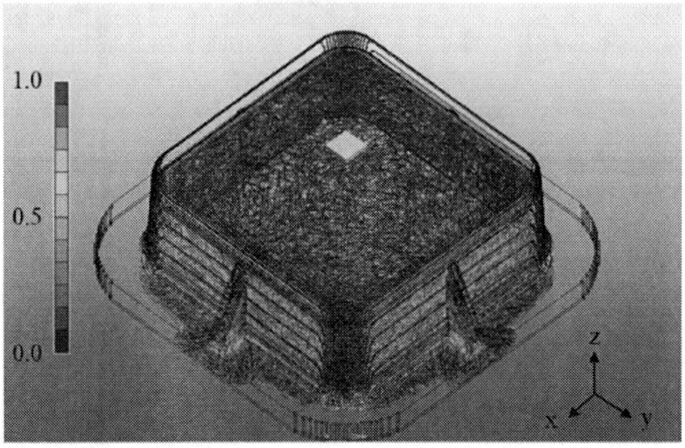

Figure 6. Process simulation result shows the alignment of the fibers referenced to z-direction (scale factor 1) with a strong z-alignment at the side walls.

Several hybrid structures using this testing tool with pressing conditions, including pressure, tool temperature, press cycle time, and weight of SMC as well as aluminum sandwiches with different densities of the aluminum foam were manufactured. The manufactured hybrid structure using an optimal condition is shown in Figure 7. It was manufactured in a one stroke compression molding using the testing tool. The cross section in Figure 7 (b) shows the aluminum sandwich status in the hybrid structure after compression molding. The area marked with a red oval is the sensor cavity that was filled with the SMC in the test sample. The best process performance was achieved for a pressure of 45 bar at a temperature of nearly 120 °C. The cycle time was 15 min, with potential of being decreased.

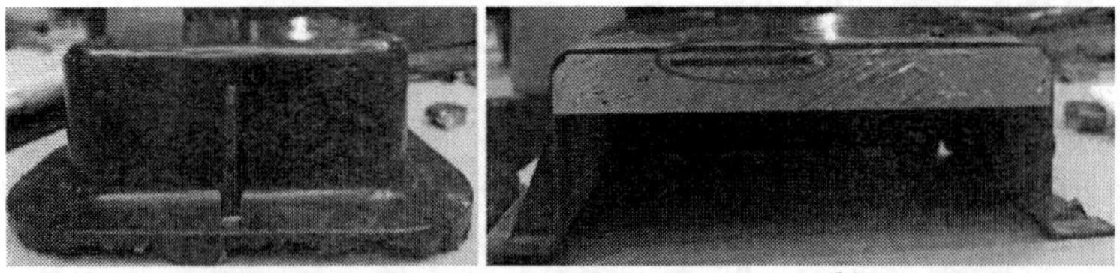

(a) (b)

Figure 7. FRP-aluminum hybrid structure manufactured using the testing tool (a) and cross section with sensor cavity filled with SMC (oval mark) in the process tests (b).

3.1.4. Sensor integration. The sensors were adapted for the battery housing and integrated into the SMC manufacturing process. A test sample is shown in Figure 8. During the first integration, it was also possible to gain insights into practical implementation, such as ensuring contact and fixing the position of the sensors as well as cable routing from the floor area. When handling the sensors, it also became clear that they require an insulating layer in the conductive SMC material. Glass fiber mats were used for this, which at the same time protect the wire structure from damage.

The sensor excitation experiments were carried out on a test stand press. A metal stamp was chosen for stimulation. In the test, the stamp position and the occurring strains were measured directly on the press and compared with a reference sensor on top. The tests were carried out in a path-controlled manner and the sensor data was recorded with a measuring amplifier.

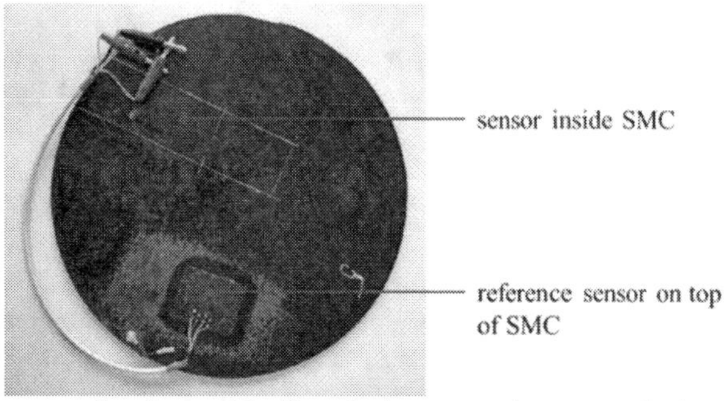

Figure 8. Test sample plate with directly integrated sensors during SMC-forming.

3.2. Tool to manufacture the Battery tray

3.2.1. Tool concept. The upper and lower unit of the SMC-tool shown in Figure 9 is equipped with several heating elements to minimize temperature deviation along the heated tool area to ± 10 °C. The temperature of 12 individual heating zones is constantly monitored with thermocouples and controlled by a control unit. Insulating plates cover the outer wall and separate the tool from the die shoe to decrease heat losses and to increase temperature uniformity.

The forming tool is designed to process three different materials in one production step. Therefore, the tool is scaled to compensate the geometrical change of the individual materials, caused by thermal shrinkage, when the component cools to room temperature. The tolerance of the gap between the two units of the tool is set to + 0.05 mm to ensure a good sealing performance and simultaneously enabling the exiting of air from the cavity.

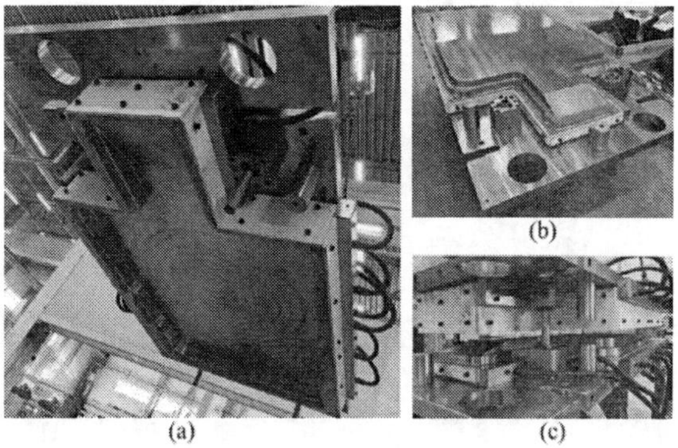

Figure 9. Tool upper unit (a), tool lower unit (b) and tool assembled (c).

Furthermore, several milled pockets are integrated in the wall area of the tool for in-mold sensor integration, see Figure 10. In total eight sensors are positioned prior to forming and are then over molded by the SMC. This tool function enables a reduction of manufacturing steps since the sensors are integrated during the SMC process instead at subsequent process steps.

Figure 10. Tool side view with milled pockets for sensor integration (oval mark).

3.2.2. Process description. The tool is heated to 145 °C before the forming takes place. To start with the manufacturing process the aluminum foam component and sensors are placed inside the tool, see Figure 11 (a). Important to mention is that the aluminum foam must be preheated to the working temperature to decrease cycle time.

The SMC is placed on top of the aluminum foam component with 80 - 90 % coverage of the surface, which is indicated in Figure 11 (b). The tool is used is a servo hydraulic press which makes it possible to vary, control and monitor the entire forming process with extremely high precision.

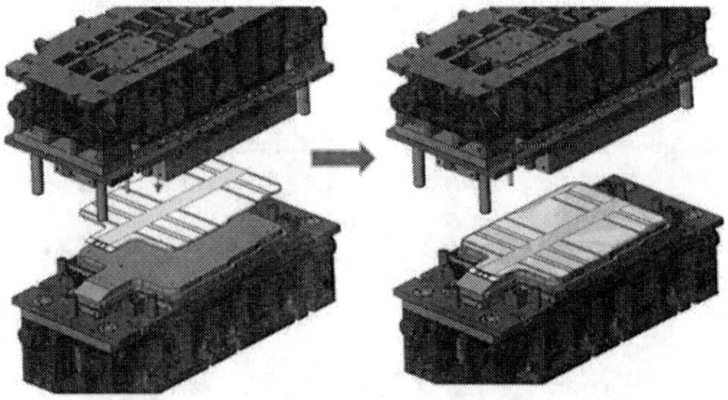

(a) (b)

Figure 11. Placement of aluminum foam part on the tool (a) and SMC indicated in yellow (b).

4. Summary and further goals

The aim was to develop a new design for a battery housing with improved mechanical performance and functional integration for passive temperature control and its production process. This was achieved by a setup based on a design with closed cell aluminum foam sandwiches containing an aluminum and SMC cover sheet. The battery housing is designed to withstand certain impacting forces from e.g. a bollard from underneath and a pole from the side in the event of a crash. For increased thermal functionality the PCM paraffin is infiltrated into the aluminum foam.

A corresponding manufacturing process for the battery lower tray was developed. The process includes the main steps: Manufacturing the upper aluminum part with profiles including the foaming process and the one stroke molding of the SMC (forming, joining and curing) onto this structure to complete the sandwich and with this the battery lower tray.

The next step is the manufacturing of the demonstrator in scale 1:1 to prove the feasibility of the concept and analyze the optimization potential. This includes the manufacturing of the foam parts, as wells as the forming and joining with the SMC. Afterwards the PCM will be infiltrated into the foam. The last step is the validation of the simulation results via experimental tests. Further goals include the partial merging of the two now separated processes. By using the rest heat of the foaming the pre-heating of the aluminum foam for the combination with the SMC can be reduced or completely left out. With this way the energy input and production time can be reduced significantly. In the aspired demonstrator the cooling channels are already included, see Figure 3. In further developments the aim is to include demonstrator cooling flaps, e.g. made of shape memory alloys, as well.

The next generation of this battery tray includes further research regarding the usage of more sustainable materials, like bio-based PCM, natural fibers (e. g. hemp, flax) in combination with thermoplastic resin systems under regard of the mechanical properties to simplify the matter of repair and recycling. For this also a comparative consideration of the impacts on the environment shall be included.

Acknowledgments
The authors from Germany gratefully thank the Federal Ministry for Economic Affairs and Climate Action BMWK and the German Federation of Industrial Research Associations AiF for funding (funding codes: ZF4016241PO9 and ZF4696702PO9). The authors from Sweden gratefully thank the Swedish financier, VINNOVA, in this project (3rd German-Swedish Call for proposals for joint SME R&D projects, funding code: 2019-05235).

References

[1] https://insideevs.de/news/512411/meistverkaufte-elektroautos-januar-mai-2021/, view: 16.03.2023
[2] https://www.euroncap.com/en/vehicle-safety/the-ratings-explained/adult-occupant-protection/lateral-impact/side-pole/, view: 16.03.2023
[3] Choudhari V G, Dhoble A S, Sathe T M 2020: A review on effect of heat generation and various thermal management systems for lithium ion battery used for electric vehicle *Journal of Energy Storage 32*, **101729**
[4] Feng X, Ren D, HE X, Ouyang M 2020: Mitigating Thermal Runaway of Lithium-Ion Batteries *Joule 4 Nr. 4* **743–770**
[5] Hipke T, Lange G, Poss R 2007 *Taschenbuch für Aluminiumschäume. 1. Auflage. Düsseldorf, Aluminium-Verlag*, **3-4**

IOP Conf. Series: Materials Science and Engineering 1284 (2023) 012047 doi:10.1088/1757-899X/1284/1/012047

Simulation of deep drawing of dual phase steel sheets incorporating tribological behaviour predicted by advanced friction modelling

Archit Shrivastava and D Ravi Kumar

Department of Mechanical Engineering, Indian Institute of Technology Delhi, New Delhi, India

E-mail: archit.shri25@gmail.com

Abstract. In a deep drawing process, friction at the sheet metal-die contact interface affects the material flow in the flange as well as at the die entry affecting product quality significantly. The accuracy of predicted drawing force and thinning in deep drawing simulation strongly depends on friction modelling. The Coulomb friction model with a constant coefficient of friction is generally assumed in simulations. In reality, the actual friction coefficient depends on process variables such as local contact pressure, relative sliding velocity, and strain in the sheet material. In this work, an approach is presented for deep drawing simulation of dual phase (DP600) steel sheets incorporating the effect of these variables on the friction coefficient. Strip drawing tests under conditions similar to that exist in the flange area during deep drawing have been carried out. Deep drawing simulations have been performed using two friction models taking into consideration the variation of friction coefficient with respect to pressure and velocity. The predicted punch force and thickness variation are validated using deep drawing experiments. A considerable improvement in the accuracy of predicted results has been observed when compared to a constant coefficient of friction.

1. Introduction

Deep drawing is one of the frequently used sheet metal forming processes to manufacture parts with moderate to large depth for mass production in the automotive, packaging and many other industries. In this process, a blank is radially drawn into a die cavity using a punch under the influence of blank holding force between the die and the binder [1]. The drawing force and the quality of parts are heavily influenced by the tribological conditions that exist at the interface of the blank and the tools [2]. In processes like deep drawing, where the blank slides over the tool surface, a low friction coefficient is desired [3].

In sheet metal forming industries, numerical simulation is now essential to predict the parameters required to form a successful component. This facilitates a significant reduction in time and effort during the try-out phase [4]. Numerous examples of research findings on the accuracy of simulations can be found in the literature, including analyses of the effects of material modelling, boundary conditions, and tribological factors. Over the past 20 years, material modelling has been the focus of a reasonably large number of researchers [5-7], while limited work has been reported on the tribological effects in the deep drawing process. Defining a constant friction coefficient based on the coulomb law is the most common practice in simulation. Another enhanced version of coulomb law incorporates the pressure and velocity dependent variation of friction coefficient. Alaitz Zabala [8] demonstrated the complexities of a tribological system and its effect on forming simulations and therefore, it is necessary to incorporate the effect of process variables on the friction coefficient. Using some advanced friction models [9], the real time friction coefficient value can be predicted in the simulation instead of assuming a constant friction coefficient value for all the elements of a blank. TriboForm model in AutoForm software was employed for formability analysis of some autobody parts [10 - 13]. It computes friction coefficients for a predefined set of process parameters, including local contact pressure, relative sliding velocity, plastic strain in the sheet material, and interface temperature.

In the present work, strip drawing tests were carried out to investigate the effect of pressure and velocity on the friction coefficient value of DP600 steel sheets under lubricated condition. The obtained friction coefficient is then used to determine constants in Pressure Velocity Dependent Friction (PVDF) model. The swift bottom cup test is simulated incorporating different friction models (constant friction coefficient, PVDF model, and TriboForm model) to predict drawing load and thinning. Experiments were carried out to validate the predicted results from the simulation.

2. Methodology

2.1. Mechanical properties

In this work, Dual phase (DP600) high strength steel sheets of 1.4 mm thickness were used. The properties were determined by performing uniaxial tensile tests as per ASTM E8M [14]. The standard mechanical properties, coefficients of Swift hardening law and plastic strain ratio values of the sheet are shown in table 1.

Table 1. Mechanical properties of DP600 sheet.

Material	YS (MPa)	UTS (MPa)	Total elongation (%)	Coefficients of Swift hardening law			Plastic strain ratio		
				ε_0	n	K (MPa)	R_0	R_{45}	R_{90}
DP600	405	648	19	0.0046	0.172	1054	0.96	0.71	0.93

2.2. Strip drawing experiments

To determine the friction coefficient (μ) value in conditions similar to those exist in the flange region during deep drawing, strip drawing experiments were conducted at different normal pressures and draw speeds. Figure 1 depicts the schematic of the strip drawing experiments in which a predefined normal force is applied on the strip using the upper and lower dies. The strip is then pulled from one end using a jaw with a displacement of 200 mm. The friction coefficient is determined using coulomb law as shown in equation (1).

$$\mu = \frac{F}{2*N} \quad (1)$$

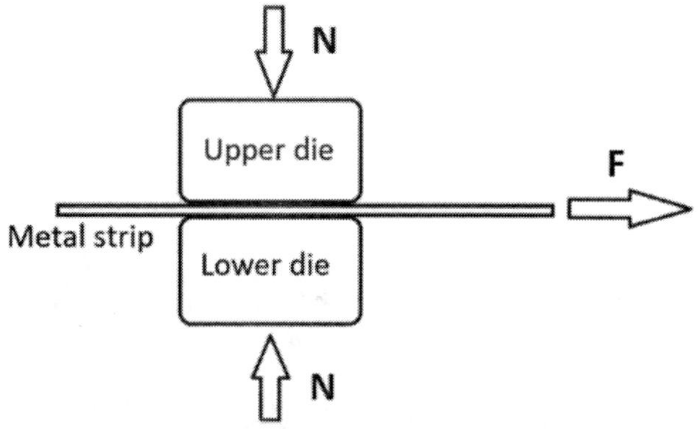

Figure 1. Schematic diagram of strip drawing experiments.

The tribological system is a combination of characteristics of the sheet material, lubricant type, and the die material. Strip drawing dies of contact area of 50 mm*25 mm were fabricated using die steel (D2) with the surface finish similar to that of deep drawing dies. The surfaces of the dies were hardened and polished. The die surface roughness was measured using Talysurf and it was found to be $R_a = 0.3 \pm 0.05$ μm. The initial surface roughness of the sheet was found to be $R_a = 0.9 \pm 0.05$ μm. Mineral oil-based lubricant was used to pre-lubricate the sheet material. An average lubricant amount of 2.0 g/m^2 was applied on the specimen. The strip drawing experiments were conducted at normal pressures of 4 MPa, 6 MPa and 8 MPa, and three sliding speeds (1 mm/s, 5 mm/s and 10 mm/s). The load displacement data was obtained and the average friction coefficient was calculated over a sliding length of 50 mm to 100 mm.

2.3. *Deep drawing simulation*

Deep drawing simulations for drawing 100 mm diameter circular blanks with a 50 mm flat bottom cylindrical punch were carried out by AutoForm software. The FE model used for the simulation is shown in figure 2. The elasto-plastic shell (EPS-5) elements of size 1.0 mm were used for meshing the blank with five integration points. The refining mesh option with a maximum refinement level of 7 was used for better accuracy in the predicted results.

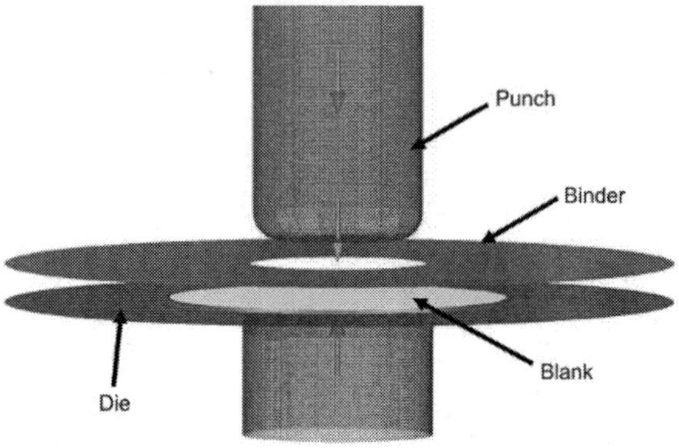

Figure 2. FE model used in simulation of Swift flat bottom cup test.

2.3.1. Material Model. The material model consists of flow curve and yield surface as shown in figure 3a and figure 3b. Swift hardening law is used to extrapolate the flow stress to larger strains. The Barlat89 material model is used to incorporate the effect of anisotropy in the yielding behavior and it requires plastic strain ratios in 0^0, 45^0 and 90^0 with respect to the rolling direction as input. The forming limit curve (figure 3c) of the material was experimentally determined using Nakazima method [15] and it was provided as input for the failure criterion.

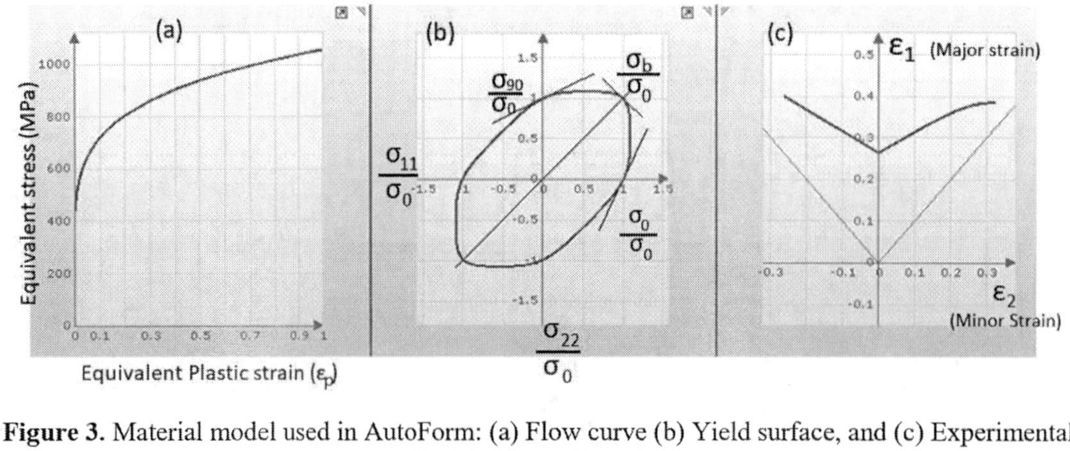

Figure 3. Material model used in AutoForm: (a) Flow curve (b) Yield surface, and (c) Experimental Forming limit diagram.

Using the above material model, deep drawing simulations were carried out at an initial blank holding pressure of 4 MPa. A constant punch speed of 1 mm/s was used in the simulation. To define the friction at the tool-blank interface, two friction models were used in addition to constant friction coefficient. These models are explained in the following section.

2.3.2. Friction models. The following two models are used in this study.
(a) Pressure Velocity Dependent Friction (PVDF) model. Enhanced coulomb law incorporates the dependence of pressure and velocity on friction coefficient as shown in equation (2).

$$\mu_{eff} = \mu\left(\frac{p}{p_{ref}}\right)^{e-1} - \left(a * \ln\left(\frac{\max(v_{rel}, v_{ref})}{v_{ref}}\right)\right) \tag{2}$$

Where, μ - Base friction coefficient, p_{ref} – Reference pressure, e – Pressure exponent, a – Velocity factor, v_{ref} -Reference velocity, v_{rel} -Relative velocity

The constants of PVDF model were determined using strip drawing experiments performed at different normal pressures (4 MPa, 6 MPa, and 8 MPa) and draw velocities (1 mm/s, 5 mm/s and 10 mm/s). The reference pressure and reference velocity of 4 MPa and 1 mm/s were selected respectively. The base friction coefficient is the friction coefficient obtained at the reference pressure and the reference velocity. The constants used in equation (2) are shown in table 2.

Table 2. Constants of PVDF model and inputs used in TriboForm module.

PVDF model					TriboForm Model		
Base friction coefficient	Reference pressure p_{ref} (MPa)	Pressure exponent	Relative velocity v_{rel} (mm/s)	Velocity factor	Average Surface roughness of the sheet (R_a) (μm)	Average Tool surface roughness (R_a) (μm)	Amount of the lubricant (g/m^2)
0.183	4	0.953	1	0.003	0.9	0.4	2

(b) TriboForm Model. In TriboForm module, the friction model for dual-phase steel and tool steel lubrication system was used. Draw oil was selected as the lubricant. The average surface roughness of the sheet, tool surface roughness and the amount of the lubricant are defined as shown in table 2. The friction model is affected by local contact pressure, sheet material strain, relative sliding velocity, and interface temperature. In this model, the volume of the lubricant entrapped in non-contacting surface pockets is determined using the calculated deformation of sheet asperities. This data is then used to compute the fluid film thickness (h) and hydrodynamic pressure distribution (P_{lub}) (i.e. the load carrying capacity of the lubricant). To solve P_{lub}, FE approach with hydrodynamic contact elements and additional pressure degrees of freedom was used. Based on the obtained hydrodynamic pressure distribution, the viscous shear stress (τ_{lub}) at the fluid-solid interface is calculated [9]. The coefficient of friction is obtained at the fluid-solid interface as shown in equation (3).

$$\mu = \frac{\tau_{asp}+\tau_{lub}}{P_{nom}} \tag{3}$$

Using the above models, a local and time dependent friction coefficient is computed at each node for each increment, and used in the finite element model calculation.

2.4. Deep drawing experiments
Swift flat bottom cup test was performed on circular blank of Dual Phase steel sheet (DP600) of 1.4 mm thickness using a 50 mm diameter flat bottom cylindrical punch. Blanks of diameter 100 mm were used in the experiments and hence the draw ratio was 2.0. Figure 4 depicts a schematic diagram of the deep

drawing process. The tests were carried out on a servo-hydraulic press with a constant punch speed of 1 mm/s and initial blank holding pressures of 4 MPa. The cups were drawn to a depth of 35 mm. The punch load and displacement were measured using a load cell and an LVDT respectively.

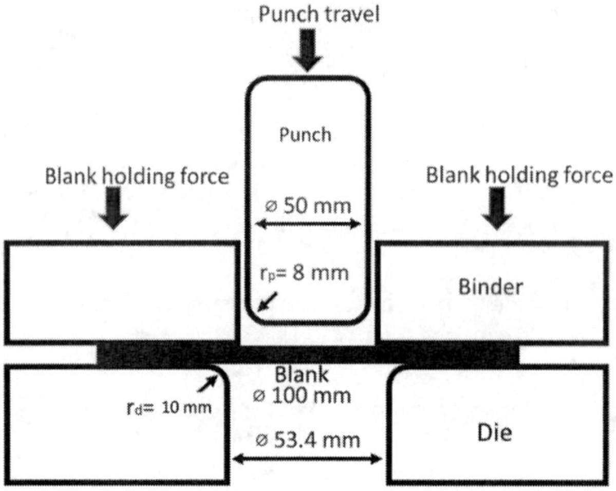

Figure 4. Schematic of experiment setup of deep drawing experiments.

3. Results and discussion

The predicted load and thinning from deep drawing simulation are compared with experimental results in this section.

3.1. Comparison of load-displacement curves

The predicted load–displacement curves from the numerical simulations are compared with the experimental results at a draw depth of 35 mm. Figure 5 shows the load-displacement curves obtained from the experiments and the simulation using different friction models. The resistance against deformation in the flange region leads to an increase in the drawing force. The error in the prediction of maximum load is 11% on using a constant friction coefficient, whereas it decreases to 7% and 6% using the TriboForm model and PVDF model respectively as seen in figure 5.

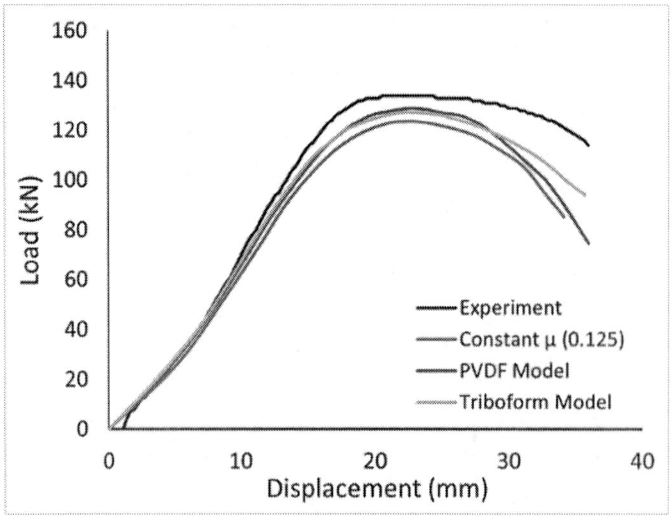

Figure 5. Comparison of load-displacement curves predicted using a constant μ, TriboForm model, and PVDF model with the experimental curve.

The PVDF model and TriboForm model improved the accuracy of prediction by incorporating the actual friction coefficient that exists in the contact zone. TriboForm model also considers the effect of flattening of surface asperities and strain induced in the material on the friction coefficient. Since the blank holding pressure and strain induced in the blank increases to a very high value when a very less area of the blank is present in the flange region. Therefore, the rate of decrease in punch load is more accurately predicted by the TriboForm model as shown in figure 5.

3.2. Comparison of thickness variation

Figure 6a shows an experimentally drawn cup. R and D represent the lines along which the thickness was measured at an interval of 2 mm for plotting the thickness variation. The predicted results for the same depth of draw (35 mm) are shown in figure 6b. Figure 7 shows the thickness distribution (along R and D lines on the cups) drawn at initial blank holding pressures of 4 MPa. Thinning occurs in the bottom and near the punch corner due to stretching, whereas thickening occurs in the flange region due to circumferential compressive stress. Minimum thickness at the punch corner radius and thinning in the bottom region due to stretching are accurately predicted by the PVDF model and Triboform model than using a constant friction coefficient value.

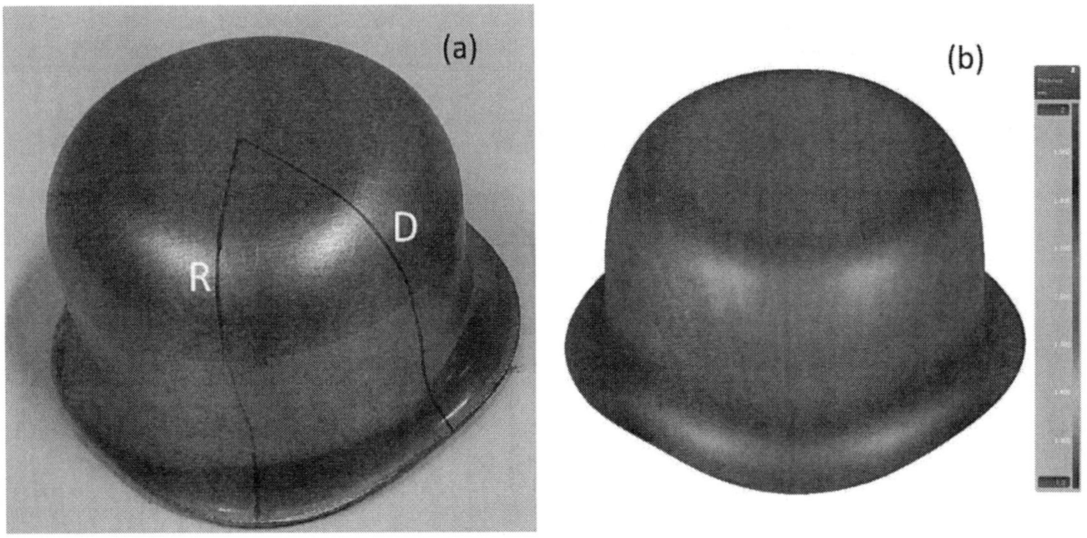

Figure 6. (a) Thickness distribution path in experimentally drawn cup (b) Simulation results for thickness distribution.

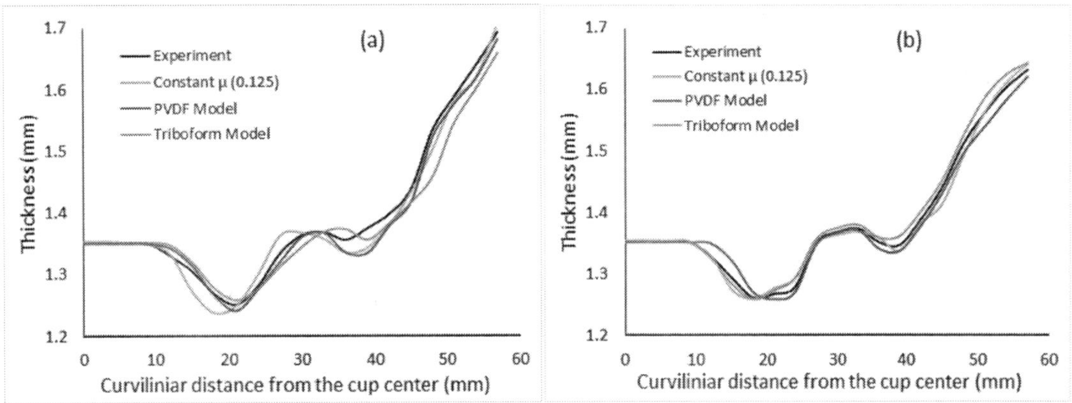

Figure 7. Thickness distribution in deep drawn cup using constant friction coefficient, PVDF model, TriboForm model and experimental (a) along R curve (b) along D curve.

The PVDF model and TriboForm model accounts for variation of friction coefficient with the increase in blank holding pressure in the flange during drawing. This incorporates the changes in flow of material in the flange region. An error of 3% is observed in the predictions of minimum thickness using constant friction coefficient. Whereas, the minimum thickness predictions by TriboForm's and PVDF's are accurate to within 1% of the experimental value.

The advantage of using TriboForm model is extended for prediction of surface roughness, and lubricant amount. Figure 8 shows the predicted average surface roughness (R_a) and lubrication amount present during drawing at an initial blank holding pressure of 4 MPa. The average surface roughness (R_a) of the experimentally drawn cup is measured in three regions i.e. cup bottom (0.9 ± 0.05 m), cup wall (0.3 ± 0.05 μm) and flange region (0.2 ± 0.05 μm). The predicted roughness values are in good agreement with the experimental observations.

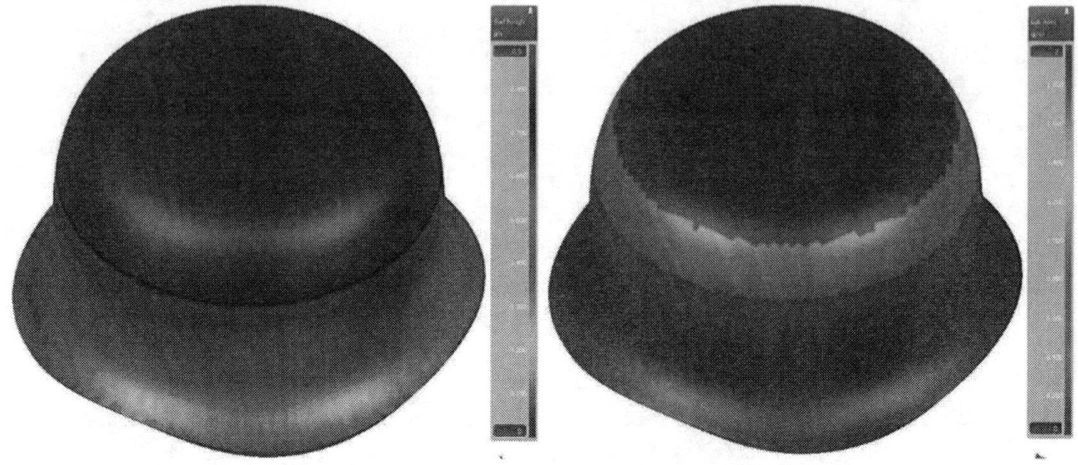

Figure 8. Predicted results of (a) surface roughness (b) lubricant amount available using TriboForm.

4. Conclusions

This work demonstrates the importance of advanced friction models in the numerical simulation of DP600 steel. The dependence of friction coefficient on pressure and velocity is studied through strip drawing experiments. The results of which were incorporated in numerical and experimental investigation of deep drawing. The following are the main findings.

The error in the prediction of maximum load is 11% on using a constant friction coefficient, whereas it decreases to 7% and 6% using the TriboForm model and PVDF model respectively as seen in figure 5. The PVDF model and TriboForm model improved the accuracy of prediction by incorporating the actual friction coefficient that exists in the contact zone.

The minimum thickness predictions by TriboForm's and PVDF model are accurate to within 1% of the experimental value. Whereas an error of 3% in the predictions is observed using constant friction coefficient.

TriboForm model can be used for prediction of surface roughness, and lubricant amount present during drawing. The predicted roughness values by the triboform model were in good agreement with the experimentally drawn cup.

Acknowledgments

Archit Shrivastava would like to acknowledge the assistantship received from the Ministry of Human Resource Development (MHRD), Government of India for this research work. Authors would like to acknowledge the support received from the Tata Steel, R&D, Jamshedpur for conducting this work.

References

[1] Altan T and Tekkaya AE 2012 *Sheet metal forming: processes and applications* ASM International.
[2] Ludwig M, Müller C and Groche P 2010 *Key Eng. Mater.* **438** pp 171–178.
[3] Prakash V and D Ravi Kumar 2022 *Adv. Mater. Process. Technol.* **8** (1) pp 1044-1057.
[4] Tatipala S, Pilthammar J, Sigvant M, Wall J and Johansson CM 2018 *IOP Conf. Ser. Mater. Sci. Eng.* **418** 012111.
[5] Barlat F, Yoon SY, Lee SY, Wi MS and Kim JH 2020 *Int. J. Solids Struct.* **202** pp 947–62.
[6] Banabic D, Kuwabara T, Balan T, Comsa D and Julean D 2003 *Int. J. Mech. Sci.* **45** (5) pp 797- 811.
[7] Yoshida F, Hamasaki H and Uemori T 2015 *Int. J. Plast.* **75** pp170–88.
[8] Alaitz Zabala, Eneko Sáenz de Argandoña, Daniel Cañizares, Iñigo Llavori, Nagore Otegi, Joseba Mendiguren 2022 *Tribol. Int.* **165** 107259.
[9] Sigvant M, Hol J, Chezan T and van den Boogaard T 2015 *Proc. of FTF - Advanced constitutive models in sheet metal forming* ETH Zurich, Switzerland pp 135-142.
[10] Hol J, Wiebenga JH, Stock J, Wied J, Wiegand K and Carleer B 2016 *J. Phys. Conf. Ser. (Bristol, UK, 4-9 September)* **734** 032091.
[11] Sigvant M, Pilthammar J, Hol J, Wiebenga JH, Chezan T, Carleer B and van den Boogaard AH 2016 *J. Phys. Conf. Ser.* **734** 032090.
[12] Hol J, Wiebenga JH and Carleer B. 2017 *J. Phys. Conf. Ser.* **896** 012026.
[13] Sigvant M, Pilthammar J, Hol J, Wiebenga JH, Chezan T and Carleer B 2018 *IOP Conf. Ser. Mater. Sci. Eng.* **418** 012093.
[14] E8/E8M-16a, A. 2016 *Standard test methods for tension testing of metallic materials* Tech. rep. ASTM International.
[15] Affronti E, Weidinger M and Merklein M 2018 *IOP Conf. Ser.: Mater. Sci. Eng.* **418** 012047.

Numerical modelling of shear cutting using particle methods

Olle Sandin[1], Juan Manuel Rodríguez Prieto[1,2], Samuel Hammarberg[1], Daniel Casellas[1,3]

[1] Division of Solid Mechanics, Luleå University of Technology, Luleå, Sweden
[2] School of Applied Sciences and Engineering, EAFIT University, Medellin, Colombia
[3] Eurecat, Centre Tecnològic de Catalunya, Unit of Metallic and Ceramic Materials, Manresa, Spain

E-mail: olle.sandin@ltu.se

Abstract. The use of Advanced High Strength Steel (AHSS) allows for lightweighting of sheet steel components, with maintained structural integrity of the part. However, AHSS grades show limitations in edge crack resistance, primarily influenced by sheared edge damage introduced by the shear cutting process. Numerical modelling of the shear cutting process can aid the understanding of the sheared edge damage, thus avoiding unforeseen edge cracking in the subsequent cold forming. However, the extreme deformations of the blank during the shear cutting process are likely to cause numerical instabilities and divergence using conventional Finite Element modelling. To overcome these challenges, this work presents the use of a particle-based numerical modelling method called the Particle Finite Element Method (PFEM). PFEM accurately solves some of the challenges encountered in shear cutting with the standard Finite Element method, such as large deformation, angular distortions, generation of new boundaries and presents an efficient way of transfer historical information from the old to the new mesh, minimising the results diffusion. The present work shows prediction of cut edge morphology of AHSS using a PFEM modelling scheme, where the numerical results are verified against experiments. With these results, the authors show new possibilities to obtain accurate numerical prediction of the shear cutting process, which promotes further advances in prediction of edge damaged related to shear cutting of AHSS.

1. Introduction

Today, advanced high strength steel (AHSS) grades are the preferred materials in the automotive industry. The high strength of AHSS enables lightweighting of structural parts, with the additional benefit of low production costs compared to other lightweight materials, e.g., aluminium and carbon fibre reinforced structures. However, AHSS grades have limitations concerning crack resistance and formability, partly due to the presence of edge cracks. The edge cracking phenomena has been thoroughly investigated over the last decades. In the works of Konieczny and Henderson [1], Thomas [2], Sigvant et al. [3] and Frómeta et al. [4] the detrimental influence from the shear cutting process on the edge cracking resistance of AHSS grades has been investigated experimentally. Further insights of the phenomena may be gained through numerical modelling of the shear cutting process.

Numerous articles describing numerical shear cutting models are found in the literature and generally consists of a Finite Element model using a ductile damage criteria for defining material failure. In the work by Hambli et al. [5] and Klingenberg and Singh [6] such an approach was adopted for axisymmetric models, whereas Wang et al. [7] extends the modelling to three dimensions. However, numerical modelling of shear cutting, using the finite element method (FEM), generates mesh distortion, causing numerical instabilities, as the blank material undergoes severe deformation in the zones where roll-over and burnish formation is present.

Numerical methods for addressing the issue of mesh distortion includes the Aribtrary Lagrangian Eulerian (ALE) formulation and re-meshing techniques. The ALE formulation was used by Brokken et al. [8] and in recent work by Sahli et al. [9] in order to overcome troublesome mesh distortion. In the ALE formulations, the blank material may move independently to the blank mesh, thus allowing for large material deformation without causing skewed and distorted elements. Meanwhile, the blank mesh maintains tracking of the material boundary and interfaces. Re-meshing was applied by Thipprakmas et al. [10] and Basak et al. [11] and generally involves creation of a new mesh based on the prior deformation of the mesh, where the results are mapped on to the new configuration. A combination of ALE and re-meshing was adopted by Pätzold et al. [12] to benefit from both techniques. Unfortunately, both ALE and re-meshing have drawbacks. According to Cremonesi et al. [13], ALE is limited in extreme and unpredictable deformations, while the results mapping utilised in re-meshing introduce numerical diffusion of the solution. To solve the issues of mesh distortion and results diffusion, the present article presents an alternative modelling approach using the Particle Finite Element Method (PFEM) for modelling the shear cutting process. PFEM was first presented by Idelsohn et al. [14] as an approach for avoiding mesh distortion in large deformation problems by combining conventional FEM and particle based methods. By doing this, PFEM combines the accuracy and numerical efficiency of mesh-based Lagrangian techniques and the robustness of particle methods concerning large deformations. This article presents a PFEM model of shear cutting in CP1000HD of 1.5mm thickness, a common AHSS grade for structural parts in the automotive industry. With this, the author's aims to present an efficient and robust numerical tool to predict the shear cutting process and aid the understanding of sheared edge damage of AHSS.

2. Particle Finite Element Method (PFEM)

The PFEM method was originally developed by Idelsohn [14] for modelling of incompressible flows with free-surfaces and fluid structure interaction, but its efficiency in large deformation extended its area of use to non-linear solid mechanics. The Lagrangian governing equations will vary depending on the physical problem to be modelled, but the fundamental features of PFEM according to Cremonesi et al. [13] are the following:

- Treat the mesh nodes as physical particles.
- Store all results and historical data at the nodes/particles.
- Solve the governing equations using a Lagrangian Finite element framework.
- Perform re-meshing using Delaunay Tessellation, defining the nodes/particles connectivity.
- The α-shape method regenerates internal and external boundaries.

This approach means that PFEM benefits from FE-methods as the particles interact using a finite element mesh, providing an efficient solution of the governing equations. When the mesh is overly distorted the connectivity is deleted, but the position and historical values of the particles are kept. A new mesh is then generated based on the deformed configuration. Keeping the position of the particles between re-meshing minimises the results diffusion, which is otherwise a common challenge of conventional re-meshing algorithms. Figure 1 shows how the PFEM steps are performed during modelling of the shear cutting process. Figure 1 shows first how the blank geometry is formed by a cloud of particles at t_0. Next, the initial blank volume is defined and the blank volume is then discretisised by connection of the blank particles using Delaunay Triangulation. The same steps are performed at t_n and t_{n+1} as modelling of the shear cutting process is progressing. At high strain level, new particles are created and inserted according to the principles described by Rodriguez et al. [15] based on an plastic strain energy distribution criteria. This step is schematically shown in Figure 1 at t_{n+1}. The accuracy of PFEM in non-linear solid mechanics is shown in work by Oñate et al. [16], that displays how various industrial forming processes may be modelled using PFEM. The work by Rodriguez et al. [15, 17, 18] specifically applies PFEM techniques in modelling of orthogonal cutting and presents precise representation of chip formation, which is unobtainable using FEM.

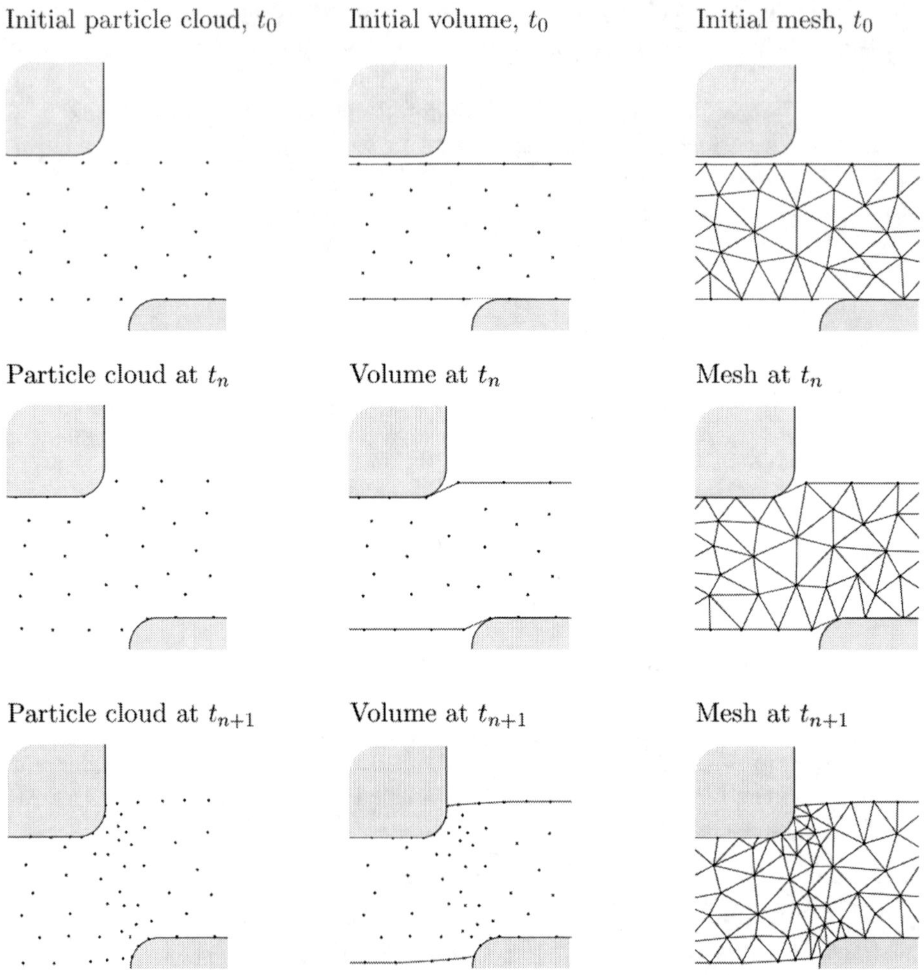

Figure 1: The PFEM steps as performed during modelling of the shear cutting process. The image shows the formation of the particle cloud, definition of the particle volume and the mesh connectivity for initial configuration t_0 and at times t_n and t_{n+1}

3. Constitutive modelling

The blank material, i.e. CP1000HD, was modelled using an elasto-plastic constitutive routine assuming strain hardening. Hardening was represented by a Power-Law plasticity model and its parameters were determined by curve fitting against experimental data of a hydraulic bulge test (ISO standard 16808 [19]). The bulge test was used as it enables the measurement of the plastic deformation behaviour well beyond the point of necking and up until failure for the

IOP Conf. Series: Materials Science and Engineering 1284 (2023) 012048 doi:10.1088/1757-899X/1284/1/012048

CP1000HD material. The experimentally determined hardening curve and the fitted power-law model is presented in Figure 2. The Power-Law expression does not provide a perfect fit of the experimental data. However, the current PFEM implementation only supports the Power-Law plasticity model which is a limiting factor at this stage. Implementation of additional plasticity models is an ongoing work.

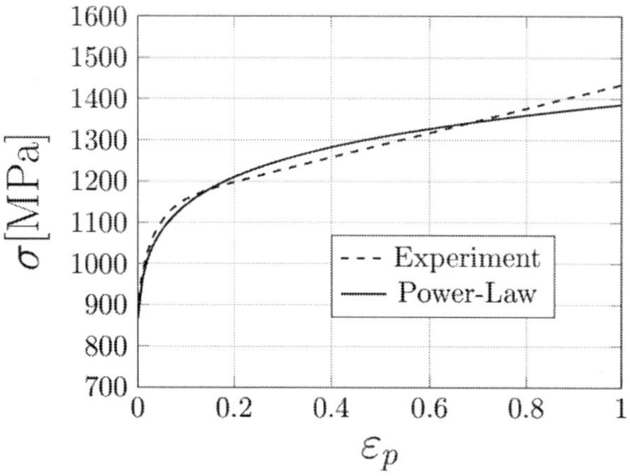

Figure 2: Plastic hardening of CP1000HD from hydraulic bulge testing, extrapolated to $\varepsilon_p = 1$ for true stress σ, and the Power-Law plasticity fit used in the PFEM model.

Table 1: CP1000HD power law parameters.

Parameter	K	n	ε_y
Value	1385	0.08491	0.0041

In this work, the fracture behaviour of the CP1000HD material was modelled using a phenomenological stress-state dependent damage and failure model. The damage and failure model implemented in the PFEM shear cutting code is based on the *Generalized Incremental Stress State Dependent Damage Model* (GISSMO), developed by Neukamm et al. [20] and Basaran et al. [21]. GISSMO is commonly used for modelling of ductile failure in high strength metals, as presented in the work of Omer et al. [22] and Pérez Caro et al. [23]. In the fracture model, instability and failure are prescribed separately in terms of stress triaxiality η, the normalised Lode angle parameter $\bar{\theta}$ and effective plastic strain ε_{eff}. The stress triaxiality and normalised Lode angle parameter are expressed in Equations (1) and (2), where s_{ij} refers to the deviatoric part of the Cauchy stress tensor.

$$\eta = \frac{\sigma_{kk}}{3\sqrt{\frac{3}{2}s_{ij}s_{ij}}} \tag{1}$$

$$\bar{\theta} = \frac{1}{3}\arccos\left(\frac{3\sqrt{3}}{2}\frac{\frac{1}{3}s_{ij}s_{jk}s_{ki}}{\left(\frac{1}{2}s_{mn}s_{mn}\right)^{3/2}}\right). \tag{2}$$

Material damage is accounted for through a damage parameter D, which accumulates according to Equation (3), Here, a value of $n = 1$ means linear damage accumulation and $n > 1$ gives an exponential damage accumulation rate. When $D = 1$, failure occurs.

$$\dot{D} = \frac{n}{\varepsilon^{f}_{eff}\left(\eta, \bar{\theta}\right)} D^{\left(1-\frac{1}{n}\right)} \dot{\varepsilon}^{p}_{eff} \tag{3}$$

Similarly, Equation (4) defines the material instability parameter F. Here, the $\varepsilon^{i}_{eff}(\eta, \bar{\theta})$ defines the stress-state dependent instability strain.

$$\dot{F} = \frac{n}{\varepsilon_{i}(\eta, \bar{\theta})} F^{\left(1-\frac{1}{n}\right)} \dot{\varepsilon}^{p}_{eff} \tag{4}$$

$F = 1$ corresponds to the point of effective plastic strain where coupling between the stress tensor and the damage parameter is initiated. At this point, the load bearing capacity of the material is degraded according Equation (5), describing the relationship between the stress tensor, the damage parameter D and the critical damage value D_{crit}. The parameter D_{crit} is the accumulated damage value of the material at the time F equals 1. The parameter m defines the material degradation rate.

$$\sigma^{*} = \sigma \left(1 - \left(\frac{D - D_{crit}}{1 - D_{crit}}\right)^{m}\right) \tag{5}$$

In this work, experimental values of failure strain and instability strains were determined through tensile testing of four different specimen geometries that were designed to deform and fail at designated stress states. The tensile specimen geometries were designed by Sjöberg et al. [24] and their geometries are schematically shown in Figure 3.

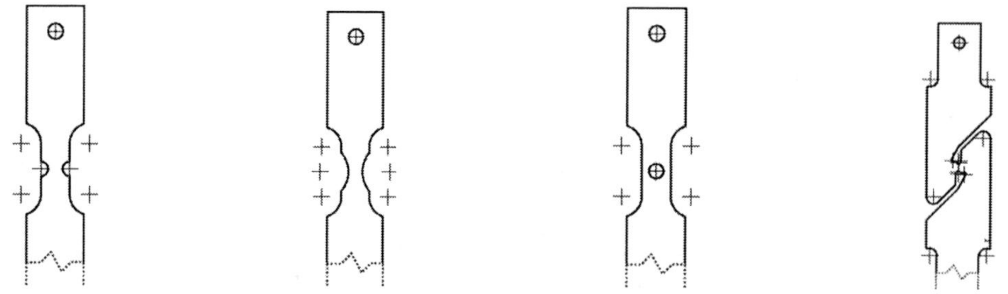

(a) R3.75. **(b)** R15. **(c)** Hole. **(d)** Shear.

Figure 3: Tensile specimen geometries.

Digital image correlation (DIC) enabled strain field measurement of the tensile specimen shown in Figure 3 up until failure. The strain field data was used for calculating the failure strain and the corresponding stress triaxiality for each of the tensile specimen. The experimentally determined failure strain data were then used for calibration of a Modified Mohr-Coulomb failure locus, as described by Bai and Wierzbicki [25] and shown in Equation (6), reduced for von Mises plasticity. In Equation (6), the parameters C_1 to C_3 where fitted against the experimental failure strain data assuming plane stress, where Equation (7) relates the normalised Lode angle parameter $\bar{\theta}$ to the stress triaxiality η for the plane stress assumption.

$$\varepsilon^{f}_{eff}(\eta, \bar{\theta}) = \left\{ C_2 \left[\sqrt{\frac{1 + C_1^2}{3}} cos\left(\frac{\bar{\theta}\pi}{6}\right) + C_1 \left(\eta + \frac{1}{3} sin\left(\frac{\bar{\theta}\pi}{6}\right)\right)\right]\right\}^{-\frac{1}{C_3}}, \tag{6}$$

$$\bar{\theta} = 1 - \frac{2}{\pi} \cos^{-1}\left(-\frac{27}{2}\eta\left(\eta^2 - \frac{1}{3}\right)\right) \tag{7}$$

Table 2 presents the calibrated MMC parameters for the CP1000HD material, while Figure 4 shows the corresponding MMC failure surface. Stress-state analysis of the shear cutting process show that deformation and failure occurs at $\bar{\theta} \approx 0$. Consequently, for numerical simplicity the shear cutting model presented in this work only used a failure curve defined by $\varepsilon^f_{eff}(\eta, 0)$ of the MMC failure surface.

Table 2: MMC parameters.

Parameter	C_1	C_2	C_3
Value	0.0918	1.6337	0.2313

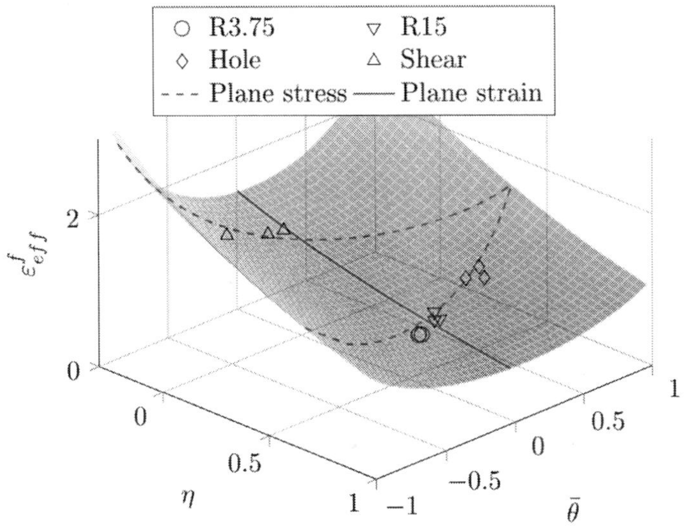

Figure 4: MMC failure locus of CP1000HD, where the plane stress failure curve (dashed) and the curve defined by $\varepsilon^f_{eff}(\eta, 0)$ (solid) is plotted along with the experimental tensile failure points.

A cut-off value at $\varepsilon^f_{eff}(\eta < 0, 0)$ was assigned, meaning that no failure occur in the material for negative triaxialities. What this means in terms of numerical modelling, is that the formation of the burnish zone is formed by a material flow rather than compressive material failure. This statement contrasts the work by i.e. Bao and Wierzbicki [26] that sets the cut-off value for $\varepsilon^f_{eff}(\eta < -1/3)$. Meanwhile, Zhao et al. [27] could see that the large hydrostatic pressure at negative triaxilities suppressed the formation and growth of voids during burnish formation in fine-blanking. In the work by Zhao et al., a punch penetration of approximately 70% of the blank thickness was required for creating enough void growth and coalescence through shearing in order to cause final fracture of the blank. At this penetration depth, the hydrostatic pressure was significantly reduced thus accelerating the tearing of the blank material. As conventional blanking processes involves larger clearances than fine-blanking, thus shorter burnish zones, the assumption of a cut-off value at $\eta < 0$ seems reasonable as no substantial shearing void growth is considered present for such short punch penetration depths.

4. PFEM shear cutting model

The PFEM shear cutting model consisted of axisymmetric representation according to the schematic configuration shown in Figure 5, where the blank, punch tool, die tool and blank tool are highlighted. The tools were modelled with a linear elastic material model, whereas the blank part was assigned the constitutive relations described in Section 3. The numerical model was solved with a vectorised MATLAB program using an implicit solution scheme. A penalty based node-to-surface contact algorithm was used between tools and blank. In the present work, a non-linear finite element framework is implemented for solving the governing equations and equations of motion. The model uses mixed displacement/pressure FE-formulation according to the work of Chiumenti et al. [28] and Saracibar et al. [29] for applying pressure stabilisation in near incompressible deformation of triangular elements. This formulation prevents volumetric locking of triangular elements subjected to large plastic deformation. Detailed model geometry and re-meshing setting are presented in Table 4.

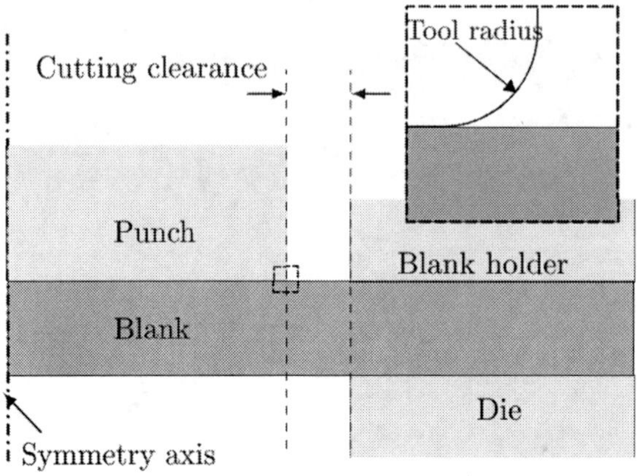

Figure 5: Schematic image of the shear cutting configuration, showing the blank and tool parts and a close-up on the tool edge. The symmetry axis and cutting clearance are highlighted as well.

Table 3: PFEM model parameters, sectioned as *Geometry-* and *Re-meshing-* parameters.

Geometry		**Re-meshing**	
Blank thickness	1.5 [mm]	Smallest elem. length	0.01 [mm]
Punch radius	5 [mm]	Re-mesh frequency	10
Die radius	5.134 [mm]	Curvature	0.01
Cutting clearance	8.93 [%]	α	1
Tool edge radius	30 [μm]	-	-

5. Results and discussion

In Figure 6 the cut edge morphology of the experiment and the results of the PFEM shear cutting process are presented. In the figure, the roll-over-, shear- and fracture surfaces for the experiment and PFEM are highlighted and compared. The close correlation of the experimental and numerical cut edge morphologies indicates that the damage and failure model initiate and propagate material failure in an accurate manner.

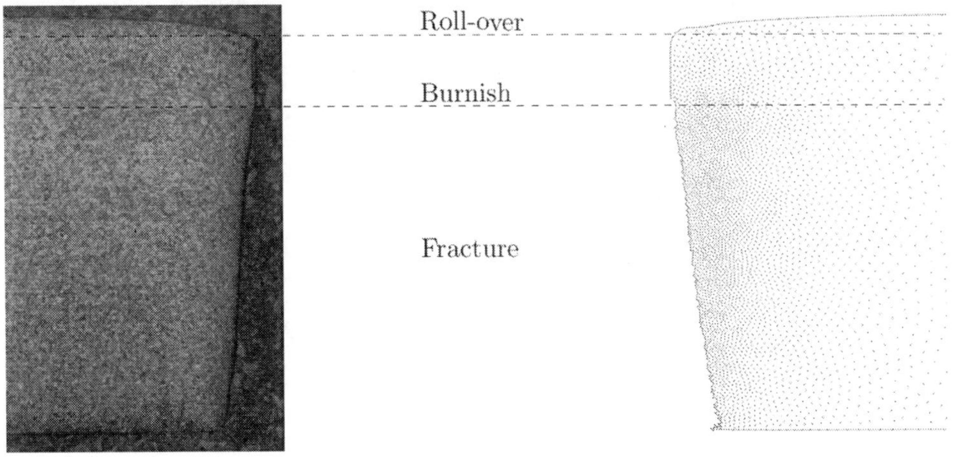

Figure 6: Cut-edge morphology from experiment and PFEM modelling of CP1000HD of 8.93% cutting clearance

Figure 7 shows the experimental and PFEM punch displacement versus force results and the crack initiation before the final fracture process. The experimental punch displacement results are corrected in order to account for the elastic deformation of the punch and machine compliance.

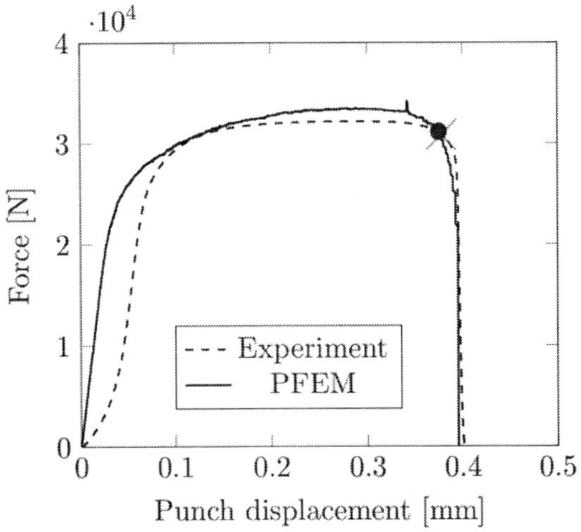

(a) Force - displacement results.

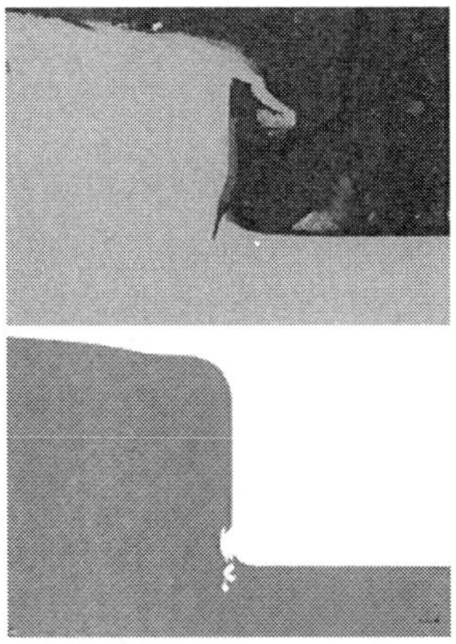

(b) Crack initiation in experiment (top) and numerical modelling (bottom).

Figure 7: Force - displacement results from punching experiment (dashed) and numerical modelling (solid) for CP1000HD with 8.93% cutting clearance. The point of experimental crack initiation is marked with ×, while the numerical fracture initiation is marked with •.

Comparing the force - displacement curves in Figure 7 a) shows that punch displacement and peak force are similar, while the exact distribution differs due to the mismatch between experimental plastic hardening response. Another possible source of result dissimilarity is the experimental tool compliance, which was linearly corrected but in fact have a non-linear behaviour. However, what is seen in Figure 7 b) is that both experimental and PFEM crack initiation occurs at a similar position along the force displacement curve and the cracks show similar angle and size. Considering Figure 6 and Figure 7, it appears that the PFEM implementation successfully model the shear cutting process as both numerical cut edge morphology predictions and force - displacement output correlates well with experiments. Further validation of the PFEM shear cutting model would require simulation of a wide range of cutting clearances and investigation of, for instance, the width of the shear affected zone and residual stress distribution to assure the the sheared edge damage is captured correctly.

6. Conclusions

Based on the presented results, following conclusions can be drawn:

- The presented PFEM model allows for accurate simulation of the shear cutting process, including robust representation of severe material deformation during burnish formation as well as ductile material failure during the fracture process.
- The PFEM connectivity algorithm ensured well-structured elements throughout the simulation, guaranteeing adequate element aspect ratio and avoiding numerical instability. Maintaining well-shaped elements was of extra importance during the burnish formation, as this process otherwise would generate overly distorted elements. By preserving well-shaped elements, it was ensured that material failure occurred under correct stress states.
- The accuracy and robustness of the PFEM modelling approach in the formation of burnish surface suggest that the method additionally would serve as an effective numerical tool for fine-blanking applications. In fine-blanking, punching penetration of 50-70% of the blank thickness is common before fracture, meaning that a majority of the cut edge morphology consists of burnish formation.
- The PFEM modelling approach ensures low results diffusion during re-meshing and allows for high level of preservation of information along the shear affected zone. Such information would be relevant to model further forming operations (as hole expansion or flange stretching) or predicting part performance (as fatigue behaviour).

Acknowledgments

The authors are grateful for the contribution by Dr. David Frómeta and Sergi Parareda at Eurecat for providing experimental results and voestalpine Stahl GmbH for providing the sheet steel investigated.

References

[1] Konieczny A and Henderson T 2007 *SAE Tech. Pap. 2007-01-0340* 1–12

[2] Thomas D J 2013 *J. Fail. Anal. Prev.* **13** 451–462

[3] Sigvant M, Falk J and Pilthammar J 2017 *J. Phys. Conf. Ser.* **896**

[4] Frómeta D, Tedesco M, Calvo J, Lara A, Molas S and Casellas D 2017 *J. Phys. Conf. Ser.* **896** 012102

[5] Hambli R and Potiron A 2000 *J. Mater. Process. Technol.* **102** 257–265

[6] Klingenberg W and Singh U P 2003 *J. Mater. Process. Technol.* **134** 296–302

[7] Wang K, Greve L and Wierzbicki T 2015 *Int. J. Solids Struct.* **71** 206–218

[8] Brokken D, Brekelmans W A M and Baaijens F P 1998 *J. Mater. Process. Technol.* **83** 192–199

[9] Sahli M, Roizard X, Colas G, Assoul M, Carpentier L, Cornuault P H, Giampiccolo S and Barbe J P 2020 *18th International Conference Metal Forming*

[10] Thipprakmas S, Jin M, Tomokazu K, Katsuhiro Y and Murakawa M 2008 *J. Mater. Process. Technol.* **198** 391–398

[11] Basak S, Kim C, Jeong W, Jung Y I and Lee M G 2022 *Int. J. Mech. Sci.* **219** 107109

[12] Pätzold I, Stahl J, Golle R and Volk W 2023 *J. Mater. Process. Technol.* **313** 117872

[13] Cremonesi M, Franci A, Idelsohn S and Oñate E 2020 *Arch. Comput. Methods Eng.* **27** 1709–1735

[14] Idelsohn S R, Onate E and Del Pin F 2004 *Int. J. Numer. Methods Eng.* **61** 964–989

[15] Rodriguez Prieto J M, Carbonell J M, Cante J C, Oliver J and Jonsén P 2018 *Comput. Mech.* **61** 639–655

[16] Oñate E, Franci A and Carbonell J M 2014 *Comput. Mech.* **54** 85–107

[17] Rodriguez J M, Carbonell J M, Cante J C and Oliver J 2016 *Int. J. Numer. Methods Eng.* **107** 733–785

[18] Rodriguez J M, Larsson S, Carbonell J M and Jonsén P 2022 *Comput. Part. Mech.* **9** 709–733

[19] ISO 16808 2014 Metallic materials — Sheet and strip — Determination of biaxial stress-strain curve by means of bulge test with optical measuring systems

[20] Neukamm F, Feucht M, Haufe A and Roll K 2008 *10th International LS-DYNA Users Conference*

[21] Basaran M, Wölkerling S D, Feucht M, Neukamm F and Weichert D 2010 *LS-DYNA Anwenderforum*

[22] Omer K, Ten Kortenaar L, Butcher C, Worswick M J, Malcolm S and Detwiler D 2017 *Int. J. Impact Eng.* **103** 12–28

[23] Pérez Caro L, Schill M, Haller K, Odenberger E L and Oldenburg M 2020 *Int. J. Mater. Form.* **13** 15–28

[24] Sjöberg T, Marth S, Kajberg J and Häggblad H Å 2017 *Eur. J. Mech. A/Solids* **66** 279–286

[25] Bai Y and Wierzbicki T 2008 *Int. J. Plast.* **24** 1071–1096

[26] Bao Y and Wierzbicki T 2005 *Eng. Fract. Mech.* **72** 1049–1069

[27] Zhao P J, Chen Z H and Dong C F 2016 *J. Mater. Process. Technol.* **236** 16–25

[28] Chiumenti M, Valverde Q, Agelet de Saracibar C and Cervera M 2004 *Int. J. Plast.* **20** 1487–1504

[29] Agelet de Saracibar C, Chiumenti M, Valverde Q and Cervera M 2006 *Comput. Methods Appl. Mech. Eng.* **195** 1224–1251

IOP Conf. Series: Materials Science and Engineering 1284 (2023) 012049 doi:10.1088/1757-899X/1284/1/012049

Evaluation of anisotropic fracture in AA1050-O sheet using uncoupled ductile fracture models with Hill48 plasticity

Aishwary Gupta[1], Yong Hou[1], Hariharan K[2], Heung Nam Han[1], Myoung-Gyu Lee[1]

[1]Department of Materials Science and Engineering,Seoul National University, South Korea
[2] Department of Mechanical Engineering, Indian Institute of Technology Madras, India

E-mail: aishwary@snu.ac.kr

Abstract. Uncoupled ductile fracture models have gained popularity for accurate prediction of fracture in ductile metallic sheets due to their ease of implementation. Accurate description of anisotropic fracture behavior is an active research field for sheet metal forming. In the present study, two isotropic ductile fracture models, i.e., modified Mohr-Coulomb and Hosford-Coulomb are evaluated for an O-tempered aluminum alloy (AA1050-O) sheet. The plastic anisotropy is modeled with a non-associated Hill48 plasticity model. In order to verify the models, experiments are conducted for a 1.2 mm thick AA1050-O sheet under various loading conditions, such as uniaxial tension, in-plane shear, and plane strain tension at room temperature. For evaluating the anisotropic ductile fracture, tensile tests are carried out in 15° intervals to the rolling direction using digital image correlation technique. The predicted fracture limit strains by the two uncoupled fracture models are compared with experimental results to evaluate the accuracy of these models.

1. Introduction

Accurate modeling of failure is one of the key issues in metal forming. Failure modes in metal forming are mainly classified into three types: necking, ductile fracture, and wrinkling. Ductile fracture modeling is broadly classified into two types of fracture models i.e. coupled and uncoupled fracture models. Coupled fracture models consists of Gurson porous plasticity models [1] which describe the growth of voids and its effect on plastic behaviour and Continuum Damage mechanics based fracture models where a scalar damage variable is derived from a thermodynamic dissipation potential[2]. Uncoupled fracture models are empirical models proposed for predicting ductile fracture initiation, in which the elastic and plastic behavior is assumed to be unaffected by the evolution of damage. These uncoupled damage models have gained wide popularity due to their ease of implementation. In the present work, two uncoupled fracture models i.e. Hosford-Coulomb and Modified Mohr-Coulomb have been evaluated with the help of Finite Element Analysis using Hill-48 non associative plasticity. For evaluating anisotropy in ductile fracture, uniaxial tensile tests in 15 degree intervals to rolling direction have been conducted and the fracture strains from the predicted models have been compared with experimental results.

IOP Conf. Series: Materials Science and Engineering 1284 (2023) 012049 doi:10.1088/1757-899X/1284/1/012049

2. Experiments and Methodology

Commercially pure aluminium alloy (AA1050-O) sheet of 1.2 mm thickness is considered in this study. A series of mechanical tests were conducted to record the mechanical behaviour of the material till fracture. Material response under different loading conditions was achieved by varying the specimen geometry. The specimen geometries were chosen such that it provides a wide range of stress states. Five types of specimen geometry were chosen, uniaxial tensile (UT), notch specimen (NT), center hole (CH), plane strain (PS) and in-plane shear specimen (SH) as shown in Fig 1. All the mechanical tests were performed using a (Instron 5942) universal testing machine (UTM) with stereo-type digital image correlation (3D-DIC). The uniaxial tensile tests were carried out at a strain of $10^{-3} sec^{-1}$ and all other tests were conducted at constant crosshead speed of 0.5 mm/min. For evaluating the anisotropic ductile fracture uniaxial tensile (UT) were carried out in 15 degree intervals to the rolling direction. The engineering stress vs engineering strain curve for RD, DD and TD direction is shown in Fig 2.

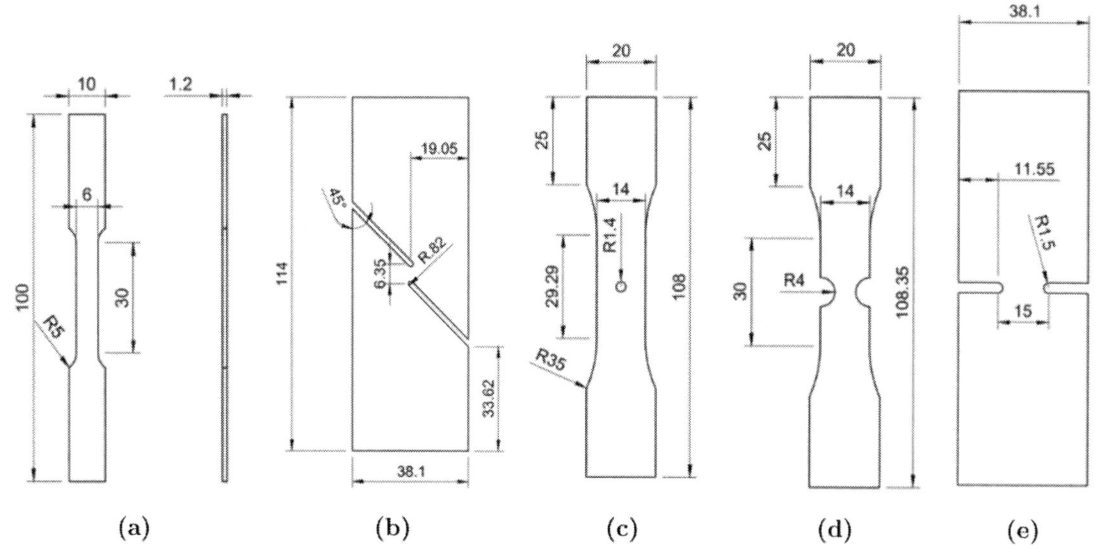

Figure 1: Schematic of specimen geometries for (a) UT (b) SH (c) CH (d) NT (e) PS specimens. (All dimensions are in mm)

3. Constitutive modelling and calibration of fracture models

3.1. Plasticity model and Hardening law

In the present work, Hill-48 yield criteria with non-associative flow rule has been used. The yield function and flow potential are denoted by f and g, and are given in Eq 1-2.

$$f = \bar{\sigma}_\sigma - \sigma_y(\bar{\epsilon}^p) \leq 0 \tag{1}$$

$$g = \bar{\sigma}_r - \sigma_y(\bar{\epsilon}^p) \leq 0 \tag{2}$$

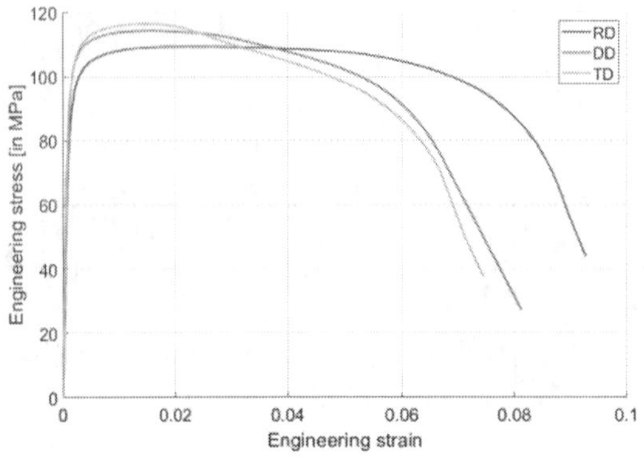

Figure 2: Engineering stress vs Engineering strain curve for AA1050-O Aluminum alloy in RD,DD and TD.

Where,

$$\bar{\sigma}_\sigma = \sqrt{\frac{1}{2}[F_\sigma(\sigma_{22}-\sigma_{33})^2 + G_\sigma(\sigma_{33}-\sigma_{11})^2 + H_\sigma(\sigma_{11}-\sigma_{22})^2] + L_\sigma\sigma_{23} + M_\sigma\sigma_{13} + N_\sigma\sigma_{12}} \quad (3)$$

$$\bar{\sigma}_r = \sqrt{\frac{1}{2}[F_r(\sigma_{22}-\sigma_{33})^2 + G_r(\sigma_{33}-\sigma_{11})^2 + H_r(\sigma_{11}-\sigma_{22})^2] + L_r\sigma_{23} + M_r\sigma_{13} + N_r\sigma_{12}} \quad (4)$$

The parameters for yield function and flow potential are estimated using Eq 5-12 and are calibrated using uniaxial tensile data in RD, DD and TD direction. For simplicity, the biaxial yield stress has been considered as average of RD, TD and DD yield stresses ($\sigma_b = (\sigma_0 + 2\sigma_{45} + \sigma_{90})/4$). The r values are estimated with help of longitudinal strain and width strain using virtual extensometers in DIC ($r = \epsilon_w/(-(\epsilon_w + \epsilon_l))$). The calibrated parameters of AA1050-O aluminum alloy are given in Table 1 .

$$F_\sigma = \sigma_0^2/\sigma_{90}^2 - 1 + \sigma_0^2/\sigma_b^2 \quad (5)$$

$$G_\sigma = 1 - \sigma_0^2/\sigma_{90}^2 + \sigma_0^2/\sigma_b^2 \quad (6)$$

$$H_\sigma = 1 + \sigma_0^2/\sigma_{90}^2 - \sigma_0^2/\sigma_b^2 \quad (7)$$

$$N_\sigma = 4\sigma_0^2/\sigma_{45}^2 - \sigma_0^2/\sigma_b^2 \quad (8)$$

$$F_r = 2r_0/(r_{90}(1+r_0)) \quad (9)$$

$$G_r = 2/(r_{90}(1+r_0)) \quad (10)$$

$$H_r = 2r_0/(1+r_0) \quad (11)$$

$$N_r = 2(r_0 + r_{90})(r_{45} + 0.5)/(r_{90}(1+r_0)) \quad (12)$$

$$L_\sigma = L_r = M_\sigma = M_r = 3 \quad (13)$$

Table 1: Calibrated coefficients for Hill-48 plasticity

F_σ	G_σ	H_σ	N_σ	F_r	G_r	H_r	N_r
0.806878	2.806878	0.962620	2.697941	0.378798	0.695748	0.304251	1.082069

The strain hardening behaviour is modelled using the combined Swift and Voce hardening law as given in Eq.14.

$$\sigma_y(\epsilon_p) = QK(\epsilon_o + \epsilon_p)^n + (1 - Q)[\sigma_y + A(1 - e^{-B\epsilon_p})] \tag{14}$$

Here, σ_y denote the yield stress, A and B are fitting constants. The values of K, ϵ_o, n are estimated by fitting the experimental data using Swift hardening law till the necking strain. Q is a post necking hardening parameter which was estimated by iterative numerical simulations of uniaxial tension test by minimizing the error between the predicted and experimental load-displacement curve till the onset of fracture. The calibrated hardening parameters are given in Table 2

Table 2: Calibrated coefficients of the hardening law

K(MPa)	ϵ_o	n	Q	A	B	σ_y
141.0736	0.002707	0.055	0.73	11.15	220.7	101.9138

3.2. Fracture models

An arbitrary stress state can be represented using η and $\bar{\theta}$. Significant work has been reported, defining the relationship between the fracture strain and stress state [3–7]. η and $\bar{\theta}$ are defined as:

$$\eta = \frac{\sigma_1 + \sigma_2 + \sigma_3}{3\sigma_{vM}} \tag{15}$$

$$\bar{\theta} = 1 - \frac{2}{\pi} cos^{-1}\left(\frac{27}{2}\frac{J_3}{\bar{\sigma}_{vM}}\right) \qquad \text{where } -1 \leq \bar{\theta} \leq 1 \tag{16}$$

where σ_1,σ_2,σ_3 are the principal stresses ($\sigma_1<\sigma_2<\sigma_3$), σ_{vM} is the equivalent stress and J_3 is devatoric third stress invariant.

A damage variable D is introduced as a state variable given by eq.17.

$$D = \int_0^{\bar{\epsilon}_f} \frac{d\bar{\epsilon}}{\bar{\epsilon}_f(\eta, \bar{\theta})} \tag{17}$$

where $d\bar{\epsilon}$ denotes equivalent plastic strain increment, $\bar{\epsilon}_f$ represents the equivalent plastic strain at the onset of fracture. $D = 1$ marks the onset of fracture and the corresponding strain $\bar{\epsilon}_f$ is referred to as fracture strain. The use of damage variable ensures strain path dependence on the onset of fracture[8].

3.3. Modified Mohr-Coulomb Model (MMC)

The classical Mohr-Coulomb criteria assumes that failure occurs when the normal and shear stresses on an arbitrary plane reaches a critical value, given by:

$$max(\tau + c_1\sigma_n) = c_2 \tag{18}$$

Here, τ and σ_n are shear and normal stress respectively, on an arbitrary fracture plane . $c_1 \geq 0$ is called the friction coefficient which control the fracture plane orientation and $c_2 > 0$ is called as shear resistance [9]. For $c_1 = 0$, Eq.(18) reduces to maximum shear stress criterion. Supposing proportional loading, stress based Mohr-Coulomb model can be reformulated in the $(\bar{\epsilon}, \bar{\theta}, \eta)$ space, assuming η and $\bar{\theta}$ dependent work hardening law as proposed by [9]. Considering further simplifying assumptions[10], MMC fracture locus is given by:

$$\bar{\epsilon}_f(\eta, \bar{\theta}) = \left\{ \frac{K}{c_2} \left[c_3 + \frac{\sqrt{3}}{2-\sqrt{3}}(1-c_3)(sec\frac{\bar{\theta}\pi}{6} - 1) \right] \left[\sqrt{\frac{1+c_1^2}{3}} cos\frac{\bar{\theta}\pi}{6} + c_1(\eta + \frac{1}{3} sin\frac{\bar{\theta}\pi}{6}) \right] \right\}^{\frac{-1}{n}} \tag{19}$$

Here K is the strength coefficient and n is work hardening exponent, which are deduced from the Swift hardening law. MMC model is calibrated for three damage parameters, c_1 and c_2 and c_3, which controls the shape of the fracture locus.

3.4. Hosford–Coulomb (HC) model

Mohr and Marcadet [11] proposed Hosford-Coulomb fracture criterion to model ductile fracture initiation for advanced high strength sheets under proportional loading. Hosford-Coulomb (HC) fracture model assumes localization of deformation in a narrow zone and the localization criterion can be given as:

$$\left\{ \frac{1}{2}((\sigma_1 - \sigma_2)^a + (\sigma_2 - \sigma_3)^a + (\sigma_3 - \sigma_1)^a) \right\}^{1/a} + c(\sigma_1 + \sigma_3) = b \tag{20}$$

Using Lode angle parameter dependent functions,

$$f_1[\bar{\theta}] = \frac{2}{3}cos[\frac{\pi}{6}(1-\bar{\theta})] \quad ; f_2[\bar{\theta}] = \frac{2}{3}cos[\frac{\pi}{6}(3+\bar{\theta})] \quad ; f_3[\bar{\theta}] = -\frac{2}{3}cos[\frac{\pi}{6}(1+\bar{\theta})] \tag{21}$$

the principal stress may be expressed in $(\eta,\bar{\theta},\bar{\epsilon})$ space, as given by:

$$\sigma_1 = \bar{\sigma}(\eta + f_1) \quad ;\sigma_2 = \bar{\sigma}(\eta + f_2) \quad ;\sigma_3 = \bar{\sigma}(\eta + f_3) \tag{22}$$

Using Eqs. 22, in the Eq.(20) the stress at the onset of fracture (σ_f) can be formulated as:

$$\bar{\sigma} = \sigma_f(\eta, \bar{\theta}) = \frac{b}{[\frac{1}{2}((f_1 - f_2)^a + (f_2 - f_3)^a + (f_1 - f_3)^a)]^{\frac{1}{a}} + c(2\eta + f_1 + f_3)} \tag{23}$$

Furthermore, strain at onset of fracture $(\bar{\epsilon}_f)$ is obtained by taking inverse of isotropic hardening law, which can be expressed as, $\bar{\epsilon}_f = k^{-1}\bar{\sigma}_f$. Parameter a controls the Lode angle sensitivity of the fracture strain, and the dependency of triaxiality is controlled by the friction parameter c. The parameter b is the strain to onset of fracture under uniaxial tension.

3.5. Calibration of Fracture models

The loading history denoted as $\eta_i(\epsilon_p)$ for triaxiality and $\bar{\theta}_i(\epsilon_p)$ for Lode angle parameter of each calibration experiment (i) i.e. UT, SH and NT tests is evaluated with help of FEA simulation. For the set of parameters $[\Omega]$ to be optimized for the fracture model, the fracture strain $\bar{\epsilon}_f^i = \bar{\epsilon}_f^i(\Omega)$ for each loading case i is calculated according to the Eq.(24):

$$\int_0^{\bar{\epsilon}_f} \frac{d\bar{\epsilon}}{\bar{\epsilon}_f(\eta, \bar{\theta})} = 1. \tag{24}$$

The following minimization problem (Eq.25 is solved with help of simplex algorithm using Matlab to obtain the final optimized set of parameters $[\Omega]$ for each fracture model.

$$\Omega = \min_{\Omega} \sum_{i} |\epsilon_f^i[\Omega] - \epsilon_{exp}^i| \quad \forall \quad i \in [UT, SH, NT] \tag{25}$$

Where ϵ_{exp}^i is eq. plastic strain at onset of fracture for ith experiment. Complete loading histories, i.e. Triaxiality ratio and Lode angle parameter have been used for the calibration of fracture models by inverse analysis using FEA simulation[12]. It is assumed that the location of onset of fracture coincides with the location of the highest equivalent plastic strain in each specimen geometry. Thus, the critical element is selected at the location of the highest equivalent plastic strain, as shown in fig. 3. The loading histories of critical element for each specimen geometry can be seen in the fig. 4. The 3D Calibrated fracture locus for MMC and HC models are shown in Fig. 5-6.

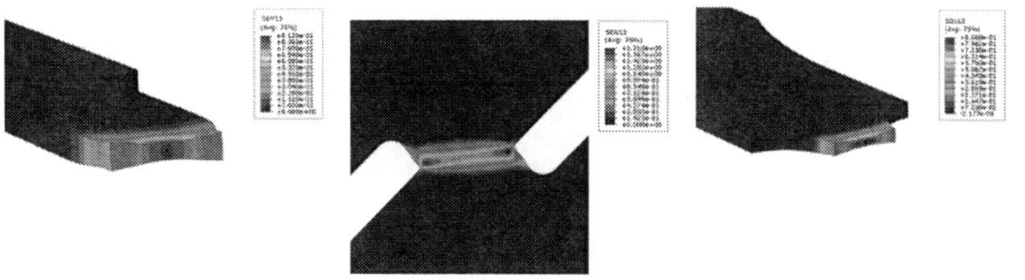

(a) (b) (c)

Figure 3: Plastic strain distribution and location of critical element for (a) UT (b) SH (c) NT specimens (The black dot marked symbol represents the location of critical element).

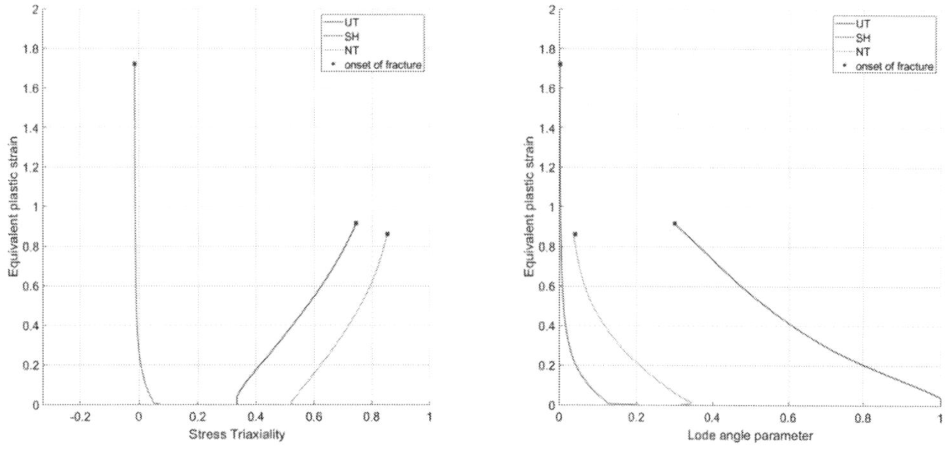

(a) (b)

Figure 4: Variation of Stress triaxiality and Lode angle parameter.

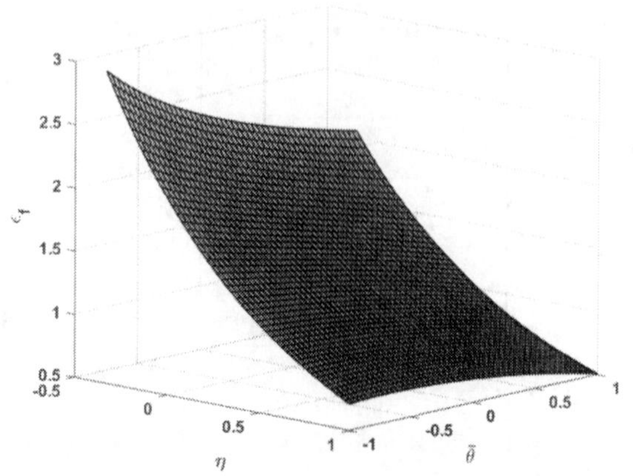

Figure 5: Representation of modified Mohr-Coulomb (MMC) model in the ($\bar{\epsilon}$,η, $\bar{\theta}$) space.

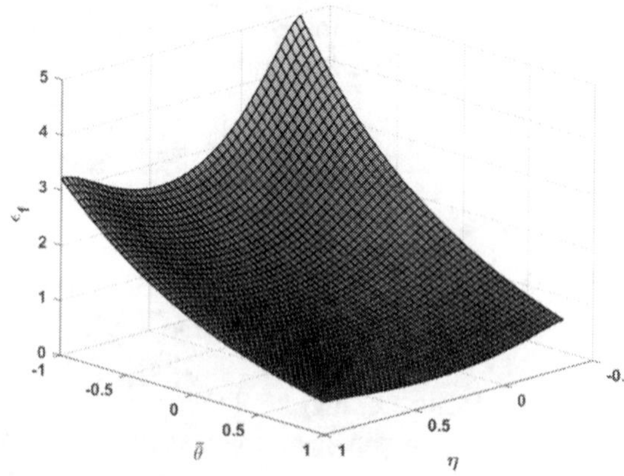

Figure 6: Representation of Hosford–Coulomb (HC) model in the ($\bar{\epsilon}$,η, $\bar{\theta}$) space.

The calibrated parameters for HC and MMC fracture models are tabulated in Table 3.

Table 3: Calibrated parameters for fracture models

a	b	c	c_1	c_2	c_3	C_1	C_2
1.70	141.012	0.02755	0.028956	71.6038	0.8640	1.6232	0.9087

4. Finite Element Model

Finite element (FE) simulation of the specimens shown in Fig. 1 were performed with ABAQUS \Explicit software. Three-dimensional continuum elements (C3D8R) were used to mesh the sheet specimens. Elasticity modulus of 71.659 GPa and Poisson's ratio of 0.33 has used to model the elastic response of AA1050-O Aluminum alloy. The MMC and HC damage criteria is implemented as a user defined subroutine (VUMAT) in ABAQUS software. The effective plastic strain, continuum damage variable (D), stress triaxiality (η) and Lode angle parameter ($\bar{\theta}$) are defined as state variables in the VUMAT. The onset of fracture for each FE simulation is assumed when the damage variable (D) reaches unity. The mesh size has been chosen based on a mesh sensitivity analysis; an element size of 0.05 mm was used near the critical region. Around five through thickness elements have been chosen for each specimen geometry.

5. Results and Discussion

As explained earlier, Finite element simulations for various specimen geometries were performed with the calibrated fracture model. The Experimental and simulated Force-Displacement curves of Shear, Notch, Centre Hole and Plane strain specimen along with experimental and simulated fracture points are shown in Fig 7-10. It can be seen that both HC and MMC fracture model are able to predict the fracture displacements well for rolling direction. The distribution of damage variable D at the onset of fracture has been shown for various specimen geometries in Fig 11.

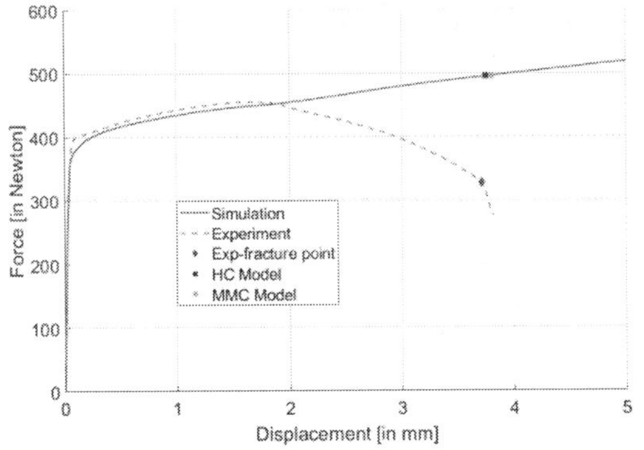

Figure 7: Force vs Displacment for Shear specimen.

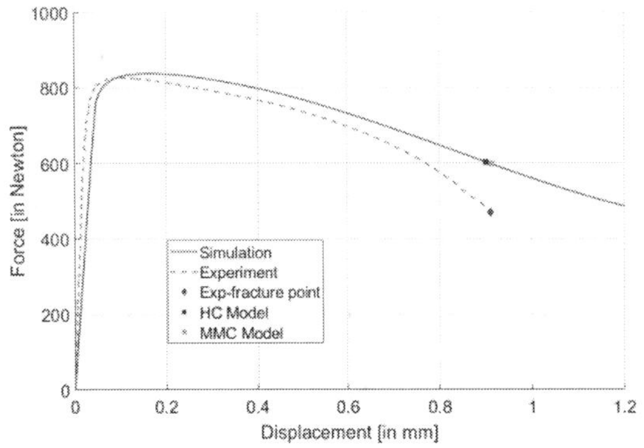

Figure 8: Force vs Displacment for Notch specimen.

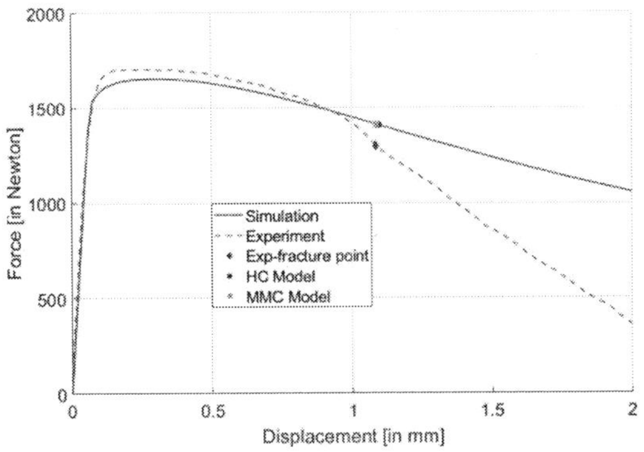

Figure 9: Force vs Displacment for Centre Hole specimen.

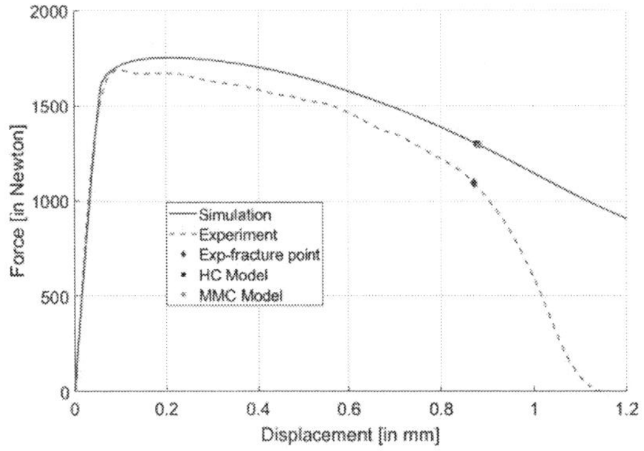

Figure 10: Force vs Displacment for Plane Strain specimen.

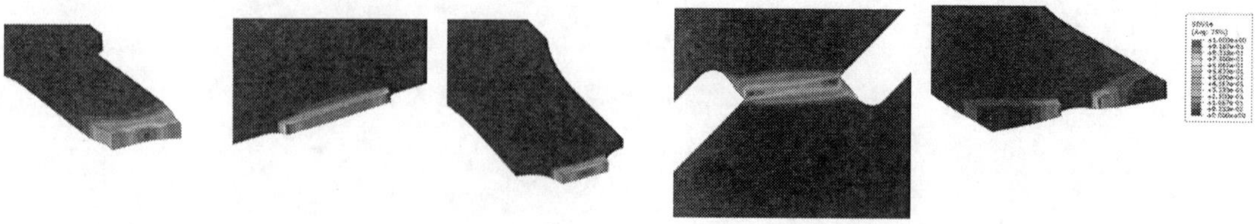

(a) (b) (c) (d) (e)

Figure 11: Contour plot of Damage variable D (SDV 14) at onset of fracture for (a) UT (b) PS (c) NT (d) SH (e) CH specimens.

The experimental Force-Displacement curve for uniaxial tests along various orientation of rolling direction is shown in Fig 12. The simulated Force-Displacement along with fracture points for the same is shown in Fig 13. It is seen that for rolling direction the experimental and simulated fracture points of both MMC and HC model show excellent match. However, for angles 30-90, both MMC and HC fracture model show significant deviation from experimental fracture points. The equivalent plastic strain at the onset of fracture for all uniaxial tensile specimen has been shown in Fig 14.

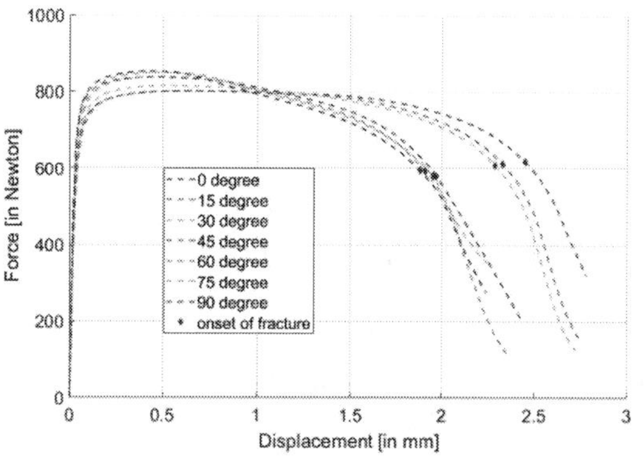

Figure 12: Experimental Force vs Displacment for uniaxial tensile specimens along various directions.

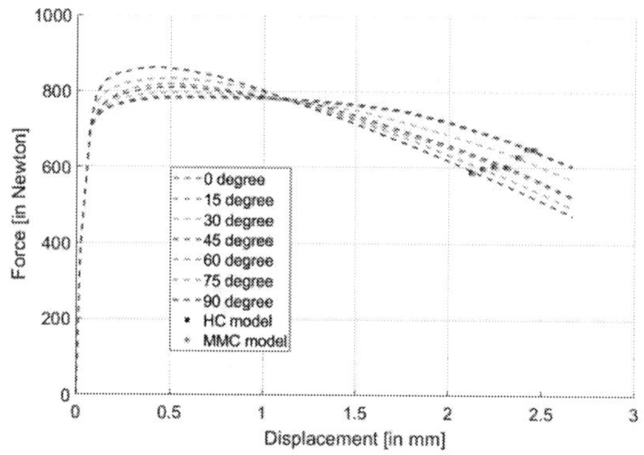

Figure 13: Simulated Force vs Displacment for uniaxial tensile specimens along various directions.

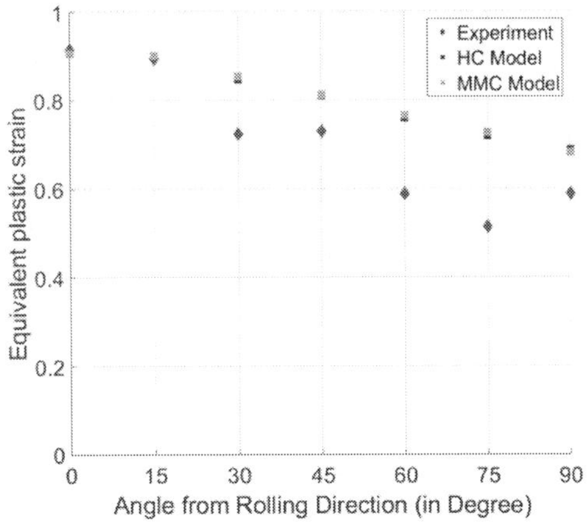

Figure 14: Equivalent plastic strain at onset of fracture at the critical element for UT tests

6. Conclusion

In the present study anisotropy in fracture has been evaluated for two fracture models i.e. Hosford-Coulomb and Modified Mohr-Coulomb fracture model using Hill-48 non associated plasticity model. It has been seen that both MMC and HC model are able to predict the fracture very well in RD for various specimen geometries. Through evaluation of fracture in uniaxial tesnile test at 15 degree variation to rolling direction, it has been seen that both the uncoupled isotropic fracture models are unable to predict the anisotropy in fracture. For better prediction of fracture in various directions anisotropic uncoupled fracture models should be used for AA1050-O aluminum alloy, which will be addressed in future work.

References

[1] Hosford W F 1972 *Journal of Applied Mechanics* **39** 607–609 ISSN 0021-8936

[2] Bonora N 1997 *Engineering Fracture Mechanics* **58** 11–28 ISSN 0013-7944 URL `https://www.sciencedirect.com/science/article/pii/S001379449700074X`

[3] Rice J 1967 *Fatigue crack propagation*

[4] Rice J and Rosengren G 1968 *Journal of the Mechanics and Physics of Solids* **16** 1–12 ISSN 0022-5096

[5] Rice J and Tracey D 1969 *Journal of the Mechanics and Physics of Solids* **17** 201–217 ISSN 0022-5096

[6] Gurson A L 1977 *Journal of Engineering Materials and Technology* **99** 2–15 ISSN 0094-4289 (*Preprint* http://arxiv.org/abs/https://asmedigitalcollection.asme.org/materialstechnology/article-pdf/99/1/2/5788715/2_1.pdf:https://asmedigitalcollection.asme.org/materialstechnology/article-pdf/99/1/2/5788715/2_1.pdf) URL `https://doi.org/10.1115/1.3443401`

[7] Bao Y and Wierzbicki T 2004 *International Journal of Mechanical Sciences* **46** 81–98 ISSN 0020-7403 URL `https://www.sciencedirect.com/science/article/pii/S0020740304000360`

[8] Prasad K, Gupta A, Krishnaswamy H, Chakkingal U, Banerjee D K and Lee M G 2023 *Friction* **11** 820–835 ISSN 2223-7704 URL `https://doi.org/10.1007/s40544-022-0698-2`

[9] Bai Y and Wierzbicki T 2008 *International Journal of Plasticity* **24** 1071–1096 ISSN 0749-6419 URL `https://www.sciencedirect.com/science/article/pii/S0749641907001246`

[10] Bai Y and Wierzbicki T 2010 *International Journal of Fracture* **161** 1–20

[11] Mohr D and Marcadet S J 2015 *International Journal of Solids and Structures* **67-68** 40–55 ISSN 0020-7683 URL `https://www.sciencedirect.com/science/article/pii/S0020768315000700`

[12] Bharti S, Gupta A, Krishnaswamy H, Panigrahi S and Lee M G 2022 *CIRP Journal of Manufacturing Science and Technology* **37** 499–517 ISSN 1755-5817

Measurement and prediction methods for forming limit strain of stretch bending deformation in sheet metal forming

Y Taguchi[1], Y Maeda[1], Y Maeda[2]

[1]Application Technology Center, Technical Development Group, Kobe Steel, LTD.1-5-5 Takatsukadai, Nishi-ku, Kobe, HYOGO 651-2271 Japan
[2]Materials Research Laboratory, Technical Development Group, Kobe Steel, Ltd., 1-5-5 Takatsukadai, Nishi-ku, Kobe, Hyogo 651-2271 Japan

Taguchi.yuki@kobelco.com

Abstract. In this study, a new test method using protruding punch was proposed, that can separate bending and stretching deformations. In this test, the blank makes simultaneous contact with the punch head and shoulder to ensure a constant bending angle during the test. After the simultaneous contact, increasing the punch stroke increases the amount of stretch deformation without changing the bending deformation. In addition, this test method can control the inflow of materials by controlling the BHF, and it is possible to evaluate cracks during stretch bending deformation even for materials with low forming limits. This test can control the distribution of strain through the thickness. As a result, the measurable surface formation limit strain in this test was higher than FLC with uniform strain along the thickness. It is suggested that this is because the strain gradient through the thickness delayed the timing of necking occurrence. There is a linear relationship between this strain gradient and the maximum principal strain on the sheet surface, which can be applied to the prediction of stretch-bending cracks in FEM analysis.

1. Introduction

High-strength steel sheets have been used to reduce the weight of automobile parts, but cracking in press forming has become a problem because of their lower elongation than conventional mild steel sheets. Cracking in press forming can be avoided by utilizing FEM analysis, but it is difficult to predict the forming limit strain under non-uniform deformation conditions with strain gradients, such as bending and stretch flange. The forming limit strain of bending deformation has long been evaluated in tests without stretch deformation, such as V-bending tests [1, 2, 3]. In press forming, however, stretch and bending deformations occur at the same time, and cracking occurs as a result of their mixing. One kind of stretching and bending test has been performed with the V-punch method, in which both ends of the specimen are fixed. In this test, both bending and stretching strain increase at the same time with the punch stroke, and the strain gradient through thickness is not constant. For this reason, special tools to control the stretching or bending strains have been proposed [4]. However, these special tools are complex, and it is difficult to obtain the stretch-bending limit strain. In addition, there is no established technique for predicting stretch-bending other than the test method. In particular, there are many reports suggesting the shape of the forming limit as a product design criterion for predicting bending cracks, but there are not many examples showing how to predict bending cracks properly in FEM analysis [3].

In this paper, a new stretch-bending test method that controls individual strains is proposed. Using this method, the stretch-bending forming limit strain was estimated. As an evaluation function applicable to FEM analysis, a prediction of stretch-bending forming strain was proposed.

2. Stretch-bending test method with separate bending strain and tension

2.1. *New stretch-bending test method*

The schematic illustration of the new stretch-bending test is shown in Fig. 1. In this test, a protruding punch is used to separate bending and stretch deformations. In the beginning, a low BHF was used because the specimen moved into the folder. At this stage, the stretching and bending deformations simultaneously increase with punch stroke. If the shoulder of the protruding punch makes contact with the specimen, the bending angle at the top of the punch is geometrically fixed. (Fig. 1(b)). After the specimen contacts the punch shoulder, it was switched to high BHF until cracks occurred (Fig. 1(c)). In this test, the bending strain can be determined by the punch head R, the stretching strain can be determined by the punch stroke, and various bending strains can be evaluated by changing the punch head R. In addition, by the control of BHF, low formability materials can be kept forming until the prescribed bending strain, and the stretch-bending limit strain can be obtained regardless of the material properties.

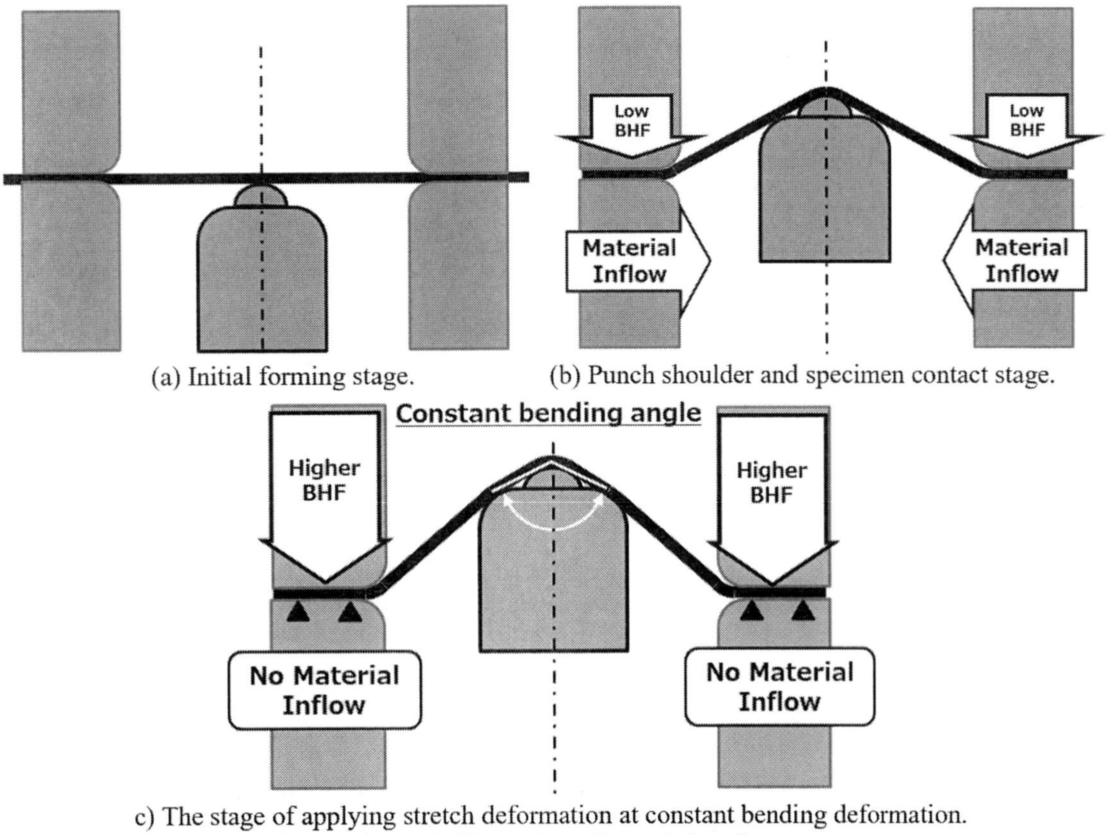

(a) Initial forming stage. (b) Punch shoulder and specimen contact stage.

c) The stage of applying stretch deformation at constant bending deformation.

Fig. 1 Schematic illustration of stretch-bending test.

2.2. *Experimental method*

In the stretch-bending test, to change the bending strain, 3 types of punch head R (R3.0, 5.0, and 10.0 mm) were designed as shown in Fig. 2. The stretch-bending test should be performed under plane strain conditions, but when the punch head R is large, the deformation mode is close to the uniaxial stretch mode. So the R5.0 and R10.0 mm punches create an additional steps in the width direction. These steps are designed to avoid compressive strain along the width and the deformation is controlled under plane strain. The specimen was a rectangular shape with a width of 30 mm×250 mm, thickness of 1.4 mm. The BHF was formed at 5 tons with the material flowing to a 30 mm punch stroke where the specimen fit the punch, then switched to 50 tons of BHF to constrain the material.

The forming limit strain in stretch and bending deformations was also measured and compared with that in stretch-bending deformations. Uniaxial, plane strain, and biaxial stretch deformation were evaluated by the Marciniak test, and bending deformation was evaluated by the crush bending test in which pressure was applied from the top of the specimen after U-bending.

In these experiments, the forming limit strains were calculated from the strain measured by DIC (ARAMIS 6M). The forming limit strains in the Marciniak test, and the stretch-bending test were calculated by a time-dependent method. The forming limit strain in the crushing bending test was defined as the average strain of the cross sections in the width direction when the bending surface cracked. The two materials used in these tests have similar stretch strengths, but different elongation and lambda values (Table 1).

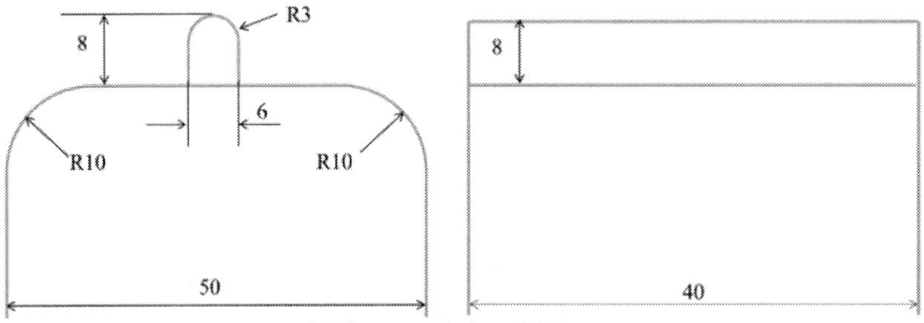

a) The punch head R3mm

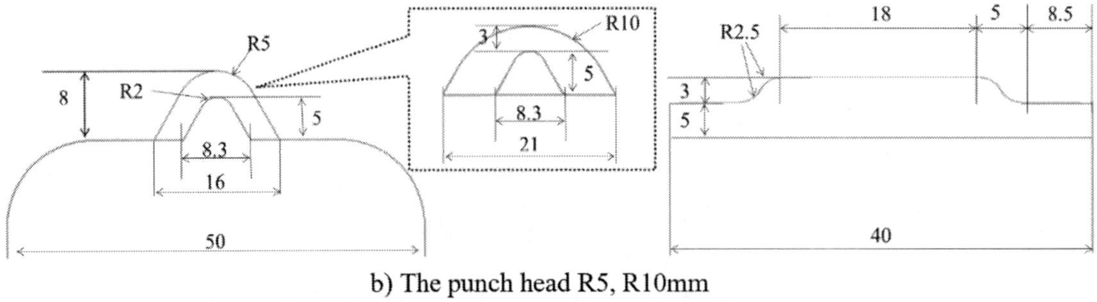

b) The punch head R5, R10mm

Fig. 2 Two types of punch of stretch-bending test.

Table 1. Material properties of 980 grade steel.

Material	Thickness / mm	YS / MPa	TS / MPa	EL / %	λ / %
A-Type	1.4	760	1006	15.8	57
B-Type	1.4	662	1049	17.7	27

2.3. Examination of stretch-bending test conditions by FEM analysis

In the stretch-bending test, it can be inferred that the ratios of bending deformation and stretch deformation varies in the punch head R and the thickness strain has various values, but it cannot be measured in the experiment. Therefore, the thickness strain was investigated by FEM analysis (LS-DYNA 9.3.0: Solid mesh). The material used A-Type, and the material model assumed isotropic hardening by extrapolating the stress-strain relationship from uniaxial tensile test.

The maximum principal strain of the bending surface and the thickness-strain gradient at each punch head R were calculated by FEM analysis and the results are shown in Fig. 3. The thickness-strain gradient was defined as the strain difference between the bending surface and inside of the plate divided by the thickness. In punches R5.0 and 10.0 mm, it was found that the thickness-strain gradient remained constant after BHF switching and the strain on the bending surface increased. In the punch R3.0mm, the thickness-strain gradient changed after switching to BHF, but remained constant after fitting to the punch surface. These results show that this new method with a constant bending angle can evaluate the forming limit strain of stretch-bending deformations with the thickness-strain gradient corresponding to the punch head R.

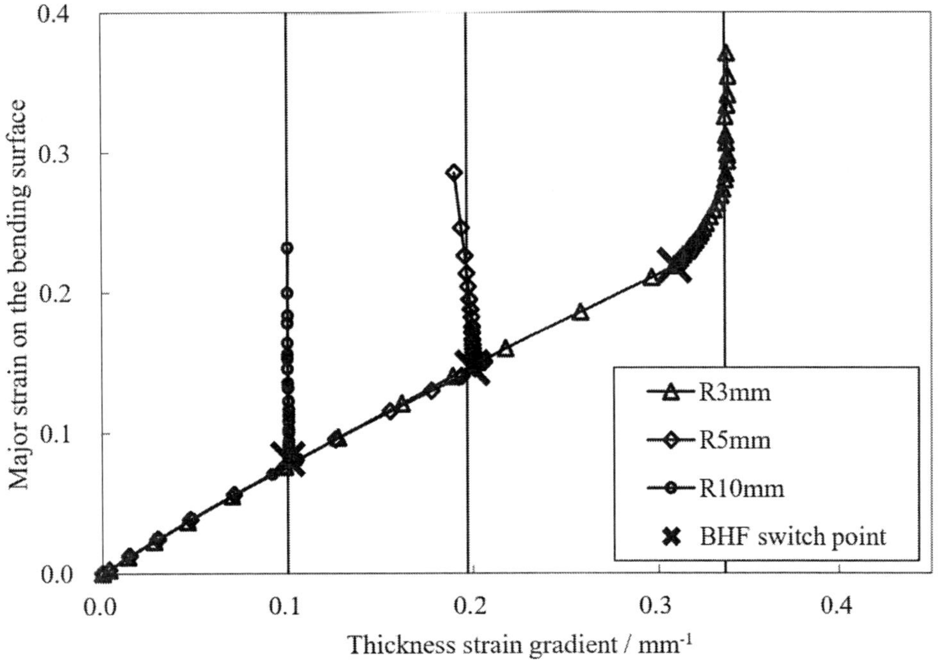

Fig. 3 Relationship between maximum principal strain and thickness-strain gradient of the A-Type material.

3. Forming limit strain in various tests

Based on the results up to the previous section, the forming limit strain (maximum principal strain on the bending surface) was determined in various tests. Fig. 4 shows the forming limit strains from various tests. By providing steps in the width direction of the punch, it was possible to calculate the forming limit strain by the plane strain ratio. The forming limit strain increased in the order of the Marciniak test, stretch-bending and crush-bending regardless of the material. It was assumed that the forming limit strain increased with increasing bending strain. In addition, in the stretch-bending test, the forming limit strain at the plane strain ratio could be collected for all punch R shapes, and it was assumed that the step applied in the punch width direction provided sufficient restraint in the width direction.

IOP Conf. Series: Materials Science and Engineering 1284 (2023) 012050 doi:10.1088/1757-899X/1284/1/012050

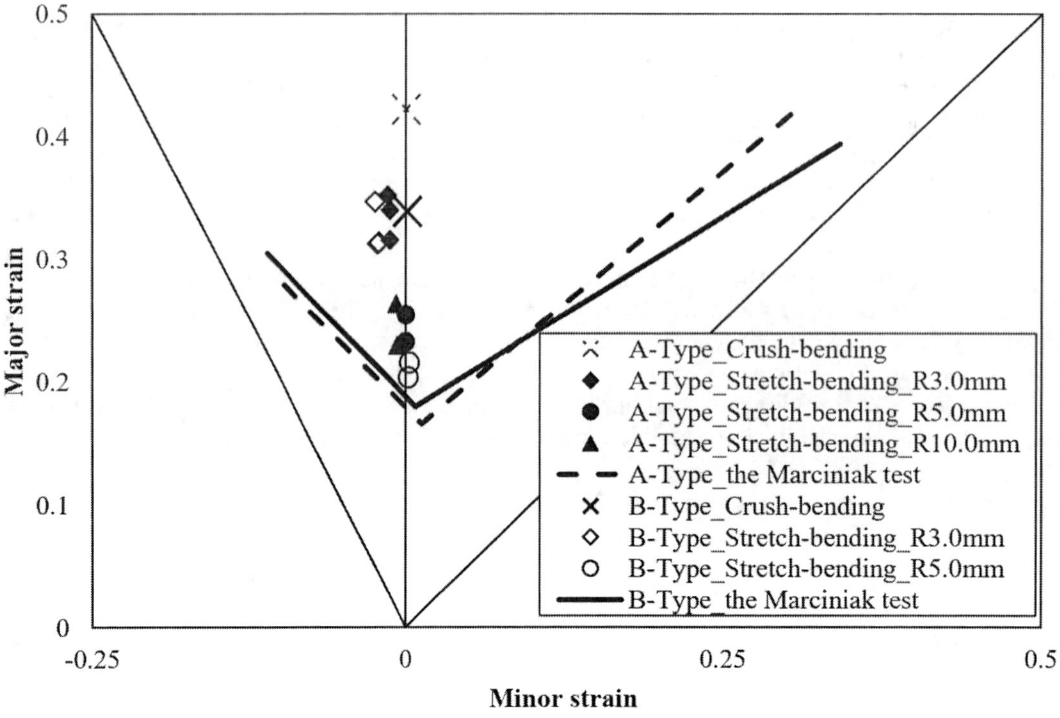

Fig. 4 Forming limit strain by various tests.

4. Construction of bending crack prediction focusing on strain gradient through thickness

The relationship between the forming limit strain and the thickness-strain gradient is shown in Fig. 5. Since the Marciniak test only involves stretch deformation, the thickness-strain gradient was set to 0. The thickness-strain gradient of the stretch-bending test was calculated by FEM analysis. The thickness strain gradient in the crush bending test was calculated by dividing the forming limit strain on the bending surface by the thickness/2, assuming that the maximum principal strain on the inside surface is the same value of compressive strain as that on the bending surface.

It was found that the forming limit strain increases almost linearly with increasing thickness-strain gradient, regardless of the material. In conventional stretch deformation, local necking occurs and leads to failure, but in crush bending, failure occurs without necking. This suggests that the thickness-strain gradient affects the delay of necking development. This makes it possible to uniformly predict forming limit strain of the stretch-bending from the maximum principal strain on the bending surface and the thickness-strain gradient, and to apply this method to FEM analysis. And, the dependence of the thickness-strain gradient on the forming limit strain depends on material properties, with higher forming limits for material B in the low strain gradient region and higher forming limits for material A in the high strain gradient region. This suggests that the forming limit strain of the bending deformation varies not only by elongation but also by factors such as λ, which contribute to local ductility.

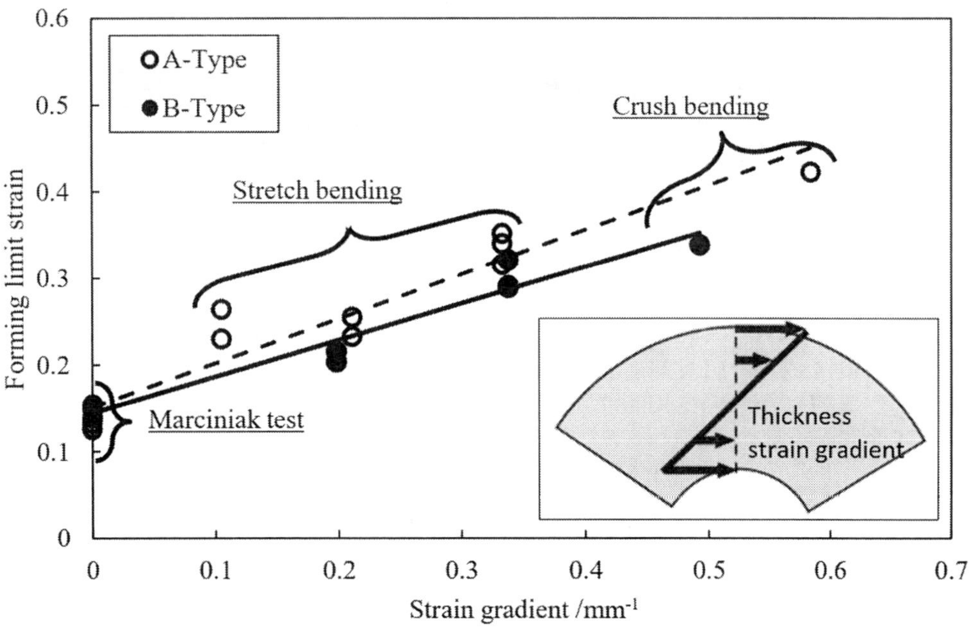

Fig. 5 Forming limit strain by thickness-strain gradient at the bending surface.

5. Conclusion

1) A new stretch-bending test method has been developed that can easily evaluate bending and stretch deformations separately only by bending the punch.
2) In this method, various thickness-strain gradients can be evaluated by changing the punch R shape.
3) It was found that the forming limit strain was larger in the order of the Marciniak test < stretch-bending < crush bending, and the forming limit strain increased with the bending strain.
4) The bending strain is proportional to the thickness-strain gradient, and a method to predict forming limit strain of the bending deformation by the thickness-strain gradient was developed.

References

[1] D.K.Uko et al.: Mining & Metal, 70-781(1977), 127-134.
[2] M.Y. Demeri et al.: J. Appl. Metal working, 70(1977), 127-134.
[3] M. Yoshida et al.: KOBE STEEL ENGINEERING REPORTS, 58(2008), 46-50.
[4] R. Yonebayashi et al.: Proceedings of Plasticity Technical Conference, 63(2012), 423-424.

Research on the evolving yielding behaviour and micro-mechanism in biaxial deformation of DP1180 sheet based on crystal plasticity finite element model

Bingying Zhou[1], Fei Han[2], Xiangdong Wu[1]*, and Min Wan[1]

[1]704#, School of Mechanical Engineering and Automation, Beihang University, Beijing 100191, China
[2]R&D Center, Baoshan Iron & Steel Co., Ltd., Shanghai 201900, China

E-mail: wuxiangdongbuaa@163.com

Abstract. Dual-phase steels (DP steels) consist of ferrite and hard martensitic phase. For the high strength and good formability, they are widely used in automotive applications. Compared with the DP steels with lower strength, DP1180 is more special for the volume fraction of martensite is higher than that of ferrite. In this paper, the evolving plastic yielding behaviours and insight into their microscale mechanisms of DP1180 were studied by biaxial tension experiments and crystal plasticity finite element method. The representative volume element (RVE) model of DP1180 taking the dual phase fraction, grain size and texture into consideration is established and the relevant parameters of crystal plasticity are reverse calibrated by the uniaxial tensile test stress strain curve. The simulated biaxial tension tests were conducted by the established RVE model to investigate the yielding micro-mechanisms of DP1180 steel, and the obtained yield loci matches well with the test results. The simulation results show that the texture has a significant effect on the contours of the yield loci. The soft ferrite grains weaken the kinetic constraints to the polycrystalline matrix and further free the rotation and plastic deformation of the neighbouring grains. This study thus provides a comprehensive understanding of the effect of the microstructure (crystal structure, texture and dual phase fraction) on the yielding behaviour of DP1180 steel as well as a method of virtual experiment to get the yield loci of metallic materials.

1. Introduction

Advanced high-strength steel (AHSS) is one of the most important materials which can satisfy both the lightweight and crash safety of the automobile. Because of the multiphase microstructures, such as ferrite, martensite, bainite, or austenite, the AHSS generally possesses both high strength and good plasticity. However, because of the complex microstructures, the deformation behavior of AHSS gets complicated and hard to predict. In order to study the deformation mechanism of AHSS, the crystal plastic finite element method (CPFEM) could be well used. Compared with the traditional macroscopic phenomenological method, CPFEM could better describe the physical essence of plastic deformation in crystalline materials, such as slip and dislocation. It could be used to analyze the effect of grain size, shape, orientation, texture, and deformation incompatibility on mechanical properties. Meanwhile, the high-resolution full-field crystal plastic simulation has been put forward as a virtual laboratory. This innovative virtual testing method may overcome the limitations of traditional physical experimental methods, and reduce the cost of mechanical properties testing and the development cycle of new material [1, 2].

The dual-phase (DP) steel is the first generation of AHSS containing ferrite and martensite phases. Studying the deformation mechanism of DP steel by CPFEM is essential for establishing the cross-scale relationship between the macro-level and meso-level. Tasan et al. [3] obtained the microstructure and local strain of DP steels based on the electron backscatter diffraction (EBSD) and the microscopic digital image correlation (DIC), and built the physically based microstructure crystal plastic models. Liu et al. [4] proposed a representative volume element (RVE) evaluation standard that fully considered the phase volume fraction, grain size, shape, and texture, and summarized a complete set of methods and processes for the geometric modelling and parameter calibration of DP steel. Cai et al. [5] established the polycrystalline RVE model based on CPFEM and conducted biaxial tensile simulations to analyze the influence of microstructure on yield loci of DP590, DP780, and DP980 steel. Li et al. [6] studied the plastic behaviors of DP980 steel by CPFEM, and summed up that the anisotropy behaviors was mainly caused by the heterogeneous microstructure. Rahimi et al. [7] obtained the yield strength and hardening index of ferrite and martensite in DP1180 steel by combining the reverse solution and nanoindentation test, and found that the mechanical properties of individual phases were quite different from those reported in other literature. Men et al. [8] studied the effect of grain orientation and grain size distribution on the yield loci evolution of SUS304 stainless steel foils.

For ultra-high strength DP steel with strength larger than 1000 MPa and the volume fraction of martensite exceeding ferrite, such as DP1180 steel, little literature has been reported. In this paper, uniaxial and biaxial tensile tests were conducted and the experimental yield loci were obtained by the plastic work per unit volume principle. The microstructure and texture of DP1180 steel was obtained by the scanning electron microscope (SEM) and EBSD. The statistical microstructure-based RVE model based on CPFEM was established. The uniaxial and biaxial tensile simulations were conducted and compared with the experiments. The simulation results agree well with the experiments. The microstructure evolution of DP1180 under different loading conditions was studied.

2. Material and experimental procedure

2.1. Microstructure characterization

The material used in this work was the DP1180 steel sheet produced by BaoSteel, Inc, and the thickness was 1.4 mm. The grain distribution map of DP1180 steel was obtained with the help of SEM, shown in Figure 1(a).

Based on the SEM, the volume fraction of ferrite and martensite in DP1180 steel was determined to be 66.72% and 33.28%. The threshold of BC value between ferrite and martensite phases was further obtained, so as to achieve the purpose of separating ferrite and martensite grains in EBSD [6, 9]. The region with smaller BC values corresponds to the hard martensite particles and the region with larger BC values to the soft ferrite. The BC map of DP1180 steel is shown in Figure 1(b). The separated ferrite and martensite are shown in Figure 1(c-d). The grain size of ferrite and martensite along the RD and TD directions was obtained by the linear intercept method. And Figure 1(e) provides the histograms of grain size distribution and the log-normal fitting curves of ferrite and martensite. The average grain size of ferrite and martensite is 1.03 μm and 0.61 μm, respectively. Figure 1(f) is the RVE model generated via the open-source code DREAM.3D with the data of the log-normal fitting curves as input. The RVE model contains 15 ferrite grains and 109 martensite grains, with 30*30*30 grid elements. The orientation distribution function (ODF) of ferrite and martensite calculated from the EBSD data as sections through the reduced Euler space for cubic-orthorhombic symmetry is shown in Figure 2(a-b).

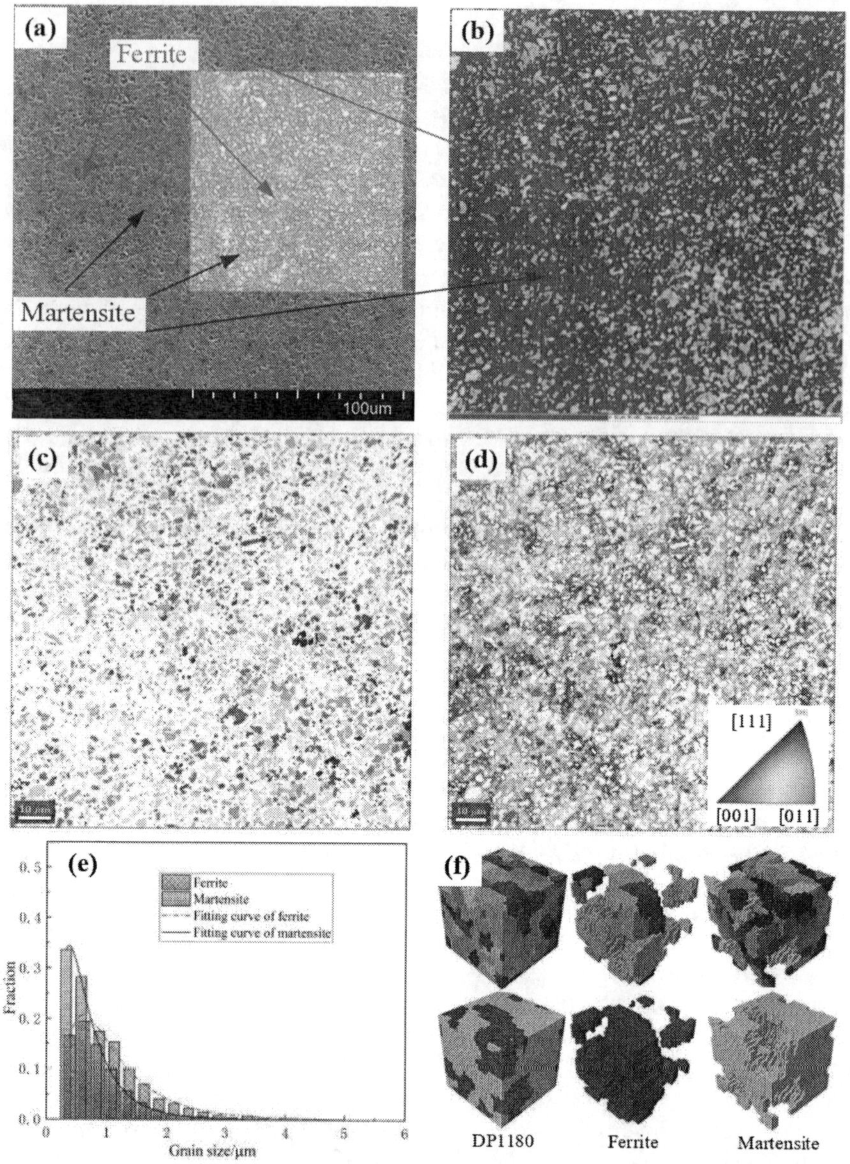

Figure 1. The microstructure characterization of ferrite and martensite in DP1180 steel: (a) the SEM figure, (b) the BC map, (c) the IPF figure of ferrite, (d) the IPF figure of martensite, (e) the grain size frequency, and (f) the RVE model.

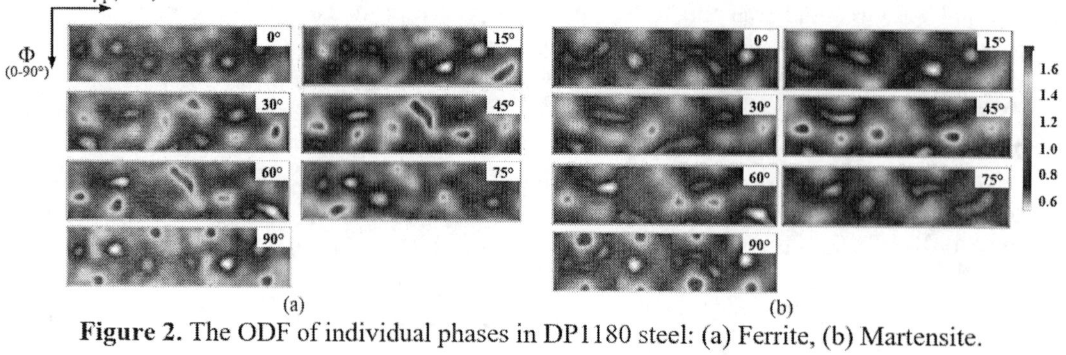

Figure 2. The ODF of individual phases in DP1180 steel: (a) Ferrite, (b) Martensite.

2.2. *Uniaxial and biaxial tensile tests*
According to the Chinese standard GB/T 228.1-2010 (ISO 6892-1:2009), the uniaxial tensile tests for DP1180 steel sheet along 0°, 45°, and 90° to the rolling direction (RD) were conducted by the MTS test machine at speed of 1.5 mm/min, corresponding to the strain rate of 0.001 s-1, as shown in Figure 3(a).

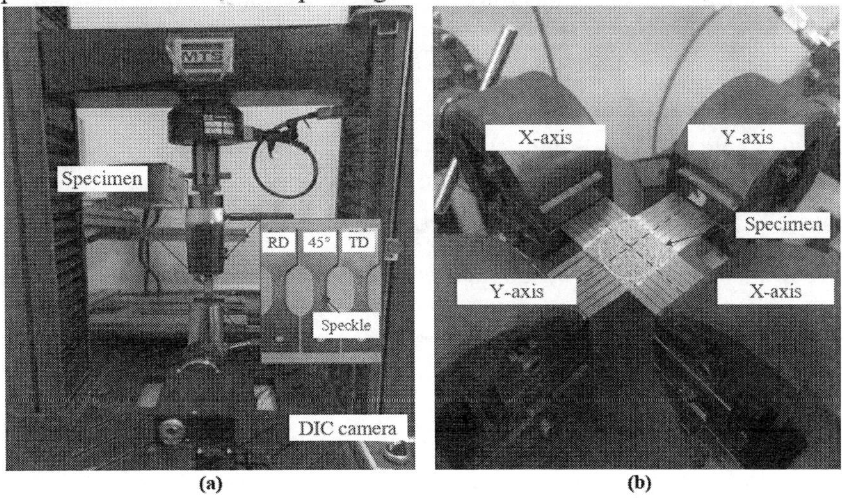

Figure 3. The tensile tests of DP1180 steel: (a) uniaxial tension; (b) biaxial tension

A digital image correlation (DIC) system was used to measure the full-field strain. The biaxial tensile tests were conducted under linear loading path to obtain the yield loci in the normal stress space. The cruciform specimens were used to lower the shear stress in the middle area, as shown in Figure 3(b).

3. CPFEM modelling

3.1. *Crystal plasticity theory*
A continuum crystal plasticity (CP) model was utilized to model the slip deformation of the active slip systems and the rotation of the crystal lattice. The CP model was programmed as UMAT subroutine by Pro. Huang [10].

The total deformation $\boldsymbol{F} = \boldsymbol{F}^{e}\boldsymbol{F}^{p}$, and the elastic deformation $\boldsymbol{F}^{e}$ contains crystal lattice stretching R^{e} and lattice distortion U^{e}. And $\boldsymbol{F}^{p}$ can be described in a rate form:

$$\dot{\mathbf{F}}^{p} = \mathbf{L}^{p} \cdot \mathbf{F}^{p} \tag{1}$$

where $\mathbf{L}^{p}$ is the plastic velocity gradient, which is related to the slip activities:

$$\boldsymbol{L}^{p} = \sum_{\alpha=1}^{N} \dot{\gamma}^{(\alpha)} \boldsymbol{m}^{(\alpha)} \otimes \boldsymbol{n}^{(\alpha)} \tag{2}$$

where $\dot{\gamma}^{(\alpha)}$ is the shear rate on the slip system α, $\boldsymbol{m}^{(\alpha)}$ and $\boldsymbol{n}^{(\alpha)}$ are the unit vectors defining the slip direction and the slip plane normal of α. And $\dot{\gamma}^{(\alpha)}$ can be expressed as follows:

$$\dot{\gamma}^{(\alpha)} = \dot{\gamma}_0^{(\alpha)} \left(\frac{\tau^{(\alpha)}}{g^{(\alpha)}} \right)^{n} \operatorname{sgn}(\tau^{(\alpha)}) \tag{3}$$

The values of $\dot{\gamma}_0^{(\alpha)}$ and n are set as 0.001 and 20. $g^{(\alpha)}$ represents the strain hardening strength of the slip system at a certain moment, and it can be expressed as:

$$\dot{g}^{(\alpha)}(\gamma) = \sum_{\beta} h_{\alpha\beta} \left| \dot{\gamma}^{(\beta)} \right| \tag{4}$$

where $h_{\alpha\beta}$ depends on the following formula:

IOP Conf. Series: Materials Science and Engineering 1284 (2023) 012051 doi:10.1088/1757-899X/1284/1/012051

$$\begin{cases} h_{\alpha\alpha} = h_0 \operatorname{sech}^2 \left| \dfrac{h_0 \gamma}{\tau_{\mathrm{sat}}^{\alpha} - \tau_0^{\alpha}} \right|, \alpha = \beta \\ h_{\alpha\beta} = q h_{\alpha\alpha}, \alpha \neq \beta \end{cases} \tag{5}$$

where h_0 is the reference hardening modulus, τ_0^{α} is the critical value of the initial resolved shear stress, τ_{sat}^{α} is the saturated resolved shear stress. q is the coefficient describing the relative strength of self-hardening and latent hardening, and it is $1 \leq q \leq 1.4$ in general.

3.2. Identification of the CP parameters

Based on the work done by Tasan et al. [3] and Li et al. [6], the CP constitutive parameters for DP steel were further determined. Uniaxial tensile load was applied to the RVE model in RD and TD, respectively. The average flow stress curves were obtained by the volume average method via a python script. The comparisons between the experiments and the CPFEM simulations are shown in Figure 4. The cloud atlas of true stress and strain is shown in Figure 5, and it indicted that the stress in martensite was much bigger than that in ferrite, while the strain is opposite.

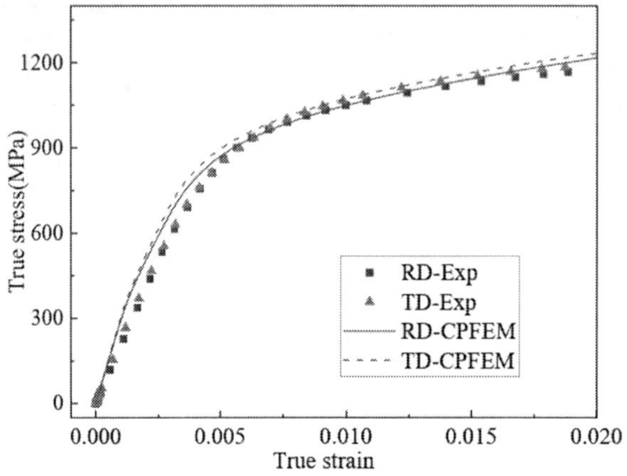

Figure 4. The comparison between the experimental and CPFEM predicted flow stress curves in both RD and TD

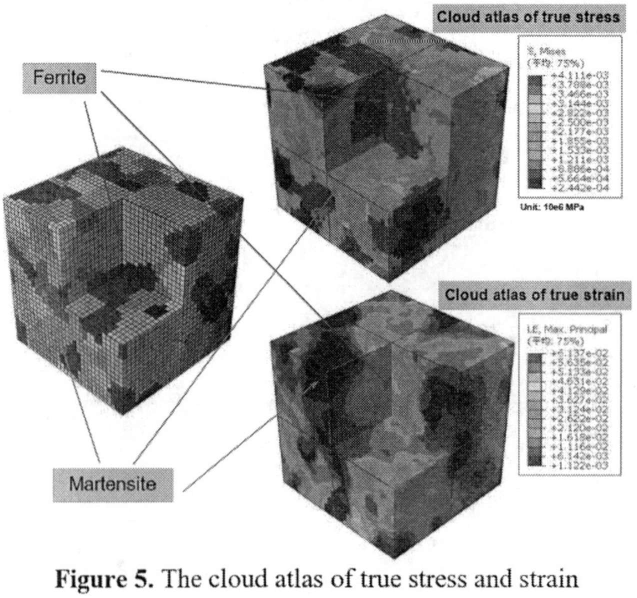

Figure 5. The cloud atlas of true stress and strain

4. Result and discussion

4.1. The comparation of biaxial tensile stress-strain curves

Based on the CP-RVE model considering grain size and orientation, and the calibration of CP constitutive parameters by uniaxial tensile tests in RD and TD, the deformation behaviours of DP1180 steel under complex loading conditions has been further studied. The biaxial tension with the ratios of 4:0, 4:1, 4:1, 4:2, 4:3, 4:4, 3:4, 2:4, 1:4, 0:4 was applied to the CP-RVE model. The strain in the central area of cruciform specimen was calculated by DIC. The biaxial tensile stress-strain curves in RD and TD were obtained, as shown in Figure 6(a-b). It can be seen that the results of CPFEM is slightly higher than the experiments, which may be because the CP-RVE model has not taken the damage of crystal boundaries and intra-crystalline into consideration. And the load conditions applied to the model have made it hard to keep a constant strain rate during the tensile process. In general, the CP-RVE model can well predict the macro-mechanical response of DP1180 steel under complex loading conditions.

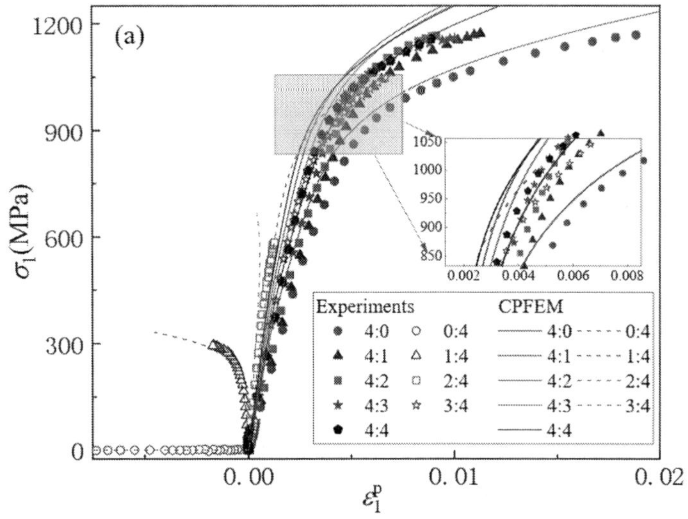

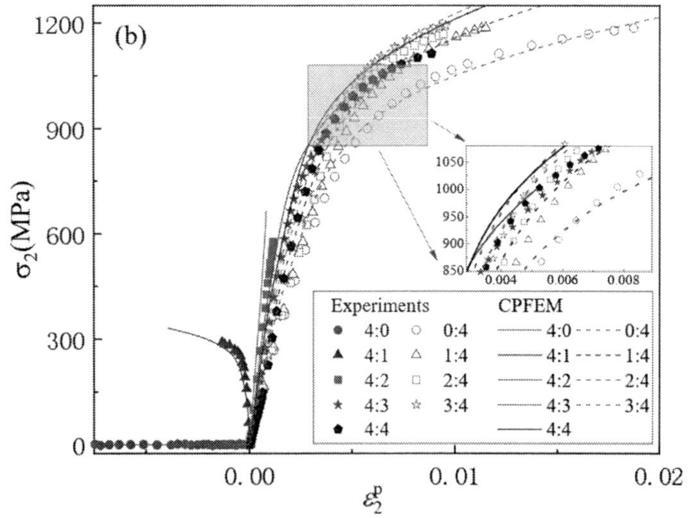

Figure 6. The comparison between the experimental and CPFEM predicted flow stress curves in: (a) rolling direction; and (b) transverse direction

IOP Conf. Series: Materials Science and Engineering 1284 (2023) 012051 doi:10.1088/1757-899X/1284/1/012051

4.2. The prediction of yield loci

The yield loci were obtained by the plastic work per unit volume principle [11]. Figure 7 shows the yield loci of test and predictions by Mises, Hill48, and Yld2000-2d yield criteria. It can be seen that Yld2000-2d has the highest prediction accuracy, followed by Mises, and Hill48 is the worst. Figure 8 shows the yield loci of predictions by Yld2000-2d and CPFEM, and it can be seen that the prediction accuracy of CPFEM is a little worse than that of Yld2000-2d. In order to quantitatively evaluate the accuracy of predictions to the yield loci of tests, an error function [11] has been applied, and the results are show in Figure 9. It can be clearly seen that the prediction accuracy of CPFEM ranked second only to Yld2000-2d, and better than Mises and Hill48 yield criteria.

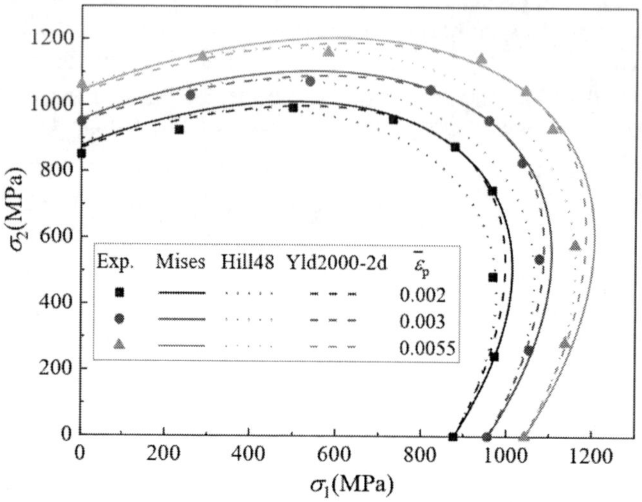

Figure 7. Yield loci predicted by experiments and general yield criterions

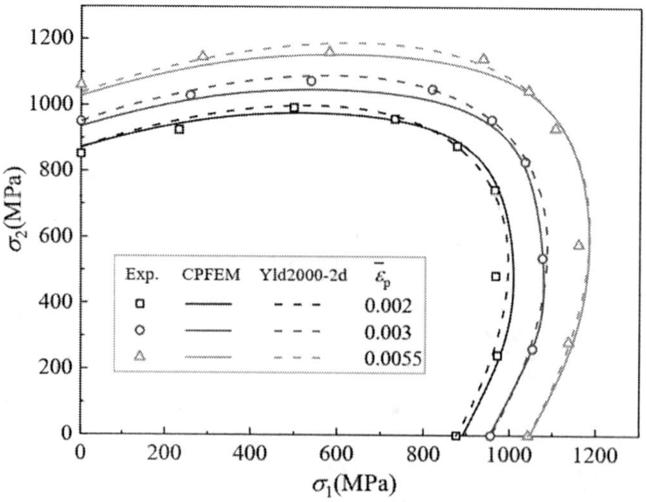

Figure 8. Yield loci predicted by experiments and CPFEM

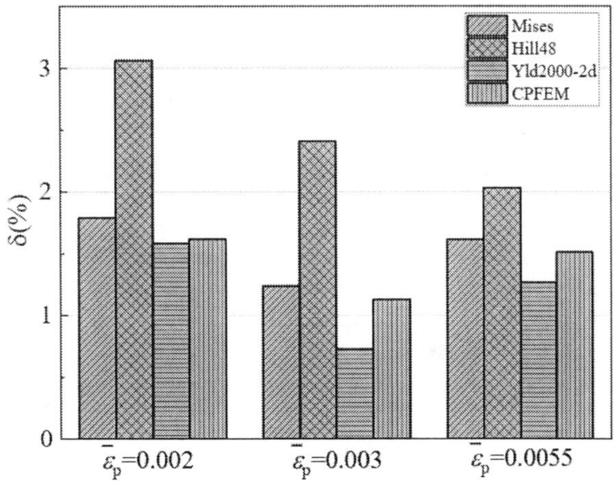

Figure 9. Errors of predictions by general yield criterions and CPFEM

5. Conclusion

This research characterizes the yield loci of DP1180 steel sheet through the multiscale method. The microstructure data was obtained by SEM and EBSD. Then the RVE model considering grain size and crystal orientation was established. Based on the uniaxial tensile tests on RD and TD, the CP constitutive parameters were calibrated. The biaxial tensile condition was applied to the above CP-RVE model, and the biaxial tensile stress-strain curves and yield loci based on CPFEM were obtained. Compared with the experimental yield loci, the prediction accuracy of CPFEM and general yield criteria for DP1180 steel was compared. It was found that Yld2000-2d had the highest prediction accuracy, followed by CPFEM and Mises, and Hill48 is the worst.

Acknowledgements

The authors would like to acknowledge the financial supports by the National Natural Science Foundation of China (Grant No.51875027), and the State Key Laboratory of Development and Application Technology of Automotive Steels (Baosteel Group, Grant No. 2021090602).

References

[1] Zhang H, Diehl M, Roters F, et al. 2016 *Int. J. Plast.* 80: 111-38
[2] Ma H, Li Y, Zhang H, et al. 2022 *Scientific. Reports.* 12(1)
[3] Tasan C C, Diehl M, Yan D, et al. 2014 *Acta. Mater.* 81: 386-400
[4] Liu W, Lian J, Aravas N, et al. 2020 *Int. J. Plast.* 126: 102614
[5] Cai Z Y, Meng B, Wan M, et al. 2020 *Int. J. Plast.* 129: 102707
[6] Li Q, Zhang H, Chen F, et al. 2020 *Int. J. Mech. Sci.* 176
[7] Rahimi R M, Bahr D F 2019 *Mater. Sci. En. A.* 756: 328-35
[8] Men M, Meng B 2022 15(3): 1140.
[9] Choi S-H, Kim E-Y, Woo W, et al. 2013 *Int J Plast.* 45: 85-102
[10] Huang Y. A User-Material Subroutine Incorporating Single Crystal Plasticity in the ABAQUS Finite Element Program, F, 1991
[11] Cai Z, Wan M, Liu Z, et al. 2017 *Int. J. Mech. Sci.* **126**: 79-94

IOP Conf. Series: Materials Science and Engineering 1284 (2023) 012052 doi:10.1088/1757-899X/1284/1/012052

An artificial neural network approach on crystal plasticity for material modelling in macroscopic simulations

L Martinitz* and C Hartmann

Chair of Metal Forming and Casting, Technical University of Munich, Walther-Meissner-Str. 4, Garching, 85748, Bavaria, Germany

lukas.martinitz@tum.de

Abstract. Anisotropy plays a significant role in engineering, especially in the field of sheet metal forming. This particular characteristic stems mainly from the crystallographic structure of the metals and the influence of the rolling process, inducing preferred orientations of the grains. In this context, the crystal plasticity theory plays an important role as it accounts for the anisotropic nature of the elastic tensor and the orientation dependencies of the crystallographic deformation mechanisms. Despite the advantages and capabilities, the integration of the crystal plasticity theory in macro simulations is hindered by high computational costs. A novel approach aims to rectify this problem through the application of machine learning. Therefore, this work investigates the machine learning of crystal plasticity simulations, whereby the DAMASK simulation kit package is used both as a benchmark for quality and costs as well as for providing a data basis for the training and testing of the neural networks. A phenomenological material model for an AA5083 aluminium alloy provides the training data for a neural network study, testing different input parameters as well as network setups.

1. Introduction

Anisotropy as a property can be found in a multitude of fields of study, ranging from physics, computer imagery to engineering and materials science. The basic definition is that a considered property of a respective material exhibits a directional dependency.

Crystal plasticity as a field of study models the mechanical behaviour of metals on the microscopic scale considering the direction or orientation dependent deformation mechanism in polycrystalline structures. Accordingly, this physics-based approach provides excellent means to model the inherent mechanical anisotropy of metals. [1, 2]

Multi scale simulations offer the possibility to integrate, e.g. crystal plasticity models into macro scale simulations on the component level, and therefore increase the quality and reliability of said models. With the emergence of holistic approaches for the design of products, such as the Integrated Computational Materials Engineering (ICME) [3] or the Materials Genome Initiative [4], multi scale simulations come more and more into focus, for example to develop new materials or to model suitable manufacturing routes.

This work focuses on a potential hybrid approach combining the prevalent concept of machine learning and crystal plasticity for multi scale simulation. The focus lies on the machine learning of crystal plasticity and the capability of neural networks to reproduce its predictions.

2. State of the art

Data driven approaches, such as machine learning (ML), are becoming more and more widespread in a variety of research areas, from natural language processing to computer vision, economics and engineering. In the case of material science and computational engineering, two main directions can be distinguished.

On the one hand machine learning is used to analyse and classify the micro structure of materials. Chowdhury et al. applied ML for microstructure recognition using a 2D convolutional neural network (CNN) to distinguish and identify micro structural images with respect to their dendrite structures [9]. They propose an approach that involves using image segmentation techniques to extract micro structure features, followed by applying machine learning algorithms to recognise the micro structure. By comparing the performance of several machine learning algorithms, including support vector machines, random forests and convolutional neural networks, they demonstrate the effectiveness of their approach on various microstructure data sets, which highlights the potential of image-driven machine learning methods for micro structure recognition and their importance in advancing materials science research.

In a similar context, Lubbers et al. again used 2D-CNN to reconstruct micro structural images using unsupervised detection and learning of low-dimensional quantitative, compact descriptions of micro structure [15]. The basic approach is to extract low-dimensional representations of micro structural arrangements and features by CNNs. The authors show that their approach is effective for various micro structures and that these low-dimensional set of features can be used for further tasks, such as microstructure classification and reconstruction.

On the other hand, machine learning models are used as surrogate alternatives to model the material-structure-property relationship that describes the transition from micro structure to component-scale properties. Here, one major focus is the reduction of computational costs [10]. In this field of study, Cecen et al. combined a 3D-CNN with spatial statistics to link microstructure and material properties [8]. The 3D-CNN are used to learn the linkage by analysing the micro structure of materials as 3D images and predicting their macroscopic properties. This approach involves training a model using a large data set of different micro structures labelled with their corresponding properties. Based on that training, the model can be used to predict the properties of an unknown material based on its micro structure. The 3D-CNNs were able to efficiently analyse large data sets of complex micro structures and predict their properties. The applications of this technique lies, for example, in the design of new materials with desired properties. Yuan et al. proposed a machine learning framework, which uses a random forest model fed with data from a viscoplastic self-consistent crystal plasticity model in order to obtain the stress strain curve and the final texture of a given micro structure under load [19]. This approach involves creating a reduced-order model using machine learning algorithms, which are trained on data generated by crystal plasticity simulations. The reduced-order model is then used to predict the macroscopic mechanical behaviour of new materials enabling a straightforward ICME analysis. Ali et al. applied an artificial neural networks (ANN) to achieve similar results, whereby they were able to predict both stress strain curves and texture evolution [5]. They present a novel approach that uses ANNs to predict the mechanical behavior of polycrystalline metals at the microscale level. The authors trained their ANNs with data generated by a crystal plasticity finite element method (CPFEM) model, which is widely used to study the mechanical behaviour of polycrystalline materials. The ANNs were trained to predict the stress and strain responses of polycrystalline metals based on their micro structural features such as grain size, orientation and shape. Beniwal et al. explore the use of deep learning techniques for predictive modelling of the material-structure-property linkage [6]. A 2D-CNN is fed micro structure images of a dual phase material, whereby the stress evolution was related to volume fraction and morphology of the phases. The network was able to predict a yield stress like quantity, in the context of isotropic elastic plastic mechanics. The absence of human interaction for feature extraction and selection is noteworthy. Pandey and Pokharel were able to predict detailed 3D texture evolution during uniaxial tension for given strain increments using deep ANNs [16]. In their approach, an ANN is trained to predict the deformation of a crystal given its initial configuration and the external loads applied. The ANN is trained on a data set of pre-simulated crystal deformations. The resulting model can accurately predict crystal deformation for unknown initial configurations under uniaxial loading conditions.

3. Problem formulation and solution approach

As made clear in the previous section, AI and ML provide novel approaches to material science particularly in the context of modern material development initiatives. Besides these applications, a combination of ML and CP may open up new possibilities in terms of modelling anisotropy, especially in the field of sheet metal forming.

The theory of crystal plasticity is particularly well suited for the simulation of anisotropy, as it takes the tensorial nature of the deformation into account, namely the anisotropy of the elastic stiffness tensor and the orientation dependent activation of the considered deformation mechanisms like twinning or dislocation slip. One drawback of CP simulation on the macro scale is the computational cost connected to it.

The proposed approach aims to reduce computation time of multi scale simulations, similar to surrogate models, whereby the 'on'-time of the simulation is reduced, through computations done in the 'off'-time, accelerating the actual simulation process.

Therefore, the aim of the work is to use machine learning in order to reduce computational costs. In conjunction with this goal is the hypothesis that a neural network trained with suitable features and data can replicate the behaviour of the considered crystal plasticity model.

4. Objectives and general approach

The main objective of the present work is to train neural networks with data generated by crystal plasticity simulations in order to replicate the simulated behaviour using the trained network. The successfully trained network is then to replace CP simulations e.g. in multi scale simulations as illustrated by the following figure 1.

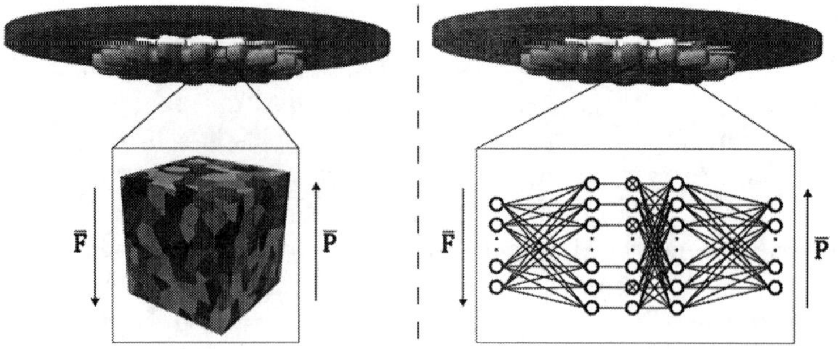

Figure 1. Integration of NN in multi scale simulations

In comparison to state of the art works as presented in section 2 the proposed approach aims to provide predictions for both texture evolution and mechanical responses, but with a focus on mechanical properties. Furthermore, it is important to emphasise that the present work maps 3D output variables, tensors, and does not resort to averaged and comparative variables, which exceeds the capabilities demonstrated by the state of the art.

The first step of the proposed approach comprises data generation based on a crystal plasticity simulation, preprocessing of the data and a preliminary sensitivity analysis with regards to the characteristics of the representative volume element. Following this, different network architectures are designed and implemented. In the last step the obtained data is again processed and used to train and test the networks capability to correctly predict the mechanical responses of the material under different loads.

5. Data Generation

The Düsseldorf Advanced Material Simulation Kit (DAMASK) provides the basis for the data generation [16]. DAMASK is a modular-hierarchical simulation software developed by the Max-Planck Institute für Eisenforschung (MPIE), which has been applied to a number of problems regarding the properties of crystalline materials. The implemented hierarchy is necessary in order to model the hierarchy of scale levels of crystalline materials. The numeric solvers provide the mathematical means to solve the system of partial differential equations, describing the considered boundary problem. Partitioning and homogenisation schemes connect the anisotropic material behaviour on the crystalline level with the global variables. The elastoplastic decomposition in conjunction with elasticity and plasticity sub-modules, i.e. the constitutive laws, allow for the computation of the mechanical response of the material.

Each representative volume element (RVE), which describes e.g. an integration point of a macro scale simulation, encompasses a certain number of constituents or grains with a specific orientation of the basic crystal lattice each in order to represent the material in both the composition in a statistical sense and the material response of the material point with sufficient quality.

The DAMASK software represents this material point model with a material file containing information about the phases, homogenisation, and the applied material model. The geometry file provides the basis for the grid solver, referencing the material in the aforementioned material file. The load file contains information on the applied solver and the loadstep, with details regarding temporal discretisation and the boundary conditions.

5.1. Material model

The AA5083 aluminium alloy is chosen for the simulations, characterised by its fcc crystal structure. The phenomenological model used in this case considers dislocations as the only deformation mechanism, as twinning plays an minor role in materials with high stacking fault energies like aluminium and their alloys.

The deformation gradient is multiplicatively decomposed into an elastic and plastic part,

$$\mathbf{F} = \mathbf{F}_e \mathbf{F}_P \,, \tag{1}$$

$\mathbf{F}_e$ denoting the rotation and elastic distortion of the crystal lattice and $\mathbf{F}_P$ representing the inelastic shear deformation on the slip planes.

The stress response of the material is computed using the generalised Hook's Law,

$$\mathbf{S} = \mathbb{C} : \mathbf{E}, \tag{2}$$

with $\mathbf{S}$ being the Second-Piola Kirchoff stress, $\mathbf{E}$ being the Green–Lagrange strain and $\mathbb{C}$ the fourth order anisotropic elastic tensor.

The plastic velocity gradient is composed of the shear contributions of the slip systems α,

$$\mathbf{L}_p = \sum_{\alpha=1}^{n} \dot{\gamma}^\alpha \, \mathbf{m}^\alpha \otimes \mathbf{n}^\alpha \,, \tag{3}$$

with $\dot{\gamma}^\alpha$, the slip rate of the slip system, the slip direction $\mathbf{m}^\alpha$ and the normal to the slip plane $\mathbf{n}^\alpha$, constituting the Schmidt tensor.

In the context of the chosen material model the shear rates of the slip systems α are computed as,

$$\dot{\gamma}^\alpha = \dot{\gamma}_0^\alpha \left|\frac{\tau^\alpha}{\xi^\alpha}\right|^n \operatorname{sgn}(\tau^\alpha)\,, \beta = 1, ..., 12 \tag{4}$$

with $\dot{\gamma}_0^\alpha$ being the reference slip rate, τ^α denoting the resolved shear stress, n determining the rate sensitivity and ξ^α being the slip resistance of the respective slip system.

The slip resistance itself evolves asymptotically from an initial value ξ_0 to a saturation value ξ_∞ following,

$$\dot{\xi}^\alpha = h_0|\dot{\gamma}^\beta|\left|1 - \frac{\xi^\beta}{\xi_\infty^\beta}\right|^a \operatorname{sgn}\left(1 - \frac{\xi^\beta}{\xi_\infty^\beta}\right) h^{\alpha\beta} \text{ with } \beta = 1, ..., 12\,, \tag{5}$$

whereby the h_0 is the self hardening coefficient, a the hardening exponent, and $h^{\alpha\beta}$ is components of the slip interaction matrix, with a value of 1.0 for coplanar slip systems and 1.4 for non-coplanar slip systems, respectively.

The parameter and constants values are summarised in table 1 and were computed using a Nelder-Mead algorithm.

Table 1. Material parameters of AA5083 aluminium alloy.

C_{11}	C_{12}	C_{44}	$\dot{\gamma}_0$	ξ_0	ξ_∞	h_0	a	n
103 GPa	49 GPa	53 GPa	1.0×10^{-3}	35 MPa	138 MPa	989 MPa	2.25	20

5.2. Sensitivity analysis

For the considered case a sensitivity analysis of two critical variables, namely the number of grains per RVE and the spatial resolution was conducted. The main focus was the influence of said variables on the crystal plasticity simulation results.

The number of grains has a direct influence on the number of features of the artificial neural networks (cf. table 3), whereas the resolution, i.e. the number of points in the spatial directions of the RVE has a substantial influence on simulation quality, speed and results file size.

In this context, a total number of 28 simulations were carried out, whereby the number of grains is altered in steps from 20 to 100 grains per RVE, with a changing spatial resolution of 16 and 32 grid points in each axis direction. Each single simulations is assigned the same geometry file, but different grain orientations. The considered load case generates uniaxial tension in X-direction, applied as a rate of deformation gradient $\dot{\mathbf{F}}$ of $23e - 3$ for 10 seconds.

The obtained results for the true strain and true stress are averaged across three single simulations per alternative, leading to the results as depicted by figure 2. The figure 2 shows the anticipated mechanical behaviour comparable to a basic tensile test, being indicative of the saturation type hardening of the polycrystalline matter.

The results are stable and show no particular sensitivity to the chosen number of grains or resolution. The maximum average difference in stress response is 11 MPa, with a maximum relative difference of 4% with regards to the average stress across all simulations.

The choice of 50 grains per RVE with a spatial resolution of 16 points is therefore well founded, as it is firstly a direct mean to control here reduce, the number of features of the neural networks without compromising the quality of the simulation data and secondly in order

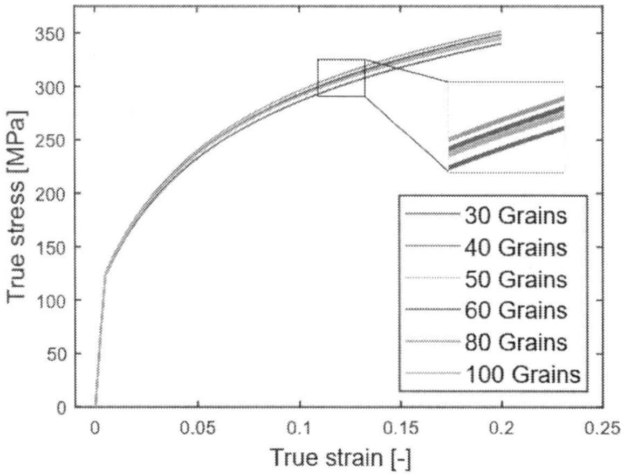

Figure 2. Results of sensitivity analysis for uniaxial tension case.

to speed up the data generation and results file size, easing data handling and postprocessing. An increase of features in particular would likely lead to a decrease of information density of the data sets, hampering the training process of the networks.

5.3. Simulation setup

Based on the results of the sensitivity analysis, the simulation was set up as follows. For each of the simulations a geometry file with a spatial resolution of 16 points per dimension and a total of 50 grains is used, accordingly with the results of the sensitivity analysis. The chosen resolution of 16 points in conjunction with the chosen number of 50 grains per RVE, results in an average number of $\approx$ 80 grid points or elements per grain, providing a high resolution for each grain. The material file is generated based on the phenomenological material model presented earlier, a total of 50 grains in accordance with the geometry file, whereby the grains orientations are generated randomly. The load cases are set up to generate basic load cases, being composed of uniaxial loading in the three spatial orientations, i.e. tension and compression, and simple shear loading in the the respective spatial orientations, leading to a total of 12 load cases. For the simulations the load was applied defining the rate of deformation gradient $\dot{\mathbf{F}}$ during the load step and the total time of the load case. For the considered case the spectral grid solver is chosen, due to the advantages regarding computation time in comparison with FEM solvers.

Table 2. Simulation parameters.

Simulation parameter	Value
Solver	spectral grid solver
Grains per RVE	50
Spatial resolution	16 x 16 x 16
Orientations	random
Boundary condition	$\dot{\mathbf{F}}$
Max deformation	25%
Min deformation	1,5%

6. Neural Network Architectures

The applied artificial neural networks (ANN) are characterised as multilayer perceptrons (MLPs). A MLP is a feedforward neural network, whereby the input is passed forward through the network. The basic perceptron first described by Rosenblatt, modelled after the structure of human neurons, represent the basic building element of MLPs. The simplest MLP consists of three layers, the input layer, a hidden layer and an output layer. Each layer equals a perceptron, with linear mappings and except for the case of the input function a nonlinear activation function.

Specifically, MLPs with the following structure were used. An input layer supplies the network with the corresponding features, followed by hidden layers consisting of a linear fully connected layer, a nonlinear activation function in the form of the ReLu function and a dropout layer. Finally, the computed results are output. The basic structure of the networks is illustrated in figure 3, whereby the box encloses the hidden layer described above.

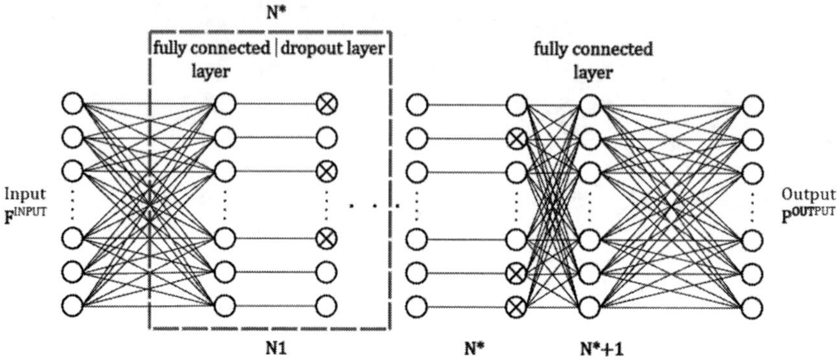

Figure 3. Multilayerperceptron architecture with dropout layer

In the course of the network study, a number of different network configurations are used. Shallow networks, i.e. a MLP with one hidden layer, are used as those offer sufficient approximation capabilities for many problems. Furthermore, deep networks with 3 and 5 layers each are applied, since deeper networks have proven to be especially useful for more complex problems due to their good approximation capabilities of complex relations.

6.1. Data preprocessing

The homogenised deformation gradient in tensor form of the RVE and the initial orientations of the grains as quaternions were defined as features or inputs of the networks. The observations or outputs are the homogenised tensor stress response of the RVE in form of the First-Piola-Kirchhoff stress and the orientation resulting from the deformation. The resulting input and output vectors are concatenated, flattened and have final dimensions of (number of observations x number of features (209)), defining the number of input and output neurons of the neural network architectures.

Due to the difference of the features with respect to their magnitude and the associated risk of exploding/vanishing gradients, a normalisation of the in- and output data is applied before training the ANNs. To prevent data leakage, the entire data set is initially split into training, validation and test data sets following a 60/20/20 split. The total number of files per data set is summarised in table 3 Subsequently, all data sets are normalised to a value range of [0,1] using the coefficients used on the training data sets.

6.2. Training

During training an exhaustive sweep of the parameters in table 4 was conducted. Each combination of hyperparameters was trained using the ADAM optimiser. In contrast to a basic gradient descend algorithm, the ADAM algorithm achieves faster and more stable convergence, using the stochastic first and second moments in order to adapt the basic updating scheme of an ordinary gradient descend implementation.

Table 3. Network design specific hyperparameters.

INPUT variables	Dimension
Deformation gradient(t)	(3x3)
Initial orientations	(50x4)
OUTPUT variables	
First-Piola-Kirchhoff stress	(3x3)
Resulting orientations	(50x4)
Data sets	**Size**
Training data	162.000
Validation data	54000
Test data	54000

Table 4. Network training specific hyperparameters.

Hyperparameter	Value
Initial learning rate	$1e-3$
Grad.DecayFactor β_1	0.9
Squ.Grad.DecayFactor β_2	0.999
Denominator offset ϵ	1e-8
Mini batch size	[64, 128]
Max Epochs	[30]
Dropout rate	[20%, 40%]
Hidden layers	[1, 3, 5]
Neurons/hidden layer (Num_{neu})	[105, 209, 2090]

The ADAM algorithm in combination with Mini batching is applied for training the networks. Two different initial values for the network learning rate are used in combination with different batch sizes. For the dropout rate a fixed value of 0.4 is set. The weights are initialised according to the approach of Glorot for neural network, whereas the bias values are initialised to zero.

During training the loss, i.e. the squares of the difference between predicted and observed values, and the root mean square error (RMSE), i.e. the square root of the mean of the squares of the difference, are metrics used to asses the quality of the respective networks.

7. Results

From the total number of 18 combinations of hyperparameters and network designs the most promising network architecture is chosen based on the RMSE and loss function metrics of the training process. Table 5 shows the chosen configuration and its respective hyperparameter configuration.

In order to provide an engineering metric, the outputs of the test data set are used to compute stress strain curves for specific test cases. The test load cases only represent only a part of the total test data set, but are indicative of the results of the networks.

Table 5. Tested network configurations.

Hidden layers	Num_{neu}	Dropout rate	Minibatch size	Max Epochs	Loss	RMSE
1	2090	20%	128	30	0.56	0.157

In order to generate the stress strain curves the output values of the networks are used to compute the true strain $\overline{\gamma}$, true stress $\overline{\sigma}$ and their respective equivalent values using the following relationships,

$$\overline{\gamma} = \frac{1}{2} ln(\overline{\mathbf{F}}^{\mathrm{T}}\,\overline{\mathbf{F}})\,,\ \overline{\gamma}_{eq} = \sqrt{\frac{2}{3}\boldsymbol{\gamma}^{D} : \boldsymbol{\gamma}^{D}}$$
$$\overline{\sigma} = \frac{1}{det(\overline{\mathbf{F}})}\,\overline{\mathbf{P}}\,\overline{\mathbf{F}}^{\mathrm{T}}\,,\ \overline{\sigma}_{eq} = \sqrt{\frac{3}{2}\boldsymbol{\sigma}^{D} : \boldsymbol{\sigma}^{D}}\,, \tag{6}$$

where $\boldsymbol{\gamma}^{D}$ and $\boldsymbol{\sigma}^{D}$ represent the deviatoric parts of the tensors.

The network is able to capture the general deformation behaviour, both with respect to the elastoplastic transition and the magnitude of the stress, as depicted in figure 4. The general trend of the stress strain curves, i.e. the saturation type material model used for the crystal plasticity model is captured with good quality. The elastic part of the deformation is modelled correctly with regards to the slope of the graph, though the prediction of the transition from elastic to plastic deformation shows shortcomings of the network. Shear cases are modelled with a high degree of accuracy, both qualitiatively and quantitatively. With regards to uniaxial loads the network struggles with modelling the mechanical behaviour. As the uniaxial load cases were split between tension and compression evenly, the information density of the data set was half the density for shear load cases, explaining the deficiencies of the network for those particular load cases.

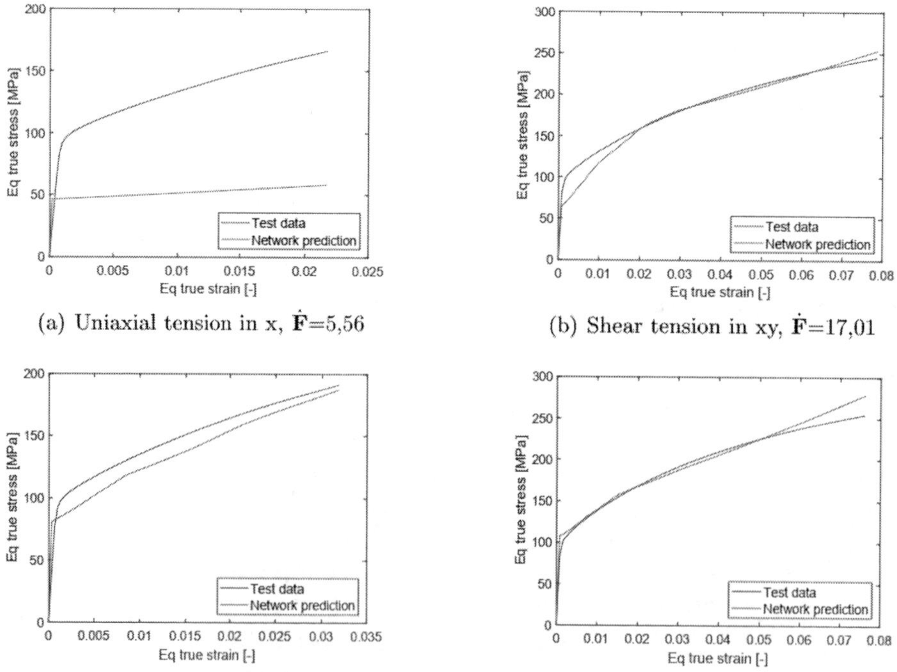

(a) Uniaxial tension in x, $\dot{\mathbf{F}}$=5,56

(b) Shear tension in xy, $\dot{\mathbf{F}}$=17,01

(c) Shear tension in zy, $\dot{\mathbf{F}}$=9,19

(d) Shear tension in yz, $\dot{\mathbf{F}}$=10,99

Figure 4. Predicted strain stress curves

8. Conclusion

During this study machine learning was applied to the subject of crystal plasticity simulations. The advantages of trained neural networks lie in their adaptability and low computational costs, when compared to sophisticated FEM or FFT simulations. The trained deep NNs are able to predict a full 3D mechanical response with reasonable accuracy.

The considered problem was first analysed with regards to suitable features for the neural networks. A sensitivity analysis showed minimal differences in the obtained simulation results considering different numbers of grains. Subsequently, the chosen phenomenological material model for the AA5083 aluminium alloy was used in conjunction with the DAMASK simulation kit in order to generate both training and test data for the networks. The data was postprocessed to extract the selected features and different network designs with their respective hyperparameters configurations were trained. The networks were subsequently tested and examined with respect to their quality and ability to predict stress responses. The selected features proved to be able to correctly predict the elastoplastic deformation of the RVE using the NNs. The implemented approach is flexible with regards to the types of load cases used and a training of the existing networks with different more complex load cases and or orientations is possible.

Based on the shown performance during the evaluation of the predicted stress strain curves, the stated hypothesis was confirmed. The neural network trained with suitable features and data, can replicate the behaviour of the considered crystal plasticity simulation. Although the networks capabilities were specifically assessed only with regards to the mechanical quantities, the assumption, that the crystallographic quantities are equally well predicted, due to the strong linkage of mechanical outputs and crystallographic variables within the theory of crystal plasticity is reasonable.

Further research could be conducted concerning the application of experimental orientation data gained from EBSD analysis, whereby an orientation density function could be used represent realistic material structures in the CP simulation. Furthermore, the expansion of the repertoire of load cases as indicated earlier could be analysed. Further networks studies centred around the application of different network types, like convolutional neural networks or transformers, to crystal plasticity simulations represent future research topics.

References

[1] Roters F, Eisenlohr P, Hantcherli L, Tjahjanto D, Bieler T and Raabe D 2010 *Acta Mater.* **58** 1152–1211.
[2] Roters F 2010 *Crystal plasticity finite element methods in materials science and engineering* (Weinheim: Wiley-VCH).
[3] Allison J 2011 *JOM* **63** 15–18.
[4] Liu Z 2014 *Sci. Bull.* **59** 1619–1623.
[5] Ali U, Muhammad W, Brahme A, Skiba O and Inal K 2019 *Int. J. Plast.* **120** 205-19.
[6] Beniwal A, Dadhich R and Alankar A 2019 *Mater.* **8** 205-19.
[7] Bhadeshia HKDH 1999 *ISIJ Int.* **39** 966-979.
[8] Cecen A, Dai H, Yabansu YC, Kalidindi SR and Song L 2018 *Acta Mater.* **146** 76-84.
[9] Chowdhury A, Kautz E, Yener B and Lewis D 2016 *Comput. Mater. Sci.* **123** 176-187.
[10] Hartmann C 2023 *TMS 152nd Ann. Meeting & Exh. Suppl. Proc.(San Diego)* (Cham:Springer) 576–583.
[11] Herriott C and Spear AD 2020 *Comput. Mater. Sci.* **175** 109599.
[12] Kalidindi SR 2020 *J. Appl. Phys.* **128** 041103.
[13] Kalidindi SR, Brough DB, Li S, Cecen A, Blekh AL, Congo FYP and Campbell C 2016 *MRS Bull.* **41** 596-602.
[14] Koenuma K, Yamanaka A, Watanabe I and Kuwabara T 2020 *Mater. Trans.* **61** 2276-2283.
[15] Lubbers N, Lookman T and Barros K 2017 *Phys. Rev. E* **96** 052111.
[16] Pandey A and Pokharel R 2021 *Scr. Mater.* **193** 1-5.
[17] Prakash A and Lebensohn RA 2009 *Modelling Simul. Mater. Sci. Eng.* **17** 064010.
[18] Yamanaka A, Kamijyo R, Koenuma K, Watanabe I and Kuwabara T 2020 *Mater. Des.* **195** 108970.
[19] Yuan M, Paradiso S, Meredig B and Niezgoda SR 2018 *Integr. Mater. Manuf. Innov.* **7** 214-230.

IOP Conf. Series: Materials Science and Engineering 1284 (2023) 012053 doi:10.1088/1757-899X/1284/1/012053

Displacement driven simulation and material calibration based on digital image correlation Part I - Theory

C Liebold[1], C Ilg[2], A Haufe[2], T Usta[1], V Sreenivasa[2]

[1] DYNAmore GmbH - an Ansys Company, Industriestr. 2, 70565 Stuttgart, GER
[2] DYNAmore GmbH - an Ansys Company, MCC - Material Competence Center, Kolumbusstraße 47, 70771 Leinfelden-Echterdingen, GER

E-mail: christian.liebold@ansys.com

Abstract.
Standard material characterization methods for subsequent finite element analysis rely on the simulation utilizing ideally discretized specimen that represent its geometrical characteristics as in the real experiment. Since many material models hold a set of parameters which is dependent on the discretization, e.g. the mesh size, especially when failure prediction for crash analysis is relevant, various mesh sizes are utilized during this calibration process. Optical measurement systems accompanying the experimental tests have become a standardized procedure, at least for a 2-dimensional (or specimen top-surface) measurements. Investigated calibration methods relying on the results of these optical measurements either include the derived strain field into the characterization process (full-field calibration), determine the crack path and iteratively derive plasticity curves (stepwise-modelling approach) or establish virtual fields based on the principal of virtual work allowing to solve for the scalar unknowns of the underlying chosen constitutive equation (virtual fields method). In the presented work, we directly use displacements measured using a GOM/ARAMIS system, mapped as boundary conditions for the nodal displacements of differently discretized specimen to establish a material characterization scheme being as close to the experiment as possible.

1. Introduction

The calibration of material data to properly describe the stress response of a specific material under various loading conditions in finite element analysis (FEA) is still a challenging task. Not only that new steel grades or aluminum alloys are investigated by the material manufacturers to enhance certain characteristics of their product such as stiffness, strength, or maximum strain to failure, but also FEA software tools add new physically or non-physically motivated material models to their libraries, allowing users to improve the predictability of their process- or structural analysis with these new materials. Since depending on the model size various element lengths are utilized, it is often necessary to repeat the calibration process several times, e.g. for 0.5 mm – 10.0 mm meshes, in order to properly capture mesh size dependent damage and fracture parameters.

Experimental test campaigns for material characterization are nowadays often accompanied by measurements of the displacement field of the specimen, but current research investigates the possibility to consider 3-dimensional optical measurements utilizing a second camera system

IOP Conf. Series: Materials Science and Engineering 1284 (2023) 012053 doi:10.1088/1757-899X/1284/1/012053

or even computer tomography to enhance the predictability of the chosen phenomenological material model. Quite commonly, the GOM/ARAMIS optical measurement system is used for this purpose. Prior to the test, a randomized speckle pattern is sprayed onto the specimen, allowing the underlying image processing software to create triangulated facets which are used internally to calculate strain fields. Previous publications ([1]) have investigated how these locally varying strain fields can be used to improve accuracy of the parameter identification process comparing measured and simulated stress-strain responses during the optimization. Since a larger area of the specimen is used as optimization target instead of only comparing to the force response, this method is called Full-Field Calibration (FFC). Another method utilizing measured strain fields based on the Principal of Virtual Work is the so-called Virtual Fields Method (VFM) proposed by [2]. Within a selected area of interest, virtual fields are created, allowing to establish scalar equations which together with a selected constitutive model help to obtain the unknown material parameters. In [3], a piecewise modelling techniqueis proposed, identifying the integration path for force calculation from the last measured state prior to specimen failure. Yield stress vs. effective plastic strain curves can then be derived iteratively based on the deformation gradient and the derived strains at each experimental time step and an underlying constitutive equation.

Within this work, we investigate the possibility to combine finite element data mapping with material calibration. Therefore, we transfer the displacement field measured with the GOM/ARAMIS system onto finite element meshes with different element sizes as nodal boundary condition. This allows us to simulate the behavior of the specimen as realistic as possible for several mesh sizes while deriving material parameters through an optimization process. This contribution is split into two parts: the first dealing with the theoretical background and implementation of the image processing and finite element data mapping into the software Envyo®, the second with the parameter identification using LS-OPT®, LS-DYNA®, and a modified test specimen, which provides a broader range of various stress states than the currently used standardized tensile specimen ([4]).

2. Theory and Implementation

In a first step, the measured GOM/Aramis data has to be read by the mapping software Envyo. The data is provided in an encrypted `*.xml` file format, holding information about the nodal positions at each measured time step and connectivity of the nodes, creating facets used by the software to calculate resulting strain fields on the specimen. These facets, which are usually created through mesh triangulation, are considered as triangular finite elements by the mapping software Envyo, so that the standard natural coordinates ξ, η and shape functions $N_i(\xi, \eta)$ and their partial derivatives $\frac{\partial N_i}{\partial \xi}$ and $\frac{\partial N_i}{\partial \eta}$ as provided in eq. (1) are applicable [5].

$$
\begin{aligned}
&N_1 = 1 - \xi - \eta \qquad && \frac{\partial N_1}{\partial \xi} = -1. \qquad && \frac{\partial N_1}{\partial \eta} = -1. && \\
&N_2 = \xi && \frac{\partial N_2}{\partial \xi} = 1. && \frac{\partial N_2}{\partial \eta} = 0. \qquad && \text{with } 0. \leq \xi, \eta \leq 1. \\
&N_3 = \eta && \frac{\partial N_3}{\partial \xi} = 0. && \frac{\partial N_3}{\partial \eta} = 1. &&
\end{aligned} \tag{1}
$$

In order to transfer the displacement field from the GOM/Aramis measurement to the finite element model, a standard closest point projection (CPP) method as commonly used in contact mechanics and described in detail in [6] has been implemented into the utilized mapping software. Given that only a measurement of the top surface has been carried out and that the data transfer is being realized for shell only meshes in a first step, a differentiation between *elliptic*, *hyperbolic*, *parabolic*, and *planar* points is being neglected. Therefore, the distance r between the target point and the source surface as illustrated in fig. 1 has to be minimized:

$$\|\boldsymbol{r}(\xi,\eta)\| = \|\boldsymbol{p}_{SRC} - \boldsymbol{p}_{TAR}\| \to min. \tag{2}$$

This can be solved, e.g. using a standard Newton-Raphson scheme given in its generalized formulation as:

$$\boldsymbol{x}_{n+1} := \boldsymbol{x}_n - \boldsymbol{J}(\boldsymbol{x}_n)^{-1}\boldsymbol{f}(\boldsymbol{x}_n) \tag{3}$$

where $\boldsymbol{J}(\boldsymbol{x}) = \boldsymbol{f}'(\boldsymbol{x}) = \left(\frac{\partial f_i}{\partial x_j}(\boldsymbol{x})\right)_{ij}$ represents the Jacobian matrix of $\boldsymbol{f}(\boldsymbol{x})$. As is a common procedure, the Jacobian is not inverted but instead, the linear equation system

$$\boldsymbol{J}(\boldsymbol{x}_n)\Delta\boldsymbol{x}_n = -\boldsymbol{f}(\boldsymbol{x}_n) \tag{4}$$

is being solved so that the next step of the iteration can be calculated as:

$$\boldsymbol{x}_{n+1} = \boldsymbol{x}_n + \Delta\boldsymbol{x}_n. \tag{5}$$

This procedure is being carried out for all nodes of the target finite element mesh, assigning a source facet and corresponding natural coordinates to each target node. For the case that the CPP fails in a first instance, the target normal as average of the normals of the connected elements of the node is used instead of the source normal for another trial. If that trial fails as well, a standard closest point search is carried out, assigning the closest node on the source mesh to the target node. A bucket sort algorithm is implemented allowing to speed-up the search process due to a restricted number of trials based on geometrically assigned bucket identification numbers.

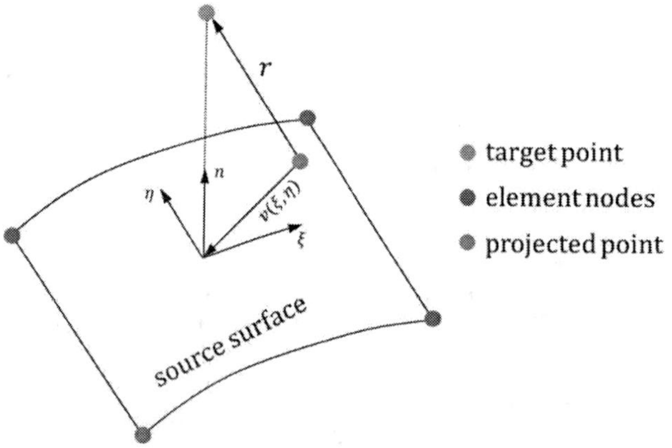

Figure 1. Illustration of the closest point projection (CPP) method utilized to determine natural coordinates on GOM facets for displacement calculation of target nodes [6].

Once all target nodes have been assigned either an element and corresponding projected natural coordinates, or a nearest neighbor node, the software will iterate through all the stages, calculating the displacement in x- and y-direction of each identified position on the source side and storing them with respect to time as a curve for each target node. As a final step, the identified displacement curves will be assigned to the nodes using the `*BOUNDARY_PRESCRIBED_MOTION` keyword and all nodes, elements, displacement over time curves, and assigned boundary conditions will be written in the ascii - based LS-DYNA input format.

3. Validation

To validate the implemented routines, severel tasks have been carried out prior to utilizing the proposed approach for a material calibration procedure. The basic idea behind all validation tests is to ensure that the qualitative comparison between input and output are within an acceptable range. Three different scenarios have been chosen to evaluate the implemented routines: a transfer from finite element simulation stages to a finite element simulation model utilizing the same mesh (*FE-FE initialization*), a transfer from a standard GOM specimen to a discretization with different element sizes (*mesh size dependent initialization*), and a transfer from a rather complicated specimen to a mesh with arbitrary element distributions (*arbitrary specimen initalization*). All models described in the following subsections have been subject to a small amount of stiffness damping (20%), fully integrated shell elements (`ETYP=16`), and a material card for a DP800 HD sheet metal material utilizing the isotropic elastic-plastic phenomenological description implement in `*MAT_PIECEWISE_LINEAR_PLASTICITY` (`*MAT_024`). Besides the applied displacements, nodes are restricted in their z-displacement and their rotational degrees of freedom. All validation simulations have been carried out utilizing LS-DYNA mppR13.1 double precision version.

3.1. FE-FE initialization

As shown in fig. 2, the nodes of both the reference simulation model and the simulation carried out with the transferred nodal displacements are comparable and the difference, which is slightly visible on the right side of the specimen is negligible. In fig. 3, the resulting cross section forces are plotted, showing that there is a small divergence after the point of uniform elongation.

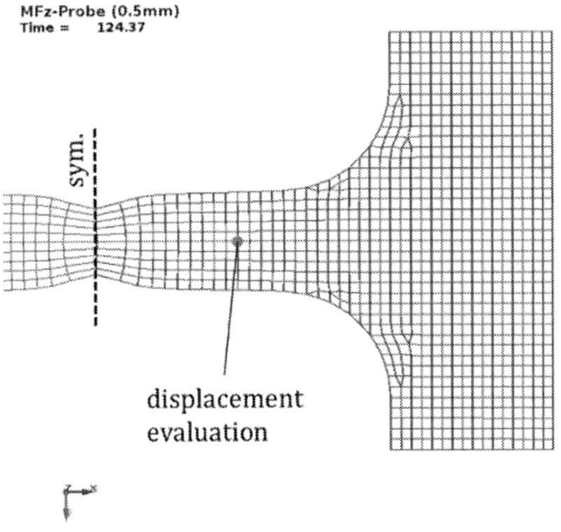

Figure 2. Comparison of the last stage of the original simulation with the simulation based on applied nodal displacements for each time step. The symmetry plane, also serving as the plane calculating the cross section forces and the point to evaluate nodal displacements are highlighted.

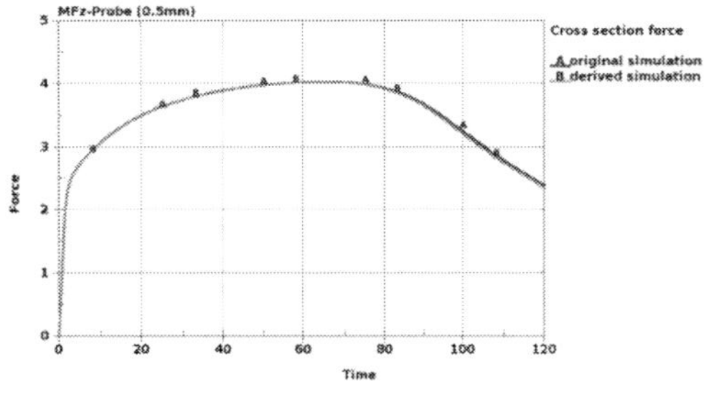

Figure 3. Comparison of the force calculated by the original simulation model and the derived simulation model.

The validity of the approach is underlined by the comparison of the nodal displacements of evaluation points to the right and left of the cross-section plane, as highlighted in fig. 2. In fig. 4, it can be observed that after the point of necking, there is a small divergence of the nodal x-displacements from the derived simulation model to the original, especially for the displacement of the node on the right side of the specimen, whereas there seems to be no visible difference of the nodal displacements of the evaluation points to the left of the cross section plane. Looking at the y-displacements, it can be observed that there are almost no displacements in this direction (order of $E-06$). Nevertheless, the way the nodal displacements reach their plateau is slightly different, which may be due to numerical reasons or because of the damping and boundary conditions applied to the derived model.

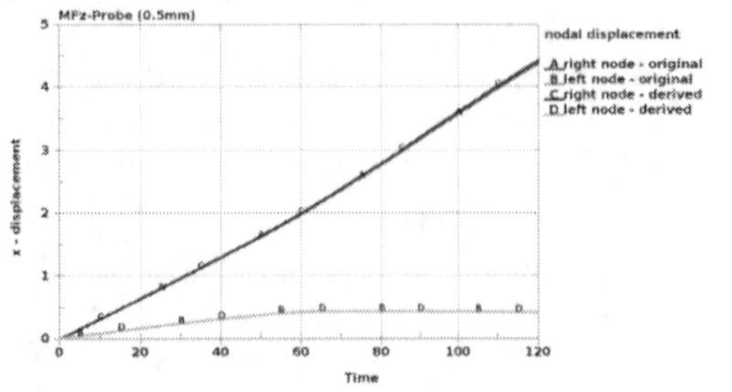

Figure 4. Comparison of the x-displacements over time between the original and the derived model.

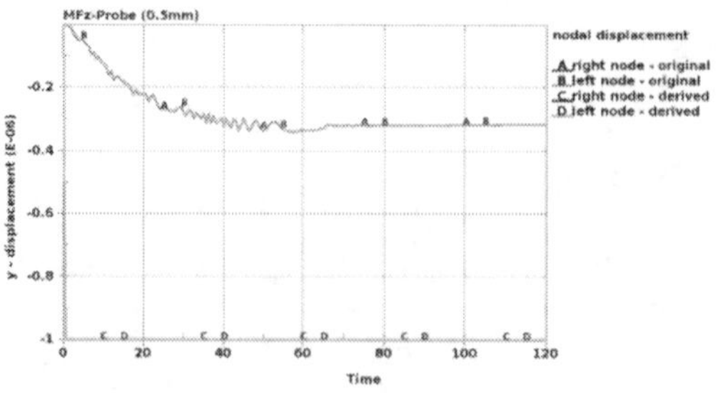

Figure 5. Comparison of the y-displacements over time between the original and the derived model.

It can be summarized that based on the *FE-FE initialization* approach, the validity of the implementation has been proven so that it can be proceeded with the investigation of the element size dependency.

3.2. Mesh size dependent initialization

As a next validation step, displacements measured using a GOM/ARAMIS system on a tensile specimen have been transferred onto simplified tensile specimen utilizing quadrilateral elements with mesh sizes from 0.25 mm to 20 mm, resulting in a range or 1 to up to 80 elements across the width of the specimen. The minimum edgle length of the triangulated GOM measurement is given as ~0.327 mm, the maximum edge length is ~0.720 mm.

Figure 6 shows the final state of the simulated tensile specimen with different mesh lengths, with the resulting plastic strain in comparison to the model simulated utilizing the facets derived by the GOM/ARAMIS system. It can be stated that for mesh sizes between 0.25 mm and 5.0 mm, the crack path running through the specimen from bottom left to top right is visible, for larger element sizes, only an average plasticity value is calculated. Furthermore, nodal values which have been assigned a closest node transormation due to non-matching geometries can be identified visually, since lower plastic strains and characteristic stripes of similar strains are being formed.

To ensure that these outliers and the mesh size do not significantly influence the results, the cross section forces for all mesh sizes, a simulation utilizing the GOM/ARAMIS facets directly and the experimentally measured force curve are plotted in fig. 7. All simulation models have the tendency to reach a higher force level than the experimentally measured force over time curve. Generally, the plastic zone is reached a bit later compared to the experiment. The simulation utilizing the GOM/ARAMIS facets shows significant scatter of the force values, similar to the simulations carried out utilizing a 0.25 mm and a 0.5 mm mesh. Mesh sizes being larger than the utilized facet size result in smoother force over time curves. This can be explained by an automatically introduced smoothing when transferring the displacements to larger mesh sizes since the scatter which is observed in any experiment will be somewhat neglected.

Given that the force levels derived with the different mesh sizes are within the same range, it can be stated that the method is valid for different mesh sizes as well, even though some enhancements towards smoothing should be considered. Therefore, the method will be validated using a rather complicated specimen with a randomized generated mesh.

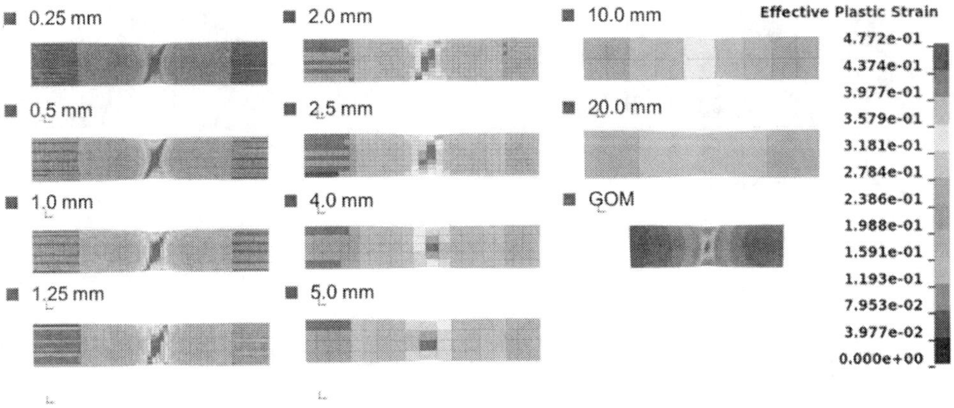

Figure 6. Final state of simulations carried out with transferred displacements from a GOM/ARAMIS system to differently discretized simplified specimen.

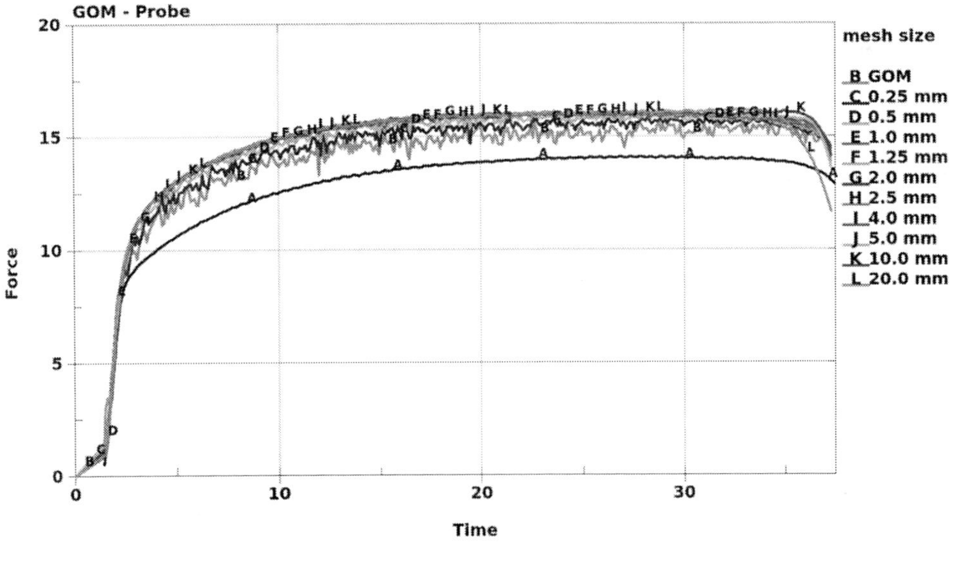

Figure 7. Resulting forces from experimental tests (A), simulation utilizting the GOM derived facets (B) and for different element lengths (C) - (L).

3.3. Arbitrary specimen initialization

As a final validation step, simulation results of a specimen proposed by [7] utilizing the GOM mesh directly and displacements mapped onto a different mesh are compared. As shown in fig. 8, both von Mises equivalent stress and the plastic strain are reproduced in a comparable way. At elements where not all nodes could be projected onto the GOM/ARAMIS facets, the scatter observed at the regular specimen from fig. 6 is also visible, especially in the areas close to holes and at the specimen boundaries. This might also explain the difference in the derived cross section forces, which tend to be slightly higher for the mapped specimen compared to the GOM/ARAMIS based model.

Therefore, it can be concluded that the introduced method is valid for arbitrary, 2-dimensional geometries, utilizing different discretizations such as triangular and quadrilateral elements.

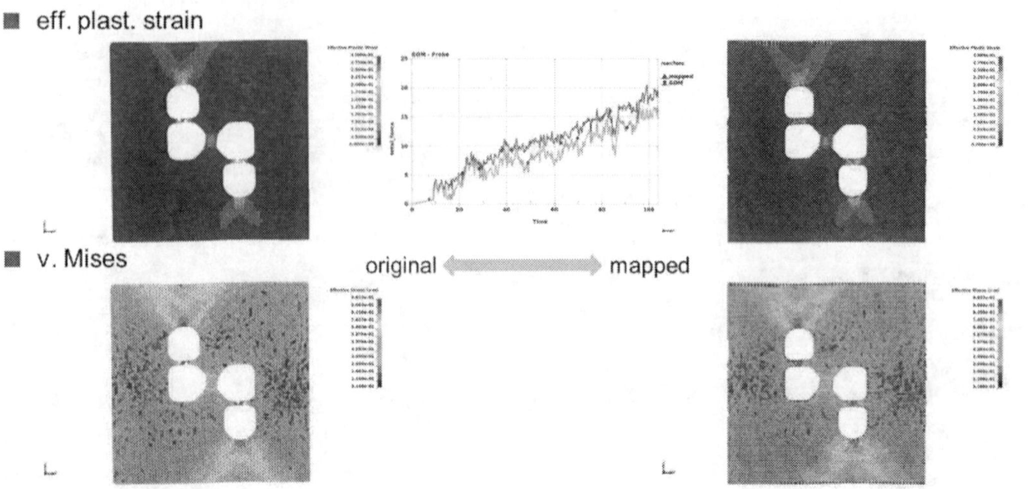

Figure 8. Comparison of simulation results utilizing the GOM/ARAMIS displacements directly on the created facets (l) and with displacements mapped onto a different mesh (r). The top row shows the comparison of the effective plastic strain, the bottom row shows the von Mises equivalent stress. The graph in the middle displays the resulting cross section forces.

Material calibration relying on the proposed method will be presented in [4], based on a specimen structure derived from the one presented above and introduced in [8].

4. Conclusion and Outlook

A method to transfer displacement fields measured using a GOM/ARAMIS system onto finite element meshes has been introduced. Boundary conditions which can be derived using shape function interpolation after utilizing the closest point projection procedure have been applied directly onto the nodes and simulations have been performed for different element sizes and rather complicated specimen structures to validate the applicability of the implemented mapping routines. Further activities utilizing this method and presented by [4] shall demonstrate its ability to increase accuracy during material calibration for finite element material models and a significant speed-up compared to the full-field calibration (FFC) method where the resulting strains have to be compared as a result from each iteration. The direct transfer of the displacements as it is propsed in this work will avoid that step since equivalent displacement fields within the experiment and the simulation model will lead to similar strain fields. Using the displacements derived from experimental testing within the model setup process is expected to increase the accuracy of the calibration results. Furthermore, the method can be extended to larger components and is an important contribution towards the linkage between measured data and simulation input and might help to overcome iterative procedures when calibrating mesh size dependent parameters such as failure strains or fracture toughnesses. Until now, such data transfer is usually carried out between different simulation stages - adding measurement results from experiments for material data calibration is therefore another important step to bring simulation closer to reality.

Acknowledgments

This work has been carried out within the research project AIMM (Artifical Intelligence for Material Models) funded by the Federal Ministry for Economic Affairs and Climate Action

(BMWK) under the grant number 19 | 20024*A*.

References

[1] Ilg C, Haufe A, Witowski K, Koch D and Liewald M 2019 *Proc. 38th Int. Deep-Drawing Research Group Conf.* (Enschede, NL)

[2] Grédiac M, Pierron F, Avril S and Toussaint E 2006 The Virtual Fields Method for Extracting Constitutive Parameters from Full-Field Measurements: a Review *Strain* **6** 233-53

[3] Marth S, Häggblad H, Oldenburg M and Östlund R 2016 Post necking characterisation for sheet metal materials using full field measurement *J. Mate. Proc. Tech.* **238** 315-24

[4] Ilg C, Liebold C, Haufe A, and Liewald M 2023 *Proc. 42nd Int. Deep-Drawing Research Group Conf.* (Lulea, SWE)

[5] Bathe K J 1996 *Finite Element Procedures* (New Jersey: Prentice-Hall Inc.)

[6] Konyukhov A and Schweizerhof K 2008 On the solvability of closest point projection procedures in contact analysis: Analysis and solution strategy for surfaces of arbitrary geometry *Comput. Meth. Appl. Mech. Eng.* **191** 3045-56

[7] Williams B and Simha C 2019 *Proc. 38th Int. Deep-Drawing Research Group Conf.* (Enschede, NL)

[8] Ilg C, Haufe A, Sreenivasa V, Karadogan C and Liewald M 2023 *Proc. 14th Int. Conf. on the Tech. of Plast.* (Mandelieu-La Napoule)

IOP Conf. Series: Materials Science and Engineering 1284 (2023) 012054 doi:10.1088/1757-899X/1284/1/012054

An overview of Methods for Simulating Sheet Metal Forming with Elastic Dies

J Pilthammar[1,2], M Sigvant[1,2], M S Islam[2], M Schill[3], S Sjöblom[2], V Sjöblom[2], M Lind[2]

[1] Volvo Cars Dept. 81110 Strategy Development, Olofström, Sweden
[2] Department of Mechanical Engineering, Blekinge Institute of Technology, Karlskrona, Sweden
[3] Dynamore Nordic, Linköping, Sweden

E-mail: johan.pilthammar@volvocars.com

Abstract. Sheet metal forming (SMF) simulations are traditionally carried out with rigid active forming surfaces. This means that the elasticity and dynamics of presses and die structures are ignored. The only geometries of the tools included in the simulations are the active forming surfaces. One reason for this simplification is the large amount of computational power that is required to solve finite element (FE) models that incorporates elastic stamping dies. Another reason is the lack of die CAD models before the later stages of stamping projects. Research during the last couple of decades indicated potential large benefits when including elastic dies in SMF simulations. For example, for simulating die try-out or for Digital Twins of presses and dies. Even though the need and potential benefits of elastic dies in simulations are well known it is not yet implemented on a wide scale. The main obstacles have been lacking data on presses and dies, long simulation times, and no standardized implementation in SMF software. This paper presents an overview of existing methods for SMF simulations with elastic dies and discuss their respective benefits and drawbacks. The survey of methods shows that simulation models with elastic tools will be needed for detailed analyses of forming operations and also for purposes like digital twins. On the other hand, simplified and robust models can be developed for non-FEA users to carry out simple one-step compensation of tool surfaces for virtual spotting purposes. The most promising and versatile method from the literature is selected, modified, and demonstrated for industrial sized dies.

1. Introduction

Sheet metal forming (SMF) simulations are traditionally carried out with rigid forming surfaces. This means that the elasticity and dynamics of presses and die structures are ignored. One reason for this simplification is the large amount of computational power that is required to solve finite element (FE) models that incorporates elastic stamping dies and presses. Another reason is the lack of die CAD models before the later stages of stamping projects. Previous research during the last couple of decades indicated large benefits if elastic dies and presses are included in SMF simulations [1-8]. The benefits of simulations with elastic tools are also seen in related manufacturing techniques using dies and presses. Some examples outside SMF are developments of processes and dies in High Pressure Die Casting [9], Thin Profile Extrusions [10], Multi-Point Sandwich Forming [11], Clinching [12], and Forging [13-14]. Even though the need and potential benefits of elastic dies in simulations are well known it is not yet implemented on a wide scale.

The first purpose of this paper is to present an overview of existing methods for SMF simulations with elastic dies and discuss their respective benefits and drawbacks. The survey of methods shows that simulation models with elastic tools will be needed for highly detailed analyses of forming operations, and for purposes like training digital twins. On the other hand, simplified models can be developed for non-FEA users to carry out simple one-step compensation of tool surfaces for virtual spotting purposes.

The second purpose of this paper is to select and demonstrate the most promising method for SMF simulations with elastic dies method from the literature. Modifying the selected method is key to enabling simulations for industrial-sized dies, otherwise it is impossible to solve models of industrial sized stamping dies with a reasonable amount of time and computational power.

The method selected and demonstrated herein has one significant advantage: it can realistically represent both the dynamic and elastic behavior of the cushion and blankholder system. This is further demonstrated in [15]. There are no other existing methods available to represent these behaviors which can have a profound impact on the forming operation.

Models throughout this paper are created in LS-PrePost V4.6.0 and solved with the explicit FE software LS-Dyna R11.1.0 [16]. Keywords and setup of models throughout this paper are referencing this software.

2. Methods for SMF Simulations with Elastic Dies

SMF simulations with elastic tools can be very tedious and time-consuming work. Here, several existing methods are presented and discussed. First, it must be mentioned that forming dies are mounted in presses that have elastic and dynamic behaviors. Predicting die deformations is challenging because it depends on the full press structure which is challenging to include in a FE model. Full representations of presses will result in very large FE models. Presses can however be represented by various measuring and modeling techniques [6, 8, 17-21]. Measuring and characterizing a press cushion is more challenging. It is however easier to model based on CAD data of the cushion structure [22-23]. The elasticity of the blankholder and cushion system can be accurately described by including a model of the press cushion in single action presses or blankholder plates in double action presses.

2.1. Discretization with solid elements

Discretization of the entire die and press structure with solid elements is not a viable way for forming analyses on an industrial scale. This approach uses brute force by adding an enormous number of solid elements to achieve a high resolution of the forming surfaces. This makes the models computationally costly [5-7, 22]. Further, they require access to die CAD data with updated forming surface geometries throughout stamping die project. However, this technique can be used on small er dies to study the influence of the elastic forming tools [1, 4]. The important first steps in method development and motivation for using elastic tools were taken with this method.

2.2. Uncoupled FE models

In [24] a method was implemented of first carrying out a structural FE-simulation of the forming tools in the FE software ABAQUS. The simulation was based on scanning data of existing forming dies past die try-out. The die surfaces and elastic tool parts were loaded with forming forces before exporting geometries to the SMF simulation software AutoForm. The method consists of pressing the die parts against each other, also used in [25], in structural simulation software. Then transferring the deformed die shapes to the forming surfaces in AutoForm. The use of sub-modeling in ABAQUS was recommended to transfer the deformed shapes. This approach showed in [24] that die deformations can have a large influence on the forming process, even though the resulting die deformations are frozen at a single point in time. Die deformations and dynamics throughout the press stroke are not included in the simulation. Manually running two or more software and transferring data between them is not desired.

Uncoupled models are not fully suited for SMF simulations with elastic tools, even though they were used to demonstrate the influence of elastic dies on the forming operation. The method can be good enough to get a sense of how much die surface compensation is needed for virtual spotting.

2.3. *Reduction techniques for DOFs*

In [4] and [7], different methods for reducing the degrees of freedom (DOF) in tool structures are presented. The first method is called static condensation, or super elements, and is available in most finite element codes. Parts of the model are presolved, then only nodes on forming surfaces and other contact areas are retained. Static condensation is shown in [4] to not be a viable way for the reduction of sheet metal forming simulations. The method is only suitable when the retained degrees of freedom are on the order of a magnitude smaller than the initial amount of degrees of freedom. In SMF simulations a very large amount of segments have to be retained on the forming surfaces.

A second method suggested in [4] is deformable rigid bodies, or modal method, where the deformation of the structure is estimated by a linear combination of so-called modes. For a cross die, presented in [4], this approach is increasing the solver time only by a few percent when deformable tools are added to the forming simulation. The method needs presolving which adds extra steps during the implementation of the model. So, unless the solving for modes and model setup is easy and simplified for the FE user, the methods presented in the following sections might be more suited for SMF simulations with elastic dies. However, Industry 4.0, Machine Learning, Digital Twins, etc. are areas that are growing fast. There, a large amount of data is needed for the data-driven models. FE models might need to be continuously running to give stamping plants decision support. For this, reducing the number of DOFs might be valuable and worth the extra effort.

2.4. *Coupled FE models*

In [6, 26-29], the forming simulation based on the geometry of the active surface of the die is performed in PAM-Stamp. Then a separate elastic structural simulation model of the die, punch, and blankholder was created. Contact forces from the forming simulation were used to estimate the displacements in a structural FE-model, then sending back updated surfaces for continuing the forming operation. This exchange of contact forces and displacements between forming simulation and structural simulation is performed through coupling steps in the simulation.

This method yields sound and realistic results that are confirmed in real try-out conditions. Especially if it is coupled with realistic estimations of press behavior based on experimental measurements [27]. To estimate a sound structural behavior and avoid unrealistic oscillations in tool deformations the coupling steps between the forming model and the structural model must be performed with high frequency.

One possible drawback of this method could be the modeling of the cushion and blankholder in single action presses. The deformations of the tools and press parts are calculated in an implicit structural model where boundary conditions are needed on any structure to prevent rigid body movements from causing divergence in the solver. In a realistic case where the cushion is not in contact with its guiding rails, only a mapped pressure on the top and point force underneath are loading the cushion structure. In [6], the bottom surfaces of all the cushion pins are constrained in all directions when deformations are calculated; it is mentioned in [27] that work is ongoing to create a universal substitute model for the cushion system based on experimental measurements. These limitations in blankholder modeling limit the use of this modeling technique e.g., robustness analysis, digital twins, and process control applications. There, blankholder forces, deformations, shimming, tilting, etc. are very important parameters and need to be realistically modeled.

2.5. *Shell elements connected to solid elements*

In [7] the idea is raised that the tool geometry probably does not have to be resolved by a very fine mesh to estimate the global tool deformations. On the other hand, the forming surfaces need a fine resolution. It is therefore suggested to, in one simulation model, connect shells representing the forming surfaces to a solid element mesh of the die structure. An example is demonstrated by simulating the forming of an s-rail geometry where the forming surfaces are connected to the tool structure by tied offset contacts in LS-Dyna.

3. Selecting and modifying a method for SMF simulations with elastic dies on industrial-sized dies.

Reduced FE models, described in chapter 2.3, is highly dependent on the model from which it is originally reduced. It also requires extra preprocessing and solving steps before the model is completed. The method is best suited when solver time is critical, e.g., large robustness analyses or for building digital twins. But even then, the model could be too slow, metamodeling is needed for production control when responses are needed within minutes or even seconds.

	Simulation of small forming dies, eg. lab tools.	**Feasibility simulations in car projects.**	**Robustness simulations.**	**Virtual Spotting and Die Tryout**	**Use the model for inline process control (Digital Twin)**	**COMMENT**
Simulations with rigid dies. (Current Industrial and Academic standard)	OK if die deformations can be ignored.	Die elasticity is ignored in the early phases of feasibility.	Missing effects of elastic deformations.	Missing effects of elastic deformations.	Missing effects of elastic deformations.	Only reliable when the influence of elastic deformations can be ignored.
Solid elements for tool discretization. (Chapter 2.1)	Fast enough for small dies.	No CAD data of solids early in SMF projects.	Requires too much computational power	Requires too much computational power	Requires too much computational power	Theoretically the most accurate model. Limited by computational power for simulations of larger dies.
Uncoupled FE models. (Chapter 2.2)	Not required, the method in chapter 2.1 is OK.	No CAD data of solids early in SMF projects.	No deformations during the stroke.	OK for one-step spotting, but the methods below are more advanced and reliable.	No deformations during the stroke.	The method is missing dynamic effects and continuous deformations of the die. The method was however a valuable step on the path toward the methods described in Chapters 2.3-2.5.
Reduced FE models. (Chapter 2.3)	Not required, the method in chapter 2.1 is OK.	No CAD data of solids early in SMF projects.	Highly dependent on the reduced original FE-model.	Highly dependent on the reduced original FE-model.	Highly dependent on the reduced original FE-model.	Will be important when the simulation time is very critical, but the properties of the model will depend on the model on which it is based.
Coupled FE models. (Chapter 2.4)	Not required, the method in chapter 2.1 is OK.	No CAD data of solids early in SMF projects.	Limited in blankholder and cushion modeling.	Reliable for punch and matrix deformations.	Limited in blankholder and cushion modeling.	Fully elastic SMF simulations. There are currently unresolved limitations of blankholder and cushion modeling. This limits the use to blankholder spotting, robustness simulations, and digital twins.
Shell elements connected to solid elements. (Chapter 2.5)	Not required, the method in chapter 2.1 is OK.	No CAD data of solids early in SMF projects.	Reliable, might require long simulation times for robustness analyses.	Includes all important dynamic and elastic effects.	Reliable predictions, powerful if it can be used as a basis for metamodels.	Fully elastic SMF simulations. Reliably represents elastic press and die deformations and dynamics.

Table 1. Comparison of identified models for sheet metal forming simulations with elastic dies.

Coupled FE models, described in chapter 2.4, is a full working model for SMF simulations with elastic dies. The models are solved within a reasonable time and seem to yield good accuracy for virtual spotting of punch and matrix. There is however one identified drawback of the model: the blankholder and the cushion system will not be realistically constrained. This is because the blankholder and cushion need to be artificially constrained to avoid rigid body motions when blankholder deformations are calculated. This happens because the deformations are calculated in a separate FE model where they should be loaded solely by forces from the blank and the cushion cylinders. In theory, this will work, but the slightest numerical imbalance will cause rigid body motions and non-convergence of the FE-model. This is solved by artificially locking nodes on the blankholder and/or cushion solids, often the bottom of the cushion pins restricting deformations and rotations.

The preferred method by the authors is a method originally proposed by Haufe et al. in [7] where a shell mesh is coupled to a solid structural mesh. This idea to include both forming and structural behavior in the same model seems very attractive, both from a preprocessing and a solver perspective. This method has none of the previously mentioned drawbacks of extra preprocessing steps or numerical issues during solving. However, in its´ current form it is impossible to solve models on an industrial scale. Important modifications of the method to enable this is presented in this paper.

Since 2015, research within Swedish and international research projects is developing methods for SMF simulations with elastic dies [4, 22, 30] where this simulation method is further adopted in LS-Dyna. The goal has been to define a method that makes it possible to include all desired parts in a single FE-model. The model shall be solved continuously, with realistic boundary conditions, without any mapping or simplifications between different codes of FE-models. Modeling should also be quick and easy, and geometries fast to update. The simulations presented in subsequent chapters of this paper are based on this approach. The models were found to be relatively straightforward to set up in LS-Dyna, but when it was implemented for industrial-sized dies and simulations some further modifications and considerations were needed that are presented in this chapter. When applying the method to larger body panels it was discovered that the solver cost for the contact algorithm is large. A major issue in running forming models in an explicit code with elastic tools is the dynamic behavior of the large and heavy mass scaled die parts. Density is therefore artificially reduced to minimize the dynamic oscillations while keeping the structural stiffness intact. It is also important to distribute the model in a smart way across the computing cores during the solving process.

Based on literature and the previous discussion, table 1 compares methods identified for SMF simulations with elastic dies.

3.1. NULL material for forming surfaces

Forming surfaces in an FE model needs a fine resolution, which will generate a lot of very small solid elements if only solid elements are used. The meshing becomes complicated and solver times extremely long. The solution proposed by Haufe et.al. [7] is to cover a coarse solid mesh with a finer shell mesh. A flexible and nimble way to utilize this strategy is to cover a solid mesh with *MAT_NULL shell elements [16], which is a material lacking any mechanical properties. It will act as an infinitely thin blank covering the coarse solid elements and representing the active forming surfaces, not influencing time steps and mass scaling. During design iterations, this method enables the same solid mesh to be used repeatedly while only swapping out the forming surfaces.

The shell elements representing the forming surfaces are connected to the solid elements in the die structure by a tied contact with offset. The contact type for connecting the shell elements to the solid structure is *CONTACT_TIED_NODES_TO_SURFACE_CONSTRAINED_OFFSET [16].

Two contact parameters are critical in ensuring that the NULL elements are tied to the solid structure [30]: The depth of the contact search shall be large enough so that each node finds a solid element segment to follow, this can be adjusted by setting negative values for the SST/MST parameter on the contact card. PARMAX should also be used to extend the contact segments in the segment-based projection. Otherwise, there is a risk that some nodes will not find a contact segment on the solid.

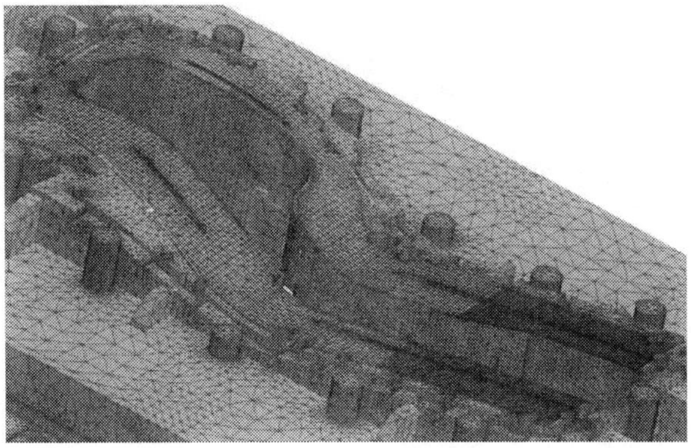

Figure 1. Example of a solid blankholder covered with NULL elements

3.2. Reduced density of solid parts

Explicit SMF simulations are conducted with very high velocities, often in the range of 500-2000 mm/s. This will lead to very large dynamic effects if large elastic bodies are used in the models to represent elastic stamping dies. The elastic bodies would start to bounce against each other and vibrate a lot during the simulation, see figure 2. One solution to this issue is to run the simulations with small velocities that will reduce mass scaling but lead to very long simulation times.

Another way is to artificially lower the density of the elastic bodies, this allows for the simulations to run at velocities normally used in SMF simulations but with small dynamic effects, still maintaining a physically sound representation of the elastic response. Figure 2 depicts the movement of the ram driving the blankholder with normal density and mass scaling, and when the density of blankholder parts is reduced by a factor of 100. The blankholder is closing with 2000 mm/s, thereafter, applying a blankholder force of 1900 kN.

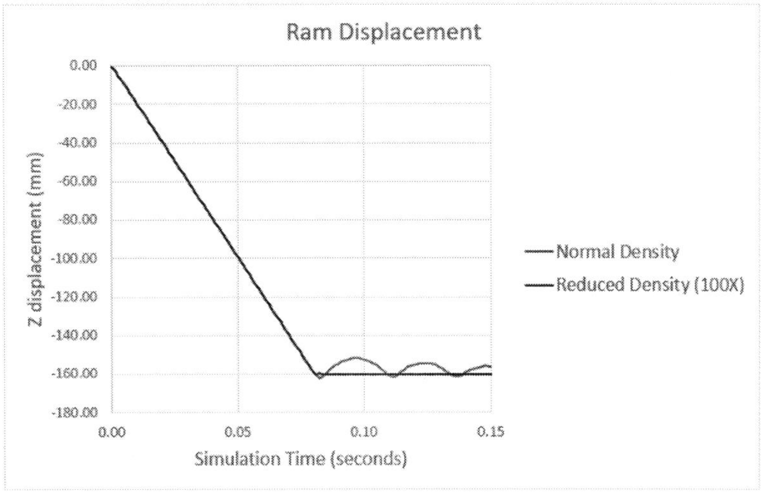

Figure 2. Ram displacement for blankholder with normal and reduced density.

3.3. Model Decomposition

An important aspect when solving SMF models with elastic dies is the decomposition of the models. Decomposition distributes the model across cores on the cluster that is solving the model. Here it is important to select a method that distributes the calculations for elements, material, and contacts as evenly as possible across the cores. A bad decomposition can lead to a few cores doing a lot of calculations while others are idle. Poor ways of performing the decomposition are e.g., to divide the model along the x, y, or z-axis in the model. Decomposition along the height of a tool is illustrated in figure 3A below. This can lead to most contact calculations, and elastoplastic material calculations are carried out on very few cores. Meanwhile, most of the cores quickly carry out the calculations for the elastic elements in the tool structure and then sit idle. An attractive way of performing decomposition for SMF simulations with elastic dies is *CONTROL_MPP_DECOMPOSITION_CONTACT_DISTRIBUTE which distributes the model across the cores based on contact interfaces selected by the user. It is recommended to use the contact interface between the blank and the die which will yield a good distribution based on the large number of elements in this interface, this is shown in figure 3B. Using *CONTROL_MPP_DECOMPOSITION_CONTACT_DISTRIBUTE decreases the simulation time by approximately one third for the presented model, from 3 to 2 hours.

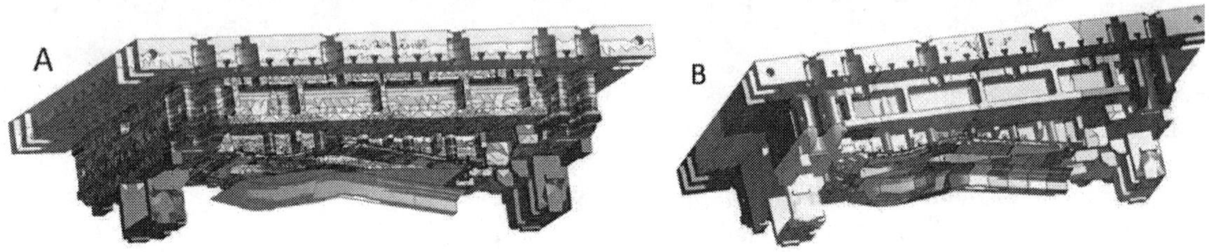

Figure 3. A: Decomposition along the height of the die, for 24 cores. Each core represented by a color. B: Decomposition by contact interface between die and blank.

3.4. *Scaling analysis*

Standard forming analyses in LS-Dyna with rigid tools are normally conducted with 4-8 cores for most car body parts. Simulations with elastic tools are found to benefit from more cores. For the specific part simulated in this example an optimum number of cores seems to be around 20. All simulations here are run with 24 cores, unless other specified. A comparison is therefore done with the rigid die simulation using 24 cores. The result can be seen in figure 4. A normal simulation with rigid dies takes <1 hour. Adding elastic tools increases the simulation time by a couple of hours.

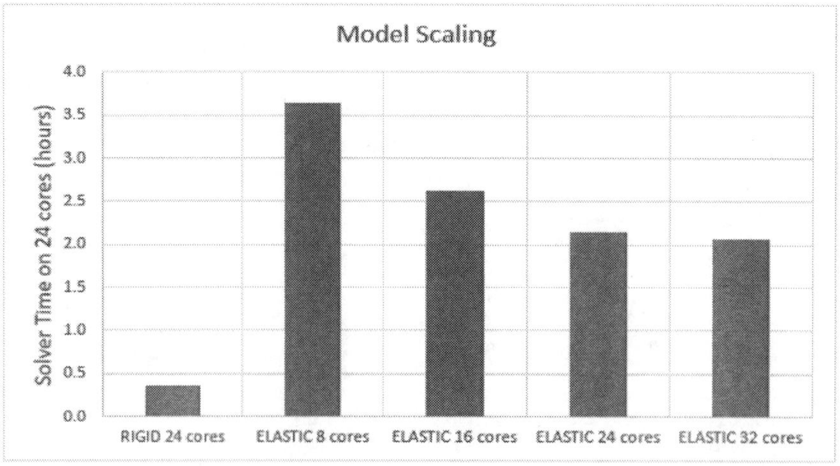

Figure 4. Scaling of FE model. The number of hours required to solve the presented model is plotted versus the number of cores.

3.5. Summary of methods for short simulation times

For a fast and nimble modeling approach, it is efficient to use a coarse solid mesh for the tool structure together with NULL shell elements representing the forming surfaces. If templates are created for LS-PrePost it takes <1 hour in LS-PrePost to set up forming simulations with elastic dies. It is only the meshes, together with material parameters, forces, and velocity curves that need to be replaced.

There is basically just one major difference in preprocessing compared to standard simulations with elastic dies: preparation and meshing of the die solids. For a full-sized stamping die this will take at least an hour to complete. The authors found the possibilities of meshing within the CAD software Catia V5 to be very efficient in meshing the die structures for use in LS-PrePost. It is very important to have standardized and parameterized CAD files for an efficient work flow. For example, it is not desired to include every small drilling hole in the die structure in the meshed die body. There can be hundreds of them in each forming die and removing them individually would be very time consuming and tedious. It is key to have a CAD model where it is possible to deactivate groups of features with a few clicks in the CAD software. This is fortunately common practice in the car industry today.

Artificially reducing the density of the elastic tool parts is the single most important step for reducing simulation times. This singlehandedly makes it possible to reduce the simulations time to a few hours using 10-20 cores. This should be compared to simulation times of tens of hours using hundreds of cores as in [23]. Selecting a reliable decomposition method, and a good number of cores, can then further cut the simulation times by more than 50% [22, 30].

Depending on the desired level of detail in the simulation it is also possible to save time by selecting less detailed material models, fewer integration points in the shell elements, a lower frequency of updates in the contact algorithms, etc. Reducing the detail level in the blank and contacts will still yield very similar results for the tool deformations, even if the response of the blank becomes less detailed. However, it is the described meshing technique, distribution across cores, and density reduction of the tools that mainly influences the simulation time. So, there are no limitations to reach the same accuracy for the blank as in simulations with rigid surfaces.

The focus of this modeling is to describe the global deformations of the die structures. The accuracy of describing these deformations is sufficient due to a large number of elements across the die structures resolving these deformations. However, these models should not be used for studying the effect of local deformations within the die. This is due to the underlying die structure being meshed with a coarse mesh that is not capturing these small deformations, for example inside the draw beads.

4. Conclusion

None of the existing methods in the literature for SMF with elastic dies are fully mature for simulations of industrial sized dies. Either they are missing important features or are too simplified. Or, as for the method presented in this paper, the method is promising but computationally very demanding and requires several modifications to become feasible for use in industry and academia. Important modifications of the selected method in an explicit FE code are to artificially reduce the density of the die solids, and distribute the model in an efficient way across the solver cores. The scaling of mass is not the common mass scaling used in explicit FE codes. Here, the mass of the die solids is reduced by a factor of 100. This cancels unrealistic dynamic effects arising from high velocities in explicit FE simulations. The scaling is still preserving the elastic behavior of the solid structures.

After modifications, the selected method for SMF with elastic dies is deemed to be fast, robust, and powerful in the sense that it provides an easy way for modeling. The method consists of using explicit time integration where the tool parts are meshed with a coarse solid mesh. The forming surfaces are then modeled with a finer shell mesh and attached to the coarse tool mesh by tie contacts with offset, originally proposed by Haufe et al. in [7].

It is easy to update the model by replacing separate parts such as die solids or forming surfaces.

Die deformations are calculated continuously throughout the simulation in one single model.

There is an increase in preprocessing and simulation time compared to using rigid tools, but industrial dies can now be modeled within an hour and solved within a working day.

One of the main criteria in favor of the selected approach is that it can realistically model the blankholder and cushion system in single action presses. This is very important for realistic results and there is no known solution for this if other methods are used. However, the elasticity of the bolster and ram needs to be represented reliably. Methods to measure and represent the bolster and ram have therefore been suggested in the literature. The existing methods all seem promising and will hopefully converge to a standard method within the SMF industry and academia.

The proposed method for SMF simulations with elastic dies can also act as a base model for robustness analyses or digital twins. Carrying out a robustness analysis, through parameter study, is feasible with the current method. It might be worth the effort to reduce the model by the model reduction techniques presented in chapter 2.3. But the most important argument for using these models in parameter studies is that elastic tools can yield a very different result compared to simulations with rigid forming surfaces, exemplified by the model in chapter 4.2.

A close topic of parameter studies is to create digital twins for inline production control where decisions are needed in seconds. In this case, it is needed to build meta-models that represent the behavior of the real production system. An interesting approach is to combine twins based on data and twins based on numerical models into hybrid twins [31, 32].

Models are now available for reliable virtual tryout and production support. The next frontier in the research area of SMF with elastic dies is probably related to topics such as digital factories controlled and monitored by smart digital systems.

Acknowledgements
This research received funding within the project “Reduced Lead Time through Advanced Die Structure Analysis” financed by the Swedish Innovation Agency Vinnova within the “Strategic Vehicle Research and Innovation (FFI) Programme”. Grant number: 2016-03324.

This research received funding within the project “Advanced numerical solutions and experimental deflection measuring devices for an efficient stamping tool cambering (CAMBER)” financed by Vinnova within the SMART EUREKA Cluster program in Advanced Manufacturing. Grant number: Vinnova, 2018-03331.

References

[1] Neto D M, Coër J, Oliveira M C, Alves J L, Manach P Y and Menezes L F 2016 Numerical analysis on the elastic deformation of the tools in sheet metal forming processes. *International Journal of Solids and Structures*

[2] Burkart M et al. 2020 Compensation of elastic die and press deformations during sheet metal forming by optimizing blank holder design *IOP Conf. Ser.: Mater. Sci. Eng. 967 012074*

[3] Grossmann K et al. 2010 Static compensation for elastic tool and press deformations during deep drawing *Prod. Eng. Res. Devel. 4:157–164*

[4] Lingbeek RA 2008 Virtual tool reworking. New strategies in die design using finite element forming simulations *Enschede, The Netherlands, ISBN 978-90-77172-38-4*

[5] Pilthammar J 2020 Towards Virtual Tryout and Digital Twins: Enhanced Modeling of Elastic Dies, Sheet Materials, and Friction in Sheet Metal Forming *Blekinge Institute of Technology, Karlskrona Sweden ISBN 978-91-7295-406-9*

[6] Zgoll F et.al. 2019 Virtual die spotting: Compensation of elastic behavior of forming presses *IOP Conf. Ser.: Mater. Sci. Eng. 651 012021*

[7] Haufe A, Roll K and Bogon P 2008 Sheet Metal Forming with Elastic Tools in LS-DYNA *Numisheet 2008*

[8] Tehel R et.al. 2019 Effective FE models for simulating the elasto-mechanical characteristics of forming machines *The International Journal of Advanced Manufacturing Technology*

[9] Müller S, Müller A, Rothe F. et al. 2018 An Initial Study of a Lightweight Die Casting Die Using a Modular Design Approach *Inter Metalcast 12*

[10] Biba N et al. 2012 The Model for Coupled Simulation of Thin Profile Extrusion *Key Engineering*

Materials

[11] Zhang Q et al. 2006 Numerical simulation of deformation in multi-point sandwich forming *International Journal of Machine Tools & Manufacture*

[12] Hiller M et al 2017 Influence of tool elasticity on process forces and joint properties during clinching with rotational tool movement *J. Phys.: Conf. Ser. 896 012116*

[13] Lu X and Balendra R 2001 Finite element simulation for die-cavity compensation *Journal of Materials Processing Technology*

[14] Lee Y et al. 2002 Analysis of the elastic characteristics at forging die for the cold forged dimensional accuracy *Journal of Materials Processing Technology*

[15] Pilthammar J et.al. 2023 Three Industrial Cases of Sheet Metal Forming Simulations with Elastic Dies *In Proceedings of IDDRG 2023*

[16] LS-Dyna Manual R11.0 2018 *Livermore Software Technology Corporation (LSTC)*

[17] Pilthammar J et.al. 2017 Characterizing the elastic behaviour of a press table through topology optimization. *Journal of Physics: Conference Series (Vol. 896, No. 1, p. 012068). IOP Publishing.*

[18] Pilthammar J et.al. 2021 New press deflection measuring methods for the creation of substitutive models for efficient die cambering *IOP Conf. Ser.: Mater. Sci. Eng.* 1157 012076

[19] Grossmann K et.al. 2014 FE forming models including press behavior allow for realistic computation of blankholder forces *Proceedings of IDDRG 2014 Conference*

[20] Müller P et al 2019 Experimental analysis of the elastic boundary conditions of press machines for modelling the deep-drawing process *The International Journal of Advanced Manufacturing Technology*

[21] Abbasi F et.al. 2022 Substitutive Press-Bolster and Press-Ram Models for the Virtual Estimation of Stamping-Tool Cambering *Materials*

[22] Allesson S 2019 Sheet Metal Forming Simulations with Elastic Dies: Emphasis on Computational Cost *Blekinge Institute of Technology*

[23] Lind M and Sjöblom V 2018 Industrial Sheet Metal Forming Simulation with Elastic Dies *Blekinge Institute of Technology*

[24] Pilthammar J, Sigvant M and Kao-Walter S 2018 Introduction of elastic die deformations in sheet metal forming simulations *International Journal of Solids and Structures*

[25] Del Pozo D et.al. 2008 Prediction of press/die deformation for an accurate manufacturing of drawing dies *Int J Adv Manuf Technol*

[26] Song Y J et al. 2021 Numerically efficient sheet metal forming simulations in consideration of tool deformation *International Journal of Automotive Technology*

[27] Zgoll F et.al. 2021 Virtual die spotting: Advanced setup for coupling of forming and structure simulation *IOP Conf. Series: Materials Science and Engineering*

[28] Choi K Y et.al. 2013 Sheet metal forming simulation considering die deformation *International Journal of Automotive Technology*

[29] Vrolijk M et.al. 2018 A study with ESI PAM-STAMP® on the influence of tool deformation on final part quality during a forming process *AIP Conference Proceedings. Vol. 1960. No. 1. AIP Publishing LLC*

[30] Schill et.al. 2019 Simulation of Sheet Metal Forming using Elastic Stamping Dies *12th European LS-DYNA Conference*

[31] Chinesta F et.al. 2020 Virtual, digital and hybrid twins: a new paradigm in data-based engineering and engineered data *Archives of computational methods in engineering*

[32] Havinga J et.al. 2020 Exploiting data in smart factories: real-time state estimation and model improvement in metal forming mass production *International Journal of Material Forming*

IOP Conf. Series: Materials Science and Engineering 1284 (2023) 012055 doi:10.1088/1757-899X/1284/1/012055

Three Industrial Cases of Sheet Metal Forming Simulations with Elastic Dies

J Pilthammar[1,2], M Sigvant[1,2], M S Islam[2], M Schill[3], S Sjöblom[2], V Sjöblom[2], M Lind[2]

[1] Volvo Cars Dept. 81110 Strategy Development, Olofström, Sweden
[2] Department of Mechanical Engineering, Blekinge Institute of Technology, Karlskrona, Sweden
[3] Dynamore Nordic, Linköping, Sweden

E-mail: johan.pilthammar@volvocars.com

Abstract. Previous research and experience points to many advantages if sheet metal forming is simulated with elastic dies. Some areas that are enabled by simulations with elastic dies are virtual spotting, improved digital twins, and improved production support. A promising method was selected from the literature, and after important modifications it is deemed to be fast and robust for simulating industrial sized dies. The method consists of meshing die solids with a coarse mesh to represent the structural behaviour of the die. The forming surfaces are then represented by a fine shell mesh connected to the solid mesh by tied contacts with an offset. With additional modifications to reduce solver time this yields a robust and flexible way of modelling sheet metal forming with elastic dies. There is an increase in preprocessing and simulation time compared to using rigid tools, but industrial dies can now be modeled within an hour and solved within a working day. It is also easy to update the model by replacing separate parts such as die solids or forming surfaces. One of the main criteria in favor of the selected approach is the realistic modeling of blankholder and cushion systems. In this paper simulations of three industrial cases are demonstrated: one case of virtual die spotting and two cases of production support. The three cases demonstrate the importance and potential of using elastic dies during virtual die tryout, production support, and for cases like digital twins and production control.

1. Introduction

Previous research and experience points to many advantages if sheet metal forming is simulated with elastic dies [1-8]. Some areas that are enabled by simulations with elastic dies are virtual spotting, improved digital twins, and improved production support. However, simulations with elastic dies have previously been very demanding and computational costly. In [9] an overview is presented of simulation methods incorporating elastic dies. This paper presents three industrial cases simulated by the recommended, and further developed, method in [9]. The method consists of meshing die solids with a coarse mesh to represent the structural behaviour of the die. The forming surfaces are represented by a fine shell mesh connected to the solid mesh by tied contacts with an offset. The surface mesh is present just to represent the geometry of the forming surfaces, it consists of a NULL material without any mechanical properties. Apart from the meshing strategy the density of the die solids is artificially reduced to minimize dynamic effects, and the distribution of the model across solver cores is controlled to minimize the solver time. All of this generates a flexible and robust simulation method and also enables a realistic representation of blankholders and hydraulic cushions.

The first case in this paper demonstrates a virtual tryout of a blankholder. In the second case, chapter 2.2, the importance of blankholder deformations is demonstrated. Effects are found that cannot be seen with rigid or constrained blankholder models. The same is true in the example in chapter 2.3 where hydraulic cushions start to tilt during forming operations. There are no other existing methods available to represent these dynamics which can have a profound impact on the forming operation process control and die development process.

Models throughout this paper are created in LS-PrePost V4.6.0 and solved with the explicit FE software LS-Dyna R11.1.0 [16]. Keywords and setup of models throughout this paper are referencing this software.

2. Three Case Studies: Applications of SMF with Elastic Dies

2.1. Case 1: Virtual try-out of a double-action blankholder

In this first case, the focus is on a double-action blankholder. Double action blankholders are notoriously known for large deformations. This is because the forces are applied by an outer ram, then transduced by a plate to a forming area that can be quite far from the ram.

2.1.1. Step 1: Simulation with rigid tools. This case shows an outer rear door manufactured in a double-action press. Symmetry is utilized in the model, two doors are manufactured in each stroke. Once the blankholder is closed displacement control is turned off, instead, a blankholder force of 1300kN is applied.

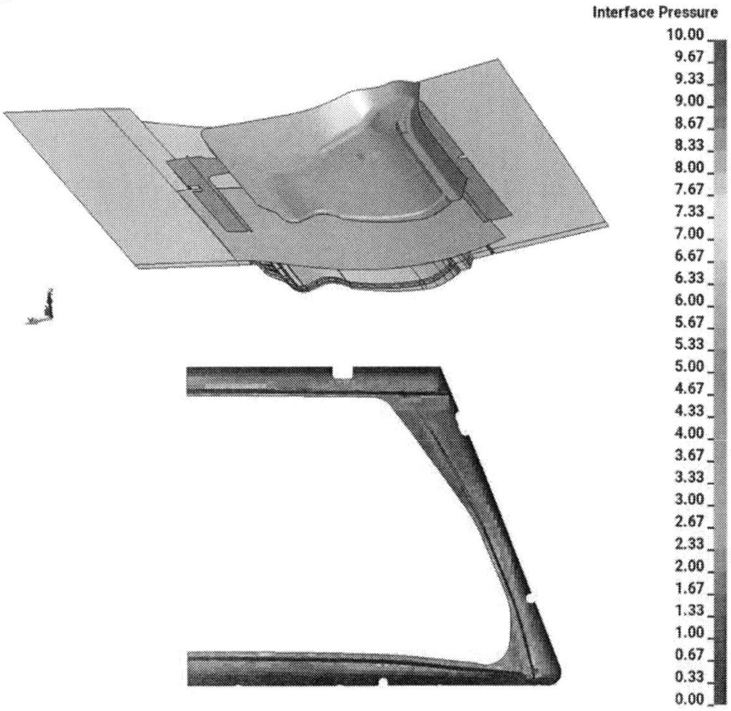

Figure 1. Blankholder pressure at blankholder closing

The initial step in the virtual die try-out is to run a standard simulation with rigid tools. The user then selects which point in time to use as a reference for the tool deformation. Here a point is selected just at the beginning of the stroke when the punch reaches the blank, this will give the tool deformation at a point that is called blankholder closing. The pressure at this point is what we later would like to achieve when the die is compensated. It can be discussed what point in time that should be chosen, currently this point was selected because the tool shop starts with blankholder spotting at this point of the stroke.

The simulation with rigid tools is a standard forming simulation in LS-Dyna. The blank material is a CR210BH simulated with *MAT_BARLAT_YLD2000. The blank has a thickness of 0.7mm. Tool velocity of the moving parts is 2000 mm/s, with a timestep for the mass scaled solution of 1.2e-6 seconds. Contact between sheet and tool is facilitated by *CONTACT_FORMING_ONE_WAY_SURFACE_TO_SURFACE. The initial element size of the sheet is 10mm, consisting of Belytschko-Tsay elements (mainly square 4-node elements, with some triangle elements). Adaptivity with three refinement levels is included in the model. The tools are meshed with a maximum elements size of 15mm, the minimum element size is 1mm, and a max deviation of 0.1mm.

2.1.2. Step 2: Deforming the blankholder and blankholder plate. The resulting pressure is converted to a pressure load, which is easily done by exporting results from LS-PrePost and changing the format in software such as Excel or MATLAB. A simulation is then conducted with just the NULL surfaces, the solid blankholder, the blankholder plate, and the ram which is held fixed. This gives a deformation that corresponds to the amount of compensation that is needed in the virtual die try-out. The blankholder is modeled as nodular iron with Young´s modulus of 173 GPa and a Poisson´s ratio of 0.25.

Presented simulations with elastic blankholder consist of the same shell meshes for the forming surfaces as previous. The outer ram in the double-action press is modeled as a rigid surface. The solid blankholder is meshed with tetrahedron elements in Catia V5 and transferred to LS-PrePost. The minimum element size is 10mm, with maximum element sizes of around 50mm. The maximum allowed deviation from geometric surfaces (sag in Catia), is set to 0.1mm. The blankholder is in contact with the ram and can separate except around the attachment slots where the blankholder is fastened to the ram, there groups of nodes on the blankholder are rigidly connected to the ram. The ram is driving the blankholder at 2000 mm/s.

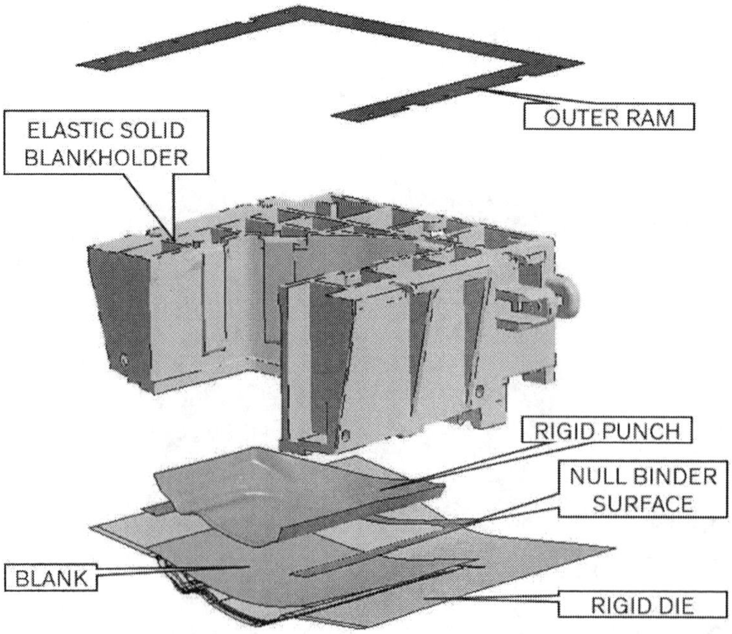

Figure 2. Exploded view of the model with elastic blankholder.

IOP Conf. Series: Materials Science and Engineering 1284 (2023) 012055 doi:10.1088/1757-899X/1284/1/012055

2.1.3. Step 3: Compensating the blankholder surface. The resulting deformation in Step 2 is exported and used to compensate the NULL shell elements. Each node on the NULL surfaces is moved the calculated distance in Figure 3, in the opposite direction. The resulting pressure distribution at blankholder closing with uncompensated surfaces is visualized in the top part of Figure 4. Contact is lost in the middle of the blankholder when no compensation is made. Using the compensated forming surfaces, depicted in Figure 4, shows that pressure in the middle of the tool increases again. The distribution doesn´t fully return to the one in Figure 1, this can be due to several reasons as slight non-linearities in the model and other elastic and dynamic effects. More compensation loops can probably achieve a converging solution towards the pressure in Figure 1. However, the compensation has a significant impact on the blankholder pressure distribution. There is a large difference in pressure in the corners of the blankholder, areas where a lot of grinding traditionally is applied. It is estimated in the tool shop at Volvo Cars that this type of virtual try-out cuts the spotting time by at least 50%, sometimes much more.

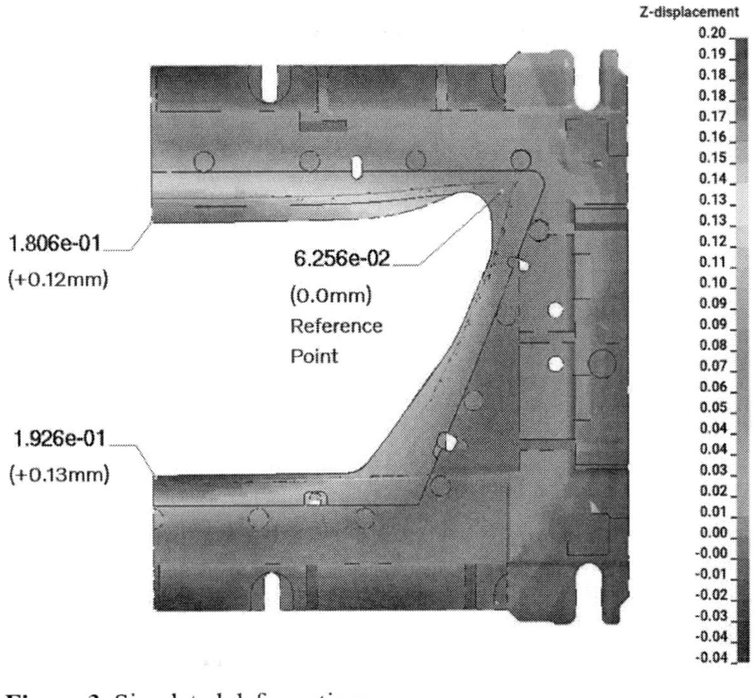

Figure 3. Simulated deformations

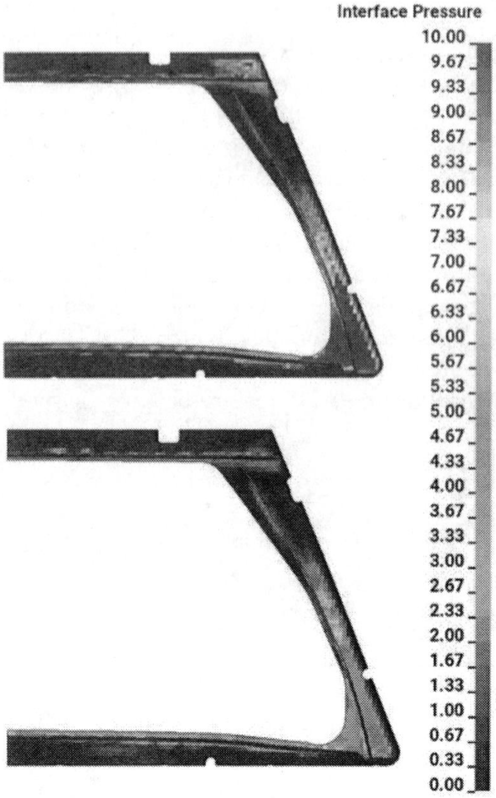

Figure 4. Pressures with elastic blankholder. Before and after compensation.

2.2. Case 2: Production support through simulations for a part non-responsive to changes in blankholder force

Issues were reported for an inner structural car part manufactured from 0.7mm thick CR4 steel. Different types of issues were present from time to time in the circled areas in Figure 5, wrinkles, cracks, etc. However, these issues shouldn´t be impossible to solve by adjusting the blankholder force. In this specific case, it was found that it was near impossible to get any response in these areas by adjusting the blankholder force. Standard SMF simulations predicted that an increased blankholder force would yield a relatively evenly distributed linear response: higher strains and less draw-in around the whole part. This was not observed in the real die.

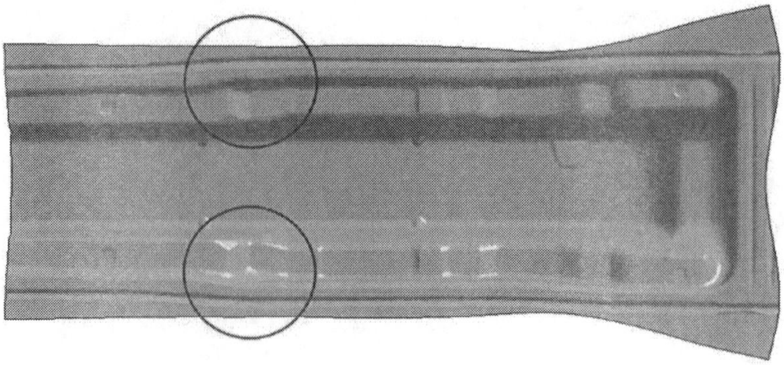

Figure 5. Areas with issues in the car part.

IOP Conf. Series: Materials Science and Engineering 1284 (2023) 012055 doi:10.1088/1757-899X/1284/1/012055

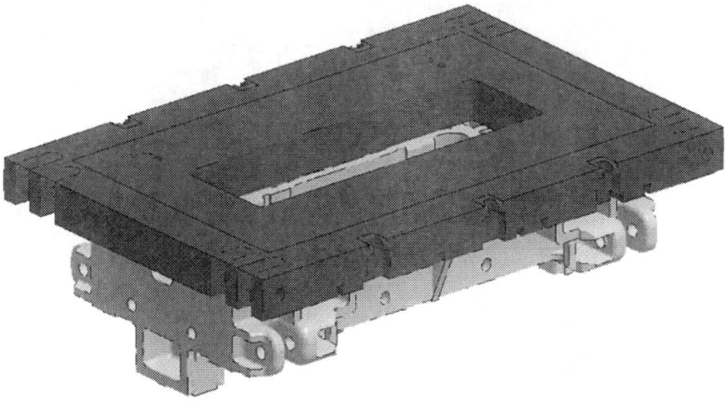

Figure 6. SMF model with elastic blankholder and blankholder plate.

For a deeper analysis of the problem, SMF simulations with elastic tool parts were utilized. A double-action die was used for producing this part, from experience it was known that the blankholder deformations can have quite a large impact on the produced part. The blankholder forces are bending the blankholder, thereby reducing contact pressure in the middle of the blankholder. Blankholder and blankholder plate were therefore modeled as elastic structures in these simulations. The setup of the model is the same as in the previous case 1, just different tool geometries and an added blankholder plate in steel (brown in Figure 6). Figure 7 depicts the difference in blankholder pressure distribution when an elastic blankholder is used in the simulation. Nominal force is simulated as well as a simulation where the force is increased by 20%. Increasing the blankholder force distorts the blankholder pressure by moving it outwards towards the short edges of the blank. In Figure 8, the difference in major strain between the two simulations with different force levels is visualized. On average the strains are increasing throughout the part. However, the simulation with an elastic blankholder shows a very minor influence on the problematic area. This confirms what was observed in the stamping plant, that it is almost impossible to influence this area of the part by adjusting blankholder force. This happens because a deforming blankholder increases the pressure at blank edges when the blankholder force is increased. It is also lowering the pressure in the middle of the blank. Right around the problematic area barely anything happens, this is a neutral area with neither increase nor decrease in the blankholder pressure. Shimming or adjusting the blankholder geometry is thereby the only options available to influence the part here. These options are both more time-consuming and expensive than just adjusting the blankholder force.

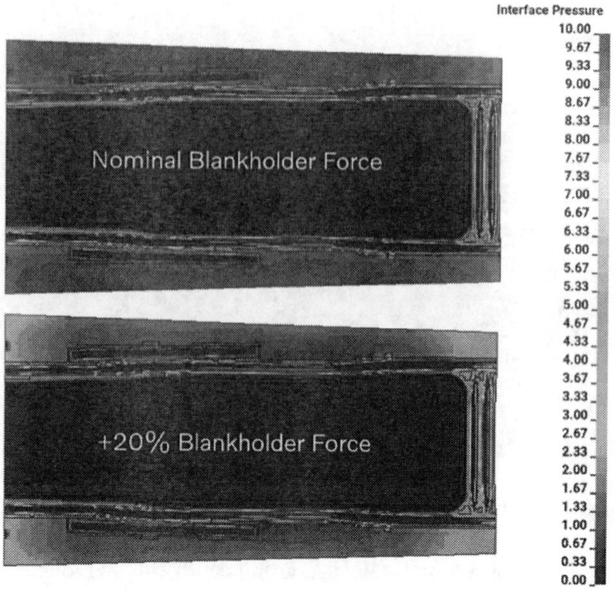

Figure 7. Predicted blankholder pressures with elastic blankholder and blankholder plate.

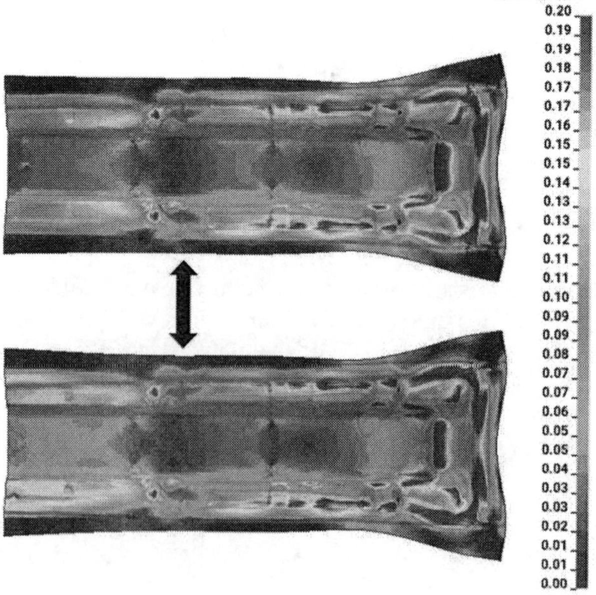

Figure 8. Differences in major strain between nominal and increased blankholder force.

2.3. Case 3: A single die in a dual cushion press

Single action press lines can have more than one hydraulic cushion in the forming press. Two cushions could be used for producing two parts in one press with one press stroke. By using two cushions the blankholder pressure can be controlled individually for the two parts. There is also the possibility to place a single larger die in the press. However, a die that isn´t large enough risks tilting the hydraulic cushions inwards towards the center of the press. This will severely distort the blankholder pressure and risk damaging the cushion guiding rails.

With the method selected and presented in this paper for SMF simulations with elastic dies, it is possible to simulate these kinds of cases. The different parts in the model will hold each other in place with this method, preventing rigid body motions when deformations are calculated. It is a major strength of this approach compared to others presented in the literature survey. The case here with two cushions and one die might not be very common in real life, but similar issues like finding an evenly distributed blankholder pressure are critical in any forming die. Asymmetric part shapes in SMF are very common causing challenges in die design and predicting cushion settings. These kinds of simulations can help set individual cushion cylinder pressures, optimize die design, optimum placement of the die on the press table, etc.

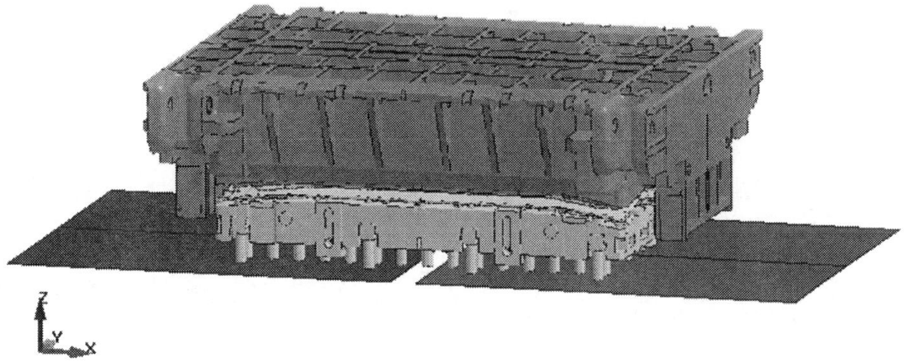

Figure 9. Simulation model of a single die in a press with two hydraulic cushions.

With the selected method for simulating SMF with elastic dies, it is possible to simulate e.g., the model presented in Figure 9. Here the tool parts are modeled as nodular iron. The ram is assumed to be rigid. The two cushions are modeled as two rigid surfaces, but they are free to move in the z-direction and rotate around the x- and y-axes. This is not possible with the method where two models are coupled, presented in chapter 2.4. With the method presented in chapter 2.4, it is possible to realistically simulate the deformations of matrix and punch because these parts are constrained in space in each coupling step. However, the boundary conditions needed for the cushion system make it impossible to realistically calculate the displacements and deformations of the blankholder and cushions. When the approach of coupled models is used the forming forces are extracted from an SMF model and then applied to a structural FE model to calculate and update the shape of the forming surfaces before forming continues. The cushions will in this case be loaded by a hydraulic force from underneath and by the contact pressure between the blank and blankholder from the top. There are no possibilities to realistically constrain this structure in space, both rigid body moments and rotations will cause the model to diverge and thereby making it impossible to solve. See the example of realistic boundary conditions in Figure 10. An example of how the pressure on the blank becomes distorted is seen in Figure 11, here a very slight tilting of the hydraulic cushions places all the cushion forces on the outermost row of blankholder pins, marked as red in the schematic Figure 10.

In conclusion, with coupled models, the blankholder needs to be artificially constrained. For example, by locking all the bottom nodes of the blankholder pins. This prevents both internal blankholder deformations and realistic cushion tilting. Using the method suggested in this paper realistic boundary conditions can be utilized, the blank with support of the matrix will hold the blankholder in a correct position.

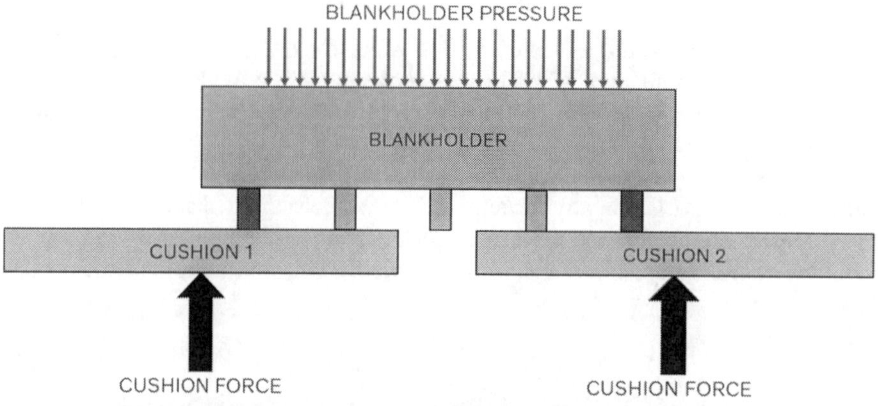

Figure 10. Blankholder and cushion system with realistic boundary conditions.

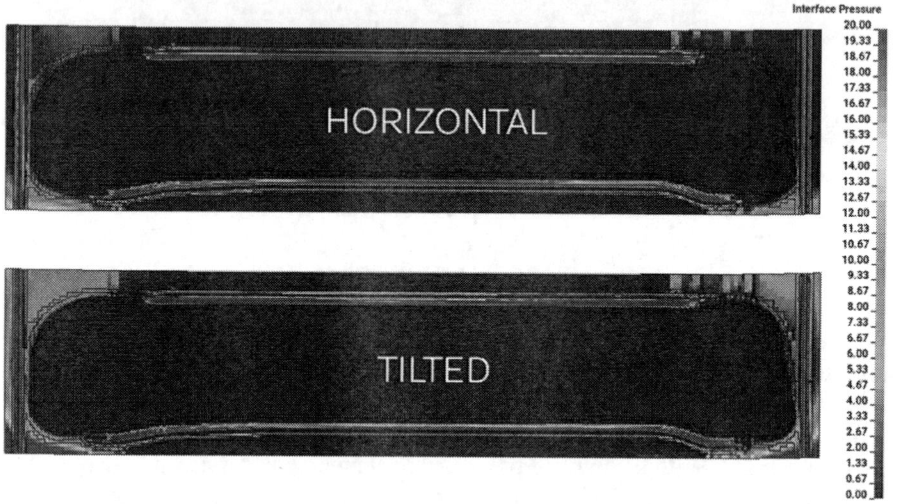

Figure 11. Distortion of blankholder pressure due to tilting of the two cushions.

3. Conclusion

A try-out was demonstrated for a double-action blankholder in chapter 2.1. The method is straightforward to set up and solve, and it yields reliable results. It is estimated in the tool shop at Volvo Cars that this type of virtual try-out cuts the spotting time by at least 50%, sometimes much more. Considering that hundreds of hours can be spent spotting each die this is a significant step forward. Some reflections on die spotting are that punch and die should be compensated at the bottom of the stroke since that is the position with maximum force and final geometry of the part. However, it will be important to gain practical experience for what position during the stroke that should be used for blankholder surface compensation. Here blankholder closing was used, other options could be considered such as blankholder closing or an average of the deformations during the stroke.

Apart from the die try-out the suggested method can be used for analyzing existing dies as in [10-12]. This has the potential to be very helpful during production support where scanned die surfaces can easily be included by the demonstrated method. The demonstrated cases of production support in chapters 2.2 and 2.3 of this paper demonstrate more benefits of being able to simulate sheet metal forming with elastic die parts and press dynamics. Apart from production support this will also be important creating digital twins for process analysis and control. Many of the behaviors seen in these simulations cannot be represented with rigid forming surfaces and will not be represented in these digital twins unless elastic dies are use.

Acknowledgements
This research received funding within the project "Reduced Lead Time through Advanced Die Structure Analysis" financed by the Swedish Innovation Agency Vinnova within the "Strategic Vehicle Research and Innovation (FFI) Programme". Grant number: 2016-03324.

This research received funding within the project "Advanced numerical solutions and experimental deflection measuring devices for an efficient stamping tool cambering (CAMBER)" financed by Vinnova within the SMART EUREKA Cluster program in Advanced Manufacturing. Grant number: Vinnova, 2018-03331.

References
[1] Neto D M, Coër J, Oliveira M C, Alves J L, Manach P Y and Menezes L F 2016 Numerical analysis on the elastic deformation of the tools in sheet metal forming processes. *International Journal of Solids and Structures*
[2] Burkart M et al. 2020 Compensation of elastic die and press deformations during sheet metal forming by optimizing blank holder design *IOP Conf. Ser.: Mater. Sci. Eng. 967 012074*
[3] Grossmann K et al. 2010 Static compensation for elastic tool and press deformations during deep drawing *Prod. Eng. Res. Devel. 4:157–164*
[4] Lingbeek RA 2008 Virtual tool reworking. New strategies in die design using finite element forming simulations *Enschede, The Netherlands, ISBN 978-90-77172-38-4*
[5] Pilthammar J 2020 Towards Virtual Tryout and Digital Twins: Enhanced Modeling of Elastic Dies, Sheet Materials, and Friction in Sheet Metal Forming *Blekinge Institute of Technology, Karlskrona Sweden ISBN 978-91-7295-406-9*
[6] Zgoll F et.al. 2019 Virtual die spotting: Compensation of elastic behavior of forming presses *IOP Conf. Ser.: Mater. Sci. Eng. 651 012021*
[7] Haufe A, Roll K and Bogon P 2008 Sheet Metal Forming with Elastic Tools in LS-DYNA *Numisheet 2008*
[8] Tehel R et.al. 2019 Effective FE models for simulating the elasto-mechanical characteristics of forming machines *The International Journal of Advanced Manufacturing Technology*
[9] Pilthammar J et.al. 2023 An overview of Methods for Simulating Sheet Metal Forming with Elastic Dies *Proceedings of IDDRG23*
[10] Abbasi F et.al. 2022 Substitutive Press-Bolster and Press-Ram Models for the Virtual Estimation of Stamping-Tool Cambering *Materials*
[11] Allesson S 2019 Sheet Metal Forming Simulations with Elastic Dies: Emphasis on Computational Cost *Blekinge Institute of Technology*
[12] Vrolijk M et.al. 2018 A study with ESI PAM-STAMP® on the influence of tool deformation on final part quality during a forming process *AIP Conference Proceedings. Vol. 1960. No. 1. AIP Publishing LLC*

Displacement based Simulation and Material Calibration based on Digital Image Correlation Part II - Application

C Ilg[1], C Liebold[2], V Sreenivasa[1], A Haufe[2], C Karadogan[3] and M Liewald[3]

[1] DYNAmore GmbH – an Ansys Company, MCC – Material Competence Center, Kolumbusstraße 47, 70771 Leinfelden-Echterdingen, GER
[2] DYNAmore GmbH – an Ansys Company, Industriestraße 2, 70565 Stuttgart, GER
[3] Institute for Metal Forming Technology (IFU), University of Stuttgart, Holzgartenstraße 17, 70174 Stuttgart, GER

E-mail: Christian.ilg@dynamore.de

Abstract. An alternative approach to calibrate material models is proposed. The entire displacement field of a specimen measured with a DIC system was converted into boundary conditions by the Envyo® mapping software. These boundary conditions were imposed on all nodes in the corresponding LS-DYNA® simulation model. In this way a digital twin of the specimen geometry and the experimental kinematics was obtained. Considering the local stress equilibrium over the entire specimen and the equilibrium of global force integrated over many hypothetical cross sections, LS-OPT® was used to iteratively determine the isotropic model parameters of the tested material. A new tensile specimen geometry is introduced to get the most out of a single test. The findings suggest that, at present, convergence in iterative parameter optimization cannot be attained without prior filtering and smoothing of the DIC input data in time and space, despite the utilization of the chosen material model.

1. Introduction

Finite element analysis (FEA) has become an indispensable tool for predicting the mechanical behavior of structures. One of the most important inputs for a reliable prediction are accurate material models. Calibration of material models for metallic materials presents a number of challenges involving all types of nonlinearities, directional, history, and rate dependencies of material behavior. Optical deformation measurement systems that employ digital image correlation (DIC) techniques are often used to calibrate material models as they can capture the full-field deformation of a tested specimen. Avril et al. [1] wrote a valuable review on DIC techniques. Techniques such as Finite Element Model Updating (FEMU) do not require a complete full-field measurement of the specimen, but rather utilize optimization tools to minimize the difference between the predicted and measured values. Güner et al. [2] successfully employed this approach to fit yield locus parameters using three waisted tensile specimens. Brosius et al. [3] introduced an approach for calculating stresses based on measured full-field strain data. This approach requires a definition of both yield locus and flow curve. Stresses are computed using the associated flow rule of plasticity. The integral of stresses at continuous sections must equal the global force at any time and at any cross-section of the specimen if the parameters for the yield locus and flow curve are correct. An inverse optimization strategy is also required here. Ilg et al. [4] and Stander et al. [5] proposed the full-field calibration (FFC) approach, which combines the force values with the strain curves and uses the difference between the multi-history curves from the experiment and the FEMU iterations as the optimization basis for parameter fitting.

In this paper, we suggest a different method for calibrating material models using DIC systems. Our approach involves using the entire displacement field measured by the DIC system as a boundary condition in FEA. This method provides several advantages compared to the approaches mentioned earlier. With this method, the local consistency of the internal force balance at the vicinity of each node can be utilized as a constraint in the optimization process. Moreover, the traditional constraint of comparing the global force with the integral of stresses at different continuous sections remains a target function that can be utilized by the optimization environment.

The remaining sections of the paper will provide a detailed description of the proposed approach and showcase its effectiveness. The research also introduces a new tensile specimen geometry and an unordinary way of testing. The results of this study in the field of material model calibration will help improve the accuracy of FEA predictions.

2. Methodology

To demonstrate the strengths of the method, the first step was to develop a novel tensile specimen geometry that can induce heterogeneous stress states to make the deformation field sensitive to as many material parameters as possible. To achieve this, various cutouts in the specimen geometry, their arrangement, and size were iteratively tested and optimized using the LS-DYNA solver and the LS-OPT optimization software [6]. The new specimen geometry, as shown in Figure 2, is intended to represent concurrently various stress states during the test.

The new specimen geometry was cut out from a DP800 sheet with a thickness of 1.5 mm by waterjet cutting. Tensile tests were conducted using a universal tensile testing machine equipped with a digital image correlation (DIC) system to measure the full displacement field of the specimen during deformation. The measured global force values were used to calibrate the von Mises material model (MAT_24 in LS-DYNA). All these steps are explained below in subsections.

2.1. Specimen geometry optimization

The objective of the optimization was to find a specimen geometry that provides high plastic strains under as many different stress states as possible. The diversity of different stress states can be well represented by the stress triaxiality, which is defined by the ratio of hydrostatic stress and von Mises equivalent stress (1):

$$\eta = \frac{\sigma_m}{\bar{m}} \qquad (1)$$

This information - the evolution of plastic strains and stress triaxiality – on a simulated specimen can be extracted by LS-PrePost® for each shell element over the entire simulation. Subsequently, this data can be processed through a Python script and can be used as optimization criterion in LS-OPT. Figure 1a shows such a graph of plastic strain vs triaxiality for the optimized sample geometry. The color scale represents the frequency of data points in this area – red meaning many data points, blue on the contrary few data points. The colored pixel data is then converted into black and white pixel data (see Figure 1b) and so the proportion of the black area to the white unfilled area can be calculated. This way, the percentage occupancy of the zone of interest in the graph can be calculated. This occupancy value is then used as the optimization criterion in LS-OPT.

IOP Conf. Series: Materials Science and Engineering 1284 (2023) 012056 doi:10.1088/1757-899X/1284/1/012056

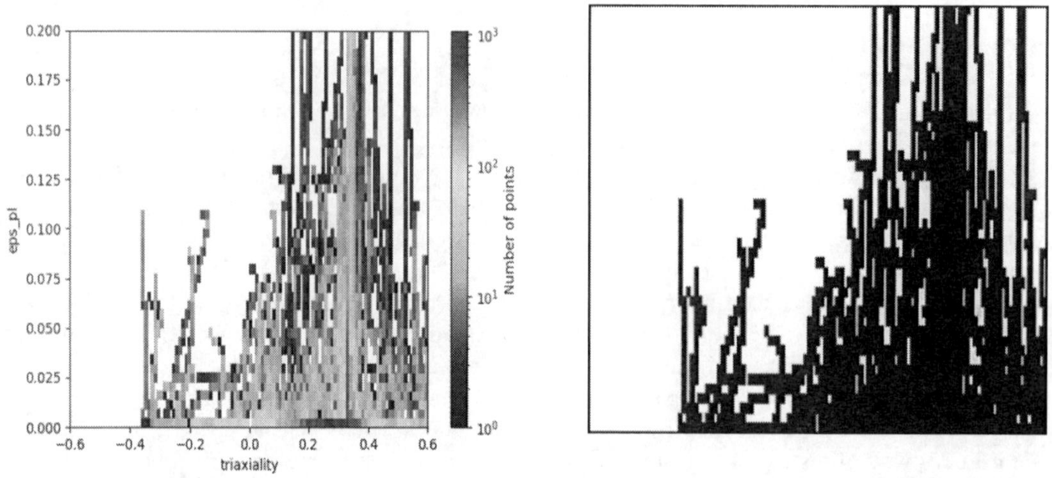

a. With intensity scale. **b.** Black-white plot used for optimization

Figure 1. Plastic strain vs triaxiality diagrams used for specimen geometry optimization.

Scripting in LS-PrePost was employed to generate parameterized specimen geometries with cutouts. These parameters were then passed on to LS-OPT to be optimized using the previously discussed criteria. During the so called *Sequential with Domain Reduction* - metamodel based optimization process, several possible tensile specimen geometries were evaluated, and it was found that specimen with L-shaped cutouts result in high plastic strains and a wide distribution in stress triaxiality. Figure 2 shows the optimized sample geometry with 8 L-shaped cut-outs modelled with a course mesh.

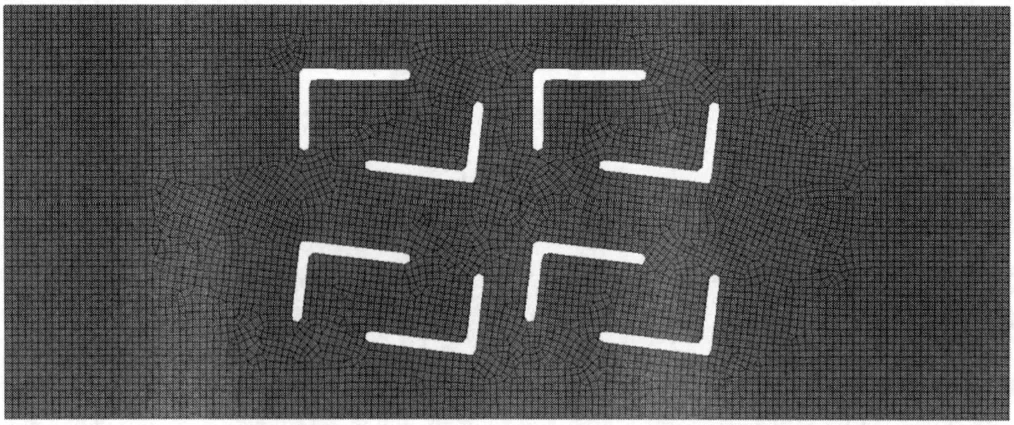

Figure 2. Optimized sample geometry with 8-L-shaped cut-outs modelled with a course mesh.

Figure 3 shows the distribution of plastic strains during the loading of the sample. Locally, plastic strains of up to approximately 0.7 can be achieved. Figure 4 shows the distribution of stress triaxialities. Areas colored in blue correspond to a pressure-like loading of the shell elements in their plane, while areas colored in red represent a biaxial loading.

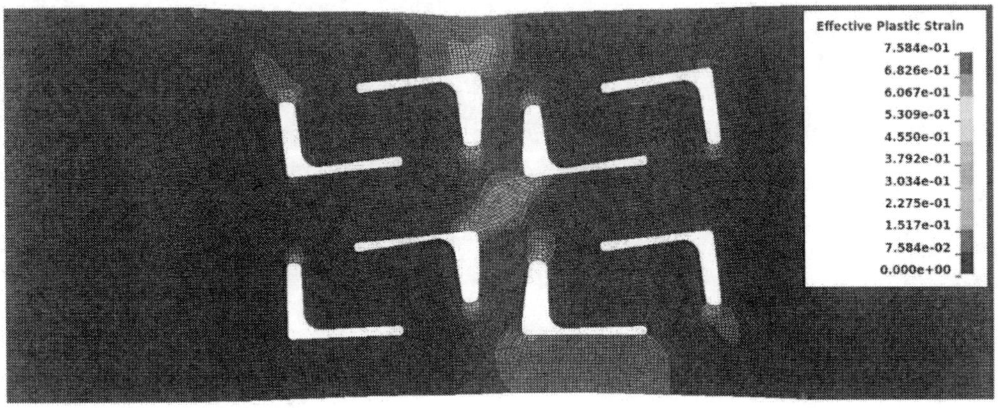

Figure 3. Distribution of plastic strains during the loading of the sample.

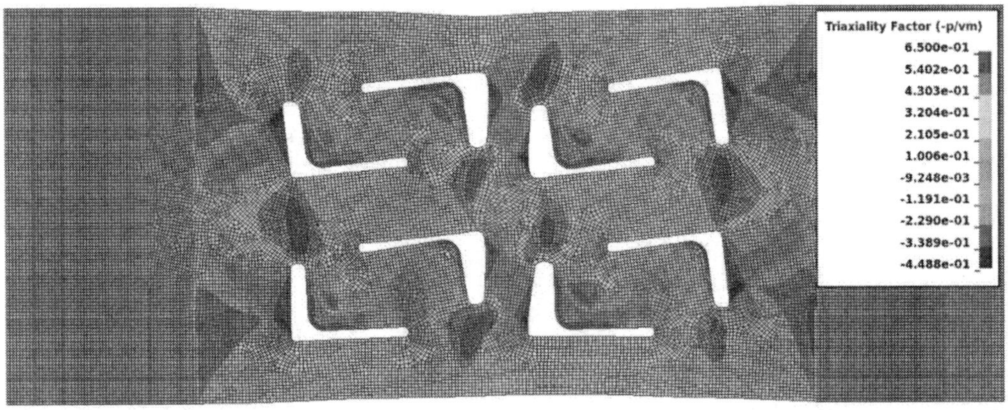

Figure 4. Distribution of stress triaxialities during the loading of the sample.

2.2. Material parameter identification strategy using the novel specimen geometry

The iterative inverse parameter identification strategy is performed using LS-OPT to find the material model parameters. The theory and software development for this method is presented by Liebold et. al. 2023 [7]. By using the measured displacement field during the tensile test, a strategy for the experimental extrapolation of the flow curve parameters will first be developed. This will be achieved by the optimization of the flow curve entered in the von Mises material model implemented as MAT_24 in LS-DYNA [8]. This specific material model is based on J2 elasto-plasticity in combination with strain rate and temperature dependency of flow curves typically represented in tabulated form. The initial estimation of the flow curve is based on the true stress versus true plastic strain diagram [4]:

$$\sigma_y(\varepsilon_{pl})=A-Be^{(-c\varepsilon_{pl}^{n})} \tag{2}$$

Preserving the experimental yield stresses up to diffuse necking and providing C^1-continuity only the following yield stress values are calculated by the Hockett-Sherby hardening law. In this extrapolation the parameters *A* and *B* in (2) are excluded from the variable parameter list of the optimization and only the c and n are used as variables to be optimized. Therefore, the numerical optimization procedure (based on metamodels built with the feed-forward neural network method) implemented in LS-OPT considers the parameters *c* and *n* as optimization variables. Figure 5 depicts schematically the creation of the flow curve, where the blue curve represents a typical engineering stress-strain curve with an elastic segment at the beginning, obtained from an experimental tensile test. The black curves above demonstrate the various shapes of the resulting flow curves with different combinations of parameters. The figure demonstrates the conversion into true stress-strain values by removal of the elastic strains, and the calculation of the true stresses for the phase of uniform elongation, and the subsequent C^1-continuous extrapolation using the Hockett-Sherby approach as explained above.

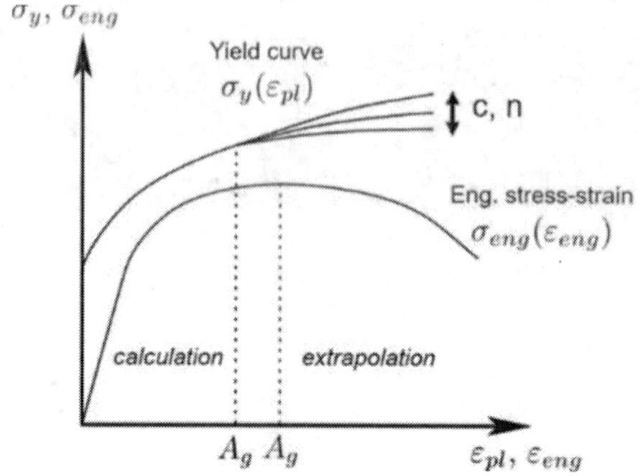

Figure 5. Relationship between the engineering stress-strain curve and the flow curve with a C1-continuous extrapolation using Hockett-Sherby hardening law.

The two parameters *c* and *n* used in the modified Hockett-Sherby extrapolation approach of the yield curve are optimization variables; they are both constant within the discretized domain. After each iteration a comparison between the experimental global force and simulative forces of different cross sections trough the FE mesh were made. The goal of the optimization was to find a yield curve which is able to provide identical global tensile force-time curves from the simulation compared to the experimental counterpart.

Figure 6 shows the measured displacement field from the experiment converted into nodal displacements as boundary conditions in LS-DYNA input deck – building the digital twin of the displacement field in the LS-DYNA environment. Using the direct conversion, the mesh contains only triangular shell elements, actual node-coordinates of which correspond to the ones that are obtained from DIC measurements. The graph also shows the defined digital cross sections of the finite element package where the global forces in the simulation are computed. These integrated forces are used for the optimization procedure thereafter.

Figure 6. The digital twin: measured displacement field from the experiment converted into a LS-DYNA input deck. The mesh contains only triangle shell elements where the nodal coordinates correspond to the ones, which are obtained from the optical displacement computation. The digital cross sections are depicted as black circles.

Using a sufficiently close initial guess for the yield surface and the flow curve, the explained optimization approach above should yield the two variable parameters of the Hockett-Sherby parameters c and n converged to their optimal values. In a broader sense this leads to the conclusion that the model that is set up shall contain enough degrees of freedom within its spatial and physical representation of the problem to find a suitable set of parameters that allows correlation to the experiment. Clearly, this imposes, that an optimization run with an isotropic constitutive model may fail for a material with orthotropic behavior.

3. Optimization result

The comparison of strain fields (Figure 7 and Figure 8) during the simulation shows that the vector displacement fields specified as boundary conditions in the simulation match those of the experiment. Even missing measurement spots (elements) are simulated accordingly since the nodal values are transferred properly.

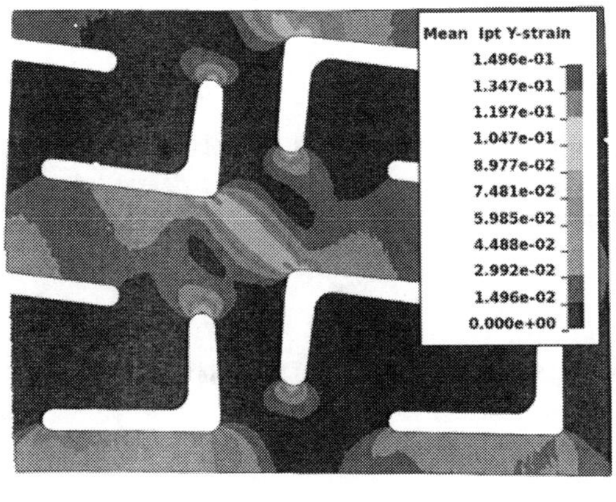

Figure 7. Distribution of the y-strains in the experiment.

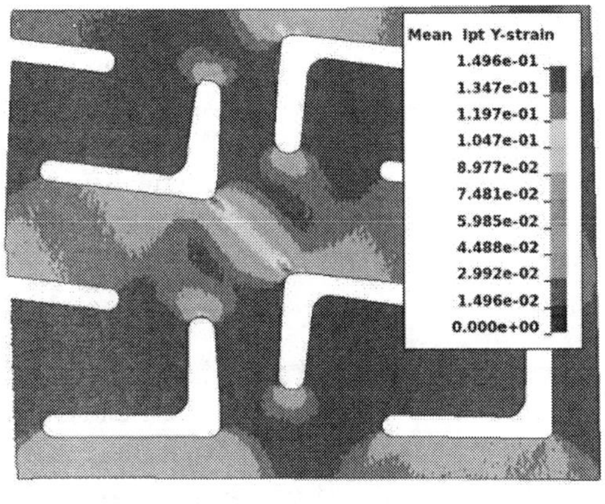

Figure 8. Distribution of the y-strains in the simulation.

The limits for the variables c and n were chosen such that non-physical flow curves are avoided. However, such values are targeted during the optimization process. A comparison of the global forces obtained from the simulation and the experiment shows – unfortunately – poor agreement.

To understand the cause of the poor optimization result, a simplified example using an A80 tensile test specimen is considered: Figure 9 shows the part of the FE mesh of the displacement field measured during the experiment from the gauge section, which was transferred to the simulation as explained in

IOP Conf. Series: Materials Science and Engineering 1284 (2023) 012056 doi:10.1088/1757-899X/1284/1/012056

the previous sections. The considered five different cross sections are depicted; each one is used to determine the global force during the simulation. The calculated forces are shown in the diagram in Figure 10. The black curve represents the force-time curve as recorded in the experiment. The orange curves show the force-time curves obtained from the considered cross sections. Despite the identical displacement field reproduced in the simulation, the resulting force curves exhibit a strong zigzag pattern. Nevertheless, the average value of force is relatively well reproduced for the phase of uniform elongation, where the flow curve is accurate. After uniform elongation, however, the simulated curves deviate significantly from the experimental curve.

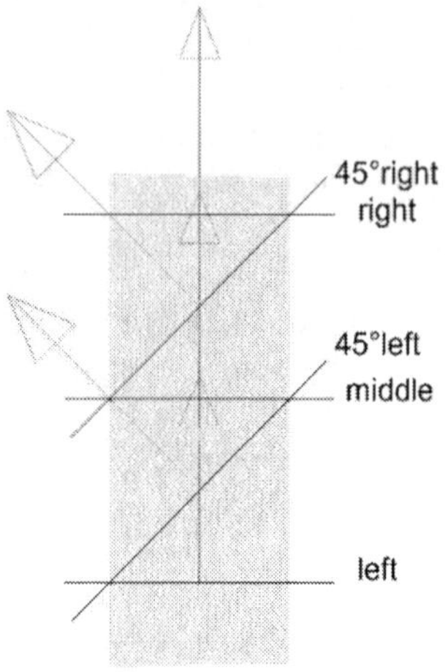

Figure 9. The digital twin of an A80 tensile test specimen.

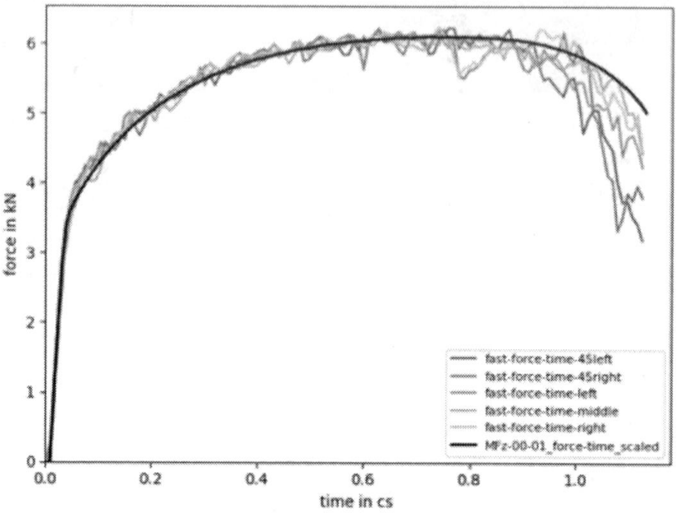

Figure 10. Force-time curve of the experiment in black and force-time curves obtained from the cross sections of the digital twin in orange.

3.1. Reasons for the deviations

Liebold et al. [7] has shown that the method, when based on simulative data only, works as expected. This leads to the conclusion that the possible causes for the poor results may be attributed the noisy test data, the simplified constitutive model, or the spatial discretization. The following sections discuss the possible causes in more detail.

3.1.1. Material model

The plasticity material model used in the simulation used a von Mises yield criterion, which – by definition – only represents isotropic material behaviour. In the present case, 7his limited capability of the material model may be a contributing factor to poor simulation results, especially in cases where the real material exhibits even the slightest anisotropic behaviour. To improve the accuracy of the simulation results, more advanced constitutive models that are able to account for the anisotropic behaviour should be used. His topic will be addressed in future work.

3.1.2. Noise in test data

Another factor that may lead to the observed poor simulation results is seen in the noise of the experimentally measured DIC data at the surface only. Experimental optical measurements are rarely perfect and may have inevitable errors and fluctuations, one of which can be seen in the fact that the measurement is on the surface, the spatial discretization represent the mid-surface of the structure. If these errors are not correctly accounted for, especially past necking, it is assumed that they lead eventually to the inaccurate simulation results in the shown digital twin. Here we are using the DIC data unmodified, while filtering and smoothing in time and space may be a first step to circumvent the shown problems. For instance, it was observed that the noise in the displacement measurements lead to inaccurate estimates of stress values, which in turn lead to incorrect predictions of the subsequent material behaviour.

Another factor that can lead to poor simulation results is noise in the experimental measurement data. Experimental optical measurements are rarely perfect and may have inevitable errors and fluctuations. If these errors are not correctly accounted for, they can lead to inaccurate simulation results in the digital twin which is using this information without any filtering of the data. For example, noise in displacement measurements can lead to inaccurate estimates of stress values that can lead to incorrect predictions of material behaviour.

3.1.3. Spatial discretization
Furthermore, it is assumed that the shell discretization is the cause for another factor that contributes to the shown poor optimization results. The chosen classical 5-parameter shell formulation (Reissner-Mindlin-kinematic, bi-linear ansatz functions, straight transversal fibres, stresses in thickness direction remain zero) was not capable of capturing the stress state in the thickness direction, which is seen as necessary for correctly modelling the localization w.r.t. the experimental data. This limitation results in inaccurate simulation results, particularly in cases where the deformation results in non-straight fibers and subsequently the section exhibits non-constant stresses across the thickness. To improve the accuracy of the simulation, alternative shell formulations that are able to account for non-linear stress and strain states in the thickness direction should be considered. Clearly, these formulations are rather expensive and rarely used in engineering practise.

4. Conclusion
In the present study, a parameter-identification procedure was carried out for the yield curve extrapolation according to a modified Hockett-Sherby extrapolation based on the optical full-field measurement. The procedure is able to build a digital twin by using local displacement measurements during the test and using this information as boundary conditions for a corresponding FE mesh.

Furthermore, we present a new tensile test specimen geometry designed to enable more accurate and representative measurements of the strain field under complex loading conditions, including uniaxial tension, plane strain, and equibiaxial tension.

The optimization of the variables used in the modified Hockett-Sherby extrapolation approach based on the new tensile test specimen shows a poor agreement regarding the experiment. The results obtained can be attributed to multiple factors, including the simplified constitutive model, noise in the experimental measurement data, and inadequate spatial discretization. The use of the J2-plasticity model, which only represents isotropic material behavior, may have contributed to inaccuracies in the simulation results. Noise in experimental measurement data, if not correctly accounted for, may also lead to inaccurate simulation results. Additionally, the chosen shell formulation for spatial discretization was not capable of capturing stress in the thickness direction at all, which may result as well in inaccurate simulation results. To improve the accuracy of the method, more advanced constitutive models that account for anisotropic behavior, DIC data smoothing and filtering techniques in space and time to fight the noise in the measurement data, and alternative shell formulations or the use of multiple solid elements across the sheet metal thickness should be considered. These topics will be addressed in future work.

Acknowledgments
This work has been carried out within the research project AIMM (Artifical Intelligence for Material Models) funded by the Federal Ministry for Economic Affairs and Climate Action (BMWK) under the grant number 19 | 20024A.

References

[1] Avril S et al. 2008 *Overview of identification methods of mechanical parameters based on full-field measurements* Experimental Mechanics 48(4) pp 381–402

[2] Güner A, Soyarslan C, Brosius A and Tekkaya A E 2012 *Characterisation of anisotropy of sheet metals employing inhomogeneous strain fields for Yld2000-2D yield function* Int. Journ. of Solids and Structures vol 49 pp 3517–3527

[3] Brosius A, Küsters N and Lenzen M 2018 *New method for stress determination based on digital image correlation data* CIRP Annals 67(1) pp 269–272

[4] Ilg C, Haufe A, Witowski K, Koch D and Liewald M 2019 *Parameter Identification using Full-Field Calibration (FFC)* vol 38th International Deep-Drawing Research Group Conference (Enschede)

[5] Stander N, Witowski K, Ilg C, Haufe A, Helbig M and Koch D 2017 *Application of Digital Image Correlation to Material Parameter Identification using LS-OPT* vol 11th European LS-DYNA Conference (Salzburg)

[6] Ilg C, Haufe A, Sreenivasa V, Karadogan C and Liewald M 2023 *Parameter Identification applying Full-Field Calibration (FFC) Techniques* vol 14th Int. Conf. on the Technology of Plasticity (Mandelieu-La Napoule)

[7] Liebold C, Ilg C, Haufe A, Usta T and Sreenivasa V 2023 *Displacement based simulation and material calibration based on digital image correlation Part I – Theory* vol 42nd Int. Deep-Drawing Research Group Conf. (Lulea)

[8] Livermore Software Technology 2020 *LS-DYNA Keyword User's Manual* vol 2 R12

IOP Conf. Series: Materials Science and Engineering 1284 (2023) 012057 doi:10.1088/1757-899X/1284/1/012057

Development of a new cyclic shear test setup for characterizing thin metallic foils

C Karadogan[1], M Beck[1*], P Cyron[1], K R Riedmüller[1] and M Liewald[1]

[1] Institute for Metal Forming Technology, University of Stuttgart, Holzgartenstr. 17, D-70174 Stuttgart

E-mail: maxim.beck@ifu.uni-stuttgart.de

Abstract. Recent developments aimed at reducing cumulative CO_2 emissions in the energy sector and e-mobility are leading to an increase in the production volumes of formed components made from thin metallic foils used for alternative energy supply concepts. Thus, components such as bipolar plates in fuel cells and a variety of parts in e-motors and batteries are being manufactured in ever higher quantities and thinner material thicknesses. As a consequence, this leads to increased challenges regarding the feasibility and robustness of required production processes. Finite Element Analysis in fact do represent a useful tool in this context to predict and to optimize expected outcome of forming processes virtually at an early stage of development. The prediction accuracy of such simulation codes thereby significantly depends on the precision of modelling the sheet metal material behavior, which is derived from special characterization methods. However, classical approaches to mechanical characterization usually tend to fail in case of thin metallic foils, especially when the structural stability of the specimen becomes an important factor for the validity of the characterization test. The cyclic shear test, for example, is used to determine the hardening behavior of the material to be characterized, but appear unsuitable for metallic foil investigations unless a special anti-wrinkling device is used. Current anti-wrinkling devices proposed in literature must be attached directly to the deforming area of the specimen and cause further weaknesses. Attaching such devices to prevent the specimen from wrinkling requires remarkable skills of the operator to avoid deforming of the specimen even before testing. Against this background, this paper presents a novel experimental setup, which increases the structural stability of the cyclic shear specimen and thus prevents wrinkling. Structural stability of specimen is enhanced by curving the gauge area of the specimen to drastically suppress the tendency to wrinkle. As a proof of concept FE-simulations with LS-Dyna were performed in this study to verify this novel idea and to design a suitable experimental setup.

1. Introduction

The growing demand for e-mobility, microelectronics, and renewable energy technologies has led to a significant increase in production volumes of thin and miniaturized components made from metallic foils having thicknesses below 100 μm. Such small thicknesses pose significant challenges to the feasibility and robustness of the forming process, as thin metallic foils are very responsive to defects caused by wrinkling or tearing during forming [1-3]. In addition, the phenomenon of springback until today remains a major issue in achieving dimensional accuracy of components formed from thin metal foils [4-6]. In modern fuel cells, for instance, thin stainless-steel foils are used to be formed into metallic bipolar plates. The sections of flow channels on these formed plates must meet very precise dimensional tolerances within a few micrometers in both vertical and lateral directions [4,7]. Otherwise, problems may arise when joining the plates in a subsequent step, resulting in decreased performance and efficiency of the entire fuel cell stack.

Finite element simulations can be used to obtain virtual results of such complex and miniaturized forming processes and thus to optimize them in early stages of development. However, the accuracy of these simulation results relies heavily on the realistic modelling of material's mechanical behavior obtained from material characterization procedures [1,8]. In order to improve prediction quality of finite element simulations, especially regarding dimensional accuracy affected by springback of formed component, it is advisable to consider reverse loading sequences in addition to monotonic stress-strain curves when determining mechanical properties of ultra-thin metallic foils [9]. The reverse loading behavior then can be considered in elaborate phenomenological material models such as the Yoshida-Uemori kinematic hardening [10] or the Chaboche model [11-12] to account for material hardening and the Bauschinger effect in the forming simulations. Apart from the cyclic-tension-compression test, which is impractical for thin foils, the reverse loading behavior can be observed in a cyclic in-plane shear test [13-14] or in a cyclic in-plane torsion test [15].

Since, the cyclic in-plane torsion test requires a special equipment, the cyclic in-plane shear test was used in the scope of this study to characterize the reverse loading behavior of the thin metallic foils considered. Typically, the cyclic in-plane shear test (hereafter this will be named as cyclic shear test) is realized as shown in Figure 1. To reduce the adverse effect of stress inhomogeneities near the free edges, a narrow and long gauge section is loaded under cyclic in-plane shear. Assuming that the clamped regions remain as rigid-body and ignoring the inhomogeneities at the free edges, the true strain and true stress are calculated quite simply using equations (1) and (2) and plotted as in Figure 1c. In order to capture the real shear strain achieved within the gauge, it is suggested to measure the strain distribution using a DIC measurement system. More importantly, the stress state in the gauge section must ideally be in-plane, which is not the case with wrinkles. As a result, the occurrence of wrinkling in the gauge section is a crucial issue that needs to be addressed.

$$\varepsilon_{v.Mises}^{plastic} = \sqrt{1/3}\,\varepsilon_{shear}^{plastic} \quad \text{with} \quad \varepsilon_{shear}^{plastic} = \gamma_{xy} - \tau_{xy}/G \quad \text{where}\ \gamma_{xy} = \frac{u}{w} \tag{1}$$

$$\sigma_{v.Mises} = \sqrt{3}\tau_{xy} \quad \text{with} \quad \tau_{xy} = F/tL \tag{2}$$

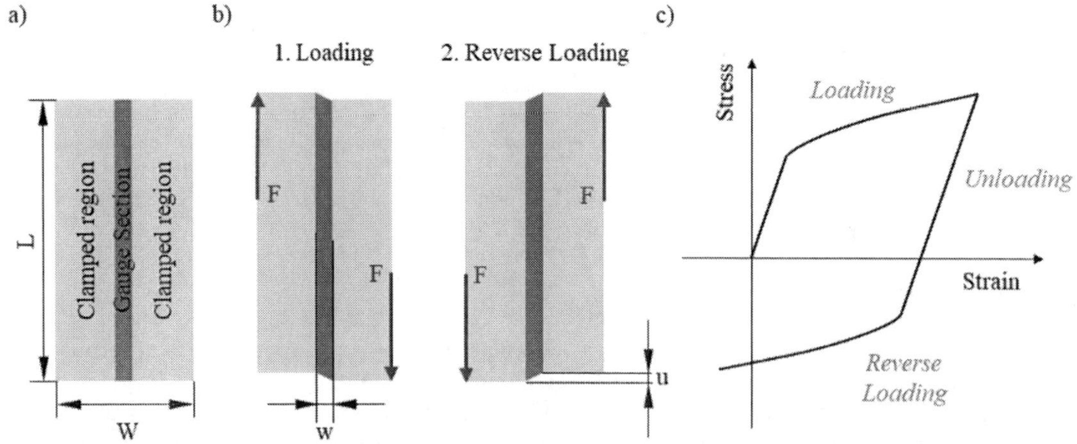

Figure 1. a) Schematic rectangular cyclic shear test specimen. L = Sample length and W = Sample width. b) Sample in loading & reverse loading conditions. w = width of gauge section. c) Schematic stress-strain curve obtained by shear testing.

Conventional shear test setups with a wider gauge section (unclamped deformation area) are unsuitable for characterizing thin metal foils, since foils are prone to wrinkling or tearing due to their low structural stability. [1-3]. If wrinkling occurs, the stress state is no longer purely in-plane shear, which means that the strains and stresses calculated using equations (1) and (2) would be inaccurate. To ensure the structural stability of the metal foils, the width of the measurement area should be reduced in proportion to the foil thickness. However, this reduction in width in turn would result in being too small to be used for conventional optical DIC measurement systems.

These problems motivated the investigation of new cyclic shear test setups as presented in previous research, mostly addressing the problem of wrinkling in the measurement area [9, 16]. One solution that has been proposed to eliminate wrinkles during the simple-shear-test is to apply a normal pressure in the thickness direction of specimen using anti-wrinkling devices. For this purpose, Pham et al. [9] used transparent pressure pads made of transparent plastics, which applied a normal pressure onto the surface of the metallic foil during cyclic shear testing. According to the authors, this method on the one side effectively eliminated wrinkles while still allowing for visual strain measurements to be taken through transparent pressure pads. On the other side, the tiny anti-buckling device needs to be fixed onto the specimen with a controlled torque. Then, the specimen, along with the attached anti-buckling device, needs to be carefully mounted into the clamps of the shear test setup without damaging or deforming the foil unintentionally. Although the authors have demonstrated that friction between clamping device and foil has a negligible effect on the measured forces and can be ignored, this document delivers insufficient information about the friction's impact on the stochastic pattern applied on the specimen surface used for optical strain measurement objectives [9]. It is believed that rubbing between the specimen and the pressure-pad may slightly distort the stochastic pattern.

In the view of this state of the art, this paper introduces a new test setup for conducting cyclic shear tests on thin metal foils. The proposed method enhances the structural stability of cyclic shear specimens in the gauge section, thus preventing wrinkling during the test without requiring an additional anti-buckling device. Structural stability is improved here by slightly curving the gauge section of the foil sample by positioning a cylindrical surface behind it, suppressing the tendency to wrinkle even at larger gauge widths and allowing for conventional optical DIC measurement. This eliminates the need to use an anti-buckling device, and thus the detrimental effect of high friction on the measured force and rubbing on the stochastic pattern.

Finite element simulations using LS-Dyna were conducted to validate this novel idea. Initially, the simulations were aimed at reproducing the occurrence of wrinkling in conventional shear tests to obtain valid simulation settings. These simulations employed a material model for austenitic stainless steel (1.4404 / 316L) with a thickness of 100 µm, as described in [4]. With the valid simulation settings in place, it was then demonstrated that a slight curvature of the foil with a cylindrical support behind could effectively prevent wrinkling in the gauge section. Moreover, an analysis of the resulting forces and displacements revealed that the slight bending stresses induced in the specimen disclose no discernible effect on the measured forces, and therefore, no adverse impact on the test results would be expected.

2. Modelling and numerical study

To eliminate considerable wrinkling that occurs during conventional shear tests of thin metallic foils, a new test setup was used in the present investigations that slightly curves the shear specimen and thus increases its structural stability without any additional anti-wrinkling device pressing on the whole gauge section. The hypothesis that wrinkling of thin metallic foils can be prevented by slightly curved fixturing during shear testing was validated through a simulation study performed for a 2×30 mm gauge section (Figure 1) on a specimen made of stainless-steel (1.4404 / 316L) foil of 0.1 mm thickness. Thereby, two simulations were conducted with LS-Dyna, one with the conventional approach using a flat shear sample and the second one with the proposed approach using a specimen bent (without stretching) in a slight curvature of $r_c = 5$ mm with rigid cylindrical support from behind the gauge section. Such supporting surface from one side is necessary for bending and also for keeping the specimen in its bent shape during testing. The simulation setups used for these two kinds of shear tests are shown in Figure 2, whereby the used material model MAT-36 was based on data from [4]. Simulations were conducted using shell elements of type 16, incorporating 11 integration points in the thickness direction.

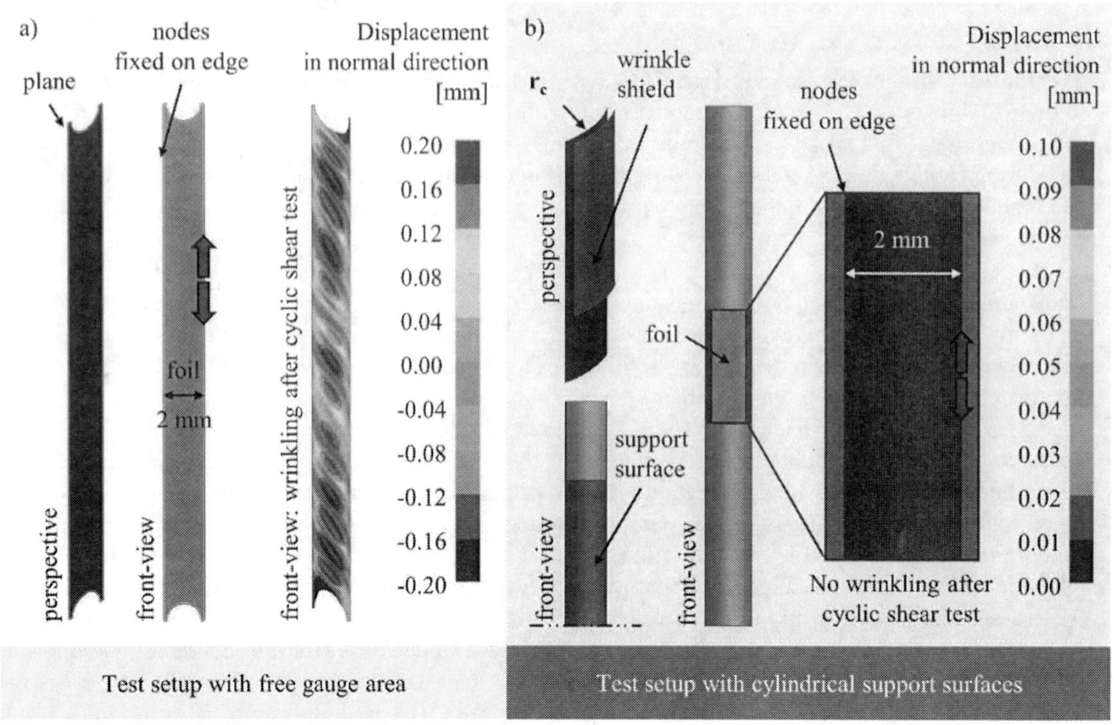

Figure 2. a) Simulation model for the conventional shear test setup with a free gauge and wrinkling in the measuring area after testing. b) Simulation model with the proposed approach of using a curved specimen in the gauge section showing no wrinkling after testing of $\gamma = 0 \rightarrow 0.3 \rightarrow -0.3$ (γ = shear strain)

In Figure 2, the thin stainless-steel foils are depicted in brown color. Both virtual test setups are exposed to the same global cyclic shear deformation. Regarding chosen width of the gauge section, the cyclic motion up and down first deforms the specimen up to a shear strain value $\gamma = 0.3$ and then in the reverse direction up to a shear strain value of $\gamma = -0.3$ (where $\gamma_{xy} = 2\varepsilon_{xy}$). The conventional testing approach (Figure 2.a) shows severe wrinkles that can distort intended stress state no longer in-plane shear and may prevent the measurement of shear deformation with a DIC system in real tests. The new test setup as depicted in Figure 2b has a cylindrical supporting edge of $r_c = 5$ mm on the back of the gauge section. In the simulation, the specimen must be slightly curved on this cylindrical edge and fixed in the flat clamps on the left and right of this cylindrical support. In doing so, the specimen must not be stretched so that the contact between the specimen and the cylindrical support edge remains slightly touched. Note that this bending is simulated as an initial step in the FE simulation before any shear deformation. Thus, the effects of the stress distribution that arise due to bending are considered over the 11 integration points used in the thickness direction. The front surface of the specimen is free from any tool pressure. Instead, cylindrical shields (depicted in blue in Figure 2) are situated a short distance away from the specimen surface above and below the gauge section, leaving ample space in the middle of the gauge open for DIC measurement. These cylindrical shields are designed to prevent the growth of wrinkles that typically develop at the top and bottom of the gauge section by limiting their height to an uncritical maximum value.

Extensive simulation studies carried out in the scope of this study showed that preventing the growth of wrinkles in the upper and lower areas of the specimen (not necessarily used for the evaluation of shear strain by the DIC system) can prevent further propagation of wrinkles into the center area of specimen. This is where the actual measurement of the representative shear strain value is assumed to be valid for the whole gauge of the specimen. In the FE models according to Figure 2, nodes outside the gauge section, which are fixed by clamping jaws in the real tests, are locked or moved in the corresponding directions by means of nodal boundary conditions in order to induce the shear deformation. More precisely, in simulation the left side of the shear specimen is fixed, whereas the right side of the specimen is moved vertically, thus inducing a cyclic shear deformation in the measurement area of the specimen.

3. Results and discussion

The simulation results, depicted in Figure 2a), show that excessive wrinkling occurs in the flat specimen tested conventionally. In contrast, no wrinkling can be observed in the measurement area of the slightly curved specimen with a curvature radius of $r_c = 5$ mm as shown in Figure 2 b). This results from the increased structural stability of loaded cross section of specimen against wrinkling in case of a shear deformation imposed in the shown direction on the specimen. The use of wrinkle-shields is justified with the following argumentation. The progression of the wrinkles during application of cyclic shear deformation as shown in Figure *3*, originating from the top and bottom free edges of the gauge section, can be hindered by using restraining surfaces that constrain the height of the initial wrinkles to a minimum without applying pressure on the specimen surface in the gauge section. In this way, wrinkles in the gauge section are eliminated by local and slight contact, which is advantageous compared to the case of using an anti-buckling device. The success of the proposed approach gets recognizable from the wrinkle-free gauge section of the specimen achieved using the proposed testing setup (see Figure 2 b and Figure 3).

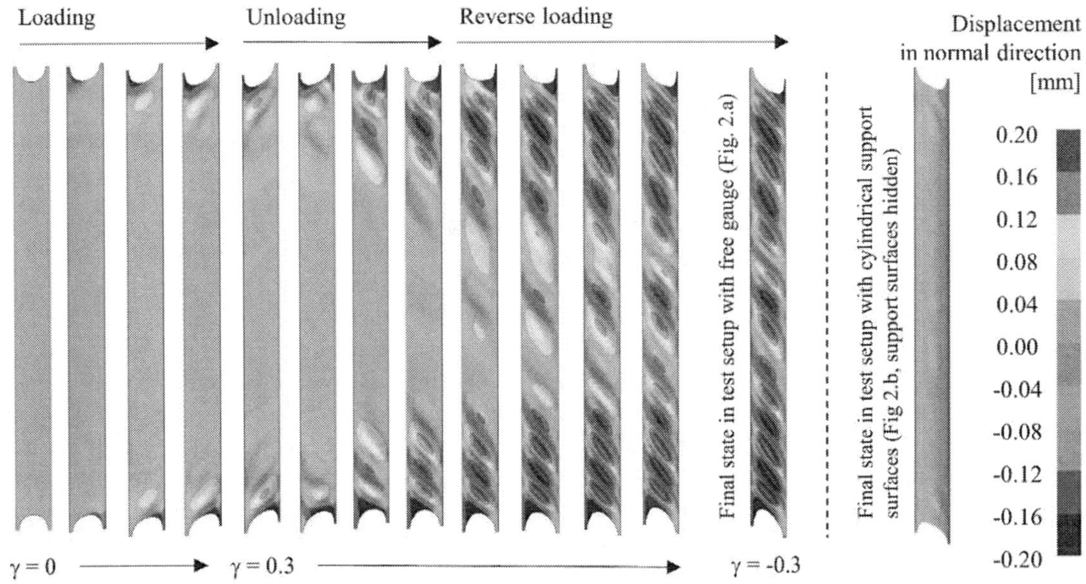

Figure 3. Wrinkling formation in upper and lower areas of the specimen and their propagation in the unconstrained gauge section of Figure 2a). Comparison of final states of the setups of Figure 2a) and b) shows significant improvement of the novel test setup

In order to ensure a neglectable impact of slightly curved cross-section of specimen and with this on the measurement and evaluation of shear-stress and -strain, forces and displacements from both the planar and curved shear testing procedure were compared. A reference simulation was performed with a planar measurement area in which wrinkling is prevented by constraining emerging displacements of the nodes perpendicular to the specimen surface. This case represents the ideal case in the following.

A comparison between the resulting forces in the idealized flat specimen, in the conventional flat specimen with wrinkling and in the newly developed curved specimen is shown in Figure 4. It can be stated that there is negligible deviation between the force profiles of the ideal case and the curved samples. In contrast to the conventional planar sample (highlighted in red in Figure 4), the curved sample provides a more precise depiction of the ideal scenario, since force values are not distorted by wrinkling. Furthermore, analysis of shear strains γ, which were calculated based on the displacement in the specimen plane, also show no significant differences between the curved and the idealized sample.

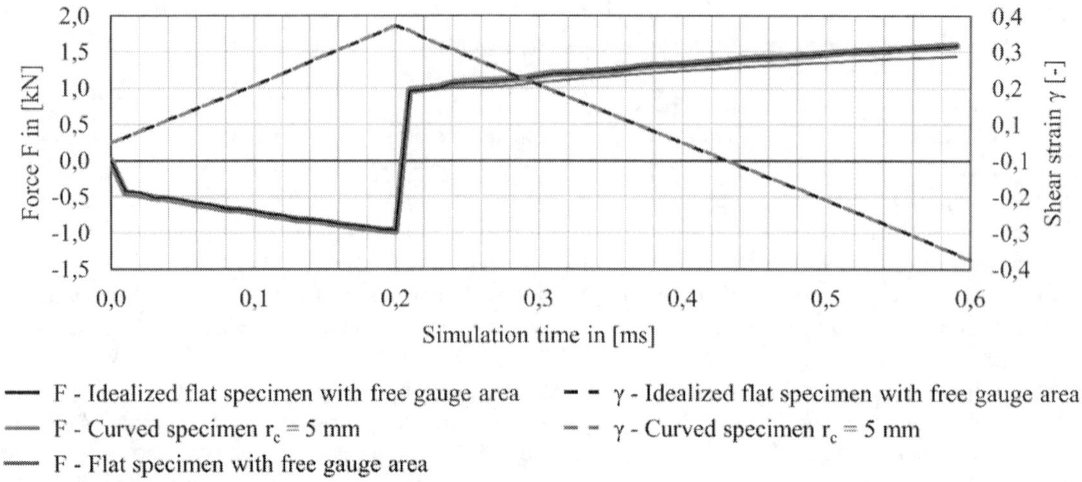

Figure 4. Comparison between the calculated force curves and shear strain curves of idealized flat specimen, the curved specimen with a radius of r_c = 5 mm and the flat specimen with free gauge section

Results gained by this confirm the validity of the suggested design of test device dedicated to thin foils in preventing the formation of wrinkles by slightly curving of specimen in the gauge section. This approach proves to be a practical and effective way to conduct cyclic shear tests on thin foils, as it ensures that the recorded force values and displacements remain accurate and undistorted.

4. Conclusion

This paper presents a novel approach for performing cyclic shear tests on thin metallic foils. By introducing a slight curvature in the gauge section of the specimen, the formation of undesired wrinkles can be prevented, thus ensuring reliable measurement and determination of shear stress and strain in the measurement area. To validate this approach, finite element simulations were conducted using LS-Dyna code, serving as a proof of concept. The simulations of conventional shear tests showed that wrinkling generally starts at the free edges of the gauge section and spreads inward. Curving the gauge section into a slight cylindrical shape using a supporting edge and preventing possible wrinkling near the free edges through so-called wrinkle shields allows a wrinkle-free measurement area that can be measured via DIC systems. The critical comparison of the forces calculated for the cyclic deformation showed that the bending stresses resulting from the cylindrical bending of the specimen are found negligible compared to the detrimental effects of the wrinkles on the measured forces as depicted in Figure 4. Future work will focus on further investigations with varied bending radii as well as with different foil materials of various thicknesses. To validate proposed concept, a suitable testing setup will be established for experimental investigations.

Acknowledgments
This work was supported within the "DFG-Fraunhofer Transfer Programme", established by the German Research Foundation (DFG-Project 460294948). Authors feel grateful for the financial support. In addition, the authors would like to thank the Working Group *Foil-Forming* established at the Institute for metal forming technology, University of Stuttgart for their support.

References

[1] Vollertsen F, Biermann D, Hansen HN, Jawahir IS, Kuzman K 2009 Size effects in manufacturing of metallic components. *CIRP Ann Manuf Technol* **58** pp. 566–587

[2] Neto D M, Oliveira M C, Alves J L, Menezes L F 2019 Numerical study on the formability of metallic bipolar plates for proton exchange membrane (PEM) fuel cells *Metals* **9** 810

[3] Hu Q, Zhang D, Fu H, Huang K 2014 Investigation of stamping process of metallic bipolar plates in PEM fuel cell - numerical simulation and experiments. *Int. J. Hydrogen Energy* **39** pp. 13770–13776.

[4] Beck M, Riedmüller K R, Liewald M, Bertz A, Aslan M J, Carl D 2023 Investigation on the influence of geometric parameters on the dimensional accuracy of high-precision embossed metallic bipolar plates *Production at the Leading Edge of Technology WGP 2022 pp.* 427-438

[5] Bauer A 2020 Experimentelle und numerische Untersuchungen zur Analyse der umformtechnischen Herstellung metallischer Bipolarplatten *(Chemnitz: Technische Universität Chemnitz).*

[6] Adzima F, Balan T, Manach P Y 2020 Springback prediction for a mechanical micro connector using CPFEM based numerical simulations *Int J Mater Form* **13** pp. 649–659

[7] Porstmann S, Wannemacher T, Drossel W G 2020 A comprehensive comparison of state-of-the-art manufacturing methods for fuel cell bipolar plates including anticipated future industry trends *J. Manuf. Process.* **60** pp. 366–383

[8] Vollertsen F, Schulze Niehoff H, Hu Z 2006 State of the art in micro forming *Int J Mach Tools Manuf* **46** pp. 1172–1179

[9] Pham C, Adzima F, Coër J et al. Anti-buckling device for ultra-thin metallic sheets under large and reversed shear strain paths 2017 *Experimental Mechanics* **57** pp 593–602

[10] Yoshida F, Uemori T 2003 A model of large-strain cyclic plasticity and its application to springback simulation *Int. J. Mater. Sci.* **45**(10) pp. 1687–1702

[11] Chaboche JL 1986 Time-independent constitutive theories for cyclic plasticity *Int J Plast* **2** (2) pp. 149-188

[12] Chaboche JL 1989 Constitutive equations for cyclic plasticity and cyclic viscoplasticity *Int J Plast* **5** (3) pp. 247-302

[13] Bouvier S, Haddadi H, Levée P, Teodosiu C 2006 Simple shear test: experimental techniques and characterization of the plastic anisotropy of rolled sheets at large strains *Journal of Materials Technology* **172** pp. 96-103

[14] Yin Q et al. 2014 An experimental and numerical investigation of different shear test configurations for sheet metal characterization *International Journal of Solids and Structures*, **51** (5) pp. 1066–1074

[15] Brosius A, Yin Q, Güner A, Tekkaya A E 2011 A new shear test for sheet metal characterization, *Steel Research International* **82** (4) pp. 323–328

[16] Zhang P, Pereira M P, Abeyrathna B, Rolfe B F, Wilkosz D E, Weiss M 2019 Improving the shear test to determine shear fracture limits for thin stainless-steel sheet by shape optimisation, *International Journal of Mechanical Sciences* **164**

IOP Conf. Series: Materials Science and Engineering 1284 (2023) 012058 doi:10.1088/1757-899X/1284/1/012058

Impact of elastic modulus evolution on springback prediction

X Lemoine[1*] and J M Devin[2]

[1]Global R&D ArcelorMittal Maizières, voie Romaine, BP30320, F-57283 Maizières-lès-Metz, France
[2]Global R&D ArcelorMittal Montataire, 1 Route de Saint-Leu, F-60160 Montataire, France

E-mail: xavier.lemoine@arcelormittal.com

Abstract. Today, the hardening Yoshida-Uemori model is implemented in the most used FEA stamping codes (Autoform®, Pamstamp®, LS-Dyna®...) with occasionally some variations. This forming material card is requested regularly by the car manufacturers. The Yoshida-Uemori model is a combined hardening model coupled with the elastic evolution modulus which depends on the equivalent plastic strain. According to the experimental tests conducted and their identification protocol, the set of parameters improves the springback prediction compared to the prediction with a purely isotropic hardening model. The numerical analysis shows that the elastic modulus evolution has an influence on springback prediction of the same order as the combined hardening (isotropic and kinematic). The adjusting parameters governing the evolution of the elastic modulus are identified thanks to hysteresis loops conducted between 0.2% of plastic deformation and uniform elongation (Uel). For each loop, there are several ways to define an unloading elastic module: the chord of loop; the slope deducted from the linear regression on unloading path on a range. According to the choice of range, the value differs. What is the impact of these choices on the springback prediction of a stamped part?

1. Introduction

The Yoshida-Uemori model [1-3] is a combined hardening model coupled with the elastic evolution modulus which depends on the equivalent plastic strain. The Yoshida-Uemori model is implemented in FEA stamping software and allows a better prediction of springback. ArcelorMittal is now ready to produce accurate Yoshida-Uemori material cards for its products, considering the various implementations of this model in the most used stamping softwares PamStamp®, LS-DYNA® and Autoform®. The methodology for identifying the Yoshida-Uemori parameters was presented last year at the IDDRG2022 conference, based on reversed shear test and hysteresis loops conducted between 0.2% of plastic deformation and uniform elongation (Uel) [4]. For elastic evolutive modulus model, the aim of the present study is to confirm the choices made for the parameters' identification, the impact on springback prediction and that the value of the elastic modulus is well in adequacy with the final state of stress after springback.

2. Several definitions and identification

The adjusting parameters governing the evolution of the elastic modulus are identified thanks to hysteresis loops conducted between 0.2% of plastic deformation and uniform elongation (Uel) every 1% on ISO 12.5x50 uniaxial tensile specimens (Figure 1a). The strain is measured with a high-resolution extensometer (Class = 0.5) as per Annex G of ISO 6892-1 standard.

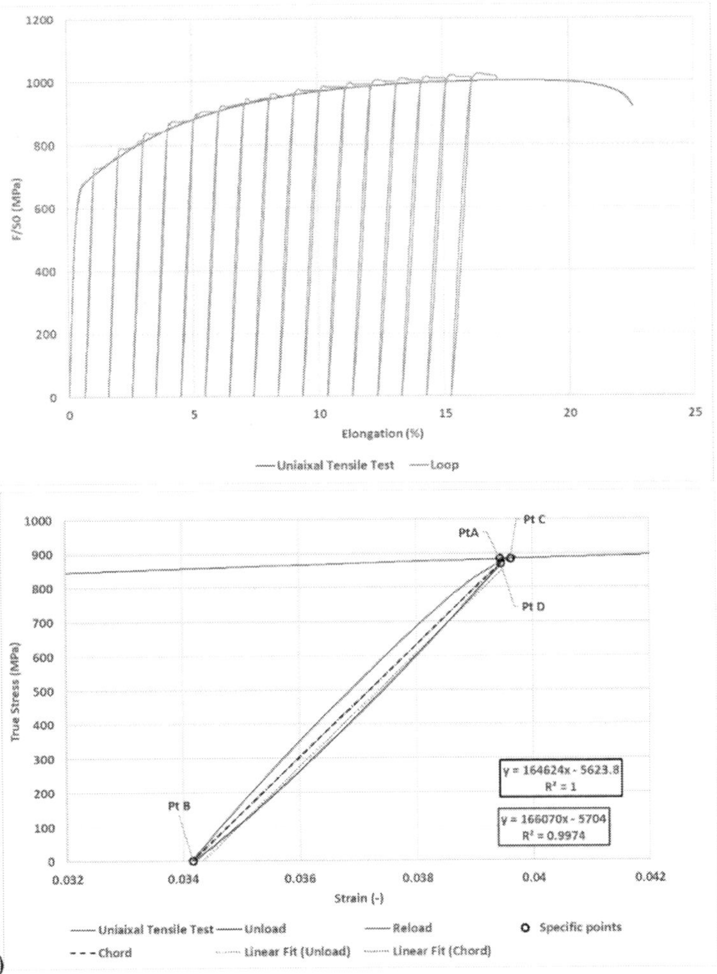

a)
b)

Figure 1. Determination of the pre-strain dependence of the elastic modulus for Fortiform® 980
a) Hysteresis loop (orange curve) conducted every 1% until the end of the uniform elongation.
b) the elastic modulus following two definitions: Linear regression on unload path (red curve), chord slope (black curve).

For each unloading/reloading loop, there are several possible to consider the elastic moduli:

- deduction from the linear regression on unloading path between the points A (maximal stress at the begin of the loop) and B (minimal stress of loop) (red curve in figure 1b) [4].
- chord slope (segment [DB]) (black dotted curve in figure 1b) [5].
- average of instantaneous modulus resulting from a polynomial regression of 2nd degree on unloading path between the points A and B (or in a sub-range of the unloading path: for example, the middle third).

These four definitions lead to a difference of elastic modulus of about 6 GPa (figure 2).

The two adjusting parameters E_a and ξ of equation (1) governing the evolution of the elastic modulus in the Yoshida-Uemori model are fitted according to a gradient method (red curve on figure 3) in order to get the better correspondence between the equation and the experimental results derived from hysteresis loops tests (blue points).

$$E(p) = E_0 - (E_0 - E_a)\left(1 - exp(-\xi p)\right) \quad (1)$$

where E_0 is initial Young modulus; E_a and ξ are two material parameters, p is equivalent plastic strain. The parameters identified for each definition are presented in table 1.

IOP Conf. Series: Materials Science and Engineering 1284 (2023) 012058 doi:10.1088/1757-899X/1284/1/012058

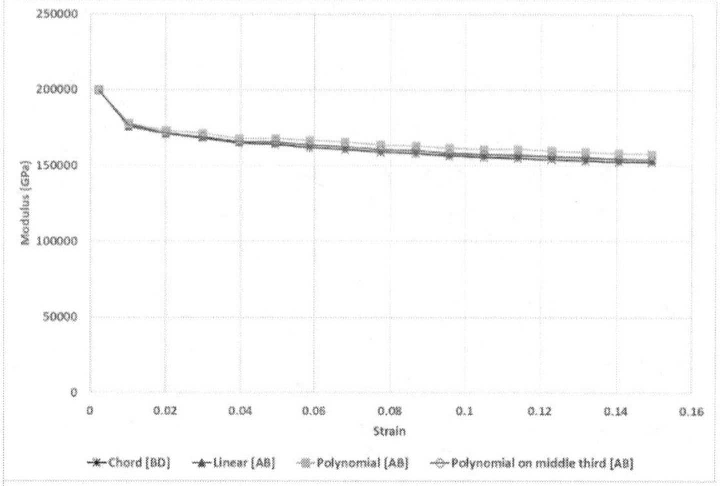

Figure 2. Evolution of elastic modulus for 4 definitions.

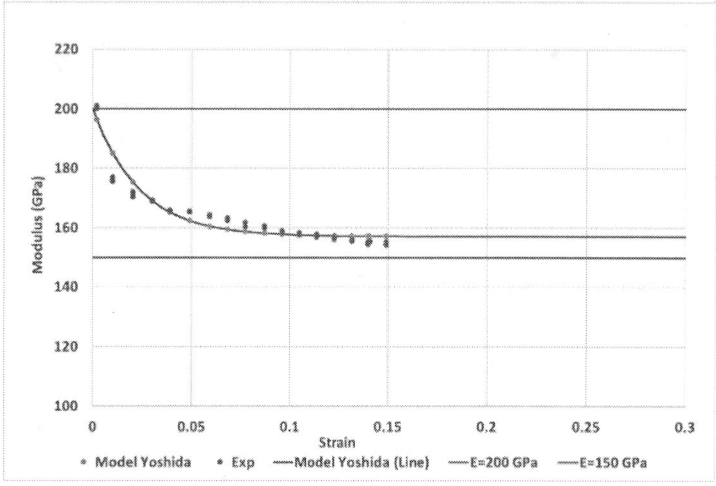

Figure 3. Experimental points (blue dots) versus Elastic modulus evolution model (red line) for a Fortiform® 980 steel ($E_0 = 200GPa, E_a = 157GPa$ and $\xi = 42.85$)

Table 1. Parameters identified for 4 definitions (Fortiform® 980 steel), $E_0 = 200GPa$.

	Chord [DB]	Linear regression [AB]	Average of instantaneous modulus	
			[AB]	Middle third of [AB]
E_a (GPa)	154	157	160	156
ξ	38.94	42.85	42.13	40.82

3. Springback applications

For the Fortiform® 980 steel grade, springback predictions of 9 parts have been realized with PamStamp®2022 considering three configurations of elastic modulus: one configuration with the elastic modulus evolution and two others with a constant elastic modulus (either 200 GPa or 150 GPa) (Figure 3).

For most of the parts, the springback prediction is very different between these three cases (table 2). The level of the strains at end of stamping has been analysed for each integration point of each element of the part mesh, and for the 9 parts (table 3). The results are discussed below.

Table 2. Comparison of springback prediction for 9 parts between Elastic modulus evolution and constant modulus (Fortiform® 980 steel).

	Gap Evolutive E / E= 200 GPa		Gap Evolutive E / E= 150 GPa	
	Max / Min distance (mm)	% of part outside [-1; +1]	Max / Min distance (mm)	% of part outside [-1; +1]
U-Form (Low BHF)	+6.7 / -0.1	57 %	+0.1 / -3.1	42 %
U-Form (High BHF)	+4.7 / -0.1	54 %	+0.2 / -0.7	0 %
Inner side-sill part	+3.3 / -2.4	32 %	+6.4 / -5	43 %
Roof arch reinforcement	+2.8 / -0.9	15 %	+1 / -1.7	10 %
Rear side-rail part	+4.2 / - 1.9	15 %	+1.5 / -1.7	1 %
Door lateral beam	+0.8 / -2.3	14 %	+0.7 / - 3.5	24 %
A pillar part	+2.6 / -1.6	11 %	+3.2 / -1.8	4 %
Floor reinforcement part	+3.3 / -1.0	7 %	+1.5 / -2.6	3 %
Front side-rail part	+0.9 / -0.5	0 %	+1 / -1.7	3 %

Table 3. % of Integration points by equivalent plastic deformation range (Fortiform® 980 steel). Four ranges are defined ≤ 0.01,]0.01;0.05],]0.05;0.1[, ≥ 0.1 corresponding respectively at elastic modulus (GPa) equal to ≥ 185,]185; 162],]162;158[,]158;157[

		Integration points				
			Equivalent plastic strain			
		Number in part	≤ 0.01	]0.01;0.05]	]0.05;0.1[	≥ 0.1
U-Form (Low BHF)	Evolutive modulus	388340	27 %	3 %	14 %	56 %
U-Form (Low BHF)	E = 200 GPa	388340	27 %	3 %	12 %	58 %
U-Form (Low BHF)	E = 150 GPa	388340	27 %	3 %	16 %	54 %
U-Form (High BHF)	E = 200 GPa	397085	7 %	8 %	5 %	80 %
Inner side-sill part	E = 200 GPa	272745	22 %	35 %	24 %	19 %
Roof arch reinforcement	E = 200 GPa	832780	17 %	29 %	23 %	31 %
Rear side-rail part	E = 200 GPa	639485	1 %	16 %	18 %	65 %
Door lateral beam	E = 200 GPa	428940	8 %	18 %	26 %	48 %
A pillar part	E = 200 GPa	714015	25 %	41 %	17 %	17 %
Floor reinforcement part	E = 200 GPa	681635	9 %	37 %	27 %	27 %
Front side-rail part	E = 200 GPa	178725	55 %	17 %	12 %	16 %

IOP Conf. Series: Materials Science and Engineering 1284 (2023) 012058 doi:10.1088/1757-899X/1284/1/012058

For a large part of the integration points (55%) of the Front side-rail part, the strains are lower than 0.01, therefore an elastic modulus higher than 185 GPa is applied for most of the integration points. Hence the difference between evolutive modulus and the 2 constant modulus scenarios is lowest for 200 GPa.

On the other hand, for the U-Form (High BHF (Blank-Holder Force)) strains area higher than 0.1 for 80% of the integration points, therefore an elastic modulus lower than 158 GPa is applied for most of the integration points. Hence the difference between evolutive modulus and the 2 constant modulus scenarios is lowest for 150 GPa.

Between the three configurations studied, for U-Form (Low BHF), the distribution of equivalent plastic strains and the major principal stresses are almost identical (table 3 and figure 4).

At the equilibrium state of a stamped part after springback, the stress state is not zero at all integration points (figure 4). Less than 5% of the integration points are close to zero (table 4).

Table 4. Specific information State of Integration points (Fortiform® 980 steel).

		Integration points	
		Number in part	Stress state after springback close to 0 (± 5 MPa)
U-Form (Low BHF)	Evolutive modulus	388340	1.3 %
U-Form (Low BHF)	E = 200 GPa	388340	1.2 %
U-Form (Low BHF)	E = 150 GPa	388340	1.5 %
U-Form (High BHF)	E = 200 GPa	397085	0.1 %
Inner side-sill part	E = 200 GPa	272745	0.6 %
Roof arch reinforcement	E = 200 GPa	832780	0 %
Rear side-rail part	E = 200 GPa	639485	2.5 %
Door lateral beam	E = 200 GPa	428940	1.1 %
A pillar part	E = 200 GPa	714015	4.8 %
Floor reinforcement part	E = 200 GPa	681635	2.8 %
Front side-rail part	E = 200 GPa	178725	2.8 %

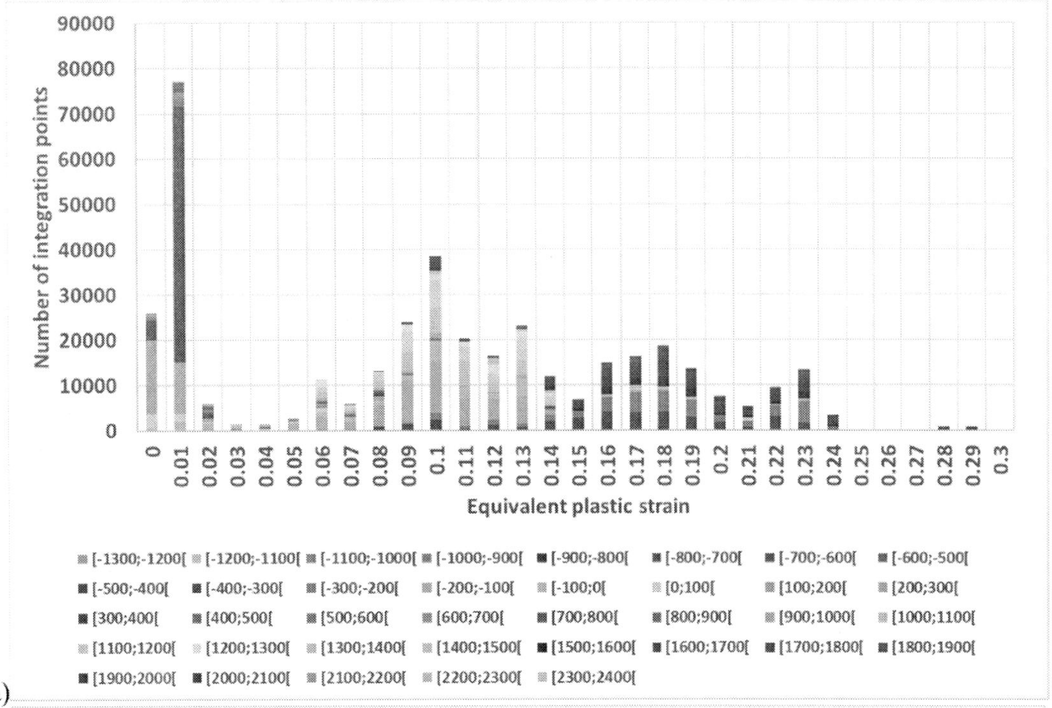
Number of integration points
Equivalent plastic strain
[-1300;-1200[[-1200;-1100[[-1100;-1000[[-1000;-900[[-900;-800[[-800;-700[[-700;-600[[-600;-500[
[-500;-400[[-400;-300[[-300;-200[[-200;-100[[-100;0[[0;100[[100;200[[200;300[
[300;400[[400;500[[500;600[[600;700[[700;800[[800;900[[900;1000[[1000;1100[
[1100;1200[[1200;1300[[1300;1400[[1400;1500[[1500;1600[[1600;1700[[1700;1800[[1800;1900[
[1900;2000[[2000;2100[[2100;2200[[2200;2300[[2300;2400[
a)

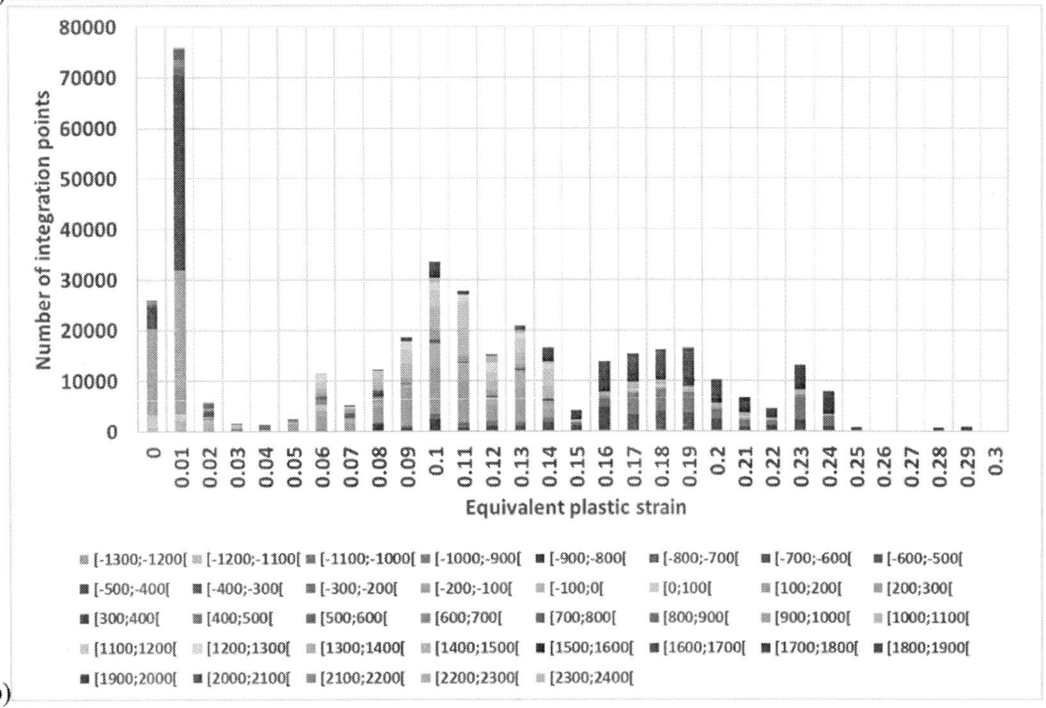
Number of integration points
Equivalent plastic strain
[-1300;-1200[[-1200;-1100[[-1100;-1000[[-1000;-900[[-900;-800[[-800;-700[[-700;-600[[-600;-500[
[-500;-400[[-400;-300[[-300;-200[[-200;-100[[-100;0[[0;100[[100;200[[200;300[
[300;400[[400;500[[500;600[[600;700[[700;800[[800;900[[900;1000[[1000;1100[
[1100;1200[[1200;1300[[1300;1400[[1400;1500[[1500;1600[[1600;1700[[1700;1800[[1800;1900[
[1900;2000[[2000;2100[[2100;2200[[2200;2300[[2300;2400[
b)

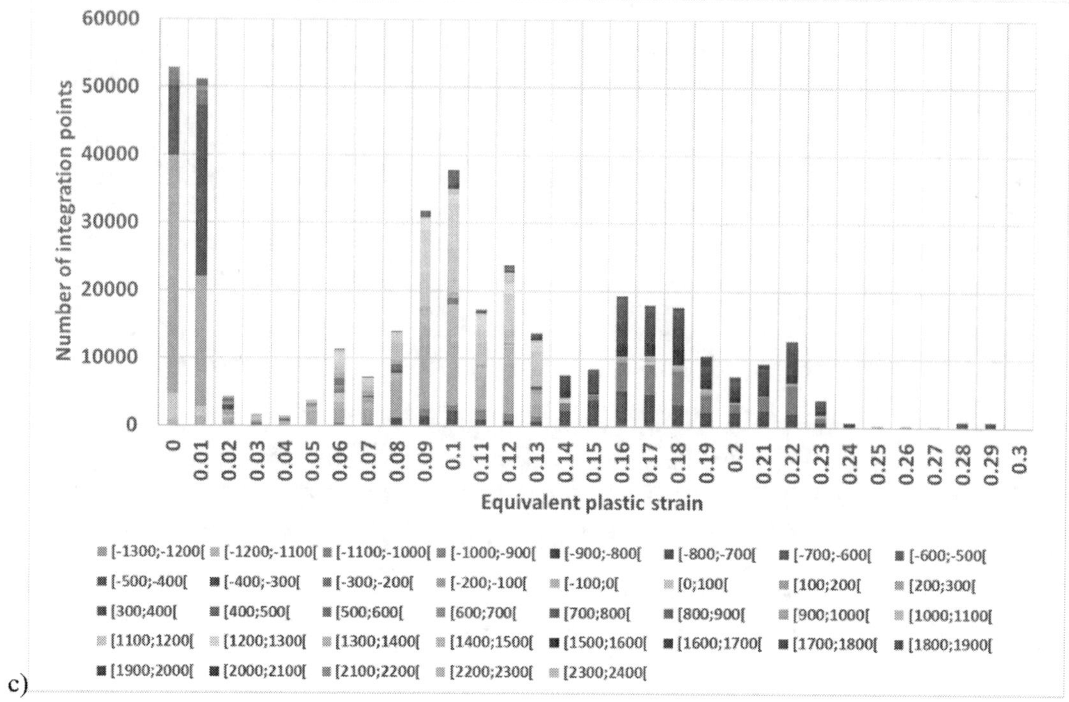

Figure 4. Stress state after springback in U-Form (Low BHF) for Fortiform® 980: Evolutive modulus (a); Constant modulus of 200 GPa (b) and of 150 GPa (c). The stress ranges indicated in the legend are for the major main stress.

4. Which elastic modulus model to consider according to the stress state after springback?

For the adjusting parameters identification, the unloading of hysteresis loops from flow curve to a value close to zero (± 5 MPa) are used (figure 1).

The elastic modulus has been determinate at different stress ranges and for 4 loops at different minimum stresses: 0; 200; 400 and 600 MPa (figure 5 and table 5). Starting from the same point on the flow curve, larger is the range, lower is the value of the elastic modulus. On the same range, the modulus value is same whatever the minimum stress of the loop.

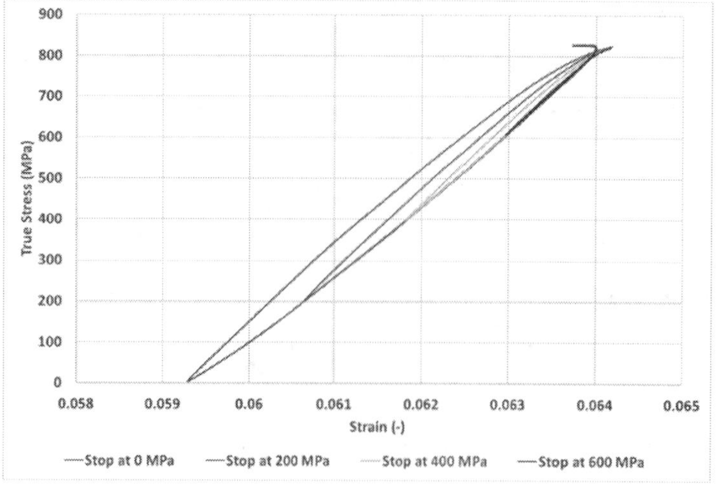

Figure 5. For TRiP800 steel, Loops at different minimum stresses: 0; 200; 400 and 600 MPa.

Table 5. Elastic modulus determinate by linear regression [AB] for loops at different minimum stresses (TRiP800 steel).

	Elastic Modulus (GPa) for			
	[800 – 600]	[800 – 400]	[800 – 200]	[800 – 0]
Stop at 0 MPa	214	195	184	173
Stop at 200 MPa	214	196	186	
Stop at 400 MPa	214	195		
Stop at 600 MPa	212			

For the numerical simulation of springback, a first solution is to realize two steps. The first step is a simulation with elastic modulus of 210 GPa to obtain the final stress state. The second step is a simulation with elastic modulus according to the equivalent strain before springback and the final stress state of the first step. The advantage of this solution is that it is not necessary to implement a specific elastic model in FEA code for the springback step.

A second solution is to implement a full loops model in FEA code as proposed by [6-8].

5. Conclusions

For each loop, there are several ways to define an unloading elastic module: the chord of loop; the slope deducted from the linear regression on unloading path on a range. Four definitions proposed lead to a difference of elastic modulus of about 6 GPa.

For a Fortiform® 980 steel, springback predictions of 9 parts with PamStamp®2022 in three configurations of elastic modulus confirm the interest of using Yoshida-Uemori model.

Two solutions are proposed to improve the elastic modulus evolutive model according to the stress state after springback in addition to the cumulative plastic deformation. One of these two solutions requires a law implementation in FEA code, the other does not.

References

[1] Yoshida F and Uemori T 2002 A model of large-strain cyclic plasticity describing the Bauschinger effect and workhardening stagnation *Int. J. Plast.* **18** 661

[2] Yoshida F, Uemori T and Fujiwara K 2002 Elastic–plastic behavior of steel sheets under in-plane cyclic tension–compression at large strain *Int. J. Plast.* **18** 633

[3] Yoshida F and Uemori T 2003 A model of large-strain cyclic plasticity and its application to springback simulation *Int. J. Mech. Sci.* **45** 1687

[4] Lemoine X, Chichery Q, Devin J.M. and Duroux P 2022 Identification Protocol of Yoshida-Uemori Hardening Model *Proc. of IDDRG 2022* (Lorient France)

[5] Lajarin S F, Filho R A C, Rebeyka C J, Nikhare C P and Marcondes P V P 2020 Numerical study on variation of chord modulus on the springback of high-strength steels *Int J Adv Manuf Technol* **106** 4707-4713

[6] Placidi F, Lemoine X, Till E and Georgi H 2007 Assessment and prediction of spring-back behaviour for high-strength steels *RFCS EUR22835 Final report*

[7] Sun L and Wagoner R H 2011 Complex unloading behavior: Nature of the deformation and its consistent constitutive representation *Int. J. Plast.* **27** 1126–1144

[8] Torkabadi A 2018 Toward an accurate springback prediction experiments and modeling *PhD report*

IOP Conf. Series: Materials Science and Engineering 1284 (2023) 012059 doi:10.1088/1757-899X/1284/1/012059

Prediction of texture-induced plastic anisotropy in AA6014-T4 aluminium sheets utilising two different crystal plasticity-based constitutive models

A Wessel[1,2], E S Perdahcıoğlu[3], A Butz[1], T van den Boogaard[3] and W Volk[2]

[1] Fraunhofer Institute for Mechanics of Materials IWM, Woehlerstrasse 11, 79108 Freiburg, Germany
[2] Chair of Metal Forming and Casting, Technical University of Munich, Walther-Meissner-Strasse 4, 85748 Garching, Germany
[3] Chair of Nonlinear Solid Mechanics, Faculty of Engineering Technology, University of Twente, Drienerlolaan 5, 7522 NB Enschede, The Nederlands

E-mail: alexander.wessel@iwm.fraunhofer.de

Abstract. Two crystal plasticity-based constitutive models that differ with respect to the flow rule (rate-dependent/rate-independent) and hardening law (phenomenological/physical-based), among other aspects, are compared with each other. To this end, both crystal plasticity-based constitutive models were deployed within a finite element framework to predict the texture-induced plastic anisotropy of an AA6014-T4 aluminium alloy considering uniaxial loading at 0°, 15°, 30°, 45°, 60°, 75° and 90° with respect to the rolling direction. The results of the stress-strain curves, the normalised yield stresses and the r-values demonstrate that both crystal plasticity-based constitutive models provide comparable results. Also, the experimental r-values were predicted with reasonable accuracy. Differences with respect to the experimental normalised yield stresses are discussed and were most likely caused by an additional direction-dependent mechanism.

1. Introduction

Information on plastic anisotropy in sheet metals is important for manufacturing sheet metal parts and for designing the necessary forming operations by using finite element simulations in particular. This plastic anisotropy is primarily caused by a preferred crystallographic texture, which develops during the rolling process of the sheet metal. To characterise plastic anisotropy of sheet metals, various experiments, such as uniaxial tensile tests in different directions, hydraulic bulge tests or plane strain tension tests, are typically applied [1, 2]. An alternative approach to analyse the plastic anisotropy of sheet metals are virtual experiments. Virtual experiments, or rather crystal plasticity simulations, utilise a crystal plasticity-based constitutive model to predict the plastic anisotropy based on the crystallographic texture. Thus, loading conditions that cannot be realised experimentally can be studied. For instance, Barlat et al. [3] performed virtual experiments on AA2090-T3 and AA6111-T4 aluminium sheets to characterise the plastic anisotropy regarding shear with respect to the thickness direction. In addition, there are many more examples in the literature [4-6], where virtual experiments were utilised to analyse the plastic anisotropy of sheet metals. However, most of these studies use slightly dissimilar crystal plasticity-based constitutive models that differ with respect to their mathematical formulation as well as their numerical implementation.

This study analyses the effect of two crystal plasticity-based constitutive models that differ with respect to the flow rule (rate-dependent/rate-independent) and hardening law (phenomenological/physical-based), among other aspects. To this end, both formulations are deployed to predict the texture-induced plastic anisotropy of an AA6014-T4 aluminium sheet. The results are compared with experimentally determined uniaxial tensile test in different directions with respect to the rolling direction (RD).

2. Material characterisation

2.1. *Electron backscatter diffraction*

AA6014-T4 aluminium sheets (Trademark Advanz™ 6F - e170) with a thickness of 1.0 mm were supplied by Novelis Switzerland SA. The crystallographic texture of these aluminium sheets was characterised by an electron backscatter diffraction (EBSD) measurement of the longitudinal cross-section. The EBSD measurement was conducted in a Zeiss Sigma 300 scanning electron microscope (SEM) equipped with an EBSD system by EDAX using APEX™ for data recording and OIM Analysis™ 8.6 for data processing. An area of roughly 2.5 mm x 0.9 mm using a hexagonal grid with a step size of 3.0 µm was scanned with an accelerating voltage of 20 kV. The EBSD data was analysed using the MATLAB toolbox MTEX 5.7.0 [7]. Only measurement points with a confidence index greater than 0.1 were considered for post-processing, as recommended by Field [8].

2.2. *Uniaxial tensile tests*

The mechanical material behaviour of AA6014-T4 was characterised by uniaxial tensile tests at 0°, 15°, 30°, 45°, 60°, 75° and 90° with respect to RD. Uniaxial tensile tests were performed on a ZwickRoell Kappa 50 DS uniaxial testing machine. The specimens were manufactured by water jet cutting and had a gauge length of 80 mm and gauge width of 20 mm in accordance with DIN EN ISO 6892-1, test piece type 2. All tensile tests were carried out until fracture using a constant engineering strain rate of 0.002 1/s. During the experiment, the change in the gauge length was measured in the longitudinal and transverse directions of the specimen using two tactile extensometers with accuracy class 0.5 according to EN ISO 9513. Three identical samples were tested for each direction. All uniaxial tensile tests were carried out two to three months after their final heat treatment at Novelis Switzerland SA.

3. Crystal plasticity modelling

3.1. *Crystal plasticity-based constitutive models*

The crystal plasticity-based constitutive models utilised in this study were available as UMAT user subroutines for the commercial finite element software Abaqus/Standard 2021. The first crystal plasticity-based constitutive model is based on the numerical framework presented in Kalidindi et al. [9] and is fully described in Pagenkopf [10]. A detailed summary of the second crystal plasticity-based constitutive model is given in Aşık et al. [11]. Hence, the following section focuses on the common fundamental equations of both formulations and their differences with respect to the flow rule (rate-dependent/rate-independent) and hardening law (phenomenological/physical-based).

The basis of both crystal plasticity-based constitutive models is the multiplicative decomposition of the deformation gradient into an elastic and plastic part:

$$\mathbf{F} = \mathbf{F}_\mathrm{e}\mathbf{F}_\mathrm{p}. \tag{1}$$

While the elastic part of the deformation gradient $\mathbf{F}_\mathrm{e}$ describes the reversible stretching and rotation of the crystal lattice, the plastic part of the deformation gradient $\mathbf{F}_\mathrm{p}$ defines the irreversible deformation due to crystallographic slip. Due to the multiplicative decomposition of the deformation gradient, the following applies to the velocity gradient:

$$\mathbf{L} = \dot{\mathbf{F}}\mathbf{F}^{-1} = \left(\dot{\mathbf{F}}_\mathrm{e}\mathbf{F}_\mathrm{p} + \mathbf{F}_\mathrm{e}\dot{\mathbf{F}}_\mathrm{p}\right)\mathbf{F}_\mathrm{p}^{-1}\mathbf{F}_\mathrm{e}^{-1} = \mathbf{L}_\mathrm{e} + \mathbf{F}_\mathrm{e}\mathbf{L}_\mathrm{p}\mathbf{F}_\mathrm{e}^{-1}. \tag{2}$$

The plastic part of the velocity gradient $\mathbf{L}_\mathrm{p}$ is defined as the sum of the shear rates $\dot{\gamma}$ acting on every slip system α:

$$\mathbf{L}_\mathrm{p} = \sum_{\alpha=1}^{n} \dot{\gamma}^{\alpha}\, \mathbf{m}^{\alpha} \otimes \mathbf{n}^{\alpha}. \tag{3}$$

The unit vectors $\mathbf{m}^{\alpha}$ and $\mathbf{n}^{\alpha}$ are the slip direction and the slip plane normal of the slip system, respectively. For face-centred cubic materials 12 slip systems, crystallographically called $\{111\}\ \langle 110 \rangle$, are considered in both crystal plasticity-based constitutive models.

3.1.1. Flow rule. With respect to the flow rule, the first crystal plasticity-based constitutive model utilises a rate-dependent approach. To this end, the plastic shear rate $\dot{\gamma}$ is derived by

$$\dot{\gamma}^{\alpha} = \dot{\gamma}_0 \left|\frac{\tau^{\alpha}}{\tau_\mathrm{c}^{\alpha}}\right|^{1/m} \mathrm{sign}(\tau^{\alpha}) \tag{4}$$

as a power law-type equation, where $\dot{\gamma}_0$ and m are the reference shear rate and the rate sensitivity of slip, respectively. The resolved shear stress is calculated by Schmid's law

$$\tau^{\alpha} = (\mathbf{C}_\mathrm{e}\mathbf{S}) : (\mathbf{m}^{\alpha} \otimes \mathbf{n}^{\alpha}) \tag{5}$$

with

$$\mathbf{C}_\mathrm{e} = \mathbf{F}_\mathrm{e}^\mathrm{T}\mathbf{F}_\mathrm{e}. \tag{6}$$

The term $\mathbf{C}_\mathrm{e}\mathbf{S}$ is also known as the Mandel stress tensor, with $\mathbf{S}$ denoting the stress tensor in the intermediate configuration. The slip direction $\mathbf{m}^{\alpha}$ and the slip plane normal $\mathbf{n}^{\alpha}$ of a slip system are defined in the intermediate configuration. In Equation (4), the critical resolved shear stress τ_c^{α} of a slip system α evolves according to a phenomenological hardening law, which is described in Section 3.1.2.

The second crystal plasticity-based constitutive model uses a rate-independent formulation of the flow rule. Thus, crystallographic slip only occurs when the resolved shear stress τ^{α} is equal to the slip resistance τ_f^{α}:

$$\phi^{\alpha} = \tau^{\alpha} - \tau_f^{\alpha} \leq 0. \tag{7}$$

Here, the resolved shear stress τ^{α} of a slip system α is defined on the base of the Cauchy stress tensor $\boldsymbol{\sigma}$ in the current configuration as

$$\tau^{\alpha} = \boldsymbol{\sigma} : (\tilde{\mathbf{m}}^{\alpha} \otimes \tilde{\mathbf{n}}^{\alpha}). \tag{8}$$

In contrast to Equation (5), the slip direction $\tilde{\mathbf{m}}^{\alpha}$ and the slip plane normal $\tilde{\mathbf{n}}^{\alpha}$ of a slip system are given in the current configuration.

3.1.2. Hardening law. The hardening law for the critical resolved shear stress τ_c^{α} in Equation (4) of the rate-dependent crystal plasticity-based constitutive model is described by a phenomenological approach. As defined in Lebensohn et al. [12], the hardening rate of the critical resolved shear stress $\dot{\tau}_c^{\alpha}$ is given by

$$\dot{\tau}_\mathrm{c}^{\alpha} = \frac{\mathrm{d}\bar{\tau}^{\alpha}}{\mathrm{d}\Gamma} \sum_{\beta=1}^{n} q^{\alpha\beta} \left|\dot{\gamma}^{\beta}\right| \tag{9}$$

with the extended Voce type hardening law of Tomé et al. [13]:

$$\bar{\tau}^{\alpha} = \tau_0 + (\tau_1 + \theta_1 \Gamma)\left[1 - \exp\left(-\frac{\Gamma\theta_0}{\tau_1}\right)\right]. \tag{10}$$

The quantities τ_0, τ_1, θ_0 and θ_1 are material-dependent parameters and are identical for all slip systems. While τ_0 and θ_0 describe the initial yield stress and initial hardening rate, the asymptotic hardening behaviour for large strains is characterised by τ_1 and θ_1. In Equations (9) and (10), Γ is the accumulated plastic shear strain over all slip systems n, which is expressed in terms of

$$\Gamma = \int_0^t \sum_{\alpha=1}^{n} |\dot{\gamma}^\alpha| \mathrm{d}t. \tag{11}$$

Interaction between two different slip systems α and β is incorporated by the interaction matrix $q^{\alpha\beta}$. The interaction matrix represents the latent hardening behaviour of a crystal. For self and coplanar slip systems, the entries of the interaction matrix are set to one. All other entries of the interaction matrix are specified by the parameter q.

The second crystal plasticity-based constitutive model utilises a physical-based hardening law. Therefore, a Taylor-type hardening model [14] is employed

$$\tau_f^\alpha = \tau_0 + \mu b \sqrt{\sum_\beta \tilde{q}^{\alpha\beta} \rho^\beta}, \tag{12}$$

where τ_0, μ and b are the lattice friction, the shear modulus, and the length of the Burgers vector, respectively. The variable ρ^β is the total dislocation density of a slip system β and is governed in accordance with Perdahcıoğlu et al. [15] as a linear ordinary differential equation

$$\dot{\rho}^\alpha = \frac{\dot{\gamma}^\alpha}{\gamma_s} (\rho_s - \rho^\alpha), \tag{13}$$

with ρ_s and γ_s denoting the saturation dislocation density and saturation shear rate, respectively. In the initial state, the total dislocation density ρ^α is defined by the initial dislocation density ρ_0. The variable $\tilde{q}^{\alpha\beta}$ in Equation (12) is the interaction matrix. In contrast to the first crystal plasticity-based constitutive model, this interaction matrix distinguishes six interaction types following Franciosi and Zaoui [16] and Kubin et al. [17]: self, coplanar, collinear, orthogonal, glissile and sessile.

3.2. Representative volume element

The representative volume element (RVE) for AA6014-T4 was generated by using the free software package Neper 3.5.2 [18]. It has a cubic shape with an edge length of 1.0 and contains 1000 grains. Each grain was assigned a crystallographic orientation, which was obtained from a reconstruction of an experimentally obtained orientation density function (ODF) using the MATLAB toolbox MTEX 5.7.0 [7]. The RVE was meshed by 40x40x40 linear hexahedral elements of type C3D8. As suggested by Schmidt [19], periodic boundary conditions were applied for constraining the RVE.

3.3. Crystal plasticity simulations

All crystal plasticity simulations were performed using the commercial finite element software Abaqus/Standard 2021. In that respect, two different kinds of crystal plasticity simulations were carried out: First, crystal plasticity simulations of the uniaxial tensile tests at 0° with respect to RD were performed to identify the crystal plasticity parameters describing the hardening behaviour for each crystal plasticity-based constitutive model. Thus, the hardening parameters were adjusted to match the experimental stress-strain curve. To this end, a reverse engineering approach using the commercial software LS-OPT 6.0 was used. Second, uniaxial tensile tests at 15°, 30°, 45°, 60°, 75° and 90° with respect to RD were simulated to predict the texture-induced plastic anisotropy. These simulations were performed on the base of a texture rotation, i.e. for each simulation, the RVE was loaded in RD with the texture rotated at the respective angle. As the grains of the studied RVE had an aspect ratio of 1.0, this approach is seen as reasonable.

4. Results

4.1. Experimental characterisation

The ODF in Figure 1 (a) shows a non-uniform distribution of crystallographic orientations. The maximum intensity amounts to 13 multiples of a random density (MRD). Areas with higher intensity are associated to the crystallographic orientation {001} <100> or rather the cube texture component.

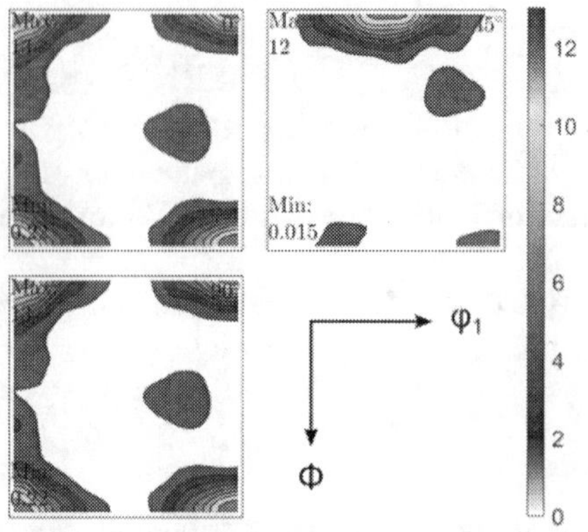

Figure 1. Orientation density function (ODF) of AA6014-T4 in the longitudinal cross-section shown as φ_2-sections from 0° to 90° in steps of 45° through the reduced Euler space.

The representative stress-strain curves of the uniaxial tensile tests at 0°, 15°, 30°, 45°, 60°, 75° and 90° with respect to RD in Figure 2 show a different hardening behaviour with respect to the angle considered. Therefore, AA6014-T4 exhibits a direction-dependent plastic material behaviour. As illustrated in Figure 5, the yield stresses were highest at 0° and lowest at 60° with respect to RD. The r-values were lowest at 45° with respect to RD.

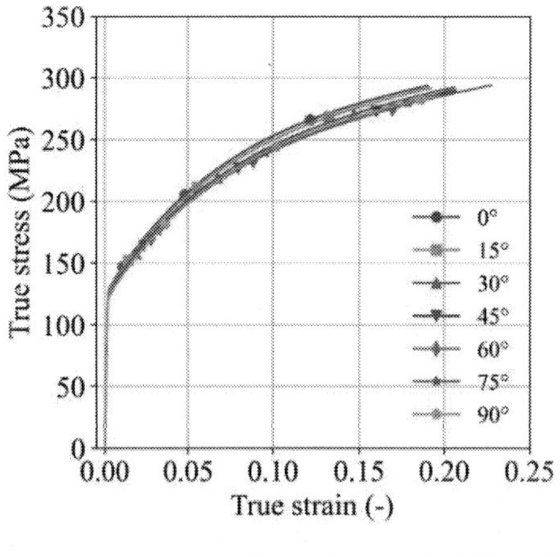

Figure 2. Stress-strain curves of uniaxial tensile tests at 0°, 15°, 30°, 45°, 60°, 75° and 90° with respect to the rolling direction (RD). One of three repetitions for each direction is illustrated.

4.2. Microstructure model

The RVE for AA6014-T4 as generated according to Section 3.2 is depicted in Figure 3 (a). The results of the parameter identification for the hardening parameters of both crystal plasticity-based constitutive models are illustrated in Figure 3 (b) in the form of the corresponding stress-strain curves in RD. Both formulations match the experimentally determined stress-strain curves in RD with good accuracy.

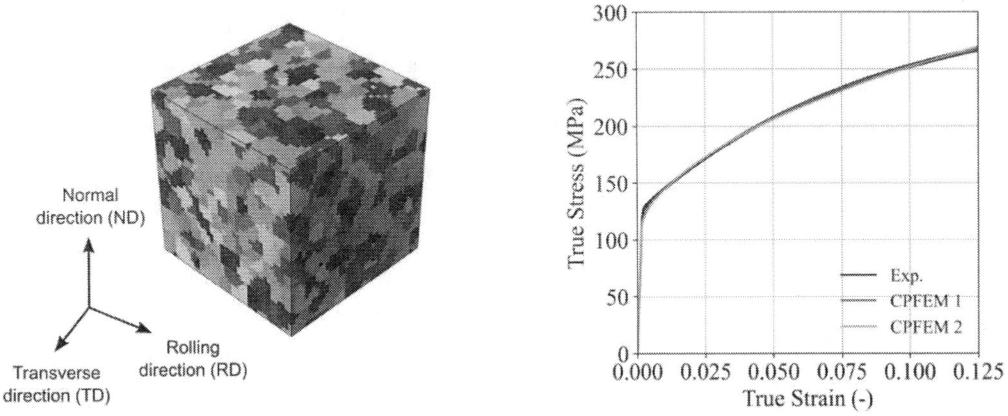

(a) (b)

Figure 3. (a) Representative volume element (RVE) for AA6014-T4. Each set of equally coloured finite elements represents one grain of the microstructure. (b) Stress-strain curves as predicted by both crystal plasticity-based constitutive models in comparison with experimental data in RD.

All crystal plasticity parameters as identified by the reverse engineering approach as well as taken from the literature are summarised in Table 1 and Table 2 for both constitutive models. The parameters C_{11}, C_{12} and C_{44} are the elastic constants describing the fourth-order elasticity tensor for cubic crystals.

Table 1. Parameters as representative for AA6014-T4 utilised for the first crystal plasticity-based constitutive model (rate-dependent, phenomenological hardening law).

C_{11} [a] (MPa)	C_{12} [a] (MPa)	C_{44} [a] (MPa)	$\dot{\gamma}_0$ [b] (-)	m [c] (-)	τ_0 (MPa)	τ_1 (MPa)	θ_0 (MPa)	θ_1 (MPa)	q [d] (-)
106750	60410	28340	0.001	0.0125	48.45	50.01	265.18	6.52	1.4

[a] Reference [20].
[b] Reference [4-5].
[c] Reference [6, 9].
[d] Reference [21].

Table 2. Parameters used for the second crystal plasticity-based constitutive model (rate-independent, physical-based hardening law) representing AA6014-T4.

C_{11} [a] (MPa)	C_{12} [a] (MPa)	C_{44} [a] (MPa)	b [b] (mm)	τ_0 (MPa)	ρ_0 (mm^{-2})	ρ_s (mm^{-2})	γ_s (-)	$\tilde{q}^{\alpha\beta}$ [c] (-)
106750	60410	28340	2.86e-07	18.0	1.00e7	6.89e8	0.21	

[a] Reference [20].
[b] Reference [15, 22].
[c] Reference [17].

4.3. *Prediction of plastic anisotropy*

Results of the stress-strain curves at 15°, 30°, 45°, 60°, 75° and 90° with respect to RD as predicted by both crystal plasticity-based constitutive models are illustrated in Figure 4. Both formulations show similar results. With respect to the experimental results, there are differences depending on the direction of the uniaxial tensile test. In Figure 4 (a), the difference between the experimental and predicted stress-strain curves at 15° with respect to RD is rather small. For larger angles, this difference increases to its maximum at 45°/60° with respect to RD before decreasing again. The best match between experimental and predicted stress-strain curves is given at 90° with respect to RD.

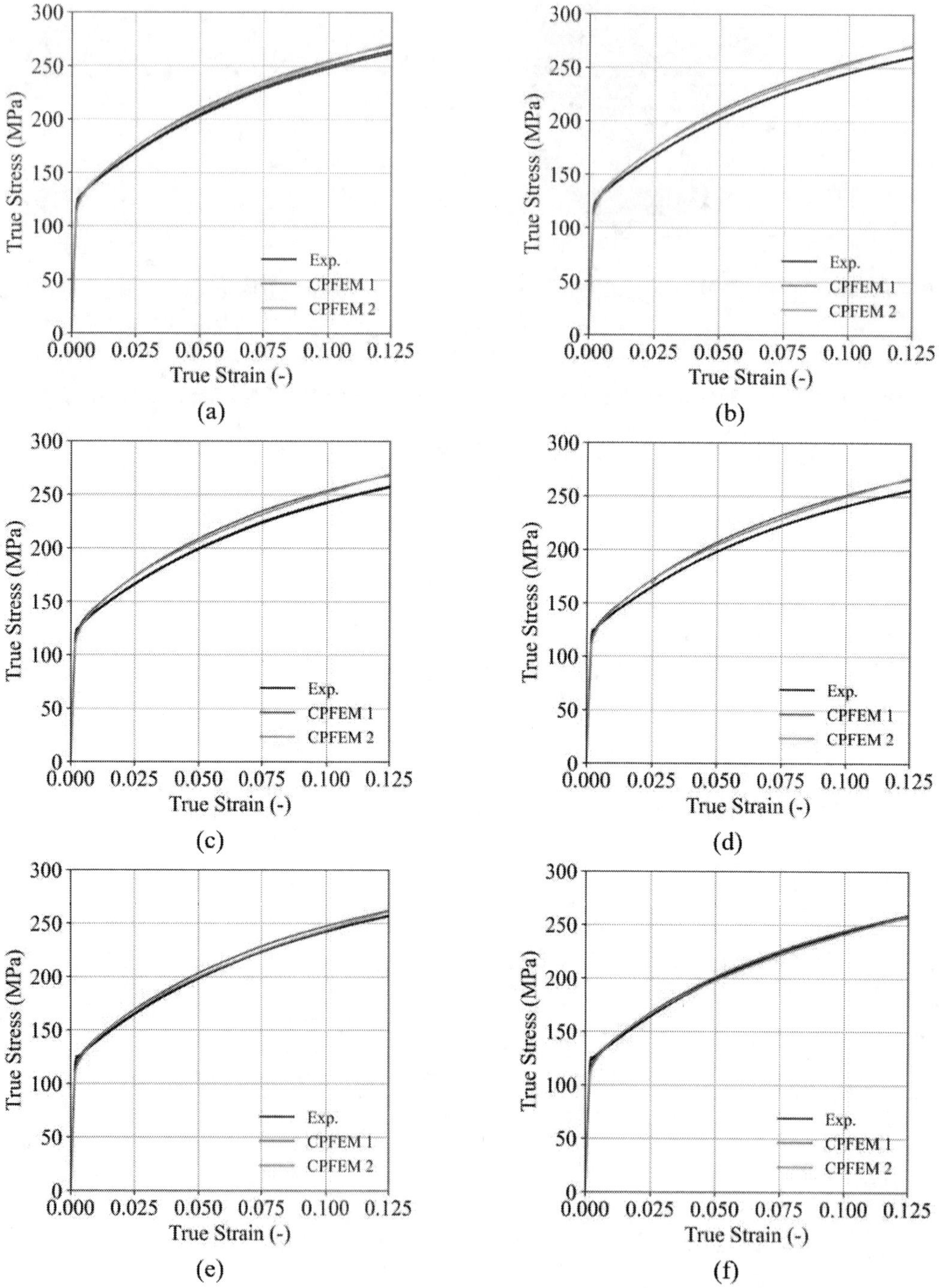

Figure 4. Stress-strain curves at (a) 15°, (b) 30°, (c) 45°, (d) 60°, (e) 75° and (f) 90° with respect to RD as predicted by both crystal plasticity-based models and compared with experimental data.

The normalised yield stresses were determined at a specific plastic work of 15.49 MPa, which corresponds to 0.08 true plastic strain and are shown in Figure 5 (a). In accordance with the results of the stress-strain curves in Figure 4, both crystal plasticity-based constitutive models are in good agreement. Again, there are differences with respect to the experiment. In the experiment, the normalised yield stress decreases from an angle at 0° to its minimum at 60° with respect to RD. In contrast to this, both crystal plasticity-based constitutive models predict a rather complementary behaviour with a maximum normalised stress at 30° with respect to RD. The results of the r-values in Figure 5 (b), which were analysed between 0.1 and 0.175 true plastic strain, demonstrate that both formulations can predict the experimental r-values with good accuracy.

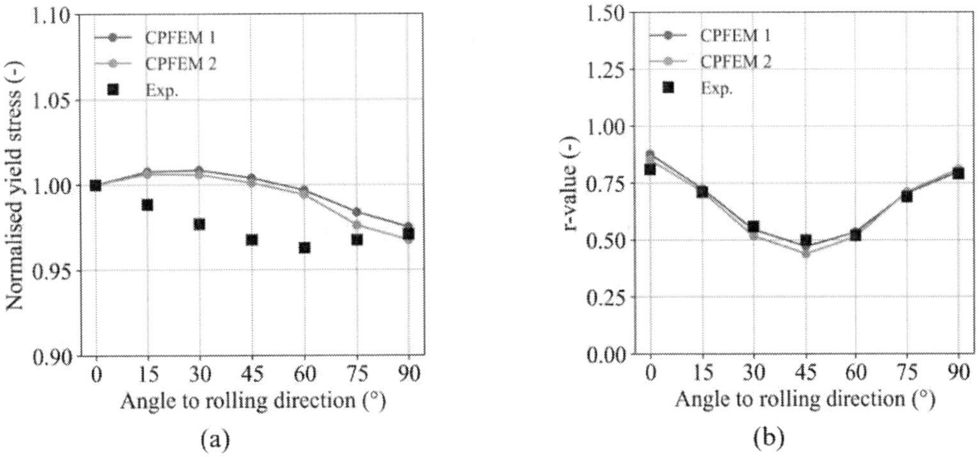

Figure 5. Experimental (a) normalised yield stresses and (b) r-values in comparison with results of both crystal plasticity-based constitutive models. Yield stresses were determined considering a specific plastic work of 15.49 MPa, which corresponds to a true plastic strain of 0.08, while r-values were analysed between 0.1 and 0.175 true plastic strain.

5. Discussion

The crystallographic and mechanical results in Section 4.1 indicate that the AA6014-T4 aluminium alloy exhibits a pronounced plastic anisotropy, which is in general agreement with data available in the literature. For example, Yoshida et al. [23] present uniaxial tensile test results at 0°, 15°, 30°, 45°, 60°, 75° and 90° with respect to RD for a non-specified AA6XXX aluminium alloy in T4 heat treatment condition. Both the normalised yield stresses and r-values reported are in good agreement with the experimental results generated in this study. In that respect, normalised yield stresses had their minimum at 60° with respect to RD, while the r-values were lowest at 45° with respect to RD. In addition, Kuwabara et al. [24] also performed uniaxial tensile test at 0°, 22.5°, 45°, 67.5° and 90° with respect to RD for aluminium sheets made out of AA6016-T4. Here, both normalised yield stresses as well as r-values were lowest at 45° with respect to RD.

With respect to the crystal plasticity simulations in Section 4.2 and Section 4.3, both crystal plasticity-based constitutive models are flexible enough to reproduce the experimental stress-strain curves in RD and show a rather similar behaviour. In addition, also the uniaxial stress-strain curves at 15°, 30°, 45°, 60°, 75° and 90° with respect to RD, the normalised yield stresses as well as r-values predicted by both crystal plasticity-based constitutive models are fairly similar. Overall, differences between both crystal plasticity-based constitutive models are rather small and seem to be negligible. With respect to the experimental data, the r-value is predicted with high accuracy for all directions. Also, the uniaxial stress-strain curve at 90° with respect to RD in Figure 4 (f) as well as the corresponding normalised yield stress in Figure 5 (a) are in good agreement with the experiment. In contrast to this, there are slightly higher differences between the crystal plasticity simulations and the experimental yield stresses at 15°, 30°, 45°, 60° and 75° with respect to RD. The highest differences appear at 45° with respect to RD. Overall, these differences may still be reasonable, but they lead to a kind of opposite curve progression. Similar results for AA6XXX aluminium alloys in T4 heat treatment condition, i.e. crystal plasticity simulations predict a different curve progression for the normalised yield stress with respect to the experiment, were already reported by several authors. Examples are given in Yoshida et al. [23], Gawad et al. [25], Hama et al. [26], Pagenkopf [10], Hirsiger [27] and Habraken et al. [28].

A possible explanation for this difference between crystal plasticity simulations and experiments can also be found in the literature. Kuwabara et al. [24] utilised uniaxial tensile tests at 0°, 22.5°, 45°, 67.5° and 90° with respect to RD for an AA6016 aluminium alloy in T4 and O heat treatment conditions. While the normalised yield stress at 45° with respect to RD was lowest for the AA6016-T4 aluminium alloy, it was highest for AA6016-O. At the same time, the r-values for both aluminium alloys were nearly similar. According to Kuwabara et al. [24], the main difference between both aluminium alloys was the existence of GP-zones in the case of the T4 heat treatment condition. Thus, it was concluded that for AA6016-O the plastic anisotropy was controlled by crystallographic texture alone, whereas that for AA6016-T4 was governed by crystallographic texture and GP-zones. Also, Yoshida et al. [23] pointed out that the difference between experimental results and crystal plasticity simulations for a non-specified AA6XXX-T4 aluminium alloy may be caused by GP-zones or rather GP-zone related effects. Moreover, in the early 70s, Hosford and Zeisloft [29] already suggested that precipitates – GP-zones are an early stage of precipitation – in AA2XXX aluminium alloys can affect the texture-induced plastic anisotropy. Hence, there seems to be some evidence that the plastic anisotropy of the AA6014-T4 aluminium alloy studied is controlled by crystallographic texture as well as GP-zone/precipitate related effects. As both crystal plasticity-based constitutive models predict the plasticity anisotropy of AA6014-T4 aluminium alloy based on crystallographic texture alone, some differences with respect to experiments – specifically for the normalised yield stress at 45° with respect to RD – appear. In that respect, it is assumed that an extension of the crystal plasticity-based constitutive models regarding the effect of GP-zones/precipitates could improve the prediction accuracy of the crystal plasticity approach with respect to age-hardenable aluminium alloys.

6. Conclusions

In this study, two crystal plasticity-based constitutive models that differ with respect to the flow rule (rate-dependent/rate-independent) and hardening law (phenomenological/physical-based), among other aspects, were utilised to predict the texture-induced plastic anisotropy for an AA6014-T4 aluminium alloy. In summary, the following conclusions can be drawn from the present work:

- Both crystal plasticity-based constitutive models are suitable to predict the texture-induced plasticity anisotropy for an AA6014-T4 aluminium alloy. In particular, the r-values are predicted with high accuracy.
- Differences with respect to the prediction of the normalised yield stresses are observed in both crystal plasticity models. These differences are most likely caused by GP-zone/precipitate related effects that are not included in either crystal plasticity-based constitutive model yet.

Acknowledgement

The authors gratefully acknowledge funding from the Federal Ministry for Economic Affairs and Climate Action via the German Federation of Industrial Research Associations – AiF (Arbeitsgemeinschaft industrieller Forschungsvereinigungen e.V.) within the scope of the programme for Industrial Collective Research (Industrielle Gemeinschaftsforschung, IGF), grant numbers 19707 N and 21466 N. The authors would also like to thank Jürgen Timm from Novelis Switzerland SA for providing the AA6014-T4 aluminium sheets. A. Wessel would like to acknowledge the Fraunhofer-Gesellschaft for funding his 5-month research stay at the University of Twente via the Fraunhofer International Mobility (FIM) programme.

References

[1] Banabic D, Barlat F, Cazacu O and Kuwabara T 2010 *Int. J. Mater. Form.* **3** 165–89
[2] Banabic D, Barlat F, Cazacu O and Kuwabara T 2020 *Int. J. Mater. Form.* **13** 749–87
[3] Barlat F, Aretz H, Yoon J W, Karabin M E, Brem J C and Dick R E 2005 *Int. J. Plast.* **21** 1009–39
[4] Zhang K, Holmedal B, Hopperstad O S, Dumoulin S, Gawad J, Van Bael A and Van Houtte P 2015 *Int. J. Plast.* **66** 3–30
[5] Zhang H, Diehl M, Roters F and Raabe D 2016 *Int. J. Plast.* **80** 111–38
[6] Wessel A, Morand L, Butz A, Helm D and Volk W 2022 *Preprint* arXiv:2211.00090
[7] Bachmann F, Hielscher R and Schaeben H 2010 *Solid State Phenom.* **160** 63–8
[8] Field D P 1997 *Ultramicroscopy* **67** 1–9
[9] Kalidindi S R, Bronkhorst C A and Anand L 1992 *J. Mech. Phys. Solids* **40** 537–69
[10] Pagenkopf J 2018 *PhD Thesis* Fraunhofer Verlag, Stuttgart
[11] Aşık E E, Perdahcıoğlu E S and van den Boogaard T 2020 *Materials* **13** 1795
[12] Lebensohn R A, Tomé C N and CastaÑeda P P 2007 *Philos. Mag.* **87** 4287–322
[13] Tome C, Canova G R, Kocks U F, Christodoulou N and Jonas J J 1984 *Acta Metall.* **32** 1637–53
[14] Taylor G I 1934 *Proc. R. Soc. Lond. A* **145** 362–87
[15] Perdahcıoğlu E S, Soyarslan C, Aşık E, van den Boogaard T and Bargmann S 2018 *Materials* **11** 1425
[16] Franciosi P and Zaoui A 1982 *Acta Metall.* **30** 1627–37
[17] Kubin L, Devincre B and Hoc T 2008 *Acta Mater.* **56** 6040–9
[18] Quey R, Dawson P R and Barbe F 2011 *Comput. Methods Appl. Mech. Eng.* **200** 1729–45
[19] Schmidt I 2011 *Comput. Mech.* **48** 579–90
[20] Frederikse H P R 2014 *CRC Handbook of Chemistry and Physics* ed Hayne W M (Boca Raton: CRC Press)
[21] Kocks U F 1970 *Metall. Mater. Trans. B* **1** 1121–43
[22] Hansen N and Huang X 1998 *Acta Mater.* **46** 1827–36
[23] Yoshida K, Yamazaki Y and Nakanishi H 2021 *Metals* **11** 1979
[24] Kuwabara T, Mori T, Asano M, Hakoyama T and Barlat F 2017 *Int. J. Plast.* **23** 164–86
[25] Gawad J, Banabic D, Van Bael A, Comsa D S, Gologanu M, Eyckens P, Van Houtte P and Roose D 2015 *Int. J. Plast.* **75** 141–69
[26] Hama T, Yagi S, Tatsukawa K, Maeda Y, Maeda Y and Takuda H 2021 *Int. J. Plast.* **137** 102913
[27] Hirsiger S 2020 *PhD Thesis* ETH Zurich, Zurich
[28] Habraken A M, Aksen T A, Alves J L, Amaral R L, Betaieb E, Chandola N, Corallo L, Cruz D J, Duchêne L, Engel B, Esener E, Firat M, Frohn-Sörensen P, Galán-López J, Ghiabakloo H, Kestens L A I, Lian J, Lingam R, Liu W, Ma J, Menezes L F, Nguyen-Minh T, Miranda S S, Neto D M, Pereira A F G, Prates P A, Reuter J, Revil-Baudard B, Rojas-Ulloa C, Sener B, Shen F, Van Bael A, Verleysen P, Barlat F, Cazacu O, Kuwabara T, Lopes A, Oliveira M C, Santos A D and Vincze G 2022 *Int. J. Mater. Form.* **15** 61
[29] Hosford W F and Zeisloft R H 1972 *Metall. Mater. Trans. B* **3** 113–21

Substitutive models of press deflections for efficient numerical die cambering

P Ottosson[1]*, J Pilthammar[2,3], D Wiklund[1], T Skåre[1] and M Sigvant[2,3]

[1] Division of Materials and Production, RISE AB, Vällaregatan 30, SE-293 38 Olofström, Sweden
[2] Volvo Cars, Dept. 81110 Strategy Development, Olofström, Sweden
[3] Blekinge Institute of Technology, Valhallavägen, 371 41 Karlskrona, Sweden

E-mail: peter.ottosson@ri.se

Abstract. Cost and time for stamping die tryouts are significant within the car industry. A major contributing factor is that elastic deflections of stamping dies and presses are usually not considered during the virtual die design and forming simulation phase. Active surfaces of stamping dies are only cambered based on previous experiences of tool types and stamping presses. However, almost all stamping dies and presses are unique, and available experiences are not valid for new sheet materials. This leads to component deviations and often several loops of tool adjustments are needed. Previously partners within the SMART Advanced Manufacturing research project CAMBER have developed advanced deflection measuring devices to quantify the elastic deformations of stamping presses. Using these measurements, cambering methodologies can be utilized in sheet metal forming simulations. In this paper numerical substitutive stamping press models are described which are capable of compensating for measured stamping press dynamics. The result show that a numerical compensated tool can improve the contact by over 80% compared to the corresponding contact without compensation.

1. Introduction

Cost and time for stamping die tryout are significant within the car industry, and elastic deflections of stamping dies and presses are most commonly not considered during the virtual die design and forming simulation phase. Because of this, stamping dies are only cambered based on previous experiences of tool types and stamping presses. However, almost all stamping dies and presses are unique, and available experiences are not valid for new sheet materials. This creates extensive problems when a toolmaker has optimised a tool through a spotting process. The toolmaker's stamping press has unique press dynamics and consequently problems arise when the tool is moved to a production unit with stamping presses with completely different characteristics. This means that a new adjustment process must conducted and sometimes with delays of several weeks. The same problem could also arise when tools need to be shifted between different lines within a production unit.

Several efforts have been carried out in order to reduce the number of try-out loops and time of the spotting processes. Numerical method for automatic contact pressure detection based on the blue pattern have been developed [1]. The objective is automation of the stamping die spotting process and stamping optimising contact patterns for the forming surfaces. Research efforts have also been aimed at developing measurement technology to detect stamping press deflections, both portable solutions and imbedded sensors in the presses [7]. The deflection strain has been measured by optical fibers with Fiber

Bragg Gratings (FBG). Research have also been focusing on cambering strategies and methods. Important breakthroughs in recent years enabling the cambering methodology consists of efficient simulation strategies for full scale simulations with elastic stamping dies [2-4]. The different simulation methods have in common that they, apart from blank and elastic die, need to include numerical representations of the elastic stamping press structure, e.g. [5-6]. The disadvantage is that they can be very computationally expensive when large sections of the stamping press are modeled. They are also often based on CAD data, which can create an idealized representation of reality. Consequently, the objective of this research work has been to generate a substitutive stamping press model which is computationally efficient and that the substitutive model can be calibrated with measurement data from real stamping presses.

2. Method

The activities to achieve numerical stamping die cambering is outlined below. Purpose of the implemented die compensation algorithm has up to now mainly been aimed at spring back compensation of sheet metal parts whereas here the objective is to reduces stamping die rework effort during try-out and ultimately to account for differences between try-out and production facilities.

2.1. Stamping press deflection measurements

Stamping press deflection measurements needs to be undertaken for the press at hand under loading conditions relevant to some generic forming situation in a sufficiently simplified manor. As described in [7] slender beams equipped with either fiber sensors, conventional strain sensors or linear displacement sensors can be used to obtain the deflections under well-defined loads measured by calibrated load cells.

Figure 1. The figure to the left show sensor beams temporarily mounted on a stamping press table [7]. The figure to the right shows the details of a sensor beam mounted on the press table optical fibers with Fiber Bragg Gratings (FBG).

Rotational stiffness associated with tilt was measured according to DIN55189-1 and DIN55189-2. Additionally input for calibration of substitutive models measurement data from fiber sensors and dial gauges have been used.

2.2. Substitutive model

To take stamping press deformations into account in FE modelling of sheet metal forming processes the influence from the press needs to be modelled somehow. Modelling the entire stamping press structure based on e.g., CAD models is one way to do this but with the drawback of creating large models that requires significant computational resources.

Another way to go about this is to use substitutive models that focuses on modelling the major influencing contributions in a simplified manor as compared to using full models of the stamping press. This approach also has the benefit over using nominal CAD models that they can account for deviations in stamping press performance from age, damage, material, and manufacturing deviations from the intended and ideal situation as present in a simulation model based on CAD-data.

Here the substitutive stamping press model is intended to mimic the most important features in the press as seen from the tooling perspective, namely the ability of the press table and ram to bend, deform and tilt, all of which influence the tooling via contact interfaces. To account for this, both table and ram is made of shell elements of various thicknesses in a framework resembling actual press tables. To the left in Figure 2 this substitutive press model is shown where the press table with underlying shell element framework is visible for both press table and ram. Both ram and cushion can tilt, a function that is provided by additional sets of shell element supporting the ram and blank holder. The tilt itself comes from use of joints (*CONSTRAINED_JOINT_STIFFNESS_GENERALIZED in LS-Dyna) in the finite element model that enables angular deviation simultaneously in two directions. The amount of tilt is determined by a non-linear resisting moment as function of the rotation angle and at the same time the degree of freedom in the forming direction is unconstrained. Surfaces of shell elements that supports the tilt function is shown in Figure 2 as yellow surfaces. A complete tool set for process simulation using the substitutive press model is shown in the right part of Figure 2.

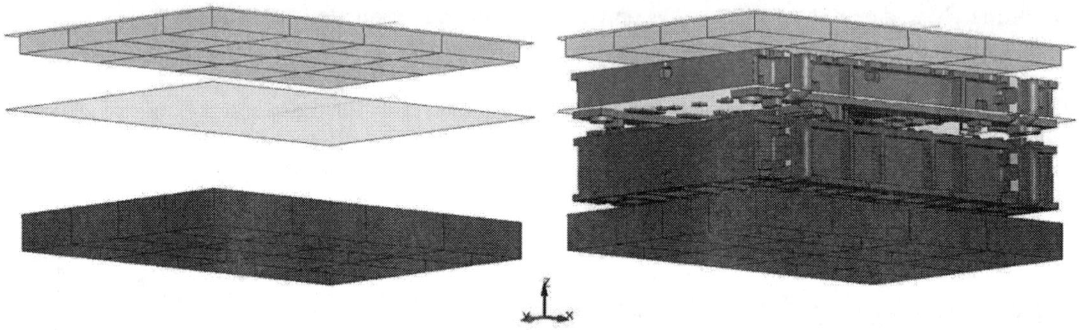

Figure 2. Substitutive press model (left) with ram (light grey), bolster (dark grey) and built in tilt function (yellow). Complete tool set model for process simulation using the substitutive press model (right).

2.3. *Optimisation of stamping press table layout*

Stamping press measurements can be accounted in a substitutive model by an optimization of the model with respect to e.g., the measured deflections that gives a response in the substitutive model which in a global sense resembles the physical press behavior in a FE simulation. It is then also possible to use process simulations with substitutive models to camber the forming tools to account for occurring deformations during forming in a certain stamping press.

Optimization of the substitutive model is done with the software LS-OPT using the measured deflection in the ram and bolster as described above. The response is evaluated with the composite function for curve matching implemented therein as responses in a D-Optimal sampling for a quadratic polynomial meta-model. The optimization objective is the minimization of the mean squared error of deflections with respect to the responses. An example of results from an optimisation based on stamping press measurements can be seen in Figure 3 where one iteration varying the associated design variables in terms of shell element thicknesses is shown. Each line in the diagram represents one design iteration and the data to match is represented by the crosses, hence it can be seen that among the selected designs there is an optimal solution for this design variable. It is obviously possible to also vary other influencing factors such as Young's modulus to achieve a desired behaviour corresponding to measurements. In Figure 4 the optimized press table with underlying shell element framework is visible. Here the colour scale indicates the deformations in z-direction when the press table is subjected to the calibration load however the deformations are exaggerated.

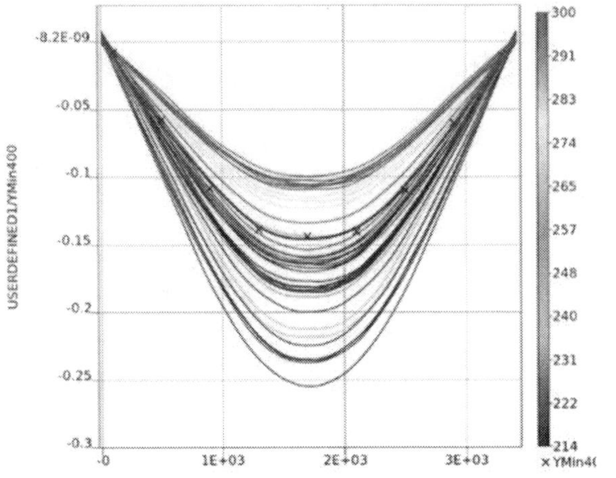

Figure 3. Output from one iteration of the optimisation of table stiffness. Cross symbols indicate response function from discretized press measurements, republished from [7].

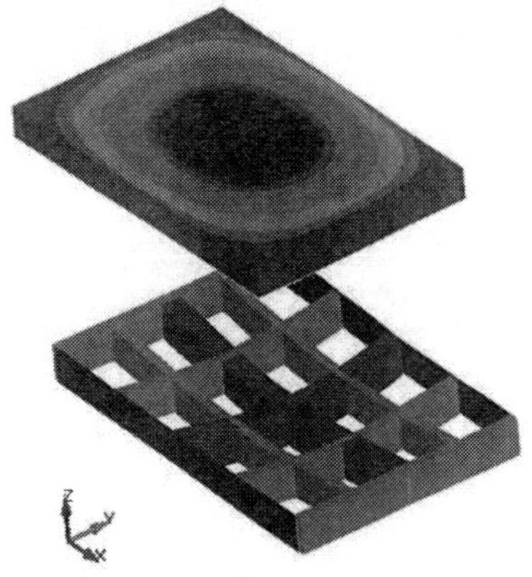

Figure 4. Exploded view of bolster with various thicknesses for the supporting framework. Deformations are exaggerated. Colour scale indicates the deformations in z-direction, republished from [7].

2.4. Stamping press tool modelling for stamping die cambering

The stamping press tooling needs to be modelled with enough detail to capture its fundamental deformation behaviour. A complex forming tool includes hundreds of components. To limit the time for simulation of the forming process it is necessary simplify press tooling taking main active features into account. At the same time, special care not to omit guiding pins and other features devised to control the tooling motions and that might influence the outcome in terms of contact surface deformations.

In this study a tool set to be used for forming of a physical part in the press characterised by press deflection measurements has been selected. One forming station in this tool is in focus, hence other stations in the tool has been omitted to being able to focus the computational resources on the major forming step which constitutes the challenge with stamping die spotting. In Figure 5 a cross section of the elastic tool model is displayed. To the left in Figure 5 the entire tools is seen together with the substitutive model of the press table the ram and the cushion for the blank holder and to the right a close up of the forming station is shown.

Tetrahedron elements are used to discretize the deformable tool body with a mesh resolution optimised for numerical speed rather than high resolution of local stresses. This means that small geometrical features not relevant to the major deformations can be meshed with low resolution.

The tooling interface with the substitutive press model corresponding to the press table and ram surfaces respectively is managed by means of numerical tied contacts. A simplified way of setting this up is to use tied contact of the tooling attachment to the substitutive press model. If more detail is required a regular contact definition between tool and press model can be used in combination with constrained nodal rigid bodies to model discrete clamping points between tool and press model.

The stamping die surface needs to have a resolution sufficient to shape the part geometry. That includes radii and other local features such as embossments. This causes mesh size gradients causing challenges in terms of mesh qualities for the deformable tool model. As long as deformations are small enough not to cause major element distortions these shortcomings are manageable.

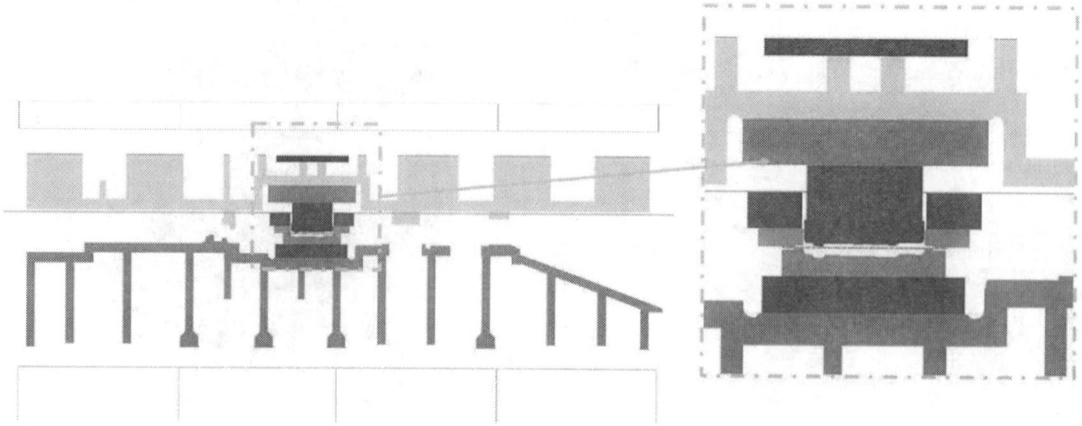

Figure 5. Cross section of elastic model of simplified tool set together with the substitutive press model with a close-up on the forming station at hand.

2.5. Numerical forming simulation of stamping press operation

The input for cambering stamping press tooling is a numerical simulation of the forming process at hand in combination with a substitutive press model based on press deflection measurements as described above that enables the capture of elastic deflections of stamping dies and presses. Forming simulation and corresponding elastic deformation in the tool and substitutive press model are integrated into one simulation eliminating the need for coupling of separate models. A benefit from using deformable tool surfaces also in the forming stage is that it enables the tooling surface to deform according to local thinning of the blank in the forming as well.

To keep simulation time reasonable the tool model needs to be simplified apart from the features most relevant for the deflection during forming operation. Simplifications needs to be undertaken without compromising the tool surface that is relevant for the geometry of the formed part. In addition to this the simulation is set up using mass scaling to enforce a larger explicit timestep than determined by the element size. The validity of this with respect to the dynamics is justified by review of the kinetic energy in relation to the total energy since the former is a minute fraction of the total energy.

The part being formed in this study is made of 1.4 mm thick DP600 material modelled by Barlat 3 parameter material model in combination with Belytschko-Tsay shell elements with five integration points through thickness. Mesh adaptivity is not invoked, the element size in the blank is sufficiently refined from the beginning. Automatic surface to surface contact between tool and blank with a Coulomb friction of 0.125 is used.

2.6. Compensation of tool and stamping press deflections

Spring back compensation of tool surfaces has been available in LS-Dyna for some time, recently under the keyword *INTERFACE_COMPENSATION_3D. The compensation algorithm was initially devised for compensation of spring back to reduce or eliminate tooling rework to achieve accurate part shape within tolerances after forming operations. The part shape deviations after forming are compensated by the algorithm by modifying the rigid tooling and it usually takes a few iterations to converge to deviations within a reasonable tolerance.

Here however we are dealing with elastically deformable three-dimensional elements in the tooling. Thus, it is necessary to associate the tooling surfaces with corresponding shell elements that can be used in the numerical compensation process. These shells, corresponding to the contact surface are used for the compensation by mirroring the coordinates of the deformed shape in the corresponding coordinates

IOP Conf. Series: Materials Science and Engineering 1284 (2023) 012060 doi:10.1088/1757-899X/1284/1/012060

of the undeformed shell. This works as long as the surface deformations are small enough not to violate element quality of the underlying tooling element. A difference with respect to regular spring back compensation is the input for the cambering, here we are looking at the maximum deformation occurring during the entire punch stroke and not the shape after elastic unloading as with usual spring back compensation.

Various options are available for the execution of compensation e.g., accelerator to speed up convergence, mesh smoothing and refinement, and scale factors to determine the amount of compensation which can contribute to reduce the need for iterations.

Several configurations for the cambering have been tested. Performing one iteration only, but with a scale factor of 1.5 turned out to be a good choice in this particular study. This is associated to both the simulation time and manual labour associated with a restart using updated geometry for the tooling surface after running the compensation algorithm. In the first iteration for compensation of the deformations it is necessary to determine a proper zero point for the deformed press and tool model as a pure offset of the tool surface is not of interest for the cambering. Here only the tool surface is compensated, i.e., the bulk tool material is left in its original shape.

3. Results

A substitutive press model was generated and calibrated as described above and simulation of the forming process using this setup has been undertaken. From this simulation a spotting image was generated which can be seen in **Figure 6**. Here a comparison with a spotting image from a manufactured sample using this very tooling and press is also displayed. The spotting image from the numerical simulation is based on the interface pressures from the simulation by setting the range to a maximum of 0.5 MPa, corresponding to what is denoted as "very soft contact" in [1].

A fairly good correlation, albeit not perfect can be seen in this comparison. For the physical sample approximately 60% of the blank has no or insufficient contact with the tool. For the corresponding simulation, the area without contact is even higher, 70%. Two major differences can be noted. On one hand for the embossments in the part better contact is achieved in the physical sample which might be attributed to sensitivity of the numerical contact in these narrow regions of the geometry. On the other hand, the blank holder pressure appears to be different in the simulation than in the physical sample since poor contact is indicated in this spotting image in these regions. Anyhow the correspondence between simulation and experiment is determined sufficient to proceed with the compensation algorithm.

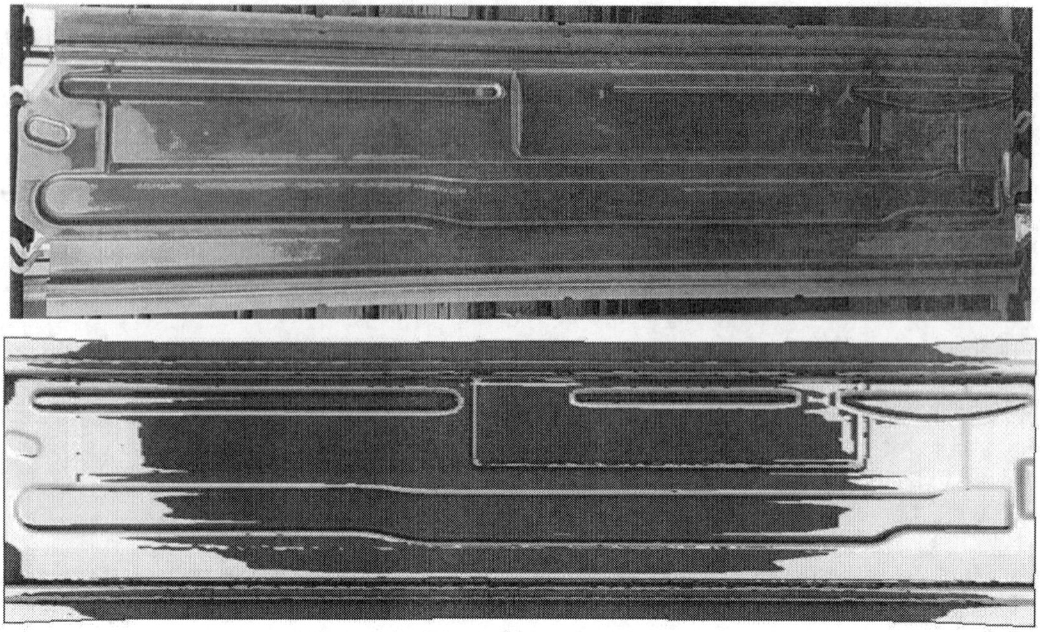

Figure 6. Spotting image of formed physical blank (top) regrettably the top picture is taken at an inclined angel with respect to the object. Corresponding numerical result (bottom) where colour range is set to 0.5 MPa corresponding to "very soft contact".

For the compensation, the deviations of the tooling surface from the original simulation are used as input, see Figure 7. By using the maximum deflections during the punch stroke in combination with the described compensation algorithm a cambered tool surface was generated and used for a subsequent simulation of the forming process to generate an updated spotting image as seen in Figure 8.

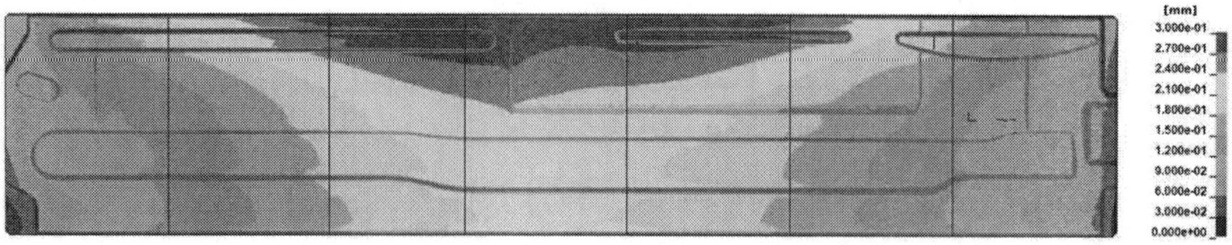

Figure 7. Maximum tool surface deflections in the forming process used for the cambering.

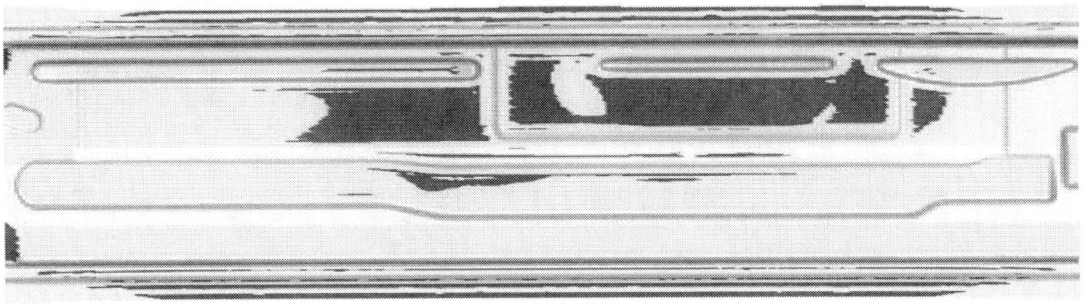

Figure 8. Spotting image after compensation of tool surface. Colour range is set to 0.5 MPa corresponding to "very soft contact".

Instead of using several iterations a scale factor was applied for the compensation. After some attempts it was found that a scale factor of 1.5 was preferable in this specific case. The resulting spotting image of this setting is displayed in Figure 8 which can be compared with the photo on top in Figure 6. A significant improvement was achieved leaving minor part of the blank without full contact with the tool. The area of the part not in contact with the tool is reduced by 80% compared to the simulation before cambering of the tool surface and slightly more than 70% when compared to the physical sample top in Figure 6. However, during the subsequent forming simulation using the compensated tool surface it was observed that a slight loss of blank holding occurred causing an increase in draw in of the blank (3 mm). At the same time the punch force required to close the tool properly increased by 7%. Thus, process adjustments might have to accommodate the virtual cambering to yield an even better outcome. In the process of virtual cambering care must also be taken not to adversely affect the final part shape with respect to tolerance requirements.

4. Discussion and conclusions

The objective of this research work has been to generate a substitutive stamping press model which is computationally efficient that can be calibrated with measurement data from real stamping presses. The result show that the spring back compensation algorithm implemented in LS-Dyna can be used for adjusting tool surface to take stamping press and tool deformations into account to minimise the effort for rework of tool and stamping die during try-out for production. This however rests on thorough measurement of stamping press deformations and associated substitutive modelling to be used in the process simulation for the input to the compensation algorithm. At the same time virtual cambering has the benefit of reduce the amount of manual expertise judgement and possibly also the labour involved in try-out processes since 80% reduction of area of part not in full contact was achieved in this study. The described attempt to improve the spotting image by means of cambering the tool surface demonstrated the ability but was unfortunately not pursued until the spotting was perfect.

In this work the cambering was performed on the entire forming surface to come in contact with the blank. As described in [8] elastic tool deformations can be divided into local and global and significant effects on local deformations can be attributed to locally high pressures especially in forming situations where e.g., thickening effects come into play. This will set entirely different demands on the tool modelling in terms of discretisation, mesh quality and computational resources. The geometry in this case however is a quite shallow u-shape thus not a deep drawing process and hence can be expected to have a more limited effect from e.g., thickening on the stamping die radii. In cases when more local variations of compensation are desired, variants of compensation strategies for instance by using locally varying scale factors in the compensation can be worked out.

A scale factor was used for the compensation algorithm instead of several iterations as used in regular spring back compensation. This has the benefit of saving both computational and labor resources in updating and handling the consecutive simulations necessary when performing several iterations.

When the compensation algorithm is used to reduce spring back, convergence is fast and most often only a few iterations are required to reach tolerances. The reason behind the strategy to use scale factor is the fact that deviations for the cambering in general are smaller than in the case of spring back compensation. Compensation of small deviations has been seen to result in a kind of oscillation were results flip from one side to the other as a result of consecutive iterations, hence it was judged to be productive to start off with a scale factor that sets the compensation off sufficiently to give a clear effect.

A more difficult aspect to possibly consider is taking the dynamic loading into account such as not only using the maximum deflection throughout the stroke but also looking at different phases of the forming process to study if a compromise of different loading situations can by useful for cambering. In a real complex situation this might even involve some means of active cambering.

Here the punch was subjected to compensation whereas the die was left unchanged. This choice was based on the practice in the tooling workshop to perform adjustments on the punch only. But it could also be valuable to investigate variants of the strategy e.g., whether the punch, the die or both are preferable for the cambering once the physical try-out is significantly reduced by virtual cambering. Physical verification of this virtual cambering is also a desired result to complete this work.

Acknowledgement
This work has been conducted under the CAMBER European research project; a project funded under the SMART EUREKA CLUSTER on Advanced Manufacturing programme. **Figure 3** and **Figure 4** republished from [7] with permission under creative commons 3.0.

References

[1] Zabala A, Llavori I, Sáenz de Argandoña E and Mendiguren J, Towards the automation of the die spotting process: Contact blue pattern decryption, Journal of Manufacturing Processes, Volume 58, 2020, Pages 1285-1296, ISSN 1526-6125,. DOI:10.1016/j.jmapro.2020.09.022.

[2] Haufe, A., Roll, K., and Bogon, P. (2008). Sheet metal forming simulation with elastic tools in LS-DYNA. In Proceedings of the 7th International Conference, NUMISHEET.

[3] Pilthammar, J, Towards Virtual Tryout and Digital Twins: Enhanced Modeling of Elastic Dies, Sheet Materials, and Friction in Sheet Metal Forming. Diss. 2020 Blekinge Tekniska Högskola.

[4] Zgoll, F., et al. "Virtual die spotting: Compensation of elastic behavior of forming presses." IOP Conference Series: Materials Science and Engineering. Vol. 651. No. 1. IOP Publishing, 2019, doi:10.1088/1757-899X/651/1/012021

[5] Tehel, R, Päßler T, and Bergmann M, "Effective FE models for simulating the elasto-mechanical characteristics of forming machines." The International Journal of Advanced Manufacturing Technology 106.7 (2020): 3505-3514, DOI:10.1007/s00170-019-04832-6

[6] Barton, G, Barth, C, Wohlmuth, M, Schafstall, H and Tannert, M, Praxis-orientated modeling of tool & machine elasticity in metal forming processes. Proc. 10th ICTP, 2011, 1270-1275.

[7] Pilthammar J et al New press deflection measuring methods for the creation of substitutive models for efficient die cambering 2021 IOP Conf. Ser.: Mater. Sci. Eng. 1157 012076 DOI 10.1088/1757-899X/1157/1/012076

[8] Larsson M. and Wiklund D. Numerical Study of Local Elastic Tool Deformation in Sheet Metal Forming Simulation (2022) Minerals, Metals and Materials Series DOI: 10.1007/978-3-031-06212-4_59

IOP Conf. Series: Materials Science and Engineering 1284 (2023) 012061 doi:10.1088/1757-899X/1284/1/012061

Impact of the process velocities on the quality of free-form bent parts

Lorenzo Scandola, Jeremias Tschannerl, Daniel Maier, Matthias Konrad Werner, Christoph Hartmann and Wolfram Volk

Chair of Metal Forming and Casting, TUM School of Engineering and Design, Technical University of Munich, Walther-Meissner-Str. 4, 85748 Garching, Germany

E-mail: lorenzo.scandola@tum.de

Abstract. The geometrical accuracy of the bent components as well as their surface quality are crucial parameters to evaluate the feasibility of a bent geometry in free-form bending. In the process the deformation of the part is obtained by displacing and rotating a movable bending head with respect to the fixed holder with a given relative velocity with respect to the feed velocity. High process velocities have a beneficial effect on the realisation of bending radii and angles, but induce high local solicitations on the semi-finished part, which can result in surface defects like indentations and buckles. In this work the impact of the bending head velocity as well as of the feed velocity on the bend's stability and the arising of indentations is investigated. Experimental tests with different head and feed velocity are conducted on mild, moderate and severe bends keeping the same kinematics. The resulting bending radii, angles and surface defects are evaluated. The results allow to identify the parameters having the highest influence on the quality of bent profiles and to propose a compromise between geometrical stability and surface quality.

1. Introduction

During the last years the free-form bending process has gained increasing interest as a suitable process for the production of 3D-bent components. Especially in the automotive field, current trends tend towards the use of high-strength materials with decreasing thickness and complex 3D-geometries, for which the full potential of free-form bending can be exploited. In comparison to traditional bending processes such as rotary-draw-bending [1], only a single tool is required in the process to reproduce arbitrary bending radii and angles along varying bending planes. Although smaller radii can still be achieved with rotary-draw-bending [2], the flexibility of free-form bending allows to reduce the tooling costs and the manufacturing time for complex parts. Similarly to three-roll push bending [3], the process is kinematics-based and the deformation of the part is realised through the relative motion of the bending head with respect to the fixed holder [4]. Different machine configurations have been proposed for realising the free-form bending principle. The first investigation were carried out by Murata [5], who invented the MOS bending machine. Other configurations are the torque superposed spatial bending (TSS) [6] and the Hexabend machine [7]. Recent studies involved the development of simulation models by Gantner [8], as well as the investigation of free-form bending as preforming step for the hydroforming process [9]. In addition, a broad experimental basis has been proposed by Guo on copper [10] and aluminum [11] semi-finished parts.

Although most of the actual investigations focus on the geometrical accuracy and the manufacturing of the parts in the required tolerance range, the surface quality of the parts is also an important quality criterion. In this paper, the impact of the bending head and the feed velocity on the geometrical stability as well as the surface defects on free-form bent components are investigated. First, the impact of the process velocities on the bending radii and angles is investigated. Successively, the resulting surface defects are quantified and correlated with the used process velocities. The results allow to introduce guidelines for the achievement of defect-free parts with the desired bending features.

2. Bending head kinematics for simple bends

In order to design the free-form bending process, the most crucial step is to generate the kinematics of the bending head. Through the control of the translation u and rotation a of the bending head with respect to the fixed holder, a curvature profile can be induced on the part. The relation between the target radius and the required translation degree-of-freedom u^* is experimentally determined with the machine calibration, for which different fitting curves can be used [12]. On the other hand, the achieved bending angle is determined by the feeding of the tube while the head reaches the target position. This results in the trapezoidal kinematics profile depicted in Fig. 1 (left).

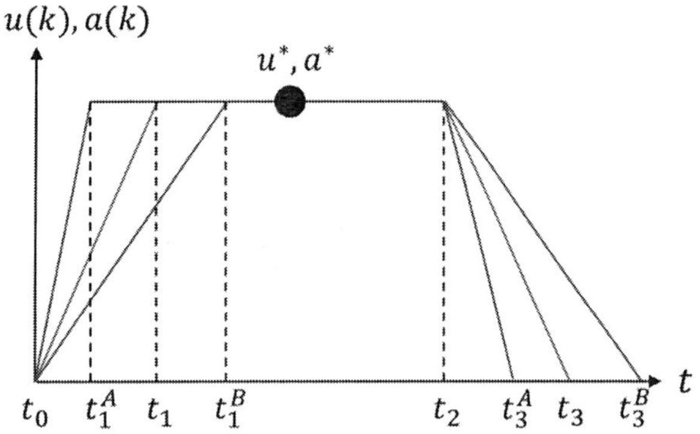

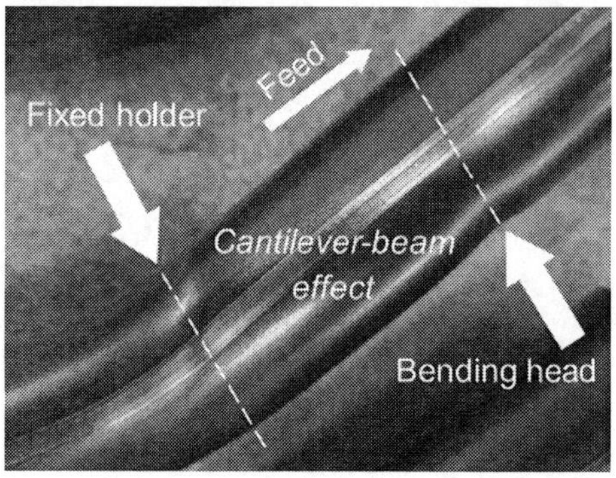

Figure 1. Kinematic profile for the realisation of a single bent (left); Cantilever beam - effect (right)

At the time t_0 the bending head starts moving towards the target position, represented by the translation u^*. Concurrently, the rotation a^* takes place. Between the time t_0 and t_1 the *run-in* transition takes place and the bending head reaches the target position. This can be achieved with different relative velocities of the head with respect to the holder, namely the run-in kinematics A and B, resulting respectively in a faster (red) and slower (blue) transition path with regard to the reference kinematics (green). In the time interval $t_1 - t_2$ the bending head remains in the *hold* position and most of the bending angle is realised. At the time t_2, the *run-out* transition takes place, which can also be realised with variable velocities. Finally, at t_3 the bending head is back in the original position and the bent operation is completed.

The arising of surface defects is typical of the transition zones. In particular during the *run-in*, the tube is loaded by the bending head and at the same time constrained by the holder in the opposite direction. This induces a *cantilever beam-effect* on the part. In this transition zone, the deformation of the part resembles that of a cantilever beam and results in an indentation at the contact points with no curvature development, which is more pronounced with higher velocities. On the other side, slower transition paths induce a lower holding time of the bending head in the constant zone, causing a smaller bending angle.

IOP Conf. Series: Materials Science and Engineering 1284 (2023) 012061 doi:10.1088/1757-899X/1284/1/012061

In the following, the sensitivity of the process to the arising of defects and the resulting bending radii and angles depending on the velocity parameters combined in Fig. 2 are quantitatively investigated. First the impact of the bending head velocity in the run-in with constant feed velocity is addressed. In addition, the impact of different feed velocity at constant bending head velocity is evaluated. Successively, the resulting indentations are considered. The results and discussion are presented in the following sections.

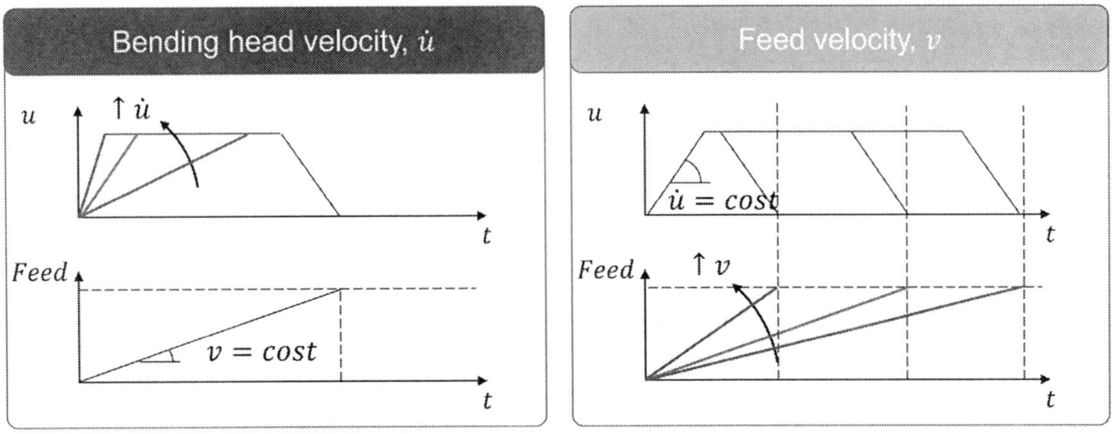

Figure 2. Investigated parameters influencing the surface defects and the bent geometry

3. Experimental setup

In order to evaluate the impact of the relative process velocities on the free-form bends, the experimental plan in Table 1 is carried out. This consists of three different run-in velocity, namely 20 %, 40 % and 80 % of the constant feed velocity of 50 mm/s. The tests are done for three radii ranges, namely moderate, mild and severe bending.

Table 1. Experimental plan for the investigation of the bending head relative velocity in the run-in

		Mild bent	Moderate bent	Severe bent
R_{tgt}	(mm)	$10 \cdot D = 337$	$5 \cdot D = 168.5$	$3 \cdot D = 101.1$
T_{tgt}	(deg)	50	60	90
$\dot{u}$	(%)	20, 40, 80	20, 40, 80	20, 40, 80
$\dot{u}$	(mm/s)	10, 20, 40	10, 20, 40	10, 20, 40
v	(mm/s)	50	50	50

The impact of the feed velocity is evaluated by conducting tests at 10 mm/s, 50 mm/s and 100 mm/s with the constant run-in velocity of 40% of the feed, namely 20 mm/s as shown in Table 2. Also in this case the tests are done for all radii ranges.

The tests are conducted on round tubes made of steel S235, with a nominal yield strength of 235MPa. The outer diameter is 33.7 mm and the thickness 2 mm. Every test is carried out 3 times in order to ensure statistical robustness, resulting in a total of 54 experiments. The tests are performed on a $6-DOFs$ free-form bending machine by J. Neu [13] available at the department.

Table 2. Experimental plan for the investigation of the feed velocity

		Mild bent	Moderate bent	Severe bent
R_{tgt}	(mm)	$10 \cdot D = 337$	$5 \cdot D = 168.5$	$3 \cdot D = 101.1$
T_{tgt}	(deg)	50	60	90
$\dot{u}$	$(\%)$	40	40	40
$\dot{u}$	(mm/s)	20	20	20
v	(mm/s)	10, 50, 100	10, 50, 100	10, 50, 100

4. Impact of velocities on the geometrical stability

4.1. Bending head relative velocity

The effect of the bending head velocity during the run-in is shown in Fig. 3. It can be seen, that considering a standard error of $3 \cdot \sigma$ the obtained bending radius remains constant independently on the relative velocity. This confirms, that the radius is solely a function of the target translation and is not influenced by the velocity and time in which it is reached. On the contrary, the bending angle decreases strongly with decreasing run-in velocity, Fig. 3 (left). The effect is particularly evident for severe bents, which require bigger translation values, resulting in angle differences of up to 20°. Although for the case of mild radii the angle fluctuations hide this effect, the trend is clearly identified in the moderate and severe radii ranges. In addition, it can be observed that the uncertainties are on the same order of magnitude for the radii as well as for the angles. Nevertheless, the scattering of the result decreases with smaller radii. This is a consequence of the non-linear relation between the obtained radius and required translation.

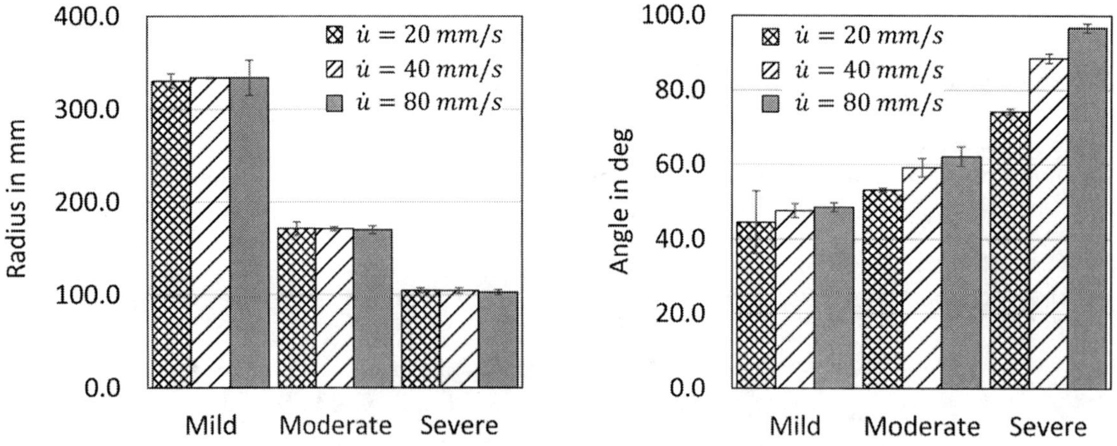

Figure 3. Influence of the relative bending head velocity on the bending radius (left) and angle (right)

4.2. Feed velocity

The impact of the feed velocity on the bends is shown in Fig. 4. It can be seen, that no relevant differences arise in the obtained radii. This is an additional proof that the bending radius correlates only with the bending head translation and has no time-dependency with the absolute feed velocity. As opposed to in the case of the bending head velocity, the feed velocity shows no influence in the resulting bending angle. This can be ascribed to the fact that a faster feed velocity only shrinks or expands the trapezoidal kinematic profile of the bent, but does not modify its shape.

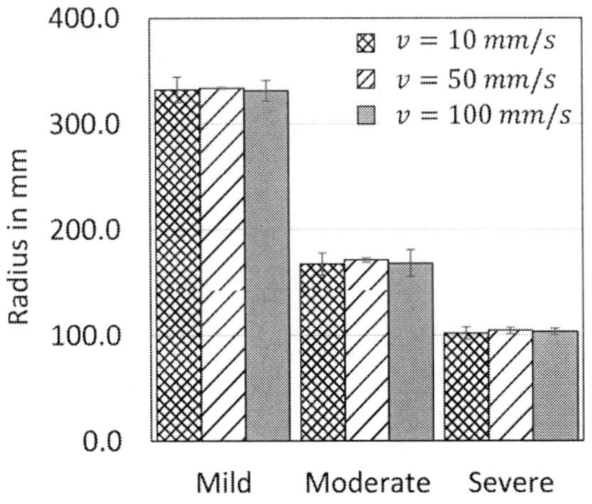

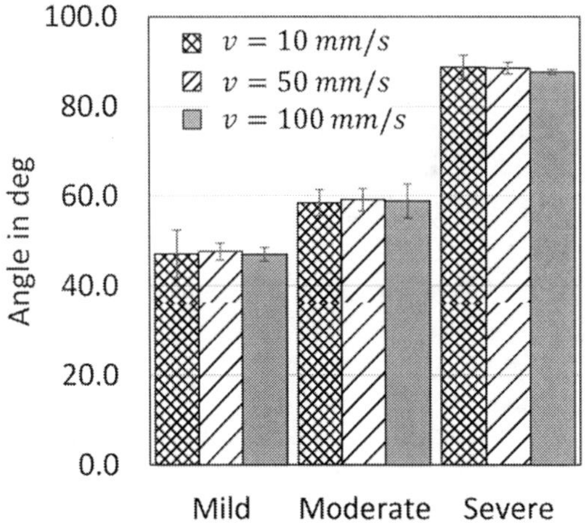

Figure 4. Influence of the feed velocity on the bending radius (left) and angle (right)

4.3. Discussion

The explanation of the different impact of the bending head and feed velocities can be explained as follows, and is schematically depicted in Fig. 5. The bending angle correlates with the integral of the bending head translation over time, namely:

$$A = \int_{t_0}^{t_3} u \cdot dt \tag{1}$$

which for constant feed velocity can be expressed in terms of the arc-length s as:

$$A = \frac{1}{v} \cdot \int_{t_0}^{t_3} u \cdot ds \tag{2}$$

and considering the run-in, hold and run-out transitions:

$$A = \frac{1}{v} \cdot \int_{t_0}^{t_1} u \cdot ds + \frac{1}{v} \cdot \int_{t_1}^{t_2} u \cdot ds + \frac{1}{v} \cdot \int_{t_2}^{t_3} u \cdot ds \tag{3}$$

.

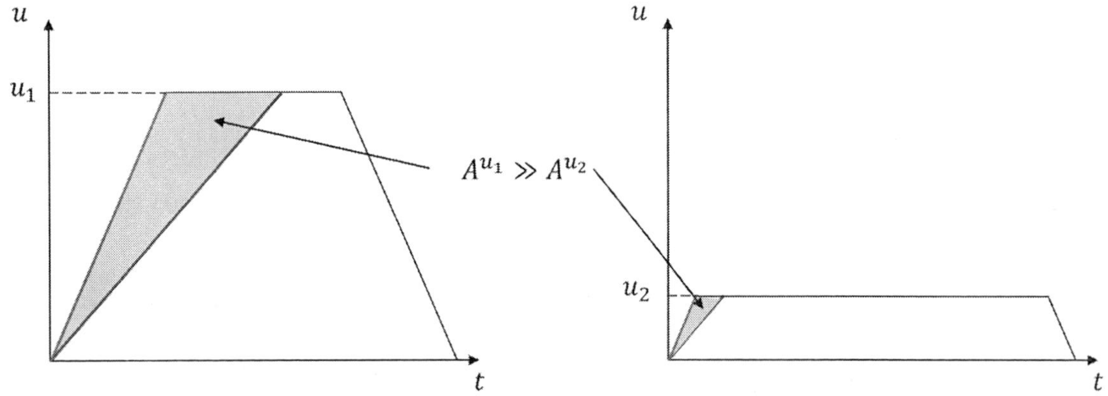

Figure 5. Bending angle sensitivity to the relative velocity of the bending head

Since changing the run-in velocity means displacing the time t_1, Eq. (3) shows that a decrease of the bending head velocity (blue in Fig. 5) increases the required time for reaching the target position and causes a decrease of the subtended area A, and so of the bending angle. In addition, the smaller the radius, the more pronounced is this effect, since smaller radii correlate with higher bending head translations u, as shown in Fig. 5 where $A^{u_1} \gg A^{u_2}$. On the other hand, Eq. (3) implies that a variation in the feed velocity does not modify the resulting bending angle, since the time extremes t_0, t_1, t_1 and t_3 are scaled accordingly.

5. Impact of velocities on the surface defects

5.1. Bending head relative velocity

The effect of the bending head relative velocity on the surface quality of the part is shown in Fig. 6. In order to quantify the arising surface defects, the indentation induced by the movement of the bending head, *head indentation*, as well as the one caused on the opposite side of the profile by the holder, *holder indentation*, are considered. These are calculated as normal deviations from the target geometry at the investigated location.

It can be observed, that the smaller the radius, the more prominent the indentation on the profile becomes. Especially for velocities equal to 40% and 80% of the feed, the transition from mild to moderate and to severe radii results approximately in a factor 2 increase in the head and holder marks respectively. Nevertheless, lower bending head relative velocities have a beneficial effect on the surface quality. This effect is particularly evident for moderate and severe radii, where the head indentation can be markedly reduced up to 50% by setting a velocity equal to 20% of the feed. This hold true also for the holder indentation which follows a similar trend to that of the head and stays in the same order of magnitude. If adopting as criterion that indentations smaller than 0.5 *mm* are not considered as surface defect, it can be concluded that relative bending head velocities around 20% allow to manufacture parts with no surface defects in the mild and moderate radius range. For the severe radius range the reduction of velocity is still not enough to obtain a defect-free component.

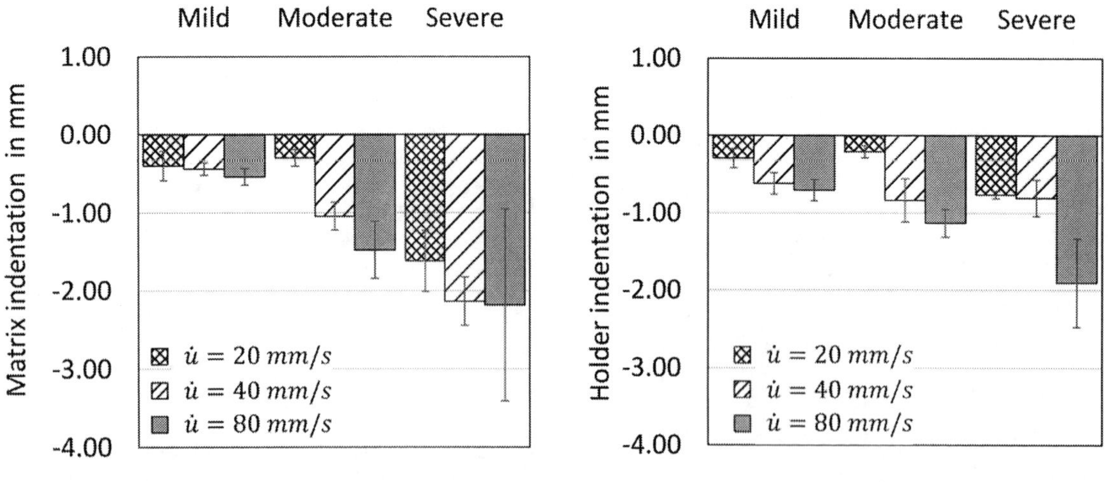

Figure 6. Influence of the relative bending head velocity on the head indentation (left) and the holder indentation (right)

5.2. Feed velocity

The influence of the feed velocity on the indentation is shown in Fig. 7. The resulting head indentations, Fig. 7 (left), confirm that the most impacting factor is the target radius range, namely the bending head translation. Severe radii show a mean head indentation of about 2 mm, while moderate and mild assess respectively on 1 mm and 2 mm respectively. In addition, it should be noted that the severe radii result show a higher dispersion with respect to the other datasets. This is due to the more stochastic distribution of the buckles on the tube for higher bending head translations, which occur as an instability phenomenon when the mandrel cannot support the part anymore. In addition, the semi-finished part tolerances contribute to the scattering of the results, as the outer diameter tolerance affects the value of the measured indentation. Nevertheless, the feed velocity does not affect the results significantly, since for every radius range the mean values of feed velocity of $10m/s$, $50mm/s$ and $100mm/s$ remain on the same values considering the standard error. The same behaviour is observed for the holder indentation. Here, it can be additionally concluded, that not only the feed velocity but also the radius range are not relevant factors. The greatest effect on the holder indentation is the relative bending head velocity, as shown in Fig. 8 (right).

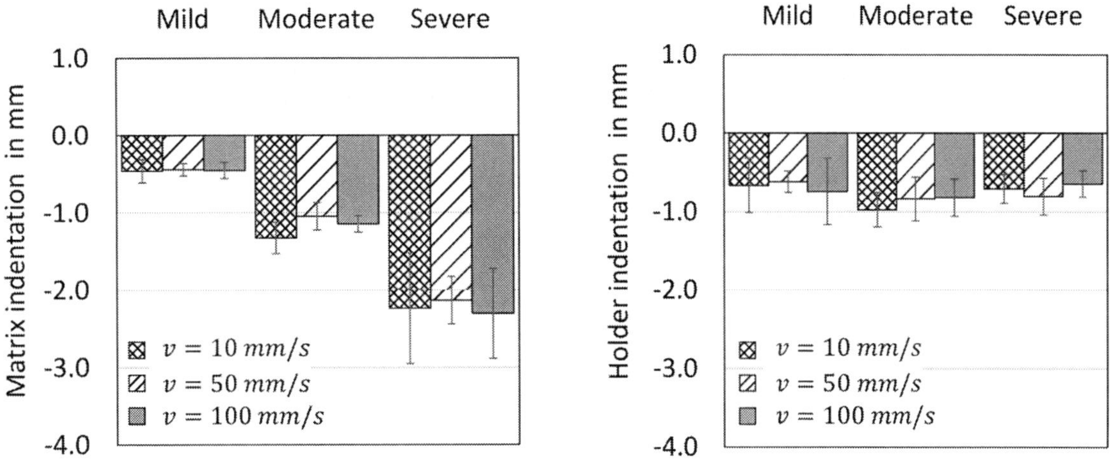

Figure 7. Influence of the feed velocity on the head indentation (left) and the holder indentation

5.3. Discussion

In order to understand the impact of the process velocities on the surface quality, the indentation profiles on the head and holder sides during the process are commented. When the bending begins, the bending head is translated upwards and hits the tube from the bottom, point A in Fig. 8 (above, left). Concurrently, the fixed holder opposes the part deformation and exerts a force on the tube, point D in Fig. 8 (below, left). After the bending head reached the target position, the part is continuously fed through the head and hits again its lower edge, point B in Fig. 8 (above, left), before entering the constant bending zone. Finally, the head begins the run-out and the bending operation is over, point C in Fig. 8 (above, left).

The effect of the relative bending head velocity during the whole bending operation is shown in Fig. 8 (above, right). Low relative velocities, such as 20% of the feed (blue), do not result in appreciable head indentations at the point A, while higher velocities correlate with deeper indentations. In addition, the head indentation at point B is wider spread, the lower the adopted velocity. Similar conclusions can be drawn for the holder indentations. On this surface defects, the impact of higher relative velocities is more evident, whereas the width of the peaks is not modified.

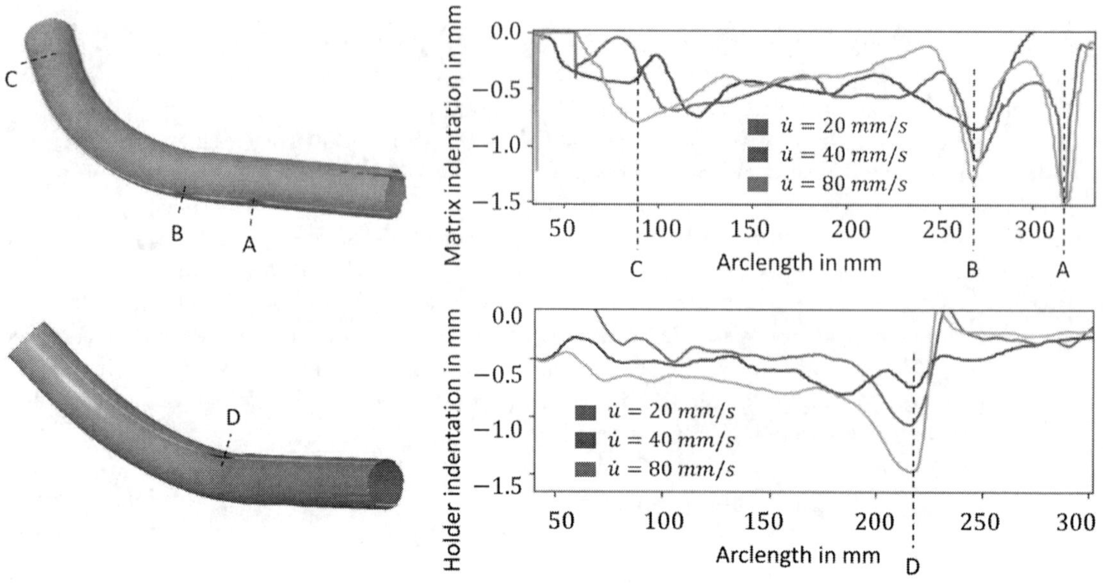

Figure 8. Sensitivity of indentation profile to bending head velocity on head (above) and holder (below)

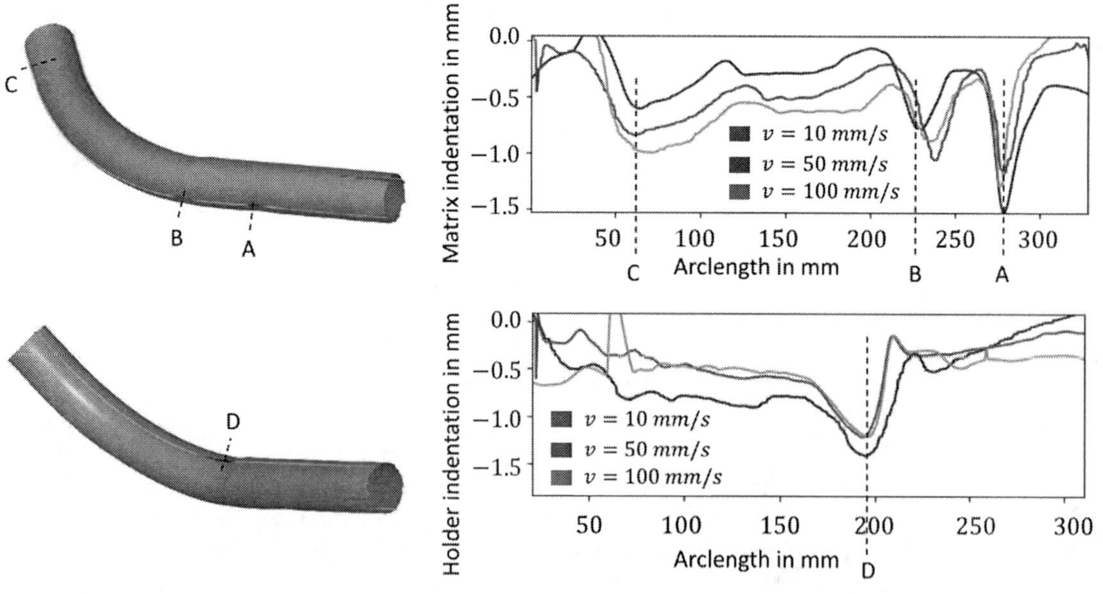

Figure 9. Sensitivity of indentation profile to feed velocity on head (above) and holder (below)

With respect to the feed velocities, the indentation profiles in Fig. 9 on the head (above, right) and the holder (below, right) are obtained. It can be observed, that the head indentation profiles are very similar in shape. The head indentation, point A, achieves similar results for every feed velocity considering the standard error. The second peak, point B, is slightly displaced for the case $v = 10mm/s$, but their widths are comparable. This confirms, that a higher feed velocity shrinks the kinematics profile only, without changing its ground shape. Finally, the run-out peak, point C, is not affected by adopting higher feed velocities. Also the holder indentation profile is similar for the three velocities. The reference indentation, point D, is equally deep for every test, confirming the results of Fig. 7 (right).

6. Summary and conclusions

The aim of this work was to investigate the impact of the process velocities on the geometrical accuracy of the part as well as on the surface quality of the bent components. The investigated process parameters are the relative bending head velocity, namely 20%, 40% and 80% of the feed, as well as the feed velocity, taken as $10mm/s$, $50mm/s$ and $100mm/s$. These have been investigated performing a total of 54 tests tubes made of stees $S235$ of diameter $33.7mm$ and $2mm$ thickness. In order to evaluate the geometrical accuracy, the resulting bending radius and angle have been considered, while the surface quality has been evaluated by measuring the head as well as the holder indentation on the tube, due to the *cantilever beam-effect*. The result shows that:

- The bending radius only depends on the bending head translation and rotation, and does not correlate with the process velocities in the case of the investigated material. In general, this holds true if the material under investigation is not strain-rate sensitive.
- The bending angle decreases strongly with decreasing run-in velocities and is not sensitive on the feed velocity.
- Bigger radii show higher scattering in the radius and angle results.
- The smaller the radii the deeper the head and holder indentations.
- Relative bending head velocity of 20% allow for defect-free parts in the mild and moderate radius range.
- Smaller radii show higher dispersion in the indentation results because of the semi-finished part tolerances and the stochastic of buckling instability.

Therefore, it is suggested as guideline to use a relative bending head velocity around 20% where the surface quality is the first quality criterion. In the case the geometrical accuracy is more important and small transitions angles have to be performed, higher velocities can be used. Nevertheless, velocities in the range of 80% will result in indentations in the range of $2mm$. As a compromise, a relative velocity of 40% can be suggested. Regarding the feed, the medium velocity of $50mm/s$ can be generally suggested as a guideline, turning to $100m/s$ if fast manufacturing is required and to $10mm/s$ for critical bents. This works contributes to a wider aim to build up a phenomenological model for free-form bending, with the aim of predicting the bent properties in terms of geometry and surface quality by addressing the bending kinematics. Further investigations are planned for semi-finished parts consisting of different cross-sections as well as materials.

Acknowledgments

Acknowledgments The authors would like to thank the company Hörmann Automotive GmbH for funding the project and the company J. Neu GmbH for providing the bending machine and technical support.

References

[1] VDI-Gesellschaft Produktion und Logistik. *VDI 3430: Rotationszugbiegen von Profilen.* Berlin, 2014.

[2] Bernd Engel, Christian Gerlach, and Stefan Cordes. “Biegemomentenabschätzung des Dornbiegeverfahrens”. In: *MM - Maschinenmarkt* (2008).

[3] Andrea Ghiotti et al. "Springback measurement in three roll push bending process of hollow structural sections". In: *CIRP Annals* 66.1 (2017), pp. 289–292. ISSN: 00078506. DOI: 10.1016/j.cirp.2017.04.119.

[4] M. K. Werner et al. "Motion profile calculation for freeform bending with moveable die based on tool parameters". In: *Proceedings of the 24th International Conference on Material Forming*. Ed. by ULiège Library. 2021. ISBN: 978-2-87019-302-0. DOI: 10.25518/esaform21.1879.

[5] Makoto Murata. "Tube Bending of New Generation by MOS Bending Machine". In: *1st International Tube and Profile Bending Conference*. Ed. by TU Dortmund. Institut für Umformtechnik und Leichtbau, 2011, pp. 9–18.

[6] Matthias Hermes. *Neue Verfahren zum rollenbasierten 3D-Biegen von Profilen*. Dortmunder Umformtechnik. Shaker, 2012. ISBN: 9783844007930.

[7] Michael Hoffmann. *HexaBend – Freiformbiegen auf einer parallelkinematischen Biegemaschine*. DOI: 10.2314/gbv:769920128. URL: https://www.tib.eu/suchen/id/TIBKAT:769920128/.

[8] P. Gantner et al. "The Development of a Simulation Model and the Determination of the Die Control Data for the Free-Bending Technique". In: *Proceedings of the Institution of Mechanical Engineers, Part B: Journal of Engineering Manufacture* 221.2 (2007), pp. 163–171. ISSN: 0954-4054. DOI: 10.1243/09544054JEM642.

[9] N. Beulich, J. Spoerer, and W. Volk. "Sensitivity analysis of process and tube parameters in free-bending processes". In: *IOP Conference Series: Materials Science and Engineering* 651.1 (2019), p. 012031. ISSN: 1757-8981. DOI: 10.1088/1757-899X/651/1/012031.

[10] Xunzhong Guo et al. "Free-bending process characteristics and forming process design of copper tubular components". In: *The International Journal of Advanced Manufacturing Technology* 96.9-12 (2018), pp. 3585–3601. ISSN: 0268-3768. DOI: 10.1007/s00170-018-1788-1.

[11] Xunzhong Guo et al. "U-R relationship prediction method for aluminum alloy circular tube free-bending process based on sensitivity analysis of material parameters". In: *The International Journal of Advanced Manufacturing Technology* 99.5-8 (2018), pp. 1967–1977. ISSN: 0268-3768. DOI: 10.1007/s00170-018-2614-5.

[12] L. Scandola et al. "Optimization of the Calibration Process in Freeform Bending Regarding Robustness and Experimental Effort". In: *Production at the Leading Edge of Technology*. Ed. by Mathias Liewald et al. Lecture Notes in Production Engineering. Cham: Springer International Publishing, 2023, pp. 170–178. ISBN: 978-3-031-18317-1. DOI: 10.1007/978-3-031-18318-8{\textunderscore}18.

[13] J. Neu GmbH. *Technische Daten: 6-Achs-Technologie*. Ed. by J. Neu GmbH. 2018. URL: http://www.neu-gmbh.de/site/de/produkte/biegen/nissin/technische-daten.php.

IOP Conf. Series: Materials Science and Engineering 1284 (2023) 012062 doi:10.1088/1757-899X/1284/1/012062

Study of the Influence of part features on springback and its robustness

Chao Niu[1], **Zhili Hu**[2], **Jian Zhang**[3], **Fei Han**[4]

[1] Researcher, Baoshan Iron & Steel Co., Ltd., Shanghai, China;
[2] Associate professor, School of Autometive Engineering, Wuhan University of Technology, Wuhan, China;
[3] Researcher, Baoshan Iron & Steel Co., Ltd., Shanghai, China;
[4] Lead Researcher, Baoshan Iron & Steel Co., Ltd., Shanghai, China;

E-mail: niuchao@baosteel.com

Abstract. The influence of part geometry on twist springback was studied by a A-pillar inner plate. The part structural parameters was divided into cross-section and longitudinal axis, and the response surface model was established to analyze the influences of various structural parameters on the twist springback. The results show that the straight line 1, straight line 3, angle 2 and angle 4 of the cross section have a significant effect on the twist springback. In addition, the twist springback increases continuously with increasing of total length and ratio of arc to straight line and the decreasing of radius in the longitudinal axis. After adding features through the angular edges along the longitudinal direction, the maximum springback at the part end drops from -10.92mm to -2.934mm. The springback robustness analysis of this part were also analyzed, the results show that the maximum springback is strongly related to material yield strength, and the robustness is improved greatly after adding part features.

1. Introduction

In recent years, with the increasingly strict regulations on energy conservation and environmental protection, lightweight technology has been developing rapidly in various industrial fields. As one of the lightweight materials, high-strength steel is widely used in aerospace and automobile fields by virtue of its high strength and light quality. However, in the application of automobile body parts, serious springback and its rules is difficult to master [1]. At present, a large number of scholars are studying the influence of stamping process and stamping mold on springbcak, and some scholars have begun to study the relationship between the structure of parts and its springback. Chetan et al. performed numerical simulations of longer 3D channels and predicted springbacks at different cross sections, and found that different cross sections produce different springbacks, and the influence of flange springback is large [2]. HongzhouLi et al. introduced a torsional rail to have a deep understanding of the influence of design parameters and process parameters such as cross-section width transition ratio, corner, tensile bar depth, material strength grade and so on on the twist springback [3]. LiXue et al. developed a simple C-shaped guide rail and proposed a reasonable method to calculate the twist springback of long members. They found that some areas of the radius of the mold to control the bending moment to minimize the twist springback, and preliminarily concluded the relationship between the radius ratio of the mold and the twist springback [4-5]. At present, the research on the influence of the geometric structure of parts on the springback is limited to simple parts, and it cannot provide effective guidance for the design of complex parts such as automobile body parts.

In this paper, we take a typical A pillar inner plate as an example and establish the mathematical model of the part springback and section parameters to obtain the relationship between the part structure and the torsion springback; Further, we make the sensitivity analysis of the material properties, process parameters, and find out the possible scheme to improve the robustness of the parts through the structural optimization to reduce the fluctuation of the part springback.

2. Definition of part parameters and CAE model establishment

2.1. Part structure analysis

The A pillar inner panel is a typical slender and beam part, as shown in Figure 1. According to the structural form, they can be decomposed into transverse section, longitudinal axis of space and local features. The longitudinal axis of space is a three-dimensional curve, which as shown in Figure 2 can be divided into circular arc segment and straight line segment. The parameters are the total length, arc vs line ratio and radius, where the total length is the sum of straight length and arc length, and the arc vs line ratio is the ratio of arc length and straight length. The cross-sectional structure is shown in Figure 3 with 13 parameters including 5 lines, 4 clips and 4 rounded corners.

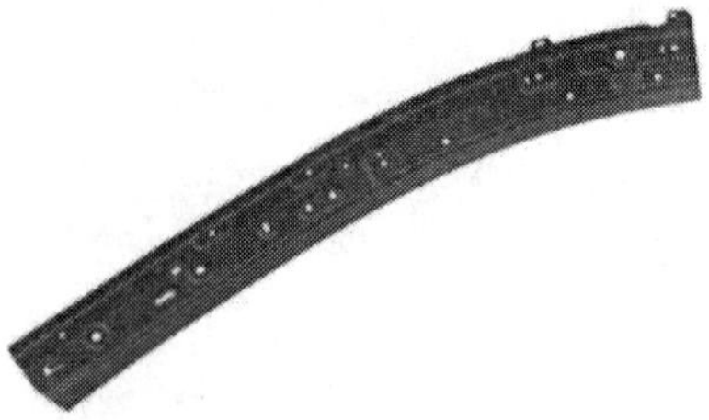

Figure 1. A pillar inner panel

The parameters of total length, radius, arc ratio and the range of 13 structural parameters in the cross section of the A pillar inner panel are shown in Table 1. This value is obtained from the measurement of the actual parts of multiple models. The line 4 and circle corner 1 vary less, which are set as invariants, and the values are 6mm and 5mm.

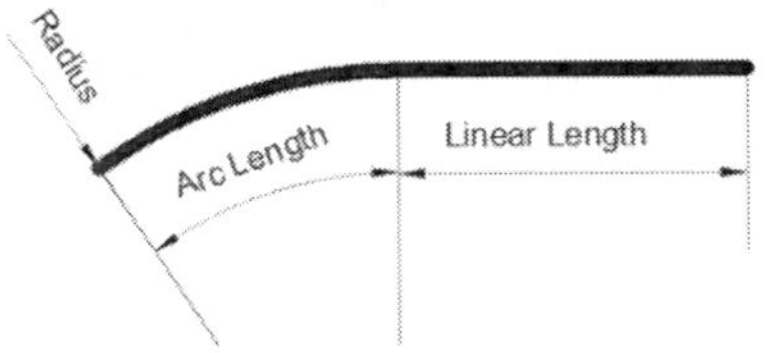

Figure 2. Longitudinal axis of the part structure features

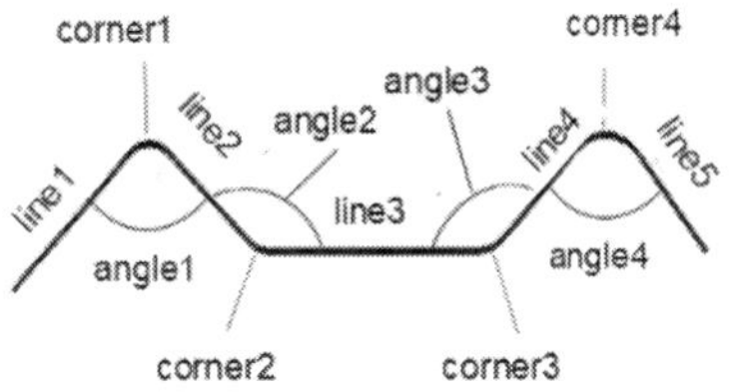

Figure 3. Cross section of the part structure features

Table 1. Range of structure parameters

Type		Min value	Max value
Longitudinal axis parameters	Total length	918	720
	Arc vs line ratio	0.64	1.3
	radius	924	1270
cross section parameters	Line1	14	20
	Line2	14	28
	Line3	20	35
	Line4	5	7
	Line5	13	16
	Angle1	90	103
	Angle2	134	170
	Angle3	100	130
	Angle4	103	150
	Corner1	4	6
	Corner2	9	17
	Corner3	4	10
	Corner4	4	10

2.2. Definition of twist springback

The springback mode of this part is mainly space twist. In this paper, the method used by Hongzhou Li[3] et al. The twist springback angle of the section can be calculated indirectly by measuring the ratio of the springback displacement of the torsion of the workpiece t and the distance between two points L on the section. The torsion springback is:

$$\theta = \frac{t}{L} = \frac{[d+(a-c)]-b}{L} \quad (1)$$

Where a and b are the original part positions, c and d are the positions after springback, a and e are the part positions after translation along N which is the normal direction of the line 3. after translation, angle 1 of ab and cd should nearly coincide.

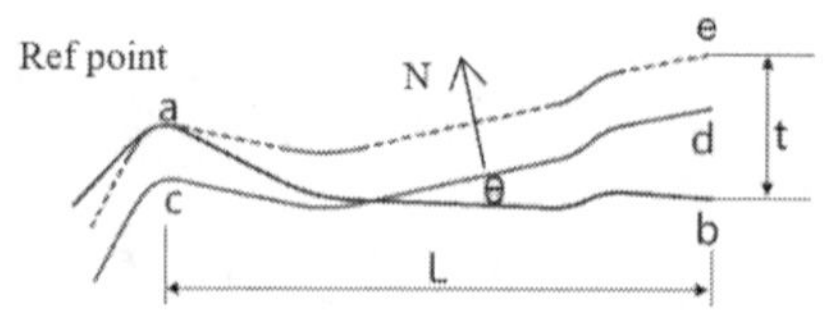

Figure 4. Schematic diagram of measurement of twist springback

2.3. *Finite element modelling*

In order to analyze the relationship between part characteristics and twist springback, this paper adopts the way of constructing virtual samples, that is, designs parts with different structural parameters, and obtains the twist springback value through CAE calculation and then conducts parameter analysis. This paper using the finite element software Autoform for springback analysis of A pillar inner panel, the material is the third generation steel QP1180, and the thickness is 1.2mm, where the material properties are shown in table 2. The FEA model of the drawing process is shown in figure 5. For the material model, material hardening model uses mixed Hockett-Sherby model and Swift model hybrid model, and BBC for the yield model . Table 3 shows the model parameters.

Table 2. QP1180CR Material property parameters

Yield strength/MPa	Tensile strength/MPa	Uniform elongation%	Fracture elongation%	R-value	N-value
1000	1201	13.75	17.8	0.748	0.11

Table 3. The parameters of the material model

Hardening model	ε0		σi	m	σsat	C	a	p	α
	0.052		664	0.18	1511	1717	3.67	0.64	0.25
Yield model	R0	R45	R90	σ0	σ45	σ90	Rb	σb	M	Biax
	0.8	0.8	0.9	1005	1009	1042	0.83	1007	6	1.03

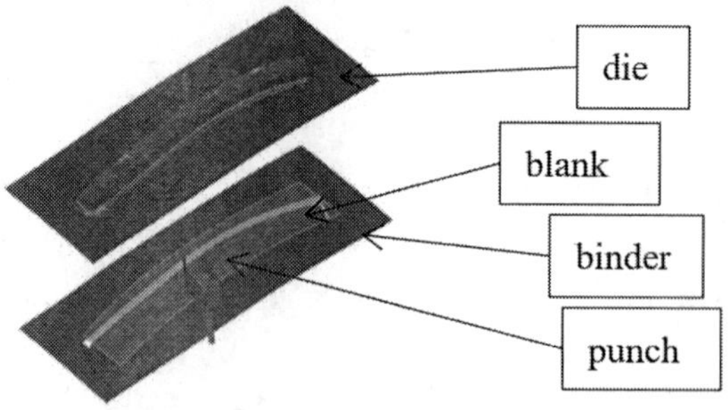

Figure 5. The FEA model

3. Analysis of the relationship between twist springback with part characteristics

3.1. *The influence of cross-section structural parameters on twist springback*

According to the variation range of each parameter in Table 1, the method of DOE factor design is first selected to determine the significance of the factor [7]. Table 4 shows the experimental design scheme

IOP Conf. Series: Materials Science and Engineering 1284 (2023) 012062 doi:10.1088/1757-899X/1284/1/012062

with 11 parameters as variables and twist springback.

The second-order factor analysis of each factor was performed with twist springback as the response, and the effect map of each variable was obtained as shown in figure 6. Here the confidence is 0.1, which is 90% confidence on the accuracy of Pareto graph analysis. From the figure, among the 13 structural parameters in the cross section significantly influence the interaction of angle 2, line 3, line 1 and angle 4. In addition, compared with other influence factors, the significance of these four structural parameters is much greater than the other factors.

Table 4. The DOE factor design table and the springback value

No.	Line 1	Line 2	Line 3	Line 5	Angle 1	Angle 2	Angle 3	Angle 4	Corner 2	Corner 3	Corner4	Twist Springback
1	14	28	35	13	90	134	130	150	17	4	4	0.124
2	20	28	20	13	90	170	130	103	17	10	4	0.345
3	20	28	35	16	103	170	130	150	17	10	10	0.124
4	20	14	35	16	90	134	130	103	9	4	10	0.174
5	14	28	35	16	90	170	100	103	9	4	4	0.125
6	20	14	20	16	103	170	100	103	17	4	4	0.187
7	14	14	35	13	103	170	130	103	9	10	4	0.174
8	14	28	20	13	103	170	100	150	9	4	10	0.244
9	14	14	20	16	90	170	130	150	9	10	10	0.213
10	14	14	35	16	103	134	100	150	17	10	4	0.138
11	20	14	20	13	103	134	130	150	9	4	4	0.179
12	20	14	35	13	90	180	100	150	17	4	10	0.078
13	14	14	20	13	90	134	100	103	17	10	10	0.121
14	14	28	20	16	103	134	130	103	17	4	10	0.117
15	20	28	20	16	90	134	100	150	9	10	4	0.131
16	20	28	35	13	103	134	100	103	9	10	10	0.108

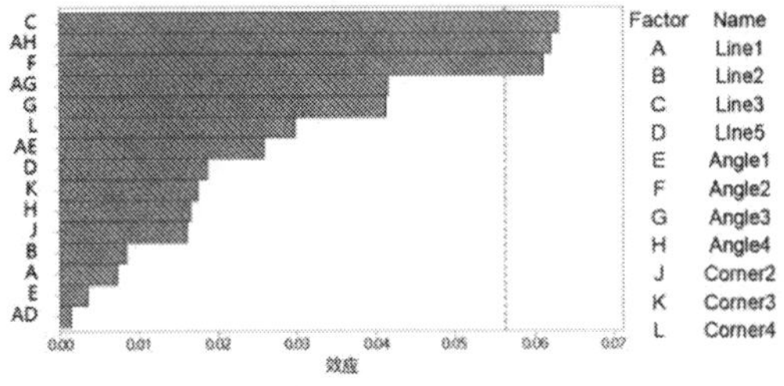

Figure 6. The pareto plot of the twist springback

The response surface trial design is an optimization technique that combines mathematical and statistical methods to evaluate the relationship between multiple factors [8] by modeling and analyzing multiple factors affecting the relevant response values. For the factors with other non-significant effects in the

IOP Conf. Series: Materials Science and Engineering 1284 (2023) 012062 doi:10.1088/1757-899X/1284/1/012062

cross section, the median value was taken as the fixed value, and the response surface model was established with the four significant influence factors as variables. The experimental scheme and simulation data are shown in Table 5. The analysis gives a second-order polynomial mathematical model of cross-sectional factors on twist springback Y as shown in Equation 2.

ANOVA(Analysis of Variance) with the regression equation, and the results of the model analysis are shown in Table 6, the correlation coefficient of the model R2=0.9503, indicating that 4.97% of twist springback is not explained by this model[9-10]. The P value of the model is less than 0.0001 and the F value is 44.8, indicating that the model is highly significant. In one term, x, z, and v had a significant effect on the model; in the second term, z * z and v * v had a significant effect on the model; x * x and y * y were not significant; in interaction terms, y * z and x * d had a significant effect on the results; and the rest were not significant on the results.

Table 5. 4-factor Experimental design and simulation data

No.	Line 1	Line 3	Angle 2	Angle 4	Twist springback	No.	Line 1	Line 3	Angle 2	Angle 4	Twist springback
1	17	27.5	152	126.5	0.159	17	14	20	170	150	0.124
2	20	35	170	150	0.226	18	20	20	170	103	0.137
3	20	35	134	103	0.131	19	17	27.5	152	126.5	0.137
4	17	27.5	152	126.5	0.136	20	14	35	170	150	0.125
5	17	27.5	152	126.5	0.137	21	20	35	170	103	0.156
6	20	20	134	150	0.131	22	11	27.5	152	126.5	0.073
7	14	35	170	103	0.089	23	17	27.5	152	126.5	0.137
8	14	35	134	103	0.063	24	14	35	134	150	0.074
9	20	20	170	150	0.199	25	17	27.5	152	126.5	0.137
10	20	35	134	150	0.145	26	17	12.5	152	126.5	0.159
11	23	27.5	152	126.5	0.188	27	17	42.5	152	126.5	0.146
12	14	20	134	103	0.084	28	14	20	134	150	0.069
13	17	27.5	152	79.5	0.105	29	17	27.5	188	126.5	0.192
14	17	27.5	152	126.5	0.137	30	20	20	134	103	0.107
15	14	20	170	103	0.143	31	17	27.5	152	173.5	0.067
16	17	27.5	116	126.5	0.047						

$$Y=0.091-0.0150x-0.01235y+0.00386z+0.00036v-0.000196x*x+0.000066y*y-0.000014z*z-0.000023v*v+0.000422x*y+0.000013x*z+0.000140x*v-0.000014y*z+0.000028y*v+0.000017z*v \quad (2)$$

Where x represents line 1, y represents line 3, z represents angle 2, and v represents angle 4.

By solving equation (2), the maximum estimate of twist springback is 0.215, each parameter is x=20mm, y=35mm, z=170°, v=150°; the minimum estimate is 0.050, the parameter is x=14mm, y=30.8mm, z=134° and v=150°. Using the two sets of parameters to establish parts model and simulation analysis, measure the springback value and substitution formula (1) the maximum twist springback value is 0.209, the minimum twist springback value is 0.052, the actual measured value and predicted value error is 2.87% and 4.00%, indicating that the response surface model is accurate and reliable, and can reflect the influence of cross-section structure parameters on the twist springback.

Table 6. Analysis of variance of the regression equation

	Freedom	AdjSS	F-value	P-value
Model	14	0.0833	21.85	<0.0001
x	1	0.0120	43.83	<0.0001
y	1	0.0004	1.54	0.232
z	1	0.0510	184.7	<0.0001
v	1	0.0056	20.58	<0.0001
x*x	1	0.0006	2.08	0.168
y*y	1	0.0006	2.38	0.143
z*z	1	0.0049	17.93	0.001
v*v	1	0.0085	31.41	<0.0001
xy	1	0.0001	0.43	0.520
xz	1	0.0000	0.11	0.746
xv	1	0.0000	0.11	0.743
yz	1	0.0014	5.19	0.037
yv	1	0.0000	0.17	0.689
zv	1	0.0003	1.15	0.300
Pure error	16	0.0044		
R-Sq				95.03
AdjR-Sq				90.68

3.2. The influence of longitudinal axis parameters on twist springback

According to the variation range of each parameter in Table 1, a response surface model of 3 factors was established with twist springback as the response, and the test scheme and simulation results are shown in table 7.

The response surface analysis of the above data gives a polynomial model of twist springback and longitudinal axis 3 factors. The mathematical expression of the model is shown in Equation 2:

$$Y=-0.165- 0.000297\,m- 0.120\,n+ 0.000644\,j+ 0.000001\,m*m+ 0.0130\,n*n+ 0.000000\,j*j+ 0.000344\,m*n-0.000001\,m*j- 0.000129\,n*j \quad (3)$$

Where m represents the total length, n the arc vs line ratio and j the radius.

The ANOVA for this response surface model is shown in table 8, and the P value of the model less than 0.001 indicates that the quadratic term model is highly significant. The correlation coefficient of the model was 99.02%, indicating that 0.98% of the model cannot be explained by the model. As shown by the ANOVA of each of the variables of the regression equation, the P value of all primary terms is less than 0.001, indicating that the three one-time variables have extremely significant effects on the twist springback. In the quadratic term, only the m * m is less than 0.0001, and the other two terms have P values greater than 0.05, indicating that only m * m has a significant effect on the results, and the other quadratic terms are not significant for the results. The mn terms and mj terms function significantly in the interaction effect. In addition to the primary term, the quadratic term and interaction also significantly affect the results, indicating that the twist springback is not simply linear with the total length, arc vs line ratio and radius.

Table 7. Response surface test scheme and simulation results

No.	Total length/ mm	Arc vs line ratio	Radius /mm	Twist spring back	No.	Total length/ mm	Arc vs line ratio	Radius /mm	Twist spring back
1	819	0.97	806	0.129	11	985	0.97	1097	0.157
2	918	0.64	924	0.126	12	819	0.97	1097	0.106
3	720	0.64	924	0.073	13	819	1.52	1097	0.125
4	653	0.97	1097	0.088	14	819	0.97	1097	0.104
5	819	0.42	1097	0.090	15	819	0.97	1097	0.105
6	819	0.97	1097	0.105	16	918	1.3	924	0.214
7	819	0.97	1097	0.103	17	819	0.97	1388	0.086
8	819	0.97	1097	0.106	18	720	0.64	1270	0.082
9	918	0.64	1270	0.108	19	720	1.3	1270	0.096
10	720	1.3	924	0.086	20	918	1.3	1270	0.136

Table 8. ANOVA with longitudinal axis regression models

	freedom	AdjSS	F-value	P-value
Model	9	0.12197	112.53	<0.0001
m	1	0.03185	264.44	<0.0001
n	1	0.05738	476.47	<0.0001
j	1	0.02484	206.31	<0.0001
m*m	1	<0.00001	0.05	0.030
n*n	1	0.00590	49.02	0.615
j*j	1	0.00027	2.21	0.605
mn	1	0.00121	10.05	0.012
mj	1	0.00001	0.1	0.003
nj	1	0.00013	1.07	0.071
Pure error	5	0.00005		
R-Sq				94.42
AdjR-Sq				90.14

The above quadratic polynomial model is solved, and when the total length m=918mm, the arc vs line ratio n=1.30 and the radius j=924mm is obtained, the maximum twist springback value is 0.199; when the total length is 720mm, the arc vs line ratio is 0.64, and the radius is 924mm, the minimum twist springback value is 0.073. After modeling and simulation verification, the measured twist springback simulation values are 0.209 and 0.072. The error between the predicted value and the simulation value is 4.78% and 1.30%, indicating that the response surface model of the longitudinal axis is accurate and reliable, which can reflect the influence of the longitudinal axis on the twist springback to a certain extent.

4. Analysis and countermeasures of springback robustness

According to the above analysis results, in order to study the influence of the springback, the parts are designed according to the longitudinal axis and cross section structure parameters of the maximum twist springback. Also the springback analysis is based on the Autoform-sigma module considering the above process route for the robustness of the design parameters[11~13] in Table 9, where the maximum and minimum value range from the nominal value by ±10%. When do the springback simulation, three fixed boundary points are defined here to constrain the rigid displacement, where the springback are nearly zero here, which shows in figure 7. In Figure 8, it can be seen that the maximum springback value is at both ends of the parts, while the upper and lower positions springback are in opposite directions, the maximum is 4.246mm and the minimum is -10.92mm; Figure 9 shows that the absolute value of cpk is in an unacceptable state. Figure 10 shows the springback sensitivity of the variables. And it can be seen by the bar chart that different areas show different sensitivity of the variables, while the areas where springback are large are associated with material yield strength. That is to say, in the subsequent mass production, the fluctuation of material performance has a great probability of causing part geometry instability.

Table 9. Range of each variable for the robustness analysis

Variables	Binder Force (KN)	Friction Coefficient	R value	Yield Stress(MPa)	Tensile Strength (MPa)
Nominal value	1200	0.15	0.825	1005	1226

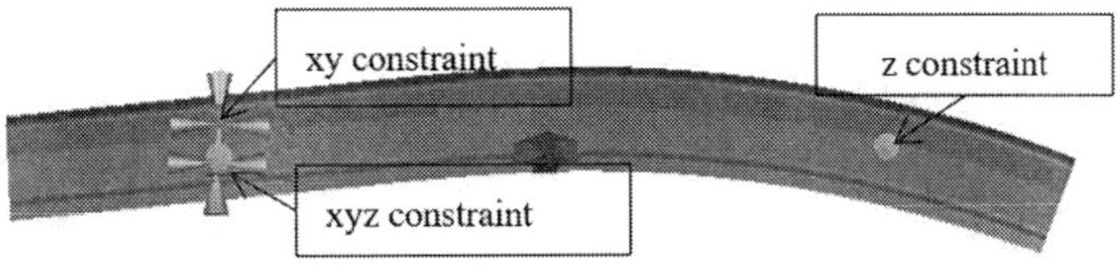

Figure 7. Constrain point definition for springback simulation

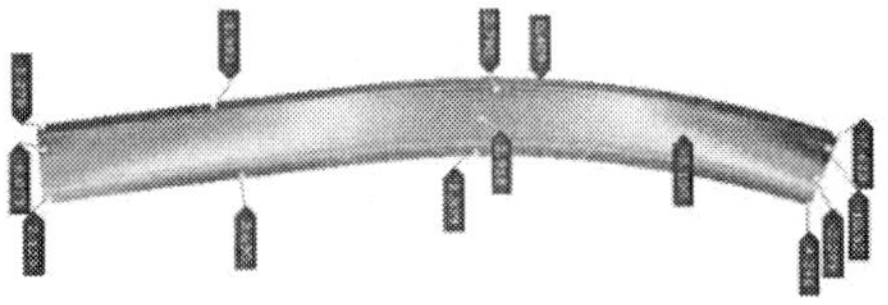

Figure 8. Springback Value (nominal)

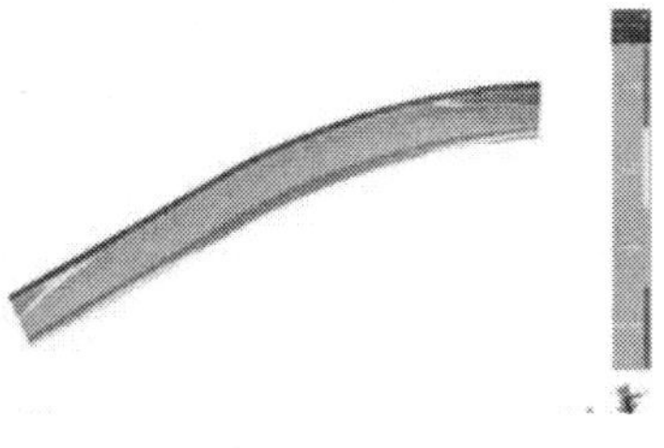

Figure 9. Evaluation of Robustness (CPK)

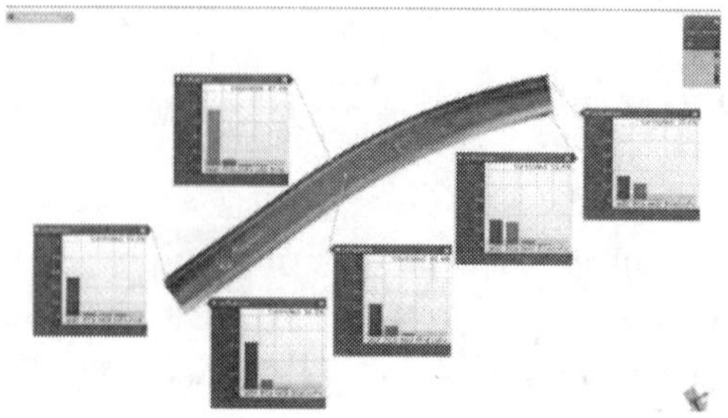

Figure 10. Results of the springback sensitivity analysis for each variable

In order to reduce the sensitivity of the parts springback to various parameters, the rigidity of the parts is considered to reduce the release of residual stress after unloading, thus reducing the springback value and fluctuation range. In order to verify the effect, this paper tries to arrange the slender features throughout the whole cross section along different longitudinal positions of the parts. What needs to be explained is this paper does not conduct a special study on the position of the feature arrangement. The actual design of the features should be considered to ensure the assembly and formability of the parts. Figure 8 shows the springback analysis results after adding features. According to the analysis results of in figure 11, the maximum springback value compared with Figure 8 decreases from 4.246mm to 1.821mm, and the minimum value decreases from-10.92mm to-2.934mm, indicating that the control effect of springback by adding features is obvious. Further, the analysis of springback robustness in figure 12 shows that the end area with unacceptable robustness is significantly improved and now is in a reliable state, indicating that the effect of springback value and fluctuation can be reduced by adding features.

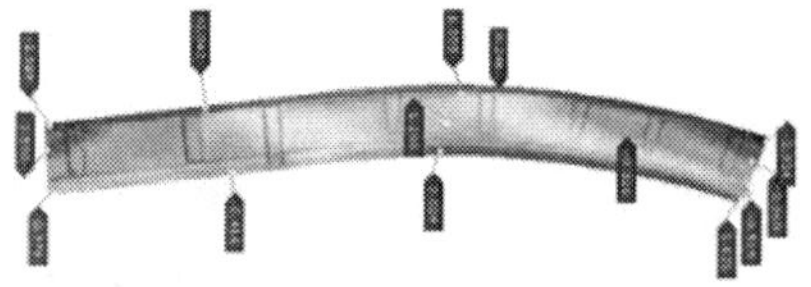

Figure 11. Springback Value (nominal) of optimized part

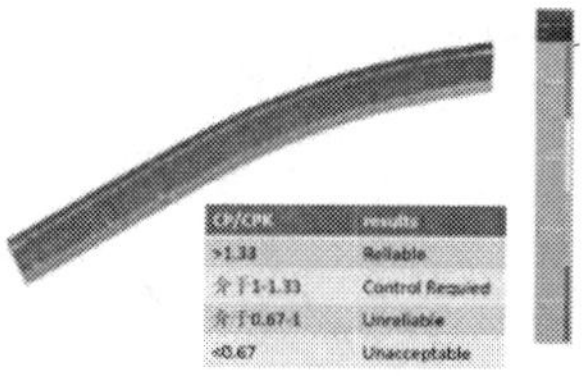

Figure 12. Robustness evaluation (CPK) of optimized part

5. Conclusions

In this paper, the relationship between the part structure and the twist springback is built by a mathematical model of the part springback and structure parameters; Also the sensitivity analysis of springback with material properties, process parameters was finished to find out the possible scheme to improve the robustness of the parts.

(1) A typical A pillar inner panel is resolved into cross-section and longitudinal axis features. Through factor analysis and response surface analysis result, the cross section parameters of line 1, line 3, Angle 2 and angle 4 have a significant influence on the twist springback, and the total length, arc vs line ratio and radius influence the twist springback. By solving the model at extreme value, the corresponding cross section parameters of the maximum and minimum springback can be obtained.

(2) After adding features through the angular edges along the longitudinal direction, the maximum springback at the part end drops from -10.92mm to -2.934mm. The robustness analysis results of part

springback show that the maximum springback is strongly related to material yield strength, and the robustness is improved greatly after adding part features. It can be used as an effective measure to improve the mass production of ultra high strength steel parts due to material performance fluctuations.

References

[1] Paweł Bałon, Andrzej Świątoniowski, et al. Springback Compensation for a Vehicle's Steel Body Panel[J]. AIP Conference Proceedings, 2017.

[2] Chetan, Nikhare P . Prediction of Springback in Long Channels[J]. Journal of Metals: English edition, 2015 (12 issues): 1525-1531.

[3] Hongzhou Li, Guangyong Sun, Guangyao Li, et al.. On twist springback in advanced high-strength steels[J]. Materials & Design, 2011, 32(6):3272-3279.

[4] Xin Xue, Juan Liao, Gabriela Vincze, et al.. Twist springback characteristics of dual-phase steel sheet after non-axisymmetric deep drawing[J]. International Journal of Material Forming, 2015, 10(2):1-6.

[5] Xue X, Liao J, Vincze G, et al. Modelling and sensitivity analysis of twist springback in deep drawing of dual-phase steel[J]. 2016, 90:204-217.

[6] Lu Jinhuo, Zhao Zijian, Liang Guogui. Design method of main section of car body [J]. Automotive Technology, 2006 (9): 9-13.

[7] Tao Xiong, Wang Yunzhi, Li Ying, et al. Based on TFT-LCD switching [J]. LCD and Display, 2018 (2): 123-128.

[8] Helga Jordão, António Jorge Sousa, Carvalho M T . Optimization of wet shaking table process using response surface methodology applied to the separation of copper and aluminum from the fine fraction of shredder ELVs[J]. Waste Management, 2015, 48:366-373.

[9] Samuel O D , Okwu M O . Comparison of Response Surface Methodology (RSM) and Artificial Neural Network (ANN) in modelling of waste coconut oil ethyl esters production[J]. Energy Sources Part A Recovery Utilization and Environmental Effects, 2018:1-13.

[10] Yang Zhe, CAI Huinong, Ni Hui, et al. Optimization of inosine shake flask fermentation medium and culture conditions by response to surface method[J]. Chinese Food Journal, 2014,14 (3): 86-93.

[11] WANG Wu rong, YOU Jie, ZHANG Xin peng,LI Shu hui,LIN Zhong qin. Robust Design for Characteristic Part Stamping Based on Statistic Analysis of Variation of Material's Property[J]. Shanghai Jiaotong Univ.(Sci.) , 2011, 45(01): 1-0005.

[12] Wu Lei;Jiang Haomin;Wang Chen;Li Guangyao. Effect of Material Properties on Springback of High Strength Steel Parts [J], 2009, 20(11): 0-1292.

Process optimisation and robustness analysis for HFQ process

M Mohamed[1], D Szegda[1], J Swift[1], O Gaines[1], D Ling[2], M Sonntag[3] and A Güner[3]
[1] Impression Technologies Limited, Coventry, CV5 9PF, UK
[2] AutoForm Engineering B.V., Netherlands
[3] AutoForm Engineering, Germany

E-mail: m.mohamed@impression-technologies.com

Abstract. Lightweight materials and lightweight technologies have gained significant attention from all EV vehicle manufacturers. Aluminium has been commonly used as a lightweight replacement for steel in the automotive industry for many years. High-strength aluminium alloys have high specific strength, and their formability increases significantly at elevated temperatures. New manufacturing technologies incorporating elevated temperature forming have emerged recently, allowing forming of high and ultrahigh-strength aluminium alloys. One such technology is Hot Form Quench (HFQ)® which is a process that combines high formability with virtually no springback for (ultra)-high-strength aluminium alloys. The thermal expansion of aluminium alloys is approximately twice that of steel, leading to significant material contraction during in-die quenching. This is especially pronounced for large or long components. Therefore, robustness analysis of the process parameters can be used to optimise the process parameters and enable a capable production of parts meeting dimensional specification requirements. Such analysis can be performed using AutoForm-Sigma solver and this paper presents a simulation methodology developed in AutoForm and its use to verify the accuracy of the thermal distortion prediction for one of the safety cell structure components. A sensitivity analysis has been implemented to optimise the process parameters to improve manufacturing robustness.

1. Introduction

The use of lightweight materials, such as aluminium alloys, is increasing significantly, particularly for automotive applications. Although aluminium alloys have many advantages compared to steel, their formability is lower at room temperature and when used for manufacturing complex shapes, particularly from sheet, poses limitations and creates additional design challenges [1]. To overcome the formability problem, particularly for high-strength aluminium alloys, hot stamping processes have been developed. For example, in the process of hot stamping and cold die quenching (HFQTM), an aluminium alloy is heated to its solution heat treatment temperature (or close to it) prior to forming stage. It is then press formed in cooled dies thereby being quenched rapidly to retain a super-saturated solid solution [2]. Springback is the main source of shape distortion and geometric tolerance noncompliance in cold sheet metal forming. This major drawback for cold forming is a phenomenon that occurs when the forming force applied to the blank is released. The deformed material tends to partially return to its original shape as an attempt to release the internal elastic residual stresses. At elevated temperatures, the material response becomes viscoplastic, i.e., rate dependent, whilst the process is inherently non-isothermal due to contact with cold tools and ambient air [1-2]. Moreover, the quenching phase is affected by mechanical boundary conditions applied through contact which is characterised by contact pressure distribution leading to non-uniform cooling and, hence, non-uniform thermal contraction [3].

The complex interactions between thermal and mechanical conditions can often lead to difficulties in the accurate simulation of the process leading to inaccuracies in the prediction of thermo-mechanical distortion. Without accurate distortion prediction, effective tool compensation is not possible. The finite element method has become an essential tool for process design and specification. Complex interrelations between the mechanical, thermal, and microstructural fields should be considered when implementing a realistic FE model for the hot stamping process to receive a more precise prediction of the resulting component properties [4]. For hot stamping, accurate process data and material card were input to a simulation environment such as Autoform and used to design compensated tool faces, which will allow thermal mechanical distortion to be controlled.

2. Design of Experimental (DOE)

This section presents the validation approach of the FE simulation setup for HFQ process. This work creates a Design of Experiments (DOE) with a combination of process parameters (forming temperature, forming speed, and quenching force and time) for a high-strength A-pillar panel made in AA6082 aluminium alloy as shown in Table 1. The DOE was designed in two groups, A and B, to get different levels of thermal mechanical distortion. Group A includes parts 1 to 4 which were used to validate the simulation model by comparing them with corresponding pressed part scans. Group B includes Parts 5 to 8 which were used to evaluate the thermal distortion by comparing pressing scans with CAD-designed surface. Process parameters for the trials were recorded and the different conditions are summarised in Figure 1.

Table 1. DOE for A-Pillar trails.

Part No		Temperature (°C)	Speed (mm/sec)	Quenching-Force (ton)	Quenching - Time (sec)
Group A	1	525	250-79	250	8
	2	525	79-79	50	30
	3	460	250-79	50	8
	4	460	79-79	50	30
Group B	5	525	250-79	250	30
	6	525	79-79	50	2
	7	460	250-79	250	30
	8	460	79-79	50	2

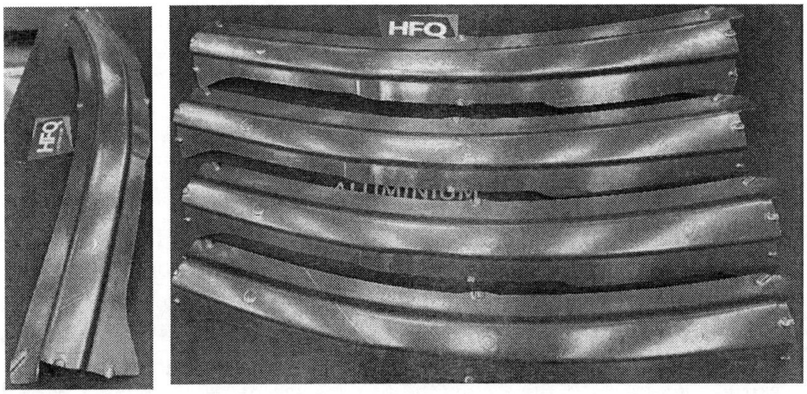

Figure 1. Pressed A-Pillar panels with different conditions.

3. HFQ simulation model validation

In order to simulate the HFQ process as accurately as possible within the Autoform software [4], the entirety of the process line setup was produced. This included solutionising temperature of the blank, transfer time, effect of gravity after the blank is located on the tool, holding by binders, stamping and quenching and then air cooling and thermo-mechanical distortion (labelled as 'springback') as shown in Figure 2. Within these stages, there are a larger number of variables to be quantified to ensure that the expected forming environment, and its effects, are captured. The defined variables include ambient temperature, heat transfer coefficients, tool temperatures, press speeds, press forces, and time delays within the tool for robot automation. The material model used in the simulation was the 6082 isotropic lookup table and more details were published by Mohamed et al 2014 [2]. The effect of heat transfer coefficient, HTC, was investigated by Foster, et al. 2008 [5]. The Coulomb model was used with a constant coefficient of friction set at 0.3. The meshing setting was defined in AutoForm using an initial maximum master element size of 10 with a refinement level of 5. The 3D FE model with boundary conditions is shown in Figure 3.

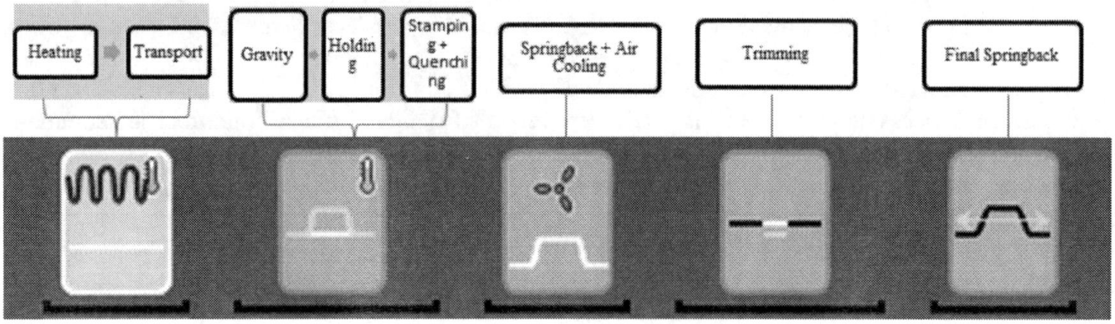

Figure 2. AutoForm HFQ simulation stages.

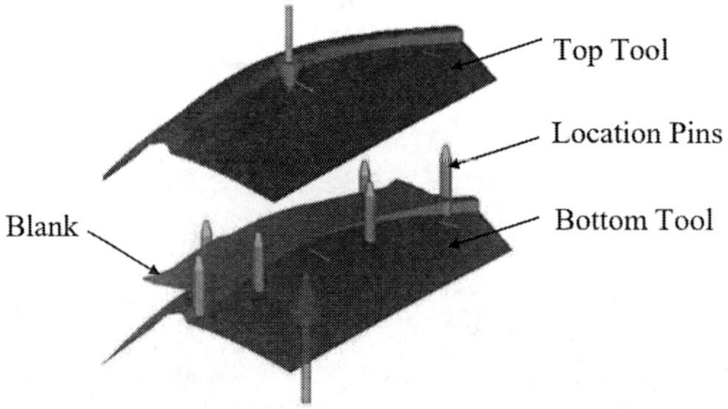

Figure 3. FE process model built in AutoForm.

As previously mentioned, the thermomechanical distortion is strongly influenced by the quenching stage and is assumed to be driven by the temperature gradients in the panel. This hypothesis is based on the uneven material shrinkage due to varying temperatures at different locations of the panel.

For an accurate and detailed comparison of shape distortion between the simulation and experiment (scanned panel), Figure 4 (right) shows the four sections (200 mm between sections) that have been cut to examine the agreement of the two parts at different locations within the parts. The overlay of the scan of the pressed part (solid shell line) with the top surface of the simulated part shows the deviation between the two surfaces. Figures 5, 6, 7 and 8 present the deviation (distance between the two surfaces) for the four conditions outlined in Table 1.

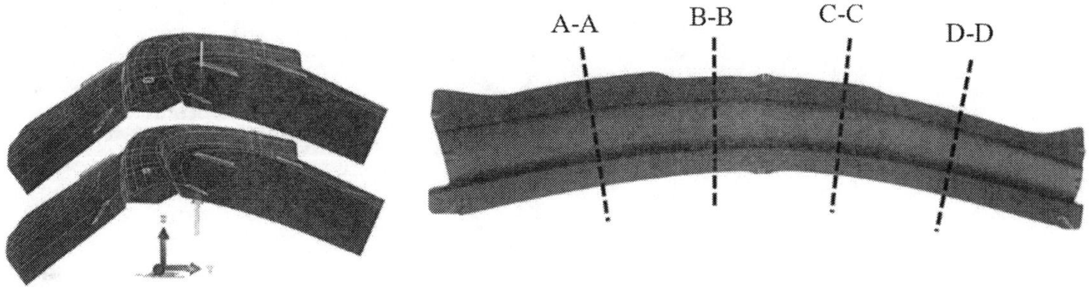

Figure 4. left Tools, right sections that are used to evaluate simulation results (200 mm between sections).

Comparing the simulated and pressed (scanned) parts reveals that a very good agreement between the simulation-predicted shape distortion and the actual part scan was achieved, as can be seen in Figures 4, 5, 6 and 7. The maximum and minimum deviations between the two surfaces are within ±1mm for all conditions.

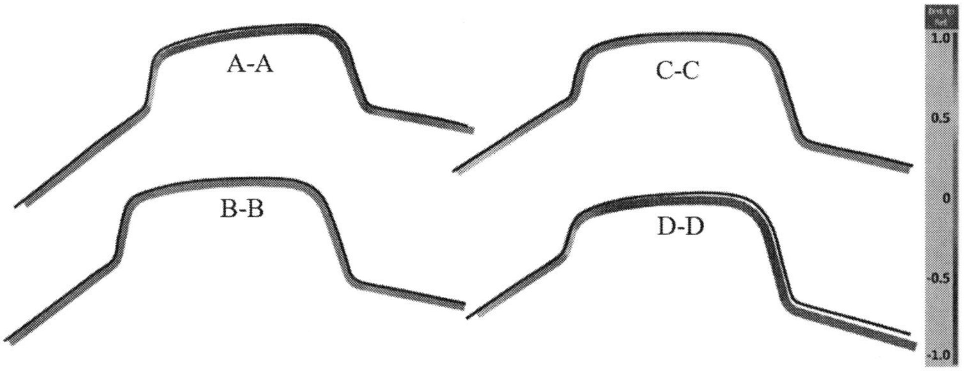

Figure 5. Part No 1 – Distance (mm) to Reference (pressed part).

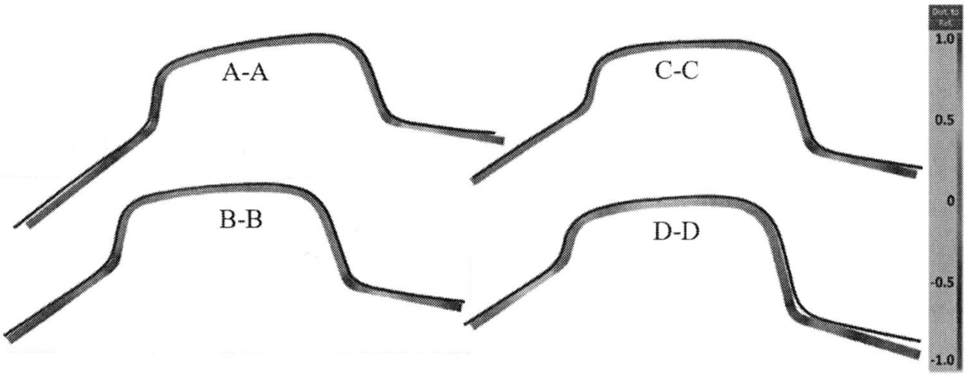

Figure 6. Part No 2 - Distance (mm) to Reference (pressed part).

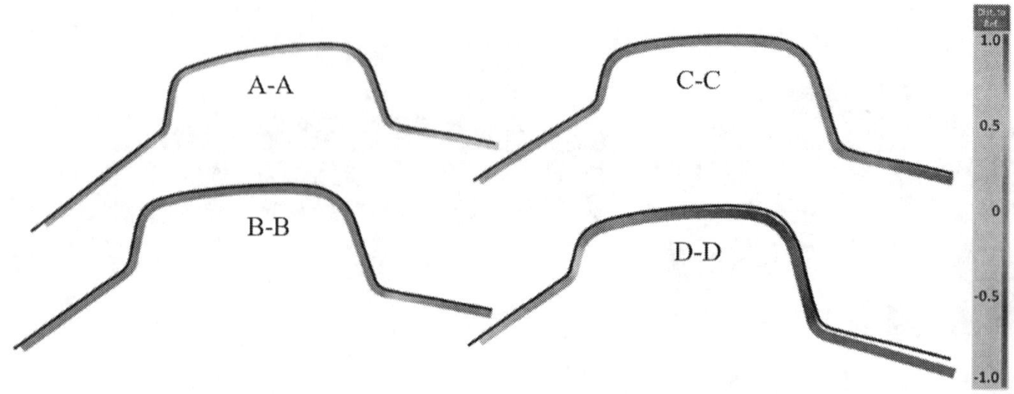

Figure 7. Part No 3 – Distance (mm) to Reference (pressed part).

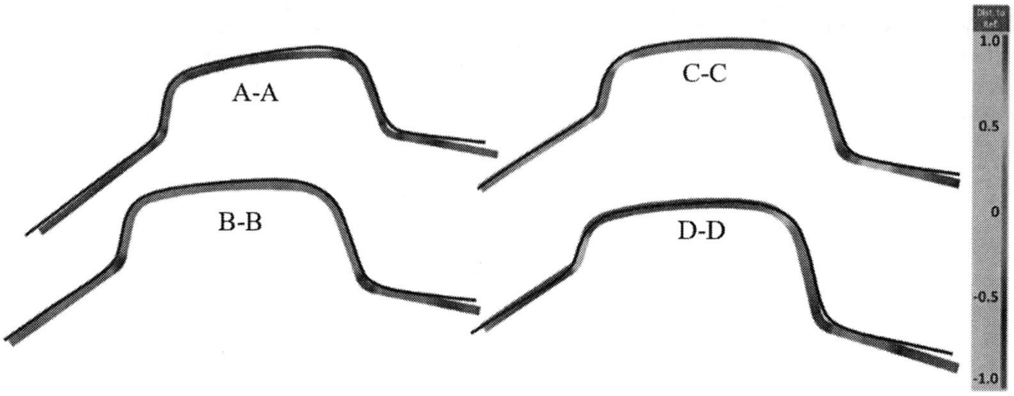

Figure 8. Part No 4 – Distance (mm) to Reference (pressed part).

4. Thermal distortion evaluation

The accurate process parameters and material data together with the dedicated simulation methodology established in AutoForm for the prediction of thermos mechanical distortion was verified and validated against press trials for an A-Pillar component. This paper shows the evaluation of thermos mechanical distortion of the part relating to the process parameters. Parts 5, 6, 7 and 8 were used to compare with the design CAD file. Parts 5 and 6 had a forming temperature of 525°C and a short quench with low quenching force, while parts 7 and 8 had a forming temperature of 460°C and a long quench with high quenching force. The thermos-mechanical distortion becomes particularly obvious when a short quenching time and low quenching force give rise to significant temperature gradients and residual stresses. This is apparent in Figures 9 and 10 which show the temperature gradient in the part at forming temperatures of 525°C and 460°C respectively. This gradient causes uneven stress distribution in the part after quenching which leads to distortion.

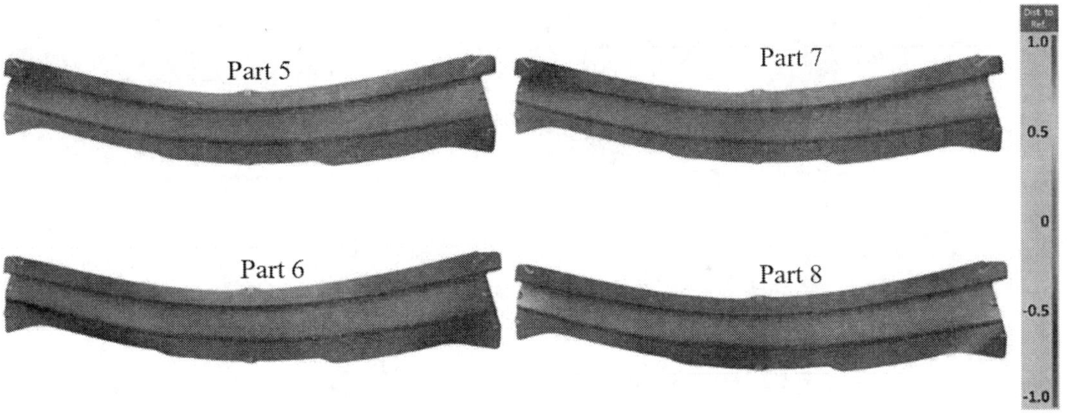

Figure 9. Distance (mm) to Reference (CAD-designed).

To explore the effect of quenching parameters on the thermal mechanical distortion further, examining the temperature and stress distributions on the parts with different conditions was essential. Figure 10 shows the temperature distributions after Quenching in °C for Parts 5, 6 (left) and parts 7, 8 (right). It is obvious that the temperature is evenly distributed and reaches room temperature for parts 5 and 7 at forming temperatures of 525°C and 460°C respectively. The reason behind these results is the high quenching force and long quenching time which are the recommended settings for the HFQ process. On the other hand, when process conditions deviate from the recommended settings an uneven temperature distribution occurs and room temperature is not reached after quenching as shown in parts 6 and 8 in Figure 10. In the case of even temperature distribution (parts 5, 7), the major and minor stresses are nearly uniformly distributed on the part which leads to minimum thermal mechanical distortion as shown in Figure 11. Conversely, in the case of non-uniform temperature distribution (parts 6, 8), a non-uniform stress distribution appears, and the distortion can increase significantly [6].

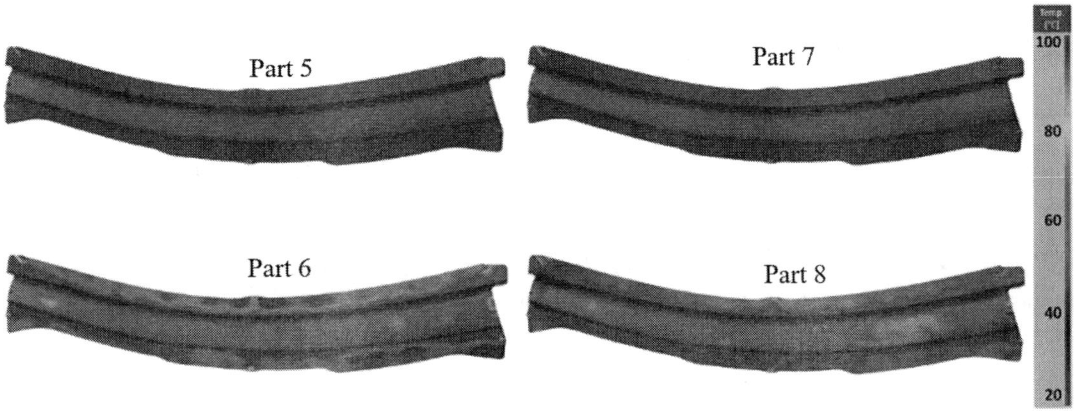

Figure 10. Temperature after Quenching in °C for Parts 5, 6 (left) and parts 7, 8 (right).

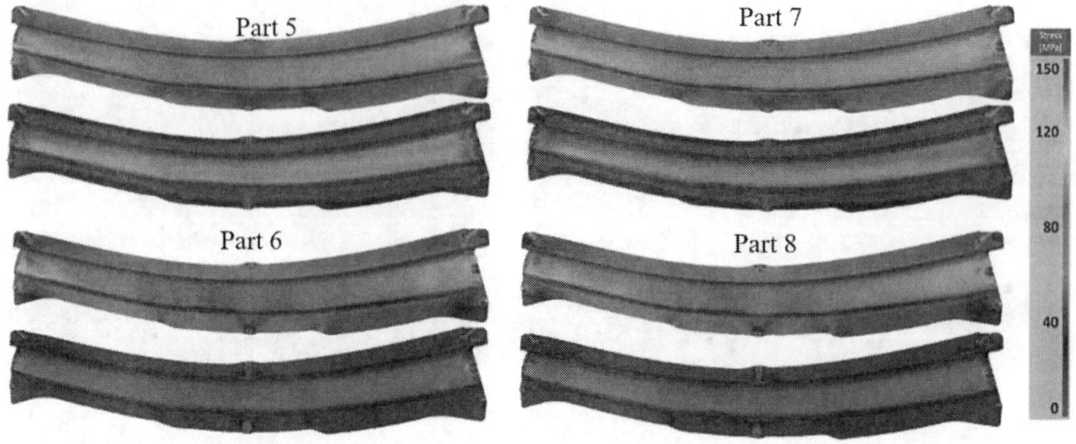

Figure 11. Major Stress (upper figures) Minor Stress (lower figures) in MPa for Parts 5, 6 (left) and parts 7, 8 (right).

5. Robustness analysis

The results in the previous sections show that the temperature and stress distributions at the end of quenching play a significant role in the springback behaviour of the parts. Two key process parameters which control temperature and stress distribution are identified as the quenching force and the quenching time. This means that process engineers need to optimize the combination of these parameters in order to end up with the part in geometrical tolerances. At the same time, they must minimize the cycle times to increase production efficiency in terms of stroke per minute. This optimization problem can be solved in press shop by a conventional trial-and-error iterative approach which would cost time and money. A more efficient solution would be to decide on the optimum set of parameters during the engineering phase.

For that purpose, a metamodel-based sensitivity analysis has been performed by using AutoForm-Sigma®[7]. In this study, the quenching force and quenching time have been varied between 50-2500 KN and 0-30 s. respectively. Different combinations of these parameters were simulated to construct a metamodel which gives the springback response of the part as a function of quenching time and force. This analysis has been performed for two distinctive cases: Part temperature before forming (after furnace and transfer to the press) at 525°C. and at 460°C. As a first step, the total range of calculated springbacks were evaluated which can be seen in Figure 12. The results show that the expected springback variation as a result of all parameter variations is in the range of 0-3.0 mm depending on the measurement location on the part. Generally, the four corners of the part show the maximum springback. It is also seen that the case with a 525°C temperature before forming is more severe in terms of springback. This is an expected behaviour due to the fact that the temperature distribution at the end of quenching will be more inhomogeneous due to higher starting temperatures.

IOP Conf. Series: Materials Science and Engineering 1284 (2023) 012063 doi:10.1088/1757-899X/1284/1/012063

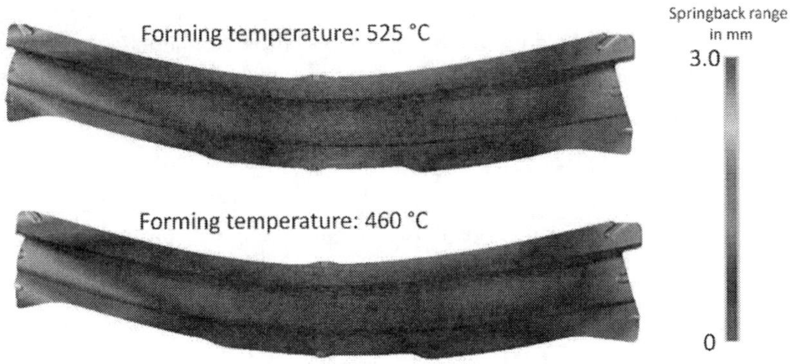

Figure 12. The total range of variation in springback as a result of the sensitivity analysis.

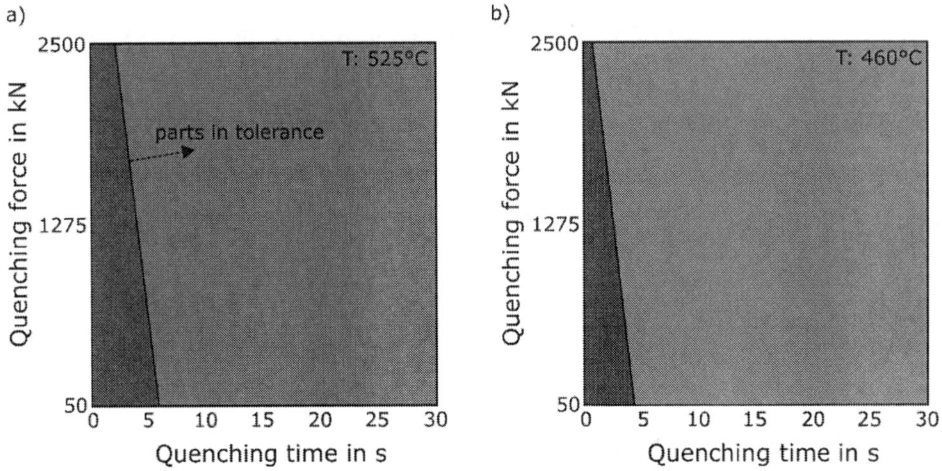

Figure 13. Process windows regarding quenching time and force for a) 525°C forming temperature and b) 460°C forming temperature. The green area represents the cases where the springback is below ± 1.0 mm.

In the next step, process windows for the two cases were calculated which can be found in Figure 13. These process windows are calculated automatically by using the information coming from the constructed metamodel. In order to define a process window, a definition of parts in tolerance is needed. In this study, springback values less than ± 1.0 mm is used to define “good parts”. It can be seen that quenching times more than about 6 seconds always result in parts within tolerance. If the quenching force is increased, the quenching time can be reduced to 2 seconds. The process window of the 525°C case is slightly smaller, needing 1-2 seconds more quenching time to ensure parts in tolerance than the case with a 460°C forming temperature.

6. Conclusion

This paper presented the validation of the dedicated simulation methodology established in AutoForm for the prediction of thermal mechanical distortion which may arise during HFQ forming of aluminium parts. The process parameters and material play a crucial role in controlling the thermal mechanical distortion for the 6082 A-Pillar panel. It was demonstrated that part distortion depends strongly on process conditions and when process conditions deviate from the recommended settings distortion can increase significantly. This becomes particularly obvious when low quenching force and short quenching time give rise to significant temperature gradients and residual stresses. The simulation approach discussed here was tested against different test cases with very different process parameters and very good correlation was found in all cases. This confirms the model's versatility and wide range of applicability spectrum. A metamodel-based sensitivity analysis has been performed using AutoForm-Sigma® to determine the process windows regarding quenching time and force showing the area which represents cases where the springback was below ± 1.0 mm.

References

[1] Mohamed M 2010 An investigation of Hot Forming Quench process for AA6082 aluminium alloys (PhD Thesis, Imperial College London)

[2] Mohamed Mohamed, Jianguo Lin, Alistair Foster, Trevor Dean, John Dear, 2014, A New Test Design for Assessing Formability of Materials in Hot Stamping, Procedia Engineering, Volume 81, Pages 1689-1694.

[3] Mohamed M, Foster A , Lin J, Balint D and Dean T D 2012 Investigation of deformation and failure features in hot stamping of AA6082: Experimentation and modelling *Int. J. of Machine Tools and Manuf* **53** pp 27-38

[4] Mori K, Bariani P.F, Behrens B.-A, Brosius A, Bruschi S, Maeno T, Merklein M, Yanagimoto J, 2017, Hot stamping of ultra-high strength steel parts, CIRP Annals, Volume 66, Issue 2, Pages 755-777.

[5] Foster A, Lin J, Mohamed M, 2008, An investigation of lubrication and heat transfer for a sheet aluminium heat, form-quench (HFQ) process - Proc. 12th Conf. Metal Forming.

[6] Szegda D, Mohamed M and Ziane M 2021 On thermal compensation of Hot-Form-Quench stamping die *IOP Conf. Ser.: Mater. Sci. Eng.* **1157** 012087

[7] AutoForm User Guide 2022.

IOP Conf. Series: Materials Science and Engineering 1284 (2023) 012064 doi:10.1088/1757-899X/1284/1/012064

Measurement of Skid Line Formation and Formation Mechanism of Automotive Outer Panel

Y Kishigami[1], T Minote[1], M Urabe[2], Y Tamai[1], K Suzuki[3], M Maeda[3] and S Takahashi[3]

[1]Forming Technology Research Dept., JFE Steel Corporation, 1 Kawasaki-cho, Chuo-ku, Chiba 260-0835, Japan.
[2]Forming Technology Research Dept., JFE Steel Corporation, DAINAGOYA BUILDING27F, 3-28-12, Meieki, Nakamura-ku, Nagoya-city, Aichi 450-6427, Japan.
[3]College of Industrial Technology, Nihon University, 1-2-1 Izumicho, Narashino 275-8575, Japan.

E-mail: Y-kishigami@jfe-steel.co.jp

Abstract. Appearance of an automobile is one of the appealing points to customers. Character ridgelines in outer panels such as hood and door panels play an important role to make automobiles attractive, and a sharp character ridgeline which has a ridge of a small curvature radius is much more preferred to a blunt character ridgeline. However, a skid line, a kind of appearance defect, is apt to generate along the character ridgeline in press forming of a sharp character ridgeline. Although it is well-known that skid lines are caused by inhomogeneous tension, the mechanism of skid line generation was not yet fully understood. In this study, the skid line defect is quantitatively evaluated by draw forming experiment using a newly devised die which can measure a panel shape at any time during press forming. The mechanism of skid line formation was clarified by experimental results. The results of the shape measurement suggest that the ridge shape in a panel formed on the punch ridge should move to a flat area of the punch. A wavy shape should be generated there immediately before the ridge shape is flattened at the bottom dead center of the die. Then the wavy shape should grow to a skid line through springback.

1. Introduction

Recently, more complex automotive outer panel shapes have been required by the diversification of automotive appearance design [1]. A frequent concern among appearance features is the sharpness of the character ridgelines, which are ridge or hollow lines on parts such as fenders, doors, and other parts. This is because car designers often require a smaller corner radius for character ridgelines. [2].

However, sharp character ridgelines often cause difficulties during press forming. In particular, the sensitive surface defect called "skid line," which appears more often with a sharper character ridgeline, i.e., with a smaller character radius, is an especially serious issue which deteriorates auto appearance. As a practical issue, more time is necessary to adjust the forming tools to form a sharper character ridgeline [3].

Imbalance of tensions during press forming process are broadly admitted cause of a skid line. An example is shown in Figure 1. A skid line is a visually observable surface distortion band which is caused by passing through on the punch radius [4]. Moreover, as shown in Figure 1. (b), it is said to occur when the forces acting on the material from both sides of the ridge of a character ridgeline are not balanced with the character ridgeline on the bottom on the character ridgeline of the punch [5].

The method of increasing the radius of curvature of the character ridgeline reduces appearance attractiveness. Controlling the inflow of material from outside the parts [6] increases the amount of material used and costs more. Both measures are difficult to use in industrial applications. In order to resolve these difficulties, it is necessary to clarify material movement, and the process of skid line formation during press forming of a character ridgeline.

About critical conditions and mechanism of skid line formation, there are some reports known such as skid line limit diagram which consists of tensile and back forces [4], and a study that a skid line should be a groove caused by partial thickness reduction [7]. The experiment conducted in these studies, making steel strip slide around on a cylindrical tool under tension, is significantly different from actual production, press forming during which a panel is sandwiched between punch and die. Hence this experimental method in the past studies cannot be applicable to an investigation of the mechanism of skid line formation.

There are two reported examples of material behavior measurement during forming between a punch and die. One is by ultrasonic flaw detector installed in the die, by which the change of contact state between the die and panel was detected using the amplitude ratio of reflected and transmitted ultrasonic waves [8]. The other is by semiconductor strain gauge embedded in a die, by which a bent angle of material was measured based on the strain of the die during forming [9]. In both of the experiments, the shape of a panel can be estimated when the panel is in contact with the die, but it cannot be done when there is a gap between the panel and die.

Therefore, in this study, the skid line defect is quantitatively evaluated by draw forming experiment using a newly devised die which can measure a panel shape at any time during press forming. The mechanism of skid line formation was clarified based on the experimental results.

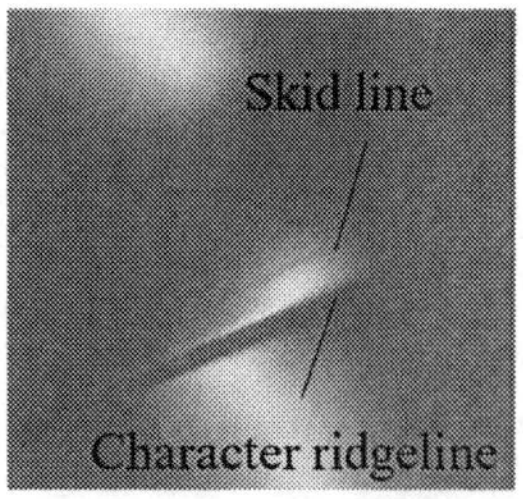

(a) Example of skid line.

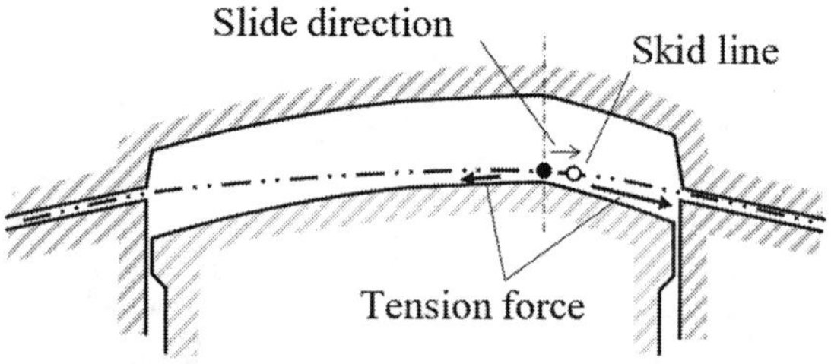

(b) Formation behaviour of skid line.

Figure 1. Skid line and imbalance of tension forces during press forming.

2. Reproduction of skid line by press forming experiment

In order to reproduce the same skid line formation as in actual production, an experiment with a punch and a die which can form a typical character ridgeline of the automobile door was carried out.

2.1. *Shapes and structures of punch and die*

Several items of 5 automobile doors were measured in order to determine the punch shape for the experiment. The Measurement items consist of the curvature radius of the character ridge, the pinch angle, and the vertical and horizontal curvature radii, as shown in Figure 2. The results measured along the outside surfaces of the door panels are listed in Table 1. The curvature radius of a character ridgeline is much smaller than those of the door panel in the vertical and horizontal directions. Therefore, the experimental punch top is made up of a ridge with a pinch angle of 150 degree and two flat areas on the both sides of the ridge as illustrated in Figure 3. The corner radius of the experimental punch was 3mm the same as car model a (Table 1) in which a skid line was observed.

In general, door panel outers are draw-formed under tension in mass production [10]. For this reason, draw-forming was adopted in this study as seen in Figure 3. The skid line is caused by the imbalance in tensions from both sides of the character ridgeline. The asymmetric tensions were attained by difference in blank holder forces (BHF) between upstream and downstream blank holders in this study since the punch, die, and blank holders have symmetric shapes, as seen in Figure 3. The BHF was controlled by the number of gas springs as shown in Figure 4. The downstream blank holder with higher BHF has more gas springs than the upstream blank holder with lower BHF.

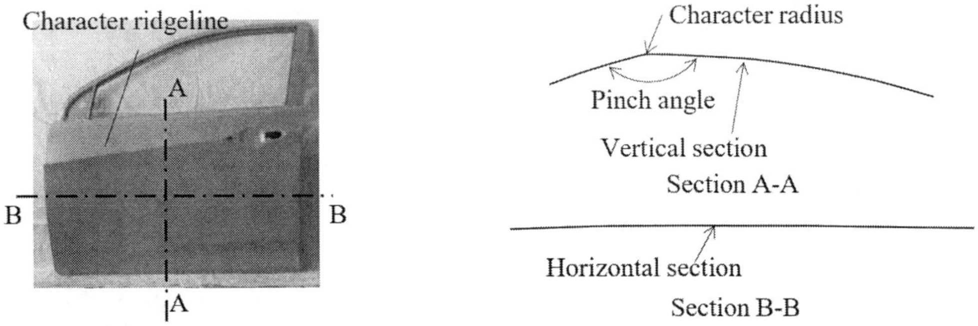

Figure 2. Measured items of door panel profile.

Table 1. Measured results of door shapes.

Car type	Thickness (mm)	Character radius (mm)	Pinch angle (°)	Curvature radius (mm)	
				Vertical	Horizontal
a	0.70	3	153	210	24500
b	0.75	8	157	580	25000
c	0.60	9	170	480	1300
d	1.10	10	159	1200	460
e	0.65	30	168	3300	3700

2.2. Experimental conditions

Mechanical properties of the tested mild steel are shown in Table 3 where YP means 0.2% proof stress, TS dose tensile strength, and EL does total elongation. Blanks of 340x80x0.7 were draw-formed using a 980kN hydraulic press machine with tools of Figure 3 under upstream and downstream BHF of Table 2. The downstream BHF was fixed at 22.5kN using 6 gas springs, on the other hand, the upstream BHF was changed from 3.8kN (1 gas cylinder) to 22.5kN (6 gas cylinders). The test material near the character ridge slightly moves in the direction from upstream to downstream depending on the difference between the BHF during press forming. Three experiments were carried out under one set of conditions listed in Table 2. The effects of the difference between the upstream and downstream BHF were investigated on skid line formation.

2.3. Experimental results

Figure 5 shows the appearance of a formed panel. The width of a formed panel did not depend on the upstream BHF. The formed panels in the vicinity of their character ridgeline after whetstone polishing are shown in Figure 6. The areas in the panels which were polished with the whetstone looks white in Figure 6, and the dark area between the white areas has a hollow shape. The hollow shape is noticeable under an upstream BHF of 3.8 and 7.5kN, not noticeable under 11.3kN, and disappears under 15.0kN or more. The hollow shape is weakened by the increase of an upstream BHF.

The outer surface profile was measured with a two-dimensional triangulation laser displacement laser (LJ-G080, KEYENCE) along the measurement line in Figure 5. Measured profiles in the downstream

IOP Conf. Series: Materials Science and Engineering 1284 (2023) 012064 doi:10.1088/1757-899X/1284/1/012064

area under upstream BHF of 7.5 and 22.5kN are shown in Figure 7 (b) on the *Y'*-*Z'* coordinate plane of Figure 7 (a). In the case of an upstream BHF of 7.5kN, there are two outstanding peaks of a character ridgeline and a skid line, and a hollow shape between them. In the case of 22.5kN, there is just one prominent peak of a character ridgeline, and no apparent hollow shape. These measured profiles are in good agreement with the appearances of panels in Figure 6.

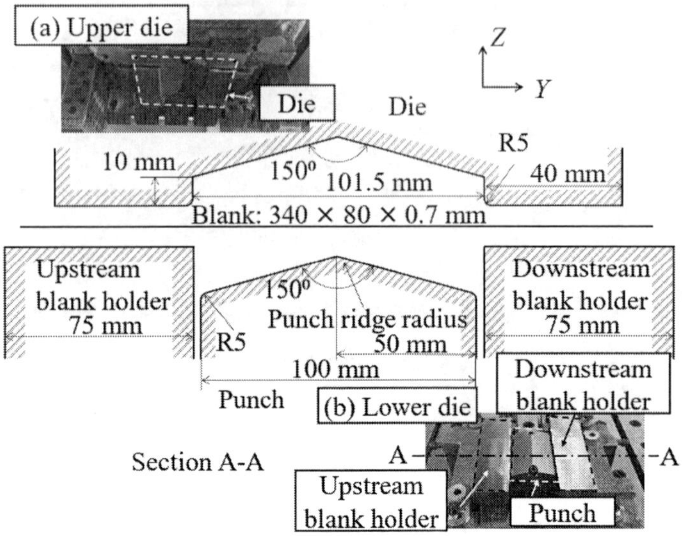

Figure 3. Configuration of punch, die, and blank holders.

(a) Upstream BHF 22.5 kN.

(b) Upstream BHF 7.5 kN.

Figure 4. Number and position of gas springs.

Table 2. Upstream and downstream BHF.

Upstream BHF (kN)	Downstream BHF (kN)
3.8	22.5
7.5	
11.3	
15.0	
18.8	
22.5	

Table 3. Mechanical properties of test steel.

YP (MPa)	TS (MPa)	EL (%)
162	296	51

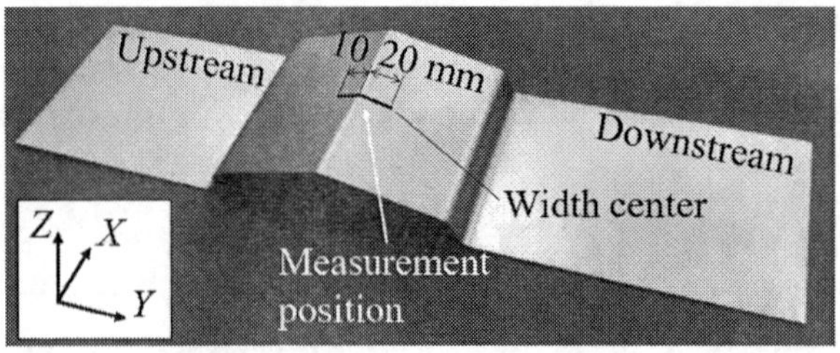

Figure 5. Shape of formed panel and line of profile measurement.

IOP Conf. Series: Materials Science and Engineering 1284 (2023) 012064 doi:10.1088/1757-899X/1284/1/012064

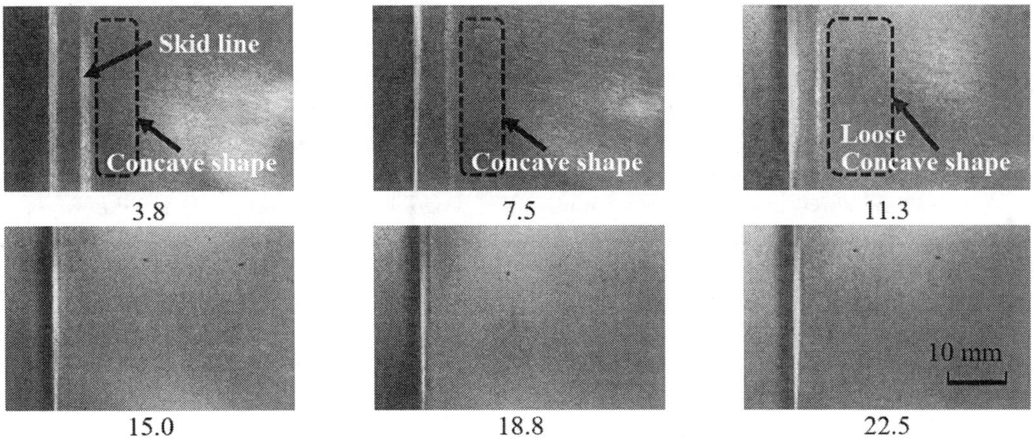

Figure 6. Vicinity of character ridgeline of the formed panel in each upstream BHF (kN).

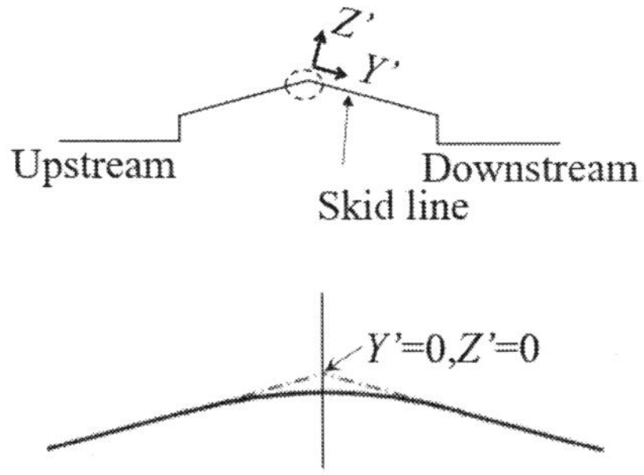

(a) *Y'*-*Z'* coordinate system.

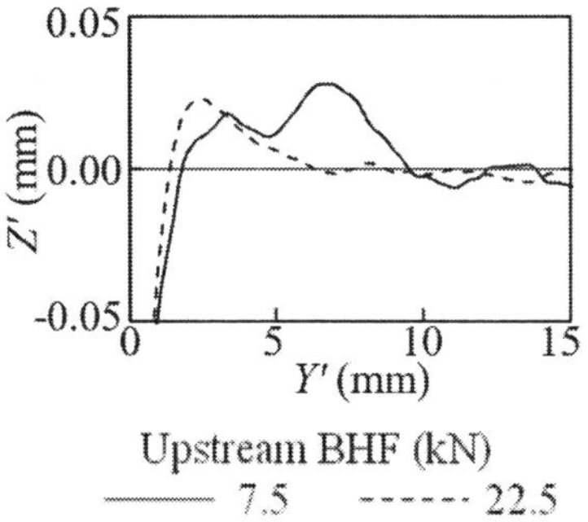

(b) Profiles of formed panel.

Figure 7. Profiles of formed panel in the vicinity of the character ridgeline.

3. In situ measurement of panel profile during press forming

3.1. Examination of die structure

During press forming, the contact between a panel and the die surface begins just before the bottom dead center (BDC), and it is considered that the panel is not in contact with the die surface in most of the process of skid line formation. Furthermore, the panel is considered to be in plane strain status during press forming because no width contraction was detected in the press formed panels. In this case, the shape of a panel during press forming can be grasped by measuring two-dimensional shape at the width center.

A new die for this experimental press forming was developed fitted with a triangular type laser displacement sensor, KEYENCE LJ-G080. Figure 8 (a) illustrates the range measured by the laser sensor. The new die has a slit hole of 8mm width and 38mm length as seen in Figure 8 (b). The dimensions of the slit hole were determined from the two viewpoints. An enough space was secured for the laser displacement sensor, and the *Y* required for the measurement of shown in Fig. 8 (a) ranged from 0.0 mm to 14.3 mm. An irradiated laser is seen in Figure 8 (c).

3.2. Effect of slit hole in die

The slit hole in the die might affect the final shape of a formed panel. As illustrated in Figure 9 (a), panel profiles along *Y*-axis were measured at the center at the directly below the slit hole and at the location of 20mm outer from the center, not below the slit hole. The position where *Y* is 0 is the center of the punch and the position where Z is 0 is the punch apex. The curvature radius of a formed panel ridge at center is larger than that at 20mm outer from the center as shown in Figure 9 (b). That is probably because the formed panel did not fit with the punch ridge at BDC due to the slit hole in the die. The panel at 20mm from the center was pressed firmly between the punch and die with the BDC. However, for the evaluation of skid line formation, a panel shape needs to be evaluated not by profile but by change in curvature. Figure 10 (a) shows the curvature change evaluation method in this study, the same method commonly used to evaluate surface distortion in mass production of automotive panels in Japan. A center height, s, was measured in the span, l, of 5mm. s can be an index of the curvature radius. s gets closer to zero as the curvature radius increases. Corresponding to the panel profiles in Figure 9 (a), s in the middle of the rigid line area is smaller at the center than at 20mm outer from the center as seen in Figure 10 (b). But the distributions of s along these two measured lines in Figure 10 (b) are similar in the flat area where a skid line arises. Therefore, the effect of the slit hole can be regarded as enough small on skid line formation and the newly developed die was used in this study.

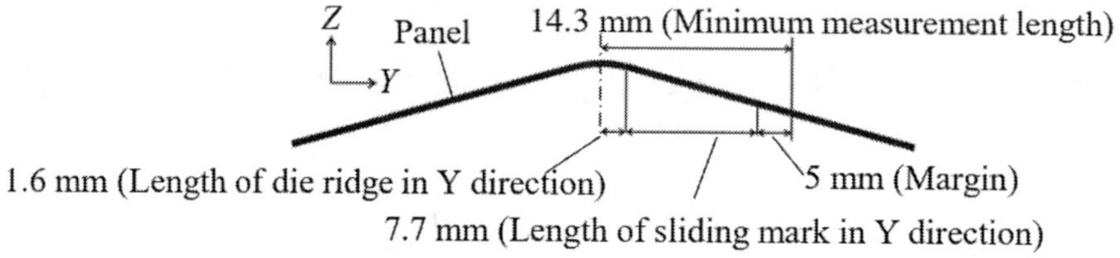

(a) Necessary range of measurement.

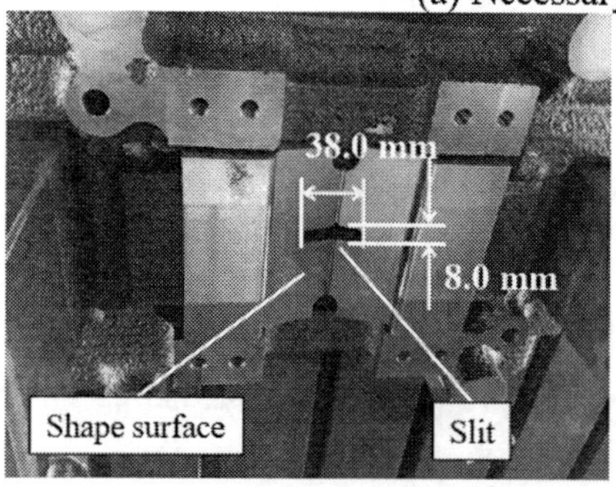

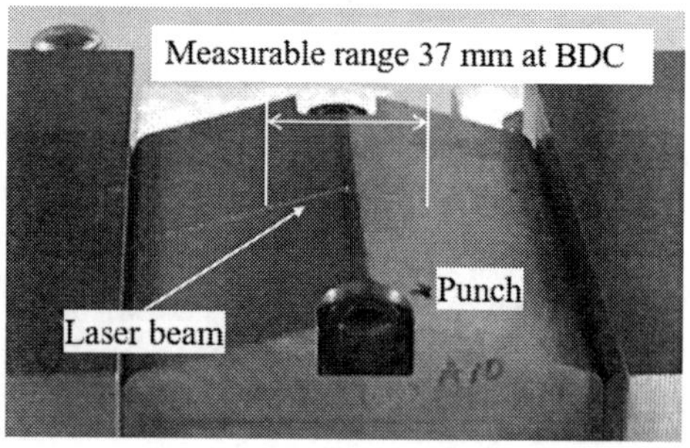

(b) Slit shape. (c) State of laser irradiation.

Figure 8. In situ measurement of panel shape during press forming.

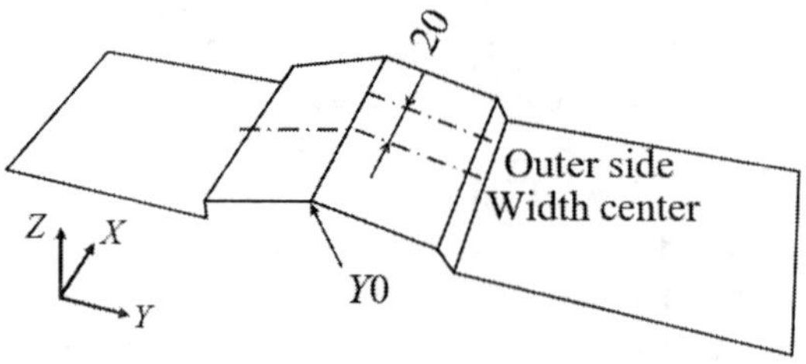

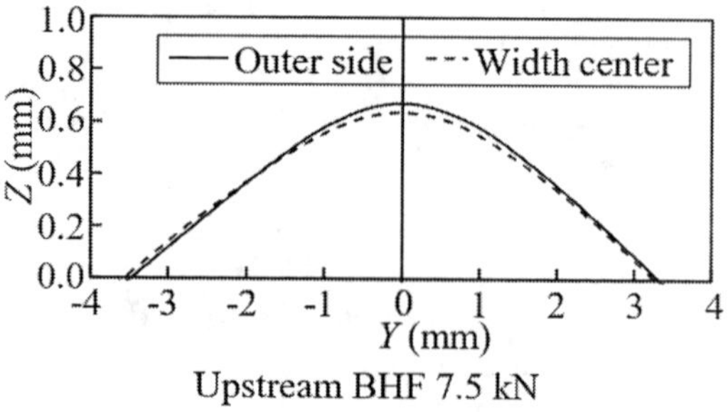

(a) Measurement position of the profile. (b) Measured profiles in Y-direction.

Figure 9. Profile of formed panel at width center and outer side.

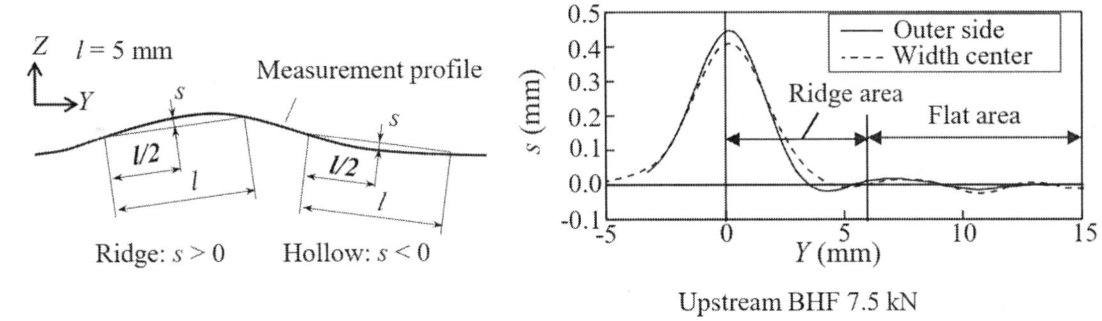

(a) Curvature radius evaluation method and definitions of center height (s) and span (l). (b) Measured center height distribution at center and at 20mm from center.

Figure 10. Evaluation method of curvature change.

3.3. *In situ measurement of panel shape*

Figure 11 shows panel profiles measured at some moments during press forming with the laser displacement meter in the die. A measured panel profile depends on the upstream BHF. As seen in Figure 11 (a), the peak of a profile measured before BDC is on the right of the line of Y is 0. Under conditions where the upstream BHF 7.5 kN was lower than the downstream BHF 22.5 kN, the peak was formed at the initial stage on the punch ridge, and then moved downstream because of the imbalance of tensions. On the other hand, in the case the upstream and downstream BHF are the same 22.5kN, the peak of the measured panel is at Y is 0 at every measured moment as shown in Figure 11 (b).

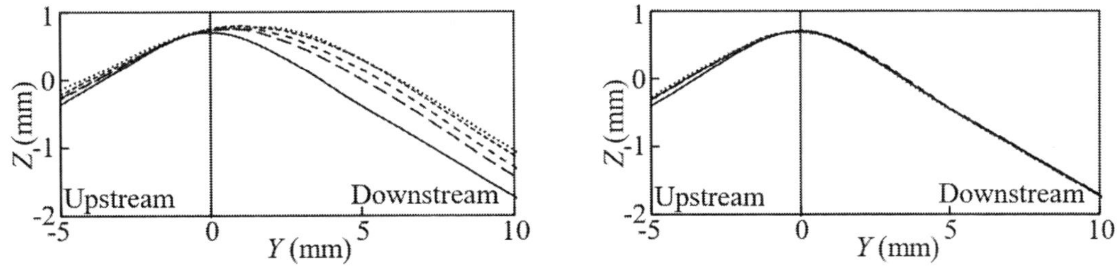

(a) Upstream BHF 7.5 kN. (b) Upstream BHF 22.5 kN.

Die position from BDC (mm)

5.0 3.0 1.0 0.3 0.0

Figure 11. Profiles of panels obtained by in situ measurement during press forming under downstream BHF of 22.5 kN.

4. Discussion

4.1. *Process of skid line formation during forming*

In this section, the process of skid line formation is discussed based on Figure 11.

Under the upstream BHF of 22.5kN (Figure 11 (b)) in the case no skid line appeared, the panel wraps around the punch ridge at its peak and has no apparent hollow profile at the die position of 5mm before BDC, and then its profile has almost no change until BDC.

Under an upstream BHF of 7.5kN (Figure 11 (a)) in the case a skid line appeared, the peak of the panel profile at the die position of 5mm before BDC is away from the punch ridge. The peak was formed on the punch ridge at the initial stage of press forming, moved to the area where the punch shape is flat because of the difference between upstream and downstream BHF, and flattened on the flat area of the punch at BDC. That seems to be the process of skid line formation during press forming under the imbalance of tensions due to the difference between upstream and downstream BHF. (1) initial ridge formation on the punch ridge, (2) movement of the formed ridge to the area where the punch is flat, and (3) flattening of the ridge at BDC.

4.2. Origin of skid line

The process of skid line formation during press forming mentioned in section 4.1 is examined in detail focusing on the flat area in Figure 10 (b) in this section. Profiles of s in the flat area, measured around BDC and after springback, are shown in Figure 12, under the conditions of an upstream BHF of 7.5kN, and of downstream BHF of 22.5kN. There is no irregularity in the flat area at 3.0mm before BDC (Figure 12 (a)). A hollow profile evolves around Y=10 in the flat area at 1.0mm before BDC (Figure 12 (b)). If the amount of waviness, Δs, is defined as shown in Figure 12 (d), Δs is 0.049mm at 0.07mm before BDC, 0.023mm at BDC, and 0.030mm after springback. The location of the uneven shape, a ridge and a hollow, does not change before and after springback. These results suggest that a wave shape should arise in the flat area just before BDC, and then it should grow to a skid line through springback.

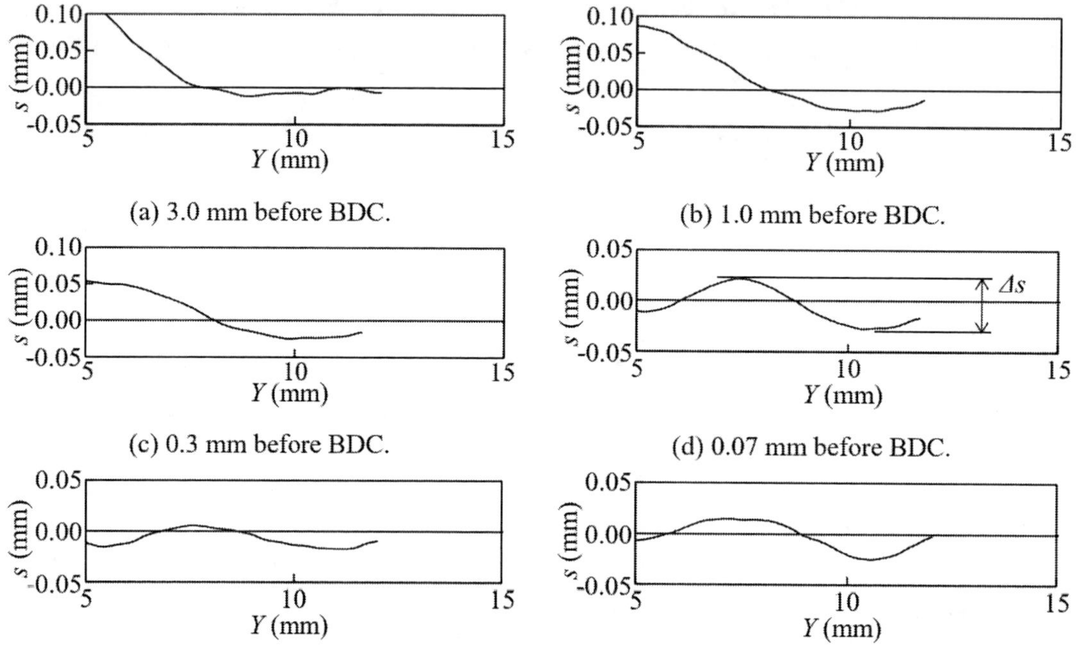

(a) 3.0 mm before BDC. (b) 1.0 mm before BDC.
(c) 0.3 mm before BDC. (d) 0.07 mm before BDC.
(e) BDC. (f) After springback.

Figure 12. Change of center height distribution in flat area of formed panel under upstream BHF of 7.5kN and downstream BHF of 22.5kN.

4.3. Effect of tension on skid line

Figure 13 shows a peak which was formed on the punch ridge and then moved downstream. The relationship between curvature radius at 5mm before BDC defined in Figure 13 and upstream BHF is shown in Figure 14, under the downstream BHF of 22.5kN. The curvature radius increases with the increase of the upstream BHF. When the peak formed at punch ridge at initial stage moves down stream, it is unbent by tensions from both sides. The higher the upstream BHF, the higher the tensions. And the higher tensions seem to cause the larger curvature radius as seen in Figure 14. In other words, skid line reduction. These results suggest that the skid line should be weakened by high tensions.

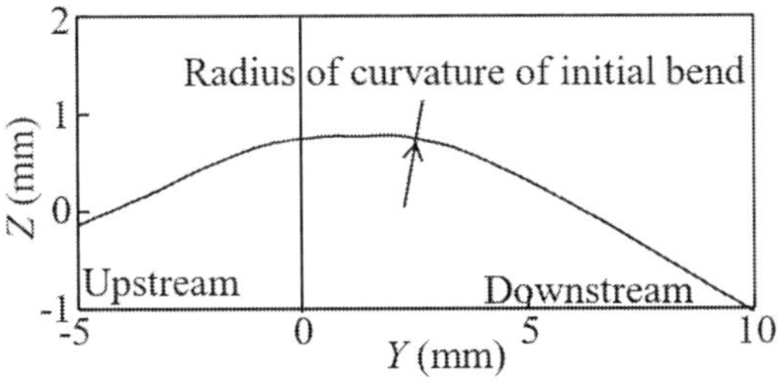

Figure 13. Peak which was formed on punch ridge at initial stage and moved downstream (under upstream BHF of 7.5kN, downstream BHF of 22.5kN, at 5mm before BDC).

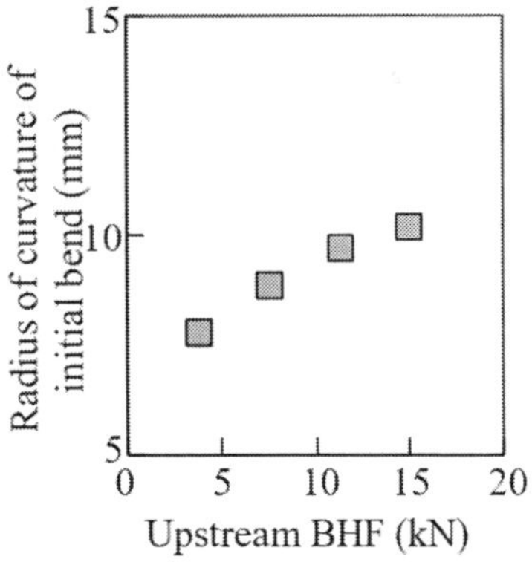

Figure 14. Relationship between curvature radius at 5mm before BDC defined in Figure 13 and upstream BHF under downstream BHF of 22.5kN.

5. Conclusions

In this study, the following conclusions were obtained about the skid line by draw forming experiment.

1) An in situ panel profile measurement method during press forming was established using a newly devised die fitted with a laser displacement meter. Panel profiles during press forming could be measured at some moments in the process of skid line formation.

2) The results of this study suggests that the skid line formation should consist of four steps. (1) a ridge formation on the punch ridge at the initial stage of press forming, (2) the movement of the ridge to the flat area of the punch, (3) the formation of a wavy shape on the flat area of the punch in the vicinity of BDC, and (4) the growth of the wavy shape to a skid line through springback.

3) The curvature radius of the ridge area which was formed on the punch ridge at the initial stage and then moved to the flat area of the punch, increased with the increase of the upstream BHF, in other words, the ridge shape became smoother. That could reduce the waviness at BDC, i.e. the skid line.

Based on the clarified mechanism of the skid line, it is considered that blunting the shape of the ridge forming section of the punch that has moved to the flat area is effective as a countermeasure against the skid line. Future studies will focus on countermeasures in the press forming process.

References

[1] Ichijo N, Iwata N, Iwata T, Mita T and Tsutamori H 2013 *J. Jpn. Soc. Technol. Plast.* **54**-628 421-5
[2] Nishimura Y, Kono T and Kawata K 2010 *Mazda Technical Review* **28** 23-7
[3] Hong Y, Sriram S, Min K, Liang H and Evangelos L 2017 *SAE Technical Papers* March
[4] Wanintradul C, Gurumurthy G.T, Smith L M, Du C, Geng L, Zhou D J, Hsiung C K, Chen J and Feng C 2011 *The 8th International Conference and Workshop on Numerical of 3D Sheet Metal Forming Processes NUMISHEET 2011 AIP Conf. Proc.* 1383 557-64
[5] Qiaomin L, Yuqi L, Gui L, Zhibing Z and Ting D 2014 *Advanced Materials Research* 842 500-4
[6] Li X, Cheng Y, Liu G and Jiang G 2011 *Forming & Stamping Technology* 2011-03
[7] Hayashi H and Furubayashi T 1979 *Proc., 1979 Jpn. Spring. Conf. Technol. Plast.* 249-52
[8] Hagino N, Endou J, Katoh S, Okudera S, Maruyama M, Kubota M and Murata C 2010 Steel research internetional **81**-9 674-7
[9] Koyama J and Yang M 2010 *J. Jpn. Soc. Technol. Plast.* **51**-596, 60-4
[10] Iwata N, Iwata T and Ichijo N 2018 *J. Jpn. Soc. Technol. Plast.* **59**-686 101-13.

Effect of strain mode and blank orientation on waviness for paint appearance of automotive outer panels

Y G An[1], L Carless[1], A R Chezan[1], F Schouten[1], M Appelman[1], M Toose[1] and B vanVeldhuizen[2]

[1] Research and Development, Tata Steel Europe, 1970 CA IJmuiden, The Netherlands
[2] Sector Automotive, Tata Steel Europe, 1970 CA IJmuiden, The Netherlands

Email: yuguo.an@tatasteeleurope.com

Abstract. Paint appearance is an important requirement for outer automobile panels. Since the last decade automotive manufacturers started reducing paint thickness to address environmental concerns. As a result, the surface roughening of steel substrate due to stamping becomes a critical factor for paint appearance. Although much research work has been conducted on surface roughening, limited research has been focused on waviness and stamping conditions in the pressing of highly formable sheet steels. In this work, the effect of strain mode and blank orientation on waviness evolution has been investigated. It was found that strain mode has a strong influence on waviness, in addition to strain magnitude. Also, the relative orientation of the applied major strain and the direction of the roughness profile relative to the rolling direction affects waviness. When surface waviness is compared at the same surface major strain, the waviness is highest in equibiaxial stretch, followed by plane strain tension in the rolling direction, and then plane strain tension in the transverse diction. When waviness is plotted against thickness strain, the waviness in plane strain tension stretched in the rolling direction is highest, independent of the roughness profile measurement direction. The waviness in equibiaxial stretching is almost the same as that in plane strain tension stretched in the transverse direction.

1. Introduction

Paint appearance is an important requirement for outer automobile panels. In the last decade, automotive manufacturers have started introducing primer-less technology in order to reduce energy consumption and CO_2 emission [1]. Some paint layers may be combined or wet-on-wet application without intermediate drying used, with the result that total paint thickness is reduced for these modern, state-of-the-art paint systems. This means that the influence of the steel substrate, in particular the surface waviness after stamping, has become more pronounced in comparison with systems that use a thick primer layer. On the other hand, more complicated exterior designs with sharper streamlines are used in the exterior panels. Such sharper ridge lines are usually accompanied with increased plastic deformation, leading to increased surface roughening and inferior paint appearance. All of these factors lead to more stringent demands on the surface topography of sheet steel, not only in the as-received state, but also after stamping. A top view of round stretched products after painting is shown in Figure 1. The products have different surface waviness after stamping, with decreasing waviness from left to right. The left one has the highest surface waviness and the largest distortion of the light grid, while the right one the least distortion. These images show that a low surface waviness improves paint appearance remarkably.

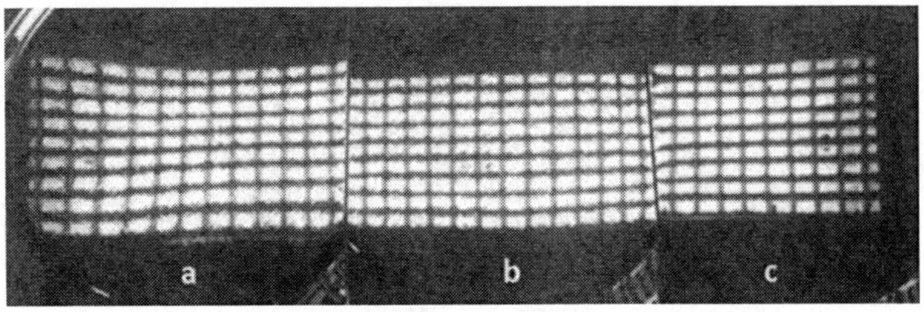

Figure 1. Top view of three round stretched products after painting. The appearance is influenced strongly by the surface waviness of steel sheet after stamping. (a) Highest waviness to (c) Lowest waviness.

Surface roughening is a kind of unevenness on the sheet surface at mesoscopic scale. The most commonly known surface defect related to this phenomenon is orange peel [2]. Many studies have been conducted on aluminium and steel sheets. It has been generally reported that surface roughening linearly increases with plastic strain [3,4,5] and is affected not only by press forming conditions such as the strain ratios or strain path, but also by material factors such as crystal grain size [6], crystal orientation [6,7] and texture.[8,9]. Some research work indicated that forming limits in the stretching state could be improved considerably by removing deformation-induced surface asperities during the forming process for sheet metals, especially for thin materials [3,10,11]. Although a substantial body of literature exists on surface roughening, there appears to have been little investigation on highly formable interstitial free sheet steels, which are mainly used for automotive outer panels. Some research work indicated that the surface roughening of IF steels depends upon the strain level as well as the strain ratio. Increased strain causes an increase in roughness, with strain paths close to plane strain exhibiting the highest rate of roughening [12,13]. Most outer panels, such as the fender and hood, are pressed at a typical strain of approximately 5%. At such low strain levels, the surface roughening of IF steels is less pronounced due to the fine grain microstructure in comparison to aluminium alloys. However, despite the fact that IF steels show minimal forming-induced roughening on a very fine scale, as paint systems have become more sensitive to the substrate beneath the paint, it has become clear that highly formable IF steels do not always lead to good paint appearance. This is due to the introduction of surface disturbances on a waviness scale during deformation, with a wavelength typically around a few millimeters. For IF steels, the waviness increase due to deformation can be significant and this is detrimental to final paint appearance. The purpose of the present work is to quantify the effect of strain state and blank orientation on the surface waviness of a highly formable galvanised sheet steel. Understanding and quantifying this effect would help automotive engineers be able to (1) incorporate waviness effects during the part design phase and (2) define appropriate stamping processes to keep waviness within acceptable tolerances that correspond with desired paint appearance levels.

2. Experimental procedures

2.1. Material

The material used in the present study was cold-rolled and galvanized highly formable IF steel sheet. The averaged mechanical properties and the plastic anisotropy r-value measured in the tensile test are listed in Table 1. The material has a well-developed {111} texture as shown in Figure 1. The average grain size was 11.8μm.

Table 1. The averaged mechanical properties and the r-value of the material

Rp (MPa)	*Rm* (MPa)	*Ag* (%)	*A* (%)	*n*	*r*
162.8	290.7	23.5	46.5	0.23	2.13

Rp: yield strength, *Rm*: tensile strength, *Ag*: uniform elongation, *A*: total elongation, *n*: work hardening exponent, *r*: plastic anisotropy.

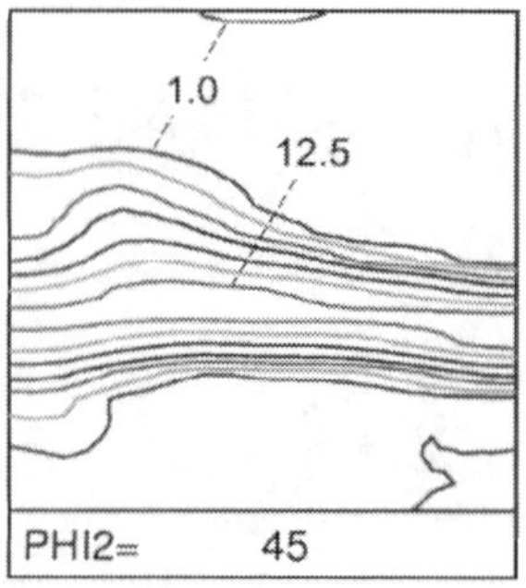

Figure 2. The ODF at φ_2= 45° section with contour levels of 1.0 _2.0 _3.5 _5.0 _7.5 _10 _12.5 _15.

2.2. Set up of biaxial stretching and plane strain tension

Equibiaxial stretching was performed using the Marciniak test [14]. Unlike the conventional Marciniak method that employs a driving blank during forming operation, a recessed cylindrical flat headed punch was used. This ensures that the deformation zone used for waviness measurement does not come into contact with the centre of punch [15]. The punch diameter was 100mm with a die radius of 10mm. Rectangular blanks, with a dimension of 140mmx140mm, were cut in the rolling direction and transverse direction. The samples were degreased and cleaned from debris. Samples with surface scratches were rejected. The stretching was carried out using a universal Erichsen deep drawing machine. The punch speed was 1 mm/s. Mineral oil Quaker N6130 was used as lubricant. The blank was fully clamped and stretched to an equibiaxial surface strain of different magnitudes.

Plane strain tension was achieved using similar tooling as for the equibiaxial stretching, but the contacting surface of the blank holder was only partly serrated, as illustrated in Figure 3(a). A dog-bone shaped blank was used, in which the parallel part of the blank is not constrained and can flow towards the die cavity. The dimension of the blank was varied to find the appropriate blank size to achieve a strain state close to plane strain. At the same time, a large uniform strain area was also required in order to be able to measure the roughness profile, from which the waviness is calculated. It was found that a blank width of 180mm was the ideal dimension for the test, as shown in Figure 3 (b). For each strain level, one blank in the rolling direction (RD) and one in the transverse direction (TD) was stretched. The first blank has a major strain in the RD, and the second one has a major strain in the TD. The test was carried out on the same deep drawing machine. The punch speed was 1 mm/s. As lubricant, Teflon foil oiled with Quaker N6130 was used.

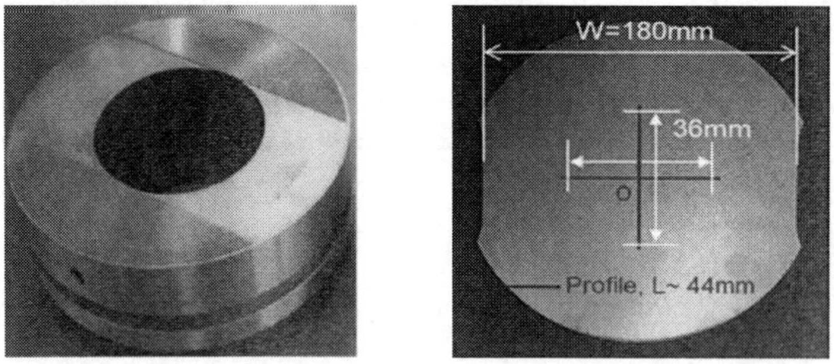

(a) (b)

Figure 3. Blank holder partly serrated (a), and the blank geometry for the plane strain tension (b).

2.3. Surface roughness profile measurement and waviness calculation

The surface roughness profiles on the stretched cup and on the blank in the as-received state before stretching were measured using a tactile BMT Expert System in the rolling direction and transverse direction. For equibaixial stretching, the major strain direction makes no difference because the deformation is geometry dominant and strains in all directions are equal. However, the waviness

evolution with straining does differ due to material anisotropy. In this study, two roughness profiles, one in the rolling direction and one in the transverse direction, were measured to evaluate the waviness evolution. The sampling length for the roughness profile was approximately 75mm.
For plane strain tension, the situation is more complex than in equibiaxial stretching, as illustrated in Figure 4. The waviness evolution depends on both the major strain direction, or stretching direction, and the direction of the roughness profile measurement. For each major strain direction, there are two directions in which the roughness profile is measured. In this investigation, material was stretched under plane strain in both the rolling direction and the transverse direction. For each stretched sample, roughness profiles along the rolling direction and transverse direction were measured, as illustrated in Figure 4. The sampling length was approximately 44mm. Because the strain distribution is uniform in the middle of the blank, a profile length of approximately 36mm was used for the waviness calculation, as illustrated in Figure 3(b). As with equiaxial stretching, the samples were also measured in the as-shipped state before stretching.

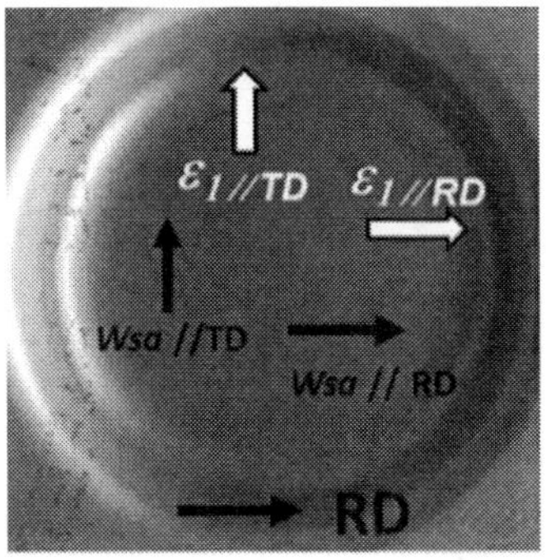

Figure 4. Illustration of surface major strain and the direction of roughness profile on the blank. ε_1 and *Wsa* refer to the major strain and roughness profile directions, respectively. The rolling direction is marked on the blank.

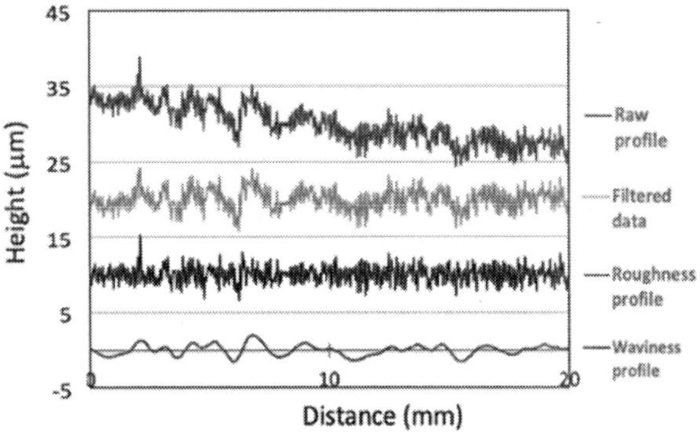

Figure 5. An example of primary measured profile and roughness and waviness profiles after filtering.

The waviness is the result of repetitive or non-repetitive macro geometrical surface irregularities in sheet metal. To characterize the shape of a product, the roughness profile is often filtered using higher order polynomials. Such modified data is further filtered, yielding a new signal in which high-frequency signals are included and low-frequency signals are filtered away. The utilized filter's boundary, or so-called cut off value, is 5mm in this work. Figure 5 illustrates the measured surface profile, filtered data after form correction, and the roughness and waviness profile after filtering. The waviness parameter, *Wsa*, is the arithmetic mean of the absolute ordinate values *Zw(x)* within the evaluation length. It is calculated using the following formula according to the VDA standard [16].

IOP Conf. Series: Materials Science and Engineering 1284 (2023) 012065 doi:10.1088/1757-899X/1284/1/012065

$$Wsa = \frac{1}{Lm}\int_{0}^{Lm} |Zw(x) - Zm| dx$$

Here the evaluation length, *Lm*, is the distance in the direction of the *X*-axis used for assessing the profile parameters under evaluation. *Zm* is the mean line of waviness that separates the enclosed area of the waviness profile into two equal areas below and above the mean line.

3. Experimental results

3.1. Strain distribution in plane strain tension

The major and minor strain distributions in the center of the blank, measured along the stretching direction in the RD, and transverse to the stretching direction, TD, were plotted in Figure 6. The results indicate that the strain state is close to plane strain and that its distribution is quite uniform in the center of the blank for both measured directions. With a target major strain set to 0.10, an averaged major strain and minor strain, and their standard deviations within a profile length of 36mm, are listed in Table 2. There is a negligible difference in the major strain between the two directions, with a deviation smaller than 0.002. The minor strain is 0.008 with a standard deviation of 0.001, very close to plane strain state within the whole profile length. The major strains in both directions are quite uniform, so the calculated waviness based on the roughness profile measured in the area is representative of this strain state and strain level.

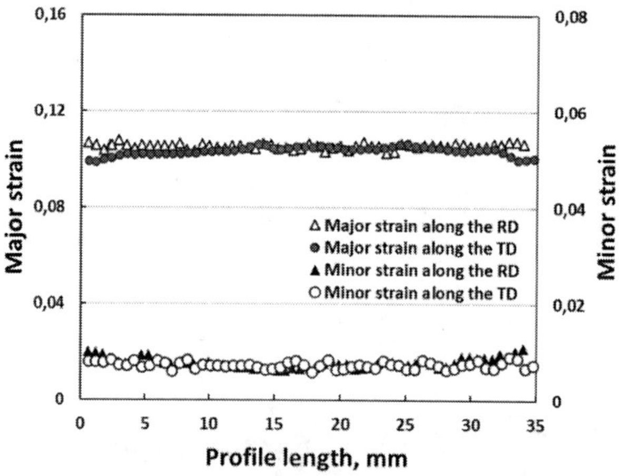

Figure 6. The major and minor strain distributions in the centre of the blank stretched along the rolling direction. The major and minor strain profiles were measured along the RD and the TD.

Table 2. Averaged major strain and minor strain and the standard deviation within a profile length of 36mm for profiles measured in the RD and TD.

Strain profile direction	Major strain	Minor strain
RD	0.106 ± 0.001	0.008 ± 0.001
TD	0.104 ± 0.002	0.008 ± 0.001

3.2. Waviness and strain relationship in biaxial stretching and plane strain tension

The waviness as a function of major strain in equibiaxial stretching was plotted in Figure 7(a). For the equibiaxial stretching state, the major strain direction makes no difference because the strains in any direction, inclusive the RD and the TD, are equal. However, an anisotropy in the evolution of waviness with straining was observed. As shown in Figure 7(a) there is a clear difference in waviness between the rolling direction and the transverse direction. At strains below 0.03, the difference is not significant. However, the difference increases with increasing strain, in particular in the rolling direction.

For plane strain tension, there are two waviness profiles for stretching in the rolling direction, and two waviness profiles for stretching in the transverse direction as shown in Figure 7 (b). Both the stretching direction, or the major strain direction, and the direction for the profile measurement have a strong influence on the waviness. Firstly, the stretching direction plays an important role. The results indicate that the waviness increases more rapidly with straining when the stretching direction is along the RD than it does along the TD. Secondly, for the same stretching direction, the direction of the surface profile also plays an role, as also found in equibiaxial stretching. The waviness measured along the RD is greater than that along the TD for the same stretching direction. For the investigated material, the waviness in plane strain tension can be ranked in a decreasing order at the same surface major strain as follows;

$$Wsa_{\text{RD-RD}} > Wsa_{\text{RD-TD}} > Wsa_{\text{TD-RD}} > Wsa_{\text{TD-TD}}$$

Here the first and second direction refer to the stretching direction and surface profile direction, respectively.

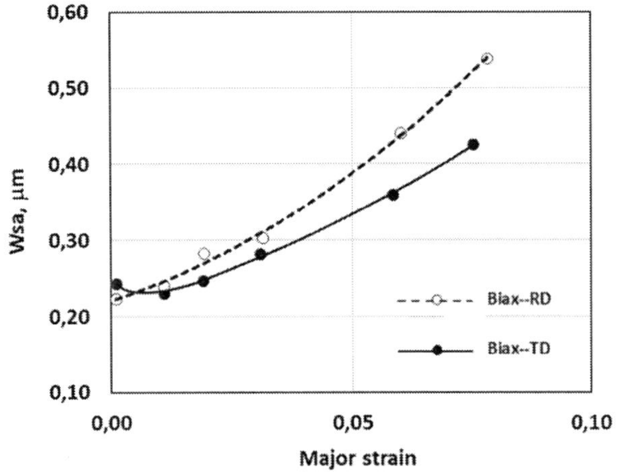

(a)

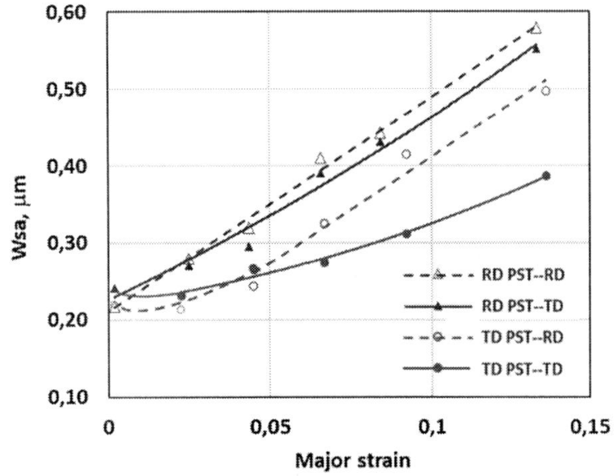

(b)

Figure 7. Waviness as a function of surface major strain in equibiaxial stretching (a), and plane strain tension (b). The direction before PST refers to the stretching direction, and the direction after refers to the direction for the measured profile.

4. Discussion

Waviness is a regular height differences of a surface that causes scattering of light. It stands out from roughness because the wavelength is high and an increased waviness degrades the aesthetics of a formed component after painting. The waviness parameter, *Wsa*, as specified in the standard correlates well with paint appearance. The effect of strain state and major strain direction on waviness at the same blank orientation is compared and discussed.

4.1. Waviness as a function of surface major strain

Equibiaxial stretching and plane strain tension are two typical strain states that cover the strain modes for most outer panels, so it is interesting to study the waviness evolution with straining at these two conditions. The waviness measured and calculated along the rolling direction for these two strain modes is shown in Figure 8(a). At the same surface major strain, the waviness in the equibiaxial stretching state is the highest, followed by the plane strain tension stretched in the rolling direction, and then the plane strain tension stretched in the transverse direction. In addition, the waviness increase with straining is more rapid in the equibiaxial stretching than in the plane strain tension.

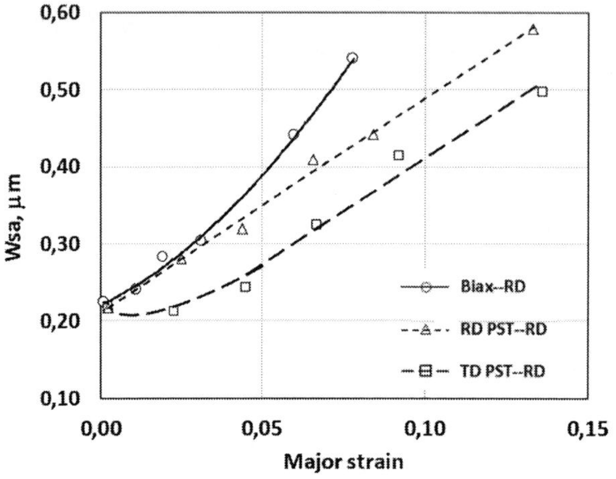

(a)

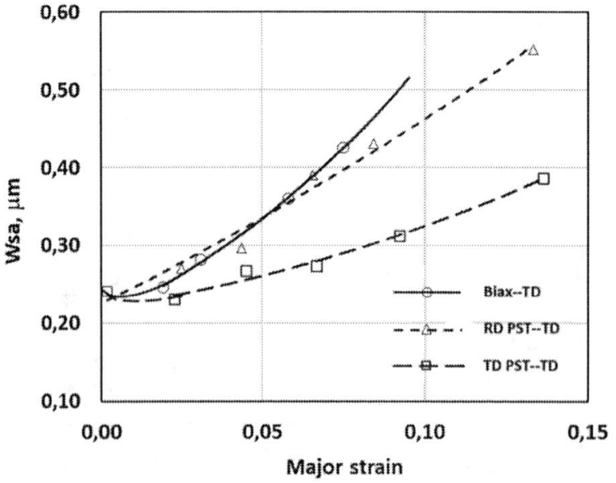

(b)

Figure 8. Waviness as a function of surface major strain at different strain state and different stretching direction. The waviness is along the rolling direction (a), and along the transverse direction (b).

The waviness measured and calculated along the transverse direction for the equibiaxial stretching and plane strain tension is shown in Figure 8(b). At the same surface major strain, the waviness in the equibiaxial stretching state is still the highest, followed by the plane strain tension in the rolling direction and in the transverse direction as the trend observed in Figure 8(a). However, the difference between the equibiaxial stretching and plane strain tension in the rolling direction is relatively small. Meanwhile, the difference in waviness between plane strain tension in the RD and TD is significant. Another interesting phenomena is that at small strains, the waviness is quite stable, or even some trend to decrease with straining for the sample stretched in the transverse direction.

Table 3. Estimated waviness at surface major strain of 5%

	Strain mode		
Waviness profile	Equi-biaxial	PST ε_1 // RD	PST ε_1 //TD
Wsa in RD	0.39	0.34	0.28
Wsa in TD	0.34	0.34	0.27

Large area fractions of the surface of typical outer panels like the hood, roof, or fender show surface major strain levels of approximately 5%. The waviness at 5% for each stretching condition is estimated and listed in Table 3. For the waviness along the rolling direction, clearly the highest waviness occurs at equibiaxial stretching state, followed by the plane strain tension in the rolling direction, and then the plane strain tension in the transverse direction. For the waviness along the transverse direction, the waviness at equibiaxial stretching is of similar magnitude as the plane strain tension in the rolling direction, both are substantially larger than the waviness in the plane strain tension in the transverse direction. In a brief summary, when the waviness is compared at the same magnitude of surface major strain, the equibiaxial stretching is the worst strain state, while the plane strain tension in the transverse direction is the most favorable one for waviness.

4.2. Waviness as a function of thickness strain

Many interrelated aspects of material performance are important when designing a car and in the selection of sheet materials. These include paint appearance, dent resistance and formability. In the previous section, the evolution of waviness with surface strain in different strain states and along different directions is discussed. This information is appropriate when investigating formability and part manufacturability, so that waviness increases may also be taken into consideration. However, for dent resistance, it is important to monitor sheet thickness. Dent performance depends strongly on the final thickness and yield strength of press formed product. For this reason, it is necessary to compare the waviness evolution with thickness strain in different strain states so that this may be taken into account when designing for optimal dent performance.

The waviness measured and calculated along the rolling direction for equibiaxial stretching and plane strain tension is shown in Figure 9 (a). At the same thickness strain, the waviness in the plane strain tension stretched in the rolling direction is the highest, followed by the equibiaxial stretching and the plane strain tension stretched in the transverse direction. The difference in waviness between the equibiaxial stretching and the plane strain tension stretched in the transverse direction is relatively small. However, the waviness increases more with straining in the plane strain tension than in the equibiaxial stretching. The waviness along the transverse direction for the equibiaxial stretching and plane strain tension is shown in Figure 9 (b). The effect of strain state on the waviness in the transverse direction is quite similar to that in the rolling direction. At the same thickness strain, the waviness in plane strain tension stretched in the rolling direction is still the highest, followed by the equibiaxial stretching and the plane strain tension stretched in the transverse direction. The difference in waviness between the equibiaxial stretching and plane strain tension stretched in the transverse direction is negligible.

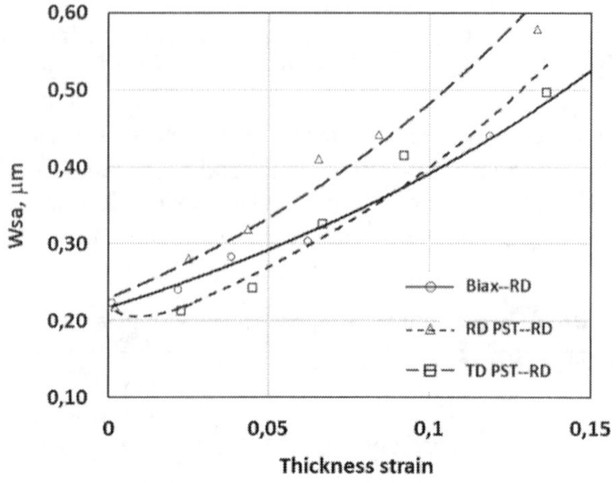

(a)

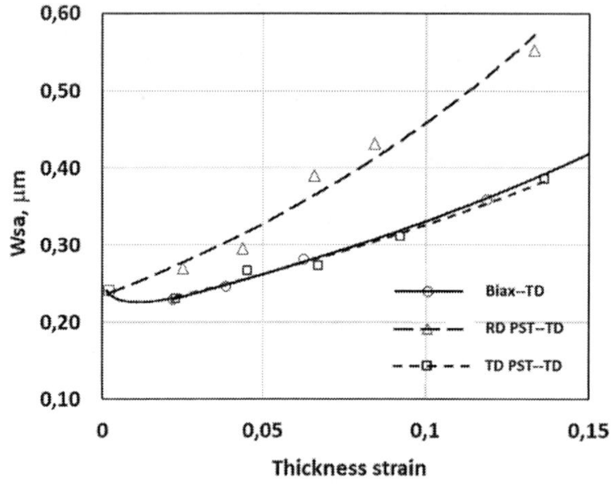

(b)

Figure 9. Waviness as a function of thickness strain at different strain state. The waviness is measured in the rolling direction (a), and in the transverse direction (b).

The waviness measured along different directions for each stretching condition at 5% thickness strain is estimated and listed in Table 4. The plane strain tension stretched in the rolling direction causes the most remarkable increase in waviness independent of the waviness profile direction. In a brief summary, when the waviness is compared at the same magnitude of thickness strain, the plane strain tension stretched in the rolling direction is the most unfavorable forming condition for waviness. The difference in waviness between the equibiaxial stretching and the plane strain tension stretched in the transverse direction is small and the waviness value is low.

Table 4. Estimated waviness at thickness strain of 5%.

	Strain mode		
Waviness profile	Equi-biaxial	PST ε_1 // RD	PST ε_1 //TD
Wsa in the RD	0.29	0.34	0.28
Wsa in the TD	0.27	0.32	0.27

IOP Conf. Series: Materials Science and Engineering 1284 (2023) 012065 doi:10.1088/1757-899X/1284/1/012065

5. Conclusions

The results and analysis presented in this paper illustrate that the strain state plays an important role in waviness change with straining in addition to strain magnitude. The blank orientation for the major strain and roughness profile measurement also have strong influence on waviness.
When waviness is compared at the same surface major strain, the equibiaxial stretching and plane strain tension in the rolling direction are unfavorable forming conditions, whereas plane strain tension stretched in the transverse direction is the most favorable condition for waviness.
When waviness is plotted against thickness strain, plane strain tension stretched in the rolling direction is the most unfavorable forming condition for waviness. The waviness in equibiaxial stretching is low and of the same magnitude as the plane strain tension stretched in the transverse direction.
The methodology presented in this paper can be used to quantify waviness effects for different material classes and to consider waviness as a design parameter for future automotive outer skin panels. Appropriate selection of the main stretching mode and major strain direction relative to the rolling direction of the blank can reduce undesired waviness increase caused by deformation during stamping, further supporting efforts to reduce paint system thickness.

References

[1] Streitberger H J Dossel K F 2008 *Automotive Paints and Coatings* (John Wiley & Sons)
[2] Wilson D V Roberts W T and Rodriguez P M B 1981 *Metall. Trans.* **12A** 1603-11
[3] Yamaguchi K Takakura N and Imatani S 1995 *J. Mater. Process. Technol.* **48** 27-34
[4] Guangnan C Huan S Shiguang H and Baudelet B 1990 *Mater. Sci. Eng.* **A128** 33-38
[5] Beaudoin A J Bryant J D and Korzekwa D A 1998 *Metall. Mater. Trans.* **29A** 2323-32
[6] Bate P S 1992 *Scripta Metall Mater* **27** 515-20
[7] Wu P D Lloyd D J Bosland A Jin J and MacEwen S R 2003 *Acta Mater.* **51** 1945-57
[8] Lee P S Piehler H R Adams B L Jarvis G Hampel H and Rollett A D 1998 *J. Mater. Process. Technol.* **80** 315-19
[9] Banovic S W and Foecke T 2003 *Metall. Mater. Trans.* **34A** 657-71
[10] An Y G 2001 *J Mater Sci Technol* **17** 249-57
[11] Bong H J Lee J 2021 *Int. J. Mech. Sci,* **191** 106066
[12] Kubo M Hama T Tsunemi Y Nakazawa Y and Takuda H 2018 *ISIJ International* **58** 704–13
[13] Wichern C M De Cooman B C Van Tyne C J 2005 *J. Mater. Process. Technol.* **160** 278-88
[14] Marciniak Z and Kuczynski K 1967 *Int. J. Mech. Sci.* **9** 609-12
[15] VDA 238-501 2020 Waviness Testing of Cold Formed Sheet Metal - Preparation of formed test specimens
[16] VDA 238-500 2020 Waviness Testing of Cold Formed Sheet Metal - Test procedure

IOP Conf. Series: Materials Science and Engineering 1284 (2023) 012066 doi:10.1088/1757-899X/1284/1/012066

Experimental investigation of the effect of W-temper forming process parameters on springback

D E Jemal, L Zsolt

Institute of Materials Science and Technology, Faculty of Mechanical Engineering and Informatics, University of Miskolc, H-3515 Miskolc, Egyetemváros, Hungary

E-mail: metjemal@uni-miskolc.hu

Abstract. High strength aluminium alloys have drawn a lot of attention because of the expanding demand for lightweight vehicle design in the automotive sector. Due to poor formability at room temperature warm and hot forming have been advised. However, warm and hot forming methods need more steps in the production process and an advanced tooling system. In contrast, since ordinary tools can be used, forming sheets at room temperature in the W-temper condition is advantageous. However, a significant problem that must be resolved during the use of this technique is the springback of supersaturated sheets and their thinning. In this study, the high strength aluminium alloy was solution heat treated, and a U-shaped bending test was carried out at different time periods between W-temper heat treatment and forming operation. As a result, it is noted that the springback is highly dependent on the time between W-temper heat treatment and forming operations, and the change in shape increases with increasing time. The change in shape after forming was also evaluated using numerical simulation aiming to validate the experimental result. To determine the material parameters of the kinematic hardening model for numerical simulation, uniaxial tension-unload tests were performed under different ageing conditions.

1. Introduction

High-strength aluminium alloys are a type of aluminium alloy that has a high strength-to-weight ratio and in commonly used in aerospace, automotive, and construction applications. They offer a combination of high strength, high hardness, and good corrosion resistance, making them an attractive option for many applications. High strength aluminium alloys are typically made by adding elements such as zinc, magnesium, copper, and manganese to aluminium to increase its strength. These alloys can be further strengthened through heat treatment processes, such as ageing, quenching, and tempering. However, high strength aluminium alloys are more prone to springback than other aluminium alloys due to their high strength and low ductility. It is the alteration of dimensions of the formed part after tool removal due to the change in the state of stresses and strains in the deformed material [1,2]. This can make forming these materials more challenging, especially for complex shapes. Furthermore, the higher strength of these alloys makes them more prone to cracking or failure under certain conditions. The most popular way to counteract this effect is to overbend stamped parts. Such faults must be identified, eliminated, or avoided, and their exact origins and relationships with other influencing elements must be studied and understood. The distribution and amplitude of stresses created during sheet thickness forming are thought to be one of the factors that significantly contribute to the generation of the springback phenomenon [1,3,4]. Accuracy of springback investigation necessitates a material model capable of accurately describing the complex material behaviour during unloading [5,6]. The friction coefficient of the contact surfaces and the blank holding force are also often influential on springback in addition to the material characteristics of the sheet. The radiuses of the tool profile and sheet thickness have an impact as well [7]. Therefore, a thorough analysis of the various ways in which these factors affect the spring-back is crucial and requires a high level of expertise with numerous variables of the experiment design matrix.

The W-temper forming process is an effective way to reduce springback during metal forming operations. The W-temper process involves heating the metal sheet to a certain temperature and then rapidly cooling it, which improves its formability and reduces the tendency for springback to occur. Several studies have investigated the relationship between various W-temper process parameters, such as temperature, cooling rate, and holding time, and their effect on springback. Gronostajski et al. [8] determined the best W-Temper condition for forming the B-pillar from 7075 aluminium alloy using both mathematical modelling and experiments. The research focused on the most important problems that arise during the cold forming of solution heat-treated sheets: the selection of optimal temperature and cooling parameters, thinning of the post-formed element, followed by the ageing and springback process. Choi et al. [9] studied the mechanical properties of WT high strength aluminium alloy and to implement them into the FE simulations using properly modelled constitutive equations. The u-draw bending test is employed to evaluate the springback in the WT forming. Anisotropic plasticity in terms of yield function and hardening law under loading path changes was highlighted in the numerical modelling of springback for potential application to the cold-forming process. Lee et al. [10] investigated the mechanical behaviour of sheets of 7075 aluminium alloy that had been W-tempered and assessed the viability of using the W-temper forming technique to produce an accurate automotive component. For this purpose, mechanical tests such as uniaxial and balanced biaxial tension tests and the Nakajima test were conducted to estimate the basic elastic and plastic mechanical properties and formability. Choi et al. [11] investigated the mechanical properties of the heat-treated Al 7xxx alloys. This study examines the strengths of each production stage. The W-temper state exhibited a strong sensitivity to the cooling technique after solution heat treatment in the high-strength 7xxx series aluminium alloy. The holding period affects the material characteristics of the W-tempered sheet. Wang et al. [12] comprehensively examine the impact of heat treatment parameters on the stamping deformation and springback of 6061 aluminium alloy sheets using three-point bending tests. In addition, by integrating it with FE numerical modelling, the mechanism of the sheets' springback was examined. Overall, the literature on this topic has demonstrated that the W-temper forming process can effectively reduce springback during metal forming operations, and that further research is needed to better understand the relationship between W-temper process parameters and springback.

In the ongoing study, the effect of W-Temper heat treatment process parameters on the springback of the AA6082 aluminium alloy was investigated. Experimental and numerical studies have been done to determine the impact of the aging time and blank holding force on springback. The numerical analysis was conducted using the commercial code AutoForm-Sigma. It is possible to adjust the heat treatment process to reduce springback while ensuring the appropriate microstructure and mechanical properties by looking at the relationship between the W-Temper process parameters and springback. A higher springback angle can be obtained by ageing over a longer period of time because decreasing the material's ductility and increasing residual stress give it more time to take on its final shape. Due to the fact that the deformation of material during the forming process is distributed uniformly throughout the thickness, the quantity of springback is reduced by a higher blank holding force.

2. W-temper heat treatments

It's important to note that the W-Temper heat treatment process can be complex and requires precise control over temperature and cooling rate to achieve the desired results. Improper heat treatment can result in cracking, distortion, or other defects that can impact the performance of the component. As a result, to get the best and optimal outcome, it is important to consult with a skilled and experienced heat treatment specialist or provider. Therefore, the W-Temper heat treatment process needs a special furnace that must be able to maintain a controlled atmosphere or vacuum and precisely control the temperature.

IOP Conf. Series: Materials Science and Engineering 1284 (2023) 012066 doi:10.1088/1757-899X/1284/1/012066

In this study, a forced warm-air heating furnace has been designed and developed as shown in Figure 1. The temperature is controlled using four thermocouples which were installed at an equal offset distance position on the barrel of the furnace to achieve a homogeneous temperature distribution.

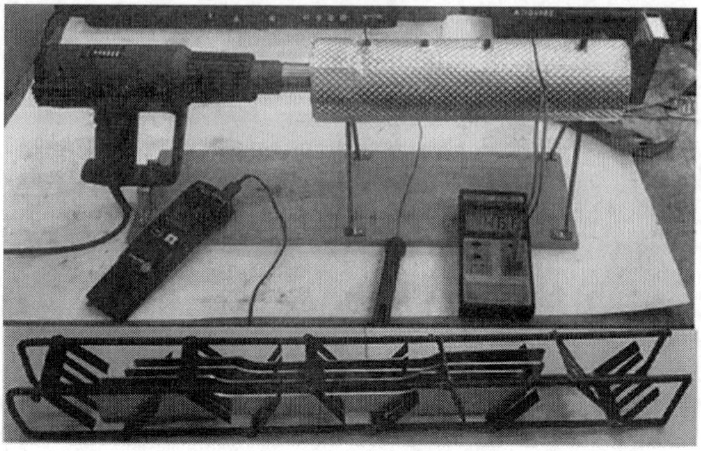

Figure 1. Forced warm-air heating furnace.

Figure 2 shows the temperature profile of the WT forming technique used in this study. The as-received sheet was subjected to a solution heat treatment (SHT). The sheet material was heated to a targeted SHT temperature and retained for some time inside the furnace to gain homogenous temperature distribution. Then, the heated material was placed in water for rapid quenching to room temperature. However, it is noted that the mechanical properties of the samples could vary depending on the holding and ageing times. A U-shaped bending test was carried out at different time periods between W-temper heat treatment and forming operation and its effect on springback analysed.

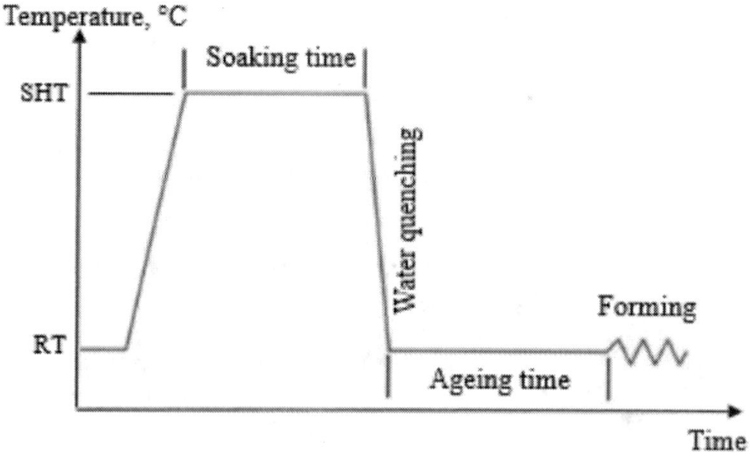

Figure 2. Temperature profile of the WT forming.

3. Material Characterization

In the following, "6082-T6," we refer to commercially produced peak-aged (T6 tempered) as-received aluminium alloy 6082 sheets. Table 1 is a list of the 6082-T6 sheet's chemical compositions.

Table 1. Chemical composition of AA6082-T6 (wt%).

Material	Fe	Si	Cu	Mn	Mg	Zn	Cr	others	Al
AA 6082-T6	0.5	0.7-1.3	0.1	0.4-1	0.6-1.2	0.2	0.15	0.15	Balanced

Because T6 tempered aluminium alloy is stronger, it is often very hard to form, which makes it fail quickly during the forming process at room temperature. In contrast, the WT sheet has been viewed as potentially being created at room temperature because of its improved formability. The sheet that was delivered was given the so-called WT heat treatment. SHT at 525 °C for 30 min, followed by water quenching to room temperature, ensures complete dissolution of the alloying components into the aluminium matrix. The "6082-WT" sheet refers to the sample that was created under the WT condition. It is known that SHT increases ductility, which is why this specific heat treatment is being researched.

3.1. Tension and unloading tests

The main advantage of WT forming is that it improves formability at room temperature, which would make traditional cold stamping possible. Then, in contrast to hot forming, which greatly reduces the flow stress, the springback following cold stamping may be a serious problem. The Bauschinger effect in the cyclic loading and the degradation of the apparent elastic modulus are reported to be factors that affect the accuracy of the springback prediction for the springback simulation [13]. From the tension-unloading tests, a significant difference was observed in the flow stress curves at 15, 30, 90, and 120-minute ageing times (the time after quenching until conducting the uniaxial tension-unloading test), as shown in Figure 3a. As shown in Figure 3b, the ageing time is an important factor in the flow curve, as is the degradation of the elastic modulus. To determine the uniaxial tension-unloading test, the test specimens with the appropriate geometry were prepared, and then the test was conducted at different aging times with 14% pre-strains according to the specifications of the MSZ EN ISO 6892-1 standard on a universal material testing machine type MTS 810.23.

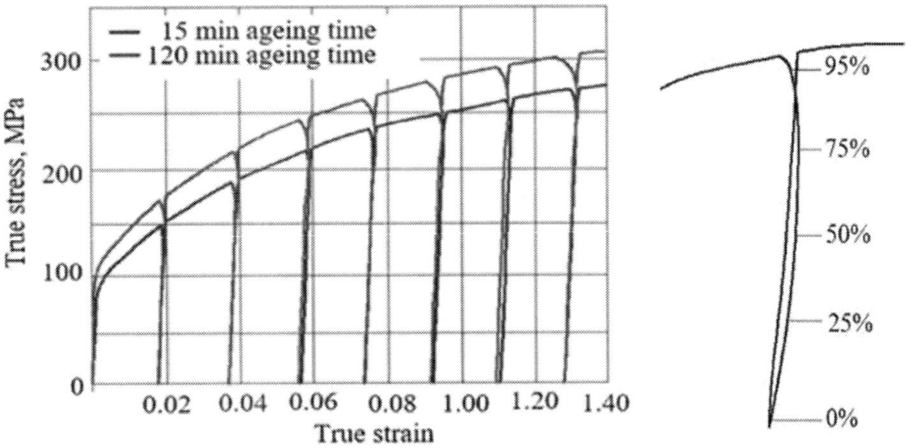

(a) (b)

Figure 3. (a) Tension-unloading test; (b) Degradation of the elastic modulus.

4. Finite Element Modelling

A U-bending test of 250x20x1 mm rectangular sheets was performed numerically using AutoForm commercial code. Accurate springback prediction needs a material model that precisely describes the complex material behaviour at loading and unloading conditions. A novel approach has been developed and implemented in the AutoForm commercial code to model the kinematic hardening behaviour of the material [14,15] as Eq. (1),

$$E_l = E_0(1 - \gamma(1 - e^{-\chi p})), \tag{1}$$

where E_l is the tangent modulus and which ordinarily falls off exponentially as comparable plastic strain p builds up; E_0 is the tangent modulus at zero plastic strain; χ is the saturation constant; γ is Young's reduction factor formulated as Eq. (2),

$$\gamma = (E_0 - E_a)/\, E_0, \tag{2}$$

where E_a is the Young's modulus at infinite plastic strain.

Transient softening rate K is expressed by the summation of linear and non-linear reverse strain as Eq. (3) [16],

$$\varepsilon_r = \varepsilon_{rl} + \varepsilon_{rn} = \frac{\sigma_r}{E_1(p)} + Karctanh^2\left(\frac{\sigma_r}{2\sigma_h(p)}\right)^2, \tag{3}$$

where $\sigma_h(p)$ is the reverse plastic strain dependent isotropic stress; σ_r is the reverse stress curve; ε_r is the total reverse strain; ε_{rl} is the linear reverse strain; and ε_{rn} is the nonlinear reverse strain. This study considered a pre-defined K in the material card of AutoForm.

The changes in the elasticity modulus during reverse loading can be determined as given by Eq. (1). The γ and χ values describing the changes in the elasticity modulus during reverse loading can be found by fitting curves to these measured points using the curve fitting module of the MATLAB program [17]. Figure. 4, shows the decrease in chord modulus as a function of pre-strain. It can be shown that as the deformation increases, the value of the average chord modulus decreases. As a result, the kinematic hardening parameters determined from the fitting curve for 15 and 120-minute ageing times were used in this study, as shown in table 2. Table 3 lists other design factors and their levels that were taken into consideration for the numerical analysis. The maximum BHF was decided because of the maximum compression force generated by the spring in the real physical experiment apparatus.

Table 2. Kinematic hardening parameters.

Aging time	γ	χ
15-minute	0.157	23.6
120-minute	0.112	28.2

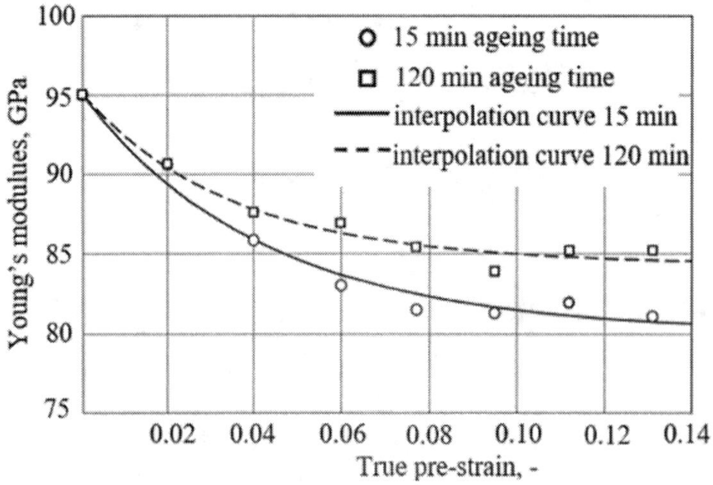

Figure 4. Reduction of elasticity modulus as a function of pre-strain.

Table 3. Design variables and their level for the simulation.

Parameters	Level
The radius of die and punch profile	5 mm
BHF	3, 5, 7 kN
μ	0.08

5. U-bending test

U-bending test experiments were conducted in lubrication conditions with the intention of validating the influence of aging time and BHF on springback parameters, which was achieved in the numerical investigation. Figure 5 shows the experimental setup of U-bending dies equipment, which is installed on a 20 kN hydraulic press machine. The punch width was 40 mm, and its half position matched the half position of the width of the sheet. The die clearance on both sides was equal to double sheet thickness. The design variables and their level for the experiment are shown in table 4.

Figure 5. U-bending test apparatus.

Table 4. Design variables and their level for the experiment.

Parameters	Level
Punch velocity	5 mm/sec
Radius of die and punch profile	5 mm
BHF	3, 5 &7 kN
Lubrication condition	oil lubrication

6. Results and discussion

After removing the tool, the springback profiles in the simulation at 15-minute ageing time and BHF=3 kN are shown in Figure 6. The amount of springback of each test was measured using NUMISHEET '93 benchmark standard [18] as it is shown in Figure 7. The 2D picture of each sample in numerical and experimental results is exported to AutoCAD commercial drafting software to measure the springback parameters.

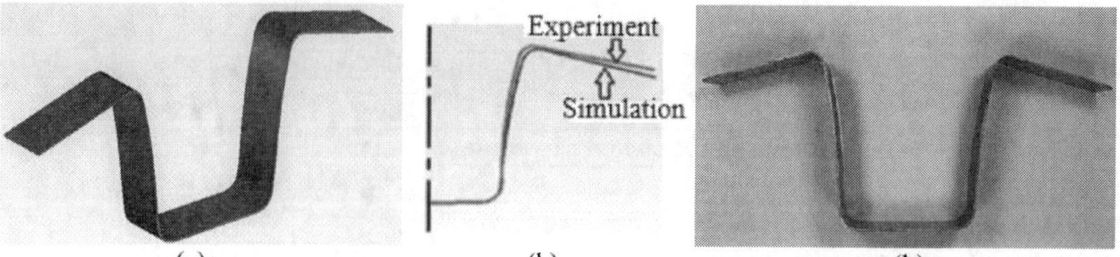

(a) (b) (b)

Figure 6. Springback profiles at 5kN BHF and 15-minute ageing time (a) 3D in simulation, (b) 2D in both experiment and simulation, (c) 3D in experiment.

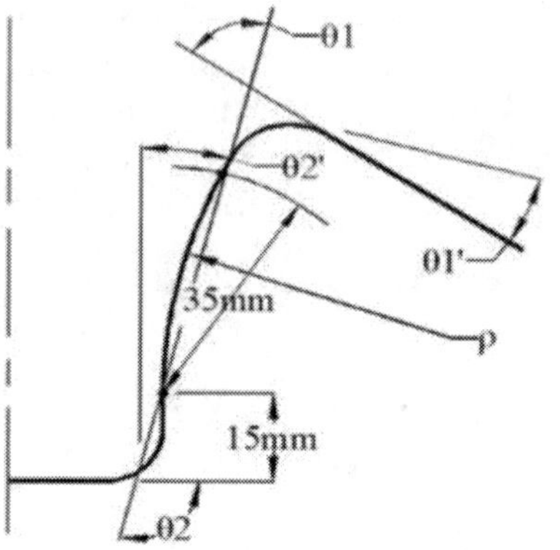

Figure 7. Springback parameters in NUMISHEET '93 benchmark.

6.1. Effect of ageing time on springback

The springback of 6082 WT aluminium alloy can be affected by its ageing time. A longer ageing time can result in a higher springback angle, as the material has more time to settle into its final shape during the aging process. However, this effect can depend on lubrication conditions and BHF. The effect of ageing time on springback profiles at 5kN BHF is shown in Figure 8a and b. Increasing the ageing time from 15 to 120 minutes resulted in a decrease of θ1 from 85.6° to 83.2° in the experiment and from 85.7° to 82.8° in the simulation. On the other hand, θ2 was increasing from 98.5° to 103° in the experiment and from 99.2° to 102.1° in the simulation. Because the material may naturally ageing to some extent, which may improve its strength while decreasing its ductility, increasing the ageing time had no further effect on springback angle. Therefore, it is important to balance this against other factors, such as production efficiency and cost, and to carefully optimize the ageing time for the specific application and material being used.

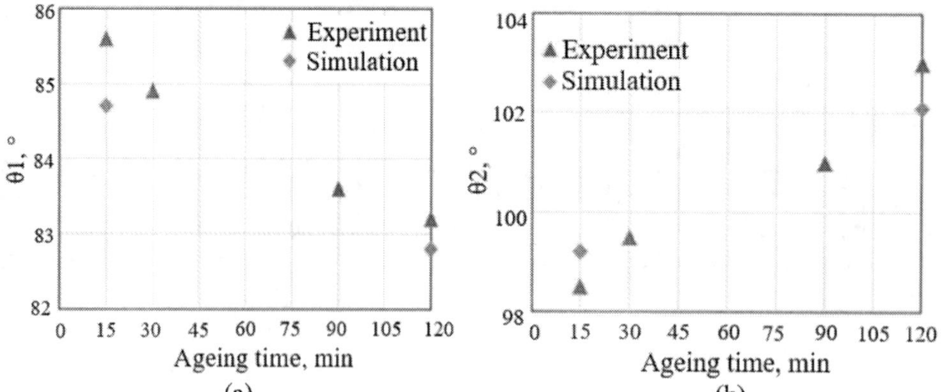

Figure 8. Effect of aging time at BHF=5kN (a) on θ1, (b) on θ2.

6.2. Effect of BHF on springback

The higher the blank holding force, the lower the amount of springback. This is because the blank holding force helps to distribute the deformation evenly throughout the sheet metal during the forming process. This results in a more uniform stress distribution and reduces residual stresses, which in turn reduces the tendency for the material to springback to its original shape after forming. However, it is important to note that increasing the blank holding force beyond a certain point can lead to other problems, such as excessive thinning or tearing of the material. Therefore, the optimal blank holding force should be determined based on the specific material properties and forming conditions.

7. Conclusions

In conclusion, the W-Temper forming of high-strength aluminium alloys are a highly effective heat treatment process that can significantly improve the mechanical properties and performance of these materials. Its benefits make it a popular choice for a wide range of industrial applications, and it is expected to continue to play a vital role in the manufacturing of high strength aluminium parts and components. However, the springback effect is a serious issue that can happen during the W-Temper forming. This is because the ageing time of high strength aluminium alloys can change the properties of the material during forming.

After the W-Temper process, the aluminium alloy is typically in a hardened state, with increased strength, hardness, and reduced ductility. If the time between the W-Temper process and forming is too long, the material may experience some degree of natural aging, which can further increase its strength and reduce its ductility. On the other hand, if the time between the W-Temper process and forming is too short, the material may not have fully stabilized, resulting in inconsistent properties during forming. The exact time between the W-Temper process and forming will depend on several factors, including the specific alloy being used, the size and complexity of the part being formed, and the types of forming processes used.

References

[1] Brabie G, Chirita B, Nanu N and Ciubotariu V 2009 Analysis of the springback and residual stresses generated by cold plastic forming in drawn round parts made from steel sheets *Metalurgia International* **14(12)** 21-27

[2] Samuel M 2000 Experimental and numerical prediction of springback and side wall curl in U-bendings of anisotropic sheet metals *J. Mater. Process. Technol.* **105(3)** 382-393

[3] Behrens B. A, Yun J. W, Schäffner C, Kroening M, Altpeter I, and Kopp M 2005 Influences on the part quality in conventional deep drawing processes *International conference on steels in cars and trucks* Wiesbaden Germany 144-151

[4] Asnafi N 2001 On springback of double-curved autobody panels *Int. J. Mech. Sci.* **43(1)** 5-37

[5] Aday A J 2019 Analysis of Springback Behavior in Steel and Aluminum Sheets Using FEM *In Annales de Chimie: Science des Materiaux* **43(2)** 95-98

[6] Yang X, Choi C, Sever N.K and Altan, T 2016 Prediction of springback in air-bending of Advanced High Strength steel (DP780) considering Young' s modulus variation and with a piecewise hardening function *Int. J. Mech. Sci.* **105** 266-272

[7] Dai H.L, Jiang H.J, Dai T, Xu W.L and Luo A.H 2017 Investigation on the influence of damage to springback of U-shape HSLA steel plates *J. Alloys Compd.* **708** 575-586

[8] Gronostajski Z, Jaśkiewicz K, Kaczyński P, Skwarski M, Polak S, Krawczyk J, Chorzępa W and Trzpis P 2022 W-temper forming of B-pillar from 7075 aluminum alloy *CIRP Annals* **71(1)** 221-224

[9] Choi Y, Lee J, Panicker S.S, Jin, H.K, Panda S.K and Lee M.G 2020 Mechanical properties, springback, and formability of W-temper and peak aged 7075 aluminum alloy sheets: Experiments and modelling *International Journal of Mechanical Sciences* **170** 105344

[10] Lee J, Bong H.J, Kim D, Lee Y.S, Choi Y and Lee M.G 2019 Application of combined W-temper and cold forming technology to high-strength aluminum alloy automotive parts *JOM* **71** 4393-4404

[11] Choi Y, Moon C and Lee M.G 2019 Experimental study on the mechanical properties of 7xxx aluminium alloy sheet under different heat treatment conditions *IOP Conference Series: Materials Science and Engineering* **651(1)** 012080
[12] Wang G, Zhu G, Kou L, Li T, Liu Z, Shang X, Jiang X and Zhu X 2019 Effect of heat treatment conditions on stamping deformation and springback of 6061 aluminum alloy sheets *Materials Research Express* **7(1)** 016512
[13] Lee J, Lee J.Y, Barlat F, Wagoner R.H, Chung K and Lee M.G 2013 Extension of quasi-plastic–elastic approach to incorporate complex plastic flow behavior–application to springback of advanced high-strength steels *International Journal of Plasticity* **45** 140-159
[14] Kubli W, Krasovskyy A, and Sester M 2008 Modeling of reverse loading effects including workhardening stagnation and early re-plastification *International Journal of Material Forming* **1(1)** 145–148
[15] Wagner L, Wallner M, Larour P, Steineder K, and Schneider R 2021 Reduction of Young's modulus for a wide range of steel sheet materials and its effect during springback simulation *IOP Conference Series: Materials Science and Engineering* **1157(1)** 012031
[16] Tisza M and Lukács Z 2015 Formability investigations of high-strength dual-phase steels *Acta Metallurgica Sinica (English Letters)* **28(12)** 1471-1481
[17] Hegedűs G 2017 Application of knowledge-based design in computer aided product development *In Vehicle and Automotive Engineering: Proceedings of the JK2016* Miskolc Hungary 109-114
[18] Makinouchi A 1993 NUMISHEET'93 *Proceedings of the 2nd International Conference Numerical Simulation of 3-D Sheet Metal Forming Processes; Verification of Simulation with Experiment* Isehara Japan

IOP Conf. Series: Materials Science and Engineering 1284 (2023) 012067 doi:10.1088/1757-899X/1284/1/012067

Influence of process and material parameters on the twist springback prediction of a panel

M C Oliveira[1], C Gomes[1], D M Neto[1], J L Alves[2] and L F Menezes

[1] CEMMPRE, ARISE, Department of Mechanical Engineering, University of Coimbra, Polo II, Rua Luís Reis Santos, Pinhal de Marrocos, 3030-788 Coimbra, Portugal

[2] CMEMS, Microelectromechanical Systems Research Unit, University of Minho, Campus de Azurém, Guimarães, 4800-058, Portugal

E-mail: marta.oliveira@dem.uc.pt

Abstract. Both advanced high strength steels and aluminum alloys gained increasing popularity because of the constant effort to meet the social pressure of reducing the vehicle weight while keeping their safety. The ratio between material strength and weight is much higher when compared to other materials but, the high springback after forming is one of the disadvantages of these materials, particularly when exhibiting twisting and side wall curl, which are the two most difficult types of springback to control and compensate in die engineering. This study analysis the influence of several material and numerical parameters on the prediction of the twist springback, using the Twist Die. Two processing conditions were considered: (i) blank-holder force that constrains the metal flow and (ii) a stake bead, which changes the forming conditions from draw to stretch forming, at the end of the process. The modelling of the elastoplastic behavior of the sheet materials uses a wide range of experimental data, for a DP980 steel and a 6xxx aluminum alloy. In addition to the effect of the sheet material on springback, the effect of the following parameters is also analyzed numerically: (i) friction coefficient; (ii) beads; (iii) modulus of elasticity; (iv) work hardening law and (v) yield criterion. All numerical simulations were performed with the finite element in-house code DD3IMP. The process parameter that most affects the twist springback is the presence of the stake bead, reducing significantly the springback. In addition, the reduction of the friction coefficient and of the elastic modulus also leads to an increase in springback. On the other hand, the yield criterion has little influence on springback of this component. The twist springback shows a different trend for the steel and the aluminum alloy that seems to be related with the different yield strength but also with the distinct gap between the punch and the die.

1. Introduction

There are several forms of springback, depending on the component geometry, which are commonly categorized into: (i) angular distortion; (ii) side wall curl and (iii) twisting. The angle between the plane of the part edge and the plane of the tool is called the angular distortion and is caused by the difference in stresses through the sheet thickness direction, created when the sheet flows along the die radius. The difference in stresses in the sheet thickness direction creates a bending moment at the radius of curvature after tool removal, which results in an angular deflection. The side wall curl corresponds to an unwanted curvature which creates challenges in assembly operations, affecting the productivity and quality of the part [1]. The main cause is the stress gradient in the thought-thickness direction, generated during the bending and unbending under tension [2]. Twisting is caused by torsional moments of two cross sections of the part and results in different rotations of the two cross sections along their axes. The torsional moment is induced by the unbalanced springback and the asymmetrical distribution of residual stresses in the wall and flange of the part [3]. Unbalanced forces are more likely to occur on asymmetrical parts, parts with wide side flanges, and in areas where there is a sudden change in cross section.

The quantification of the twist springback can be ambiguous requiring the definition of an evaluation procedure. Different methods can be applied to evaluate the rotation of each transversal section of the component, considering as reference the position before springback. One approach is to determine the line that best fits the transversal section using the least squares method, before:

$$y_{\mathrm{BS}} = \alpha_{\mathrm{BS}} x + \beta_{\mathrm{BS}}, \tag{1}$$

and after springback:

$$y_{\mathrm{AS}} = \alpha_{\mathrm{AS}} x + \beta_{\mathrm{AS}}. \tag{2}$$

The twist springback is defined as [3]:

$$\Delta\theta = \theta_{\mathrm{AS}} - \theta_{\mathrm{BS}} = \arctan\left(\alpha_{\mathrm{AS}}\right) - \arctan\left(\alpha_{\mathrm{BS}}\right). \tag{3}$$

To compensate or control the springback it is important to assess the quality of its prediction using numerical simulation tools. This was the challenge of the Numisheet 2022 Benchmark #1: Springback Prediction of Twist Die Panel [4], which considers a tool used by the Auto/Steel Partnership for several years. The benchmark committee selected a steel an aluminum alloy, since the goal was to assess the springback prediction capability for different conditions. In this context, also two processing conditions were also considered, one with a blank-holder force to constrain the material flow into the die and the other with a draw-bead like feature, known as stake bead. This feature is machined on an expansion of the punch, acting like a miniature one, like shown in Figure 1. Therefore, it contacts the metal sheet only in the last increments of the punch stroke. If the flow is stopped, the last moments of the forming process is changed from draw forming to stretch forming. The tooling adopted was the same for both materials and is referred to as the Twist Die, due to the tendency of the panel to twist during springback [4].

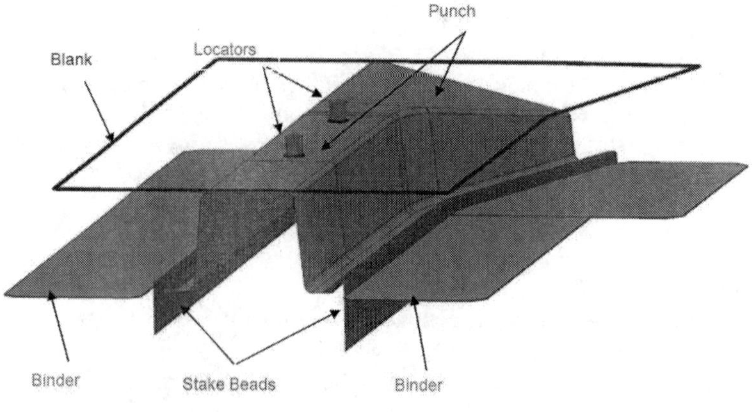

Figure 1. Tools assembly highlighting the stake bead used in the Numisheet 2022 Benchmark #1: Springback Prediction of Twist Die Panel [4],

2. Finite element model

The details concerning the finite element models are described in this section. All numerical simulations were performed with the in-house finite element code DD3IMP [5,6]. The mechanical model adopted considers that the elastic properties are described by the Hooke's law, while the plastic ones combine an associated flow rule with a hardening law and an orthotropic yield criterion. The following subsection describes the details of the constitutive models considered in this study, considering the experimental data supplied by the benchmark committee.

2.1. *Material properties and constitutive models*

The two materials selected for the benchmark are a DP980 steel, with an initial thickness of 1.4 mm, and an aluminum alloy from the 6xxx series, with an initial thickness of 1.5 mm. The benchmark committee provided the raw data (3 repetitions) of: (i) uniaxial tension-compression tests along the rolling direction (RD); (ii) hydraulic bulge test with a circular die; and (iv) uniaxial tension tests in 15° increments to the RD, from 0° to 90°. The benchmark committee also reported standard material properties, from their own analysis of the data, which include: (i) elastic parameters; (ii) initial yield stress; (iii) stress-strain relations and (iv) plastic strain ratios. Finally, the data files also include calibrations of several constitutive models, including isotropic and kinematic hardening models [4].

2.1.1. Elastic properties. According to the analysis performed, the Young's modulus of the DP980 steel presents an in-plane variation of about 7%, while the aluminum alloy show only 2%. The weighted average corresponds to 202.1 GPa and 70.6 GPa, for the steel and the aluminum, respectively. The Poisson ratio presents higher in-plane variation, of 19% for steel and 17% for the aluminum alloy. In this case, the weighted average values for the steel and the aluminum alloy are 0.299 and 0.338, respectively [4]. Despite the anisotropy observed, in this work the elastic behavior is assumed isotropic, described by average values of the elastic modulus, E, and the Poisson ratio, υ.

It is known that the elastic modulus degradation with the plastic strain is an important factor to consider in the prediction of the springback behavior [7]. In this context, the benchmark committee also provided the parameters for the model proposed by Yoshida [8], which suggests that the dependency of the elastic modulus, E_{av}, follows an asymptotic behavior, given by:

$$E_{\text{av}} = E_0 - (E_0 - E_{\text{A}})\left(1 - \exp\left(-\xi\bar{\varepsilon}^{\text{p}}\right)\right), \tag{4}$$

where E_0 is the initial value, E_{A} is the minimum one and ξ is the rate of decrease with the increase of the plastic strain, $\bar{\varepsilon}^{\text{p}}$ [8]. Figure 2 presents a comparison between the numerical model reported for the two materials (labelled EMD) by the benchmark committee [4] and values reported in other studies for steels [7,9]. The analysis shows that the model suggested and adopted in this study leads to a trend like the one reported in other studies, although the variation is slightly higher for the DP980 steel under analysis than the one considered in [9].~

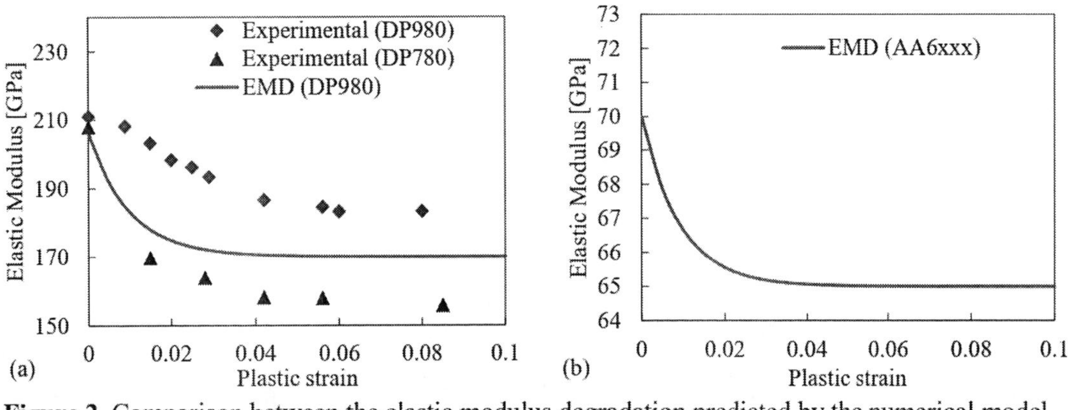

Figure 2. Comparison between the elastic modulus degradation predicted by the numerical model (EMD) and the one reported in other studies for: (a) DP980 steel and (b) 6xxx aluminum alloy.

2.1.2. Plastic properties. The stress-strain curves highlight the different material behavior, with the steel presenting a much higher flow stress and lower ductility; the former only for the uniaxial tensile tests. Both materials show a negligeable variation of the in-plane distribution of the flow stress. For the steel, the yield stress presents 6% of variation, while for the aluminum alloy it is 7%. The analysis of the uniaxial tension-compression test results shows that the steel presents a higher Bauschinger effect. Regarding the plastic strain ratios, both materials present an orthotropic behavior. For steel, the r-value varies between 0.8 and 1.6, between 0° and 60° to RD. On the hand, the r-value of the aluminum alloy varies only between 0.6 and 0.8 [4].

The hardening law parameters were obtained by minimizing the difference with the experimental data from the uniaxial tension test at RD. For the DP980 steel, the best results were obtained with the Swift law:

$$Y = K\left(\varepsilon_0 + \bar{\varepsilon}^{\mathrm{p}}\right)^n, \tag{5}$$

where Y is the flow stress and K, ε_0 and n are the materials parameters. On the other hand, for the 6xxx series aluminum alloy, the best results were attained with the Voce law:

$$Y = Y_0 + \left(Y_{\mathrm{Sat}} - Y_0\right)\left[1 - \exp\left(-C_{\mathrm{Y}}\bar{\varepsilon}^{\mathrm{p}}\right)\right], \tag{6}$$

where Y_0 is the initial yield stress and C_{Y} is the rate of increase to the saturation value, Y_{Sat}. These isotropic hardening laws were combined with the Armstrong and Frederick kinematic law, given by:

$$\mathrm{d}\mathbf{X} = C_{\mathrm{X}}\left[\frac{X_{\mathrm{Sat}}}{\bar{\sigma}}\left(\boldsymbol{\sigma}' - \mathbf{X}\right) - \mathbf{X}\right]\mathrm{d}\bar{\varepsilon}^{\mathrm{p}}, \tag{7}$$

where $\mathbf{X}$ is the back stress tensor, $\boldsymbol{\sigma}'$ is the deviatoric component of the Cauchy stress tensor, $\bar{\sigma}$ is the equivalent stress and C_{X} and X_{Sat} are material parameters. The parameters for the combined isotropic and kinematic hardening law were identified by minimization in the error in the description of the uniaxial tension-compression experimental test results. The parameters for the hardening laws are presented in Table 1. Figure 3 presents the comparison between the experimental and numerical results obtained for the tension-compression test, considering only isotropic hardening, or combined (labelled + KH). While the inclusion of the kinematic component has a negligeable impact for the aluminum alloy, which shows a low Bauschinger effect, for the DP980 steel the results are clearly improved. Nevertheless, the hardening behavior is still overestimated, when compared with the experimental one, highlighting the limitations of the model adopted.

Table 1. Hardening law parameters.

AA6xxx		Y_0 [MPa]	Y_{Sat} [MPa]	C_{Y}		
	Voce	176.9	404.3	10.0		
		Y_0 [MPa]	Y_{Sat} [MPa]	C_{Y}	X_{Sat} [MPa]	C_{X}
	Voce + KH	172.72	395.23	5.19	35.50	199.89
DP980		K [MPa]	ε_0	n		
	Swift	1349.7	0.0005	0.085		
		K [MPa]	ε_0	n	X_{Sat} [MPa]	C_{X}
	Swift + KH	1037.4	0.026	0.104	230.96	72.15

Three yield criteria were adopted: von Mises, Hill'48 [10] and Barlat'91 [11]. The parameters adopted for the orthotropic yield criteria are the ones supplied by the benchmark committee [4] and are listed in

Table 2. Figure 4 presents the comparison of the yield surfaces obtained for the three yield criteria, in the σ_{RD} - σ_{TD} plane (assuming that all other stress components are null). The results show that both orthotropic yield criteria lead to an identical description of the r-values. For the DP980 steel, this means that both underestimate the higher value at 60°. As expected, both yield criteria are unable to describe the almost isotropic behavior of the yield stresses, with the Hill'48 leading to a higher amplitude of variation when compared with the Barlat'91. For the DP980 steel, both yield criteria overestimate the yield stress for the transverse direction (TD), with a higher value predicted by Hill'48, as shown in Figure 4 (a). On the other hand, for the aluminum alloy the yield stress at TD is underestimated by both yield criteria, particularly by the Hill'48 (see Figure 4 (b)). Figure 4 also highlights that the differences between both orthotropic yield criteria are more accentuated for the aluminum alloy, which is related with the value recommended for the exponent m for FCC and BCC materials (see Table 2).

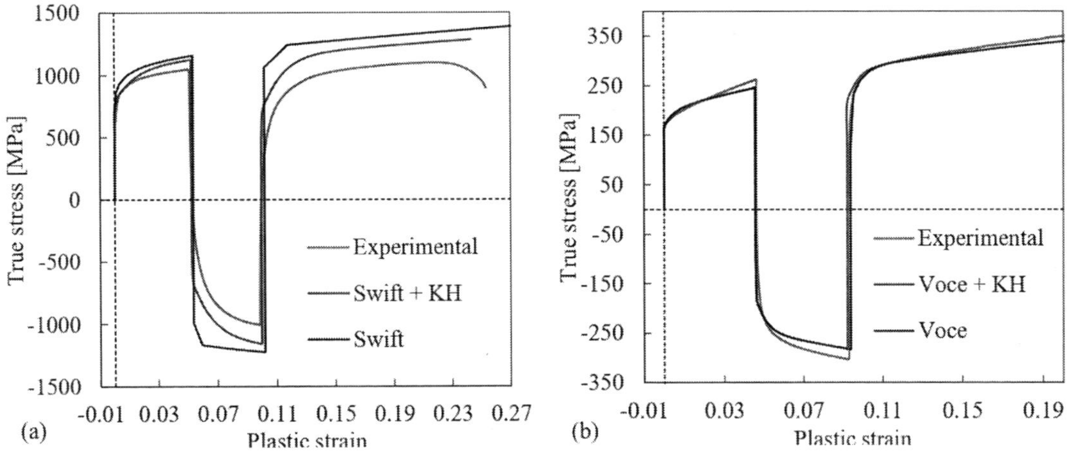

Figure 3. Comparison between the experimental and numerical results of the tension-compression test for: (a) DP980 steel and (b) 6xxx aluminum alloy.

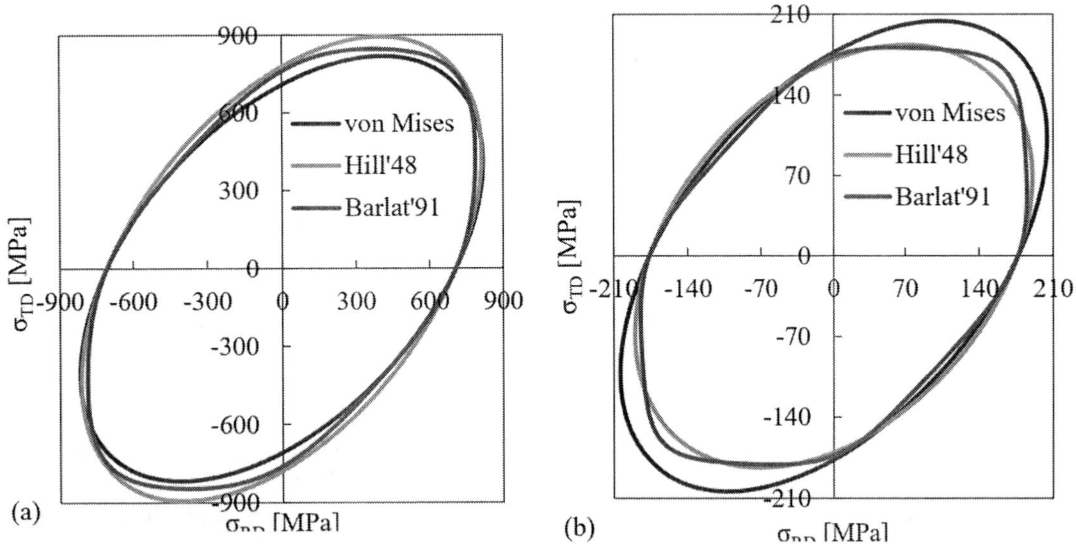

Figure 4. Comparison between the yield surfaces predicted by the three yield criteria adopted for: (a) DP980 steel and (b) 6xxx aluminum alloy.

Table 2. Orthotropic yield criteria parameters.

AA6xxx		F	G	H	N	$L=M$
	Hill'48	0.6906	0.6159	0.3841	1.6526	1.5
		c_1	c_2	c_3	c_6	$c_4=c_5$
	Barlat'91 (m=8)	1.1046	1.0650	0.9316	1.0300	1.0
DP980		F	G	H	N	$L=M$
	Hill'48	0.3807	0.5531	0.4469	1.6809	1.5
		c_1	c_2	c_3	c_6	$c_4=c_5$
	Barlat'91 (m=6)	0.9046	1.0410	0.9580	1.041877	1.0

2.2. Spatial and time discretization

The tools are assumed as rigid and modelled with Nagata patches [6]. Figure 5 (a) presents the blank geometry, including the holes used to maintain its position relatively to the punch. The blank was discretized with solid 8-node finite elements, combined with a selective reduced integration technique. The in-plane mesh combines unstructured zones, in the areas submitted to low plastic strain levels, and structured ones for the other. Special care was taken in the selection of the finite element in-plane size for the regions that flow into the die. Two layers of elements were used through the thickness, leading to a total of 86382 elements.

To facilitate the modelling, the kinematic conditions of the tooling are different from the experimental ones. The die and the blank-holder are assumed as fixed while the punch moves with the same relative displacement has in the experimental setup. The punch displacement is discretized in a minimum of 500 increments. The springback is performed in a single increment, by simply removing the influence of all tools simultaneously. Figure 5 (b) show the positions of the points with imposed boundary conditions during springback. Point E is fixed while F can only move in the Ox direction and G is only constrained in the Oz direction. To assure consistent springback measurements, five sections were defined, using parallel planes distributed along the component's length, as shown in Figure 5 (b).

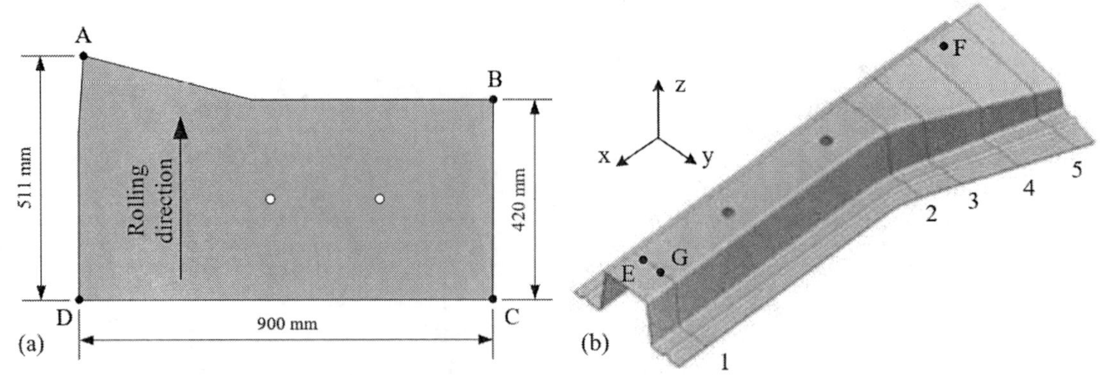

Figure 5. (a) Blank geometry and (b) Points with boundary conditions and reference sections for springback analysis.

3. Results and discussion

Numerical simulations were performed for both materials considering: (i) isotropic and orthotropic yield criteria; (ii) elastic modulus with or without degradation; (iii) a high (0.15) and a low (0.01) value for the friction coefficient; (iv) isotropic or combined hardening laws; and (v) punch with or without draw-bead. The reference result for the DP980 steel corresponds to the numerical simulation considering the

Hill'48 yield criterion, a constant value for the elastic modulus of 206 GPa, the Swift isotropic hardening law, the higher friction coefficient value, and the tools with draw-bead. For the 6xxx aluminum alloy, the reference result corresponds to the numerical simulation considering the Hill'48 yield criterion, a constant value for the elastic modulus of 70 GPa, the Voce isotropic hardening law, the higher friction coefficient value, and the tools with draw-bead. Simulations were performed changing only one parameter, as presented in Table 3, for both materials.

Figure 6 presents the evolution of the punch force with its displacement, with and without draw-bead. For both materials, the presence of the draw-bead leads to a strong increase of the force in the last 10 mm of punch displacement. The increase is approximately of 11 times for steel and of 18 times for the aluminum alloy. Although not shown here, the effect of the yield criteria is negligeable during the drawing stage, for both materials. For the aluminum alloy, in the stretching stage the von Mises leads to an even higher increase of the force, since the yield surface is more expanded (see Figure 4 (b)). The reduction of the friction coefficient leads to a decrease of the stretching force of approximately 25% and 22%, for the steel and the aluminum alloy, respectively. The combined hardening law also leads to a decrease of the maximum value of force, in this case of 6% and 4%, for the steel and the aluminum alloy, respectively. The elastic modulus has a negligible effect on this process parameter.

Table 3. Description of the parameter changed in each simulation in relation to the reference one.

DP980		AA6xxx	
Description	**Label**	**Description**	**Label**
Reference	Hill'48_E206_μ=0.15	Reference	Hill'48_E70_μ=0.15
Barlat'91 (see Table 2)	Barlat'91	Barlat'91 (see Table 2)	Barlat'91
Isotropic plastic behavior	von Mises	Isotropic plastic behavior	von Mises
Constant friction of 0.01	μ=0.01	Constant friction of 0.01	μ=0.01
Constant E of 170 GPa	E170	Constant E of 65 GPa	E65
EMD (see Figure 2 (a))	EMD	EMD (see Figure 2 (b))	EMD
Hardening law (see Figure 3 (a))	Swift + KH	Hardening law (see Figure 3 (b))	Swift + KH
Tool without draw-bead	wo/bead		wo/bead

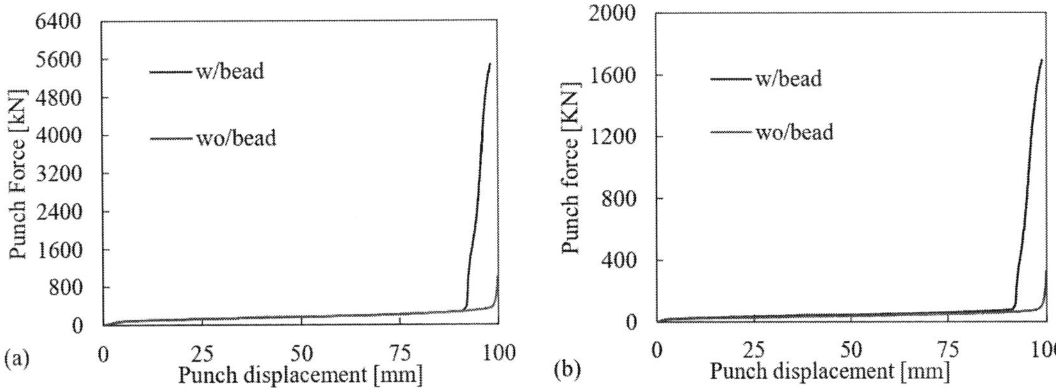

Figure 6. Comparison of the punch force evolution with or without draw-bead for: (a) DP980 steel and (b) 6xxx aluminum alloy.

The draw-in was evaluated for the corners of the blank (see Figure 5 (a)) highlighting that point A always presents a lower value than the other points, as a consequence of the lower length of the vertical wall of the component in that section. Moreover, the differences in the draw-in predicted indicates slightly lower values for the steel. For both materials, the friction coefficient is the more influential factor, with an impact of approximately 4% of increase. Figure 7 presents the comparison between the five sections selected for analysis (see also Figure 5 (b)), before and after springback, predicted with the reference conditions, for both materials. For the steel, the springback corresponds to a counterclockwise rotation of the right part of the component. On the other hand, the aluminum alloy presents a clockwise rotation of the same zone. Globally, the springback is lower for the aluminum alloy. Since the behavior of the different sections is quite similar, the analysis of the influence of the different parameters is performed only for SEC5.

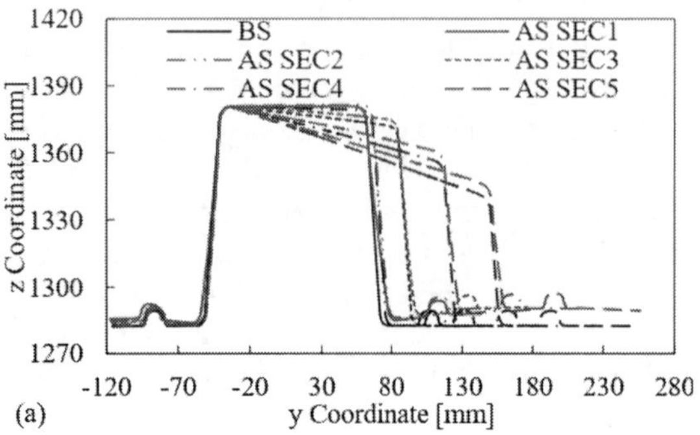

Figure 7. Comparison between the different section before (BS) and after (AS) springback (reference results) for: (a) DP980 steel and (b) 6xxx aluminum alloy.

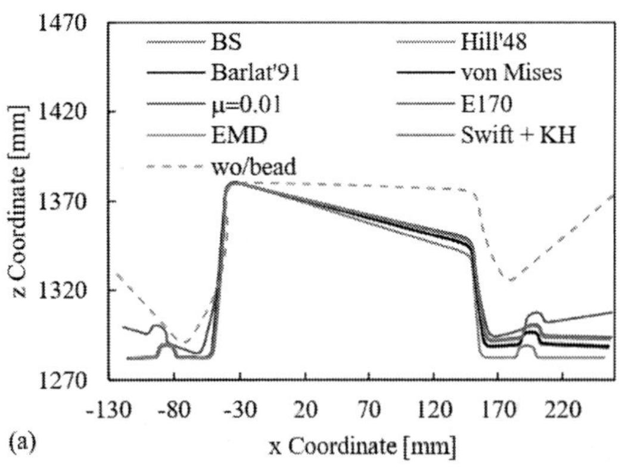

Figure 8. Comparison of SEC5 before (BS) and after (AS) springback for: (a) DP980 steel and (b) 6xxx aluminum alloy.

Figure 8 presents SEC5 before and after springback, predicted with the different parameters under analysis, for both materials. The results show the small influence of the yield criteria, for both materials. Regarding the elastic modulus, it effect is negligible for the aluminum alloy while for steel is perceptible, on the right side of the component. Nevertheless, it should be mentioned that the range of values used for steel is significantly higher (see Figure 2). The reduction of the friction coefficient leads to the increase of springback, for both materials, since the reduction of the friction forces leads to lower plastic strain levels, particularly along the vertical wall. It is interesting to note that the effect is negligible for the right side of the component, for the aluminum alloy. This seems to relate with its low impact in the length of the flange before springback. The friction coefficient presents a stronger impact for the steel, because the forming forces are 3 times higher for this material than for the aluminum alloy (see Figure 6). The combined hardening law leads to an increase of the springback predicted for steel and a reduction for the aluminum alloy, with a similar impact, despite the fact that the Bauschinger effect is more clear for the DP980 (see Figure 3). The simulation performed without draw-bead leads to a strong springback for both materials, with a particular increase of the side wall curvature, as a result for the lower plastic strain levels attained along the vertical walls. For the aluminum alloy, the top region of the part also changes its curvature, indicating a lower level of stretching.

The rotation due to springback of each transversal section was evaluated using Eq. (3), considering only the points located in the linear zone of the top part of the component to perform the linear fit. Figure 9 presents the distribution of the rotation angle for the five different sections, for each combination of parameters. For both materials, the rotation angle is almost zero in SEC1 (x=360 mm), i.e., the top part of the component suffers no rotation. This is a consequence of the boundary conditions adopted, which restrain the vertical movement of points E and G (see Figure 5 (a)). Figure 9 highlights the strong effect of the draw-bead on the reduction of the twist springback, which is reduced from at least 10° for tools without draw-bead to at most 3° for the situation with draw-bead. In this case the trend is similar for both materials. When using the tool with draw-bead, Figure 9 also confirms the reduced effect of the yield criteria, for both materials, and of the elastic modulus, for the aluminum alloy. For the DP980 steel, the friction coefficient, the elastic modulus, and the kinematic hardening present similar effects on the twist springback. On the other hand, for the aluminum alloy, the friction coefficient is the parameter with higher impact on twist springback.

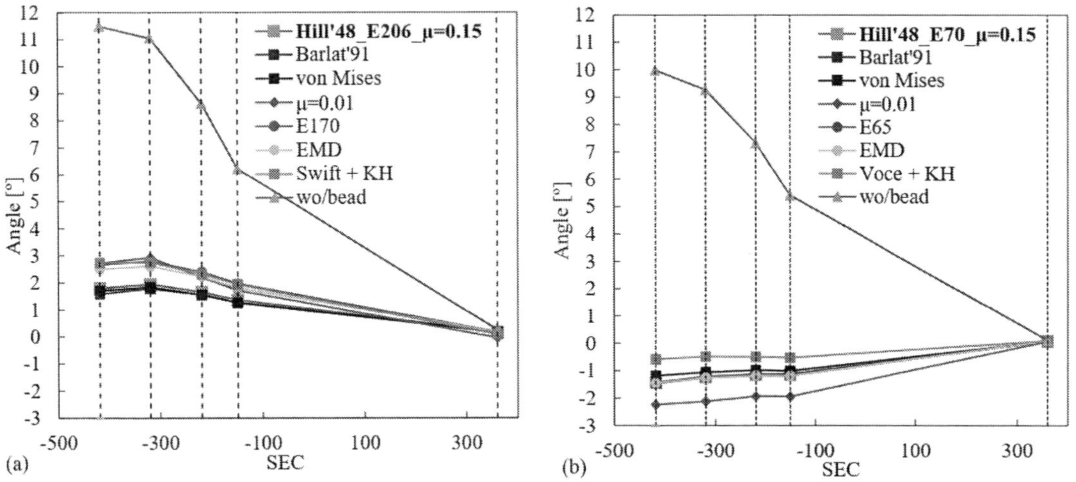

Figure 9. Effect of each parameter on the torsion angle evaluated on the upper surface of each of the 5 sections of the component for: (a) DP980 steel and (b) 6xxx aluminum alloy.

The fact that the twist springback is larger for steel in comparison with the aluminum alloy is in good agreement with the experimental measurements [4]. Nevertheless, it should be mentioned that the same tool was used to deform a steel and an aluminum alloy, which present quite different values of yield strength but also different values of initial thickness. In fact, the steel blank is 1.4 mm thick, while the aluminum alloy is 1.5 mm. The difference in the thickness values is compensated in the punch displacement, when using or not draw-beads, to assure the final calibration of the part. Nevertheless, the gap between the punch and the die during the draw forming cannot be compensated, i.e., it is higher for the steel. This seems to influence the twist springback of the part, when using the stake bead, because the steel component presents a counter-clockwise rotation of the right part of the component, while the aluminum presents a clockwise rotation of the same zone. The gap seems to be affecting the range of the through-thickness plastic zone differently on both sides of the transversal sections. For the DP980, in the region of the component with higher depth, the higher yield stress of DP980 steel and the lower gap between the punch and the die seem to result in a higher range of the through-thickness plastic zone. This can help explaining the lower sensitivity of the left region of the component to the material parameters studied, as shown in Figure 8 (a). On the other hand, for the aluminum alloy the sensitivity to the parameters under analysis seems to be more uniformly distribution in all section, although the gap between the die and the punch is smaller.

4. Conclusions

In this study, the effect of the following parameters in the twist springback were analyzed numerically: (i) friction coefficient; (ii) beads; (iii) elastic modulus; (iv) work hardening law, (v) yield criterion and (vi) sheet material. The results show that the process parameter that most affects the twist springback is the presence of the stake bead, reducing significantly the springback. In addition, the reduction of the friction coefficient and of the elastic modulus also leads to an increase in springback. On the other hand, the yield criterion has little influence on the springback of this component. The kinematic hardening also influences the springback, presenting different trends for both materials. The combined hardening law leads to an increase of the springback predicted for the steel and a reduction for the aluminum alloy. These results can be related with the differences in the hardening model, since the one adopted for the DP980 steel presents a clear Bauschinger effect. For the tool with draw-bead, whatever the parameter under analysis the steel component presents a counter-clockwise rotation of the right part of the component, while the aluminum presents a clockwise rotation of the same zone. Therefore, the difference trend seems to result from the different gap between the punch and the die, during the draw forming.

Acknowledgements
The authors gratefully acknowledge the financial support of the Portuguese Foundation for Science and Technology (FCT) under projects with reference UIDB/00285/2020 and LA/P/0112/2020.

References

[1] Ayres R A 1984 SHAPESET: A Process to Reduce Sidewall Curl Springback in High-Strength Steel Rails *Journal of Applied Metalworking volume* **3** 127–34
[2] Wagoner R H and Li M 2007 Simulation of springback: Through-thickness integration *Int J Plast* **23** 345–60
[3] Dezelak M, Stepisnik A, Pahole I and Ficko M 2014 Evaluation of twist springback prediction after an AHSS forming process *International Journal of Simulation Modelling* **13** 171–82
[4] Qin Z G, Stoughton T, Anderson D, Zhou D J, McCarty E, Smith E and Park P 2020 *NUMISHEET BENCHMARK 1 — Springback Prediction of Twist Die Panel*
[5] Menezes L F and Teodosiu C 2000 Three-dimensional numerical simulation of the deep-drawing process using solid finite elements *J Mater Process Technol* **97** 100–6
[6] Neto D M, Oliveira M C and Menezes L F 2015 Surface Smoothing Procedures in Computational Contact Mechanics *Archives of Computational Methods in Engineering* **24** 37–87
[7] Xue X, Liao J, Vincze G, Pereira A B and Barlat F 2016 Experimental assessment of nonlinear elastic behaviour of dual-phase steels and application to springback prediction *Int J Mech Sci* **117** 1–15
[8] Yoshida F, Uemori T and Fujiwara K 2002 Elastic-plastic behavior of steel sheets under in-plane cyclic tension-compression at large strain *International Journal of Plasticity* **18** 639–59
[9] Lajarin S F, Nikhare C P and Marcondes P V P 2018 Dependence of plastic strain and microstructure on elastic modulus reduction in advanced high-strength steels *Journal of the Brazilian Society of Mechanical Sciences and Engineering* **40**
[10] Hill R 1948 A Theory of the Yielding and Plastic Flow of Anisotropic Metals *Proceedings of the Royal Society of London A: Mathematical, Physical and Engineering Sciences* **193**
[11] Barlat F, Lege D J and Brem J C 1991 A six-component yield function for anisotropic materials *Int J Plast* **7** 693–712

Deformation Prediction of Titanium Alloy in Incremental Sheet Forming with Anisotropic and Asymmetric Yield Function

M Kim[1] **and N Park**[1]

[1] Korea Institute of Industrial Technology (KITECH), Incheon, South Korea

E-mail: nspark@kitech.re.kr

Abstract. The present study is concerned with the prediction of deformation in incremental sheet forming (ISF) for a Ti6Al4V sheet. To investigate anisotropic and asymmetric behaviors, uniaxial tension and compression tests were conducted along the rolling direction (RD), diagonal direction (DD), and transverse direction (TD). A representative yield function for a HCP material, the Cazacu–Plunkett–Barlat (CPB06) yield criterion, was selected for constitutive modeling to take into account anisotropy/asymmetry-induced distortional yielding of the titanium alloy. The chord modulus degradation was confirmed through uniaxial loading–unloading tests. The ISF tests were carried out using a single contact-point tool following a truncated pyramid tool path. To predict the deformation during the ISF, finite element analyses were conducted with the ABAQUS explicit and the vectorized user-subroutine (VUMAT). The performance of various constitutive models was evaluated based on the comparison of deformed shapes and distribution of thickness from the experiment and the FEA.

1. Introduction

Titanium alloy Ti6Al4V is widely selected in structural design due to its specific-strength, resistance to corrosion, machinability, and etc. [1]. As geometries of aerospace parts are large and complicated, it is hard to adopt traditional sheet metal forming processes. To take account of this issue, incremental sheet forming (ISF) offers a promising solution due to its flexibility from a die-less forming system for diversity of product geometries. One of simple ISF processes is the single point ISF that employs a stylus without die-tooling required. Because, the stylus contacts one side of the sheet for forming and the other side is free in the single point ISF, there would be a limitation in the dimensional accuracy [2]. Many studies have been conducted to improve it by analyzing the deformation mechanism and formability regarding process-factors such as tool path [3], feed rate [4], and form-angle [5].

Most researches in numerical simulation of the ISF process have been carried out utilizing shell elements and relatively simple yield functions (e.g., von Mises and Hill48) with isotropic hardening considering the computational costs [6, 7]. The performance of the simple yield functions to predict deformed shape or springback may be limited because von Mises and Hill48 yield functions are developed to describe isotropic and in-plane anisotropic yielding, respectively, in case of symmetric material behavior. However, asymmetric material behavior of the titanium alloy, which have varying flow stress levels in tension and compression, requires an asymmetric yield function [8]. To predict the springback more accurately, the elastic modulus change must be considered as a function of the plastic strain during the simulation [9].

This paper addresses the deformation prediction of the Ti6Al4V with the thickness of 1.3 mm in the ISF process by utilizing the numerical simulation with an anisotropic/asymmetric yield function and a model for the elastic modulus degradation. To handle the advanced technique, the previous work [10] are resummarized especially focusing on the performance of constitutive models.

2. Material characterization

Material is a titanium alloy sheet of Ti6Al4V (1.3 mm thickness). Uniaxial tension, uniaxial compression, hydraulic bulge, and loading-unloading tests were conducted to observe mechanical responses under different loading conditions. For the uniaxial tension and compression tests, specimens were prepared along three orientations 0°, 45° and 90° from the rolling direction (RD). The dimensions of specimens and the correction procedure for the data in the compression tests were referred to the previous works [10, 23]. Figure 1 shows the obtained mechanical responses of the Ti6Al4V. As shown in figure 1(a), the titanium alloy sheet shows anisotropic and asymmetric behavior in flow stress with the strength differential (SD) effect [12]. Figure 1(b) depicts R-value in the uniaxial tension test in different orientations. Since a 3D-DIC system was utilized to measure the strain data during the tension test while a laser extensometer was used during the compression test, the R-values were observed in only tension tests.

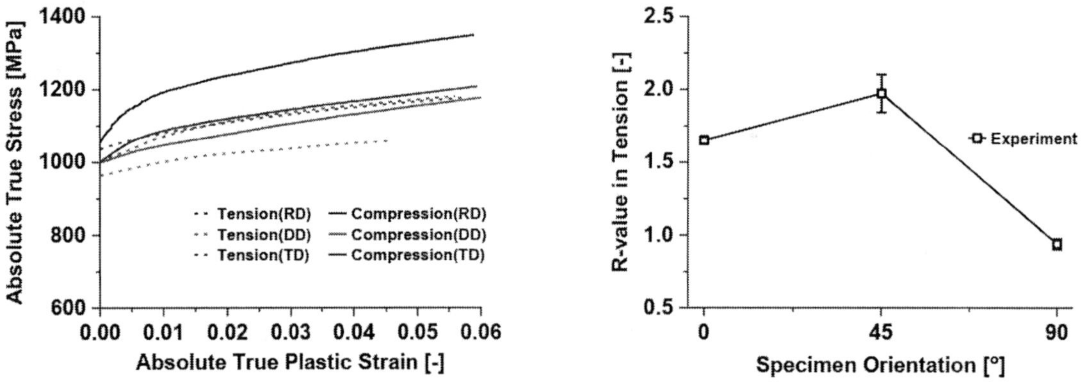

(a) (b)

Figure 1. Mechanical responses of Ti6Al4V sheet: (a) flow stress; (b) R-value.

The hydraulic bulge test [10] was additionally performed to obtain the biaxial tension flow stress. To characterize the anisotropic and the asymmetric yielding of the Ti6Al4V, an anisotropic and asymmetric yield function, CPB06ex2 [8], was employed. The CPB06ex2 is an extended model from the CPB06 [12] to capture more complex yielding behavior by adding the numbers of liner transformations, i.e., the CPB06ex2 includes two different linear transformations. The CPB06ex2 is described in Eq. (1) and eq. (2).

$$f=\left(|\Sigma_1'|-k'\Sigma_1'\right)^a+\left(|\Sigma_2'|-k'\Sigma_2'\right)^a+\left(|\Sigma_3'|-k'\Sigma_3'\right)^a+\left(|\Sigma_1''|-k''\Sigma_1''\right)^a+\left(|\Sigma_2''|-k''\Sigma_2''\right)^a+\left(|\Sigma_3''|-k''\Sigma_3''\right)^a \quad (1)$$

$$\Sigma'=\boldsymbol{C}'\cdot\boldsymbol{S}',\ \ \Sigma''=\boldsymbol{C}''\cdot\boldsymbol{S}'', \quad (2)$$

Here, a is the exponent while k' and k" are the asymmetry parameters. The C' and C", are the linear transformation tensors while S' and S" are the deviatoric stress tensor. The CPB06ex2 was calibrated with the experimental data at a plastic work of 30 MJ/m^3, stress (tension, compression and biaxial tension) and R-value (tension and biaxial tension), by solving the optimization problem [10]. The calibrated parameters in the CPB06ex2 are listed in Table 1.

Table 1. Anisotropic/Asymmetric parameters of CPB06ex2 (a=8).

k'	C'_{11}	C'_{12}	C'_{13}	C'_{22}	C'_{23}	C'_{33}	C'_{44}	C'_{55}	C'_{66}
-0.0461	0.1143	-2.5164	-1.6785	-0.3274	-2.5931	-0.1187	1	1	2.7252
k''	C''_{11}	C''_{12}	C''_{13}	C''_{22}	C''_{23}	C''_{33}	C''_{44}	C''_{55}	C''_{66}
0.2085	0.1143	-0.1646	-2.1423	1.0588	-1.1104	0.0381	1	1	0.9132

To characterize the degradation of the elastic modulus, the loading-unloading in uniaxial tension tests were performed along the RD. The degradation model [9] was employed to fit the experimental data. The model is described in eq. (3). E_0 and E_a are the initial and saturated elastic modulus, respectively while q is the decaying parameter. The calibrated parameters in the degradation model are listed in Table 2.

$$E = E_0 - (E_0 - E_a)\{1 - \exp(-q\varepsilon_p)\} \tag{3}$$

Table 2. Parameters in the degradation model.

E_0 [GPa]	E_a [GPa]	q [-]
107.54	95.77	26.84

The calibration results are graphically demonstrated in figure 2. As shown in the figure 2(a), the CPB06ex2 well-describes both anisotropic and asymmetric yielding of the Ti6Al4V. Figure 2(b) depicts the degradation of the elastic modulus regarding the plastic deformation. The modulus exponentially decreases as the material deforms further.

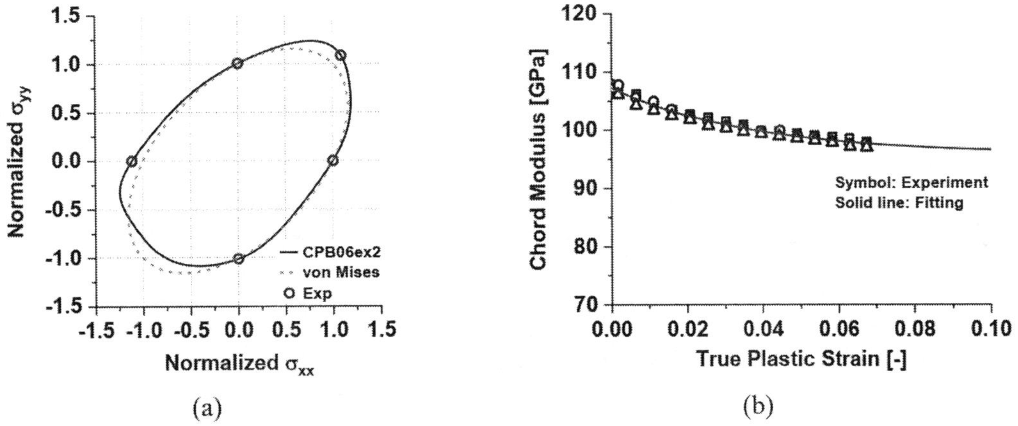

Figure 2. Material characterization: (a) Yield behavior; (b) Modulus degradation.

3. Incremental sheet forming

The ISF test was performed by using a 3-axis CNC system [10] in order to make a truncated pyramid. Figure 3 shows the ISF testing system, the specimen dimensions, and the tool path of the stylus. A dome type stylus tool with a diameter of 10 mm was used to form the pyramid with the wall angle of 30°. The tool speed and the step size were set to 10 mm/s and 0.5 mm/loop, respectively. To reduce the friction between the stylus and the specimen, a lubricant oil was sufficiently utilized as shown in figure 3.

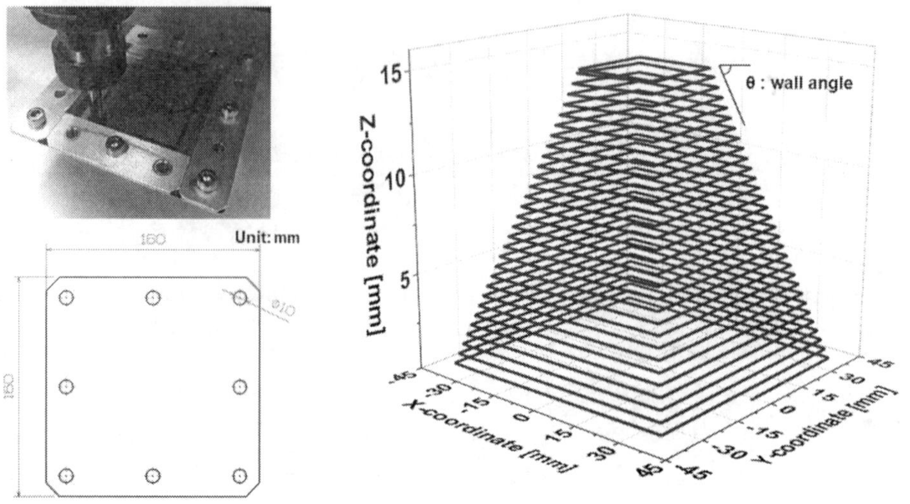

Figure 3. Increment sheet forming of Ti6Al4V.

4. Deformation prediction and model performance

Finite element (FE) models were designed by using the ABAQUS explicit with two different constitutive model: case (1) von Mises; case (2) CPB06ex2 with the elastic modulus degradation. The FE models were composed of 105,296 solid elements with the reduced integration (C3D8R). The details of the FE models are described in the previous work [10]. The FE simulations of the ISF process and the unloading were conducted to compare the model performance in both cases. For the case (2), the vectorized UMAT and the UMAT were constructed for obtaining the forming and the springback results, respectively. Figure 4 shows the distributions of the equivalent stress, the equivalent plastic strain, and the elastic modulus at the end of the ISF process. The levels of the equivalent stress in the cases (1) and (2) are comparable but the distributions differ from each other. The difference in the equivalent plastic strain level is noticeable due to the difference in the constitutive model. The degradation of the elastic modulus is well-described in the case (2).

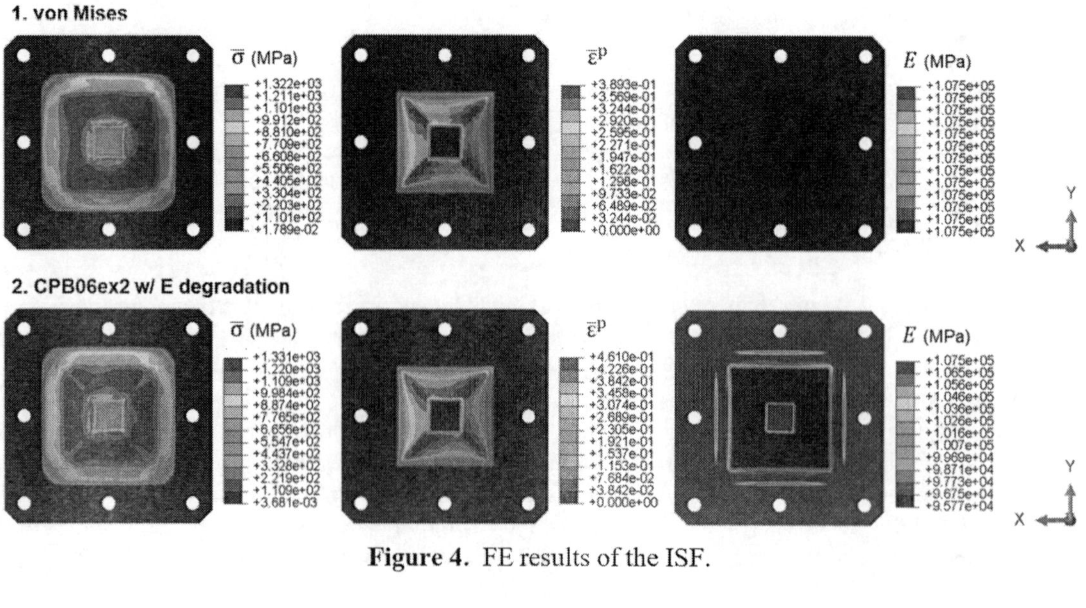

Figure 4. FE results of the ISF.

The surface profile of the truncated pyramid from the experiment and the FE simulation were evaluated as shown in figure 5. As demonstrated in figure 5(a), both predicted profiles from the FE simulations in cases of (1) and (2) show a good agreement with that from the experiment when the profiles were evaluated from the center of the pyramid to 50 mm from the center in X-direction. However, near the edge part of the specimen is predicted with a small amount of the variation in the case (2). Since the amount of the springback is quite remarkable near the edge compared to the center of the specimen, the case (2) which includes the modulus degradation, predicts better than the case (1). The quantitative evaluation was performed using Eq. 4 where Z_i^{EXP} and Z_i^{FEM} indicates Z-coordinates of surface profiles and n is the total number of data. Total amount of the difference in the case (2) is much smaller than that in the case (1).

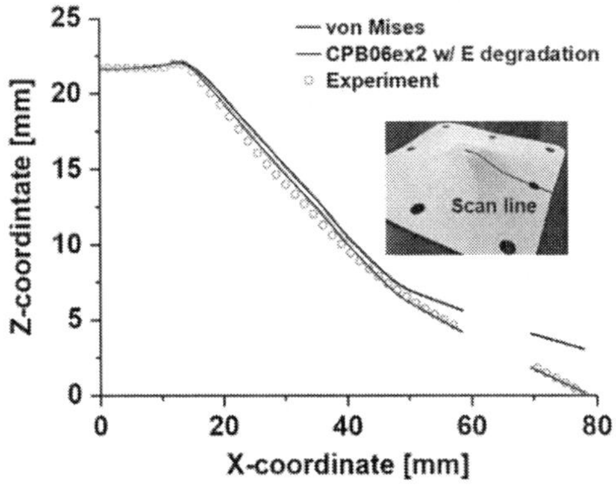

(a)

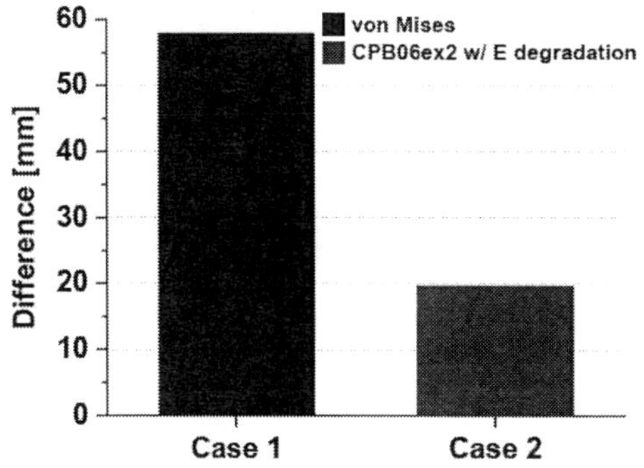

(b)

Figure 5. Deformation prediction: (a) Surface profiles; (b) Quantitative evaluation.

$$D = \sqrt{\sum_{i=1}^{n} \left(Z_i^{EXP} - Z_i^{FEM} \right)^2} \quad (4)$$

Thickness distribution in the truncated pyramid from the center to the edge was evaluated in the same manner for the comparison of the surface profile as shown in figure 6. The FE results show the case (1) and (2) are matched with the experimental results well. However, the performance of the case (2) is slightly better in predicting the thickness change.

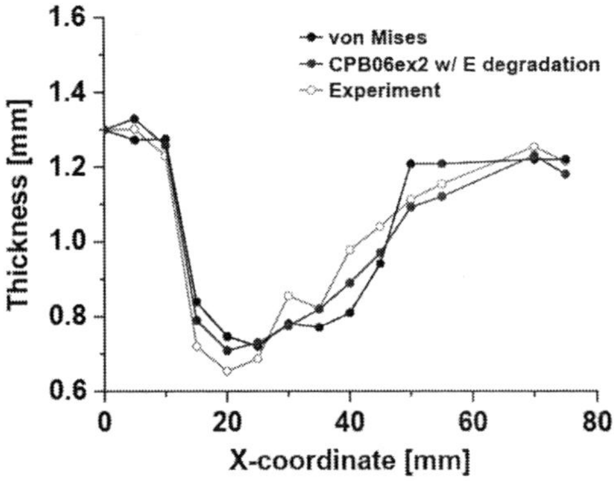

(a)

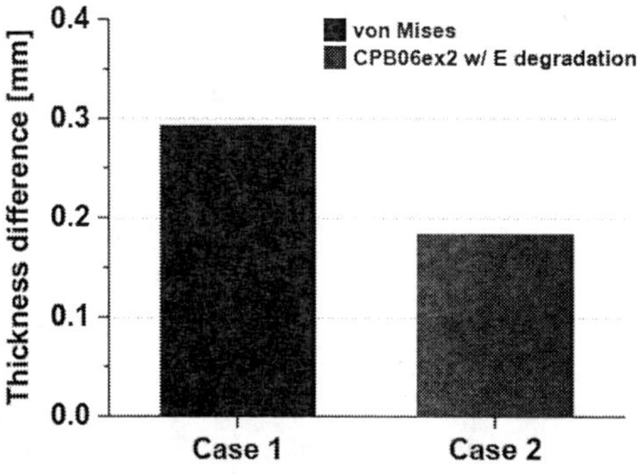

(b)

Figure 6. Thickness evaluation: (a) Distribution; (b) Quantitative comparison.

Furthermore, the forming force was investigated to evaluate the model performance from the experiment and the FE simulations as shown in figure 7. Since the limitation of the load measuring system in the CNC machine, only the maximum value during the ISF process was extracted and depicted in figure 7(a) as a reference data. The forming forces from the FEM were extracted from the reaction force in Z-axis applied to the stylus tool. The forming force in the case (2) which is CPB06ex2 with the modulus degradation, is closer to the experimental data. The level of the forming force in the case (2) is larger than that in the case (1). The reason why the predicted force in the case (2) is larger could be explained by employing the difference between stress levels in the plane-strain tension (PST) region of the yield functions. As shown in figure 7(b), the PST part in the CPB06ex2 always predicts higher levels of the stress. Because the wall region of the truncated pyramid, which is the majority of the deformed part, undergoes the deformation under the PST condition.

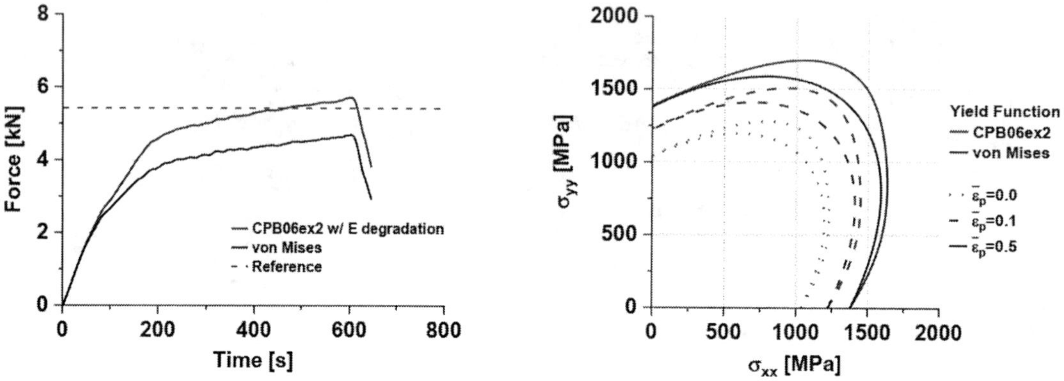

(a) (b)

Figure 7. Effect of yield function on prediction of forming force: (a) Force; (b) Yield behavior.

5. Concluding remarks

This paper investigates the deformation prediction of the Ti6Al4V truncated pyramid fabricated by the ISF. The performance of two different constitutive models, von Mises (isotropic case) and CPB06ex2 (anisotropic/asymmetric case), was evaluated through the comparison of the predicted surface profiles, thickness profile, and the forming force. The CPB06ex2 with the elastic modulus degradation which is the advanced model better predicts the experimental results. The simplest constitutive model, von Mises, are still useful when the critical part is the deformed shape from the center to 50 mm in X-direction of the specimen. For the prediction of the springback tendency and the forming force, the advanced model must be employed to obtain better FE results.

References

[1] Oleksik V, Trzepieciński T, Szpunar M, Chodoła Ł, Ficek D, Szczęsny I 2021 Single-Point Incremental Forming of Titanium and Titanium Alloy Sheets *Materials* **14** 6372

[2] Behera AK, de Sousa RA, Ingarao G, Oleksik V 2017 Single point incremental forming: An assessment of the progress and technology trends from 2005 to 2015 *Journal of Manufacturing Process* **27** 37-62

[3] Lu B, Chen J, Ou H, Cao J 2013 Feature-based tool path generation approach for incremental sheet forming process *Journal of Materials Processing Technology* **213** 1221-1233

[4] Pereira Bastos RN, Alves de Sousa RJ, Fernandes Ferreira JA 2016 Enhancing time efficiency on single point incremental forming processes *International Journal of Material Forming* **9** 653-662

[5] Uheida EH, Oosthuizen GA, Dimitrov DM, Bezuidenhout MB, Hugo PA 2018 Effects of the relative tool rotation direction on formability during the incremental forming of titanium sheets *International Journal of Advanced Manufacturing Technology* **96** 3311-3319

[6] De Sena JIV, Guzmán CF, Duchene L, Habraken AM, Behera AK, Duflou J, Valente RAF, Alves de Sousa RJ 2016 Simulation of a two-slope pyramid made by SPIF using an adaptive remeshing method with solid-shell finite element *International Journal of Material Forming* **9** 383-394

[7] Eyckens P, Belkassem B, Henrard C, Gu J, Sol H, Habraken AM, Duflou JR, Van Bael A, Van Houtte P 2011 Strain evolution in the single point incremental forming process: digital image correlation measurement and finite element prediction *International Journal of Material Forming* **4** 55-71

[8] Plunkett B, Cazacu O, Barlat F 2008 Orthotropic yield criteria for description of the anisotropy in tension and compression of sheet metals *International Journal of Plasticity* **24** 847-866

[9] Yoshida F, Uemori T, Fujiwara K 2002 Elastic–plastic behavior of steel sheets under in-plane cyclic tension–compression at large strain *International Journal of Plasticity* **18** 633-659

[10] Kim M, Lee H, Park N 2022 Evaluation of deformation for titanium alloy sheet in single point incremental forming based on asymmetric yield function *International Journal of Material Forming* **15** 66

[11] Kim M, Huh H 2019 Asymmetric hardening behavior of az31b magnesium alloy sheet with large strain at various strain rates *International Journal of Automotive Technology* **20** 645-653

[12] Cazacu O, Plunkett B, Barlat F 2006 Orthotropic yield criterion for hexagonal closed packed metals *International Journal of Plasticity* **22**, 1171-1194

IOP Conf. Series: Materials Science and Engineering 1284 (2023) 012069 doi:10.1088/1757-899X/1284/1/012069

Artificial intelligence-based springback compensation of EV motor component

Hyunsung Choi[1], Yongnam Kwon[1], Joon Ho Cho[2], Jeong Whan Yoon[3,4]

[1] Korea Institute of Materials Science, Center of Aerospace Materials, Changwon 51508, Republic of Korea

[2] Metal Forming Technology Team, LG Electronics Inc, LG-ro, Jinwi-myeon, Pyeongtaek-si, Gyronggi-do, Republic of Korea

[3] Department of Mechanical Engineering, Korea Advanced Institute of Science and Technology, Daejeon 34141, Republic of Korea

[4] School of Engineering, Deakin University, Waurn Ponds, VIC 3220, Australia

E-mail: h.choi@kims.re.kr, j.yoon@kaist.ac.kr

Abstract. The hairpin is a crucial component of the drive motor of an electric vehicle, consisting of a copper coil and a very thin enamel layer. It is manufactured using successive forming processes due to its complex shape. After the forming processes, springback is inevitably observed, which causes critical issues in fabricating the drive motor. Compensating springback is a time and cost-inefficient process, and a new process is required for a cost-efficient forming process. In this research, an intelligent forming process has been proposed for springback compensation of hairpin forming by applying Artificial Intelligence to the forming process. The key parameters for the AI are the dimensions of the copper coil, including bobbin information, material property variance in a coil bobbin, punch stroke, and amount of springback. As a preliminary design, the data set has been constructed through the finite element simulation. For the springback compensation, punch stroke was optimized with the trained AI and optimization algorithm. The results show that AI could play a key role in solving the manufacturing issue.

1. Introduction

For the efficiency of the drive motor of Electric Vehicle, it is important to minimize copper loss. One of the methods to reduce the copper loss is hairpin winding due to its tight slot coverage and higher thermal conductance. The hairpin consists of copper and very thin enamel layer and is formed into its designed shape through the successive bending processes. The springback of hairpin after the bending processes has been a major issue in fabricating of the drive motor. In a metal forming, there has been many researches for the sprinback compensation in terms of constitutive modeling, finite element modeling [1-11], tool and specimen geometry optimization [12-17], and the application of artificial intelligence [18-19]. There has been many trials to effectively compensate springback in a metal forming, but it is limited to explicitly relate the process parameters and springback. In this research, the Artificial Intelligence (AI) has been applied to compensate springback. The data set of the AI was constructed through the finite element simulation. The input parameters of AI are variation of material properties and cross-sectional dimension in a coil bobbin and punch stroke. The output parameters of AI are eight springback parameters. For the accuracy of the FE simulation, material and process properties has been characterized with newly designed tensile specimen, nano-indentation and inverse method. The accuracy of the FE simulation was identified with the experimental results. The trained AI was incorporated with the optimization algorithm for the optimized punch stroke that minimize springback of hairpin forming.

2. Finite element simulation

2.1. Material properties

Since hairpin consists of copper coil which has 2 mm by 3 mm cross section that coated with 0.1 mm very thin enamel layer, it is necessary to propose new method to characterize the material properties of copper and enamel layer.

2.1.1. Material properties of copper. For the properties of Elastic modulus, yield stress, ultimate tensile strength, the new specimen was designed as shown in figure 1. Since the 0.1 mm enamel layer is coated on the copper, the enamel layer in 20 mm gage section was eliminated and additional machining was performed to ensure the fracture in gage section. To investigate variation of material properties in a coil bobbin, the specimens were manufactured at the initial, middle, and last parts of the coil bobbin. For the reproducibility in a repeated tests, the tests were conducted three times for each part. The results of the test are indicated in figure 2 where the properties of the last parts show the deviate from others.

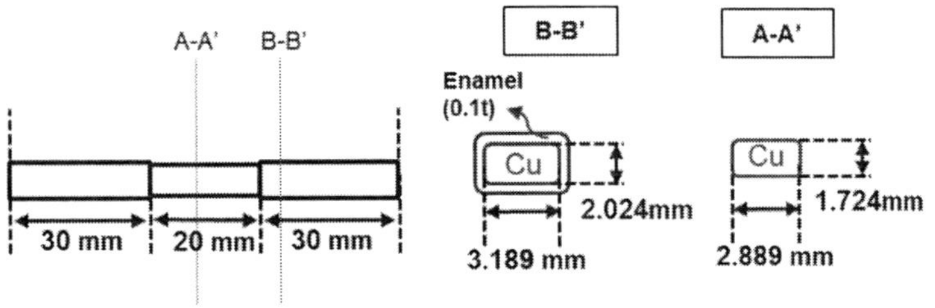

Figure 1. The designed specimen for the characterization of material properties of copper [20].

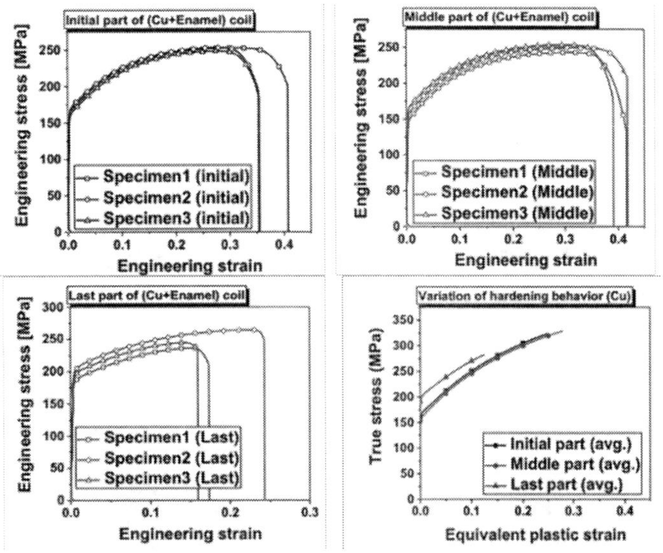

Figure 2. The results of tensile test for the variation of material properties of copper [20].

2.1.2. Material properties of enamel layer. Since it is hard to obtain the material properties such as elastic modulus, yield stress, and ultimate tensile strength of the 0.1 mm thin enamel layer, the nano-indentation and inverse method was applied by matching the load-depth curve of simulation with the experimental results. The 9 tests were performed and obtained Elastic Modulus is indicated in Table 1. The two successive inverse analysis was performed to identified the amount of wear of nanoindenter tip and the yield stress of enamel layer where the enamel layer was assumed as elastic perfectly-plastic material for the hardening behaviour. The finite element modelling of the nano-indentation is indicated in figure 3 and the results of inverse method are indicated in figure 4.

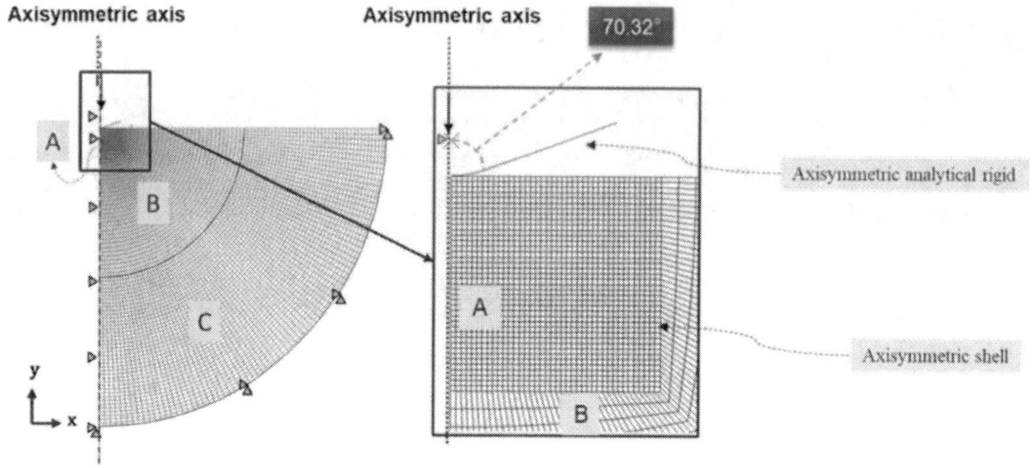

Figure 3. The boundary conditions of 2D finite element simulation of nanoindentation (Both A and B area represent enamel layer and C represents copper) [20].

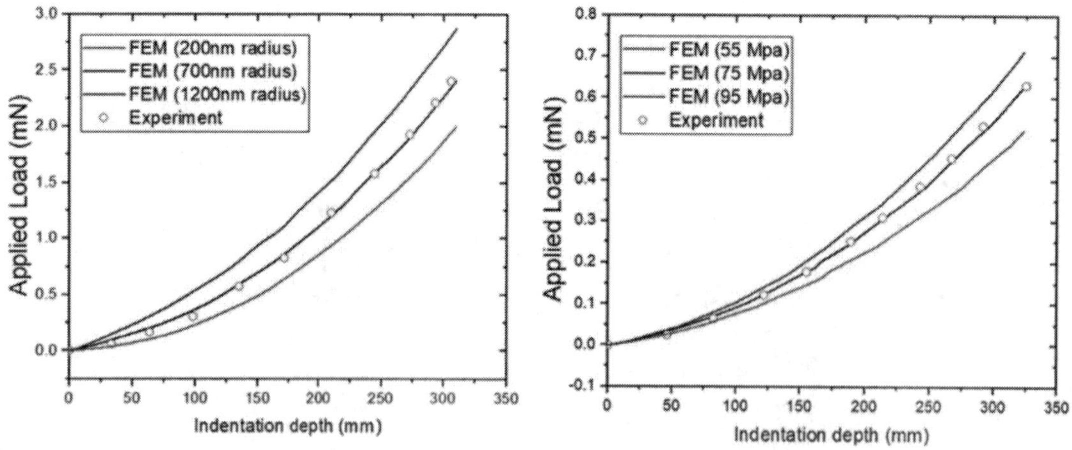

Figure 4. The results of the inverse method for the calibration of (a) tip geometry and (b) yield stress of enamel layer [20].

Table 1. The results of nano-indentation for the elastic modulus of enamel layer (Unit: GPa) [20].

	Test 1	Test 2	Test 3	Test 4	Test 5	Test 6	Test 7	Test 8	Test 9	Avg.
Enamel	5.0	5.1	5.1	5.1	5.2	5.0	5.0	5.0	5.0	5.05

2.2. Finite element modelling of hairpin forming

The dataset for the AI was constructed through FE simulation so that it is important to secure the accuracy of the FE simulation by modelling all the forming processes and comparing with the experimental results. For the material modelling, the isotropic hardening and elastic-perfectly plastic model were used for the copper and enamel, respectively. The coefficients and the properties of the material models are well explained in [20]. For the hairpin forming, three successive bending process which are V-shape, U-shape, and R-shape were modelled by using ABAQUS DYNAMIC/IMPLICIT solver. For each process punch return step is included to accurately model the real manufacturing environment. The copper and enamel layer were modelled as 8 node linear brick element with 0.4 mm element average size. In addition, tie constraint was imposed between copper and enamel layer. The finite element models are indicated in figure 5.

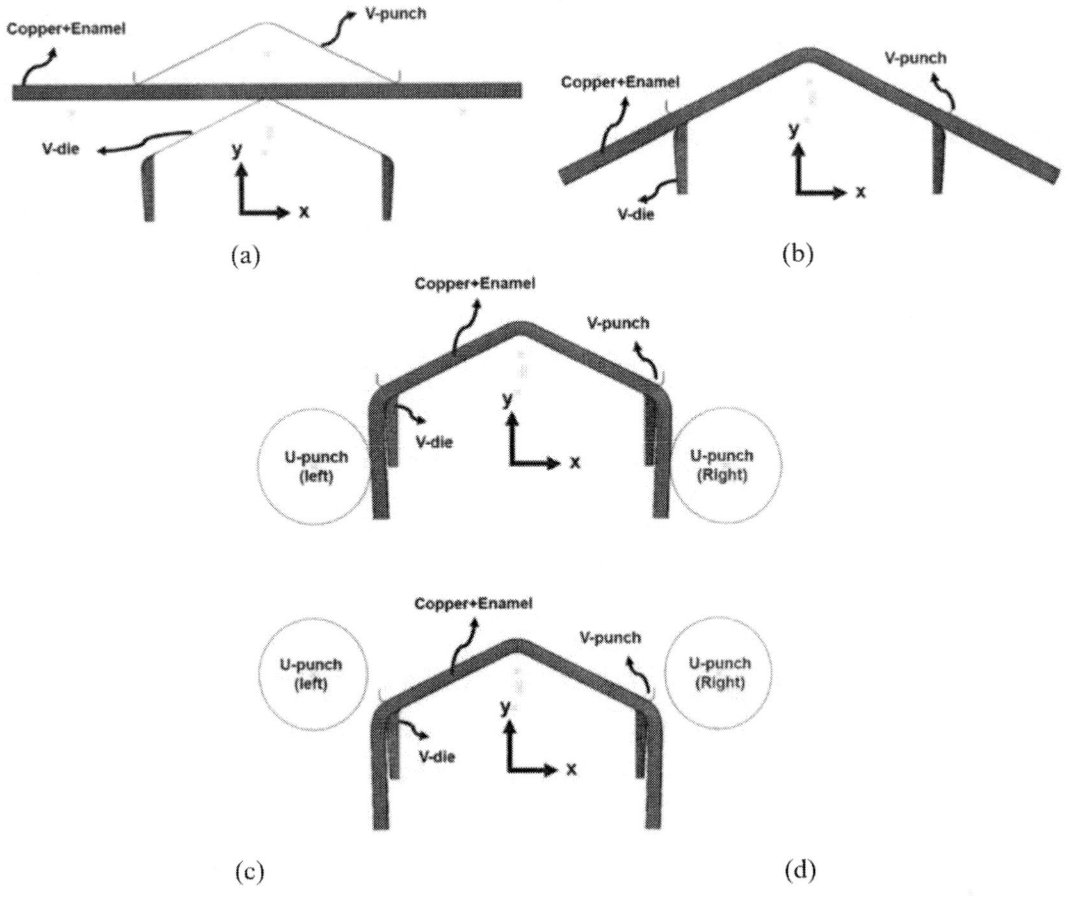

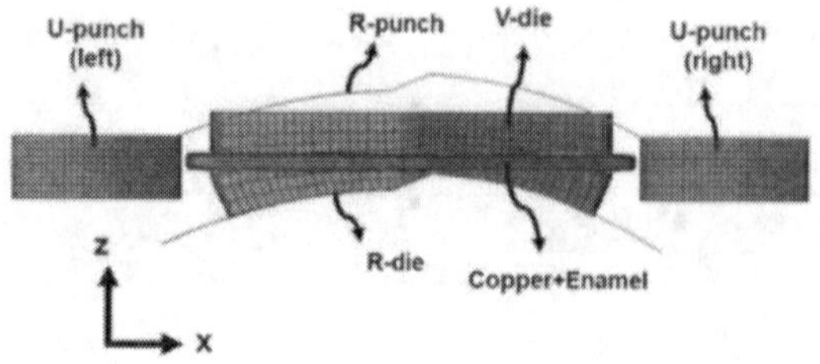

(e)

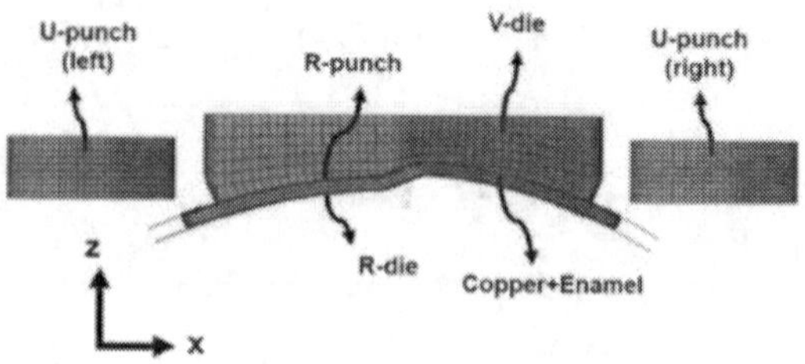

(f)

Figure 5. Finite element model (a) initial stage, (b) V-shape forming, (c) U-punch forming, (d) U-punch return, (e) R-shape forming (initial), and (f) R-shape forming [20].

After the forming simulation, springback simulation was conducted with gravity condition. For the accuracy of the FE simulation, the eight springback parameters are defined as shown in figure 6. Note that after the simulation, to eliminate the rigid body motion, coordinate transformation was applied to the node (P1 to P5) [20]. The comparison of springback are indicated in figure 7 where black and red bar stand for the results from experiment and simulation, respectively. As shown in the figure, the simulation shows a good agreement with experimental results. Therefore, it is reasonable to use the FE simulation for the dataset of the AI.

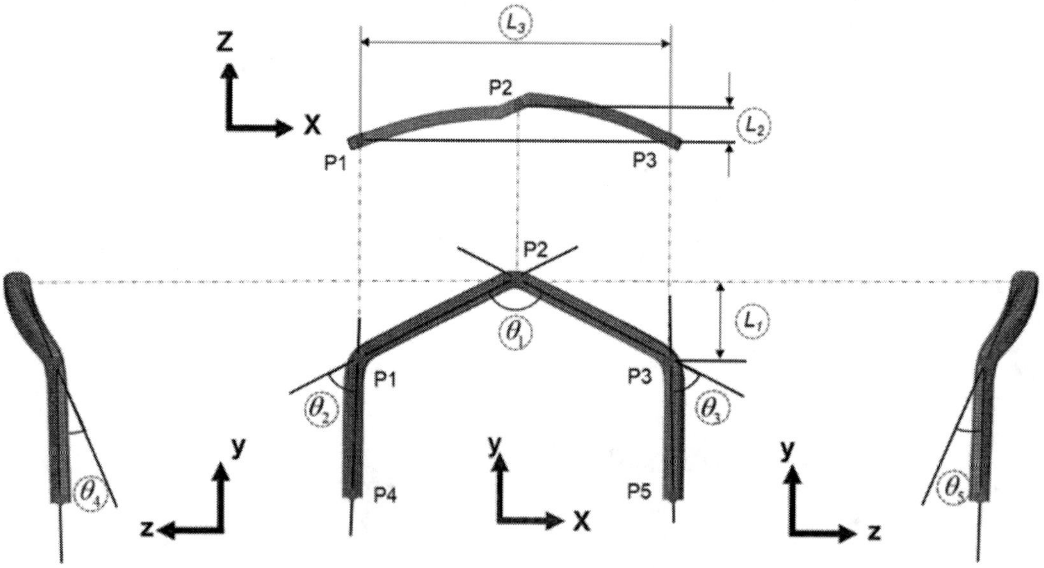

Figure 6. The springback parameters for hairpin forming.

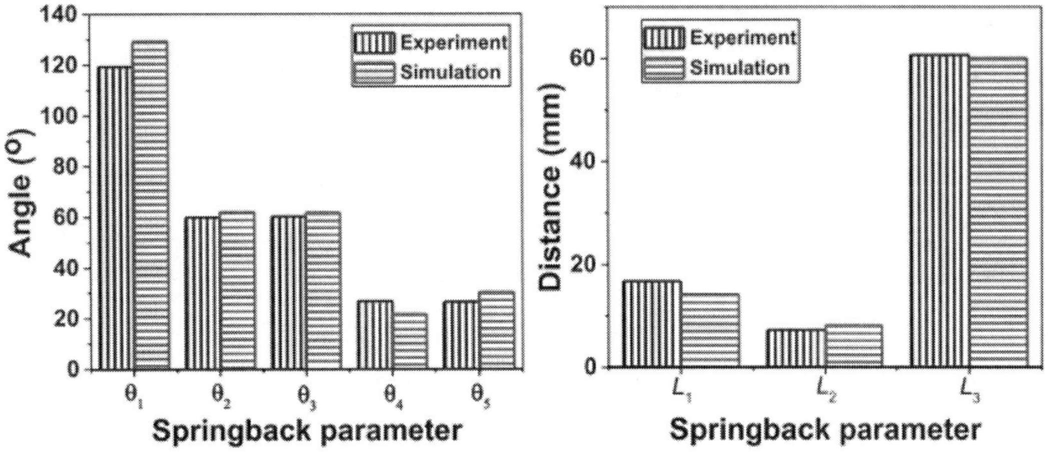

Figure 7. The comparison of springback.

3. Artificial Intelligence and springback compensation

In this research, deep neural network was adopted that consists of input, output, and multiple hidden layers. The inputs are variation of dimension of copper, variation of material properties of copper, and r-punch stroke. In addition, the outputs are eight springback parameters. Each hidden layer has set of neurons and each neuron receives input from the neurons in the preceding layer. The output of each neuron in the hidden layer defined as follows:

$$y_k = \phi(\sum_{j=1}^{m} w_{kj} x_j + b_k) \tag{1}$$

where ϕ is activation function and w_{kj} is the weight tensor, b_k is the basis vector, and x_j is the output of the preceding layer. The Rectified Linear Unit was used for the activation function. The training was performed with minimization of cost function defined as follows:

$$R^2(y, \hat{y}) = 1 - \frac{\sum_{i=1}^{n} (y_i - \hat{y})^2}{\sum_{i=1}^{n} (y_i - \overline{y})^2} \tag{2}$$

where $\hat{y}_i$ and y_i are predicted output of the ith sample and corresponding true value, respectively. In addition,

$$\overline{y} = \frac{1}{n} \sum_{i=1}^{n} y_i \tag{3}$$

The schematic of the ANN model is indicated in figure 8(a) which consists of six hidden layers and 128 nodes per hidden layer. For the training of the AI, 160,000 data were prepared by using FE simulation with data augmentation where dataset is divided into training and validation set as shown in figure 8(b).

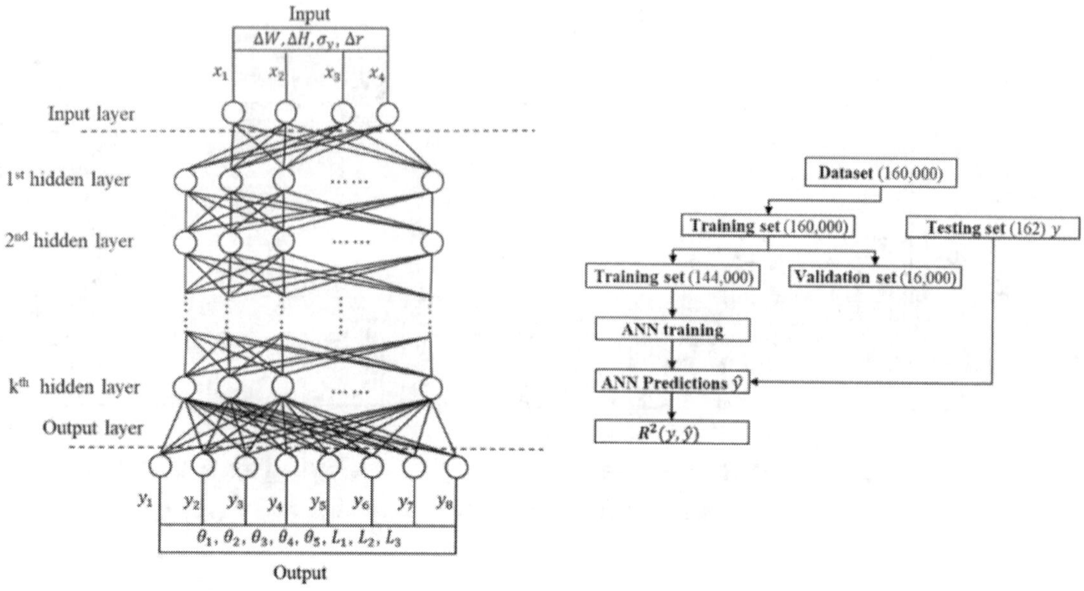

(e) (f)

Figure 8. (a) The schematic of the ANN structure and (b) dataset preparation [20].

The performance of the ANN model is indicated in figure 9 where the trained ANN model shows a good prediction for the eight springback parameter with the inputs from validation dataset. With the trained ANN model, the r-punch stroke was obtained by incorporated with optimization solver for the compensation of springback. The procedure to obtain r-punch stroke is indicated in flow chare as shown inf figure 10(a). The example of the optimization process is indicated in figure 10(b). With respect to the random input given by the optimization solver, the ANN model predicts the eight springback parameter and find the optimum r-punch stroke that minimize springback in hairpin forming. Note that the iteration was completed after 16 iterations and it took about 6.6s [20].

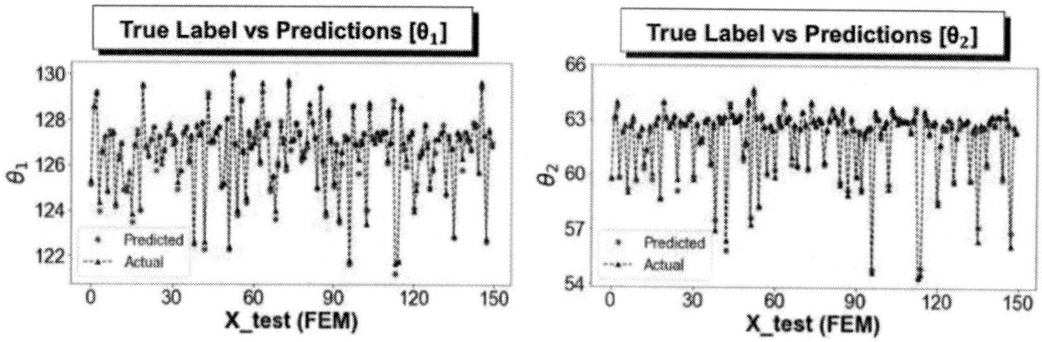

IOP Conf. Series: Materials Science and Engineering 1284 (2023) 012069 doi:10.1088/1757-899X/1284/1/012069

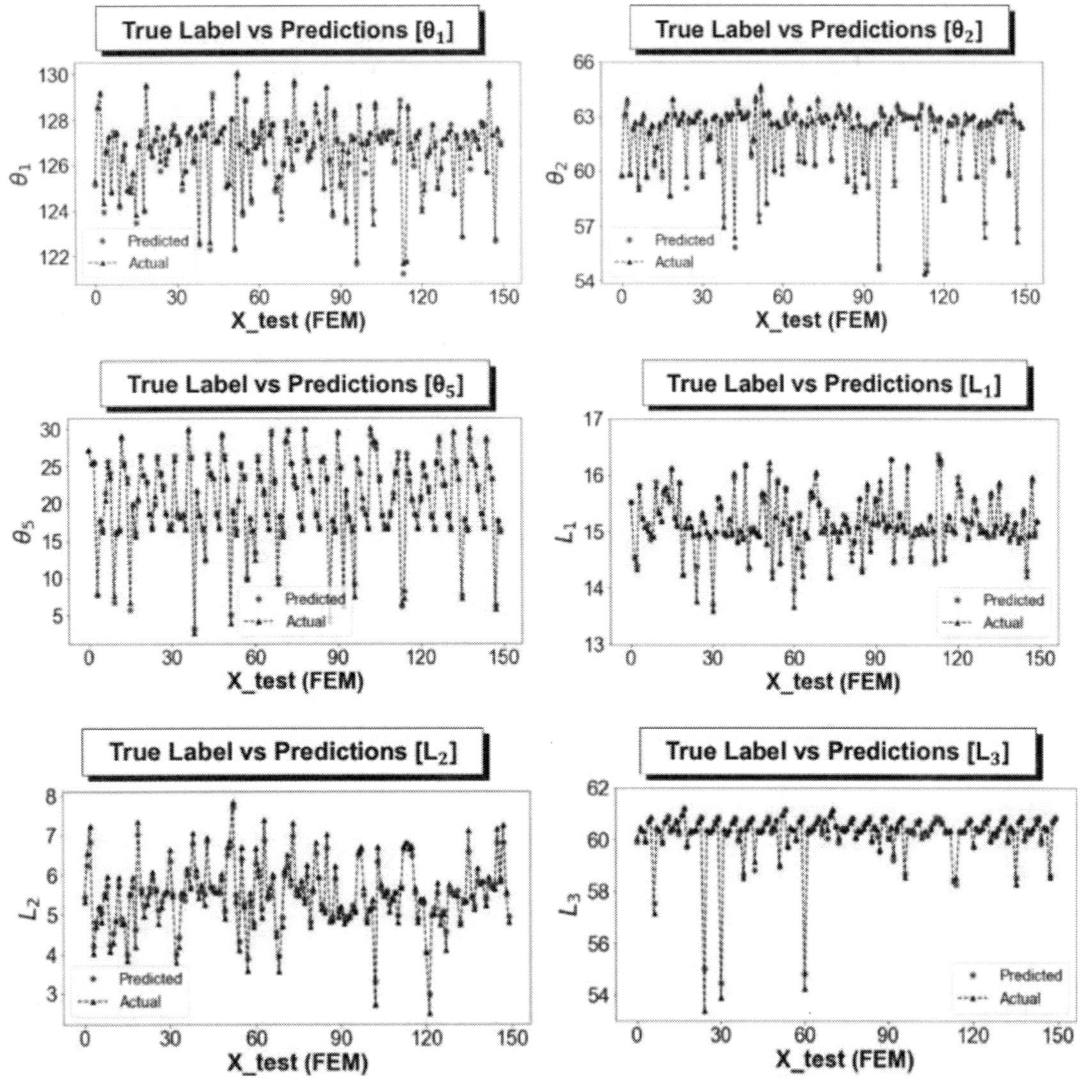

Figure 9. The springback parameters prediction with validation dataset [20].

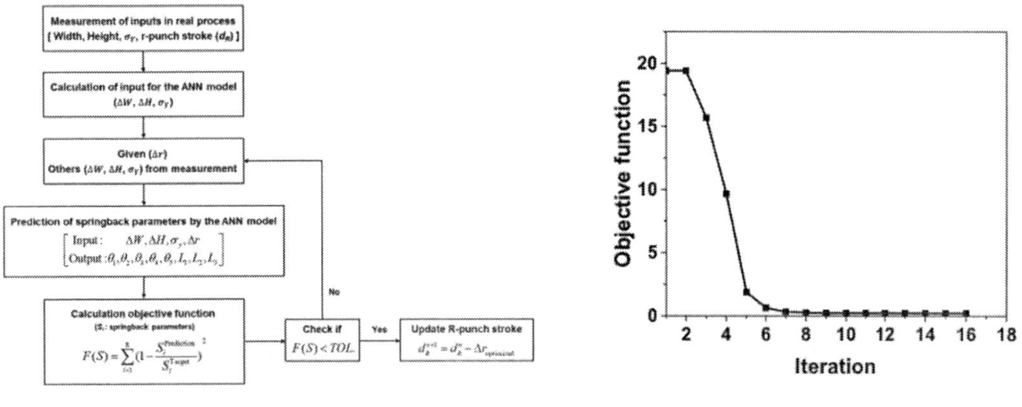

Figure 10. Schematic of optimization process for the springback compensation [20].

4. Conclusion

In this research, the new method for the springback compensation was proposed by using the Artifical Neural Network method. The variation of the dimension, material properties were considered as an input of the ANN model. In addition, the punch stroke was optimized to minimize springback in the hairpin forming process. The dataset was prepared by using the FE simulation and data augmentation. To secure the accuracy of the FE simulation, material characterization with the newly designed specimen and inverse method with nanoindentation. The FE results shows a higher accuracy that could be used for the training the ANN model. The results showed that by incorporating the ANN model with the optimization solver, the springback compensation could be effectively conducted.

References

[1] Joo G, Huh H and Kwon J 2019 *J Mater Process Technol* **270** pp 365–379
[2] Choi HS, Yoon JW, Lee JS and Kim GH 2019 *In: Key Engineering Materials Trans Tech Publ* pp 267–274
[3] Sumikawa S, Ishiwatari A and Hiramoto J 2017 *J Mater Process Technol* **241** pp 46–53
[4] Eggertsen P-A and Mattiasson K 2011 *IntJ Mater Form* **4** pp103–120
[5] Ghaei A, Green D and Aryanpour A 2015 *Mater Des* **88** pp 461–470
[6] Taherizadeh A, Green DE and Yoon JW 2015 *Int J Solids Struct* **69** pp 370–382
[7] Yoshida F and Uemori T 2003 *Int J Mech Sci* **45** pp 687–1702
[8] Barlat F, Vincze G, Grácio J, Lee MG, Rauch E and Tomé C 2014 *Int J Plast* **58** pp 201–218
[9] Chen S, Liao J, Xiang H, Xue X and Pereira AB 2021P *J Mater Process Technol* **287** 116224
[10] Badr OM, Rolfe B, Zhang P and Weiss M 2017 *Int J Mech Sci* **128** pp 389–400
[11] Lee JY, Lee JW, Lee MG and Barlat F 2012 *Int J Solids Struct* **49** pp 3562–3572
[12] Karafllis A and Boyce M 1992 *J Mater Process Technol* **32** pp 499–508
[13] Karafllis AP and Boyce MC 1996 *Int J Mach Tools Manuf* **36** pp 503–526
[14] Gan W and Wagoner R 2004 *Int J Mech Sci* **46** pp 1097–1113
[15] Lingbeek R, Huetink J, Ohnimus S, Petzoldt M and Weiher J 2005 *J Mater Process Technol* **169** pp 115–125
[16] Yang XA and Ruan F 2011 *Int J Mech Sci* **53** pp 399–406
[17] Mole N, Cafuta G and Štok B 2014 *J Mater Process Technol* **214** pp 1673–1685
[18] Palau T, Kuhn A, Nogales S, Böhm H, Rauh A and Ag MB 2012 *In: 6th European Congress on Computational Methods in Applied Sciences and Engineering*
[19] Jang DP, Fazily P and Yoon JW 2021 *Int J Plastic* **138** 102919
[20] Choi H, Fazily P, Park J, Kim Y, Cho JH, Kim J and Yoon JW 2022 *International Journal of Material Forming* **15(3)** 22

IOP Conf. Series: Materials Science and Engineering 1284 (2023) 012070 doi:10.1088/1757-899X/1284/1/012070

Formed developables - a new class of sheet metal parts

J Bhuva[1], M Weiss[1], B Rolfe[2], J Jiao[1], M Paulino[2]

[1] Institute for Frontier Materials, Deakin University, Waurn Ponds, Pigdons Rd., Geelong, VIC 3216, Australia
[2] School of Engineering, Deakin University, Waurn Ponds, Pigdons Rd., Geelong, VIC 3216, Australia

E-mail: jhbhuva@deakin.edu.au

Abstract. Sheet metal parts are broadly classified as *stamped* parts versus *folded developables*. While stamped parts are parts manufactured using conventional sheet metal forming techniques, folded developables are curved developable connections produced by the virtue of the classical theory of folded developables. However, folding is not suitable for parts where material integrity cannot be compromised at the fold lines. In such cases, forming techniques can be used to introduce a distributed curvature and avoid material self-intersection at the curved fold axis. This paper combines the two genres to introduce a third class of parts called the *formed developables*. A developable shaped, thin sheet metal of very high strength and reasonable inextensibility is formed by flange drawing over a die radius. The die tool establishes the required curved bend axis, and the die radius is small enough to introduce a distributed curvature. A developable connection is successfully formed where the two connected corrugations are mirror images of each other. While no tearing or wrinkling is observed, experimental strain measurements highlight significant surface strains in the transverse direction while marginal stretching strains are observed in the forming direction. A trend of equal and opposite transverse strains strongly suggests that the major deformation in this geometrical process, and in-turn the shape of the final part, is driven by the involved geometrical parameters.

1. Introduction

A high strength-to-weight ratio alongside the ability to be mass-produced at comparatively lower costs makes sheet metal an attractive raw material for several industries[1]. Its high ductility presents the advantage of forming sheet metal parts of various shapes that are commonly found in the automotive and transportation industries, domestic appliances, building materials and so on [1]. While plenty of sheet metal parts constitute of adjoining surfaces that are flat, making it effortless to form [2], a lot of applications involve connecting curved surfaces like corrugated sheets used in roofing, external body panels of mopeds and internal automotive parts like the B-pillar, and rail carbody parts, to name a few. In some cases, like in the case of corrugations, foldability of the developable shape [3] of the part can be used as an advantage to minimize forming effort and defects. With the application of industrial origami on the rise (Figure 1-c), the ability to achieve a curved developable connection by drawing a developable blank adds significant value. A potential example of this is illustrated in Figure 1-c.

IOP Conf. Series: Materials Science and Engineering 1284 (2023) 012070 doi:10.1088/1757-899X/1284/1/012070

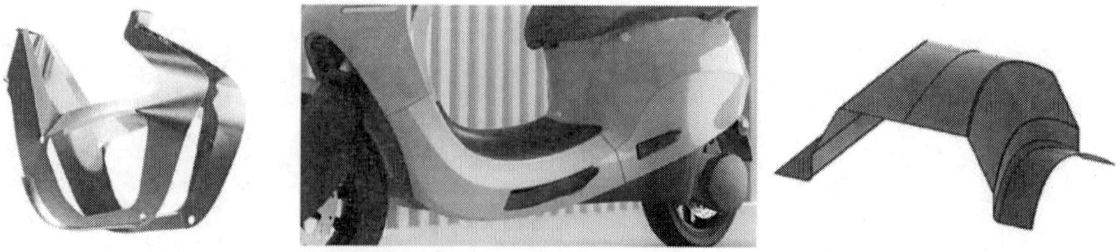
(a) (b) (c)

Figure 1. (a) Body of Stilride © - an industrial origami scooter [4], (b) Ola Electric © moped [5], (c) an example of developable autobody panel part

Unlike forming a flat sheet however, developable shapes involve forming over a curved bend axis where bending becomes increasingly complex as deformation occurs away from the bend as well, like in contour flanging [1]. In contour flanging, the major deformation occurs in the plane of the flange and can cause splitting or wrinkling of the flange in the case of stretch or shrink flanging respectively [6]. This shear distortion in the plane of the sheet, or a change in the sheet thickness like in deep drawing [7], is what differentiates *stamped* sheet metal parts from *folded-developable* parts [8, 9]. The classical theory of folded developables [3] postulates that unlike in contour flanging, a curved developable can only be connected to another curved developable along a fold line that is also curved in plane development. The two developables of this curved folded-developable will have opposing curvatures.

The findings of the theory of folded developables [3] have been extensively used to produce sheet metal parts with curved surfaces [9-13]. Researchers have employed innovative folding techniques for making developable connections with sheet metal. However, finite sheet thickness entails material self-intersection at the fold lines which hinders the geometry of folded-developable parts. This is because *folding* causes localization of the radius of curvature at the fold line as opposed to *bending* over a distributed curvature [14], as illustrated in Figure 2-a and b respectively. Therefore, to facilitate curved crease folding of sheet metal, material is removed at the fold lines usually by grooving [11, 13] or perforation [9, 12].

To avoid compromising material integrity at the curved fold line, a distributed curvature must be introduced at the bend axis to connect the two curved developables. A potential method to achieve this is by *flange drawing* a developable shaped blank over a die tool with a finite radius, where the blank is traditionally clamped between a blank holder and a die, and the punch draws the sheet to form a flange.

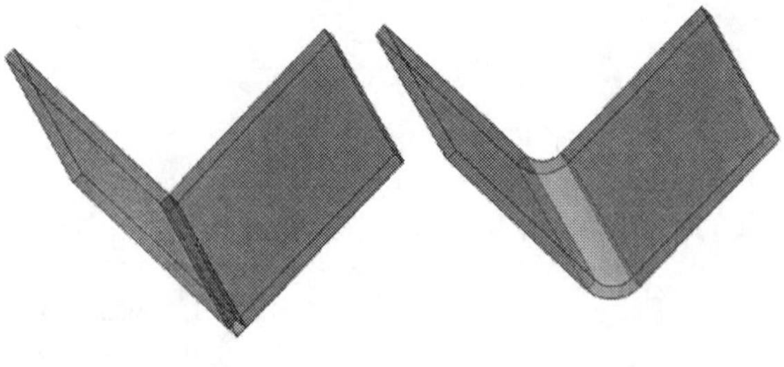
(a) (b)

Figure 2. (a) localized curvature in folding causing material self-intersection, (b) distributed curvature in bending over a small radius.

This paper attempts to assess the potential and understand the process of forming a folded-developable part using the conventional sheet metal forming technique *flange drawing* [15]. The flange drawing process causes shear deformation in the plane of the sheet which is not the case with classical folded-developables [8]. Thus, the drawn flange, even though produced by virtue of the theory of folded developables [3], can be expected to show surface strains. Following the theoretical recommendation

IOP Conf. Series: Materials Science and Engineering 1284 (2023) 012070 doi:10.1088/1757-899X/1284/1/012070

[3], in this study, a thin sheet made of very high strength steel and reasonable inextensibility is used, and the shape of the blank is a generalized cylindrical developable or *corrugation*. By means of visual inspection and gauging the extent of deformation, this paper demonstrates a novel process and provides a foundation for developing future forming processes for curved developable connections. The class of developable parts manufactured by using draw forming techniques can thus be termed as *formed-developables*.

2. Material properties and blank profile

The blank material used for this research was roll formed corrugated steel sheet of material grade *G550* and thickness of 0.42 mm. The corrugation shape is a cylindrical developable with a planar section sandwiched between alternating concave and convex sections (Figure 3-b). A single sheet comprises of ten complete corrugations and the overall dimensions are shown in Figure 3-a.

Material properties were determined using an Instron tensile testing machine with a 30 kN load cell. The engineering stress-strain characteristic of the material is obtained by tracking marked points in the sample using a video extensometer. Samples were obtained from the flat sections (Figure 3-b) of three different sheets and CNC machined to ASTM E8 subsize [16]. Testing speed was 1.5 mm/min. The tensile stress-strain curve of the material is shown in Figure 3-c and the key material parameters are listed in Table 1. The steel exhibits an ageing effect with the upper yield strength of about 700 MPa. The yield strength of the material is 685 MPa, equal to the flow stress. A total elongation of 1.4% shows that the material is reasonably inextensible, and ideal for testing the theory.

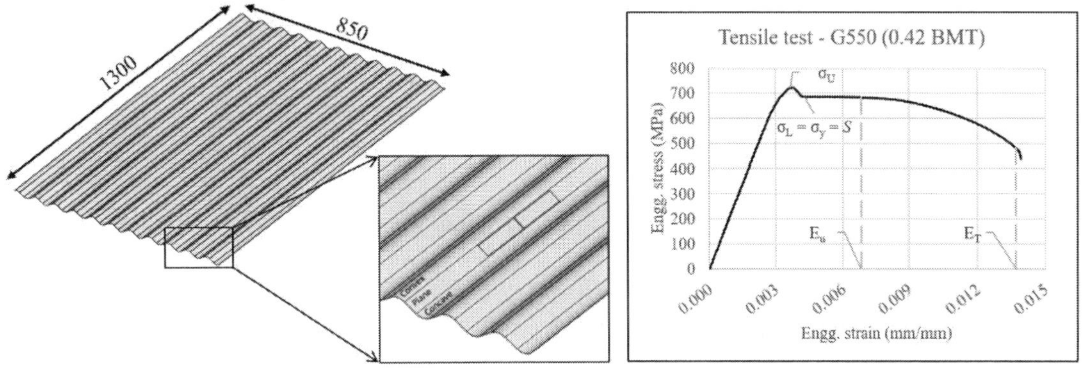

(a) (b) (c)

Figure 3. (a) Overall dimensions (in mm) of a typical sample, (b) corrugation profile with location for obtaining tensile test samples, (c) stress-strain characteristic of the material.

Table 1. Mechanical properties

Material	G550 (0.42BMT)
Young's modulus (GPa)	212
Upper yield strength (σ_U) (MPa)	721
Lower yield strength (σ_L) (MPa)	685
Yield strength (σ_y) (MPa)	685
Uniform elongation (E_u) (%)	0.7
Total elongation (E_T) (%)	1.4

3. The forming process

The forming process used to produce the formed-developable is flange drawing. The process uses a set of clamps, a die tool, and a support block, conforming with the blank shape. The corrugated sheet is inserted from one side, clamped on the other side, and flanged over the die tool as the set of clamps move vertically upwards. The clamping force is high enough to prevent the sheet from sliding. A schematic of the process is illustrated in Figure 4. No back tension is applied in the process. The die radius is 2 mm with a tool gap of 1.1 mm, and the maximum stroke length is 260 mm.

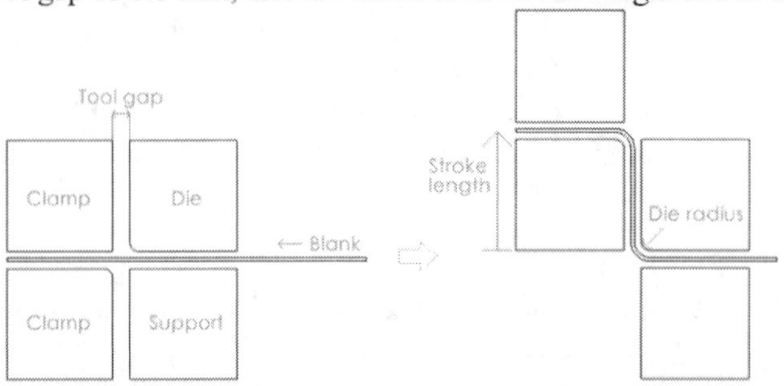

Figure 4: Typical flange drawing process

The patented design of the die tool [17] establishes the required curved bend axis by virtue of which, the concave sections of the blank corrugations invert to convex sections and the convex sections to concave in the flange. The forming cycle produces a 'Z' shaped developable part wherein the cross-section profile of the flange is a mirror reflection of that of the blank, illustrated in Figure 5-b.

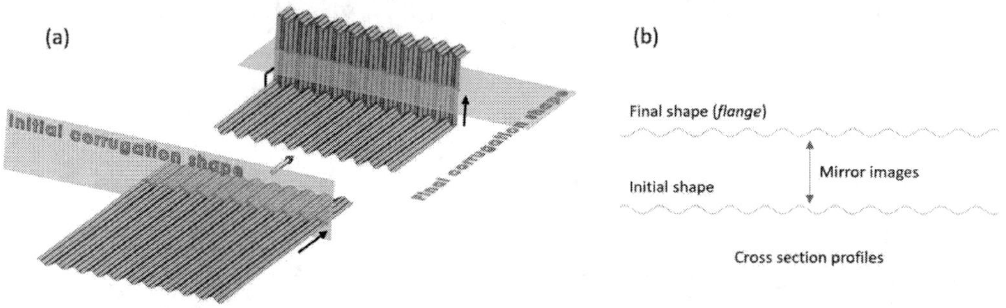

Figure 5.(a) flat corrugated sheet flanged to a 'Z' section; (b) shape of the flange is a mirror image of the initial corrugation shape

4. Strain measurements

The extent of deformation in the sheet was determined by measuring surface strains that developed in the forming process using the Autogrid® strain measurement system. To etch the grid on an initially curved surface, a 110 mm x 110 mm patch was cut out from the original stencil. To check the reduction in the grid size of the stencil patch, strains were measured after etching the cut-out patch on a flat sheet. Further, the patch was carefully rolled and affixed onto the corrugated sheet (Figure 6-a). To allow air bubbles to exit during etching, one side of the patch along the rolling direction was not taped. The grid center was aligned with the center of the flange as shown in Figure 6-b and etched on the side of the sheet that does not come into contact with the die radius during forming.

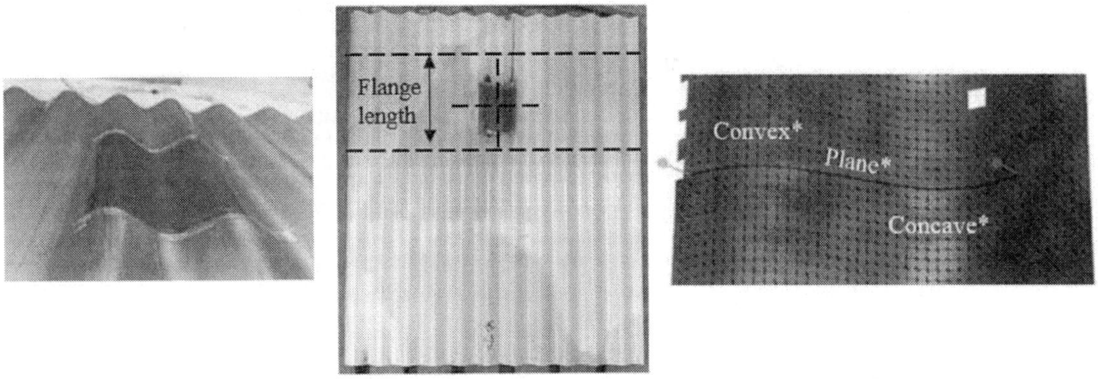

(a) (b) (c)

Figure 6. (a) grid patch rolled onto a corrugated sheet sample, (b) grid positioning and alignment, (c) transverse path for obtaining strain pattern

After etching, the quality of the grid was inspected visually, and the biaxial compression of the grid because of being cut-out from the perfectly stretched stencil, was measured prior to forming. To obtain strains developed in the forming process, these strains were added to the ones measured after forming. Strains were obtained along a transverse path as demonstrated in Figure 6-c. Three samples were analyzed to ensure concurrence of results.

5. Results and discussions

5.1. No tearing or wrinkling observed

In the forming process, the geometry of the die radius forces material in the initially concave sections of the corrugated sheet, to converge in the circumferential direction, as it becomes a part of the drawn flange. Conversely, material in the initially convex section diverges over the bend radius. Hence, there is alternating tensile and compressive deformation in the plane of the flange, in the circumferential direction. Even with a very high yield strength close to 700 MPa alongside a total elongation of just 1.4%, there was no tearing or wrinkling observed in the flange of the formed corrugated samples. This highlights the possibility of forming developable connections using advanced to ultra-high strength steels, without failure. Thus, geometrical forming opens avenues for producing high strength sheet metal parts that consists of curved surfaces.

5.2. Quality of the grid before forming

As the grid patch was etched on the curved surface of the corrugated sheets by manually rolling a cut-out of the stencil, the quality of the grid was assessed visually and found to be acceptable for further experiments. Figure 7 presents the average major and minor strains obtained for three samples, prior to forming. A similar strain pattern including magnitudes, was observed for the grid when etched on to a flat sheet (Figure 7). The results highlight that the grid underwent a biaxial compression as it was released from the stretched and framed stencil.

IOP Conf. Series: Materials Science and Engineering 1284 (2023) 012070 doi:10.1088/1757-899X/1284/1/012070

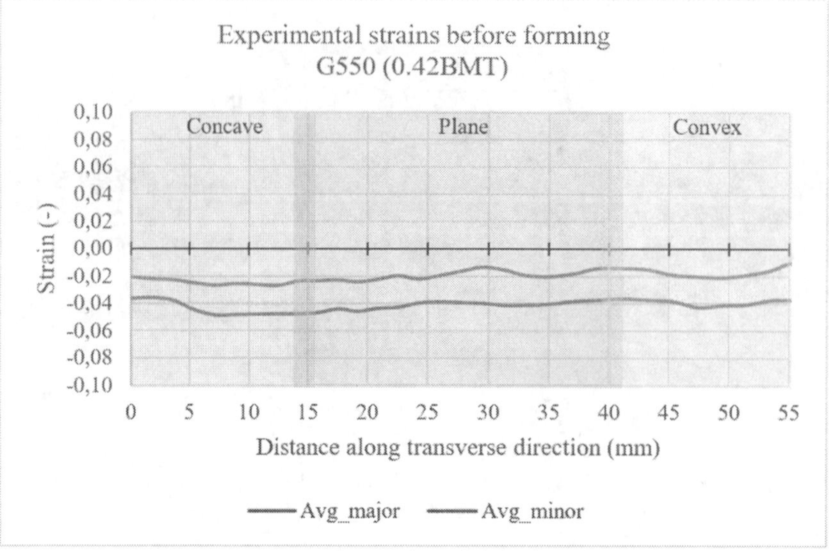

Figure 7. Experimental strains before forming

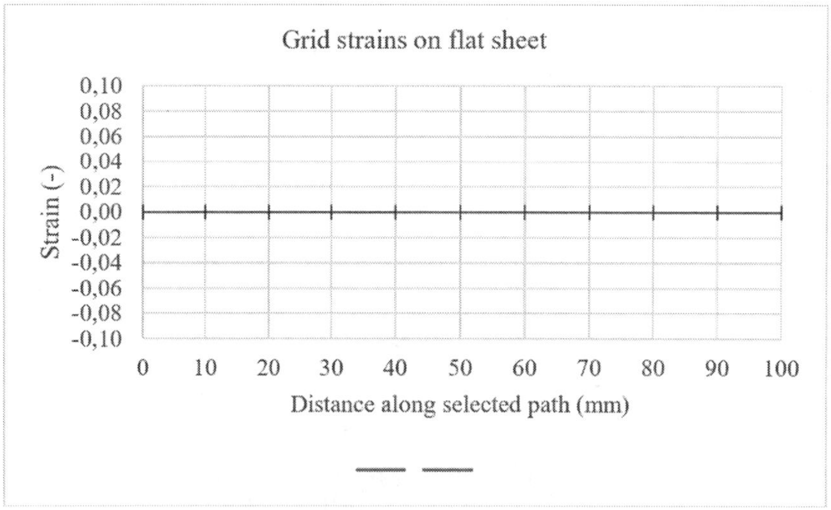

Figure 8. Grid strains on a flat sheet

5.3. Strains in the sheet after forming

Strains are extracted along a transverse path, on the non-contact side of the sheet, as illustrated in Figure 9-a and b. The pre-forming biaxial compression of the grid was accounted for by subtracting the average of the major and minor strains along a selected path before forming from the strains obtained along the same path after the sheet was flanged. This approximation was justified as the Autogrid® produces directional strains and there was no significant trend found in the transverse and longitudinal directions before forming. The final transverse and longitudinal strains obtained this way are presented in Figure 10.

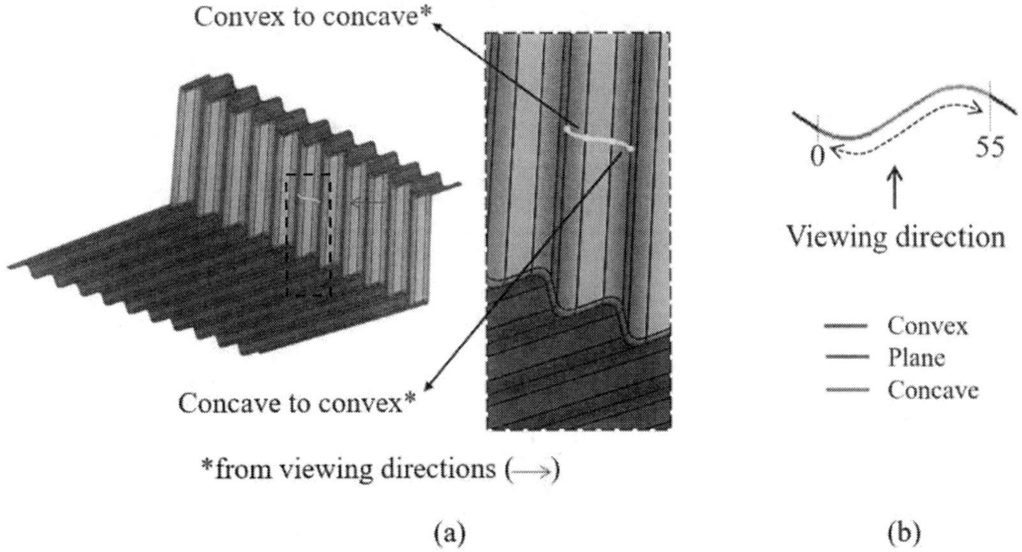

Figure 9. (a) Strain measurement path on the non-contact side of the flange, (b) measuring distance along transverse direction (mm)

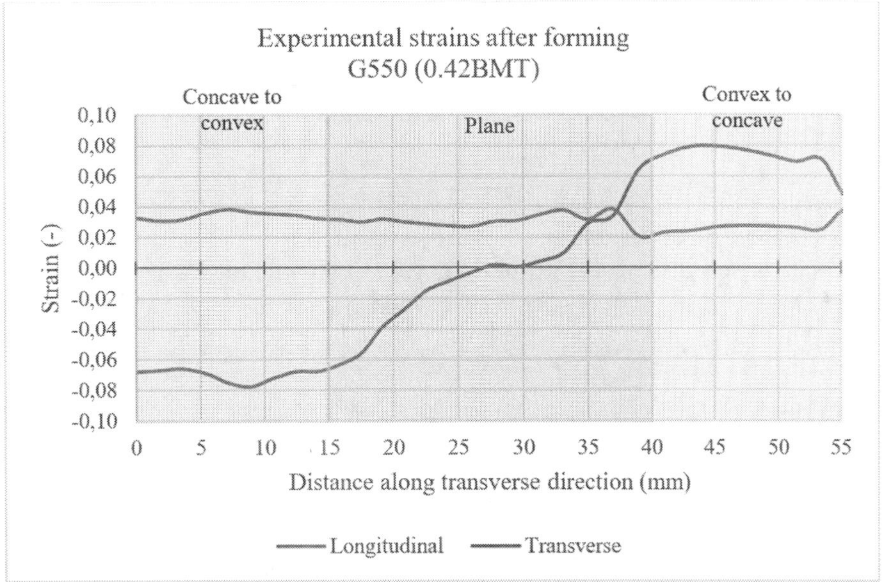

Figure 10. Experimental strains after forming

5.3.1. Transverse strains

Figure 10 shows alternating tensile and compressive strains in the convex and the concave sections of the formed developable. The high magnitude of surface strains points towards the primary difference between *folded* versus *formed-developables.* The sheet metal drawing process leads to shear deformation in the plane of the flange causing membrane strains, otherwise absent in the case of folded developables [8].

The maximum tensile and compressive strains are found to be equal and opposite and are close to the geometrical peaks of the respective curved sections (Figure 9-b and Figure 10). The strain values gradually reduce away from the geometrical peaks. This trend is governed by the doubly curved geometry of the bend radius which changes from a converging region at the concave to convex transition, to a diverging region at the convex to concave transition. The equal magnitudes of the transverse strains indicates that deformation is influenced more by geometrical and process parameters like the blank shape, the bend radius, and the bend angle.

5.3.2. *Longitudinal strains*

The sheet undergoes a stretching deformation in the longitudinal direction. Autogrid® strain measurements illustrated in Figure 10 show an average tensile strain of 3% across the selected path, in the longitudinal or forming direction. This is almost double the total elongation of the material found from the tensile test results (Figure 3-c). Hence, a higher formability is achieved for the high strength steel sheets from a geometrically compliant forming process. The ±1% variation of the longitudinal strain along the selected transverse path can be attributed to the assumption of equal biaxial reduction in the grid size of the stencil patch considered by taking the average of major and minor strains prior to forming.

The strain results highlight that the primary deformation mode in the geometrical forming of the corrugation developable is plane stress. While the deformation type for the transition from a concave to convex section is classical drawing, the convex section undergoes biaxial stretching while forming into the concave shape. Thus, it is likely that the material undergoes thinning in the latter case, causing an uneven material thickness across the developable shape. Change in thickness and the overall deformation on the shape of the formed developable must be studied to better inform the novel sheet metal forming process.

In conventional flange drawing of a flat sheet [15], the major deformation is in the longitudinal/forming direction, and the associated springback causes a shape defect often termed as sidewall curling [18, 19]. However, in the novel process, major deformation occurs in the transverse direction, orthogonal to the direction of drawing. The release of internal stresses in the transverse direction may therefore play a major role in the shape of the final part and could lead to unique shape defects. Further research must be conducted to link the deformation in the process to shape of the final part.

6. Conclusions

Following conclusions can be drawn from the current research,

- Combining the classical theory of folded developables with the traditional sheet metal forming flange drawing process a developable connection is successfully produced with thin sheet metal of very high strength and reasonable inextensibility The class of parts manufactured by such a combination are termed as *formed-developables*.
- The paper presents a unique, modified methodology to capture strains in a forming process where the blank sheet is curved.
- The novel forming process results in significant strains in the plane of the sheet, yet formed parts showed no tearing or wrinkling.
- While the sheet stretches in the longitudinal direction, higher strains are obtained in the transverse direction. Thus, unlike classical plane strain flange drawing, the major deformation is orthogonal to the forming direction, and is mainly governed by the doubly curved geometry of the bend radius.

Acknowledgements

The authors would like to thank Deakin University Postgraduate Research Scholarships (DUPRS) and Australian Research Council Linkage Program (ARCLP) scheme for financial support alongside FormFlow Pty. Ltd. for providing material and equipment for conducting experiments.

References

[1] Z. Marciniak, J. Duncan, and J. Hu, *Mechanics of sheet metal forming*. Elsevier, 2002.
[2] A. E. Moghadam, "Flexible roll forming of the variable depth profiles," Deakin University, 2018.

[3] J. P. Duncan and J. Duncan, "Folded developables," *Proceedings of the Royal Society of London. A. Mathematical and Physical Sciences,* vol. 383, no. 1784, pp. 191-205, 1982.
[4] "Stilride: Technology." Stilride. [Online]. https://www.stilride.com/pages/technology (accessed May 10, 2023).
[5] "Ola Electric: Home." Ola Electric. [Online]. https://olaelectric.com/ (accessed May 10, 2023).
[6] C.-T. Wang, G. Kinzel, and T. Altan, "Failure and wrinkling criteria and mathematical modeling of shrink and stretch flanging operations in sheet-metal forming," *Journal of Materials Processing Technology,* vol. 53, no. 3-4, pp. 759-780, 1995.
[7] A. I. Zaid, "Deep drawing mechanism, parameters, defects and recent results: state of the art," in *IOP Conference Series: Materials Science and Engineering*, 2016, vol. 146, no. 1: IOP Publishing, p. 012009.
[8] R. G. Templer, "Computer aided modelling of sheet metal forming," ResearchSpace@ Auckland, 1994.
[9] M. Schneider and M. Liewald, "Development of a Folding Tool for Miuri-Structures," in *IOP Conference Series: Materials Science and Engineering*, 2018, vol. 418, no. 1: IOP Publishing, p. 012120.
[10] J. S. Chalfant, "Analysis and design of developable surfaces for shipbuilding," Massachusetts Institute of Technology, 1997.
[11] G. Epps and S. Verma, "Curved Folding: Design to fabrication process of RoboFold," *Shape Modeling International 2013,* vol. 75, 2013.
[12] V. A. Raducanu, V. D. Cojocaru, and D. Raducanu, "Structural Architectural Elements Made of Curved Folded Sheet Metal," 2016.
[13] E. Erdine *et al.*, "Curved-crease folding and robotic incremental sheet forming in design and fabrication," in *Proceedings of the Symposium on Simulation for Architecture and Urban Design*, 2019, pp. 1-8.
[14] C. Lauff *et al.*, "Differentiating bending from folding in origami engineering using active materials," in *International Design Engineering Technical Conferences and Computers and Information in Engineering Conference*, 2014, vol. 46377: American Society of Mechanical Engineers, p. V05BT08A040.
[15] S.-W. Lee and Y.-T. Kim, "A study on the springback in the sheet metal flange drawing," *Journal of Materials Processing Technology,* vol. 187, pp. 89-93, 2007.
[16] A. E8/E8M-15, "Annual Book of ASTM Standards," 2015: ASTM West Conshohocken, PA.
[17] J. A. Duncan, NZ), Dingle, Matthew (Torquay, AU), "Sheet junction," United States Patent Appl. D885613, 2020. [Online]. Available: https://www.freepatentsonline.com/D885613.html
[18] F. Pourboghrat and E. Chu, "Springback in plane strain stretch/draw sheet forming," *International Journal of Mechanical Sciences,* vol. 37, no. 3, pp. 327-341, 1995.
[19] M. Samuel, "Experimental and numerical prediction of springback and side wall curl in U-bendings of anisotropic sheet metals," *Journal of Materials Processing Technology,* vol. 105, no. 3, pp. 382-393, 2000.

IOP Conf. Series: Materials Science and Engineering 1284 (2023) 012071 doi:10.1088/1757-899X/1284/1/012071

Innovative experimental setup for the investigation of material models with regard to strain hardening behavior

L Maier*, M Eder, R Norz, W Volk

Chair of Metal Forming and Casting, Technical University of Munich, Walther-Meissner-Strasse 4, 85748 Garching, Germany

E-mail: lorenz.maier@utg.de

Abstract. Sheet metal formed parts are often produced in multistage processes. The multi-directional deformations that occur in this process pose a challenge for sufficiently accurate modeling of the material behavior. In this paper, we present an experimental methodology and its potential that serves as a basis for a comprehensive validation of material models with respect to multi-directional deformation. Therefore, we present the experimental results of pre-strained DP600 specimens, which we afterwards subjected by the MUC-Test. The MUC-Test has already been established as a methodology for validating material models under proportional loading. By using pre-strained specimens, this methodology can be extended to non-linearities in the load history. For the purpose of the paper, we pre-strain the sheet material in three conditions: uniaxial tension, plane strain and equibiaxial tension. Thus, we show that the MUC-Test is suitable for the representation of multi-directional strain histories and useful for comprehensive validation of material models.

1. Introduction

In the production of deep drawing parts, it is essential to investigate numerically the components at an early stage and during the development process in order to define tool geometries and test processes digitally [1]. The use of a suitable material model is essential for the quality of the simulation results [2]. It has long been proven that the complexity of the material model has an influence on the simulation result [3]. The typical multi-stage production steps in sheet metal forming usually also involve a non-linear strain path, which the simulation must take into account in terms of formability [4]. Especially for ferritic-martensitic dual phase (DP) steels, the modeling is not trivial due to the complex microstructure and their elastic-plastic deformation behavior [5].

In this paper, we present an experimental setup by which specific pre-strains can be introduced in a sheet metal. Subsequently, these pre-strained samples are prepared for a further investigation by a material validation test, the so-called MUC-Test. This allows the generation of a complex strain distributions starting from the different pre-strain state. We compare the difference between pre-strained and non-pre-strained specimens of a DP600 steel. This presents an experimental setup through which material models can be investigated with respect to complex strain path history.

2. State of the Art

2.1. Material Card Validation

In order to validate material models as free of influence as possible, a appropriate test setup is necessary [6]. On the one hand, one wants to keep frictional influences as low as possible, since not a friction model but the material model is to be investigated. In addition, the model is to be used for complex geometries. Accordingly, no specific strains are to be investigated, but rather the broadest possible strain distribution in the range relevant for the deep drawing process. It is also advantageous for validation not only to be able to compare an initial and final state, but also to be able to carry out continuous validation during the forming process. [2]

Various experiments have been established for the validation of material models. One example is the cross-die, in which a wide range of strain states is achieved in the specimen material [7]. Here, the major and minor strains after a forming process are compared with the simulation. In this test setup, however, friction dominates and validation is only possible at the end of the forming process [2]. Another validation test, which we will also investigate further in this paper, is the so-called MUC-Test [8].

The MUC-Test is an experimental setup that generates complex strain states during forming by using a butterfly-shaped tool geometry. By examining three specimen geometries, strain ranges in the predominantly uniaxial, plane strain or biaxial state range can be investigated. In addition to these different geometries, three rolling directions are also investigated. This allows a full validation of the yield locus with a small amount of geometries. During the test, an optical measuring system records the deformation of the sheet material on the upper side of the sheet.

The use of a digital twin allows the experimental setup to be examined by an FEM simulation. The material model to be validated is used. By comparing the major and minor strain over the entire punch stroke as well as the punch force of experiment and simulation, a validation of the material model can be carried out [2].

2.2. Strain-path Dependent Hardening

An important factor affecting the outcome of the forming process is the strain path dependence, which refers to the relationship between the direction of deformation and the resulting mechanical properties of the metal [6].

Forming during the production of a sheet metal component can take place in different directions, e.g. by tension, compression or shear. This results in different strain paths which affect the mechanical properties of the formed component. There are many approaches for generating these strain paths experimentally and simulating them [5], [6]. Due to the complexity of generating different strain paths that deviate from the uniaxial tension, few is found in the literature here. Most deal with uniaxial pre-strained blanks and subsequent testing of cross tensile, tensile, Nakajima or plane strain specimens in different rolling directions [9], [10].

Moreover, the strain path dependence is not only affected by the direction of deformation but also by the magnitude and distribution of the deformation. For instance, if the deformation is non-uniform, with some areas of the sheet experiencing higher strains than others, the resulting strain path can be significantly different from a uniformly deformed sheet. Nevertheless, understanding the strain path dependence is critical for predicting the mechanical properties of complex formed parts accurately [11].

3. Experimental Setup

3.1. Pre-strain Conditions

To achieve different pre-strains, we used a specially manufactured tool and frequently used by us for this purpose on a hydraulic press [12]. This tool is shown in Figure 1. The material flow in this tool we can control by using specimens of different widths and by the drawing bead [10]. For the series of tests presented in this paper, we created three different pre-strain conditions. We lubricated the initial blanks with a deep drawing oil in the area of tool contact.

Figure 1. Deep drawing tool for generating different pre-strain states in sheet metal materials: a) punch and blank holder, b) die [12].

In order to assess the strain state following a forming process, we utilized the GOM Argus system. The blank was marked with evenly spaced dots via an etching process, and following the forming process, the optical deformation of these points was measured. This approach enabled us to evaluate whether the desired pre-strain was attained based on the specimen geometry and tool used. After achieving the pre-strain, specimens were then cut from the blank for use in subsequent testing. Figure 2 shows the location of a specimen cut from the blank in the rolling direction, with a specimen width of 70 mm, while specimens with widths of 110 mm and 230 mm were also cut. Two rolling directions (0°, and 90°) were prepared for the subsequent MUC-Test, resulting in a total of 18 different experimental setups that varied according to the degree of pre-straining. We have induced the pre-strain into the material in rolling direction.

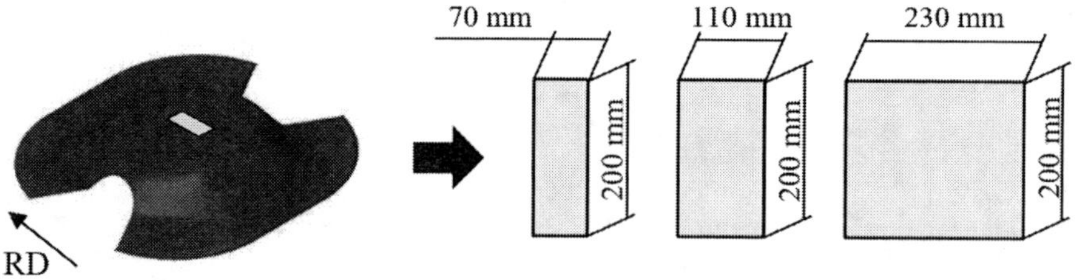

Figure 2. Specimen position of a 70 mm MUC-Test specimen at a plane strain pre-strained sheet and the size of the three geometries for the MUC-Test. Pre-strained by the utilized pre-strain tool shown in Figure 1.

3.2. MUC-Test

These 18 pre-post-strained samples we are going to compare in this paper with nine samples of the MUC-Test, which did not pass the pre-straining procedure. For this purpose, we cut out of the initial sheet specimens with a width of 70 mm, 110 mm and 230 mm and a length of 200 mm in the three rolling directions. For each combination, we performed three experiments.

In the next step, we prepared a white primer on all prepared specimens and applied a black stochastic pattern to them. This serves as preparation for the optical measurement system, more precisely the digital image correlation (DIC) system GOM Aramis 4M, which registers the displacement of the facets on the sheet material top surface through this pattern during the MUC-Test. The punch, the blankholder and the die of the material card validation tool are shown in Figure 3. Due to the butterfly shape of the drawing bead and the geometry of the punch, a complex strain distribution is obtained in this second deformation step.

Figure 3. Validation tool of the MUC-Test for installation in a hydraulic testing machine with (left to right) punch, blankholder and die.

A combination of deep-drawing foil (two layers) and lanolin (three layers) is applied on the punch to reduce friction between punch and specimen. After closing the blankholder, a blankholder force of 400 kN is applied. In conjunction with a draw bead, slippage of the specimen is prevented.

After the clamping process, the punch moves upwards at a speed of 1.0 mm/s and deforms the material. The displacement of the sheet surface is recorded by the DIC system at a measuring frequency of 10 Hz. The two cameras observe the process through the opening in the die. This allows us to calculate major and minor strains with high temporal resolution. The area of the specimen top surface that can be recorded by this DIC system during the complete forming process is shown in Figure 4.

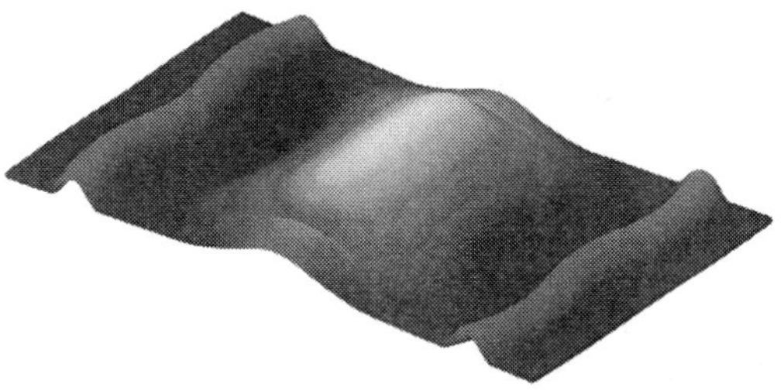

Figure 4. The area visible to the DIC system through the opening of the die is highlighted in color for a 110 mm specimen. In this area we can determine the strains of the specimen surface [2].

4. Results

4.1. Pre-straining Results

First, we generated the individual pre-strain states with the tool from Figure 1. Different specimen widths were used for this purpose. The resulting strain state after the first forming step can be seen in Table 1. The specimen passes through a linear strain path during the forming process up to the strain point shown in Figure 5. In order to obtain a sufficient post-deformation strain after the first forming step and avoiding a failure at the beginning of the second followed up forming process, we performed comparative deformations according to

$$\varepsilon_{v,M} = \sqrt{\frac{2}{3} \cdot (\varepsilon_1^2 + \varepsilon_2^2 + \varepsilon_3^2)} \tag{1}$$

between $\varepsilon_{v,M} = 0.098$ and $\varepsilon_{v,M} = 0.121$, as presented in Figure 5. This is about half the uniform elongation achieved during the uniaxial tensile test when the specimen was locally necked. It can be seen in the figure that the selected specimen geometries are very well suited for uniaxial, plane and biaxial pre-strain. Here we achieved the objective of obtaining strain states relevant for forming applications through the first forming process.

Table 1. Logarithmic strain states of the three different pre-strains investigated.

	Uniaxial strain	Plane strain	Biaxial strain
Major Strain [-]	0.114	0.083	0.067
Minor Strain [-]	-0.050	0.003	0.054
von Mises Strain [-]	0.114	0.098	0.121

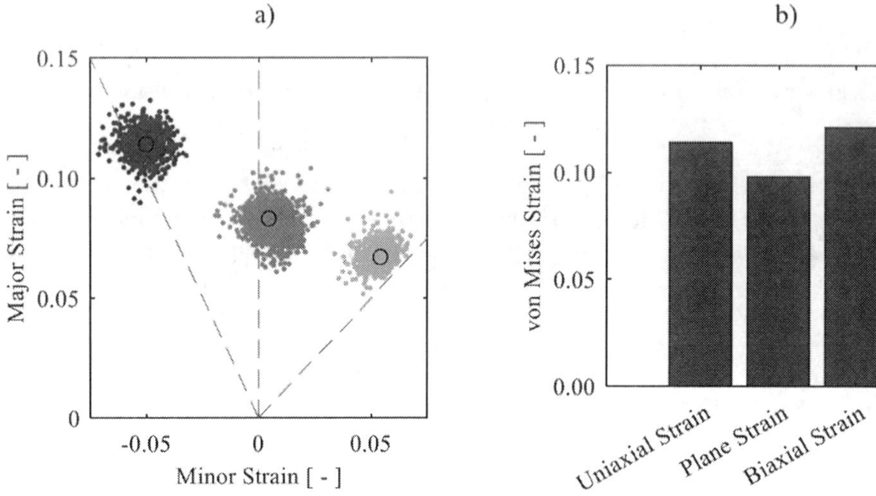

Figure 5. a) Logarithmic strain states and b) the according equivalent value of the three different pre-strains after the first forming process with the tool from Figure 1 for the area from which we cut the specimen for the MUC-Test.

4.2. MUC-Test Results

The strain evolution of the MUC-Test, which is recorded by the DIC system, runs continuously during the tests. The results we show in this chapter show the deformation in major and minor strain at 80 % of the punch stroke at the failure of the tested specimen. However, the results are available over the entire test and can be evaluated over the complete punch stroke. In the following figures with pre-strained specimens, the initial pre-strain value corresponds to the mean values of the results from the GOM Argus measurements from Table 1 and Figure 5. When using this point, we have marked it in the respective figure.

Figure 6 shows the strain distribution of the three investigated specimen geometries of the MUC-Test in two rolling directions. Under a), the rolling direction is at 90° to the 200 mm specimen length. Under b) at 0° to the 200 mm specimen length. We first consider the 70 mm specimen, which predominantly follows a uniaxial strain. Here it can be seen that the 0° specimen has less negative minor strain. Also, compared to the 90° specimen, the W-shape is less pronounced in the high major strain range. For the 110 mm specimen geometry, it can be seen that at 80 % of the punch travel at 0°, there is a higher major strain. Comparing the two 230 mm specimens, only a marginal difference can be detected

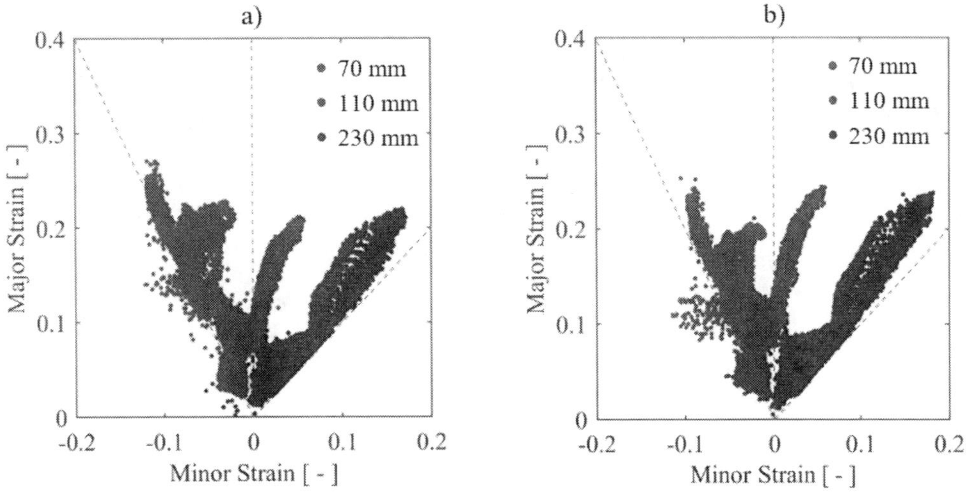

Figure 6. Strain distribution of three MUC-Test experiments at 80 % of punch stroke; a) rolling direction at 90° to the 200 mm specimen length, b) rolling direction at 0° to the 200 mm specimen length.

With regard to the pre-strained specimens, we start with the discussion of the uniaxial pre-strain. Here, Figure 7 shows the results of the second forming step, the MUC-Test. The results are again from the time step at 80 % of the punch stroke. Under a) we find the strain distribution of the three specimen geometries with rolling direction at 90° to the specimen length. Under b) the results are shown at 0° to the specimen length. Here, a clear difference in the V-shape can be seen for the 70 mm specimen. Also, at 0°, significantly higher major strains were obtained for the 110 mm specimen width. The same tendency can be seen with the 230 mm specimen width.

Besides this finding, we can also observe that the shape of the strain distribution changes for the 110 mm specimen geometry. The arm, which developed positively in the minor strain, widens significantly. The negative minor strain disappears completely and the 110 mm specimen geometry experiences only a positive minor strain change starting from the pre-strain state from the first forming step.

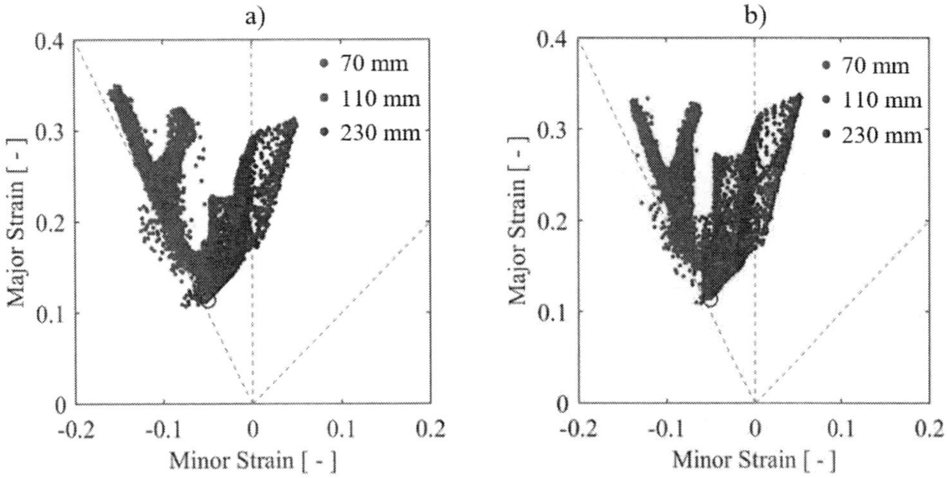

Figure 7. Strain distribution of the three different specimen of the MUC-Test with the uniaxial pre-strain from Figure 5; a) rolling direction in 90° to the 200 mm specimen length, b) rolling direction in 0° to the 200 mm specimen length.

Figure 8 shows the punch force over punch travel until failure of the specimens. The specimens shown were tested in the rolling direction. All three specimen widths are shown, one set without pre-strain and one set with uniaxial pre-strain up to the pre-strain from Table 1. The pre-strengthening of the material by the first forming process is clearly visible. This is accompanied by the fact that the specimens fail at much lower punch travel.

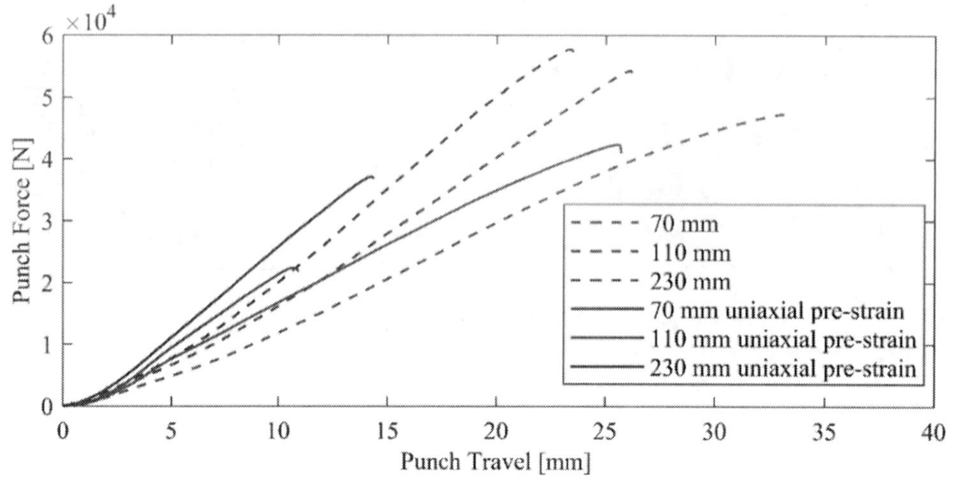

Figure 8. Punch force over punch travel of the non-pre-strained specimens compared to the uniaxial pre-strained specimens - all specimens shown were examined in the rolling direction to the 200 mm specimen length.

As in Figure 6 for the 0° specimens, we can also observe for the plane pre-strained specimens in Figure 9 that we achieve less negative minor strain with this specimen orientation for the 70 mm specimen. The higher major strain range with this specimen geometry behaves similarly to the 90° specimen orientation. An expression of the W-shape, as we can see in Figure 6, does not occur here.

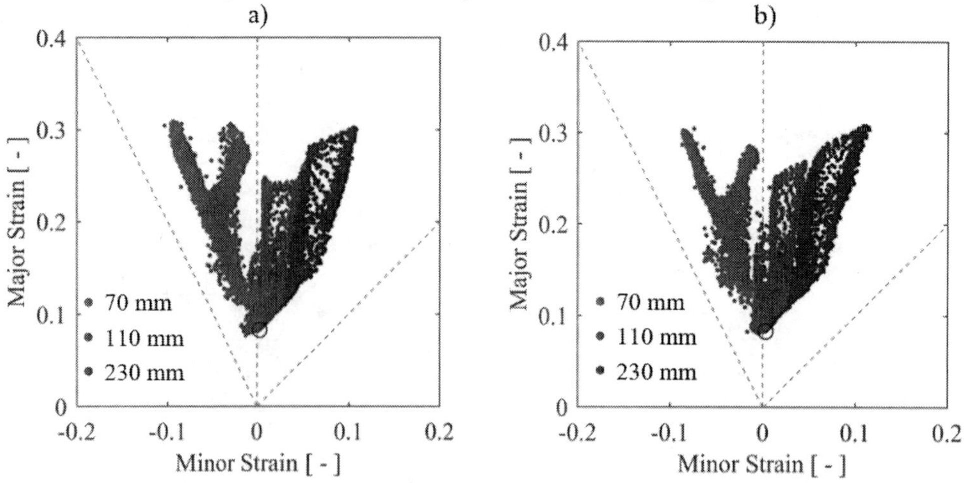

Figure 9. Strain distribution of the three different specimen of the MUC-Test with the plane pre-strain a) rolling direction in 90° to the 200 mm specimen length, b) rolling direction in 0° to the 200 mm specimen length.

Due to the pre-strain, the 70 mm specimen follows a V-shape. This is also the case with the uniaxial pre-strained specimens from Figure 7. The largest difference in strain distribution can be seen in the 110 mm specimen geometry. Here, a clear visible deviation in the upper region of the strain distribution can be seen.

For the biaxial pre-strained specimens, the 90° and the 0° MUC-Test experiments are shown in Figure 10. Slight differences in the major strain are evident between these two experimental tests. Most noticeable in this pre-strain is the 110 mm specimen geometry. Here, the arm assumes an L-shape in the higher major strain region. This shape could not be detected in the other experimentally investigated pre-strains.

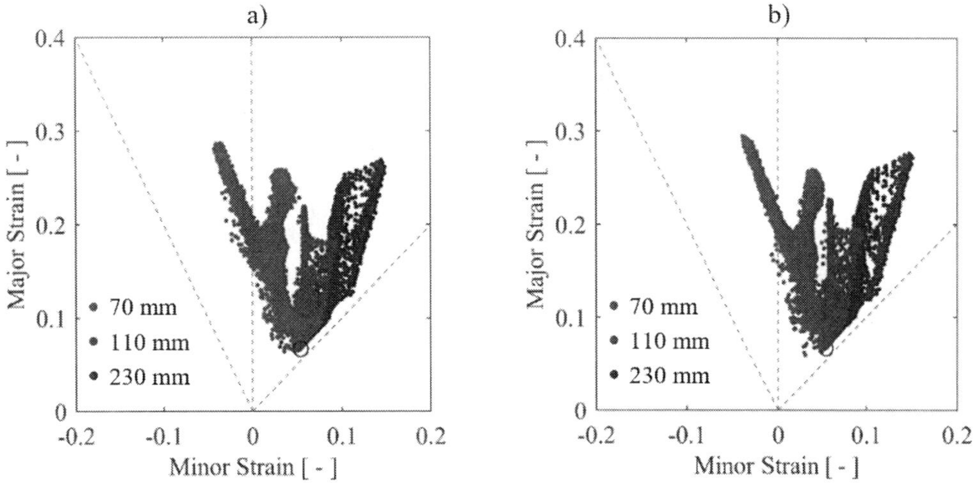

Figure 10. Strain distribution of the three different specimen of the MUC-Test with the biaxial pre-strain a) rolling direction in 90° to the 200 mm specimen length, b) rolling direction in 0° to the 200 mm specimen length.

5. Conclusion

We have presented a two-step forming process that can be applied to the experimental investigation for strain paths. We measured the introduced pre-strain by a GOM Argus system. The von Mises pre-strain has been between $\varepsilon_{v,M} = 0.098$ and $\varepsilon_{v,M} = 0.121$. Uniaxial, plane and biaxial pre-strains were applied. Subsequently, we performed an MUC-Test with the pre-strained material. Three different specimen widths were tested, and each in two orientations relative to the rolling directions. During the MUC-Test, the deformation was measured by a GOM Aramis 4M system.

The results in comparison with non-pre-strained and in the different orientations to the rolling direction per pre-strain show a significant difference. Thus, in this paper we were able to show that the pre-strain condition at the 70 mm specimen width compared to the non-pre-strained MUC-Test specimens has an influence at all the pre-strains investigated. The 110 mm specimen width not only shows a difference compared to the non-pre-strained specimens. Furthermore, the shape of the strain distribution in the major and minor strain diagram changes depending on the pre-strain. The influence of the 230 mm specimens was the weakest in this study.

With regard to the required punch forces, we were able to determine that the setup presented is suitable for the validation of material models. We were able to determine a considerable difference in the course of the force-displacement curves, which must also be represented in a material model validation by a digital twin. With the presented experimental setup, a validation test for a material model can be performed, where strain path dependence can be investigated.

Acknowledgements
The authors would like to thank the German Research Foundation (Deutsche Forschungsgemeinschaft, DFG) for the financial support under the grant number 455388465.

References

[1] Lee S-Y, Kim J-M, Kim J-H, Barlat F, Validation of homogeneous anisotropic hardening model using non-linear strain path experiments *International Journal of Mechanical Sciences* 2020, *183*, 105769

[2] Eder M, Gruber M, Volk W, Validation of material models for sheet metals using new test equipment *International Journal of Material Forming 2022*, 15

[3] Banabic D, Comşa D, Sester M, Selig M, Kubli W, Mattiasson, K, Sigvant, M, Influence of constitutive equations on the accuracy of prediction in sheet metal forming simulation *Numisheet* 2008

[4] Dhara S, Taylor S, Figiel Ł, Hughes D et al., Development of a novel testing methodology for in-situ microstructural characterisation during continuous strain path change *Materials Characterization* 2022, 183

[5] Chongthairungruang B, Uthaisangsuk V, Suranuntchai S, Jirathearanat S, Experimental and numerical investigation of springback effect for advanced high strength dual phase steel *Materials & Design* 2012, 39, 318–328

[6] Aung T L, Ma N, Okitsu Y, Hayashi S et al., Measurement and modelling of strain-path dependent anisotropic hardening behaviors of high strength steels subjected to pre-strains *Journal of Materials Research and Technology* 2023, 23, 451–465

[7] Doig M, Isik K, Clausmeyer T, Heibel S, Richter H, Tekkaya A E, Material Characterization and Validation Studies for Modeling Ductile Damage during Deep Drawing *Procedia Engineering* 2017, 183, 77–82

[8] Eder M, Gruber M, Manopulo N, Volk W, Potentials for material card validation using an innovative tool *IOP Conference Series: Material Science and Engineering* 2021, 1157, 12067

[9] Cheng Z, Zhang C, Wang M, Wang Y, Fang S, Liang C, Wenchao S, Bo L, Forming limit analysis of AA5052 sheet under bilinear strain path by designing a novel large-scale gripping fixture *Journal of Materials Research and Technology* 2023, 22, 2425–2439

[10] Norz R, Vitzthum S, Volk W, Failure behaviour of various pre-formed steel sheets with respect to the mechanical grain boundary properties *International Journal of Material Forming* 2022

[11] Das B, Singh A, Kumar P S, Low cycle fatigue performance of DP600 steel under various pre-straining paths *International Journal of Fatigue* 2020, 132, 105331

[12] Weinschenk A, Volk W, FEA-based development of a new tool for systematic experimental validation of nonlinear strain paths and design of test specimens *20th International ESAFORM Conference on Material Forming* 2017

Optimised parametrisation of the MMC fracture model using AE and FE analysis

E Stockburger*, H Wester and B-A Behrens

Institute of Forming Technology and Machines, Leibniz Universität Hannover, An der Universität 2, 30823 Garbsen, Germany

E-mail: stockburger@ifum.uni-hannover.de

Abstract. To save resources in component production, finite element (FE) simulations are state of the art during process design to prevent expensive tool modifications. For a high accuracy of a FE simulation, realistic material data is required. Unlike for flow behaviour, the determination of the fracture behaviour is not standardised. Conventionally, tensile tests with different geometries are performed, monitored with digital image correlation (DIC) and fracture is determined on the surface. However, for many specimens the initiation of fracture is assumed to occur inside the material prior to a macroscopic fracture on the surface. Therefore, various tensile tests with butterfly specimens were performed for a HCT980X steel and monitored with a DIC as well as an acoustic emission analysis (AEA) system. The displacement at fracture was evaluated conventionally for fracture on the specimen's surface based on the DIC and unconventionally for fracture inside the specimen by AEA. FE models of the butterfly tests were created with fracture displacements of both evaluation methods as boundary conditions. Using the numerically determined stress state and fracture strain from the butterfly models, the MMC fracture model was parametrised for both evaluation methods and compared by applying both models to a FE simulation of a forming process.

1. Introduction

For the characterisation of material fracture different experimental tests are currently performed and tracked using digital image correlation (DIC) [1]. The fracture initiation is identified based on a macroscopic surface crack, which is subjectively determined with the images of an optical measurement system. However, for many materials and stress states the fracture initiation takes place inside the specimen before a surface crack is visible [2]. As shown in figure 1, after reaching the maximum sustainable load while forming, the specimen first starts to neck. With further forming voids start to nucleate inside the material. These microscopic cracks grow and coalesce until a macroscopic crack and fracture of the specimen occur. Therefore, using DIC to estimate material fracture can lead to inaccuracies. A further issue using DIC for fracture initiation is that the onset of fracture is partly difficult to identify, in particularly for tests using shear specimens. The acoustic emission analysis (AEA) offers a good possibility to overcome these difficulties. In prior investigations, the onset of fracture inside the material during forming could be investigated [3]. Therefore, within this work an AEA system is applied for the characterisation of material fracture for a more accurate and clearer prediction of fracture initiation.

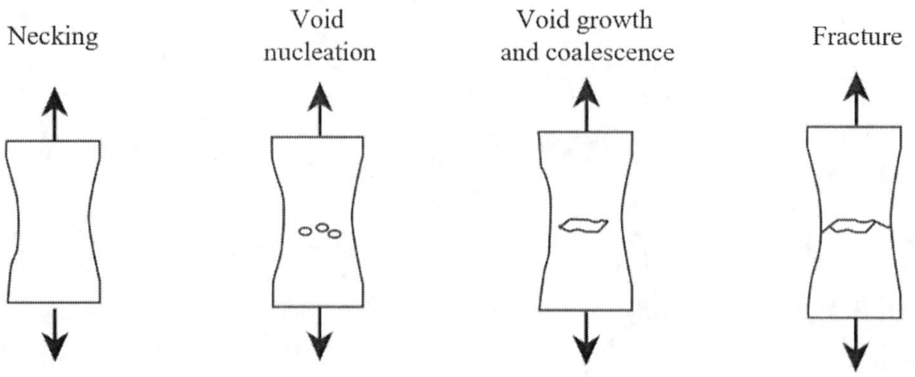

Figure 1. Schematic representation of the fracture development in metallic materials during uniaxial tensile load.

In this paper, fracture characterisation tests with butterfly specimens are performed. The applied test rig with DIC was prior shown by the authors in [4] and extended by an AEA system in this research. The butterfly tests were numerically modelled with the displacement until fracture as a boundary condition, determined from the DIC and the AEA measurements. The DIC models were analysed on the surface nodes of the specimen, while the AEA models were analysed inside the specimen. Two data sets of plastic strain at fracture, stress triaxiality as well as normalised Lode angle were created and the Modified Mohr-Coulomb (MMC) fracture model was fitted for the DIC and for the AEA evaluation method [5]. Both fracture models were used in a forming simulation and compared regarding their difference in fracture prediction.

2. Methodology

2.1. Experimental fracture characterisation and fracture modelling

A HCT980X sheet steel in 1.5 mm thickness was used for the presented investigations. The steel consists of a ferritic matrix with martensitic clusters. To characterise the fracture behaviour of the HCT980X, butterfly tests were conducted. The applied test rig is shown in figure 2 (A), which also displays the butterfly specimen. The outer part of the specimen was produced by water jet cutting and the inner part, which has the shape of a horizontal number eight, was produced by milling on both sides. In the rest rig, the butterfly specimen can be rotated under different loading angles with intervals of 15.5°, in order to create different stress states in the specimen. The tests were performed in a tensile testing machine with a forming speed v_B of 0.02 mm/s at room temperature. Tested loading angles ranged from -3° to 90°. While testing, the forming force was measured by the tensile testing machine and the acoustic emissions of the butterfly specimens by the AEA. To measure the exact displacement of the butterfly specimen in x- and y-direction, the tests were also monitored with DIC. A more detailed description of the test setup is presented by Stockburger et al. in [6].

The specimen fracture was evaluated and compared using two evaluation methods. The first method is based on the DIC, where the material fracture is optically detected based on a surface crack, as it is currently state of the art [7]. The second method is based on the AEA, where the material fracture is determined by identification of clear burst signals. In figure 2 (B) the forming force and amplitude over the testing time is plotted for a butterfly test using both evaluation methods. As for this example, the material fracture is determined earlier by the AEA evaluation using burst signals than by the DIC evaluation using the surface crack as an indicator. The evaluation methods are described in further detail for the steel HCT780C in [8]. Based on the two evaluation methods, the experiments for the seven loading angles were evaluated similarly for their displacements in x- and y-direction at the point of material fracture.

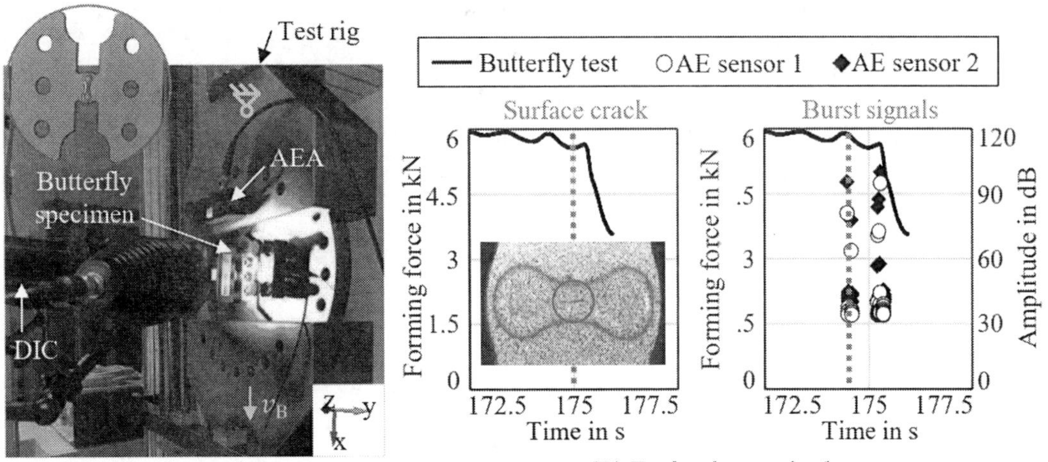

(A) Experimental set-up (B) Evaluation methods

Figure 2. Experimental set-up of the butterfly tests (A) and DIC as well as AEA evaluation methods for material fracture (B).

The MMC fracture model from Bai and Wierzbicki [9] can be parametrised using the butterfly specimens. The strain at fracture according to the simplified MMC model can be written as

$$\varepsilon_{\mathrm{f,MMC}} = \left[\left[\left(\frac{1+c_1{}^2}{3}\right)^{1/2} \cdot \cos\frac{\pi\bar{\theta}}{6} + c_1 \cdot \left(\eta + \frac{1}{3}\sin\frac{\pi\bar{\theta}}{6}\right)\right] \cdot c_2\right]^{-1/n}, \tag{1}$$

where η is the stress triaxiality, $\bar{\theta}$ is the normalised Lode angle and c_1, c_2 as well as n are material specific parameters. For plane strain states the normalised Lode angle $\bar{\theta}$ becomes zero and for plane stress states the relation

$$\bar{\theta} = 1 - \frac{2}{\pi} \cdot arccos\left(-\frac{27}{2} \cdot \eta \cdot \left[\eta^2 - \frac{1}{3}\right]\right) \tag{2}$$

is given between the normalised Lode angle $\bar{\theta}$ and the stress triaxiality η.

2.2. Numerical modelling of the butterfly tests and of the B-pillar deep drawing

Since the stress triaxiality and the normalised Lode angle are experimental not measurable quantities, the butterfly tests were numerically modelled in LS-PrePost. The associated model with boundary conditions is shown in figure 3 (A).

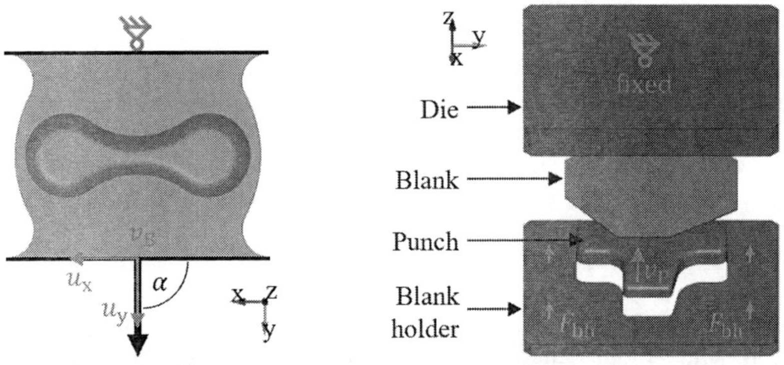

(A) Butterfly (B) B-pillar

Figure 3. Geometry with boundary conditions for the butterfly tests (A) and for the B-pillar deep drawing (B).

For an efficient numerical simulation only a reduced geometry of the butterfly test was used. Furthermore, specimen's symmetry in z-direction was considered, thus it was modelled as half geometry. For the discretisation of the geometry, hexahedron elements (constant stress solid) with an edge length of 0.1 mm were used. The nodes of the specimen's front face were fixed in all directions as well as rotations, while a displacement was applied to the node set on the direct opposite. For each loading angle and evaluation method, the corresponding experimental displacement in x-direction u_x and y-direction u_y from table 1 was set as boundary condition.

Table 1. Average fracture displacements in x- and y-directions of the butterfly tests for the tested HCT980X.

Displace-ment	Evaluation method	-3°	12.5°	28°	43.5°	59°	74.5°	90°
u_x	DIC	0.019	0.136	0.229	0.275	0.355	0.219	0.006
	AEA	0.026	0.136	0.225	0.275	0.353	0.219	0.005
u_y	DIC	1.198	0.834	0.645	0.489	0.499	0.281	0.188
	AEA	1.224	0.832	0.626	0.489	0.497	0.279	0.181

The displacements were adjusted by an angle α according to the seven loading angles. As in the experiments, the forming speed v_B was 0.02 mm/s. The material behaviour of the HCT980X was modelled as elastic-plastic. For the Poisson's ratio, density and young's modulus the values of 0.3, 7.85e-09 ton/mm³ and 2.1e5 MPa, respectively, were used. For the flow behaviour the yield criterion YLD2000 from Barlat et al. [10] and the flow curve extrapolation approach from Hockett and Sherby [11] were applied. The corresponding material characterisation and material modelling of the HCT980X was priorly published by the authors in [12]. The rolling direction was set to y-direction, as in the experiments. The model was calculated implicit with a parallel multi-frontal sparse solver from LS-Dyna version R12. Based on the 14 simulation models, the characteristic values of the stress triaxiality development $\eta(t)$ as well as the normalised Lode angle development $\bar{\theta}(t)$ and the plastic strain at fracture ε_f were determined for fitting the parameters of the MMC fracture model for the two evaluation methods. The term "plastic strain at fracture" in this context represents the plastic strain in the numerical simulation present at the displacements at fracture determined with the two evaluation methods in the experiments.

To compare the impact of the MMC fracture models, a numerical model of deep drawing component abstracted from the lower part of a B-pillar was created. Figure 3 (B) shows an exploded view of the 3D model with its boundary conditions. The model included a die, a 1.5 mm thick blank, a punch and a blank holder. The die was fixed in all spatial directions. The dimensions of the blank were 380 mm x 460 mm and the rolling direction was set into x-direction. The blank holder was only movable in z-direction and had a blank holder force F_{bh} of 1000 kN. No spacing was used for the forming, so the blank holder and the die were closed completely. The punch was also only movable in z-direction and had a forming speed v_P of 15 mm/s. The forming temperature was set to 20 °C. A surface-to-surface contact with penalty method was applied between the tools and the blank. Coulomb's law of friction with a friction coefficient of 0.1 were picked to represented realistic friction conditions [13]. The geometries of the tools were discretised in LS-PrePost with rectangular-shaped shell elements (Belytschko-Tsay) and the blank with hexahedron solid elements (Constant Stress). An element edge length of 2 mm was used for the tools and 1.5 mm for the blank. The tools were model as rigid bodies with Poisson's ratio of 0.3, density of 7.85e-09 ton/mm³ and young's modulus of 2.1e5 MPa. According to the material modelling of the butterfly tests, the blank was modelled elastic-plastic deformable using the material data for HCT980X, but extended by the GISSMO damage model

$$D_{\mathrm{GISSMO}} = \sum_i \left(\frac{\Delta\varepsilon_i}{\varepsilon_f}\right)^m \quad (3)$$

with a damage exponent m of 2 and the MMC fracture models from both evaluation methods. The model was calculated implicit with a parallel multi-frontal sparse solver from LS-Dyna version R12. With this approach, two numerical models were generated, which allowed to compare the impact of the DIC and AEA evaluation methods on the MMC fracture model.

3. Results

3.1. Experimental-numerical butterfly tests

In order to validate the numerical models of the butterfly tests, the numerical force-displacement curves were compared to the experimental curves. Figure 4 (A) shows the force-displacement curves for loading angles from 12.5° to 90°. It can be stated that the numerical curves correspond to the experimental curves sufficiently enough. In general, the maximum force reduces from a loading angle of 90° to 12.5°, while the maximum displacement increases.

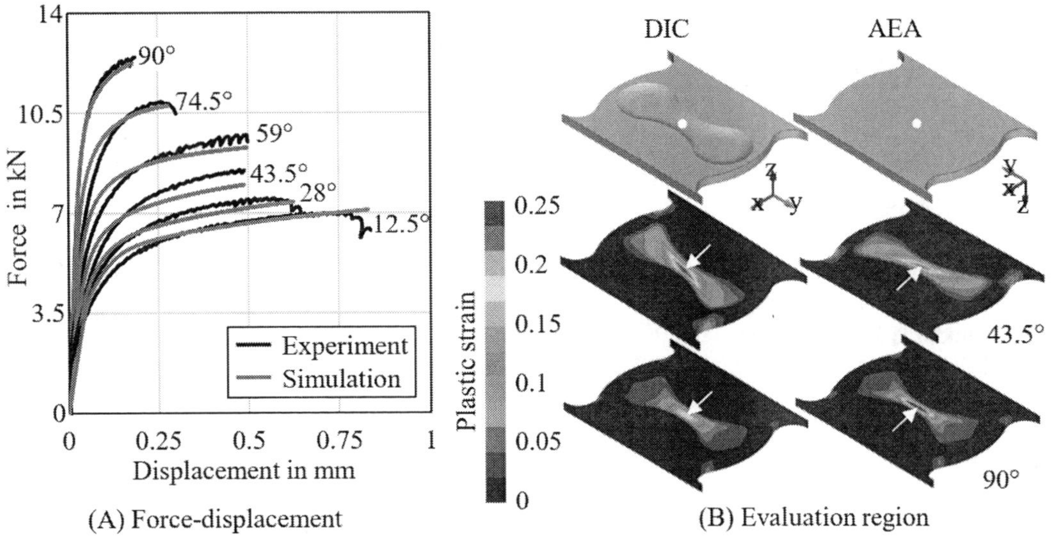

Figure 4. Comparison of the experimental force-displacement-curves to the numerical ones (A) and specimen regions for the evaluation methods (B).

Based on the numerical models of the butterfly tests, parameter sets for the parametrisation of the MMC fracture model are determined. Therefore, the butterfly models were evaluated as shown in figure 4 (B). Based on the experiments, the butterfly models were evaluated in the middle of the specimen surface for the DIC evaluation. For the AEA method, the butterfly models were evaluated inside the centre of the specimen, which is on the opposite side of the half geometry. For each of the 14 simulation models the plastic strain, stress triaxiality and normalised Lode Angle were determined as functions of time.

The plastic strain-stress triaxiality (A) and the plastic strain-normalised Lode angle curves (B) are shown in figure 5. Similar courses of the curves for both evaluation methods are visible for the loading angles from -3° up to 28° regarding both the stress triaxiality and the normalised Lode angle. However, a difference in the course is recognisable for higher loading angles starting at 43.5°. The curves of the AEA evaluation method start and finish at higher stress triaxialities than the curves of the DIC method. Regarding the plastic strain, both evaluation methods lead to similar values up to a loading angle of 59°, while the plastic strain from the AEA evaluation method is below the plastic strain from the DIC evaluation method. For higher loading angles, the plastic strain is higher for the AEA than for the DIC evaluation method, which indicates a high gradient of the plastic strain over the sheet thickness for these loading angles.

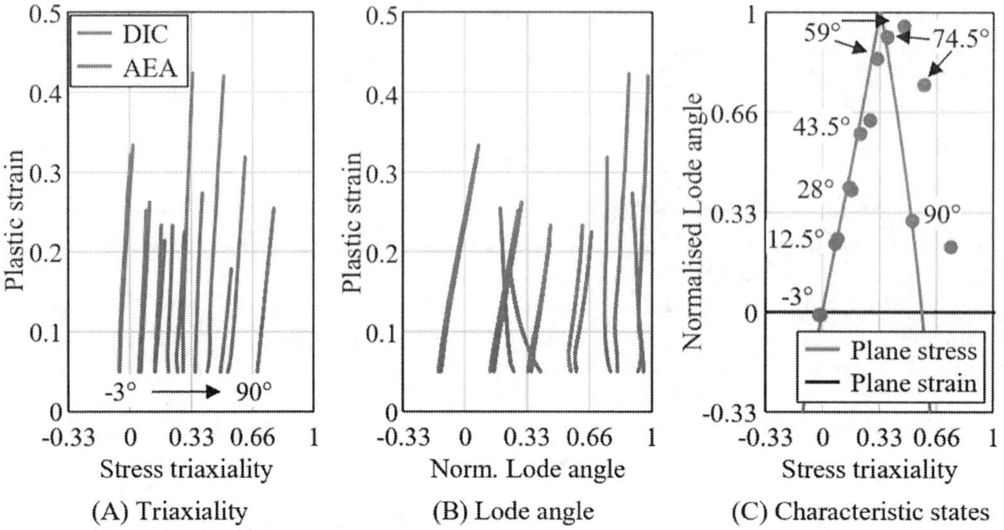

Figure 5. Plastic strain over stress triaxiality (A) as well as over normalised Lode angle development (B) and characteristic stress states (C) for the DIC as well as the AEA evaluation method.

To achieve the parameter sets for the parametrisation of the MMC fracture models, first the characteristic stress states were determined. For this reason, the area-weighted centroids of the strain-stress triaxiality curves were calculated as in [14] and the area-weighted centroids of the plastic strain-normalised Lode Angle curves as in [15]. The area-weighted centroids are displayed in figure 5 (C) for the seven loading angles and both evaluation methods in a stress triaxiality-normalised Lode angle plane. Here, the differences in the stress states of both evaluation methods for loading angles above 28° are clearly recognisable. The stress state of the AEA evaluation method differs significantly more from the plane stress state than of the DIC evaluation method beginning at the loading angle of 43.5° and is shifted to higher stress triaxialities. Finally, to complete the parameter sets for the parametrisation, the maxima of the plastic strain curves were defined as the plastic strains at fracture.

3.2. Parametrisation and evaluation of the MMC fracture models

Using the parameter sets plastic strain at fracture, characteristic stress triaxiality as well as characteristic normalised Lode Angle of the DIC and AEA evaluation methods, the parameters of the MMC fracture model were determined with the method of least squares. The parameters of the MMC fracture models for each evaluation method are summarised in table 2.

Table 2. Parameters of the MMC fracture models for both evaluation methods.

Parameter	c_1	c_2	n
DIC	0.1099	2.0074	0.1298
AEA	0.0161	2.4112	0.2373

A 3D plot of the fracture surfaces is displayed in figure 6 (A) and (B) for both evaluation methods. Here, the influence of the stress state on the parametrisation is clearly visible. The slope of the DIC

MMC fracture model is much lower than the slope of the AEA MMC fracture model. For a better comparability, the plane stress (C) and plane strain state (D) with the fracture parameter set is shown for both evaluation methods in figure 6. Although the fracture parameter sets of 43.5°, 59°, 74.5° and 90° from the AEA evaluation method are not located on the plane stress function, those parameter sets are also plotted for completeness.

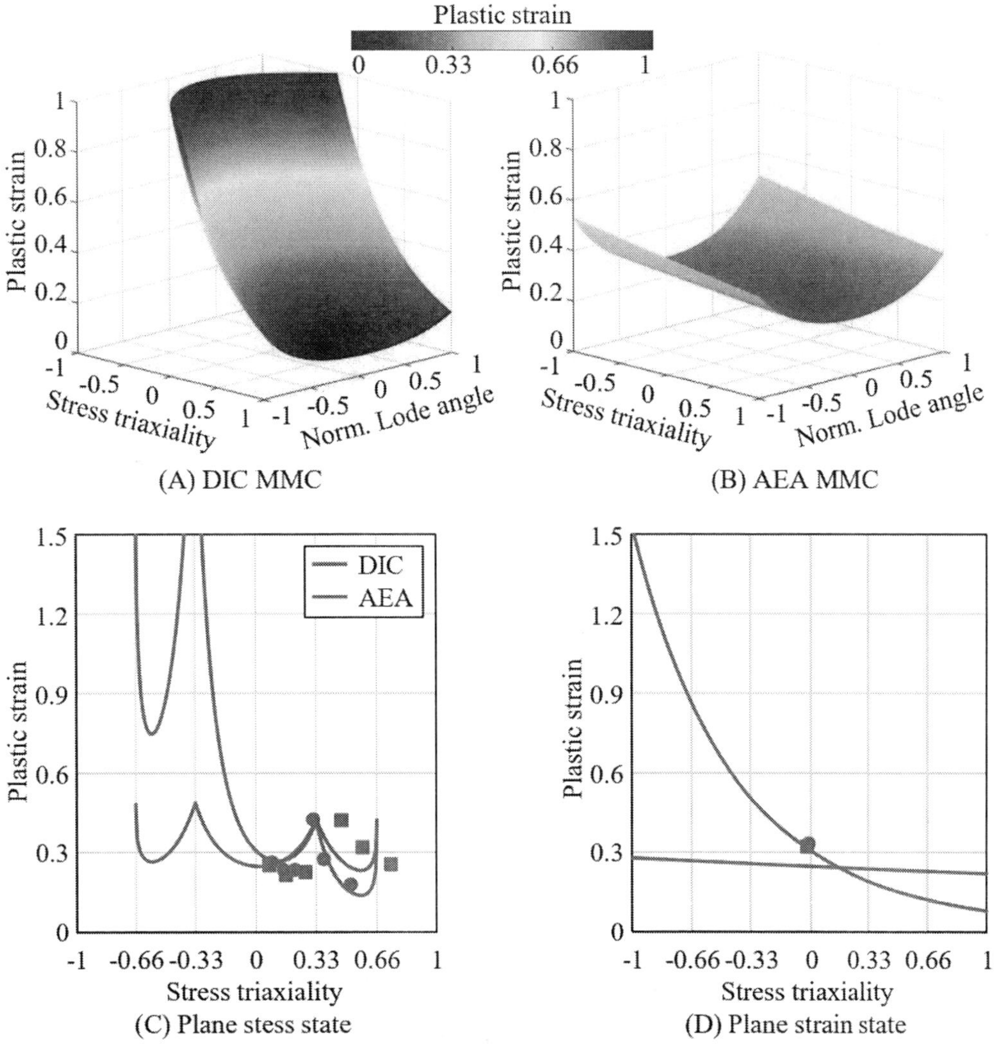

Figure 6. 3D plot of the DIC (A) as well as the AEA MMC fracture model (B) and plane stress (C) as well as plane strain states (D) of the MMC fracture models.

In general, the AEA MMC fracture model proceeds much flatter than the DIC MMC fracture model in both stress states. Therefore, the DIC MMC fracture model runs well above the AEA MMC fracture model up to a stress triaxiality of approximately 0.1. In the further course the DIC MMC fracture model is below the AEA MMC fracture model. The reason for these progressions of the models is caused by two reasons. While the plastic strain for the loading angle of -3° is similar for both evaluation methods, the difference is significantly higher for the loading angle of -90°. This leads to the different slopes in both fracture models. Through the shift of the characteristic stress triaxialities of the AEA evaluation method to higher stress triaxialities, the effect of the plastic strain on the lower slope of the AEA MMC fracture model is even increased.

To compare the impact of the MMC fracture models, the damage variable was evaluated for both B-pillar simulations, at the point in time when the damage variable reached a value of one. This is at a punch displacement of 21.33 mm for the DIC and 26.24 mm for the AEA MMC fracture model. The punch displacement and therefore the drawing depth of the part is much smaller for the B-pillar using the DIC MMC fracture model than of the AEA model. This is because the AEA MMC fracture model is higher than the DIC MMC fracture model for the stress states occurring in the B-pillar and especially for the critical stress states such as plane strain tension. The contour plot of the damage variable is illustrated in figure 7 for both models.

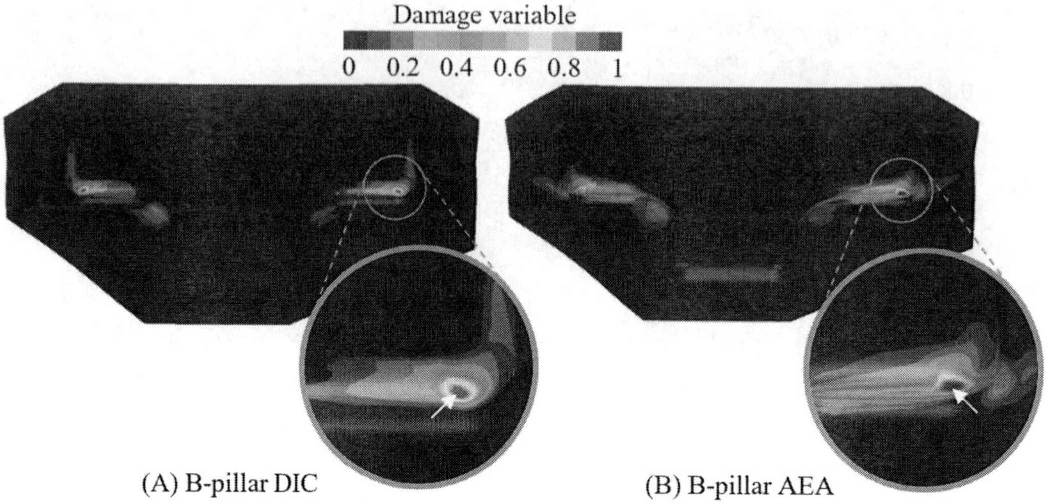

Figure 7. Contour plot of the damage variable for the B-pillar simulation using the DIC (A) and the AEA MMC fracture model (B).

It is clearly recognisable, that the damage is accumulated more locally when creating a high gradient of the damage variable using the DIC MMC fracture model, while it is spread broader in the simulation using the AEA MMC fracture model. This can be well explained by the different slopes of the two fracture models. Since the slope of the DIC MMC fracture model is much steeper than of the AEA MMC fracture model, the material fracture is reached rather for a certain stress state. Due to a flat course of the AEA MMC fracture model, fracture can be reached for more stress states at a similar point in time.

To further evaluate the fracture behaviour, the development of the plastic strain, stress triaxiality and normalised Lode angle was analysed for the failed element as indicated in figure 7 and plotted in figure 8. In figure 8 (A) the stress state of the failed element is shown for the B-pillar simulation using the DIC and the AEA MMC fracture model. The stress state proceeds along the plane stress state for both models and both elements fail at stress triaxialities of approximately 0.5 as well as at normalised Lode angles of 0.1 to 0.3, hence at a stress state between uniaxial and plane strain tension. The reason for the lower drawing depth of the part using the DIC MMC fracture model can be seen well in the plastic strain development in figure 8 (B). Because of the steep slope of the DIC MMC fracture model, the material fracture is reached much earlier than of the AEA model, although the stress state development of the failed elements is very similar. So far, first forming experiments with the B-pillar tool were performed, at which the punch displacement at fracture showed values above 26 mm. Therefore, it already can be concluded that the AEA MMC fracture model leads to a more precise fracture prediction. Nevertheless, this assumption needs to be validated with more experimental investigations, which are planned in the near future.

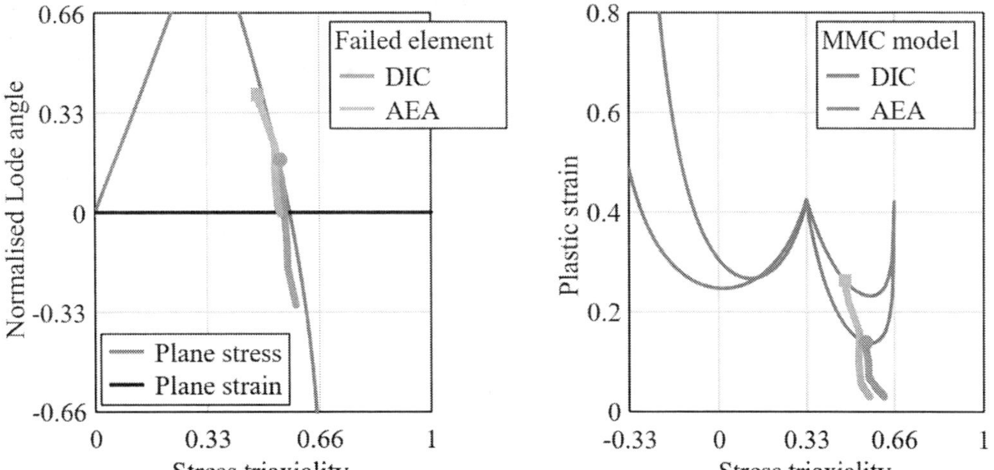

(A) Stress state (B) Plane stess state

Figure 8. Development of the stress state (A) and plane stress state (B) for the failed element of the B-pillar simulation using DIC and AEA MMC fracture models.

4. Summary

To characterise the material fracture behaviour for numerical process design, tensile tests are often performed and analytical fracture models are parametrised based on experimental data. For this purpose, the tests are usually monitored using DIC and the onset of fracture is determined optically by a macroscopic crack on the specimen surface. To create data for parametrisation, the plastic strain before fracture at the specimen surface is used. However, for some materials and specimen geometries the material fracture initiates already inside the specimen before a macroscopic crack is visible on the specimen's surface. Therefore, in this research butterfly tests with specimens in different loading angles were performed and monitored with DIC as well as AEA systems. The fracture of the specimen was determined based on each evaluation method and different fracture displacements were ascertained. The obtained displacements at fracture were used as boundary conditions in FE models of the butterfly tests. First, the numerical models of the butterfly tests were compared to the experiments. A validation was carried out by comparison of experimental and numerical force displacement curves. This showed a good agreement and thus, the numerical models were considered to be suited for further evaluation of the fracture parameter sets. Subsequently, the fracture parameter sets consisting of the plastic strain at fracture, characteristic stress triaxiality and characteristic normalised Lode angle were evaluated for the DIC and AEA evaluation method. It was found that the evaluation of the parameter sets inside the specimen with the AEA evaluation method resulted in a variance from the plane stress state compared to the DIC evaluation method. Finally, two MMC fracture models were parametrised and used in two simulations of a B-pillar forming process. The AEA MMC fracture model showed a difference regarding the onset of fracture and the distribution of the damage variable compared to the DIC MMC fracture model. First experimental forming tests showed a higher agreement of the AEA MMC fracture model for the prediction of material fracture, which will be analysed by more experiments in the future.

Acknowledgement

The authors are much obliged to the German Research Foundation (Deutsche Forschungsgemeinschaft, DFG) for the financial support of project 385276585. Furthermore, the authors would like to thank voestalpine Stahl GmbH for providing the material under investigation.

References

[1] Kohl D and Merklein M 2021 Alternative characterization method for the failure behavior of sheet metals derived from Nakajima test *IOP Conf. Ser.: Mater. Sci. Eng.* **1157** 012046 doi.org/10.1088/1757-899X/1157/1/012046

[2] Rice JR and Tracey DM 1969 On the ductile enlargement of voids in triaxial stress fields *J. Mech. Phys. Sol.* **17(3)** pp 201-217 doi.org/10.1016/0022-5096(69)90033-7

[3] Chuluunbat T, Lu C, Kostryzhev A and Tieu K 2015 Investigation of X70 line pipe steel fracture during single edge-notched tensile testing using acoustic emission monitoring *Mater. Sci. Eng. A* **640** pp 471-479 doi.org/10.1016/j.msea.2015.06.030

[4] Behrens B-A, Rosenbusch D, Wester H and Althaus P 2022 Comparison of three different ductile damage models for deep drawing simulation of high-strength steels *IOP Conf. Ser.: Mater. Sci. Eng.* **1238** 012021 doi.org/10.1088/1757-899X/1238/1/012021

[5] Bai Y and Wierzbicki T 2008 A new model of metal plasticity and fracture with pressure and Lode dependence *Int. J. Plas.* **24(6)** pp 1071-1096 doi.org/10.1016/j.ijplas. 2007.09.004

[6] Stockburger E, Vogt H, Wester H, Hübner S and Behrens B-A 2023 Evaluating Material Failure of AHSS Using Acoustic Emission Analysis *Mat. Res. Proc.* **25** doi.org/10.21741/9781644902417-47

[7] ISO 12004-2:2021 *Metallic materials - Determination of forming-limit curves for sheet and strip Part 2: Determination of forming limit curves in the laboratory* 2021 Beuth Verlag GmbH doi.org/10.31030/3212338

[8] Stockburger E, Wester H and Behrens B-A 2023 Failure modelling of CP800 using acoustic emission analysis *Appl. Sci.* **13** 4067 doi.org/10.3390/app13064067

[9] Bai Y and Wierzbicki T 2010 Application of extended Mohr-Coulomb criterion to ductile fracture *Int. J. Fract.* **161(1)** doi.org/10.1007/s10704-009-9422-8

[10] Barlat F, Brem JC, Yoon JW, Chung K, Dick RE, Lege DJ, Pourboghrat F, Choi SH and Chu E 2003 Plane Stress Yield Function for Aluminum Alloy Sheets - Part 1: Theory *Int. J. Plast.* **19** doi.org/10.1016/S0749-6419(02)00019-0

[11] Hockett JE and Sherby OD 1975 Large strain deformation of polycrystalline metals at low homologous temperatures *J. Mech. Phys. Solids* **23(2)** pp 87-98 doi.org/10.1016/0022-5096(75)90018-6

[12] Behrens B-A, Brunotte K, Wester H and Dykiert M 2021 Fracture characterisation by butterfly-tests and damage modelling of advanced high strength steels *Key Eng. Mat.* **883** pp 294-302 doi.org/10.4028/www.scientific.net/kem.883.294

[13] Grüner M, Merklein M 2014 Determination of friction coefficients in deep drawing by modification of Siebel's formula for calculation of ideal drawing force *Prod. Eng. Res. Devel.* **8** pp 577-584 https://doi.org/10.1007/s11740-014-0551-1

[14] Bao Y and Wierzbicki T 2004 On fracture locus in the equivalent strain and stress triaxiality space *Int. J. Mech. Sci.* **46(1)** pp 81-98 doi.org/10.1016/j.ijmecsci.2004.02.006

[15] Cao T-S, Gaillac A, Montmitonnet P and Bouchard P-O 2013 Identification methodology and comparison of phenomenological ductile damage models via hybrid numerical-experimental analysis of fracture experiments conducted on a zirconium alloy *Int. J. Sol. and Struc.* **50(24)** pp 3984-3999 doi.org/10.1016/j.ijsolstr.2013.08.011

Fracture toughness to assess the effect of trimming on the fatigue behaviour of high-strength steels for chassis parts

S Parareda[1], D Frómeta[1], D Casellas[1,2], D Gustafsson[2], E Olsson[2], R Munier[3], A Mateo[4]

[1] Eurecat, Centre Tecnològic de Catalunya, Unit of Metallic and Ceramic Materials, 08243 Manresa, Spain
[2] Division of Solid Mechanics, Luleå University of Technology, 971 87 Luleå, Sweden
[3] ArcelorMittal Maizières Research & Development, BP 30320, F-57283 Maizières-les-Metz Cedex, France
[4] CIEFMA – Department of Materials Science and Engineering, EEBE, Universitat Politècnica de Catalunya-BarcelonaTech, 08019 Barcelona, Spain

Email: sergi.parareda@eurecat.org

Abstract. High-strength steels are widely used in vehicle body-in-white, offering a good balance between crashworthiness and lightweight design. The increased requirements of heavier electric vehicles, in terms of fatigue resistance and crashworthiness, highlight that chassis parts have remarkable lightweighting potential. However, applying these grades in chassis parts is not straightforward, as the forming processes, like trimming, may introduce surface defects that compromise the fatigue resistance of the component. This work presents a material selection strategy for the applicability of high-strength steels in chassis parts of electric vehicles. The proposed approach allows the evaluation of the key parameters of the chassis parts in a simple way. The crash performance is evaluated through fracture toughness using the essential work of fracture (EWF) methodology. The method is applied to thin high-strength steel sheets employing double-edge notched tensile specimens (DENT). On the other hand, fatigue performance is investigated in terms of fatigue resistance for both notched and unnotched specimens. The results for different complex-phase and dual-phase steels show a good agreement between the EWF and the fatigue notch factor. The method could help apply high-strength steel to chassis parts, as designers will have a tool to focus the expensive fatigue tests on the best material candidates.

1. Introduction

Using high-strength steels in car body-in-white (BiW) is a well-established strategy for lightweighting. Although the utilization of these materials has been focused on the BiW, they can also be used in chassis parts, which represent over 20% of the structural weight of the car. This approach becomes more and more relevant in the battery electric vehicles, as the heavy batteries increase their structural requirements. Besides the required exceptional crashworthiness behaviour, the chassis parts must deal with cyclic loads that can compromise the structural integrity of the vehicle. An improved fatigue resistance will help to reduce the weight of the vehicle by gauge reduction, which also contributes to the use of less material. This goal should be easily achieved by using high-strength steels with a yield strength close to (or even above) 1000 MPa, as it is well-accepted that an increasing linear trend exists between yield strength and fatigue resistance. However, when a defect is present in the material, this linear relationship is not valid anymore, and an increase in fatigue resistance is harder to obtain.

Fatigue resistance is a local plastic deformation occurring at high stressed points of the material, either in an intrinsic defect, as an inclusion or discontinuity in the microstructure, or an extrinsic defect, as a notch in the surface. When dealing with chassis parts, the extrinsic defects are related to the manufacturing operations like trimming or punching. It has been shown that the edge morphology depends on the material, shearing clearance and punch geometry, including the wear of the tool [1]. The edge geometry becomes crucial when dealing with local formability, as a lousy edge quality could lead to edge-cracking issues. These barriers were addressed through new approaches, such as fracture toughness to understand better the fracture behaviour of new materials and propose solutions to avoid fracture issues during part forming or develop materials with high cracking resistance. Previous works showed that tougher materials are more tolerant to the edge damage induced in sharing [2]. Additionally, improving the edge quality is beneficial for local formability, as shown in the hole expansion tests [3]. Similarly, improving the edge geometries by reducing the fracture zone or the burr is proven to be effective in enhancing fatigue resistance. A clear example is the utilization of the double-shearing strategy [4]. In this case, the consecutive shearing process removes a thin layer of damaged material, leaving a smoother edge with less damage than after the first cutting operation. Post-forming operations, like sandblasting or shot peening, also improve surface quality by smoothing the edge surface, rounding the edge corners, and introducing compressive residual stresses [5]. The coining processes that plastically deform the edge surface can also introduce these beneficial residual stresses. Although the mentioned operations can improve fatigue resistance, fatigue strength is still reduced because of the shearing process, which becomes critical in materials with high tensile strength.

Many factors affect the fatigue resistance of sheared edges, so developing solutions to improve the fatigue behaviour is time-consuming. Accordingly, some works addressed the development of rapid testing methodologies to shorten the testing time. Regarding the mechanical behaviour of sheared edges, the use of the material property that informs about damage tolerance, i.e. the fracture toughness, could bring valuable information to understand their fatigue resistance. Thus, this work proposes to combine the rapid testing of the fatigue limit in smooth specimens with fracture toughness values to estimate the fatigue notch factor of high-strength steels. The correlation between these properties could become a powerful material selection tool for fatigue applications of chassis parts with sheared or trimmed edges.

2. Materials and experimental procedure

The three materials studied in this work belong to the Advanced High Strength Steels (AHSS) family. The materials supplied by ArcelorMittal are two Complex Phase (CP) and a Dual Phase (DP) steel. HR CP800 SF is a hot rolled (HR) steel with a microstructure composed of bainitic-ferritic matrix with martensite and austenite islands, specially designed for a very good Stretch Flangeability (SF). Higher tensile strength is achieved by the HR CP980 SF thanks to the complex microstructure composed of tempered martensite and ferrite, combined with upper bainite and retained austenite islands, as shown in figure 1. Both CP steels exhibit good formability, while the HR DP600 steel perfectly balances formability and strength by combining a soft ferrite matrix with martensite islands. This combination leads to an outstanding strain hardening and elongation at fracture. The rough chemical composition of the investigated materials is given in table 1.

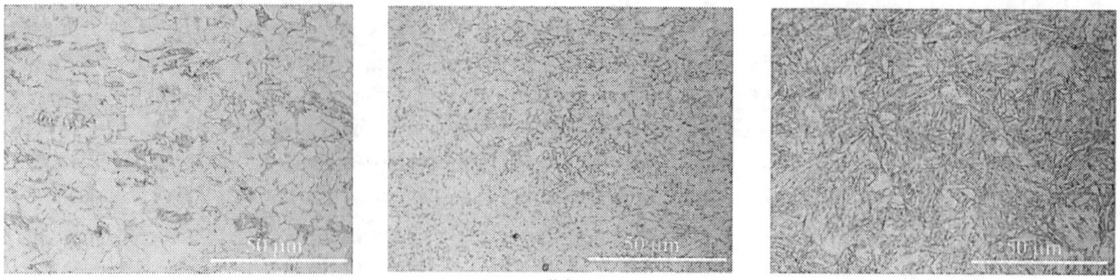

(a) (b) (c)

Figure 1. Microstructure of: a) HR DP600, b) HR CP800 SF, and c) HR CP980 SF.

Table 1. Chemical composition of the investigated materials in weight percentage. Balance is Fe.

Steel grade	C	Si	Mn	Cr	Al
HR DP600	≤ 0.10	≤ 0.30	≤ 1.50	≤ 0.90	≤ 0.06
HR CP800 SF	≤ 0.18	≤ 1.00	≤ 2.20	≤ 1.00	0.015-1.2
HR CP980 SF	≤ 0.20	≤ 1.00	≤ 2.20	≤ 1.00	0.015-1.2

Conventional tensile tests were carried out in an universal testing machine equipped with an extensometer and according to ISO6892-1 standard. The tensile specimens were machined at 90° concerning the rolling direction (RD). The specimen type 2 geometry from ISO 6892 with an L_0 of 80 mm was used. The tensile results are summarized in table 2.

Table 2. Mechanical properties in terms of yield strength (σ_{YS}), ultimate tensile strength (σ_{UTS}) and elongation at fracture (A_{80}, gauge length 80 mm).

Steel grade	t [mm]	σ_{YS} [MPa]	σ_{UTS} [MPa]	A_{80} [%]
HR DP600	4.3	439	622	21
HR CP800 SF	3.4	778	835	17
HR CP980 SF	3.5	957	1034	13

2.1. *Fatigue tests*

Fatigue behaviour was evaluated using hourglass specimens following the ISO 1099 standard. Specimens with a radius of 100 mm were machined by spark erosion at 90° concerning the RD and edge polished to a specular finish or trimmed. The gap between the punching and the die in the trimming tool was 0.3 mm, which means a trimming clearance of 7% for the DP600, 8.8% for the HR CP800 SF and 8.6% for the HRCP980 SF. This range of clearances is commonly used in the chassis parts. A drawing of the specimen geometry is shown in figure 2. The fatigue limit (σ_f) was defined at 10^6 cycles, $2 \cdot 10^6$ cycles. It was evaluated using the staircase or up-and-down method proposed by Dixon-Mood [6] and the *stiffness method* for the polished and trimmed conditions. At least 15 specimens were tested at room temperature for the conventional fatigue test, using a stress ratio of R = 0.1 and a frequency of 30 Hz. The obtained fatigue limit will be compared with results obtained by the rapid testing approach to check its accuracy.

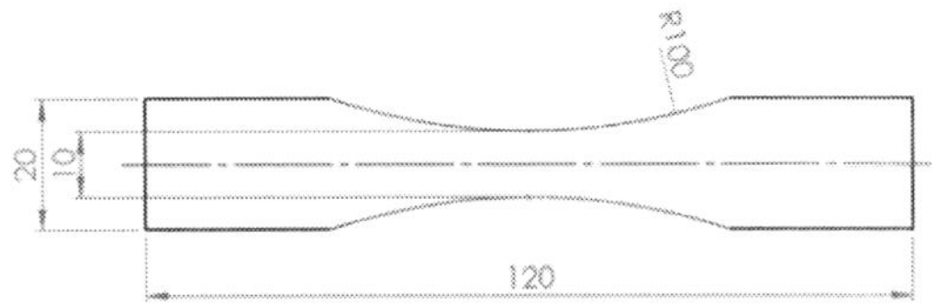

Figure 2. Hourglass fatigue specimen (K_t = 1.03) with the polished edge (in red colour). Dimensions are expressed in mm.

The *stiffness method* relies on the measurement of the fatigue damage through the determination of the specimen inelastic strains ($\Delta\varepsilon_p$) after each successive fatigue block [7]. The $\Delta\varepsilon_p$ is then plotted against the maximum stress (σ_{max}) of each block. The σ_f was determined as the interception of the fitting line to the x-axis, as schematically shown in figure 3. The length of the fatigue blocks was 6000 cycles, with increasing stress of 25 MPa between each one. For the $\Delta\varepsilon_p$ determination, the stress was set at $\sigma_{YS}/2$. At least three specimens were tested for each material and condition, all at room temperature and using a stress ratio of R = 0.1. Tests were performed in a servo-hydraulic testing machine MTS 322 Test Frame, at a frequency of 30 Hz, equipped with a Digital Image Correlation (DIC) system GOM Aramis SRX for $\Delta\varepsilon_p$ measurements using a virtual extensometer with a length of 40 mm.

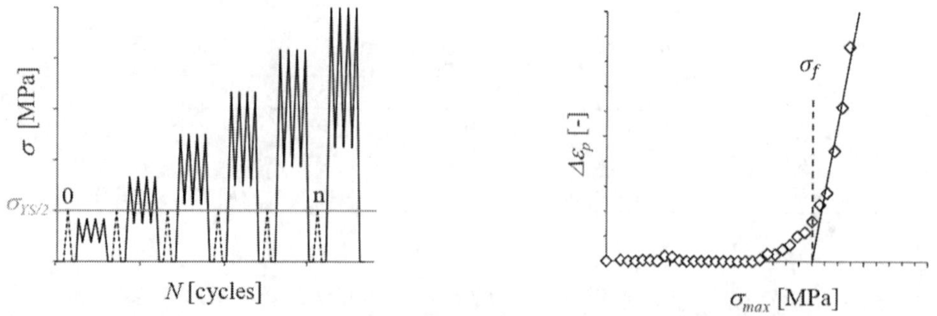

Figure 3. Schematic representation of the stepwise loading test and inelastic strain range versus maximum stress to determine the fatigue limit.

After the fatigue test, the total strain evolution rate ($d\Delta\varepsilon_t/dN$) is analyzed to identify the coalescence of microcracks to a macrocrack. Only the $\Delta\varepsilon_p$ prior to the macrocrack coalescence is used to determine the fatigue damage (D) according to:

$$D = \frac{\Delta\varepsilon_p - \Delta\varepsilon_{p0}}{\Delta\varepsilon_{pf} - \Delta\varepsilon_{p0}} \quad (1)$$

where $\Delta\varepsilon_p$ is the current plastic strain measured for each fatigue block, $\Delta\varepsilon_{p0}$ is the initial inelastic strain, and $\Delta\varepsilon_{pf}$ corresponds to microcrack coalescence inelastic strain.

2.2. *Fracture toughness tests*

Fracture toughness was evaluated through the Essential Work of Fracture (EWF) methodology based on Elasto-Plastic Fracture Mechanics (EPFM). The method, described in the CWA protocol [8], consists of loading in tension a range of Double Edge Notched Tensile (DENT) specimens with different ligament lengths (l_0). The specimens shown in figure 4 were manufactured by wire erosion and fatigue pre-cracked in a resonance testing machine Rumul Testronic. By testing those specimens up to fracture, it is possible to determine the specific essential work of fracture (w_e) by plotting the specific total work of fracture (w_f) against l_0 as represented in figure 4. The w_f is obtained by integrating the area under the load vs displacement curve for each tested specimen (W_f) and dividing it by the fractured area (l_0t). Equation 2 defines w_e as the intercept and the specific non-essential work of fracture (w_pb) as the slope. Besides separating the w_f the method also allows the separation of the energy required for crack initiation and propagation. The specific work for fracture initiation (w_e^i) is independent of the ligament length. It is determined as the mean value of the specific total work of fracture to crack onset (w_f^i), which is used to define the w_e^i as illustrated in figure 4. The measured w_e and w_e^i are not intrinsic material properties, as they depend on the sheet thickness.

$$\frac{W_f}{l_0 t_0} = w_f = w_e + w_p b l_0 \quad (2)$$

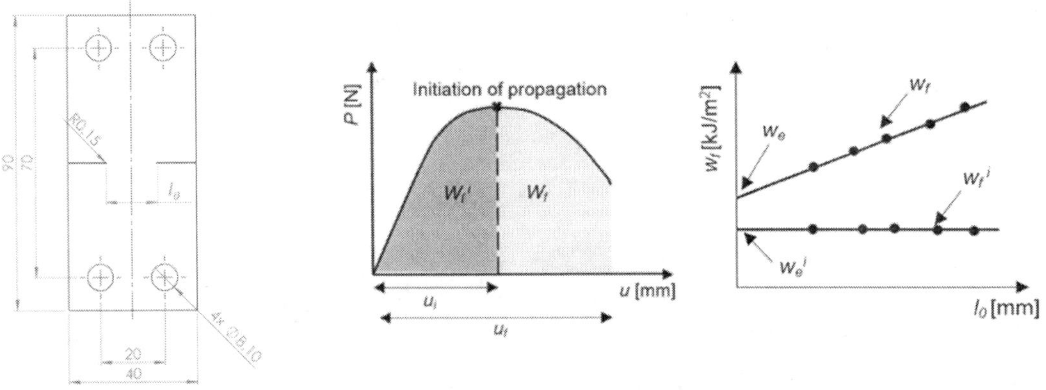

Figure 4. DENT specimen used to determine the EWF (l_0 refers to the different ligament lengths): Schematic representation of the determination of W_f and W_f^i for a given ligament length. Plot of w_f vs l_0, the intercept gives the specific work of fracture, w_e.

At least ten specimens per material were tested in a 250 kN universal testing machine Instron 5585H, equipped with a digital video extensometer. The crosshead testing speed was set at 1 mm/min. The load-displacement measurements were carried out by a video extensometer using initial marks separated by 50 mm. The crack initiation was detected using a high-resolution video camera synchronized with the testing machine. The w_f^i was determined for at least one specimen of each ligament length, and the w_e^i was determined as the mean value.

3. Results and discussion

Figure 5 shows a cross-section of the trimmed specimens for each steel. Sheared edges exhibit the four well-known regions obtained by a shearing process [9]: rollover, smooth or burnish zone, fracture zone and burr (Figure 5). Although the trimming clearance was almost the same, the shape of the trimmed edges was slightly different. The HR CP980 SF presents a longer burnish zone and less rollover than HR CP800 SF and HR DP600.

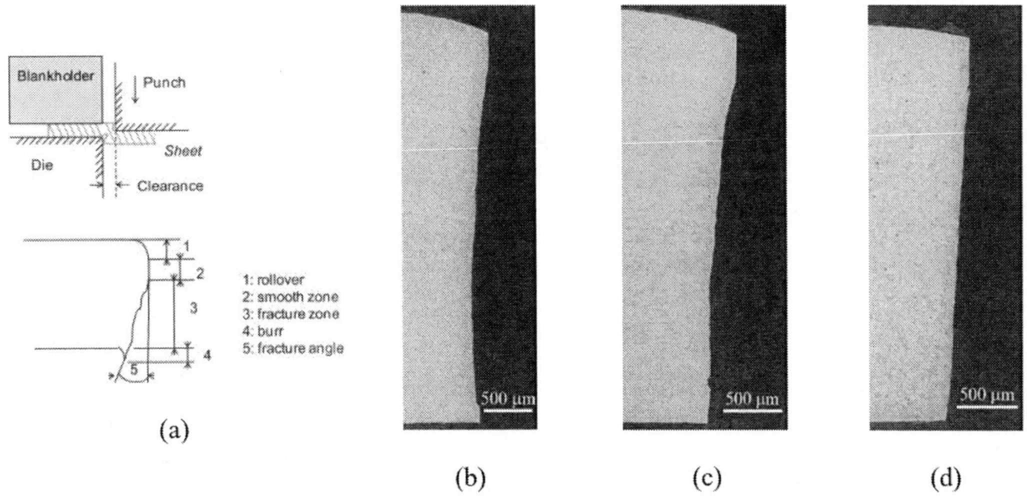

Figure 5. a) Sheared edge profile characteristics and analysis for: b) HR DP600, c) HR CP800 SF, and d) HR CP980SF.

These differences in the shear edge, related to the material properties, may affect the fatigue resistance as the burr, the defects in the fracture zone or the transition between the burnish to fracture

IOP Conf. Series: Materials Science and Engineering 1284 (2023) 012073 doi:10.1088/1757-899X/1284/1/012073

zone act as stress concentration points [4]. Such locally higher stresses induce the nucleation of cracks that propagate throughout the specimen, as observed in figure 6. The fatigue resistance of the polished specimens is driven by the defects or roughness of the as-rolled surface, while the defects in the trimmed edge govern the fatigue behaviour of trimmed specimens. Although the trimmed edge morphology and the resulting defects are similar in all the investigated steels, their effect on the fatigue resistance is different, as observed in the *stiffness test* data shown in figure 7. As expected, the polished conditions can withstand higher stress, i.e. more fatigue blocks, during the rapid fatigue tests than the trimmed specimens.

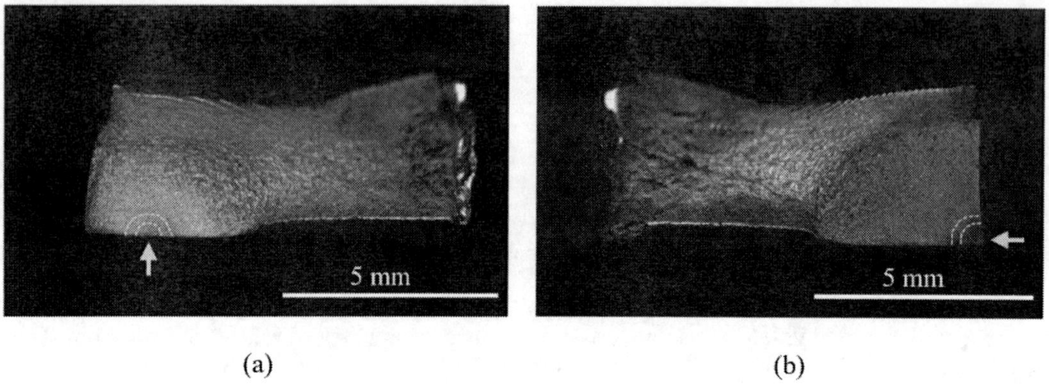

Figure 6. Fatigue crack initiation and propagation for the HR DP600 in: a) polished, and b) trimmed specimens.

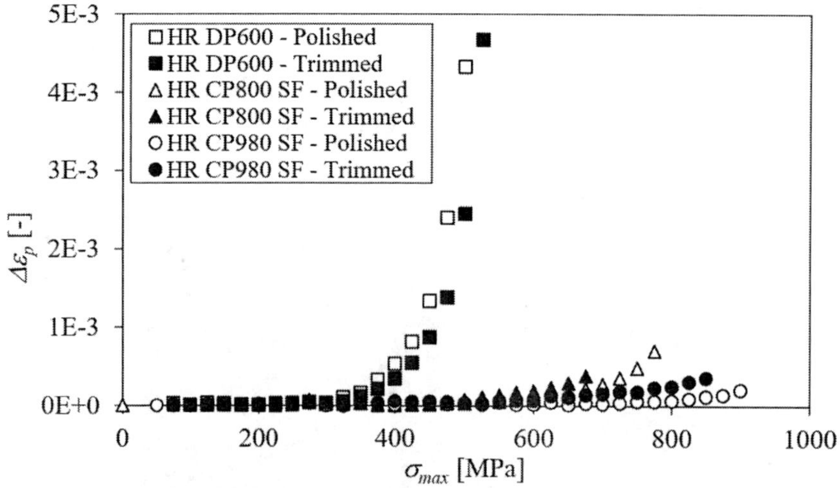

Figure 7. Inelastic strain range versus maximum stress curves from the stiffness method for the polished and trimmed conditions and investigated materials.

As summarized in table 3, the HR CP980 SF exhibits the best fatigue resistance for the polished condition, while the HR DP 600 shows the lowest one. The results obtained by the conventional staircase tests (σ_f) and *stiffness method* ($\sigma_{f\text{-}SM}$) are in good agreement for all the conditions. However, the fatigue resistance of trimmed specimens clearly decreases with respect to the polished conditions in both CP steels compared with the HR DP600. This behaviour is well described by the fatigue strength reduction factor (k_f), and calculated as the ratio between the fatigue resistance of the sheared to polished edges at 10^6 cycles:

$$k_f = \frac{\sigma_{f-Trimmed}}{\sigma_{f-Polished}} \tag{3}$$

Traditionally, it is accepted that σ_f increases by increasing the tensile strength (σ_{UTS}), as shown in figure 8a. However, although it is commonly known that high-strength steels are sensitive to pre-existing defects [10], this relationship does not give enough information for material selection when dealing with high-strength materials and defects. Instead, this information is found in the ability of the material to resist the propagation of damage, as the defects in the trimmed edges in the present work. This behaviour is rationalized by damage tolerance, which is closely linked to fracture toughness, i.e. a tough material better tolerates the pre-existing defects, while a material with a lower toughness is more sensitive to them. Damage tolerance has been conventionally related to σ_{UTS} and ductility, but in high-strength materials, ductility and fracture toughness do not show the same trend [2,11]. In AHSS, fracture toughness, evaluated in the frame of fracture mechanics, better represents the resistance of pre-existent defects. Similarly, fracture toughness expressed in terms of w_e^i (crack propagation resistance at crack initiation) and w_e (overall fracture toughness) can be used to assess the fatigue behaviour of pre-existing defects as it shows a good linear trend with the k_f factor (Figure 8b). Although only material data for the three investigated materials is available the observed trend path the way to further investigations with more materials of different microstructures, strengths, and thicknesses. Materials with low sensitivity to surface defects, i.e. with high damage tolerance, present a k_f close to 1 and have a high w_e value. On the contrary, materials with lower w_e are more sensitive to surface defects, i.e. poor damage tolerance, as the k_f is far below 1. The same trend is observed by plotting the k_f against the w_e^i. Both w_e^i and w_e are good indicators of damage tolerance for AHSS sheets.

Such a good correlation between the fatigue notch sensitivity, in terms of k_f, and the fracture toughness, in terms of w_e and w_e^i, means that the effect of trimming on fatigue can be readily determined by experimentally simple tests, i.e. with a fatigue test in the polished condition and the EWF.

Table 3. Fatigue and fracture toughness results for the three investigated materials: fatigue limit by the conventional tests (σ_f) and from the stiffness method ($\sigma_{f\text{-}SM}$) at R = 0.1 for each edge condition, fatigue strength reduction factor (k_f), fracture toughness (w_e) and fracture toughness at crack initiation (w_e^i).

Steel grade	Edge condition	σ_f [MPa]	$\sigma_{f\text{-}SM}$ [MPa]	k_f	w_e^i [kJ/m²]	w_e [kJ/m²]
HR DP600	Polished	502 ± 5	465 ± 10	0.98	285 ± 29	481 ± 67
	Trimmed, Cl = 7%	470 ± 5	454 ± 20			
HR CP800 SF	Polished	622 ± 9	619 ± 16	0.73	155 ± 28	299 ± 56
	Trimmed, Cl = 8.8%	426 ± 19	449 ± 11			
HR CP980 SF	Polished	643 ± 30	678 ± 14	0.68	107 ± 18	141 ± 45
	Trimmed, Cl = 8.6%	429 ± 11	464 ± 67			

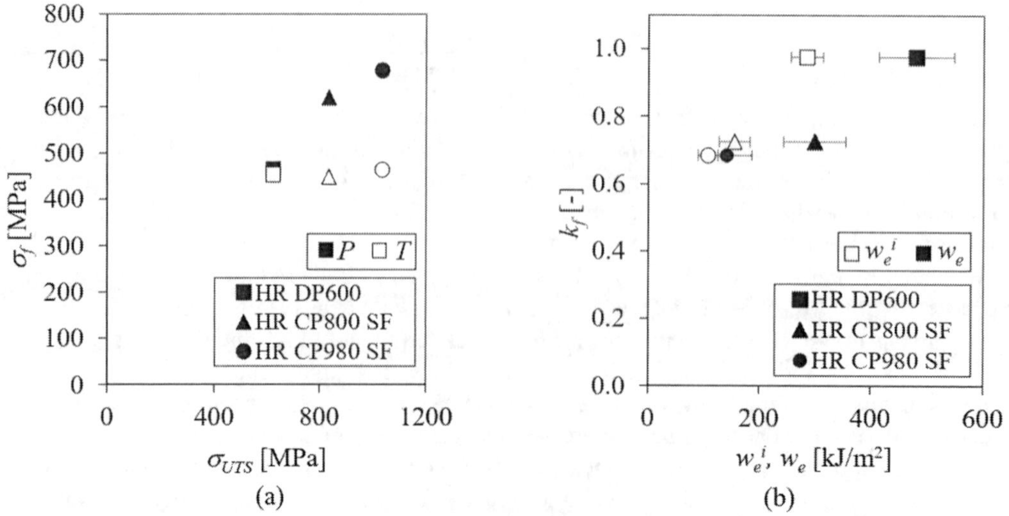

Figure 8. a) Fatigue resistance of polished (P) and trimmed (T) specimens against tensile strength (σ_{UTS}) for the investigated AHSS between 3 to 4mm. b) Fatigue strength reduction factor (k_f) as a function of specific work for fracture initiation (w_e^i) and specific essential work of fracture (w_e).

4. Summary and conclusions

The results obtained in this work show that fracture toughness can be used to assess the fatigue strength reduction of AHSS. The following conclusions can be drawn:

- The fatigue resistance can be readily determined through the stiffness method, in a quick and reliable way, for both polished and sheared specimens.
- The fatigue strength reduction due to the sheared edge can be expressed as a fatigue strength reduction factor, k_f, which shows an excellent correlation to the fracture toughness measured in terms of the EWF. It means that tougher materials have lower fatigue notch sensitivity.
- The proposed approach allows to rapidly determine the influence of the trimmed edge on the fatigue resistance of AHSS through a rapid test in the polished conditions and the fracture toughness obtained with the EWF method.

Acknowledgements

The research leading to these results has received funding from the European Union's Horizon 2020 innovation action programme under grant agreement No 101006844 – Fatigue4Light project. This work was financially supported by the Catalan Government through the funding grant ACCIÓ-Eurecat (Project PRIV - RapidFat). A.M. acknowledges the Agència de Gestió d'Ajuts Universitaris i de Recerca (AGAUR) del Departament de Recerca i Universitats de la Generalitat de Catalunya for recognising the Microstructural Design and Advanced Manufacturing of Materials Group as a consolidated research group number 2021 SGR 01053.

References

[1] Han S, Chang Y, Wang C Y and Dong H 2022 *J. Mater. Process. Technol.* **299** 117377
[2] Frómeta D, Lara A, Grifé L, Dieudonné T, Dietsch P, Rehrl J, Suppan C, Casellas D and Calvo J 2021 *Metall. Mater. Trans. A Phys. Metall. Mater. Sci.* **52** 840–56
[3] Gläsner T, Sunderkötter C, Plath A, Bednarz M, Volk W, Hoffmann H and Golle R 2014 *Key Eng. Mater.* **611–612** 1294–307
[4] Stahl J, Pätzold I, Golle R, Sunderkötter C, Sieurin H and Volk W 2020 *J. Manuf. Mater. Process. 2020, Vol. 4, Page 52* **4** 52
[5] Parareda S, Lara A, Sieurin H, Darmas H and Casellas D 2018 *MATEC Web Conf.* **165** 1–6

[6] Dixon W J and Mood A M 1948 *J. Am. Stat. Assoc.* **43** 109–26
[7] Parareda S, Casellas D, Lara A and Mateo A 2022 *Int. J. Fatigue* **156** 106643
[8] EUROPEAN COMMITTEE FOR STANDARDIZATION 2021 *Test method for determination of the essential work of fracture of thin ductile metallic sheets* (Brussels: CEN)
[9] The Association of German Engineers 1994 *VDI 2906-2: Quality of cut faces of (sheet) metal parts after cutting, blanking, trimming or piercing; general introduction, characteristic values, materials* (Berlin: Beuth)
[10] Lara A, Picas I and Casellas D 2013 *J. Mater. Process. Technol.* **213** 1908–19
[11] Frómeta D, Parareda S, Lara A, Molas S, Casellas D, Jonsén P and Calvo J 2020 *Eng. Fract. Mech.* **229** 106949

IOP Conf. Series: Materials Science and Engineering 1284 (2023) 012074 doi:10.1088/1757-899X/1284/1/012074

Cruciform tension-shear test for sheet metal: Evaluation of methods for calculating plastic work

S Sato[1]*, M Kim[2], J Ha[3], Y P Korkolis[2], T Kuwabara[4]

[1] Department of Mechanical Systems Engineering, Institute of Engineering, Tokyo University of Agriculture and Technology, 2-24-16 Naka-cho, Koganei-shi, Tokyo 184-8588, Japan

[2] Integrated Systems Engineering, The Ohio State University, Columbus, OH, 43210, USA

[3] Mechanical Engineering, University of New Hampshire, Durham, NH, 03824, USA

[4] Division of Advanced Mechanical Systems Engineering, Institute of Engineering, Tokyo University of Agriculture and Technology, 2-24-16 Naka-cho, Koganei-shi, Tokyo 184-8588, Japan

*E-Mail: s229931u@st.go.tuat.ac.jp

Abstract. The tensile-shear test specimen devised by Kim et al. (Kim M, Ha J, Bonica S and Korkolis YP 2021 Proc. 13th Int. Conf. Technol. Plast. pp. 1961-1967) is used to evaluate methods for calculating the plastic work per unit volume of a metal sheet subjected to combined tension and shear. The deformation of the tensile-shear specimen is analysed using finite element analysis (FEA). In tension-shear deformation, a material element undergoes rigid body rotation as well as in-plane deformation. Therefore, methods for calculating the stress components and the incremental strain components with respect to the material coordinate system and the spatial coordinate system are discussed. It is confirmed that the plastic work per unit volume does not depend on the coordinate system adopted in the calculation as long as the same coordinate system is used to evaluate the stress components and strain increment components.

1. Introduction

In order to accurately predict and prevent the formation of defects during press forming, accurate sheet forming simulations are required. A material model that can reproduce the deformation behaviour of sheet metal under biaxial stress conditions is crucially important for accurately predicting forming defects such as wrinkling and fracture [1-3]. Biaxial tensile test methods typically use cruciform specimens [4, 5] and tubular specimens made of sheet metal [6-8] to determine the yield locus in the first quadrant of the $\sigma_{xx} - \sigma_{yy}$ stress space, where the x and y axes are along the rolling direction (RD) and the transverse direction (TD), respectively, of the sheet metal. Combined tension-shear (TS) testing methods [9, 10] are used to determine the yield locus in the $\sigma_{xx} - \sigma_{yy} - \sigma_{xy}$ stress space.

Recently, Kim et al. [11,12] developed a cruciform TS specimen to measure the flow stresses under combined tension and shear. The specimen is a combination of a general cruciform specimen [4, 5] and a simple shear specimen [13]. It has four arms, two of which are used to apply tension and the other two are used to apply shear. The dimensions of the gage-section are optimised to minimize the variation of normal and shear stresses in the test section and to find the difference with the average stresses (i.e., force divided by the current area) so that the measured forces can be used to calculate the stresses in experiments [11]. The optimised specimen design was tested using finite element analysis (FEA) with various force ratios to confirm its performance [12].

In a material testing experiment using a cruciform TS specimen, it is necessary to determine the contours of plastic work in the $\sigma_{xx} - \sigma_{yy} - \sigma_{xy}$ stress space and to derive an appropriate material model for the test sample. In order to calculate the plastic work from the stress-strain data obtained using the cruciform TS specimen, the stress and strain components should be defined with respect to the same coordinate system. However, the material coordinate system rotates with respect to the spatial coordinate system in a TS experiment. Moreover, the true stress components are defined with respect to the spatial coordinate system in the experiment, whereas the strain components are defined with respect to the material coordinate system when a digital image correlation (DIC) strain measurement system is used. Therefore, in order to correctly calculate the plastic work, the stress and/or strain components must be transformed considering the rotation of the material coordinate system. In this study, using FEA, we discuss the calculation of the plastic work performed by a material element subjected to tension and shear.

2. FEA of combined tension-shear material tested using cruciform TS specimen

Figure 1 shows the geometry and dimensions of the cruciform TS specimen used in this study. It was originally developed by Kim et al. [11]. The *x*, *y*, and *z* axes are along the horizontal, vertical, and thickness directions of the material, respectively. A biaxial tensile testing machine is used to simultaneously apply biaxial loads F_x and F_y in the *x* and *y* directions of the cruciform TS specimen, respectively. The size of the test section is 6.5 mm in the *y* direction and 0.5 mm in the *x* direction.

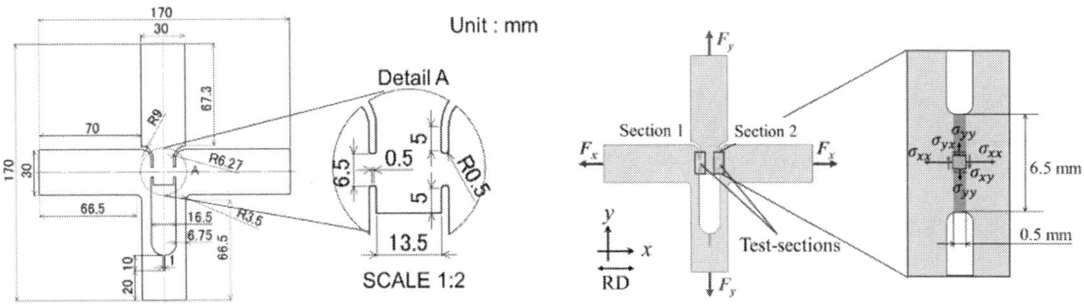

Figure 1. Cruciform TS specimen. (a) Geometry of test section and (b) stress components in test section.

The FEA of the cruciform TS specimen was performed using static implicit software Abaqus/ Standard 2017. Only one half of the specimen was analysed due to symmetry. The mesh elements for the test section had a size of 0.1 mm square, as shown in Figure 2. The rounded edge of the test section was divided into finer elements because of the stress concentration in that region. The sheet thickness was 1.0 mm. Five integration points were used in the thickness direction. The von Mises yield criterion was used as the basis of the material model and Swift's hardening law was used as the work hardening equation:

$$\sigma = 573(0.00171 + \varepsilon_0^{\mathrm{p}})^{0.277} \text{ (MPa)} \tag{1}$$

This equation was based on uniaxial tensile test data for a cold-rolled mild steel sheet.

The boundary conditions were as follows. The edge of the lower arm was fixed and that of the arm in the *x* direction was held such that there was no rotation around the *z* axis. F_x and F_y were applied to the edges of the arms in the *x* and *y* directions, respectively. The load applied ratios were $F_x : F_y = 1:0$, 1:1, 1:2, 1:4, and 0:1. The plastic work per unit volume was calculated for the element at the centre of the test section, as shown in Figure 2 (see inset).

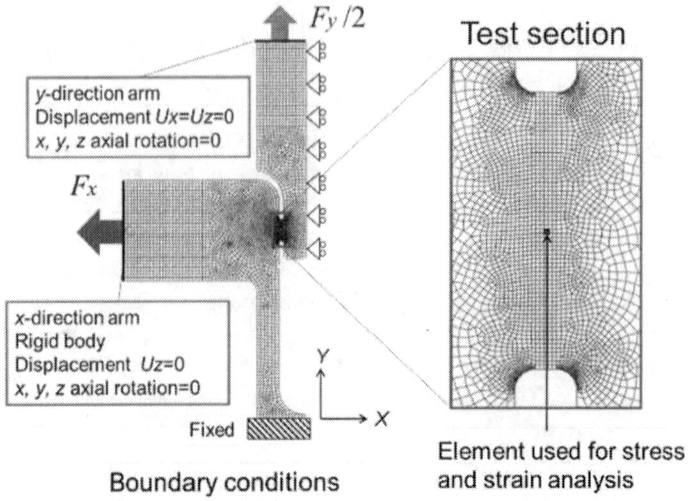

Figure 2. Mesh configuration and boundary conditions for FEA of cruciform TS specimen.

3. Calculation of plastic work

3.1. Calculation of strain and stress

The deformation gradient **F** was used to calculate the strain and stress. **F** is defined as $\mathrm{d}\mathbf{x} = \mathbf{F} \cdot \mathrm{d}\mathbf{X}$, where $\mathrm{d}\mathbf{X}$ and $\mathrm{d}\mathbf{x}$ are infinitesimal vectors in the undeformed and deformed configurations, respectively. First, we explain the calculation of **F**. Consider a unit element subjected to deformation in the *x*-*y* plane, as shown in Figure 3. The three nodes of the element are A, B, and C, and the in-plane displacements of each node in the x and y directions are $u_{\mathrm{I}x}$ and $u_{\mathrm{I}y}$ (I = A, B, and C), respectively. Then, the components of **F** are given by

$$\mathbf{F} = \begin{bmatrix} \dfrac{u_{\mathrm{B}x} - u_{\mathrm{A}x}}{\Delta X} + 1 & \dfrac{u_{\mathrm{C}x} - u_{\mathrm{A}x}}{\Delta Y} & 0 \\ \dfrac{u_{\mathrm{B}y} - u_{\mathrm{A}y}}{\Delta X} & \dfrac{u_{\mathrm{C}y} - u_{\mathrm{A}y}}{\Delta Y} + 1 & 0 \\ 0 & 0 & F_{33} \end{bmatrix} \tag{2}$$

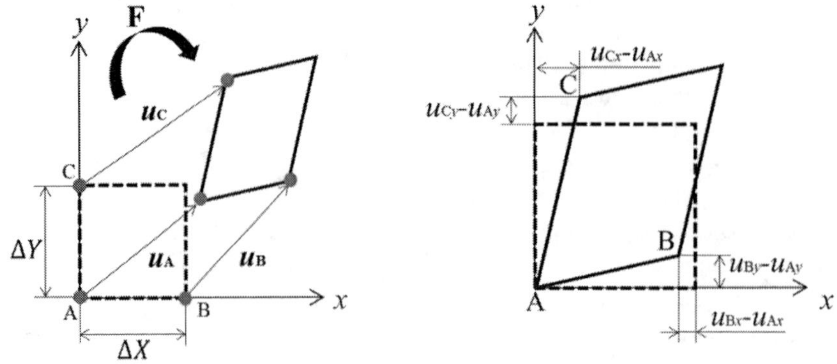

Figure 3. Schematic diagrams of calculation of components of **F** using displacement components.

In this equation, $F_{13} = F_{31} = F_{23} = F_{32} = 0$ because no shear deformation occurs in the z direction. F_{33} is calculated by assuming a constant volume (i.e., det $\mathbf{F} = 1$) as

$$F_{33} = \frac{1}{F_{11}F_{22} - F_{12}F_{21}} \tag{3}$$

$\mathbf{F}$ can be expressed as $\mathbf{F} = \mathbf{RU}$ using right-polar decomposition, where $\mathbf{U}$ and $\mathbf{R}$ are the right-stretch tensor and rotation tensor, respectively, and are given as

$$\mathbf{U} = \sqrt{\mathbf{F}^{\mathrm{T}}\mathbf{F}} \tag{4}$$

$$\mathbf{R} = \mathbf{F}\mathbf{U}^{-1} \tag{5}$$

$\mathbf{U}$ represents the elongation component on the material coordinate system (hereafter, referred to as the material frame, MF). The logarithmic strain $\boldsymbol{\varepsilon}'_{\mathrm{L}}$ with respect to the MF is calculated as

$$\boldsymbol{\varepsilon}'_{\mathrm{L}} = \ln \mathbf{U} \tag{6}$$

The stress tensor $\boldsymbol{\sigma}'$ with respect to the MF was obtained from the FEA.

The stress tensor $\boldsymbol{\sigma}$ with respect to the reference frame (RF) is calculated using $\mathbf{R}$ as

$$\boldsymbol{\sigma} = \mathbf{R}\boldsymbol{\sigma}'\mathbf{R}^{\mathrm{T}} \tag{7}$$

Similarly, the logarithmic strain $\boldsymbol{\varepsilon}_{\mathrm{L}}$ with respect to the RF is calculated as $\boldsymbol{\varepsilon}_{\mathrm{L}} = \mathbf{R}\boldsymbol{\varepsilon}'_{\mathrm{L}}\mathbf{R}^{\mathrm{T}}$.

Figure 4(a) shows the true stress-logarithmic strain component curves for $F_x : F_y = 1:2$ for the RF and MF calculated using the FEA with equation (1). It can be clearly seen that these curves differ significantly depending on the coordinate system. This can be explained by the finite rotations that the test sections experience. The rotation angle of the MF at the final stage was 28.9°, as shown in Figure 4(b).

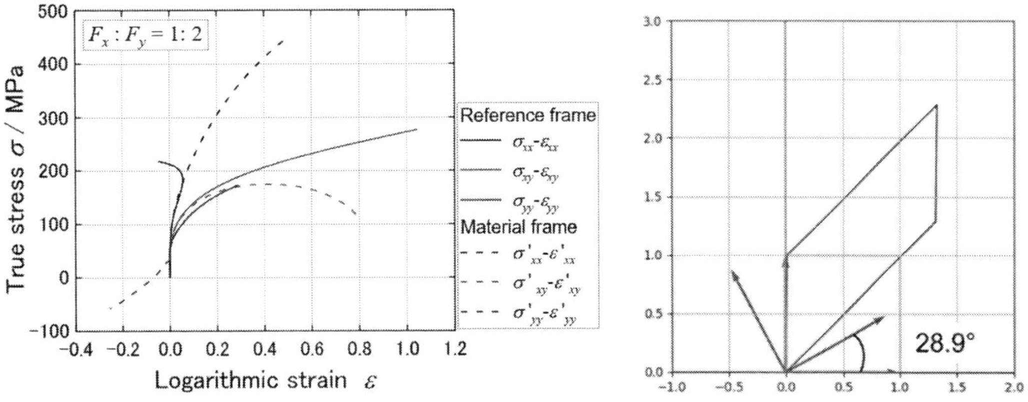

(a) (b)

Figure 4. FEA results for (a) true stress – logarithmic strain curves for $F_x : F_y = 1:2$ obtained using cruciform TS specimen shown in Figure 2, and (b) final shape of square grid and rotation angle of MF.

3.2. *Method for calculating plastic work per unit volume*

In biaxial stress tests, the contours of plastic work per unit volume are measured in the stress space to determine an appropriate material model (phenomenological yield criterion). For this measurement, it is necessary to calculate the plastic work per unit volume, W^{p}. In this study, we calculate W^{p} using two methods. One method uses the plastic strain rate tensor $\mathbf{D}^{\mathrm{p}}$ and the other uses the logarithmic strain increment tensor $d\boldsymbol{\varepsilon}^{\mathrm{p}}_{\mathrm{L}}$.

The following method is used for calculating $\mathbf{D}^{\mathrm{p}}$. First, the velocity gradient tensor $\mathbf{L}$ is obtained from $\mathbf{L} = \dot{\mathbf{F}} \cdot \mathbf{F}^{-1}$, where $\dot{\mathbf{F}}$ is the rate of $\mathbf{F}$. In this study, $\dot{F}(t)$ was calculated as $\dot{\mathbf{F}}(t) = (\mathbf{F}(t) - \mathbf{F}(t-1))/\Delta t$, where Δt is the time increment. The symmetric part of $\mathbf{L}$ can be calculated as $\mathbf{D} = (\mathbf{L} + \mathbf{L}^{\mathrm{T}})/2$ and the asymmetric part of $\mathbf{L}$ gives the spin tensor $\mathbf{W} = (\mathbf{L} - \mathbf{L}^{\mathrm{T}})/2$. The Jaumann rate for the Cauchy stress can be calculated as $\tilde{\boldsymbol{\sigma}} = \dot{\boldsymbol{\sigma}} - \mathbf{W} \cdot \boldsymbol{\sigma} + \boldsymbol{\sigma} \cdot \mathbf{W}$. Using Hooke's law, the elastic strain rate tensor $\mathbf{D}^{\mathrm{e}}$ can be obtained as

$$\begin{bmatrix} D_{xx}^{\mathrm{e}} \\ D_{yy}^{\mathrm{e}} \\ D_{xy}^{\mathrm{e}} \end{bmatrix} = \begin{bmatrix} 1/E & -\nu/E & 0 \\ -\nu/E & 1/E & 0 \\ 0 & 0 & 2(1+\nu)/E \end{bmatrix} \begin{bmatrix} \tilde{\sigma}_{xx} \\ \tilde{\sigma}_{yy} \\ \tilde{\sigma}_{xy} \end{bmatrix} \tag{8}$$

where E and ν are Young's modulus and Poisson's ratio, respectively. The shear modulus G is calculated as $G = E/2(1+\nu)$. Then, $\mathbf{D}^{\mathrm{p}}$ is obtained as $\mathbf{D}^{\mathrm{p}} = \mathbf{D} - \mathbf{D}^{\mathrm{e}}$.

Denoting the W^{p} calculated using $\mathbf{D}^{\mathrm{p}}$ as W_1^{p}, we obtain

$$W_1^{\mathrm{p}} = \int \boldsymbol{\sigma} : \mathbf{D}^{\mathrm{p}} dt \tag{9}$$

where $\mathbf{A} : \mathbf{B} = \mathrm{tr}(\mathbf{A} \cdot \mathbf{B}^{\mathrm{T}})$.

In the method for calculating W^{p} using the logarithmic plastic strain rate, W^{p} is calculated as the area under the true stress-plastic strain curve. The strain increment with respect to the MF was calculated as $d\boldsymbol{\varepsilon}'_{\mathrm{L}} = \boldsymbol{\varepsilon}'_{\mathrm{L}}(t) - \boldsymbol{\varepsilon}'_{\mathrm{L}}(t-1)$. Next, the elastic strain increment tensor $d\boldsymbol{\varepsilon}'^{\mathrm{e}}_{\mathrm{L}}$ was obtained from $d\varepsilon'^{\mathrm{e}}_{\mathrm{L}\alpha\beta} = d\sigma'_{\alpha\beta}/E'_{\alpha\beta}$ (no sum on α and β), where $E'_{\alpha\beta}$ $(\alpha,\ \beta = x, y)$ is the slope of the $\sigma'_{\alpha\beta} - \varepsilon'_{\mathrm{L}\alpha\beta}$ curve in the elastic range. Thus, $d\boldsymbol{\varepsilon}'^{\mathrm{p}}_{\mathrm{L}}$ can be calculated as $d\boldsymbol{\varepsilon}'^{\mathrm{p}}_{\mathrm{L}} = d\boldsymbol{\varepsilon}'_{\mathrm{L}} - d\boldsymbol{\varepsilon}'^{\mathrm{e}}_{\mathrm{L}}$. Therefore, denoting the W^{p} calculated using $d\boldsymbol{\varepsilon}'^{\mathrm{p}}_{\mathrm{L}}$ as W_2^{p}, we obtain

$$W_2^{\mathrm{p}} = \int \boldsymbol{\sigma}' : d\boldsymbol{\varepsilon}'^{\mathrm{p}}_{\mathrm{L}} \tag{10}$$

$d\boldsymbol{\varepsilon}^{\mathrm{p}}_{\mathrm{L}}$ with respect to the RF is obtained as $d\boldsymbol{\varepsilon}^{\mathrm{p}}_{\mathrm{L}} = \mathbf{R} \cdot d\boldsymbol{\varepsilon}'^{\mathrm{p}}_{\mathrm{L}} \cdot \mathbf{R}^{\mathrm{T}}$. Denoting the W^{p} calculated using $d\boldsymbol{\varepsilon}^{\mathrm{p}}_{\mathrm{L}}$ as W_3^{p}, we obtain

$$W_3^{\mathrm{p}} = \int \boldsymbol{\sigma} : d\boldsymbol{\varepsilon}^{\mathrm{p}}_{\mathrm{L}} \tag{11}$$

To check whether W^{p} was correctly calculated using the physical values measured on the MF and RF, we chose $\boldsymbol{\sigma}$ and $d\boldsymbol{\varepsilon}^{\mathrm{p}}_{\mathrm{L}}$ as those based on the RF and $\boldsymbol{\sigma}'$ and $d\boldsymbol{\varepsilon}'^{\mathrm{p}}_{\mathrm{L}}$ as those based on the MF. As the reference data for $W^{\mathrm{p}}, W_4^{\mathrm{p}} = \int \bar{\sigma} \overline{d\varepsilon}^{\mathrm{p}}$ was used, where $\bar{\sigma}$ and $\overline{d\varepsilon}^{\mathrm{p}}$ are the equivalent stress and equivalent plastic strain increment, respectively. Table 1 summarizes the calculation methods for dW^{p}.

Table 1 Calculation methods for dW^{p}

Equation	Remark
$dW_1^{\mathrm{p}} = \boldsymbol{\sigma} : \mathbf{D}^{\mathrm{p}}$	$\boldsymbol{\sigma}$ and $\mathbf{D}^{\mathrm{p}}$: RF
$dW_2^{\mathrm{p}} = \boldsymbol{\sigma}' : \mathbf{d}\boldsymbol{\varepsilon}'^{\mathrm{p}}_{\mathrm{L}}$	$\boldsymbol{\sigma}'$ and $\mathbf{d}\boldsymbol{\varepsilon}'^{\mathrm{p}}_{\mathrm{L}}$: MF
$dW_3^{\mathrm{p}} = \boldsymbol{\sigma} : \mathbf{d}\boldsymbol{\varepsilon}^{\mathrm{p}}_{\mathrm{L}}$	$\boldsymbol{\sigma}$ and $\mathbf{d}\boldsymbol{\varepsilon}^{\mathrm{p}}_{\mathrm{L}}$: RF
$dW_4^{\mathrm{p}} = \bar{\sigma} \cdot \overline{d\varepsilon}^{\mathrm{p}}$	$\bar{\sigma}$: Equivalent stress $\overline{d\varepsilon}^{\mathrm{p}}$: Equivalent plastic strain increment

The calculation results for W^{p} obtained using the five methods given in Table 1 are shown in Figure 5. For F_x:F_y=1:0, the test section is in plane strain tension and there is no rotation of the coordinate system. Therefore, it is natural that the calculation results obtained using the three methods, W_1^{p} to W_3^{p},

IOP Conf. Series: Materials Science and Engineering 1284 (2023) 012074 doi:10.1088/1757-899X/1284/1/012074

are identical to that for W_4^{p}. Moreover, for F_x: F_y=1:1, 1:2, 1:4, and 0:1, the calculation results for W_1^{p} to W_3^{p}, are identical to that for W_4^{p}. This verifies that the methods for W_1^{p} to W_3^{p} also give the correct W^{p} when the coordinate system rotates.

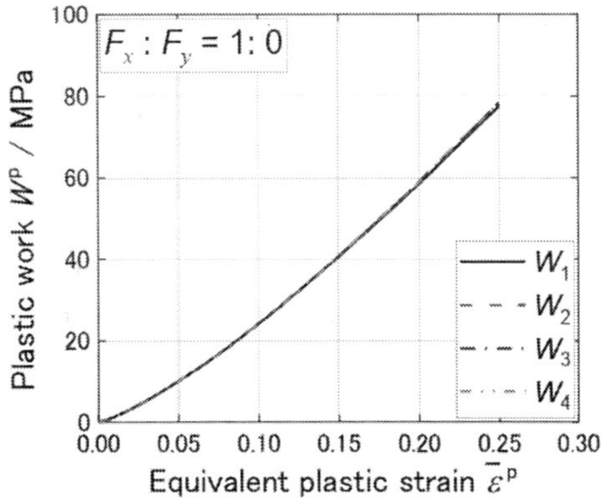

(a) F_x: F_y=1:0 (plane strain tension)

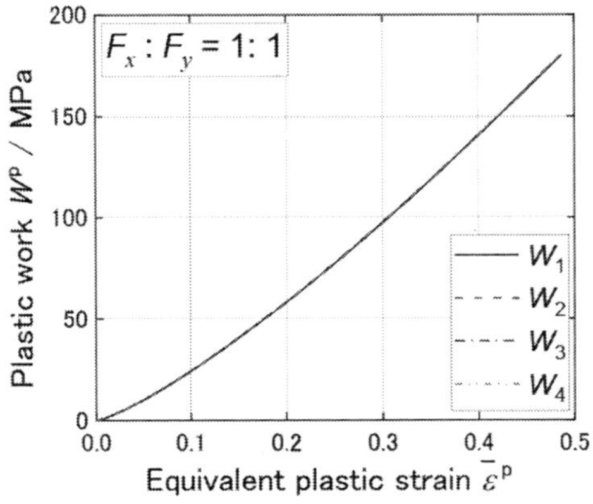

(b) F_x: F_y=1:1

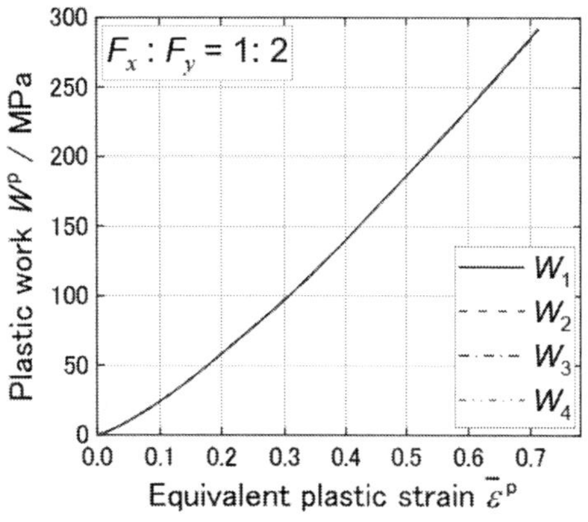

(c) F_x: F_y=1:2

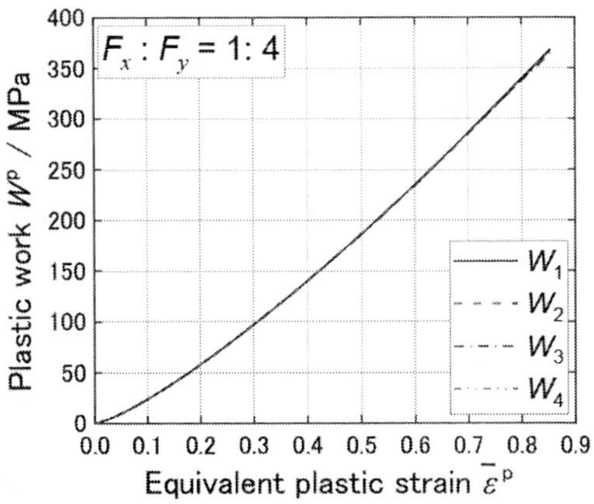

(d) F_x: F_y=1:4

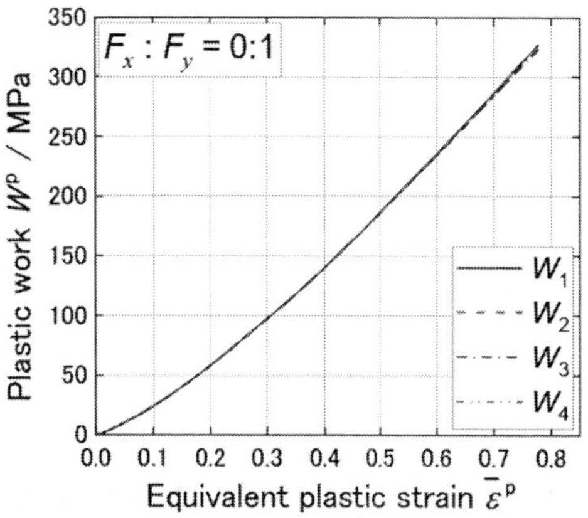

(e) F_x: F_y=0:1 (simple shear)

Figure 5. Comparison of $W^{\mathrm{p}} - \bar{\varepsilon}^{\mathrm{p}}$ curves calculated using methods given in Table 1.

IOP Conf. Series: Materials Science and Engineering 1284 (2023) 012074 doi:10.1088/1757-899X/1284/1/012074

In the experiment using the cruciform TS specimen, the stress tensor $\boldsymbol{\sigma}_{\mathrm{E}}$ is obtained as

$$\boldsymbol{\sigma}_{\mathrm{E}} = \begin{bmatrix} \sigma_{xx} & \sigma_{xy} \\ \sigma_{yx} & \sigma_{yy} \end{bmatrix} = \begin{bmatrix} \frac{F_x}{A} & \frac{F_y}{2A} \\ \frac{F_y}{2A} & \frac{F_x}{2A} \end{bmatrix} \tag{12}$$

Where A is the current cross-sectional area of the stress measurement section. A can be calculated as equation (13), using the deformation gradient tensor $\mathbf{F}$ measured at the middle of the stress measurement section and assuming that the deformation of the stress measurement section is uniform:

$$A = wt = w_0 F_{22} \times t_0 F_{33} = w_0 F_{22} \times \frac{t_0}{F_{11}F_{22} - F_{12}F_{21}} \tag{13}$$

Where w_0 and t_0 are the initial length in the y-direction and thickness of the stress measurement section, respectively. $\mathbf{F}$ is determined as the average in the central area, 1 x 0.5 mm^2, of the stress measurement section, see **Figure 6**. The area is divided into eight squares with a side of 0.25 mm, and $\mathbf{F}$ is calculated for each square using the displacement of the corner points of each square measured using the DIC.

The length of the stress measurement section in the y-direction, 6.5 mm, is sufficiently larger than that in the x-direction, 0.5 mm; therefore, it is assumed that $d\varepsilon_{yy} = 0$, and that $\sigma_{yy} = \sigma_{xx}/2$ by assuming the Levy-von Mises equation. In the TS test, we will apply the plastic work calculation W_1^{p} or W_4^{p} using $\mathbf{F}$ and $\boldsymbol{\sigma}_{\mathrm{E}}$ with respect to RF.

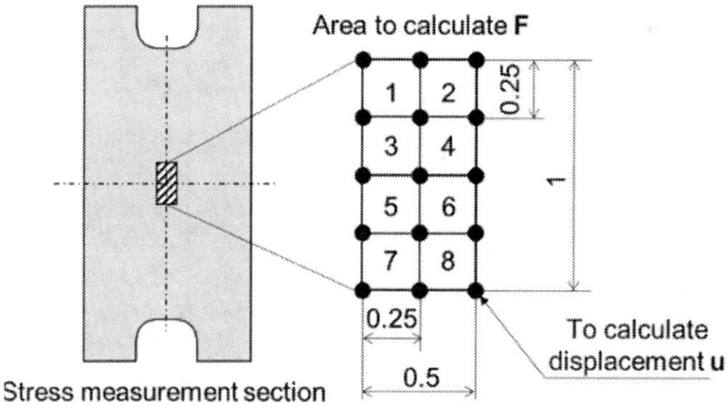

Figure 6. Schematic diagram of the region for which the deformation gradient tensor **F** is calculated in the experiment.

4. Conclusion

Methods for calculating the plastic work per unit volume of a metal sheet subjected to combined tension and shear were evaluated. Such methods are critical for properly calibrating the anisotropic yield criterion. In tension-shear deformation, a material element undergoes rigid body rotation as well as in-plane deformation. Therefore, methods for calculating the stress components and the incremental strain components with respect to the MF and RF were discussed. It was confirmed that the plastic work per unit volume does not depend on the coordinate system adopted in the calculation as long as the same coordinate system is used to evaluate the stress components and strain increment components. In the experiments using a cruciform TS specimen, the stress components are evaluated with respect to the RF. When a digital image correlation system is used for the measurement, the strain components are evaluated with respect to the MF. Therefore, when calculating the plastic work, it is necessary to match the coordinate system for evaluating the stress and strain increments.

References

[1] Kuwabara T 2007 Advances in experiments on metal sheets and tubes in support of constitutive modeling and forming simulations. *Int. J. Plast.* **23** 385–419.

[2] Banabic D, Barlat F, Cazacu O and Kuwabara T 2010 Advances in anisotropy and formability. *Int. J. Mater. Form.* **3** 165–89.

[3] Banabic D, Barlat F, Cazacu O and Kuwabara T 2020 Advances in Anisotropy of plastic behaviour and formability of sheet metals. *Int. J. Mater. Form.* **13** 749–87.

[4] Kuwabara T, Ikeda S and Kuroda T 1998 Measurement and analysis of differential work hardening in cold-rolled steel sheet under biaxial tension. *J. Materials Process. Technol.* **80/81** 517–23.

[5] Deng N, Kuwabara T and Korkolis YP 2018 On the non-linear unloading behavior of a biaxially loaded dual-phase steel sheet. *Int. J. Mech. Sci.* **138–139** 383–97.

[6] Kuwabara T and Sugawara F 2013 Multiaxial tube expansion test method for measurement of sheet metal deformation behavior under biaxial tension for a large strain range. *Int. J. Plast.* **45** 103–18.

[7] Hakoyama T and Kuwabara T 2015 Effect of biaxial work hardening modeling for sheet metals on the accuracy of forming limit analyses using the Marciniak-Kuczynski approach. In: Altenbach H, Matsuda T and Okumura D (Eds.) *From Creep Damage Mechanics to Homogenization Methods.* Springer, pp. 67–95.

[8] Kuwabara T, Mori T, Asano M, Hakoyama T and Barlat F 2017 Material modeling of 6016-O and 6016-T4 aluminum alloy sheets and application to hole expansion forming simulation. *Int. J. Plast.* **93** 164–86.

[9] Mohr D and Oswald M 2008 A New Experimental Technique for the Multi-axial Testing of Advanced High Strength Steel Sheets. *Exp. Mech.* **48** 65–77.

[10] Dunand M, Maertens AP, Luo M, Mohr D 2012 Experiments and modeling of anisotropic aluminum extrusions under multi-axial loading – Part I: Plasticity. *Int. J. Plast.* **36** 34–49.

[11] Kim M, Ha J, Bonica S and Korkolis Y P 2021 Proc. 13th Int. Conf. Technol. Plast. pp. 1961-1967.

[12] Kim M, Ha J and Korkolis Y P 2022 *Proc. NUMISHEET2022* pp. 389–397.

[13] Roth CC and Mohr D 2016 Ductile fracture experiments with locally proportional loading histories. *Int. J. Plast.* **79** 328–54.

A uniform method for the characterization of shear fracture strain for high-strength steel sheets under quasi-static and crash loading

S Klitschke[1], S Westhäuser[2], M Schneider[2], A Heitzmann[1], A Pearce[3], H Friebe[4]

[1] Fraunhofer Institute for Mechanics of Materials IWM, Freiburg, Germany
[2] Salzgitter Mannesmann Forschung GmbH, Salzgitter, Germany
[3] Instron GmbH, High Wycombe, UK
[4] Carl Zeiss GOM Metrology GmbH, Braunschweig, Germany

E-mail: silke.klitschke@iwm.fraunhofer.de

Abstract. In the conflict between lightweight design and crash safety of high-strength steel components failure models for crash simulation have to be calibrated precisely. To predict shear failure strains in a wide strain rate range shear experiments must be performed from quasi-static up to crash-relevant strain rates of about 100 s^{-1}. In particular, asymmetric notched shear tensile specimens have proven useful for this purpose in previous investigations. Using this type of shear specimen, a characterization concept for shear loading in a wide strain rate range is presented for high strength steel sheets of different strength, ductility and sheet thickness in this contribution. Within a round robin test of different laboratories the influence of the degrees of freedom in the clamping area and the free clamping length on the kinematic of the specimens and on the deformation and failure behavior in the shear zone are investigated in detail. The main outcome is that the disadvantages of a fix clamping in terms of shear deformation, shear fracture and homogeneity of the shear zone can be compensated by a large free clamping length. An evaluation method based on digital image correlation (DIC) is presented, which enables an operator- and laboratory-independent determination of shear strains up to failure.

1. Introduction

In order to satisfy continuously increasing regulations on passenger car bodies, further improvement of crash performance still proves to be one of the major technological challenges of modern automotive engineering. Facing this goal it becomes crucial to represent the material behavior under high strain rates and different complex loading situations realistically in modelling. High strength steels are widely used in strength-relevant and energy absorbing components, especially in automotive structures because of their high strength and ductility. Though, high strength steels show a tendency to sudden failure without previous necking or observable instability, especially when it comes to shear failure under high strain rates. Several investigations in the past have shown, that for different metallic materials shear failure strains are significantly lower under crash loading than under quasi-static loading based on a gauge length scale of about 100 µm and larger [1-6]. Therefore shear failure can be seen as a critical loading state in terms of crash worthiness and this effect should be considered in failure prediction for crash simulation.

For a reliable detection of strain rate effects on fracture strain under shear loading a systematic experimental procedure with suitable shear specimens is essential and should be useful for quasi-static shear tests as well as for shear tests under high strain rates. Notched shear tensile specimens have been seen as a promising option for this purpose in the past and numerous specimen geometries have been presented in the literature, examples are [1, 3, 6, 7]. In the vast majority asymmetrical specimen geometries are preferred because they show reduced lateral forces in the shear zone compared to the symmetric options. Frequently proposed specimen geometries are the Peiers/Verleysen type with horizontal introduction of the notches [1], the Lademo type with nearly V-notches [8] and the ASTM type with diagonal notches [9], which is standardized for quasi-static shear test for aluminum sheet alloys. Nevertheless, it is still a challenge to avoid failure initiation in the notch region and to provide shear failure. In literature exists a consensus to design the geometry of the shear zone dependent on the material behavior, especially the hardening behavior, to reduce previous notch failure and to evaluate reliable shear failure strains [6, 10, 11]. Shear specimen geometries for the characterization of fracture initiation has to be clearly distinguished to the characterization of yield initiation [12].

Beside the design of the specimen geometry also the clamping of the specimen has a significant impact on the deformation and failure behavior. Previous works indicate that a fix clamping with no rotational and translational degrees of freedom could lead to early failure in the notch region [13]. As this clamping type is frequently used profound investigations concerning this topic are important. A further point is the fact that for highspeed testing shorter specimens are used compared to tests under quasi-static loading to reduce oscillations in the force signal [1, 6, 10, 14, 15]. There are still investiga-tions necessary concerning the influence of the free clamping length of asymmetrical notched shear tensile specimens for fracture characterization under shear loading.

In this contribution investigations for an optimized characterization concept for the determination of shear fracture strains in a wide strain rate range is presented. As geometry asymmetrical notched shear tensile specimens with diagonal notches are chosen because they are widely used, they are near to the standardized ASTM-specimen and easy to extract out of sheets. The presented studies focus on the main topics:

- Experiments under various clamping types and free clamping lengths L_f for quasi-static and highspeed loading
- Definition of an evaluation strategy for determining shear strains by optical methods
- Analyses of the influence of the clamping situation and the free clamping length L_f on the shear deformation in the shear zone

2. Experimental investigations

An experimental testing campaign is carried out. Quasi-static tests are performed to investigate the influence of the clamping and the free clamping length on the local shear deformation behavior. Therefore the kinematic of the specimen is analyzed in detail. To cover various clamping situations, fix clamping types with no freedom of movement are used and clamping types with two or more bolts to provide rotational and translational movement. Also different free clamping lengths are investigated. The high-rate tests are performed at crash relevant strain rates using clamping types with rotational and translational movement of the specimens. The tensile orientation was perpendicular to the rolling direction. To cover different material ductilities a dual phase steel CR590Y980T-DP-GI, a microalloyed steel CR300LA-GI and a bainitic multiphase steel HR660Y760T-CP are considered, Table 1.

Table 1. Mechanical properties perpendicular to the rolling direction.

Material	Sheet thickness (mm)	Yield strength $R_{p0,2}$/ R_{eL} (MPa)	Tensile strength R_m (MPa)	Hardening value n (-)
CR590Y980T-DP-GI	1.5	683	1007	0.13
CR300LA-GI	1.5	390	461	0.15
HR660Y760T-CP	4.0	756	825	0.10

IOP Conf. Series: Materials Science and Engineering 1284 (2023) 012075 doi:10.1088/1757-899X/1284/1/012075

2.1. Experimental procedure

The quasi-static experiments are performed at different laboratories using hydraulic and electrical testing machines. Four different types of clamping (cl1, cl2, cl3 and cl4) are investigated, see Figure 1. Concerning the freedom of horizontal and rotational movement cl1 shows no degrees of freedom, cl2 and cl3 show rotational degrees of freedom and cl4 shows additional translational degrees of freedom.

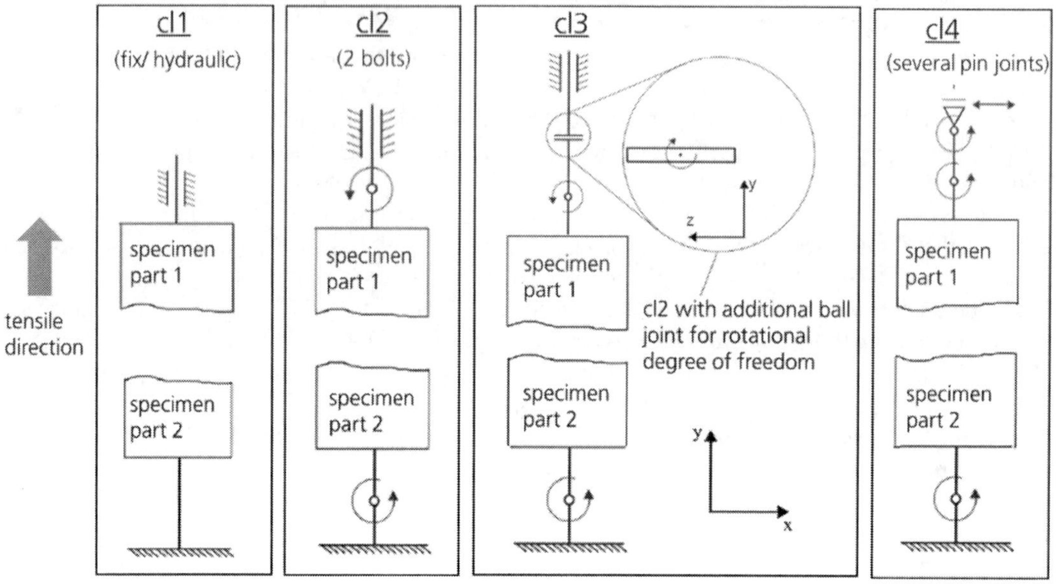

Figure 1. Variations of specimen clamping cl1, cl2, cl3 and cl4

The specimens are tested under displacement control at test speeds between 0.04 - 0.1 mm/s. The specimen geometries for the dual phase steel and the microalloyed steel were chosen based on the recommendations in [6] with a notch radius of R = 1 mm and a negative notch overlap based on the material hardening behavior, see Figure 2a.). For the bainitic steel with a hardening value up to 0.1 a specimen geometry was chosen with a notch overlap of about -10 °, which is recommended in [6]. The tests with fix clamping were performed with a free clamping length of 62 mm and 165 mm, see Figure 2b.). For the other clamping variations cl2, cl3 and cl4, a free clamping length of 62 mm was used.

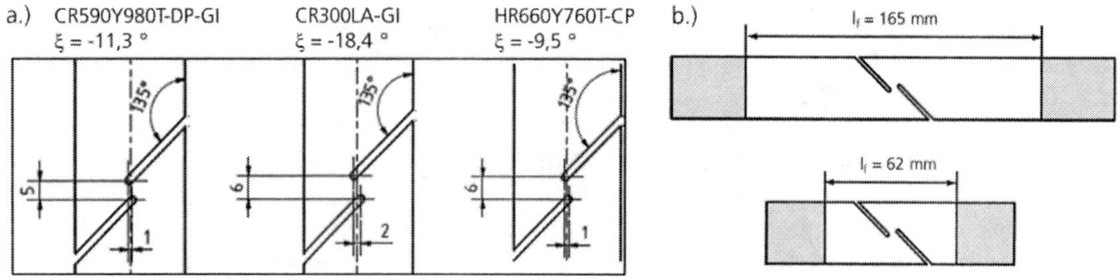

Figure 2. a.) Material dependent geometry of the shear zone; b.) free clamping length

The force was measured with suitable load cells like universal load cells or piezo load cells. The displacement was measured with optical extensometers for the original gauge length of 15 mm. The kinematic of the specimens are observed by 2D video recording of at least 55 mm length of the specimen. The local strains in the shear zone were determined by 2D and partly 3D video recording of the shear zone and evaluation of the displacement field by digital image correlation (DIC). Therefore, a random speckle pattern is applied to the specimen surface before testing. The optical resolution was specified as 100 – 200 pixel/mm allowing a local strain gauge length in the range of 0.1 mm to 0.2 mm with the correlation software GOM ARAMIS [17]. The video recording was performed with a frame rate of at least 10 frames /second.

The high-rate tests are performed using highspeed testing machines with a clamping type cl4 and using the same material dependent specimen geometries shown in Figure 2a.) with a free clamping length L_f of 62 mm. The test speeds of 25 mm/s, 250 mm/s and 2.5 m/s were chosen, because they lead to strain rates in the order of 1 s^{-1}, 10 s^{-1} and 100 s^{-1} respectively. The force was measured by piezo and quasi-local load cells in the clamping region [4, 14]. As it was done for the quasi-static tests the displacement was measured with optical extensometers for the initial gauge length of 15 mm.

2.2. Results

The specimens with a short free clamping length of 62 mm and tested with a fix clamping show a tendency of earing, especially for the most ductile microalloyed steel CR300LA-GI, see Figure 3a.). Earing is observed, if the fractured shear zone is positioned from the middle part of the upper notch to the middle part of the lower notch. In this part of the notch a tensile stress state occurs [6]. Therefore a fractured shear zone with earing is a hint of fracture initiation under tensile loading. The specimens with a free clamping length of 165 mm show much less earing although they are tested with a fix clamping, see Figure 3b.). Also specimens tested with rotational and horizontal mobility do not show any significant earing, see Figure 3c.). This fracture position is necessary for shear fracture initiation. The specimens tested under high strain rates do not show earing.

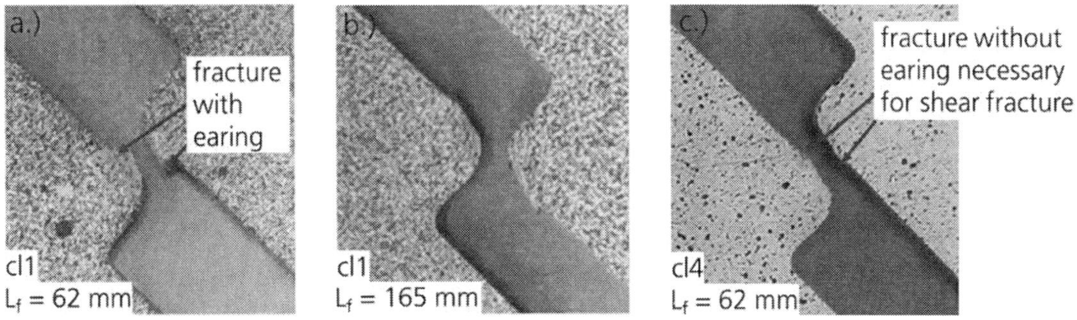

Figure 3. Shear zone of tested specimens for CR300LA-GI with two different clamping methods and two different free clamping lengths L_f

The measurement curves for the engineering stress, as the measured force related to the initial cross section, are shown dependent on the displacement in Figure 4. The red curves, representing the results for fix clamping cl1 contain the results for the specimens with investigated free clamping lengths. For the microalloyed steel and the dual phase steel the specimens tested with a fix clamping show moderately larger forces compared to the other clamping situations. This is not the case for the bainitic multiphase steel. For clamping types with various degrees of freedom, cl2 to cl4, there is no influence of the clamping on the stress level visible. An influence of the clamping situation on the displacement at fracture is only observable for the microalloyed steel, which shows the largest displacement at fracture for the fix clamping cl1. For the other two materials there is no influence of the clamping on the displacement at fracture evaluated. Summarizing the influence of the clamping type on the deformation and fracture behavior there seems to be a difference between the fix clamping and the other clamping types with various degrees of freedom.

The results for the high-rate tests are shown for the dual phase steel and the microalloyed steel in Figure 5. The results for the test speed v = 25 mm/s and partly for v = 2.5 m/s are taken from [6]. The results for quasi-static loading, for 250 mm/s and partly for v = 2.5 m/s are new results evaluated for the same batches within the presented studies. For both materials the displacement to fracture decreases with increasing test speed for the complete investigated strain rate range. These results might show a continuous decrease of failure strain with increasing strain rate. Though, local strain analysis up to fracture are necessary to investigate this topic in detail and are not in the scope of this contribution.

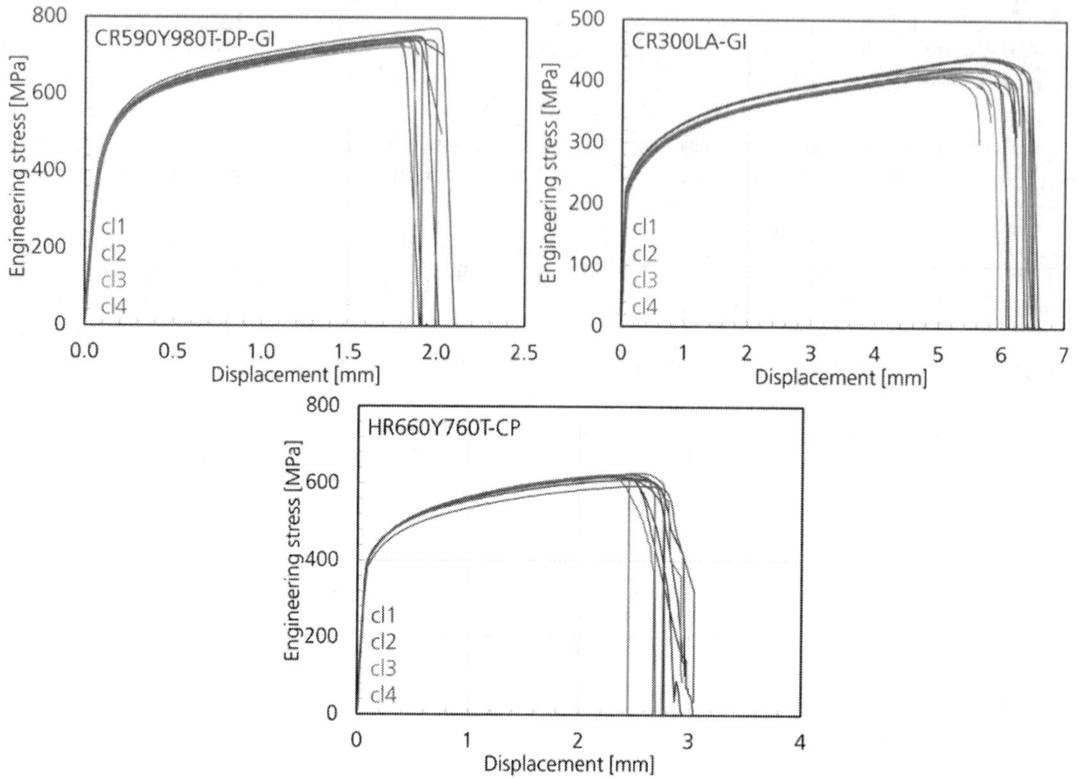

Figure 4. Influence of the clamping type on the stress-displacement curves for quasi-static shear tests for CR590Y980T-DP-GI, CR300LA-GI and HR660Y760T-CP

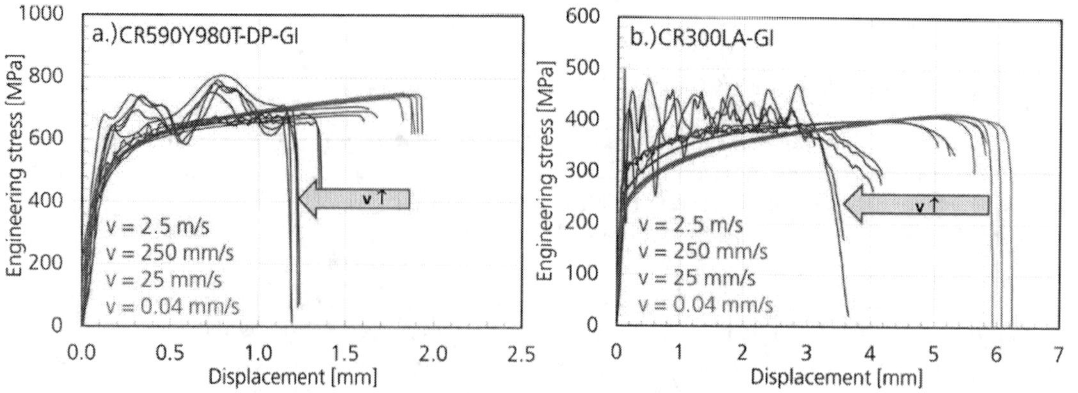

Figure 5. Influence of the test speed on the stress-displacement curves for CR590Y980T-DP-GI and CR300LA-GI determined by different laboratories, the results for 25 mm/s are taken from [6]

3. Evaluation method for shear strains with optical strain measurement (DIC)

For the determination of reproducible and comparable shear strains up to fracture using digital image correlation (DIC) an evaluation process is proposed in this chapter. As the characteristic strain gauge length L_0 as well as the choice of the relevant evaluation points influence the evaluated shear strain results significantly they have to be defined precisely.

3.1. Determination of the characteristic strain gauge length L_0

With a too large strain gauge length, the strains in small shear bands will be underestimated [18]. The strain gauge length is influenced by the optical resolution of the images, which is controlled by the size of the detector (pixel field) and the measured object (mm^2). Dependent on the method of strain computation defined by the used DIC software the facet size and the facet gap are the important parameters influencing the strain gauge length. It has to be emphasized that the characteristic strain gauge length is a value gained by a complex calculation within the software and is not identical with the facet gap.

A suitable characteristic strain gauge length depends on the material properties, which influence the localization behavior of the material. In [12] the hardening value of the material is proposed to define the maximal allowed strain gauge length. For high strength steels investigated in this contribution a characteristic strain gauge length of $L_0 \sim 0.2$ mm was used. This size could be achieved by the involved laboratories with their video cameras and DIC systems. The characteristic strain gauge length should lead to multiple evaluation points in the region of maximal strain immediately before fracture, see Figure 6a.). It should be mentioned that the maximal strain before fracture, evaluated in this way, is still a conservative approach for the failure strain.

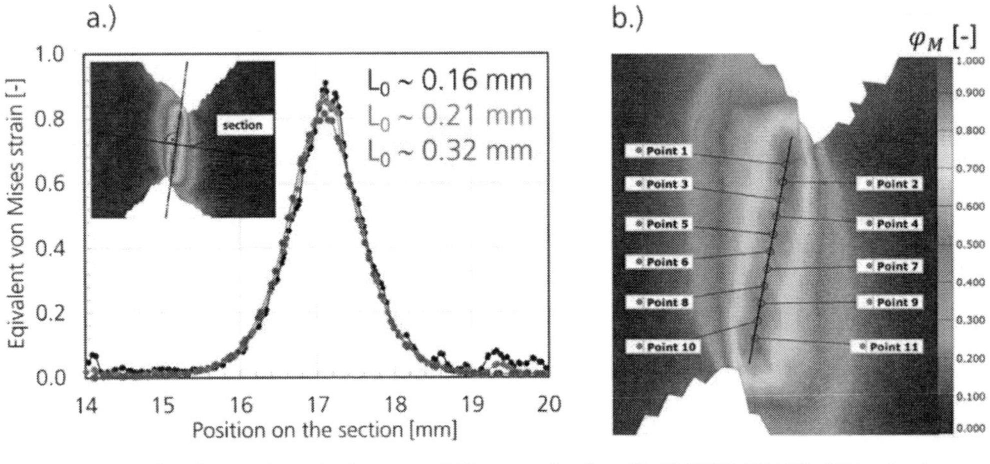

Figure 6. Evaluations of equivalent von Mises strain for CR590Y980T-DP-GI in the last image before fracture: a.) Section perpendicular to the localized zone and strain along the section line for three spatial resolutions; b.) Evaluation points along the shear band

3.2. The choice of relevant evaluation points

For an evaluation of robust shear strain paths and shear failure strain values by DIC it is helpful to determine average values considering multiple points in the central part of the shear zone. In this contribution it is proposed to define at least 10 points in the last image before fracture. These points should be positioned along a line following the maximal strains in the shear band, see Figure 6b.). For further evaluations like major, minor and equivalent von Mises strain the average values for all selected points should be taken.

For materials with a large shear failure strain it is often useful, to use the previous stage as reference stage for the correlation. The standard correlation procedure, in contrast, refers to the initial stage. Using the previous stage as reference stage it is mandatory to check at least the evaluated strain result with those determined by the standard correlation procedure.

IOP Conf. Series: Materials Science and Engineering 1284 (2023) 012075 doi:10.1088/1757-899X/1284/1/012075

4. Influence parameters on the shear deformation behavior

4.1 Influence of the clamping type and the free clamping length on the local shear deformation

The evaluation of the local deformation behavior in the shear zone is presented exemplarily for the dual phase steel, because of its low ductility with evaluable DIC images up to fracture compared to the two other investigated materials. There is a significant influence of the clamping type and the free clamping length on the strain distribution in the shear zone and on the strain paths observed:

- For the short specimen with a fix clamping the strain localization before fracture looks uneven and shows three partly separated localized zones along the shear length, see Figure 7a.). The corresponding strain paths for the different evaluation points also show a large scatter leading to a non-linear shape of the averaged strain path, see Figure 8a.).
- For the short specimen with rotational and limited horizontal mobility (cl3) a sufficiently uniform strain localization along the shear zone is observed, see Figure 7b.). The corresponding strain paths for the different evaluation points show much less scatter and a linear averaged strain path is evaluated, see Figure 8b.).
- For the long specimen with fix clamping again a mostly uniform strain localization can be seen in Figure 7c.). The corresponding strain paths in Figure 8c.) are nearly comparable to those evaluated for the short specimen with clamping variation cl3.

The averaged strain paths up to fracture for the short specimens are shown in Figure 8d.) for the different clamping situations.

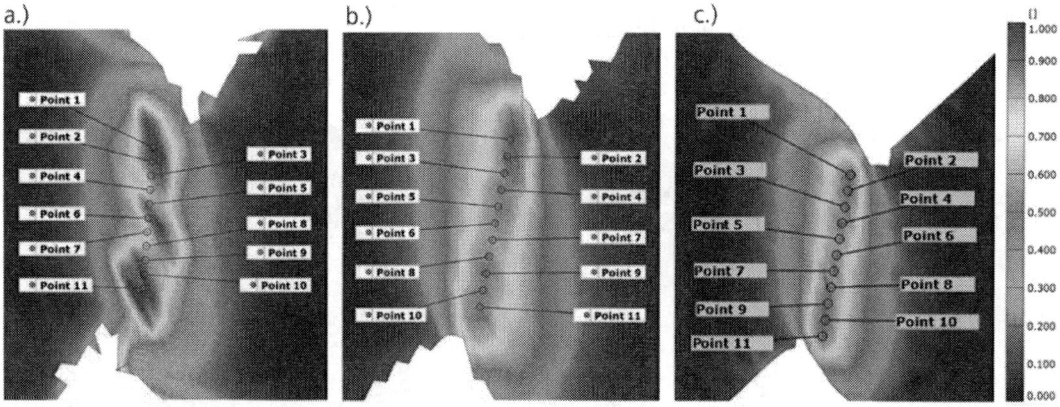

Figure 7. Equivalent von Mises strain field in the last image before fracture evaluated by DIC for CR590Y980T-DP-GI: a.) cl1 and L_f = 62 mm; b.) cl3 and L_f = 62 mm; c.) cl1 and L_f = 165 mm

All clamping variations with at least rotational degrees of freedom show linear strain paths, see Figure 8d.). The strain to fracture scatters moderately but does not show a significant dependency on the clamping type. For the fix clamping with the long specimen the strain path also shows a linear shape in contrast to the short specimen, see Figure 8d.).

For the microalloyed steel and the bainitic multiphase steel the local strain evaluations up to fracture partially show loss of facets caused by larger strains to fracture. Nevertheless this negative effect on the strain paths for short specimens with fix clamping is also observable but less pronounced. Pro-bably the fix clamping with short specimens does not allow a relative movement of the two shearing parts of the specimen to each other. This is necessary to avoid overlayed tensile stress components necessary for a pure shear stress state. Therefore the influence of the clamping type and the free clamping length on the kinematic of the specimen is analyzed in the following chapter.

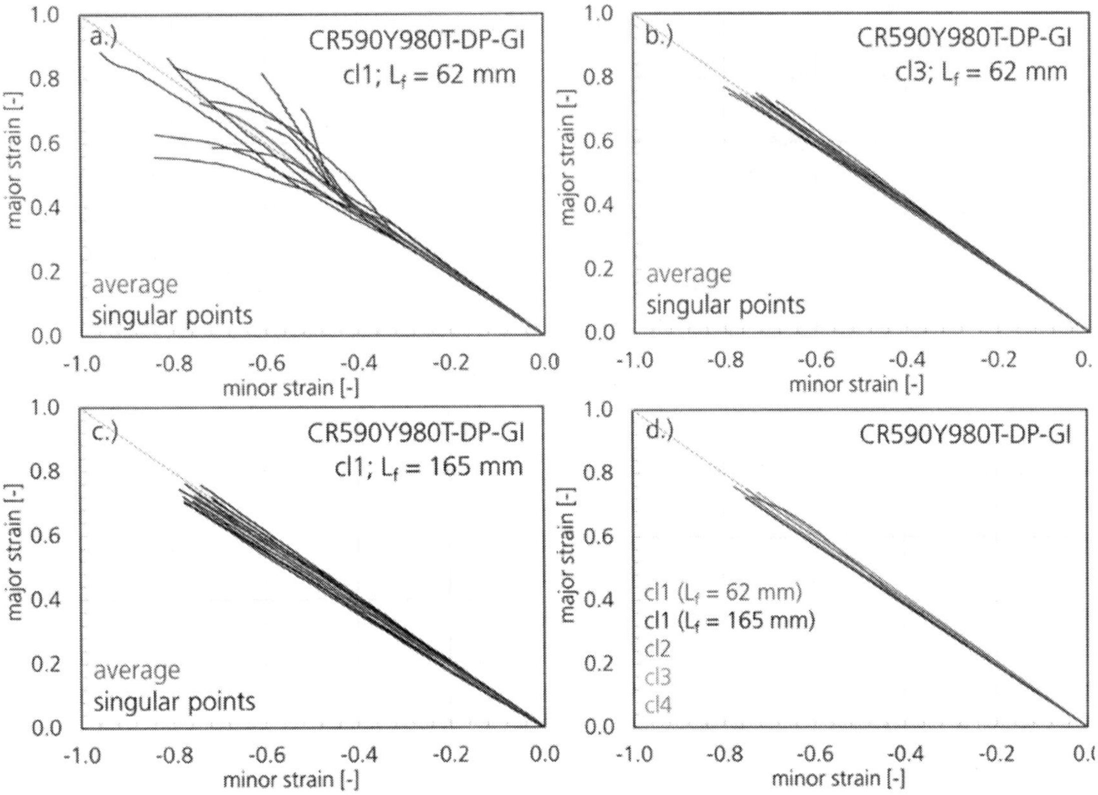

Figure 8. Influence of the clamping type and the free clamping length on the strain paths in the shear zone for CR590Y980T-DP-GI, singular and average values for a.) cl1, $L_f = 62$ mm; b.) cl3, $L_f = 62$ mm; c.) cl1, $L_f = 165$ mm and in d.) average values for all clamping combinations

4.2 Influence of the clamping type and the free clamping length on the kinematic of the specimen

The relative horizontal movement of the two parts of the specimen is analyzed for three pairs of facets, (3/4), (5/6) and (7/8), each in a vertical distance of 55 mm symmetrical to the shear zone, see Figure 9a.) [19]. The pairs of facets are positioned in a horizontal distance of $a_x = 13.5$ mm. The relative horizontal movement of the two facets of each pair, $\Delta x_{ij} = x_j - x_i$, is determined using the horizontal coordinates of the facets i and j, x_i and x_j (i = 3, 5, 7 and j = 4, 6, 8).

For the microalloyed steel the displacement dependent on the relative horizontal movement of each pair of facets Δx_{ij} is shown in Figure 9b.). In the beginning of the deformation the specimen parts are moving apart, which can be seen by positive values of Δx_{ij}. This is due to the negative notch overlap which leads to a compression stress state in the beginning of the deformation. During the deformation procedure the initial compression stress state changes into a shear-tension stress state [6]. At that point the specimen parts begin to get again closer to each other up to negative values of Δx_{ij}. It is obvious that with horizontal translatoric degrees of freedom the relative horizontal movement of the two specimen parts is larger compared to rotatoric degrees of freedom. Also for the specimens with a large free clamping length under a fix clamping large horizontal mobility is observed, see the black curves in Figure 9b.). The kinematic of the long specimens under fix clamping cl1 resembles those with short specimens and rotational and horizontal movement cl4.

The other two materials show a comparable tendency concerning the influence of the clamping type and the free clamping length on the relative horizontal movement but less pronounced, see Figure 9c.) and Figure 9d.). This is due to their lower fracture strains compared to the microalloyed steel.

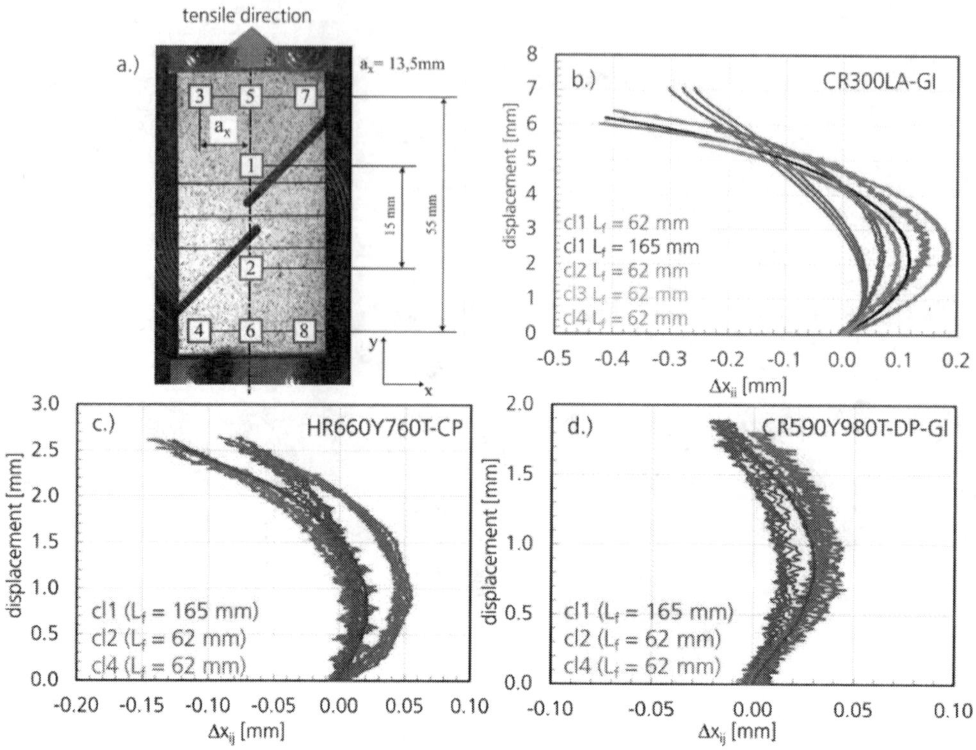

Figure 9. a.) Evaluation points for the kinematic analyzes; Influence of the clamping type and free clamping length on the relationship between relative horizontal movement Δx_{ij} and displacement for b.) CR300LA-GI, c.) HR660Y760T-CP and d.) CR590Y980T-DP-GI [19]

5. Summary and conclusion

The influence of the clamping type and the free clamping length on the shear deformation and damage behavior and on the kinematics of asymmetrical notched shear tensile specimen for fracture characterization is investigated. As materials, a dual phase steel CR590Y980T-DP-GI, a microalloyed steel CR300LA-GI and a bainitic multiphase steel HR660Y760T-CP are selected. Asymmetrical notched shear tensile specimens with diagonal notches are used with a material dependent negative notch overlap. For the evaluation of shear strains with digital image correlation a robust evaluation method is proposed. The following conclusions can be drawn from these studies:

- The evaluation of shear strains by DIC should be performed with an appropriate characteristic strain gauge length and using average strain values from a selection of evaluation points.
- The degrees of freedom concerning the specimen clamping and the free clamping length of the specimen have an impact on the shear deformation behavior in the shear zone.
- Fixed clamping using specimens with a short free clamping length of about 62 mm leads to nonlinear strain paths in the shear region and partly subsequent previous failure under tensile loading in the notch region. Using fixed clamping with specimens with a large free clamping length, e.g in the size of about 165 mm, this disadvantage can be balanced.
- Short specimens, which are usually used for the characterization under crash loads, should be tested with a clamping providing horizontal mobility. For the investigated materials a horizontal mobility in the range of about 0.3 mm in a distance of 55 mm should be sufficient.
- The investigations indicate a continuous decrease of failure strain with increasing strain rate in the range between 0.001 s^{-1} and 100 s^{-1}. Though, local strain analysis up to fracture according to the proposed method are necessary to investigate this topic in detail. Such studies are actually in process.

Acknowledgement
We gratefully acknowledge the German Federal Ministry for Economic Affairs and Energy for funding the project WipaNORMetall. Much thanks to all associated partners for supporting this project.

References
[1] Peirs J, Verleysen P, Degrieck P, 2010, Novel technique for static and dynamic shear testing of Ti6Al4V sheet, *Exp Mech* **52**(7) pp 729-741. doi: 10.1007/s11340-011-9541-9
[2] Böhme W, Memhard D, Brand M, Siegele D, 2012, Failure behavior of crash-relevant welded aluminium joints, *Weld. Cut.* **11**(1) pp 42-48
[3] Roth CC, Mohr D, 2015, *Experimental Investigation on Shear Fracture at High Strain Rates, Lugano, Switzerland, Sept. 2015,* EPJ web of Conferences **94**, 01078
[4] Klitschke S, Böhme W, 2016, Deformation and damage behavior of lightweight steels at highrate multiaxial loading, *Mater. Test.* **58** (3) pp 173-181, doi: 10.3139/ 120.110836
[5] Abedini A, Butcher C, Worswick MJ, 2017, Fracture Characterization of Rolled Sheet Alloys in Shear Loading: Studies of Specimen Geometry, Anisotropy and Rate Sensitivity, *Exp Mech* **57** pp 75-88
[6] Klitschke S, Trondl A, Huberth F, 2019, *Influence of strain rate on deformation and failure behavior of sheet metals under shear loading, 12th Europ. LS-Dyna Conf. 2019, Koblenz, Germany*
[7] Rahmaan T, Abedini A, Butcher C, Pathak N, Worswick MJ, 2017, Investigation into shear stress, localization and fracture behaviour of DP600 and AA5182-O sheet metal alloys under elevated strain rates, *Int. J. Impact Eng.* **108** pp 303-321.
[8] Achani D, Lademo OG, Engler O, Hopperstad OS, 2011, Evaluation of constitutive models for textured aluminium alloys using plane-strain tension and shear tests, *Int. J. Mater. Form.* **4** pp 227-241
[9] ASTM Standard B 831-14, 2014, Standard Test method for Shear testing of Thin Aluminum Alloy Products. ASTM Standards, West Conshohocken, PA 19428-2959, United States
[10] Roth CC, Mohr D, 2018, Determining the strain to fracture for simple shear for a wide range of materials, *Int. J. Mech. Sci.* **149** pp 224-240
[11] Narayanan A, Abedini A, Weinschenk A, Worswick MJ, Butcher C, Evaluation of simple shear test geometries for constitutive characterization using virtual experiments, 2021, IOP Conf. Ser.: *Mater. Sci. Eng.* 1157
[12] Gesamtverband der Aluminiuindustrie e.V. (GDA), 2021, Determination of plastic material properties from the shear tensile test (yield limit), GDA Technical Fact Sheet
[13] Klitschke S, Huberth F, 2019, *Einheitliches Prüfkonzept für die Schercharakterisierung metallischer Blechwerkstoffe in einem weiten Dehnratenbereich*, Tagung Werkstoffprüfung Dez. 2019 Neu-Ulm, Germany
[14] Böhme W, 2008, Dynamische Werkstoffkennwerte für die Crashsimulation, *Mater. Test.* **50** (4) pp 199-205
[15] ISO 26203-2, 2011, Metallic materials – Tensile testing at high strain rates – Part 2: Servohydraulic and other test systems, Beuth Verlag
[16] Trondl A, Klitschke S, Böhme W, Sun DZ, 2015, Verformungs- und Versagensverhalten von Stählen für den Automobilbau unter crashartiger mehrachsiger Belastung, FAT-Schriftenreihe 283-VDA
[17] http://www.gom.com/de/messsysteme/aramis.html
[18] Behrens BA, Bouguecha A, Vucetic M, Peshekhodov I, 2013, *An experimental-numerical method to characterize formability of sheet metals in a wide range of stress states with the help of a tensile-shear test on new butterfly specimens*, International Scientific and technical conference on advanced metal materials and technologies, June 2013 St. Petersbourg, Russia.
[19] Heitzmann A, Untersuchungen zum Scherversagen hochfester Blechwerkstoffe in einem weiten Dehnratenbereich, 2023, *MSc Thesis, Karlsruhe Institute of Technology KIT*

Bendability assessment of advanced high strength steels thanks to artificial neural networks

M Teaca[1], P Dietsch[1], L Durrenberger[1], K Tihay[1] and S Oueslati[2]

[1]Global R&D ArcelorMittal Maizières, voie Romaine, BP30320, F-57283 Maizières-lès-Metz, France
[2]Arts et Métiers Institute of Technology, Campus Metz

E-mail: mihaela.teaca@arcelormittal.com

Abstract. Thanks to their very important strength in addition to their considerable formability and ductility, the newly developed Advanced High Strength Steel (AHSS) allow carmakers to save mass on the vehicles they produce. However, before industrial implementation by OEMs, any grade has to be approved. The bending test defined in VDA238-100 [1] test specification is a key indicator of the material ductility and is increasingly requested in the scope of their homologation. The test consists in bending a square sample of sheet material with a knife. The sample is supported by two rolls. The aim of this test is to determine the maximum angle reachable by the sample before it breaks. We can also derive the maximum strain reachable by the material before fracture initiation. Mainly 2 different angles are defined: the angle after springback, that can be measured manually and the angle at maximum load, which is to be considered in the VDA238-100. The test specification gives instructions on the determination of the bending angle. Nevertheless, the approach is purely geometric and only relies on the parameters of the test. No influence of the tested material is taken into account. This approach can be improved because for AHSS that are very ductile and have a bending angle over 100°, the proposed formula can provide unphysical results. For this reason, the innovative proposed approach is to calculate the bending angle at maximal force and to determine the fracture strain thanks to Artificial Neural Networks, one of the most promising tools for resolution of engineering problems. This new approach considers a large panel of materials tested experimentally or numerically and significantly improves the prediction.

1. Introduction

The bendability (usually characterized using a bend angle and fracture strain) provides critical insight into the plane strain fracture limit, corresponding to the lowest ductility of sheet materials. These values are a relevant criterion for the evaluation of steels crash ductility.

Some experimental work has been performed to understand the influence of various parameters, such as punch stroke, rolls distance, punch diameter on calculated bending angle [2],[3]. It has been shown that the punch stroke and diameter increase with bending angle unlike the rolls distance.

A numerical approach has been chosen [4] to predict the evolution of the bending angle before springback, allowing to predict the angle corresponding to another sheet thickness.

Furthermore, the bend angle at fracture is not directly transferable into numerical models to predict fracture as it is not a strain measurement [5]. To address these limitations in the VDA238-100 test specification, digital image correlation (DIC) has been used to relate the bend angle with the major strain for the considered materials [5] [6] [7] [8], [9].

During the last 20 years, researchers explored the potential of artificial neural networks (ANNs) as an analytical alternative to conventional methods [10] which are often limited by strict assumptions of linearity, normality, homogeneity, and variable independence [11]. ANN approach has proved to be a helpful tool for the engineers to determine springback and predict the final part shape in air bending [12] and for springback prediction in the free cylindrical bending of metallic sheet [11].
An analytical formula is employed to estimate the bend angle before spring back based on the thickness, punch displacement, when the load at Fmax is reached. Any mathematical bending formula using the punch displacement can only deliver an approximate bending angle value [2]. This approach needs to be improved because for ductile AHSS having a bending angle over 100°, the proposed formula can provide unphysical result (negative springback).
The aim of the present study is to predict the bending angle before spring back, to solve the issue of the negative values of spring back and predict fracture strain by combining experimental data, numerical simulation and ANN approach. DIC will be also used for validation of fracture strain model.

2. Experimental procedure and test results

2.1. Test equipment
The test is performed with a traction compression machine equipped with the specific tools. The bending device consists of two rolls of diameters 30 mm each (Figure 1). The rolls are mounted on an adjustable mechanism which allows the operator to manually adjust the distance to be twice the thickness of the specimen plus 0.5 mm.

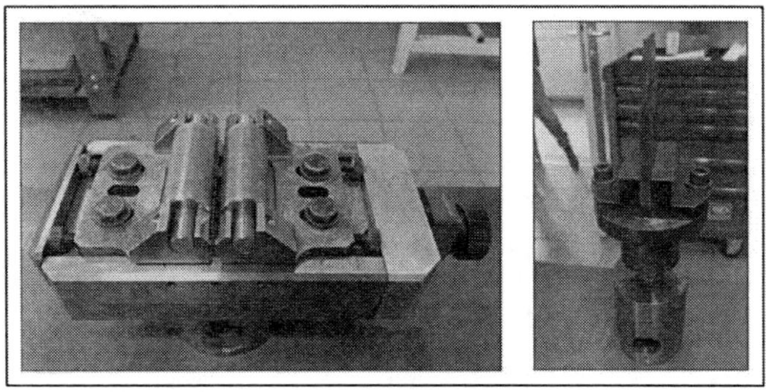

Figure 1. Bending set-up and punch

2.2. Test Procedure
In general, the test specimens are 60x60 mm^2 square. The samples are machined to avoid crack initiation from the edges. Mostly, the rolling and transverse directions are characterized. The sheet is put on two rotary rolls; then, the punch goes down and stops when the force drops of 60N, corresponding to the crack initiation on the bend.
The assumption that crack initiate at the max force measured during bending isn't too many cases valid for all steel grades and cannot be universally trusted [9], [13], [14], [15].

2.3. Different obtained angles
In Figure 2(a) is reported the definition of the different bending angles that could be obtained or measured, respectively:

1. α_{ApRE}: Angle after spring back, measured manually at the end of the test and after the load relief. In this case only the plastic strain is considered. This irreversible strain is actually due to the sliding of crystallographic plans.
2. α_{AvRE}: Angle before spring back, measured manually after the force drop and while maintaining the punch load (Fmax – 60N). The operator uses a digital level mounted on a magnetic shim to ensure the assembly shown in figure 2(b).

IOP Conf. Series: Materials Science and Engineering 1284 (2023) 012076 doi:10.1088/1757-899X/1284/1/012076

3. **α_{Fmax}**: Angle at maximum load, calculated according to the formula extracted from the VDA 238-100 last revised in 2017. The norm gives instructions on how to perform the test and how to determine the bending angle. In a matter of fact, this value obtained from the formula derives from a purely geometric approach. No material influence is taken into account.
4. Spring back (SB) is defined as the difference between the angle underload (Fmax-60N) and the angle after unloading (α_{ApRE}).

However, for materials with high bending angles, the measured angle after spring back is higher than the angle at maximum load, which contradicts experimental evidence where ($\alpha_{Fmax} > \alpha_{ApRE}$).

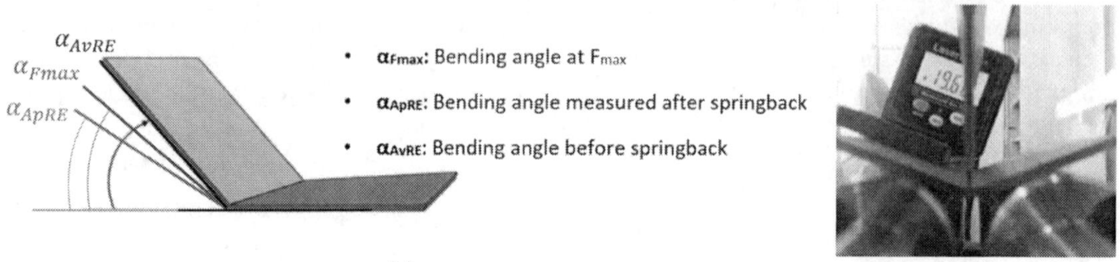

(a) (b)

Figure 2. (a) Different obtained bending angles, (b) Measuring of bending angle before spring back

2.4. Fracture strain
Using DIC coupled with bending tests, makes possible to measure the strain at the outer surface. In this set-up the rolls are fixed, and only the punch moves up down. A mirror is used to visualize the sample with the camera as illustrated below.

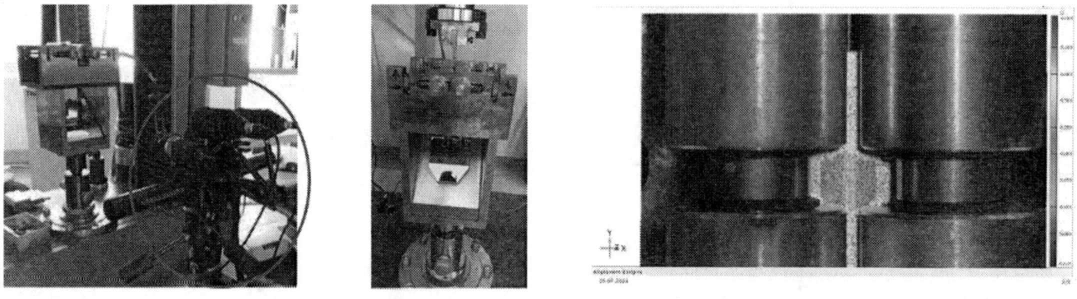

Figure 3. Fracture strain measurement using DIC method

3. Addressed issues
The geometrical conditions imposed by the VDA238-100 (test specification) for the calculated angle includes only the plastic proportion and doesn't take into account the elastic curvature between the rolls figure 4. The calculations give smaller bending angles as highlighted in figure 5. To overcome this issue, it is suggested in the standard that the influence of horizontal stiffness and vertical stiffens can be minimized by means of a correction factor when applicable.
Therefore, regarding both the oddities faced in experimental and analytical methods of bending angle obtention and the time-consuming procedure of making FE simulations and corrections for each steel grade, a machine learning approach is suggested. The proposed idea is to develop an algorithm that takes as input simulation and experimental data, used it to learn, adapt and gives us the wanted outputs: fracture strain and angle before spring back.

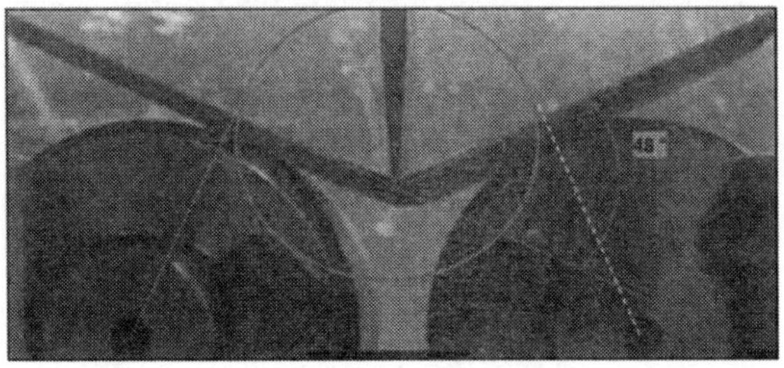

Figure 4. Experimental measurement is different from angle defined in VDA238-100 (test specification)

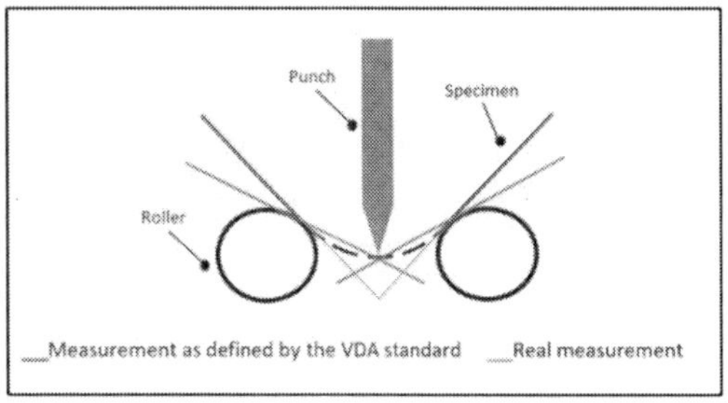

Figure 5. Bent sample at the end of the test, still under load Fmax-60N

4. ANN implementation

In order to achieve our above mentioned objectives, both models were developed into 5 parts following the data flow shown in figure 6:

- Data processing: created, import, clean, normalize and define outputs and potential inputs.
- Feature importance: define the importance of each and decide inputs.
- Splitting the data base: get train and test sets.
- Building and training the model: chose hyper parameters and train the model (train set).
- Evaluating the model: predict and score (test set).

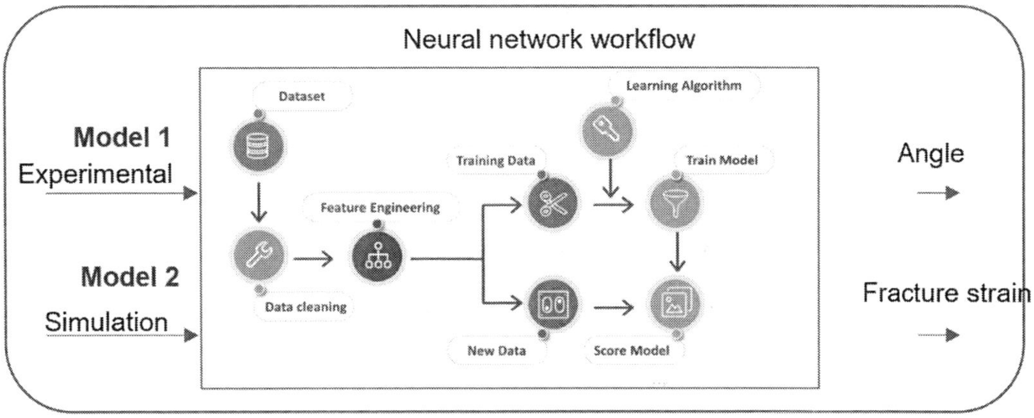

Figure 6. Data workflow in the models

4.1. Data processing

An algorithm of machine learning needs to be trained with a large amount of data to generate an accurate predictive model. The more data is collected, the more accurate the prediction of the model can be. For data collection, 2 methods can be used:

1. Experimental approach. Laser Liner mounted for bending angle measurements is mandatory
2. Finite element approach. The main advantage of this approach is the ability to consider a large number of materials (realistic or not). More than 4000 simulations were performed with tensile strengths between 350 MPa and 3000 MPa, yield stresses between 400 MPa and 1500MPa and uniform elongation from 3% to 30%.

The differences existed between considered data base from simulation and experimental data. In the simulations, the considered punch stroke is the displacement at punch while in experimentally the punch stroke is the displacement of the crosshead of the machine. In this case correction factors were needed for every new prediction. Furthermore, these corrections must be determined according to the tests steel grade and the real value of punch stroke. A correction must be applied on experimental data in order to

IOP Conf. Series: Materials Science and Engineering 1284 (2023) 012076 doi:10.1088/1757-899X/1284/1/012076

calculate the experimentally the displacement of the punch. This correction should take into account the machine stiffness.
Different methods of ANN application for fracture and bending angle is given figure 7. For fracture strain data base from numerical simulation has been used (due to the lack of experimental measurements) whereas for bending angle experimental data has been considered in the learning process of our ANN.

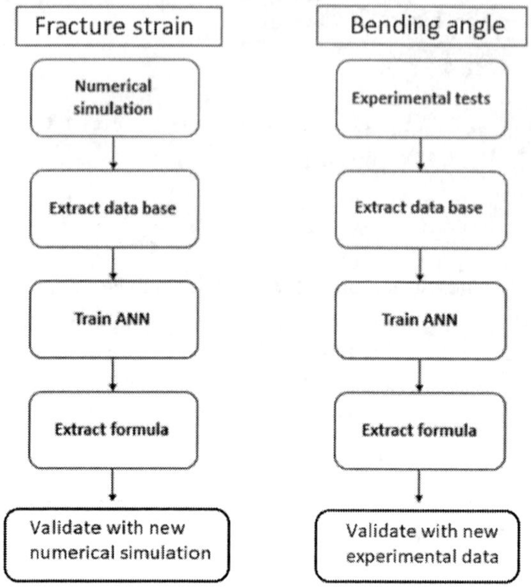

Figure 7. Data procesing

4.1.1. Finite element approach
The choice of the appropriate laws introduced to the Finite Element (FE) code is highly related to the precision of the obtained results. For the VDA 238 - 100 test specification, in order to represent the work hardening that takes place in the simulations, a combined Swift-Voce hardening model with a weighting coefficient α is used. However, the anisotropy of the material is neglected.Regarding the test geometry, only half of the specimen was represented with Abaqus software with CPE4 elements. The elements used to represent the specimen were 2D elements dedicated for 3D structures under plane strain hypothesis. The fracture strain is measured at the extrados figure 8 for at least 10 values of punch displacement during test.

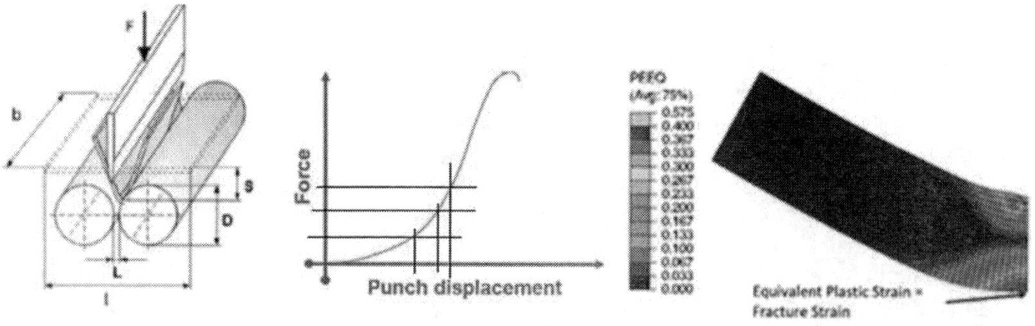

Figure 8. VDA 238-100 test specification: illustration of numerical approach for failure strain determination

4.2. Feature importance

This step allows us to attribute a score parameter to assess which ones of the input vector have the high effect on the outputs and which must be kept as inputs for our models (figure 9).

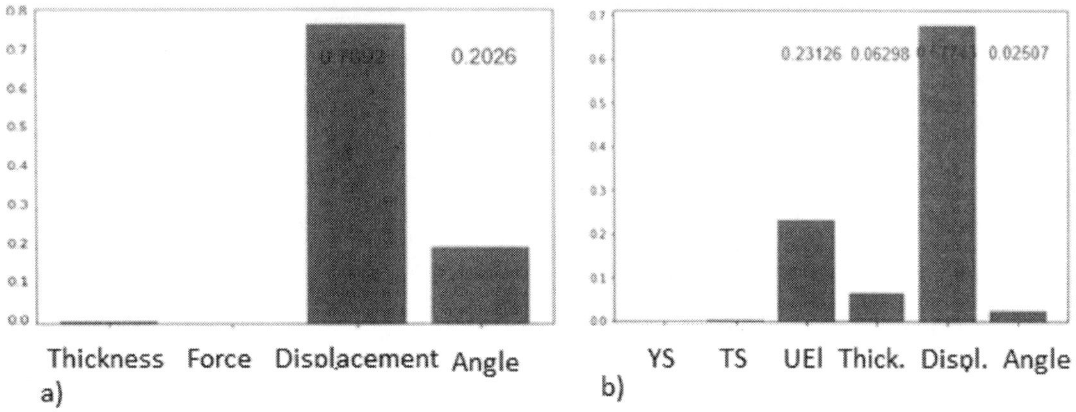

Figure 9. Feature importance a) Bending angle b) Fracture strain Where YS- Yield Strength , TS- Tensile Strength , UEl- Uniform Elongation, Thick- thickness, Displ-Displacement

The relevant inputs that are used in the ANN :

- For bending angle: Displacement and Measured angle after SB
- For fracture strain: Uniform Elongation UEl, displacement and thickness

4.3. Splitting the data

The last step before starting to design the model is to split our data bases into train set and test set. The train set is used for the training of the model and the test one is used for scoring the model at the end. In our case we chose to use 80% of the data for training and 20% of the data for the scoring.

4.4. Building and training the model

Model were built with following parameters (table 1).

Table 1. Hyper parameters

Hyper Parameter	Model 1 Bending angle	Model 2 Fracture strain
Number of hidden layers	2	2
Number of neurons per layer	30	20
Activation function	Sigmoïd	Sigmoïd
Batch size	50	50
Number of epochs	400	500
Optimizer	Adam	Adam

Then, the training is made over browsing the train set over 400 - 500 epochs. Figure 10 represents an example the loss curve of fracture strain. It decreases from a loss value of 0.9075 until it converges on error value of 0.011 at the end of the training.

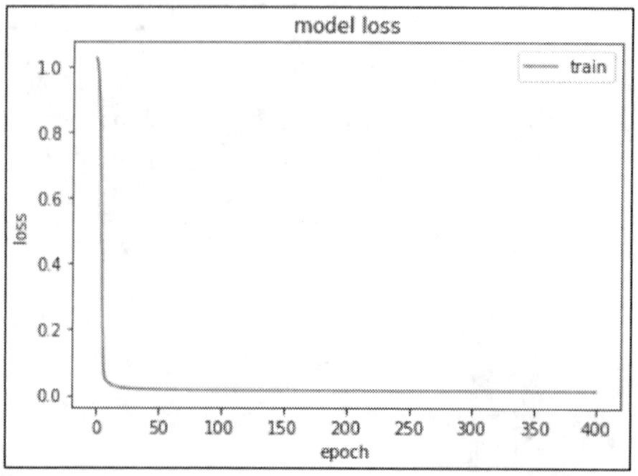

Figure 10. Visualization of error (Model 1: Bending angle)

4.5. Evaluating model

The evaluation of the model is possible thanks the remaining 20% of the data. Scikit Learn library offers a predefined function that calculates the mean absolute error (MAPE) according to equation (1). And then the accuracy is calculated as shown in equation (2).

$$\text{MAPE} = \frac{1}{n}\sum_{k=1}^{n}\left|\frac{Y_{REAl}-Y_{PRED}}{Y_{REAl}}\right| \tag{1}$$

$$\text{Accuracy} = 100\% - \text{MAPE} \tag{2}$$

The accuracy of our models was calculated as follow:

- Model 1 Bending angle: 95. %
- Model 2 Fracture strain: 92 %

5. Results

5.1. Results for bending angle

A first model based on ANN was built to determine bending angle at Fmax-60N and relate the bend angle with the displacement and angle after springback. A good correlation between neural network model and experimentally bending angle given by using laser liner mounted was found and illustrated by figure 11.

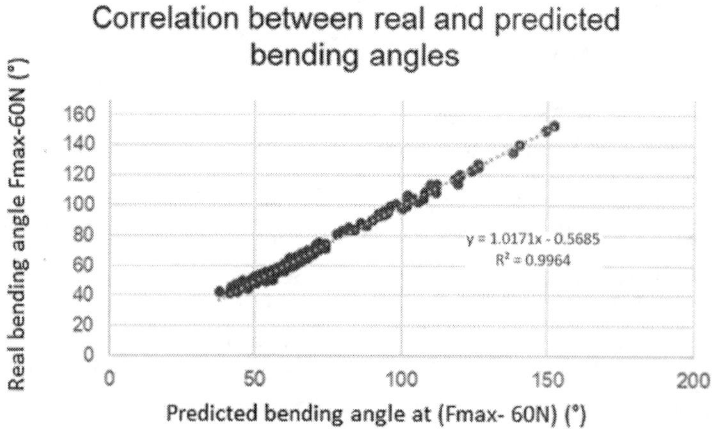

Figure 11. Neural network estimations are in good agreement with experimental angle measured

In figure 12 a comparison is done between spring back observed in the experimental test, spring back predicted by ANN and the difference between (α_{Fmax}- α_{ApRE}) all plotted against the measured angle under load (Fmax-60N).

It could be noticed that for high bending angles (>100°), the measured angle after spring back α_{ApRE} is higher than the angle at maximum load α_{Fmax} which contradicts experimental evidence where (α_{Fmax}> α_{ApRE}). Furthermore, the predicted spring back values are closer to the actual values obtained in the experimental test than those calculated using the VDA238-100 formula.

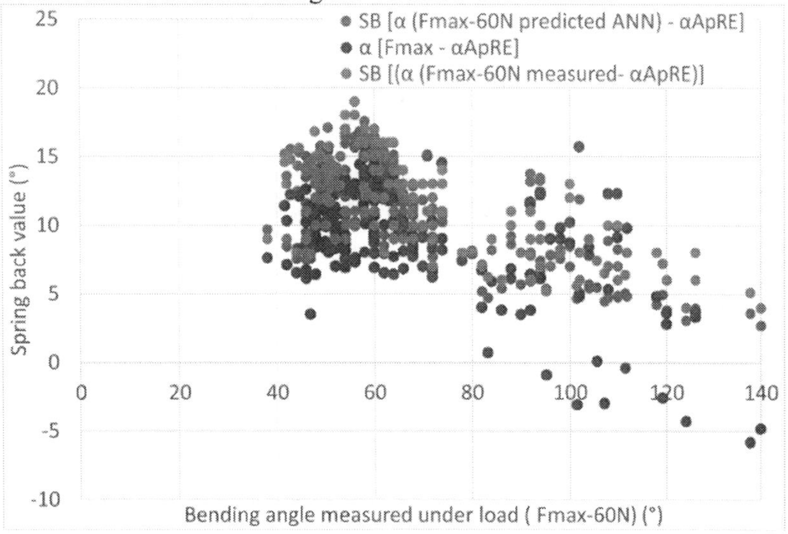

Figure 12. With ANN, predicted values are positive like experimental ones

5.2. Results for fracture strain

A second model based on ANN was built to determine fracture strain, to relate the bending angle with the fracture strain and to replace the numerical simulation currently used to determine these values. The developed neural networks should consider as input the punch force displacement curve obtained in a VDA238-100 test specification and the mechanical characteristics. A good correlation between neural network model and numerical simulation was found and reported in figure 13.

The numerical model is compared with the experimental data obtained with DIC as shown in Figure 14. The ANN obtained model agrees reasonably well with the experimental data.

Nevertheless, the model coming from ANN overestimates the values of experimental data, this is due to ANN inputs learning, based on FE results and not one experimental data. To improve model ANN a larger number of the experimental test needs to be done in order to provide a proper statistical evaluation.

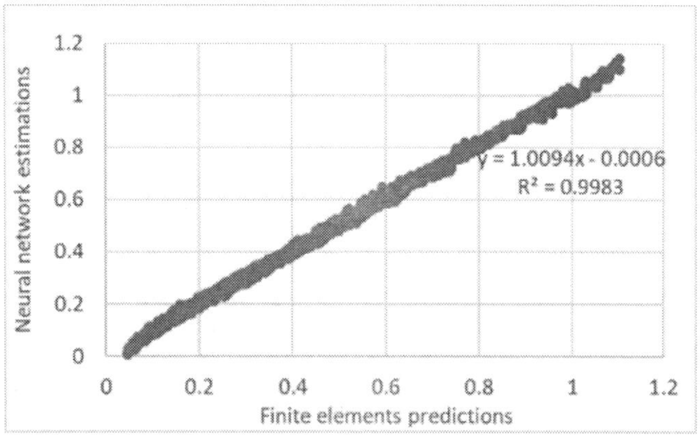

Figure 13. Neural network estimations are in good agreement with finite elements predictions

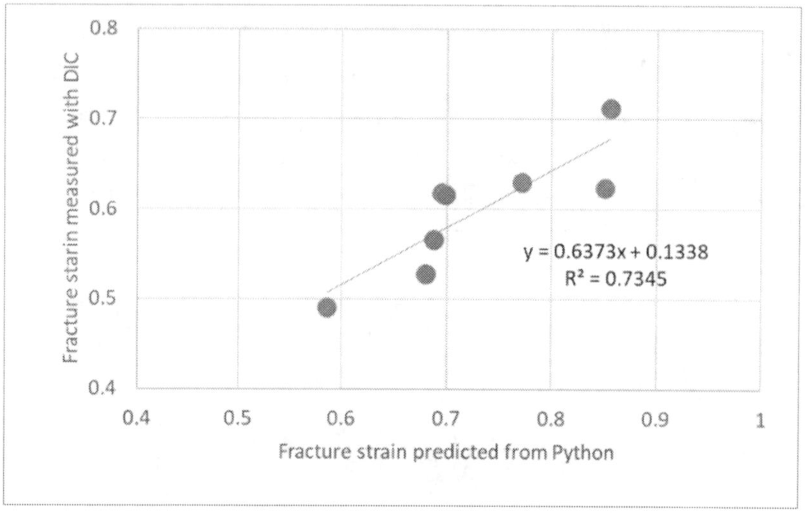

Figure 14. Neural network estimations are in agreement with experimental data

6. Conclusions

In order to assess predictive ductility an approach using neural network was developed to enable the determination of fracture strain and bending angle value before springback.

Finally, the results given by the models ANN were satisfying.

- A first model was developed to relate the VDA238-100 bending angle before spring back with displacement and measured angle after spring back. The required data are collected from experimental tests. Model gave good correlation with experimental values, much better and more reliable than formula proposed by VDA238-100 test specification
- A second model was built to estimate fracture strain as a substitution to finite element simulations which takes much more time and resources. The developed ANN considers as input, the thickness, the punch displacement curve obtained in a VDA238-100 test specification and the mechanical characteristics measured by uniaxial tensile test. A good correlation between neural network model and numerical simulation was found. Moreover, the numerical model responses agree reasonably well with the experimental data obtained with DIC.

Nevertheless, direct online optical bending angle measurement, is therefore recommended to obtain these values only from experimental test.

References

[1] VDA 238–100 2017. Test Specification: Plate Bending Test for Metallic Materials
[2] Larour P, Hackl B, Leomann F 2013 Sensitivity analysis on the calculated bending angle in the instrumented bending test, *Proc. Int. Conf. IDDRG* (Zurich: Switzerland) p 317-322
[3] Larour P, Hackl B, Leomann F, Benedykt K 2012 Bending angle calculation in the instrumented three-point bending test *Proc. Int. Conf. IDDRG (Mumbai)*
[4] Durrenberger L, Dietsch D 2018 Bending angle correction regarding sheet thickness *Proc. Int. Conf. IDDRG (Canada)*
[5] Cheong K, Omer Butcher JC, George R and Dykeman J 2017 Evaluation of the VDA 238-100 Tight Radius Bending Test using Digital Image Correlation Strain Measurement *Conf. Series: Journal of Physics*
[6] Cheong K, Butcher C and Dykeman J 2018 The Influence of the Through-Thickness Strain Gradients on the Fracture Characterization of Advanced High Strength Steels SAE Int J Mater Manuf, p 541-552

[7] Arolaa M-A, Kaijalainena K, Kestib A, Pokkaa V, Larkiolaa AP J 2017 Digital image correlation and optical strain measuring in bendability assessment of ultra-high strength structural steels *Journal of Physics 36th IDDRG Conference – Materials Modelling and Testing for Sheet Metal Forming*

[8] Noder J, Abedini A, Butcher C 2020 Evaluation of the VDA 238–100 Tight Radius Bend Test for Plane Strain Fracture Characterization of Automotive Sheet Metals *Experimental Mechanics* vol 60, p 787–800

[9] Noder J, Dykeman JC, Butcher C 2021 New Methodologies for Fracture Detection of Automotive Steels in Tight Radius Bending: Application to the VDA 238–100 V-Bend Test *Exp Mech* volume 61, pp 367–394

[10] Hanna M A, Ural D, and Saygili G, 2007 Evaluation of liquefaction potential of soil deposits using artificial neural networks. *Engineering Computations, 24, 5*

[11] Serban FM, Grozav S, Ceclan V, Turcu 2014 A Artificial Neural Networks Model for Springback Prediction in the Bending Operations *https://doi.org/10.17559/TV-20141209182117*

[12] Senol O, Esat V, Darendeliler H 2014 Springback Analysis in Air Bending Process through Experiment based Artificial Neural Networks 11th International *Conference on Technology of Plasticity* (Nagoya: Congress Center)

[13] Wagner L, Larour P, Dolzer D, Leomann F, and Suppan C 2020 Experimental issues in the instrumented 3 point bending VDA238-100 test *Materials Science and Engineering* p 967

[14] Boul I, Lang M, Mees G, Sponem F, Sturel T and Garat X 2007 Bending test and bendability of Ultra High Strength Steels. *Proc. Int. Conf. IDDRG (Győr)*

[15] Seemuang N, Panich S and Slatter T 2017 Bendability evaluation of sheet metals in three-point bending test by using acoustic emission features. *J Appl Sci 16 (2),* pp 15-22.

Experimental characterization of forming behavior of 3rd GEN AHSS

E Champolivier [1*], D Brancherie [1], P Feissel [1], S Gaied [2] and J-F Canourgues [2]

[1]Roberval Laboratory, Université de Technologie de Compiègne, France
[2] ArcelorMittal Global R&D Montataire, France

* elise.champolivier@utc.fr

Abstract. Third generation of Advanced High Strength Steels (3rd GEN AHSS) for cold stamping combine superior formability and ductility at higher strengths. Their combination with the Laser Welded Blanks (LWB) technology is one answer to improve safety and weight reduction of vehicules for automakers. In this study, a Nakazima test is used to evaluate the forming behavior of LWB based 3rd GEN AHSS and to estimate their Forming Limit. Tests equipped with a Digital Image Correlation (DIC) system are carried out for several materials combinations in order to understand the influence of the weld on LWB formability. The experimental results are used to determine the critical area during this forming test and a dedicated FLC for each material combination. The location of the onset of failure is investigated, from optical observations, DIC data as well as modeling. It appears that different failure locations are obtained depending on both AHSS combination and strain paths. In order to improve the accuracy of the determination of the onset of the failure, a method is proposed to evaluate the failure risk in base metal and in the vicinity of the weld. The improvement of failure prediction will be very helpful to analyze part formability and provide design recommendations.

1. Introduction

In recent years, car manufacturers have faced challenges to improve safety and weight reduction of their vehicules. Several solutions have been developed to meet the higher performance requirements of body structures including the Laser Welded Blanks technology. This technology consists in butt-welding steel sheets before stamping to obtain a vehicle's part. Thickness combinations and material location can then be optimized to reduce the weight of parts while maintaining or improving safety performances.

New solutions based on this technology are in development by integrating Advanced High Strength Steels (AHSS). The development of these new solutions leads to adapt the formability prediction.

In industry, the most common forming criteria is the Forming Limit Curve (FLC) which gives the maximum admissible principal strains of a sheet before the onset of necking as a function of the deformation path. Regarding Laser Welded Blanks (LWB), in literature, most of the authors agree with the fact that their formability is reduced by comparison with monolithic sheets [1, 2, 3, 4]. Furthermore, experience in industry shows that FLC, defined for monolithic sheets, cannot be applied in the case of LWB due to particular phenomena induced by the joint and its Heat Affected Zone (HAZ). Various types of failure can be observed in the vicinity of the joint depending on the loading direction with respect to the weld: failures perpendicular to the weld in the case of longitudinal loading and failures parallel to the welds in regions close to the weld in the case of transverse loading [3, 5, 6, 7]. Thus, the FLC criterion needs to be adapted to be used for LWB.

The use of AHSS rises another challenge. During the welding process, a hardness drop appears in the HAZ of AHSS welds. Due to this softened zone in the HAZ, the prediction of the forming behavior of LWB could be different [8, 9, 10, 11] and the formability of LWB with AHSS could be affected under specific loading cases [1, 9, 10].
It is therefore necessary to understand the effect of this specific area on formability. This could facilitate the design of new parts, in particular the choice of the location of the weld.

In this study, a method is proposed to evaluate the formability of LWB made of AHSS. It is based on both experiments and computations for the analysis of various tests and aims at proposing a model at both the weld and the sheet scales. In this framework, this paper focuses on a specific Nakazima test on LWB. The aim is to understand the influence of the material interaction, weld and HAZ on the forming behavior of LWB, in particular by investigating failure locations and strain map prior to failure under severe forming conditions. Failure criteria for AHSS is then proposed based on the experimental results.

2. Methodology

2.1. Nakazima test

The Nakazima test is currently used to determine the Forming Limit Curve (FLC) of monolithic sheet metals. It consists in deforming a blank with a hemispherical punch until failure. Figure 1 shows a schematic diagram of the test.

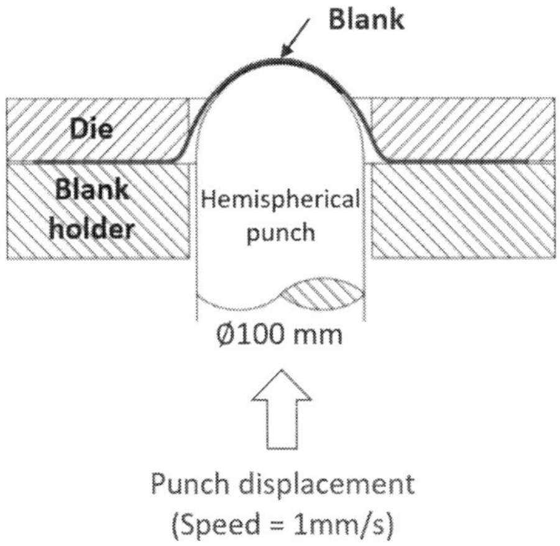

Figure 1. Schematic diagram of the Nakazima test

The experimental tool is equipped with a Digital Image Correlation (DIC) system in order to measure the strain distribution during the tests. This test is defined by the ISO 12004-2 standard [12] which provides the protocol modalities and the method to determine FLC from the DIC measurements. By varying the geometry of the samples, various strain paths can be generated in the blank to evaluate its forming capability (figure 2).

2.2. Design of experiment

Various steel combinations based on a 3rd GEN AHSS with varying grades and/or thickness are investigated. The 3rd GEN steel used in the study was supplied by ArcelorMittal and will be

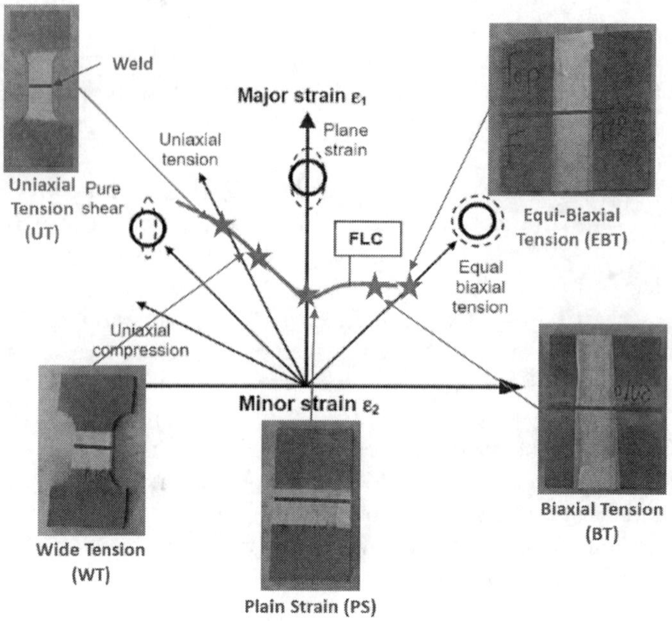

Figure 2. Strain paths on strain diagram and corresponding sample geometry.

refered in the following by its trade name: Fortiform®. A dedicated welding process developped by ArcelorMittal is used to avoid Brittle Liquid Metal embrittlement and brittle behavior of the weld. The assessed modalities in this work are presented in table 1 and denominations used in the table are defined in figure 3. The "Comment" column in the table specifies the weakest material of each assembly defined as the lowest value of the product "thickness by maximum tensile stress". Basically, the failure occurs in the weakest material. For each modality, the weld is positioned in the direction parallel to the sheet rolling direction. In the following, the materials of the blanks are denoted as the base metals (BM).

Table 1. Investigated modalities Fortiform ®based LWB.

Modality	Material A	Th. A (mm)	Material B	Th. B (mm)	Comment
1	Fortiform ®1180	1.23	Fortiform ®1180	1.23	Mechanical equilibrium
2	Fortiform ®1180	1.23	DP600	1.2	Weakest material : DP600

5 strain paths (therefore 5 sample geometries) and 5 specimens for each combination of materials are tested (figure 2).
For each specimen, the punch force and corresponding displacement are measured as well as the strain field during the test (figures 4 and 5). Failure location analysis is first carried out by visual inspection of broken specimens.

3. Experimental results and discussion

Results of the comparison between the 2 considered modalities are presented in this section. Modality 1 is a same grade/same thickness LWB. This is the most critical case for the weld and

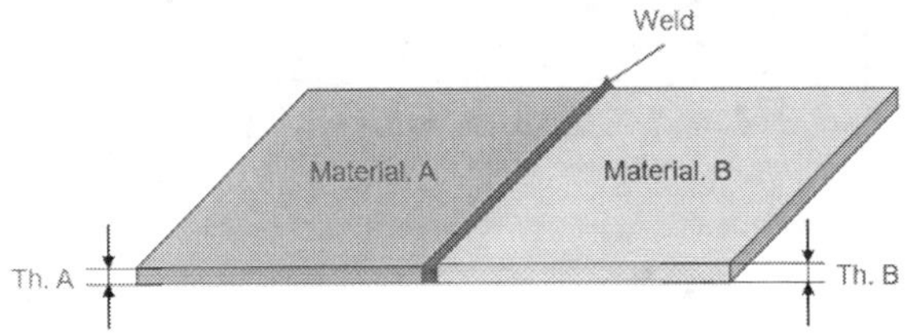

Figure 3. Definition of the denominations used in Table 1

Figure 4. Experimental set-up

Figure 5. Experimental set-up with DIC device

its HAZ as there is a mechanical balance on each side of the weld. In modality 2, a mechanical interaction is introduced by the use of a base metal with different mechanical properties.

3.1. Micro-hardness study

Micro-hardness measurements were firstly performed to analyze the local properties in each zone including weld, HAZ and base metal. Figure 6 presents the superposition of the micro-hardness profiles on a line orthogonal to the weld in the section of a specimen.

A drop of hardness can be observed in the HAZ of Fortiform ®1180 in both modalities. This observation is in accordance with the literature where HAZ softening is observed in AHSS welds. For modality 2, DP600 hardness is lower than Fortiform ®1180 BM and there is no hardness drop on the DP600 HAZ.

3.2. Failure types during Nakazima test

In this paper, 3 types of failure are mentioned: failures occuring in the BM far from the weld, failures occuring in the weld and failures ocuring in the HAZ or a few millimeters from the weld in the BM. This last type of failure is due to the mechanical interaction between Base Metals and the presence of the weld. It will be refered as *interaction zone* failure.

Figures 7 and 8 show a summary of the failure types observed for modalities 1 and 2 respectively.

IOP Conf. Series: Materials Science and Engineering 1284 (2023) 012077 doi:10.1088/1757-899X/1284/1/012077

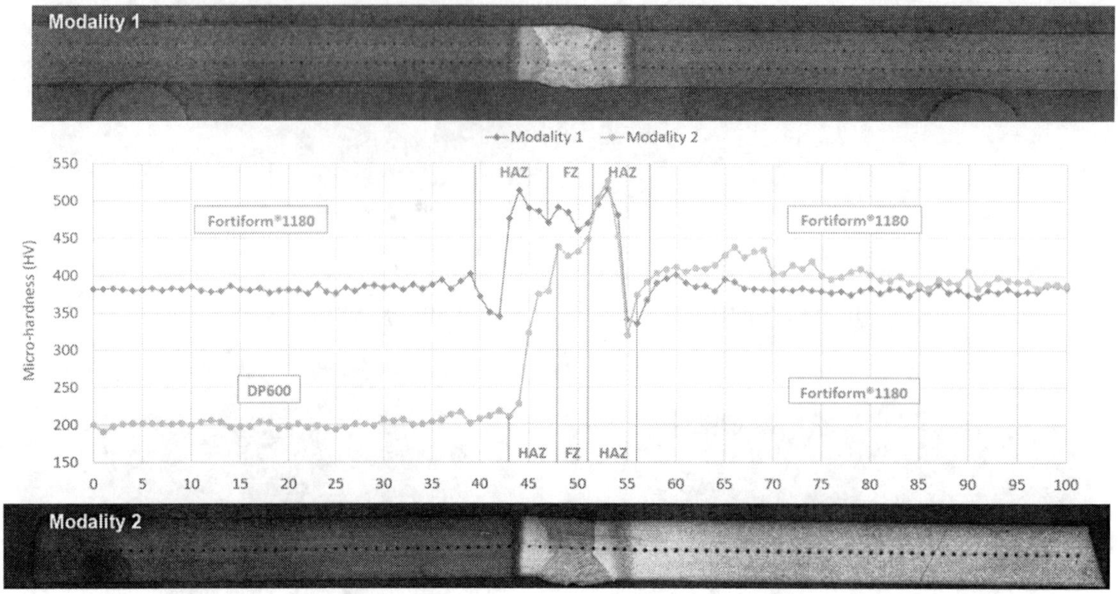

Figure 6. Micro-hardness profiles transverse to the weld for modality 1 and 2.

For all the modalities, various failure locations are observed depending on the strain path. For modality 1 (figure 7), 3 types of failure are observed. For Uniaxial Tension (UT) specimens, all the failures occur in the BM far from the weld. The Wide Tension (WT) path is a transition point as mixed failure types are obtained showing the competition between interaction zone and BM behavior. Then, from WT to Biaxial Tension (BT), only interaction zone failures are observed. Finally, specimens in Equi-Biaxial Tension (EBT) show mixed failure types demonstrating competitive behavior between interaction zone and Fusion Zone properties.

For modality 2 (figure 8), all the failures appear on DP600 side as expected since DP600 is the weakest material of the assembly. Failures are located on the BM far from the weld for the strain paths on the left of the deformation diagram. For Plain Strain (PS), Biaxial Tension and Equi-Biaxial Tension failures occur in the BM a few millimeters from the weld in the "interaction zone".

Interaction zone failures are observed for modality 1 at the location corresponding to the hardness drop in the hardness profile. Observed failures demonstrate the weakest mechanical behavior of the interaction zone with respect to the BM depending on the strain path. For modality 2, no drop of hardness is observed in DP600 HAZ. The hardness of DP600 is also lower than the minimum hardness value measured in the Fortiform ®1180 on the other side of the weld. Therefore, the interaction zone is located in the weakest material is the DP600.

3.3. Strain field analysis

For each LWB specimen, the FLC calculation is based on an internal method used by ArcelorMittal. This method is inspired from the ISO12004-2 standard which describes the calculation method for FLC of monolithic sheets. For more details on the ISO12004-2 method, readers are invited to refer to [12]. To be applied to LWB, an additional step is added to the standard method before the exclusion of the points located in the necking region. The measured strain profiles perpendicular to the failures are symmetrized with respect to the crack to build theorical strain profiles before failures similar to the ones that could be obtained for an equivalent

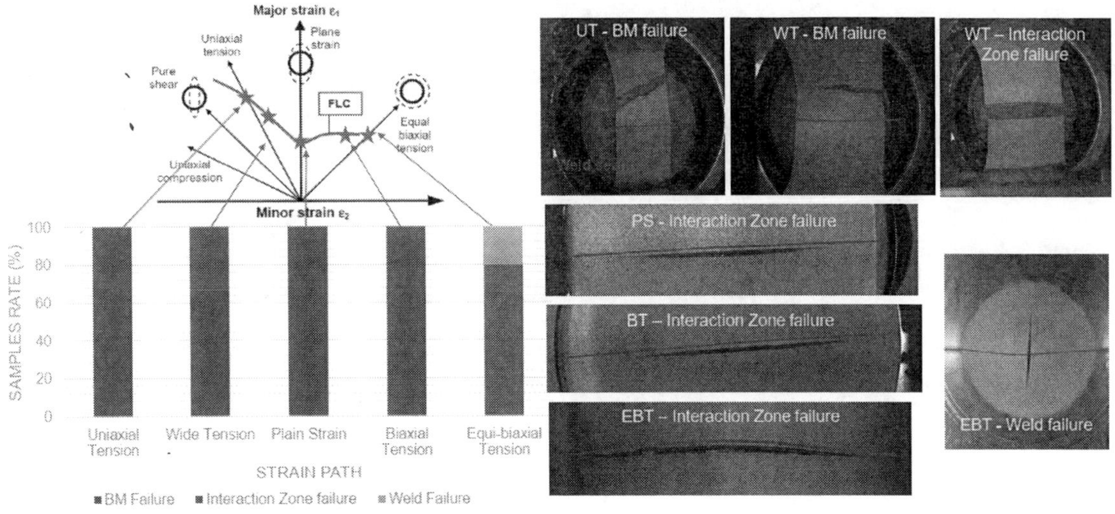

Figure 7. Failure types in function of the strain path for modality 1.

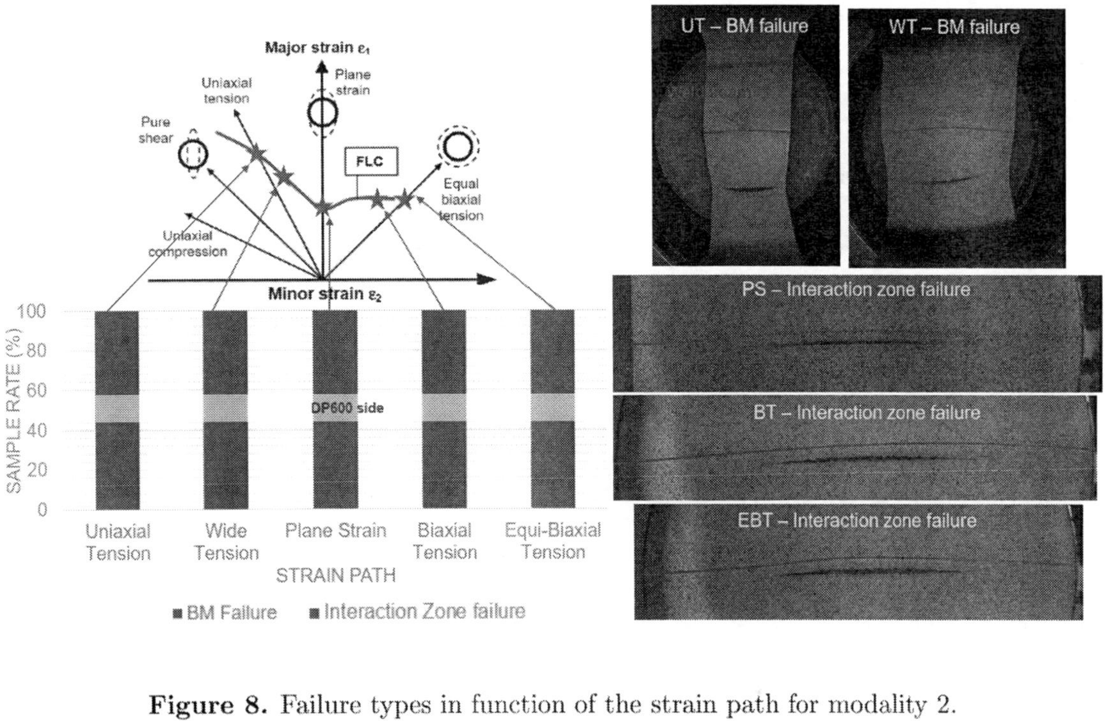

Figure 8. Failure types in function of the strain path for modality 2.

monolithic sheet. These profiles are then treated as described in the ISO12004-2 standard to determine the forming limit points of the specimens.

The forming limit points obtained for the 2 studied modalities are shown on figures 9 and 10.

The plots show that BM failures occur on the left side of the forming limit diagram (when the minor strain is negative) and that interaction zone failures occur in expansion for both assemblies under study. It can also be noted that interaction zone failures occur at lower strain levels than BM failures.

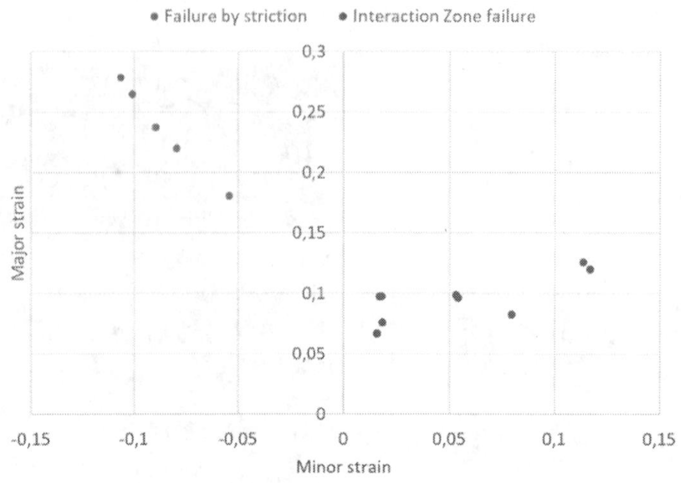

Figure 9. Forming limit diagram - modality 1.

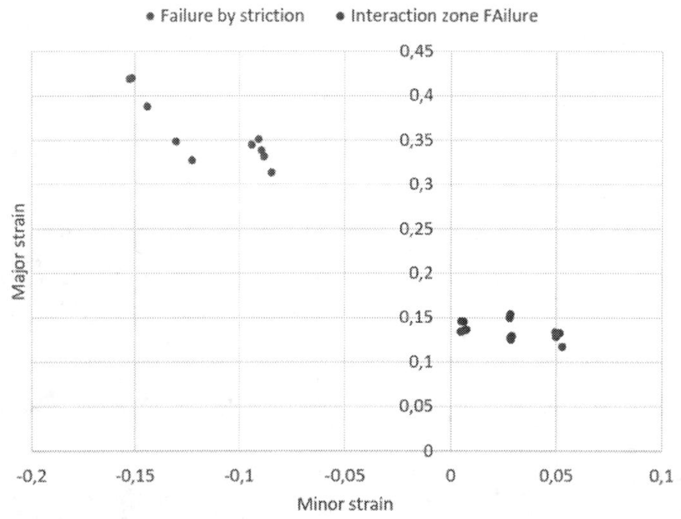

Figure 10. Forming limit diagram - modality 2.

4. Modeling of LWB FLC

Figures 11 and 12 show the forming limit points of each modality superimposed on the FLC of the BM (weakest material). These FLC are obtained from a forming prediction model, usually called ArcelorMittal V9.2, developped by Cayssials [13, 14] and validated for a wide range of steels.

The plots show that BM failures are well predicted by the FLC of the weakest BM in the assembly in both cases.

In the case of interaction zone failures, the formability is reduced compared to monolithic BM. A previous study [15] showed that forming behavior of LWB was mainly depending on the mechanical interaction between Base Metals in case of failure in interaction zone. Based on experimental data and modeling, a specific interaction criterion was defined by the authors to predict the forming behavior of the area in the vicinity of the weld on the weakest BM of the LWB

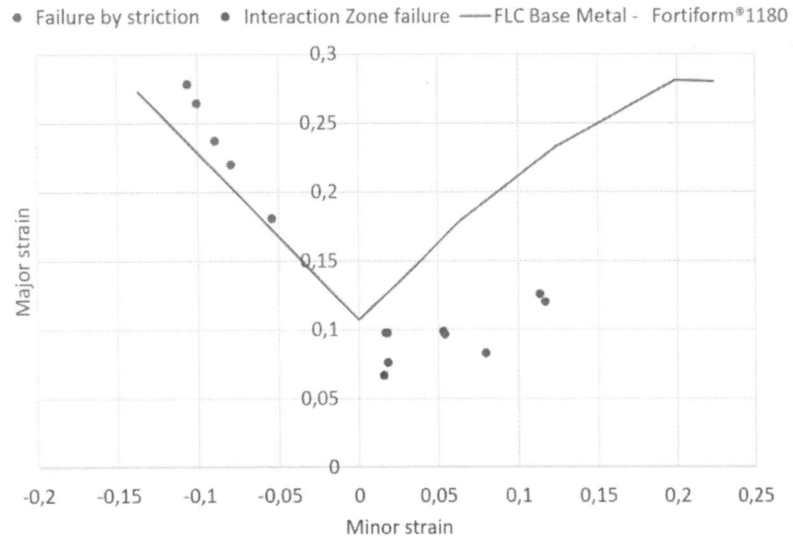

Figure 11. Experimental forming limit points and existing forming limit criterion - modality 1.

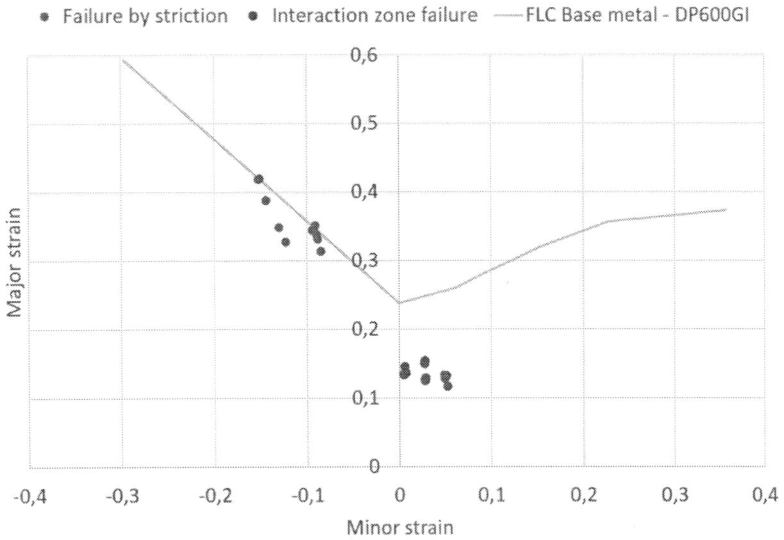

Figure 12. Experimental forming limit points and existing forming limit criteria - modality 2.

(this region corresponds to the interaction zone referred to above). This study was performed on LWB based on mild steels. In the present study, the first experimental results show the presence of failures in the interaction zone with the use of 3rd GEN AHSS too. The interaction zone model could therefore be used to predict the forming behavior of this area. However, the previous model of interaction zone FLC, developped for mild steels, is not suitable for LWB based on 3rd GEN AHSS. Therefore, this model needs to be extended for 3rd GEN AHSS. The first elements of the construction of this model will be presented during the conference.

5. Conclusion

In this study, Nakazima tests are performed in order to investigate the formability of 3rd GEN AHSS based LWB under various strain paths. Results show that various types of failures are obtained depending on the strain path: BM, interaction zone and weld failures. It can be noted that the failure in interaction zone occur at lower strain levels than BM failures.

Experimental results demonstrate that the base metal forming limit curve could be used if the failure occurs far from the weld. However, if failure occurs in the interaction zone a dedicated FLC for LWB has to be used. In a previous study, a forming limit criteria for the interaction zone of mild steels based LWB was developped. First results show that this model needs to be adapted in the case of 3rd GEN AHSS use. The purpose of the work to be followed is to complete the study on the specific forming prediction model for the interaction zone of LWB to apply it in the case of 3rd GEN AHSS. The improvement of failure prediction will be very helpful to analyze part formability and provide design recommendations.

References

[1] Ono M, Yoshitake A and Ohmura M 2002 Laser weldability of high strength steel sheets in fabrication of tailor welded blanks *NKK TECHNICAL REVIEW* **86** 13–8

[2] Baysore K, Williamson M S, Adonyi Y and Milian J L 1995 Laser beam welding and formability of tailored blanks *Weld. Res. Suppl.* 345–52

[3] Saunders F I and Wagoner R H 1996 Forming of tailor-welded blanks *Metall. Mater. Trans.* A **27** 2605–16

[4] Abbasi M, Ketabchi M, Ramazani A, Abbasi M and Prahl U 2012 Investigation into the effects of weld zone and geometric discontinuity on the formability reduction of tailor welded blanks *Comput. Mater. Sci.* **59** 158–64

[5] Zadpoor A A, Sinke J and Benedictus R 2007 Mechanics of tailor welded blanks : an overview *Key Eng. Mater.* **344** 373–82

[6] Panda S K, Kumar D R, Kumar H and Nath A K 2007 Characterization of tensile properties of tailor welded IF steel sheets and their formability in stretch forming *J. Mater. Process. Technol.* **183** 321–32

[7] Gaied S, Cretteur L and Schmit F 2013 Modeling of laser-welded blank formability *Rev. métall.* **110** 199–203

[8] Jia Q, Guo W, Li W, Zhu Y, Peng P and Zou G 2016 Microstructure and tensile behavior of fiber laser-welded blanks of DP600 and DP980 steels *J. Mater. Process. Technol.* **236** 73–83

[9] Farabi N, Chen D L, Li J, Zhou Y and Dong S J 2010 Microstructure and mechanical properties of laser welded DP600 steel joints *Mater. Sci. Eng.* A **527** 1215–22

[10] Gong H, Wang S, Knysh P and Korkolis Y P 2016 Experimental investigation of the mechanical response of laser-welded dissimilar blanks from advanced- and ultra-high-strength steels *Mater. Des.* **90** 1115–23

[11] Bandyopadhyay K, Panda S K, Saha P, Baltazar-Hernandez V H and Zhou Y N 2016 Microstructures and failure analyses of DP980 laser welded blanks in formability context *Mater. Sci. Eng.* A **652** 250–63

[12] International Organization for Standardization 2008 ISO/DIS 12004-2:2008. Metallic materials — Sheet and strip — Determination of forming-limit curves Part 2: Determination of forming-limit curves in the laboratory

[13] Cayssials F 1998 *Sheet Metal Forming Beyond 2000 : Proc. of the 20th Biennial [IDDRG] Congress (Brussels)* (Liege: CRM) p 443–54

[14] Cayssials F and Lemoine X 2005 *Proc. of the 24th IDDRG Congress* (Besançon)

[15] Gaied S, Roelandt J M, Pinard F, Schmit F and Balabane M 2009 Experimental and numerical assessment of tailor-welded blanks formability *J. Mater. Process. Technol.* **209** 387–395

IOP Conf. Series: Materials Science and Engineering 1284 (2023) 012078 doi:10.1088/1757-899X/1284/1/012078

Dimensional precision and wear of a new approach for prototype tooling in deep drawing

K Holzer[1], L Maier[1], V Böhm[1] and W Volk[1]

[1]Chair of Metal Forming and Casting, Technical University of Munich, Walther-Meissner-Strasse 4, 85748 Garching, Germany

E-mail: katja.holzer@utg.de

Abstract. In this work, we present and evaluate a new approach for prototype tooling in deep drawing based on direct polymer additive tooling. With fused filament fabrication (FFF) a PLA shell is printed additively. Afterwards, this is filled with ultra-high performance concrete (UHPC). UHPC is characterized by its higher strength properties compared to conventional concrete materials, which makes the material feasible for forming applications. Two configurations of these hybrid UHPC polymer additive are possible: either the PLA shell is in contact with the sheet metal during forming or UHPC. The hybrid UHPC polymer additive tooling approach has the potential to be more cost-efficient for small series. The dimensional precision and wear of such hybrid tools is evaluated using a standard cup geometry. A test series of 30 cups with sheet metal DX56+Z with 1 mm thickness was drawn with the hybrid tools as well as with a polymeric tool and a conventional steel tool. The dimensional precision and wear of the prototype tools was evaluated optically.

1. Introduction

A suitable technology for the economic production of deep drawn parts in low quantities is needed. As the market is facing shortened product life cycles, an uncertain demand and pressure to innovate to meet the customer's desire for individualization.

An approach to reduce the production costs of prototypes, spare parts and individual parts is using tools made from milled polymers or casted resins [1–3]. With the latest improvements and industrialization of additive technologies, directly printed polymeric tools for sheet metal drawing become feasible. These direct polymer additive tooling are mainly produced by fused filament fabrication (FFF). The most common filament type used is polylactide (PLA). [4, 5]

Direct polymer additive tooling made from PLA and CF-reinforced polyamide (PA) [6] as well as pure PA [7] are used to produce a standard cup geometry from different deep drawing materials. The dimensional precision of the drawn cups can be improved by introducing shrinkage and springback allowances in the tool design [6]. Additionally, considerable degradation of the PA tooling was observed from a small batch of 30 cups drawn [7]. Drawing more complex geometries the tolerances with direct polymer additive tooling are rarely met, as shown by Durgun when drawing a tailgate lock from S355MC with a FFF tool made from polycarbonate [8]. The same result is seen by Schuh, Bergweiler et al. drawing a demonstrator part with near to reality geometric features from car bodies with a FFF tool made from PLA [9]. According to Nakamura, Mori et al. [10], the dimensional accuracy of sheets bent with a polymeric tools is lower than that fabricated with steel tools, due to their low stiffness. The results deteriorate with increasing flow stress of the sheet metal. They found that increasing the tool stiffness by reinforcement improves the dimensional accuracy of the bent sheets. This seems to be true for geometrically more complex parts as well, as stated by Schuh, Bergweiler et al. [9]. Therefore, there is a need for the optimization in the development of tool design to achieve the necessary dimensional tolerances.

An alternative approach to reduce the production costs of deep drawn parts is the use of ultra-high performance concrete (UHPC). UHPC is characterized by a small particle size of the binder and the use of additives to improve packing density. This results in an accelerated hydration reaction and higher strength properties compared to conventional concrete materials. [11] Schwartzentruber, Bournazel et al. [12] used high performance concrete tools with a gel coat for the production of sheet metal parts for prototype cars. Furthermore, UHPC dies have been successfully used for sheet metal hydroforming by Kleiner et al. [13] and Ritter [14, 15]. The friction conditions between UHPC and sheet metal could be improved such that the properties of the drawn parts were of good quality.

In this work, we propose the use of deep drawing tools made out of a combination of additive layered (FFF) PLA and UHPC. These hybrid UHPC polymer additive tools promise improvement by combining the contrasting properties of the materials. UHPC, with an elastic modulus between 45 – 55 GPa [11], increases the overall tool stiffness compared to direct polymer additive tooling, as PLA build up by FFF shows a modulus of 1.34 – 3.23 GPa depended on the layer thickness [7]. On the other hand, the PLA part absorbs stress peaks. This reduces the risk of brittle failure prone to UHPC. Additionally, hybrid UHPC polymer additive tools can reduce the tooling costs even further compared to direct polymer additive tooling, as UHPC is a cheap, castable material reducing the overall additive manufacturing time of the tools.

2. Approach and Tools

2.1. Hybrid UHPC polymer additive tooling

The hybrid UHPC polymer additive tooling approach consists of the combination of a PLA shell build up by FFF filled with UHPC. Two configurations of these hybrid UHPC polymer additive are possible as shown in Figure 1. Either the PLA shell is in contact with the sheet metal during forming or UHPC.

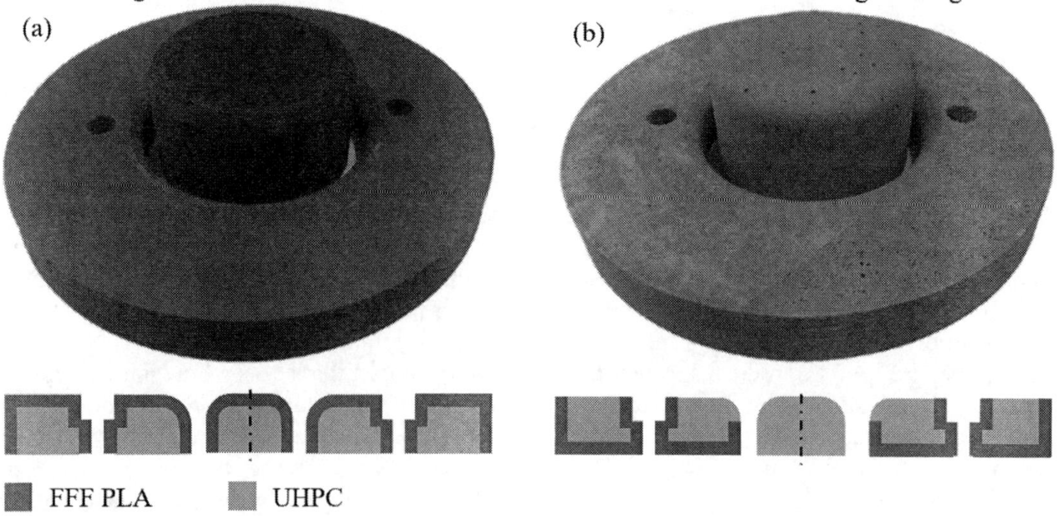

Figure 1. Hybrid UHPC polymer additive tooling photograph and scheme with (a) FFF PLA in contact (b) UHPC in contact with sheet metal.

2.1.1. FFF PLA in contact. For hybrid UHPC polymer additive tooling with FFF PLA in contact with the sheet metal, a shell of PLA is build up by FFF directly in the shape of the target geometry of the formed part. The shell used in this work has a thickness of 2 mm. The UHPC is mixed accordingly and

IOP Conf. Series: Materials Science and Engineering 1284 (2023) 012078 doi:10.1088/1757-899X/1284/1/012078

filled into the shell. No additional tooling or aid is needed. The tools are afterwards placed in saturated atmosphere at room temperature for 24 h. After the curing the UHPC reached around 80 % of its final strength. Final strength is reached after 28 days. Due to the low melting point of PLA a heat treatment, to reach final strength in 2 days, is not possible [16]. The filling side of the tools is grinded down manually to a flat surface. The tools were stored at room temperature.

2.1.2. UHPC in contact. For hybrid UHPC polymer additive tooling with UHPC in contact with the sheet metal, a shell of PLA is build up by FFF giving the shape according to the connection to the tool frame. The shape of the target geometry of the formed part is given by an additional formwork. This can be achieved by a FFF PLA shell that gets removed after curing of UHPC or by using a reusable formwork. In this work the indirect rapid tooling approach was used for casting near-net-shape tools [17]. The curing, demolding, grinding and storage of the tools is done as for the first configuration.

2.2. Reference tools
Within the scope of this work the configurations of hybrid UHPC polymer additive tooling will be compared to conventional steel tooling as well as direct polymer additive tooling in deep drawing tests of a standard cup geometry, Figure 2. The conventional steel tool is used for Marciniak cup test and forms the basis for the geometric specification of all tools used in this work [18]. The polymer additive tooling is build up by FFF in PLA with a shell thickness of 2 mm and a 80 % triangle infill according to [19].

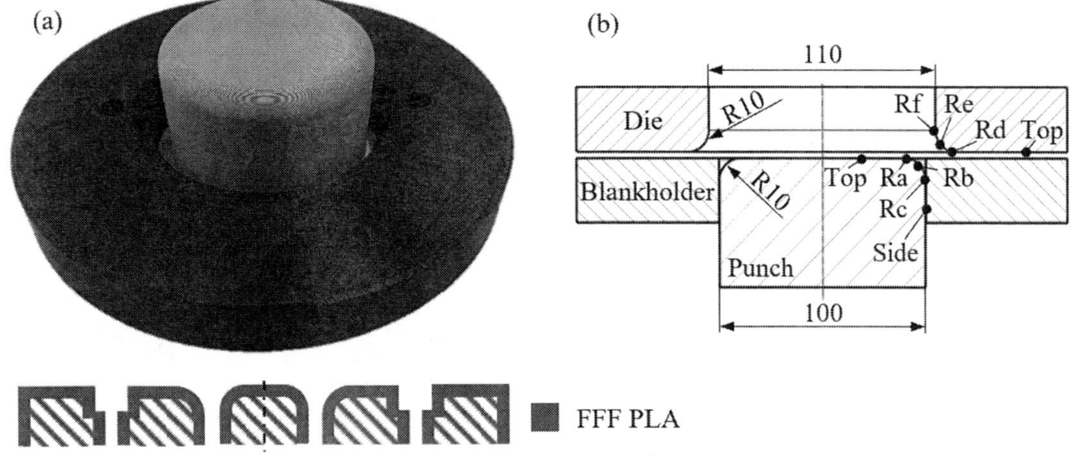

Figure 2. (a) Direct polymer additive tooling photograph, scheme and (b) geometric specifications and measurement points of tools used.

3. Materials and Methods

3.1. Materials

3.1.1. PLA. The filament used is PLA, by Das Filament, Emskirchen, Germany and has a diameter of 1.75 mm. To process it by fused filament fabrication (FFF) a Chiron Anycubic, Shenzhen, Hongkong is used. It has a maximum print volume of 400 × 400 × 450 mm3 in X-, Y-, and Z-axis, respectively. The nozzle diameter, which determines the layer resolution is 0.4 mm. The printing accuracy of the Chiron is 0.05 to 0.3 mm, with a position accuracy of 12.5 μm in the X- and Y and 2 μm in Z-axis. The printing parameters for the PLA parts used is given in Table 1.

3.1.2. UHPC. In this work the UHPC used consists of binder, mixing water and a superplasticizer. As binder the ready-to-use cement Nanodur Compound 5941 from Dyckerhoff GmbH, Wiesbaden, Germany, is used. The superplasticizer ADVA Flow 375 (BV/FM) from GCP Germany GmbH, Lügde,

Germany, was added to the mixing water to prevent the formation of agglomerates and improves workability of the fresh concrete. Each constituent is weighed with a tolerance of ± 1 mass percent before mixing. Mixing is done by hand for at least 12 minutes using a commercial mortar bucket and trowel. When a homogeneous consistency with dilatant behavior is obtained the UHPC is poured into the formwork. The added mass per volume of the constituents is given in Table 2.

Table 1. Printing parameters of PLA in FFF.

Parameter	Value
Layer height (mm)	0.2
Shell thickness (mm)	2
Nozzle temperature (°C)	195
Build plate temperature (°C)	60
Print speed (mm/s)	50

Table 2. Composition of UHPC.

Binder kg/m³	Water kg/m³	Superplasticizer kg/m³
2100.0	316.0	30.0

3.1.3. Sheet Metal. The sheet metal used to evaluate the tooling approaches in deep drawing tests is DX56+Z (CR 4, 1.0963) with a thickness of 1 mm and hot dip galvanizing. The chemical composition and mechanical properties of this soft steel for cold forming are shown in Table 3 and Table 4.

Table 3. Chemical composition of DX56+Z in weight percentage.

C max.	Si max.	Mn max.	P max.	S max.	Ti max.	Al min.	Cu max.
0.06	0.50	0.40	0.025	0.020	0.030	0.010	0.02

Table 4. Mechanical properties of DX56+Z.

Yield strength MPa	Ultimate strength MPa	Failure strain %
120 - 180	260 - 350	≥ 39

3.2. Experimental Setup and Tool Design

To perform the deep drawing experiments a BUP sheet metal testing machine with a maximum test load of 100 kN by ZwickRoell, Ulm, Germany, is used. The dimensions of the tooling is given in Figure 2 (b). The punch is attached to the bottom by a centering device and a pin to act as a twist lock. The blankholder is hold in place by a circumferential groove. The die is fixed to the top by two screws. A rotational blank with diameter of 200 mm of DX56+Z is used for lubricant free deep drawing of 30 cups with each tooling approach. The forming parameters and resulting maximum drawing force are given in Table 5.

Table 5. Forming parameters and resulting maximum drawing force.

Tooling	Forming velocity mm/s	Clamping pressure N/mm²	Drawing depth mm	Max. drawing force kN
Steel	15	1.4	40	94.07
FFF	15	0.7	40	90.36
Hybrid PLA in contact	15	0.7	40	87.87
Hybrid UHPC in contact	15	1.4	40	92.31

3.3. *Determining of precision and tool wear*

Precision and wear or the tools is assessed using the ATOS Compact Scan / 300 of Carl Zeiss GOM Metrology GmbH, Braunschweig, Germany. The optical scanning device has an accuracy of 0.017 mm in a measuring volume of 600 x 500 x 420 mm^3. The geometric deviations are evaluated by surface comparison with a CAD model fitted by a marking in model and part to ensure equal positioning at the beginning and after 30 drawing operations. A detailed evaluation is done at measurement points at eight positions radially and then averaged. For the punch five and for the die four measurement points are taken, these are given in Figure 2 (b). At the punch the point Top is located 10 mm from the center. Measurement point Side is located at the cylindrical surface 30 mm from the top. The points Ra, Rb and Rc are located at the punch radius inlet, radius center and radius outlet, respectively. The same applies to points Rd, Re and Rf on the die radius. The measurement point Top on the die is located on a circle with radius 87.5 mm from the center, still in the clamping area of the sheet metal. Precision is determined by the deviation from the CAD model in the initial state. Tool wear is defined as the negative change in deviation from initial to the state after the maximum number of strokes achieved.

4. Results

The dimensional precision and wear of the punch for direct polymer additive tooling (referenced as FFF), hybrid UHPC polymer additive tooling with FFF PLA in contact (referenced as PLA) and with UHPC in contact with sheet metal (referenced as UHPC) are given in Figure 3. For the die the same information is given in Figure 4. The left side shows the initial state of the tools, the right side mirrored the state after the maximum number of strokes achieved.

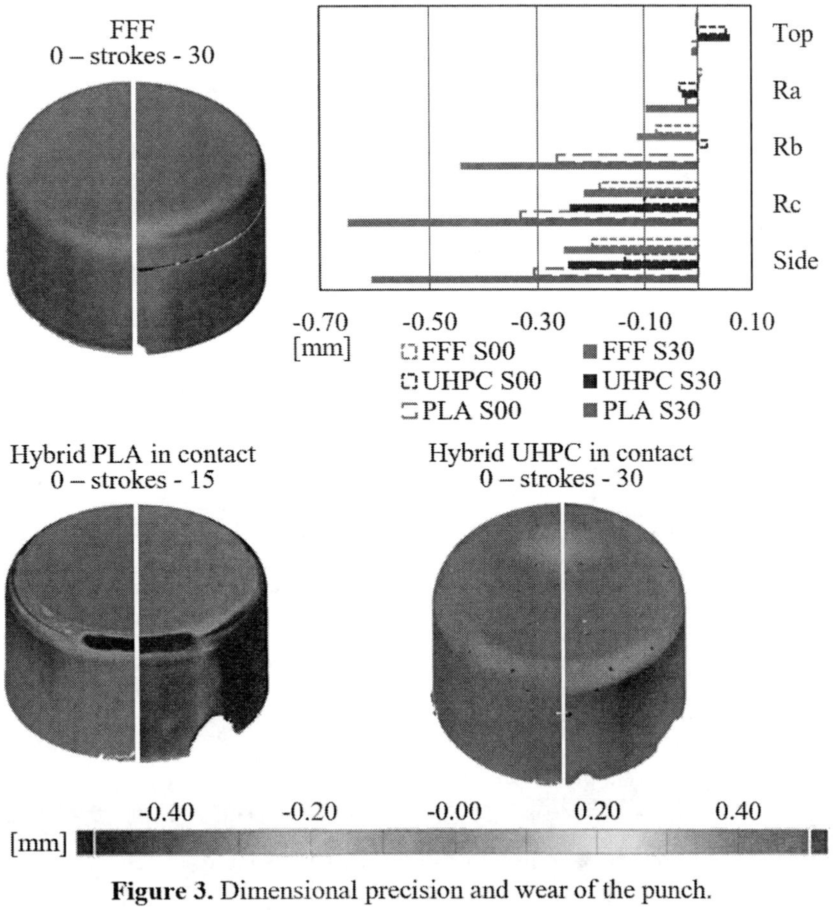

Figure 3. Dimensional precision and wear of the punch.

The direct polymer additive tooling punch shows an initial printing precision of min. -0.1 mm. The biggest wear can be seen at the radius center (Rb) at -0.08 mm after 30 strokes. The standard deviation at the measurement points is a max. of 0.1 mm with a mean value of 0.06 mm. After four strokes, the edge layer on the punch chipped, this did not affect the further deep drawing.

The hybrid UHPC polymer additive tooling punch with FFF PLA in contact shows an initial printing precision of min. -0.3 mm. The greatest wear can be seen all around the cylinder side with max. -0.3 mm. At the radius center (Rb) a reduction of -0.2 mm after 15 strokes compared to the initial state can be observed. The standard deviation at the measurement points is a max. of 0.15 mm with a mean value of 0.1 mm.

The hybrid UHPC polymer additive tooling punch with UHPC in contact shows an initial precision of min. -0.1 mm. A milled component was molded, the accuracy of which is not included here. The biggest wear can be seen at the side at -0.14 mm. At the radius center (Rb) a reduction of -0.02 mm after 30 strokes can be seen. The standard deviation at the measurement points is a max. of 0.1 mm with a mean value of 0.08 mm.

The direct polymer additive tooling die shows an initial printing precision of max. +0.4 mm. The biggest wear can be seen at the radius center (Re) at -0.13 mm after 30 strokes. The standard deviation at the measurement points is a max. of 0.05 mm with a mean value of 0.04 mm.

The hybrid UHPC polymer additive tooling die with FFF PLA in contact shows an initial printing precision of max. +0.4 mm. The greatest wear can be seen at the radius center (Re) at -0.9 mm after 10 strokes. After 15 strokes the die failed at the screw hole. A measurable reduction of the quality of the formed cup could already be seen after 10 strokes. The experiments were not continued. The standard deviation at the measurement points is a max. of 0.2 mm with a mean value of 0.13 mm.

The hybrid UHPC polymer additive tooling die with UHPC in contact failed after two strokes showing several radial cracks. The experiments were continued using a steel die.

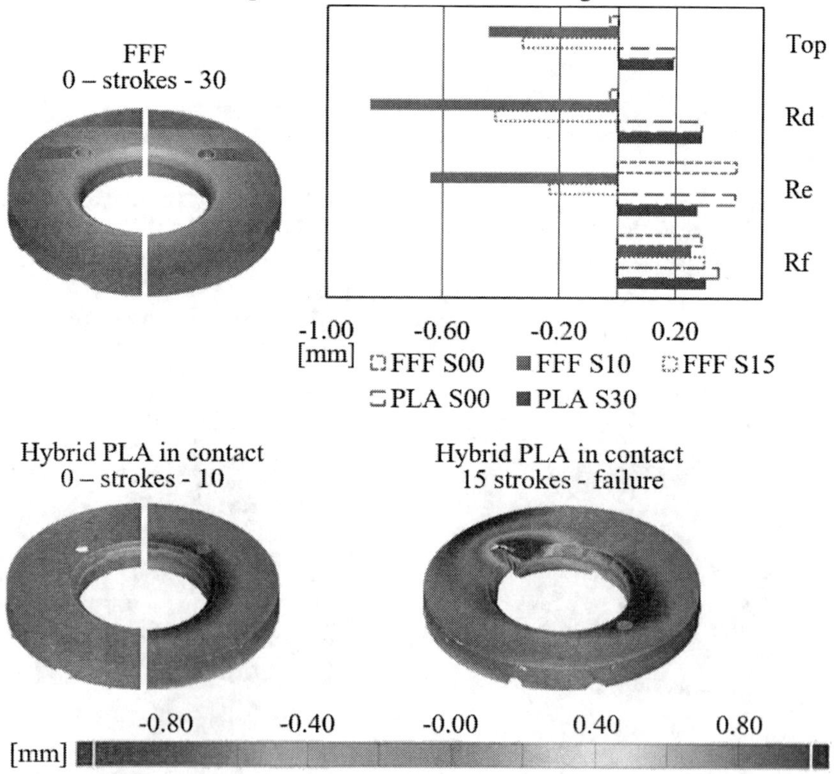

Figure 4. Dimensional precision and wear of the die.

IOP Conf. Series: Materials Science and Engineering 1284 (2023) 012078 doi:10.1088/1757-899X/1284/1/012078

5. Conclusion

PLA is the cheapest material to be fabricated by additive methods. The here used material is priced at 20 € per kg. Direct polymer additive tools are easy to fabricate and use. The blankholder pressure can be reduced and the PLA surface results in a good surface quality of the parts in dry forming. Minor defects of the tools do not affect the forming result. The wear of the tools is noticeable especially at the drawing radius of punch and die.

The fabrication cost of prototype tools can be reduced further by using ultra-high performance concrete (UHPC) at 1.73 € per kg for Nanodur Compound 5941. The application of hybrid UHPC polymer additive tooling still poses challenges in durability due to the brittle behavior of UHPC. The hybrid UHPC polymer additive tooling with PLA in contact shows the same advantages as direct polymer additive tooling due to the PLA surface but shows high wear and breakage. This can be attributed to the high deformation degree due the low thickness of the PLA shell. Disadvantageous of this approach is also the low initial precision due to the orientation at printing. To reduce the filling area the drawing radius of the punch and die were placed facing the printing platform. Contrary to the printing position used for direct additive tools with the drawing radius on top.

The hybrid UHPC polymer additive tooling with UHPC in contact shows the least wear at the punch of all approaches presented here. The quality of the drawn parts is comparable with the results archived with dry forming of steel tools. The zinc coating of the sheet is damaged in areas contacted with the drawing radius and small wrinkles are leveled. It has to be noted that the pores in the UHPC surface press down on the sheet metal in areas of high pressure like the drawing radius and the blankholder.

This study provides a new tooling concept for the economic production of deep drawn parts in low quantities as well as insights about the use of UHPC as material for deep drawing tools. Further work should be focused on optimizing the fastening of UHPC tooling to the tool frame, as the screw holes used in this study introduced a major weakening to the UHPC. Additionally, an optimization of the proportions of PLA and UHPC for best performance and cost in both configurations of hybrid UHPC polymer additive tooling could push the approach forward.

References

[1] Frank C 1999 Kunststoff als Werkzeugwerkstoff für das Tiefziehen von Feinblechen *Dissertation* Fachbereich Maschinenbau, Universität Hannover

[2] Witulski J, Trompeter M, Tekkaya A E and Kleiner M 2011 High wear resistant deep drawing tools made of coated polymers *CIRP Annals* **60** 311–4

[3] Kolbe J 2012 Thermisch beschichtete, faserverstärkte Polymerwerkzeuge für die Umformung höherfester Blechwerkstoffe Dissertation Institut für Umformtechnik und Leichtbau, Technische Universität Dortmund

[4] Gebhardt A 2011 *Understanding additive manufacturing: Rapid prototyping, rapid tooling, rapid manufacturing* (München: Hanser)

[5] Schuh G, Bergweiler G, Fiedler F, Bickendorf P and Colag C 122019 A Review on Flexible Forming of Sheet Metal Parts *2019 IEEE International Conference on Industrial Engineering and Engineering Management (IEEM) 2019 IEEE International Conference on Industrial Engineering and Engineering Management (IEEM) (Macao, Macao, 15.12.2019 - 18.12.2019)* (IEEE) pp 1221–5

[6] Bergweiler G, Fiedler F, Shaukat A and Löffler B 2021 Experimental Investigation of Dimensional Precision of Deep Drawn Cups Using Direct Polymer Additive Tooling *JMMP* **5** 1–16

[7] Frohn-Sörensen P, Geueke M, Engel B, Löffler B, Bickendorf P, Asimi A, Bergweiler G and Schuh G 2022 Design for 3D Printed Tools: Mechanical Material Properties for Direct Polymer Additive Tooling *Polymers* **14** 1694

[8] Durgun I 2015 Sheet metal forming using FDM rapid prototype tool *Rapid Prototyping Journal* **21** 412–22

[9] Schuh G, Bergweiler G, Fiedler F, Bickendorf P and Schumacher P 12142020 Small Series Production and Geometric Analysis of Sheet Metal Car Body Parts Using Forming Tools Made of Fused Filament Fabricated PLA *2020 IEEE International Conference on Industrial Engineering and Engineering Management (IEEM) 2020 IEEE International Conference on Industrial Engineering and Engineering Management (IEEM) (Singapore, Singapore, 14.12.2020 - 17.12.2020)* (IEEE) pp 156–60

[10] Nakamura N, Mori K and Abe Y 2020 Applicability of plastic tools additively manufactured by fused deposition modelling for sheet metal forming *Int J Adv Manuf Technol* **108** 975–85

[11] Fehling E, Schmidt M, Walraven J C, Leutbecher T and Fröhlich S 2014 *Ultra-high performance concrete UHPC: Fundamentals - design - examples (BetonKalender)* (Berlin: Ernst & Sohn)

[12] Schwartzentruber A, Bournazel J-P and Gacel J-N 1999 Hydraulic concrete as a deep-drawing tool of sheet steel *Cement and Concrete Research* **29** 267–71

[13] Kleiner M, Curbach M, Tekkaya A E, Ritter R, Speck K and Trompeter M 2008 Development of ultra high performance concrete dies for sheet metal hydroforming *Prod. Eng. Res. Devel.* **2** 201–8

[14] Ritter R, Curbach M, Trompeter M and Tekkaya A E 2009 Material Behavior of an Ultra-High-Performance Concrete Forming Die for Sheet Metal Hydroforming *ACI Materials Journal* **106** 515–22

[15] Ritter R and Curbach M 2015 Material Behavior of Ultra-High-Strength Concrete under Multiaxial Stress States *ACI Materials Journal* **112** 641–51

[16] Sagmeister B (ed) 2017 *Maschinenteile aus zementgebundenem Beton (Praxis)* 1st edn (Berlin, Wien, Zürich: Beuth Verlag GmbH)

[17] Holzer K, Füchsle F, Steinlehner F, Ettemeyer F and Volk W 2023 Potential of Ultra-High Performance Fiber Reinforced Concrete UHPFRC in Metal Forming Technology *Production at the Leading Edge of Technology (Lecture Notes in Production Engineering)* ed M Liewald *et al* (Cham: Springer International Publishing) pp 407–15

[18] Marciniak Z and Kuczyński K 1967 Limit strains in the processes of stretch-forming sheet metal *International Journal of Mechanical Sciences* **9** 609–20

[19] Schuh G, Bergweiler G, Bickendorf P, Fiedler F and Colag C 2020 Sheet Metal Forming Using Additively Manufactured Polymer Tools *Procedia CIRP* **93** 20–5

IOP Conf. Series: Materials Science and Engineering 1284 (2023) 012079 doi:10.1088/1757-899X/1284/1/012079

Proposal of a New Tool for Pre-Straining Operations of Sheet Metals and an Initial Investigation of CR4 Mild Steel Formability

A Barlo[1,*], M Sigvant[1,2], M S Islam[1], Ll Pérez[3], E Olofsson[4], M Al-Fadhli[5], Q T Pham[1], J Pilthammar[1,2], E-L Odenberger[3]

[1]Dept. of Mechanical Engineering, Blekinge Institute of Technology, Karlskrona, Sweden
[2]Volvo Cars Dept. 81110 Strategy Development, Olofström, Sweden
[3]RISE Research Institutes of Sweden, Component Manufacturing Unit, Olofström, Sweden
[4]Volvo Cars Dept. 81253 Commodity Hang on Parts, Gothenburg, Sweden
[5]Volvo Cars Dept. 81151 Product & Process Structures, Olofström, Sweden

E-mail: Alexander.Barlo@bth.se

Abstract. With the increased focus on reducing carbon emissions in the automotive industry, more advanced materials are introduced to reduce the vehicle weight, and more complex component geometries are designed to both satisfy customer demands and to optimize the vehicle aerodynamically. With the increase in component complexity, the strain paths produced during the forming operation of car body components often display a highly non-linear behavior which makes the task of failure prediction during the manufacturing feasibility studies more difficult. Therefore, CAE engineers need better capabilities to predict failure induced by strain path nonlinearity. This study proposes a new tool designed for creating bi-linear strain paths, by performing a pre-strain of a sheet large enough to cut out Nakajima specimens to perform the post-straining in any direction. From five pre-straining tests the tool present a stable pre-straining operation with a uniform strain field in a radius of 100 [mm] from the centre, corresponding to the region of interest of a Nakajima specimen. From the five pre-strained samples, different Nakajima specimens are cut transverse and longitudinal to the rolling direction and a failure prediction approach in an alternative, path independent evaluation space was used to predict the onset of necking with promising results.

1. Introduction

With an increased focus on carbon emission reduction in today's society, more and more advanced material grades of sheet metals are introduced to the market. This, combined with the consumers increasing demand for complex products, creates a number of challenges in the sheet metal forming industry. One already existing challenge amplified by these demands is the presence of non-linear strain paths during the forming process. For several decades, this particular issue has been targeted by multiple authors, including Stoughton [1], Volk and Suh [2], and Mattiason et al. [3]. With the recent changes driven by demand, the severity of the path non-linearity also increases. Recently, both Norz et al. [4] and Barlo et al. [5] tested different numerical approaches to predict failure in an automotive component with a critical element displaying a strain path with a radical change in the loading direction - both not successfully. This paper proposes a new tool geometry for the pre-straining of sheets and sets out to investigate an alternative evaluation space for the prediction of failure caused by non-linear strain paths. The tool consists of three parts - a punch, a binder, and a die. The punch has a size of 598 × 398 [mm] with rounded edges and a shoulder radius of 25 [mm]. The die has an opening of 603.7 × 403.5 [mm] with a die radius of 10 [mm]. To restrain the material during the forming operation, draw beads with a bead height of 6 [mm] and shoulder radius of 4 [mm] are used. An illustration of the entire tool and a picture of the punch can be found in Figure 1.

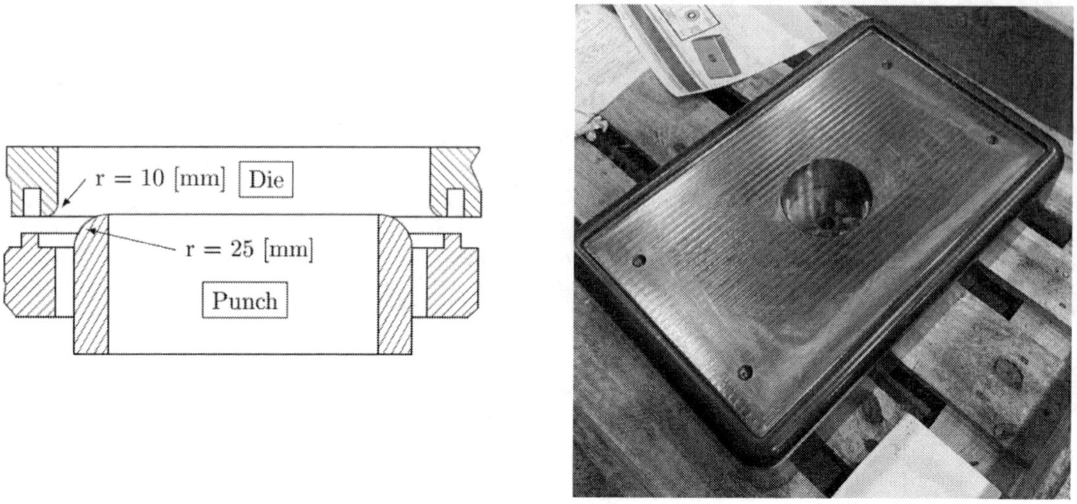

Figure 1: Cross-section of entire tool geometry and the manufactured punch.

The tool is designed to provide a pre-straining with a uniform strain distribution in the centre of the sheet in a circular area with a diameter of 200 [mm]. With the uniform strain distribution in this area, the full range of Nakajima specimens according to ISO-12004 [6] can be cut from the blank allowing for the determination of a pre-strain Forming Limit Curve (FLC). This process is outlined in Figure 2.

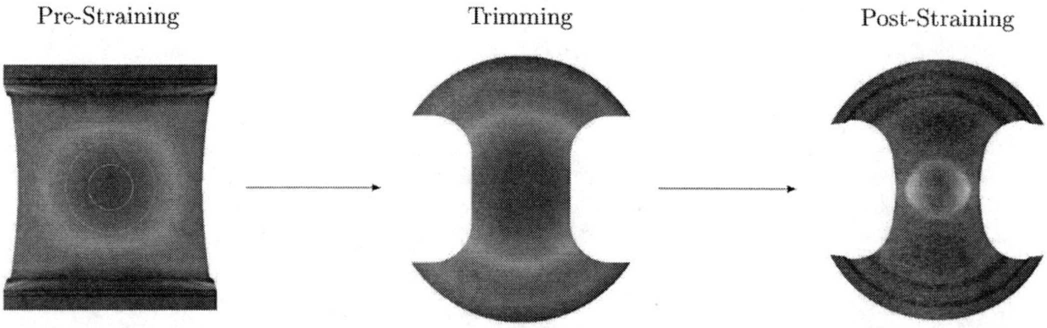

Figure 2: Proceedure of the bi-linear straining operation exemplified by the pre-straining of the 480 × 550 [mm] blank and post-straining of a Nakajima 75 sample. The circles presented on the pre-straining outlines the 200 [mm] diameter of the Nakajima full width blank, and the 100 [mm] diameter of the Nakajima punch serving as the region of interest for the DIC analysis.

2. Experimental Work

To perform the initial investigation, five rectangular 480 × 550 [mm] blanks were cut to obtain straining transverse to the rolling direction. For the pre-straining, the cut blanks were stretch to a punch depth of 62.6 [mm] with a ram velocity of approximately 22 [mm/s]. To reduce the friction in the forming operation, a suitable lubrication system was applied between the tool parts and blank. For the determination of the pre-straining levels, ARGUS measurements were performed, where the average major and minor strains of a Ø100 [mm] circle in the centre of the formed blank (see inner circle highlighed on Figure 2) were determined as the pre-straining levels, and can be found in Table 1.

For the post-straining, five different Nakajima test geometries were cut from the pre-strained blanks: 75, 100, 125, and 200 samples transverse to the rolling direction, and a 125 sample longitudinal to the rolling direction. The Nakajima tests were performed according to the ISO 12004-2 standard [6], and the strain fields were captured by a 3D ARAMIS DIC system. The determined mode of failure for the Nakajima post-straining samples was the onset of necking, and the method proposed by Sigvant et al. [7] was used to determine the necking strain limits. The determined limit strains for the post-straining operations of the 5 samples can be found in Table 1.

Table 1: Pre- and post-straining strain values. PS_XX test numbers indicate pre-straining, and the Nakajima sample size are presented followed by either T for transverse or L for longitudinal indicating the rolling direction relative to the major straining direction for the post-straining operations.

Pre-Straining			Post-Straining		
Test No	ε_2^{pre}	ε_1^{pre}	Specimen	ε_2^{post}	ε_1^{post}
PS_01	-0.197	0.306	75T	-0.065	0.164
PS_02	-0.198	0.306	100T	-0.008	0.082
PS_03	-0.193	0.300	125T	0.005	0.059
PS_04	-0.197	0.306	125L	0.002	0.053
PS_05	-0.192	0.297	200T	0.161	0.168

Based on the strain values presented, a new FLC can be generated based on the pre-strained samples. Figure 3 presents the new FLC based on the four Nakajima specimens tested in the transverse direction. For the case of the 125L specimen, this experiment was performed to replicate the strain path investigated by Norz et al. [4] and Barlo et al. [5], where a radical change in strain direction is seen. For the strain path presented in Figure 3 representing the 125L sample, the change in loading direction results in the minor strain of the post-straining operation being added to the major strain of the pre-straining and the major strain of the post-straining to the minor strain of the pre-straining.

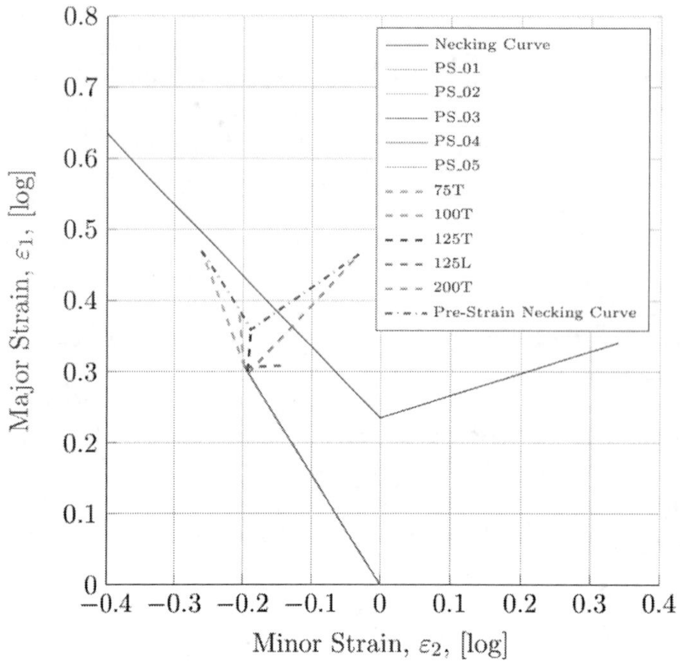

Figure 3: Forming Limit Curve generated from the pre-strained Nakajima specimens. The specimen 125L is not used as a data point in the limit curve, as the testing direction is different from the pre-straining.

The experimental results presented in Table 1 and Figure 3 shows that the presented tool produce stable pre-straining to the tested geometry. Therefore, an initial investigation of the ability to predict non-linear failure in the specimens tested can be performed.

3. Failure Prediction of Bi-Linear Strain Paths

For the initial investigation of failure prediction, the methodology proposed by Zeng et al. [8] for prediction of failure due to non-linear strain paths was used. In this methodology, a transformation of the standard forming limit diagram to a new evaluation space defined by the material flow direction α and the effective plastic strain $\bar{\varepsilon}^p$ is performed. In this initial investigation, an assumption of linear and proportional deformation ($d\varepsilon_i = \varepsilon_i$) is made for the pre- ($0 \rightarrow \varepsilon_i^{pre}$) and post-straining ($\varepsilon_i^{pre} \rightarrow \varepsilon_i^{post}$) strain paths.

In order to apply the methodology, inverse modeling of the experiments was performed, as the effective plastic strain cannot be obtained from DIC. For the inverse modeling, Finite Element models were built in AutoForm™ R10 using the BBC2005 material model. The input parameters for the material model can be found in Table 2, and the yield surface and hardening curve can be found in Figure 4.

Table 2: BBC2005 material input for the VDA239 CR4 mild steel. The exponent $M = 2k$ in the BBC2005 deviates from the crystallographic determined values, with the presented valued having been determined through inverse modeling to obtain a more accurate principal strain prediction [9].

r_0	r_{45}	r_{90}	r_b	σ_0/σ_0	σ_{45}/σ_0	σ_{90}/σ_0	σ_b/σ_0	M
1.805	1.336	1.876	0.982	1	1.0214	0.9959	1.1938	4.5

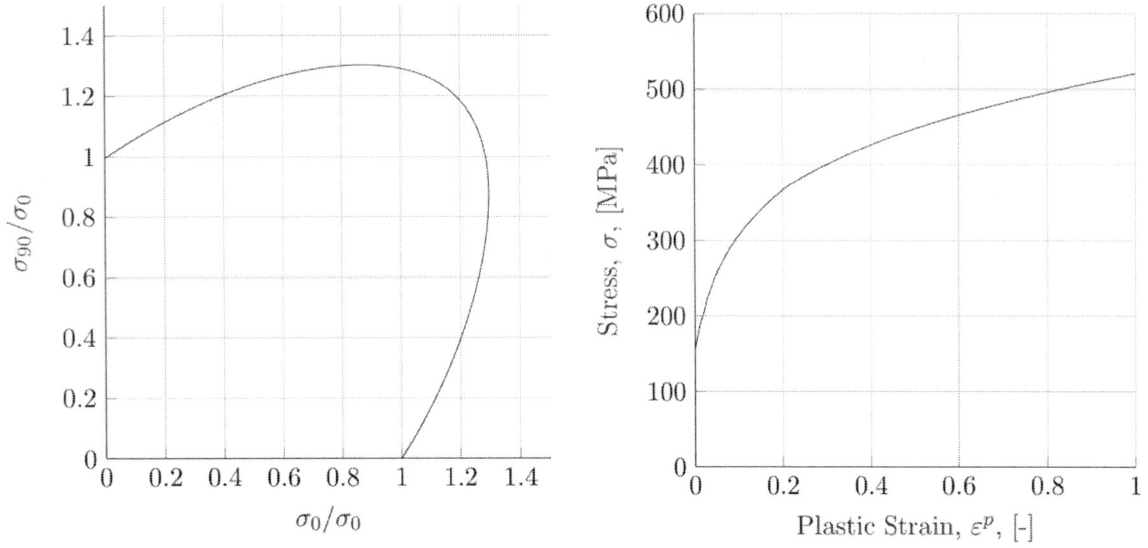

Figure 4: Yield surface and hardening curve of the VDA239 CR4 mild steel.

To replicate the friction conditions in the pre-straining experiment caused by the lubrication system, a mixed approach is taken using both Coulomb and Triboform® friction models. For the forming of the Nakajima specimens, an identical approach to eliminate friction was taken. Table 3 presents the different friction models applied to different areas of the model.

Table 3: Friction model setup of Finite Element Model.

Tool Part	Friction Model	Friction Coefficient, μ, [-]
Die	Coulomb	0.15
Binder	Coulomb	0.15
Punch	Coulomb	0.01
Die Radius	TriboForm	-

With the described setup there are deviations between the experimental and simulated force-displacement responses, see Figure 5. For the pre-straining operation, there is an acceptable fit between the experiments and simulation up to 30 [mm] of punch displacement. After this, the simulation starts to drift, and overestimates the maximum force by approximately 20 [kN]. For the post-straining operation, the fit of the simulation varies between the Nakajima samples. An acceptable fit is seen for both the 75T and 100T samples, a slightly less accurate fit is seen for the 125T and 125L samples, and finally an even less accurate fit for the 200T sample. For this initial investigation, these fits are however deemed acceptable, as for the majority of cases in the post-straining (except the 200T sample) the major part of the deviation occurs after the point determined to be the onset of necking.

With the calibrated models, the effective plastic strain value $\bar{\varepsilon}^p$ was determined by identifying the punch displacement in the simulation corresponding to the onset of necking identified in the experimental analysis of the post-straining operation. In the determined step, the element with the maximum major strain was identified and corresponding effective plastic strain values were found. To assess the quality of the investigated method, a comparison to the Advanced Formability option in AutoForm™ R10 (based on the Generalized Forming Limit Concept (GFLC) proposed by Volk and Suh [2]) was made. The Advanced Formability assessment

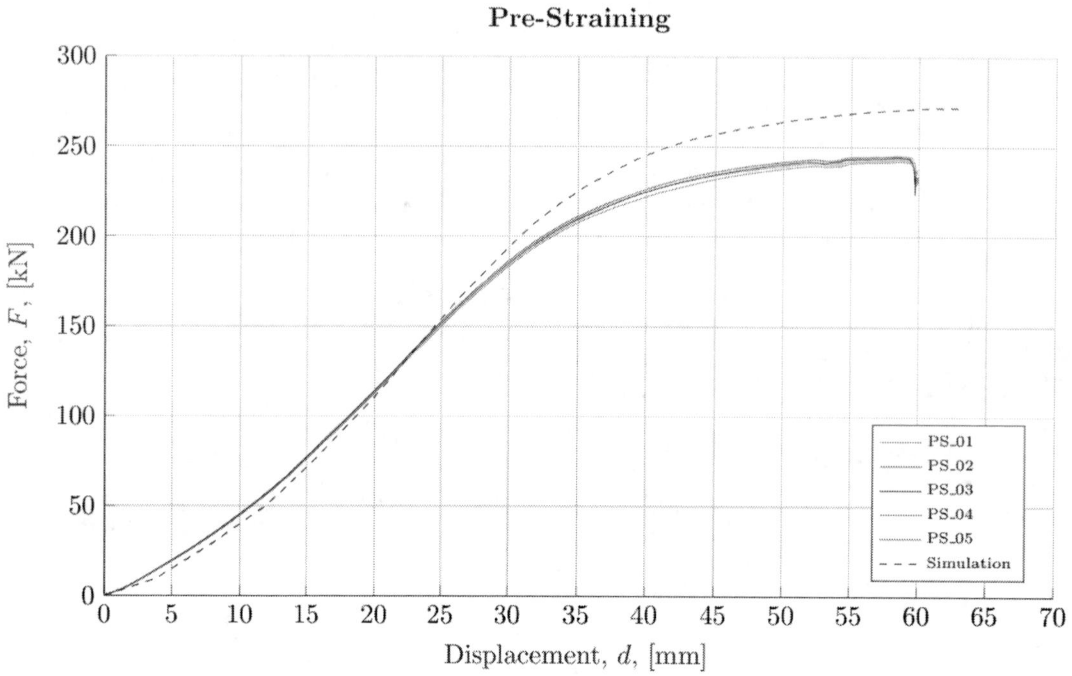

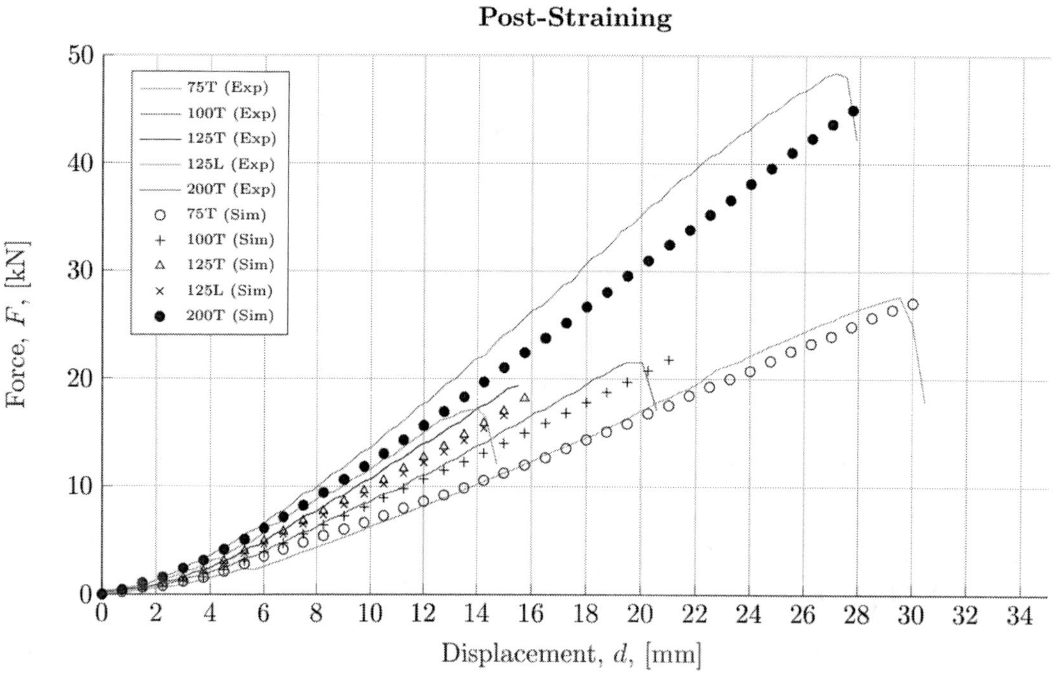

Figure 5: Comparison of experimental and simulated force-displacement curves for the pre- and post-straining operations.

was performed a step *after* where the effective plastic strain was determined corresponding to roughly a difference of 0.75 [mm] in punch displacement. Lastly, the material flow direction α was determined as the relationship between the minor and major strain increment from the end of the pre-straining operation to the end of the post-straining increment, with the exception of the 125L sample. For this sample, the α-value is determined as the ratio between major and minor strain due to the radical turn of the strain path, practically meaning that for the post-straining, major strain becomes minor and vice versa. For the determination of the material flow direction, the experimental values are used. The values identified for the different samples for the material flow direction, effective plastic strain, and Advanced Formability state can be found in Table 4.

Table 4: Values of material flow direction (α), effective plastic strain ($\bar{\varepsilon}^p$) and AutoForm™ R10 Advanced Formability. The effective plastic strain and Advanced Formability values are obtained from simulation, and the material flow direction is calculated based on experimental values. 'Excessive Thinning' is defined according to Volvo Cars standard.

	Pre-Strain	**75T**	**100T**	**125T**	**125L**	**200T**
α	-0.644	-0.396	-0.103	0.085	0.038	0.958
$\bar{\varepsilon}^p$	0.293	0.457	0.364	0.368	0.341	0.637
Advanced Formability	Safe	Safe	Safe	Safe	Safe	Excess Thinning

For the investigated approach, a transformation of the FLC into the alternative evaluation space is needed. The work presented in [5] investigated the same VDA239 CR4 mild steel and how sensitive the transformation was to the choice of material model. It was concluded that the transformation could not be decoupled from the choice of material model used in the simulation, why a transformation based on the BBC2005 material model was needed for this study, and the curve presented in [5] was adopted. Combining the curve with the information presented in Table 4, the failure evaluation in the alternative evaluation space displayed in Figure 6 can be created.

The results presented in Figure 6 provide an accurate failure prediction for four out of the five pre-strained Nakajima tests. For this initial investigation, the method fails to predict the failure for the 200T Nakajima sample, which is captured by the Advanced Formability functionality in AutoForm™ R10. The four pre-strained Nakajima samples accurately predicted (75T, 100T, 125T, and 125L) were all predicted safe by the Advanced Formability functionality. At first glance, the investigated methodology seems to be an improvement. This statement seems to be further strengthened by the fact that the Advanced Formability functionality does not account for load path changes with respect to the rolling direction, and this exact phenomenon was captured by the investigated approach.

4. Discussion

The investigated approach for failure prediction of the bi-linear strain paths proved to perform well for four out of the five pre-strain Nakajima specimens. For the 200T specimen geometry, the Advanced Formability functionality in AutoForm™ R10 predicted the onset of necking, however this is most likely a false positive. The first indication of failure occurs at a punch depth of 16.5 [mm] which is approximately 3.75 [mm] before the failure is determined in the experimental setup. Observing Figure 3 it can be found that the new necking limit for the pre-strain biaxial point is located above the original FLC, which might have let to the false positive identification of failure in the 200T sample.

As for the investigated approach's inability to predict the failure for the 200T sample, several reasons could be causing this. One obvious area of improvement that must be dealt

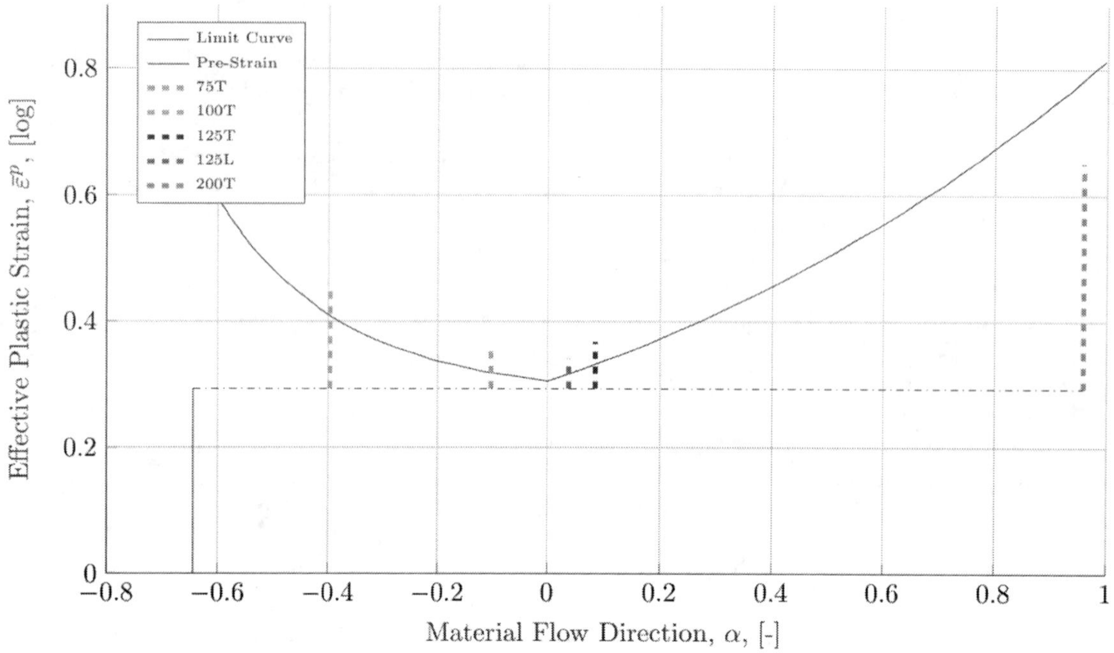

Figure 6: Failure assessment of the five pre-strained Nakajima samples in the alternative evaluation space proposed by Zeng et al. [8].

with in further investigations is a more accurate fit between the experimental and simulated force-displacement curves. Particularly the 200T sample displays a sub-par fit between the experimental and simulated curve which potentially impacts the strain predictions of the simulation. To obtain a better agreement between the experimental and simulated curve, several parameters could be investigated such as modifying the hardening curve (i.e. perform a power law extrapolation), the implementation of a strain rate sensitive hardening model, or modifications to the biaxial point of the yield locus.

In the presented work, the material flow direction α determined for each Nakajima sample was based on the experimentally obtained values. This choice was made as the material card applied to the Finite Element model was a slightly modified library card. This also means that the presented approach is in its current form not ready to be implemented in code as a post-processing technique, but further investigation into the method with more advanced models must be made.

5. Conclusion

The presented work introduced a new tool geometry for pre-straining of sheets. The work presented the results of five pre-strained samples that shows a stable pre-straining process in a single direction. The design of the tool allowed for a subsequent trimming operation producing Nakajima specimens of various width for the characterization of necking limits.

Following the successful pre-straining operation, five Nakajima tests were performed on sample widths of 75, 100, 125, and 200 transverse to the rolling direction, and a sample with width 125 along the rolling direction of the sheet. In an effort to accurately predict the failure in the pre-strained Nakajima tests, an approach using an alternative path independent evaluation space was used. This evaluation yielded an accurate prediction of failure in four of the five Nakajima tests, most notably an accurate prediction of the 125 sample in the longitudinal direction, with a radical change in the loading direction.

Acknowledgments
Open Access funding was provided by Blekinge Institute of Technology. This study was also funded by VINNOVA in the Sustainable Production sub-program within Vehicle Strategic Research and Innovation (FFI) program (grant number 2020-02986).

References
[1] Stoughton, T. 2000 A general forming limit criterion for sheet metal forming *Int. J. Mech. Sci.* **42** pp. 1-27.
[2] Volk, W. and J. Suh 2013 Prediction of formability for non-linear deformation history using generalized forming limit concept (GFLC) *AIP Conference Proceedings* **1567** pp. 556-561.
[3] Mattiasson, K., J. Jergéus, and P. DuBois 2014 On the prediction of failure in metal sheets with special reference to strain path dependence *Int. J. Mech. Sci* **88** pp. 175-191.
[4] Norz, R., N. Manopulo, M. Sigvant, A. R. Chezan and W. Volk 2022 Prediction of necking initiation in case of abrubt changes in the loading direction *The Minerals, Metals & Materials Series: NUMISHEET 2022* pp. 617-625.
[5] Barlo, A., M. Sigvant, N Manopulo, M. S. Islam. and J. Pilthammar 2022 Failure prediction of automotive components utilizing a path independent forming limit criterion *Key Engineering Materials* **926** pp. 906-916.
[6] International Standard Organization 2008 Metallic Materials - sheet and strip - Determination of forming-limit-curves (ISO 12004-2).
[7] Sigvant, M., K. Mattiasson and M. Larsson 2008 The definition of incipient necking and its impact on experimentally or theoretically determined forming limit curves *IDDRG 2008 Conference Proceedings.*
[8] Zeng, D., L. Chappuis, Z. C. Xia and X. Zhu 2009 A path independent forming limit criterion for sheet metal forming simulations *SAE Int. J. Mater. Manu.* Vol. 1, No. 1, pp. 809-817.
[9] Pilthammar, J., D. Banabic and M. Sigvant 2021 BBC05 with non-integer exponent and ambiguities in nakajima yield surface calibration *Int. J. Mater. Form.* **14**, pp. 577-592.

Investigation of tool damage in shear cutting of ultra-high strength sheet steels

Thorsten Beier [1], Eiji Iizuka [2], Toyohisa Shimmiya [3], Hiroto Miyake [3]

[1] thyssenkrupp Steel Europe AG, Eberhardstraße 12, 44145 Dortmund, Germany
[2] JFE Steel Corporation, 3-28-12, Meieki, Nakamura-ku, Nagoya-city, Aichi, 450-6427 Japan
[3] JFE Steel Corporation, 1, Kokanchyo, Fukuyama-city, Hiroshima, 721-8510 Japan

thorsten.beier@steeleurope.com

Abstract. Proportions of high- and ultra-high strength steels are constantly increasing in modern car bodies. In the press shop there are uncertainties concerning the application of steels with strength higher than 1000MPa particularly concerning shear cutting processes. To support the steel user regarding a tool material and coating and the wear behavior, long-term shear cutting test had been carried out. The effects of various shear cutting conditions on tool damage were investigated by continuous shear cutting test with an amount of cutting cuts up to 100,000 times. Tested sheet materials were 780, 1'180 and 1'470MPa grade steels. The changed cutting conditions were sheet steel strength, tool material, cutting clearance, and rake angle. Measurements by laser and confocal microscopes were carried out in order to reveal tool damage mechanisms. The features of tool damage were classified as wear, chipping, and plastic deformation. The amount of plastic deformation rises with the increase of sheet steel strength. Not only plastic deformation but also chipping in the tested tool was observed in the cutting test of 1180MPa sheet steel by standard tool steel (DIN 1.2379) after 40'000 cuts. Large plastic deformation in the tested tool was observed in a cutting test of 1470MPa sheet steel by high-speed steel (DIN 1.3343) after 100'000 cuts although no chipping was detected. Under a rake angle of 5°, serious chipping and tool wear were observed especially near the final sheet-tool contact position. Regarding clearance, the amount of plastic tool deformation increased with the decrease of the clearance. This effect of the clearance can be explained by vertical force and by sheet rotation during cutting.

1. Introduction and state of the art

Proportions of high- and ultra-high strength steels are constantly increasing in modern car bodies. In the press shop there are uncertainties concerning the application of steels with strength higher than 1'000MPa particularly concerning shear cutting processes. To support the steel user regarding tool material, coating or the wear behavior, long-term shear cutting test had been carried out.

Issues of cutting tools are also serious in mass production with high strength steels. Shinmiya et al. conducted experiments of continuous trimming press after drawing using DP980 steel sheets in which trimming dies with different blank holding conditions were used and investigated the influence of blank holding condition on the damages in cutting blades. In the results, chips occurred in the edge of cutting blades early in the case of using the trimming die with poor blank holding condition. On the other hand, in the case of using the die manufactured based on drawing sample shape after springback deformation calculated by FEM, which has good blank holding condition, chips were not observed after 3'000 cuts [1]. Akyürek et al. conducted continuous piercing tests up to 8'000 cuts using 400 MPa grade steel sheets and investigated the effect of hardness of cutting tools and cutting die clearance on the punch edge and face wear. They demonstrated that the wear of cutting tools decrease with increasing of hardness of cutting tools and cutting die clearance [2]. Won et al. proposed the design criteria for an aspect ratio of the punching pin to prevent buckling failure by experimental and numerical analyses. They also reported that the maintenance of a blank holder force was important to prevent stripping failure [3]. Cui et al. examined the mechanisms of AISI D2 steel trim die wear and their effects on plastic deformation and fracture behavior of the sheared edge of DP980 steel sheet.

In recent automotive, 1'500 MPa steel sheet have been applied to structural parts which were manufactured by cold pressing process. Therefore, it is important to understand damage behavior of cutting tools in the case of using 1'500 MPa steel sheets as already described by tool steel suppliers [7]. Figure 1 illustrates the known wear phenomena beside tool breakage. In this study, long-term shear cutting tests were carried out and investigated the effect of various cutting conditions and material strength.

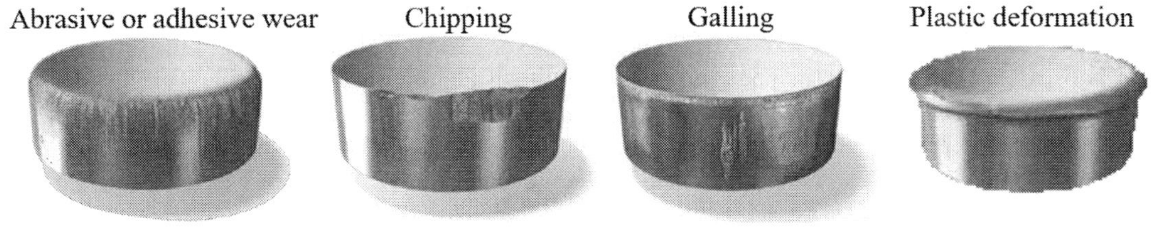

Figure 1: tool wear phenomena for cutting high strength steels [7]

2. Test procedure, used material

Especially looking at sheet material strength higher than 1'000 MPa there are concerns regarding tool wear and maintenance intervals. For this reason, two AHHS grades with tensile strength >1'200 MPa are used in this study and as reference a dual-phase steel of the 800 MPa class. Table 1 shows the mechanical properties of the tested steel sheet materials.

Table 1: mechanical properties test material according to DIN EN ISO6892-1

	t [mm]	Y.S. [N/mm²]	T.S. [N/mm²]	U.E. [%]	T.E. [%]	n
JFE-CA 1470	1.4	1293	1547	3.1	6.2	0.035
CR780Y1180T-DP -GI	1.5	850	1235	6.8	12.5	0.073
CR440Y780T-DP -GI	1.5	468	795	12.1	19.1	0.117

For the test procedure split coil material with a width of 50 mm was used. The test set-up is schematically shown in figure 2a. It consists of a coiler, a leveling device, a lubrication unit and the cutting tool itself. The cutting tool had a modular buildup so cutting parameters like clearance, tool steel, rake angle or the gap between pad and die cut be adjusted (figure 2b). In this case the pad is distance controlled, not force controlled. With a minimum amount of lubrication and a feeding rate of 10-20mm a straight cut is performed.

IOP Conf. Series: Materials Science and Engineering 1284 (2023) 012080 doi:10.1088/1757-899X/1284/1/012080

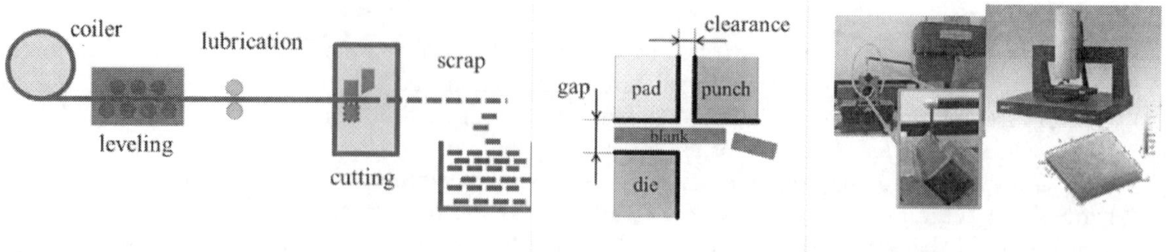

a) b) c)

Figure 2: a) schematic test set-up b) layout cutting tool c) measuring equipment.

To investigate tool wear evaluation long-term studies are done, for every parameter setting up to 100'000 cuts are performed without any rework of the active tool surfaces. After a maximum of 20'000 cuts the die is dismounted and surveyed before continuing the test run. The shape of the die is evaluated by a tactile surface tester or a non-tactile 3D-confocal sensor (figure 2c). Table 2 gives an overview of the complete test program and the combination of material and cutting parameters.

Beside the steel sheet material and the cutting parameters, the tool steel material plays an important role when looking at wear evolution. To quantify the improvement of the use of advance tool steel materials two different tool materials are used in this work for punch and die. The in the press shop well known 1.2379 as a reference and the advanced material 1.3244. Both materials are used for the active cutting tool elements (punch, die) and are heat treated with a hardness of HRC60 +2.

The material 1.2379 is a ledeburitic cold work tool steel with high wear resistance and good cutting strength. A high resistance to abrasive and adhesive wear and high-pressure resistance are given. Due to the good mechanical properties, this steel can be used in many different applications. The 1.3244 is a cobolt alloyed powder metallurgy high speed steel with excellent compressive strength and good abrasive wear resistance.

Beside the investigated parameters there are more parameters in press shop applications which influence wear evolution like inclined surfaces, tool coatings, lubricants, cutting geometry or the use of pads. Additionally, the tool design maybe restricted e.g. to available press forces or noise development

Table 2: overview test program

#	materials	tool steel	clearance' [%]	rake angle [°]	gap [mm]
A1	CR440Y780T-DP -GI	1.2379 HRC60 +2	12		t
A2	CR440Y780T-DP -GI	1.2379 HRC60 +2	20		t
A3	CR440Y780T-DP -GI	1.2379 HRC60 +2	28	0	t
B1	CR780Y1180T-DP -GI	1.2379 HRC60 +2	12		t
B2	CR780Y1180T-DP -GI	1.2379 HRC60 +2	20		t
B3	CR780Y1180T-DP -GI	1.2379 HRC60 +2	28		t
B5	CR780Y1180T-DP -GI	1.2379 HRC60 +2	20	5	t
D2	JFE-CA – 1470	1.3244 HRC60+2	20		t
D3	JFE-CA - 1470	1.3244 HRC60+2	28	0	t
D8	JFE-CA - 1470	1.3244 HRC60+2	20		3*t

3. Test evaluation and analysis

The influence on wear development of sheet material strength, clearance and rake angle were the some of the aims in the present study. The results and findings are the focus in the following chapters.

3.1. Influence of sheet material strength

To compare and quantify the amount of wear in dependency of the sheet material strength the tactile measurements of the centre area of the die are compared in Figure 3. The comparison is done for the cutting clearance of 20% regarding sheet thickness. Significant differences are getting visible around 40'000 cuts. Cutting the CR440Y780T-DP -GI only a small increase of the tool edge radii of ~0.01mm is measurable. Here the standard tool material 1.2379 is used.

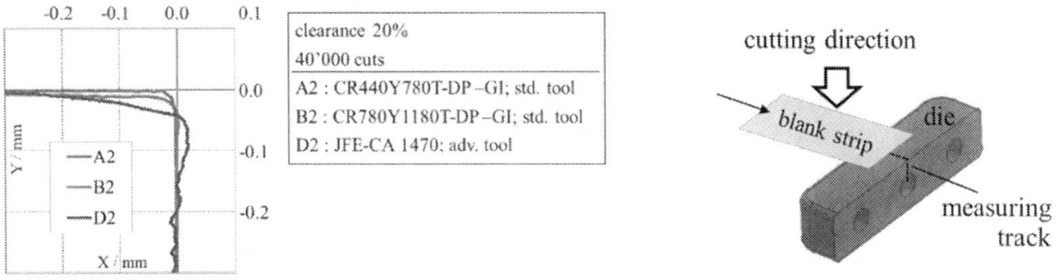

Figure 3: shape of cutting die after 40'000 cuts for 3 different sheet material strength

Looking at the measurement of the tool surface for cutting CR780Y1180T-DP -GI wear on upper surface and at the edge is getting obvious. For the material with the highest strength in this project the JFE-CA 1470 wear is overlayed by plastic deformation. When looking at the standard tool material 1.2379 beside wear also chipping was observed for number of cuts >40'000. This phenomenon was not observed using the advanced tool material 1.3244.

These observation leads to the conclusion that for steel sheet with strength ≥1'200MPa advanced tool steels can be recommended and for even higher strength is needed to prevent plastic deformation. As shown by Woestmann et al. [8] the radii of cutting tools have also a negative influence on stretch flange formability, so edge crack can occur for worn tools. Also, plastic deformation reduces the nominal clearance which can also have a negative impact on stretch flange formability.

3.2. Evaluation on influence of clearance

As already mentioned, cutting clearance plays an important role in stretch flange formability. The effect on tool wear development is shown in figure 4 and discussed in the following.

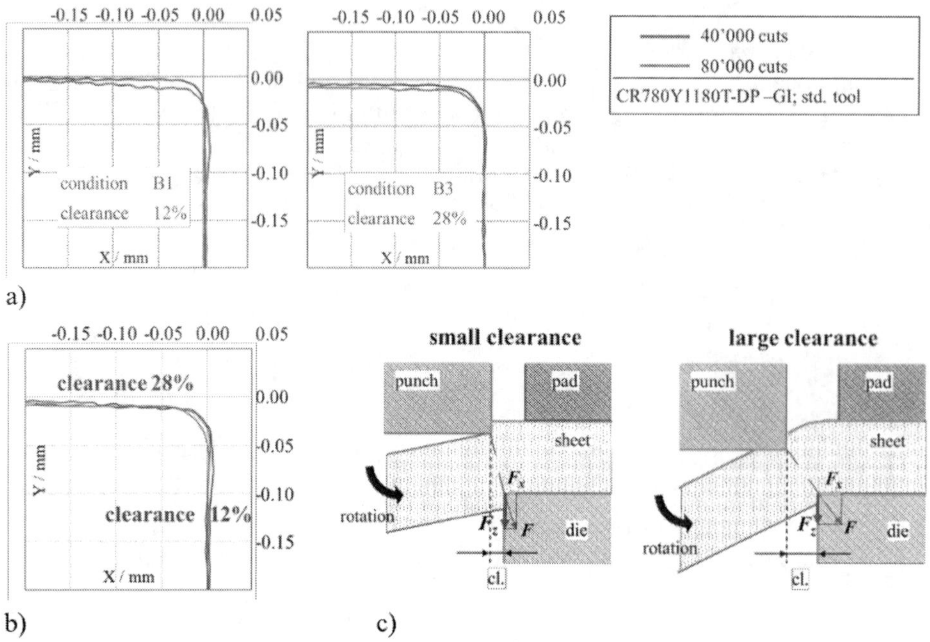

Figure 4: CR780Y1180T-DP -GI; 1.2379 std. tool steel; a) tool wear development; b) variation of clearance; test run B1, B2, B3; 80'000 cuts c) analytic cutting forces.

Figure 4a shows the development of wear for the CR780Y1180T-DP -GI. There are two facts getting obvious when looking at the tactile surface measurements. Again, like examined in the test runs in chapter 4.1 below 40'000 cuts only less wear is observed. For low cutting clearance of 12% when performing cuts up to 80'000 plastic deformation occurs in combination with wear. The higher clearance of 28% shows only wear (figure 4b). Analytic investigation of the loading condition explains this behavior. The rotation of the sheet during the cutting process dependents on the clearance. Figure 4c shows an overstated sketch of this relation. The vertical force component *Fz* is higher for a small clearance and encourages plastic deformation. For larger a larger clearance the horizontal force component *Fx* is higher which may increase wear development.

The findings to use higher clearances for many DP-steels [8] for better stretch flange formability leads also to the benefit of reduced plastic deformation. From experiments of the study with CR780Y1180T-DP -GI this recommendation is verifiable.

3.3. Influence of rake angle

In press shop application rake angle between upper and lower cutting tools is used with aim to reduce e.g. cutting forces and noise. The influence on tool wear development is discussed in the following. For CR780Y1180T-DP -GI, a cutting clearance of 20% and the standard tool steel wear development is shown in figure 5. The surfaces measurements for 40'000 and 80'000 cuts of three areas on the die are plotted. The regions of interest are the area S1 where first contact between blank and tools occur, the middle stage S2 of the process and final stage S3 of the separation process. For the area S1 due to high pressure on the die plastic deformation could be observed for a higher number of cuts (>40'000). In the middle area wear is the dominant effect. At the final stage of the cutting process chipping is obvious. Additional transverse forces and large material translation lead to additional pressure which encourage chipping.

From this observation a minimal rake angle with respect to press shop needs is recommended for CR780Y1180T-DP -GI.

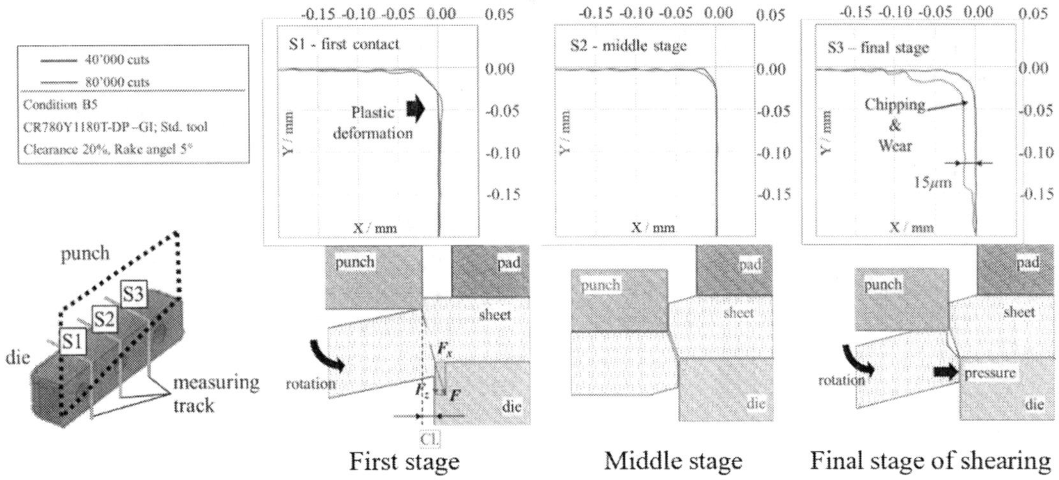

Figure 5: CR780Y1180T-DP -GI; 1.2379 std. tool steel; 20% cutting clearance; influence of rake angle on tool wear development

4. Conclusions and outlook

In sum tool wear and damage of cutting tools such as plastic deformation, wear and chipping become severe in higher material strength condition. Table 3 sums up the conclusions and recommendations which are made from the described investigations

Table 3: Recommendation for cutting ultra high strength steel

cutting process influencing parameter	result	recommendation
Material strength	For CR780Y1180T -DP -GI with standard tool materials, chipping was observed in addition to plastic deformations starting from 40´000 cuts.	Advanced tool material
	For JFE-CA 1470 with advanced tool material large plastic deformations were observed without chipping in 100´000 cuts.	Higher hardness of cutting tool
Clearance	For CR780Y1180T-DP -GI, plastic deformation wasn't detected for higher clearance condition.	Higher clearance
Rake angle	By using a rake angel of 5°, serious chipping and wear was observed especially in the final contact position for CR780Y1180T-DP -GI	Lower rake angle
Gap	The results for higher gap of 3*t between pad and die were almost the same for JFE-CA 1470 until 20k cuts.	

References

[1] T. Shinmiya, K. Higai1, Y. Yamasaki and T. Inazumi: Proceedings of IDDRG 560-565. (2012)
[2] F. Akyürek, K. Yaman, Z. Tekiner: Arab. J. Sci. Eng. 42, 4683–4692. (2017)
[3] C. Won, S. Lee, J. Seo, S. H. Park, J. Yoon: Int. J. Adv. Manuf. Technol. 94, 73–83. (2018)
[4] Z. Cui, S. Bhattacharya, D.E. Green, A.T. Alpas: Wear 426–427, 1635–1645. (2019)
[5] H. Lia, X. Wub, G. Lic, D. Zhoud: J. Mater. Process. Technol. 285, 116787. (2020)
[6] N. Mizuta, Y. Hayashida, K. Yamamoto: Proceedings of 2021 JSAE Annual Congress (Spring) 20215111. (2021)
[7] Technical broschure Uddeholm "Automotive tooling solutions"; 2017; www.uddeholm.co.uk
[8] Woestmann, S; Beier, T; Linnepe, M; Sikora, S; Forming Potential of Current Dual-Phase Steel Developments with Strengths of 800 MPa; 7th International Conference on Accuracy in Forming Technology ICAFT, Chemnitz, Germany, 2–3 November 2022

Optimization of the physical scaling approach through the adaption of the inverted membrane stresses

P Zimmermann[1]*, A Birkert[1], M Nowack[1] and T Kraft[2]

[1] Center for Metal Forming and Car Body Manufacturing, Faculty of Mechanics and Electronics, Heilbronn University of Applied Sciences, Heilbronn, Germany
[2] inigence gmbh, Bretzfeld, Germany

E-mail: philipp.zimmermann@hs-heilbronn.de

Abstract. Dimensional deviations of stamped car body components caused by elastic springback still represent a significant problem for the development of the stamping tools. First, the springback causes deviations of the part in vertical direction to the part surface due to bending stresses. Second, the surface area of the part contracts (and also bends) due to elastic membrane stresses. The precise manifestation of the contraction depends both on the part stiffness and on the – usually heterogeneous – distribution of the membrane stresses. Usually, the resulting deformations cannot adequately be compensated by globally homogeneous scaling approaches. In order to carry out a locally correct compensation of the stamping tools, Birkert et al. have recently presented a physical scaling approach based on the inverted membrane stresses gained from the part in the closed die to compensate the active die surfaces. This approach delivered significantly better results but left potential for improvement. It is shown in the present paper how the existing approach and thus the scaling values can further be improved by adapting the inverted membrane stresses in an appropriate way. This is done by comparing the amount of strain changes during spring-back with those of the scaling process in a first step and adjusting the compensation stresses, in a second step, in such a way that the considered strain amounts are approximately identical.

1. Introduction

During the production of sheet metal parts, elastic springback of the formed parts occurs when the forming tools are opened. Thus, a naive tool design relying on a replication of the desired target geometry leads to manufactured parts significantly deviating from this target. This is observed in the continuous process and can, more or less accurately, be simulated by modern numerical simulation packages like *AutoForm Forming* or *LS-Dyna* for example.

From a physical perspective, the reason for these deviations are elastic membrane and bending stresses being induced in the sheet metal when the die is closed. When the die is opened again the release of bending stresses causes bending and/or twisting while, additionally, an inhomogeneous membrane stress distribution – the membrane stresses are mainly tensile stresses – leads to a shrinkage of the formed parts. Because the tensile stresses – being decisive for springback in the manufacturing process of sheet metal parts – predominantly occur in the first forming stage, i.e. in the drawing stage, the above statement particularly refers to the draw shell. The different types of dimensional deviations, namely bending and twisting on the one hand and shrinkage of the drawshell on the other, are handled in practice by different methods. The shrinkage of the drawshell is compensated by an increase in size of the die, see Section 2. In contrast, bending and twisting, which can be characterized and quantified by the normal distance of the part from the target geometry, are compensated by "overbending" of the die surfaces according to the displacement adjustment method [1] or variants [2-12] of the same.

In [13] a stress-based scaling approach was presented leading for given examples to an improved local scaling of the draw shell compared to the state of the art. As a positive side effect, the approach led to a reduction of dimensional deviations (in normal direction) from the bending and twisting of the draw shell so that the subsequent overbending of the die faces simplified considerably. However, it also became apparent that this approach still has potential for improvement.

2. State-of-the-art scaling approaches

In practice, global scaling approaches are still often used today for the scaling of the die surfaces. In such a process the die surfaces are increased uniformly, i.e. geometrically similar, in all surface directions by a constant factor. The mechanical material parameters of the part, such as Young's modulus and the biaxial yield stress, can be used to appropriately estimate the global scaling factor. This procedure can easily be improved by using different scaling factors for the axes directions, typically leading to better results, at least for geometrically simple parts such as a roof or a door planking, because planking parts usually underlie significantly different strain changes during the deep drawing process in their main and side curvature direction. The main and side curvature direction is defined in [14].

However, in order to enable a highly accurate compensation of the contraction behaviour of the draw shell, which is locally varying in the general case, a new scaling approach was presented in [13]. In the suggested process, the membrane stresses before springback are inverted to generate local strain changes in the part that counteract the springback. The displacement vector field generated in this way is then used to scale the tool surfaces locally different. From the example shown in [13] it is evident that this approach potentially compensates for the contraction effects much better than any global scaling approach could; with the enormous additional benefit that unwanted plastic deformations are drastically reduced when the blankholder is closed in the subsequent operation.

The results from the investigation in [13] demonstrate that optimal local scaling values should approximately correspond to the local contraction values during springback with reversed sign in order to achieve area/length dimensions of the draw shell close to those of the target geometry. However, due to nonlinearities the part structure does not react with exactly the same inverse strain to an inverse membrane stress of the same amount; thus, in the general case the stresses used for scaling have to be adjusted in an iterative procedure to achieve the required local scaling values.

3. Optimization of the existing stress-based scaling approach

For a better understanding of the stress-based scaling it is helpful to visualize the respective stress and strain states in a uniaxial stress-strain diagram (Figure 1). As the stress-based scaling approach uses the results of the simulated, FE-based manufacturing process and is a FE-based approach itself, the physical values mentioned in the further text always refer to the discrete values of the elements. In the diagram on the left-hand side (Figure 1a) the previous approach and the related problems are visualized. In the first step (0→1), the drawshell is formed from the flat sheet blank using an unscaled and uncompensated draw die (zero die). The present part geometry in the closed die corresponds to the target geometry (part zero geometry) and is represented in the diagram by the strain state ε_{bSB_0}. When opening the die (1→2) the part changes its geometry due to the removal of the external forces. The contraction of the part is represented by the strain change (contraction value) $\Delta\varepsilon_{SB\_0}$. The sprung-back part geometry now exhibits the strain state ε_{aSB_0}.

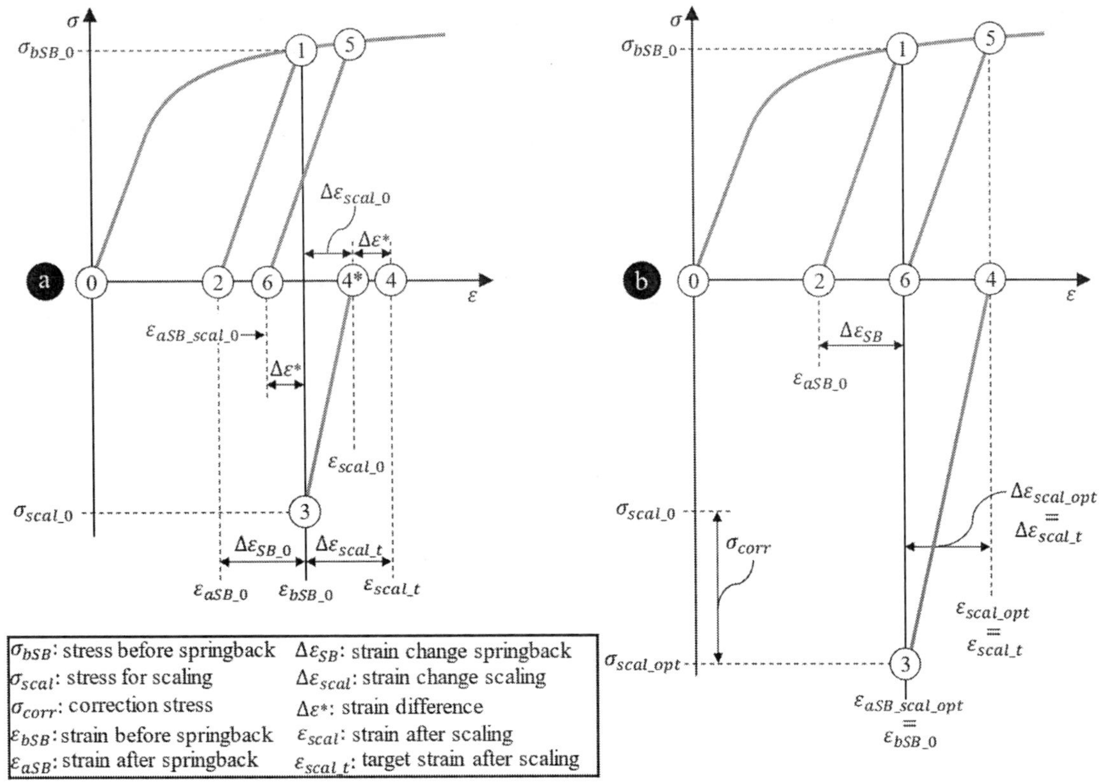

Figure 1. Workflow to optimize the physical scaling approach.

For scaling, the present membrane stresses before springback σ_{bSB_0} are inverted without changing their amount (1→3). The resulting stresses $\sigma_{scal\_0}(= -\sigma_{SB\_0})$ are now applied to the target part geometry and an "inverse springback" is calculated (3→4*). The resulting deformation $\Delta\varepsilon_{scal\_0}$ can now be utilized along the lines of the previous scaling approach to scale the tool geometries of the drawing stage. A scaled draw shell is now formed with the tools scaled in this way (0→5). During the opening of the tools the workpiece contracts as a result of springback (5→6). If a strain difference $\Delta\varepsilon^*$ between the sprung-back part (6) and the target geometry remains, the optimized approach can be used to adjust the present scaling value for a considered element $\Delta\varepsilon_{scal\_0}$.

From the results shown in [13], it can be concluded that the contraction values during springback $\Delta\varepsilon_{SB}$ do not change significantly due to the scaling itself. Therefore, the required local strain changes (local scaling values) $\Delta\varepsilon_{scal_t}$ occurring during scaling should have the same amount as the contraction values occurring during springback with an inverted sign (3→4). The resulting strain difference $\Delta\varepsilon^*$ between the sprung-back part (formed with scaled tools) and the target geometry can thus be determined (in sufficiently good approximation) even without forming simulation by comparing the existing scaling value $\Delta\varepsilon_{scal_0}$ with the required scaling value $\Delta\varepsilon_{scal_t}$.

To further reduce the strain difference $\Delta\varepsilon^*$, the stress used for scaling σ_{scal_0} must be adjusted. In the diagram (Figure 1a) the present strain at scaling ε_{scal_0} is lower than the required ε_{scal_t}. To increase the strain, a negative correction stress σ_{corr} must be added to the present scaling stress (Figure 1b). On the other hand, if the present strain value is higher than the required strain value, a positive correction stress would have to be added. The amount of the required correction stress can be calculated based on the Young's modulus (for shear strain from the shear modulus) of the part material and the existing strain difference $\Delta\varepsilon^*$. Thus, the calculation of the adjusted scaling stress is given by equation (1).

$$\sigma_{scal\_i+1} = \sigma_{scal\_i} + \left(\varepsilon_{scal\_i} - \varepsilon_{scal_t}\right) \cdot E \tag{1}$$

If there is still an unacceptably high strain difference $\Delta\varepsilon^*$ after the first stress adjustment, the process is repeated until this difference falls below a given threshold. In each iteration the respective adjusted stress from the previous iteration is adjusted with the present strain difference.

For simplification, the described procedure was considered in the one-dimensional case, which means that the dependency of the strains on different axial directions was neglected. Considering real material behaviour, strain along one axis direction also causes transverse strain. This relationship is characterized by the material-specific Poisson's ratio μ. To enable adjustment of the strain value related to one axis without altering the transverse strain, the Poisson's ratio in the material model was set to zero (μ=0).

Figure 1b shows the process with optimized scaling stress σ_{scal_opt}. The utilized strain value ε_{scal_opt} in the scaling procedure corresponds to the required strain value ε_{scal_t} here. After forming the part with the tools scaled in this way (1→5) and subsequent springback (5→6), the dimensions (strain values) of the part $\varepsilon_{aSB\_scal\_opt}$ should match with those of the target geometry ε_{bSB_0}.

4. Comparison of the results of the optimized approach with those of the existent approach

To validate the effectiveness of the new approach, it was tested on a car body structural part. For comparison two manufacturing processes were simulated and evaluated, one without any scaling and one with global scaling. The FE software *AutoForm Forming R10* was used to simulate the manufacturing process. Figure 2 shows the manufacturing process of the chosen structural part. In the first operation the draw shell is formed from a shaped blank in the drawing operation (D20); then, the part is separated from the addendum by a virtual laser cut (T30); and finally, two flange areas are formed in the flanging operation (F40). After each of the two operations T30 and F40 the elastic springback of the part is calculated. To save computation time in the simulation, springback between the drawing stage (D20) and the virtual laser cutting (T30) is prevented.

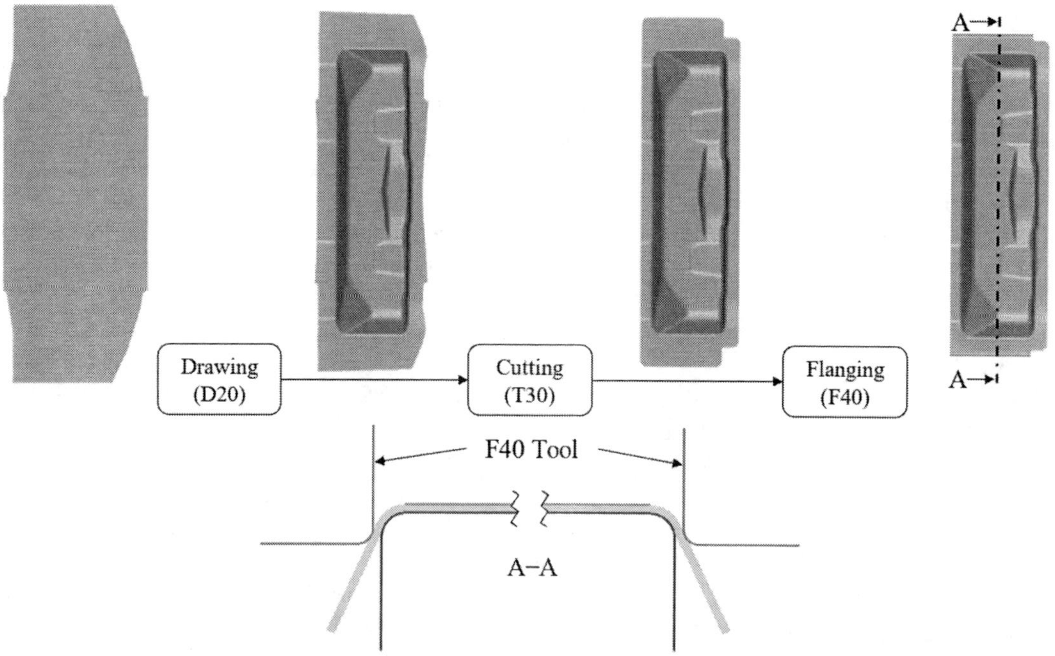

Figure 2. Manufacturing process of the reinforcement part.

Figure 3 shows the strain changes during springback after the virtual laser cutting. For a better visualization, the strain changes are shown referring to the global coordinate system. To evaluate the quality of the results in the scaling process and to adjust the stresses, the strain changes referred to the local element coordinate systems are used there. Both the strain changes along the considered axes and the corresponding shear changes vary significantly along the part surface. To compensate these contraction effects as accurately as possible, the local scaling method according to [13] is required. Global scaling approaches fail in this case because, for example, a positive scaling factor would further increase the area/length error in those regions where a positive strain change occurs. In addition, a compensation for the shear changes in the global scaling approaches is not provided.

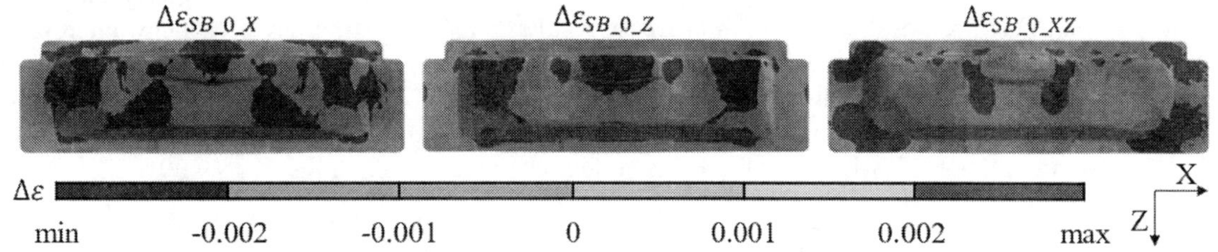

Figure 3. Local strain change during springback after drawing and cutting.

In the following, at first the convergence behaviour when adjusting the stresses in the physical scaling approach with respect to the reduction of the strain differences is shown. Subsequently, the scaling vector fields from chosen iteration loops are used to generate locally scaled tool geometries and to generate and evaluate the corresponding simulation setups. These results are then compared to the results of the simulation setups without scaling and with global scaling.

4.1. Convergence behaviour by adapting the inverted membrane stresses

As already described, the adjustment of the membrane stresses should reduce and ideally eliminate the strain differences (difference between present and desired strains). Figure 4 shows the over all elements averaged absolute strain differences ($\overline{\Delta\varepsilon *}$) along thirty iteration loops. The strain values here refer to the local coordinate systems of the FE elements. The average of all three strain variables decreases significantly at the beginning. However, the decrease reduces from iteration to iteration. This behaviour suggests that it is not possible to eliminate the strain differences completely. A useful compromise must be made between the effort required for the iteration loops and the increase in the quality of the results.

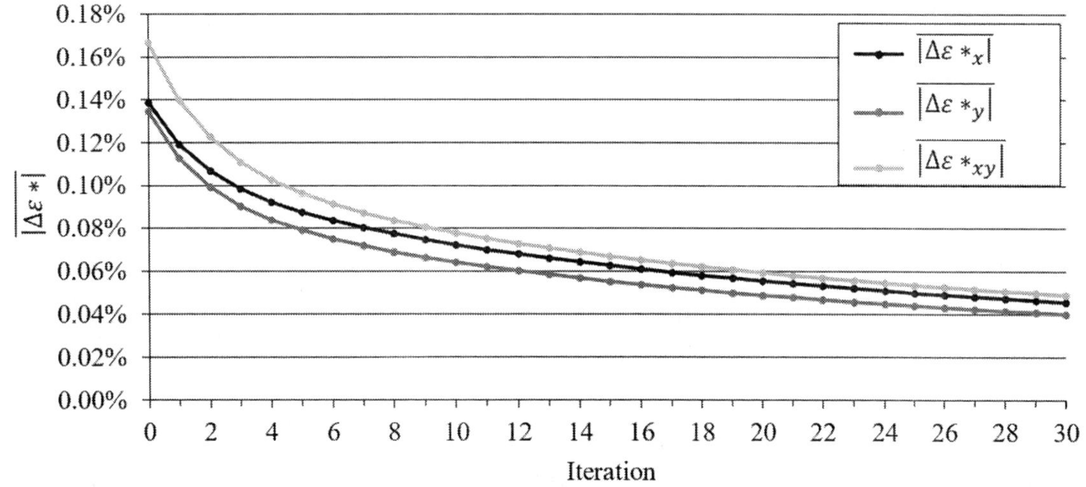

Figure 4. Development of the strain change differences during the optimization of the membrane stresses.

4.2. Effects of differently scaled tools

An inadequate scaling results in area/length differences between the sprung-back part geometry and the part target geometry. The consequence of this (besides the dimensional deviations themselves) is that the part can be plastically deformed during the closing of the blankholders in the subsequent operations. To evaluate these unwanted plastic deformations, the *AutoForm* evaluation variable *Plastic Strain Rate* (PSR) is evaluated in the last time step when closing the blankholders in the subsequent operation. The results are shown in Figure 5. In the simulation setup without scaling of the drawshell (No Scaling), nearly a quarter of the part surface has a PSR value above 0.001 (approximately no plastic deformation). In the simulation setup with global scaling, the drawshell was increased with a constant scaling factor of +0.18%. This factor was calculated based on the biaxial yield stress and the Young's modulus of the part material. However, global scaling in this case results in an even larger area with unwanted plastic deformation. If the stress-based local scaling approach without optimizing the membrane stresses (Local Scaling Opt 0) is applied, a significant reduction of the area with unwanted plastic deformation can be seen. The best result is reached in the simulation setup where the stresses for local scaling were adjusted in thirty iteration loops. Nearly 98% of the part surface here has a PSR of less than 0.001.

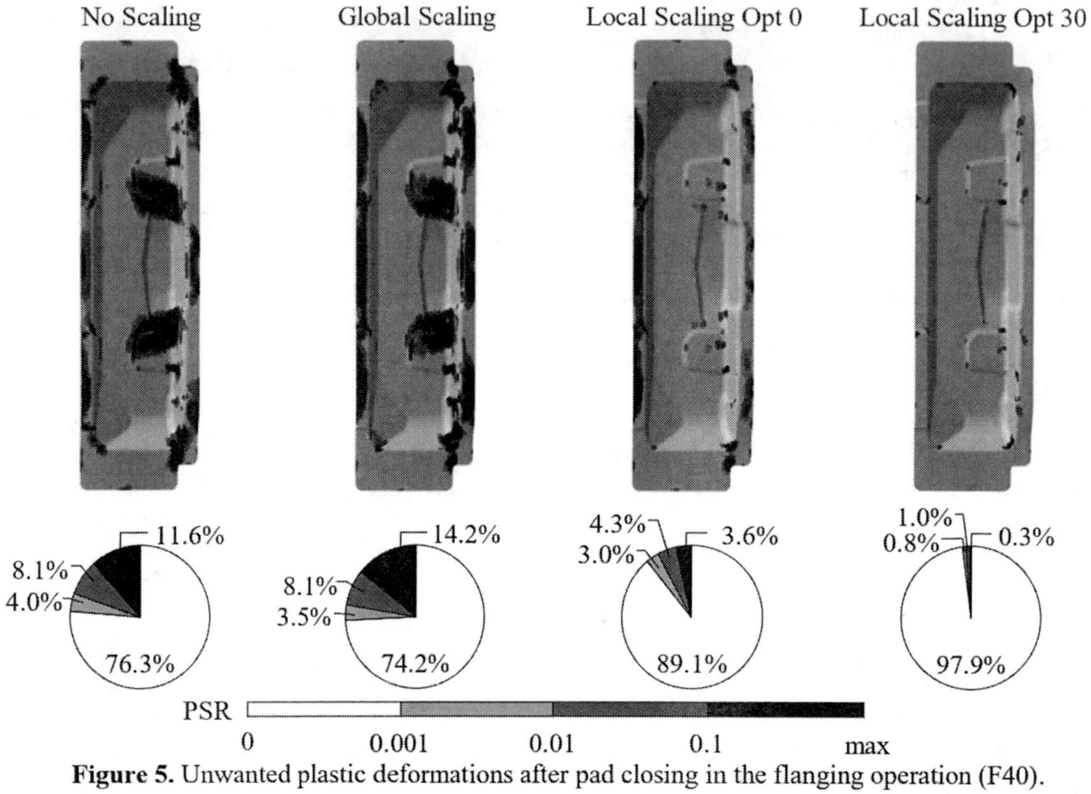

Figure 5. Unwanted plastic deformations after pad closing in the flanging operation (F40).

The unwanted plastic deformations have an impact on the springback behaviour and thus also on the dimensional accuracy of the part after the respective operation. To quantify this impact, in the simulation setups the sprung-back part geometries after drawing and cutting are compared with the part geometries after flanging. For this study, however, only the closing of the blankholders without flanging was simulated in the flanging operation F40. Thereby the isolated impact of the unwanted plastic deformations during blankholder closing on the dimensional accuracy of the part can be identified. The dimensional accuracy is quantified by the average dimensional deviation, which is defined as deviation in normal direction averaged over all elements, whereby the element size is taken into account in the averaging. For the calculation of the average PSR the PSR of all elements is averaged and here also the element size is taken into account in the averaging. Figure 6 shows that a high average PSR in the simulation setups without (No scaling) or with global scaling (Global scaling) results in a relatively high average dimensional deviation. In the simulation setups with local scaling (Opt 0 - Opt30), the average

IOP Conf. Series: Materials Science and Engineering 1284 (2023) 012081 doi:10.1088/1757-899X/1284/1/012081

PSR is significantly lower, which also results in smaller dimensional deviations between both the arriving and leaving part in the flanging operation F40.

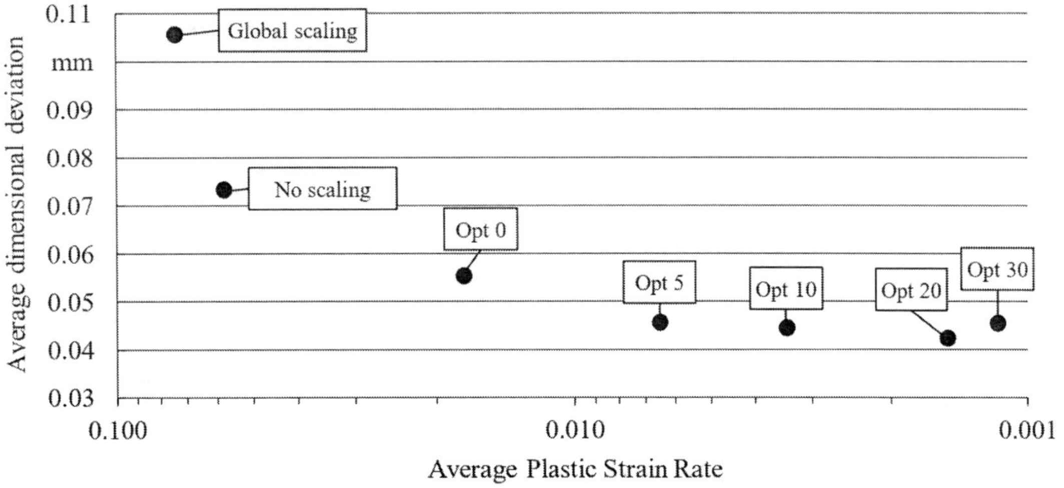

Figure 6. Correlation between average plastic strain rate and average dimensional deviation with different scaling approaches.

Another meaningful aspect in the context of scaling is the required blankholder force in the flanging operation F40, which is needed to close the blankholders completely. Figure 7 shows the average PSR and the corresponding required blankholder force of the differently scaled simulation setups. In the simulation setup without scaling (No scaling) a high average PSR occurs, which causes a high blankholder force. With the locally scaled simulation setups (Opt 0 - Opt 30), the average PSR decreases with the number of optimization loops, leading to a reduction in blankholder forces. Only the result of the simulation setup with global scaling (Global scaling) does not correlate with those of the other setups. Here, there is a higher average PSR value than in the setup without scaling (no-scaling) and in the setup with local scaling without optimization of the stresses (Opt 0). However, the blankholder force there is lower than in the other two setups. This means that not only the average unwanted plastic deformations have an impact on the amount of the blankholder forces but also the elastic deformation components and the stiffness of those areas that are unwanted deformed.

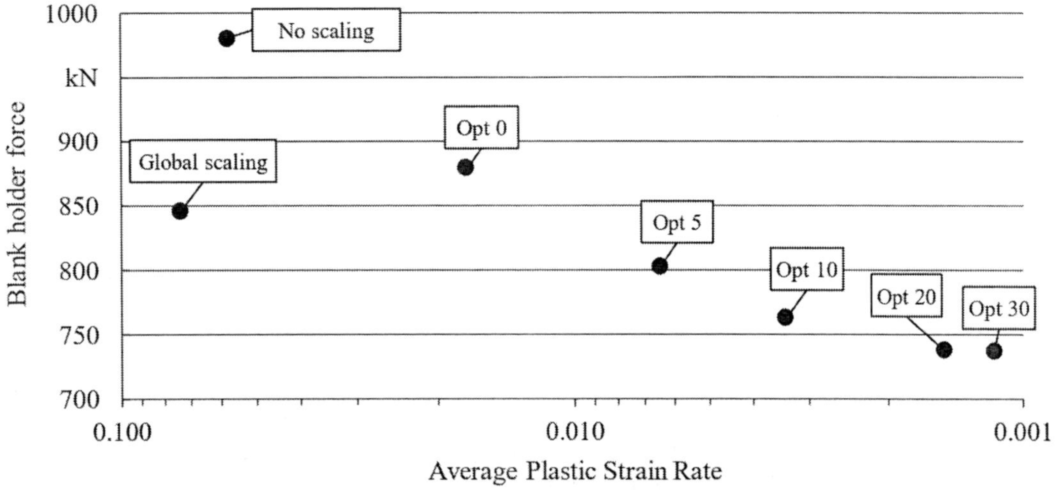

Figure 7. Correlation between average plastic strain rate and blank holder force with different scaling approaches.

5. Conclusion

In the present paper a modified approach to optimize the stress-based local scaling [13] is suggested. The local (element-wise) strain values after springback from the unscaled manufacturing process are used to define target strains (scaling values). If strain values after scaling deviate from the corresponding target values, correction stresses are calculated for the respective elements and element directions using the strain differences and Young's modulus of the part material. These correction stresses are then used to adjust the stress values for the scaling. If strain differences are still too high after adjusting the stresses once, the procedure can be repeated iteratively until the strain differences fall under a given threshold.

To validate the effectivity of the new approach, it was tested on the virtual manufacturing process of a car body structural part. The results were then compared to those from simulation setups without any and with global scaling. First, the average strain differences were evaluated when the stresses were iteratively optimized over thirty iterations. The resulting behaviour demonstrates that the strain differences can be significantly reduced with this approach but probably not completely eliminated. This means, a useful compromise must be found between the effort required for the iteration loops and the increase in the quality of the results.

The evaluation of the unwanted plastic deformations occurring when closing the blankholders in the flanging operation showed, that the area fraction with a plastic strain rate below 0.001 (approximately no plastic deformation) could be increased from 76.3% (without scaling) respectively 74.2% (global scaling) to 97.9% (local scaling with thirty optimization loops). Furthermore, the impact of the unwanted plastic deformations on the dimensional accuracy of the part was evaluated. Here, it was shown that less unwanted plastic deformation leads to smaller dimensional deviations, as expected. Finally, the impact of the unwanted plastic deformations on the closing force of the blankholders in the flanging operation was evaluated. Here, it was shown that less unwanted plastic deformations usually correlate with a lower blankholder force. However, the blankholder force also depends on other factors, such as the elastic deformation or the local distribution of the unwanted plastic deformations.

In future work it will be investigated how to make the stress adjustment more efficient to reduce the number of iteration loops required for stress optimization.

References

[1] Gan W and Wagoner RH 2004 Die design method for sheet springback *Int. J. Mech. Sci.* **46**(7) 1097-113

[2] Karafillis A and Boyce M 1992 Tooling design in sheet metal forming using springback calculations *Int. J. Mech. Sci.* **34** 113-31

[3] Yang X and Ruan F 2011 A die design method for springback compensation based on Displacement adjustment *Int. J. Mech. Sci.* **53** 399-406

[4] Birkert A, Hartmann B and Straub M 2017 New method for springback compensation for the stamping of sheet metal components *J. Phys.: Conf. Ser.* **896** 012067

[5] Weiher J, Rietman B, Kose K, Ohnimus S and Petzoldt M 2004 Controlling springback with compensation strategies *AIP Conf. Proc.* **712** 1011-15

[6] Lingbeek R, Huétink J, Ohnimus S, Petzoldt M and Weiher J 2005 The development of a finite elements based springback compensation tool for sheet metal products *J. Mater. Proc. Tech.* **169** 115-25

[7] Cheng HS, Cao J and Xia ZC 2007 An accelerated springback compensation method *Int. J. Mech. Sci.* **49** 267–79

[8] Cafuta G, Mole N and Štock B 2012 An enhanced displacement adjustment method: springback and thinning compensation *Mater.& Design* **40** 476-87

[9] Liao J, Xue X, Zhou C, Barlat F and Gracio JJ 2013 A springback compensation strategy and applications to bending cases *steel research international* **84**(5) 463-72

[10] Li G, Liu Y, Du T and Tong H 2014 Algorithm research and system development on geometrical springback compensation system for advanced high-strength steel parts *Int. J. Adv. Manuf. Tech.* **70** 413–27

[11] Siswanto WA, Anggono AD, Omar B and Jusoff K 2014 An alternate method to springback compensation for sheet metal forming *Sci. Wor. J.* **2014** 301271
[12] Zhang Z, Wu J, Zhang S, Wang M, Guo R and Guo S 2016 A new iterative method for springback control based on theory analysis and displacement adjustment *Int. J. Mech. Sci.* **105** 330-9
[13] Birkert A, Zimmermann P, Hartmann B, Scholle M and Straub M 2020 New physical scaling approach to compensate the part contraction in drawing operations *IOP Conf. Ser.: Mater. Sci. Eng.* **967** 012041
[14] Birkert A, Straub M and Haage S 2013 *Umformtechnische Herstellung komplexer Karosserieteile* (Berlin/Heidelberg: Springer-Verlag) pp 232-3

Pressure Distributions in Deep Drawing with a Passive Cardanic Blank Holder

A Breunig[1] and P Groche[1]

[1] Institute for Prodution Engineering and Forming Machines, Technical University of Darmstadt, Otto-Berndt-Straße 2, 64287 Darmstadt, Germany

E-mail: breunig@ptu.tu-darmstadt.de

Abstract. Deep drawing processes are strongly influenced by the blank holder forces, as well as their local distribution. In current deep drawing tools, this pressure distribution is dominated by the compliance of the specific machine the tool is mounted on. Therefore, each tool must be manually polished to enable the desired distribution of the blank holder pressure. This regularly causes problems when transferring tools from one machine to another, for example from a try-out press to a regular press line. As both machines have unique compliances, the pressure distribution on the flange of the drawn part changes. Additional manual spot grinding is needed to achieve the desired pressure distribution on the flange. To counteract this problem, the use of a passive cardanic blank holder is proposed in this paper. The cardanic blank holder can freely tilt around two axes, which leads to an auto-leveling effect of the blank holder on the sheet, resulting in a more uniform blank holder pressure. This effect is demonstrated experimentally using pressure sensitive film on the same tool with two different blank holders, one equipped with a conventional blank holder, the other with a cardanic blank holder. To demonstrate transferability, the experiments are repeated on multiple single acting presses. The contribution will present and discuss the resulting pressure distributions.

1. Introduction

Even in times of Industry 4.0, manual work-in of the active surfaces of deep drawing tools (i.e. punch, blank holder and die) is required. This takes place during the try-out phase, where skilled workers manually grind the active tool surfaces to achieve the desired pressure distribution for contact with the sheet. This is one of the most time-consuming steps in the production, accounting for approximately 32% of the throughput-time [1]. Therefore, a reduction in try-out time, would result in significant cost savings. While the work-in time can be reduced via surface treatments, such as machine hammer peening [2], the focus of this paper is placed on the influence of different blank holder systems, as they directly influence the material-flow in the flange and thus the properties of the final part: In the beginning of the deep drawing process, the material in the flange has to overcome the static friction between the sheet and blank holder (BH) and between sheet and die respectively. Increased retention forces can lead to a super-imposed stretch forming which can trigger unwanted necking and ultimately lead to premature failure [3].

In an attempt to remedy these effects both active and passive approaches have been investigated. Active approaches rely on feedback from the process which is used as a control input for actuators such as gas springs [4], hydraulic cylinders, [5], active draw-beads, [6] or piezo elements in the blank holder [7]. While the aforementioned approaches have a positive influence on the process, they lead to complex tools.

In contrast to active tools, passive approaches aim to improve the pressure distribution without the need for external actuation. They rely on a predefined elastic deformation of the tool which is achieved through the use of e.g. rubber elements [8], hydro-elastic pockets [9] or a blank holder with a predefined compliance [10].

The approaches mentioned before were carried out on conventional forming machines with only one degree of freedom in the ram. A novel forming machine with two additional degrees of freedom was developed at the Institute of Production Engineering and Forming Machines [11]. Compared to conventional presses, this 3D Servo Press (3DSP) enables the ram to superimpose the vertical motion with two independent tilting movements. In order to make these additional degrees of freedom accessible during a deep drawing process, a special deep drawing tool was developed [12]. With the use of this new forming tool with a cardanic blank holder, the authors previously demonstrated a machine based closed loop control of the drawing process [13]. While the use of an active-cardanic blank holder proved beneficial, its use in closed loop control is currently limited to a 3DSP.

Since cardanic suspensions in general are used in various different fields for their self-leveling capabilities in the likes of navigational equipment and camera stabilization, it poses the question how a passive-cardanic blank holder influences the pressure distribution in deep drawing.

It should be noted that more recently, a deep drawing tool with the punch mounted on a cardanic cross was theorized by [14]. This would offer degrees of freedom similar to the ones of a 3DSP on tool basis, however no use in practice has been reported so far. Furthermore, the blank holder is a conventional one.

In contrast to the previous works, the present paper will focus on the pressure distribution resulting from the use of a cardanic blank holder and aims to answer the following research questions:

- How does the pressure distribution between work piece and blank holder differ between a rigid and a cardanic blank holder?
- How does a cardanic blank holder influence the transferability between forming machines?

2. Methodology

2.1. Forming Tool

The forming tool used is similar to the tone in previous works by the authors, [12], [13] and the overall schematic is shown in Figure 1.

The punch carrier is coupled with the press through a ball joint, which transfers the vertical motion of the ram to the punch itself. The ball joint is needed to decouple the vertical motion of the ram from the additional degrees of freedom provided by the 3DSP. For more detail, please refer to [12]. Furthermore, the vertical motion of the press leads to a compression of eight equidistantly spaced springs, which generate a force acting on the blank holder. Note that the blank holder force (BHF) is coupled with the displacement of the ram, so that the BHF is linearly increasing throughout the forming operation.

The cardanic blank holder is shown in Figure 2: The blank holder carrier is divided up into three parts: An outer frame, which is guided by sliding bearings to ensure accurate relative positioning of the blank holder with regard to die and punch. Within the outer frame, there is a secondary inner frame which forms the first degree of freedom of the cardanic joint together with the outer frame (θ_y). The inner frame guides the actual mount for the blank holder plates. The axis of rotation θ_x is perpendicular to θ_y, thus completing the cardanic suspension.

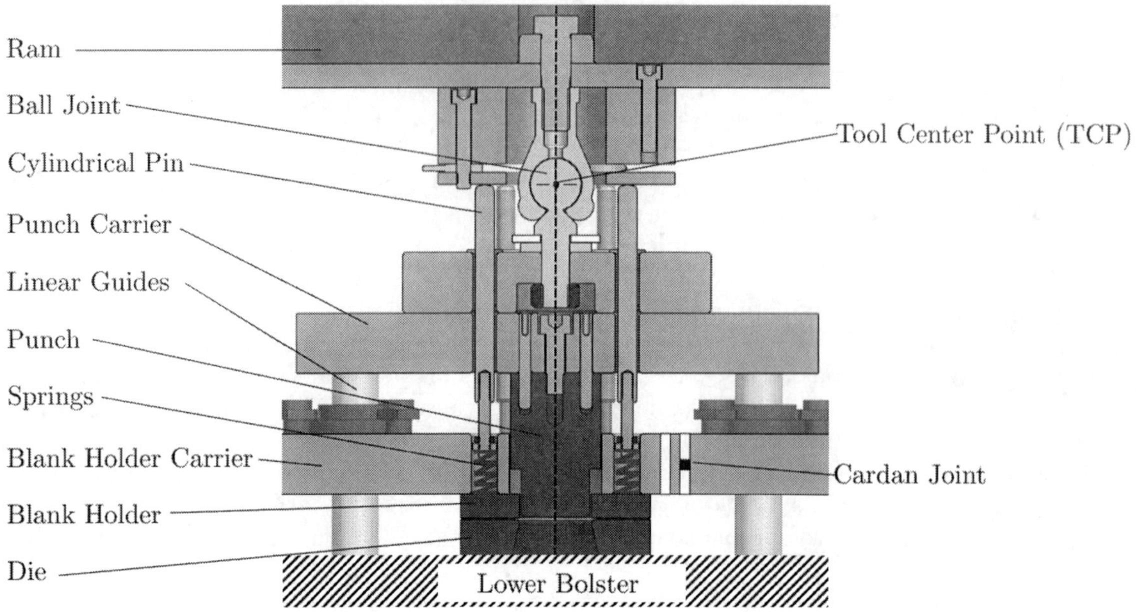

Figure 1. Section of Drawing Tool.
Left side of Blank Holder Carrier: conventional Blank Holder; **Right** side: Cardanic Blank Holder.

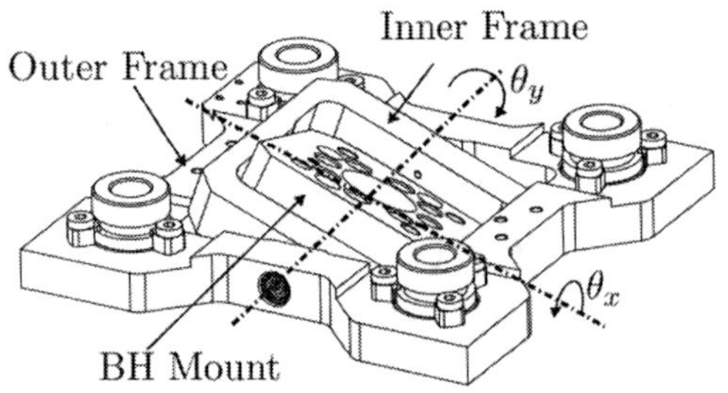

Figure 2. Cardanic blank holder carrier (Tilting highly exaggerated to demonstrate motion).

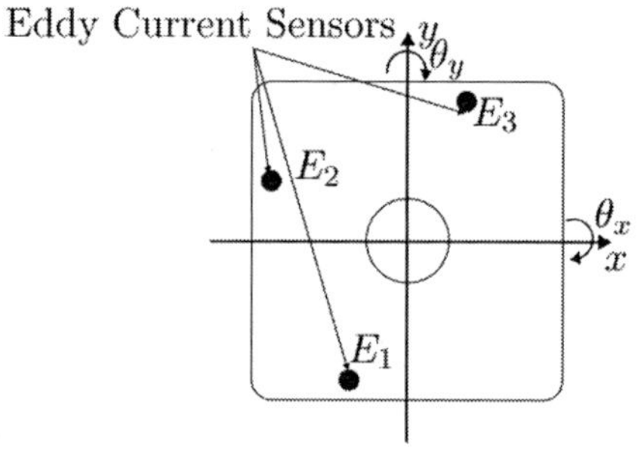

Figure 3. Position of eddy current sensors on blank holder and coordinate system.

2.2. Forming Machines
The experiments are performed on a 3D Servo Press (3DSP), however, the tilitng motion will not be used and instead the pose of the ram will be kept horizontal. This emulates a conventional press, where the ram cannot be tilted. The second machine used for the experiments is a combined tensile compression test machine (ZwickRoell AllroundLine 100 kN). Both machines can be considered single acting presses, as only the ram is driven and no additional actuators (e.g. an additional draw bed) are available.

2.3. Data Acquisition
The resulting pressure distributions are recorded using pressure sensitive film (Fujifilm Prescale type LLLW). For later evaluation of the pressure sensitive film, the air temperature and the humidty is recorded using a thermohygrometer (testo 608-H1). In order to determine the

Figure 4. Evaluation procedure for pressure sensitive film.

pressure distribution on the flange of the part, the punch is removed from the tool before loading the part. This step is necessary, as the punch would interfere with the pressure sensitive film. The pressure distributions are determined as "momentary pressures" as defined by Fujifilm: The force is increased for 5 seconds, held for 5 seconds and is then released.

Furthermore, three piezoelectric force transducers (Kistler 9012-A) are used to capture the total BHF. The self weight of the blank holder (approx. 45 N) is added to the recorded force data, resulting in the overall load.

Lastly, the blank holder is equipped with three eddy current sensors (MicroEpsilon ES08) as shown in Figure 3 which continuously monitor the distance between the blank holder and the die in three points. Since the position of all three sensors are known, this information can be used to determine the current pose of the blank holder relative to the die.

All force data is sampled at 100 Hz. At the 3DSP the integrated programmable logic controller is used, while a QuantumX MX840A by HBK is used at the tensile testing machine.

2.4. Data Processing

2.4.1. Determination of Blank Holder Pose In order to evaluate the tilting angle of the blank holder about the x- and y-axis, the coordinates of the eddy-current-sensors are assumed to be on the same plane (i.e. the surface of the blank holder). With known locations for each eddy-current-sensor in x_n, y_n and the measured distance z_n, the location vector for each sensor can be expressed as a position vector:

$$\vec{E_n} = \begin{pmatrix} x_n \\ y_n \\ z_n \end{pmatrix} \tag{1}$$

With this, two vectors $\vec{V_1} = \vec{E_1} - \vec{E_2}$ and $\vec{V_2} = \vec{E_3} - \vec{E_2}$ which make up the plane are constructed. A normal vector $\vec{N}$ to the plane is determined by calculating the cross-product $\vec{V_1} \times \vec{V_2}$. The individual values of $\vec{N} = [N_1, N_2, N_3]$ can be used to determine the angles of the plane with regard to the x-axis (θ_x) and the y-axis (θ_y).

$$\theta_x = -\arctan\left(\frac{N_2}{N_3}\right) \qquad \theta_y = \arctan\left(\frac{N_1}{N_3}\right) \tag{2}$$

Note that the x- and y-position of eddy-current-sensors are assumed to be identical to their original positions in the non-tilted case. This is valid, since only very small tilting angles are expected, so that $\lim_{\theta \to 0} x \cos(\theta) \approx x$.

2.4.2. Evaluation of Pressure Distribution Data The resulting pressure prints are scanned at 600 DPI and later on processed using a custom Python script. Within the script, the colored scans are first translated to a grayscale image and the corresponding pressures are subsequently mapped on the image (Figure 4).

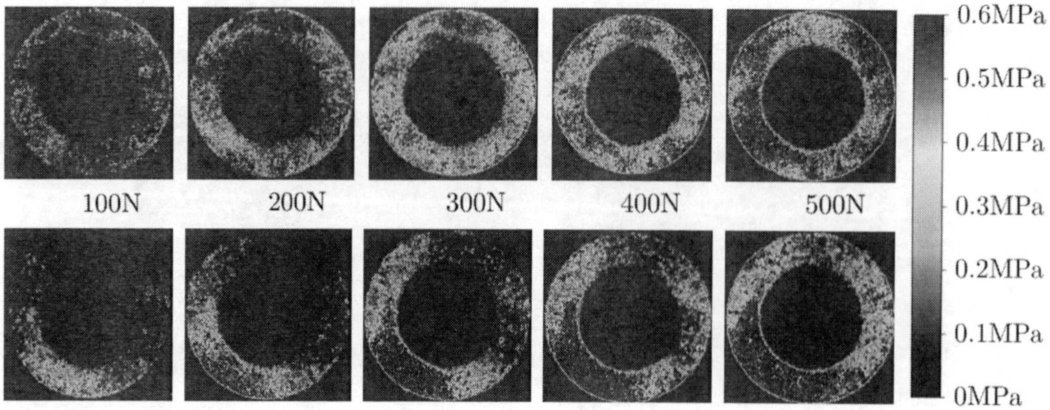

Figure 5. Pressure Distribution on 3DSP.
Top row: cardanic blank holder; **bottom row:** rigid blank holder.

3. Results

In the beginning of a deep drawing process, the blank is at a stand still. Additionally to the actual forming forces required, the static coefficient of friction has to be overcome in order to allow for a flange draw-in. In order to investigate the initial pressure distribution between blank holder and sheet, different levels of BHF are applied to a blank with an initial diameter of $D_0 = 37$ mm and the resulting pressure-distributions are presented in the following sections.

3.1. 3D-Servo-Press

The resulting pressure distributions obtained by using the 3DSP are presented in Figure 5. It can be seen that a rigid blank holder (bottom row) leads to a localized pressure in the bottom left of the image. This localization is present during the first loading stage with 100 N and slowly expands towards the top right with an increase in blank holder force. However, even at a constant load of 500 N, the blank holder pressure is concentrated on the bottom left, with little contact in the top right section.

The use of a cardanic blank holder (top row) leads to an even distribution of the blank holder pressures on the flange. Even at low forces (100 N), the blank holder makes contact with the blank in the top right corner, which was not the case with conventional blank holder. With an increase in force, the pressure increases proportionally throughout the whole flange, while maintaining an almost constant spatial distribution.

The resulting tilt angles for the experiments performed on the 3DSP are presented in Figure 6. For the cardanic blank holder, both tilting angles hardly change with increasing blank holder force. A positive rotation about the pitch-axis (θ_x) and a negative rotation about the roll-axis θ_y represent slight tilt towards the bottom left.

The same overall behavior can be observed in the conventional BH. However, the magnitude of the tilt angles is higher than the ones observed in the cardanic case. With an increase of overall blank holder force, the pitch angle decreases, while the roll angle increases. The values appear to converge towards the tilt angles observed in the cardanic case. This describes a tilting motion towards the top right of the images, hence matching the results observed in the pressure distributions.

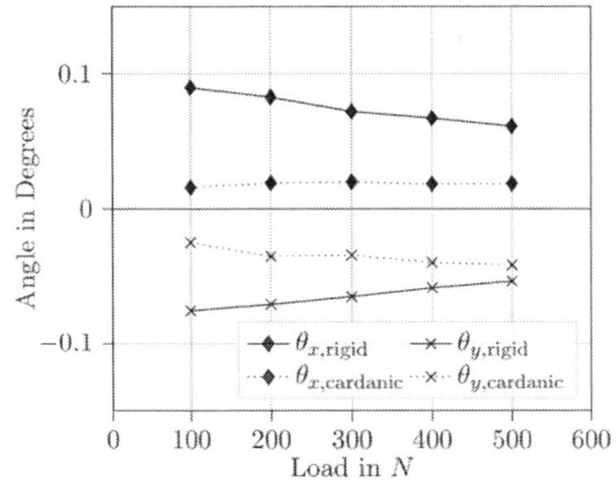

Figure 6. BH Angles over applied force for 3DSP.

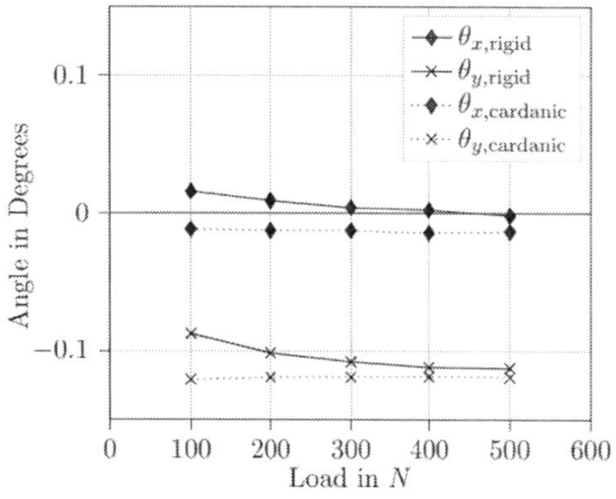

Figure 7. BH Angles over applied force for Tensile Machine.

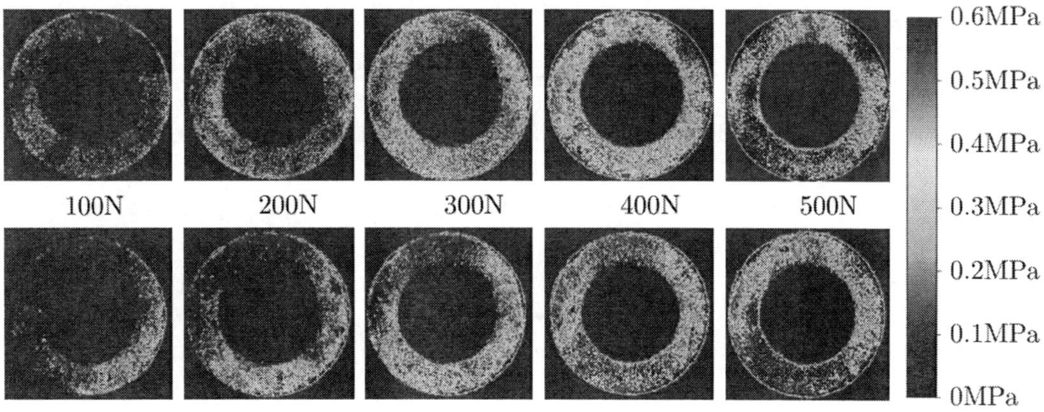

Figure 8. Pressure Distribution on Tensile-Testing machine
Top row: cardanic blank holder; **bottom row:** ridigd blank holder

3.2. Tensile Testing Machine

Next, the pressure distributions as they occur on the tensile-testing-machine are presented in Figure 8.

As was previously observed on the 3DSP, the use of a conventional blank holder (bottom row) leads to a localization of pressure, however this time starting on the bottom right and slowly expanding towards the top left. Even at higher forces, more pressure is applied on the bottom of the flange than on the top of the flange.

The cardanic blank holder (top row) leads to more evenly distributed pressures throughout the whole part, even at low applied forces. The spatial distribution of the blank holder remains almost identical, while the overall pressure increases with increasing force.

The tilt angles shown in Figure 7 remain essentially constant in case of the cardanic blank holder. With a slightly negative tilt around the pitch-axis, as well as the roll-axis, the blank holder is slightly tilted towards the top right corner of the images. The conventional one is initially tilted to the bottom right (positive pitch-angles and negative roll angle), and begins to tilt towards the top left corner. As observed on the 3DSP, the pose of the conventional blank holder gradually converges towards the pose of the cardanic blank holder.

IOP Conf. Series: Materials Science and Engineering 1284 (2023) 012082 doi:10.1088/1757-899X/1284/1/012082

4. Discussion

We showed that the passive-cardanic blank holder has self-leveling capabilities, which result in a more evenly distributed blank holder pressure throughout the flange. This can attributed to the tilt-angles provided by the cardanic joint: While the motion of the conventional blank holder is constricted and higher forces are needed to establish contact between the active surfaces and the sheet, the cardanic blank holder can tilt freely and establish contact much more evenly and at lower loads. Furthermore, this indicates that there is an optimal pose for the blank holder.

Comparing the pressures between the two machines used in the study, we found that a conventional blank caused a localization of pressures towards the bottom in both cases. The direction of the initial contact varied from bottom left on the 3DSP to bottom right on the second machine. Since the springs in the tool are much more compliant than the machines used, this difference cannot be directly attributed to the difference in stiffness of the presses but to secondary effects, such as the parallelism of punch and lower bolster or operator error when moving the tool between machines.

While the root causes cannot be determined for certain, the cardanic blank holder establishes a much more even contact with the sheet on both machines. This shows that the use of cardanic blank holders has the potential to increase the robustness of forming proesses and improving transferability between two machines.

The results presented in the present paper were achieved using flat active surfaces on both parts of the tool. Commercial blank holder systems oftentimes make use of the draw beads to influence the material flow locally. While the authors expect that the use of a cardanic blank holder will also have a positive effect on those forming tools, further research should be focused on this aspect. Furthermore, the effectiveness for non-axis-symmetrical parts needs to be investigated.

5. Conclusion

The manual work-in of deep drawing tools makes up a significant amount of time in the overall production process. This issue has previously been addressed using active and passive blank holder systems.

In the present study, we used pressure sensitive film to investigate the influence of a passive-cardanic blank holder on the initial pressure distribution on an unformed blank. We demonstrated that the use of a cardanic blank holder can aid the uniform distribution of blank holder forces, and thus has the potential to reduce the work-in time deep drawing tool, while also increasing the transferability between forming machines.

Acknowledgments

The presented investigations were funded by the Deutsche Forschungsgemeinschaft (DFG, German Research Foundation) – Project number 386415239. The authors thank the DFG for supporting this project.

References

[1] Meyerhoff M 1998 *NC-Programmierung für das kraftgesteuerte Bandschleifen von Freiformflächen* (*Fortschrittberichte VDI : Reihe 2, Fertigungstechnik* vol 467) (Düsseldorf: VDI-Verl.) ISBN 978-3-18-346702-0
[2] Oechsner M, Wied J and Stock J 2014 *Advanced Materials Research* **966–967** 397–405 ISSN 1662-8985
[3] Lange K 1986 *Journal of Applied Metalworking* **4** ISSN 0162-9700
[4] Rittmeier S 2007 *Systemunterstützte Umformung* (Göttingen: Cuvillier) ISBN 978-3-86727-419-7
[5] Barthau M and Liewald M 2017 *Procedia Engineering* **207** 66–71 ISSN 18777058
[6] Siegert K, Liewald M, Blaich C, Großmann K and Kauschinger B 2007 *wt Werkstatttechnik online* 781–791
[7] Neugebauer R, Mainda P, Kerschner M, Drossel W G and Roscher H J 2011 *AIP Conference Proceedings* **1353** 189–194 ISSN 0094-243X

[8] Sommer N 1986 *Niederhalterdruck und Gestaltung des Niederhalters beim Tiefziehen von Feinblechen* (*EFB-Forschungsbericht* no 31) (Hannover: EFB) ISBN 978-3-86776-106-2
[9] Stock G 1996 *Alternative Verfahren Zur Verbesserung Des Tiefziehergebnisses Auf Einfachwirkenden Pressen* (*Fortschritt-Berichte VDI Reihe 2, Fertigungstechnik* no 380) (Düsseldorf: VDI-Verl) ISBN 978-3-18-338002-2
[10] Doege E and Elend L E 2001 *Journal of materials processing technology* **111** 182–187
[11] Groche P, Scheitza M, Kraft M and Schmitt S 2010 *CIRP Annals* **59** 267–270 ISSN 00078506
[12] Breunig A, Hoppe F and Groche P 2019 Localized Blank-Holder Pressure Control in Cup Drawing through Tilting of the Ram *Proceedings of NUMIFORM 2019: The 13th International Conference on Numerical Methods in Industrial Forming Processes* (Portsmouth, NH, USA)
[13] Groche P, Breunig A, Chen K, Molitor D A, Ha J, Kinsey B L and Korkolis Y P 2022 *CIRP Annals* ISSN 0007-8506
[14] Paunoiu V, Teodor V, Baroiu N, Morosanu G A and Epureanu A 2022 *MATEC Web of Conferences* **368** 01022 ISSN 2261-236X

IOP Conf. Series: Materials Science and Engineering 1284 (2023) 012083 doi:10.1088/1757-899X/1284/1/012083

Novel tool steel for application in hot stamping

J Knippenberg, P Niederhofer, H-G Krull and S Richert

Deutsche Edelstahlwerke Specialty Steel & Co. KG

E-mail: jonas.knippenberg@swisssteelgroup.com

Abstract. Hot stamping is a commonly used procedure to manufacture components from high strength steels, which are used for parts of car bodies. Against the background of production efficiency and costs, there is a number of factors which determine the performance of tools used in hot stamping, e.g. cycle time in production and lifetime of the tools. Main requirement is a high wear resistance, while additional properties such as high thermal conductivity can be considered being beneficial. Many steel grades, which are currently applied in hot stamping tools (e.g. hot work tool steel 1.2367) contain high amounts of alloying elements which gives them good mechanical properties, particularly at high temperatures. However, these steels in turn suffer from a rather low thermal conductivity and only moderate wear resistance. Optimization of the balance of mechanical properties, wear resistance, and thermal conductivity can lead to a better general performance of hot working tool steels used in hot stamping applications. In this study, a novel tool steel is characterized with special focus on properties relevant for hot stamping tools. Results are compared to those of reference hot work tool steels. Wear tests that were conducted in the framework of this paper include laboratory setups like ASTM G65 / G75 and high-temperature strip-drawing test setups to simulate the conditions of the process closer to reality. Results show that the novel tool steel is characterized by an increased wear resistance in combination with a higher thermal conductivity compared with common hot work tool steels. Furthermore it is approved that a wide range of desired properties can be adjusted in this alloy via heat treatment. The favourable combination of properties makes the novel tool steel suitable for application in hot stamping tools.

1. Introduction

1.1. Hot stamping

Against the background of global climate change by anthropogenic exhaust of carbon dioxide, efforts are being made by automotive industries to limit and reduce fuel consumption of vehicles and thus CO_2 emissions. One way to meet these goals is the usage of ultra-high strength steels such as 22MnB5 (1.5528) which may possess a tensile strength up to 1.500 MPa [1]. Higher tensile strengths allow for the usage of thinner components to achieve the same or even higher body-in-white rigidity and crash safety but also decreasing weight of vehicles and thus fuel consumption [2].

However high mechanical properties also place very high requirements in forming technology. For example conventional cold forming processes need very high bending forces to reach a necessary effect in forming steel sheets which reduces tool lifetime. Another major problem is the so called springback effect which obstructs accurate manufacturing of parts and is therefore not wanted [3].

The production process of hot stamping describes a special hot forming procedure for body-in-white parts which has been established in car part manufacturing since several years. A rising trend on hot stamped parts produced in automotive industry can be observed. [4]. This way ultra-high strength steel sheets require only one production step to be formed and hardened. The results are fully martensitic production parts without the need of another heat treatment [5]. At first steel sheets of 1.5528 need to be heated to appropriate hardening temperatures. After a short transfer to the press the part is formed and immediately quenched. Fast heat transfer out of the workpiece is enabled via cooling channels inside of the tool. Because of forming at elevated temperatures problems in cold formability mentioned before are circumvented. Followed by that the main advantages of this forming process lie in produced parts having a reduced or totally removed springback effect as well as significantly reduced forming forces. It becomes clear that this production process is most suitable for manufacturing crash-relevant automobile parts from ultra-high strength steels, such as A- or B- pillars. [1].

However there are also disadvantages resulting from this kind of production method. To prevent the steel sheets surface from scaling and corrosion an additional protective layer is needed. The most commonly used type of protective layer is Al-Si-coating which is applied in a solid state. During annealing this kind of layer forms into a liquid state and protects the steel sheet from oxidation whilst giving good weldability as well as coatability after hot stamping. This is a necessity due to welding and painting of body-in-white parts. Also other types of protective layers do exist but are not used in serial production due to them being in experimental state (sol-gel layers) or introducing other technical problems (Fe-Zn layers) [6]. Elevated process temperatures also demand higher resistance to the tool materials in terms of thermal load and increased wear. Currently there are no lubricants available that can withstand these high process temperatures used in hot stamping, resulting in higher friction and thus more wear. Also corrosive attacks of cooling media located inside the cooling channels of the tools have been reported [7]. Summed up, all elements mentioned above reduce the overall lifetime of tools used in hot stamping.

1.2. Tool steels for hot stamping

It becomes clear that materials for hot stamping need to meet several requirements. First of all a high and evenly distributed hardness is of great importance but also a high resistance to wear and high thermal conductivity of the tool material has been identified as important. The latter in particular can be seen as crucial to enable sufficiently fast quenching to achieve the desired martensitic microstructure. Also higher thermal conductivity may contribute to reduce the cycle time of the process because the workpieces can be cooled faster and productivity rises.

Conventionally used hot working tool steels have high amounts of alloying elements which cause high hardness even in large cross sections. The standardized alloy 1.2367 (X38CrMoV5-3) shall be used as an example, since it is widely applied in hot stamping tools. On the other side a high amount of alloying elements, especially those elements which accumulate interstitially or substitutionally, are known to reduce thermal conductivity. Chromium and silicon seem to have the largest influence while the elements carbon, molybdenum, and manganese show a lower effect on thermal conductivity [8]. Wilzer et al. showed that the described effect can be reduced when the detrimental alloying elements are present in precipitated form, for example as carbides [9]. As a result amounts of elements which have little to no affinity to form carbides should be kept as low as possible. In addition there needs to be balanced amounts of carbide forming alloying elements in order to achieve a thermal conductivity as high as possible.

As the process of hot stamping has been established in automotive manufacturing since nearly 10 to 15 years there have also been efforts in optimizing tool steels for this specific process. As first concepts only focused on high thermal conductivity alone, impairments in terms of wear resistance and achievable hardness may occur. Consequently a balance between thermal conductivity, wear resistance and hardenability needs to be found to further improve tool steels for hot stamping [10]. Furthermore it has been confirmed that needed physical properties of hot working tool steels as thermal conductivity can be adjusted via heat treatment methods. As explained by Wilzer et al. [11] thermal conductivity of tool steels is not only dependent on composition but also on the state of heat treatment the material is in. In general, steels tempered at lower temperatures show high hardness values but also comparatively low thermal conductivity coefficients. Compared to higher tempering temperatures the overall hardness is reduced in favor of gaining toughness but also due to formation of carbides and decreased amount of alloying elements in the steel matrix a higher thermal conductivity is reached.

Based on the known facts and more research effort aiming at the effects of the alloying system of a tool steel on the important properties, a hot work tool steel especially designed to meet the mentioned characteristics has been developed and standardized as 1.2383 (X45 MoV14-15, trade name Thermodur 2383 Supercool) [12] [13]. This steel grade was developed for balancing the characteristics mentioned above containing hardness, wear resistance, hardening behavior and thermal conductivity.

1.3. *Novel tool steel development*

Based on steel grade 1.2383 further modifications had been done. The modifications aimed at achieving a higher resistance to wear in combination with a thermal conductivity which is more suitable for the application in hot stamping. It is noted that the top priority in this study is wear resistance followed by thermal conductivity in importance. For determining both properties of the new steel two reference materials were also investigated in terms of thermal conductivity. Furthermore this study focusses on the results of comprehensive wear testing of the new tool steel compared to conventionally used steels for hot stamping applications.

2. Experimental

2.1. *Investigated steel grades*

The established tool steel 1.2367 is applied in this study as reference material. A second steel grade 1.2383 which has been developed with thermal conductivity as first priority is integrated here as well. This steel shows excellent values in thermal conductivity but performs not optimal in terms of wear resistance. The novel tool steel mentioned in section 1.3 is named Thermodur HSP Superclean. Further information on the development of the alloy is given in [14]. An overview of all steel grades used in this study and their chemical composition is given in table 1.

Table 1. A composition of investigated steel grades in mass.-%

	C	Si	Cr	Mo	V	Ni	Mn
1.2367	0.37	0.3	5.0	3.0	0.6	--	--
1.2383	0.45	--	--	1.5	1.5	0.9	0.9
HSP	0.65	--		Σ 4.0		0.5	0.5

Materials 1.2367 and 1.2383 were manufactured via common industrial processes including melting in an electric arc furnace (EAF), secondary metallurgical treatment, ingot casting, electro-slag re-melting, forging and suitable heat treatment processes. Some of the results of Thermodur HSP were derived from experimental state, where the alloy was manufactured on a laboratory scale featuring induction melting, ingot casting and rolling. Since the development process, HSP was manufactured on an industrial scale as well, some results derived from this material were provided as well. The manufacturing route here is the same as for 1.2367 and 1.2383. All specimen which were used in the upcoming tests were either cut from the laboratory scale ingot or from the transition area of large blocks.

Two different heat treatment stages for the new alloy were chosen to perform the following wear resistance measurements. For referential purposes the hardness annealing curve for Thermodur HSP is presented in figure 1. Necessary heat treatments were conducted in vacuum furnaces using nitrogen as

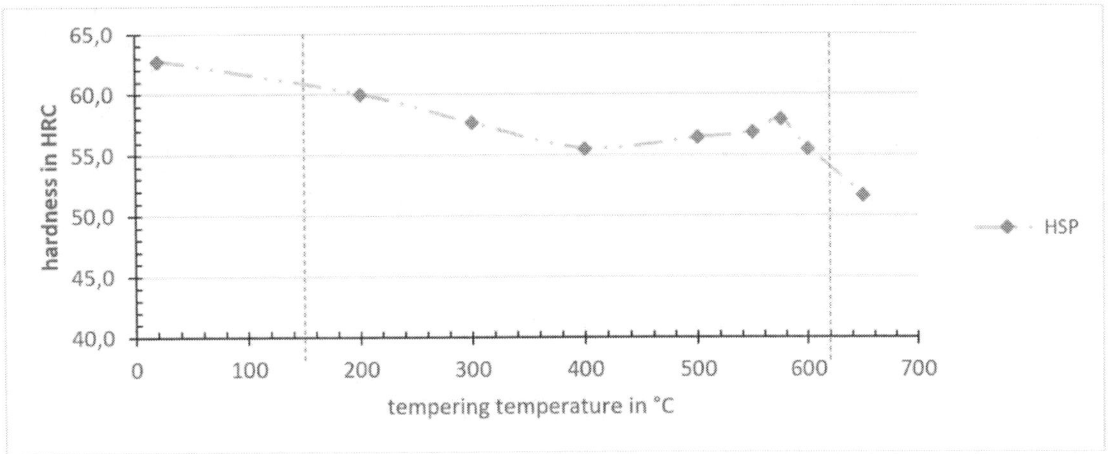

Figure 1: hardness annealing curve for HSP, hardened at T = 1100 °C, as presented in section 2.1. Dashed lines indicating both tempering temperatures chosen for further research (150 °C / 620 °C)

quenching medium. Further measurements of hardness values after different tempering stages were taken to determine suitable heat treatment conditions.

2.2. Testing methods

2.2.1. Thermal conductivity

Thermal conductivity values were determined via measuring three different physical material properties. Density was measured via buoyancy method and dilatometry, for specific heat capacity a differential scanning calorimeter was used and the temperature conductivity was measured from laser flash analysis. After that, calculation of thermal conductivity could be done via multiplication of the individual properties.

2.2.2. Strip drawing tests

To specify wear resistance strip drawing tests were performed and compared. A specifically designed test rig was used for this testing method which is able to simulate tribological processes occurring in hot stamping closer to reality than standardized wear resistance tests could do. Analysis of cause-effect relations between the parameters tool and workpiece temperature, contact pressure and drawing velocity for different workpiece and tool materials can be concluded [14]. Counter body material in all strip drawing tests were steel sheets of AlSi-coated 22MnB5. Used test parameters were a contact pressure of 5 MPa and a drawing velocity of 50 mm/s. Three experiments were conducted on each material. Evaluated properties were the measurement of the resulting strip surface topography using confocal microscopy and determining profile amplitude heights.

2.2.3. Abrasive wear testing

For comparing purposes wear resistance measurements from standardized tests were also considered. The first test series was conducted according to ASTM G65-15 with the use of dry sand as abrasive material. Method A was used and the volume loss was measured. ASTM G75 was also considered using a standardized slurry and determining again volume loss.

3. Results and discussion

3.1. Thermal conductivity

Now the results from measurements of thermal conductivity values described in section 2.2.1 will be presented. An overview of thermal conductivity values at room temperature is given in figure 2. The effect of heat treatment on the material explained in section 1.2 can be seen directly here: in all cases the low tempered variants of all materials show a reduction in thermal conductivity. It also stands out that the modified variant can be placed between both reference materials for the high tempered state in terms of thermal conductivity (1.2383: 41,5 W/m*K; HSP: 34 W/m*K; 1.2367: 28 W/m*K). The high tempered variants are ranked different (1.2383: 40 W/m*K, 1.2367: 25 W/m*K, HSP: 19 W/m*K). Manufacturing of HSP on an industrial scale resulted in similar values of thermal conductivity (33,5 W/m*K, while HSP manufactured on laboratory scale shows 34 W/m*K). As thermal conductivity was proofed to only have limited impacts on the overall performance of tool steels in hot stamping (see section 1.2) the novel tool steel can be rated as a good compromise.

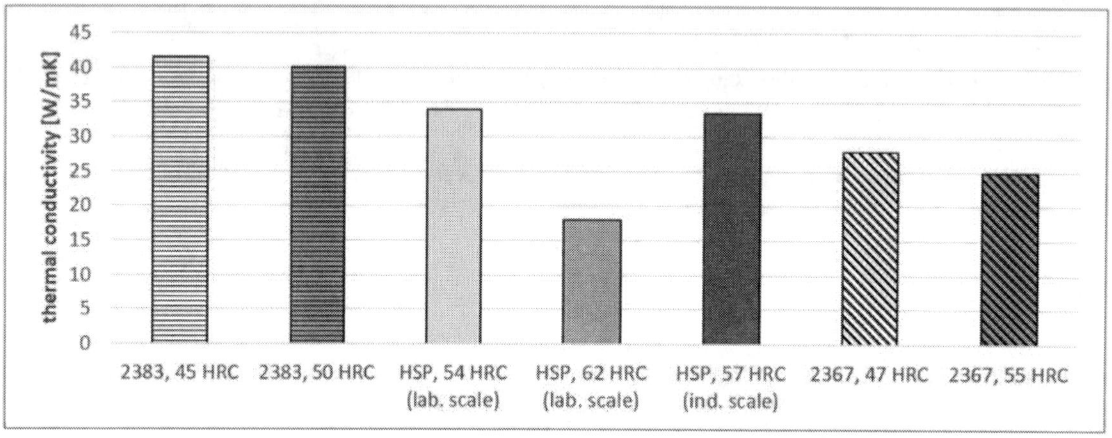

Figure 2: thermal conductivity values at room temperature (20 °C)

3.2. Wear resistance: strip drawing tests

Results from measuring the surface topography via confocal microscopy (see section 2.2.2) are presented in figure 3. Thermodur HSP in the former described two different heat treatment stages (resulting in different hardness values between 62 HRC and 52 HRC) after conducting the strip drawing

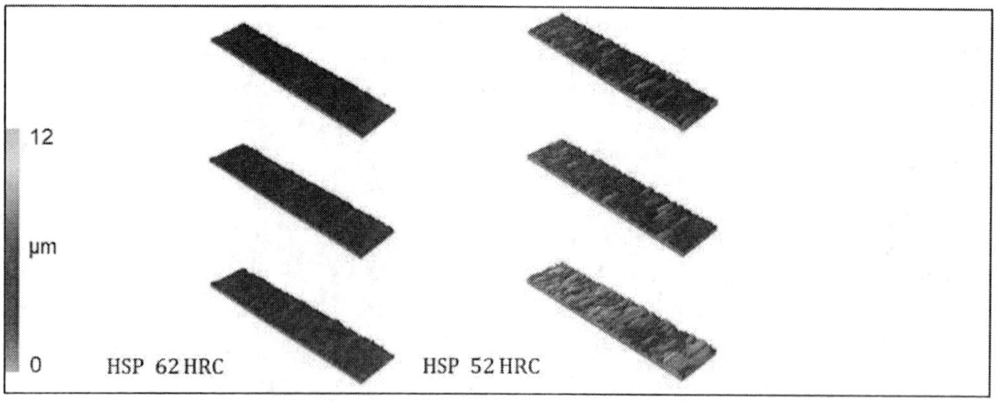

Figure 3: topography of used tool surfaces made from HSP and tempered at 150 °C (left) and 620 °C (right) after strip drawing tests

tests are shown. It becomes clear, that the modified alloy in its lower tempered state has a much higher resistance to adhesive wear than its high tempered variant. This is due to the much smoother appearing surface of the left strips as can be observed via the scale on the left side.

In order to better evaluate the wear resistance of the compared materials the profile amplitudes which are presented in figure 4 may also be considered. Here it is shown that the reference materials are also influenced by adhesive wear. 1.2383 seems to have a slightly lower tendency to adhesion than 1.2367. Also the strip drawing tests indicate that the high tempered version of HSP has a resistance to adhesive wear which is comparable to the reference materials; the low tempered variant is influenced much more. All strips have been grinded before starting the experiments, so the heights of all amplitudes at the start of each experiments was around 3 μm. As already mentioned above the dominant wear mechanism is adhesion which means that a higher profile amplitude in figure 5 indicates an increasing amount of adhesive wear at the tools surface. Here it is shown that reference 1.2367 has in fact the highest profile amplitudes from all investigated steel grades, followed by 1.2383 and HSP in high tempered condition. The lowest profile amplitudes is achieved by HSP in low tempered condition.

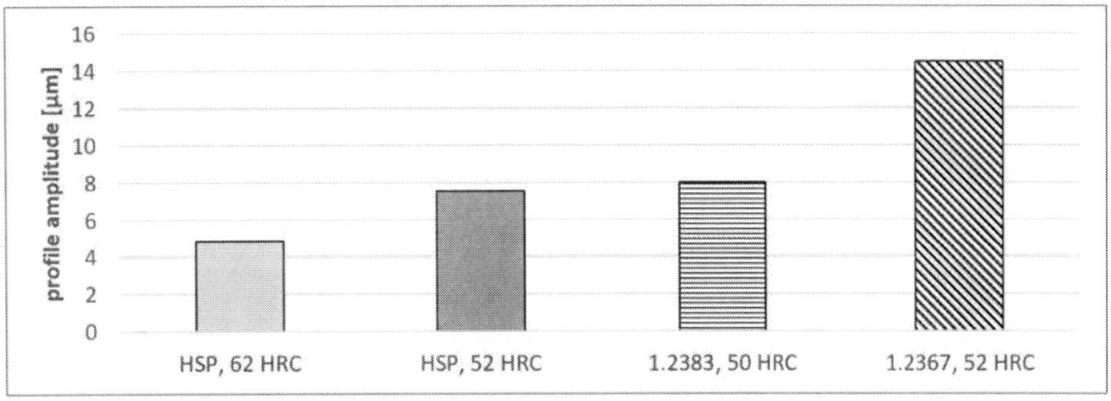

Figure 4: profile amplitudes of tool surfaces after strips after drawing tests

3.3. *Wear resistance: G65 / G75*

In terms of abrasive wear tests according to ASTM G65 and ASTM G75 have also been taken into account (see section 2.2.3). The experimental results originate from [15] and are presented in figure 5 and figure 6. It becomes clear that 1.2367 shows very high amount of volume loss to due abrasive attacks in both test cases. 1.2383 has significantly lower volume losses in G65 but even higher losses under G75 testing conditions. Also the following facts have to be considered: the abrasive medium in wear resistance tests according to ASTM G65 is sand, which means rather coarse abrasive elements. In this case large primary carbides, which are present in 1.2383 may help reduce mass loss. The abrasive elements used under G75 testing conditions are much smaller and therefore have a larger tendency to attack steel matrix between the carbides which has a lower resistance to wear. A certain comparison can therefore only be rated as ambiguous. Still the new variant HSP performs better than the reference materials, since they combine favorable carbide size and distribution with a high hardness of the steel matrix. Under both testing conditions HSP reaches good results while the material in low tempered condition benefits from an even lower volume loss due to abrasive wear than the high tempered variant. Comparing the hardness values of both this fact seems reasonable as higher hardness gives a better resistance to abrasive wear.

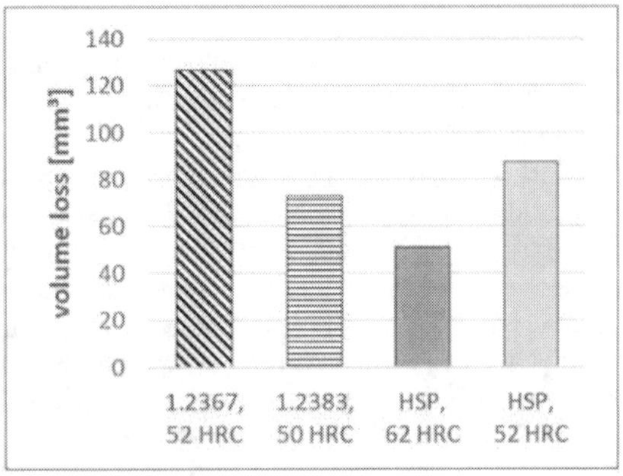

Figure 5: volume loss values of all materials resulting from **ASTM G65**

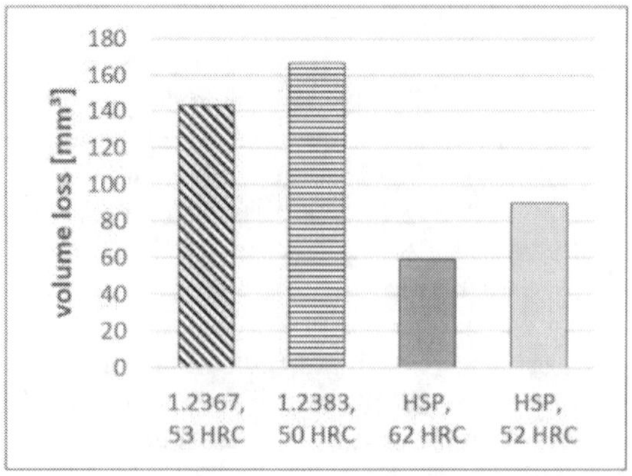

Figure 6: volume loss values of all materials resulting from **ASTM G75**

4. Conclusion

A number of experiments regarding thermal conductivity and resistance to wear stresses both abrasive an adhesive had been performed on a selection if materials including two conventional alloys for reference (1.2367 and 1.2383) and one experimental, novel tool steel alloy (Thermodur HSP Superclean). The experiments included measurement and calculation of thermal conductivity as well as strip drawing tests and experiments according to ASTM G65 and ASTM G75. Considering all results from this experiments above, it can be stated that the novel tool steel shows better properties against mechanical wear than conventional alloys. When compared to the steel grade 1.2383 thermal conductivity is less but still larger than compared to conventional alloys such as 1.2367. Combined with the better resistance against both abrasive and adhesive wear the novel tool steel can be characterized as a suitable solution for hot stamping applications.

The results of the wear resistance experiments were verified by various industrial partners in additional tests not described here. Manufacturing this steel grade in larger dimensions for the successful implementation of this material in the market and is topic of currently running investigations.

Acknowledgements
The strip drawing tests which results have been used in this paper had been performed at the Friedrich Alexander Universität (FAU) at Erlangen which is thankfully acknowledged. Also parts of the experiments were conducted within the project P1228 “Tool-side impact on the tribological conditions within hot stamping” funded by the Research Association for Steel Application (Forschungsvereinigung Stahlanwendung e. V. FOSTA), which is also gratefully acknowledged.

References

[1] Hasegawa K., Kaneko K., Seto K., 2013 “Cold-Rolled and Galvannealed (GA) High Strength Steel Sheets for Automotive Cabin Structure”, *JFE Technical Report* **18** pp. 80–88.

[2] K. Mori, S. Saito, S. Maki, 2008 “Warm and hot punching of ultra high strength steel sheet”, *CIRP Annals – Manufacturing Technology*, **57**, pp. 321-324

[3] K. Mori, P. Bariani, B. Behrens, A. Brosius, S. Bruschi, T. Maeno, M. Merklein and J. Yanagimoto, 2017, "Hot stamping of ultra-high strength steel parts," *CIRP Annals - Manufacturing Technology*, **66**, pp. 755-777

[4] Habert H, 2015 “Products in Hot Stamped Boron Steel”, *Capital Market Day 2015*

[5] K. Mori, D. Ito, 2009, “Prevention of oxidation in hot stamping of quenchable steel sheet by oxidation preventive oil”, *CIRP Annals – Manufacturing Technology*, **58**, pp. 267-270

[6] Grigorieva R, Drillet P, Mataigne JM, Redjaïmia, 2011, "Phase Transformations in the Al-Si Coating during the Austenitization Step", *Solid State Phenomena* **172–174**, pp. 784–790
[7] R. Rahn and I. Schruff, "The Selection of Tool Steels for Hot-Stamping Tools with Respect to Increased Loads," *5th International Conference on Hot Sheet Metal Forming of High-Performance Steel* **CHS2**, pp. 659-668
[8] Y. Terada, K. Ohkubo, T. Mohri and T. Suzuki, 2002, "Effects of Alloying Additions on Thermal Conductivity of Ferritic Iron," *ISIJ International*, **no. 3**, pp. 322-324
[9] J. Wilzer, S. Weber, C. Escher and W. Theisen, 2014, "Werkstofftechnische Anforderungen an Presshärtewerkzeuge am Beispiel der Werkzeugstähle X38CrMoV5-3, 30MoW33-7 und 60MoCrW28-8-4," *HTM J. Heat Treatm. Mat.*, **6**, pp. 325-332
[10] J. Wilzer, C. Escher, M. Kotzian, S. Weber and W. Theisen, 2016, "Optimierte Eigenschaften von Werkzeugstahl für Presshärtewerkzeuge," *HTM J. Heat Treatm. Mat.*, **no. 1**, pp. 35 - 42
[11] S. Hafenstein, E. Werner, J. Wilzer, W. Theisen, S. Weber, C. Sunderkötter and M. Bachmann, 2015, "Influence of Temperature and Tempering Condition on Thermal Conductivity of Hot Work Tool Steels for Hot Stamping Applications," *Steel Research International*, **no. 12**, pp. 1628-1635
[12] E. Meurisse, C. Ernst and W. Bleck, 2012, "Improvement of thermal conductivity of hot-work tool steels by alloy design and heat treatment," *Proceedings of the 9th International Tooling Conference*, pp. 215-226
[13] E. Meurisse, 2015, Untersuchung der Wärmeleitfähigkeit von Warmarbeitsstählen, *Dissertation Aachen RWTH*
[14] P. Schwingenschlögl, J. Tenner and M. Merklein, 2018, "Tribological Behavior of Different Tool Steels and Surface Properties under Hot Stamping Conditions," *Key Engineering Materials*, **767**, pp. 212-219
[15] P. Niederhofer , K. Eger , P. Schwingenschlögl , M. Merklein, 2019, „Properties of tools steels for application in hot stamping", *steel research international*, **91**, 1900422

IOP Conf. Series: Materials Science and Engineering 1284 (2023) 012084 doi:10.1088/1757-899X/1284/1/012084

Prediction strategy for the forming and stuffing of non-circular tube

M Stennei[1], T Clausmeyer[1], J Grodotzki[1], S Rosenthal[1], M Stiglmair[2] and A. E. Tekkaya[1]

Institute of Forming Technology and Lightweight Components, Germany
Faurecia Emissions Control Technologies Germany GmbH, Germany

E-mail: markus.stennei@iul.tu-dortmund.de

Abstract. The manufacturing of exhaust filter systems for cars typically comprises the forming of a tube and the later stuffing of the formed product with a ceramic monolith and an elastomer mat for filtration. It is shown, that for oval tubes the stuffing process leads to inhomogeneous deformation of the final product wherefore it must be considered when designing the tools for the forming step. The high number of iterations for the tool design can hence be reduced. Additionally, under certain conditions, the necessity of simulating the forming process is shown to no longer be given. The reason for this is that during the stuffing step, no plastic deformation is introduced in the tubes. Consequently, the forming history, including local strain and residual stress distributions, do not need to be explicitly considered. A simplified methodology is presented based on these findings, which can find the required cross-section geometry of the tube entering the stuffing process. The goal of the previous forming process therefore is to produce exactly that geometry, which can be done using conventional forming process planning approaches.

1. Introduction

In automotive engineering, tubes are installed as structural components with high specific bending stiffness or for transporting fluids (fuel lines or exhaust system). The variation of tube cross-sections (round, oval or rectangular) allows an application-oriented design of tube structures. Especially in the exhaust tract, components based on tubes with different cross sections are used, e.g. as catalysts. For the production or further processing of tube cross-sections, forming processes such as expanding are commonly used. A division can be made between active-media-based and mechanical forming operations. Both process classes can create tubes with constant and variable cross-sections.

Kamaliev et al. [1] presented the possibility of production of rectangular cross-section tubes by applying internal pressure at elevated temperatures.

As a mechanical expansion process, Fischer et al. [2] considered the flaring of a tube by a conical punch. For this purpose, the authors derive a semi-analytical model to predict the punch force. The same process was analyzed by Karrech and Seibi [3] in terms of sheet thickness distribution. For this purpose, they established the dependence on the punch inclination and derived design proposals. Also Avalle and Scattina [4] present a simplified model to predict the relation to the pipe diameter and thickness was established. Further, Daxner [5] considered the forming limits in flaring.

The previous publications show that flaring has been widely analyzed in terms of process forces and limits. However, the shape accuracy of the tubes after flaring was not analyzed in this detail. An approach to this was provided by Yin et al. [6] which expanded oval tube cross-sections into round tubes by using mechanical tools. For this purpose, they established an analytical approach to determine the springback. However, this model cannot be used for rectangular geometries. Muschik [7] showed, that in the manufacturing process of catalysts the distance between tube and inserted monoliths has an influence on the product quality. Therefore, the springback after releasing the forming load of a non-round cross-section must be understood in detail to create parts with high geometric accuracy as requested by many OEMs. In this paper, tubes with two different cross-sections are modelled numerically and validated through experimental processing. Furthermore, own investigations have been shown that assembly, referred as stuffing in the following, also has an influence on the shape accuracy of the final part. For this reason, stuffing is investigated in more detail as part of the process chain. The knowledge gained will then be used to present a new method that converts the undesirable deformation into a desired change in shape.

2. Tube forming process

A typical industrial process to produce oval catalysts (figure 1), as investigated in this paper, can be subdivided into the following process steps:

(i) **Blanking:** Cutting rectangular sheets from coils or larger blanks
(ii) **Welding:** Roll-bending and subsequent welding to produce circular tubes
(iii) **Preforming:** Creating a small oval cross-section from the circular tube by hydraulic presses
(iv) **Calibration:** Expanding to establish the target geometry with inner tools
(v) **Endforming:** Expanding of tube ends (connection areas)
(vi) **Stuffing:** Inserting an elastomer-wrapped ceramic monolith

The process steps of calibration, endforming and stuffing (iii – vi) are discussed the next sections. The influence of blanking and welding are considered negligible due to the low amount of forming work introduced. For this purpose, the process kinematics, material characterization and numerical modelling are presented. After, the numerical results are validated with the aid of experimental studies.

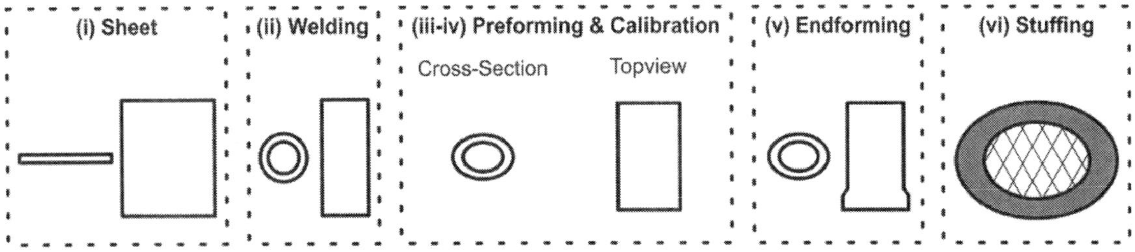

Figure 1. Industrial process chain to manufacture oval catalytic converters from initially plane sheets

2.1. Process kinematics of forming steps

In preforming, the tube is formed into the raw shape by hydraulic presses, whereby the forming process starts by closing a positive and negative die. Own initial investigations have shown that the small degrees of deformation introduced in this process are in most cases negligible.

In the mechanical expansion process steps of calibration and endforming, a mandrel with a polygonal cross-section, known as a cathedral, is pushed into a number of segments (figure 2 a). The contact surfaces A_i have an angle α_i to the axial directions Z, wherefore the radial displacement dR of the segments i can be approximated by the equation (1). The constant Si includes the mechanical tolerances and the tool clearance.

$$dR_i = \mathrm{dZ} \cdot \tan \alpha_i - S_i \tag{1}$$

The exact number of segments varies typically between 4 and 12 depending on the complexity of the part. The occurring inhomogeneous strain distributions lead to complex springbacks. For localization of the forming zone during expansion of the tube ends (endforming), fixed and pressure-controlled counter die segments are used in addition to the internal tools.

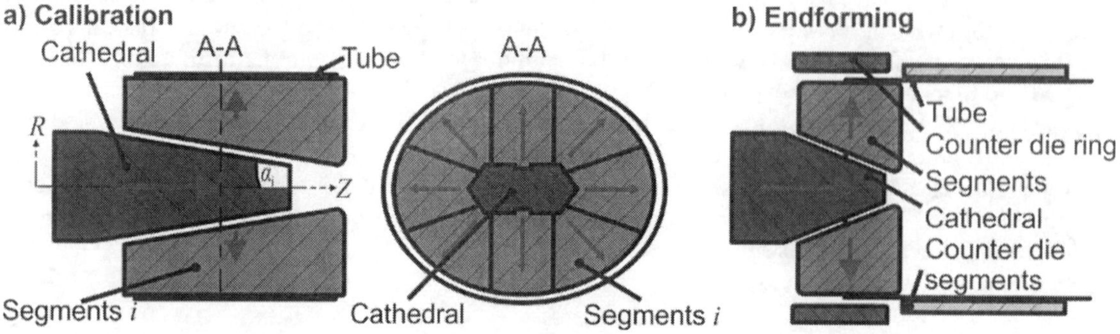

Figure 2. Sections view of a calibration (a) and endforming tool (b)

2.2. *Process kinematics of forming steps*

Stuffing is the process of encasing the ceramic monoliths with an elastomer mat. The initial height of the mat is between 8.7 and 12 mm, depending on the area density of the mat. The wrapped monolith is then inserted into the tube with the use of a funnel. After inserting, resulting pressure between the mat and the tube causes a change in the tube geometry, cf. figure 3 Expansion.

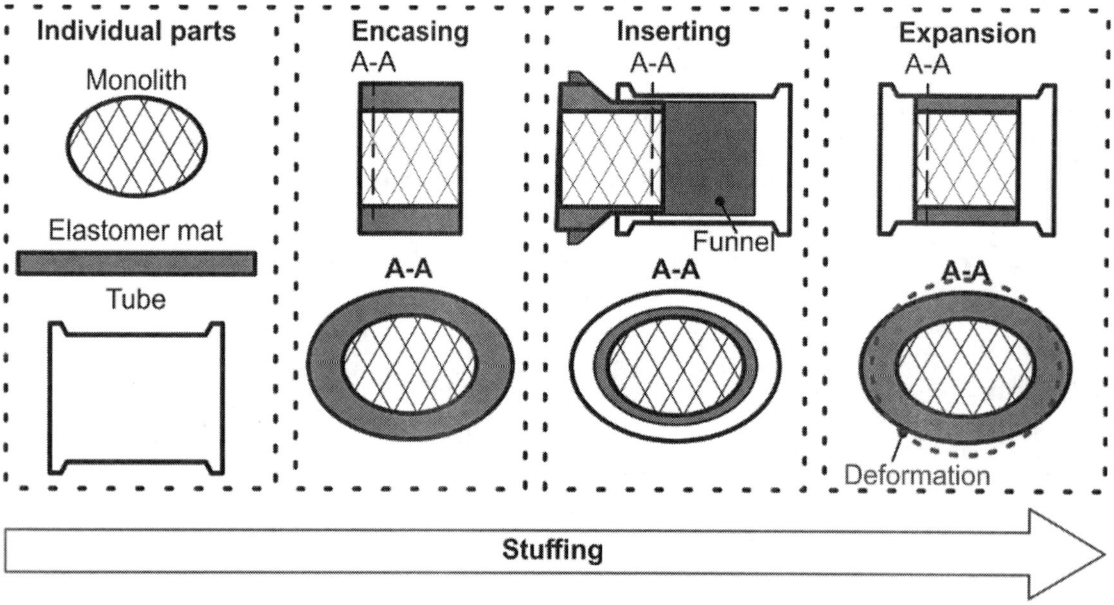

Figure 3. Process kinematic during the stuffing process

2.3. *Material characterization of stainless steel*

The exhaust system of a car is constantly exposed to hot, corrosive exhaust gases. As a result of their ability to prevent corrosion, stainless steels are used to make components for exhaust systems. Circular tubes made of stainless steel 1.4509 (EN 10088-2) sheets are used in the current study. On a Zwick 250 universal testing machine operating at room temperature, uniaxial tension tests in accordance with DIN 6892 are carried out to obtain engineering stress-strain curves from a 1.5 mm thick sheet (as-received), a pre-formed tube, a calibrated tube, and a tube after endforming to evaluate the impact of the forming operations on the material behavior. The specimens were cut out with a laser. The tensile specimens had a gauge length of 80 mm taken parallel to the axial direction. The side of the tube opposite the weld seam is where the specimens are taken from.

In figure 4a, the difference between the as-received sheet and the pre-formed tube is small; specifically, the initial yield stresses and tensile strength are distanced by less than 5% and 4%, respectively. After calibration and endforming, there's not much variation in the stress-strain curve of the tube specimens (0.5% in stress). In comparison to the as-received sheet, the specimens taken from the calibrated or endformed tube show an initial yield stress that is significantly higher (by about 20%). It's interesting to note that there is only a 2% difference in flow stress between the as-received sheet's and the endformed's tensile strengths. This is explained by the fact that a significant amount of the material's formability was consumed in the processes before the tensile test. The uniform elongation of the end-formed tube is 16%. The as-received sheet is 24 %.

2.4. Material characterization of elastomer mat

The stress-strain behavior of the elastomer mats is determined using compression tests on a universal testing machine from Zwick. The specimen size is 50 x 50 mm. During the test, the mats are compressed at a compression speed of 10 mm/s in the thickness direction from the initial height h_0 = 8.7 mm to different target heights h_1. The tested compression speed corresponds to the maximum travel speed of the testing machine. In the real process, compression speeds of approx. 40 mm/s occurred during insertion, depending on the geometry and mat thickness. The investigated target heights are selected according to the expected gap distribution between monolith and tube. For the analysis of the expected relaxation, the compression states have been kept constant for 120 s using fixed displacement control.

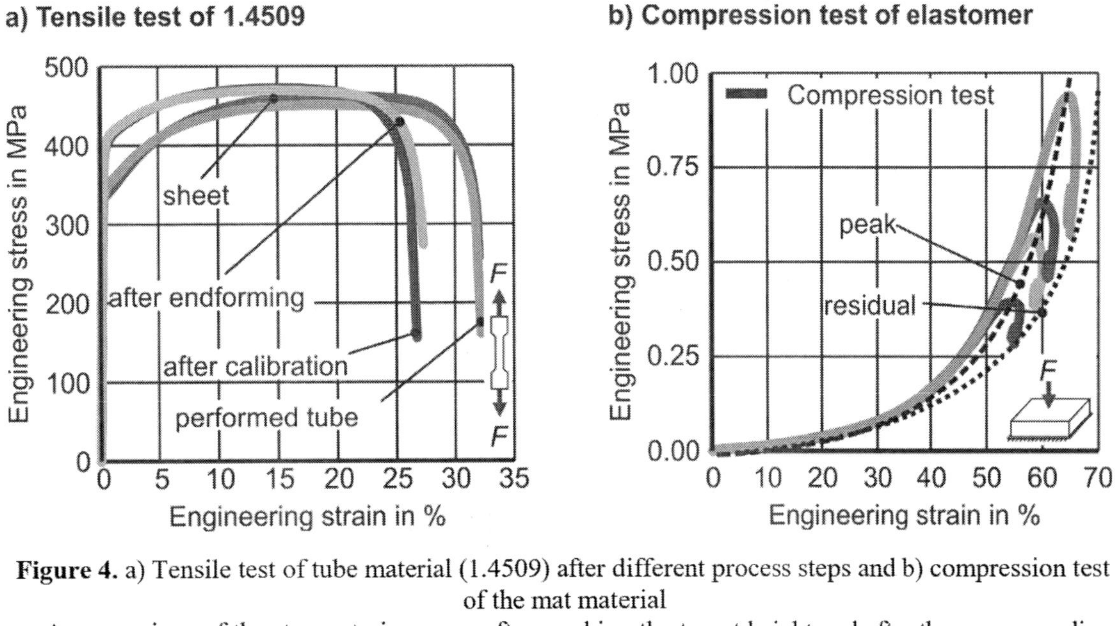

Figure 4. a) Tensile test of tube material (1.4509) after different process steps and b) compression test of the mat material

A comparison of the stress-strain curves after reaching the target height and after the corresponding holding time shows a drop in stress (figure 4 b). These pressures will be referred to as peak and residual pressures in the following. Polynomial functions are derived from the individual pressure values, which calculate the peak and residual pressures as a function of the technical strain.

3. Numerical investigation of the process chain

In this section, the previously explained manufacturing process is numerically modelled and validated. In addition, the deformation during stuffing is examined in more detail. Because the results of the tensile tests from chapter 1.3 do not show a significant difference between the initial sheets and the pre-formed tubs, the preforming is also neglected in the numerical modelling.

3.1. Numerical modelling

Two different simulation programs are used for the numerical representation of process steps 4-6. The mechanical expansion is modeled by the software TubeXpert from AutoForm. The software allows the separate investigation of process steps 4 and 5, where the tube is modeled by shell elements with a triangular mesh [8]. The tools are modeled as rigid (figure 5 a)

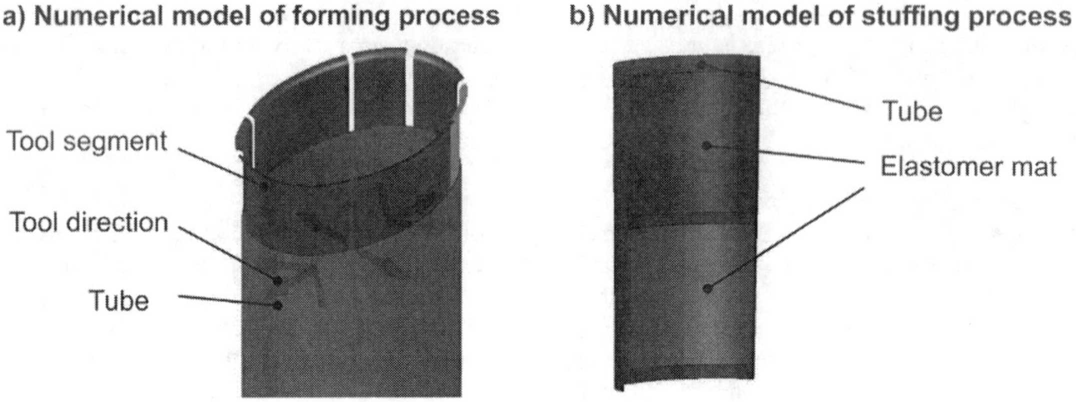

Figure 5. Numerical modelling of a) expansion process and b) stuffing process

The tool kinematics can be defined by vectors. The average calculation time of both process steps combined is about 15 min. An automatic remeshing during the calculation is performed. A similar simulation with Abaqus requires 4 hours, thus TubeXpert can model the mechanical process part in a time-efficient and user-friendly way.

TubeXpert does not offer a module to describe the stuffing process, but the tube can be exported to LS-DYNA after the 5th process step (endforming) with the forming history. For this reason, the stuffing is modeled numerically in LS-DYNA. During stuffing, the tube is also mapped by shell elements. But the mats are modeled with volume elements. In contrast to the real process, there is no compression of the mat by using funnels. The mats are derived from the monolith outline geometry, fixed at the inner shell surface, and then compressed by a radial displacement. After the target height is reached, the tube is activated, and the re-expansion of the mat is enabled. The numerical solver determines the static equilibrium between tube expansion and mat pressure using an implicit calculation. The resulting displacement can be considered as expansion due to stuffing (figure 6 c).

3.2. Validation of Numerical Simulation

Clausmeyer et. al [9] showed that the expansion of tubes with round and oval cross-sections can be predicted with sufficient accuracy in the TubeXpert software.

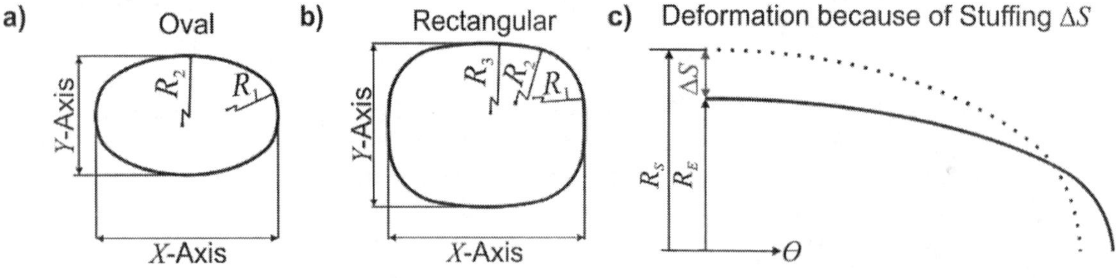

Figure 6. Tube cross-section of an a) oval shape, a b) rectangle tube and the explanation of calculation of the stuffing deformation

In the following, this paper focuses on the evaluation of tubes with strongly non-circular cross-sections. For this purpose, on the one hand an oval tube and on the other hand a rectangular tube with rounded corners are considered (figure 6 a-b). For validation, the contour of the tubes was measured by laser on the outside before and after each process step. A deviation between simulation and experiment of less than 0.2 mm is considered sufficiently accurate. The exact evaluation takes place at 0 and 90°. The average contour deviation is then calculated. To validate stuffing, the shape changes because of mat pressure are compared between experiment and simulation (figure 6 c).

3.2.1. Oval shape The oval cross-section has a ratio of 1.25 between the long and short axes. The geometry is defined with two different radii R_i (figure 6 a). For validation, contour measurements have been carried out on different tube heights. The following results refer to a cut at a height of 75 mm. The height corresponds to the centre of the monolith. The deviation between the simulation and the experiment after mechanical expansion is calculated as an average of 0.12 mm. On the X axis there is a deviation of 0.03 mm. The deviation on the Y-axis is slightly larger at 0.12 mm.

Table 1. Evaluation of XY-Data after mechanical expanding and during stuffing for Experiment (Exp.), Simulation (Sim.) and Deviation (Dev.) in mm.

Shape	Location	XY-Data after Expanding			Expanding during Stuffing		
		Exp.	Sim.	Dev.	Exp.	Sim.	Dev.
Oval	*X*-Axis	78.56	78.44	+0.12	-0.36	-0.35	-0.01
	Y-Axis	52.29	52.31	+0.03	+0.74	+0.65	+0.09
Rectangular	*X*-Axis	101.42	101.48	+0.06	+0.30	+0.41	-0.11
	Y-Axis	81.53	81.45	+0.08	+1.16	+1.29	-0.13

During stuffing, a bounce occurs on the short axis. This behavior (+0.74 mm in the experiment) can be predicted with a deviation of 0.09 by the simulation (+0.65 mm). The collapse (-0.36 mm) at the long axis can be predicted with an accuracy of 0.01 mm by the simulation (-0.35 mm).

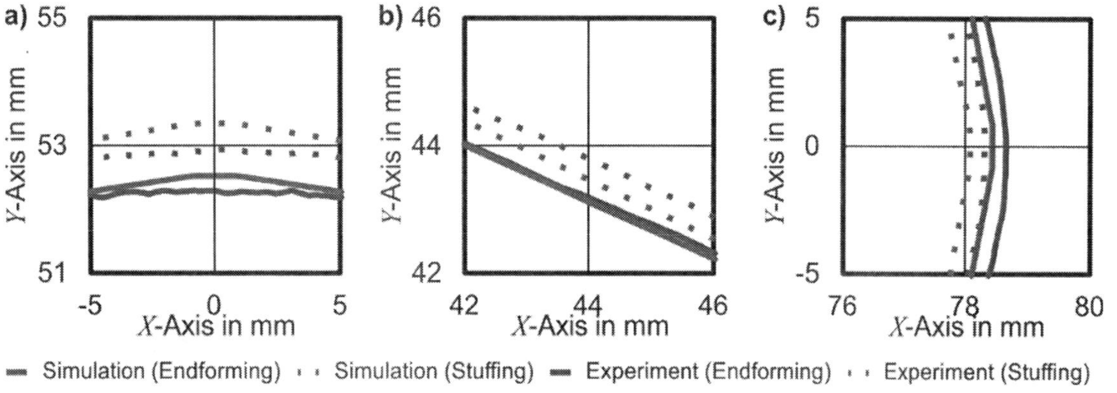

Figure 7. Plots of oval shape after endforming and after stuffing in a) 0°, b) 45° and c) 90°

3.2.2. Rectangular shape The rectangular cross-section has a ratio of 1.5 between the long and short axes. The geometry is defined with three different radii R_i, (figure 6 b). A different mat is also used for this tube geometry. It has an initial height of 10 mm. Analogous to the procedure from section 1.4, residual pressure curves were determined and stored in the simulation. In comparison, the pressure values are slightly lower than shown in figure 4. For validation, contour measurements have been carried out on different tube heights. Compared to the oval cross-section geometry, greater deviations occur between the simulation and the experiment. The average deviation is 0.16 mm. On the axes, there was a deviation of 0.06 mm on the X axis and 0.08 mm on the Y axis. It should be added that the experiments had a scatter (0.15) in the axis dimensions. However, the deformation after stuffing can be predicted in both axis directions with an accuracy of less than or equal to 0.2, so that the desired tolerance can be maintained.

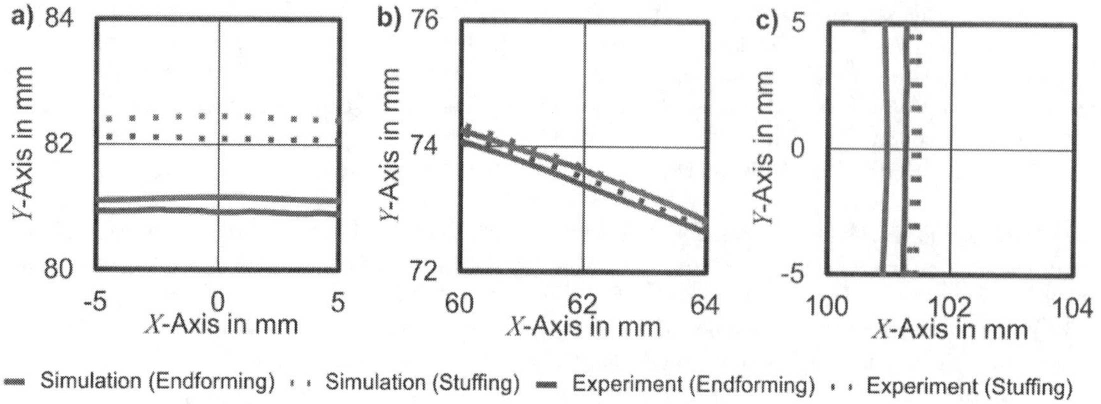

Figure 8. Plots of rectangular shape after endforming and after stuffing in a) 0°, b) 45° and c) 90°

3.3. Validation of Numerical Simulation

Based on the validation results above, the effect of mechanical expansion on the stuffing process is investigated in this section. For this purpose, the stuffing of the oval and rectangular tube is simulated with and without including the forming history.

The comparison of the simulations shows that the deformation of the stuffing is purely elastic. Furthermore, the simulation without forming history calculates the same results as already shown above. This means that under certain conditions the deformation because of the stuffing is only dependent on the initial shape of the tube, the monolith geometry, and the mat pressure. These conditions can be summarized as follows, which will be explained in more detail below:

- Isotropic material behavior
- Stress below the original initial yield stress Y
- Pure tensile stress

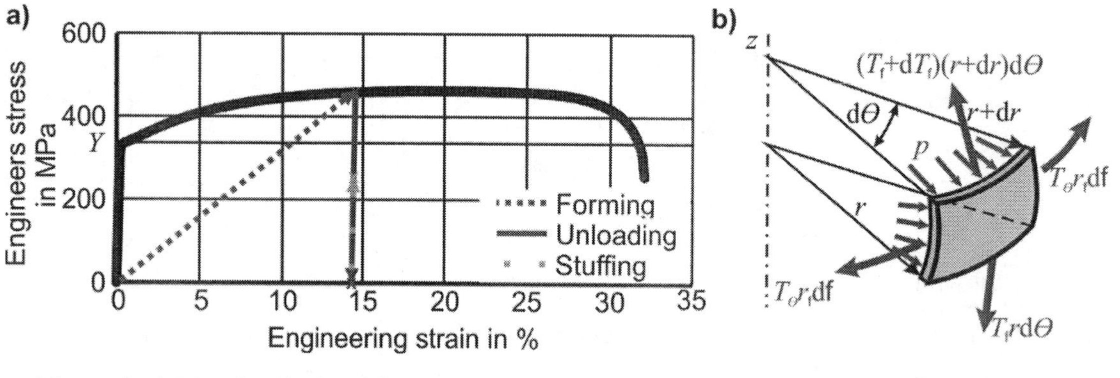

Figure 9. a) Mapping the local forming history because of expanding on a schematic stress-strain curve and b) the basic of the membrane theory [10]

The neglect of strength variations along the circumferential direction (isotropic material behavior) enables the mapping of the local forming history (expansion) through different strain states on a stress-strain curve (figure 9 a). The insertion of the mat applies a pressure to the tube in the thickness direction. Using the membrane theory according to Marciniak et. al [10], a tensile stress in the circumferential direction results. The maximum permissible mat pressure for all mats considered is in the range of 2 MPa.

$$\frac{p}{t} = \frac{\sigma_\theta}{r_\theta} + \frac{\sigma_\phi}{r_\phi} \tag{2}$$

Given an equivalent radius of curvature of approx. 150 mm (estimated upward) and a sheet thickness of min. 1 mm, the result is, according to equation 2, a maximum tensile stress of 300 MPa. This value is below the initial yield stress Y according to figure 4 a, therefore any strain state during stuffing is independent of the plastic strain hardening. Further, there is no load reversal. A reduction of the Youngs modulus due to the Bauschinger effect does not have to be considered.

4. Designing of new tools

After presenting and numerically modelling the manufacturing process chain in the previous chapters, this knowledge is applied to improve the design of new tube geometries. Normally, the design of new tools for mechanical expansion is an iterative process. At the beginning, the tool is manufactured according to the target tube geometries. Subsequently, catalysts are produced according to the process chain (figure 1) and then a variance analysis is carried out. If the deviations are too large, the tool must be reworked. The adjustments are usually based on the experience of toolmakers. Due to the complex deformation during the stuffing of non-round tubes, three or more iteration loops are often necessary. Since each iteration loop increases production costs, a simulation-based software algorithm for tooling design is desired and presented below.

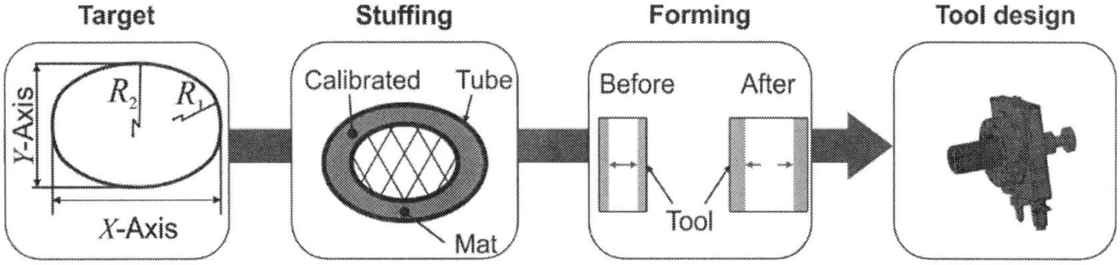

Figure 10. Approach to reduce tooling iterations

4.1. *Software algorithm*

The aim of the software solution was to automatically change the geometry by adjusting the specific contour parameters. The choice of parameters should be flexible, enabling, for example, the radii or axial dimensions to be modified according figure 6 a. Furthermore, a simple tracking of the changed parameters should be possible. For this reason, the Catia software was selected for the creation of CAD models. The CAD program allows the generation of parameterized models, which means that parts as well as assemblies can be modified by changing defined values.

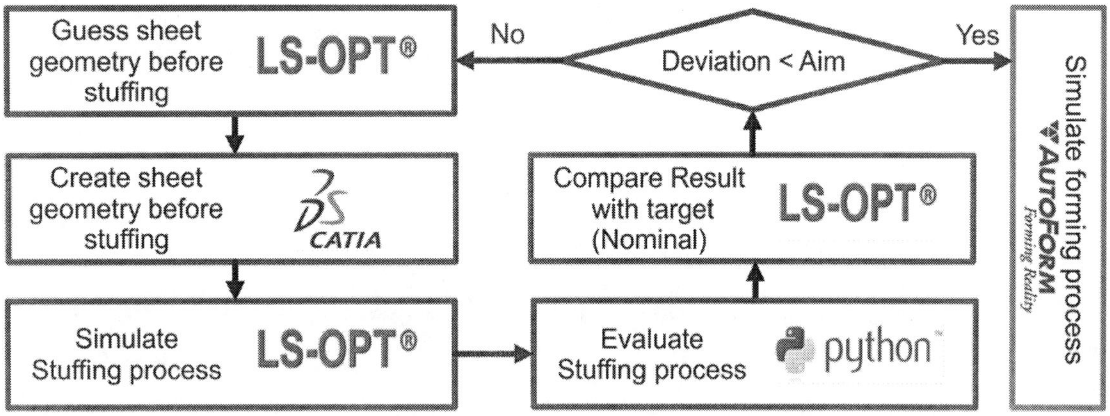

Figure 11. Software algorithm for the iterative determination of an ideal pre-stuffing geometry

For the automatic, iterative adaptation of the tool geometry, an interface between Catia and the simulation model in LS-DYNA has been developed. This allows for the modification of the parameters in Catia and the subsequent export and conversion of the CAD data into simulation data. Finally, this interface was integrated into the optimization software LS-OPT, which allows the graphical analysis of an optimization problem and the selection of different solution strategies. In this case, the software was used to minimize the deviation between the nominal and actual geometry after stuffing. The full algorithm is shown in the figure 11.

4.2. Results of a new type of tool design

In this section, the presented algorithm is applied to a fictitious geometry. The target cross section of the fictitious geometry corresponds to figure 6 a, such that the cross section is defined by two radii and the axis dimensions. The exact parameters are listed in table 2. The axial length of the tube is 228 mm. In the imaginary assembly, a monolith of axial length 200 mm is inserted in the center of the tube. Its contour was generated by offsetting the tube geometry by 4 mm. A mat with the same mechanical properties as the oval contour (figure 4 b) was chosen for wrapping. The thickness of the mat in the uncompacted state is 8.7 mm.

In the following, only the geometry before stuffing is discussed. The consideration of the springback for the creation of the tool geometry after figure 10 is not presented. However, the AutoForm TubeXpert software offers a corresponding module, which can be used to generate the tool surface taking springback into account, if the source and target tube geometry (Catia CAD data) are known.

Table 2. Nominal dimension of the fictitious geometry

X-Axis [mm]	*Y*-Axis [mm]	R_1 [mm]	R_2 [mm]	Sheet thickness [mm]	Tube length [mm]	Mat thickness [mm]	Mat length [mm]	Mat weight [g]
70.5	49	42	103.5	1.2	228	8.7	200	140

For the results, a full-factorial test matrix with 11 values per parameter was used for simplicity. Alternative optimization algorithms allow a better resolution of the parameters but require a certain mathematical understanding and training. The number of simulations in the selected matrix corresponds to 1331 with 3 independent parameters. The average calculation time per simulation was about 20 min. The computer was able to run 6 simulations in parallel, resulting in a total calculation time of about 36 hours.

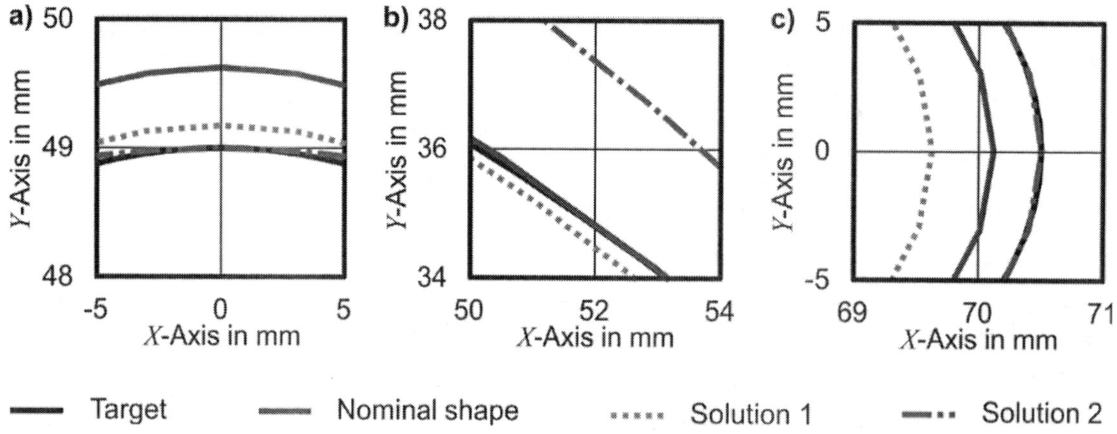

Figure 12. Results after stuffing with different tube cross sections before stuffing

The evaluation of the optimization tool shows that a reduction of the deviation between nominal and actual geometry is possible by adjusting the geometry before stuffing (figure 12). Stuffing with nominal initial geometry shows a large deviation between nominal/actual value. It is also shown that an ideal solution is not obtained. There are deviations either in the area of the axes (Solution 1) or in the area of the 45° (Solution 2). Nevertheless, the new software provides useful suggestions for the design of new tool designs and can thus be used alongside the toolmakers' empirical values. The application of a full-factorial test matrix allows the subsequent determination of weighting factors, whereby certain areas, such as the axial dimensions, can be focused on during the selection. An ideal shape could not be determined with a full-factorial test matrix, but a reduction of the tool iteration is nevertheless realistic due to the experience gained.

5. Summary

In this paper, a manufacturing process of components for the automotive exhaust tract was presented. The process consists of mechanical expansion of tubes and assembly, where a tube is stuffed with a monolith that is covered by an elastic mat. Numerical models were presented for both process steps to predict the forming behavior and the contour of the tube. These individual simulations were performed with the numerical solvers AutoForm TubeXpert and LS-DYNA and validated with experimental tests. The tube contour could be predicted with an accuracy of ± 0.2 mm. It was found, that in many circumstances, the forming history does not need to be taken into account when analyzing the stuffing process. Hence, a new approach for the tool design of the forming step is proposed: based on the desired final product geometry, the iterative simulation of the stuffing process predicts the required geometry of the tube prior to stuffing. This geometry is in turn the result of the initial tube forming process. The tools which yield the desired geometry, can be found using conventional software-based approaches. This procedure helps to reduce developments costs due to multiple iterations of the tool design.

Acknowledgements
The Institute of Forming Technology and Lightweight Construction gratefully acknowledges the support of Faurecia Emissions Control Technologies Germany GmbH as part of the Research Center in Metals Processing (ReCIMP). The authors would like to thank Uwe Beckmann from AutoForm Engineering Deutschland GmbH for the continuous support of this study.

References

[1] Kamaliev M, Kolpak F and Tekkaya A E 2022 Isothermal high temperature pneumoforming of stainless steel tubes at low pressure levels *CHS²* p. 616–623
[2] Fischer F, Rammerstorfer F and Daxner T 2006 Flaring - An analytical approach *International Journal of Mechanical Sciences* **11** p. 1246–1255
[3] A. Karrech and A. Seibi 2010 Analytical model for the expansion of tubes under tension; *Journal of Materials Processing Technology* **210** Vol. 2 (2010) S. 356–362
[4] Avalle M and Scattina A 2018 Experiment based modeling of the mechanical expansion of tubes for the construction of heat exchangers *Procedia Structural Integrity* p. 130–144
[5] Daxner T, Rammerstorfer F and Fischer F 2005 Instability phenomena during the conical expansion of circular cylindrical shells *Computer Methods in Applied Mechanics and Engineering* **21-24** p. 2591–2603
[6] Zhao J, Yin J, Ma R and Ma L 2011 Springback equation of small curvature plane bending *Sci. China Technol. Sci.* **9** p. 2386–2396
[7] Muschik C 2018 Oval Canning for Substrates of Exhaust Aftertreatment Systems for Commercial Vehicles *MTZ Worldw* **1** p. 46–49
[8] AutoForm Engineering GmbH Webpage presenting AutoForm TubeXpert 2023
[9] Clausmeyer T, Gutknecht F, Hassan H, Kaya D, Stiglmair M, Tadeu F and Stennei M 2021 Expansion of oval tubes: prediction and experiment *ESAFORM 2021*
[10] Marciniak Z, Duncan J and Hu S. 2002 Mechanics of Sheet Metal Forming

Wear characteristics of the tool in the cold stamping process of ultra high strength steel sheets by establishing a novel wear test method based on the progressive die

J Bang[1], M Kim[1], G Bae[1], J Song[1], M G Lee[2]

[1] Korea Institute of Industrial Technology, 156, Gaetbeol-ro, Yeonsu-gu, Incheon, 21999, South Korea
[2] Department of Materials Science and Engineering, Seoul National University, 1, Gwanak-ro, Gwanak-Gu, Seoul 08826, South Korea

E-mail: jhsong@kitech.re.kr

Abstract. Demands for lightweighting and crashworthiness of the vehicle body have increased. For this purpose, advanced high strength steel to the automobile body was regarded as a solution with respect to manufacturing cost and impact energy absorption. However, increasing the strength of the automotive steel sheet lead to a diversity of problems caused by the tool wear due to higher forming load than that of commonly used steel sheets. Hence, systematic wear experiment and evaluation methods are required to quantitatively and qualitatively evaluate the tool wear amount in the sheet metal forming process. In this study, a methodology is proposed to quantitatively evaluate the wear of sheet metal forming tool based on the experimental results. In order to carry out a systematic wear test and save the time and cost, the progressive die set was designed to be suitable for wear test. The designed testing machine can simultaneously test four types of punches made under various tooling conditions such as materials, shapes, and coatings. Through the measurement of the wear depth, roughness, and surface imaging of the punch and product roughness, which represent the wear characteristics, quantitative evaluation methods for tool wear in the sheet metal forming process are established. By referring to the wear test results, it is confirmed that it is appropriate to analyze the reasonable tool wear characteristics by using the proposed methodology for quantifying the tool wear in the sheet metal forming process.

1. Introduction

Recently, there has been a steady increase in demand for eco-friendly cars such as electric and hydrogen vehicles. As a result, research on improving crash performance and lightweight design for extended driving range is actively being conducted to ensure the protection of batteries, fuel cells, and passenger safety. Lightweighting of the car body can be effectively achieved through the application of lightweight materials [1]. The most commonly used lightweight materials for car bodies are aluminum alloys, magnesium alloys, advanced high-strength steel (AHSS), glass fiber composites, and carbon-fiber reinforced composites (CFRP). Among them, AHSS is the most economical and has excellent crash energy absorption performance [2-5]. However, the forming of AHSS sheets results in earlier tool wear due to higher forming loads compared to conventional steel sheets. Tool wear is an important factor that needs to be managed during the forming process of AHSS sheets as it leads to various issues such as increased maintenance costs for tools, decreased product quality, and decreased productivity. Various studies have been conducted to improve the wear resistance of tools for efficient management of tools in the forming process. Pereira et al. [6] studied the effect of machining errors of tools on wear and found that even small shape errors have a significant effects on the lifespan of tools. Podgornik and Jerina [7] revealed that the surface roughness of tools affects the wear resistance of coated tools. Gonzalez-Pociño et al. [8] investigated the effect of temperature changes in heat treatment processes such as quenching, tempering, and nitriding on adhesive wear. Ghiotti, Bruschi, and Cora [9-10] compared the wear resistance of various coatings using a wear test method that imitates the forming process. These studies have identified the effect of various tool production variables on tool wear and effectively improved wear resistance by controlling these variables. However, these studies only qualitatively compared the effect of each tool variable on tool wear, and did not reflect the effect of plastic deformation in the actual forming process on wear, making it difficult to apply to the evaluation and prediction of tool wear in actual forming processes. This study quantitatively evaluated the wear of tools in the sheet metal forming process. The punches, which were the subject of the wear test, were made of K340 tool steel and coated with PVD CrN. They were designed to mimic the tool shape used in the forming process of automotive body parts. Progressive dies were used to enable continuous and efficient experiments. The wear behavior and characteristics of the punches were evaluated by quantitatively measuring the wear depth and surface roughness to determine the degree of wear on the punches. Such research can greatly contribute to improving the quality of the forming process and reducing costs.

2. Setup of wear test

2.1. Punch design
Contact pressure and sliding distance are factors that have a significant impact on the tool wear life in the sheet metal forming process [11-12]. Tool wear generally occurs on the curvature of the tool where contact pressure and sliding distance are concentrated. The punch used in the wear test refers to the curved shape of the tool, which is a vulnerable area where wear occurs in the forming process of automotive parts. The punch is designed to concentrate the contact pressure on the curved part of the punch when forming a blank, as shown in Fig. 1, and the curvature radius is 3.0 mm. The punch is made in an insert type to facilitate measurement and replacement. The punch stroke is 35.0 mm. A detailed description of the punch shape was referenced from the paper of Bang et al. [13].

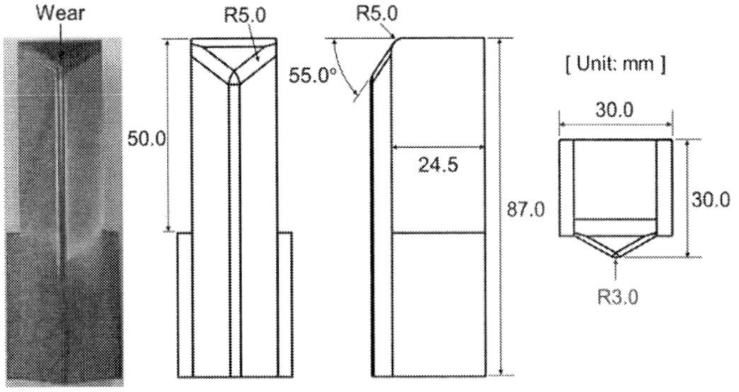

Figure 1. Punch shape used in the wear test [13].

2.2. Design of continuous wear test
The wear test of the tools for cold stamping process requires a significant amount of time, materials and labor. Therefore, in this study, a progressive die was used to enable continuous and efficient wear test. The progressive die is designed to include multiple subdivided metal processing operations in a single tool set, and the sheet metal is processed through each operation step by step to obtain the final product

shape. Consequently, with a constant operating speed, continuous mass production of the product is possible, which is why in this study, the progressive die was installed in a mechanical press, and a coil with a width of 120 mm was continuously fed to perform the wear test.

Fig. 2(a) schematically illustrates the metalworking process of the progressive die used in the test. The designed metalworking process includes piercing, punching, stamping, and parting. The press speed applied in the test was 15 strokes per minute (spm), which is commonly used in the forming process of automotive parts. Fig. 2(b) is the experimental sample, and it shows each of the metalworking operations. For detailed descriptions on the experimental design and each process, please refer to the paper by Bang et al. [13].

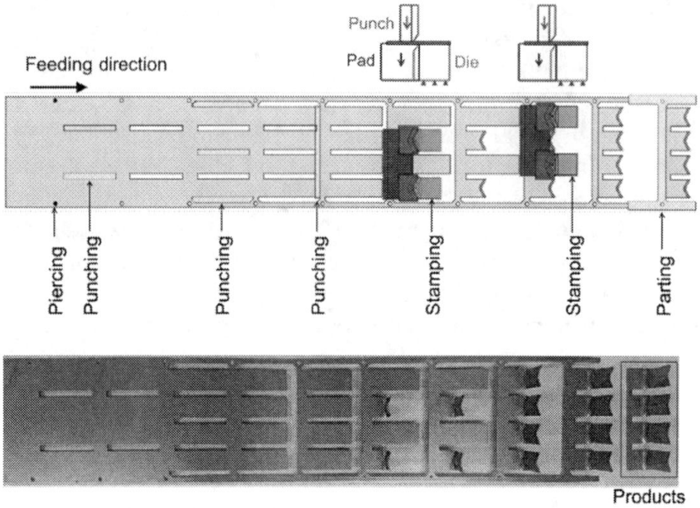

(a) Schematic illustration
(b) Steel strip
Figure 2. Layout of the designed progressive die [13].

2.3. Material properties of tool and steel sheet

This study used uncoated TRIP1180 steel sheet with a thickness of 1.2 mm, and the tensile material properties can be found in Table 1. The punch used was made of BÖHLER K340 tool steel. Heat treatment was performed to form a martensite microstructure to harden the tool steel. It was hardened at 650, 850, 1030°C for 80, 120, and 200 minutes, respectively, in a vacuum, followed by air cooling. To reduce brittleness caused by quenching, the punch was tempered twice at 500°C for 200 minutes. The punch surface was coated with CrN using physical vapor deposition (PVD) method.

Table 1. Tensile properties of TRIP1180 1.2t.

			Swift hardening law $\sigma = k(\varepsilon_0 + \varepsilon)^n$			R-value		
Thickness (mm)	YS (MPa)	UTS (MPa)	k (MPa)	ε_0	n	0°	45°	90°
1.2	933	1196	1584.6	0.0036	0.0858	0.7382	0.9786	0.8671

2.4. Measurement methods of wear

Damage to the punch surface during sheet metal forming leads to changes in surface profile. Therefore, to analyze the gradual wear behavior of the punch, the wear depth was quantitatively measured with respect to number of stamping hits. The surface profile of the punch curvature radius where wear was expected to occur due to concentrated contact pressure and sliding was measured. The surface roughness and imaging of the punch radius were examined to identify the wear characteristics, and to examine the wear track. In this regard, the roughness and imaging of the punch radius were observed using a confocal laser scanning microscope (CLSM; LEXT OLS3100 manufactured by OLYMPUS). The surface roughness of the punch radius was expressed as the arithmetical mean height of a surface (Sa). Sa is the average of the absolute point heights within the measured area, as follows:

$$Sa = \frac{1}{A}\iint_A |Z(x,y)|dxdy \tag{1}$$

Where A is the measured area, and x and y represent the coordinates. Z is the absolute point height within the measured area.

The specific measurement conditions and methods were referenced from the paper by Bang et al. [13]. The surface profile and roughness of the punch radius were first measured before the wear test. In order to track the changes in wear information as the wear test progressed, the surface profile and roughness were measured at the same region. The wear depth was calculated from the difference in surface profile before and after the wear test. The surface profile and roughness were measured after 5,000 and 10,000 stamping hits. The wear test was stopped if wear occurred on the punch.

3. Test results

3.1. Wear depth

Fig. 3 shows the measurement results of wear depth on the punch radius ranging from 55 to 90°. The wear depth measured at 5,000 and 10,000 stamping hits was close to zero on all punch radius range, indicating that the deviation was considered as measurement error, hence it was confirmed that no wear was found on the punch radius at 5,000 and 10,000 stamping hits. After 14,000 stamping hits, wear was observed on the punch radius from 85 to 90°, and the maximum wear depth was 5.52 μm at 89°. It was observed that the wear depth sharply increased in a specific region at 14,000 stamping hits.

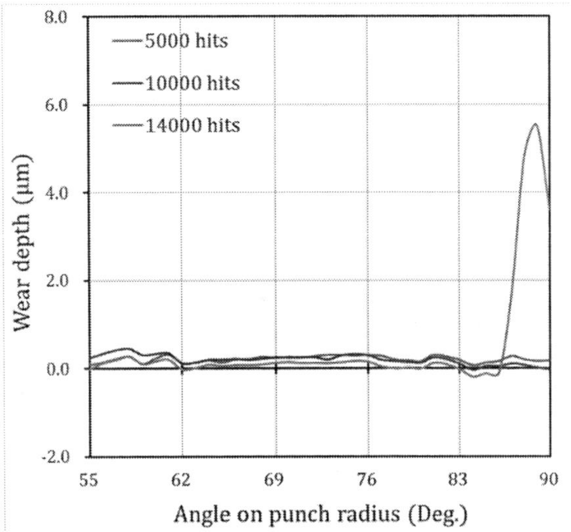

Figure 3. Wear depth of punch with respect to the number of stamping hits.

3.2. Surface roughness and imaging

Due to the defects on the punch surface, the condition of the punch surface deteriorates over time. To characterize the wear characteristics and mechanism of the progressively deteriorating wear track, surface roughness and imaging of the punch radius were observed. The surface roughness of the as-produced exhibits a maximum of 0.32 μm at 86° and is less than 0.30 μm in other measured areas, as shown in Fig. 4. The average roughness of the as-produced is 0.23 μm. The roughness measured at 5,000 and 10,000 stamping hits also has a similar value to that of the as-produced. As shown in Fig. 5, no severe wear tracks were found after 5,000 and 10,000 stamping hits, and micro scratches were found on the entire measured area after 10,000 stamping hits. After 14,000 stamping hits, the surface roughness increases rapidly when the measuring angle changes from 83 to 90°, and the maximum roughness is 1.07 μm at 90° (Fig. 4). As can be seen in the surface condition after 14,000 stamping hits in the Fig. 5, a severe wear track is confirmed from 83 to 90°. The wear depth results are consistent with the surface roughness and imaging results.

According to Bang et al.'s paper [13], the wear mechanism of PVD CrN coated tools in the sheet metal forming process is attributed to fretting wear. Therefore, the wear in this study was confirmed to be due to fretting wear.

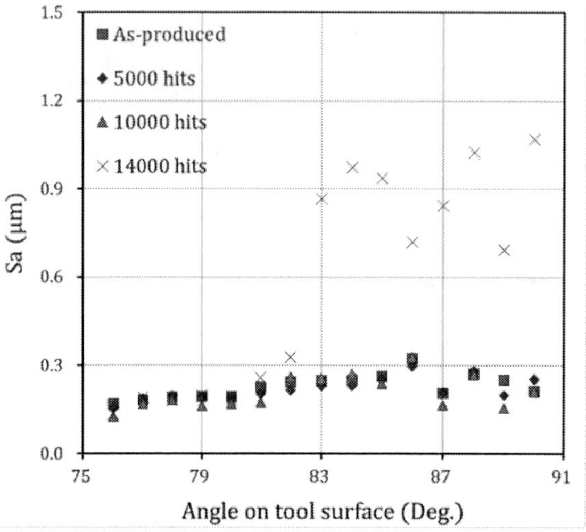

Figure 4. Roughness of punch radius with respect to the number of stamping hits.

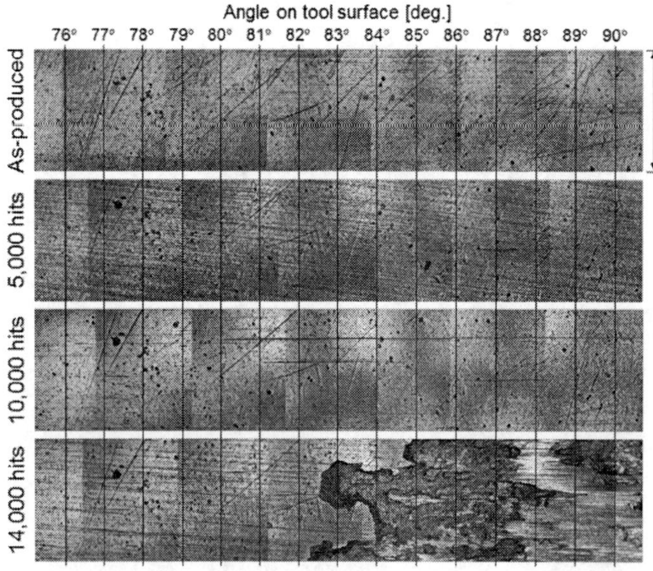

Figure 5. Imaging of punch radius with respect to the number of stamping hits.

4. Conclusions

This paper proposes an experimental method for quantitatively evaluating tool wear in the sheet metal forming process, and confirms the tool wear characteristics and behaviors through the wear tests. The punch used in the wear test is made of K340 tool steel and coated with PVD CrN. Because tool wear tests in the forming process require a significant amount of time, material, and labor, the test was performed using a progressive die proposed by Bang et al. [13]. The tool surface is clear before tool wear occurs, but is was confirmed that wear occurs rapidly due to the fretting wear mechanism.

Acknowledgements
This study has been conducted with the support of MOTIE R&D project (20018447).

References

[1] Li Y, Lin Z, Jiang A and Chen G 2003 *Mater. Des.* **24** 177
[2] Jambor A and Beyer M 1997 *Mater. Des.* **18** 203
[3] Cheon S S, Lee D G Jeong K S 1997 *Compos. Struct.* **38** 29
[4] Taub A I and Luo A A 2015 *MRS Bull.* **40** 1045
[5] Safari H, Nahvi H and Esfahanian M 2018 *Int. J. Crashworthiness* **23** 645
[6] Pereira M P, Weiss M, Rolfe B F and Hilditch T B 2013 *Int. J. Mach. Tools Manuf.* **66** 44
[7] Podgornik B and Jerina J 2012 *Surf. Coat. Technol.* **206** 2792
[8] Gonzalez-Pociño A, Alvarez-Antolin F and Asensio-Lozano J 2019 *Materials* **12** 2831
[9] Ghiotti A and Bruschi S 2011 *Wear* **271** 2454
[10] Cora Ö N, Namiki K and Koç M 2009 *Wear* **267** 1123
[11] Bang J, Park N, Song H, Kim H G, Bae G and Lee M G 2021 *Metals* **11** 306
[12] Pereira M P, Yan W and Rolfe B F 2010 *Wear* **268** 1275
[13] Bang J, Kim M, Song J, Kim H G, Bae G and Lee M G 2022 *Metals and Mat. Int.* **29** 327

Optimization of wear-, fracture- and corrosion performance of Vanadis 6 steel through cryogenic treatment and tempering

P Jurči[1], V Yarasu[1], I Dlouhý[2], A Bartkowska[3]

[1]Slovak University of Technology, Faculty of Materials Science and Technology in Trnava, Bottova 25, 917 24 Trnava, Slovakia
[2]Institute of Physics of Materials, CEITEC-IPM, Czech Academy of Sciences, Zizkova 22, 61662 Brno, Czech Republic
[3]Institute of Materials Technology, Faculty of Mechanical Engineering; Poznan University of Technology, ul. Piotrowo 3, 61-138 Poznan, Poland

E-mail: p.jurci@seznam.cz

Abstract. Vanadis 6 cold work tool steel was vacuum austenitized, room temperature quenched, differently cryogenically treated (-75, -140 and -196 °C for different durations) and double tempered, at temperatures of either 170 or 530 °C. Galling resistance was assessed by pin-on-disc testing of differently treated specimens against CuSn bronze, at wide ranges of sliding velocities and loads. The obtained results show that cryogenic treatments combined with low temperature tempering improves anti-galling performance of the steel. The best tribological properties were obtained when high-temperature tempering was combined with cryogenic treatment at -196 °C, due to the maximum extent of microstructural alteration generated by this kind of treatment. The improvement in wear performance is accompanied with slight but undisputable increase in corrosion resistance of the steel, at almost no affected toughness as compared with the state after conventional heat treatment. In conclusion, the obtained results imply an opportunity to simultaneously improve wear performance, corrosion resistance and toughness of the examined steel, by choosing a proper combination of cryogenic treatment and tempering.

1. Introduction

Deep drawing is a widely used forming process to produce different industrial parts, office and home appliances, automotive bodies, kitchen utensils etc. It belongs to one of the most energy-efficient production techniques, based on its high material utilisation. However, there is a variety of possible defects in deep drawing among them the galling belongs to the most frequently occurring ones [1]. The galling source is based on the forming technology itself. The tools (i.e. punch and blank holders) load and slide over the workpiece generating the required product geometry. Sliding and loading the workpiece surface results in flattening of the initial surface roughness. At the micro scale, tool asperities slide through the flattened workpiece surface causing wear in the softer workpiece, which may result in material transfer from sheet to tool. In other words, galling is a kind of adhesive wear involving cold welding of sheet material fragments to the die. After severe galling the tool surface will cause scratches or scoring on the blanks and finally the tool must be replaced by other one.
A number of trials focusing to the reduction or complete elimination of galling have been made recently. These trials involved the use of hard coatings [2] or duplex-coatings on tool steels [3], utilization of

ceramic tools [4] or laser surface structuring [5]. Despite promising laboratory- or field tests, however, none of these treatments has gained a wider industrial use. Another opportunity of how to reduce the galling is the proper choice of tool steel. The AISI D2 grade is very popular in cold forming applications but the grades such as Sleipner or Calmax are also being increasingly used. Moreover, newly developed steel grades were introduced into real applications recently. Liljengren et al. [3] have reported that Vancron 40 outperformed all of tested steel grades in laboratory testing with deep drawing of steel sheets used in automotive. Also, the proper heat treatment is important to provide the steels the best possible wear performance. It is highly desirable to maintain/produce as much as possible carbides in steel since the adhesive wear resistance is generally improved by increased amounts of these particles [6].

The cryogenic treatment (CT) has been established as a procedure suitable for getting an increment in carbides number in cold work- and high speed steels. Recent studies [7,8] revealed manyfold increased number of "extra" carbides in AISI D2 and Vanadis 6 steels. Unfortunately, these studies were either focused to effects of CT on steels microstructures or their main goal was to examine the mechanical properties and abrasive wear resistance. There is only one comprehensive study available over the last decade where variations in anti-galling properties of S390 Microclean steel due to CT in liquid nitrogen have been investigated [9]. A decreased friction coefficient and correspondingly greatly increased galling resistance of CT steel has been reported. These phenomena were attributed to 1-2 HRC hardness increment as well as overall, although very small microstructural refinement.

Improvement of tribological performance due to cryogenic treatment is counterbalanced by significant toughness reduction in many cases, as reported for AISI D2 [10] or AISI D3 [11] steels after CT and low-temperature tempering. Similar results were published also by Kumar for AISI D3 steel. Only Rhyim [12] pointed out that the toughness of the AISI D2 may also be improved, when high-temperature tempering is followed the CT. There were made trials with determination of toughness after cryogenic treatments of powder metallurgy Vanadis 6 steel and the principal outcome was that the toughness might not be inevitably worsened by applying this kind of treatment [13].

The corrosion resistance of cryogenically treated cold-work tool steels has been examined by only few researchers. Uygur at el. [14] reported worsened corrosion behaviour of cryogenically treated AISI D3 steel in 3.5% NaCl solution. However, also opposite result, > 50% improvement in corrosion resistance of AISI D3 steel in borate buffer with pH 10, was reported [15].

From the literature review it follows that even though attempts were made to reduce the galling through surfacing of existing tool materials or through replacing them by newly developed ones none of the trials has brought a decisive advantage into sheet metal working practice. The current contribution is an attempt to find an alternative way in minimization of galling, by using the cryogenic treatments of tool steel.

2. Material and experimental procedures

The experimental material was PM cold work tool steel Vanadis 6 with nominally (in wt.%) 2.1 %C, 1.0 %Si, 0.4 %Mn, 6.8 %Cr, 1.5%Mo, 5.4 %V and Fe as balance, soft annealed to a hardness of 284 HV 10. Samples of the experimental steel, with dimensions of 10 x 18 x 70 mm (wear testing), 10 x 10 x 55 mm (fracture toughness) and 15 x 15 x 6 mm (corrosion testing) were machined, fine grinded and finished up to the surface roughness of $R_a = 0.04$ μm.

In the first stage the effect of austenitizing temperature (T_A) has been examined. The specimens were austenitized at 1000, 1025, 1050 or 1075 °C in a vacuum furnace, held at the final temperature for 30 min. and nitrogen gas quenched (5 bar pressure). After that the samples were immediately moved into the tempering furnace where they were tempered in two cycles, each of them carried out at the temperature of 530 °C for 2 h, and air-cooled.

Based on the first results the T_A of 1050 °C was chosen for further investigations. After the quenching the specimens were subjected to cryogenic treatments with lowest temperature of -196, -140, or -75°C, and for 17h each, and re-heated to the room temperature. Both cooling and heating rates in cryogenic cycles were 1 °C/min. Immediately after reaching the room temperature the specimens were subjected to tempering treatments. These treatments were low- (170 °C) or high (530 °C) temperature, and consisted of two cycles, each of them for 2 h.

The microstructures were studied by the scanning electron microscopy (SEM) coupled with energy-dispersive spectroscopy (EDS). The same method has also been used for examinations of specimens surfaces after wear testing. The anti-galling performance of the tool steel was expressed by the percentage of surface that was covered by counterpart material after the testing. Twenty-five randomly acquired SEM micrographs were used from each tested specimen for determination of the area fraction covered with bronze, and the mean values were then considered for further examinations. The retained austenite amounts were determined following the ASTM standard [16]. Vickers hardness (HV 10) was determined as per the standard operating procedure of ASTM E384-17 [17].

Dry sliding wear tests have been performed with a CSM pin-on-disc tribometer. Balls with a diameter of 6 mm made of CuSn6 bronze (74 HV 10) were used as counterparts. The tests were conducted at an ambient temperature (25 °C) and in laboratory air, at a relative humidity of 40 – 50 %. To obtain sufficient statistical approach experiments were carried out with 3 samples for each combination of heat treatment parameters. For the first stage of investigations the normal load was 5 N, the sliding velocity of 0.064 ms^{-1} and the total sliding distance of 100 m were used. However, the experimental design was significantly extended in the optimization stage.

For this stage, the Taguchi design has been adopted to minimize the number of measurements. The total sliding distance was 100 m but the loading (1, 5, or 10 N) and sliding velocities (0.064, 0.128, or 0.1885 m^{s-1}) were the testing variables. The experiments were carried out using a four-factor mixed level design with an L18 orthogonal array. The MINITAB-19 statistical analysis software, which is extensively used in many fields of engineering study, was used for the analysis.

For fracture toughness (K_{IC}) determination, heat treated specimens were pre-cracked. Both the pre-crack preparation and the tests were carried out at room temperature according to the ISO 12137 standard [18]. For the determination of fracture toughness, the samples were loaded in three-point bending with a roller span of 40 mm, and at a loading rate of 0.1 $mm.min^{-1}$. Five samples were tested for each investigated condition, to ensure the statistical reliability of the obtained data.

Corrosion resistance examinations have been carried out by the potentiodynamic polarisation measurements (TAFEL). The 3.5 (in mass %) NaCl water solution was used as the corrosive medium. For the corrosion tests, the temperature of the solution was kept constant at 22 °C. The potentiodynamic polarisation measurements were carried out within the range of potentials of −1.5 to 1.5 V, and at a scan rate of 1 $mV.s^{-1}$. The corrosion rate for each specimen was estimated according to the ASTM G 102-89 standard [19]. More details of experimental works are reported elsewhere [20].

3. Results and discussion

Figure 1 shows the microstructures by using different T_A followed with 530 °C tempering. The matrix of the steel is tempered martensite in all the specimens. The second microstructural feature is the presence of different carbides, namely eutectic and secondary phases. The total carbide percentages were 20, 18, 15 and 14 vol.% for the specimens austenitized at 1000, 1025, 1050 and 1075 °C, respectively. The retained austenite was not detected; the reason is that high-temperature tempering eliminates this phase almost completely. The hardness values were 722 ± 5, 727 ± 6, 754 ± 4 and 803 ± 16 HV 10 for the specimens austenitized at 1000, 1025, 1050 and 1075 °C, respectively.

Determination of covered area percentage with counterpart material showed insignificant differences with respect to the T_A, **Figure 2.** Therefore, and also with respect to the steel supplier´s recommendation, the T_A of 1050 °C has been used for experiments with cryogenic treatments.

An example of the obtained microstructures by -140 °C CT is in **Figure 3.** They consist of the matrix and three different carbides, namely eutectic carbides, secondary carbides and small globular carbides (ECs, SCs and SGCs). Recent investigations have revealed that cryogenic treatments do not alter the number of both the ECs and SCs [8]. Alternatively, the number of SGCs is substantially increased, with the maximum after -140 °C cryogenic treatment, **Figure 4.** After low-temperature tempering (170 °C) the steel also contains some amounts of retained austenite. These amounts were determined to be 5.5,

4.3 and 2.1 vol.% after cryogenic treatment at -75, -140 and -196 °C, respectively. High-temperature tempered (at 530 °C) steel does not contain any significant amounts of retained austenite.

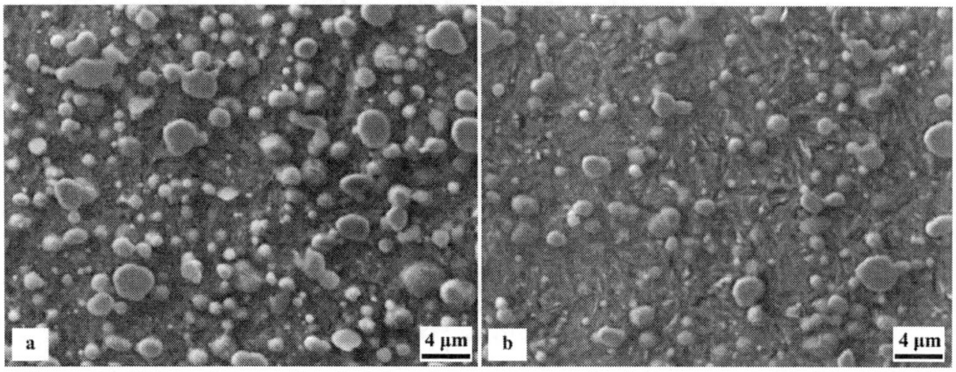

Figure 1. SEM micrographs showing the as-tempered microstructures (2x530 °C/2 h) of differently austenitized steel, a) 1000 °C, b) 1075 °C.

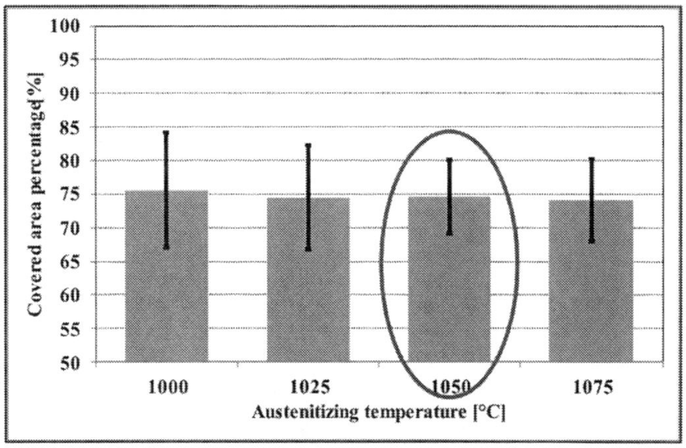

Figure 2. Covered area percentage (with counterpart material) for differently austenitized steel specimens.

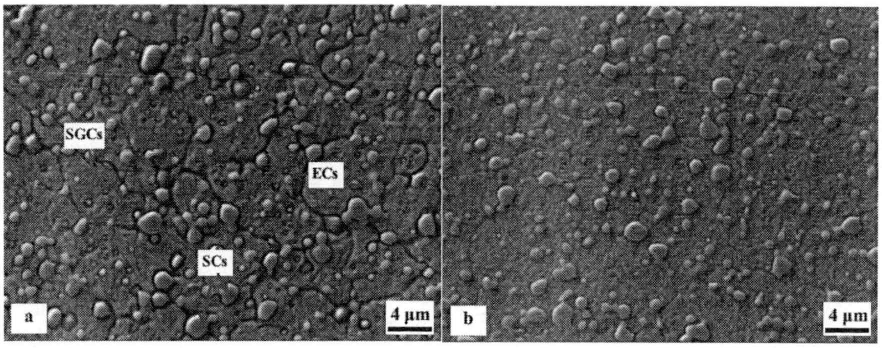

Figure 3. SEM micrographs showing the obtained microstructures by austenitizing at 1050 °C, -140 °C CT and tempering: a) at 170 °C, b) at 530 °C.

The hardness of differently treated specimens is in **Figure 5.** If the steel was low-temperature tempered then the hardness ranged between 821 and 854 HV 10 for different cryogenic treatments. While, the hardness was reduced to 754 – 772 HV 10 after high-temperature tempering. Mentioned hardness variations can be referred to much more extensive martensite softening after 530 °C tempering along with slight decrease in population density of SGCs. On the other side, high temperature tempered material contains very low amount of soft retained austenite; however, this fact cannot fully compensate the effects of martensite softening along with reduction of SGCs.

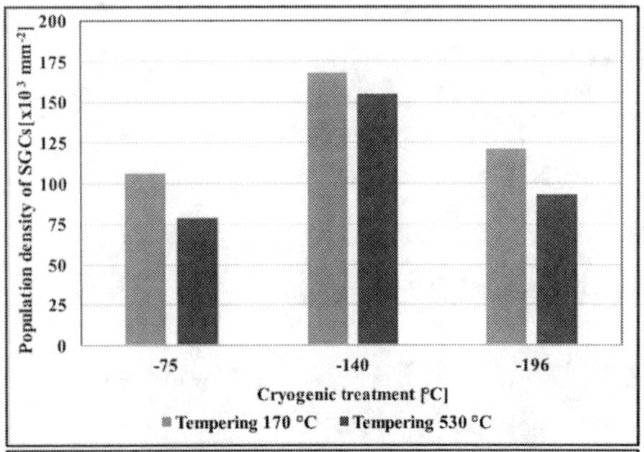

Figure 4. Population density of SGCs in differently cryogenically treated and tempered specimens. Note that the population density of SGCs in conventionally heat treated specimens did not exceed $40x10^3$ mm^{-2}.

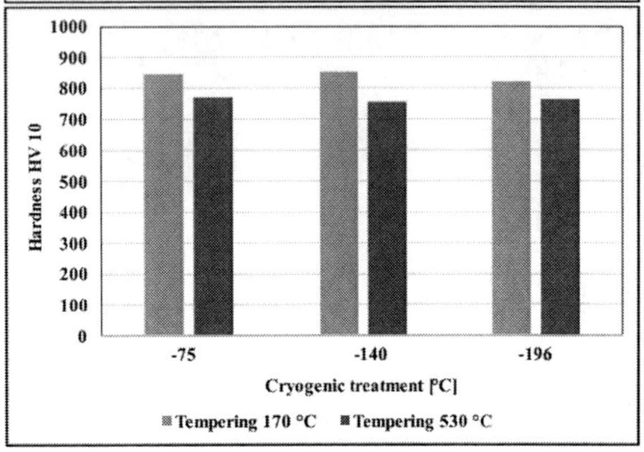

Figure 5. Mean hardness HV 10 values of differently cryogenically treated and tempered specimens.

The extent of counterpart´s adhesion raises rapidly with increasing load, **Figure 6**. While only small tracks of adhesion are visible after testing at 1 N the surface is covered with bronze by approx. 50% after testing at 5 N and, the extent of coverage reaches around 90% after testing at 10 N.
After the testing the obtained data of adhesion extent were assessed for normality testing. The results are shown in **Figure 7**. The estimated p-value is a criterion for data normality; it is seen that the p-value is greater than 0.05, suggesting that the data is normal.
The mean counterpart´s adhesion extent along with standard error values for all the experiments are **Table 1.** There are two general trends clearly visible: 1) The adhesion extent increases with increasing load (see also **Figure 6**), 2) Tempering at 530 °C produces lower adhesion extent that what is achieved by tempering at 170 °C. In addition, by comparing the obtained results after tempering at 530 °C and testing by using a 5 N load with the adhesion values obtained in the first stage (effect of austenitizing temperature, **Figure 2**) it is seen that CTs generally reduce the adhesion. To find the optimal conditions the adhesion behaviour of counterpart material is further addressed using Taguchi analysis and ANOVA results.
The main effects plot for SN ratios is in **Figure 8**. The graph shows that adhesion of counterpart´s material has a strong negative linear relationship with the tempering temperature; as the tempering temperature increases the adhesion level decreases. This confirms the visual inspection of the results given in **Table 1**. Also, the load has a strong positive linear relationship; the adhesion extent increases with increasing load. These two factors have, according to the results in **Table 2**, statistically most significant effect on the adhesion as their P values are less than 0.05. CT at -75 °C and -140 °C manifest no significant difference in counterpart´s adhesion extent but CT at -196 °C shows an improvement of the adhesion extent. And finally, the effect of sliding speed is almost minimal. One can summarize that the cryogenic treatment at -196 °C followed by tempering at 530 °C are the optimal processing

parameters for the Vanadis 6 steel when tested against CuSn6. The minimum adhesion extent can then be obtained at 1 N normal load combined with sliding speed of 0.1885 ms^{-1}.

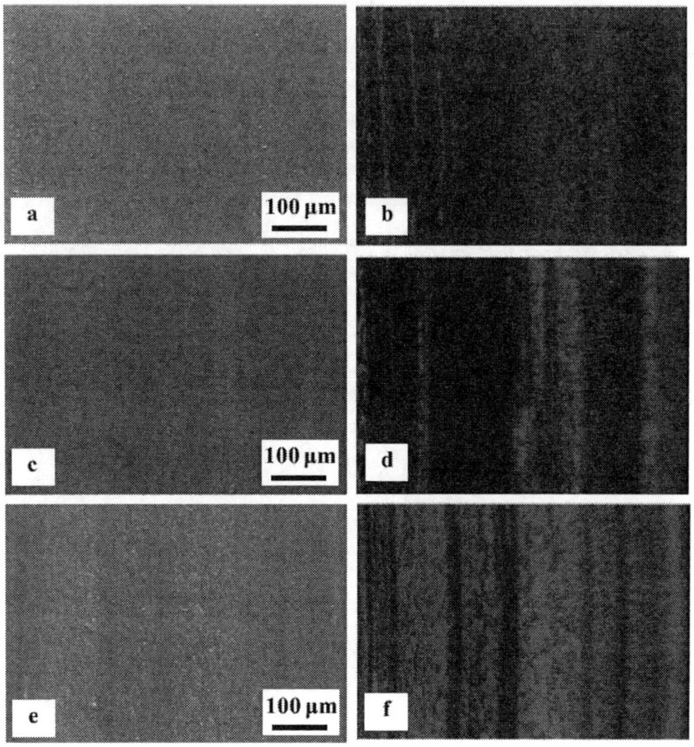

Figure 6. Plan-view SEM micrographs (a, c, e) and corresponding EDS maps of copper (b, d, f) showing the extent of counterpart´s adhesion after testing at 1 N (a, b), 5 N (c, d), and 10 N (e, f).

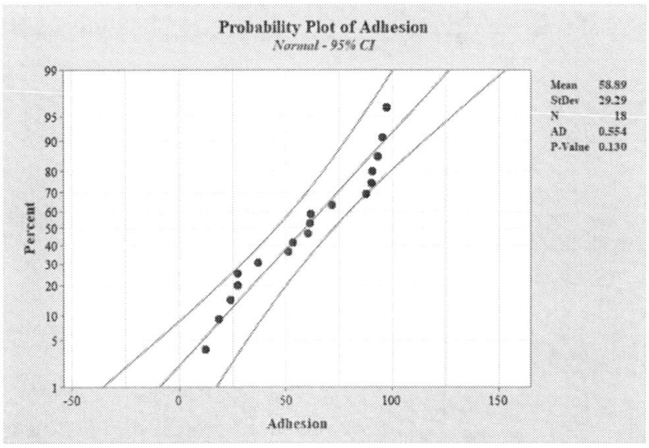

Figure 7. Normal probability plot at 95% CI for extent of adhesion.

Figure 9 is a compilation of CoF dependences on sliding distance for differently CT specimens. The CoF increases very rapidly over the running-in stages for all the specimens suggesting great extent of counterpart´s material transfer being taken place already in early stage of experiments. In the steady state, the CoF characterizes the bronze-to-bronze sliding rather than steel-to-bronze sliding contact. The lower CoF in experiments with 10 N load can be explained as follows: The actual free surface always contains local asperities where the contact pressure reaches very high values. The contact pressure values are higher at 10 N leading than at 1 N loading. One can thus expect more extensive flattening of existing surface asperities in the case of higher loading, thus leading to lower CoF.

Figure 10 shows the mean fracture toughness values for differently treated specimens. It is seen that CTs followed with low-temperature tempering worsen this property, with the maximum reduction after -196 °C cryogenic treatment. Only treatment at -140 °C provides the steel comparable K_{IC}. In contrast, CTs followed with high-temperature tempering improve the fracture toughness slightly (-196 °C) or more markedly (-75 or -140 °C). Lower K_{IC} of cryogenically treated and low-temperature tempered specimens as compared with conventionally heat treated ones can be simply explained by reduction of retained austenite (soft and ductile phase), which is only partly compensated by enhanced number of SGCs (note that these carbides may serve as crack propagation arrestors [13]).

Table 1. The extent of counterpart material transfer for all 18 experiments.

Tempering (°C)	Cryotreatment (°C)	Sliding speed (ms^{-1})	Load (N)	Adhesion (%)	Std.E
170	-75	0.064	1	37.0	5.0
170	-75	0.128	5	90.9	1.8
170	-75	0.1885	10	93.2	0.6
170	-140	0.064	1	61.0	2.7
170	-140	0.128	5	90.3	1.6
170	-140	0.1885	10	97.3	0.5
170	-196	0.064	5	71.7	2.2
170	-196	0.128	10	95.2	0.7
170	-196	0.1885	1	18.6	1.0
530	-75	0.064	10	60.3	3.1
530	-75	0.128	1	27.2	1.3
530	-75	0.1885	5	51.0	4.5
530	-140	0.064	5	27.4	2.5
530	-140	0.128	10	88.0	0.8
530	-140	0.1885	1	23.8	1.3
530	-196	0.064	10	61.4	4.4
530	-196	0.128	1	12.5	0.7
530	-196	0.1885	5	53.3	1.6

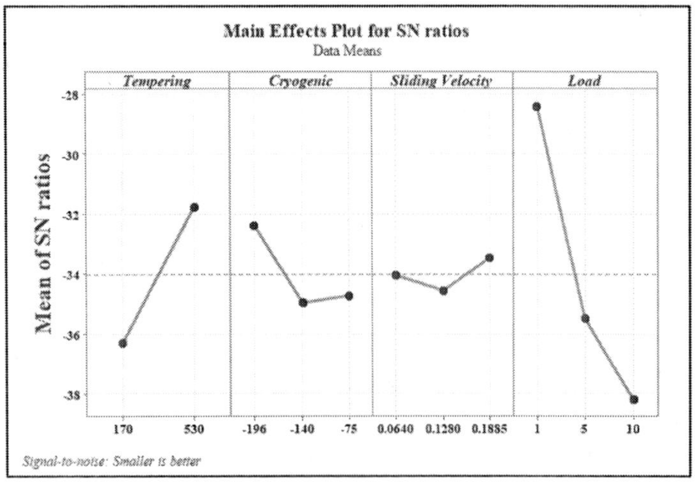

Figure 8. Main effects plot for the SN ratios of counterpart material adhesion.

Table 2. Analysis of variance (ANOVA) table for SN ratios for adhesion (%).

Source	DF	Seq SS	Adj SS	Adj MS	F	P	%C
Tempering temperature	1	92.071	92.071	92.071	10.27	0.009	18.0
Cryogenic treatment	2	24.271	24.271	12.136	1.35	0.302	4.7
Sliding velocity	2	3.657	3.657	1.828	0.2	0.819	0.7
Load	2	302.303	302.303	151.152	16.87	0.001	59.1
Residual error	10	89.621	89.621	8.962			
Total	17	511.924					

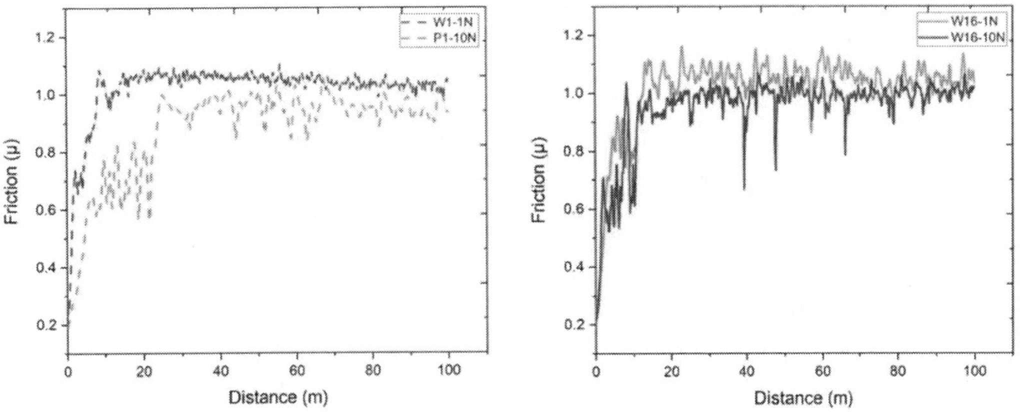

Figure 9. Dependence of friction coefficient on sliding distance for differently cryogenically treated and tempered specimens: W1 – cryogenic treatment at -75 °C + tempering at 170 °C, W1 – cryogenic treatment at -75 °C + tempering at 170 °C, W16 – cryogenic treatment at -196 °C + tempering at 530 °C.

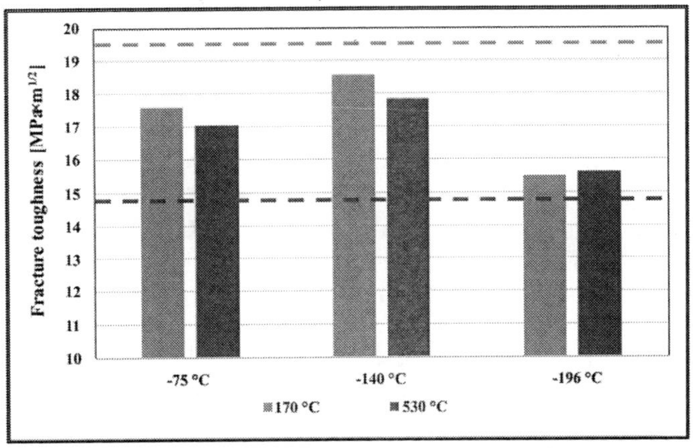

Figure 10. Mean values of fracture toughness for differently cryogenically treated and tempered Vanadis 6 steel. Dashed lines denote the fracture toughness values for conventionally treated steel after tempering at 170 °C (green line) and 530 °C (red line).

The fracture surfaces of specimens tempered at 170 °C contain a great number of micro-voids and holes, **Figure 11a,c**. The presence of these features corresponds to the extraction of SGCs from the fracture surface during the crack propagation, which relates to the decohesion at the carbide/matrix interfaces (DCs), and with (micro)plastic deformation (MPLD, PLD). But, the character of the crack propagation is semi-cleavage (SCL) or cleavage (CL) in the rest of the matrix. Cleavage fracture is also a typical crack propagation manner for coarser carbides (FCs). In specimens tempered at 530 °C lower amount of micro-plastically matrix has been observed, **Figure 11b,d** as compared with 170 °C tempered specimens. This may be associated with extensive precipitation of nano-sized carbides in the matrix of conventionally quenched steel [8], which makes the steel less amenable to plastic deformation. Moreover, conventionally heat treated steel contains much lower amount and population density of SGCs, which serve as nucleation sites for decohesive fracture mechanism. In CT specimens the application of higher tempering temperatures is always associated with decrease in SGCs count, thus with decrease in number of nucleation sites for decohesive fracture mechanism. This is of why the cryogenically treated and 530 °C tempered specimen (**Figure 11d**) manifests much more pronounced cleavage fracture than the one after 170 °C tempering, **Figure 11b**. One can thus summarize that tempering at 170 °C results always in fracture toughness reduction of CT specimens (albeit only slight in the case of treatment at -140 °C). In contrast, high-temperature tempering induces slight or moderate increase in fracture toughness as compared with conventionally treated steel.

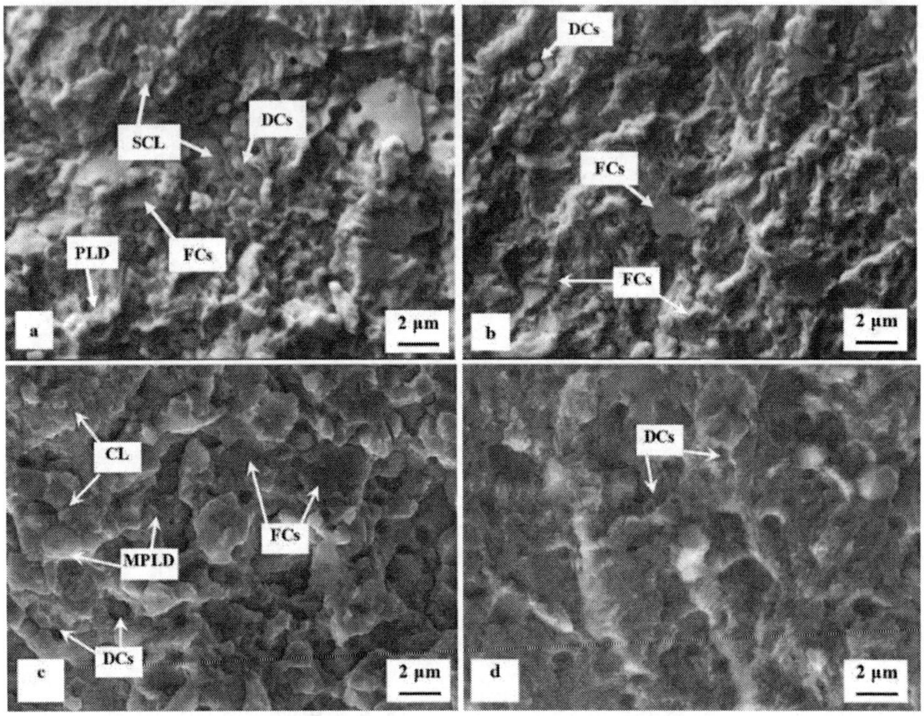

Figure 11. Detail SEM micrographs of the obtained fracture surfaces after K_{IC} testing, a) conventionally quenched and tempered at 170 °C, b) conventionally quenched and tempered at 530 °C, c) CT at -140 °C and tempered at 170 °C, d) CT at -140 °C and tempered at 530 °C. Legend: PLD, MPLD (micro) plastic deformation, SCL – semi-cleaved regions, CL – fully cleavage regions, DCs – decohesive carbides, FCs – fractured carbides.

Figure 12 shows the potentiodynamic polarization curves for differently treated Vanadis 6 steel. The corrosion potentials of cryogenically treated specimens (dotted, dot-dashed and dashed lines) are always higher (more anodic) as compared to the material state after CHT. The treatment at –140 °C produced the highest E_{corr}. Also it is evident that low-temperature tempering provides the material higher corrosion potential than high-temperature tempering. Mentioned variations are reflected in corrosion rate values, **Figure 13**. The main difference in corrosion rate is produced by different tempering treatments; high-temperature tempering always produces higher corrosion rate. The application of CTs reduces the

IOP Conf. Series: Materials Science and Engineering 1284 (2023) 012086 doi:10.1088/1757-899X/1284/1/012086

corrosion rate as compared with the CHT specimens; among differently cryogenically treated specimens the -140 °C treatment provides the steel the lowest corrosion rate.

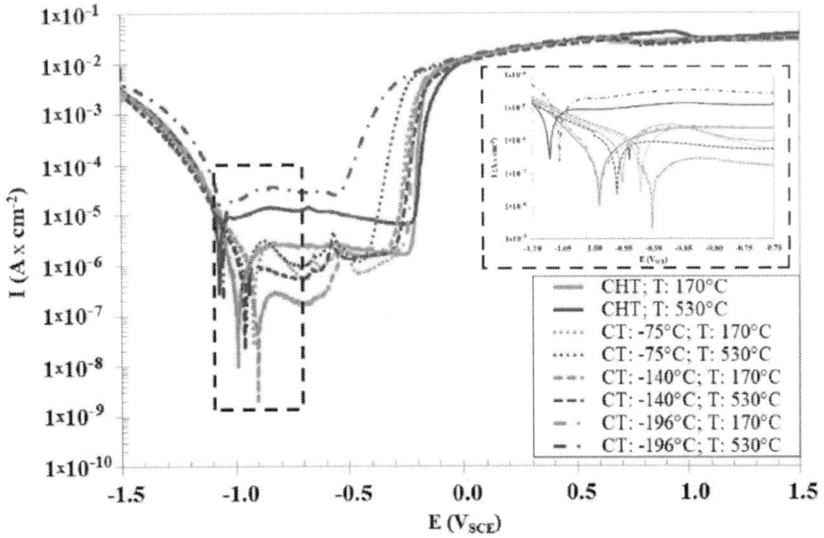

Figure 12. Potentiodynamic polarization curves of differently treated Vanadis 6. Legend: CHT–conventional heat treatment, CT–cryogenic treatment, T–tempering.

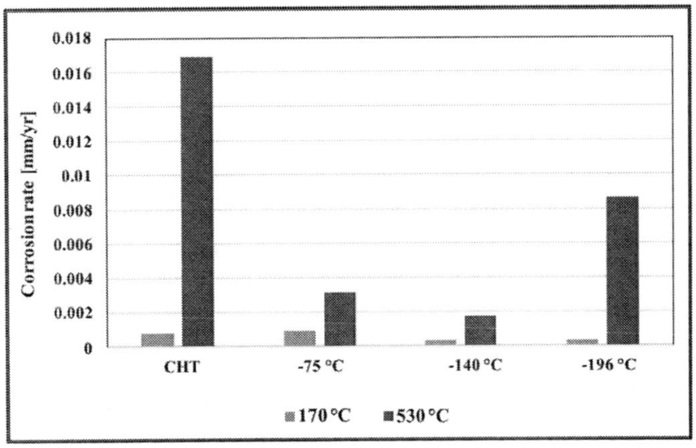

Figure 13. Corrosion rate of differently treated specimens made of Vanadis 6 steel.

Variations in corrosion properties are closely related to different microstructures produced by the analysed treatment strategies. Effect of tempering temperature, for instance, can be explained by much more extensive precipitation of special alloyed carbides at 530 °C than at 170 °C [8, 20]. Precipitation is a thermally activated process; thus, the precipitates differ from the matrix in terms of their chemistry. Therefore, they may act as galvanic cells that make the steel more sensitive to corrosion. Improvement of corrosion behaviour due to application of cryogenic treatment is mainly related to the presence of SGCs in microstructures. These carbides are formed at cryogenic temperature and their chemistry is almost the same as that of the matrix [8]. Therefore, they do not act as galvanic cells; on the contrary, carbides always manifest more noble behaviour than either ferrite of martensite. The more carbides are in the microstructure the higher exposed surface area is covered by this more noble phase. This phenomenon is considered as the most reliable explanation of corrosion resistance improvement by cryogenic treatments.

Conclusions

A model steel Vanadis 6 was subjected to different heat treatment strategies involving austenitizing temperatures, cryogenic treatments and tempering regimes. Anti-galling resistance, fracture toughness and corrosion resistance have been studied. The main outcomes of the research works can be summarized as follows:

1) The level of austenitizing temperature does not affect the anti-galling performance of the material.
2) Cryogenic treatment moderately improves the anti-galling resistance of the steel; the effect is more pronounced when the steel is high-temperature tempered. However, the counterpart´s material transfer can only be reduced but not completely eliminated by cryogenic treatments.
3) Fracture toughness is more (-196 or -75 °C) or less (-140 °C) worsened when cryogenic treatment is followed by low-temperature tempering; in the contrary, general improvement of this property is obtained for high-temperature tempered steel.
4) Corrosion resistance of the examined steel is generally slightly improved by cryogenic treatments.

Acknowledgement

The authors acknowledge that the paper is a result of experiments realized within the project VEGA 1/0112/20.

References

[1] Zaid A I O 2016 *IOP Conf. Series: Materials Science and Engineering* **146** 012009
[2] Robertson J 2002 *Mater. Sci. Eng.* **R37** 129
[3] Liljengren M, Kjellsson K, Johansson T and Asnafi N 2006 *In: proc. of the Annual Conference of the International Deep Drawing Research Group (IDDRG)* Porto, Portugal, 19-21 June 2006 597 – 603
[4] Kataoka S, Murakawa M, Aizawa T and Ike H 2004 *Surf. Coat. Technol.* **177–178** 582
[5] Gachot C, Rosenkranz A, Hsu S M and Costa H L 2017 *Wear* **372–373** 21
[6] Gaard A, Krakhmalev P and Bergström J 2009 *Wear* **267** 387
[7] Das D, Dutta A K and Ray K K 2010 *Mater. Sci. Eng.* **A527** 2182
[8] Ďurica J, Ptačinová J, Dománková M, Čaplovič L, Čaplovičová M, Hrušovská L, Malovcová V and Jurči P 2019 *Vacuum* **170** 108977
[9] Leskovšek V and Podgornik B 2012 *Mater. Sci. Eng.* **A531** 119
[10] Das D, Sarkar R, Dutta A K and Ray K K 2010 *Mater. Sci. Eng.* **A528** 589
[11] Kumar S, Nahraj M, Bongale A and Khedkar N K 2019 *Int. J. Mater. Eng. Innov.* **10** 98
[12] Rhyim Y M, Han S H, Na Y S and Lee, J H 2006 *Solid State Phenom.* **118** 9
[13] Jurči P, Ďurica J, Dlouhý I, Horník J, Planieta R and Kralovič D 2019 *Metall. Mater. Trans.* **50A** 2413
[14] Uygur I, Gerengi H, Arslan Y and Kurtay M 2015 *Materials Research* **18** 569
[15] Jovičevič-Klug P, Kranjec T, Jovičevič-Klug M, Kosec T and Podgornik B 2021 *Materials* **14** 6357
[16] ASTM E975-13: *Standard Practice for X-Ray Determination of Retained Austenite in Steel with Near Random Crystallographic Orientation*. In *ASTM Book of Standards, vol. 3.01*; ASTM: West Conshohocken, PA, USA, 2004
[17] American Society for Testing and Materials, ASTM E384-17: *Test Method for Microindentation Hardness of Materials*, ASTM International
[18] ISO 12137: 2010. *Metallic materials – determination of plane strain fracture toughness*
[19] ASTM G 102: *Standard Practice for Calculation of Corrosion Rates and Related Information from Electrochemical Measurements*. In *ASTM Book of Standards*; ASTM: West Conshohocken, PA, USA, 1994
[20] Jurči P, Bartkowska A, Hudáková M, Dománková M, Čaplovičová M, Bartkowski D 2021 *Materials* **14** 3759

IOP Conf. Series: Materials Science and Engineering 1284 (2023) 012087 doi:10.1088/1757-899X/1284/1/012087

Friction and lubrication modelling in sheet metal forming: Influence of local tool roughness on product quality

A T M Domhof[1], D. Waanders[1], J Hol[1] *, M Sigvant[2,3]

[1] TriboForm Engineering, Hengelosestraat 500, 7521 AN, Enschede, the Netherlands
[2] Volvo Cars dept. 81110 Strategy Development, Olofström, Sweden
[3] Department of Mechanical Engineering, Blekinge Institute of Technology, Karlskrona, Sweden

E-mail: j.hol@triboform.com

Abstract. To improve the accuracy of forming simulations, advanced friction models are increasingly used in the industry. These models account for the physical properties of the sheet, tool and lubrication and describe the tribological conditions during the forming operation. One of the main influencing factors on the tribology system, and therefore the friction coefficient, is the surface roughness of the tools. Until now, it is often assumed all tools have the same uniform surface roughness. In reality however, the tool might have different surface conditions dependent on the type and location of the tool. That is, the blank holder might be differently polished than the punch, and sharp radii might have a different tool roughness compared to flat areas. This paper investigates a significant number of tool measurements from different tool sets from Volvo Cars, and quantifies the effect of local surface conditions on product quality.

1. Introduction

Sheet metal forming is a manufacturing technology widely used within the automotive industry. When in production, sheet metal formed parts are created at a high rate with consistent quality. To acquire a part of high quality, knowledge on the friction and lubrication conditions is essential. In fact, friction conditions can be more influential than the variation of sheet material properties or stamping process settings [1]. The friction conditions between the forming tools and the sheet material differs locally and over time. More specifically, these conditions depend on the local pressure, sliding velocity, interface temperature, plastic strain of the sheet material, the type and amount of lubrication and the surface topography of both the sheet and the stamping tools [2, 3]. For the latter parameter, it is often assumed that the surface roughness is the same for all tools and uniform across the tool geometry. Most of the times, the tools are polished manually and often performed by multiple polishers. Polishing is therefore not done equally for all tools and across their surfaces. Curved surfaces, for instance, are usually more thoroughly polished than flat surfaces. This paper therefore studies the influence of the tool type; die, blank holder and punch, as well as the influence of the location on the tool on the surface roughness. This is done through a statistical analysis of a significant amount of surface measurements from different tool sets from Volvo Cars. The influence of the tool surface roughness variations on part quality is subsequently shown on a door inner part from Volvo Cars by making use of AutoForm.

To be able to study the influence of the tool roughness on the simulation results in AutoForm, the TriboForm friction model is used [4, 5]. The statistical analysis of the tool surface measurements is performed in Section 2. In Section 3, the statistical results are applied to a simulation of a rear door inner part from Volvo Cars. Finally, the conclusions and recommendations are given in Section 4.

2. Statistical analysis of surface measurements

This section contains information on the performed measurements and the surface characteristics considered.

2.1. Specifications

Over a period from 2015 to 2022, surface measurements have been performed on a significant number of production and try-out tools at Volvo Cars. In this study, a total of 100 surface measurements are analyzed corresponding to polished and uncoated cast iron tools used in cold sheet metal forming processes. These measurements are taken of each of the tool types shown in Figure 1. Of the total amount of surface measurements, 39 are from dies, 41 from blank holders and 20 from punches. The surface measurements are taken from different locations at both flat and radii areas. Figure 2 shows an example of a die of a rear door inner part and the locations at which surface measurements have been taken.

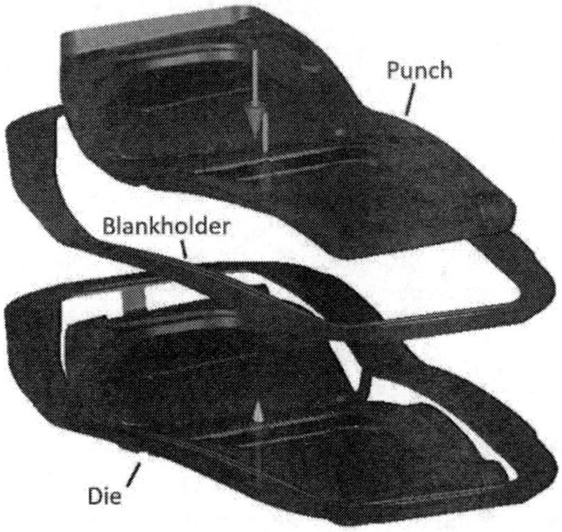

Figure 1. Tool set of a rear door inner part in AutoForm.

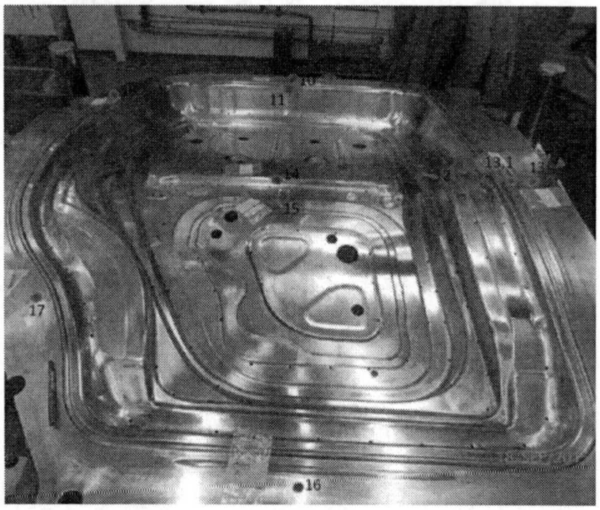

Figure 2. Example of surface measurement locations of a rear door inner part.

The surface characterization was performed using a surface replication technique in combination with the NanoFocus µsurf system to perform 3D surface measurements. Before the scanned 3D surface data is used, the data is processed (straightened and filtered) as described in [4].

From the surface measurements, the Sa-parameter is used as a measure for the surface roughness, which is the arithmetic mean height of the surface and is commonly used in the industry. Additionally, the radius of each surface is determined. This is done by fitting a second order polynomial through the rough surface data. Subsequently, the radius is computed in each direction of the surface, with the smallest found radius taken as the radius of the surface. Based on the computed radius, a distinction is made between flat and curved surface measurements.

2.2. Results

Figure 3 shows the Sa-roughness values of the total batch of measurements. On the left, the Sa-roughness value is plotted for each measurement. Within the box plot, both the median and the mean value are shown by respectively the green and the black line. The right side shows the probability density function of Sa-roughness values of all surface measurements, including the median and the mean value as thick horizontal lines.

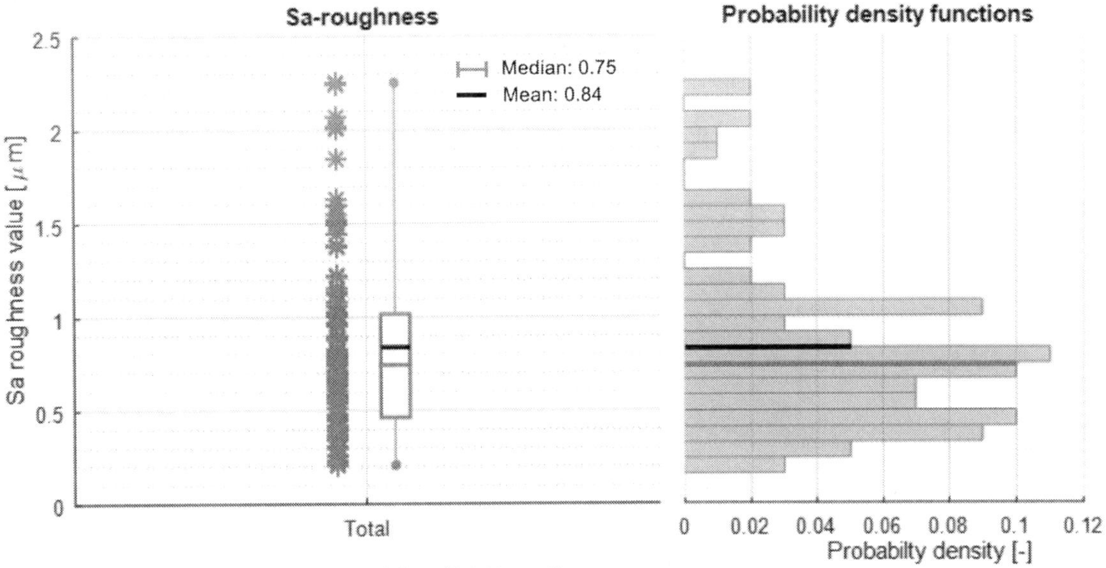

Figure 3. The Sa-roughness plotted for all 100 surface measurements.

From the probability density function it is shown that the Sa-roughness distribution is skewed and contains one-sided outliers with high roughness values. The median value is used within this study as the value shows a higher robustness against outliers. The median, with a value of 0.75 μm, is found in an area with a higher density of measurements compared to the mean, which has a value of 0.84 μm.

Splitting the total set of measurements into measurements belonging to the die, punch and blank holder, allows to see differences in surface roughness between the three tool types. The Sa-roughness values of each tool type are presented in Figure 4.

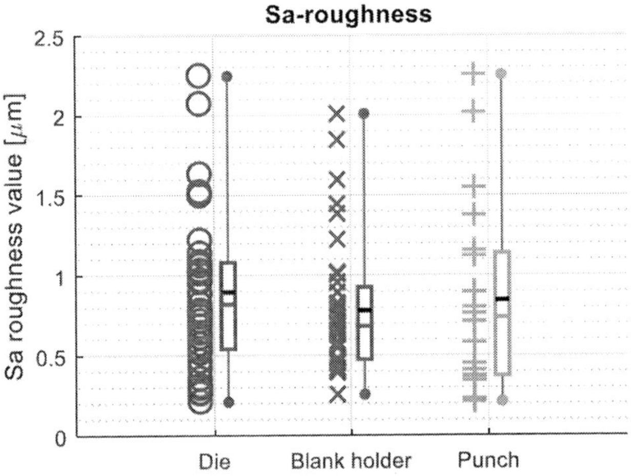

Figure 4. The Sa-roughness distributions of the die, blank holder and punch.

Table 1. The median and mean Sa-roughness value in μm corresponding to each tool type.

	Median	Mean
Die	0.82	0.90
Blank holder	0.68	0.78
Punch	0.74	0.84

The spread of Sa-roughness values is the largest for the punch, followed by the die and then the blank holder, as displayed by the height of the box plots. The median and mean values of the die are higher than those of the blank holder and the blank holder has higher values compared to the punch, as can also be seen in Table 1.

As seen in Figure 4, there exists a significant spread in surface roughness values for each of the tool types. This demonstrates that the tools have different surface conditions, which might depend on the location on the tool. That is, the blank holder might be differently polished than the punch, and sharp radii might have a finer tool roughness compared to flat areas. For this reason, a distinction is made between flat and curved areas as shown in Figure 5, where the radius of each measurement is plotted against the Sa-roughness. The surfaces are divided into two groups, one with measurements on curved surfaces and one with measurements from flat surfaces. Surfaces with a radius smaller than 20 mm are considered curved and measurements with a larger radius are considered flat. The dashed line in Figure 5 shows how the measurements are divided.

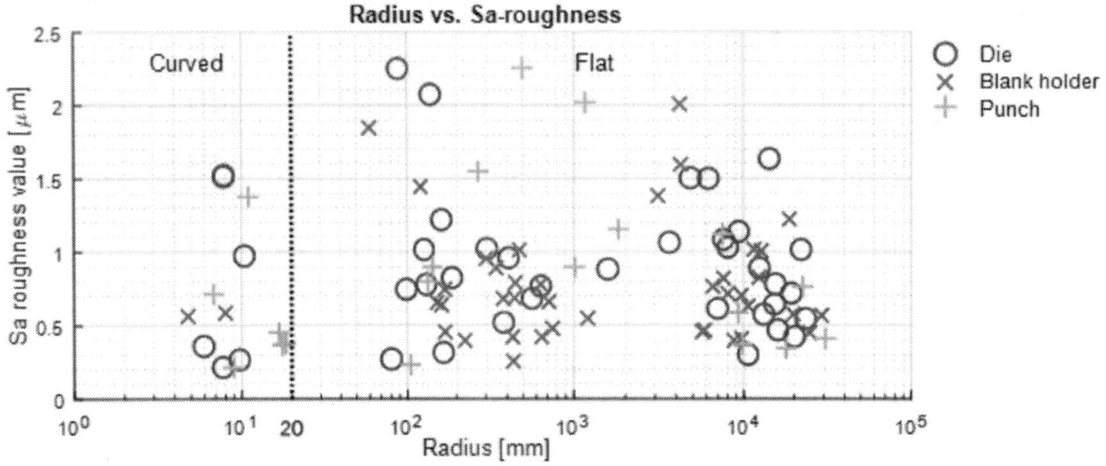

Figure 5. The Sa-roughness plotted against the radius for all surface measurements of each tool type.

In total, fourteen surface measurements have been performed on curved surfaces. Six measurements correspond with the die, two measurements belong to the blank holder and the remaining six belong to the punch.

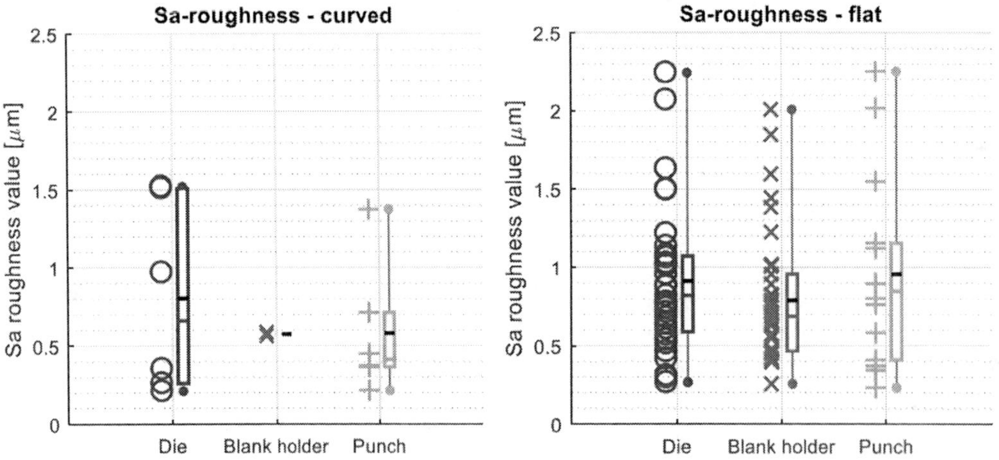

(a) Curved surface measurements **(b)** Flat surface measurements

Figure 6. The Sa-roughness plotted against the radius for all surface measurements of each tool type in case of flat measurements (a) and curved measurements (b).

IOP Conf. Series: Materials Science and Engineering 1284 (2023) 012087 doi:10.1088/1757-899X/1284/1/012087

In Figure 6, the Sa-roughness of the curved and flat measurements are plotted separately for each tool type. From Figure 6a, it can be noticed that the measurements corresponding with the die show a large spread and that the mean and median values of the die are relatively high, as can also be seen in Table 2. This is caused by two outlying measurements with an Sa-roughness value close to 1.5 μm. The six curved punch measurements show less spread, with a relatively low median value of 0.41 μm. Of the set of blank holder measurements, only two are considered curved, resulting in equal mean and median values.

The Sa-roughness values corresponding with flat measurements are shown in Figure 6b. The mean and median values corresponding with the punch are higher than those of the die. Also, the spread of points is the highest for the punch. The measurements that belong to the blank holder on average have the lowest Sa-roughness values.

From Figure 6a and 6b, it is shown that the Sa roughness is lower for curved surfaces than for flat surfaces, which is also presented in Table 2. The largest difference is found for the punch measurements, with the median value being more than a factor two larger for flat surfaces.

Table 2. The median and mean Sa-roughness values in μm of each tool type, separating curved and flat surface measurements.

	Median		Mean	
	Curved	Flat	Curved	Flat
Die	0.66	0.82	0.80	0.91
Blank holder	0.58	0.69	0.58	0.79
Punch	0.41	0.85	0.58	0.96

3. Numerical analysis

To demonstrate the effect of different roughness regimes on product quality, numerical analyses are performed on a rear door inner part from Volvo Cars. The tribology system present, consists of a mild steel (CR3-ZM) with an average sheet roughness of 1.6 μm. The sheet is uniformly lubricated with Fuchs Anticorit PL3802 39S with an amount of 0.8 g/m^2. The tool material is a nodular cast iron (GGG70) with a surface roughness depending on the applied roughness regime.

Four different cases are evaluated, see Table 3. Case 1 uses Coulomb friction with a fixed value of the friction coefficient of 0.15. The other three cases use TriboForm friction, making use of the characteristics of the real tribological system. Case 2 utilizes a uniform surface roughness distribution for all tool types, using the median value of 0.75 μm. Case 3 uses tool type dependent roughness and uses the median values shown in Table 1, using all measurements independent of the surface radius. In the last case, Case 4, also radius-dependency is taken into account, using the summarized results shown in Table 2 corresponding to the median values. In this case, surfaces with a radius smaller than 20 mm are considered to be curved and use a different tool roughness value compared to flat areas.

Table 3. The four analyzed cases and their corresponding tool roughness values in μm.

Case	Description	Die		Blank holder		Punch	
1	**Coulomb friction 0.15**	-		-		-	
2	**TF, tool independent**	0.75		0.75		0.75	
3	**TF, tool dependent**	0.82		0.68		0.74	
		Curved	Flat	Curved	Flat	Curved	Flat
4	**TF, tool & radius dependent**	0.66	0.82	0.58	0.69	0.41	0.85

Figure 7 shows the difference in friction coefficients between Coulomb and TriboForm friction using three different values for the Sa-roughness of the tools. Here, the dependency of TriboForm friction on the pressure and the amount of strain in the sheet is shown, whereas Coulomb friction remains constant. Due to the relatively low lubrication amount, the velocity dependency of the current tribology system is minor. TriboForm friction shows that with an increase in pressure of the tool on the sheet, the friction coefficient decreases. Also, the friction coefficient decreases with increasing the strain in the sheet material. Furthermore, with an increasing tool-roughness, the friction coefficient increases as well, which is most significant for low pressure and strain values. The three friction models converge for higher values of pressure and strain.

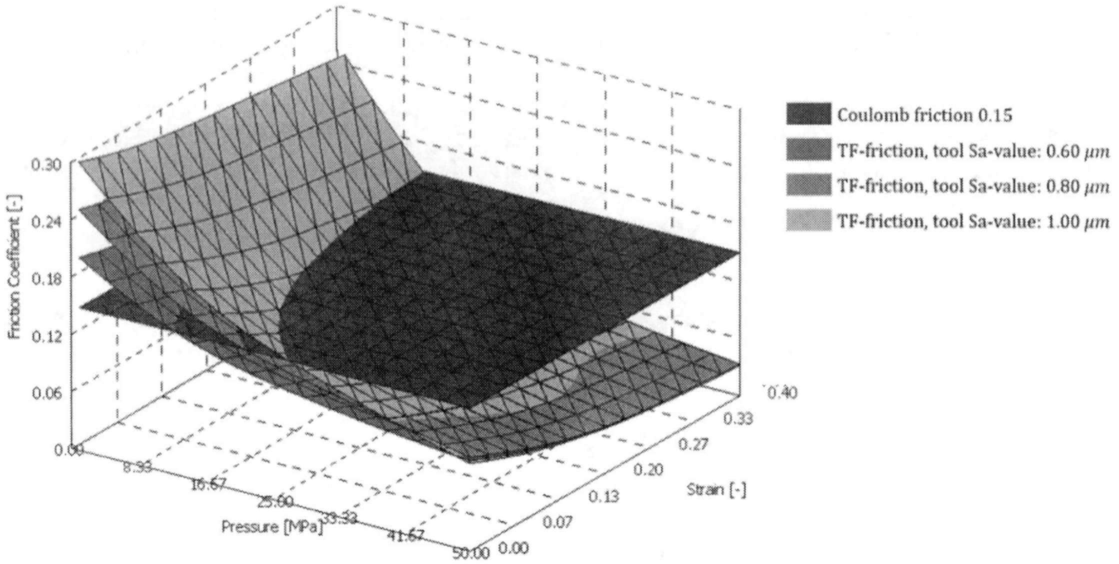

Figure 7. Comparison of Coulomb friction with TriboForm friction for different tool roughness values at a sliding velocity of 50 mm/s and a temperature of 20 °c.

The simulation results of the four cases are compared at the bottom end of the closing. Comparison is done using Advanced Formability (see Figure 8) in the AutoForm R10 software. Non-critical simulation results are expected from try-out and production of the rear door inner part, indicated by Volvo Cars.

The simulation results of Case 1 using Coulomb (Figure 8a), show that the part fails at multiple locations, which are encircled in red and are mostly found on the left hand side of Figure 8a. The fractures are clearly visible from the FLD, showing that Coulomb is too conservative in predicting the quality of the part.

The results of Case 2, using TriboForm friction with average tool roughness for all tools, are shown in Figure 8b. The FLC is not violated and hence no fractures are predicted. However, the part is still considered critical as excessive thinning and risk of splits occur at the encircled locations.

For Case 3, where each tool type has a different Sa-roughness value, the results are displayed in Figure 8c, which are similar to the results of Case 2. This is as expected, as the roughness values of each tool type are comparable and do not deviate much from the median value used in Case 2 (see Table 3).

The results of Case 4, which distinguishes between flat and radii areas on the tool surface, are shown in Figure 8d. The part becomes less critical with a reduction in excessive thinning areas and risk of splits. Overall, the parts become more safe with respect to Cases 2 and 3 and critical areas only occur at sharp radii locations.

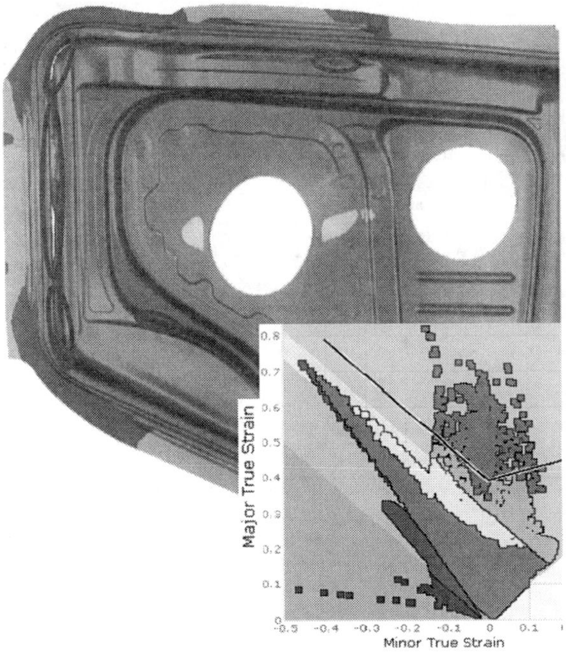

(a) Case 1 – Coulomb friction 0.15

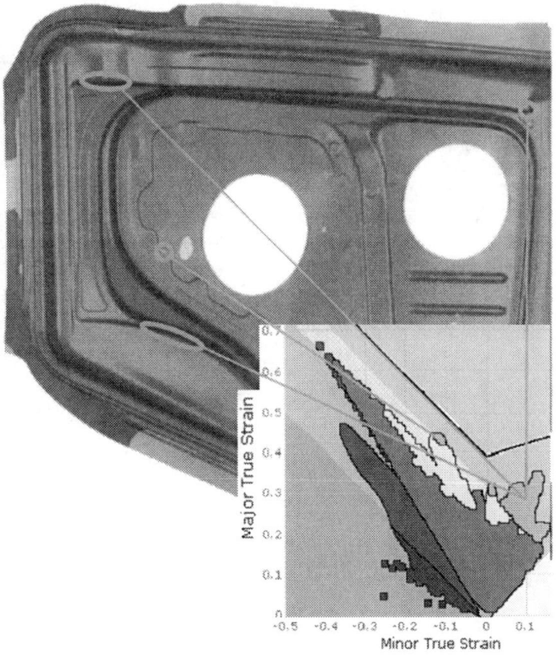

(b) Case 2 – TF, tool independent

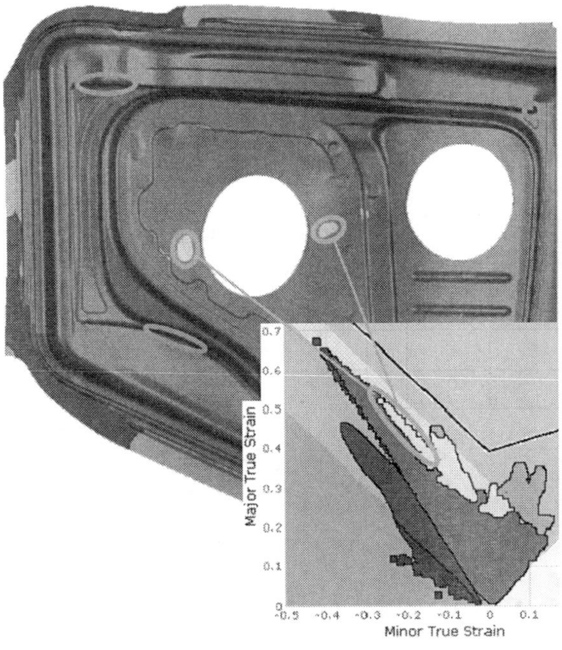

(c) Case 3 – TF, tool dependent

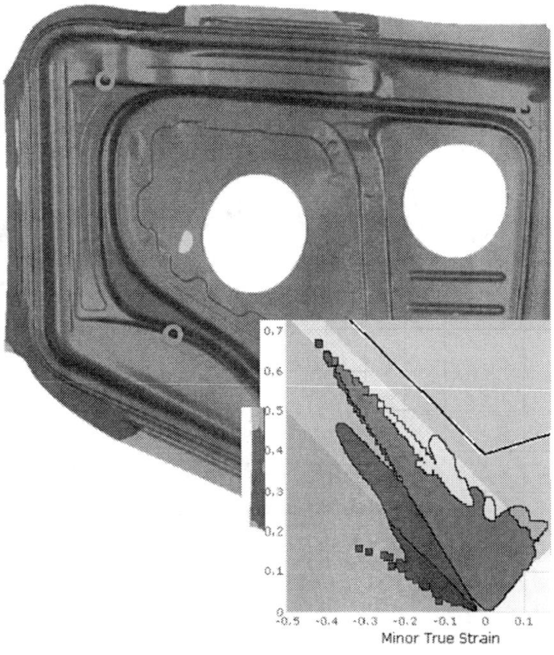

(d) Case 4 – TF, tool & radius dependent

Figure 8. Advanced formability results of all four cases at the end of drawing, showing the formability across the part as well as the FLD.

The significant difference between the results using Coulomb friction and TriboForm friction can be explained by Figures 9 and 10, which show the frictional behaviour of Cases 1 and 2 across the part geometry.

IOP Conf. Series: Materials Science and Engineering 1284 (2023) 012087 doi:10.1088/1757-899X/1284/1/012087

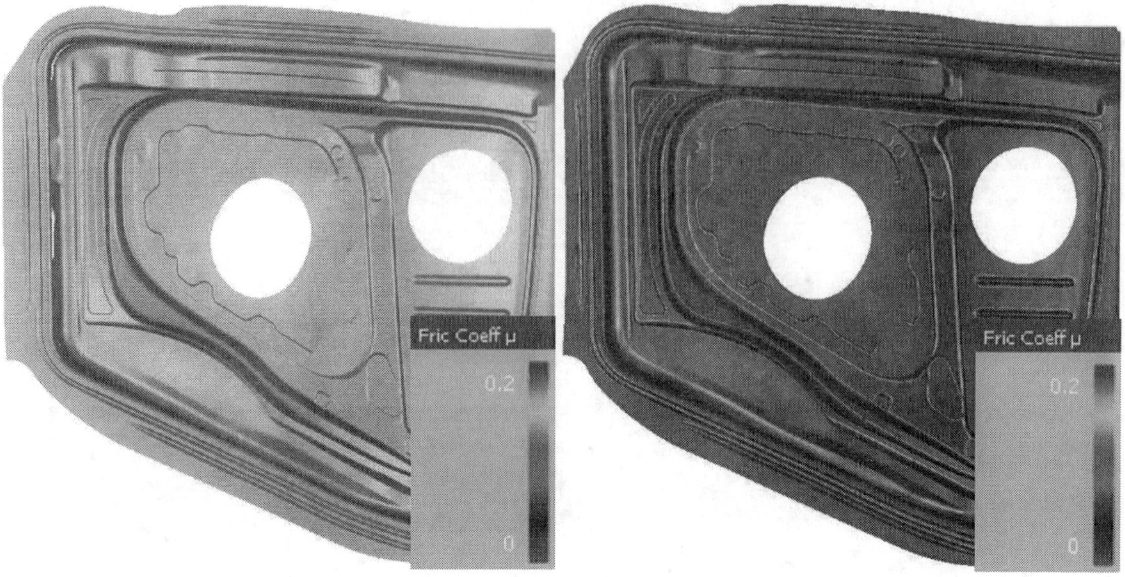

(a) Case 1 – Coulomb friction 0.15 (b) Case 2 – TF, tool independent

Figure 9. The friction coefficient across the part surface in contact with the punch for Cases 1 and 2.

On the left, Figure 9a, the uniform friction coefficient of Case 1 is shown (Coulomb friction 0.15). On the right, Figure 9b, the friction coefficient is shown for Case 2, which varies in accordance with the friction model shown in Figure 7. That is, regions with high values of pressure and/or strain show a lower friction coefficient. As can be seen, these regions are mainly found on curved regions such as the drawbeads. Furthermore, it can be observed that flat areas have a relatively high friction coefficient, which is displayed in red. The friction coefficient is not computed for areas which have not been in contact with the tool and are displayed in grey.

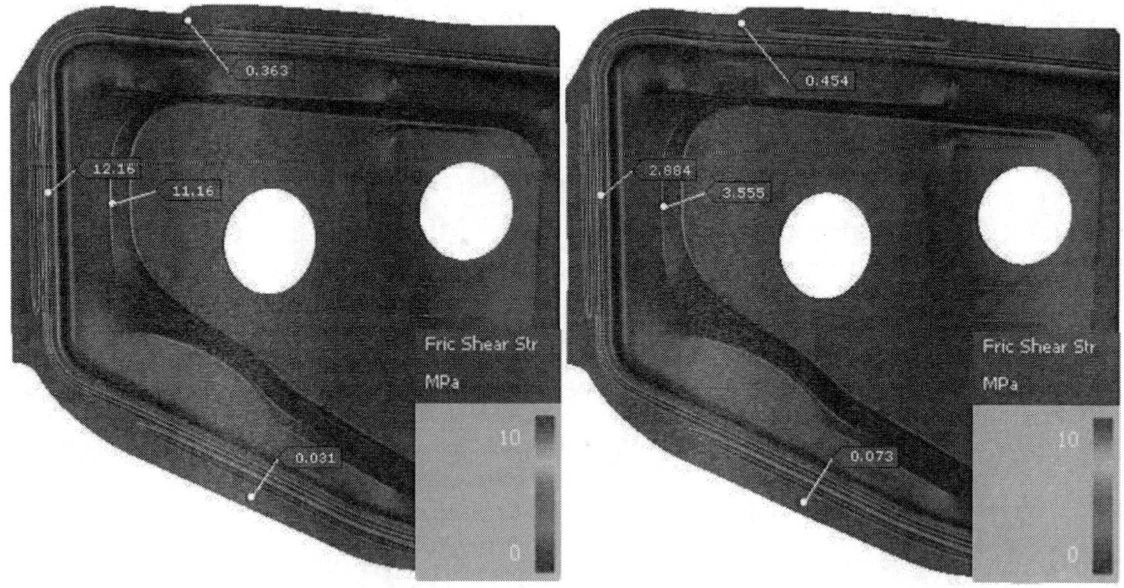

(a) Case 1 – Coulomb friction 0.15 (b) Case 2 – TF, tool independent

Figure 10. The friction shear stress in MPa across the part surface for Cases 1 and 2 at 20 mm prior to closing, when the first splits appear for Case 1.

Figure 10 shows the friction shear stress for Cases 1 and 2 across the part geometry 20 mm prior to closing, when the first splits appear for Case 1. The friction shear stress is computed as the friction coefficient multiplied with the normal pressure. As can be seen, curved regions of the part correspond to high friction shear stress values. These regions therefore have a relatively large impact on the forming process, despite having relatively low friction coefficient values.

Using Coulomb friction, the friction coefficient (fixed value of 0.15), and therefore also the friction shear stress, is larger in curved regions compared to TriboForm Friction, see Figure 9. On the other hand, the friction coefficient in flat sheet areas is lower compared to TriboForm friction, but because of relatively low contact pressures at these areas, the effect on shear stress is relatively low, see Figure 10. Overall, this leads to a more critical simulation, resulting in splits, as seen in Figure 8a.

Comparing the results from the three cases using TriboForm, it is shown that Case 4, with radius dependency, shows the least critical results. By lowering the tool-roughness in curved regions, the friction coefficient decreases, resulting in a reduction of the friction shear stress at these areas. As radii area have the highest contribution to the friction shear stress, these also have the highest impact on the part quality. The results of Case 4 are expected to give the most accurate simulation results and are in line with the expectations from try-out trials at Volvo Cars.

4. Conclusions

From the statistical analysis it was shown that the tools have a Sa-roughness value of 0.75 and 0.84 μm (median and mean, respectively). Furthermore, a slight difference in surface roughness values between tool types was found. A larger difference was found between curved and flat areas. However, the data set of curved measurements was relatively small and more measurements are required to perform a proper statistical analysis. To extend the study, additional surface measurements should be evaluated and care should be taken regarding the location on the tool the measurement is performed. That is, the measurements should be taken in areas in contact with the sheet and not from e.g. offsetted tool areas or non-bearing areas.

From the numerical results it was concluded that Coulomb overpredicts frictional shear stresses leading to a too conservative simulation. Using TriboForm and differentiating in Sa-roughness values between curved and flat areas, results becomes less critical and in line with expectations from try-out results at Volvo Cars. The difference in simulation results of tool type independent and dependent TriboForm friction is small, which is due to the slight difference in surface roughness values of the different tool types.

References

[1] Tatipala S, Pilthammar J, Sigvant M, Wall J and Johansson C M 2018 Introductory study of sheet metal forming simulations to evaluate process robustness *IOP Conf. Ser. Mater. Sci. Eng.* **418**

[2] Grueebler R and Hora P 2009 Temperature dependent friction modeling for sheet metal forming *Int. J. Mater. Form.* **2** 251–4

[3] Ludwig M, Müller C and Groche P 2010 Simulation of dynamic lubricant effects in sheet metal forming processes *Key Eng. Mater.* **438** 171–8

[4] Hol J, Meinders V T, de Rooij M B and van den Boogaard A H 2014 Multi-scale friction modeling for sheet metal forming: the boundary lubrication regime *Tribol. Int.* **81** 112–28

[5] Hol J, Meinders V T, Geijselaers H J M, and van den Boogaard A H 2015 Multi-scale friction modeling for sheet metal forming: the mixed lubrication regime *Tribol. Int.* **85** 10–25

IOP Conf. Series: Materials Science and Engineering 1284 (2023) 012088 doi:10.1088/1757-899X/1284/1/012088

Experimental investigation of the effect of lubricant change on automotive component

Carvalho, L A; Zsolt L

Institute of Materials Science and Technology, Faculty of Mechanical Engineering and Informatics, University of Miskolc, H-3515 Miskolc, Egyetemváros, Hungary

E-mail: de.carvalho.lucas.alexandre@student.uni-miskolc.hu, metzsolt@uni-miskolc.hu

Abstract. Manufacturing problems resulting from lubricant supplier changes are relatively common in automotive press shops. Simulations using computer-aided design (process and tool) detect a significant proportion of manufacturing problems before tools are physically produced. In modern forming codes, an enhanced Coulomb model is used to numerically model the friction phenomena associated with lubrication. In the laboratory condition, determining model parameters through physical measurement is challenging. As a result, it became necessary to design a device that could be utilized in industrial environments. In this study, we present the prediction of a manufacturing problem resulting from a change in the lubricant of an industrial sheet metal part. The measuring device has been developed based on the parameters of the friction model of the lubricant materials. The developed measuring device may be used to determine the pressure and velocity dependent friction model parameters. The measuring device can only measure a limited range of pressures and velocities, so the results are extended to a wider range using a mathematical method. The results are used to demonstrate the effect of lubricant material changes on the forming process using the AutoForm code.

1. Introduction

Due to the high material utilization, the metal forming process is already among the most energy efficient production methods. Currently, lubricants are used in all categories of metal forming processes to minimize friction between workpieces and forming tools together with the reduction of tool loads and other purposes. However, manufacturing problems resulting from lubricant supplier changes are relatively common in automotive press shops. Those problems are a result of the difficulty in characterizing the lubricant performance for a certain application. An accurate and practical estimation of friction coefficients to satisfy the tribological characteristics for metal forming simulations is a complex task. Completely distinct friction values can be achieved under identical load conditions for different tribological systems.

Lubricant industries need solid data for the development of effective boundary lubricants. Historically, not much attention has been given to the investigation of tribological behavior in the boundary regime. The tribological behavior of the related parts (sheet, tool and lubrication) at high contact pressures, long forming geometries and low levels of lubrication still need attention. Lubricant requirements extend well beyond enhancing formability. These requirements are often the limiting factor for the development of new lubricants. [1]

Other important aspect to be considered in the nature of friction in a particular tribosystem including stainless steel and tool interaction is the role that oxide layers and contaminant films on the surfaces have in it [5]. If the contact pressure is low enough, contaminant films (mainly oxides) separate the surfaces and the adhesion is low. As the pressure rises the films are broken and the contact points and adhesion occurs. This possibility also depends on the surface roughness. [2] A complete mapping of the actual contact area as a function of normal pressure, bulk deformation, material properties, and surface conditions appears to continue to need investigation. Finally, it is essential to work more closely with actual asperity sizes rather than modeling them, and it is expected that in the future, improvements to current FE codes will frequently come from advances in friction modeling. []

Computer-aided design (processes and tool) have an important role in detecting a significant portion of manufacturing problems before tools are produced. In modern forming codes, an enhanced Coulomb model is used to numerically model friction phenomena associated with lubrication. In the laboratory condition, determining the model parameters through physical measurement is challenging. As a result, it becomes necessary to design a device that can be utilized in industrial environments.

Previous studies [2]–[4] indicate that it is better to use a strip drawing test rather than abstract testing principles like the pin-on-disc tribometer considering the complexity of microscopic friction mechanisms. For the assessment of friction behavior the flat contact area located under the blank holder is of great interest. This fact can be connected to the influence on forming results caused by the friction forces that act in these large areas. The strip drawing test consists of pressing a metal strip between the tool and the stage, which is simultaneously drawn by the stage fuse moves. Both the applied normal load and the tangential force which opposes the movement are continuously measured. The measurement can be made via a load cell, allowing friction coefficient values to be computed. [5], [6]

Different parameters can affect the coefficient of friction. The contact pressure is one of the most relevant parameters and should be considered in macro and micro scales. The topography of the contact surface is submitted to flattening of each asperity due to the increase of contact pressure. As a consequence, the contact geometry changes generating a change on the coefficient of friction. As the material flows to the cavity of the die, the contact pressure applied to the sheet will increase as the flange area decreases. This is a very important base of the sheet metal forming. Beside this fact, an heterogeneous contact pressure distribution can be also found as a result of the change in thickness in different areas of the sheet during the flow of the material. The effective coefficient of friction can then be expressed as a function of pressure. The velocity dependent friction model takes into account the decrease of friction coefficient due to the increase of the relative velocity between the tool and sheet. [7], [8]

The major relevance of friction in simulations is the fact that the occurring friction forces act as a substantial additional resistance. Thus, friction directly influences material flow and stress distribution. These friction influences affect the simulation result "part quality" in terms of thickness distribution, thinning and draw-in as well as springback. [4]

In this study, we present the investigation of a manufacturing problem resulting from a change in the lubricant of an industrial sheet metal part. The measuring device has been developed based on the parameters of the friction model of the lubricant materials. The developed measuring device is used to access the friction coefficient in order to determine the pressure and velocity dependent friction model parameters. Due to the limited range of pressures and velocities of the measuring device the results are extended to wider range using a mathematical method. The results are used to demonstrate the effect of lubricant material changes on the forming process using the AutoForm code.

2. Industrial problem

The current paper discusses a real problem caused by the change of a lubricant supplier of an automotive company. The figures 1 and 2 illustrate the problem faced by the industrial partner. It is indicated on the pictures both failures, in the produced part and in the die radius.

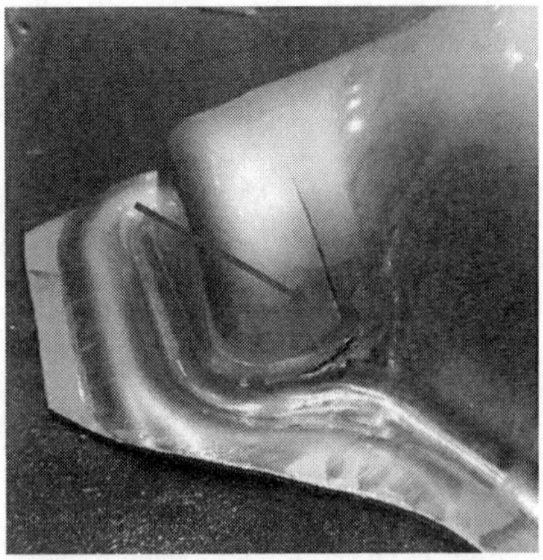

Figure 1. Cracking area on the surface of industrial workpiece.

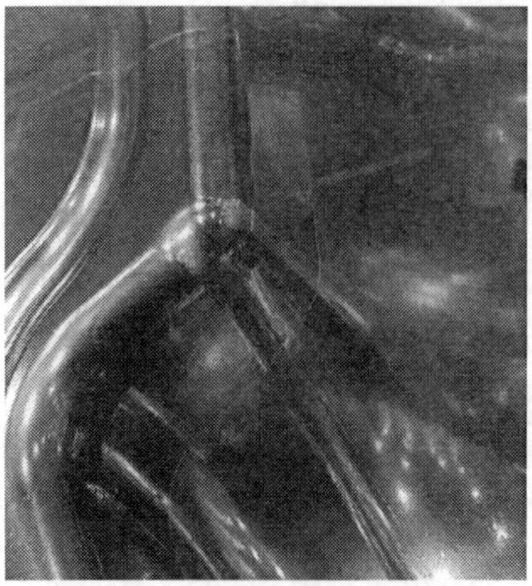

Figure 2. Removed coating at the tool radius.

In figure 1 it can be seen a crack as a consequence of the lubricant change on the workpiece. This failure can be associated to the lubrication change due to the increase of friction forces between the sheet metal and the forming tool. Higher friction can result in excessive stress and strain in the metal part leading to crack. Figure 2 shows wear in the tool radius characterized by the removal of coating in this area. Adhesion is one of the most common causes of wear on forming tools. It occurs when during the process the sheet metal sticks to the surface of the tool due to frictional forces.

Both indicated failures are located in the same contact area. Galling is a sever type of adhesive wear caused by high frictional forces and localized welding that can cause damage to both the forming tool and the sheet metal. The current method comes into this scenario to help predicting failures associated to the lubricant replacement by proposing an improvement on realistic friction description to provide the required input variables for the enhanced Coulomb models on AutoForm.

3. Materials and methods

3.1. Numerical model

As a requirement to protect the intellectual property of the industrial partner a similar shape has been based on the original part has been designed for this study. The dummy workpiece shown in figure 3 preserves a similar geometry and allow the investigation of the area of interest of this study.

Figure 3. Shape of the dummy workpiece.

For the material behavior description of the stainless-steel sheet metal, the Swift hardening curve is expressed in equation (1) and the material parameters are in Table 1.

$$\sigma = C(\varepsilon_0 + \varepsilon)^n \tag{1}$$

Table 1. Material parameter inputs for hardening curve.

ε_0	n	C
0.05	0.53	1637 MPa

3.2. Friction model on AutoForm

Among the friction models available on AutoForm, pressure and velocity dependent enhanced Coulomb friction models are the models considered for the development of this research. The effective pressure dependent coefficient of friction *(μ_{eff})* in AutoForm is shown in the equation (2). For this model, the effective friction coefficient is a function of two important parameters, the p_{ref} indicating the reference pressure, and the exponent *(e-1)* where *e* refers to the pressure exponent.

$$\mu_{eff} = \mu(\frac{p}{p_{ref}})^{(e-1)} \tag{2}$$

The velocity dependent friction model accounts for the decrease in friction coefficient as the relative velocity between the tool and the sheet increases. Equation (3) is the AutoForm enhanced Coulomb model to calculate the effective velocity dependent coefficient of friction, μ_{eff},

$$\mu_{eff} = \mu - a\,ln(\frac{max\, v_{rel}, v_{ref}}{v_{ref}}) \tag{3}$$

For this approach it is considered that if the relative velocity is smaller than the reference velocity, the effective friction coefficient is equal to the base friction coefficient μ.

A combination of both enhanced models is stated in the equation. (4).

$$\mu_{eff} = \mu(\frac{p}{p_{ref}})^{(e-1)} - a\,ln(\frac{max\, v_{rel}, v_{ref}}{v_{ref}}) \tag{4}$$

Due to the limitations of the velocity range achieved by the current device this analysis will be focused on the pressure dependent friction model.

3.3. Physical measurement of friction

There is currently no established technique for estimating the coefficient of friction. This is caused by a diversity of geometry on the tool-material interface, as well as the presence of varying stress and strain states in different areas and different moments during the sheet metal forming process. There are recent efforts [5], [9] in order to design and validate an appropriated method to access the friction coefficient for the sheet metal forming.

Based on the strip drawing test and the current available devices in both the literature and the industry, the friction measurement rig created at the Institute of Material Science and Technology at the University of Miskolc is shown in figure 4. For the friction measurement the normal force is previously determined, is applied and visually controlled by the dynamometer. The applied pressure is transferred to the tool and the normal force is calculated using the nominal contact area between tool and metal strip. The friction force is controlled by the Universal Testing Machine.

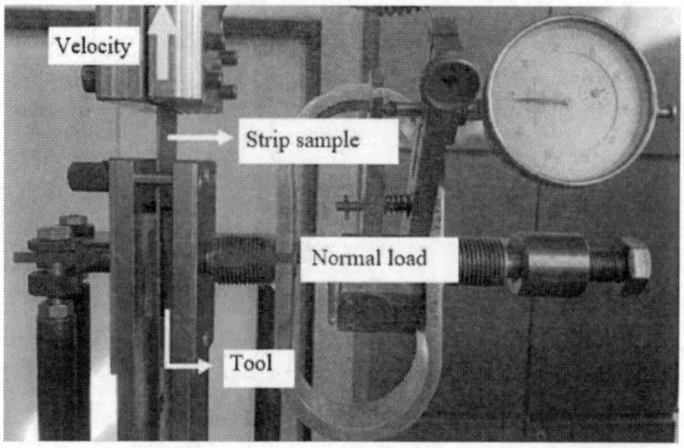

Figure 4. Detailed pictured evidencing the functionality of the new rig.

With the new rig assembled to the Universal Testing Machine two measurements are made. First, the pressure between the tools is taken. Secondly, the fille containing the pulling force (friction force) by the Universal Testing Machine. After the measurements, the normal force is calculated considering the contact area between tool and metal strip and the results are applied to equation 5. The calculation of the friction coefficient (μ) is based on the well-known Amontos-Coulomb friction model. F_R and F_N denote the shear and normal forces, respectively.

$$\mu = \frac{F_R}{F_N} \tag{5}$$

The Table 2 presents the chemical composition of the material of the strip samples (35x300mm).

Table 2: Chemical composition of AISI 304 Stainless Steel

Component	C	Cr	Mn	Ni	P	S	Si	Fe
(wt%)	0.08	20.0	2.00	10.5	0.045	0.03	1.00	Balanced

The physical measurements were conducted following design of experiments (DOE). For the analyses both petroleum-based lubricants were tested. Table 3 presents the lubricants.

Table 3. Lubricant labeling.

Lubricant	**Description**
A	Previously mineral standard lubricant.
B	New mineral lubricant provided by the new supplier.

As the final output of the proposed method the points obtained with the physical measurement are plotted in MATLAB [10] and an interpolated curve is generated using the relevant enhanced friction equation. The specific variables are then defined and used as an input for the enhanced friction model in AutoForm.

4. Results

4.1. Material surface characterization

The surface roughness of the metal strip sample was characterized before and after the friction measurements [11]. Figure 5 shows the comparison between both surfaces. On figure 5 (a) the material as received from the company. On figure 5 (b) the material with a lower roughness where it can be noticed the flattening of the surfaces caused by the contact with the tool.

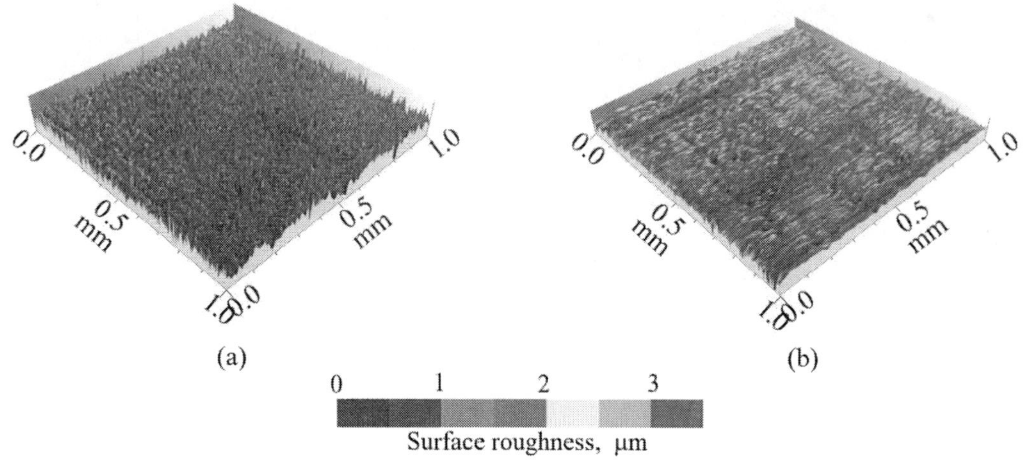

Figure 5. Material surface characterization.

4.2. *Physical measurement results*

The results of the physical friction measurements are shown in figure 6.

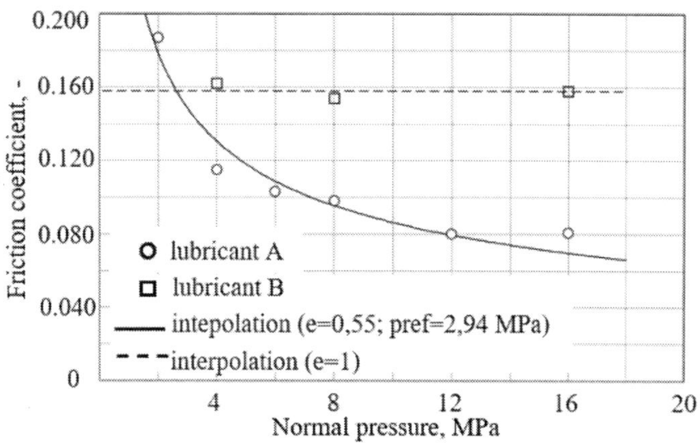

Figure 6. Friction measurement results.

In figure 6 it can be seen both tested lubricants, the interpolation curves based in equation 4 and the calculated variables (e and p_{ref}) to be used as an input for the pressure dependent friction model in AutoForm. First interesting finding from the physical measurements is that the tribosystem using lubricant A shows a pressure exponent of 0.55 indicating the bigger influence of pressure. On the other hand, the new alternative lubricant shows a pressure exponent equal to 1, where the constant friction behavior results in a bigger friction resistance even for higher pressure.

4.3. *Numerical modelling results*

Figure 7 shows the simulation results for both tribosystems. The friction shear stress results show that higher shear stress was found using lubricant B.

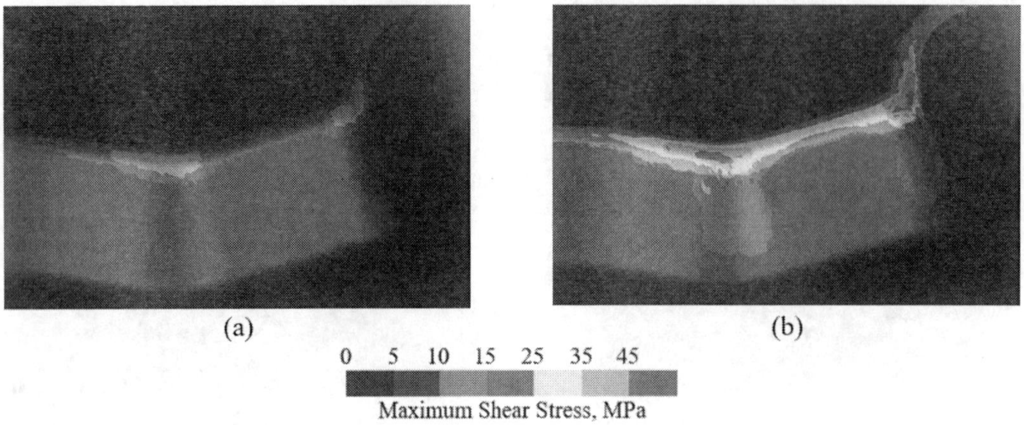

Figure 7. The distribution of Maximum Shear Stress on the critical part of die radius with Lubrication A (a) and Lubrication B (b).

The results of the simulations have shown max shear stress of 13 MPa for lubricant A in contrast to 50 MPa for lubricant B. The change in lubrication can modify the frictional forces between the sheet metal and the forming tools. The friction share stress cannot exceed the shear yield stress of the material. If the friction becomes too high, it can cause excessive shear stress on the sheet metal. As a physical interpretation, this more severe stress strain conditions can lead the sheet metal to the development of cracks.

5. Conclusions

The results of this study indicate the potential of the presented friction measurement device as an advantageous tool to predict failures in sheet metal forming. By using the new device to access the friction coefficient, we were able to identify after simulation results the higher shear stress on the same part of die radius where the failure has occurred on the real part. Therefore, these results can be used to identify the potential failure points in the forming process at an early stage. In the current scenario the current method could be used for the study of the new lubricant before its implementation in the production line. Overall, the new friction measurement device provided valuable insights for the sheet metal forming process by facilitating the analyses of the lubrication strategy for the process improvement and higher quality parts.

References

[1] G. Dalton and G. Dalton, 'New Friction Model for Sheet Metalformin', vol. 110, no. 2001, pp. 56–61, 2023.
[2] J. Coello, V. Miguel, C. Ferrer, A. Calatayud, and A. Martínez, 'Study of the Frictional Behavior of Austenitic Stainless Steel Sheet Under Flat-Die Friction Testing', 2007, no. May.
[3] L. Figueiredo, A. Ramalho, M. C. Oliveira, and L. F. Menezes, 'Experimental study of friction in sheet metal forming', Wear, vol. 271, no. 9–10, pp. 1651–1657, 2011, doi: 10.1016/j.wear.2011.02.020.
[4] J. Filzek, M. Ludwig, and P. Groche, 'Improved FEM Simulation of Sheet Metal Forming with Friction Modelling using Laboratory Tests', in IDDRG Conference Proceedings, 2011, p. 10.
[5] A. F. Tavares, A. P. O. Lopes, E. A. Mesquita, D. T. Almeida, J. H. C. Souza, and H. L. Costa, 'Effect of transfer layers on friction and wear mechanisms in strip drawing tests of commercially coated forming tools', Wear, no. February, p. 203733, 2021, doi: 10.1016/j.wear.2021.203733.
[6] B. Vadavadagi, K. U. Yazar, P. Kulkarni, and K. Narasimhan, 'Friction studies with different lubricants in low carbon steel strips: major strain predictions with microstructural inputs', IDDRG - Fr., pp. 486–490, 2014.

IOP Conf. Series: Materials Science and Engineering 1284 (2023) 012088 doi:10.1088/1757-899X/1284/1/012088

[7] I. Gil, J. Mendiguren, L. Galdos, E. Mugarra, and E. Saenz de Argandoña, 'Influence of the pressure dependent coefficient of friction on deep drawing springback predictions', Tribol. Int., vol. 103, pp. 266–273, 2016, doi: 10.1016/j.triboint.2016.07.004.

[8] Y. S. Kim, M. K. Jain, and D. R. Metzger, 'Determination of pressure-dependent friction coefficient from draw-bend test and its application to cup drawing', Int. J. Mach. Tools Manuf., vol. 56, pp. 69–78, 2012, doi: 10.1016/j.ijmachtools.2011.12.011.

[9] L. Kirkhorn, K. Frogner, M. Andersson, and J. E. Ståhl, 'Improved tribotesting for sheet metal forming', Procedia CIRP, vol. 3, no. 1, pp. 507–512, 2012, doi: 10.1016/j.procir.2012.07.087.

[10] G. Hegedűs, 'Newton's method based collision avoidance in a CAD environment on ball nut grinding', Int. J. Adv. Manuf. Technol., vol. 84, no. 5, pp. 1219–1228, 2016, doi: 10.1007/s00170-015-7796-5.

[11] K. Csaba, Felhő ; János, 'Investigation of the Topography of Machined Surfaces', Appl. Mech. Mater., pp. 412–417, 2014.

AUTHOR INDEX